Elements
of Chemical
Reaction
Engineering

Fourth Edition

BALZHISER, SAMUELS, AND ELIASSEN   Chemical Engineering Thermodynamics
BEQUETTE   Process Control: Modeling, Design, and Simulation
BEQUETTE   Process Dynamics
BIEGLER, GROSSMAN, AND WESTERBERG   Systematic Methods of Chemical Process
   Design
BROSILOW AND JOSEPH   Techniques of Model-based Control
CONSTANTINIDES AND MOSTOUFI   Numerical Methods for Chemical Engineers
   with MATLAB Applications
CROWL AND LOUVAR   Chemical Process Safety: Fundamentals with Applications,
   2nd edition
CUTLIP AND SHACHAM   Problem Solving in Chemical Engineering with Numerical
   Methods
DENN   Process Fluid Mechanics
ELLIOT AND LIRA   Introductory Chemical Engineering Thermodynamics
FOGLER   Elements of Chemical Reaction Engineering, 4th edition
HIMMELBLAU AND RIGGS   Basic Principles and Calculations in Chemical
   Engineering, 7th edition
HINES AND MADDOX   Mass Transfer: Fundamentals and Applications
PRAUSNITZ, LICHTENTHALER, AND DE AZEVEDO   Molecular Thermodynamics
   of Fluid-Phase Equilibria, 3rd edition
PRENTICE   Electrochemical Engineering Principles
SHULER AND KARGI   Bioprocess Engineering, 2nd edition
STEPHANOPOULOS   Chemical Process Control
TESTER AND MODELL   Thermodynamics and Its Applications, 3rd edition
TURTON, BAILIE, WHITING, AND SHAEIWITZ   Analysis, Synthesis, and Design
   of Chemical Processes, 2nd edition
WILKES   Fluid Mechanics for Chemical Engineers, 2nd edition

Elements of Chemical Reaction Engineering

Fourth Edition

H. SCOTT FOGLER

Ame and Catherine Vennema Professor
of Chemical Engineering
The University of Michigan, Ann Arbor

Prentice Hall Professional Technical Reference

Upper Saddle River, NJ • Boston • Indianapolis • San Francisco
New York • Toronto • Montreal • London • Munich • Paris • Madrid
Capetown • Sydney • Tokyo • Singapore • Mexico City

The publisher offers excellent discounts on this book when ordered in quantity for bulk purchases or special sales, which may include electronic versions and/or custom covers and content particular to your business, training goals, marketing focus, and branding interests. For more information, please contact:

    U.S. Corporate and Government Sales
    (800) 382-3419
    corpsales@pearsontechgroup.com

For sales outside the U.S., please contact:

    International Sales
    international@pearsoned.com

Visit us on the Web: www.prenhallprofessional.com

Library of Congress Cataloging-in-Publication Data

Fogler, H. Scott.
    Elements of chemical reaction engineering / H. Scott Fogler—4th ed.
        p.   cm.
    Includes bibliographical references and index.
    ISBN 0-13-047394-4 (alk. paper)
      1. Chemical reactors. I. Title.

    TP157.F65 2005
    660'.2832—dc22   2004060140

ISBN 0-13-047394-4

Text printed in the United States on recycled paper at Courier in Westford, Massachusetts.
Second printing, May 2006

Dedicated to the memory of

Professors

Giuseppe Parravano
Joseph J. Martin
Donald L. Katz

of the University of Michigan
whose standards and lifelong achievements
serve to inspire us

Contents

4   ISOTHERMAL REACTOR DESIGN   143

11   *EXTERNAL DIFFUSION EFFECTS ON HETEROGENEOUS REACTIONS*   **757**

12   *DIFFUSION AND REACTION*   **813**

13   *DISTRIBUTIONS OF RESIDENCE TIMES FOR CHEMICAL REACTORS*   867

Preface

> The man who has ceased to learn ought not to be
> allowed to wander around loose in these dangerous
> days.
>
> M. M. Coady

A. The Audience

This book and interactive CD-ROM is intended for use as both an undergraduate-level and a graduate-level text in chemical reaction engineering. The level will depend on the choice of chapters and CD-ROM *Professional Reference Shelf* (PRS) material to be covered and the type and degree of difficulty of problems assigned.

B. The Goals

B.1. To Develop a Fundamental Understanding of Reaction Engineering

The first goal of this book is to enable the reader to develop a clear understanding of the fundamentals of chemical reaction engineering (CRE). This goal will be achieved by presenting a structure that allows the reader to solve reaction engineering problems **through reasoning** rather than through memorization and recall of numerous equations and the restrictions and conditions under which each equation applies. The algorithms presented in the text for reactor design provide this framework, and the homework problems will give practice at using the algorithms. The conventional home problems at the end of each chapter are designed to reinforce the principles in the chapter. These problems are about equally divided between those that can be solved with a calculator and those that require a personal computer and a numerical software package such as Polymath, MATLAB, or COMSOL.

To give a reference point as to the level of understanding of CRE required in the profession, a number of reaction engineering problems from the California Board of Registration for Civil and Professional Engineers—Chemical Engineering Examinations (PECEE) are included in the text.[1] Typically, these problems should each require approximately 30 minutes to solve.

Finally, the CD-ROM should greatly facilitate learning the fundamentals of CRE because it includes summary notes of the chapters, added examples, expanded derivations, and self tests. A complete description of these *learning resources* is given in the "The Integration of the Text and the CD-ROM" section in this Preface.

B.2.   To Develop Critical Thinking Skills

A second goal is to enhance critical thinking skills. A number of home problems have been included that are designed for this purpose. Socratic questioning is at the heart of critical thinking, and a number of homework problems draw from R. W. Paul's six types of Socratic questions[2] shown in Table P-1.

TABLE P-1.   SIX TYPES OF SOCRATIC QUESTIONS

(1) *Questions for clarification:* Why do you say that? How does this relate to our discussion?   "Are you going to include diffusion in your mole balance equations?"
(2) *Questions that probe assumptions:* What could we assume instead? How can you verify or disprove that assumption?   "Why are you neglecting radial diffusion and including only axial diffusion?"
(3) *Questions that probe reasons and evidence:* What would be an example?   "Do you think that diffusion is responsible for the lower conversion?"
(4) *Questions about viewpoints and perspectives:* What would be an alternative?   "With all the bends in the pipe, from an industrial/practical standpoint, do you think diffusion and dispersion will be large enough to affect the conversion?"
(5) *Questions that probe implications and consequences:* What generalizations can you make? What are the consequences of that assumption?   "How would our results be affected if we neglected diffusion?"
(6) *Questions about the question:* What was the point of this question? Why do you think I asked this question?   "Why do you think diffusion is important?"

---

[1] The permission for use of these problems, which, incidentally, may be obtained from the Documents Section, California Board of Registration for Civil and Professional Engineers—Chemical Engineering, 1004 6th Street, Sacramento, CA 95814, is gratefully acknowledged. (Note: These problems have been copyrighted by the California Board of Registration and may not be reproduced without its permission).

[2] R. W. Paul, *Critical Thinking* (Santa Rosa, Calif.: Foundation for Critical Thinking, 1992).

Scheffer and Rubenfeld[3,4] expand on the practice of critical thinking skills discussed by R. W. Paul by using the activities, statements, and questions shown in Table P-2.

TABLE P-2.   CRITICAL THINKING SKILLS[2,3]

Analyzing: separating or breaking a whole into parts to discover their nature, function, and relationships "I studied it piece by piece." "I sorted things out."
Applying Standards: judging according to established personal, professional, or social rules or criteria "I judged it according to...."
Discriminating: recognizing differences and similarities among things or situations and distinguishing carefully as to category or rank "I rank ordered the various...." "I grouped things together."
Information Seeking: searching for evidence, facts, or knowledge by identifying relevant sources and gathering objective, subjective, historical, and current data from those sources "I knew I needed to look up/study...." "I kept searching for data."
Logical Reasoning: drawing inferences or conclusions that are supported in or justified by evidence "I deduced from the information that...." "My rationale for the conclusion was...."
Predicting: envisioning a plan and its consequences "I envisioned the outcome would be...." "I was prepared for...."
Transforming Knowledge: changing or converting the condition, nature, form, or function of concepts among contexts "I improved on the basics by...." "I wondered if that would fit the situation of ...."

I have found the best way to develop and practice critical thinking skills is to use Tables P-1 and P-2 to help students write a question on any assigned homework problem and then to explain why the question involves critical thinking.

   More information on critical thinking can be found on the CD-ROM in the section on *Problem Solving*.

B.3.   To Develop Creative Thinking Skills

The third goal of this book is to help develop creative thinking skills. This goal will be achieved by using a number of problems that are open-ended to various degrees. Here the students can practice their creative skills by exploring the example problems as outlined at the beginning of the home problems of each

---

[3] Courtesy of B. K. Scheffer and M. G. Rubenfeld, "A Consensus Statement on Critical Thinking in Nursing," *Journal of Nursing Education, 39*, 352–9 (2000).

[4] Courtesy of B. K. Scheffer and M. G. Rubenfeld, "Critical Thinking: What Is It and How Do We Teach It?" *Current Issues in Nursing* (2001).

chapter and by making up and solving an original problem. Problem P4-1 gives some guidelines for developing original problems. A number of techniques that can aid the students in practicing and enhancing their creativity can be found in Fogler and LeBlanc[5] and in the *Thoughts on Problem Solving* section on the CD-ROM and on the web site *www.engin.umich.edu/~cre*. We will use these techniques, such as Osborn's checklist and de Bono's lateral thinking (which involves considering other people's views and responding to random stimulation) to answer add-on questions such as those in Table P-3.

TABLE P-3.   PRACTICING CREATIVE THINKING

(1) Brainstorm ideas to ask another question or suggest another calculation that can be made for this homework problem.
(2) Brainstorm ways you could work this homework problem incorrectly.
(3) Brainstorm ways to make this problem easier or more difficult or more exciting.
(4) Brainstorm a list of things you learned from working this homework problem and what you think the point of the problem is.
(5) Brainstorm the reasons why your calculations overpredicted the conversion that was measured when the reactor was put on stream. Assume you made no numerical errors on your calculations.
(6) "What if…" questions: The "What if…" questions are particularly effective when used with the *Living Example Problems* where one varies the parameters to explore the problem and to carry out a sensitivity analysis. For example, *what if someone suggested that you should double the catalyst particle diameter, what would you say?*

One of the major goals at the undergraduate level is to bring students to the point where they can solve complex reaction problems, such as multiple reactions with heat effects, and then ask "What if…" questions and look for optimum operating conditions. One problem whose solution exemplifies this goal is the Manufacture of Styrene, Problem P8-26. This problem is particularly interesting because two reactions are endothermic and one is exothermic.

(1) Ethylbenzene $\rightarrow$ Styrene + Hydrogen: *Endothermic*
(2) Ethylbenzene $\rightarrow$ Benzene + Ethylene: *Endothermic*
(3) Ethylbenzene + Hydrogen $\rightarrow$ Toluene + Methane: *Exothermic*

To summarize Section B, it is the author's experience that both critical and creative thinking skills can be enhanced by using Tables P-1, P-2, and P-3 to extend any of the homework problems at the end of every chapter.

C. The Structure

The strategy behind the presentation of material is to build continually on a few basic ideas in chemical reaction engineering to solve a wide variety of problems. These ideas, referred to as the *Pillars of Chemical Reaction Engineering*,

---

[5] H. S. Fogler and S. E. LeBlanc, *Strategies for Creative Problem Solving* (Upper Saddle River, N.J.: Prentice Hall, 1995).

are the foundation on which different applications rest. The pillars holding up the application of chemical reaction engineering are shown in Figure P-1.

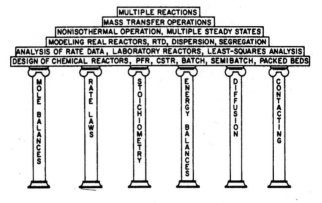

Figure P-1   Pillars of Chemical Reaction Engineering.

From these Pillars we construct our CRE algorithm:

Mole balance + Rate laws + Stoichiometry + Energy balance + Combine

With a few restrictions, the contents of this book can be studied in virtually any order after students have mastered the first four chapters. A flow diagram showing the possible paths can be seen in Figure P-2.

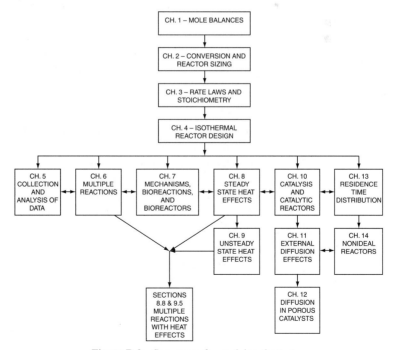

Figure P-2   Sequences for studying the text.

Table P-4 shows examples of topics that can be covered in a graduate course and an undergraduate course. In a four-hour undergraduate course at the University of Michigan, approximately eight chapters are covered in the following order: Chapters 1, 2, 3, 4, and 6; Sections 5.1–5.3; and Chapters 7, 8, and parts of Chapter 10.

TABLE P-4.   UNDERGRADUATE/GRADUATE COVERAGE OF CRE

Undergraduate Material/Course	*Graduate Material/Course*
Mole Balances (Ch. 1)	Short Review (Ch. 1–4, 6, 8)
Smog in Los Angeles Basin (PRS Ch. 1)	Collision Theory (PRS Ch. 3)
Reactor Staging (Ch. 2)	Transition State Theory (PRS Ch. 3)
Hippopotamus Stomach (PRS Ch. 2)	Molecular Dynamics (PRS Ch. 3)
Rate Laws (Ch. 3)	Aerosol Reactors (PRS Ch. 4)
Stoichiometry (Ch.3)	Multiple Reactions (Ch. 6):
Reactors (Ch. 4):	Fed Membrane Reactors
Batch, PFR, CSTR, PBR,	Bioreactions and reactors (Ch. 7, PRS 7.3, 7.4,
Semibatch, Membrane	7.5)
Data Analysis: Regression (Ch. 5)	Polymerization (PRS Ch. 7)
Multiple Reactions (Ch. 6)	Co- and Counter Current Heat
Blood Coagulation (SN Ch. 6)	Exchange (Ch. 8)
Bioreaction Engineering (Ch. 7)	Radial and Axial Gradients in a PFR
Steady-State Heat Effects (Ch. 8):	COMSOL (Ch. 8)
PFR and CSTR with and without	Reactor Stability and Safety (Ch. 8, 9, PRS 9.3)
a Heat Exchanger	Runaway Reactions (PRS Ch. 8)
Multiple Steady States	Catalyst Deactivation (Ch. 10)
Unsteady-State Heat Effects (Ch. 9)	Residence Time Distribution (Ch. 13)
Reactor Safety	Models of Real Reactors (Ch. 14)
Catalysis (Ch. 10)	Applications (PRS): Multiphase Reactors,
	CVD Reactors, Bioreactors

The reader will observe that although metric units are used primarily in this text (e.g., $kmol/m^3$, J/mol), a variety of other units are also employed (e.g., lb/ft^3). This is intentional! We believe that whereas most papers published today use the metric system, a significant amount of reaction engineering data exists in the older literature in English units. Because engineers will be faced with extracting information and reaction rate data from older literature as well as the current literature, they should be equally at ease with both English and metric units.

The notes in the margins are meant to serve two purposes. First, they act as guides or as commentary as one reads through the material. Second, they identify key equations and relationships that are used to solve chemical reaction engineering problems.

D. The Components of the CD-ROM

The interactive CD-ROM is a novel and unique part of this book. The main purposes of the CD-ROM are to serve as an enrichment resource and as a professional reference shelf. The home page for the CD-ROM and the CRE web site (*www.engin.umich.edu/~cre/fogler&gurmen*) is shown in Figure P-3.

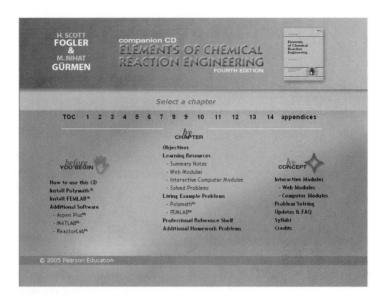

Figure P-3   Screen shot of the home page of the CD-ROM.

The objectives of the CD-ROM are threefold: (1) to facilitate the learning of CRE by interactively addressing the *Felder/Solomon Inventory of Learning Styles*[6] in the Summary Notes, the additional examples, the Interactive Computing Modules (ICMs), and the Web Modules; (2) to provide additional technical material for the professional reference shelf; (3) to provide other tutorial information, examples, derivations, and self tests, such as additional thoughts on problem solving, the use of computational software in chemical reaction engineering, and representative course structures. The following components are listed at the end of most chapters and can be accessed from each chapter in the CD-ROM.

- **Learning Resources**

    The Learning Resources give an overview of the material in each chapter and provide extra explanations, examples, and applications to reinforce the basic concepts of chemical reaction engineering. The learning resources on the CD-ROM include the following:

    Summary Notes

    Web Modules

    1. *Summary Notes*

        The Summary Notes give an overview of each chapter and provide on-demand additional examples, derivations, and audio comments as well as self tests to assess each reader's understanding of the material.

    2. *Web Modules*

        The Web Modules, which apply key concepts to both standard and nonstandard reaction engineering problems (e.g., the use of wetlands to degrade toxic chemicals, cobra bites), can be loaded directly from the CD-ROM.

---

[6] *http://www.ncsu.edu/felder-public/ILSdir/styles.htm*

Additional Web Modules are expected to be added to the web site (*www.engin.umich.edu/~cre*) over the next several years.

3. *Interactive Computer Modules* (ICMs)

Interactive

Computer Modules

Students have found the Interactive Computer Modules to be both fun and extremely useful to review the important chapter concepts and then apply them to real problems in a unique and entertaining fashion. In addition to updating all the ICMs from the last edition, two new modules, *The Great Race* (Ch. 6) and *Enzyme Man* (Ch. 7), have been added. The complete set of 11 modules follows:

- Quiz Show I (Ch. 1)
- Reactor Staging (Ch. 2)
- Quiz Show II (Ch. 3)
- Murder Mystery (Ch. 4)
- Tic Tac (Ch. 4)
- Ecology (Ch. 5)

- The Great Race (Ch. 6)
- Enzyme Man (Ch. 7)
- Heat Effects I (Ch. 8)
- Heat Effects II (Ch. 8)
- Catalysis (Ch. 10)

Solved Problems

4. *Solved Problems*

A number of solved problems are presented along with problem-solving heuristics. Problem-solving strategies and additional worked example problems are available in the *Problem Solving* section of the CD-ROM.

- **Living Example Problems**

A copy of Polymath is provided on the CD-ROM for the students to use to solve the homework problems. The example problems that use an ODE solver (e.g., Polymath) are referred to as "living example problems" because students can load the Polymath program directly onto their own computers in order to study the problem. Students are encouraged to change parameter values and to "play with" the key variables and assumptions. Using the Living Example Problems to explore the problem and asking "What if..." questions provide students with the opportunity to practice critical and creative thinking skills.

Living Example Problem

- **Professional Reference Shelf**

This section of the CD-ROM contains

1. Material that was in previous editions (e.g., polymerization, slurry reactors, and chemical vapor disposition reactors) that has been omitted from the printed version of the fourth edition

2. New topics such as *collision and transition state theory, aerosol reactors, DFT,* and *runaway reactions*, which are commonly found in graduate courses

Reference Shelf

3. Material that is important to the practicing engineer, such as details of the industrial reactor design for the oxidation of SO_2 and design of spherical reactors and other material that is typically not included in the majority of chemical reaction engineering courses

- **Software Toolbox on the CD-ROM**

Polymath. The Polymath software includes an ordinary differential equation (ODE) solver, a nonlinear equation solver, and nonlinear regression. As with previous editions, Polymath is included with this edition to explore the example problems and to solve the home problems. A special Polymath web site (*www.polymath-software.com/fogler*) has been set up for this book by Polymath authors Cutlip and Shacham. This web site provides information on how to obtain an updated version of Polymath at a discount.

*COMSOL.** The COMSOL Multiphysics (referred to throughout the book as COMSOL) software includes a partial differential equation solver. This edition includes a specially prepared version of COMSOL on its own CD-ROM. With COMSOL the students can view both axial and radial temperature and concentration profiles. Five of the COMSOL modules are:

- Isothermal operation
- Adiabatic operation
- Heat effects with constant heat exchange fluid temperature
- Heat effects with variable heat exchanger temperature
- Dispersion with Reaction using the Danckwerts Boundary Conditions (two cases)

As with the Polymath programs, the input parameters can be varied to learn how they change the temperature and concentration profiles.

Instructions are included on how to use not only the software packages of Polymath, MATLAB, and COMSOL, but also on how to apply ASPEN PLUS to solve CRE problems. Tutorials with detailed screen shots are provided for Polymath and COMSOL.

- **Other CD-ROM Resources**
  FAQs. The Frequently Asked Questions (FAQs) are a compilation of questions collected over the years from undergraduate students taking reaction engineering.

  Problem Solving. In this section, both critical thinking and creative thinking are discussed along with what to do if you get "stuck" on a problem.

  Visual Encyclopedia of Equipment. This section was developed by Dr. Susan Montgomery at the University of Michigan. Here a wealth of photographs and descriptions of real and ideal reactors are given. The students with visual, active, sensing, and intuitive learning styles of the Felder/Solomon Index will particularly benefit from this section.

  Reactor Lab. Developed by Professor Richard Herz at the University of California at San Diego, this interactive tool will allow students not only to test their comprehension of CRE material but also to explore different situations and combinations of reaction orders and types of reactions.

  Green Engineering Home Problems. Green engineering problems for virtually every chapter have been developed by Professor Robert Hesketh at Rowan University and Professor Martin Abraham at the University of Toledo and can be found at *www.rowan.edu/greenengineering*. These problems also accompany the book by David Allen and David Shonnard, *Green Chemical Engineering: Environmentally Conscious Design of Chemical Processes* (Prentice Hall, 2002).

Green engineering

E.   The Integration of the Text and the CD-ROM

E.1.   The University Student

There are a number of ways one can use the CD-ROM in conjunction with the text. The CD-ROM provides enrichment resources for the reader in the form of

---

* The name *FEMLAB* was changed to *COMSOL Multiphysics* on July 1, 2005.

interactive tutorials. Pathways on how to use the materials to learn chemical reaction engineering are shown in Figure P-4. The keys to the CRE learning flow sheets include primary resources and enrichment resources:

In developing a fundamental understanding of the material, students may wish to use only the primary resources without using the CD-ROM (i.e., using only the boxes shown in Figure P-4), or they may use a few or all of the inter-active tutorials in the CD-ROM (i.e., the circles shown in Figure P-4). How-ever, to practice the skills that enhance critical and creative thinking, students are strongly encouraged to use the *Living Example Problems* and vary the model parameters to ask and answer "What if…" questions.

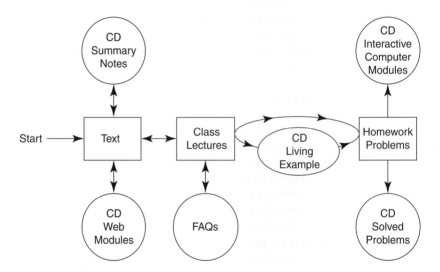

Figure P-4   A Student Pathway to Integrate the Class, the Text, and the CD.

Note that even though the author recommends studying the Living Example Problems before working home problems, they may be bypassed, as is the case with all the enrichment resources, if time is short. However, class testing of the enrichment resources reveals that they not only greatly aid in learning the material but also serve to motivate students through the novel use of CRE principles.

E.2.   For the Practicing Engineer

A figure similar to Figure P-4 for the practicing engineer is given in the CD-ROM Appendix.

F.   The Web

The web site (*www.engin.umich.edu/~cre*) will be used to update the text and the CD-ROM. It will identify typographical and other errors in the first and later printings of the fourth edition of the text. In the near future, additional material will be added to include more solved problems as well as additional Web Modules.

G.   What's New

Pedagogy. The fourth edition of this book maintains all the strengths of the previous additions by using algorithms that allow students to learn chemical reaction engineering through logic rather than memorization. At the same time it provides new resources that allow students to go beyond solving equations in order to get an intuitive feel and understanding of how reactors behave under different situations. This understanding is achieved through more than sixty interactive simulations provided on the CD-ROM that is shrink wrapped with the text. The CD-ROM has been greatly expanded to address the Felder/Solomon Inventory of Different Learning Styles[7] through interactive Summary Notes and new and updated Interactive Computer Modules (ICMs). For example, the Global Learner can get an overview of the chapter material from the Summary Notes; the Sequential Learner can use all the ⬤ **Derive** ⬤ hot buttons; and the Active Learner can interact with the ICM's and use the ⬤ **Self Test** ⬤ hot buttons in the Summary Notes.

A new pedagogical concept is introduced in this edition through expanded emphasis on the example problems. Here, the students simply load the Living Example Problems (LEPs) onto their computers and then explore the problems to obtain a deeper understanding of the implications and generalizations before working the home problems for that chapter. This exploration helps the students get an innate feel of reactor behavior and operation, as well as develop and practice their creative thinking skills. To develop critical thinking skills, instructors can assign one of the new home problems on troubleshooting, as well as ask the students to expand home problems by asking a related question that involves critical thinking using Tables P-1 and P-2. Creative thinking skills can be enhanced by exploring the example problems and asking "what if. . ." questions, by using one or more of the brainstorming exercises in Table P-3 to extend any of the home problems, and by working the open-ended problems. For example, in the case study on safety, students can use the CD-ROM to carry out a post-mortem on the nitroanaline explosion in Example 9-2 to learn what would have happened if the cooling had failed for five minutes instead of ten minutes. Significant effort has been devoted to developing example and home problems that foster critical and creative thinking.

---

[7] *http://www.ncsu.edu/felder-public/ILSdir/styles.htm*

New Material.

Bioreaction Engineering. The greatest expansion of material is in the area of bioreaction engineering. New material has been added on

- Tissue Engineering
- Pharmacokinetics
- Blood Coagulation
- DNA Lab-On-A-Chip
- Methanol Poisoning

- Enzyme Kinetics
- Cell Growth
- Ethanol Metabolism
- Transdermal Drug Delivery
- RTD of Arterial Blood Flow in the Eye

There is a bioreaction engineering home problem in virtually every chapter. Bio-related web modules include physiological-based-pharmacokinetic (PBPK) models of ethanol metabolism, of drug distribution, and of venomous snake bites by the Russels' viper and the cobra.

Chemical Reaction Engineering. There is a greater emphasis on the use of mole balances in terms of concentrations and molar flow rates rather than conversion. It is introduced early in the text so that these forms of the balance equations can be easily applied to membrane reactors and multiple reactions, as well as PFRs, PBRs, and CSTRs.

The partial differential equation solver COMSOL is included to allow students to see 2-D axial and radial temperature and concentration profiles in reactors with heat effects and dispersion. Other new material includes

- Microreactors
- Side-Fed-Membrane Reactors
- Laminar Flow Reactors
- The Advanced Reactor Safety Screening Tool (ARSST)
- Runaway Reactions
- Counter Current Heat Exchangers in Plug Flow Reactors
- Tubular Reactors with Axial and Radial Gradients (COMSOL)
- The Use of RTD to Troubleshoot and Diagnose Faulty Reactor Operation

Expanded material includes collision theory, transition state theory, molecular dynamics, and molecular chemical reaction engineering (DFT) to study rate constants. Many of the new home problems reflect this wide range of application.

The fourth edition contains more industrial chemistry with real reactors and real reactions and extends the wide range of applications to which chemical reaction engineering principles can be applied (i.e., cobra bites, medications, ecological engineering). However, all intensive laws tend often to have exceptions. Very interesting concepts take orderly, responsible statements. Virtually all laws intrinsically are natural thoughts. General observations become laws under experimentation.

H. Acknowledgments

There are so many colleagues and students who contributed to this book that it would require another chapter to thank them all in an appropriate manner. I again acknowledge all my friends, students, and colleagues for their contributions to the first, second, and third editions (see Introduction, CD-ROM). For the fourth edition, I give special recognition as follows.

First of all, I thank my colleague Dr. Nihat Gürmen who coauthored the CD-ROM and web site. His creativity and energy had a great impact on this project and really makes the fourth edition of this text and associated CD-ROM special. He has been a wonderful colleague to work with.

Professor Flavio F. de Moraes not only translated the third edition into Portuguese, in collaboration with Professor Luismar M. Porto, but also gave suggestions, as well as assistance proofreading the fourth edition. Dr. Susan Montgomery provided the *Visual Encyclopedia of Equipment* for the CD-ROM, as well as support and encouragement. Professor Richard Herz provided the *Reactor Lab* portion of the CD-ROM. Dr. Ed Fontes, Anna Gordon, and the folks at COMSOL provided a special version of COMSOL Multiphysics to be included with this book. Duc Nguyen, Yongzhong Liu, and Nihat Gürmen also helped develop the COMSOL material and web modules. These contributions are greatly appreciated.

Elena Mansilla Diaz contributed to the blood coagulation model and, along with Nihat Gürmen, to the pharmacokinetics model of the envenomation of the Fer-de-Lance. Michael Breson and Nihat Gürmen contributed to the Russell's Viper envenomation model, and David Umulis and Nihat Gürmen contributed to the alcohol metabolism. Veerapat (Five) Tantayakom contributed a number of the drawings, along with many other details. Senior web designers Nathan Comstock, Andrea Sterling, and Brian Vicente worked tirelessly with Dr. Gürmen on the CD-ROM, as well as with web designers Jiewei Cao and Lei He. Professor Michael Cutlip, along with Professor Mordechai Schacham, provided Polymath and a special Polymath web site for the text. Brian Vicente took major responsibility for the solution manual, while Massimiliano Nori provided solutions to Chapters 13 and 14. Sombuddha Ghosh also helped with the manual's preparation and some web material.

I would also like to thank colleagues at the University of Colorado. Professor Will Medlin coauthored the Molecular Reaction Engineering Web Modules (DFT), Professor Kristi Anseth contributed to the Tissue Engineering Example, and Professor Dhinakar Kompala contributed to the Professional Reference Shelf R7.4 Multiple Enzymes/Multiple Substrates.

I also thank my Ph.D. graduate students—Rama Venkatesan, Duc Nugyen, Ann Piyarat Wattana, Kris Paso, Veerapat (Five) Tantayakom, Ryan Hartman, Hyun Lee, Michael Senra, Lizzie Wang, Prashant Singh, and Kriangkrai Kraiwattanawong—for their patience and understanding during the period while I was writing this book. In addition, the support provided by the staff and colleagues at the departments of chemical engineering at University College London and the University of Colorado while I finished the final details of the text is greatly appreciated. Both are very stimulating and are great places to work and to spend a sabbatical.

The stimulating discussions with Professors Robert Hesketh, Phil Savage, John Falconer, D. B. Battacharia, Rich Masel, Eric McFarland, Will Medlin, and Kristi Anseth are greatly appreciated. I also appreciate the friendship and insights provided by Dr. Lee Brown on Chapters 13 and 14. Mike Cutlip not only gave suggestions and a critical reading of many sections but, most importantly, provided continuous support and encouragement throughout the course of this project.

Don MacLaren (composition) and Yvette Raven (CD-ROM user interface design) made large contributions to this new edition. Bernard Goodwin (Publisher) of Prentice Hall was extremely helpful and supportive throughout.

There are three people who need special mention as they helped pull everything together as we rushed to meet the printing deadline. Julie Nahil, Full-Service Production Manager at Prentice Hall, provided encouragement, attention to detail, and a great sense of humor that was greatly appreciated. Janet Peters was not only a meticulous proofreader of the page proofs, but also added many valuable editorial and other comments and suggestions. Brian Vicente put out extra effort to help finish so many details with the CD-ROM and also provided a number of drawings in the text. Thanks Julie, Janet, and Brian for your added effort.

Laura Bracken is so much a part of this manuscript. I appreciate her excellent deciphering of equations and scribbles, her organization, and her attention to detail in working with the galley and copy edited proofs. Through all this was her ever-present wonderful disposition. Thanks, Radar!!

Finally, to my wife Janet, love and thanks. Without her enormous help and support the project would never have been possible.

HSF
Ann Arbor

 For updates on the CD and new and exciting applications, see the web site:

www.engin.umich.edu/~cre
or
www.engin.umich.edu/~cre/fogler&gurmen

For typographical errors, click on Updates & FAQ on the Home page to find

www.engin.umich.edu/~cre/byconcept/updates/frames.htm

Mole Balances   1

> The first step to knowledge
> is to know that we are ignorant.
>
> Socrates (470–399 B.C.)

The Wide Wide Wild World of Chemical Reaction Engineering

How is a chemical
engineer different
from other
engineers?

Chemical kinetics is the study of chemical reaction rates and reaction mechanisms. The study of chemical reaction engineering (CRE) combines the study of chemical kinetics with the reactors in which the reactions occur. Chemical kinetics and reactor design are at the heart of producing almost all industrial chemicals such as the manufacture of phthalic anhydride shown in Figure 1-1. It is primarily a knowledge of chemical kinetics and reactor design that distinguishes the chemical engineer from other engineers. The selection of a reaction system that operates in the safest and most efficient manner can be the key to the economic success or failure of a chemical plant. For example, if a reaction system produces a large amount of undesirable product, subsequent purification and separation of the desired product could make the entire process economically unfeasible.

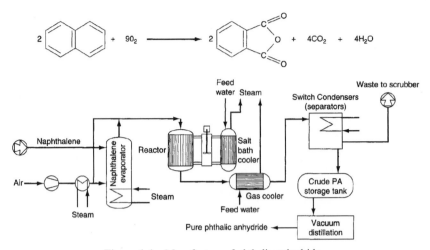

Figure 1-1   Manufacture of phthalic anhydride.

The Chemical Reaction Engineering (**CRE**) principles learned here can also be applied in areas such as waste treatment, microelectronics, nanoparticles and living systems in addition to the more traditional areas of the manufacture of chemicals and pharmaceuticals. Some of the examples that illustrate the wide application of CRE principles are shown in Figure 1-2. These examples include modeling smog in the L.A. basin (Chapter 1), the digestive system of a hippopotamus (Chapter 2), and molecular CRE (Chapter 3). Also shown is the manufacture of ethylene glycol (antifreeze), where three of the most common types of industrial reactors are used (Chapter 4). The CD-ROM describes the use of wetlands to degrade toxic chemicals (Chapter 4). Other examples shown are the solid-liquid kinetics of acid-rock interactions to improve oil recovery (Chapter 5); pharmacokinetics of cobra bites and of drug delivery (Chapter 6); free radical scavengers used in the design of motor oils (Chapter 7), enzyme kinetics, and pharmacokinetics (Chapter 7); heat effects, runaway reactions, and plant safety (Chapters 8 and 9); increasing the octane number of gasoline (Chapter 10); and the manufacture of computer chips (Chapter 12).

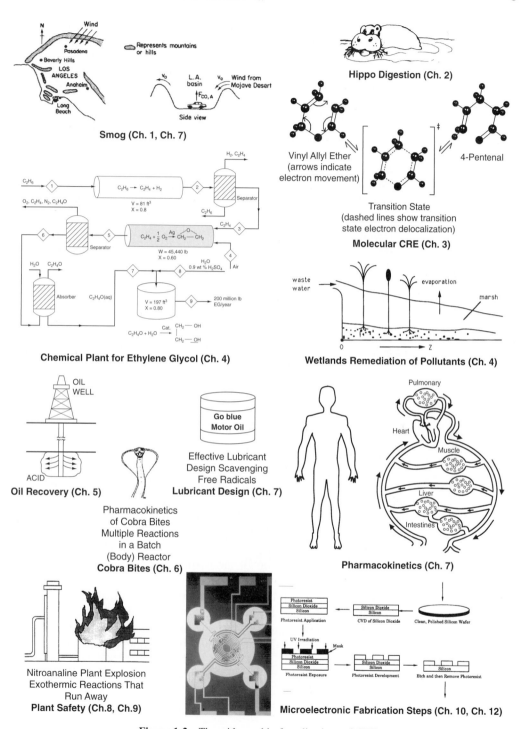

Figure 1-2   The wide world of applications of CRE.

Overview—Chapter 1. This chapter develops the first building block of chemical reaction engineering, mole balances, that will be used continually throughout the text. After completing this chapter the reader will be able to describe and define the rate of reaction, derive the general mole balance equation, and apply the general mole balance equation to the four most common types of industrial reactors.

Before entering into discussions of the conditions that affect chemical reaction rate mechanisms and reactor design, it is necessary to account for the various chemical species entering and leaving a reaction system. This accounting process is achieved through overall mole balances on individual species in the reacting system. In this chapter, we develop a general mole balance that can be applied to any species (usually a chemical compound) entering, leaving, and/or remaining within the reaction system volume. After defining the rate of reaction, $-r_A$, and discussing the earlier difficulties of properly defining the chemical reaction rate, we show how the general balance equation may be used to develop a preliminary form of the design equations of the most common industrial reactors: batch, continuous-stirred tank (CSTR), tubular (PFR), and packed bed (PBR). In developing these equations, the assumptions pertaining to the modeling of each type of reactor are delineated. Finally, a brief summary and series of short review questions are given at the end of the chapter.

1.1 The Rate of Reaction, $-r_A$

The rate of reaction tells us how fast a number of moles of one chemical species are being consumed to form another chemical species. The term *chemical species* refers to any chemical component or element with a given *identity*. The identity of a chemical species is determined by the *kind, number,* and *configuration* of that species' atoms. For example, the species nicotine (a bad tobacco alkaloid) is made up of a fixed number of specific atoms in a definite molecular arrangement or configuration. The structure shown illustrates the kind, number, and configuration of atoms in the species nicotine (responsible for "nicotine fits") on a molecular level.

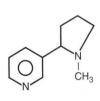

Nicotine

Even though two chemical compounds have exactly the same number of atoms of each element, they could still be different species because of different configurations. For example, 2-butene has four carbon atoms and eight hydrogen atoms; however, the atoms in this compound can form two different arrangements.

$$
\begin{array}{c}
\text{H} \qquad\qquad \text{H} \\
\diagdown \qquad\quad \diagup \\
\text{C}=\text{C} \\
\diagup \qquad\quad \diagdown \\
\text{CH}_3 \qquad\quad \text{CH}_3
\end{array}
\qquad \text{and} \qquad
\begin{array}{c}
\text{H} \qquad\qquad \text{CH}_3 \\
\diagdown \qquad\quad \diagup \\
\text{C}=\text{C} \\
\diagup \qquad\quad \diagdown \\
\text{CH}_3 \qquad\quad \text{H}
\end{array}
$$

cis-2-butene   *trans*-2-butene

As a consequence of the different configurations, these two isomers display different chemical and physical properties. Therefore, we consider them as two different species even though each has the same number of atoms of each element.

When has a chemical reaction taken place?

We say that a *chemical reaction* has taken place when a detectable number of molecules of one or more species have lost their identity and assumed a new form by a change in the kind or number of atoms in the compound and/or by a change in structure or configuration of these atoms. In this classical approach to chemical change, it is assumed that the total mass is neither created nor destroyed when a chemical reaction occurs. The mass referred to is the total collective mass of all the different species in the system. However, when considering the individual species involved in a particular reaction, we do speak of the rate of disappearance of mass of a particular species. *The rate of disappearance of a species, say species A, is the number of A molecules that lose their chemical identity per unit time per unit volume through the breaking and subsequent re-forming of chemical bonds during the course of the reaction.* In order for a particular species to "appear" in the system, some prescribed fraction of another species must lose its chemical identity.

There are three basic ways a species may lose its chemical identity: decomposition, combination, and isomerization. In *decomposition*, the molecule loses its identity by being broken down into smaller molecules, atoms, or atom fragments. For example, if benzene and propylene are formed from a cumene molecule,

A species can lose its identity by decomposition, combination, or isomerization.

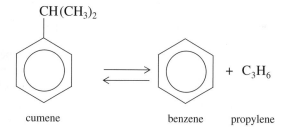

cumene   benzene   propylene

the cumene molecule has lost its identity (i.e., disappeared) by breaking its bonds to form these molecules. A second way that a molecule may lose its species identity is through *combination* with another molecule or atom. In the example above, the propylene molecule would lose its species identity if the reaction were carried out in the reverse direction so that it combined with benzene to form cumene. The third way a species may lose its identity is through *isomerization,* such as the reaction

$$
\underset{\displaystyle CH_2{=}C{-}CH_2CH_3}{\overset{\displaystyle \overset{CH_3}{|}}{}} \longrightarrow \underset{\displaystyle CH_3C{=}CHCH_3}{\overset{\displaystyle \overset{CH_3}{|}}{}}
$$

Here, although the molecule neither adds other molecules to itself nor breaks into smaller molecules, it still loses its identity through a change in configuration.

To summarize this point, we say that a given number of molecules (e.g., mole) of a particular chemical species have reacted or disappeared when the molecules have lost their chemical identity.

The rate at which a given chemical reaction proceeds can be expressed in several ways. To illustrate, consider the reaction of chlorobenzene and chloral to produce the insecticide DDT (dichlorodiphenyl-trichloroethane) in the presence of fuming sulfuric acid.

$$CCl_3CHO + 2C_6H_5Cl \longrightarrow (C_6H_4Cl)_2 CHCCl_3 + H_2O$$

Letting the symbol A represent chloral, B be chlorobenzene, C be DDT, and D be H_2O we obtain

$$A + 2B \longrightarrow C + D$$

The numerical value of the rate of disappearance of reactant A, $-r_A$, is a positive number (e.g., $-r_A = 4$ mol A/dm^3·s).

What is $-r_A$?

> The rate of reaction, $-r_A$, is the number of moles of A (e.g., chloral) reacting (disappearing) per unit time per unit volume (mol/dm^3·s).

$A + 2B \rightarrow C + D$
The convention

$-r_A = 4$ mol A/dm^3·s
$r_A = -4$ mol A/dm^3·s
$-r_B = 8$ mol B/dm^3·s
$r_B = -8$ mol B/dm^3·s
$r_C = 4$ mol C/dm^3·s

The symbol r_j is the rate of formation (generation) of species j. If species j is a reactant, the numerical value of r_j will be a negative number (e.g., $r_A = -4$ moles A/dm^3·s). If species j is a product, then r_j will be a positive number (e.g., $r_C = 4$ moles C/dm^3·s). In Chapter 3, we will delineate the prescribed relationship between the rate of formation of one species, r_j (e.g., DDT[C]), and the rate of disappearance of another species, $-r_i$ (e.g., chlorobenzene[B]), in a chemical reaction.

Heterogeneous reactions involve more than one phase. In heterogeneous reaction systems, the rate of reaction is usually expressed in measures other than volume, such as reaction surface area or catalyst weight. For a gas-solid catalytic reaction, the gas molecules must interact with the solid catalyst surface for the reaction to take place.

What is r_A'?

The dimensions of this heterogeneous reaction rate, r_A' (prime), *are the number of moles of A reacting per unit time per unit mass of catalyst* (mol/s·g catalyst).

Most of the introductory discussions on chemical reaction engineering in this book focus on homogeneous systems.

The mathematical definition of a chemical reaction rate has been a source of confusion in chemical and chemical engineering literature for many years. The origin of this confusion stems from laboratory bench-scale experiments that were carried out to obtain chemical reaction rate data. These early experiments were batch-type, in which the reaction vessel was closed and rigid; consequently, the ensuing reaction took place at constant volume. The reactants were mixed together at time $t = 0$ and the concentration of one of the reactants, C_A, was measured at various times t. The rate of reaction was determined from the slope of a plot of C_A as a function of time. Letting r_A be the rate of formation of A per unit volume (e.g., mol/s·dm³), the investigators then defined and reported the chemical reaction rate as

$$r_A = \frac{dC_A}{dt} \tag{1-1}$$

Summary Notes

However this "definition" is wrong! It is simply a mole balance that is only valid for a constant volume batch system. Equation (1-1) will not apply to any continuous-flow reactor operated at steady state, such as the tank (CSTR) reactor where the concentration does not vary from day to day (i.e., the concentration is not a function of time). For amplification on this point, see the section "Is Sodium Hydroxide Reacting?" in the Summary Notes for Chapter 1 on the CD-ROM or on the web.

Definition of r_j

In conclusion, Equation (1-1) is not the definition of the chemical reaction rate. We shall simply say that r_j *is the rate of formation of species j per unit volume*. It is the number of moles of species j generated per unit volume per unit time.

The rate law does not depend on the type of reactor used!!

The rate equation (i.e., rate law) for r_j is an algebraic equation that is solely a function of the properties of the reacting materials and reaction conditions (e.g., species concentration, temperature, pressure, or type of catalyst, if any) at a point in the system. The rate equation is independent of the type of reactor (e.g., batch or continuous flow) in which the reaction is carried out. However, because the properties and reaction conditions of the reacting materials may vary with position in a chemical reactor, r_j can in turn be a function of position and can vary from point to point in the system.

What is $-r_A$ a function of?

The chemical reaction rate law is essentially an algebraic equation involving concentration, not a differential equation.[1] For example, the algebraic form of the rate law for $-r_A$ for the reaction

$$A \longrightarrow products$$

may be a linear function of concentration,

$$-r_A = kC_A$$

or, as shown in Chapter 3, it may be some other algebraic function of concentration, such as

$$-r_A = kC_A^2 \tag{1-2}$$

or

---

[1] For further elaboration on this point, see *Chem. Eng. Sci.*, 25, 337 (1970); B. L. Crynes and H. S. Fogler, eds., *AIChE Modular Instruction Series E: Kinetics*, 1, 1 (New York: AIChE, 1981); and R. L. Kabel, "Rates," *Chem. Eng. Commun.*, 9, 15 (1981).

$$-r_A = \frac{k_1 C_A}{1 + k_2 C_A}$$

For a given reaction, the particular concentration dependence that the rate law follows (i.e., $-r_A = kC_A$ or $-r_A = kC_A^2$ or ...) must be determined from experimental observation. Equation (1-2) states that the rate of disappearance of A is equal to a rate constant k (which is a function of temperature) times the square of the concentration of A. By convention, r_A is the rate of formation of A; consequently, $-r_A$ is the rate of disappearance of A. Throughout this book, the phrase *rate of generation* means exactly the same as the phrase *rate of formation*, and these phrases are used interchangeably.

1.2 The General Mole Balance Equation

To perform a mole balance on any system, the system boundaries must first be specified. The volume enclosed by these boundaries is referred to as the *system volume*. We shall perform a mole balance on species j in a system volume, where species j represents the particular chemical species of interest, such as water or NaOH (Figure 1-3).

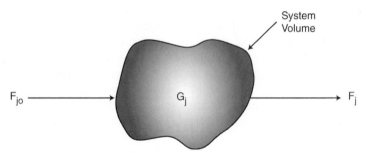

Figure 1-3   Balance on system volume.

A mole balance on species j at any instant in time, t, yields the following equation:

$$\begin{bmatrix} \text{Rate of flow} \\ \text{of } j \text{ into} \\ \text{the system} \\ \text{(moles/time)} \end{bmatrix} - \begin{bmatrix} \text{Rate of flow} \\ \text{of } j \text{ out of} \\ \text{the system} \\ \text{(moles/time)} \end{bmatrix} + \begin{bmatrix} \text{Rate of generation} \\ \text{of } j \text{ by chemical} \\ \text{reaction within} \\ \text{the system} \\ \text{(moles/time)} \end{bmatrix} = \begin{bmatrix} \text{Rate of} \\ \text{accumulation} \\ \text{of } j \text{ within} \\ \text{the system} \\ \text{(moles/time)} \end{bmatrix}$$

In   –   **Out**   **+ Generation**   **= Accumulation**

$$F_{j0} \quad - \quad F_j \quad + \quad G_j \quad = \quad \frac{dN_j}{dt} \qquad (1\text{-}3)$$

where N_j represents the number of moles of species j in the system at time t. If all the system variables (e.g., temperature, catalytic activity, concentration of the chemical species) are spatially uniform throughout the system volume, the rate of generation of species j, G_j, is just the product of the reaction volume, V, and the rate of formation of species j, r_j.

$$G_j = r_j \cdot V$$

$$\frac{moles}{time} = \frac{moles}{time \cdot volume} \cdot volume$$

Suppose now that the rate of formation of species j for the reaction varies with the position in the system volume. That is, it has a value r_{j1} at location 1, which is surrounded by a small volume, ΔV_1, within which the rate is uniform: similarly, the reaction rate has a value r_{j2} at location 2 and an associated volume, ΔV_2 (Figure 1-4).

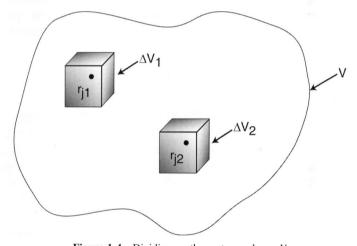

Figure 1-4   Dividing up the system volume, V.

The rate of generation, ΔG_{j1}, in terms of r_{j1} and subvolume ΔV_1, is

$$\Delta G_{j1} = r_{j1}\, \Delta V_1$$

Similar expressions can be written for ΔG_{j2} and the other system subvolumes, ΔV_i. The total rate of generation within the system volume is the sum of all the rates of generation in each of the subvolumes. If the total system volume is divided into M subvolumes, the total rate of generation is

$$G_j = \sum_{i=1}^{M} \Delta G_{ji} = \sum_{i=1}^{M} r_{ji}\, \Delta V_i$$

By taking the appropriate limits (i.e., let $M \rightarrow \infty$ and $\Delta V \rightarrow 0$) and making use of the definition of an integral, we can rewrite the foregoing equation in the form

$$G_j = \int^V r_j \, dV$$

From this equation we see that r_j will be an indirect function of position, since the properties of the reacting materials and reaction conditions (e.g., concentration, temperature) can have different values at different locations in the reactor.

We now replace G_j in Equation (1-3)

$$F_{j0} - F_j + G_j = \frac{dN_j}{dt} \tag{1-3}$$

by its integral form to yield a form of the general mole balance equation for any chemical species j that is entering, leaving, reacting, and/or accumulating within any system volume V.

This is a basic equation for chemical reaction engineering.

$$\boxed{F_{j0} - F_j + \int^V r_j \, dV = \frac{dN_j}{dt}} \tag{1-4}$$

From this general mole balance equation we can develop the design equations for the various types of industrial reactors: batch, semibatch, and continuous-flow. Upon evaluation of these equations we can determine the time (batch) or reactor volume (continuous-flow) necessary to convert a specified amount of the reactants into products.

1.3 Batch Reactors

When is a batch reactor used?

A batch reactor is used for small-scale operation, for testing new processes that have not been fully developed, for the manufacture of expensive products, and for processes that are difficult to convert to continuous operations. The reactor can be charged (i.e., filled) through the holes at the top (Figure 1-5[a]). The batch reactor has the advantage of high conversions that can be obtained by leaving the reactant in the reactor for long periods of time, but it also has the disadvantages of high labor costs per batch, the variability of products from batch to batch, and the difficulty of large-scale production (see Professional Reference Shelf [PRS]).

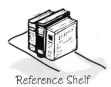

Reference Shelf

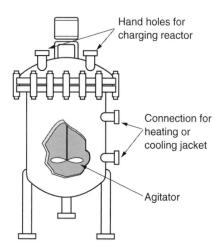

Figure 1-5(a) Simple batch homogeneous reactor. [Excerpted by special permission from *Chem. Eng., 63*(10), 211 (Oct. 1956). Copyright 1956 by McGraw-Hill, Inc., New York, NY 10020.]

Figure 1-5(b) Batch reactor mixing patterns. Further descriptions and photos of the batch reactors can be found in both the *Visual Encyclopedia of Equipment* and in the *Professional Reference Shelf* on the CD-ROM.

A batch reactor has neither inflow nor outflow of reactants or products while the reaction is being carried out: $F_{j0} = F_j = 0$. The resulting general mole balance on species j is

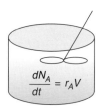

$$\frac{dN_j}{dt} = \int^V r_j \, dV$$

If the reaction mixture is perfectly mixed (Figure 1-5[b]) so that there is no variation in the rate of reaction throughout the reactor volume, we can take r_j out of the integral, integrate, and write the mole balance in the form

Perfect mixing

$$\boxed{\frac{dN_j}{dt} = r_j V} \tag{1-5}$$

Let's consider the isomerization of species A in a batch reactor

$$A \longrightarrow B$$

As the reaction proceeds, the number of moles of A decreases and the number of moles of B increases, as shown in Figure 1-6.

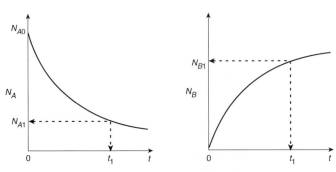

Figure 1-6 Mole-time trajectories.

We might ask what time, t_1, is necessary to reduce the initial number of moles from N_{A0} to a final desired number N_{A1}. Applying Equation (1-5) to the isomerization

$$\frac{dN_A}{dt} = r_A V$$

rearranging,

$$dt = \frac{dN_A}{r_A V}$$

and integrating with limits that at $t = 0$, then $N_A = N_{A0}$, and at $t = t_1$, then $N_A = N_{A1}$, we obtain

$$t_1 = \int_{N_{A1}}^{N_{A0}} \frac{dN_A}{-r_A V} \tag{1-6}$$

This equation is the integral form of the mole balance on a batch reactor. It gives the time, t_1, necessary to reduce the number of moles from N_{A0} to N_{A1} and also to form N_{B1} moles of B.

1.4 Continuous-Flow Reactors

Continuous flow reactors are almost always operated at steady state. We will consider three types: the continuous stirred tank reactor (CSTR), the plug flow reactor (PFR), and the packed bed reactor (PBR). Detailed descriptions of these reactors can be found in both the Professional Reference Shelf (PRS) for Chapter 1 and in the *Visual Encyclopedia of Equipment* on the CD-ROM.

1.4.1 Continuous-Stirred Tank Reactor

A type of reactor used commonly in industrial processing is the stirred tank operated continuously (Figure 1-7). It is referred to as the *continuous-stirred tank reactor* (CSTR) or vat, or *backmix reactor,* and is used primarily for liquid

What is a CSTR
used for?

phase reactions. It is normally operated **at steady state** and is assumed to be **perfectly mixed**; consequently, there is no time dependence or position dependence of the temperature, the concentration, or the reaction rate inside the CSTR. That is, every variable is the same at every point inside the reactor. Because the temperature and concentration are identical everywhere within the reaction vessel, they are the same at the exit point as they are elsewhere in the tank. Thus the temperature and concentration in the exit stream are modeled as being the same as those inside the reactor. In systems where mixing is highly nonideal, the well-mixed model is inadequate and we must resort to other modeling techniques, such as residence-time distributions, to obtain meaningful results. This topic of nonideal mixing is discussed in Chapters 13 and 14.

What reaction systems use a CSTR?

The ideal CSTR is assumed to be perfectly mixed.

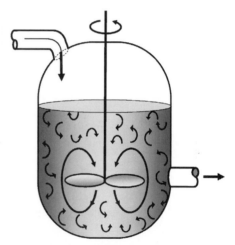

Figure 1-7(a)   CSTR/batch reactor. [Courtesy of Pfaudler, Inc.]

Figure 1-7(b)   CSTR mixing patterns. Also see the *Visual Encyclopedia of Equipment* on the CD-ROM.

When the general mole balance equation

$$F_{j0} - F_j + \int^V r_j \, dV = \frac{dN_j}{dt}$$ (1-4)

is applied to a CSTR operated at steady state (i.e., conditions do not change with time),

$$\frac{dN_j}{dt} = 0$$

in which there are no spatial variations in the rate of reaction (i.e., perfect mixing),

$$\int^V r_j \, dV = V r_j$$

it takes the familiar form known as the *design equation* for a CSTR:

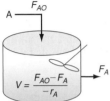

$$V = \frac{F_{j0} - F_j}{-r_j} \tag{1-7}$$

The CSTR design equation gives the reactor volume V necessary to reduce the entering flow rate of species j, from F_{j0} to the exit flow rate F_j, when species j is disappearing at a rate of $-r_j$. We note that the CSTR is modeled such that the conditions in the exit stream (e.g., concentration, temperature) are identical to those in the tank. The molar flow rate F_j is just the product of the concentration of species j and the volumetric flow rate v :

$$F_j = C_j \cdot v$$

$$\frac{\text{moles}}{\text{time}} = \frac{\text{moles}}{\text{volume}} \cdot \frac{\text{volume}}{\text{time}} \tag{1-8}$$

Consequently, we could combine Equations (1-7) and (1-8) to write a balance on species A as

$$V = \frac{v_0 C_{A0} - v C_A}{-r_A} \tag{1-9}$$

1.4.2 Tubular Reactor

In addition to the CSTR and batch reactors, another type of reactor commonly used in industry is the *tubular reactor*. It consists of a cylindrical pipe and is normally operated at steady state, as is the CSTR. Tubular reactors are used most often for gas-phase reactions. A schematic and a photograph of industrial tubular reactors are shown in Figure 1-8.

When is tubular reactor most often used?

In the tubular reactor, the reactants are continually consumed as they flow down the length of the reactor. In modeling the tubular reactor, we assume that the concentration varies continuously in the axial direction through the reactor. Consequently, the reaction rate, which is a function of concentration for all but zero-order reactions, will also vary axially. For the purposes of the material presented here, we consider systems in which the flow field may be modeled by that of a plug flow profile (e.g., uniform velocity as in turbulent flow), as shown in Figure 1-9. That is, there is no radial variation in reaction rate and the reactor is referred to as a plug-flow reactor (PFR). (The laminar flow reactor is discussed in Chapter 13.)

Figure 1-8(a)   Tubular reactor schematic.
Longitudinal tubular reactor. [Excerpted by
special permission from *Chem. Eng., 63*(10),
211 (Oct. 1956). Copyright 1956 by
McGraw-Hill, Inc., New York, NY 10020.]

Figure 1-8(b)   Tubular reactor photo.
Tubular reactor for production of Dimersol G.
[Photo Courtesy of Editions Techniq
Institute Francois du Petrol].

Plug flow–no radial variations in velocity,
concentration, temperature, or reaction rate

Reactants   **Products**

Figure 1-9   Plug-flow tubular reactor.

The general mole balance equation is given by Equation (1-4):

$$F_{j0} - F_j + \int^V r_j \, dV = \frac{dN_j}{dt} \tag{1-4}$$

The equation we will use to design PFRs at steady state can be developed in
two ways: (1) directly from Equation (1-4) by differentiating with respect to
volume V, or (2) from a mole balance on species j in a differential segment of
the reactor volume ΔV. Let's choose the second way to arrive at the differential form of the PFR mole balance. The differential volume, ΔV, shown in Figure 1-10, will be chosen sufficiently small such that there are no spatial
variations in reaction rate within this volume. Thus the generation term, ΔG_j, is

$$\Delta G_j = \int^{\Delta V} r_j \, dV = r_j \, \Delta V$$

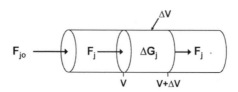

Figure 1-10   Mole balance on species j in volume ΔV.

$$\begin{bmatrix} \text{Molar flow} \\ \text{rate of species } j \\ \textit{In} \text{ at V} \end{bmatrix} - \begin{bmatrix} \text{Molar flow} \\ \text{rate of species } j \\ \textit{Out} \text{ at } (V+\Delta V) \end{bmatrix} + \begin{bmatrix} \text{Molar rate of} \\ \textit{Generation} \\ \text{of species } j \\ \text{within } \Delta V \end{bmatrix} = \begin{bmatrix} \text{Molar rate of} \\ \textit{Accumulation} \\ \text{of species } j \\ \text{within } \Delta V \end{bmatrix}$$

In   −   Out   + Generation   = Accumulation

$$F_j\big|_V \quad - \quad F_j\big|_{V+\Delta V} \quad + \quad r_j\Delta V \quad = \quad 0 \qquad (1\text{-}10)$$

Dividing by ΔV and rearranging

$$\left[\frac{F_j\big|_{V+\Delta V} - F_j\big|_V}{\Delta V} \right] = r_j$$

the term in brackets resembles the definition of the derivative

$$\lim_{\Delta x \to 0} \left[\frac{f(x+\Delta x)-f(x)}{\Delta x} \right] = \frac{df}{dx}$$

Taking the limit as ΔV approaches zero, we obtain the differential form of steady state mole balance on a PFR.

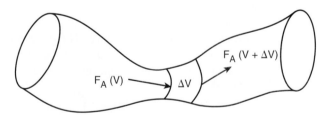

Tubular reactor

$$\boxed{\frac{dF_j}{dV} = r_j} \qquad (1\text{-}11)$$

We could have made the cylindrical reactor on which we carried out our mole balance an irregular shape reactor, such as the one shown in Figure 1-11 for reactant species A.

Picasso's reactor

$F_A (V) \longrightarrow$ ΔV $F_A (V + \Delta V)$

Figure 1-11   Pablo Picasso's reactor.

However, we see that by applying Equation (1-10) the result would yield the same equation (i.e., Equation [1-11]). For species A, the mole balance is

$$\boxed{\frac{dF_A}{dV} = r_A} \qquad (1\text{-}12)$$

Consequently, we see that Equation (1-11) applies equally well to our model of tubular reactors of variable and constant cross-sectional area, although it is doubtful that one would find a reactor of the shape shown in Figure 1-11 unless it were designed by Pablo Picasso. The conclusion drawn from the application of the design equation to Picasso's reactor is an important one: the degree of completion of a reaction achieved in an ideal plug-flow reactor (PFR) does not depend on its shape, only on its total volume.

Again consider the isomerization $A \rightarrow B$, this time in a PFR. As the reactants proceed down the reactor, A is consumed by chemical reaction and B is produced. Consequently, the molar flow rate of A decreases and that of B increases, as shown in Figure 1-12.

$$V = \int_{F_A}^{F_{A0}} \frac{dF_A}{-r_A}$$

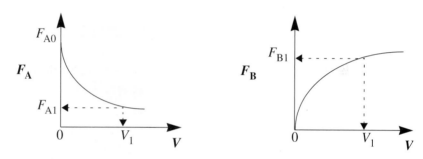

Figure 1-12   Profiles of molar flow rates in a PFR.

We now ask what is the reactor volume V_1 necessary to reduce the entering molar flow rate of A from F_{A0} to F_{A1}. Rearranging Equation (1-12) in the form

$$dV = \frac{dF_A}{r_A}$$

and integrating with limits at $V = 0$, then $F_A = F_{A0}$, and at $V = V_1$, then $F_A = F_{A1}$.

$$V_1 = \int_{F_{A0}}^{F_{A1}} \frac{dF_A}{r_A} = \int_{F_{A1}}^{F_{A0}} \frac{dF_A}{-r_A} \tag{1-13}$$

V_1 is the volume necessary to reduce the entering molar flow rate F_{A0} to some specified value F_{A1} and also the volume necessary to produce a molar flow rate of B of F_{B1}.

1.4.3 Packed-Bed Reactor

The principal difference between reactor design calculations involving homogeneous reactions and those involving fluid-solid heterogeneous reactions is that for the latter, the reaction takes place on the surface of the catalyst. Consequently, the reaction rate is based on mass of solid catalyst, W, rather than on reactor volume, V. For a fluid–solid heterogeneous system, the rate of reaction of a substance A is defined as

$$-r'_A = \text{mol A reacted/s} \cdot \text{g catalyst}$$

The mass of solid catalyst is used because the amount of the catalyst is what is important to the rate of product formation. The reactor volume that contains the catalyst is of secondary significance. Figure 1-13 shows a schematic of an industrial catalytic reactor with vertical tubes packed with catalyst.

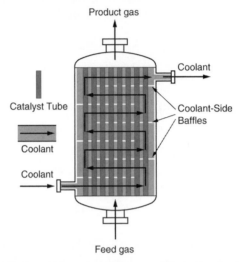

Figure 1-13 Longitudinal catalytic packed-bed reactor. [From Cropley, American Institute of Chemical Engineers, *86*(2), 34 (1990). Reproduced with permission of the American Institute of Chemical Engineers, Copyright © 1990 AIChE. All rights reserved.]

In the three idealized types of reactors just discussed (the perfectly mixed batch reactor, the plug-flow tubular reactor (PFR), and the perfectly mixed continuous-stirred tank reactor (CSTR), the design equations (i.e., mole balances) were developed based on reactor volume. The derivation of the design equation for a packed-bed catalytic reactor (PBR) will be carried out in a manner analogous to the development of the tubular design equation. To accomplish this derivation, we simply replace the volume coordinate in Equation (1-10) with the catalyst weight coordinate W (Figure 1-14).

**PBR
Mole Balance**

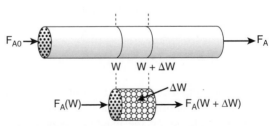

Figure 1-14 Packed-bed reactor schematic.

As with the PFR, the PBR is assumed to have no radial gradients in concentration, temperature, or reaction rate. The generalized mole balance on species A over catalyst weight ΔW results in the equation

In	**– Out**	**+ Generation**	**= Accumulation**

$$F_{A|W} \quad - \quad F_{A|(W+\Delta W)} \quad + \quad r'_A \, \Delta W \quad = \quad 0 \qquad (1\text{-}14)$$

The dimensions of the generation term in Equation (1-14) are

$$(r'_A)\,\Delta W \equiv \frac{moles\ A}{(time)(mass\ of\ catalyst)} \cdot (mass\ of\ catalyst) \equiv \frac{moles\ A}{time}$$

which are, as expected, the same dimensions of the molar flow rate F_A. After dividing by ΔW and taking the limit as $\Delta W \rightarrow 0$, we arrive at the differential form of the mole balance for a packed-bed reactor:

Use differential form of design equation for catalyst decay and pressure drop.

$$\boxed{\frac{dF_A}{dW} = r'_A} \qquad (1\text{-}15)$$

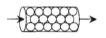

When pressure drop through the reactor (see Section 4.5) and catalyst decay (see Section 10.7) are neglected, the integral form of the packed-catalyst-bed design equation can be used to calculate the catalyst weight.

Use integral form only for no ΔP and no catalyst decay.

$$W = \int_{F_{A0}}^{F_A} \frac{dF_A}{r'_A} = \int_{F_A}^{F_{A0}} \frac{dF_A}{-r'_A} \qquad (1\text{-}16)$$

W is the catalyst weight necessary to reduce the entering molar flow rate of species A, F_{A0}, to a flow rate F_A.

For some insight into things to come, consider the following example of how one can use the tubular reactor design Equation (1-11).

Example 1–1   How Large Is It?

Consider the liquid phase *cis – trans* isomerization of 2–butene

$$\underset{\text{cis-2-butene}}{\underset{CH_3}{\overset{H}{>}} C = C \underset{CH_3}{\overset{H}{<}}} \longrightarrow \underset{\text{trans-2-butene}}{\underset{CH_3}{\overset{H}{>}} C = C \underset{H}{\overset{CH_3}{<}}}$$

which we will write symbolically as

$$A \longrightarrow B$$

The first order $(-r_A = kC_A)$ reaction is carried out in a tubular reactor in which the volumetric flow rate, v, is constant, i.e., $v = v_0$.

1. Sketch the concentration profile.
2. Derive an equation relating the reactor volume to the entering and exiting concentrations of A, the rate constant k, and the volumetric flow rate v.
3. Determine the reactor volume necessary to reduce the exiting concentration to 10% of the entering concentration when the volumetric flow rate is 10 dm^3/min (i.e., liters/min) and the specific reaction rate, k, is 0.23 min^{-1}.

Solution

1. Species A is consumed as we move down the reactor, and as a result, both the molar flow rate of A and the concentration of A will decrease as we move. Because the volumetric flow rate is constant, $v = v_0$, one can use Equation (1-8) to obtain the concentration of A, $C_A = F_A/v_0$, and then by comparison with Figure 1-12 plot the concentration of A as a function of reactor volume as shown in Figure E1-1.1.

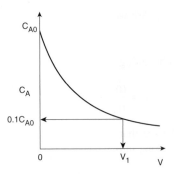

Figure E1-1.1   Concentration profile.

2. Derive an equation relating V, v_0, k, C_{A0}, and C_A.

For a tubular reactor, the mole balance on species A ($j = A$) was shown to be given by Equation (1-11). Then for species A ($j = A$) results

$$\frac{dF_A}{dV} = r_A \tag{1-11}$$

For a first-order reaction, the rate law (discussed in Chapter 3) is

$$-r_A = kC_A \tag{E1-1.1}$$

Because the volumetric flow rate, v, is constant ($v = v_0$), as it is for most liquid-phase reactions,

$$\frac{dF_A}{dV} = \frac{d(C_A v)}{dV} = \frac{d(C_A v_0)}{dV} = v_0 \frac{dC_A}{dV} = r_A \tag{E1-1.2}$$

Multiplying both sides of Equation (E1-1.2) by minus one and then substituting Equation (E1-1.1) yields

$$-\frac{v_0 dC_A}{dV} = -r_A = kC_A \tag{E1-1.3}$$

Reactor sizing

Rearranging gives

$$-\frac{v_0}{k}\left(\frac{dC_A}{C_A}\right) = dV$$

$C_A = C_{A0}\exp(-kV/v)$

Using the conditions at the entrance of the reactor that when $V = 0$, then $C_A = C_{A0}$,

$$-\frac{v_0}{k}\int_{C_{A0}}^{C_A}\frac{dC_A}{C_A} = \int_0^V dV \qquad \text{(E1-1.4)}$$

Carrying out the integration of Equation (E1-1.4) gives

$$\boxed{V = \frac{v_0}{k}\,ln\,\frac{C_{A0}}{C_A}} \qquad \text{(E1-1.5)}$$

3. We want to find the volume, V_1, at which $C_A = \frac{1}{10}C_{A0}$ for $k = 0.23$ min^{-1} and $v_0 = 10$ dm^3/min.

Substituting C_{A0}, C_A, v_0, and k in Equation (E1-1.5), we have

$$V = \frac{10\ \text{dm}^3/\text{min}}{0.23\ \text{min}^{-1}}\ ln\ \frac{C_{A0}}{0.1C_{A0}} = \frac{10\ \text{dm}^3}{0.23}ln\,10 = 100\ \text{dm}^3\ (\text{i.e., } 100\ \text{L}; 0.1\ \text{m}^3)$$

We see that a reactor volume of 0.1 m^3 is necessary to convert 90% of species A entering into product B for the parameters given.

In the remainder of this chapter we look at slightly more detailed drawings of some typical industrial reactors and point out a few of the advantages and disadvantages of each.[2]

1.5 Industrial Reactors

When is a batch reactor used?

Be sure to view actual photographs of industrial reactors on the CD-ROM and on the Web site. There are also links to view reactors on different web sites. The CD-ROM also includes a portion of the *Visual Encyclopedia of Equipment—Chemical Reactors* developed by Dr. Susan Montgomery and her students at the University of Michigan.

Links

[1] Liquid-Phase Reactions.   Semibatch reactors and CSTRs are used primarily for liquid-phase reactions. A semibatch reactor (Figure 1-15) has essentially the same disadvantages as the batch reactor. However, it has the advantages of temperature control by regulation of the feed rate and the capability of minimizing unwanted side reactions through the maintenance of a low concentration of one of the reactants. The semibatch reactor is also used for two-phase reactions in which a gas usually is bubbled continuously through the liquid.

[2] *Chem. Eng.*, *63*(10), 211 (1956). See also *AIChE Modular Instruction Series E, 5* (1984).

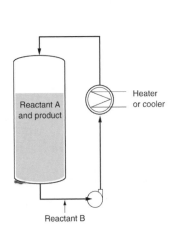

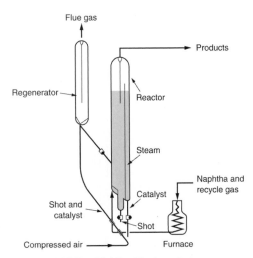

Figure 1-15(a) Semibatch reactor.
[Excerpted by special permission from
Chem. Eng., 63(10), 211 (Oct. 1956).
Copyright 1956 by McGraw-Hill, Inc.,
New York, NY 10020.]

Figure 1-15(b) Fluidized-bed catalytic reactor.
[Excerpted by special permission from *Chem.
Eng., 63*(10), 211 (Oct. 1956). Copyright 1956 by
McGraw-Hill, Inc., New York, NY 10020.]

What are the
advantages and
disadvantages of a
CSTR?

A CSTR is used when intense agitation is required. Figure 1-7(a) showed a cutaway view of a Pfaudler CSTR/batch reactor. Table 1-1 gives the typical sizes (along with that of the comparable size of a familiar object) and costs for batch and CSTR reactors. All reactors are glass lined and the prices include heating/cooling jacket, motor, mixer, and baffles. The reactors can be operated at temperatures between 20 and 450°F and at pressures up to 100 psi.

TABLE 1-1.   REPRESENTATIVE PFAUDLER CSTR/BATCH REACTOR
SIZES AND 2004 PRICES

Volume	Price	Volume	Price
5 Gallons (wastebasket)	$29,000	1000 Gallons (2 Jacuzzis)	$85,000
50 Gallons (garbage can)	$38,000	4000 Gallons (8 Jacuzzis)	$150,000
500 Gallons (Jacuzzi)	$70,000	8000 Gallons (gasoline tanker)	$280,000

The CSTR can either be used by itself or, in the manner shown in Figure 1-16, as part of a series or battery of CSTRs. It is relatively easy to maintain good temperature control with a CSTR because it is well mixed. There is, however, the disadvantage that the conversion of reactant per volume of reactor is the smallest of the flow reactors. Consequently, very large reactors are neces-

Reference Shelf

sary to obtain high conversions. An industrial flow sheet for the manufacture of nitrobenzene from benzene using a cascade of CSTRs is shown and described in the Professional Reference Shelf for Chapter 1 on the CD-ROM.

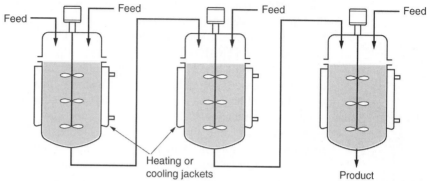

Figure 1-16   Battery of stirred tanks. [Excerpted by special permission from *Chem. Eng., 63*(10), 211 (Oct. 1956). Copyright 1956 by McGraw-Hill, Inc., New York, NY 10020.]

Links

If you are not able to afford to purchase a new reactor, it may be possible to find a used reactor that may fit your needs. Previously owned reactors are much less expensive and can be purchased from equipment clearinghouses such as Aaron Equipment Company (*www.aaronequipment.com*) or Loeb Equipment Supply (*www.loebequipment.com/*).

What are the advantages and disadvantages of a PFR?

[2] Gas-Phase Reactions.   The tubular reactor (i.e., plug-flow reactor [PFR]) is relatively easy to maintain (no moving parts), and it usually produces the highest conversion per reactor volume of any of the flow reactors. The disadvantage of the tubular reactor is that it is difficult to control temperature within the reactor, and hot spots can occur when the reaction is exothermic. The tubular reactor is commonly found either in the form of one long tube or as one of *CSTR: liquids* a number of shorter reactors arranged in a tube bank as shown in Figures *PFR: gases* 1-8(a) and (b). Most homogeneous liquid-phase flow reactors are CSTRs, whereas most homogeneous gas-phase flow reactors are tubular.

The costs of PFRs and PBRs (without catalyst) are similar to the costs of heat exchangers and can be found in *Plant Design and Economics for Chemical Engineers*, 5th ed., by M. S. Peters and K. D. Timmerhaus (New York: McGraw-Hill, 2002). From Figure 15-12 of the Peters and Timmerhaus book, one can get an estimate of the purchase cost per foot of $1 for a 1-in. pipe and $2 per foot for a 2-in. pipe for single tubes and approximately $20 to $50 per square foot of surface area for fixed-tube sheet exchangers.

A packed-bed (also called a fixed-bed) reactor is essentially a tubular reactor that is packed with solid catalyst particles (Figure 1-13). This heterogeneous reaction system is most often used to catalyze gas reactions. This reactor has the same difficulties with temperature control as other tubular reactors; in addition, the catalyst is usually troublesome to replace. On occasion, channeling of the gas flow occurs, resulting in ineffective use of parts of the reactor bed. The advantage of the packed-bed reactor is that for most reactions it gives the highest conversion per weight of catalyst of any catalytic reactor.

Another type of catalytic reactor in common use is the fluidized-bed (Figure 1-15[b]) reactor, which is analogous to the CSTR in that its contents, though heterogeneous, are well mixed, resulting in an even temperature distribution throughout the bed. The fluidized-bed reactor can only be approximately modeled as a CSTR (Example 10.3); for higher precision it requires a model of its own (Section PRS12.3). The temperature is relatively uniform throughout, thus avoiding hot spots. This type of reactor can handle large amounts of feed and solids and has good temperature control; consequently, it is used in a large number of applications. The advantages of the ease of catalyst replacement or regeneration are sometimes offset by the high cost of the reactor and catalyst regeneration equipment. A thorough discussion of a gas-phase industrial reactor and process can be found on the Professional Reference Shelf of the CD-ROM for Chapter 1. The process is the manufacture of paraffins from synthesis gas (CO and H_2) in a straight-through transport reactor (see Chapter 10).

Reference Shelf

In this chapter, and on the CD-ROM, we've introduced each of the major types of industrial reactors: batch, semibatch, stirred tank, tubular, fixed bed (packed bed), and fluidized bed. Many variations and modifications of these commercial reactors are in current use; for further elaboration, refer to the detailed discussion of industrial reactors given by Walas.[3]

Solved Problems

The CD-ROM describes industrial reactors, along with typical feed and operating conditions. In addition, two solved example problems for Chapter 1 can be found on the CD.

Closure. The goal of this text is to weave the fundamentals of chemical reaction engineering into a structure or algorithm that is easy to use and apply to a variety of problems. We have just finished the first building block of this algorithm: mole balances. This algorithm and its corresponding building blocks will be developed and discussed in the following chapters:

- Mole Balance, Chapter 1
- Rate Law, Chapter 3
- Stoichiometry, Chapter 3
- Combine, Chapter 4
- Evaluate, Chapter 4
- Energy Balance, Chapter 8

With this algorithm, one can approach and solve chemical reaction engineering problems through logic rather than memorization.

---

[3] S. M. Walas, *Reaction Kinetics for Chemical Engineers* (New York: McGraw-Hill, 1959), Chapter 11.

SUMMARY

Each chapter summary gives the key points of the chapter that need to be remembered and carried into succeeding chapters.

ST Montgomery 8/94

1. A mole balance on species j, which enters, leaves, reacts, and accumulates in a system volume V, is

$$F_{j0} - F_j + \int^V r_j \, dV = \frac{dN_j}{dt} \qquad (S1\text{-}1)$$

If, and only if, the contents of the reactor are well mixed, then a mole balance (Equation S1-1) on species A gives

$$F_{A0} - F_A + r_A V = \frac{dN_A}{dt} \qquad (S1\text{-}2)$$

2. The kinetic rate law for r_j is:

- Solely a function of properties of reacting materials and reaction conditions (e.g., concentration [activities], temperature, pressure, catalyst or solvent [if any])
- The rate of formation of species j per unit volume (e.g., mol/s·dm³)
- An intensive quantity (i.e., it does not depend on the total amount)
- An algebraic equation, not a differential equation (e.g., $-r_A = kC_A$, $-r_A = kC_A^2$)

   For homogeneous catalytic systems, typical units of $-r_j$ may be gram moles per second per liter; for heterogeneous systems, typical units of r_j' may be gram moles per second per gram of catalyst. By convention, $-r_A$ is the rate of disappearance of species A and r_A is the rate of formation of species A.

3. Mole balances on species A in four common reactors are as follows:

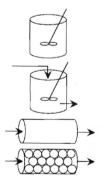

TABLE S–1 SUMMARY OF REACTOR MOLE BALANCES

Reactor	Comment	Mole Balance Differential Form	Algebraic Form	Integral Form
Batch	No spatial variations	$\dfrac{dN_A}{dt} = r_A V$		$t_1 = \displaystyle\int_{N_{A1}}^{N_{A0}} \dfrac{dN_A}{-r_A V}$
CSTR	No spatial variations, steady state	—	$V = \dfrac{F_{A0} - F_A}{-r_A}$	—
PFR	Steady state	$\dfrac{dF_A}{dV} = r_A$		$V_1 = \displaystyle\int_{F_{A1}}^{F_{A0}} \dfrac{dF_A}{-r_A}$
PBR	Steady state	$\dfrac{dF_A}{dW} = r_A'$		$W_1 = \displaystyle\int_{F_{A1}}^{F_{A0}} \dfrac{dF_A}{-r_A'}$

CD-ROM MATERIAL

Summary Notes

- **Learning Resources**
  1. *Summary Notes*
  2. *Web Material*
     A. Problem-Solving Algorithm
     B. Getting Unstuck on a Problem
        This site on the web and CD-ROM gives tips on how to overcome mental barriers in problem solving.
     C. Smog in L.A. basin

B. Getting Unstuck   **C. Smog in L.A.**

Fotografiert von ©2002 Hank Good.

  3. *Interactive Computer Modules*
     A. Quiz Show I

4.   *Solved Problems*

   A. CDP1-A_B Batch Reactor Calculations: A Hint of Things to Come

   B. P1-14$_B$ Modeling Smog in the L.A. Basin

Living Example Problem
Smog in L.A.

- **FAQ [Frequently Asked Questions]**—In Updates/FAQ icon section
- **Professional Reference Shelf**

   1. **Photos of Real Reactors**

Reference Shelf

2. **Reactor Section of the *Visual Encyclopedia of Equipment***

   This section of the CD-ROM shows industrial equipment and discusses its operation. The reactor portion of this encyclopedia is included on the CD-ROM accompanying this book.

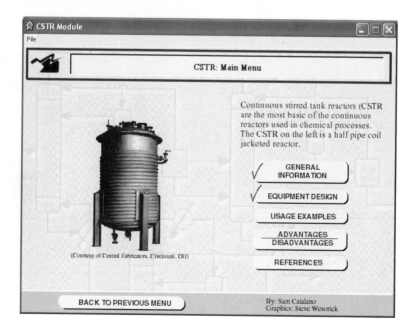

Examples of
industrial reactions
and reactors

3. **The production of nitrobenzene example problem.** Here the process flow-sheet is given, along with operating conditions.

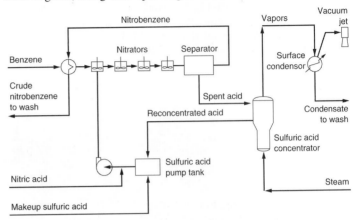

Figure PRS.A-1   Flowsheet for the manufacture of nitrobenzene.

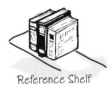

Reference Shelf

4. **Fischer-Tropsch Reaction and Reactor Example.** A Fischer-Tropsch reaction carried out in a typical straight-through transport reactor (Riser).

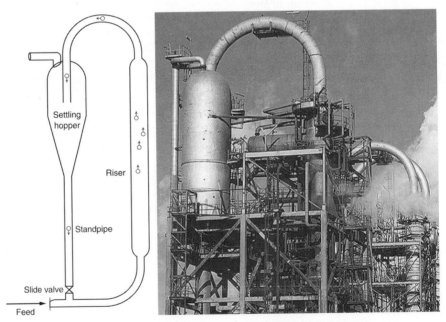

Figure PRS.B-1   The reactor is 3.5 m in diameter and 38 m tall.
[Schematic and photo courtesy of Sasol/Sastech PT Limited.]

Here photographs and schematics of the equipment along with the feed rates, reactor sizes, and principal reactions

$$nCO + 2nH_2 \longrightarrow C_nH_{2n} + nH_2O$$

are also discussed in the PRS.

QUESTIONS AND PROBLEMS

I wish I had an answer for that, because I'm getting tired of answering that question.

Yogi Berra, New York Yankees
Sports Illustrated, June 11, 1984

The subscript to each of the problem numbers indicates the level of difficulty: A, least difficult; D, most difficult.

$$A = \bullet \quad B = \blacksquare \quad C = \blacklozenge \quad D = \blacklozenge\blacklozenge$$

In each of the questions and problems below, rather than just drawing a box around your answer, write a sentence or two describing how you solved the problem, the assumptions you made, the reasonableness of your answer, what you learned, and any other facts that you want to include. You may wish to refer to W. Strunk and E. B. White, *The Elements of Style,* 4th Ed. (New York: Macmillan, 2000) and Joseph M. Williams, *Style: Ten Lessons in Clarity & Grace,* 6th Ed. (Glenview, Ill.: Scott, Foresman, 1999) to enhance the quality of your sentences.

 = Hint on the web.

Before solving the problems, state or sketch qualitatively the expected results or trends.

P1-1$_A$ (a) Read through the Preface. Write a paragraph describing both the content goals and the intellectual goals of the course and text. Also describe what's on the CD and how the CD can be used with the text and course.

(b) List the areas in Figure 1-1 you are most looking forward to studying.

(c) Take a quick look at the web modules and list the ones that you feel are the most novel applications of **CRE**.

(d) Visit the problem-solving web site, *www.engin.umich.edu/~cre/probsolv/closed/cep.htm,* to find ways to "Get Unstuck" on a problem and to review the "Problem-Solving Algorithm." List four ways that might help you in your solutions to the home problems.

P1-2$_A$ (a) After reading each page or two ask yourself a question. Make a list of the four best questions for this chapter.

(b) Make a list of the five most important things you learned from this chapter.

P1-3$_A$ Visit the web site on Critical and Creative Thinking, *www.engin.umich.edu/~cre/probsolv/strategy/crit-n-creat.htm.*

(a) Write a paragraph describing what "critical thinking" is and how you can develop your critical thinking skills.

(b) Write a paragraph describing what "creative thinking" is and then list four things you will do during the next month that will increase your creative thinking skills.

Web Hint

(c) Write a question based on the material in this chapter that involves critical thinking and explain why it involves critical thinking.

(d) Repeat (c) for creative thinking.

(e) Brainstorm a list of ways you could work problems P-XX (to be specified by your instructor—e.g., Example E-1, or P1-15$_B$) incorrectly.

P1-4$_A$ Surf the CD-ROM and the web (*www.engin.umich.edu/~cre*). Go on a scavenger hunt using the summary notes for Chapter 1 on the CD-ROM.

(a) What Frequently Asked Question (FAQ) is not really frequently asked?

(b) What [Derive] hot button leads to a picture of a cobra?

(c) What [Self Test] hot button leads to a picture of a rabbit?

(d) What [Example] hot button leads to a picture of a hippo?

(e) Review the objectives for Chapter 1 in the Summary Notes on the CD-ROM. Write a paragraph in which you describe how well you feel you met these objectives. Discuss any difficulties you encountered and three ways (e.g., meet with professor, classmates) you plan to address removing these difficulties.

(f) Look at the Chemical Reactor section of the *Visual Encyclopedia of Equipment* on the CD-ROM. Write a paragraph describing what you learned.

(g) View the photos and schematics on the CD-ROM under Elements of Chemical Reaction Engineering—Chapter 1. Look at the quicktime videos. Write a paragraph describing two or more of the reactors. What similarities and differences do you observe between the reactors on the web (e.g., *www.loebequipment.com*), on the CD-ROM, and in the text? How do the used reactor prices compare with those in Table 1-1?

ICM Quiz Show

Mole Balance	Reactions	Rate Laws
100	100	100
200	200	200
300	300	300

P1-5$_A$ Load the Interactive Computer Module (ICM) from the CD-ROM. Run the module and then record your performance number for the module which indicates your mastery of the material.

ICM Kinetics Challenge 1 Performance # _____

P1-6$_B$ **Example 1-1** Calculate the volume of a CSTR for the conditions used to figure the plug-flow reactor volume in Example 1-1. Which volume is larger, the PFR or the CSTR? Explain why. Suggest two ways to work this problem incorrectly.

P1-7$_A$ Calculate the time to reduce the number of moles of A to 1% of its initial value in a constant-volume batch reactor for the reaction and data in Example 1-1.

P1-8$_A$ What assumptions were made in the derivation of the design equation for:

(a) the batch reactor?

(b) the CSTR?

(c) the plug-flow reactor (PFR)?

(d) the packed-bed reactor (PBR)?

(e) State in words the meanings of $-r_A$, $-r'_A$, and r'_A. Is the reaction rate $-r_A$ an extensive quantity? Explain.

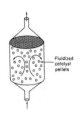

Fluidized catalyst pellets

P1-9$_A$ Use the mole balance to derive an equation analogous to Equation (1-7) for a fluidized CSTR containing catalyst particles in terms of the catalyst weight, W, and other appropriate terms. *Hint:* See margin figure.

P1-10$_A$ How can you convert the general mole balance equation for a given species, Equation (1-4), to a general mass balance equation for that species?

P1-11$_B$ We are going to consider the cell as a reactor. The nutrient corn steep liquor enters the cell of the microorganism *Penicillium chrysogenum* and is decomposed to form such products as amino acids, RNA, and DNA. Write an unsteady mass balance on (a) the corn steep liquor, (b) RNA, and (c) penicillin. Assume the cell is well mixed and that RNA remains inside the cell.

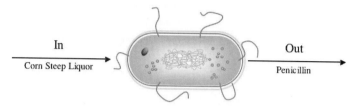

In

Corn Steep Liquor

Out

Penicillin

Penicillium chrysogenum

P1-12$_A$ The United States produced 32.5% of the world's chemical products in 2002 according to "Global Top 50," *Chemical and Engineering News*, July 28, 2003. Table P1-12.1 lists the 10 most produced chemicals in 2002.

TABLE P1-12.1.   CHEMICAL PRODUCTION

2002 Chemical	Thousands of Metric Tons	1995 Rank	2002 Chemical	Thousands of Metric Tons	1995 Rank
1. H_2SO_4	36,567	1	6. H_2	13,989	—
2. N_2	26,448	2	7. NH_3	13,171	6
3. C_2H_4	23,644	4	8. Cl_2	11,362	10
4. O_2	16,735	3	9. P_2O_5	10,789	—
5. C_3H_6	14,425	9	10. $C_2H_2Cl_2$	9,328	—

Reference: *Chemical and Engineering News*, July 7, 2003, *http://pubs.acs.org/cen/*

(a) What were the 10 most produced chemicals for the year that just ended? Were there any significant changes from the 1995 statistics? (See Chapter 1 of 3rd edition of *Elements of CRE.*) The same issue of *C&E News* ranks chemical companies as given in Table P1-12.2.

(b) What 10 companies were tops in sales for the year just ended? Did any significant changes occur compared to the 2002 statistics?

(c) Why do you think H_2SO_4 is the most produced chemical? What are some of its uses?

(d) What is the current annual production rate (lb/yr) of ethylene, ethylene oxide, and benzene?

(e) Why do you suspect there are so few organic chemicals in the top 10?

TABLE P1-12.2. TOP COMPANIES IN SALES

Rank 2002	Rank 2001	Rank 2000	Rank 1999	Rank 1995	Company	Chemical Sales [$ millions]
1	1	2	2	1	Dow Chemical	27,609
2	2	1	1	2	Dupont	26,728
3	3	3	3	3	ExxonMobil	16,408
4	5	5	6	6	General Electric	7,651
5	4	4	4	—	Huntsman Corp.	7,200
6	8	10	9	—	PPG Industries	5,996
7	9	8	10	—	Equistar Chemicals	5,537
8	7	7	—	—	Chevron Phillips	5,473
9	—	—	—	—	Eastman Chemical	5,320
10	—	—	—	—	Praxair	5,128

References:
Rank 2002: *Chemical and Engineering News,* May 12, 2003.
Rank 2001: *Chemical and Engineering News,* May 13, 2002.
Rank 2000: *Chemical and Engineering News,* May 7, 2001.
Rank 1999: *Chemical and Engineering News,* May 1, 2000.
http://pubs.acs-org/cen/

P1-13$_A$ Referring to the text material and the additional references on commercial reactors given at the end of this chapter, fill in Table P1-13.

TABLE P1.13   COMPARISON OF REACTOR TYPES

Type of Reactor	Characteristics	Kinds of Phases Present	Use	Advantages	Disadvantages
Batch					
CSTR					
PFR					
PBR					

Hall of Fame

P1-14$_B$ Schematic diagrams of the Los Angeles basin are shown in Figure P1-14. The basin floor covers approximately 700 square miles ($2 \times 10^{10}\ ft^2$) and is almost completely surrounded by mountain ranges. If one assumes an inversion height in the basin of 2000 ft, the corresponding volume of air in the basin is $4 \times 10^{13}\ ft^3$. We shall use this system volume to model the accumulation and depletion of air pollutants. As a very rough first approximation, we shall treat the Los Angeles basin as a well-mixed container (analogous to a CSTR) in which there are no spatial variations in pollutant concentrations.

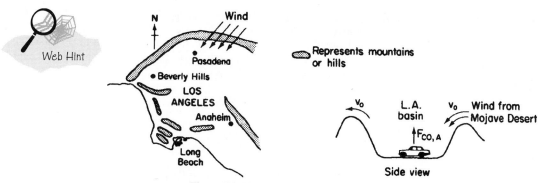

Figure P1-14   Schematic diagrams of the Los Angeles basin.

We shall perform an unsteady-state mole balance on CO as it is depleted from the basin area by a Santa Ana wind. Santa Ana winds are high-velocity winds that originate in the Mojave Desert just to the northeast of Los Angeles. Load the **Smog in Los Angeles Basin Web Module**. Use the data in the module to work part 1–14 (a) through (h) given in the module. Load the **living example polymath code** and explore the problem. For part (i), vary the parameters v_0, a, and b, and write a paragraph describing what you find.

There is heavier traffic in the L.A. basin in the mornings and in the evenings as workers go to and from work in downtown L.A. Consequently, the flow of CO into the L.A. basin might be better represented by the sine function over a 24-hour period.

P1-15$_B$ The reaction

$$A \longrightarrow B$$

is to be carried out isothermally in a continuous-flow reactor. Calculate both the CSTR and PFR reactor volumes necessary to consume 99% of A (i.e., $C_A = 0.01 C_{A0}$) when the entering molar flow rate is 5 mol/h, assuming the reaction rate $-r_A$ is:

(a) $-r_A = k$ with $k = 0.05 \dfrac{\text{mol}}{\text{h} \cdot \text{dm}^3}$ (*Ans.:* $V = 99 \text{ dm}^3$)

(b) $-r_A = kC_A$ with $k = 0.0001 \text{ s}^{-1}$

(c) $-r_A = kC_A^2$ with $k = 3 \dfrac{\text{dm}^3}{\text{mol} \cdot \text{h}}$ (*Ans.:* $V_{\text{CSTR}} = 66{,}000 \text{ dm}^3$)

The entering volumetric flow rate is 10 dm^3/h. (*Note:* $F_A = C_A v$. For a constant volumetric flow rate $v = v_0$, then $F_A = C_A v_0$. Also, $C_{A0} = F_{A0}/v_0 = [5 \text{ mol/h}]/[10 \text{ dm}^3/\text{h}] = 0.5 \text{ mol/dm}^3$.)

(d) Repeat **(a)**, **(b)**, and **(c)** to calculate the time necessary to consume 99.9% of species A in a 1000 dm^3 constant volume batch reactor with $C_{A0} = 0.5$ mol/dm^3.

P1-16$_B$ Write a one-paragraph summary of a journal article on chemical kinetics or reaction engineering. The article must have been published within the last five years. What did you learn from this article? Why is the article important?

P1-17$_B$ **(a)** There are initially 500 rabbits (x) and 200 foxes (y) on Farmer Oat's property. Use Polymath or MATLAB to plot the concentration of foxes and rabbits as a function of time for a period of up to 500 days. The predator–prey relationships are given by the following set of coupled ordinary differential equations:

$$\frac{dx}{dt} = k_1 x - k_2 x \cdot y$$

$$\frac{dy}{dt} = k_3 x \cdot y - k_4 y$$

Constant for growth of rabbits $k_1 = 0.02 \ day^{-1}$
Constant for death of rabbits $k_2 = 0.00004/(\text{day} \times \text{no. of foxes})$
Constant for growth of foxes after eating rabbits $k_3 = 0.0004/(\text{day} \times \text{no. of rabbits})$
Constant for death of foxes $k_4 = 0.04 \ day^{-1}$
What do your results look like for the case of $k_3 = 0.00004/(\text{day} \times \text{no. of rabbits})$ and $t_{\text{final}} = 800$ days? Also plot the number of foxes versus the number of rabbits. Explain why the curves look the way they do.
Vary the parameters k_1, k_2, k_3, and k_4. Discuss which parameters can or cannot be larger than others. Write a paragraph describing what you find.

(b) Use Polymath or MATLAB to solve the following set of nonlinear algebraic equations:

$$x^3 y - 4y^2 + 3x = 1$$

$$6y^2 - 9xy = 5$$

Polymath Tutorial
on CD-ROM

Summary Notes

with initial guesses of $x = 2$, $y = 2$. Try to become familiar with the edit keys in Polymath MATLAB. See the CD-ROM for instructions.
     Screen shots on how to run Polymath are shown at the end of Summary Notes for Chapter 1 on the CD-ROM and on the web.

P1-18$_C$ **What if:**
(a) the benzene feed stream in Example R1.3-1 in the PRS were not preheated by the product stream? What would be the consequences?
(b) you needed the cost of a 6000-gallon and a 15,000-gallon Pfaudler reactor? What would they be?
(c) only one operator showed up to run the nitrobenzene plant. What would be some of your first concerns?

P1-19$_A$ **Enrico Fermi (1901–1954) Problems (EEP).** Enrico Fermi was an Italian physicist who received the Nobel Prize for his work on nuclear processes. Fermi was famous for his "Back of the Envelope Order of Magnitude Calculation" to obtain an estimate of the answer through *logic* and making reasonable assumptions. He used a process to set bounds on the answer by saying it is probably larger than one number and smaller than another and arrived at an answer that was within a factor of 10.
http://mathforum.org/workshops/sum96/interdisc/sheila2.html
Enrico Fermi Problem (EFP) #1
How many piano tuners are there in the city of Chicago? Show the steps in your reasoning.
1. Population of Chicago _____
2. Number of people per household _____

Web Hint

    3.   Number of households _____

    4.   Households with pianos _____

    5.   Average number of tunes per year _____

    6.   Etc. _____

An answer is given on the web under Summary Notes for Chapter 1.

P1-20$_A$ EFP #2. How many square meters of pizza were eaten by an undergraduate student body population of 20,000 during the Fall term 2004?

P1-21$_B$ This problem will be used in each of the following chapters to help develop critical-thinking skills.

    (a) Write a question about this problem that involve critical thinking.

    (b) What generalizations can you make about the results of this problem?

    (c) Write a question that will expand this problem.

P1-22   New material for the 2nd printing the following changes/additions have been made to the 2nd printing.

NOTE TO INSTRUCTORS: Additional problems (cf. those from the preceding editions) can be found in the solutions manual and on the CD-ROM. These problems could be photocopied and used to help reinforce the fundamental principles discussed in this chapter.

Solved Problems

CDP1-A$_A$ Calculate the time to consume 80% of species A in a constant-volume batch reactor for a first- and a second-order reaction. (**Includes Solution**)

CDP1-B$_A$ Derive the differential mole balance equation for a foam reactor. [2nd Ed. P1-10$_B$]

SUPPLEMENTARY READING

1. For further elaboration of the development of the general balance equation, see not only the web site *www.engin.umich.edu/~cre* but also

    FELDER, R. M., and R. W. ROUSSEAU, *Elementary Principles of Chemical Processes*, 3rd ed. New York: Wiley, 2000, Chapter 4.

    HIMMELBLAU, D. M., and J. D. Riggs, *Basic Principles and Calculations in Chemical Engineering*, 7th ed. Upper Saddle River, N.J.: Prentice Hall, 2004, Chapters 2 and 6.

    SANDERS, R. J., *The Anatomy of Skiing.* Denver, CO: Golden Bell Press, 1976.

2. A detailed explanation of a number of topics in this chapter can be found in

    CRYNES, B. L., and H. S. FOGLER, eds., *AIChE Modular Instruction Series E: Kinetics*, Vols. 1 and 2. New York: AIChE, 1981.

3. An excellent description of the various types of commercial reactors used in industry is found in Chapter 11 of

    WALAS, S. M., *Reaction Kinetics for Chemical Engineers.* New York: McGraw-Hill, 1959.

4. A discussion of some of the most important industrial processes is presented by

    MEYERS, R.A., *Handbook of Chemicals Production Processes.* New York: McGraw-Hill, 1986.

    See also

    MCKETTA, J. J., ed., *Encyclopedia of Chemical Processes and Design.* New York: Marcel Dekker, 1976.

A similar book, which describes a larger number of processes, is

> Austin, G. T., *Shreve's Chemical Process Industries*, 5th ed. New York: McGraw-Hill, 1984.

5. The following journals may be useful in obtaining information on chemical reaction engineering: *International Journal of Chemical Kinetics, Journal of Catalysis, Journal of Applied Catalysis, AIChE Journal, Chemical Engineering Science, Canadian Journal of Chemical Engineering, Chemical Engineering Communications, Journal of Physical Chemistry,* and *Industrial and Engineering Chemistry Research.*

Links

6. The price of chemicals can be found in such journals as the *Chemical Marketing Reporter, Chemical Weekly,* and *Chemical Engineering News* and on the ACS web site *http://pubs.acs.org/cen.*

Conversion and Reactor Sizing   **2**

> Be more concerned with your character than with your
> reputation, because character is what you really are
> while reputation is merely what others think you are.
> John Wooden, coach, UCLA Bruins

Overview. In the first chapter, the general mole balance equation was derived and then applied to the four most common types of industrial reactors. A balance equation was developed for each reactor type and these equations are summarized in Table S-1. In Chapter 2, we will evaluate these equations to size CSTRs and PFRs. To size these reactors we first define conversion, which is a measure of the reaction's progress toward completion, and then rewrite all the balance equations in terms of conversion. These equations are often referred to as the design equations. Next, we show how one may size a reactor (i.e., determine the reactor volume necessary to achieve a specified conversion) once the relationship between the reaction rate, $-r_A$, and conversion, X, is known. In addition to being able to size CSTRs and PFRs once given $-r_A = f(X)$, another goal of this chapter is to compare CSTRs and PFRs and the overall conversions for reactors arranged in series. It is also important to arrive at the best arrangement of reactors in series.

After completing this chapter you will be able to size CSTRs and PFRs given the rate of reaction as a function of conversion and to calculate the overall conversion and reactor volumes for reactors arranged in series.

2.1 Definition of Conversion

In defining conversion, we choose one of the reactants as the basis of calcula-tion and then relate the other species involved in the reaction to this basis. In virtually all instances it is best to choose the limiting reactant as the basis of calculation. We develop the stoichiometric relationships and design equations by considering the general reaction

$$aA + bB \longrightarrow cC + dD \qquad (2\text{-}1)$$

The uppercase letters represent chemical species and the lowercase letters rep-resent stoichiometric coefficients. Taking species A as our *basis of calculation*, we divide the reaction expression through by the stoichiometric coefficient of species A, in order to arrange the reaction expression in the form

$$A + \frac{b}{a} B \longrightarrow \frac{c}{a} C + \frac{d}{a} D \qquad (2\text{-}2)$$

to put every quantity on a "per mole of A" basis, our limiting reactant.

Now we ask such questions as "How can we quantify how far a reaction [e.g., Equation (2-2)] proceeds to the right?" or "How many moles of C are formed for every mole A consumed?" A convenient way to answer these ques-tions is to define a parameter called *conversion*. The conversion X_A is the num-ber of moles of A that have reacted per mole of A fed to the system:

Definition of X

$$X_A = \frac{\text{Moles of A reacted}}{\text{Moles of A fed}}$$

Because we are defining conversion with respect to our basis of calculation [A in Equation (2-2)], we eliminate the subscript A for the sake of brevity and let $X \equiv X_A$. For irreversible reactions, the maximum conversion is 1.0, i.e., complete conversion. For reversible reactions, the maximum conversion is the equilibrium conversion X_e (i.e., $X_{max} = X_e$).

2.2 Batch Reactor Design Equations

In most batch reactors, the longer a reactant stays in the reactor, the more the reactant is converted to product until either equilibrium is reached or the reac-tant is exhausted. Consequently, in batch systems the conversion X is a func-tion of the time the reactants spend in the reactor. If N_{A0} is the number of moles of A initially in the reactor, then the total number of moles of A that have reacted after a time t is $[N_{A0}X]$

$$\begin{bmatrix} \text{Moles of A reacted (consumed)} \end{bmatrix} = \begin{bmatrix} \text{Moles of A fed} \end{bmatrix} \cdot \begin{bmatrix} \dfrac{\text{Moles of A reacted}}{\text{Moles of A fed}} \end{bmatrix}$$

$$\begin{bmatrix} \text{Moles of A} \\ \text{reacted} \\ \text{(consumed)} \end{bmatrix} = \quad [N_{A0}] \quad \cdot \quad [X] \qquad (2\text{-}3)$$

Now, the number of moles of A that remain in the reactor after a time t, N_A, can be expressed in terms of N_{A0} and X:

$$
\begin{bmatrix} \text{Moles of A} \\ \text{in reactor} \\ \text{at time } t \end{bmatrix} = \begin{bmatrix} \text{Moles of A} \\ \text{initially fed} \\ \text{to reactor at} \\ t = 0 \end{bmatrix} - \begin{bmatrix} \text{Moles of A that} \\ \text{have been con-} \\ \text{sumed by chemical} \\ \text{reaction} \end{bmatrix}
$$

$$[N_A] \quad = \quad [N_{A0}] \quad - \quad [N_{A0}X]$$

The number of moles of A in the reactor after a conversion X has been achieved is

$$\boxed{N_A = N_{A0} - N_{A0}X = N_{A0}(1 - X)} \tag{2-4}$$

When no spatial variations in reaction rate exist, the mole balance on species A for a batch system is given by the following equation [cf. Equation (1-5)]:

$$\frac{dN_A}{dt} = r_A V \tag{2-5}$$

This equation is valid whether or not the reactor volume is constant. In the general reaction, Equation (2-2), reactant A is disappearing; therefore, we multiply both sides of Equation (2-5) by –1 to obtain the mole balance for the batch reactor in the form

$$-\frac{dN_A}{dt} = (-r_A)V$$

The rate of disappearance of A, $-r_A$, in this reaction might be given by a rate law similar to Equation (1-2), such as $-r_A = kC_A C_B$.

For batch reactors, we are interested in determining how long to leave the reactants in the reactor to achieve a certain conversion X. To determine this length of time, we write the mole balance, Equation (2-5), in terms of conversion by differentiating Equation (2-4) with respect to time, remembering that N_{A0} is the number of moles of A initially present and is therefore a constant with respect to time.

$$\frac{dN_A}{dt} = 0 - N_{A0}\frac{dX}{dt}$$

Combining the above with Equation (2-5) yields

$$-N_{A0}\frac{dX}{dt} = r_A V$$

For a batch reactor, the design equation in differential form is

Batch reactor
design equation

$$\boxed{N_{A0}\frac{dX}{dt} = -r_A V} \tag{2-6}$$

We call Equation (2-6) the differential form of the **design equation** for a batch reactor because we have written the mole balance in terms of conversion. The differential forms of the batch reactor mole balances, Equations (2-5) and (2-6), are often used in the interpretation of reaction rate data (Chapter 5) and for reactors with heat effects (Chapter 9), respectively. Batch reactors are frequently used in industry for both gas-phase and liquid-phase reactions. The laboratory bomb calorimeter reactor is widely used for obtaining reaction rate data (see Section 9.3). Liquid-phase reactions are frequently carried out in batch reactors when small-scale production is desired or operating difficulties rule out the use of continuous flow systems.

For a constant-volume batch reactor, $V = V_0$, Equation (2-5) can be arranged into the form

$$\frac{1}{V_0}\frac{dN_A}{dt} = \frac{d(N_A/V_0)}{dt} = \frac{dC_A}{dt}$$

Constant-volume
batch reactor

(2-7)

$$\frac{dC_A}{dt} = r_A$$

As previously mentioned, the differential form of the mole balance, e.g., Equation (2-7), is used for analyzing rate data in a batch reactor as we will see in Chapters 5 and 9.

To determine the time to achieve a specified conversion X, we first separate the variables in Equation (2-6) as follows

$$dt = N_{A0}\frac{dX}{-r_A V}$$

(2-8)

Batch time t to achieve a conversion X

This equation is now integrated with the limits that the reaction begins at time equal zero where there is no conversion initially (i.e., $t = 0$, $X = 0$). Carrying out the integration, we obtain the time t necessary to achieve a conversion X in a batch reactor

$$\boxed{t = N_{A0}\int_0^X \frac{dX}{-r_A V}}$$

(2-9)

Batch Design Equation

The longer the reactants are left in the reactor, the greater will be the conversion. Equation (2-6) is the differential form of the design equation, and Equation (2-9) is the integral form of the design equation for a batch reactor.

2.3 Design Equations for Flow Reactors

For a batch reactor, we saw that conversion increases with time spent in the reactor. For continuous-flow systems, this time usually increases with increasing reactor volume, e.g., the bigger/longer the reactor, the more time it will take the reactants to flow completely through the reactor and thus, the more time to react. Consequently, the conversion X is a function of reactor volume V. If F_{A0} is the molar flow rate of species A fed to a system operated at steady state, the molar rate at which species A is reacting *within* the entire system will be $F_{A0}X$.

$$[F_{A0}] \cdot [X] = \frac{\text{Moles of A fed}}{\text{time}} \cdot \frac{\text{Moles of A reacted}}{\text{Mole of A fed}}$$

$$[F_{A0} \cdot X] = \frac{\text{Moles of A reacted}}{\text{time}}$$

The molar feed rate of A *to* the system *minus* the rate of reaction of A within the system *equals* the molar flow rate of A leaving the system F_A. The preceding sentence can be written in the form of the following mathematical statement:

$$\begin{bmatrix} \text{Molar flow rate} \\ \text{at which A is} \\ \text{fed to the system} \end{bmatrix} - \begin{bmatrix} \text{Molar rate at} \\ \text{which A is} \\ \text{consumed within} \\ \text{the system} \end{bmatrix} = \begin{bmatrix} \text{Molar flow rate} \\ \text{at which A leaves} \\ \text{the system} \end{bmatrix}$$

$$[F_{A0}] \qquad - \qquad [F_{A0}X] \qquad = \qquad [F_A]$$

Rearranging gives

$$\boxed{F_A = F_{A0}(1 - X)} \tag{2-10}$$

The entering molar flow rate of species A, F_{A0} (mol/s), is just the product of the entering concentration, C_{A0} (mol/dm^3), and the entering volumetric flow rate, v_0 (dm^3/s):

Liquid phase
$$F_{A0} = C_{A0} v_0$$

For liquid systems, C_{A0} is commonly given in terms of molarity, for example,

$$C_{A0} = 2 \text{ mol/dm}^3$$

For gas systems, C_{A0} can be calculated from the entering temperature and pressure using the ideal gas law or some other gas law. For an ideal gas (see Appendix B):

Gas phase
$$C_{A0} = \frac{P_{A0}}{RT_0} = \frac{y_{A0}P_0}{RT_0}$$

The entering molar flow rate is

$$F_{A0} = v_0 C_{A0} = v_0 \frac{y_{A0} P_0}{RT_0}$$

where C_{A0} = entering concentration, mol/dm³

y_{A0} = entering mole fraction of A

P_0 = entering total pressure, e.g., kPa

$P_{A0} = y_{A0} P_0$ = entering partial pressure of A, e.g., kPa

T_0 = entering temperature, K

$$R = \text{ideal gas constant} \left(\text{e.g., } R = 8.314 \frac{\text{kPa} \cdot \text{dm}^3}{\text{mol} \cdot K}; \text{ see Appendix B} \right)$$

The size of the reactor will depend on the flow rate, reaction kinetics, reactor conditions, and desired conversion. Let's first calculate the entering molar flow rate.

Example 2–1   Using the Ideal Gas Law to Calculate C_{A0} and F_{A0}

A gas of pure A at 830 kPa (8.2 atm) enters a reactor with a volumetric flow rate, v_0, of 2 dm³/s at 500 K. Calculate the entering concentration of A, C_{A0}, and the entering molar flow rate, F_{A0}.

Solution

We again recall that for an ideal gas:

$$C_{A0} = \frac{P_{A0}}{RT_0} = \frac{y_{A0} P_0}{RT_0} \qquad \text{(E2-1.1)}$$

where P_0 = 830 kPa (8.2 atm)

y_{A0} = 1.0 (Pure A)

T_0 = initial temperature = 500K

R = 8.314 dm³·kPa/mol·K   (Appendix B)

Substituting the given parameter values into Equation (E2-1.1) yields

$$C_{A0} = \frac{(1)(830 \text{ kPa})}{(8.314 \text{ dm}^3 \cdot \text{kPa/mol} \cdot \text{K})(500\text{K})} = 0.20 \frac{\text{mol}}{\text{dm}^3}$$

We could also solve for the partial pressure in terms of the concentration:

$$P_{A0} = C_{A0} R T_0 \qquad \text{(E2-1.2)}$$

However, since pure A enters, the total pressure and partial pressure entering are the same. The entering molar flow rate, F_{A0}, is just the product of the entering concentration, C_{A0}, and the entering volumetric flow rate, v_0:

$$F_{A0} = C_{A0}v_0 = (0.2 \text{ mol/dm}^3)(2 \text{ dm}^3/\text{s}) = (0.4 \text{ mol/s})$$

$$\boxed{F_{A0} = 0.4 \text{ mol/s}}$$

This feed rate ($F_{A0} = 0.4$ mol/s) is in the range of that which is necessary to form several million pounds of product per year. We will use this value of F_{A0} together with either Table 2-2 or Figure 2-1 to size and evaluate a number of reactor schemes in Examples 2-2 through 2-5.

Now that we have a relationship [Equation (2-10)] between the molar flow rate and conversion, it is possible to express the design equations (i.e., mole balances) in terms of conversion for the *flow* reactors examined in Chapter 1.

2.3.1 CSTR (also known as a Backmix Reactor or Vat)

Recall that the CSTR is modeled as being well mixed such that there are no spatial variations in the reactor. The CSTR mole balance, Equation (1-7), when applied to species A in the reaction

$$A + \frac{b}{a} B \longrightarrow \frac{c}{a} C + \frac{d}{a} D \tag{2-2}$$

can be arranged to

$$V = \frac{F_{A0} - F_A}{-r_A} \tag{2-11}$$

We now substitute for F_A in terms of F_{A0} and X

$$F_A = F_{A0} - F_{A0}X \tag{2-12}$$

and then substitute Equation (2-12) into (2-11)

$$V = \frac{F_{A0} - (F_{A0} - F_{A0}X)}{-r_A}$$

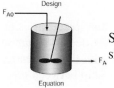

Design
Equation
Perfect mixing

Simplifying, we see the CSTR volume necessary to achieve a specified conversion X is

$$\boxed{V = \frac{F_{A0}X}{(-r_A)_{\text{exit}}}} \tag{2-13}$$

Evaluate $-r_A$ at CSTR exit. Because the reactor is *perfectly mixed*, the exit composition from the reactor is identical to the composition inside the reactor, and the rate of reaction is evaluated at the exit conditions.

2.3.2 Tubular Flow Reactor (PFR)

We model the tubular reactor as having the fluid flowing in plug flow, i.e., no radial gradients in concentration, temperature, or reaction rate.[1] As the reactants enter and flow axially down the reactor, they are consumed and the conversion increases along the length of the reactor. To develop the PFR design equation we first multiply both sides of the tubular reactor design equation (1-12) by -1. We then express the mole balance equation for species A in the reaction as

$$\frac{-dF_A}{dV} = -r_A \qquad (2\text{-}14)$$

For a flow system, F_A has previously been given in terms of the entering molar flow rate F_{A0} and the conversion X

$$F_A = F_{A0} - F_{A0}X \qquad (2\text{-}12)$$

differentiating

$$dF_A = -F_{A0}dX$$

and substituting into (2-14), gives the differential form of the design equation for a plug-flow reactor (PFR):

$$\boxed{F_{A0}\frac{dX}{dV} = -r_A} \qquad (2\text{-}15)$$

Design PFR equation

We now separate the variables and integrate with the limits $V = 0$ when $X = 0$ to obtain the plug-flow reactor volume necessary to achieve a specified conversion X:

$$\boxed{V = F_{A0} \int_0^X \frac{dX}{-r_A}} \qquad (2\text{-}16)$$

To carry out the integrations in the batch and plug-flow reactor design equations (2-9) and (2-16), as well as to evaluate the CSTR design equation (2-13), we need to know how the reaction rate $-r_A$ varies with the concentration (hence conversion) of the reacting species. This relationship between reaction rate and concentration is developed in Chapter 3.

---

[1] This constraint will be removed when we extend our analysis to nonideal (industrial) reactors in Chapters 13 and 14.

2.3.3 Packed-Bed Reactor

Packed-bed reactors are tubular reactors filled with catalyst particles. The derivation of the differential and integral forms of the design equations for packed-bed reactors are analogous to those for a PFR [cf. Equations (2-15) and (2-16)]. That is, substituting Equation (2-12) for F_A in Equation (1-15) gives

PBR design equation

$$\boxed{F_{A0}\,\frac{dX}{dW} = -r_A'}$$

(2-17)

The differential form of the design equation [i.e., Equation (2-17)] **must** be used when analyzing reactors that have a pressure drop along the length of the reactor. We discuss pressure drop in packed-bed reactors in Chapter 4.

In the *absence* of pressure drop, i.e., $\Delta P = 0$, we can integrate (2-17) with limits $X = 0$ at $W = 0$ to obtain

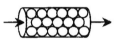

$$\boxed{W = F_{A0} \int_0^X \frac{dX}{-r_A'}}$$

(2-18)

Equation (2-18) can be used to determine the catalyst weight W necessary to achieve a conversion X when the total pressure remains constant.

2.4 Applications of the Design Equations for Continuous-Flow Reactors

In this section, we are going to show how we can size CSTRs and PFRs (i.e., determine their reactor volumes) from knowledge of the rate of reaction, $-r_A$, as a function of conversion, X. The rate of disappearance of A, $-r_A$, is almost always a function of the concentrations of the various species present. When only one reaction is occurring, each of the concentrations can be expressed as a function of the conversion X (see Chapter 3); consequently, $-r_A$ can be expressed as a function of X.

A particularly simple functional dependence, yet one that occurs often, is the first-order dependence

$$-r_A = kC_A = kC_{A0}(1 - X)$$

Here, k is the specific reaction rate and is a function only of temperature, and C_{A0} is the entering concentration. We note in Equations (2-13) and (2-16) the reactor volume in a function of the reciprocal of $-r_A$. For this first-order dependence, a plot of the reciprocal rate of reaction $(1/-r_A)$ as a function of conversion yields a curve similar to the one shown in Figure 2-1, where

$$\frac{1}{-r_A} = \frac{1}{kC_{A0}}\left(\frac{1}{1-X}\right)$$

To illustrate the design of a series of reactors, we consider the isothermal gas-phase isomerization

$$A \longrightarrow B$$

We are going to the laboratory to determine the rate of chemical reaction as a function of the conversion of reactant A. The laboratory measurements given in Table 2-1 show the chemical reaction rate as a function of conversion. The temperature was 500 K (440°F), the total pressure was 830 kPa (8.2 atm), and the initial charge to the reactor was pure A.

TABLE 2-1   RAW DATA

X	$-r_A$ (mol/m$^3 \cdot$ s)
0	0.45
0.1	0.37
0.2	0.30
0.4	0.195
0.6	0.113
0.7	0.079
0.8	0.05

If we know $-r_A$ as a function of X, we can size any isothermal reaction system.

Recalling the CSTR and PFR design equations, (2-13) and (2-16), we see that the reactor volume varies with the reciprocal of $-r_A$, $(1/-r_A)$, e.g.,

$$V = \left(\frac{1}{-r_A}\right)(F_{A0}X)\,.$$ Consequently, to size reactors, we convert the rate data in Table 2-1 to reciprocal rates, $(1/-r_A)$, in Table 2-2.

TABLE 2-2   PROCESSED DATA −1

X	0.0	0.1	0.2	0.4	0.6	0.7	0.8
$-r_A\left(\dfrac{\text{mol}}{\text{m}^3 \cdot \text{s}}\right)$	0.45	0.37	0.30	0.195	0.113	0.079	0.05
$(1/-r_A)\left(\dfrac{\text{m}^3 \cdot \text{s}}{\text{mol}}\right)$	2.22	2.70	3.33	5.13	8.85	12.7	20

These data are used to arrive at a plot of $(1/-r_A)$ as a function of X, shown in Figure 2-1.

We can use this figure to size flow reactors for different entering molar flow rates. Before sizing flow reactors let's first consider some insights. If a

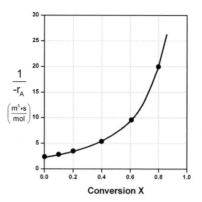

Figure 2-1 Processed data –1.

reaction is carried out isothermally, the rate is usually greatest at the start of the reaction when the concentration of reactant is greatest (i.e., when there is negligible conversion [$X \cong 0$]). Hence ($1/-r_A$) will be small. Near the end of the reaction, when the reactant has been mostly used up and thus the concentration of A is small (i.e., the conversion is large), the reaction rate will be small. Consequently, ($1/-r_A$) is large.

For all irreversible reactions of greater than zero order (see Chapter 3 for zero-order reactions), as we approach complete conversion where all the limiting reactant is used up, i.e., $X = 1$, the reciprocal rate approaches infinity as does the reactor volume, i.e.

$$A \rightarrow B + C \qquad \text{As } X \rightarrow 1, -r_A \rightarrow 0 \text{ , thus, } \frac{1}{-r_A} \rightarrow \infty \text{ therefore } V \rightarrow \infty$$

"To infinity and beyond"
—Buzz Lightyear

Consequently, we see that an infinite reactor volume is necessary to reach complete conversion, $X = 1.0$

For reversible reactions (e.g., A $\rightleftarrows$ B), the maximum conversion is the equilibrium conversion X_e. At equilibrium, the reaction rate is zero ($r_A \equiv 0$). Therefore,

$$A \rightleftarrows B + C \qquad \text{As } X \rightarrow X_e, -r_A \rightarrow 0 \text{ , thus, } \frac{1}{-r_A} \rightarrow \infty \text{ and therefore } V \rightarrow \infty$$

and we see that an infinite reactor volume would also be necessary to obtain the exact equilibrium conversion, $X = X_e$.

To size a number of reactors for the reaction we have been considering, we will use $F_{A0} = 0.4$ mol/s (calculated in Example 2-1) to add another row to the processed data shown in Table 2-2 to obtain Table 2-3.

TABLE 2-3   PROCESSED DATA –2

X	0.0	0.1	0.2	0.4	0.6	0.7	0.8
$-r_A\left(\dfrac{mol}{m^3 \cdot s}\right)$	0.45	0.37	0.30	0.195	0.113	0.079	0.05
$(1/-r_A)\left(\dfrac{m^3 \cdot s}{mol}\right)$	2.22	2.70	3.33	5.13	8.85	12.7	20
$[F_{A0}/-r_A](m^3)$	0.89	1.08	1.33	2.05	3.54	5.06	8.0

We shall use the data
in this table for the
next five Example
Problems.

Plotting $\left(\dfrac{F_{A0}}{-r_A}\right)$ as a function of X using the data in Table 2-3 we obtain the plot shown in Figure 2.2.

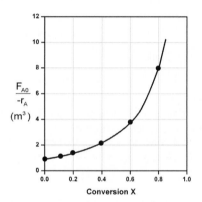

Levenspiel plot

Figure 2-2   Levenspiel plot of processed data –2.

Example 2–2   Sizing a CSTR

The reaction described by the data in Table 2-2

$$A \rightarrow B$$

is to be carried out in a CSTR. Species A enters the reactor at a molar flow rate of 0.4 mol/s.

(a) Using the data in either Table 2-2, Table 2-3, or Figure 2-1, calculate the volume necessary to achieve 80% conversion in a CSTR.

(b) Shade the area in Figure 2-2 that would give the CSTR volume necessary to achieve 80% conversion.

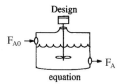

Design

equation

Solutions

(a) Equation (2-13) gives the volume of a CSTR as a function of F_{A0}, X, and $-r_A$:

$$V = \frac{F_{A0}X}{(-r_A)_{exit}} \tag{2-13}$$

In a CSTR, the composition, temperature, and conversion of the effluent stream are identical to that of the fluid within the reactor, because perfect mixing is assumed. Therefore, we need to find the value of $-r_A$ (or reciprocal thereof) at $X = 0.8$. From either Table 2-2 or Figure 2-1, we see that when $X = 0.8$, then

$$\left(\frac{1}{-r_A}\right)_{X=0.8} = 20\frac{m^3 \cdot s}{mol}$$

Substitution into Equation (2-13) for an entering molar flow rate, F_{A0}, of 0.4 mol A/s and $X = 0.8$ gives

$$V = 0.4\frac{mol}{s}\left(\frac{20\ m^3 \cdot s}{mol}\right)(0.8) = 6.4\ m^3 \tag{E2-2.1}$$

$$V = 6.4\ m^3 = 6400\ dm^3 = 6400l$$

(b) Shade the area in Figure 2-2 that yields the CSTR volume. Rearranging Equation (2-13) gives

$$V = \left[\frac{F_{A0}}{-r_A}\right]X \tag{2-13}$$

In Figure E2-2.1, the volume is equal to the area of a rectangle with a height $(F_{A0}/-r_A = 8\ m^3)$ and a base $(X = 0.8)$. This rectangle is shaded in the figure.

$$V = \left[\frac{F_{A0}}{-r_A}\right]_{X=0.8}(0.8) \tag{E2-2.2}$$

V = Levenspiel rectangle area = height × width

$$V = [8\ m^3][0.8] = 6.4\ m^3$$

The CSTR volume necessary to achieve 80% conversion is 6.4 m³ when operated at 500 K, 830 kPa (8.2 atm), and with an entering molar flow rate of A of 0.4 mol/s. This volume corresponds to a reactor about 1.5 m in diameter and 3.6 m high. It's a large CSTR, but this is a gas-phase reaction, and CSTRs are normally not used for gas-phase reactions. CSTRs are used primarily for liquid-phase reactions.

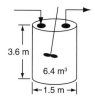

3.6 m

6.4 m³

1.5 m

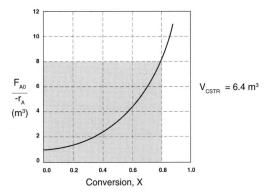

Figure E2-2.1   Levenspiel CSTR plot.

Plots of $1/-r_A$ vs. X
are sometimes
referred to as
Levenspiel plots
(after Octave
Levenspiel)

Example 2–3   Sizing a PFR

The reaction described by the data in Tables 2-1 and 2-2 is to be carried out in a PFR. The entering molar flow rate of A is 0.4 mol/s.

(a) First, use one of the integration formulas given in Appendix A.4 to determine the PFR reactor volume necessary to achieve 80% conversion.

(b) Next, shade the area in Figure 2-2 that would give the PFR the volume necessary to achieve 80% conversion.

(c) Finally, make a qualitative sketch of the conversion, X, and the rate of reaction, $-r_A$, down the length (volume) of the reactor.

Solution

We start by repeating rows (1) and (4) of Table 2-3.

TABLE 2-3   PROCESSED DATA –2

X	0.0	0.1	0.2	0.4	0.6	0.7	0.8
$[F_{A0}/-r_A](m^3)$	0.89	1.08	1.33	2.05	3.54	5.06	8.0

(a) For the PFR, the differential form of the mole balance is

$$F_{A0} \frac{dX}{dV} = -r_A \tag{2-15}$$

Rearranging and integrating gives

$$V = F_{A0} \int_0^{0.8} \frac{dX}{-r_A} = \int_0^{0.8} \frac{F_{A0}}{-r_A} dX \tag{2-16}$$

We shall use the *five point quadrature* formula (A-23) given in Appendix A.4 to numerically evaluate Equation (2-16). For the five-point formula with a final conversion of 0.8, gives for four equal segments between $X = 0$ and $X = 0.8$ with a segment length of $\Delta X = \dfrac{0.8}{4} = 0.2$. The function inside the integral is evaluated at $X = 0$, $X = 0.2$, $X = 0.4$, $X = 0.6$, and $X = 0.8$.

$$V = \frac{\Delta X}{3}\left[\frac{F_{A0}}{-r_A(X=0)} + \frac{4F_{A0}}{-r_A(X=0.2)} + \frac{2F_{A0}}{-r_A(X=0.4)} + \frac{4F_{A0}}{-r_A(X=0.6)} + \frac{F_{A0}}{-r_A(X=0.8)}\right]$$

Using values of $F_{A0}/(-r_A)$ in Table 2-3 yields

$$V = \left(\frac{0.2}{3}\right)[0.89 + 4(1.33) + 2(2.05) + 4(3.54) + 8.0]\,\text{dm}^3 = \left(\frac{0.2}{3}\right)(32.47\ \text{m}^3)$$

$$\boxed{V = 2.165\ \text{m}^3 = 2165\ \text{dm}^3}$$

The PFR reactor volume necessary to achieve 80% conversion is 2165 dm³. This volume could result from a bank of 100 PFRs that are each 0.1 m in diameter with a length of 2.8 m (e.g., see Figures 1-8(a) and (b)).

(b) The integral in Equation (2-16) can also be evaluated from the area under the curve of a plot of $(F_{A0}/-r_A)$ versus X.

$$V = \int_0^{0.8} \frac{F_{A0}}{-r_A}\,dX = \text{Area under the curve between } X = 0 \text{ and } X = 0.8$$
$$\text{(see appropriate shaded area in Figure E2-3.1)}$$

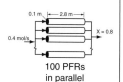

100 PFRs
in parallel

PFR

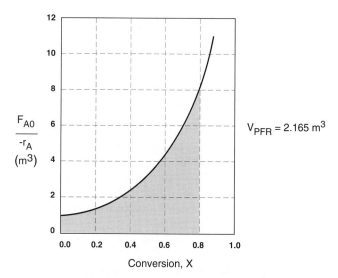

Figure E2-3.1   Levenspiel PFR plot.

The area under the curve will give the tubular reactor volume necessary to achieve the specified conversion of A. For 80% conversion, the shaded area is roughly equal to 2165 dm³ (2.165 m³).
(c) Sketch the profiles of $-r_A$ and X down the length of the reactor.

Solution

We know that as we proceed down the reactor and more and more of the reactant is consumed, the concentration of reactant decreases, as does the rate of disappearance of A. However, the conversion increases as more and more reactant is converted to product. For $X = 0.2$, we calculate the corresponding reactor volume using Simpson's rule (given in Appendix A.4 as Equation [A-21]) with $\Delta X = 0.1$ and the data in rows 1 and 4 in Table 2-3,

$$V = F_{A0} \int_0^{0.2} \frac{dX}{-r_A} = \frac{\Delta X}{3}\left[\frac{F_{A0}}{-r_A(X=0)} + \frac{4F_{A0}}{-r_A(X=0.1)} + \frac{F_{A0}}{-r_A(X=0.2)}\right]$$

$$= \left[\frac{0.1}{3}[0.89 + 4(1.08) + 1.33]\right]m^3 = \frac{0.1}{3}(6.54 \text{ m}^3) = 0.218 \text{ m}^3 = 218 \text{ dm}^3$$

$$= 218 \text{ dm}^3$$

For $X = 0.4$, we can again use Simpson's rule with $\Delta X = 0.2$ to find the reactor volume necessary for a conversion of 40%.

$$V = \frac{\Delta X}{3}\left[\frac{F_{A0}}{-r_A(X=0)} + \frac{4F_{A0}}{-r_A(X=0.2)} + \frac{F_{A0}}{-r_A(X=0.4)}\right]$$

$$= \left[\frac{0.2}{3}[0.89 + 4(1.33) + 2.05]\right]m^3 = 0.551 \text{ m}^3$$

$$= 551 \text{ dm}^3$$

We can continue in this manner to arrive at Table E2-3.1.

TABLE E2-3.1.   CONVERSION AND REACTION RATE PROFILES

X	0	0.2	0.4	0.6	0.8
$-r_A \left(\dfrac{mol}{m^3 \cdot s}\right)$	0.45	0.30	0.195	0.113	0.05
V (dm³)	0	218	551	1093	2165

The data in Table E2-3.1 are plotted in Figures E2-3.2 (a) and (b).
    One observes that the reaction rate, $-r_A$, decreases as we move down the reactor while the conversion increases. These plots are typical for reactors operated isothermally.

For isothermal reactions, the conversion increases and the rate decreases as we move down the PFR.

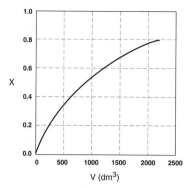

Figure E2-3.2(a)   Conversion profile.

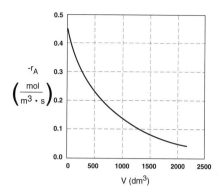

Figure E2-3.2(b)   Reaction rate profile.

Example 2–4   Comparing CSTR and PFR Sizes

It is interesting to compare the volumes of a CSTR and a PFR required for the same job. To make this comparison, we shall use the data in Figure 2-2 to learn which reactor would require the smaller volume to achieve a conversion of 80%: a CSTR or a PFR. The entering molar flow rate $F_{A0} = 0.4$ mol/s, and the feed conditions are the same in both cases.

Solution

TABLE 2-3   PROCESSED DATA –2

X	0.0	0.1	0.2	0.4	0.6	0.7	0.8
$[F_{A0}/-r_A](m^3)$	0.89	1.08	1.33	2.05	3.54	5.06	8.0

The CSTR volume was 6.4 m^3 and the PFR volume was 2.165 m^3. When we combine Figures E2-2.1 and E2-3.1 on the same graph, we see that the crosshatched area above the curve is the difference in the CSTR and PFR reactor volumes.

   For isothermal reactions greater than zero order (see Chapter 3 for zero order), the CSTR volume will usually be greater than the PFR volume for the same conversion and reaction conditions (temperature, flow rate, etc.).

   We see that the reason the isothermal CSTR volume is usually greater than the PFR volume is that the CSTR is always operating at the lowest reaction rate (e.g., $-r_A = 0.05$ in Figure E2-4.1(b)). The PFR on the other hand starts at a high rate at the entrance and gradually decreases to the exit rate, thereby requiring less volume because the volume is inversely proportional to the rate. However, for autocatalytic reactions, product-inhibited reactions, and nonisothermal exothermic reactions, these trends will not always be the case, as we will see in Chapters 7 and 8.

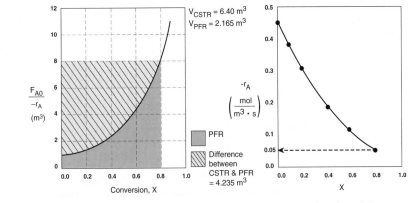

Figure E2-4.1(a) Comparison of CSTR and   **(b)** $-r_A$ as a function of X.
PFR reactor sizes.

2.5 Reactors in Series

Many times, reactors are connected in series so that the exit stream of one reactor is the feed stream for another reactor. When this arrangement is used, it is often possible to speed calculations by defining conversion in terms of location at a point downstream rather than with respect to any single reactor. That is, the conversion X is the *total number of moles* of A that have reacted up to that point per mole of A fed to the *first* reactor.

**Only valid for
NO side streams**

For reactors in series

$$X_i = \frac{\text{Total moles of A reacted up to point } i}{\text{Moles of A fed to the first reactor}}$$

However, this definition can *only* be used when the feed stream only enters the first reactor in the series and there are *no* side streams either fed or withdrawn. The molar flow rate of A at point i is equal to moles of A fed to the first reactor minus all the moles of A reacted up to point i:

$$F_{Ai} = F_{A0} - F_{A0}X_i$$

For the reactors shown in Figure 2-3, X_1 at point $i = 1$ is the conversion achieved in the PFR, X_2 at point $i = 2$ is the total conversion achieved at this point in the PFR and the CSTR, and X_3 is the total conversion achieved by all three reactors.

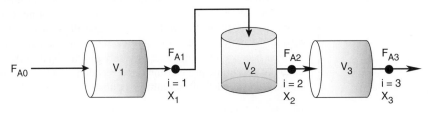

Figure 2-3 Reactors in series.

To demonstrate these ideas, let us consider three different schemes of reactors in series: two CSTRs, two PFRs, and then a combination of PFRs and CSTRs in series. To size these reactors, we shall use laboratory data that gives the reaction rate at different conversions.

2.5.1 CSTRs in Series

The first scheme to be considered is the two CSTRs in series shown in Figure 2-4.

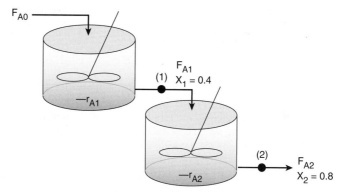

Figure 2-4   Two CSTRs in series.

For the first reactor, the rate of disappearance of A is $-r_{A1}$ at conversion X_1.
    A mole balance on reactor 1 gives

$$\text{In} - \text{Out} + \text{Generation} = 0$$

Reactor 1: $\qquad F_{A0} - F_{A1} + r_{A1}V_1 = 0 \qquad\qquad$ (2-19)

The molar flow rate of A at point 1 is

$$F_{A1} = F_{A0} - F_{A0}X_1 \qquad\qquad (2\text{-}20)$$

Combining Equations (2-19) and (2-20) or rearranging

Reactor 1

$$\boxed{V_1 = F_{A0}\left(\frac{1}{-r_{A1}}\right)X_1} \qquad\qquad (2\text{-}21)$$

In the second reactor, the rate of disappearance of A, $-r_{A2}$, is evaluated at the conversion of the exit stream of reactor 2, X_2. A mole balance on the second reactor

$$\text{In} - \text{Out} + \text{Generation} = 0$$

Reactor 2: $\qquad F_{A1} - F_{A2} + r_{A2}V = 0 \qquad\qquad$ (2-22)

The molar flow rate of A at point 2 is

$$F_{A2} = F_{A0} - F_{A0}X_2 \qquad\qquad (2\text{-}23)$$

Combining and rearranging

$$V_2 = \frac{F_{A1} - F_{A2}}{-r_{A2}} = \frac{(F_{A0} - F_{A0}X_1) - (F_{A0} - F_{A0}X_2)}{-r_{A2}}$$

Reactor 2

$$\boxed{V_2 = \frac{F_{A0}}{-r_{A2}}(X_2 - X_1)} \qquad (2\text{-}24)$$

For the second CSTR recall that $-r_{A2}$ is evaluated at X_2 and then use $(X_2 - X_1)$ to calculate V_2 at X_2.

In the examples that follow, we shall use the molar flow rate of A we calculated in Example 2-1 (0.4 mol A/s) and the reaction conditions given in Table 2-3.

Example 2–5 Comparing Volumes for CSTRs in Series

For the two CSTRs in series, 40% conversion is achieved in the first reactor. What is the volume of each of the two reactors necessary to achieve 80% overall conversion of the entering species A?

TABLE 2-3   PROCESSED DATA –2

X	0.0	0.1	0.2	0.4	0.6	0.7	0.8
$[F_{A0}/-r_A](m^3)$	0.89	1.09	1.33	2.05	3.54	5.06	8.0

Solution

For reactor 1, we observe from either Table 2-3 or Figure E2-5.1 that when $X = 0.4$, then

$$\left(\frac{F_{A0}}{-r_{A1}}\right)_{X=0.4} = 2.05 \text{ m}^3$$

then

$$V_1 = \left(\frac{F_{A0}}{-r_{A1}}\right)_{X_1} X_1 = \left(\frac{F_{A0}}{-r_{A1}}\right)_{0.4} X_1 = (2.05)(0.4) = 0.82 \text{ m}^3 = 820 \text{ dm}^3$$

For reactor 2, when $X_2 = 0.8$, then $= \left(\frac{F_{A0}}{-r_A}\right)_{X=0.8} = 8.0 \text{ m}^3$

$$V_2 = \left(\frac{F_{A0}}{-r_{A2}}\right)(X_2 - X_1) \qquad (2\text{-}24)$$

$$V_2 = (8.0 \text{ m}^3)(0.8 - 0.4) = 3.2 \text{ m}^3 = 3200 \text{ dm}^3$$

$$V_2 = 3200 \text{ dm}^3 \text{ (liters)}$$

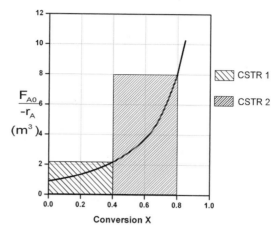

Figure E2-5.1   Two CSTRs in series.

To achieve the
same overall
conversion, the
total volume for
two CSTRs in
series is less than
that required
for one CSTR.

Note again that for CSTRs in series the rate $-r_{A1}$ is evaluated at a conversion of 0.4 and rate $-r_{A2}$ is evaluated at a conversion of 0.8. The total volume for these two reactors in series is

$$V = V_1 + V_2 = 0.82 \text{ m}^3 + 3.2 \text{ m}^3 = 4.02 \text{ m}^3 = 4020 \text{ dm}^3$$

We need only
$-r_A = f(x)$ and
F_{A0} to size
reactors.

By comparison, the volume necessary to achieve 80% conversion in **one** CSTR is

$$V = \left(\frac{F_{A0}}{-r_{A1}}\right) X = (8.0)(0.8) = 6.4 \text{ m}^3 = 6400 \text{ dm}^3$$

Notice in Example 2-5 that the sum of the two CSTR reactor volumes (4.02 m³) in series is less than the volume of one CSTR (6.4 m³) to achieve the same conversion.

Approximating a PFR by a large number of CSTRs in series

Consider approximating a PFR with a number of small, equal-volume CSTRs of V_i in series (Figure 2-5). We want to compare the *total volume* of all the CSTRs with the volume of one plug-flow reactor for the same conversion, say 80%.

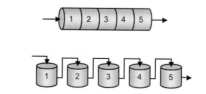

Figure 2-5   Modeling a PFR with CSTRs in series.

The fact that we can
model a PFR with a
large number of
CSTRs is an
important result.

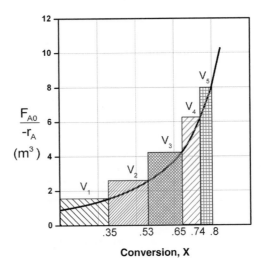

Figure 2-6   Levenspiel plot showing comparison of CSTRs in series with one PFR.

From Figure 2-6, we note a very important observation! The total volume to achieve 80% conversion for five CSTRs of equal volume in series is roughly the same as the volume of a PFR. As we make the volume of each CSTR smaller and increase the number of CSTRs, the total volume of the CSTRs in series and the volume of the PFR will become identical. *That is, we can model a PFR with a large number of CSTRs in series.* This concept of using many CSTRs in series to model a PFR will be used later in a number of situations, such as modeling catalyst decay in packed-bed reactors or transient heat effects in PFRs.

2.5.2 PFRs in Series

We saw that two CSTRs in series gave a smaller total volume than a single CSTR to achieve the same conversion. This case does not hold true for the two plug-flow reactors connected in series shown in Figure 2-7.

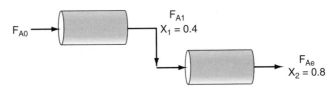

Figure 2-7 Two PFRs in series.

PRF in series   We can see from Figure 2-8 and from the following equation

$$\int_0^{X_2} F_{A0}\frac{dX}{-r_A} \equiv \int_0^{X_1} F_{A0}\frac{dX}{-r_A} + \int_{X_1}^{X_2} F_{A0}\frac{dX}{-r_A}$$

that it is immaterial whether you place two plug-flow reactors in series or have one continuous plug-flow reactor; the total reactor volume required to achieve the same conversion is identical!

The overall conversion of two PRFs in series is the same as one PRF with the same total volume.

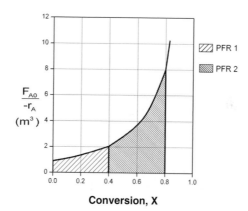

Figure 2-8   Levenspiel plot for two PFRs in series.

Example 2–6 Sizing Plug-Flow Reactors in Series

Using either the data in Table 2-3 or Figure 2-2, calculate the reactor volumes V_1 and V_2 for the plug-flow sequence shown in Figure 2-7 when the intermediate conversion is 40% and the final conversion is 80%. The entering molar flow rate is the same as in the previous examples, 0.4 mol/s.

Solution

TABLE 2-3   PROCESSED DATA –2

X	0.0	0.1	0.2	0.4	0.6	0.7	0.8
$[F_{A0}/-r_A](m^3)$	0.89	1.08	1.33	2.05	3.54	5.06	8.0

In addition to graphical integration, we could have used numerical methods to size the plug-flow reactors. In this example, we shall use Simpson's rule (see Appendix A.4) to evaluate the integrals.

Simpson's
three-point rule

$$\int_{X_0}^{X_2} f(X)\, dX = \frac{\Delta X}{3}\,[f(X_0) + 4f(X_1) + f(X_2)] \tag{A-21}$$

For the first reactor, $X_0 = 0$, $X_1 = 0.2$, $X_2 = 0.4$, and $\Delta X = 0.2$,

$$V = F_{A0} \int_0^{0.4} \frac{dX}{-r_A} = \frac{\Delta X}{3}\left[\frac{F_{A0}}{-r_A(0)} + 4\frac{F_{A0}}{-r_A(0.2)} + \frac{F_{A0}}{-r_A(0.4)}\right] \tag{E2-6.1}$$

Selecting the appropriate values from Table 2-3, we get

$$V_1 = \left(\frac{0.2}{3}\right)[0.89 + 4(1.33) + 2.05]\,m^3 = 0.551\ m^3$$

$$\boxed{V_1 = 551\ dm^3}$$

For the second reactor,

$$V_2 = F_{A0}\int_{0.4}^{0.8} \frac{dX}{-r_A}$$

$$= \frac{\Delta X}{3}\left[\frac{F_{A0}}{-r_A(0.4)} + 4\frac{F_{A0}}{-r_A(0.6)} + \frac{F_{A0}}{-r_A(0.8)}\right] \tag{E2-6.2}$$

$$= \left(\frac{0.2}{3}\right)[2.05 + 4(3.54) + 8.0]\,m^3 = \frac{0.2}{3}(24.21\ m^3) = 1.614\ m^3$$

$$\boxed{V_2 = 1614\ dm^3}$$

The total volume is then

$$V = V_1 + V_2 = 551\ dm^3 + 1614\ dm^3 = 2165\ dm^3$$

Note: This is the same volume we calculated for a single PFR to achieve 80% conversion in Example 2-4.

2.5.3 Combinations of CSTRs and PFRs in Series

The final sequences we shall consider are combinations of CSTRs and PFRs in series. An industrial example of reactors in series is shown in the photo in Figure 2-9. This sequence is used to dimerize propylene into isohexanes, e.g.,

$$2CH_3-CH=CH_2 \longrightarrow \overset{\displaystyle CH_3}{\underset{\displaystyle |}{CH_3C}}=CH-CH_2-CH_3$$

Figure 2-9   Dimersol G (an organometallic catalyst) unit (two CSTRs and one tubular reactor in series) to dimerize propylene into isohexanes. Institut Français du Pétrole process. [Photo courtesy of Editions Technip (Institut Français du Pétrole).]

A schematic of the industrial reactor system in the Figure 2-9 is shown in Figure 2-10.

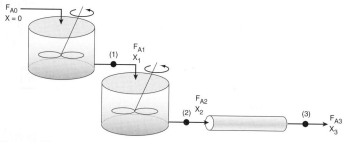

Figure 2-10   Schematic of a real system.

For the sake of illustration, let's assume the reaction carried out in the reactors in Figure 2-10 follows the same $\left(\dfrac{F_{A0}}{-r_A}\right)$ vs. X curve given by Table 2-3.

The volumes of the first two CSTRs in series (see Example 2-5) are:

Reactor 1
$$V_1 = \frac{F_{A0}X_1}{-r_{A1}}$$
(2-13)

In this series arrangement $-r_{A2}$ is evaluated at X_2 for the second CSTR.

Reactor 2
$$V_2 = \frac{F_{A0}(X_2 - X_1)}{-r_{A2}}$$
(2-24)

Starting with the differential form of PFR design equation

$$F_{A0}\frac{dX}{dV} = -r_A$$
(2-15)

Rearranging and integrating between limits, when $V = 0$, then $X = X_2$, and when $V = V_3$, then $X = X_3$.

Reactor 3
$$V_3 = \int_{X_2}^{X_3} \frac{F_{A0}}{-r_A} dX \tag{2-25}$$

The corresponding reactor volumes for each of the three reactors can be found from the shaded areas in Figure 2-11.

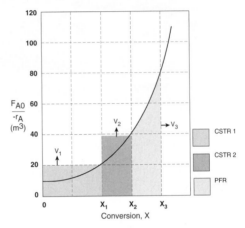

Figure 2-11   Levenspiel plot to determine the reactor volumes V_1, V_2, and V_3.

The $F_{A0}/-r_A$ curves we have been using in the previous examples are typical of those found in isothermal reaction systems. We will now consider a real reaction system that is carried out adiabatically. Isothermal reaction systems are discussed in Chapter 4 and adiabatic systems in Chapter 8.

Example 2–7   An Adiabatic Liquid-Phase Isomerization

The isomerization of butane

$$n - C_4H_{10} \;\rightleftharpoons\; i - C_4H_{10}$$

was carried out adiabatically in the liquid phase and the data in Table E2-7.1 were obtained. (Example 8.4 shows how the data in Table E2-7.1 were generated.)

TABLE E2-7.1   RAW DATA

X	0.0	0.2	0.4	0.6	0.65
$-r_A$(kmol/m$^3 \cdot$ h)	39	53	59	38	25

Don't worry how we got this data or why the $(1/-r_A)$ looks the way it does, we will see how to construct this table in Chapter 8. It is *real data* for a *real reaction* carried out adiabatically, and the reactor scheme shown in Figure E2-7.1 is used.

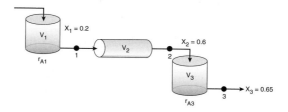

Figure E2-7.1 Reactors in series.

Calculate the volume of each of the reactors for an entering molar flow rate of n-butane of 50 kmol/hr.

Solution

Taking the reciprocal of $-r_A$ and multiplying by F_{A0} we obtain Table E2-7.2.

$$\text{E.g., at } X = 0: \quad \frac{F_{A0}}{-r_A} = \frac{50 \text{ kmol/h}}{39 \text{ kmol/h} \cdot \text{m}^3} = 1.28 \text{ m}^3$$

TABLE E2-7.2   PROCESSED DATA

X	0.0	0.2	0.4	0.6	0.65
$-r_A$ (kmol/m³ · h)	39	53	59	38	25
$[F_{A0}/-r_A]$ (m³)	1.28	0.94	0.85	1.32	2.0

(a) For the first CSTR,
 when $X = 0.2$, then $\dfrac{F_{A0}}{-r_A} = 0.94$

$$V_1 = \frac{F_{A0}}{-r_A}X_1 = (0.94 \text{ m}^3)(0.2) = 0.188 \text{ m}^3 \qquad \text{(E2-7.1)}$$

$$\boxed{V_1 = 0.188 \text{ m}^3 = 188 \text{ dm}^3} \qquad \text{(E2-7.2)}$$

(b) For the PFR,

$$V_2 = \int_{0.2}^{0.6} \left(\frac{F_{A0}}{-r_A}\right) dX$$

Using Simpson's three-point formula with $\Delta X = (0.6 - 0.2)/2 = 0.2$, and $X_1 = 0.2$, $X_2 = 0.4$, and $X_3 = 0.6$.

$$V_2 = \int_{0.2}^{0.6} \frac{F_{A0}}{-r_A}(dX) = \frac{\Delta X}{3}\left[\left(\frac{F_{A0}}{-r_A}\right)_{X=0.2} + 4\left(\frac{F_{A0}}{-r_A}\right)_{X=0.4} + \left(\frac{F_{A0}}{-r_A}\right)_{X=0.6}\right]$$

$$= \frac{0.2}{3}[0.94 + 4(0.85) + 1.32]m^3 \tag{E2-7.3}$$

$$\boxed{V_2 = 0.38 \text{ m}^3 = 380 \text{ dm}^3} \tag{E2-7.4}$$

(c) For the last reactor and the second CSTR, mole balance on A for the CSTR:

$$\text{In} - \text{Out} + \text{Generation} = 0$$

$$F_{A2} - F_{A3} + r_{A3}V_3 = 0 \tag{E2-7.5}$$

Rearranging

$$V_3 = \frac{F_{A3} - F_{A2}}{-r_{A3}} \tag{E2-7.6}$$

$$F_{A2} = F_{A0} - F_{A0}X_2$$

$$F_{A3} = F_{A0} - F_{A0}X_3$$

$$V_3 = \frac{(F_{A0} - F_{A0}X_2) - (F_{A0} - F_{A0}X_3)}{-r_{A3}}$$

Simplifying

$$\boxed{V_3 = \left(\frac{F_{A0}}{-r_{A3}}\right)(X_3 - X_2)} \tag{E2-7.7}$$

We find from Table E2-7.2 that at $X_3 = 0.65$, then $\dfrac{F_{A0}}{-r_{A3}} = 2.0 \text{ m}^3$

$$V_3 = 2 \text{ m}^3 (0.65 - 0.6) = 0.1 \text{ m}^3$$

$$\boxed{V_3 = 0.1 \text{ m}^3 = 100 \text{ dm}^3} \tag{E2-7.8}$$

A Levenspiel plot of $(F_{A0}/-r_A)$ vs. X is shown in Figure E2-7.2.

2.5.4 Comparing the CSTR and PFR Reactor Volumes and Reactor Sequencing

If we look at Figure E2-7.2, the area under the curve (PFR volume) between $X = 0$ and $X = 0.2$, we see that the PFR area is greater than the rectangular area corresponding to the CSTR volume, i.e., $V_{PFR} > V_{CSTR}$. However, if we compare the areas under the curve between $X = 0.6$ and $X = 0.65$, we see that the

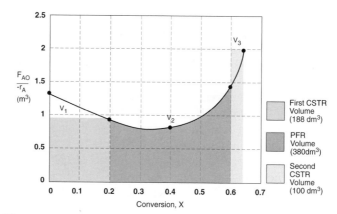

Figure E2-7.2 Levenspiel plot for adiabatic reactors in series.

Which arrangement
is best?

or

Given V and
$\left[\dfrac{1}{r_A}\ \text{vs. } X\right]$ find X

area under the curve (PFR volume) is smaller than the rectangular area corresponding to the CSTR volume, i.e., $V_{CSTR} > V_{PFR}$. This result often occurs when the reaction is carried out adiabatically, which is discussed when we look at heat effects in Chapter 8.

In the *sequencing of reactors* one is often asked, "Which reactor should go first to give the highest overall conversion? Should it be a PFR followed by a CSTR, or two CSTRs, then a PFR, or ...?" The answer is **"It depends."** It depends not only on the shape of the Levenspiel plots ($F_{A0}/-r_A$) versus X, but also on the relative reactor sizes. As an exercise, examine Figure E2-7.2 to learn if there is a better way to arrange the two CSTRs and one PRF. Suppose you were given a Levenspiel plot of ($F_{A0}/-r_A$) vs. X for three reactors in series along with their reactor volumes $V_{CSTR1} = 3$ m³, $V_{CSTR2} = 2$ m³, and $V_{PFR} = 1.2$ m³ and asked to find the highest possible conversion X. What would you do? The methods we used to calculate reactor volumes all apply, except the procedure is reversed and a trial-and-error solution is needed to find the exit overall conversion from each reactor. See Problem P2-5$_B$.

The previous examples show that *if* we know the molar flow rate to the reactor and the reaction rate as a function of conversion, *then* we can calculate the reactor volume necessary to achieve a specified conversion. The reaction rate does not depend on conversion alone, however. It is also affected by the initial concentrations of the reactants, the temperature, and the pressure. Consequently, the experimental data obtained in the laboratory and presented in Table 2-1 as $-r_A$ as a function of X are useful only in the design of full-scale reactors that are to be operated at the *identical conditions* as the laboratory experiments (temperature, pressure, initial reactant concentrations). However, such circumstances are **seldom** encountered and we must revert to the methods we describe in Chapter 3 to obtain $-r_A$ as a function of X.

Only need $-r_A = f(X)$ to size flow reactors

It is important to understand that if the rate of reaction is available or can be obtained solely as a function of conversion, $-r_A = f(X)$, or if it can be generated by some intermediate calculations, one can design a variety of reactors or a combination of reactors.

Chapter 3 shows how to find $-r_A = f(X)$.

Ordinarily, laboratory data are used to formulate a rate law, and then the reaction rate–conversion functional dependence is determined using the rate law. The preceding sections show that with the reaction rate–conversion relationship, different reactor schemes can readily be sized. In Chapter 3, we show how we obtain this relationship between reaction rate and conversion from rate law and reaction stoichiometry.

2.6 Some Further Definitions

Before proceeding to Chapter 3, some terms and equations commonly used in reaction engineering need to be defined. We also consider the special case of the plug-flow design equation when the volumetric flow rate is constant.

2.6.1 Space Time

The space time, τ, is obtained by dividing reactor volume by the volumetric flow rate entering the reactor:

τ is an important quantity!

$$\boxed{\tau \equiv \frac{V}{v_0}} \qquad (2\text{-}26)$$

The space time is the time necessary to process one reactor volume of fluid based on entrance conditions. For example, consider the tubular reactor shown in Figure 2-12, which is 20 m long and 0.2 m³ in volume. The dashed line in Figure 2-12 represents 0.2 m³ of fluid directly upstream of the reactor. The time it takes for this fluid to enter the reactor completely is the space time. It is also called the *holding time or mean residence time.*

Space time or mean residence time, $\tau = V/v_0$

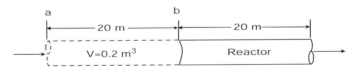

Figure 2-12   Tubular reactor showing identical volume upstream.

For example, if the volumetric flow rate were 0.01 m³/s, it would take the upstream volume shown by the dashed lines a time τ

$$\tau = \frac{0.2 \text{ m}^3}{0.01 \text{ m}^3/\text{s}} = 20 \text{ s}$$

to enter the reactor. In other words, it would take 20 s for the fluid at point a to move to point b, which corresponds to a space time of 20 s.

In the absence of dispersion, which is discussed in Chapter 14, the space time is equal to the mean residence time in the reactor, t_m. This time is the average time the molecules spend in the reactor. A range of typical processing times in terms of the space time (residence time) for industrial reactors is shown in Table 2-4.

TABLE 2-4   TYPICAL SPACE TIME FOR INDUSTRIAL REACTORS[2]

Reactor Type	Mean Residence Time Range	Production Capacity
Batch	15 min to 20 h	Few kg/day to 100,000 tons/year
CSTR	10 min to 4 h	10 to 3,000,000 tons/year
Tubular	0.5 s to 1 h	50 to 5,000,000 tons/year

Table 2-5 shows space times for six industrial reactions and reactors.

TABLE 2-5   SAMPLE INDUSTRIAL SPACE TIMES[3]

Typical industrial reaction space times

	Reaction	Reactor	Temperature	Pressure atm	Space Time
(1)	$C_2H_6 \rightarrow C_2H_4 + H_2$	PFR[†]	860°C	2	1 s
(2)	$CH_3CH_2OH + HCH_3COOH \rightarrow$ $CH_3CH_2COOCH_3 + H_2O$	CSTR	100°C	1	2 h
(3)	Catalytic cracking	PBR	490°C	20	$1 \text{ s} < \tau < 400 \text{ s}$
(4)	$C_6H_5CH_2CH_3 \rightarrow C_6H_5\ CH = CH_2 + H_2$	PBR	600°C	1	0.2 s
(5)	$CO + H_2O \rightarrow CO_2 + H_2$	PBR	300°C	26	4.5 s
(6)	$C_6H_6 + HNO_3 \rightarrow$ $C_6H_5NO_2 + H_2O$	CSTR	50°C	1	20 min

[†]The reactor is tubular but the flow may or may not be ideal plug flow.

[2] Trambouze, Landeghem, and Wauquier, *Chemical Reactors,* p. 154, (Paris: Editions Technip, 1988; Houston: Gulf Publishing Company, 1988).

[3] Walas, S. M. Chemical Reactor Data, *Chemical Engineering,* 79 (October 14, 1985).

2.6.2 Space Velocity

The space velocity (SV), which is defined as

$$SV \equiv \frac{v_0}{V} \qquad SV = \frac{1}{\tau} \tag{2-26}$$

might be regarded at first sight as the reciprocal of the space time. However, there can be a difference in the two quantities' definitions. For the space time, the entering volumetric flow rate is measured at the entrance conditions, but for the space velocity, other conditions are often used. The two space velocities commonly used in industry are the liquid-hourly and gas-hourly space velocities, LHSV and GHSV, respectively. The entering volumetric flow rate, v_0, in the LHSV is frequently measured as that of a liquid feed rate at 60°F or 75°F, even though the feed to the reactor may be a vapor at some higher temperature. Strange but true. The gas volumetric flow rate, v_0, in the GHSV is normally measured at standard temperature and pressure (STP).

$$LHSV = \frac{v_0|_{liquid}}{V} \tag{2-27}$$

$$GHSV = \frac{v_0|_{STP}}{V} \tag{2-28}$$

Example 2–8 Reactor Space Times and Space Velocities

Calculate the space time, τ, and space velocities for each of the reactors in Examples 2-2 and 2-3.

Solution

From Example 2-1, we recall the entering volumetric flow rate was given as 2 dm³/s (0.002 m³/s), and we calculated the concentration and molar flow rates for the conditions given to be $C_{A0} = 0.2$ mol/dm³ and $F_{A0} = 0.4$ mol/s.

From Example 2-2, the CSTR volume was 6.4 m³ and the corresponding space time and space velocity are

$$\tau = \frac{V}{v_0} = \frac{6.4 \text{ m}^3}{0.002 \text{ m}^3/s} = 3200 \text{ s} = 0.89 \text{ h}$$

$$SV = \frac{1}{\tau} = \frac{1}{0.89 \text{ h}} = 1.125 \text{ h}^{-1}$$

From Example 2-3, the PFR volume was 2.165 m³, and the corresponding space time and space velocity are

$$\tau = \frac{V}{v_0} = \frac{2.165 \text{ m}^3}{0.002 \text{ m}^3/s} = 1083 \text{ s} = 0.30 \text{ h}$$

$$SV = \frac{1}{\tau} = \frac{1}{0.30 \text{ h}} = 3.3 \text{ h}^{-1}$$

These space times are the times for each of the reactors to take one reactor volume of fluid and put it into the reactor.

To summarize these last examples, we have seen that in the design of reactors that are to be operated at conditions (e.g., temperature and initial concentration) identical to those at which the reaction rate data were obtained, we can size (determine the reactor volume) both CSTRs and PFRs alone or in various combinations. In principle, it may be possible to scale up a laboratory-bench or pilot-plant reaction system solely from knowledge of $-r_A$ as a function of X or C_A. However, for most reactor systems in industry, a scale-up process cannot be achieved in this manner because knowledge of $-r_A$ solely as a function of X is seldom, if ever, available under identical conditions. In Chapter 3, we shall see how we can obtain $-r_A = f(X)$ from information obtained either in the laboratory or from the literature. This relationship will be developed in a two-step process. In Step 1, we will find the rate law that gives the rate as a function of concentration and in Step 2, we will find the concentrations as a function of conversion. Combining Steps 1 and 2 in Chapter 3, we obtain $-r_A = f(X)$. We can then use the methods developed in this chapter along with integral and numerical methods to size reactors.

The CRE Algorithm
• Mole Balance, Ch 1
• Rate Law, Ch 3
• Stoichiometry, Ch 3
• Combine, Ch 4
• Evaluate, Ch 4
• Energy Balance, Ch 8

Closure

In this chapter, we have shown that if you are given the rate of reaction as a function of conversion, i.e., $-r_A = f(X)$, you will be able to size CSTRs and PFRs and arrange the order of a given set of reactors to determine the best overall conversion. After completing this chapter, the reader should be able to

a. define the parameter *conversion* and rewrite the mole balances in terms of conversion
b. show that by expressing $-r_A$ as a function of conversion X a number of reactors and reaction systems can be sized or a conversion calculated from a given reactor size
c. arrange reactors in series to achieve the maximum conversion for a given Levenspiel plot

SUMMARY

1. The conversion X is the moles of A reacted per mole of A fed.

For batch systems:
$$X = \frac{N_{A0} - N_A}{N_{A0}} \tag{S2-1}$$

For flow sytems:
$$X = \frac{F_{A0} - F_A}{F_{A0}} \tag{S2-2}$$

For reactors in series with no side streams, the conversion at point i is

$$X_i = \frac{\text{Total moles of A reacted up to point } i}{\text{Moles A fed to the first reactor}} \qquad \text{(S2-3)}$$

2. In terms of the conversion, the differential and integral forms of the reactor design equations become:

TABLE S2-1

	Differential Form	Algebraic Form	Integral Form
Batch	$N_{A0}\dfrac{dX}{dt} = -r_A V$		$t = N_{A0}\displaystyle\int_0^X \dfrac{dX}{-r_A V}$
CSTR		$V = \dfrac{F_{A0}(X_{out} - X_{in})}{(-r_A)_{out}}$	
PFR	$F_{A0}\dfrac{dX}{dV} = -r_A$		$V = F_{A0}\displaystyle\int_{X_{in}}^{X_{out}} \dfrac{dX}{-r_A}$
PBR	$F_{A0}\dfrac{dX}{dW} = -r'_A$		$W = F_{A0}\displaystyle\int_{X_{in}}^{X_{out}} \dfrac{dX}{-r'_A}$

3. If the rate of disappearance is given as a function of conversion, the following graphical techniques can be used to size a CSTR and a plug-flow reactor.

   A.   Graphical Integration Using Levenspiel Plots

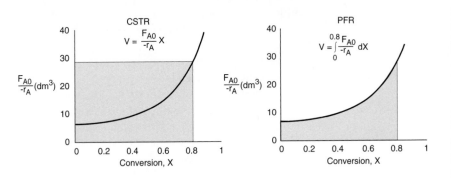

The PFR integral could also be evaluated by

   B.   Numerical Integration

   See Appendix A.4 for quatrature formulas such as the five-point quadrature formula with $\Delta X = 0.8/4$ of five equally spaced points, $X_1 = 0$, $X_2 = 0.2$, $X_3 = 0.4$, $X_4 = 0.6$, and $X_5 = 0.8$.

The integral can be evaluated by Appendix Eqn. (A-23)

$$V = \int_0^{0.8} \frac{F_{A0}}{-r_A} dX = \frac{\Delta X}{3}\left[\frac{F_{A0}}{-r(X=0)} + \frac{2F_{A0}}{-r_A(X=0.2)} + \frac{4F_{A0}}{-r_A(X=0.4)}\right.$$

$$\left. + \frac{2F_{A0}}{-r_A(X=0.6)} + \frac{F_{A0}}{-r_A(X=0.8)}\right]$$

with $\Delta X = 0.2$.

4. Space time, τ, and space velocity, SV, are given by

$$\tau = \frac{V}{v_0} \tag{S2-4}$$

$$SV = \frac{v_0}{V} \text{ (at STP)} \tag{S2-5}$$

CD-ROM MATERIALS

• **Learning Resources**
    1. Summary Notes for Chapter 2
    2. Web Module
        A. Hippopotamus Digestive System

Summary Notes

Web Modules

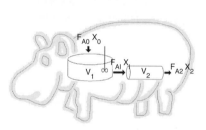

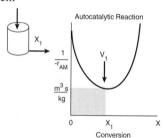

Levenspiel Plot for Autocatalytic Digestion in a CSTR

    3. Interactive Computer Modules
        A. Reactor Staging

Computer Modules

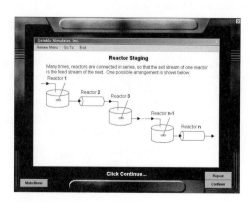

4. Solved Problems
   A. CDP2-A$_B$ More CSTR and PFR Calculations—No Memorization
• **FAQ [Frequently Asked Questions]**

• **Professional Reference Shelf**
   R2.1 *Modified Levenspiel Plots*
          For liquids and constant volume batch reactors, the mole balance equations can be modified to

$$\tau = \int_{C_A}^{C_{A0}} \frac{dC_A}{-r_A}$$

A plot of $(1/-r_A)$ versus C_A gives Figure CD2-1.

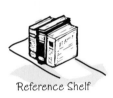

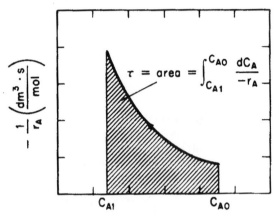

Figure CD2-1   Determining the space time, τ.

One can use this plot to study CSTRs, PFRs, and batch reactors. This material using space time as a variable is given on the CD-ROM.

QUESTIONS AND PROBLEMS

The subscript to each of the problem numbers indicates the level of difficulty: A, least difficult; D, most difficult.

A = ●   B = ■   C = ◆   D = ◆◆

P2-1$_A$ Without referring back, make a list of the most important items you learned in this chapter. What do you believe was the overall purpose of the chapter?

P2-2$_A$ Go to the web site *www.engr.ncsu.edu/learningstyles/ilsweb.html*
   (a) Take the Inventory of Learning Style test, and record your learning style according to the Solomon/Felder inventory.
          Global/Sequential_____
          Active/Reflective_____
          Visual/Verbal_____
          Sensing/Intuitive_____

Links

Web Hint

(b) After checking the web site *www.engin.umich.edu/~cre/asyLearn/ itresources.htm,* suggest two ways to facilitate your learning style in each of the four categories.

(c) Visit the problem-solving web site *www.engin.umich.edu/~cre/probsolv/ closed/cep.htm* to find ways to "Get Unstuck" when you get stuck on a problem and to review the "Problem-Solving Algorithm." List four ways that might help you in your solution to the home problems.

(d) What audio,  , from the first two chapters sounds like Arnold Schwarzenegger?

(e) What Frequently Asked Question (FAQ) would you have asked?

P2-3$_A$ **ICM Staging.** Load the Interactive Computer Module (ICM) from the CD-ROM. Run the module and then record the performance number, which indicates your mastery of the material. Your professor has the key to decode your performance number. Note: To run this module you *must* have Windows 2000 or a later version. ICM Reactor Staging Performance #_____.

P2-4$_A$ **(a)** Revisit **Examples 2-1** through **2-3.** How would your answers change if the flow rate, F_{A0}, were cut in half? If it were doubled?

(b) **Example 2-5.** How would your answers change if the two CSTRs (one 0.82 m^3 and the other 3.2 m^3) were placed in parallel with the flow, F_{A0}, divided equally to each reactor.

(c) **Example 2-6.** How would your answer change if the PFRs were placed in parallel with the flow, F_{A0}, divided equally to each reactor?

(d) **Example 2-7.** (1) What would be the reactor volumes if the two intermediate conversions were changed to 20% and 50%, respectively. (2) What would be the conversions, X_1, X_2, and X_3, if all the reactors had the same volume of 100 dm^3 and were placed in the same order? (3) What is the worst possible way to arrange the two CSTRs and one PFR?

(e) **Example 2-8.** The space time required to achieve 80% conversion in a CSTR is 5 h. The entering volumetric flow rate and concentration of reactant A are 1 dm^3/min and 2.5 molar, respectively. If possible, determine (1) the rate of reaction, $-r_A =$ _____, (2) the reactor volume, $V =$ _____, (3) the exit concentration of A, $C_A =$ _____, and (4) the PFR space time for 80% conversion.

P2-5$_B$ You have two CSTRs and two PFRs each with a volume of 1.6 m^3. Use Figure 2-2 to calculate the conversion for each of the reactors in the following arrangements.

(a) Two CSTRs in series.

(b) Two PFRs in series.

(c) Two CSTRs in parallel with the feed, F_{A0}, divided equally between the two reactors.

(d) Two PFRs in parallel with the feed divided equally between the two reactors.

(e) A CSTR and a PFR in parallel with the flow equally divided. Also calculate the overall conversion, X_{ov}

$$X_{ov} = \frac{F_{A0} - F_{ACSTR} - F_{APFR}}{F_{A0}}, \text{ with } F_{ACSTR} = \frac{F_{A0}}{2} - \frac{F_{A0}}{2} X_{CSTR},$$

$$F_{APFR} = \frac{F_{A0}}{2} (1 - X_{PFR})$$

(f) A PFR followed by a CSTR.

Member

Hall of Fame

(g) A CSTR followed by a PFR.

(h) A PFR followed by two CSTRs. Is this arrangement a good one or is there a better one?

P2-6$_A$ Read the chemical reaction engineer of hippopotamus on the CD-ROM or on the web.

(a) Write five sentences summarizing what you learned from the web module.

(b) Work problems (1) and (2) on the hippo module.

(c) The hippo has picked up a river fungus and now the effective volume of the CSTR stomach compartment is only 0.2 m^3. The hippo needs 30% conversion to survive? Will the hippo survive?

(d) The hippo had to have surgery to remove a blockage. Unfortunately, the surgeon, Dr. No, accidentally reversed the CSTR and the PFR during the operation. **Oops!!** What will be the conversion with the new digestive arrangement? Can the hippo survive?

P2-7$_B$ The exothermic reaction

$$A \longrightarrow B + C$$

was carried out adiabatically and the following data recorded:

X	0	0.2	0.4	0.45	0.5	0.6	0.8	0.9
$-r_A$ (mol/dm^3·min)	1.0	1.67	5.0	5.0	5.0	5.0	1.25	0.91

The entering molar flow rate of A was 300 mol/min.

(a) What are the PFR and CSTR volumes necessary to achieve 40% conversion? (V_{PFR} = 72 dm^3, V_{CSTR} = 24 dm^3)

(b) Over what range of conversions would the CSTR and PFR reactor volumes be identical?

(c) What is the maximum conversion that can be achieved in a 10.5-dm^3 CSTR?

(d) What conversion can be achieved if a 72-dm^3 PFR is followed in series by a 24-dm^3 CSTR?

(e) What conversion can be achieved if a 24-dm^3 CSTR is followed in a series by a 72-dm^3 PFR?

(f) Plot the conversion and rate of reaction as a function of PFR reactor volume up to a volume of 100 dm^3.

P2-8$_B$ In bioreactors, the growth is autocatalytic in that the more cells you have, the greater the growth rate

$$\text{Cells + nutrients} \xrightarrow{\text{cells}} \text{more cells + product}$$

The cell growth rate, r_g, and the rate of nutrient consumption, r_S, are directly proportional to the concentration of cells for a given set of conditions. A

Levenspiel plot of $(1/-r_S,)$ a function of nutrient conversion $X_S = (C_{S0} - C_S)/C_{S0}$ is given below in Figure P2-8.

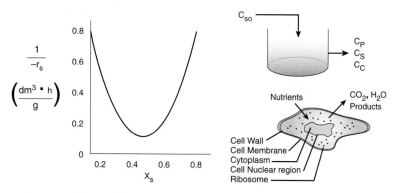

Figure P2-8$_B$ Levenspiel plot for bacteria growth.

For a nutrient feed rate of 1kg/hr with $C_{S0} = 0.25$ g/dm³, what chemostat (CSTR) size is necessary to achieve.
(a) 40% conversion of the substrate.
(b) 80% conversion of the substrate.
(c) What conversion could you achieve with an 80-dm³ CSTR? An 80-dm³ PFR?
(d) How could you arrange a CSTR and PFR in series to achieve 80% conversion with the minimum total volume? Repeat for two CSTRs in series.
(e) Show that Monod Equation for cell growth

$$-r_s = \frac{kC_SC_C}{K_M + C_S}$$

along with the stoichiometric relationship between the cell concentration, C_c, and the substrate concentration, C_s,

$$C_C = Y_{C/S}[C_{S0} - C_S] + C_{c0} = 0.1[C_{S0} - C_S] + 0.001$$

is consistent with Figure P2-8$_B$.

P2-9$_B$ The adiabatic exothermic irreversible gas-phase reaction

$$2A + B \longrightarrow 2C$$

is to be carried out in a flow reactor for an equimolar feed of A and B. A Levenspiel plot for this reaction is shown in Figure P2-9$_B$ on next page.
(a) What PFR volume is necessary to achieve 50% conversion?
(b) What CSTR volume is necessary to achieve 50% conversion?
(c) What is the volume of a second CSTR added in series to the first CSTR (**Part b**) necessary to achieve an overall conversion of 80%?
(d) What PFR volume must be added to the first CSTR (**Part b**) to raise the conversion to 80%?
(e) What conversion can be achieved in a 6×10^4 m³ CSTR and also in a 6×10^4 m³ PFR?
(f) Critique the shape of Figure P2-9$_B$ and the answers (numbers) to this problem.

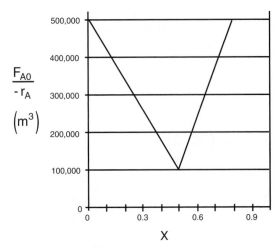

Figure P2-9$_B$ Levenspiel plot.

P2-10$_A$ Estimate the reactor volumes of the two CSTRs and the PFR shown in the photo in Figure 2-9.

P2-11$_D$ Don't calculate anything. Just go home and relax.

P2-12$_B$ The curve shown in Figure 2-1 is typical of a reaction carried out isothermally, and the curve shown in Figure P2-12$_B$ is typical of gas-solid catalytic exothermic reaction carried out adiabatically.

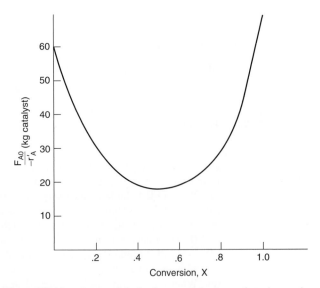

Figure P2-12$_B$ Levenspiel plot for an adiabatic exothermic reaction.

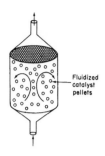

Fluidized
catalyst
pellets

(a) Assuming that you have a fluidized CSTR and a PBR containing equal weights of catalyst, how should they be arranged for this adiabatic reaction? In each case, use the smallest amount of catalyst weight and still achieve 80% conversion.

(b) What is the catalyst weight necessary to achieve 80% conversion in a fluidized CSTR?

(c) What fluidized CSTR weight is necessary to achieve 40% conversion?

(d) What PBR weight is necessary to achieve 80% conversion?

(e) What PBR weight is necessary to achieve 40% conversion?

(f) Plot the rate of reaction and conversion as a function of PBR catalyst weight, W.

Additional information: $F_{A0} = 2$ mol/s.

- **Additional Homework Problems on the CD-ROM**

CDP2-A$_B$ Use Levenspiel plots to calculate PFR and CSTR reactor volumes given $-r_A = f(X)$. **(Includes Solution)** [2nd Ed. P2-12$_B$]

CDP2-B$_A$ An ethical dilemma as to how to determine the reactor size in a competitor's chemical plant. [2nd Ed. P2-18$_B$]

CDP2-C$_B$ Uses an unusual rate law and asks how to best arrange the reactors. This problem is a good practice problem before running the ICM "Staging." (PDF) [3rd Ed. P2-7]

CDP2-D$_B$ Use Levenspiel plots to calculate PFR and CSTR volumes. (PDF)   [3rd Ed. P2-9]

CDP2-E$_A$ Use Levenspiel plots to calculate PFR and CSTR volumes.

CDP2-F$_A$ Use Levenspiel plots to calculate CSTR and PFR volumes for the reaction $A + B \rightarrow C$.

SUPPLEMENTARY READING

1. Further discussion of stoichiometry may be found in

> HIMMELBLAU, D. M., and J. D. RIGGS, *Basic Principles and Calculations in Chemical Engineering*, 7th ed. Upper Saddle River, N.J.: Prentice Hall, 2004, Chapters 2 and 6.

> FELDER, R. M., and R. W. ROUSSEAU, *Elementary Principles of Chemical Processes*, 3rd ed. New York: Wiley, 2000, Chapter 4.

2. Further discussion of the proper staging of reactors in series for various rate laws, in which a plot of -1/r_A versus X is given, is presented in

> LEVENSPIEL, O., *Chemical Reaction Engineering*, 3rd ed. New York: Wiley, 1999, Chapter 6 (pp. 139–156).

> HILL, C. G., *An Introduction to Chemical Engineering Kinetics and Reactor Design*. New York: Wiley, 1977, Chapter 8.

Rate Laws   **3**
and Stoichiometry

Success is measured not so much by the position
one has reached in life, as by the obstacles one has
overcome while trying to succeed.

Booker T. Washington

Overview. In Chapter 2, we showed that if we had the rate of reaction
as a function of conversion, $-r_A = f(X)$, we could calculate reactor volumes necessary to achieve a specified conversion for flow systems and
the time to achieve a given conversion in a batch system. Unfortunately,
one is seldom, if ever, given $-r_A = f(X)$ directly from raw data. Not to
fear, in this chapter we will show how to obtain the rate of reaction as a
function of conversion. This relationship between reaction rate and conversion will be obtained in two steps. In Step 1, Part 1 of this chapter, we
define the rate law, which relates the rate of reaction to the concentrations of the reacting species and to temperature. In Step 2, Part 2 of this
chapter, we define concentrations for flow and batch systems and develop
a stoichiometric table so that one can write concentrations as a function
of conversion. Combining Steps 1 and 2, we see that one can then write
the rate as a function conversion and use the techniques in Chapter 2 to
design reaction systems.

After completing this chapter, you will be able to write the rate of
reaction as a function of conversion for both liquid-phase and gas-phase
reacting systems.

PART 1   RATE LAWS

3.1 Basic Definitions

A *homogeneous reaction* is one that involves only one phase. A *heterogeneous reaction* involves more than one phase, and the reaction usually occurs at the interface between the phases. An *irreversible reaction* is one that proceeds in only one direction and continues in that direction until the reactants are **Types of reactions** exhausted. A *reversible reaction*, on the other hand, can proceed in either direction, depending on the concentrations of reactants and products relative to the corresponding equilibrium concentrations. An irreversible reaction behaves as if no equilibrium condition exists. Strictly speaking, no chemical reaction is completely irreversible. However, for many reactions, the equilibrium point lies so far to the product side that these reactions are treated as irreversible reactions.

The *molecularity* of a reaction is the number of atoms, ions, or molecules involved (colliding) in a reaction step. The terms *unimolecular, bimolecular,* and *termolecular* refer to reactions involving, respectively, one, two, or three atoms (or molecules) interacting or colliding in any one reaction step. The most common example of a *unimolecular* reaction is radioactive decay, such as the spontaneous emission of an alpha particle from uranium-238 to give thorium and helium:

$$_{92}U^{238} \rightarrow {}_{90}Th^{234} + {}_2He^4$$

The rate of disappearance of uranium (U) is given by the rate law

$$-r_U = kC_U$$

The true *bimolecular* reactions that exist are reactions involving free radicals such as

$$Br \bullet + C_2H_6 \rightarrow HBr + C_2H_5 \bullet$$

with the rate of disappearance of bromine given by the rate law

$$-r_{Br\bullet} = kC_{Br\bullet}C_{C_2H_6}$$

The probability of a *termolecular* reaction occurring is almost nonexistent, and in most instances the reaction pathway follows a series of *bimolecular* reactions as in the case of the reaction

$$2NO + O_2 \rightarrow 2NO_2$$

The reaction pathway for this "Hall of Fame" reaction is quite interesting and is discussed in Chapter 7 along with similar reactions that form active intermediate complexes in their reaction pathways.

3.1.1 Relative Rates of Reaction

The relative rates of reaction of the various species involved in a reaction can be obtained from the ratio of stoichiometric coefficients. For Reaction (2-2),

$$A + \frac{b}{a}B \rightarrow \frac{c}{a}C + \frac{d}{a}D \tag{2-2}$$

we see that for every mole of A that is consumed, c/a moles of C appear. In other words,

$$\text{Rate of formation of C} = \frac{c}{a} \text{ (Rate of disappearance of A)}$$

$$r_C = \frac{c}{a}(-r_A) = -\frac{c}{a}r_A$$

Similarly, the relationship between the rates of formation of C and D is

$$r_C = \frac{c}{d}r_D$$

The relationship can be expressed directly from the stoichiometry of the reaction,

$$aA + bB \rightarrow cC + dD \tag{2-1}$$

for which

$$\boxed{\frac{-r_A}{a} = \frac{-r_B}{b} = \frac{r_C}{c} = \frac{r_D}{d}} \tag{3-1}$$

Reaction
stoichiometry

or

$$\boxed{\frac{r_A}{-a} = \frac{r_B}{-b} = \frac{r_C}{c} = \frac{r_D}{d}}$$

For example, in the reaction

$$2NO + O_2 \rightleftharpoons 2NO_2$$

we have

$$\frac{r_{NO}}{-2} = \frac{r_{O_2}}{-1} = \frac{r_{NO_2}}{2}$$

If NO_2 is being formed at a rate of 4 mol/m³/s, i.e.,

$$r_{NO_2} = 4 \text{ mol/m}^3/\text{s}$$

then the rate of formation of NO is

$$2NO + O_2 \rightarrow 2NO_2$$
$$r_{NO_2} = 4 \text{ mol/m}^3/\text{s}$$
$$-r_{NO} = 4 \text{ mol/m}^3/\text{s}$$
$$-r_{O_2} = 2 \text{ mol/m}^3/\text{s}$$

$$r_{NO} = \frac{-2}{2} \, r_{NO_2} = -4 \text{ mol/m}^3/\text{s}$$

the rate of disappearance of NO is

$$-r_{NO} = 4 \text{ mol/m}^3/\text{s}$$

and the rate of disappearance of oxygen, O_2, is

$$-r_{O_2} = \frac{-1}{-2} \, r_{NO_2} = 2 \text{ mol/dm}^3/\text{s}$$

3.2 The Reaction Order and the Rate Law

In the chemical reactions considered in the following paragraphs, we take as the basis of calculation a species A, which is one of the reactants that is disappearing as a result of the reaction. The limiting reactant is usually chosen as our basis for calculation. The rate of disappearance of A, $-r_A$, depends on temperature and composition. For many reactions, it can be written as the product of a *reaction rate constant* k_A and a function of the concentrations (activities) of the various species involved in the reaction:

$$\boxed{-r_A = [k_A(T)][\text{fn}(C_A, C_B, \ldots)]} \qquad (3\text{-}2)$$

The rate law gives the relationship between reaction rate and concentration.

The algebraic equation that relates $-r_A$ to the species concentrations is called the kinetic expression or **rate law**. The specific rate of reaction (also called the rate constant), k_A, like the reaction rate $-r_A$, always refers to a particular species in the reaction and normally should be subscripted with respect to that species. However, for reactions in which the stoichiometric coefficient is 1 for all species involved in the reaction, for example,

$$1\,NaOH + 1\,HCl \rightarrow 1\,NaCl + 1\,H_2O$$

we shall delete the subscript on the specific reaction rate, (e.g., A in k_A), to let

$$k = k_{NaOH} = k_{HCl} = k_{NaCl} = k_{H_2O}$$

3.2.1 Power Law Models and Elementary Rate Laws

The dependence of the reaction rate, $-r_A$, on the concentrations of the species present, $\text{fn}(C_j)$, is almost without exception determined by experimental observation. Although the functional dependence on concentration may be postulated from theory, experiments are necessary to confirm the proposed form. One of the most common general forms of this dependence is the power law model. Here the rate law is the product of concentrations of the individual reacting species, each of which is raised to a power, for example,

$$\boxed{-r_A = k_A C_A^\alpha C_B^\beta} \tag{3-3}$$

The exponents of the concentrations in Equation (3-3) lead to the concept of *reaction order*. The **order of a reaction** refers to the powers to which the concentrations are raised in the kinetic rate law.[1] In Equation (3-3), the reaction is α *order with respect to reactant* A, *and* β *order with respect to reactant* B. The overall order of the reaction, n, is

Overall
reaction
order

$$n = \alpha + \beta$$

The units of $-r_A$ are always in terms of concentration per unit time while the units of the specific reaction rate, k_A, will vary with the order of the reaction. Consider a reaction involving only one reactant, such as

$$A \rightarrow Products$$

with a reaction order n. The units of the specific reaction rate constant are

$$k = \frac{(\text{Concentration})^{1-n}}{\text{Time}}$$

Consequently, the rate laws corresponding to a zero-, first-, second-, and third-order reaction, together with typical units for the corresponding rate constants, are:

Zero-order ($n = 0$): $\qquad\qquad -r_A = k_A$:

$$\{k\} = \text{mol/dm}^3 \cdot \text{s} \tag{3-4}$$

First-order ($n = 1$): $\qquad\qquad -r_A = k_A C_A$:

$$\{k\} = \text{s}^{-1} \tag{3-5}$$

Second-order ($n = 2$): $\qquad\quad -r_A = k_A C_A^2$:

$$\{k\} = \text{dm}^3/\text{mol} \cdot \text{s} \tag{3-6}$$

---

[1] Strictly speaking, the reaction rates should be written in terms of the activities, a_i, ($a_i = \gamma_i C_i$, where γ_i is the activity coefficient). Kline and Fogler, *JCIS,* 82, 93 (1981); ibid., p. 103; and *Ind. Eng. Chem Fundamentals* 20, 155 (1981).

$$-r_A = k_A' a_A^\alpha a_B^\beta$$

However, for many reacting systems, the activity coefficients, γ_i, do not change appreciably during the course of the reaction, and they are absorbed into the specific reaction rate:

$$-r_A = k_A' a_A^\alpha a_B^\beta = -k_A'(\gamma_A C_A)^\alpha (\gamma_B C_B)^\beta = \overbrace{\left(k_A' \gamma_A^\alpha \gamma_B^\beta\right)}^{k_A} C_A^\alpha C_B^\beta = k_A C_A^\alpha C_B^\beta$$

Third-order ($n = 3$): $-r_A = k_A C_A^3$:

$$\{k\} = (dm^3/mol)^2 \cdot s^{-1} \tag{3-7}$$

An *elementary reaction* is one that evolves a single step such as the bimolecular reaction between oxygen and methanol

$$O \bullet + CH_3OH \rightarrow CH_3O \bullet + OH \bullet$$

The stoichiometric coefficients in this reaction are *identical* to the powers in the rate law. Consequently, the rate law for the disappearance of molecular oxygen is

$$-r_{O \bullet} = k C_{O \bullet} C_{CH_3OH}$$

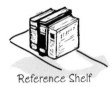

Reference Shelf

Collision theory

The reaction is first order in molecular oxygen and first order in methanol; therefore, we say both the reaction and the rate law are elementary. This form of the rate law can be derived from *Collision Theory* as shown in the Profession Reference Shelf 3A on the CD-ROM. There are many reactions where the stoichiometric coefficients in the reaction are identical to the reaction orders, but the reactions are not elementary owing to such things as pathways involving active intermediates and series reactions. For these reactions that are not elementary but whose stoichiometric coefficients are identical to the reaction orders in the rate law, we say the reaction *follows an elementary rate law*. For example, the oxidation reaction of nitric oxide discussed earlier,

$$2NO + O_2 \rightarrow 2NO_2$$

is not elementary but follows the elementary rate law

Note: the rate constant, k, is defined with respect to NO.

$$-r_{NO} = k_{NO} C_{NO}^2 C_{O_2}$$

Another nonelementary reaction that follows an elementary rate law is the gas-phase reaction between hydrogen and iodine

$$H_2 + I_2 \rightarrow 2HI$$

with

$$-r_{H_2} = k_{H_2} C_{H_2} C_{I_2}$$

Web Hint

In summary, for many reactions involving multiple steps and pathways, the powers in the rate laws surprisingly agree with the stoichiometric coefficients. Consequently, to facilitate describing this class of reactions, we say a reaction *follows an elementary rate law* when the reaction orders are identical with the stoichiometric coefficients of the reacting species for the reaction *as written*. It is important to remember that the rate laws **are determined by experimental observation!** They are a function of the reaction chemistry and not the type of reactor in which the reactions occur. Table 3-1 gives examples of rate laws for a number of reactions.

The values of specific reaction rates for these and a number of other reactions can be found in the *Data Base* found on the CD-ROM and on the web.

Where do you find rate laws?

The activation energy, frequency factor, and reaction orders for a large number of gas- and liquid-phase reactions can be found in the National Bureau of Standards' circulars and supplements.[2] Also consult the journals listed at the end of Chapter 1.

TABLE 3-1.   EXAMPLES OF REACTION RATE LAWS

A. First-Order Rate Laws

(1) $\qquad$ $C_2H_6 \longrightarrow C_2H_4 + H_2$ $\qquad$ $-r_A = kC_{C_2H_6}$

(2) $\qquad$ $\phi N = NCl \longrightarrow \phi Cl + N_2$ $\qquad$ $-r_A = kC_{\phi N = NCl}$

(3)

$$CH_2 \overset{O}{\frown} CH_2 + H_2O \overset{H_2SO_4}{\longrightarrow} \overset{CH_2OH}{\underset{CH_2OH}{|}}$$

$-r_A = kC_{CH_2OCH_2}$

(4) $\qquad$ $CH_3COCH_3 \longrightarrow CH_2CO + CH_4$ $\qquad$ $-r_A = kC_{CH_3COCH_3}$

(5) $\qquad$ $nC_4H_{10} \rightleftarrows iC_4H_{10}$ $\qquad$ $-r_n = k[C_{nC_4} - C_{iC_4}/K_C]$

B. Second-Order Rate Laws

(1) $\qquad$ (NO$_2$–Cl) $+ 2NH_3 \longrightarrow$ (NO$_2$–NH$_2$) $+ NH_4CL$ $\qquad$ $-r_A = k_{ONCB}C_{ONCB}C_{NH_3}$[†]

(2) $\qquad$ $CNBr + CH_3NH_2 \longrightarrow CH_3Br + NCNH_2$ $\qquad$ $-r_A = kC_{CNBr}C_{CH_3NH_2}$

(3)

$$\underset{A}{CH_3COOC_2H_5} + \underset{B}{C_4H_9OH} \rightleftarrows \underset{C}{CH_3COOC_4H_9} + \underset{D}{C_2H_5OH}$$

$-r_A = k[C_AC_B - C_CC_D/K_C]$

† See Problem P3-13$_B$ and Section 9.2.

Very important references, but you should also look in the other literature before *going to the lab*

[2] Kinetic data for larger number of reactions can be obtained on floppy disks and CD-ROMs provided by *National Institute of Standards and Technology (NIST)*. Standard Reference Data 221/A320 Gaithersburg, MD 20899; phone: (301) 975-2208. Additional sources are *Tables of Chemical Kinetics: Homogeneous Reactions*, National Bureau of Standards Circular 510 (Sept. 28, 1951); Suppl. 1 (Nov. 14, 1956); Suppl. 2 (Aug. 5, 1960); Suppl. 3 (Sept. 15, 1961) (Washington, D.C.: U.S. Government Printing Office). *Chemical Kinetics and Photochemical Data for Use in Stratospheric Modeling*, Evaluate No. 10, JPL Publication 92-20 (Pasadena, Calif.: Jet Propulsion Laboratories, Aug. 15, 1992).

TABLE 3-1.   EXAMPLES OF REACTION RATE LAWS (CONTINUED)

C. Nonelementary Rate Laws

(1) $CH_3CHO \longrightarrow CH_4 + CO$

$$-r_{CH_3CHO} = kC_{CH_3CHO}^{3/2}$$

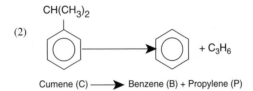

(2) Cumene (C) $\longrightarrow$ Benzene (B) + Propylene (P)

$$-r_C = \frac{k[(P_C - P_B P_P)K_P]}{1 + K_B P_B + K_C P_C}$$

D. Enzymatic Reactions (Urea (U) + Urease (E))

$$NH_2CONH_2 + Urease \xrightarrow{+H_2O} 2NH_3 + CO_2 + Urease$$

$$-r_U = \frac{kC_U}{K_M + C_U}$$

E. Biomass Reactions

Substrate (S) + Cells (C) $\rightarrow$ More Cells + Product

$$-r_U = \frac{kC_S C_C}{K_S + C_S}$$

Note: The rate constant, k, and activation energies for a number of the reactions in these examples are given in the Data Base on the CD-ROM and Summary Notes.

3.2.2 Nonelementary Rate Laws

A large number of both homogeneous and heterogeneous reactions do not follow simple rate laws. Examples of reactions that don't follow simple elementary rate laws are discussed below.

Homogeneous Reactions The overall order of a reaction does not have to be an integer, nor does the order have to be an integer with respect to any individual component. As an example, consider the gas-phase synthesis of phosgene,

$$CO + Cl_2 \rightarrow COCl_2$$

in which the kinetic *rate law* is

$$-r_{CO} = kC_{CO}C_{Cl_2}^{3/2}$$

This reaction is first order with respect to carbon monoxide, three-halves order with respect to chlorine, and five-halves order overall.

Sometimes reactions have complex rate expressions that cannot be separated into solely temperature-dependent and concentration-dependent portions. In the decomposition of nitrous oxide,

$$2N_2O \rightarrow 2N_2 + O_2$$

the kinetic *rate law* is

$$-r_{N_2O} = \frac{k_{N_2O} C_{N_2O}}{1 + k' C_{O_2}}$$

Apparent reaction orders

Both k_{N_2O} and k' are strongly temperature-dependent. When a rate expression such as the one given above occurs, we cannot state an overall reaction order. Here we can only speak of reaction orders under certain limiting conditions. For example, at very low concentrations of oxygen, the second term in the denominator would be negligible wrt 1 ($1 \gg k' C_{O_2}$), and the reaction would be "apparent" first order with respect to nitrous oxide and first order overall. However, if the concentration of oxygen were large enough so that the number 1 in the denominator were insignificant in comparison with the second term, $k' C_{O_2}$ ($k' C_{O_2} \gg 1$), the apparent reaction order would be -1 with respect to oxygen and first order with respect to nitrous oxide an overall *apparent* zero order. Rate expressions of this type are very common for liquid and gaseous reactions promoted by solid catalysts (see Chapter 10). They also occur in homogeneous reaction systems with reactive intermediates (see Chapter 7).

Important resources for **rate laws**

It is interesting to note that although the reaction orders often correspond to the stoichiometric coefficients as evidenced for the reaction between hydrogen and iodine, the rate expression for the reaction between hydrogen and another halogen, bromine, is quite complex. This nonelementary reaction

$$H_2 + Br_2 \rightarrow 2HBr$$

proceeds by a free-radical mechanism, and its reaction rate law is

$$-r_{HBr} = \frac{k_1 C_{H_2} C_{Br_2}^{1/2}}{k_2 + C_{HBr}/C_{Br_2}} \tag{3-8}$$

In Chapter 7, we will discuss reaction mechanisms and pathways that lead to nonelementary rate laws such as rate of formation of HBr shown in Equation (3-8).

Heterogeneous Reactions In many gas-solid catalyzed reactions, it historically has been the practice to write the rate law in terms of partial pressures rather than concentrations. An example of a heterogeneous reaction and corresponding rate law is the hydrodemethylation of toluene (T) to form benzene (B) and methane (M) carried out over a solid catalyst.

$$C_6H_5CH_3 + H_2 \underset{cat}{\rightarrow} C_6H_6 + CH_4$$

The rate of disappearance of toluene per mass of catalyst, $-r'_T$, follows Langmuir-Hinshelwood kinetics (Chapter 10), and the rate law was found experimentally to be

$$-r'_T = \frac{kP_{H_2}P_T}{1 + K_B P_B + K_T P_T}$$

where K_B and K_T are the adsorption constants with units of kPa^{-1} (or atm^{-1}) and the specific reaction rate has units of

$$[k] = \frac{\text{mol toluene}}{\text{kg cat} \cdot \text{s} \cdot \text{kPa}^2}$$

To express the rate of reaction in terms of concentration rather than partial pressure, we simply substitute for P_i using the ideal gas law

$$P_i = C_i RT \tag{3-9}$$

The rate of reaction per unit weight catalyst, $-r'_A$ (e.g., $-r'_T$), and the rate of reaction per unit volume, $-r_A$, are related through the bulk density ρ_b (mass of solid/volume) of the *catalyst particles* in the fluid media:

$$-r_A = \rho_b(-r'_A)$$

$$\frac{\text{moles}}{\text{time} \cdot \text{volume}} = \left(\frac{\text{mass}}{\text{volume}}\right)\left(\frac{\text{moles}}{\text{time} \cdot \text{mass}}\right)$$

In fluidized catalytic beds, the bulk density is normally a function of the volumetric flow rate through the bed.

In summary on reaction orders, they **cannot** be deduced from reaction stoichiometry. Even though a number of reactions follow elementary rate laws, at least as many reactions do not. One **must** determine the reaction order from the literature or from experiments.

3.2.3 Reversible Reactions

All rate laws for reversible reactions *must* reduce to the thermodynamic relationship relating the reacting species concentrations at equilibrium. At equilibrium, the rate of reaction is identically zero for all species (i.e., $-r_A \equiv 0$). That is, for the general reaction

$$a\text{A} + b\text{B} \; \rightleftarrows \; c\text{C} + d\text{D} \tag{2-1}$$

the concentrations at equilibrium are related by the thermodynamic relationship for the equilibrium constant K_C (see Appendix C).

Thermodynamic
Equilibrium
Relationship

$$K_C = \frac{C_{Ce}^c C_{De}^d}{C_{Ae}^a C_{Be}^b} \tag{3-10}$$

The units of the thermodynamic equilibrium constant, K_C, are K_C, are $(\text{mol/dm}^3)^{d+c-b-a}$.

To illustrate how to write rate laws for reversible reactions, we will use the combination of two benzene molecules to form one molecule of hydrogen and one of diphenyl. In this discussion, we shall consider this gas-phase reaction to be elementary and reversible:

$$2C_6H_6 \underset{k_{-B}}{\overset{k_B}{\rightleftarrows}} C_{12}H_{10} + H_2$$

or, symbolically,

$$2B \underset{k_{-B}}{\overset{k_B}{\rightleftarrows}} D + H_2$$

The specific reaction rate, k_i, must be defined wrt a particular species.

The forward and reverse specific reaction rate constants, k_B and k_{-B}, respectively, will *be defined with respect to benzene.*

Benzene (B) is being depleted by the forward reaction

$$2C_6H_6 \xrightarrow{k_B} C_{12}H_{10} + H_2$$

in which the rate of disappearance of benzene is

$$-r_{B,\text{forward}} = k_B C_B^2$$

If we multiply both sides of this equation by -1, we obtain the expression for the rate of formation of benzene for the forward reaction:

$$r_{B,\text{forward}} = -k_B C_B^2 \tag{3-11}$$

For the reverse reaction between diphenyl (D) and hydrogen (H_2),

$$C_{12}H_{10} + H_2 \xrightarrow{k_{-B}} 2C_6H_6$$

the rate of formation of benzene is given as

$$r_{B,\text{reverse}} = k_{-B} C_D C_{H_2} \tag{3-12}$$

Again, both the rate constants k_B and k_{-B} are *defined with respect to benzene!!!*

The net rate of formation of benzene is the sum of the rates of formation from the forward reaction [i.e., Equation (3-11)] and the reverse reaction [i.e., Equation (3-12)]:

$$r_B \equiv r_{B,\text{net}} = r_{B,\text{forward}} + r_{B,\text{reverse}}$$

$$r_B = -k_B C_B^2 + k_{-B} C_D C_{H_2} \tag{3-13}$$

Multiplying both sides of Equation (3-13) by -1, we obtain the rate law for the rate of disappearance of benzene, $-r_B$:

Elementary reversible
$$A \rightleftharpoons B$$

$$-r_B = k_B C_B^2 - k_{-B} C_D C_{H_2} = k_B \left(C_B^2 - \frac{k_{-B}}{k_B} C_D C_{H_2} \right)$$

$-r_A = k \left(C_A - \dfrac{C_B}{K_c} \right)$ Replacing the ratio of the reverse to forward rate law constants by the equilibrium constant, we obtain

$$\boxed{-r_B = k_B \left(C_B^2 - \frac{C_D C_{H_2}}{K_C} \right)} \tag{3-14}$$

where

$$\frac{k_B}{k_{-B}} = K_C = \text{Concentration equilibrium constant}$$

The equilibrium constant decreases with increasing temperature for exothermic reactions and increases with increasing temperature for endothermic reactions.

Let's write the rate of formation of diphenyl, r_D, in terms of the concentrations of hydrogen, H_2, diphenyl, D, and benzene, B. The rate of formation of diphenyl, r_D, **must** have the same functional dependence on the reacting species concentrations as does the rate of disappearance of benzene, $-r_B$. The rate of formation of diphenyl is

$$r_D = k_D \left[C_B^2 - \frac{C_D C_{H_2}}{K_C} \right] \tag{3-15}$$

Using the relationship given by Equation (3-1) for the general reaction

This is just stoichiometry.

$$\boxed{\frac{r_A}{-a} = \frac{r_B}{-b} = \frac{r_C}{c} = \frac{r_D}{d}} \tag{3-1}$$

we can obtain the relationship between the various specific reaction rates, k_B, k_D:

$$\frac{r_D}{1} = \frac{r_B}{-2} = \frac{k_B}{2} \left[C_B^2 - \frac{C_D C_{H_2}}{K_C} \right] \tag{3-16}$$

Comparing Equations (3-15) and (3-16), we see the relationship between the specific reaction rate with respect to diphenyl and the specific reaction rate with respect to benzene is

$$k_D = \frac{k_B}{2}$$

Consequently, we see the need to define the rate constant, k, wrt a particular species.

Finally, we need to check to see if the rate law given by Equation (3-14) is thermodynamically consistent at equilibrium. Applying Equation (3-10) (and Appendix C) to the diphenyl reaction and substituting the appropriate species concentration and exponents, thermodynamics tells us that

$$K_C = \frac{C_{\text{De}} C_{\text{H}_2\text{e}}}{C_{\text{Be}}^2} \tag{3-17}$$

At equilibrium, the rate law must reduce to an equation consistent wth thermodynamic equilibrium.

Now let's look at the rate law. At equilibrium, $-r_B \equiv 0$, and the rate law given by Equation (3-14) becomes

$$-r_B \equiv 0 = k_B \left[C_{\text{Be}}^2 - \frac{C_{\text{De}} C_{\text{H}_2\text{e}}}{K_C} \right]$$

Rearranging, we obtain, as expected, the equilibrium expression

$$K_C = \frac{C_{\text{De}} C_{\text{H}_2\text{e}}}{C_{\text{Be}}^2}$$

which is identical to Equation (3-17) obtained from thermodynamics.

From Appendix C, Equation (C-9), we know that when there is no change in the total number of moles and the heat capacity term, $\Delta C_P = 0$ the temperature dependence of the concentration equilibrium constant is

$$K_C(T) = K_C(T_1) \exp\left[\frac{\Delta H_{\text{Rx}}}{R} \left(\frac{1}{T_1} - \frac{1}{T} \right) \right] \tag{C-9}$$

Therefore, if we know the equilibrium constant at one temperature, T_1 [i.e., $K_C(T_1)$], and the heat of reaction, ΔH_{Rx}, we can calculate the equilibrium constant at any other temperature T. For endothermic reactions, the equilibrium constant, K_C, increases with increasing temperature; for exothermic reactions, K_C decreases with increasing temperature. A further discussion of the equilibrium constant and its thermodynamic relationship is given in Appendix C.

3.3   The Reaction Rate Constant

The reaction rate constant k is not truly a constant; it is merely independent of the concentrations of the species involved in the reaction. The quantity k is referred to as either the **specific reaction rate** or **the rate constant**. It is almost always strongly dependent on temperature. It depends on whether or not a catalyst is present, and in gas-phase reactions, it may be a function of total pressure. In liquid systems it can also be a function of other parameters, such as ionic strength and choice of solvent. These other variables normally exhibit much less effect on the specific reaction rate than temperature does with the exception of supercritical solvents, such as super critical water.

Consequently, for the purposes of the material presented here, it will be assumed that k_A depends only on temperature. This assumption is valid in most laboratory and industrial reactions and seems to work quite well.

It was the great Swedish chemist Arrhenius who first suggested that the temperature dependence of the specific reaction rate, k_A, could be correlated by an equation of the type

Arrhenius equation

$$k_A(T) = Ae^{-E/RT} \qquad\qquad (3\text{-}18)$$

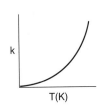

where A = preexponential factor or frequency factor
E = activation energy, J/mol or cal/mol
R = gas constant = 8.314 J/mol · K = 1.987 cal/mol · K
T = absolute temperature, K

Equation (3-18), known as the *Arrhenius equation*, has been verified empirically to give the temperature behavior of most reaction rate constants within experimental accuracy over fairly large temperature ranges. The Arrhenius equation is derived in the Professional Reference Shelf 3.A: *Collision Theory* on the CD-ROM.

Why is there an activation energy? If the reactants are free radicals that essentially react immediately on collision, there usually isn't an activation energy. However, for most atoms and molecules undergoing reaction, there is an activation energy. A couple of the reasons are that in order to react,

1. The molecules need energy to distort or stretch their bonds so that they break them and thus form new bonds.
2. The steric and electron repulsion forces must be overcome as the reacting molecules come close together.

The activation energy can be thought of as a barrier to energy transfer (from the kinetic energy to the potential energy) between reacting molecules that must be overcome. One way to view the barrier to a reaction is through the use of the *reaction coordinates*. These coordinates denote the potential energy of the system as a function of the progress along the reaction path as we go from reactants to an intermediate to products. For the reaction

$$A + BC \underset{\longrightarrow}{\rightleftarrows} A - B - C \longrightarrow AB + C$$

the reaction coordinate is shown in Figure 3-1.

Figure 3-1(a) shows the potential energy of the three atom (or molecule) system, A, B, and C, as well as the reaction progress as we go from reactant species A and BC to products AB and C. Initially A and BC are far apart and the system energy is just the bond energy BC. At the end of the reaction, the products AB and C are far apart, and the system energy is the bond energy AB. As we move along the reaction coordinate (*x*-axis) to the right in Figure 3-1(a), the reactants A and BC approach each other, the BC bond begins to break, and the energy of the reaction pair increases until the top of the barrier is reached. At the top, the *transition state* is reached where the intermolecular distances between AB and between BC are equal (i.e., A–B–C). As a result, the potential energy of the initial three atoms (molecules) is high. As the reaction proceeds

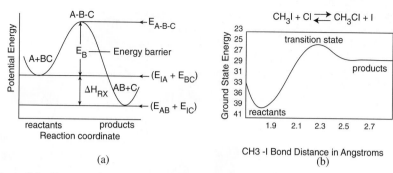

Figure 3-1 Progress along reaction path. (a) Symbolic reaction; (b) Calculated from computational software on the CD-ROM Chapter 3 Web Module.

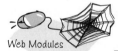

further, the distance between A and B decreases, and the AB bond begins to form. As we proceed further, the distance between AB and C increases and the energy of the reacting pair decreases to that of the AB bond energy. The calculations to arrive at Figure 3-1(b) are discussed in the CD-ROM web module, and transition state theory is discussed in the CD-ROM Professional Reference Shelf R3.2 Transition State Theory.

We see that for the reaction to occur, the reactants must overcome an energy barrier, E_B, shown in Figure 3-1. The energy barrier, E_B, is related to the activation energy, E. The energy barrier height, E_B, can be calculated from differences in the energies of formation of the transition state molecule and the energy of formation of the reactants, that is,

$$E_B = E_{fA-B-C}^{\circ} - (E_{fA}^{\circ} + E_{fB-C}^{\circ}) \qquad (3\text{-}19)$$

The energy of formation of the reactants can be found in the literature while the energy of formation of transition state can be calculated using a number of software programs such as CACHE, Spartan, or Cerius². The activation energy, E_A, is often approximated by the barrier height, which is a good approximation in the absence as quantum mechanical tunneling.

Now that we have the general idea for a reaction coordinate let's consider another real reaction system:

$$H \bullet + C_2H_6 \rightarrow H_2 + C_2H_5 \bullet$$

The energy-reaction coordinate diagram for the reaction between a hydrogen atom and an ethane molecule is shown in Figure 3.2 where the bond distortions, breaking, and forming are identified.

One can also view the activation energy in terms of collision theory (Professional Reference Shelf R3.1). By increasing the temperature, we increase the kinetic energy of the reactant molecules. This kinetic energy can in turn be transferred through molecular collisions to internal energy to increase the stretching and bending of the bonds, causing them to reach an activated state, vulnerable to bond breaking and reaction (cf. Figures 3-1 and 3-2).

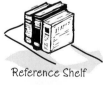
Web Modules

Web Modules

Reference Shelf

Reference Shelf

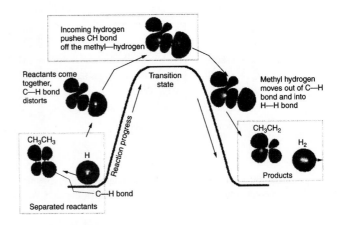

Figure 3-2 A diagram of the orbital distortions during the reaction
$$H \bullet + CH_3CH_3 \rightarrow H_2 + CH_2CH_3 \bullet$$
The diagram shows only the interaction with the energy state of ethane (the C–H bond).
Other molecular orbitals of the ethane also distort. [Courtesy of R. Masel, *Chemical Kinetics*
(McGraw Hill, 2002), p. 594.]

Reference Shelf

The energy of the individual molecules falls within a distribution of energies where some molecules have more energy than others. One such distribution is shown in Figure 3-3 where $f(E,T)$ is the energy distribution function for
the kinetic energies of the reacting molecules. It is interpreted most easily by
recognizing $(f \cdot dE)$ as the fraction of molecules that have an energy between
E and $(E + dE)$. The activation energy has been equated with a minimum
energy that must be possessed by reacting molecules before the reaction will
occur. The fraction of the reacting molecules that have an energy E_A or greater
is shown by the shaded areas in Figure 3-3. The molecules in the shaded area
have sufficient kinetic energy to cause the bond to break and reaction to occur.
One observes that, as the temperature increases, more molecules have sufficient

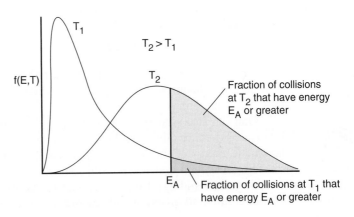

Figure 3-3 Energy distribution of reacting molecules.

energy to react as noted by an increase in the shaded area, and the rate of reaction, $-r_A$, increases.

Postulation of the Arrhenius equation, Equation (3-18), remains the greatest single step in chemical kinetics, and retains its usefulness today, nearly a century later. The activation energy, E, is determined experimentally by carrying out the reaction at several different temperatures. After taking the natural logarithm of Equation (3-18) we obtain

Calculation of the activation energy

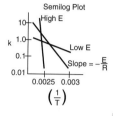

Semilog Plot

$$\ln k_A = \ln A - \frac{E}{R}\left(\frac{1}{T}\right) \tag{3-20}$$

and see that the activation energy can be found from a plot of $\ln k_A$ as a function of $(1/T)$.

Example 3–1 Determination of the Activation Energy

Calculate the activation energy for the decomposition of benzene diazonium chloride to give chlorobenzene and nitrogen:

using the information in Table E3-1.1 for this first-order reaction.

TABLE E3–1.1 DATA

k (s^{-1})	0.00043	0.00103	0.00180	0.00355	0.00717
T (K)	313.0	319.0	323.0	328.0	333.0

Solution

We start by recalling Equation (3-20)

$$\ln k_A = \ln A - \frac{E}{R}\left(\frac{1}{T}\right) \tag{3-20}$$

Summary Notes

Tutorials

We can use the data in Table E3-1.1 to determine the activation energy, E, and frequency factor, A, in two different ways. One way is to make a semilog plot of k vs. $(1/T)$ and determine E from the slope. Another way is to use Excel or Polymath to regress the data. The data in Table E3-1.1 was entered in Excel and is shown in Figure E3-1.1 which was then used to obtain Figure E3-1.2.
A step-by-step tutorial to construct both an Excel and a Polymath spread sheet is given in the Chapter 3 Summary Notes on the CD-ROM.

	A	B	C
1	k (s^{-1})	ln (k)	1/T (K^{-1})
2	0.00043	-7.75	0.00320
3	0.00103	-6.88	0.00314
4	0.00180	-6.32	0.00310
5	0.00355	-5.64	0.00305
6	0.00717	-4.94	0.00300

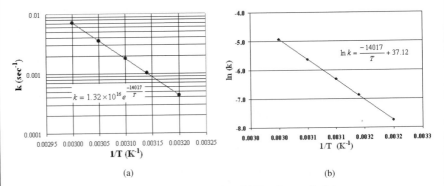

Figure E3-1.1 Excel spreadsheet.

$$k = k_1 \exp\left[\frac{E}{R}\left(\frac{1}{T_1} - \frac{1}{T}\right)\right]$$

(a) (b)

Figure E3-1.2 (a) Excel semilog plot; (b) Excel normal plot.

(a) *Graphical Solution*

Figure E3-1.2(a) shows the semilog plots from which we can calculate the activation energy. From CD-ROM Appendix D, we show how to rearrange Equation (3-20) in the form

$$\log \frac{k_2}{k_1} = \frac{-E}{2.3R}\left(\frac{1}{T_2} - \frac{1}{T_1}\right) \tag{E3-1.1}$$

Rearranging

$$E = -\frac{(2.3)(R)\log(k_2/k_1)}{1/T_2 - 1/T_1} \tag{E3-1.2}$$

To use the decade method, choose $1/T_1$ and $1/T_2$ so that $k_2 = 0.1k_1$. Then, $\log(k_1/k_2) = 1$.

When $k_1 = 0.005$: $\frac{1}{T_1} = 0.003025$ and when $k_2 = 0.0005$: $\frac{1}{T_2} = 0.00319$

Therefore,

$$E = \frac{2.303R}{1/T_2 - 1/T_1} = \frac{(2.303)(8.314 \text{ J/mol} \cdot \text{K})}{(0.00319 - 0.003025)/\text{K}}$$

$$= 116 \frac{\text{kJ}}{\text{mol}} \text{ or } 28.7 \text{ kcal/mol}$$

(b) *Excel Analysis*

The equation for the best fit of the data

$$\ln k = \frac{14,017}{T} + 37.12 \qquad \text{(E3-1.3)}$$

is also shown in Figure E3-1.2(b). From the slope of the line given in Figure 3-1.2(b)

$$-\frac{E}{R} = -14,017 \text{ K}$$

$$E = (14,017 \text{ K})R = (14,017 \text{ K})\left(8.314 \frac{\text{kJ}}{\text{mol} \cdot \text{K}}\right)$$

$$\boxed{E = 116.5 \frac{\text{kJ}}{\text{mol}}}$$

From Figure E3-1.2(b) and Equation (E3-1.3), we see

$$\ln A = 37.12$$

taking the antilog

$$A = 1.32 \times 10^{16} \text{ s}^{-1}$$

$$k = 1.32 \times 10^{16} \, exp\left[-\frac{14,017 \text{ K}}{T}\right] \qquad \text{(E3-1.4)}$$

The rate does not always double for a temperature increase of 10°C.

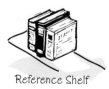

Reference Shelf

There is a rule of thumb that states that the rate of reaction doubles for every 10°C increase in temperature. However, this is true only for a specific combination of activation energy and temperature. For example, if the activation energy is 53.6 kJ/mol, the rate will double only if the temperature is raised from 300 K to 310 K. If the activation energy is 147 kJ/mol, the rule will be valid only if the temperature is raised from 500 K to 510 K. (See Problem P3-7 for the derivation of this relationship.)

The larger the activation energy, the more temperature-sensitive is the rate of reaction. While there are no typical values of the frequency factor and activation energy for a first-order gas-phase reaction, if one were forced to make a guess, values of A and E might be 10^{13} s^{-1} and 200 kJ/mol. However, for families of reactions (e.g., halogenation), a number of correlations can be used to estimate the activation energy. One such correlation is the Polanyi-Semenov equation, which relates activation energy to the heat of reaction (see Professional Reference Shelf 3.1). Another correlation relates activation energy to differences in bond strengths between products and reactants.[3] While activation energy cannot be currently predicted a priori, significant research efforts are under way to calculate activation energies from first principles.[4] (Also see Appendix J.)

One final comment on the Arrhenius equation, Equation (3-18). It can be put in a most useful form by finding the specific reaction rate at a temperature T_0, that is,

$$k(T_0) = Ae^{-E/RT_0}$$

and at a temperature T

$$k(T) = Ae^{-E/RT}$$

A most useful form
of $k(T)$ and taking the ratio to obtain

$$k(T) = k(T_0)e^{\frac{E}{R}\left(\frac{1}{T_0} - \frac{1}{T}\right)}$$

(3-21)

This equation says that if we know the specific reaction rate $k_0(T_0)$ at a temperature, T_0, and we know the activation energy, E, we can find the specific reaction rate $k(T)$ at any other temperature, T, for that reaction.

Where are we?

3.4 Present Status of Our Approach to Reactor Sizing and Design

In Chapter 2, we showed how it was possible to size CSTRs, PFRs, and PBRs using the design equations in Table 3-2 (page 99) *if* the rate of disappearance of A is known as a function of conversion, X:

$$-r_A = g(X)$$

In general, information in the form $-r_A = g(X)$ is not available. However, we have seen in Section 3.2 that the rate of disappearance of A, $-r_A$, is normally expressed in terms of the concentration of the reacting species. This functionality,

$$-r_A = [k_A(T)][\mathrm{fn}(C_A, C_B, \ldots]$$

(3-2)

$-r_A = f(C_j)$

$+$

$C_j = h_j(X)$

$\downarrow$

$-r_A = g(X)$

and then we can
design isothermal
reactors

is called a *rate law*. In Part 2, Sections 3.5 and 3.6, we show how the concentration of the reacting species may be written in terms of the conversion X,

$$C_j = h_j(X)$$

(3-22)

[3] M. Boudart, *Kinetics of Chemical Processes* (Upper Saddle River, N.J.: Prentice Hall, 1968), p. 168. J. W. Moore and R. G. Pearson, *Kinetics and Mechanisms,* 3rd ed. (New York: Wiley, 1981), p. 199. S. W. Benson, *Thermochemical Kinetics,* 2nd ed. (New York: Wiley, 1976).

[4] S. M. Senkan, *Detailed Chemical Kinetic Modeling: Chemical Reaction Engineering of the Future,* Advances in Chemical Engineering, Vol. 18 (San Diego: Academic Press, 1992), pp. 95–96.

TABLE 3-2.   DESIGN EQUATIONS

	Differential Form	Algebraic Form	Integral Form
Batch	$N_{A0} \dfrac{dX}{dt} = -r_A V$ (2-6)		$t = N_{A0} \displaystyle\int_0^X \dfrac{dX}{-r_A V}$ (2-9)
Backmix (CSTR)		$V = \dfrac{F_{A0} X}{-r_A}$ (2-13)	
Tubular (PFR)	$F_{A0} \dfrac{dX}{dV} = -r_A$ (2-15)		$V = F_{A0} \displaystyle\int_0^X \dfrac{dX}{-r_A}$ (2-16)
Packed bed (PBR)	$F_{A0} \dfrac{dX}{dW} = -r'_A$ (2-17)		$W = F_{A0} \displaystyle\int_0^X \dfrac{dX}{-r'_A}$ (2-18)

The design equations

With these additional relationships, one observes that if the rate law is given and the concentrations can be expressed as a function of conversion, *then in fact we have $-r_A$ as a function of X and this is all that is needed to evaluate the design equations.* One can use either the numerical techniques described in Chapter 2, or, as we shall see in Chapter 4, a table of integrals, and/or software programs (e.g., Polymath).

PART 2   STOICHIOMETRY

Now that we have shown how the rate law can be expressed as a function of concentrations, we need only express concentration as a function of conversion in order to carry out calculations similar to those presented in Chapter 2 to size reactors. If the rate law depends on more than one species, we must relate the concentrations of the different species to each other. This relationship is most easily established with the aid of a stoichiometric table. This table presents the stoichiometric relationships between reacting molecules for a single reaction. That is, it tells us how many molecules of one species will be formed during a chemical reaction when a given number of molecules of another species disappears. These relationships will be developed for the general reaction

$$aA + bB \underset{\longleftarrow}{\overset{\longrightarrow}{\ \ \ }} cC + dD \tag{2-1}$$

This stoichiometric relationship relating reaction rates will be used in Part 2 of Chapter 4.

Recall that we have already used stoichiometry to relate the relative rates of reaction for Equation (2-1):

$$\frac{r_A}{-a} = \frac{r_B}{-b} = \frac{r_C}{c} = \frac{r_D}{d} \tag{3-1}$$

In formulating our stoichiometric table, we shall take species A as our basis of calculation (i.e., limiting reactant) and then divide through by the stoichiometric coefficient of A,

$$A + \frac{b}{a} B \longrightarrow \frac{c}{a} C + \frac{d}{a} D \tag{2-2}$$

in order to put everything on a basis of "per mole of A."

Next, we develop the stoichiometric relationships for reacting species that give the change in the number of moles of each species (i.e., A, B, C, and D).

3.5 Batch Systems

Batch reactors are primarily used for the production of specialty chemicals and to obtain reaction rate data in order to determine reaction rate laws and rate law parameters such as k, the specific reaction rate.

Figure 3-4 shows an artist's rendition of a batch system in which we will carry out the reaction given by Equation (2-2). At time $t = 0$, we will open the reactor and place a number of moles of species A, B, C, D, and I (N_{A0}, N_{B0}, N_{C0}, N_{D0}, and N_I, respectively) into the reactor.

Species A is our basis of calculation, and N_{A0} is the number of moles of A initially present in the reactor. Of these, $N_{A0}X$ moles of A are consumed in the system as a result of the chemical reaction, leaving ($N_{A0} - N_{A0}X$) moles of A in the system. That is, the number of moles of A remaining in the reactor after a conversion X has been achieved is

$$N_A = N_{A0} - N_{A0}X = N_{A0}(1 - X)$$

We now will use conversion in this fashion to express the number of moles of B, C, and D in terms of conversion.

To determine the number of moles of each species remaining after $N_{A0}X$ moles of A have reacted, we form the stoichiometric table (Table 3-3). This stoichiometric table presents the following information:

Column 1: the particular species

Components of the stoichiometric table

Column 1: the particular species
Column 2: the number of moles of each species initially present
Column 3: the change in the number of moles brought about by reaction
Column 4: the number of moles remaining in the system at time *t*

To calculate the number of moles of species B remaining at time *t*, we recall that at time *t* the number of moles of A that have reacted is $N_{A0}X$. For every mole of A that reacts, b/a moles of B must react; therefore, the total number of moles of B that have reacted is

$$\text{moles B reacted} = \frac{\text{moles B reacted}}{\text{moles A reacted}} \cdot \text{moles A reacted}$$

$$= \frac{b}{a}(N_{A0}X)$$

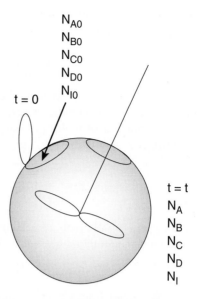

N_{A0}
N_{B0}
N_{C0}
N_{D0}
N_{I0}

t = 0

t = t
N_A
N_B
N_C
N_D
N_I

Figure 3-4   Batch reactor. (Schematic with permission by Renwahr.)

TABLE 3-3.   STOICHIOMETRIC TABLE FOR A BATCH SYSTEM

Species	Initially (mol)	Change (mol)	Remaining (mol)
A	N_{A0}	$-(N_{A0}X)$	$N_A = N_{A0} - N_{A0}X$
B	N_{B0}	$-\dfrac{b}{a}(N_{A0}X)$	$N_B = N_{B0} - \dfrac{b}{a}N_{A0}X$
C	N_{C0}	$\dfrac{c}{a}(N_{A0}X)$	$N_C = N_{C0} + \dfrac{c}{a}N_{A0}X$
D	N_{D0}	$\dfrac{d}{a}(N_{A0}X)$	$N_D = N_{D0} + \dfrac{d}{a}N_{A0}X$
I (inerts)	$\underline{N_{I0}}$	—	$N_I = N_{I0}$
Totals	N_{T0}		$N_T = N_{T0} + \left(\dfrac{d}{a} + \dfrac{c}{a} - \dfrac{b}{a} - 1\right)N_{A0}X$

Because B is disappearing from the system, the sign of the "change" is negative. N_{B0} is the number of moles initially in the system. Therefore, the number of moles of B remaining in the system, N_B, at a time t, is given in the last column of Table 3-3 as

$$N_B = N_{B0} - \frac{b}{a}N_{A0}X$$

The complete stoichiometric table delineated in Table 3-3 is for all species in the general reaction

$$A + \frac{b}{a} B \longrightarrow \frac{c}{a} C + \frac{d}{a} D \qquad (2\text{-}2)$$

Let's take a look at the totals in the last column of Table 3-3. The stoichiometric coefficients in parentheses $(d/a + c/a - b/a - 1)$ represent the increase in the total number of moles per mole of A reacted. Because this term occurs so often in our calculations, it is given the symbol δ:

$$\boxed{\delta = \frac{d}{a} + \frac{c}{a} - \frac{b}{a} - 1} \qquad (3\text{-}23)$$

The parameter δ tells us the change in the total number of moles per mole of A reacted. The total number of moles can now be calculated from the equation

$$N_T = N_{T0} + \delta N_{A0} X$$

We recall from Chapter 1 and Part 1 of this chapter that the kinetic rate law (e.g., $-r_A = kC_A^2$) is a function solely of the intensive properties of the reacting system (e.g., temperature, pressure, concentration, and catalysts, if any). The reaction rate, $-r_A$, usually depends on the concentration of the reacting species raised to some power. Consequently, to determine the reaction rate as a function of conversion X, we need to know the concentrations of the reacting species as a function of conversion.

We want
$C_j = h_j(X)$

3.5.1 Equations for Batch Concentrations

The concentration of A is the number of moles of A per unit volume:

Batch
concentration

$$C_A = \frac{N_A}{V}$$

After writing similar equations for B, C, and D, we use the stoichiometric table to express the concentration of each component in terms of the conversion X:

$$C_A = \frac{N_A}{V} = \frac{N_{A0}(1 - X)}{V}$$

$$C_B = \frac{N_B}{V} = \frac{N_{B0} - (b/a) N_{A0} X}{V}$$

$$C_C = \frac{N_C}{V} = \frac{N_{C0} + (c/a) N_{A0} X}{V} \qquad (3\text{-}24)$$

$$C_D = \frac{N_D}{V} = \frac{N_{D0} + (d/a) N_{A0} X}{V}$$

We further simplify these equations by defining the parameter Θ_i, which allows us to factor N_{A0} in each of the expressions for concentration:

$$\boxed{\Theta_i = \frac{N_{i0}}{N_{A0}} = \frac{C_{i0}}{C_{A0}} = \frac{y_{i0}}{y_{A0}}},$$

$$C_B = \frac{N_{A0}[N_{B0}/N_{A0} - (b/a)X]}{V} = \frac{N_{A0}[\Theta_B - (b/a)X]}{V}, \quad \text{with } \Theta_B = \frac{N_{B0}}{N_{A0}}$$

$$(3\text{-}25)$$

$$C_C = \frac{N_{A0}[\Theta_C + (c/a)X]}{V}, \quad \text{with } \Theta_C = \frac{N_{C0}}{N_{A0}}$$

$$C_D = \frac{N_{A0}[\Theta_D + (d/a)X]}{V}, \quad \text{with } \Theta_D = \frac{N_{D0}}{N_{A0}}$$

We need $V(X)$ to obtain $C_j = h_j(X)$.   We now need only to find volume as a function of conversion to obtain the species concentration as a function of conversion.

3.5.2 Constant-Volume Batch Reaction Systems

Some significant simplifications in the reactor design equations are possible when the reacting system undergoes no change in volume as the reaction progresses. These systems are called constant-volume, or constant-density, because of the invariance of either volume or density during the reaction process. This situation may arise from several causes. In gas-phase batch systems, the reactor is usually a sealed constant-volume vessel with appropriate instruments to measure pressure and temperature within the reactor. The volume within this vessel is fixed and will not change, and is therefore a constant-volume system ($V = V_0$). The laboratory bomb calorimeter reactor is a typical example of this type of reactor.

Another example of a constant-volume gas-phase isothermal reaction occurs when the number of moles of products equals the number of moles of reactants. The water-gas shift reaction, important in coal gasification and many other processes, is one of these:

$$CO + H_2O \rightleftharpoons CO_2 + H_2$$

In this reaction, 2 mol of reactant forms 2 mol of product. When the number of reactant molecules forms an equal number of product molecules at the *same* temperature and pressure, the volume of the reacting mixture will not change if the conditions are such that the ideal gas law is applicable, or if the compressibility factors of the products and reactants are approximately equal.

For liquid-phase reactions taking place in solution, the solvent usually dominates the situation. As a result, changes in the density of the solute **do not** affect the overall density of the solution significantly and therefore it is essentially a constant-volume reaction process. Most liquid-phase organic reactions do not change density during the reaction and represent still another case to which the constant-volume simplifications apply. An important exception to this general rule exists for polymerization processes.

For the constant-volume systems described earlier, Equation (3-25) can be simplified to give the following expressions relating concentration and conversion:

$$V = V_0$$

Concentration as a function of conversion when no volume change occurs with reaction

$$C_A = \frac{N_{A0}(1-X)}{V_0} = C_{A0}(1-X)$$

$$C_B = N_{A0}\frac{[(N_{B0}/N_{A0}) - (b/a)X]}{V_0} = \frac{N_{A0}[\Theta_B - (b/a)X]}{V_0} = C_{A0}\left(\Theta_B - \frac{b}{a}X\right)$$

$$C_C = N_{A0}\frac{[(N_{C0}/N_{A0}) + (c/a)X]}{V_0} = C_{A0}\left(\Theta_C + \frac{c}{a}X\right)$$

$$C_D = N_{A0}\frac{[(N_{D0}/N_{A0}) + (d/a)X]}{V_0} = C_{A0}\left(\Theta_D + \frac{d}{a}X\right)$$

$$(3\text{-}26)$$

To summarize for liquid-phase reactions (or as we will soon see for isothermal and isobaric gas-phase reactions with no change in the total number of moles), we can use a rate law for reaction (2-2) such as $-r_A = k_A C_A C_B$ to obtain $-r_A = f(X)$, that is,

$$-r_A = kC_A C_B = kC_{A0}^2(1-X)\left(\Theta_B - \frac{b}{a}X\right) = f(X)$$

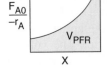

Substituting for the given parameters k, C_{A0}, and Θ_B, we can now use the techniques in Chapter 2 to size the CSTRs and PFRs for liquid-phase reactions.

Example 3–2 Expressing $C_j = h_j(X)$ for a Liquid-Phase Reaction

Soap consists of the sodium and potassium salts of various fatty acids such as oleic, stearic, palmitic, lauric, and myristic acids. The saponification for the formation of soap from aqueous caustic soda and glyceryl stearate is

$$3\text{NaOH(aq)} + (C_{17}H_{35}COO)_3C_3H_5 \longrightarrow 3C_{17}H_{35}COONa + C_3H_5(OH)_3$$

Letting X represent the conversion of sodium hydroxide (the moles of sodium hydroxide reacted per mole of sodium hydroxide initially present), set up a stoichiometric table expressing the concentration of each species in terms of its initial concentration and the conversion X.

Solution

Because we are taking sodium hydroxide as our basis, we divide through by the stoichiometric coefficient of sodium hydroxide to put the reaction expression in the form

Choosing a basis of
calculation

$$\text{NaOH} + \tfrac{1}{3}(C_{17}H_{35}COO)_3C_3H_5 \longrightarrow C_{17}H_{35}COONa + \tfrac{1}{3}C_3H_5(OH)_3$$

$$A \quad + \quad \tfrac{1}{3}B \quad \longrightarrow \quad C \quad + \quad \tfrac{1}{3}D$$

We may then perform the calculations shown in Table E3-2.1. Because this is a liquid-phase reaction, the density ρ is considered to be constant; therefore, $V = V_0$.

$$C_A = \frac{N_A}{V} = \frac{N_A}{V_0} = \frac{N_{A0}(1-X)}{V_0} = C_{A0}(1-X)$$

$$\Theta_B = \frac{C_{B0}}{C_{A0}} \quad \Theta_C = \frac{C_{C0}}{C_{A0}} \quad \Theta_D = \frac{C_{D0}}{C_{A0}}$$

TABLE E3-2.1.   STOICHIOMETRIC TABLE FOR LIQUID-PHASE SOAP REACTION

Species	Symbol	Initially	Change	Remaining	Concentration
NaOH	A	N_{A0}	$-N_{A0}X$	$N_{A0}(1-X)$	$C_{A0}(1-X)$
$(C_{17}H_{35}COO)_3C_3H_5$	B	N_{B0}	$-\tfrac{1}{3}N_{A0}X$	$N_{A0}\left(\Theta_B - \dfrac{X}{3}\right)$	$C_{A0}\left(\Theta_B - \dfrac{X}{3}\right)$
$C_{17}H_{35}COONa$	C	N_{C0}	$N_{A0}X$	$N_{A0}(\Theta_C + X)$	$C_{A0}(\Theta_C + X)$
$C_3H_5(OH)_3$	D	N_{D0}	$\tfrac{1}{3}N_{A0}X$	$N_{A0}\left(\Theta_D + \dfrac{X}{3}\right)$	$C_{A0}\left(\Theta_D + \dfrac{X}{3}\right)$
Water (inert)	I	$\dfrac{N_{I0}}{N_{T0}}$	—	$\dfrac{N_{I0}}{N_T = N_{T0}}$	C_{I0}

Stoichiometric table
(batch)

Example 3–3   What Is the Limiting Reactant?

Having set up the stoichiometric table in Example 3-2, one can now readily use it to calculate the concentrations at a given conversion. If the initial mixture consists solely of sodium hydroxide at a concentration of 10 mol/dm³ (i.e., 10 mol/L or 10 kmol/m³) and of glyceryl stearate at a concentration of 2 mol/dm³, what is the concentration of glycerine when the conversion of sodium hydroxide is **(a)** 20% and **(b)** 90%?

Solution

Only the reactants NaOH and $(C_{17}H_{35}COO)_3C_3H_5$ are initially present; therefore, $\Theta_C = \Theta_D = 0$.

(a) For 20% conversion of NaOH:

$$C_D = C_{A0}\left(\frac{X}{3}\right) = (10)\left(\frac{0.2}{3}\right) = 0.67 \text{ mol/L} = 0.67 \text{ mol/dm}^3$$

$$C_B = C_{A0}\left(\Theta_B - \frac{X}{3}\right) = 10\left(\frac{2}{10} - \frac{0.2}{3}\right) = 10(0.133) = 1.33 \text{ mol/dm}^3$$

(b) For 90% conversion of NaOH:

$$C_D = C_{A0}\left(\frac{X}{3}\right) = 10\left(\frac{0.9}{3}\right) = 3 \text{ mol/dm}^3$$

Let us find C_B:

$$C_B = 10\left(\frac{2}{10} - \frac{0.9}{3}\right) = 10(0.2 - 0.3) = -1 \text{ mol/dm}^3$$

Oops!! Negative concentration—impossible! What went wrong?

> The basis of calculation should be the limiting reactant.

Ninety percent conversion of NaOH is not possible, because glyceryl stearate is the limiting reactant. Consequently, all the glyceryl stearate is used up before 90% of the NaOH could be reacted. It is important to choose the limiting reactant as the basis of calculation.

3.6 Flow Systems

The form of the stoichiometric table for a continuous-flow system (see Figure 3-5) is virtually identical to that for a batch system (Table 3-3) except that we replace N_{j0} by F_{j0} and N_j by F_j (Table 3-4). Taking A as the basis, divide Equation (2-1) through by the stoichiometric coefficient of A to obtain

$$A + \frac{b}{a} B \longrightarrow \frac{c}{a} C + \frac{d}{a} D \qquad (2\text{-}2)$$

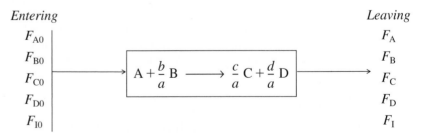

Entering   *Leaving*

F_{A0}		F_A
F_{B0}		F_B
F_{C0}	$A + \dfrac{b}{a} B \longrightarrow \dfrac{c}{a} C + \dfrac{d}{a} D$	F_C
F_{D0}		F_D
F_{I0}		F_I

Figure 3-5   Flow reactor.

TABLE 3-4.   STOICHIOMETRIC TABLE FOR A FLOW SYSTEM

Species	Feed Rate to Reactor (mol/time)	Change within Reactor (mol/time)	Effluent Rate from Reactor (mol/time)
A	F_{A0}	$-F_{A0}X$	$F_A = F_{A0}(1-X)$
B	$F_{B0} = \Theta_B F_{A0}$	$-\dfrac{b}{a}F_{A0}X$	$F_B = F_{A0}\left(\Theta_B - \dfrac{b}{a}X\right)$
C	$F_{C0} = \Theta_C F_{A0}$	$\dfrac{c}{a}F_{A0}X$	$F_C = F_{A0}\left(\Theta_C + \dfrac{c}{a}X\right)$
D	$F_{D0} = \Theta_D F_{A0}$	$\dfrac{d}{a}F_{A0}X$	$F_D = F_{A0}\left(\Theta_D + \dfrac{d}{a}X\right)$
I	$F_{I0} = \Theta_I F_{A0}$	—	$F_I = F_{A0}\Theta_I$
	F_{T0}		$F_T = F_{T0} + \left(\dfrac{d}{a} + \dfrac{c}{a} - \dfrac{b}{a} - 1\right)F_{A0}X$
			$F_T = F_{T0} + \delta F_{A0}X$

Stoichiometric table (flow)

where

$$\Theta_B = \frac{F_{B0}}{F_{A0}} = \frac{C_{B0}v_0}{C_{A0}v_0} = \frac{C_{B0}}{C_{A0}} = \frac{y_{B0}}{y_{A0}}$$

and Θ_C, Θ_D, and Θ_I are defined similarly.

3.6.1 Equations for Concentrations in Flow Systems

For a flow system, the concentration C_A at a given point can be determined from the molar flow rate F_A and the volumetric flow rate v at that point:

Definition of concentration for a flow system

$$C_A = \frac{F_A}{v} = \frac{\text{moles/time}}{\text{liters/time}} = \frac{\text{moles}}{\text{liter}} \qquad (3\text{-}27)$$

Units of v are typically given in terms of liters per second, cubic decimeters per second, or cubic feet per minute. We now can write the concentrations of A, B, C, and D for the general reaction given by Equation (2-2) in terms of their respective entering molar flow rates (F_{A0}, F_{B0}, F_{C0}, F_{D0}), the conversion X, and the volumetric flow rate, v.

$$C_A = \frac{F_A}{v} = \frac{F_{A0}}{v}(1-X) \qquad C_B = \frac{F_B}{v} = \frac{F_{B0} - (b/a)F_{A0}X}{v}$$

$$C_C = \frac{F_C}{v} = \frac{F_{C0} + (c/a)F_{A0}X}{v} \qquad C_D = \frac{F_D}{v} = \frac{F_{D0} + (d/a)F_{A0}X}{v} \qquad (3\text{-}28)$$

3.6.2 Liquid-Phase Concentrations

For liquids, volume change with reaction is negligible when no phase changes are taking place. Consequently, we can take

$$v = v_0$$

For liquids   Then

$$C_A = C_{A0}(1 - X)$$

$$C_B = C_{A0}\left(\Theta_B - \frac{b}{a}X\right)$$

Therefore, for a given rate law we have

$$-r_A = g(X)$$

$$C_A = \frac{F_{A0}}{v_0}(1 - X) = C_{A0}(1 - X)$$

$$C_B = C_{A0}\left(\Theta_B - \frac{b}{a}X\right) \quad \text{etc.}$$

(3-29)

Consequently, using any one of the rate laws in Part 1 of this chapter, we can now find $-r_A = f(X)$ for liquid-phase reactions. However, for gas-phase reactions the volumetric flow rate most often changes during the course of the reaction because of a change in the total number of moles or in temperature or pressure. Hence, one cannot always use Equation (3-29) to express concentration as a function of conversion for gas-phase reactions.

3.6.3 Change in the Total Number of Moles with Reaction in the Gas Phase

In our previous discussions, we considered primarily systems in which the reaction volume or volumetric flow rate did not vary as the reaction progressed. Most batch and liquid-phase and some gas-phase systems fall into this category. There are other systems, though, in which either V or v **do** vary, and these will now be considered.

A situation where one encounters a varying flow rate occurs quite frequently in gas-phase reactions that do not have an equal number of product and reactant moles. For example, in the synthesis of ammonia,

$$N_2 + 3H_2 \rightleftharpoons 2NH_3$$

4 mol of reactants gives 2 mol of product. In flow systems where this type of reaction occurs, the molar flow rate will be changing as the reaction progresses. Because equal numbers of moles occupy equal volumes in the gas phase at the same temperature and pressure, the volumetric flow rate will also change.

Another variable-volume situation, which occurs much less frequently, is in batch reactors where volume changes with time. Everyday examples of this situation are the combustion chamber of the internal-combustion engine and the expanding gases within the breech and barrel of a firearm as it is fired.

In the stoichiometric tables presented on the preceding pages, it was not necessary to make assumptions concerning a volume change in the first four columns of the table (i.e., the species, initial number of moles or molar feed rate, change within the reactor, and the remaining number of moles or the molar effluent rate). All of these columns of the stoichiometric table are independent of the volume or density, and they are *identical* for constant-volume (constant-density) and varying-volume (varying-density) situations. Only when concentration is expressed as a function of conversion does variable density enter the picture.

Batch Reactors with Variable Volume Although variable volume batch reactors are seldom encountered because they are usually solid steel containers, we will develop the concentrations as a function of conversion because (1) they have been used to collect reaction data for gas-phase reactions, and (2) the development of the equations that express volume as a function of conversion will facilitate analyzing flow systems with variable volumetric flow rates.

Individual concentrations can be determined by expressing the volume V for a batch system, or volumetric flow rate v for a flow system, as a function of conversion using the following equation of state:

Equation of state

$$PV = ZN_T RT \qquad (3\text{-}30)$$

in which V = volume and N_T = total number of moles as before and

T = temperature, K

P = total pressure, atm (kPa; 1 atm = 101.3 kPa)

Z = compressibility factor

R = gas constant = 0.08206 dm$^3 \cdot$ atm/mol $\cdot$ K

This equation is valid at any point in the system at any time t. At time $t = 0$ (i.e., when the reaction is initiated), Equation (3-30) becomes

$$P_0 V_0 = Z_0 N_{T0} RT_0 \qquad (3\text{-}31)$$

Dividing Equation (3-30) by Equation (3-31) and rearranging yields

$$V = V_0 \left(\frac{P_0}{P}\right) \frac{T}{T_0} \left(\frac{Z}{Z_0}\right) \frac{N_T}{N_{T0}} \qquad (3\text{-}32)$$

We now want to express the volume V as a function of the conversion X. Recalling the equation for the total number of moles in Table 3-3,

$$N_T = N_{T0} + \delta N_{A0} X \qquad (3\text{-}33)$$

where

$$\boxed{\delta = \frac{d}{a} + \frac{c}{a} - \frac{b}{a} - 1} \qquad (3\text{-}23)$$

$$\delta = \frac{\text{Change in total number of moles}}{\text{Mole of A reacted}}$$

We divide Equation (3-33) through by N_{T0}:

$$\frac{N_T}{N_{T0}} = 1 + \frac{N_{A0}}{N_{T0}}\delta X = 1 + \overbrace{\delta y_{A0}}^{\varepsilon} X$$

Then

$$\frac{N_T}{N_{T0}} = 1 + \varepsilon X \tag{3-34}$$

Relationship between δ and ε

where y_{A0} is the mole fraction of A initially present, and where

$$\varepsilon = \left(\frac{d}{a} + \frac{c}{a} - \frac{b}{a} - 1\right)\frac{N_{A0}}{N_{T0}} = y_{A0}\delta \tag{3-35}$$

$$\boxed{\varepsilon = y_{A0}\delta}$$

Equation (3-35) holds for both batch and flow systems. To interpret ε, let's rearrange Equation (3-34)

$$\varepsilon = \frac{N_T - N_{T0}}{N_{T0}X}$$

Interpretation of ε

at complete conversion, (i.e., $X = 1$ and $N_T = N_{Tf}$)

$$\varepsilon = \frac{N_{Tf} - N_{T0}}{N_{T0}}$$

$$= \frac{\text{Change in total number of moles for complete conversion}}{\text{Total moles fed}} \tag{3-36}$$

If all species in the generalized equation are in the gas phase, we can substitute Equation (3-34) with Equation (3-32) to arrive at

$$V = V_0\left(\frac{P_0}{P}\right)\frac{T}{T_0}\left(\frac{Z}{Z_0}\right)(1 + \varepsilon X) \tag{3-37}$$

In the gas-phase systems that we shall be studying, the temperatures and pressures are such that the compressibility factor will not change significantly during the course of the reaction; hence $Z_0 \cong Z$. For a batch system, the volume of gas at any time t is

Volume of gas for a variable volume batch reaction

$$\boxed{V = V_0\left(\frac{P_0}{P}\right)(1 + \varepsilon X)\frac{T}{T_0}} \tag{3-38}$$

Equation (3-38) applies only to a *variable-volume* batch reactor, where one can now substitute Equation (3-38) into Equation (3-25) to express $r_A = f(X)$. However, if the reactor is a rigid steel container of constant volume, then of course

$V = V_0$. For a constant-volume container, $V = V_0$, and Equation (3-38) can be used to calculate the gas pressure inside the reactor as a function of temperature and conversion.

Flow Reactors with Variable Volumetric Flow Rate. An expression similar to Equation (3-38) for a variable-volume batch reactor exists for a variable-volume flow system. To derive the concentrations of each species in terms of conversion for a variable-volume flow system, we shall use the relationships for the total concentration. The total concentration, C_T, at any point in the reactor is the total molar flow rate, F_T, divided by volumetric flow rate v [cf. Equation (3-27)]. In the gas phase, the total concentration is also found from the gas law, $C_T = P/ZRT$. Equating these two relationships gives

$$C_T = \frac{F_T}{v} = \frac{P}{ZRT} \tag{3-39}$$

At the entrance to the reactor,

$$C_{T0} = \frac{F_{T0}}{v_0} = \frac{P_0}{Z_0 R T_0} \tag{3-40}$$

Taking the ratio of Equation (3-40) to Equation (3-39) and assuming negligible changes in the compressibility factor, we have upon rearrangement

$$v = v_0 \left(\frac{F_T}{F_{T0}} \right) \frac{P_0}{P} \left(\frac{T}{T_0} \right) \tag{3-41}$$

We can now express the concentration of species j for a flow system in terms of its flow rate, F_j, the temperature, T, and total pressure, P.

$$C_j = \frac{F_j}{v} = \frac{F_j}{v_0 \left(\dfrac{F_T}{F_{T0}} \dfrac{P_0}{P} \dfrac{T}{T_0} \right)} = \left(\frac{F_{T0}}{v_0} \right) \left(\frac{F_j}{F_T} \right) \left(\frac{P}{P_0} \right) \left(\frac{T_0}{T} \right)$$

Use this form for membrane reactors (Chapter 4) and for multiple reactions (Chapter 6)

$$C_j = C_{T0} \left(\frac{F_j}{F_T} \right) \left(\frac{P}{P_0} \right) \left(\frac{T_0}{T} \right) \tag{3-42}$$

The total molar flow rate is just the sum of the molar flow rates of each of the species in the system and is

$$F_T = F_A + F_B + F_C + F_D + F_I + \cdots = \sum_{j=1}^{n} F_j$$

One of the major objectives of this chapter is to learn how to express any given rate law $-r_A$ as a function of conversion. The schematic diagram in Figure 3-6 helps to summarize our discussion on this point. The concentration of the key reactant, A (the basis of our calculations), is expressed as a function of conversion in both flow and batch systems, for various conditions of temperature, pressure, and volume.

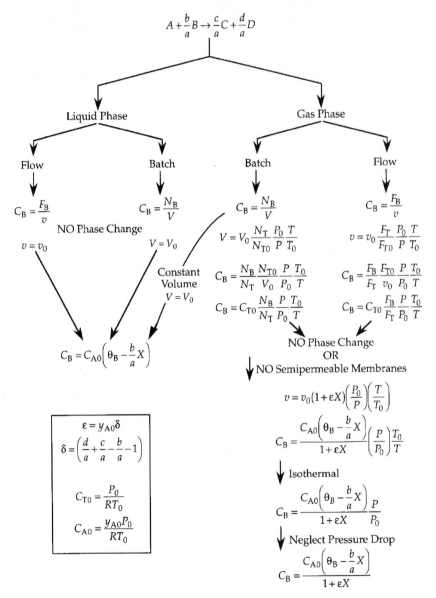

Figure 3-6   Expressing concentration as a function of conversion.

We see that conversion is not used in this sum. The molar flow rates, F_j, are found by solving the mole balance equations. Equation (3-42) will be used for measures other than conversion when we discuss membrane reactors (Chapter 4 Part 2) and multiple reactions (Chapter 6). We will use this form of the concentration equation for multiple gas-phase reactions and for membrane reactors.

Now let's express the concentration in terms of conversion for gas flow systems. From Table 3-4, the total molar flow rate can be written in terms of conversion and is

$$F_T = F_{T0} + F_{A0}\,\delta X \tag{3-43}$$

Substituting for F_T in Equation (3-41) gives

$$v = v_0 \frac{F_{T0} + F_{A0}\,\delta X}{F_{T0}} \left(\frac{P_0}{P}\right)\frac{T}{T_0}$$

$$= v_0\left(1 + \frac{F_{A0}}{F_{T0}}\,\delta X\right)\frac{P_0}{P}\left(\frac{T}{T_0}\right) = v_0(1 + y_{A0}\,\delta X)\frac{P_0}{P}\left(\frac{T}{T_0}\right) \tag{3-44}$$

Gas-phase volumetric flow rate

$$\boxed{v = v_0(1 + \varepsilon X)\frac{P_0}{P}\left(\frac{T}{T_0}\right)} \tag{3-45}$$

The concentration of species j is

$$C_j = \frac{F_j}{v}$$

The molar flow rate of species j is

$$F_j = F_{j0} + v_j(F_{A0}X) = F_{A0}(\Theta_j + v_jX)$$

where v_i is the stoichiometric coefficient, which is negative for reactants and positive for products. For example, for the reaction

$$A + \frac{b}{a}\,B \longrightarrow \frac{c}{a}\,C + \frac{d}{a}\,D \tag{2-2}$$

$v_A = -1$, $v_B = -b/a$, $v_C = c/a$, $v_D = d/a$, and $\Theta_j = F_{j0}/F_{A0}$.

Substituting for v using Equation (3-42) and for F_j, we have

$$C_j = \frac{F_{A0}(\Theta_j + v_jX)}{v_0\left((1 + \varepsilon X)\dfrac{P_0}{P}\dfrac{T}{T_0}\right)}$$

Rearranging

$$C_j = \frac{C_{A0}(\Theta_j + v_j X)}{1 + \varepsilon X}\left(\frac{P}{P_0}\right)\frac{T_0}{T} \qquad (3\text{-}46)$$

Recall that $y_{A0} = F_{A0}/F_{T0}$, $C_{A0} = y_{A0}C_{T0}$, and ε from Equation (3-35) (i.e., $\varepsilon = y_{A0}\delta$).

The stoichiometric table for the gas-phase reaction (2-2) is given in Table 3-5.

TABLE 3-5.   CONCENTRATIONS IN A VARIABLE-VOLUME GAS FLOW SYSTEM

$$C_A = \frac{F_A}{v} = \frac{F_{A0}(1-X)}{v} \quad = \frac{F_{A0}(1-X)}{v_0(1+\varepsilon X)}\left(\frac{T_0}{T}\right)\frac{P}{P_0} \quad = C_{A0}\left(\frac{1-X}{1+\varepsilon X}\right)\frac{T_0}{T}\left(\frac{P}{P_0}\right)$$

At last!
We now have
$C_j = h_j(X)$
and
$-r_A = g(X)$
for variable-volume
gas-phase reactions

$$C_B = \frac{F_B}{v} = \frac{F_{A0}[\Theta_B - (b/a)X]}{v} = \frac{F_{A0}[\Theta_B - (b/a)X]}{v_0(1+\varepsilon X)}\left(\frac{T_0}{T}\right)\frac{P}{P_0} = C_{A0}\left(\frac{\Theta_B - (b/a)X}{1+\varepsilon X}\right)\frac{T_0}{T}\left(\frac{P}{P_0}\right)$$

$$C_C = \frac{F_C}{v} = \frac{F_{A0}[\Theta_C + (c/a)X]}{v} = \frac{F_{A0}[\Theta_C + (c/a)X]}{v_0(1+\varepsilon X)}\left(\frac{T_0}{T}\right)\frac{P}{P_0} = C_{A0}\left(\frac{\Theta_C + (c/a)X}{1+\varepsilon X}\right)\frac{T_0}{T}\left(\frac{P}{P_0}\right)$$

$$C_D = \frac{F_D}{v} = \frac{F_{A0}[\Theta_D + (d/a)X]}{v} = \frac{F_{A0}[\Theta_D + (d/a)X]}{v_0(1+\varepsilon X)}\left(\frac{T_0}{T}\right)\frac{P}{P_0} = C_{A0}\left(\frac{\Theta_D + (d/a)X}{1+\varepsilon X}\right)\frac{T_0}{T}\left(\frac{P}{P_0}\right)$$

$$C_I = \frac{F_I}{v} = \frac{F_{A0}\Theta_I}{v} \qquad = \frac{F_{A0}\Theta_I}{v_0(1+\varepsilon X)}\left(\frac{T_0}{T}\right)\frac{P}{P_0} \qquad = \frac{C_{A0}\Theta_I}{1+\varepsilon X}\left(\frac{T_0}{T}\right)\frac{P}{P_0}$$

Example 3–4 Manipulation of the Equation for $C_j = h_j(X)$

Show under what conditions and manipulation the expression for C_B for a gas flow system reduces to that given in Table 3-5.

Solution

For a flow system the concentration is *defined* as

$$C_B = \frac{F_B}{v} \qquad (E3\text{-}4.1)$$

From Table 3-4, the molar flow rate and conversion are related by

$$F_B = F_{A0}\left(\Theta_B - \frac{b}{a}X\right) \qquad (E3\text{-}4.2)$$

Combining Equations (E3-4.1) and (E3-4.2) yields

$$C_B = \frac{F_{A0}[\Theta_B - (b/a)X]}{v} \qquad (E3\text{-}4.3)$$

This equation for v is only for a gas-phase reaction

Using Equation (3-45) gives us

$$v = v_0(1 + \varepsilon X)\frac{P_0}{P}\left(\frac{T}{T_0}\right) \tag{3-45}$$

to substitute for the volumetric flow rate gives

$$C_B = \frac{F_B}{v} = \frac{F_{A0}[\Theta_B - (b/a)X]}{v_0(1 + \varepsilon X)}\left(\frac{P}{P_0}\right)\frac{T_0}{T} \tag{E3-4.4}$$

Recalling $\dfrac{F_{A0}}{v_0} = C_{A0}$, we obtain

$$C_B = C_{A0}\left[\frac{\Theta_B - (b/a)X}{1 + \varepsilon X}\right]\frac{P}{P_0}\left(\frac{T_0}{T}\right)$$

which is identical to the concentration expression for a variable-volume batch reactor.

Example 3–5 Determining $C_j = h_j(X)$ for a Gas-Phase Reaction

A mixture of 28% SO_2 and 72% air is charged to a flow reactor in which SO_2 is oxidized.

$$2SO_2 + O_2 \longrightarrow 2SO_3$$

First, set up a stoichiometric table using only the symbols (i.e., Θ_i, F_i) and then prepare a second stoichiometric table evaluating numerically as many symbols as possible for the case when the total pressure is 1485 kPa (14.7 atm) and the temperature is constant at 227°C.

Solution

Taking SO_2 as the basis of calculation, we divide the reaction through by the stoichiometric coefficient of our chosen basis of calculation:

$$SO_2 + \tfrac{1}{2}O_2 \longrightarrow SO_3$$

The initial stoichiometric table is given as Table E3-5.1. Initially, 72% of the total number of moles is air containing (21% O_2 and 79% N_2) along with 28% SO_2.

$$F_{A0} = (0.28)(F_{T0})$$
$$F_{B0} = (0.72)(0.21)(F_{T0})$$
$$\Theta_B = \frac{F_{B0}}{F_{A0}} = \frac{(0.72)(0.21)}{0.28} = 0.54$$
$$\Theta_I = \frac{F_{I0}}{F_{A0}} = \frac{(0.72)(0.79)}{0.28} = 2.03$$

From the definition of conversion, we substitute not only for the molar flow rate of SO_2 (A) in terms of conversion but also for the volumetric flow rate as a function of conversion.

TABLE E3-5.1.   STOICHIOMETRIC TABLE FOR $SO_2 + \frac{1}{2}O_2 \longrightarrow SO_3$

Species	Symbol	Initially	Change	Remaining
SO_2	A	F_{A0}	$-F_{A0}X$	$F_A = F_{A0}(1-X)$
O_2	B	$F_{B0} = \Theta_B F_{A0}$	$-\dfrac{F_{A0}X}{2}$	$F_B = F_{A0}\left(\Theta_B - \dfrac{1}{2}X\right)$
SO_3	C	0	$+F_{A0}X$	$F_C = F_{A0}X$
N_2	I	$F_{I0} = \Theta_I F_{A0}$	—	$F_I = F_{I0} = \Theta_I F_{A0}$
		F_{T0}		$F_T = F_{T0} - \dfrac{F_{A0}X}{2}$

$$C_A = \frac{F_A}{v} = \frac{F_{A0}(1-X)}{v}$$

Recalling Equation (3-45), we have

$$v = v_0(1 + \varepsilon X)\frac{P_0}{P}\left(\frac{T}{T_0}\right) \qquad (3\text{-}45)$$

Neglecting
pressure
drop, $P = P_0$

Neglecting pressure drop in the reaction, $P = P_0$, yields

$$v = v_0(1 + \varepsilon X)\frac{T}{T_0}$$

If the reaction is also carried out isothermally, $T = T_0$, we obtain

$$v = v_0(1 + \varepsilon X)$$

Isothermal
operation, $T = T_0$

$$C_A = \frac{F_{A0}(1-X)}{v_0(1 + \varepsilon X)} = C_{A0}\left(\frac{1-X}{1 + \varepsilon X}\right)$$

The concentration of A initially is equal to the mole fraction of A initially multiplied by the total concentration. The total concentration can be calculated from an equation of state such as the ideal gas law. Recall that $y_{A0} = 0.28$, $T_0 = 500$ K, and $P_0 = 1485$ kPa.

$$C_{A0} = y_{A0}C_{T0} = y_{A0}\left(\frac{P_0}{RT_0}\right)$$

$$= 0.28\left[\frac{1485 \text{ kPa}}{8.314 \text{ kPa}\cdot\text{dm}^3/(\text{mol}\cdot\text{K}) \times 500 \text{ K}}\right]$$

$$= 0.1 \text{ mol/dm}^3$$

The total concentration is

$$C_T = \frac{F_T}{v} = \frac{F_{T0} + y_{A0}\delta X F_{T0}}{v_0(1+\varepsilon X)} = \frac{F_{T0}(1+\varepsilon X)}{v_0(1+\varepsilon X)} = \frac{F_{T0}}{v_0} = C_{T0} = \frac{P_0}{RT_0}$$

$$= \frac{1485 \text{ kPa}}{[8.314 \text{ kPa}\cdot\text{dm}^3/(\text{mol}\cdot\text{K})](500 \text{ K})} = 0.357\frac{\text{mol}}{\text{dm}^3}$$

We now evaluate ε.

$$\varepsilon = y_{A0}\,\delta = (0.28)(1 - 1 - \tfrac{1}{2}) = -0.14$$

$$C_A = C_{A0}\left(\frac{1-X}{1+\varepsilon X}\right) = 0.1\left(\frac{1-X}{1-0.14X}\right)\text{ mol/dm}^3$$

$$C_B = C_{A0}\left(\frac{\Theta_B - \tfrac{1}{2}X}{1+\varepsilon X}\right) = \frac{0.1(0.54 - 0.5X)}{1 - 0.14X}\text{ mol/dm}^3$$

$$C_C = \frac{C_{A0}X}{1+\varepsilon X} = \frac{0.1X}{1 - 0.14X}\text{ mol/dm}^3$$

$$C_I = \frac{C_{A0}\Theta_I}{1+\varepsilon X} = \frac{(0.1)(2.03)}{1 - 0.14X}\text{ mol/dm}^3$$

The concentrations of different species at various conversions are calculated in Table E3-5.2 and plotted in Figure E3-5.1. *Note* that the concentration of N_2 is changing even though it is an inert species in this reaction!!

TABLE E3-5.2.   CONCENTRATION AS A FUNCTION OF CONVERSION

Species		C_i (mol/dm³)				
		$X = 0.0$	$X = 0.25$	$X = 0.5$	$X = 0.75$	$X = 1.0$
SO_2	$C_A =$	0.100	0.078	0.054	0.028	0.000
O_2	$C_B =$	0.054	0.043	0.031	0.018	0.005
SO_3	$C_C =$	0.000	0.026	0.054	0.084	0.116
N_2	$C_I =$	0.203	0.210	0.218	0.227	0.236
	$C_T =$	0.357	0.357	0.357	0.357	0.357

The concentration of the inert is not constant!

We are now in a position to express $-r_A$ as a function of X. For example, *if* the rate law for this reaction *were* first order in SO_2 (i.e., A) and in O_2 (i.e., B), with $k = 200 \text{ dm}^3/\text{mol}\cdot\text{s}$, then the rate law becomes

$$-r_A = kC_A C_B = kC_{A0}^2\frac{(1-X)(\Theta_B - 0.5X)}{(1+\varepsilon X)^2} = \frac{2(1-X)(0.54 - 0.5X)}{(1 - 0.14X)^2} \qquad \text{(E3-5.1)}$$

Note: Because the volumetic flow rate varies with conversion, the concentration of inerts (N_2) is *not* constant.

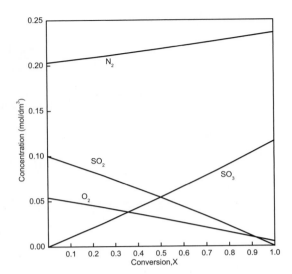

Figure E3-5.1   Concentration as a function of conversion.

Now use techniques presented in Chapter 2 to size reactors.

Taking the reciprocal of $-r_A$ yields

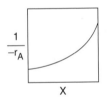

$$\frac{1}{-r_A} = \frac{0.5(1-0.14X)^2}{(1-X)(0.54-0.5X)} \qquad \text{(E3-5)}$$

We see that we could size a variety of combinations of *isothermal reactors* using the techniques discussed in Chapter 2.

Thus far in this chapter, we have focused mostly on irreversible reactions. The procedure one uses for the isothermal reactor design of reversible reactions is virtually the same as that for irreversible reactions, with one notable exception. First calculate the maximum conversion that can be achieved at the isothermal reaction temperature. This value is the equilibrium conversion. In the following example it will be shown how our algorithm for reactor design is easily extended to reversible reactions.

Need to first calculate X_e

Example 3–6   Calculating the Equilibrium Conversion

The reversible gas-phase decomposition of nitrogen tetroxide, N_2O_4, to nitrogen dioxide, NO_2,

$$N_2O_4 \; \underset{\longleftarrow}{\overset{\longrightarrow}{\rightleftharpoons}} \; 2NO_2$$

is to be carried out at constant temperature. The feed consists of pure N_2O_4 at 340 K and 202.6 kPa (2 atm). The concentration equilibrium constant, K_C, at 340 K is 0.1 mol/dm³.

(a) Calculate the equilibrium conversion of N_2O_4 in a constant-volume batch reactor.

(b) Calculate the equilibrium conversion of N_2O_4 in a flow reactor.

(c) Assuming the reaction is elementary, express the rate of reaction solely as a function of conversion for a flow system and for a batch system.

(d) Determine the CSTR volume necessary to achieve 80% of the equilibrium conversion.

Solution

$$N_2O_4 \rightleftarrows 2NO_2$$

$$A \rightleftarrows 2B$$

At equilibrium the concentrations of the reacting species are related by the relationship dictated by thermodynamics [see Equation (3-10) and Appendix C]

$$K_C = \frac{C_{Be}^2}{C_{Ae}} \tag{E3-6.1}$$

(a) Batch system—constant volume, $V = V_0$. See Table E3-6.1.

TABLE E3-6.1.   STOICHIOMETRIC TABLE

Species	Symbol	Initial	Change	Remaining
N_2O_4	A	N_{A0}	$-N_{A0}X$	$N_A = N_{A0}(1-X)$
NO_2	B	0	$+2N_{A0}X$	$N_B = 2N_{A0}X$
		$N_{T0} = N_{A0}$		$N_T = N_{T0} + N_{A0}X$

Living Example Problem

For batch systems $C_i = N_i/V$,

$$C_A = \frac{N_A}{V} = \frac{N_A}{V_0} = \frac{N_{A0}(1-X)}{V_0} = C_{A0}(1-X) \tag{E3-6.2}$$

$$C_B = \frac{N_B}{V} = \frac{N_B}{V_0} = \frac{2N_{A0}X}{V_0} = 2C_{A0}X \tag{E3-6.3}$$

$$C_{A0} = \frac{y_{A0}P_0}{RT_0} = \frac{(1)(2 \text{ atm})}{(0.082 \text{ atm} \cdot \text{dm}^3/\text{mol} \cdot \text{K})(340 \text{ K})}$$

$$= 0.07174 \text{ mol/dm}^3$$

At equilibrium, $X = X_e$, and we substitute Equations (E3-6.2) and (E3-6.3) into Equation (E3-6.1),

$$K_C = \frac{C_{Be}^2}{C_{Ae}} = \frac{4C_{A0}^2 X_e^2}{C_{A0}(1-X_e)} = \frac{4C_{A0}X_e^2}{1-X_e}$$

$$X_e = \sqrt{\frac{K_C(1 - X_e)}{4C_{A0}}} \tag{E3-6.4}$$

-math-math-math-math

We will use Polymath to solve for the equilibrium conversion and let xeb represent the equilibrium conversion in a constant-volume batch reactor. Equation (E3-6.4) written in Polymath format becomes

$$f(xeb) = xeb - [kc*(1 - xeb)/(4*cao)] \wedge 0.5$$

The Polymath program and solution are given in Table E3-6.2.

When looking at Equation (E3-6.4), you probably asked yourself, "Why not use the quadratic formula to solve for the equilibrium conversion in both batch and flow systems?" That is,

There is a Polymath tutorial in the summary Notes of Chapter 1

$$\text{Batch:} \quad X_e = \frac{1}{8}[(-1 + \sqrt{1 + 16C_{A0}/K_C})/(C_{A0}/K_C)]$$

$$\text{Flow:} \quad X_e = \frac{[(\epsilon - 1) + \sqrt{(\epsilon - 1)^2 + 4(\epsilon + 4C_{A0}/K_C)}]}{2(\epsilon + 4C_{A0}/K_C)}$$

The answer is that future problems will be nonlinear and require Polymath solutions and I wanted to increase the reader's ease in using Polymath.

TABLE E3-6.2.   POLYMATH PROGRAM AND SOLUTION FOR BOTH BATCH AND FLOW SYSTEMS

NLES Solution

Variable	Value	f(x)	Ini Guess
Xeb	0.4412598	4.078E-08	0.5
Xef	0.5083548	2.622E-10	0.5
Kc	0.1		
Cao	0.07174		
eps	1		

NLES Report (safenewt)

Nonlinear equations
[1] f(Xeb) = Xeb-(Kc*(1-Xeb)/(4*Cao))^0.5 = 0
[2] f(Xef) = Xef-(Kc*(1-Xef)*(1+eps*Xef)/(4*Cao))^0.5 = 0

Explicit equations
[1] Kc = 0.1
[2] Cao = 0.07174
[3] eps = 1

The equilibrium conversion in a constant-volume batch reactor is

$$\boxed{X_{eb} = 0.44}$$

Polymath Tutorial
Chapter 1

Summary Notes

Note: A tutorial of Polymath can be found in the summary notes of Chapter 1.

(b) Flow system. The stoichiometric table is the same as that for a batch system except that the number of moles of each species, N_i, is replaced by the molar flow rate of that species, F_i. For constant temperature and pressure, the volumetric flow rate is $v = v_0(1 + \varepsilon X)$, and the resulting concentrations of species A and B are

$$C_A = \frac{F_A}{v} = \frac{F_{A0}(1-X)}{v} = \frac{F_{A0}(1-X)}{v_0(1+\varepsilon X)} = \frac{C_{A0}(1-X)}{1+\varepsilon X} \qquad \text{(E3-6.5)}$$

$$C_B = \frac{F_B}{v} = \frac{2F_{A0}X}{v_0(1+\varepsilon X)} = \frac{2C_{A0}X}{1+\varepsilon X} \qquad \text{(E3-6.6)}$$

At equilibrium, $X = X_e$, and we can substitute Equations (E3-6.5) and (E3-6.6) into Equation (E3-6.1) to obtain the expression

$$K_C = \frac{C_{Be}^2}{C_{Ae}} = \frac{[2C_{A0}X_e/(1+\varepsilon X_e)]^2}{C_{A0}(1-X_e)/(1+\varepsilon X_e)}$$

Simplifying gives

$$K_C = \frac{4C_{A0}X_e^2}{(1-X_e)(1+\varepsilon X_e)} \qquad \text{(E3-6.7)}$$

Rearranging to use Polymath yields

$$X_e = \sqrt{\frac{K_C(1-X_e)(1+\varepsilon X_e)}{4C_{A0}}} \qquad \text{(E3-6.8)}$$

For a flow system with pure N_2O_4 feed, $\varepsilon = y_{A0}\,\delta = 1(2-1) = 1$.

We shall let Xef represent the equilibrium conversion in a flow system. Equation (E3-6.8) written in the Polymath format becomes

f(Xef) = Xef − [kc*(1 − Xef)*(1 + eps*Xef)/4/cao]^0.5

This solution is also shown in Table E3-6.2 $\boxed{X_{ef} = 0.51}$.

Note that the equilibrium conversion in a flow reactor (i.e., $X_{ef} = 0.51$), with negligible pressure drop, is greater than the equilibrium conversion in a constant-volume batch reactor ($X_{eb} = 0.44$). Recalling Le Châtelier's principle, can you suggest an explanation for this difference in X_e?

(c) Rate laws. Assuming that the reaction follows an elementary rate law, then

$$-r_A = k_A\left[C_A - \frac{C_B^2}{K_C}\right] \qquad \text{(E3-6.9)}$$

1. For a constant volume ($V = V_0$) batch system.

Here $C_A = N_A / V_0$ and $C_B = N_B / V_0$. Substituting Equations (E3-6.2) and (E3-6.3) into the rate law, we obtain the rate of disappearance of A as a function of conversion:

$-r_A = f(X)$
for a batch reactor
with $V = V_0$

$$-r_A = k_A\left[C_A - \frac{C_B^2}{K_C}\right] = k_A\left[C_{A0}(1-X) - \frac{4C_{A0}^2 X^2}{K_C}\right] \qquad (E3\text{-}6.10)$$

2. For a flow system.

Here $C_A = F_A/v$ and $C_B = F_B/v$ with $v = v_0(1 + \varepsilon X)$. Consequently, we can substitute Equations (E3-6.5) and (E3-6.6) into Equation (E3-6.9) to obtain

$-r_A = f(X)$ for a
flow reactor

$$-r_A = k_A\left[\frac{C_A(1-X)}{1+\varepsilon X} - \frac{4C_{A0}^2 X^2}{K_C(1+\varepsilon X)^2}\right] \qquad (E3\text{-}6.11)$$

As expected, the dependence of reaction rate on conversion for a constant-volume batch system [i.e., Equation (E3-6.10)] is different than that for a flow system [Equation (E3-6.11)] for gas-phase reactions.

If we substitute the values for C_{A0}, K_C, ε, and $k = 0.5$ min^{-1} in Equation (E3-6.11), we obtain $-r_A$ solely as a function of X for the flow system.

$$-r_A = \frac{0.5}{\text{min}}\left[0.072\frac{\text{mol}(1-X)}{\text{dm}^3(1+X)} - \frac{4(0.072\ \text{mol/dm}^3)^2 X^2}{0.1\ \text{mol/dm}^3(1+X)^2}\right]$$

$$-r_A = 0.0036\left[\frac{(1-X)}{(1+X)} - \frac{2.88\ X^2}{(1+X)^2}\right]\frac{\text{mol}}{\text{dm}^3 \cdot \text{min}} \qquad (E3\text{-}6.12)$$

We can now form our Levenspiel plot.

We see $(1/-r_A)$ goes to infinity as X approaches X_e.

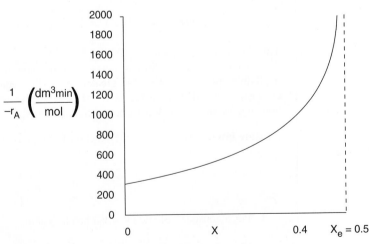

Figure E3-6.1   Levenspiel plot for a flow system.

(d) CSTR volume. Just for fun let's calculate the CSTR reactor volume necessary to achieve 80% of the equilibrium conversion of 50% (i.e., $X = 0.8X_e$) $X = 0.4$ for a feed rate of 3 mol/min.

Solution

$$-r_A = 0.0036\left[\frac{(1-0.4)}{(1+0.4)} - \frac{2.88(0.4)^2}{(1+0.4)^2}\right]$$

$$= 0.00070$$

$$V = \frac{F_{A0}X}{-r_A|_X} = \frac{F_{A0}(0.4)}{-r_A|_{0.4}} = \frac{(3 \text{ mol/min})(0.4)}{0.00070 \dfrac{\text{mol}}{\text{dm}^3 \cdot \text{min}}}$$

$$V = 1,714 \text{ dm}^3 = 1.71 \text{ m}^3$$

The CSTR volume necessary to achieve 40% conversion is 1.71 m³.

Closure. Having completed this chapter you should be able to write the rate law in terms of concentration and the Arrhenius temperature dependence. The next step is to use the stoichiometric table to write the concentrations in terms of conversion to finally arrive at a relationship between the rate of reaction and conversion. We have now completed the first three basic building blocks in our algorithm to study isothermal chemical reactions and reactors.

The CRE Algorithm
• Mole Balance, Ch 1
• Rate Law, Ch 3
• Stoichiometry, Ch 3
• Combine, Ch 4
• Evaluate, Ch 4
• Energy Balance, Ch 8

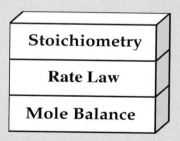

In Chapter 4, we will focus on the **combine** and **evaluation** building blocks which will then complete our algorithm for isothermal chemical reactor design.

SUMMARY

PART 1

1. Relative rates of reaction for the generic reaction:

$$A + \frac{b}{a}B \rightarrow \frac{c}{a}C + \frac{d}{a}D \qquad \text{(S3-1)}$$

The relative rates of reaction can be written either as

$$\boxed{\frac{-r_A}{a} = \frac{-r_B}{b} = \frac{r_C}{c} = \frac{r_D}{d}} \text{ or } \boxed{\frac{r_A}{-a} = \frac{r_B}{-b} = \frac{r_C}{c} = \frac{r_D}{d}} \qquad \text{(S3-2)}$$

2. *Reaction order* is determined from experimental observation:

$$A + B \longrightarrow C \qquad \text{(S3-3)}$$

$$-r_A = kC_A^\alpha C_B^\beta$$

The reaction in Equation (S3-3) is α order with respect to species A and β order with respect to species B, whereas the overall order, n, is $\alpha + \beta$. Reaction order is determined from experimental observation. If $\alpha = 1$ and $\beta = 2$, we would say that the reaction is first order with respect to A, second order with respect to B, and overall third order. We say a reaction follows an elementary rate law if the reaction orders agree with the stoichiometric coefficients for the reaction as written.

3. The temperature dependence of a specific reaction rate is given by the *Arrhenius equation*,

$$k = Ae^{-E/RT} \qquad \text{(S3-4)}$$

where A is the frequency factor and E the activation energy.

If we know the specific reaction rate, k, at a temperature, T_0, and the activation energy, we can find k at any temperature, T,

$$k(T) = k(T_0)\exp\left[\frac{E}{R}\left(\frac{1}{T_0} - \frac{1}{T}\right)\right] \qquad \text{(S3-5)}$$

Similarly from Appendix C, Equation (C-9), if we know the equilibrium constant at a temperature, T_1, and the heat of reaction, we can find the equilibrium constant at any other temperature

$$K_P(T) = K_P(T_1)\exp\left[\frac{\Delta H_{Rx}}{R}\left(\frac{1}{T_1} - \frac{1}{T}\right)\right] \qquad \text{(C-9)}$$

PART 2

4. The *stoichiometric table* for the reaction given by Equation (S3-1) being carried out in a flow system is

Species	Entering	Change	Leaving
A	F_{A0}	$-F_{A0}X$	$F_{A0}(1-X)$
B	F_{B0}	$-\left(\dfrac{b}{a}\right)F_{A0}X$	$F_{A0}\left(\Theta_B - \dfrac{b}{a}X\right)$
C	F_{C0}	$\left(\dfrac{c}{a}\right)F_{A0}X$	$F_{A0}\left(\Theta_C + \dfrac{c}{a}X\right)$
D	F_{D0}	$\left(\dfrac{d}{a}\right)F_{A0}X$	$F_{A0}\left(\Theta_D + \dfrac{d}{a}X\right)$
I	$\dfrac{F_{I0}}{F_{T0}}$	—	$\dfrac{F_{I0}}{F_T = F_{T0} + \delta F_{A0}X}$

$$\text{where} \quad \delta = \frac{d}{a} + \frac{c}{a} - \frac{b}{a} - 1$$

5. In the case of ideal gases, Equations (S3-6) and (S3-7) relate volume and volumetric flow rate to conversion.

Batch constant volume: $V = V_0$ (S3-6)

Flow systems: Gas: $v = v_0 \left(\dfrac{P_0}{P}\right)(1 + \varepsilon X)\dfrac{T}{T_0}$ (S3-7)

Liquid: $v = v_0$

For the general reaction given by (S3-1), we have

$$\boxed{\delta = \frac{d}{a} + \frac{c}{a} - \frac{b}{a} - 1}$$ (S3-8)

$$\delta = \frac{\text{Change in total number of moles}}{\text{Mole of A reacted}}$$

Definitions of δ and
and ε

$$\boxed{\varepsilon = y_{A0}\delta}$$ (S3-9)

$$\varepsilon = \frac{\text{Change in total number of moles for complete conversion}}{\text{Total number of moles fed to the reactor}}$$

6. For gas-phase reactions, we use the definition of concentration ($C_A = F_A/v$) along with the stoichiometric table and Equation (S3-7) to write the concentration of A and C in terms of conversion.

$$C_A = \frac{F_A}{v} = \frac{F_{A0}(1-X)}{v} = C_{A0}\left[\frac{1-X}{1+\varepsilon X}\right]\frac{P}{P_0}\left(\frac{T_0}{T}\right) \quad \text{(S3-10)}$$

$$C_C = \frac{F_C}{v} = C_{A0}\left[\frac{\Theta_C + (c/a)X}{1+\varepsilon X}\right]\frac{P}{P_0}\left(\frac{T_0}{T}\right) \quad \text{(S3-11)}$$

with $\Theta_C = \dfrac{F_{C0}}{F_{A0}} = \dfrac{C_{C0}}{C_{A0}} = \dfrac{y_{C0}}{y_{A0}}$

7. For incompressible liquids, the concentrations of species A and C in the reaction given by Equation (S3-1) can be written as

$$C_A = \frac{F_A}{v} = \frac{F_{A0}}{v_0}(1-X) = C_{A0}(1-X) \quad \text{(S3-12)}$$

$$C_C = C_{A0}\left(\Theta_C + \frac{c}{a}X\right) \quad \text{(S3-13)}$$

Equations (S3-12) and (S3-13) also hold for gas-phase reactions carried out at constant volume in batch systems.

8. In terms of gas-phase molar flow rates, the concentration of species i is

$$C_i = C_{T0}\frac{F_i}{F_{T0}}\frac{P}{P_0}\frac{T_0}{T} \quad \text{(S3-14)}$$

CD-ROM MATERIAL

Summary Notes

- **Learning Resource**
  1. *Summary Notes for Chapter 3*
  2. *Web Modules*
     A. Cooking a Potato
        The chemical reaction engineering is applied to cooking a potato

$$\text{Starch (crystalline)} \xrightarrow{k} \text{Starch amorphous}$$

   with

$$k = Ae^{-E/RT}$$

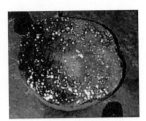

8 minutes at 400° F   12 minutes at 400° F   16 minutes at 400° F

B. Molecular Reaction Engineering

Molecular simulators (Spartan, Cerius[2]) are used to make predictions of the activation energy. The fundamentals of density functional are presented along with specific examples.

Web Modules

Molecular Modeling

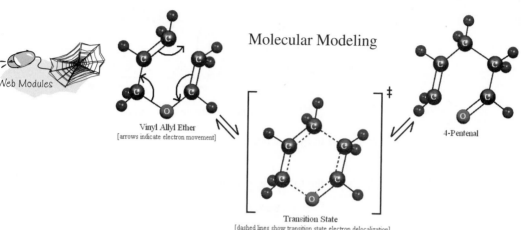

Vinyl Allyl Ether
[arrows indicate electron movement]

Transition State
[dashed lines show transition state electron delocalization]

4-Pentenal

3. *Interactive Computer Modules*
    A. Quiz Show II

Interactive

Computer Modules

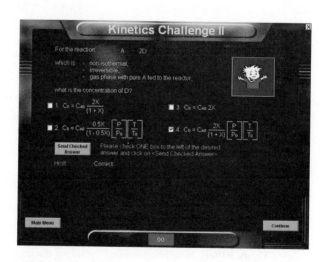

4. *Solved Problems*
    A. CDP3-A$_B$ Activation Energy for a Beetle Pushing a Ball of Dung
    B. CDP3-B$_B$ Microelectronics Industry and the Stoichiometric Table
• **FAQ [Frequently Asked Questions]**—In Updates/FAW icon section

Solved Problems

• Professional Reference Shelf

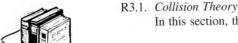

Reference Shelf

R3.1. *Collision Theory*

In this section, the fundamentals of collision theory

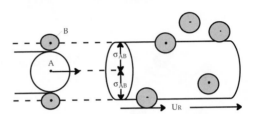

Schematic of collision cross section

are applied to the reaction

$$A + B \rightarrow C + D$$

to arrive at the following rate law

$$-r_A = \underbrace{\pi\sigma_{AB}^2\left(\frac{8\pi k_B T}{\mu\pi}\right)^{1/2} N_{Avo} e^{-E_A/RT}}_{A} C_A C_B = Ae^{-E_A/RT} C_A C_B$$

The activation energy, E_A, can be estimated from the Polanyi equation

$$E_A = E_A^\circ + \gamma_P \Delta H_{Rx}$$

R3.2. *Transition State Theory*

In this section, the rate law and rate law parameters are derived for the reaction

$$A + BC \rightleftarrows ABC^\# \rightarrow AB + C$$

using transition state theory. The Figure P3B-1 shows the energy of the molecules along the reaction coordinate which measures the progress of the reaction.

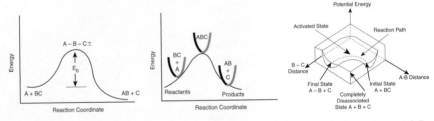

Figure P3B-1 Reaction coordinate for (a) S_{N2} reaction, and (b) generalized reaction. (c) 3-D energy surface for generalized reaction.

We will now use statistical and quantum mechanics to evaluate $K_C^\#$ to arrive at the equation

$$-r_A = \left(\frac{k_B T}{h}\right) e^{-\frac{\Delta E_0}{kT}} \frac{(q'_{ABC^\#})}{(q'_A)(q'_{BC})} C_A C_{BC}$$

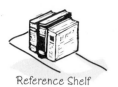

where q' is overall the partition function per unit volume and is the product of translational, vibration, rotational, and electric partition functions; that is,

$$q' = q'_T q'_V q'_R q'_E$$

The individual partition functions to be evaluated are[5]

Translation

$$q'_T = \frac{(2\pi m k_B T)^{3/2}}{h^3} = \left(\frac{9.88 \times 10^{29}}{m^3}\right)\left(\frac{m_{AB}}{1 \text{ amu}}\right)^{2/3}\left(\frac{T}{300 \text{ K}}\right)^{3/2}$$

Vibration

$$q_V = \frac{1}{1 - \exp\left(-\dfrac{hv}{k_B T}\right)}$$

$$\frac{hv}{k_B T} = \frac{hcv}{k_B T} = 4.8 \times 10^{-3}\left(\frac{\tilde{v}}{1 \text{ cm}^{-1}}\right)\left(\frac{300 \text{ K}}{T}\right)$$

Rotation

$$q_R = \frac{8\pi^2 I k_B T}{S_y h^2} = 12.4\left(\frac{T}{300 \text{ K}}\right)\left(\frac{I_{AB}}{1 \bullet \text{amu} \bullet \text{Å}^2}\right)\left(\frac{1}{S_y}\right)$$

The Eyring Equation

$$k = \left(\frac{k_B T}{h}\right)e^{\Delta S^{\#}/R}\frac{e^{-\Delta H^{\#}/RT}}{K_\gamma C_{T0}}$$

R3.3. *Molecular Dynamics*

The reaction trajectories are calculated to determine the reaction cross section of the reacting molecules. The reaction probability is found by counting up the number of reactive trajectories after Karplus.[6]

Nonreactive Trajectory

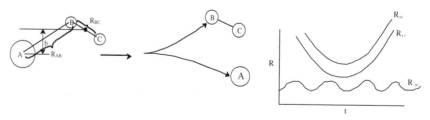

[5] R. Masel, *Chemical Kinetics*: (New York McGraw Hill, 2002), p. 594.

[6] M. Karplus, R.N. Porter, and R.D. Sharma, *J. Chem. Phys.*, *43* (9), 3259 (1965).

Reactive Trajectory

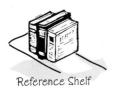

Reference Shelf

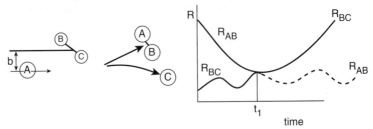

From these trajectories, one can calculate the following reaction cross section, S_r, shown for the case where both the vibrational and rotational quantum numbers are zero:

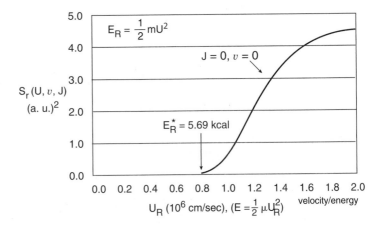

The specific reaction rate can then be calculated from first principle for simple molecules.

R3.4. *Measures Other Than Conversion*
Gas Phase
(*Note:* This topic will be covered in Chapter 4 but for those who want to use it now, look on the CD-ROM.) For membrane reactors and gas-phase multiple reactions, it is much more convenient to work in terms of the number of moles (N_A, N_B) or molar flow rates (F_A, F_B, etc.) rather than conversion.

R3.5. *Reactions with Condensation*
We now consider a gas-phase reaction in which condensation occurs. An example of this class of reactions is

$$C_2H_6(g) + 2Br_2(g) \rightarrow C_2H_4Br_2(g, l) + 2HBr(g)$$

Here we will develop our stoichiometric table for reactions with phase change. When one of the products condenses during the course of a reaction, calculation of the change in volume or volumetric flow rate must be undertaken in a slightly different manner. Plots of the molar flow rates of condensate D and the total, together with the reciprocal rate, are shown here as a function of conversion.

QUESTIONS AND PROBLEMS

The subscript to each of the problem numbers indicates the level of difficulty: A, least difficult; D, most difficult.

$$A = \bullet \quad B = \blacksquare \quad C = \blacklozenge \quad D = \blacklozenge\blacklozenge$$

Homework Problems

Web Hint

Creative Thinking

P3-1$_C$ (a) List the important concepts that you learned from this chapter. What concepts are you not clear about?

(b) Explain the strategy to evaluate reactor design equations and how this chapter expands on Chapter 2.

(c) Choose a FAQ from Chapters 1 through 3 and say why it was the most helpful.

(d) Listen to the audios 🎧 on the CD. Select a topic and explain it.

(e) Read through the Self Tests and Self Assessments for Lectures 1 through 4 on the CD-ROM. Select one and critique it. (See Preface.)

(f) Which example on the CD-ROM Lecture notes for Chapters 1 through 3 was most helpful?

(g) Which of the ICMs for the first three chapters was the most fun?

P3-2$_A$ (a) **Example 3-1.** Make a plot of k versus T. On this plot also sketch k versus $(1/T)$ for $E = 240$ kJ/mol, for $E = 60$ kJ/mol. Write a couple of sentences describing what you find. Next write a paragraph describing the activation, how it affects chemical reaction rates, and what its origins are?

(b) **Example 3-2.** Would the example be correct if water was considered an inert?

(c) **Example 3-3.** How would the answer change if the initial concentration or glyceryl sterate were 3 mol/dm³?

(d) **Example 3-4.** What is the smallest value of $\Theta_B = (N_{B0}/N_{A0})$ for which the concentration of B will *not* become the limiting reactant?

(e) **Example 3-5.** Under what conditions will the concentration of the inert nitrogen be constant? Plot Equation (E3-5.2) of $(1/-r_A)$ as a function of X up to value of $X = 0.99$. What did you find?

(f) **Example 3-6.** Why is the equilibrium conversion lower for the batch system than the flow system? Will this always be the case for constant volume batch systems? For the case in which the total concentration C_{T0} is to remain constant as the inerts are varied, plot the equilibrium conversion as a function of mole fraction of inerts for both a PFR and a constant-volume batch reactor. The pressure and temperature are constant at 2 atm and 340 K. Only N_2O_4 and inert I are to be fed.

(g) **Collision Theory**—Professional Reference Shelf. Make an outline of the steps that were used to derive

$$-r_A = A e^{-E/RT} C_A C_B$$

(h) The rate law for the reaction $(2A + B \rightarrow C)$ is $-r_A = k_A C_A^2 C_B$ with $k_A = 25(dm^3/mol)^2/s$. What are k_B and k_C?

(i) At low temperatures the rate law for the reaction $\left(\frac{1}{2}A + \frac{3}{2}B \rightarrow C\right)$ is $-r_A = k_A C_A C_B$. If the reaction is reversible at high temperatures, what is the rate law?

Kinetics Challenge II

Rate	Law	Stoich
100	100	100
200	200	200
300	300	300

P3-3$_A$ Load the Interactive Computer Module (ICM) Kinetic Challenge from the CD-ROM. Run the module, and then record your performance number for the module which indicates your mastering of the material. Your professor has

the key to decode your performance number. ICM Kinetics Challenge
Performance # _____.

P3-4$_A$ The frequency of flashing of fireflies and the frequency of chirping of crickets
as a function of temperature follow [*J. Chem. Educ.*, 5, 343 (1972) Reprinted
by permission.].

For fireflies:

T (°C)	21.0	25.00	30.0
Flashes/min	9.0	12.16	16.2

For crickets:

T (°C)	14.2	20.3	27.0
Chirps/min	80	126	200

The running speed of ants and the flight speed of honeybees as a function of
temperature are given below [Source: B. Heinrich, *The Hot-Blooded Insects*
(Cambridge, Mass.: Harvard University Press, 1993)].

For ants:

T (°C)	10	20	30	38
V (cm/s)	0.5	2	3.4	6.5

For honeybees:

T (°C)	25	30	35	40
V (cm/s)	0.7	1.8	3	?

(a) What do the firefly and cricket have in common? What are their differences?

(b) What is the velocity of the honeybee at 40°C? At –5°C?

(c) Do the bees, ants, crickets, and fireflies have anything in common? If so,
what is it? You may also do a pairwise comparison.

(d) Would more data help clarify the relationships among frequency, speed,
and temperature? If so, in what temperature should the data be obtained?
Pick an insect, and explain how you would carry out the experiment to
obtain more data. [For an alternative to this problem, see CDP3-A$_B$.]

P3-5$_B$ **Troubleshooting.** Corrosion of high-nickel stainless steel plates was found to
occur in a distillation column used at DuPont to separate HCN and water. Sulfuric acid is always added at the top of the column to prevent polymerization
of HCN. Water collects at the bottom of the column and HCN at the top. The
amount of corrosion on each tray is shown in Figure P3-5 as a function of
plate location in the column.

The bottom-most temperature of the column is approximately 125°C
and the topmost is 100°C. The corrosion rate is a function of temperature and
the concentration of an HCN–H_2SO_4 complex. Suggest an explanation for the
observed corrosion plate profile in the column. What effect would the column
operating conditions have on the corrosion profile?

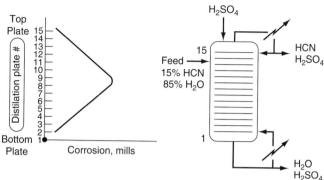

Figure P3-5

P3-6$_B$ **Inspector Sgt. Ambercromby of Scotland Yard.** It is believed, although never proven, that Bonnie murdered her first husband, Lefty, by poisoning the tepid brandy they drank together on their first anniversary. Lefty was unware she had coated her glass with an antidote before she filled both glasses with the poisoned brandy. Bonnie married her second husband, Clyde, and some years later when she had tired of him, she called him one day to tell him of her new promotion at work and to suggest that they celebrate with a glass of brandy that evening. She had the fatal end in mind for Clyde. However, Clyde suggested that instead of brandy, they celebrate with ice cold Russian vodka and they down it Cossack style, in one gulp. She agreed and decided to follow her previously successful plan and to put the poison in the vodka and the antidote in her glass. The next day, both were found dead. Sgt. Ambercromby arrives. What were the first three questions he asks? What are two possible explanations? Based on what you learned from this chapter, what do you feel Sgt. Ambercromby suggested as the most logical explanation?

[Professor Flavio Marin Flores, ITESM, Monterrey, Mexico]

P3-7$_B$ **(a)** The rule of thumb that the rate of reaction doubles for a 10°C increase in temperature occurs only at a specific temperature for a given activation energy. Develop a relationship between the temperature and activation energy for which the rule of thumb holds. Neglect any variation of concentration with temperature.

(b) Determine the activation energy and frequency factor from the following data:

k (min^{-1})	0.001	0.050
T (°C)	00.0	100.0

(c) Write a paragraph explaining activation energy, E, and how it affects the chemical reaction rate. Refer to Section 3.3 and especially the Professional Reference Shelf sections R3.1, R3.2, and R3.3 if necessary.

P3-8$_C$ **Air bags** contain a mixture of NaN_3, KNO_3, and SiO_2. When ignited, the following reactions take place:

(1) $2NaN_3 \rightarrow 2Na + 3N_2$

(2) $10Na + 2KNO_3 \rightarrow K_2O + 5Na_2O + N_2$

(3) $K_2O + Na_2O + SiO_2 \rightarrow$ alkaline silicate glass

Reactions (2) and (3) are necessary to handle the toxic sodium reaction product from detonation. Set up a stoichiometric table solely in terms of NaN_3 (A), KNO_3 (B), etc., and the number of moles initially. If 150 g of sodium azide are present in each air bag, how many grams of KNO_3 and SiO_2 must be added to make the reaction products safe in the form of alkaline silicate glass after the bag has inflated. The sodium azide is in itself toxic. How would you propose to handle all the undetonated air bags in cars piling up in the nation's junkyards.

P3-9$_A$ **Hot Potato.** Review the "Cooking a Potato" web module on the CD-ROM or on the web.

   (a) It took the potato described on the web 1 hour to cook at 350°F. Builder Bob suggests that the potato can be cooked in half that time if the oven temperature is raised to 600°F. What do you think?

   (b) Buzz Lightyear says, "No Bob," and suggests that it would be quicker to boil the potato in water at 100°C because the heat transfer coefficient is 20 times greater. What are the tradeoffs of oven versus boiling?

   (c) Ore Ida Tater Tots is a favorite of one of the Proctors/Graders in the class, Adam Cole. Tater Tots are 1/12 the size of a whole potato but approximately the same shape. Estimate how long it would take to cook a Tater Tot at 400°F? At what time would it be cooked half way through? ($r/R = 0.5$)?

P3-10$_A$ **(a)** Write the rate law for the following reactions assuming each reaction follows an elementary rate law.

   (1)
$$C_2H_6 \longrightarrow C_2H_4 + H_2$$

   (2)
$$C_2H_4 + \tfrac{1}{2}O_2 \rightarrow \overset{\displaystyle O}{\overset{\displaystyle \diagup \diagdown}{CH_2 - CH_2}}$$

   (3)
$$(CH_3)_3COOC(CH_3)_3 \rightleftharpoons C_2H_6 + 2CH_3COCH_3$$

   (4)
$$nC_4H_{10} \rightleftharpoons iC_4H_{10}$$

   (5) $CH_3COOC_2H_5 + C_4H_9OH \rightleftharpoons CH_3COOC_4H_9 + C_2H_5OH$

   (b) Write the rate law for the reaction
$$2A + B \rightarrow C$$

if the reaction (1) is second order in B and overall third order, (2) is zero order in A and first order in B, (3) is zero order in both A and B, and (4) is first order in A and overall zero order.

   (c) Find and write the rate laws for the following reactions

   (1) $H_2 + Br_2 \rightarrow 2HBr$

   (2) $H_2 + I_2 \rightarrow 2HI$

Web Hint

P3-11$_A$ Set up a stoichiometric table for each of the following reactions and express the concentration of each species in the reaction as a function of conversion evaluating all constants (e.g., ε, Θ). Then, assume the reaction follows an elementary rate law, and write the reaction rate solely as a function of conversion, i.e., $-r_A = f(X)$.

(a) For the liquid-phase reaction

$$\underset{\text{CH}_2-\text{CH}_2}{\overset{\text{O}}{\triangle}} + \text{H}_2\text{O} \xrightarrow{\text{H}_2\text{SO}_4} \underset{\text{CH}_2-\text{OH}}{\overset{\text{CH}_2-\text{OH}}{|}}$$

the initial concentrations of ethylene oxide and water are 1 lb-mol/ft³ and 3.47 lb-mol/ft³ (62.41 lb/ft³ ÷ 18), respectively. If $k = 0.1$ dm³/mol · s at 300 K with $E = 12{,}500$ cal/mol, calculate the CSTR space-time for 90% conversion at 300 K and at 350 K.

(b) For the isothermal, isobaric gas-phase pyrolysis

$$\text{C}_2\text{H}_6 \longrightarrow \text{C}_2\text{H}_4 + \text{H}_2$$

pure ethane enters the flow reactor at 6 atm and 1100 K. How would your equation for the concentration and reaction rate change if the reaction were to be carried out in a constant-volume batch reactor?

(c) For the isothermal, isobaric, catalytic gas-phase oxidation

$$\text{C}_2\text{H}_4 + \tfrac{1}{2}\text{O}_2 \longrightarrow \underset{\text{CH}_2-\text{CH}_2}{\overset{\text{O}}{\triangle}}$$

the feed enters a PBR at 6 atm and 260°C and is a stoichiometric mixture of only oxygen and ethylene.

(d) For the isothermal, isobaric, catalytic gas-phase reaction is carried out in a fluidized CSTR

$$\bigcirc +2\text{H}_2 \longrightarrow \bigcirc$$

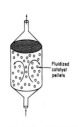

Fluidized CSTR

the feed enters at 6 atm and 170°C and is a stoichiometric mixture. What catalyst weight is required to reach 80% conversion in a fluidized CSTR at 170°C and 270°C? The rate constant is defined wrt benzene and $v_0 = 50$ dm³/min.

$$k_B = \frac{53 \text{ mol}}{\text{kgcat} \cdot \text{min } \text{atm}^3} \quad \text{at 300 K with } E = 80 \text{ kJ/mol}$$

P3-12$_B$ There were 5430 million pounds of ethylene oxide produced in the United States in 1995. The flowsheet for the commercial production of ethylene oxide (EO) by oxidation of ethylene is shown in Figure P3-12. We note that the process essentially consists of two systems, a reaction system and a separation system. Discuss the flowsheet and how your answers to P3-11 **(c)** would change if air is used in a stoichiometric feed. This reaction is studied further in Chapter 4.

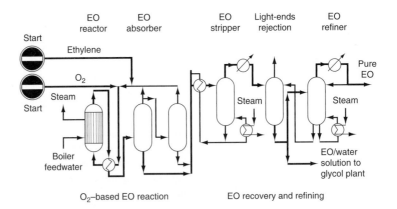

Figure P3-12   EO plant flowsheet. [Adapted from R. A. Meyers, ed., *Handbook of Chemical Production Processes, Chemical Process Technology Handbook Series*, New York: McGraw-Hill, 1983, p. 1.5-5. ISBN 0-67-041-765-2.]

P3-13$_B$ The formation of nitroanalyine (an important intermediate in dyes—called fast orange) is formed from the reaction of orthonitrochlorobenzene (ONCB) and aqueous ammonia. (See Table 3-1 and Example 9-2.)

$$\text{(NO}_2\text{, Cl benzene)} + 2NH_3 \longrightarrow \text{(NO}_2\text{, NH}_2 \text{ benzene)} + NH_4Cl$$

The liquid-phase reaction is first order in both ONCB and ammonia with $k = 0.0017$ m³/kmol · min at 188°C with $E = 11,273$ cal/mol. The initial entering concentrations of ONCB and ammonia are 1.8 kmol/m³ and 6.6 kmol/m³, respectively (more on this reaction in Chapter 9).

(a) Write the rate law for the rate of disappearance of ONCB in terms of concentration.

(b) Set up a stoichiometric table for this reaction for a flow system.

(c) Explain how parts (a) and (b) would be different for a batch system.

(d) Write $-r_A$ solely as a function of conversion. $-r_A =$ _____

(e) What is the initial rate of reaction ($X = 0$)   at 188°C? $-r_A =$ _____
                                                         at 25°C? $-r_A =$ _____
                                                         at 288°C? $-r_A =$ _____

(f) What is the rate of reaction when $X = 0.90$ at 188°C? $-r_A =$ _____
                                                         at 25°C? $-r_A =$ _____
                                                         at 288°C? $-r_A =$ _____

(g) What would be the corresponding CSTR reactor volume at 25°C to achieve 90% conversion at 25°C and at 288°C for a feed rate of 2 dm³/min
                                               at 25°C? $V =$ _____
                                               at 288°C? $V =$ _____

P3-14$_B$ Adapted from M. L. Shuler and F. Kargi, *Bioprocess Engineering*, Prentice Hall (2002). Cell growth takes place in bioreactors called chemostats.

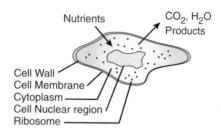

A substrate such as glucose is used to grow cells and produce a product.

$$\text{Substrate} \xrightarrow{\text{Cells}} \text{More Cells (biomass) + Product}$$

A generic molecule formula for the biomass is $C_{4.4}H_{7.3}N_{0.86}O_{1.2}$. Consider the growth of a generic organism on glucose

$$C_6H_{12}O_6 + aO_2 + bNH_3 \rightarrow c(C_{4.4}H_{7.3}N_{0.86}O_{1.2}) + dH_2O + eCO_2$$

Experimentally, it was shown that for this organism, the cells convert 2/3 of carbon substrate to biomass.

(a) Calculate the stoichiometric coefficients a, b, c, d, and e (*Hint:* carry out atom balances [*Ans:* c = 0.91]).

(b) Calculate the yield coefficients $Y_{C/S}$ (g cells/g substrate) and Y_{C/O_2} (g cells/g O_2). The gram of cells are dry weight (no water—gdw) (*Ans:* Y_{C/O_2} = 1.77 gdw cells/g O_2) (gdw = grams dry weight).

P3-15$_B$ The gas-phase reaction

$$\tfrac{1}{2}N_2 + \tfrac{3}{2}H_2 \longrightarrow NH_3$$

is to be carried out isothermally. The molar feed is 50% H_2 and 50% N_2, at a pressure of 16.4 atm and 227°C.

(a) Construct a complete stoichiometric table.

(b) What are C_{AO}, δ, and ε? Calculate the concentrations of ammonia and hydrogen when the conversion of H_2 is 60%. (*Ans:* C_{H_2} = 0.1 mol/dm³)

(c) Suppose by chance the reaction is elementary with k_{N_2} = 40 dm³/mol/s. Write the rate of reaction *solely* as a function of conversion for (1) a flow system and (2) a constant volume batch system.

P3-16$_B$ Calculate the equilibrium conversion and concentrations for each of the following reactions.

(a) The liquid-phase reaction

$$A + B \rightleftharpoons C$$

with $C_{A0} = C_{B0} = 2$ mol/dm³ and $K_C = 10$ dm³/mol .

(b) The gas-phase reaction

$$A \rightleftharpoons 3C$$

carried out in a flow reactor with no pressure drop. Pure A enters at a temperature of 400 K and 10 atm. At this temperature, $K_C = 0.25$(dm³/mol)².

(c) The gas-phase reaction in part (b) carried out in a constant-volume batch reactor.

(d) The gas-phase reaction in part (b) carried out in a constant-pressure batch reaction.

Member

Hall of Fame

P3-17$_B$ Consider a *cylindrical batch reactor* that has one end fitted with a frictionless piston attached to a spring (Figure P3-17). The reaction

$$A + B \longrightarrow 8C$$

with the rate expression

$$-r_A = k_1 C_A^2 C_B$$

is taking place in this type of reactor.

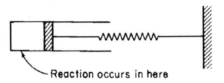

Reaction occurs in here

Figure P3-17

(a) Write the rate law solely as a function of conversion, numerically evaluating all possible symbols. (*Ans.:* $-r_A = 5.03 \times 10^{-9} [(1 - X)^3/(1 + 3X)^{3/2}]$ lb mol/ft$^3 \cdot$s.)

(b) What is the conversion and rate of reaction when $V = 0.2$ ft^3? (*Ans.:* $X = 0.259$, $-r_A = 8.63 \times 10^{-10}$ lb mol/ft$^3 \cdot$s.)

Additional information:

Equal moles of A and B are present at $t = 0$
Initial volume: 0.15 ft^3
Value of k_1: 1.0 (ft^3/lb mol)$^2 \cdot$s^{-1}
The relationship between the volume of the reactor and pressure within the reactor is

$$V = (0.1)(P) \qquad (V \text{ in ft}^3, P \text{ in atm})$$

Temperature of system (considered constant): 140°F
Gas constant: 0.73 ft$^3 \cdot$atm/lb mol$\cdot$°R

P3-18$_C$ Read the sections on Collision Theory, Transition State Theory, and Molecular Dynamics in the Professional Reference Shelf on the CD-ROM.

(a) Use collision theory to outline the derivation of the Polanyi–Semenov equation, which can be used to estimate activation energies from the heats of reaction, ΔH_R, according to the equation

$$E = C - \alpha(-\Delta H_R) \tag{P3-18.1}$$

(b) Why is this a reasonable correlation? (*Hint:* See Professional Reference Shelf 3R.1: *Collision Theory.*)

(c) Consider the following family of reactions:

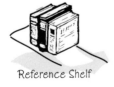

Reference Shelf

	E (kcal/mol)	$-\Delta H_R$ (kcal/mol)
H + RBr $\longrightarrow$ HBr + R	6.8	17.5
H + R'Br $\longrightarrow$ HBr + R'	6.0	20.0

Estimate the activation energy for the reaction

$$CH_3 \bullet + RBr \longrightarrow CH_3Br + R \bullet$$

which has an exothermic heat of reaction of 6 kcal/mol (i.e., $\Delta H_R = -6$ kcal/mol).

(d) What parameters vary over the widest range in calculating the reaction rate constant using transition state theory (i.e., which ones do you need to focus your attention on to get a good approximation of the specific reaction rate)? (See Professional Reference Shelf R3.2.)

(e) List the assumptions made in the molecular dynamics simulation R3.3 used to calculate the activation energy for the hydrogen exchange reaction.

(f) The volume of the box used to calculate the translational partition function for the activated complex was taken as 1 dm³. True or False?

(g) Suppose the distance between two atoms of a linear molecule in the transition state was set at half the true value. Would the rate constant increase or decrease over that of the true value and by how much (i.e., what factor)?

(h) List the parameters you can obtain from Cerius² to calculate the molecular partition functions.

Web Hint

P3-19$_D$ Use Spartan, CACHE, Cerius², Gaussian, or some other chemical computational software package to calculate the heats of formation of the reactants, products, and transition state and the activation barrier, E_B, for the following reactions:

(a) $CH_3OH + O \longrightarrow CH_3O + OH$

(b) $CH_3Br + OH \longrightarrow CH_3OH + Br$

(c) $Cl + H_2 \longrightarrow HCl + H$

(d) $Br + CH_4 \longrightarrow HBr + CH_3$

P3-20$_B$ It is proposed to produce ethanol by one of two reactions:

$$C_2H_5Cl + OH^- \longleftrightarrow C_2H_5OH + Cl^- \tag{1}$$

$$C_2H_5Br + OH^- \longrightarrow C_2H_5OH + Br^- \tag{2}$$

Use Spartan (see Appendix J) or some other software package to answer the following:

(a) What is the ratio of the rates of reaction at 25°C? 100°C? 500°C?

(b) Which reaction scheme would you choose to make ethanol? (*Hint*: Consult *Chemical Marketing Reporter* or *www.chemweek.com* for chemical prices).

[Professor R. Baldwin, Colorado School of Mines]

• **Additional Homework Problems on CD-ROM**

Temperature Effects

Solved Problems

CDP3-A$_B$ Estimate how fast a Tenebrionid Beetle can push a ball of dung at 41.5°C. (**Solution included.**)

CDP3-B$_B$ Use the Polanyi equation to calculate activation energies. [3rd Ed. P3-20$_B$]

CDP3-C$_B$ Given the irreversible rate law at low temperature, write the reversible rate law at high temperature. [3rd Ed. P3-10$_B$]

Stoichiometry

Member

Hall of Fame

CDP3-D$_B$ Set up a stoichiometric table for

$$4NH_3 + 5O_2 \longrightarrow 4NO + 6H_2O$$

in terms of molar flow rates. **(Solution included.)**

CDP3-E$_B$ Set up a stoichiometric table for the reaction

$$C_6H_5COCH + 2NH_3 \longrightarrow C_6H_5ONH_2 + NH_2Cl$$

[2nd Ed. P3-10$_B$]

CDP3-F$_B$ The elementary reaction $A(g) + B(l) \longrightarrow C(g)$ takes place in a square duct containing liquid B, which evaporates into the gas reacting with A. [2nd Ed. P3-20$_B$]

Reactions with Phase Change

CDP3-G$_B$ Silicon is used in the manufacture of microelectronics devices. Set up a stochiometric table for the reaction **(Solution included.)**

$$Si\,HCl_3(g) + H_2(g) \longrightarrow Si(s) + HCl(g) + Si_xH_yCl_z(g)$$

[2nd Ed. P3-16$_B$]

CDP3-H$_B$ Reactions with condensation between chlorine and methane. [3rd Ed. P3-21$_B$].

CDP3-I$_B$ Reactions with condensation

$$C_2H_6(g) + 2Br(g) \longrightarrow C_2H_4Br_2(l,\,g) + 2HBr(g)$$

[3rd Ed. P3-22$_B$]

CDP3-J$_B$ Chemical vapor deposition

$$3SiH_4(g) + 4NH_3(g) \longrightarrow SiN_4(s) + 12H_2(g)$$

[3rd Ed. P3-23$_B$]

CDP3-K$_B$ Condensation occurs in the gas-phase reaction:

$$C_2H_4(g) + 2Cl(g) \longrightarrow CH_2Cl_2(g,\,l) + 2HCl(g)$$

[2nd Ed. P3-17$_B$]

New Problems on the Web

CDP3-New   From time to time new problems relating Chapter 3 material to everyday interests or emerging technologies will be placed on the web. Solutions to these problems can be obtained by emailing the author.

SUPPLEMENTARY READING

1. Two references relating to the discussion of activation energy have already been cited in this chapter. Activation energy is usually discussed in terms of either collision theory or transition-state theory. A concise and readable account of these two theories can be found in

> Masel, R., *Chemical Kinetics,* New York: McGraw Hill, 2002, p. 594.
>
> LAIDLER, K. J. *Chemical Kinetics.* New York: Harper & Row, 1987, Chap. 3.

An expanded but still elementary presentation can be found in

> MOORE, J. W., and R. G. PEARSON, *Kinetics and Mechanism*, 3rd ed. New York: Wiley, 1981, Chaps. 4 and 5.

A more advanced treatise of activation energies and collision and transition-state theories is

> BENSON, S. W., *The Foundations of Chemical Kinetics.* New York: McGraw-Hill, 1960.
>
> STEINFELD, J. I., J. S. FRANCISCO, W. L. HASE, *Chemical Kinetics and Dynamics*, 2nd ed. New Jersey: Prentice Hall, 1999.

2. The books listed above also give the rate laws and activation energies for a number of reactions; in addition, as mentioned earlier in this chapter, an extensive listing of rate laws and activation energies can be found in NBS circulars:

> Kinetic data for larger number of reactions can be obtained on Floppy Disks and CD-ROMs provided by National Institute of Standards and Technology (NIST). Standard Reference Data 221/A320 Gaithersburg, MD 20899; ph: (301) 975-2208 Additional sources are Tables of Chemical Kinetics: Homogeneous Reactions, National Bureau of Standards Circular 510 (Sept. 28, 1951); Suppl. 1 (Nov. 14, 1956); Suppl. 2 (Aug. 5, 1960); Suppl. 3 (Sept. 15, 1961) (Washington, D.C.: U.S. Government Printing Office). Chemical Kinetics and Photochemical Data for Use in Stratospheric Modeling, Evaluate No. 10, JPL Publication 92-20, Aug. 15, 1992, Jet Propulsion Laboratories, Pasadena, Calif.

3. Also consult the current chemistry literature for the appropriate algebraic form of the rate law for a given reaction. For example, check the *Journal of Physical Chemistry* in addition to the journals listed in Section 4 of the Supplementary Reading section in Chapter 4.

Isothermal
Reactor Design

4

Why, a four-year-old child could understand this.
Someone get me a four-year-old child.

Groucho Marx

Tying everything
together

Overview. Chapters 1 and 2 discussed mole balances on *reactors* and the manipulation of these balances to predict reactor sizes. Chapter 3 discussed *reactions*. In Chapter 4, we combine *reactions* and *reactors* as we bring all the material in the preceding three chapters together to arrive at a logical structure for the design of various types of reactors. By using this structure, one should be able to solve reactor engineering problems by reasoning rather than by memorizing numerous equations together with the various restrictions and conditions under which each equation applies (i.e., whether there is a change in the total number of moles, etc.). In perhaps no other area of engineering is mere formula plugging more hazardous; the number of physical situations that can arise appears infinite, and the chances of a simple formula being sufficient for the adequate design of a real reactor are vanishingly small.

We divide the chapter into two parts: Part 1 "Mole Balances in Terms of Conversion," and Part 2 "Mole Balances in Terms of Concentration, C_i, and Molar Flow Rates, F_i." In Part 1, we will concentrate on batch reactors, CSTRs, and PFRs where conversion is the preferred measure of a reaction's progress for single reactions. In Part 2, we will analyze membrane reactors, the startup of a CSTR, and semibatch reactors, which are most easily analyzed using concentration and molar flow rates as the variables rather than conversion. We will again use mole balances in terms of these variables (C_i, F_i) for multiple reactors in Chapter 6.

This chapter focuses attention on reactions that are operated iso-thermally. We begin the chapter by studying a liquid-phase reaction to form ethylene glycol in a batch reactor. Here, we determine the specific reaction rate constant that will be used to design an industrial CSTR to produce ethylene glycol. After illustrating the design of a CSTR from batch date, we carry out the design of a PFR for the gas-phase pyrolsis reaction to form ethylene. This section is followed by the design of a packed bed reactor with pressure drop to form ethylene oxide from the partial oxidation of ethylene. When we put all these reactions and reactors together, we will see we have designed a chemical plant to produce 200 million pounds per year of ethylene glycol. We close the chapter by analyzing some of the newer reactors such as microreactors, membrane reactors, and, on the CD-ROM, reactive distillation semibatch reactors.

PART 1   Mole Balances in Terms of Conversion

4.1 Design Structure for Isothermal Reactors

Logic
vs.
Memorization

One of the primary goals of this chapter is to solve chemical reaction engineer-ing (CRE) problems by using logic rather than memorizing which equation applies where. It is the author's experience that following this structure, shown in Figure 4-1, will lead to a greater understanding of isothermal reactor design. We begin by applying our general mole balance equation (level ①) to a spe-cific reactor to arrive at the design equation for that reactor (level ②). If the feed conditions are specified (e.g., N_{A0} or F_{A0}), all that is required to evaluate the design equation is the rate of reaction as a function of conversion at the same conditions as those at which the reactor is to be operated (e.g., tempera-ture and pressure). When $-r_A = f(X)$ is known or given, one can go directly

Use the algorithm
rather than memo-
rizing equations.

from level ③ to level ⑨ to determine either the batch time or reactor volume necessary to achieve the specified conversion.

If the rate of reaction is not given explicitly as a function of conversion, we must proceed to level ④ where the rate law must be determined by either finding it in books or journals or by determining it experimentally in the labo-ratory. Techniques for obtaining and analyzing rate data to determine the reac-tion order and rate constant are presented in Chapter 5. After the rate law has been established, one has only to use stoichiometry (level ⑤) together with the conditions of the system (e.g., constant volume, temperature) to express con-centration as a function of conversion.

For liquid-phase reactions and for gas-phase reactions with no pressure drop ($P = P_0$), one can combine the information in levels ④ and ⑤, to express the rate of reaction as a function of conversion and arrive at level ⑥. It is now possible to determine either the time or reactor volume necessary to achieve the desired conversion by substituting the relationship linking conversion and rate of reaction into the appropriate design equation (level ⑨).

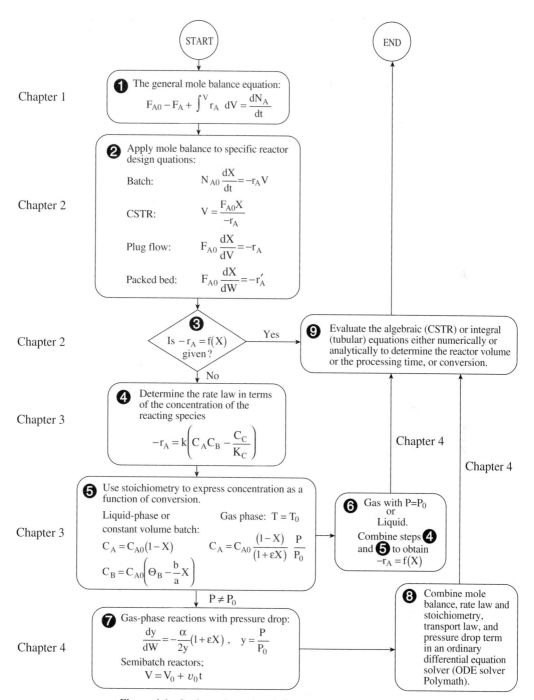

Figure 4-1   Isothermal reaction design algorithm for conversion.

For gas-phase reactions in packed beds where there is a pressure drop, we need to proceed to level ⑦ to evaluate the pressure ratio (P / P_0) in the concentration term using the Ergun equation (Section 4.5). In level ⑧, we combine the equations for pressure drop in level ⑦ with the information in levels ④ and ⑤, to proceed to level ⑨ where the equations are then evaluated in the appropriate manner (i.e., analytically using a table of integrals, or numerically using an ODE solver). Although this structure emphasizes the determination of a reaction time or reactor volume for a specified conversion, it can also readily be used for other types of reactor calculations, such as determining the conversion for a specified volume. Different manipulations can be performed in level ⑨ to answer the different types of questions mentioned here.

The structure shown in Figure 4-1 allows one to develop a few basic concepts and then to arrange the parameters (equations) associated with each concept in a variety of ways. Without such a structure, one is faced with the possibility of choosing or perhaps memorizing the correct equation from a *multitude of equations* that can arise for a variety of different combinations of reactions, reactors, and sets of conditions. The challenge is to put everything together in an orderly and logical fashion so that we can proceed to arrive at the correct equation for a given situation.

The Algorithm
1. Mole balance
2. Rate law
3. Stoichiometry
4. Combine
5. Evaluate

Fortunately, by using an algorithm to formulate CRE problems, which happens to be analogous to the algorithm for ordering dinner from a fixed-price menu in a fine French restaurant, we can eliminate virtually all memorization. In both of these algorithms, we must make choices in each category. For example, in ordering from a French menu, we begin by choosing one dish from the *appetizers* listed. Step 1 in the analog in CRE is to begin by choosing the appropriate mole balance for one of the three types of reactors shown. In Step 2 we choose the rate law (*entrée*), and in Step 3 we specify whether the reaction is gas *or* liquid phase (*cheese* or *dessert*). Finally, in Step 4 we combine Steps 1, 2, and 3 and either obtain an analytical solution or solve the equations using an ODE solver. (See the complete French menu on the CD-ROM).

We now will apply this algorithm to a specific situation. Suppose that we have, as shown in Figure 4-2, mole balances for three reactors, three rate laws, and the equations for concentrations for both liquid and gas phases. In Figure 4-2 we see how the algorithm is used to formulate the equation to calculate the *PFR reactor volume for a first-order gas-phase reaction*. The pathway to arrive at this equation is shown by the ovals connected to the dark lines through the algorithm. The dashed lines and the boxes represent other pathways for solutions to other situations. The algorithm for the pathway shown is

1. **Mole balance,** choose species A reacting in a PFR
2. **Rate law,** choose the irreversible first-order reaction
3. **Stoichiometry,** choose the gas-phase concentration
4. **Combine** Steps 1, 2, and 3 to arrive at Equation A
5. **Evaluate.** The combine step can be evaluated either
   a. Analytically (Appendix Al)
   b. Graphically (Chapter 2)

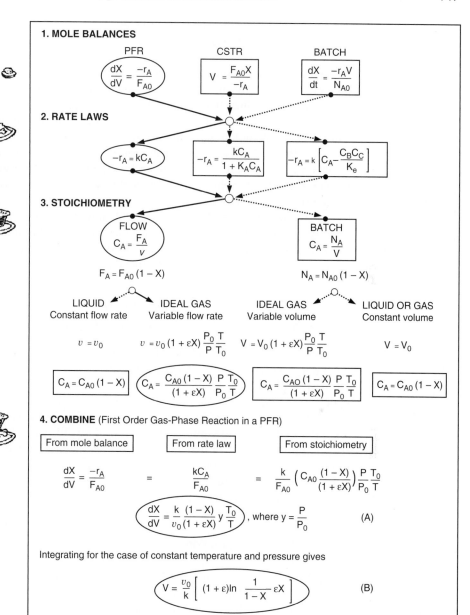

1. MOLE BALANCES

PFR

$$\frac{dX}{dV} = \frac{-r_A}{F_{A0}}$$

CSTR

$$V = \frac{F_{A0}X}{-r_A}$$

BATCH

$$\frac{dX}{dt} = \frac{-r_A V}{N_{A0}}$$

2. RATE LAWS

$$-r_A = kC_A$$

$$-r_A = \frac{kC_A}{1 + K_A C_A}$$

$$-r_A = k\left[C_A - \frac{C_B C_C}{K_e}\right]$$

3. STOICHIOMETRY

FLOW

$$C_A = \frac{F_A}{v}$$

BATCH

$$C_A = \frac{N_A}{V}$$

$$F_A = F_{A0}(1 - X)$$

$$N_A = N_{A0}(1 - X)$$

LIQUID   IDEAL GAS   IDEAL GAS   LIQUID OR GAS
Constant flow rate   Variable flow rate   Variable volume   Constant volume

$$v = v_0$$

$$v = v_0(1 + \varepsilon X)\frac{P_0}{P}\frac{T}{T_0}$$

$$V = V_0(1 + \varepsilon X)\frac{P_0}{P}\frac{T}{T_0}$$

$$V = V_0$$

$$C_A = C_{A0}(1 - X)$$

$$C_A = \frac{C_{A0}(1 - X)}{(1 + \varepsilon X)}\frac{P}{P_0}\frac{T_0}{T}$$

$$C_A = \frac{C_{A0}(1 - X)}{(1 + \varepsilon X)}\frac{P}{P_0}\frac{T_0}{T}$$

$$C_A = C_{A0}(1 - X)$$

4. COMBINE (First Order Gas-Phase Reaction in a PFR)

From mole balance

From rate law

From stoichiometry

$$\frac{dX}{dV} = \frac{-r_A}{F_{A0}} \quad = \quad \frac{kC_A}{F_{A0}} \quad = \quad \frac{k}{F_{A0}}\left(C_{A0}\frac{(1 - X)}{(1 + \varepsilon X)}\right)\frac{P}{P_0}\frac{T_0}{T}$$

$$\frac{dX}{dV} = \frac{k}{v_0}\frac{(1 - X)}{(1 + \varepsilon X)}\, y\, \frac{T_0}{T}, \quad \text{where } y = \frac{P}{P_0} \qquad \text{(A)}$$

Integrating for the case of constant temperature and pressure gives

$$V = \frac{v_0}{k}\left[(1 + \varepsilon)\ln\frac{1}{1 - X} - \varepsilon X\right] \qquad \text{(B)}$$

Figure 4-2   Algorithm for isothermal reactors.

c. Numerically(Appendix A4), or

d. Using software (Polymath)

Substitute parame-
ter values in
steps 1–4 only if
they are zero.

In Figure 4-2 we chose to integrate Equation A for constant temperature and pressure to find the volume necessary to achieve a specified conversion (or calculate the conversion that can be achieved in a specified reactor volume). Unless the parameter values are zero, we typically don't substitute numerical values for parameters in the combine step until the very end.

For the case of isothermal operation with no pressure drop, we were able to obtain an analytical solution, given by equation B, which gives the reactor volume necessary to achieve a conversion X for a first-order gas-phase reaction carried out isothermally in a PFR. However, in the majority of situations, analytical solutions to the ordinary differential equations appearing in the combine step are not possible. Consequently, we include Polymath, or some other ODE solver such as MATLAB, in our menu in that it makes obtaining solutions to the differential equations much more palatable.

We can solve the
equations in the
combine step either
1. Analytically
   (Appendix A1)
2. Graphically
   (Chapter 2)
3. Numerically
   (Appendix A4)
4. Using Software
   (Polymath).

4.2 Scale-Up of Liquid-Phase Batch Reactor Data to the Design of a CSTR

One of the jobs in which chemical engineers are involved is the scale-up of laboratory experiments to pilot-plant operation or to full-scale production. In the past, a pilot plant would be designed based on laboratory data. However, owing to the high cost of a pilot-plant study, this step is beginning to be surpassed in many instances by designing a full-scale plant from the operation of a laboratory-bench-scale unit called a *microplant*. To make this jump successfully requires a thorough understanding of the chemical kinetics and transport limitations. In this section we show how to analyze a laboratory-scale batch reactor in which a liquid-phase reaction of known order is being carried out. After determining the specific reaction rate, k, from a batch experiment, we use it in the design of a full-scale flow reactor.

4.2.1 Batch Operation

In modeling a batch reactor, we have assumed that there is no inflow or outflow of material and that the reactor is well mixed. For most liquid-phase reactions, the density change with reaction is usually small and can be neglected (i.e., $V = V_0$). In addition, for *gas-phases* reactions in which the batch reactor volume remains constant, we also have $V = V_0$. Consequently, for constant-volume ($V = V_0$) (e.g., closed metal vessels) batch reactors the mole balance

$$\frac{1}{V}\left(\frac{dN_A}{dt}\right) = r_A \tag{4-1}$$

can be written in terms of concentration.

$$\frac{1}{V}\frac{dN_A}{dt} = \frac{1}{V_0}\frac{dN_A}{dt} = \frac{d(N_A/V_0)}{dt} = \frac{dC_A}{dt} = r_A \tag{4-2}$$

Generally, when analyzing laboratory experiments, it is best to process the data in terms of the measured variable. Because concentration is the measured variable for most liquid-phase reactions, the general mole balance equation applied to reactions in which there is no volume change becomes

Used to analyze batch reaction data

$$-\frac{dC_A}{dt} = -r_A$$

This is the form we will use in analyzing reaction rate data in Chapter 5.

Let's calculate the time necessary to achieve a given conversion X for the irreversible second-order reaction

$$A \longrightarrow B$$

The **mole balance** on a constant-volume, $V = V_0$, batch reactor is

Mole balance

$$N_{A0}\frac{dX}{dt} = -r_A V_0 \tag{2-6}$$

The **rate law** is

Rate law

$$-r_A = kC_A^2 \tag{4-3}$$

From **stoichiometry** for a constant-volume batch reactor. we obtain

Stoichiometry

$$C_A = C_{A0}(1 - X) \tag{3-29}$$

The **mole balance, rate law,** and **stoichiometry** are **combined** to obtain

Combine

$$\frac{dX}{dt} = kC_{A0}(1 - X)^2 \tag{4-4}$$

To **evaluate,** we rearrange and integrate

$$\frac{dX}{(1 - X)^2} = kC_{A0}dt$$

Initially, if $t = 0$, then $X = 0$. If the reaction is carried out isothermally, k will be constant; we can integrate this equation (see Appendix A.1 for a table of Integrals used in CRE applications) to obtain

Evaluate

$$\int_0^t dt = \frac{1}{kC_{A0}} \int_0^X \frac{dX}{(1 - X)^2}$$

$$t = \frac{1}{kC_{A0}}\left(\frac{X}{1-X}\right)$$ (4-5)

This time is the reaction time t (i.e., t_R) needed to achieve a conversion X for a second-order reaction in a batch reactor. It is important to have a grasp of the order of magnitudes of batch reaction times, t_R, to achieve a given conversion, say 90%, for different values of the product of specific reaction rate, k, and initial concentration, C_{A0}. Table 4-1 shows the algorithm to find the batch reaction times, t_R, for both first- and a second-order reactions carried out isothermally. We can obtain these estimates of t_R by considering the first- and second-order irreversible reactions of the form

$$A \rightarrow B$$

TABLE 4-1.   ALGORITHM TO ESTIMATE REACTION TIMES

Mole balance	$\dfrac{dX}{dt_R} = \dfrac{-r_A}{N_{A0}} V$	
Rate law	First-Order	Second-Order
	$-r_A = kC_A$	$-r_A = kC_A^2$
Stoichiometry ($V = V_0$)	$C_A = \dfrac{N_A}{V_0} = C_{A0}(1-X)$	
Combine	$\dfrac{dX}{dt_R} = k(1-X)$	$\dfrac{dX}{dt_R} = kC_{A0}(1-X)^2$
Evaluate (Integrate)	$t_R = \dfrac{1}{k}\ln\dfrac{1}{1-X}$	$t_R = \dfrac{X}{kC_{A0}(1-X)}$

For first-order reactions the reaction time to reach 90% conversion (i.e., $X = 0.9$) in a constant-volume batch reactor scales as

$$t_R = \frac{1}{k}\ln\frac{1}{1-X} = \frac{1}{k}\ln\frac{1}{1-0.9} = \frac{2.3}{k}$$

If $k = 10^{-4}$ s^{-1},

$$t_R = \frac{2.3}{10^{-4}\text{ s}^{-1}} = 23,000 \text{ s} = 6.4 \text{ h}$$

The time necessary to achieve 90% conversion in a batch reactor for an irreversible first-order reaction in which the specific reaction rate is (10^{-4} s^{-1}) is 6.4 h.

For second-order reactions, we have

$$t_R = \frac{1}{kC_{A0}}\frac{X}{1-X} = \frac{0.9}{kC_{A0}(1-0.9)} = \frac{9}{kC_{A0}}$$

If $kC_{A0} = 10^{-3}$ s^{-1},

$$t_R = \frac{9}{10^{-3}\text{ s}^{-1}} = 9000 \text{ s} = 2.5 \text{ h}$$

We note that if 99% conversion had been required for this value of kC_{A0}, the reaction time, t_R, would jump to 27.5 h.

Table 4-2 gives the *order of magnitude* of time to achieve 90% conversion for first- and second-order irreversible batch reactions.

Estimating Reaction Times

TABLE 4-2.   BATCH REACTION TIMES

Reaction Time t_R	First-Order k (s^{-1})	Second-Order kC_{A0} (s^{-1})
Hours	10^{-4}	10^{-3}
Minutes	10^{-2}	10^{-1}
Seconds	1	10
Milliseconds	1000	10,000

Flow reactors would be used for reactions with *characteristic reaction times*, t_R, of minutes or less.

The times in Table 4-2 are the reaction time to achieve 90% conversion (i.e., to reduce the concentration from C_{A0} to 0.1 C_{A0}). The total cycle time in any batch operation is considerably longer than the reaction time, t_R, as one must account for the time necessary to fill (t_f) and heat (t_e) the reactor together with the time necessary to clean the reactor between batches, t_c. In some cases, the reaction time calculated from Equation (4-5) may be only a small fraction of the total cycle time, t_t.

$$t_t = t_f + t_e + t_c + t_R$$

Typical cycle times for a batch polymerization process are shown in Table 4-3. Batch polymerization reaction times may vary between 5 and 60 hours. Clearly, decreasing the reaction time with a 60-hour reaction is a critical problem. As the reaction time is reduced (e.g., 2.5 h for a second-order reaction with $kC_{A0} = 10^3$ s^{-1}), it becomes important to use large lines and pumps to achieve rapid transfers and to utilize efficient sequencing to minimize the cycle time.

TABLE 4-3.   TYPICAL CYCLE TIME FOR A BATCH
POLYMERIZATION PROCESS

Batch operation times

Activity	Time (h)
1. Charge feed to the reactor and agitate, t_f	1.5–3.0
2. Heat to reaction temperature, t_e	0.2–2.0
3. Carry out reaction, t_R	(varies)
4. Empty and clean reactor, t_c	0.5–1.0
Total time excluding reaction	3.0–6.0

Usually one has to optimize the reaction time with the processing times listed in Table 4-3 to produce the maximum number of batches (i.e., pounds of product) in a day. See Problems P4-6(f) and P4-7(c).

In the next four examples, we will describe the various reactors needed to produce 200 million pounds per year of ethylene glycol from a feedstock of ethane. We begin by finding the rate constant, k, for the hydrolysis of ethylene oxide to form ethylene glycol.

Example 4–1 Determining k from Batch Data

It is desired to design a CSTR to produce 200 million pounds of ethylene glycol per year by hydrolyzing ethylene oxide. However, before the design can be carried out, it is necessary to perform and analyze a batch reactor experiment to determine the specific reaction rate constant, k. Because the reaction will be carried out isothermally, the specific reaction rate will need to be determined only at the reaction temperature of the CSTR. At high temperatures there is a significant by-product formation, while at temperatures below 40°C the reaction does not proceed at a significant rate; consequently, a temperature of 55°C has been chosen. Because the water is usually present in excess, its concentration may be considered constant during the course of the reaction. The reaction is first-order in ethylene oxide.

$$\underset{A}{\underset{\text{CH}_2-\text{CH}_2}{\overset{\displaystyle \text{O}}{\diagup\!\!\diagdown}}} + \underset{+\ B}{\text{H}_2\text{O}} \xrightarrow[\text{catalyst}]{\text{H}_2\text{SO}_4} \underset{C}{\overset{\displaystyle \text{CH}_2-\text{OH}}{\underset{\displaystyle \text{CH}_2-\text{OH}}{|}}}$$

In the laboratory experiment, 500 mL of a 2 M solution (2 kmol/m^3) of ethylene oxide in water was mixed with 500 mL of water containing 0.9 wt % sulfuric acid, which is a catalyst. The temperature was maintained at 55°C. The concentration of ethylene glycol was recorded as a function of time (Table E4-1.1).

Using the data in Table E4-1.1, determine the specific reaction rate at 55°C.

TABLE E4-1.1.   CONCENTRATION-TIME DATA

Time (min)	Concentration of Ethylene Glycol (kmol/m^3)[a]
0.0	0.000
0.5	0.145
1.0	0.270
1.5	0.376
2.0	0.467
3.0	0.610
4.0	0.715
6.0	0.848
10.0	0.957

[a] 1 kmol/m^3 = 1 mol/dm^3 = 1 mol/L.

Check 10 types of homework problems on the CD-ROM for more solved examples using this algorithm.

Analysis

In this example we use the problem-solving algorithm (A through G) that is given in the CD-ROM and on the web *www.engin.umich.edu/~problemsolving*. You may wish to follow this algorithm in solving the other examples in this chapter and the problems given at the end of the chapter. However, to conserve space it will not be repeated for other example problems.

A. *Problem statement.* Determine the specific reaction rate, k_A.

B. *Sketch*:

A, B, C

Batch

C. *Identify*:

C1. Relevant theories

Mole balance: $\dfrac{dN_A}{dt} = r_A V$

Rate law: $-r_A = k_A C_A$

C2. Variables

Dependent: concentrations, C_A, C_B, and C_C

Independent: time, t

C3. Knowns and unknowns

Knowns: concentration of ethylene glycol as a function of time

Unknowns:

1. Concentration of ethylene oxide as a function of time, C_A = ?

2. Specific reaction rate, k_A = ?

3. Reactor volume, V = ?

C4. Inputs and outputs: reactant fed all at once to a batch reactor

C5. Missing information: None; does not appear that other sources need to be sought.

D. *Assumptions and approximations*:

Assumptions

1. Well mixed

2. All reactants enter at the same time

3. No side reactions

4. Negligible filling time

5. Isothermal operation

Approximations

1. Water in excess so that its concentration is essentially constant, (i.e., $C_B \cong C_{B0}$).

E. *Specification.* The problem is neither overspecified nor underspecified.

Check 10 types of homework problems on the CD-ROM for more solved examples using this algorithm.

Solved Problems

F. *Related material.* This problem uses the mole balances developed in Chapter 1 for a batch reactor and the stoichiometry and rate laws developed in Chapter 3.

G. *Use an algorithm.* For an isothermal reaction, use the chemical reaction engineering algorithm shown in Figures 4-1 and 4-2.

Following the Algorithm

Solution

Mole Balance

1. A **mole balance** on a batch reactor that is well-mixed is

$$\frac{1}{V}\frac{dN_A}{dt} = r_A \qquad\qquad \text{(E4-1.1)}$$

Rate Law

2. The **rate law** is

$$-r_A = kC_A \qquad\qquad \text{(E4-1.2)}$$

Because water is present in such excess, the concentration of water at any time t is virtually the same as the initial concentration and the rate law is independent of the concentration of H_2O. ($C_B \cong C_{B0}$.)

3. **Stoichiometry.** Liquid phase, no volume change, $V = V_0$ (Table E4-1.2):

TABLE E4-1.2. STOICHIOMETRIC TABLE

Species	Symbol	Initial	Change	Remaining	Concentration
CH_2CH_2O	A	N_{A0}	$-N_{A0}X$	$N_A = N_{A0}(1-X)$	$C_A = C_{A0}(1-X)$
H_2O	B	$\Theta_B N_{A0}$	$-N_{A0}X$	$N_B = N_{A0}(\Theta_B - X)$	$C_B = C_{A0}(\Theta_B - X)$
					$C_B \approx C_{A0}\Theta_B = C_{B0}$
$(CH_2OH)_2$	C	0	$N_{A0}X$	$N_C = N_{A0}X$	$C_C = C_{A0}X$
		N_{T0}		$N_T = N_{T0} - N_{A0}X$	

Stoichiometric Table for Constant Volume

Recall Θ_B is the initial number of moles of A to B (i.e., $\Theta_B = \dfrac{N_{B0}}{N_{A0}}$).

$$C_A = \frac{N_A}{V} = \frac{N_A}{V_0}$$

$$\frac{1}{V_0}\left(\frac{dN_A}{dt}\right) = \frac{d(N_A/V_0)}{dt} = \frac{dC_A}{dt}$$

Combining mole balance, rate law, and stoichiometry

4. **Combining** the rate law and the mole balance, we have

$$-\frac{dC_A}{dt} = kC_A \qquad\qquad \text{(E4-1.3)}$$

5. **Evaluate.** For isothermal operation, k is constant, so we can integrate this equation (E4-1.3)

$$-\int_{C_{A0}}^{C_A} \frac{dC_A}{C_A} = \int_0^t k\,dt = k\int_0^t dt$$

using the initial condition that when $t = 0$, then $C_A = C_{A0}$. The initial concentration of A after mixing the two volumes together is 1.0 kmol/m³ (1 mol/L).

Integrating yields

$$\ln \frac{C_{A0}}{C_A} = kt \tag{E4-1.4}$$

The concentration of ethylene oxide at any time t is

$$C_A = C_{A0}e^{-kt} \tag{E4-1.5}$$

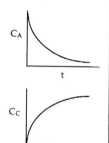

The concentration of ethylene glycol at any time t can be obtained from the reaction stoichiometry:

$$A + B \longrightarrow C$$
$$N_C = N_{A0}X = N_{A0} - N_A$$

For liquid-phase reactions $V = V_0$,

$$C_C = \frac{N_C}{V} = \frac{N_C}{V_0} = C_{A0} - C_A = C_{A0}(1 - e^{-kt}) \tag{E4-1.6}$$

Rearranging and taking the logarithm of both sides yields

$$\ln \frac{C_{A0} - C_C}{C_{A0}} = -kt \tag{E4-1.7}$$

We see that a plot of $\ln[(C_{A0} - C_C)/C_{A0}]$ as a function of t will be a straight line with a slope $-k$. Using Table E4-1.1, we can construct Table E4-1.3 and use Excel to plot $\ln(C_{A0} - C_C)/C_{A0}$ as a function of t.

TABLE E4-1.3.   PROCESSED DATA

t (min)	C_C (kmol/m³)	$\dfrac{C_{A0} - C_C}{C_{A0}}$	$\ln\left(\dfrac{C_{A0} - C_C}{C_{A0}}\right)$
0.0	0.000	1.000	0.0000
0.5	0.145	0.855	−0.1570
1.0	0.270	0.730	−0.3150
1.5	0.376	0.624	−0.4720
2.0	0.467	0.533	−0.6290
3.0	0.610	0.390	−0.9420
4.0	0.715	0.285	−1.2550
6.0	0.848	0.152	−1.8840
10.0	0.957	0.043	−3.1470

From the slope of a plot of $\ln[(C_{A0} - C_C)/C_{A0}]$ versus t, we can find k as shown on the Excel Figure E4-4.1.

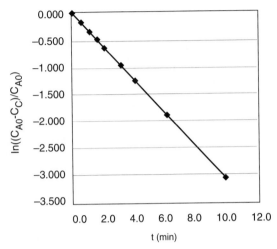

Figure E4-1.1   Excel plot of data.

$$\text{Slope} = -k = -0.311 \text{ min}^{-1}$$

$$k = 0.311 \text{ min}^{-1}$$

The rate law becomes

$$-r_A = 0.311 \text{ min}^{-1}C_A$$

Summary Notes

The rate law can now be used in the design of an industrial CSTR. For those who prefer to find k using semilog graph paper, this type of analysis is given in Chapter 4 Summary Notes on the CD-ROM. An Excel tutorial is also given in the Summary Notes for Chapter 3.

4.3 Design of Continuous Stirred Tank Reactors (CSTRs)

Continuous stirred tank reactors (CSTRs), such as the one shown here schematically, are typically used for liquid-phase reactions.

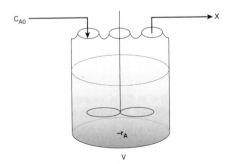

In Chapter 2, we derived the following design equation for a CSTR:

Mole balance

$$V = \frac{F_{A0} X}{(-r_A)_{\text{exit}}} \qquad (2\text{-}13)$$

which gives the volume V necessary to achieve a conversion X. As we saw in Chapter 2, the space time, τ, is a characteristic time of a reactor. To obtain the space time, τ, as a function of conversion we first substitute for $F_{A0} = v_0 C_{A0}$ in Equation (2-13)

$$V = \frac{v_0 C_{A0} X}{-r_A} \qquad (4\text{-}6)$$

and then divide by v_0 to obtain the space time, τ, to achieve a conversion X in a CSTR

$$\tau = \frac{V}{v_0} = \frac{C_{A0} X}{-r_A} \qquad (4\text{-}7)$$

This equation applies to a single CSTR or to the first reactor of CSTRs connected in series.

4.3.1 A Single CSTR

Let's consider a first-order irreversible reaction for which the rate law is

Rate law

$$-r_A = k C_A$$

For liquid-phase reactions, there is no volume change during the course of the reaction, so we can use Equation (3-29) to relate concentration and conversion,

Stoichiometry

$$C_A = C_{A0}(1 - X) \qquad (3\text{-}29)$$

We can combine mole balance Equation (4-7), the rate law and concentration, Equation (3-29) to obtain

Combine

$$\tau = \frac{1}{k}\left(\frac{X}{1 - X}\right)$$

CSTR Relationship
between space time
and conversion for a
first-order liquid-
phase reaction

Rearranging

$$X = \frac{\tau k}{1 + \tau k} \qquad (4\text{-}8)$$

We could also combine Equations (3-29) and (4-8) to find the exit reactor concentration of A, C_A,

$$C_A = \frac{C_{A0}}{1 + \tau k} \tag{4-9}$$

$$Da = \frac{-r_{A0}V}{F_{A0}}$$

For a first-order reaction, the product τk is often referred to as the reaction **Damköhler number,** Da, which is a dimensionless number that can give us a quick estimate of the degree of conversion that can be achieved in continuous-flow reactors. The Damköhler number is the ratio of the rate of reaction of A to the rate of convective transport of A at the entrance to the reactor.

$$Da = \frac{-r_{A0}V}{F_{A0}} = \frac{\text{Rate of reaction at entrance}}{\text{Entering flow rate of A}} = \frac{\text{"A reaction rate"}}{\text{"A convection rate"}}$$

The Damköhler number for a first-order irreversible reaction is

$$Da = \frac{-r_{A0}V}{F_{A0}} = \frac{kC_{A0}V}{v_0 C_{A0}} = \tau k$$

For a second-order irreversible reaction, the Damköhler number is

$$Da = \frac{-r_{A0}V}{F_{A0}} = \frac{kC_{A0}^2 V}{v_0 C_{A0}} = \tau k C_{A0}$$

$0.1 \le Da \le 10$

It is important to know what values of the Damköhler number, Da, give high and low conversion in continuous-flow reactors. A value of Da = 0.1 or less will usually give less than 10% conversion and a value of Da = 10.0 or greater will usually give greater than 90% conversion; that is, the rule of thumb is

if Da < 0.1, then $X < 0.1$

if Da > 10, then $X > 0.9$

Equation (4-8) for a first-order liquid-phase reaction in a CSTR can also be written in terms of the Damköhler

$$X = \frac{Da}{1 + Da}$$

4.3.2 CSTRs in Series

A first-order reaction with no change in the volumetric flow rate ($v = v_0$) is to be carried out in two CSTRs placed in series (Figure 4-3).

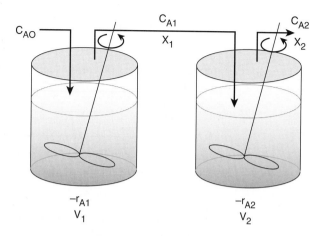

Figure 4-3   Two CSTRs in series.

The effluent concentration of reactant A from the first CSTR can be found using Equation (4-9)

$$C_{A1} = \frac{C_{A0}}{1 + \tau_1 k_1}$$

with $\tau_1 = V_1/\upsilon_0$. From a mole balance on reactor 2,

$$V_2 = \frac{F_{A1} - F_{A2}}{-r_{A2}} = \frac{\upsilon_0(C_{A1} - C_{A2})}{k_2 C_{A2}}$$

Solving for C_{A2}, the concentration exiting the second reactor, we obtain

First-order reaction

$$C_{A2} = \frac{C_{A1}}{1 + \tau_2 k_2} = \frac{C_{A0}}{(1 + \tau_2 k_2)(1 + \tau_1 k_1)}$$

If both reactors are of equal size ($\tau_1 = \tau_2 = \tau$) and operate at the same temperature ($k_1 = k_2 = k$), then

$$C_{A2} = \frac{C_{A0}}{(1 + \tau k)^2}$$

If instead of two CSTRs in series we had n equal-sized CSTRs connected in series ($\tau_1 = \tau_2 = \cdots = \tau_n = \tau_i = (V_i/\upsilon_0)$) operating at the same temperature ($k_1 = k_2 = \cdots = k_n = k$), the concentration leaving the last reactor would be

$$C_{An} = \frac{C_{A0}}{(1 + \tau k)^n} = \frac{C_{A0}}{(1 + \mathrm{Da})^n} \qquad (4\text{-}10)$$

Substituting for C_{An} in terms of conversion

$$C_{A0}(1 - X) = \frac{C_{A0}}{(1 + Da)^n}$$

and rearranging, the conversion for these n tank reactors in series will be

Conversion as a
function of the
number of
tanks in series

$$\boxed{X = 1 - \frac{1}{(1 + Da)^n} \equiv 1 - \frac{1}{(1 + \tau k)^n}}$$

(4-11)

A plot of the conversion as a function of the number of reactors in series for a first-order reaction is shown in Figure 4-4 for various values of the Damköhler

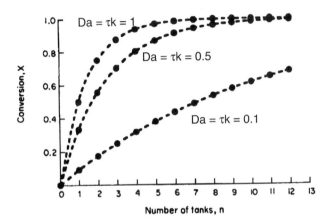

CSTRs in series

Figure 4-4 Conversion as a function of the number of tanks in series for different Damköhler numbers for a first-order reaction.

Economics number τk. Observe from Figure 4-4 that when the product of the space time and the specific reaction rate is relatively large, say, Da $\geq$ 1, approximately 90% conversion is achieved in two or three reactors; thus the cost of adding subsequent reactors might not be justified. When the product τk is small, Da ~ 0.1, the conversion continues to increase significantly with each reactor added.

The rate of disappearance of A in the nth reactor is

$$\boxed{-r_{An} = kC_{An} = k\,\frac{C_{A0}}{(1 + \tau k)^n}}$$

4.3.3 CSTRs in Parallel

We now consider the case in which equal-sized reactors are placed in parallel rather than in series, and the feed is distributed equally among each of the reactors (Figure 4-5). The balance on any reactor, say i, gives

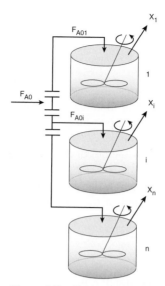

Figure 4-5   CSTRs in parallel.

the individual reactor volume

$$V_i = F_{A0i} \left(\frac{X_i}{-r_{Ai}} \right) \qquad (4\text{-}12)$$

Since the reactors are of equal size, operate at the same temperature, and have identical feed rates, the conversion will be the same for each reactor:

$$X_1 = X_2 = \cdots = X_n = X$$

as will be the rate of reaction in each reactor

$$-r_{A1} = -r_{A2} = \cdots = -r_{An} = -r_A$$

The volume of each individual reactor, V_i, is related to the total volume, V, of all the reactors by the equation

$$V_i = \frac{V}{n}$$

A similar relationship exists for the total molar flow rate, which is equally divided:

$$F_{A0i} = \frac{F_{A0}}{n}$$

Substituting these values into Equation (4-12) yields

$$\frac{V}{n} = \frac{F_{A0}}{n} \left(\frac{X_i}{-r_{Ai}} \right)$$

or

Conversion
for tanks in
parallel. Is this
result surprising?

$$V = \frac{F_{A0}X_i}{-r_{Ai}} = \frac{F_{A0}X}{-r_A} \tag{4-13}$$

This result shows that the conversion achieved in any one of the reactors in parallel is identical to what would be achieved if the reactant were fed in one stream to one large reactor of volume V!

4.3.4 A Second-Order Reaction in a CSTR

For a second-order liquid-phase reaction being carried out in a CSTR, the **combination** of the **rate law** and the **design equation** yields

$$V = \frac{F_{A0}X}{-r_A} = \frac{F_{A0}X}{kC_A^2} \tag{4-14}$$

Using our stoichiometric table for constant density $v = v_0$, $C_A = C_{A0}(1-X)$, and $F_{A0}X = v_0 C_{A0}X$, then

$$V = \frac{v_0 C_{A0} X}{kC_{A0}^2(1-X)^2}$$

Dividing by v_0,

$$\tau = \frac{V}{v_0} = \frac{X}{kC_{A0}(1-X)^2} \tag{4-15}$$

We solve Equation (4-15) for the conversion X:

Conversion for
a second-order
liquid-phase
reaction
in a CSTR

$$X = \frac{(1+2\tau kC_{A0}) - \sqrt{(1+2\tau kC_{A0})^2 - (2\tau kC_{A0})^2}}{2\tau kC_{A0}}$$

$$= \frac{(1+2\tau kC_{A0}) - \sqrt{1+4\tau kC_{A0}}}{2\tau kC_{A0}} \tag{4-16}$$

$$\boxed{X = \frac{(1+2\text{Da}) - \sqrt{1+4\text{Da}}}{2\text{Da}}}$$

The minus sign must be chosen in the quadratic equation because X cannot be greater than 1. Conversion is plotted as a function of the Damköhler parameter, τkC_{A0}, in Figure 4-6. Observe from this figure that at high conversions (say 67%) a 10-fold increase in the reactor volume (or increase in the specific reaction rate by raising the temperature) will increase the conversion only to 88%. This observation is a consequence of the fact that the CSTR operates under the condition of the lowest value of the reactant concentration (i.e., the exit concentration), and consequently the smallest value of the rate of reaction.

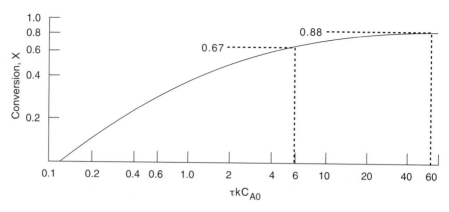

Figure 4-6   Conversion as a function of the Damköhler number $(\tau k C_{A0})$ for a second-order reaction in a CSTR.

Uses and economics

Anti-freeze

Scale-Up of Batch Reactor Data

Example 4–2   Producing 200 Million Pounds per Year in a CSTR

Close to 12.2 billion metric tons of ethylene glycol (EG) were produced in 2000, which ranked it the twenty-sixth most produced chemical in the nation that year on a total pound basis. About one-half of the ethylene glycol is used for *antifreeze* while the other half is used in the manufacture of polyesters. In the polyester category, 88% was used for fibers and 12% for the manufacture of bottles and films. The 2004 selling price for ethylene glycol was $0.28 per pound.

It is desired to produce 200 million pounds per year of EG. The reactor is to be operated isothermally. A 1 lb mol/ft³ solution of ethylene oxide (EO) in water is fed to the reactor (shown in Figure E4-2.1) together with an equal volumetric solution of water containing 0.9 wt % of the catalyst H_2SO_4. The specific reaction rate constant is 0.311 min⁻¹, as determined in Example 4-1. Practical guidelines for reactor scale-up were recently given by Mukesh[1] and by Warsteel[2].

(a)   If 80% conversion is to be achieved, determine the necessary CSTR volume.

(b)   If two 800-gal reactors were arranged in parallel, what is the corresponding conversion?

(c)   If two 800-gal reactors were arranged in series, what is the corresponding conversion?

Solution

Assumption: Ethylene glycol (EG) is the only reaction product formed.

---

[1] D. Mukesh, *Chemical Engineering,* 46 (January 2002); *www.CHE.com.*

[2] J. Warsteel, *Chemical Engineering Progress,* (June 2000).

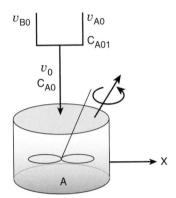

Figure E4-2.1 Single CSTR.

The specified Ethylene Glycol (EG) production rate in lb mol/min is

$$F_C = 2 \times 10^8 \, \frac{lb}{yr} \times \frac{1 \, yr}{365 \, days} \times \frac{1 \, day}{24 \, h} \times \frac{1 \, h}{60 \, min} \times \frac{1 lb \, mol}{62 \, lb} = 6.137 \, \frac{lb \, mol}{min}$$

From the reaction stoichiometry

$$F_C = F_{A0} X$$

we find the required molar flow rate of ethylene oxide to be

$$F_{A0} = \frac{F_C}{X} = \frac{6.137}{0.8} = 7.67 \, \frac{lb \, mol}{min} \quad (58.0 \, g \, mol/s)$$

(a) We now calculate the single CSTR volume to achieve 80% conversion using the CRE algorithm.

1. **Design equation**:

$$V = \frac{F_{A0} X}{-r_A} \tag{E4-2.1}$$

2. **Rate law**:

$$-r_A = kC_A \tag{E4-2.2}$$

Following the Algorithm

3. **Stoichiometry**. Liquid phase $(v = v_0)$:

$$C_A = \frac{F_A}{v_0} = \frac{F_{A0}(1-X)}{v_0} = C_{A0}(1-X) \tag{E4-2.3}$$

4. **Combining**:

$$V = \frac{F_{A0} X}{k C_{A0}(1 - X)} = \frac{v_0 X}{k(1 - X)} \qquad \text{(E4-2.4)}$$

5. **Evaluate**:

The entering volumetric flow rate of stream A, with $C_{A01} = 1$ lb mol/ft³ before mixing, is

$$v_{A0} = \frac{F_{A0}}{C_{A01}} = \frac{7.67 \text{ lb mol/min}}{1 \text{ lb mol/ft}^3} = 7.67 \frac{\text{ft}^3}{\text{min}}$$

From the problem statement $v_{B0} = v_{A0}$

$$F_{B0} = v_{B0} C_{B01} = \left(7.67 \frac{\text{ft}^3}{\text{min}}\right) \cdot \left(62.4 \frac{\text{lb}}{\text{ft}^3} \times \frac{1 \text{ lb mol}}{18 \text{ lb}}\right) = 26.6 \frac{\text{lb mol}}{\text{min}}$$

The total entering volumetric flow rate of liquid is

$$v_0 = v_{A0} + v_{B0} = 15.34 \frac{\text{ft}^3}{\text{min}} \quad (7.24 \text{ dm}^3/\text{s})$$

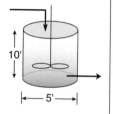

Substituting in Equation (E4-2.4), recalling that $k = 0.311$ min⁻¹, yields

$$V = \frac{v_0 X}{k(1 - X)} = 15.34 \frac{\text{ft}^3}{\text{min}} \frac{0.8}{(0.311 \text{ min}^{-1})(1 - 0.8)} = 197.3 \text{ ft}^3$$
$$= 1480 \text{ gal } (5.6 \text{ m}^3)$$

A tank 5 ft in diameter and approximately 10 ft tall is necessary to achieve 80% conversion.

(b) CSTRs in parallel. For two 800-gal CSTRs arranged in parallel (as shown in Figure E4-2.2) with 7.67 ft³/min $(v_0/2)$ fed to each reactor, the conversion achieved can be calculated by rearranging Equation (E4-2.4)

$$\frac{V}{v_0} k = \tau k = \frac{X}{1 - X}$$

to obtain

$$X = \frac{\tau k}{1 + \tau k} \qquad \text{(E4-2.5)}$$

where

$$\tau = \frac{V}{v_0/2} = \left(800 \text{ gal} \times \frac{1 \text{ ft}^3}{7.48 \text{ gal}}\right) \times \frac{1}{7.67 \text{ ft}^3/\text{min}} = 13.94 \text{ min}$$

The **Damköhler number** is

$$\text{Da} = \tau k = 13.94 \text{ min} \times \frac{0.311}{\text{min}} = 4.34$$

Substituting into Equation (E4-2.5) gives us

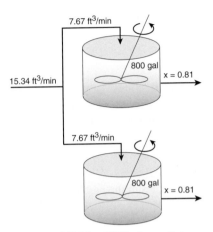

Figure E4-2.2 CSTRs in parallel.

$$X = \frac{Da}{1+Da} = \frac{4.34}{1+4.34} = 0.81$$

The conversion exiting each of the CSTRs in parallel is 81%.

(c) CSTRs in series. If the 800-gal reactors are arranged in series, the conversion in the first reactor [cf. Equation (E4-2.5)] is

$$X_1 = \frac{\tau_1 k}{1+\tau_1 k} \tag{E4-2.6}$$

where

$$\tau = \frac{V_1}{v_{01}} = \left(800 \text{ gal} \times \frac{1 \text{ ft}^3}{7.48 \text{ gal}}\right) \times \frac{1}{15.34 \text{ ft}^3/\text{min}} = 6.97 \text{ min}$$

First CSTR

The Damköhler number is

$$Da_1 = \tau_1 k = 6.97 \text{ min} \times \frac{0.311}{\text{min}} = 2.167$$

$$X_1 = \frac{2.167}{1+2.167} = \frac{2.167}{3.167} = 0.684$$

To calculate the conversion exiting the second reactor, we recall that $V_1 = V_2 = V$ and $v_{01} = v_{02} = v_0$; then

$$\tau_1 = \tau_2 = \tau$$

A mole balance on the second reactor is

In	−	Out	+	Generation		=	0
$\overbrace{F_{A1}}$	−	$\overbrace{F_{A2}}$	+	$\overbrace{r_{A2}V}$		=	0

Basing the conversion on the total number of moles reacted up to a point per mole of A fed to the first reactor,

$$F_{A1} = F_{A0}(1 - X_1) \quad \text{and} \quad F_{A2} = F_{A0}(1 - X_2)$$

Rearranging

$$V = \frac{F_{A1} - F_{A2}}{-r_{A2}} = F_{A0}\frac{X_2 - X_1}{-r_{A2}}$$

$$-r_{A2} = kC_{A2} = k\frac{F_{A2}}{v_0} = \frac{kF_{A0}(1 - X_2)}{v_0} = kC_{A0}(1 - X_2)$$

Combining the mole balance on the second reactor [cf. Equation (2-24)] with the rate law, we obtain

$$V = \frac{F_{A0}(X_2 - X_1)}{-r_{A2}} = \frac{C_{A0}v_0(X_2 - X_1)}{kC_{A0}(1 - X_2)} = \frac{v_0}{k}\left(\frac{X_2 - X_1}{1 - X_2}\right) \qquad \text{(E4-2.7)}$$

Second CSTR

Solving for the conversion exiting the second reactor yields

$$X_2 = \frac{X_1 + \text{Da}}{1 + \text{Da}} = \frac{X_1 + \tau k}{1 + \tau k} = \frac{0.684 + 2.167}{1 + 2.167} = 0.90$$

The same result could have been obtained from Equation (4-11):

$$X_2 = 1 - \frac{1}{(1 + \tau k)^n} = 1 - \frac{1}{(1 + 2.167)^2} = 0.90$$

Two hundred million pounds of EG per year can be produced using two 800-gal (3.0-m^3) reactors in series.

Conversion in the series arrangement is greater than in parallel for CSTRs. From our discussion of reactor staging in Chapter 2, we could have predicted that the series arrangement would have given the higher conversion.

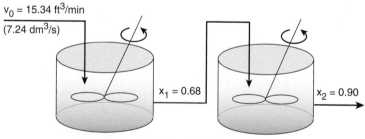

$v_0 = 15.34$ ft^3/min
(7.24 dm^3/s)

$x_1 = 0.68$

$x_2 = 0.90$

Figure E4-2.3   CSTRs in series.

The two equal-sized CSTRs in series (shown in Figure E4-2.3) will give a higher conversion than two CSTRs in parallel of the same size when the reaction order is greater than zero.

Safety considerations

We can find information about the safety of ethylene glycol and other chemicals from the World Wide Web (WWW) (Table 4-4). One source is the Vermont Safety Information on the Internet (Vermont SIRI). For example, we can learn from the *Control Measures* that we should use neoprene gloves when handling the material, and that we should avoid breathing the vapors. If we click on "Dow Chemical USA" and scroll the *Reactivity Data*, we would find that ethylene glycol will ignite in air at 413°C.

TABLE 4-4.   ACCESSING SAFETY INFORMATION

1. Type in

    http://www.siri.org/

2. When the first screen appears, click on "Material Safety Data Sheets." ("MSDS")

3. When the next page appears, type in the chemical you want to find.

    Example: Find ethylene glycol

    Then click on Enter.

4. The next page will show a list of a number of companies that provide the data on ethylene glycol.

        MALLINCKRODT BAKER
        FISHER
        DOW CHEMICAL USA
        etc.

    Let's click on "Mallinckrodt Baker." The materials safety data sheet provided will appear.

5. Scroll "ethylene glycol" for information you desire.

        1.   *Product Identification*
        2.   *Composition/Information on Ingredients*
        3.   *Hazards Identification*
        4.   *First Aid Measures*
        5.   *Fire Fighting Measures*
        6.   *Accidental Release Measures*
        7.   *Handling and Storage*
        8.   *Exposure Controls/Personal Protection*
        9.   *Physical and Chemical Properties*
        10-16. *Other Information*

4.4 Tubular Reactors

Gas-phase reactions are carried out primarily in tubular reactors where the flow is generally turbulent. By assuming that there is no dispersion and there are no radial gradients in either temperature, velocity, or concentration, we can model the flow in the reactor as plug-flow.

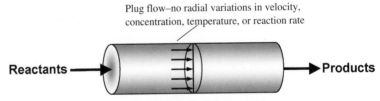

Plug flow–no radial variations in velocity, concentration, temperature, or reaction rate

Reactants ————→   ➤ Products

Figure 1-9   (Revisited) Tubular reactor.

Laminar reactors are discussed in Chapter 13 and dispersion effects in Chapter 14. The *differential form* of the PFR design equation

$$F_{A0} \frac{dX}{dV} = -r_A \tag{2-15}$$

must be used when there is a pressure drop in the reactor or heat exchange between the PFR and the surroundings. In the absence of pressure drop or heat exchange, the integral form of the *plug-flow design* equation is used,

$$V = F_{A0} \int_0^X \frac{dX}{-r_A} \tag{2-16}$$

As an example, consider the reaction

$$A \longrightarrow \text{Products}$$

for which the rate law is

Rate law
$$-r_A = kC_A^2$$

We shall first consider the reaction to take place as a liquid-phase reaction and then to take place as a gas-phase reaction.

Liquid Phase $v = v_0$

The combined PFR mole balance and rate law is

$$\frac{dX}{dV} = \frac{kC_A^2}{F_{A0}}$$

If the reaction is carried out in the liquid phase, the concentration is

Stoichiometry
(liquid phase)
$$C_A = C_{A0}(1 - X)$$

and for the isothermal operation, we can bring k outside the integral

Combine
$$V = \frac{F_{A0}}{kC_{A0}^2} \int_0^x \frac{dx}{(1-X)^2} = \frac{v_0}{kC_{A0}}\left(\frac{X}{1-X}\right)$$

This equation gives the reactor volume to achieve a conversion X. Dividing by v_0 ($\tau = V/v_0$) and solving for conversion, we find

$$X = \frac{\tau k C_{A0}}{1 + \tau k C_{A0}} = \frac{\text{Da}_2}{1 + \text{Da}_2}$$

where Da_2 is the Damköhler number for a second-order reaction.

Gas Phase

For constant-temperature ($T = T_0$) and constant-pressure ($P = P_0$) *gas-phase reactions,* the concentration is expressed as a function of conversion:

Stoichiometry
(gas phase)
$$C_A = \frac{F_A}{v} = \frac{F_A}{v_0(1 + \varepsilon X)} = \frac{F_{A0}(1 - X)}{v_0(1 + \varepsilon X)} = C_{A0}\frac{(1 - X)}{(1 + \varepsilon X)}$$

and then combining the PFR mole balance, rate law, and stoichiometry

Combine

$$V = F_{A0} \int_0^X \frac{(1 + \varepsilon X)^2}{k C_{A0}^2 (1 - X)^2} \, dX$$

The entering concentration C_{A0} can be taken outside the integral sign since it is not a function of conversion. Because the reaction is carried out isothermally, the specific reaction rate constant, k, can also be taken outside the integral sign.

For an isothermal reaction, k is constant.

$$V = \frac{F_{A0}}{k C_{A0}^2} \int_0^X \frac{(1 + \varepsilon X)^2}{(1 - X)^2} \, dX$$

From the integral equations in Appendix A.1, we find that

Reactor volume for a second-order gas-phase reaction

$$V = \frac{v_0}{k C_{A0}} \left[2\varepsilon(1 + \varepsilon) \ln(1 - X) + \varepsilon^2 X + \frac{(1 + \varepsilon)^2 X}{1 - X} \right] \qquad (4\text{-}17)$$

Using Equation (4-17), a plot of conversion along the length (i.e., volume) of the reactor is shown for four different reactions, and values of ε are given in Figure 4-7 for the same value of $[v_0/kC_{A0}]$ to illustrate the effect of volume change with reaction.

The term $\left[\dfrac{v_0}{k C_{A0}} \right]$ is the same for each reaction.

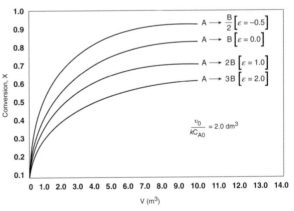

Figure 4-7   Conversion as a function of distance down the reactor.

We now look at the effect of the change in the number of moles in the gas phase on the relationship between conversion and volume. For constant temperature and pressure, Equation (3-45) becomes

$$v = v_0 (1 + \varepsilon X) \qquad (3\text{-}45)$$

Let's now consider three types of reactions, one in which $\varepsilon = 0$ ($\delta = 0$), one in which $\varepsilon < 0$ ($\delta < 0$), and one in which $\varepsilon > 0$ ($\delta > 0$). When there is no change in the number of moles with reaction, (e.g., A $\rightarrow$ B) $\delta = 0$ and $\varepsilon = 0$; then the fluid moves through the reactor at a constant volumetric flow rate ($v = v_0$) as the conversion increases.

When there is a decrease in the number of moles ($\delta < 0$, $\varepsilon < 0$) in the gas phase (e.g., 2A → B), the volumetric gas flow rate decreases as the conversion increases; for example,

$$v = v_0(1 - 0.5X)$$

Consequently, the gas molecules will spend longer in the reactor than they would if the flow rate were constant, $v = v_0$. As a result, this longer residence time would result in a higher conversion than if the flow were constant at v_0.

On the other hand, if there is an increase in the total number of moles ($\delta > 0$, $\varepsilon > 0$) in the gas phase (e.g., A → 2B), then the volumetric flow rate will increase as the conversion increases; for example,

$$v = v_0(1 + X)$$

and the molecules will spend less time in the reactor than they would if the volumetric flow rate were constant. As a result of this smaller residence time in the reactor the conversion will be less than what would result if the volumetric flow rate were constant at v_0.

The importance of changes in volumetric flow rate (i.e., $\varepsilon \neq 0$) with reaction

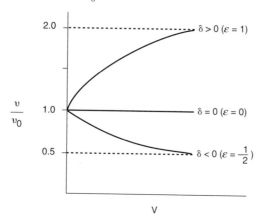

Figure 4-8   Change in gas-phase volumetric flow rate down the length of the reactor.

Figure 4-8 shows the volumetric flow rate profiles for the three cases just discussed. We note that, at the end of the reactor, virtually complete conversion has been achieved.

Example 4–3   Producing 300 Million Pounds per Year of Ethylene in a Plug-Flow Reactor: Design of a Full-Scale Tubular Reactor

The economics

Ethylene ranks fourth in the United States in total pounds of chemicals produced each year, and it is the number one organic chemical produced each year. Over 50 billion pounds were produced in 2000, and it sold for $0.27 per pound. Sixty-five percent of the ethylene produced is used in the manufacture of fabricated plastics,

The uses

20% for ethylene oxide, 16% for ethylene dichloride and ethylene glycol, 5% for fibers, and 5% for solvents.

Determine the plug-flow reactor volume necessary to produce 300 million pounds of ethylene a year from cracking a feed stream of pure ethane. The reaction is irreversible and follows an elementary rate law. We want to achieve 80% conversion of ethane, operating the reactor isothermally at 1100 K at a pressure of 6 atm.

Solution

$$C_2H_6 \longrightarrow C_2H_4 + H_2$$

Let $A = C_2H_6$, $B = C_2H_4$, and $C = H_2$. In symbols,

$$A \longrightarrow B + C$$

Because we want the reader to be familiar with both metric units **and** English units, we will work some of the examples using English units.

The molar flow rate of ethylene exiting the reactor is

$$F_B = 300 \times 10^6 \frac{lb}{year} \times \frac{1\ year}{365\ days} \times \frac{1\ day}{24\ h} \times \frac{1\ h}{3600\ s} \times \frac{lb\ mol}{28\ lb}$$

$$= 0.340 \frac{lb\ mol}{s} \left(154.4 \frac{mol}{s}\right)$$

Next calculate the molar feed rate of ethane, F_{A0}, to produce 0.34 lb mol/s of ethylene when 80% conversion is achieved,

$$F_B = F_{A0} X$$

$$F_{A0} = \frac{0.34}{0.8} = 0.425 \frac{lb\ mol}{s} \quad (402 \times 10^6\ lb/yr)$$

1. **Plug-flow design equation:**

Mole balance

$$F_{A0} \frac{dX}{dV} = -r_A \qquad (2\text{-}15)$$

Rearranging and integrating for the case of **no pressure drop** and isothermal operation yields

$$\boxed{V = F_{A0} \int_0^X \frac{dX}{-r_A}} \qquad (E4\text{-}3.1)$$

2. **Rate law:**[3]

---

[3] *Ind. Eng. Chem. Process Des. Dev., 14,* 218 (1975); *Ind. Eng. Chem., 59*(5), 70 (1967).

Rate law

$$\boxed{-r_A = kC_A}\quad \text{with } k = 0.072 \text{ s}^{-1} \text{ at 1000 K} \qquad (E4\text{-}3.2)$$

The activation energy is 82 kcal/g mol.

3. **Stoichiometry.** For isothermal operation and negligible pressure drop, the concentration of ethane is calculated as follows:

Gas phase, constant T and P:

Stoichiometry

$$\upsilon = \upsilon_0 \frac{F_T}{F_{T0}} = \upsilon_0(1 + \varepsilon X):$$

$$\boxed{C_A = \frac{F_A}{\upsilon} = \frac{F_{A0}(1-X)}{\upsilon_0(1+\varepsilon X)} = C_{A0}\left(\frac{1-X}{1+\varepsilon X}\right)} \qquad (E4\text{-}3.3)$$

$$C_C = \frac{C_{A0}X}{(1+\varepsilon X)} \qquad (E4\text{-}3.4)$$

4. We now **combine** Equations (E4-3.1) through (E4-3.3) to obtain

Combining the
design equation,
rate law, and
stoichiometry

$$V = F_{A0}\int_0^X \frac{dX}{kC_{A0}(1-X)/(1+\varepsilon X)} = F_{A0}\int_0^X \frac{(1+\varepsilon X)\,dX}{kC_{A0}(1-X)}$$

$$= \frac{F_{A0}}{C_{A0}}\int_0^X \frac{(1+\varepsilon X)\,dX}{k(1-X)} \qquad (E4\text{-}3.5)$$

5. **Evaluate.**

Since the reaction is carried out isothermally, we can take k outside the integral sign and use Appendix A.1 to carry out our integration.

Analytical solution

$$\boxed{V = \frac{F_{A0}}{kC_{A0}}\int_0^X \frac{(1+\varepsilon X)\,dX}{1-X} = \frac{F_{A0}}{kC_{A0}}\left[(1+\varepsilon)\ln\frac{1}{1-X} - \varepsilon X\right]} \qquad (E4\text{-}3.6)$$

6. **Parameter evaluation**:

Evaluate

$$C_{A0} = y_{A0}C_{T0} = \frac{y_{A0}P_0}{RT_0} = (1)\left(\frac{6 \text{ atm}}{(0.73 \text{ ft}^3\cdot\text{atm/lb mol}\cdot{}^\circ\text{R}) \times (1980^\circ\text{R})}\right)$$

$$= 0.00415 \frac{\text{lb mol}}{\text{ft}^3}\quad (0.066 \text{ mol/dm}^3)$$

$$\varepsilon = y_{A0}\delta = (1)(1+1-1) = 1$$

Oops! The rate constant k is given at 1000 K, and we need to calculate k at reaction conditions, which is 1100 K.

$$k(T_2) = k(T_1) \exp\left[\frac{E}{R}\left(\frac{1}{T_1} - \frac{1}{T_2}\right)\right]$$

$$= k(T_1) \exp\left[\frac{E}{R}\left(\frac{T_2 - T_1}{T_1 T_2}\right)\right] \tag{E4-3.7}$$

$$= \frac{0.072}{s} \exp\left[\frac{82,000 \text{ cal/g mol}(1100 - 1000) \text{ K}}{1.987 \text{ cal/(g mol·K)}(1000 \text{ K})(1100 \text{ K})}\right]$$

$$= 3.07 \text{ s}^{-1}$$

Substituting into Equation (E4-3.6) yields

$$V = \frac{0.425 \text{ lb mol/s}}{(3.07/s)(0.00415 \text{ lb mol/ft}^3)}\left[(1+1)\ln\frac{1}{1-X} - (1)X\right] \tag{E4-3.8}$$

$$= 33.36 \text{ ft}^3\left[2\ln\left(\frac{1}{1-X}\right) - X\right]$$

For $X = 0.8$,

$$V = 33.36 \text{ ft}^3\left[2\ln\left(\frac{1}{1-0.8}\right) - 0.8\right]$$

$$= 80.7 \text{ ft}^3 = (2280 \text{ dm}^3 = 2.28 \text{ m}^3)$$

It was decided to use a bank of 2-in. schedule 80 pipes in parallel that are 40 ft in length. For pipe schedule 80, the cross-sectional area, A_C, is 0.0205 ft². The number of pipes necessary is

The number of PFRs in parallel

$$n = \frac{80.7 \text{ ft}^3}{(0.0205 \text{ ft}^2)(40 \text{ ft})} = 98.4 \tag{E4-3.9}$$

To determine the concentrations and conversion profiles down the length of the reactor, z, we divide the volume Equation (E4-3.8) by the cross-sectional area, A_C,

$$z = \frac{V}{A_C} \tag{E4-3.10}$$

Equation (E4-3.9) was used along with $A_C = 0.0205$ ft², and Equations (E4-3.8) and (E4-3.3) were used to obtain Figure E4-3.1. Using a bank of 100 pipes will give us the reactor volume necessary to make 300 million pounds per year of ethylene from ethane. The concentration and conversion profiles down any one of the pipes are shown in Figure E4-3.1.

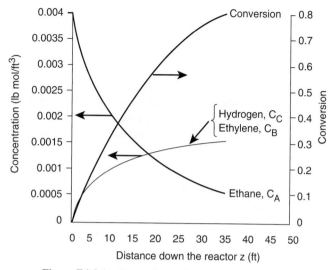

Figure E4-3.1   Conversion and concentration profiles.

4.5 Pressure Drop in Reactors

*Pressure drop is
ignored for liquid-
phase kinetics
calculations*

In liquid-phase reactions, the concentration of reactants is insignificantly
affected by even relatively large changes in the total pressure. Consequently,
we can totally ignore the effect of pressure drop on the rate of reaction when
sizing liquid-phase chemical reactors. However, in gas-phase reactions, the
concentration of the reacting species is proportional to the total pressure; consequently, proper accounting for the effects of pressure drop on the reaction
system can, in many instances, be a key factor in the success or failure of the
reactor operation. This fact is especially true in microreactors packed with
solid catalyst. Here the channels are so small (see Section 4.8) that pressure
drop can limit the throughput and conversion for gas-phase reactions.

4.5.1 Pressure Drop and the Rate Law

*For gas-phase
reactions, pressure
drop may be very
important.*

We now focus our attention on accounting for the pressure drop in the rate law.
For an ideal gas, we recall Equation (3-46) to write the concentration of reacting species i as

$$C_i = C_{A0} \left(\frac{\Theta_i + v_i X}{1 + \varepsilon X} \right) \frac{P}{P_0} \frac{T_0}{T} \qquad (4\text{-}18)$$

where $\Theta_i = \dfrac{F_{i0}}{F_{A0}}$, $\varepsilon = y_{A0}\delta$ and ν is the stoichiometric coefficient (e.g., $\nu_A = -1$, $\nu_B = -b/a$). We now must determine the ratio P/P_0 as a function of the volume, V, or the catalyst weight, W, to account for pressure drop. We then can combine the concentration, rate law, and design equation. However, whenever accounting for the effects of pressure drop, *the differential form of the mole balance (design equation) must be used*.

If, for example, the second-order isomerization reaction

$$2A \longrightarrow B + C$$

When $P \neq P_0$ one must use the differential forms of the PFR/PBR design equations.

is being carried out in a packed-bed reactor, the **differential form of the mole balance** equation in terms of catalyst weight is

$$F_{A0}\frac{dX}{dW} = -r_A' \qquad \left(\frac{\text{gram moles}}{\text{gram catalyst} \cdot \text{min}}\right) \qquad (2\text{-}17)$$

The **rate law** is

$$-r_A' = kC_A^2 \qquad (4\text{-}19)$$

From **stoichiometry** for gas-phase reactions (Table 3-5),

$$C_A = \frac{C_{A0}(1-X)}{1+\varepsilon X}\frac{P}{P_0}\frac{T_0}{T}$$

and the rate law can be written as

$$-r_A' = k\left[\frac{C_{A0}(1-X)}{1+\varepsilon X}\frac{P}{P_0}\frac{T_0}{T}\right]^2 \qquad (4\text{-}20)$$

Note from Equation (4-20) that the larger the pressure drop (i.e., the smaller P) from frictional losses, the smaller the reaction rate!

Combining Equation (4-20) with the mole balance (2-17) and assuming isothermal operation ($T = T_0$) gives

$$F_{A0}\frac{dX}{dW} = k\left[\frac{C_{A0}(1-X)}{1+\varepsilon X}\right]^2\left(\frac{P}{P_0}\right)^2$$

Dividing by F_{A0} (i.e., $v_0 C_{A0}$) yields

$$\frac{dX}{dW} = \frac{kC_{A0}}{v_0}\left(\frac{1-X}{1+\varepsilon X}\right)^2 \left(\frac{P}{P_0}\right)^2$$

For isothermal operation ($T = T_0$), the right-hand side is a function of only conversion and pressure:

Another equation
is needed
(e.g., $P = f(W)$).

$$\frac{dX}{dW} = F_1(X, P) \tag{4-21}$$

We now need to relate the pressure drop to the catalyst weight in order to determine the conversion as a function of catalyst weight.

4.5.2 Flow Through a Packed Bed

The majority of gas-phase reactions are catalyzed by passing the reactant through a packed bed of catalyst particles.

The equation used most to calculate pressure drop in a packed porous bed is the **Ergun equation**:[4,†]

Ergun equation

$$\frac{dP}{dz} = -\frac{G}{\rho g_c D_p}\left(\frac{1-\phi}{\phi^3}\right)\left[\overbrace{\frac{150(1-\phi)\mu}{D_p}}^{\text{Term 1}} + \overbrace{1.75G}^{\text{Term 2}}\right] \tag{4-22}$$

Term 1 is dominant for laminar flow, and Term 2 is dominant for turbulent flow.

---

[4] R. B. Bird, W. E. Stewart, and E. N. Lightfoot, *Transport Phenomena,* 2nd ed. (New York: Wiley, 2001), p. 200.

[†] A slightly different set of constants for the Ergun Equation (e.g., 1.8G instead of 1.75G) can be found in *Ind. Eng. Chem. Fundamentals,* 18 (1979), p. 199.

where P = pressure, lb_f/ft^2 (kPa)

$$\phi = \text{porosity} = \frac{\text{volume of void}}{\text{total bed volume}} = \text{void fraction}$$

$$1 - \phi = \frac{\text{volume of solid}}{\text{total bed volume}}$$

g_c = 32.174 $\text{lb}_m \cdot \text{ft/s}^2 \cdot \text{lb}_f$ (conversion factor)

$\quad$ = $4.17 \times 10^8 \ \text{lb}_m \cdot \text{ft/h}^2 \cdot \text{lb}_f$

(recall that for the metric system $g_c = 1.0$)

D_p = diameter of particle in the bed, ft (m)

μ = viscosity of gas passing through the bed, $\text{lb}_m/\text{ft} \cdot \text{h}(\text{kg/m} \cdot \text{s})$

z = length down the packed bed of pipe, ft (m)

u = superficial velocity = volumetric flow $\div$ cross-sectional
$\quad$ area of pipe, ft/h (m/s)

ρ = gas density, lb/ft^3 (kg/m^3)

$G = \rho u$ = superficial mass velocity, $\text{lb}_m/\text{ft}^2 \cdot \text{h}(\text{kg/m}^2 \cdot \text{s})$

In calculating the pressure drop using the Ergun equation, the only parameter that varies with pressure on the right-hand side of Equation (4-22) is the gas density, ρ. We are now going to calculate the pressure drop through the bed.

Because the reactor is operated at steady state, the mass flow rate at any point down the reactor, $\dot{m}$ (kg/s), is equal to the entering mass flow rate, $\dot{m}_0$ (i.e., equation of continuity),

$$\dot{m}_0 = \dot{m}$$

$$\rho_0 v_0 = \rho v$$

Recalling Equation (3-41), we have

$$v = v_0 \frac{P_0}{P}\left(\frac{T}{T_0}\right)\frac{F_T}{F_{T0}} \tag{3-41}$$

$$\rho = \rho_0 \frac{v_0}{v} = \rho_0 \frac{P}{P_0}\left(\frac{T_0}{T}\right)\frac{F_{T0}}{F_T} \tag{4-23}$$

Combining Equations (4-22) and (4-23) gives

$$\frac{dP}{dz} = -\frac{G(1-\phi)}{\rho_0 g_c D_p \phi^3}\left[\frac{150(1-\phi)\mu}{D_p} + 1.75G\right]\frac{P_0}{P}\left(\frac{T}{T_0}\right)\frac{F_T}{F_{T0}}$$

Simplifying yields

$$\boxed{\frac{dP}{dz} = -\beta_0 \frac{P_0}{P}\left(\frac{T}{T_0}\right)\frac{F_T}{F_{T0}}}$$

(4-24)

where β_0 is a constant down the reactor that depends only on the properties of the packed bed and the entrance conditions.

$$\beta_0 = \frac{G(1-\phi)}{\rho_0 g_c D_P \phi^3}\left[\frac{150(1-\phi)\mu}{D_P} + 1.75G\right]$$

(4-25)

For tubular packed-bed reactors, we are more interested in catalyst weight rather than the distance z down the reactor. The catalyst weight up to a distance of z down the reactor is

$$W \qquad = \qquad (1-\phi)A_c z \qquad \times \qquad \rho_c$$

$$\overbrace{\begin{bmatrix} \text{Weight of} \\ \text{catalyst} \end{bmatrix}} = \overbrace{\begin{bmatrix} \text{Volume of} \\ \text{solids} \end{bmatrix}} \times \overbrace{\begin{bmatrix} \text{Density of} \\ \text{solid catalyst} \end{bmatrix}}$$

(4-26)

where A_c is the cross-sectional area. The bulk density of the catalyst, ρ_b (mass of catalyst per volume of reactor bed), is just the product of the density of the solid catalyst particles, ρ_c, and the fraction of solids, $(1-\phi)$:

Bulk density

$$\rho_b = \rho_c (1-\phi)$$

Using the relationship between z and W [Equation (4-26)] we can change our variables to express the Ergun equation in terms of catalyst weight:

Use this form for multiple reactions and membrane reactors.

$$\frac{dP}{dW} = -\frac{\beta_0}{A_c(1-\phi)\rho_c}\frac{P_0}{P}\left(\frac{T}{T_0}\right)\frac{F_T}{F_{T0}}$$

Further simplification yields

$$\frac{dP}{dW} = -\frac{\alpha}{2}\frac{T}{T_0}\frac{P_0}{P/P_0}\left(\frac{F_T}{F_{T0}}\right)$$

(4-27)

Let $y = P / P_0$, then

$$\boxed{\frac{dy}{dW} = -\frac{\alpha}{2y}\frac{T}{T_0}\frac{F_T}{F_{T0}}}$$

(4-28)

where

$$\boxed{\alpha = \frac{2\beta_0}{A_c\rho_c(1-\phi)P_0}} \tag{4-29}$$

We will use Equation (4-28) when multiple reactions are occurring or when there is pressure drop in a membrane reactor. However, for single reactions in packed-bed reactors, it is more convenient to express the Ergun equation in terms of the conversion X. Recalling Equation (3-43) for F_T,

$$F_T = F_{T0} + F_{A0}\delta X = F_{T0}\left(1 + \frac{F_{A0}}{F_{T0}}\delta X\right) \tag{3-43}$$

and dividing by F_{T0}

$$\frac{F_T}{F_{T0}} = 1 + \varepsilon X$$

where, as before,

$$\varepsilon = y_{A0}\delta = \frac{F_{A0}}{F_{T0}}\delta \tag{3-35}$$

Differential form of Ergun equation for the pressure drop in packed beds.

Substituting for the ratio (F_T/F_{T0}), Equation (4-28) can now be written as

$$\boxed{\frac{dy}{dW} = -\frac{\alpha}{2y}(1 + \varepsilon X)\frac{T}{T_0}} \tag{4-30}$$

We note that when ε is negative, the pressure drop ΔP will be less (i.e., higher pressure) than that for $\varepsilon = 0$. When ε is positive, the pressure drop ΔP will be greater than when $\varepsilon = 0$.

For isothermal operation, Equation (4-30) is only a function of conversion and pressure:

$$\frac{dP}{dW} = F_2(X,P) \tag{4-31}$$

Two coupled equations to be solved numerically

Recalling Equation (4-21), for the combined mole balance, rate law, and stoichiometry,

$$\frac{dX}{dW} = F_1(X,P) \tag{4-21}$$

we see that we have two coupled first-order differential equations, (4-31) and (4-21), that must be solved simultaneously. A variety of software packages and numerical integration schemes are available for this purpose.

Analytical Solution.   If $\varepsilon = 0$, *or* if we can neglect (εX) with respect to 1.0 (i.e., $1 \gg \varepsilon X$), we can obtain an analytical solution to Equation (4-30) for isothermal operation (i.e., $T = T_0$). For isothermal operation with $\varepsilon = 0$, Equation (4-30) becomes

Isothermal with $\varepsilon = 0$

$$\frac{dy}{dW} = \frac{-\alpha}{2y} \qquad (4\text{-}32)$$

Rearranging gives us

$$2y\frac{dy}{dW} = -\alpha$$

Taking y inside the derivative, we have

$$\frac{dy^2}{dW} = -\alpha$$

Integrating with $y = 1$ $(P = P_0)$ at $W = 0$ yields

$$(y)^2 = 1 - \alpha W$$

Taking the square root of both sides gives

Pressure ratio **only** for $\varepsilon = 0$

$$\boxed{y = \frac{P}{P_0} = (1 - \alpha W)^{1/2}} \qquad (4\text{-}33)$$

Be sure *not* to use this equation if $\varepsilon \neq 0$ or the reaction is not carried out isothermally, where again

$$\boxed{\alpha = \frac{2\beta_0}{A_c(1 - \phi)\rho_c P_0}} \qquad (4\text{-}29)$$

Equation (4-33) can be used to substitute for the pressure in the rate law, in which case the mole balance can be written solely as a function of conversion and catalyst weight. The resulting equation can readily be solved either analytically or numerically.

   If we wish to express the pressure in terms of reactor length z, we can use Equation (4-26) to substitute for W in Equation (4-33). Then

$$y = \frac{P}{P_0} = \left(1 - \frac{2\beta_0 z}{P_0}\right)^{1/2} \qquad (4\text{-}34)$$

4.5.3 Pressure Drop in Pipes

Normally, the pressure drop for gases flowing through pipes without packing can be neglected. For flow in pipes, the pressure drop along the length of the pipe is given by

$$\frac{dP}{dL} = -G\frac{du}{dL} - \frac{2fG^2}{\rho D} \tag{4-35}$$

where D = pipe diameter, cm

$\quad\quad u$ = average velocity of gas, cm/s

$\quad\quad f$ = Fanning friction factor

$\quad\quad G = \rho u$, g/cm$^2 \cdot$ s

The friction factor is a function of the Reynolds number and pipe roughness. The mass velocity, G, is constant along the length of the pipe. Replacing u with G/ρ, and combining with Equation (4-23) for the case of constant temperature, T, and total molar flow rate, F_T, Equation (4-35) becomes

$$\rho_0\frac{P}{P_0}\frac{dP}{dL} - G^2\frac{dP}{P\,dL} + \frac{2fG^2}{D} = 0$$

Integrating with limits $P = P_0$ when $L = 0$, and assuming that f does not vary, we have

$$\frac{P_0^2 - P^2}{2} = G^2\frac{P_0}{\rho_0}\left(2f\frac{L}{D} + \ln\frac{P_0}{P}\right)$$

Neglecting the second term on the right-hand side gives upon rearrangement

$$\frac{P}{P_0} = \left[1 - \frac{4fG^2V}{\rho_0 P_0 A_c D}\right]^{1/2} = (1 - \alpha_p V)^{1/2} \tag{4-36}$$

where

$$\alpha_p = \frac{4fG^2}{A_c \rho_0 P_0 D}$$

For the flow conditions given in Example 4-4 in a 1000-ft length of $1\frac{1}{2}$-in. schedule 40 pipe ($\alpha_p = 0.0118$), the pressure drop is less than 10%. However, for high volumetric flow rates through microreactors, the pressure drop may be significant.

Example 4–4 Calculating Pressure Drop in a Packed Bed

Plot the pressure drop in a 60 ft length of $1\frac{1}{2}$-in. schedule 40 pipe packed with catalyst pellets $\frac{1}{4}$-in. in diameter. There is 104.4 lb/h of gas passing through the bed. The temperature is constant along the length of pipe at 260°C. The void fraction is 45% and the properties of the gas are similar to those of air at this temperature. The entering pressure is 10 atm.

Solution

At the end of the reactor, $z = L$ and Equation (4-34) becomes

$$\frac{P}{P_0} = \left(1 - \frac{2\beta_0 L}{P_0}\right)^{1/2} \tag{E4-4.1}$$

$$\beta_0 = \frac{G(1-\phi)}{g_c \rho_0 D_p \phi^3}\left[\frac{150(1-\phi)\mu}{D_p} + 1.75G\right] \tag{4-25}$$

Evaluating
the pressure drop
parameters

$$G = \frac{\dot{m}}{A_c} \tag{E4-4.2}$$

For $1\frac{1}{2}$-in. schedule 40 pipe, $A_c = 0.01414$ ft²:

$$G = \frac{104.4 \text{ lb}_m/\text{h}}{0.01414 \text{ ft}^2} = 7383.3 \frac{\text{lb}_m}{\text{h} \cdot \text{ft}^2}$$

For air at 260°C and 10 atm,

$$\mu = 0.0673 \text{ lb}_m/\text{ft} \cdot \text{h}$$

$$\rho_0 = 0.413 \text{ lb}_m/\text{ft}^3$$

$$v_0 = \frac{\dot{m}}{\rho_0} = \frac{104.4 \text{ lb}_m/\text{h}}{0.413 \text{ lb}_m/\text{ft}^3} = 252.8 \text{ ft}^3/\text{h} \ (7.16 \text{ m}^3/\text{h})$$

From the problem statement,

$$D_p = \tfrac{1}{4} \text{ in.} = 0.0208 \text{ ft, } \phi = 0.45$$

$$g_c = 4.17 \times 10^8 \frac{\text{lb}_m \cdot \text{ft}}{\text{lb}_f \cdot \text{h}^2}$$

Substituting these values into Equation (4-25) gives

$$\beta_0 = \left[\frac{7383.3 \text{ lb}_m/\text{ft}^2 \cdot \text{h}(1 - 0.45)}{(4.17 \times 10^8 \text{ lb}_m \cdot \text{ft}/\text{lb}_f \cdot \text{h}^2)(0.413 \text{ lb}_m/\text{ft}^3)(0.0208 \text{ ft})(0.45)^3}\right] \quad \text{(E4-4.3)}$$

$$\times \left[\frac{150(1 - 0.45)(0.0673 \text{ lb}_m/\text{ft} \cdot \text{h})}{0.0208 \text{ ft}} + 1.75(7383.3)\frac{\text{lb}_m}{\text{ft}^2 \cdot \text{h}}\right]$$

$$\beta_0 = 0.01244 \frac{\text{lb}_f \cdot \text{h}}{\text{ft} \cdot \text{lb}_m} [\overbrace{(266.9}^{\text{Term 1}} + \overbrace{12,920.8)}^{\text{Term 2}}] \frac{\text{lb}_m}{\text{ft}^2 \cdot \text{h}} = 164.1 \frac{\text{lb}_f}{\text{ft}^3} \quad \text{(E4-4.4)}$$

We note that the turbulent flow term, Term 2, is dominant.

$$\beta_0 = 164.1 \frac{\text{lb}_f}{\text{ft}^3} \times \frac{1 \text{ ft}^2}{144 \text{ in.}^2} \times \frac{1 \text{ atm}}{14.7 \text{ lb}_f/\text{in.}^2}$$

Unit Conversion
for β_0

$\dfrac{1 \text{ atm}}{\text{ft}} = 333\dfrac{\text{kPa}}{\text{m}}$

$$\boxed{\beta_0 = 0.0775 \frac{\text{atm}}{\text{ft}} = 25.8 \frac{\text{kPa}}{\text{m}}}$$

$$y = \frac{P}{P_0} = \left(1 - \frac{2\beta_0 L}{P_0}\right)^{1/2} = \left(1 - \frac{\overbrace{2 \times 0.0775}^{0.155} \text{ atm/ft} \times 60 \text{ ft}}{10 \text{ atm}}\right)^{1/2} \quad \text{(E4-4.6)}$$

$$P = 0.265P_0 = 2.65 \text{ atm} \quad (268 \text{ kPa})$$
$$\Delta P = P_0 - P = 10 - 2.65 = 7.35 \text{ atm} \quad (744 \text{ kPa}) \quad \text{(E4-4.7)}$$

Now let's use the data to plot the pressure and the volumetric flow rate profiles. Recalling Equation (4-34) for the case $\varepsilon = 0$ and $T = T_0$

$$\boxed{v = v_0\frac{P_0}{P} = \frac{v_0}{y}} \quad \text{(E4-4.8)}$$

Equations (4-34) and (E4-4.8) were used in the construction of Table E4-4.1.

TABLE E4-4.1. *P* AND *v* PROFILES

z (ft)	0	10	20	30	40	50	60
P (atm)	10	9.2	8.3	7.3	6.2	4.7	2.65
v (ft³/h)	253	275	305	347	408	538	955

For $\rho_c = 120 \text{ lb/ft}^3$

$$\alpha = \frac{2\beta_0}{\rho_c(1-\phi)A_cP_0} = \frac{2(0.0775)\text{atm/ft}}{120\ \text{lb/ft}^3(1-0.45)(0.1414\ \text{ft}^2)10\ \text{atm}}$$

$$\boxed{\alpha = 0.00165\ \text{lb}^{-1} = 0.037\ \text{kg}^{-1}}$$

The values in Table E4-4.1 were used to obtain Figure E4-4.1.

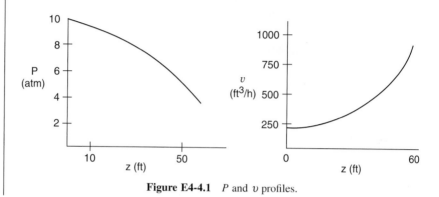

Figure E4-4.1 P and v profiles.

4.5.4   Analytical Solution for Reaction with Pressure Drop

We will first describe how pressure drop affects our CRE algorithm. Figure 4-9 shows qualitatively the effects of pressure drop on reactor design.

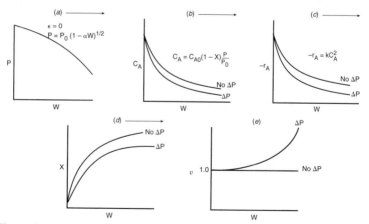

Figure 4-9   Effect of pressure drop on P **(a)**, C_A **(b)**, $-r_A$ **(c)**, X **(d)**, and v **(e)**.

These graphs compare the concentrations, reaction rates, and conversion profiles for the cases of pressure drop and no pressure drop. We see that when there is pressure drop in the reactor, the reactant concentrations and thus reaction rate for reaction (for reaction orders greater than 0 order) will always be smaller than the case with no pressure drop. As a result of this smaller reaction rate, the conversion will be less with pressure drop than without pressure drop.

Now that we have expressed pressure as a function of catalyst weight [Equation (4-33)], we can return to the second-order isothermal reaction,

$$A \longrightarrow B$$

to relate conversion and catalyst weight. Recall our mole balance, rate law, and stoichiometry.

1. Mole balance:
$$F_{A0} \frac{dX}{dW} = -r'_A \qquad (2\text{-}17)$$

2. Rate law:
$$-r'_A = kC_A^2 \qquad (4\text{-}19)$$

3. Stoichiometry. Gas-phase isothermal reaction $(T = T_0)$ with $\varepsilon = 0$. From Equation (3-45), $v = v_0/y$

$$C_A = \frac{F_A}{v} = C_{A0}(1 - X)\, y \qquad (4\text{-}37)$$

Only
for
$\varepsilon = 0$

$$y = (1 - \alpha W)^{1/2} \qquad (4\text{-}33)$$

Using Equation (4-33) to substitute for y in terms of the catalyst weight, we obtain

$$C_A = C_{A0}(1 - X)(1 - \alpha W)^{1/2}$$

4. Combining:
$$\frac{dX}{dW} = \frac{kC_{A0}^2}{F_{A0}} (1 - X)^2 \, [(1 - \alpha W)^{1/2}]^2$$

5. Separating variables:
$$\frac{F_{A0}}{kC_{A0}^2} \frac{dX}{(1 - X)^2} = (1 - \alpha W)\, dW$$

Integrating with limits $X = 0$ when $W = 0$ and substituting for $F_{A0} = C_{A0}v_0$ yields

$$\frac{v_0}{kC_{A0}} \left(\frac{X}{1 - X} \right) = W\left(1 - \frac{\alpha W}{2} \right)$$

Solving for conversion gives

$$X = \frac{\dfrac{kC_{A0}W}{v_0}\left(1 - \dfrac{\alpha W}{2}\right)}{1 + \dfrac{kC_{A0}W}{v_0}\left(1 - \dfrac{\alpha W}{2}\right)} \tag{4-38}$$

Solving for the catalyst weight, we have

Catalyst weight
for
second-order
reaction in PFR
with ΔP

$$W = \frac{1 - \{1 - [(2v_0\alpha)/kC_{A0}][X/(1 - X)]\}^{1/2}}{\alpha} \tag{4-39}$$

Example 4–5   Effect of Pressure Drop on the Conversion Profile

Reconsider the packed bed in Example 4-4 for the case where a second-order reaction

$$2A \rightarrow B + C$$

is taking place in 20 meters of a $1\frac{1}{2}$ schedule 40 pipe packed with catalyst. The flow and packed-bed conditions in the example remain the same except that they are converted to SI units; that is, $P_0 = 10$ atm = 1013 kPa, and

We need to be able
to work either
metric, S.I., or
English units.

Entering volumetric flow rate: $v_0 = 7.15$ m³/h (252 ft³/h)

Catalyst pellet size: $D_p = 0.006$ m (ca. $\frac{1}{4}$-inch)

Solid catalyst density: $\rho_c = 1923$ kg/m³ (120 lb/ft³)

Cross-sectional area of $1\frac{1}{2}$-in. schedule 40 pipe: $A_C = 0.0013$ m²

Pressure drop parameter: $\beta_0 = 25.8$ kPa/m

Reactor length: $L = 20$ m

We will change the particle size to learn its effect on the conversion profile. However, we will assume that the specific reaction rate, k, is unaffected by particle size, an assumption we know from Chapter 12 is valid only for small particles.

(a)   First, calculate the conversion in the absence of pressure drop.

(b)   Next, calculate the conversion accounting for pressure drop.

(c)   Finally, determine how your answer to (b) would change if the catalyst particle diameter were doubled.

The entering concentration of A is 0.1 kmol/m³ and the specific reaction rate is

$$k = \frac{12\,\text{m}^6}{\text{kmol}\cdot\text{kg cat}\cdot\text{h}}$$

Solution

Using Equation (4-38)

$$X = \frac{\dfrac{kC_{A0}W}{v_0}\left(1 - \dfrac{\alpha W}{2}\right)}{1 + \dfrac{kC_{A0}W}{v_0}(1 - \alpha W)} \tag{4-38}$$

For the bulk catalyst density,

$$\rho_b = \rho_c(1 - \phi) = (1923)(1 - 0.45) = 1058 \text{ kg/m}^3$$

The weight of catalyst in the 20 m of $1\frac{1}{2}$-in. schedule 40 pipe is

$$W = A_c\rho_b L = (0.0013 \text{ m}^2)\left(1058 \ \frac{\text{kg}}{\text{m}^3}\right)(20 \text{ m})$$

$$W = 27.5 \text{ kg}$$

$$\frac{kC_{A0}W}{v_0} = \frac{12\text{m}^6}{\text{kmol} \cdot \text{kg cat} \cdot \text{h}} \cdot 0.1\frac{\text{kmol}}{\text{m}^3} \cdot \frac{27.5 \ \text{kg}}{7.15 \ \text{m}^3/\text{h}} = 4.6$$

(a) First calculate the conversion for $\Delta P = 0$ (i.e., $\alpha = 0$)

$$X = \frac{\dfrac{kC_{A0}W}{v_0}}{1 + \dfrac{kC_{A0}W}{v_0}} = \frac{4.6}{1 + 4.6} = 0.82 \tag{E4-5.1}$$

$$\boxed{X = 0.82}$$

(b) Next, we calculate the conversion with pressure drop. Recalling Equation (4-29) and substituting the bulk density $\rho_b = (1 - \phi) \rho_c = 1058 \text{ kg/m}^3$

$$\alpha = \frac{2\beta_0}{P_0 A_c \rho_b} = \frac{2\left(25.8 \ \dfrac{\text{kPa}}{\text{m}}\right)}{(1013 \ \text{kPa})(0.0013\text{m}^2)\left(1058\dfrac{\text{kg}}{\text{m}^3}\right)} \tag{E4-5.2}$$

$$= 0.037 \text{ kg}^{-1}$$

then

$$\left(1 - \frac{\alpha W}{2}\right) = 1 - \frac{(0.037)(27.5)}{2} = 0.49 \tag{E4-5.3}$$

$$X = \frac{\dfrac{kC_{A0}W}{v_0}\left(1 - \dfrac{\alpha W}{2}\right)}{1 + \dfrac{kC_{A0}W}{v_0}\left(1 - \dfrac{\alpha W}{2}\right)} = \frac{(4.6)(0.49)}{1 + (4.6)(0.49)} = \frac{2.36}{3.26} \qquad \text{(E4-5.4)}$$

$$\boxed{X = 0.693}$$

We see the predicted conversion dropped from 82.2% to 69.3% because of pressure drop. It would be not only embarrassing but also an economic disaster if we had neglected pressure drop and the actual conversion had turned out to be significantly smaller.

(c)   *Robert the Worrier* wonders: *What if* we increase the catalyst size by a factor of 2? We see from Equation (E4-4.5) that the second term in the Ergun equation is dominant; that is,

$$1.75G \gg \frac{150(1 - \phi)\mu}{D_p} \qquad \text{(E4-5.5)}$$

Therefore from Equation (4-25)

$$\beta_0 = \frac{G(1 - \phi)}{\rho_0 g_c D_p \phi^3}\left[\frac{150(1 - \phi)\mu}{D_p} + 1.75G\right]$$

we have

$$\beta_0 = \frac{1.75G^2(1 - \phi)}{\rho_0 g_c D_p \phi^3} \qquad \text{(E4-5.6)}$$

We see for the conditions given by Equation (E4-4.4) that the pressure drop parameter varies inversely with the particle diameter

We will learn more about *Robert the Worrier* in Chapter 11.

$$\beta_0 \sim \frac{1}{D_p}$$

and thus

$$\alpha \sim \frac{1}{D_p}$$

For Case 2, $D_{P_2} = 2D_{P_1}$

$$\alpha_2 = \alpha_1 \frac{D_{P_1}}{D_{P_2}} = (0.037 \text{ kg}^{-1})\frac{1}{2} \qquad \text{(E4-5.7)}$$

$$= 0.0185 \text{ kg}^{-1}$$

Substituting this new value of α in Equation (E4-5.4)

$$X_2 = \frac{(4.6)\left(1 - \dfrac{0.0185(27.5)}{2}\right)}{1 + (4.6)\left(1 - \dfrac{0.0185(27.5)}{2}\right)} = \frac{3.43}{4.43}$$

$$X = 0.774$$

By increasing the particle diameter we decrease the pressure drop parameter and thus increase the reaction rate and the conversion. However, Chapters 10 and 12 explain that when interparticle diffusion effects are important in the catalyst pellet, this increase in conversion with increasing particle size will not always be the case. For larger particles, it takes a longer time for a given number of reactant and product molecules to diffuse in and out of the catalyst particle where they undergo reaction (see Figure 10-6). Consequently, the specific reaction rate decreases with increasing particle size $k \sim 1/D_P$ [see Equation (12-35)], which in turn decreases the conversion. At small particle diameters, the rate constant, k, is large, and at its maximum value, but the pressure drop is also large, resulting in a low rate of reaction. At large particle diameters, the pressure drop is small, but so is the rate constant, k, and the rate of reaction, resulting in low conversion. Thus, we see how a low conversion at both large and small particle diameters with an optimum in between. This optimum is shown in Figure E4-5.1. See Problem P4-23.

The variation

$$k \sim \frac{1}{D_p}$$

is discussed in detail in Chapter 12. Also, see Ch. 4 Summary Notes.

Summary Notes

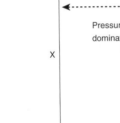

Figure E4-5.1   Finding the optimum particle diameter.

Problems with large diameter tubes

(1) Bypassing of catalyst
(2) Smaller heat transfer area

  If pressure drop is to be minimized, *why not pack the catalyst into a larger diameter tube* to decrease the superficial velocity, G, thereby reducing ΔP? There are two reasons for *not* increasing the tube diameter: (1) There is an increased chance the gas could channel and bypass most of the catalyst, resulting in little conversion (see Figures 13-2 and 13-10); (2) the ratio of the heat-transfer surface area to reactor volume (catalyst weight) will be decreased, thereby making heat transfer more difficult for highly exothermic and endothermic reactions. We now proceed (Example 4-6) to combine pressure drop with reaction in a packed bed when we have volume change with reaction and therefore cannot obtain an analytical solution.

Example 4–6 Calculating X in a Reactor with Pressure Drop

The economics

The uses

Approximately 7 billion pounds of ethylene oxide were produced in the United States in 1997. The 1997 selling price was $0.58 a pound, amounting to a commercial value of $4.0 billion. Over 60% of the ethylene oxide produced is used to make ethylene glycol. The major end uses of ethylene oxide are antifreeze (30%), polyester (30%), surfactants (10%), and solvents (5%). We want to calculate the catalyst weight necessary to achieve 60% conversion when ethylene oxide is to be made by the vapor-phase catalytic oxidation of ethylene with air.

$$C_2H_4 + \tfrac{1}{2}O_2 \longrightarrow \overset{\text{O}}{\overset{\triangle}{CH_2-CH_2}}$$

$$A + \tfrac{1}{2}B \longrightarrow C$$

Ethylene and oxygen are fed in stoichiometric proportions to a packed-bed reactor operated isothermally at 260°C. Ethylene is fed at a rate of 0.30 lb mol/s at a pressure of 10 atm. It is proposed to use 10 banks of $1\tfrac{1}{2}$-in.-diameter schedule 40 tubes packed with catalyst with 100 tubes per bank. Consequently, the molar flow rate to each tube is to be 3×10^{-4} lb mol/s. The properties of the reacting fluid are to be considered identical to those of air at this temperature and pressure. The density of the $\tfrac{1}{4}$-in.-catalyst particles is 120 lb/ft³ and the bed void fraction is 0.45. The rate law is

$$-r'_A = kP_A^{1/3}P_B^{2/3} \qquad \text{lb mol/lb cat}\cdot\text{h}$$

with[5]

$$k = 0.0141 \frac{\text{lb mol}}{\text{atm}\cdot\text{lb cat}\cdot\text{h}} \text{ at 260°C}$$

Solution

Following the Algorithm

1. **Differential mole balance**:

$$\boxed{F_{A0}\frac{dX}{dW} = -r'_A} \qquad (E4\text{-}6.1)$$

2. **Rate law**:

$$-r'_A = kP_A^{1/3}P_B^{2/3} = k(C_A RT)^{1/3}(C_B RT)^{2/3} \qquad (E4\text{-}6.2)$$

$$= kRT C_A^{1/3} C_B^{2/3} \qquad (E4\text{-}6.3)$$

The algorithm

3. **Stoichiometry.** Gas-phase, isothermal $v = v_0(1+\varepsilon X)(P_0/P)$:

$$C_A = \frac{F_A}{v} = \frac{C_{A0}(1-X)}{1+\varepsilon X}\left(\frac{P}{P_0}\right) = \frac{C_{A0}(1-X)y}{1+\varepsilon X} \quad \text{where } y = \frac{P}{P_0} \qquad (E4\text{-}6.4)$$

---

[5] *Ind. Eng. Chem.*, 45, 234 (1953).

$$C_B = \frac{F_B}{v} = \frac{C_{A0}(\Theta_B - X/2)}{1 + \varepsilon X} y \qquad (E4\text{-}6.5)$$

For stoichiometric feed $\Theta_B = \dfrac{F_{B0}}{F_{A0}} = \dfrac{1}{2}$

$$C_B = \frac{C_{A0}}{2} \frac{(1 - X)}{(1 + \varepsilon X)} y$$

For isothermal operation, Equation (4-30) becomes

$$\frac{dy}{dW} = -\frac{\alpha}{2y}(1 + \varepsilon X) \qquad (E4\text{-}6.6)$$

4. **Combining** the rate law and concentrations:

$$-r'_A = kRT_0 \left[\frac{C_{A0}(1 - X)}{1 + \varepsilon X} (y) \right]^{1/3} \left[\frac{C_{A0}(1 - X)}{2(1 + \varepsilon X)} (y) \right]^{2/3} \qquad (E4\text{-}6.7)$$

We can evaluate
the **combine** step
either

1. Analytically
2. Graphically
3. Numerically, or
4. Using software

Factoring $\left(\dfrac{1}{2}\right)^{2/3}$ and recalling $P_{A0} = C_{A0} RT_0$, we can simplify Equation (E4-6.7) to

$$\boxed{-r'_A = k'\left(\frac{1 - X}{1 + \varepsilon X}\right) y} \qquad (E4\text{-}6.8)$$

where $k' = kP_{A0}\left(\dfrac{1}{2}\right)^{2/3} = 0.63kP_{A0}$.

5. **Parameter evaluation per tube** (i.e., divide feed rates by 1000):

Ethylene: $F_{A0} = 3 \times 10^{-4}$ lb mol/s = 1.08 lb mol/h

Oxygen: $F_{B0} = 1.5 \times 10^{-4}$ lb mol/s = 0.54 lb mol/h

I = inerts = N_2: $F_I = 1.5 \times 10^{-4}$ lb mol/s $\times \dfrac{0.79 \text{ mol } N_2}{0.21 \text{ mol } O_2}$

$$F_I = 5.64 \times 10^{-4} \text{ lb mol/s} = 2.03 \text{ lb mol/h}$$

Summing: $F_{T0} = F_{A0} + F_{B0} + F_I = 3.65$ lb mol/h

$$y_{A0} = \frac{F_{A0}}{F_{T0}} = \frac{1.08}{3.65} = 0.30$$

$$\varepsilon = y_{A0}\delta = (0.3)(1 - \tfrac{1}{2} - 1) = -0.15$$

$$P_{A0} = y_{A0}P_0 = 3.0 \text{ atm}$$

$$k' = kP_{A0}\left(\frac{1}{2}\right)^{2/3} = 0.0141 \, \frac{\text{lb mol}}{\text{atm} \cdot \text{lb cat} \cdot \text{h}} \times 3 \text{ atm} \times 0.63 = 0.0266 \, \frac{\text{lb mol}}{\text{h} \cdot \text{lb cat}}$$

In order to calculate α,

Evaluating the
pressure drop
parameters

$$\alpha = \frac{2\beta_0}{A_c(1 - \phi)\rho_c P_0}$$

we need the superficial mass velocity, G. The mass flow rates of each entering species are

$$\dot{m}_{A0} = 1.08 \frac{\text{lb mol}}{\text{h}} \times 28 \frac{\text{lb}}{\text{lb mol}} = 30.24 \text{ lb/h}$$

$$\dot{m}_{B0} = 0.54 \frac{\text{lb mol}}{\text{h}} \times 32 \frac{\text{lb}}{\text{lb mol}} = 17.28 \text{ lb/h}$$

$$\dot{m}_{I0} = 2.03 \frac{\text{lb mol}}{\text{h}} \times 28 \frac{\text{lb}}{\text{lb mol}} = 56.84 \text{ lb/h}$$

The total mass flow rate is

$$\dot{m}_{T0} = 104.4 \frac{\text{lb}}{\text{h}}$$

$$G = \frac{\dot{m}_{T0}}{A_c} = \frac{104.4 \text{ lb/h}}{0.01414 \text{ ft}^2} = 7383.3 \frac{\text{lb}}{\text{h} \cdot \text{ft}^2}$$

Ah ha! The superficial mass velocity, temperature, and pressure are the same as in Example 4-4. Consequently, we can use the value of β_0 calculated in Example 4-4, to calculate α

$$\beta_0 = 0.0775 \frac{\text{atm}}{\text{ft}}$$

$$\alpha = \frac{2\beta_0}{A_c(1 - \phi)\rho_c P_0} = \frac{(2)(0.0775) \text{ atm/ft}}{(0.01414 \text{ ft}^2)(0.55)(120 \text{ lb cat/ft}^3)(10 \text{ atm})}$$

$$= \frac{0.0166}{\text{lb cat}} \quad (\alpha = 3.656 \times 10^{-5}/\text{g cat})$$

6. **Summary.** Combining Equation (E4-6.1) and (E4-6.8) and summarizing

$$\boxed{\frac{dX}{dW} = \frac{k'}{F_{A0}}\left(\frac{1 - X}{1 + \varepsilon X}\right)y} \tag{E4-6.9}$$

$$\boxed{\frac{dy}{dW} = -\frac{\alpha(1 + \varepsilon X)}{2y}} \tag{E4-6.10}$$

$$k' = 0.0266 \frac{\text{lb mol}}{\text{h lb cat}} \tag{E4-6.11}$$

$$F_{A0} = 1.08 \frac{\text{lb mol}}{\text{h}} \tag{E4-6.12}$$

$$\alpha = \frac{0.0166}{\text{lb cat}} \tag{E4-6.13}$$

$$\varepsilon = -0.15 \tag{E4-6.14}$$

We have the boundary conditions $W = 0$, $X = 0$, and $y = 1.0$, and $W_f = 60$ lb. Here we are guessing an upper limit of the integration to be 60 lb with the expectation that 60% conversion will be achieved within this catalyst weight. If 60% conversion is not achieved, we will guess a higher weight and redo the calculation.

A large number of ordinary differential equation solver software packages (i.e., ODE solvers), which are extremely user friendly, have become available. We shall use Polymath[6] to solve the examples in the printed text. With Polymath, one simply enters Equations (E4-6.9) and (E4-6.10) and the corresponding parameter values [Equations (4-6.11) through (4-6.14)] into the computer with the boundary conditions and they are solved and displayed as shown in Figures E4-6.1 and E4-6.2. Equations (E4-6.9) and (E4-6.10) are entered as differential equations and the parameter values are set using explicit equations. The rate law may be netered as an explicit equation in order to generate a plot of reaction rate as it changes down the length of the reactor, using Polymath's graphing function. The CD-ROM contains all of the MatLab and Polymath solution programs used to solve the example problems, as well as an example using ASPEN. Consequently, one can load the Polymath program directly from the CD-ROM, which has programmed Equations (E4-6.9) through (E4-6.14), and run the program for different parameter values.

It is also interesting to learn what happens to the volumetric flow rate along the length of the reactor. Recalling Equation (3-45),

$$v = v_0(1 + \varepsilon X) \frac{P_0}{P} \frac{T}{T_0} = \frac{v_0(1 + \varepsilon X)(T/T_0)}{P/P_0} \tag{3-45}$$

We let f be the ratio of the volumetric flow rate, v, to the entering volumetric flow rate, v_0, at any point down the reactor. For isothermal operation Equation (3-45) becomes

$$f = \frac{v}{v_0} = \frac{1 + \varepsilon X}{y} \tag{E4-6.15}$$

TABLE E4-6.1 POLYMATH PROGRAM

ODE REPORT (STIFF)

$$f = \frac{v_0}{v}$$

$$y = \frac{P}{P_0}$$

Differential equations as entered by the user
[1] d(X)/d(W) = -raprime/Fao
[2] d(y)/d(W) = -alpha*(1+eps*X)/2/y

Explicit equations as entered by the user
[1] eps = -0.15
[2] kprime = 0.0266
[3] Fao = 1.08
[4] alpha = 0.0166
[5] raprime = -kprime*(1-X)/(1+eps*X)*y
[6] f = (1+eps*X)/y
[7] rate = -raprime

---

[6] Developed by Professor M. Cutlip of the University of Connecticut, and Professor M. Shacham of Ben Gurion University. Available from the CACHE Corporation, P.O. Box 7939, Austin, TX 78713.

Figure E4-6.2 shows X, y (i.e., $y = P/P_0$), and f down the length of the reactor. We see that both the conversion and the volumetric flow increase along the length of the reactor while the pressure decreases. For gas-phase reactions with orders greater than zero, this decrease in pressure will cause the reaction rate to be less than in the case of no pressure drop.

Program examples Polymath, MATLAB can be loaded from the CD-ROM (see the Introduction).

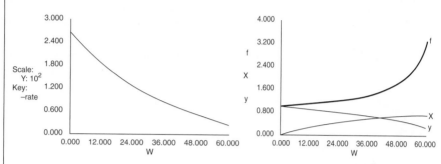

Figure E4-6.1   Reaction rate profile down the PBR.

Figure E4-6.2   Output in graphical form from Polymath.

From either the conversion profile (shown in Figure E4-6.2) or the Polymath table of results (not shown in text, but available on the CD), we find 60% conversion is achieved with 44.5-lb catalyst in each tube.

Effect of added catalyst on conversion

We note from Figure E4-6.2 that the catalyst weight necessary to raise the conversion the last 1% from 65% to 66% (3.5 lb) is 8.5 times more than that (0.41 lb) required to raise the conversion 1% at the reactor's entrance. Also, during the last 5% increase in conversion, the pressure decreases from 3.8 atm to 2.3 atm.

This catalyst weight of 44.5 lb/tube corresponds to a pressure drop of approximately 5 atm. If we had erroneously neglected pressure drop, the catalyst weight would have been found by integrating equation (E4-6.9) with $y = 1$ to give

Neglecting pressure drop results in poor design (here 53% vs. 60% conversion)

$$W = \frac{F_{A0}}{k'}\left[(1 + \varepsilon)\ln\left(\frac{1}{1-X} - \varepsilon X\right)\right] \qquad \text{(E4-6.16)}$$

$$= \frac{1.08}{0.0266} \times \left[(1 - 0.15)\ln\frac{1}{1-0.6} - (-0.15)(0.6)\right] \qquad \text{(E4-6.17)}$$

$$= 35.3 \text{ lb. of catalyst per tube (neglecting pressure drop) (16 kg/tube)}$$

Embarrassing!

If we had used this catalyst weight in our reactor we would have had insufficient catalyst to achieve the desired conversion. For this catalyst weight (i.e., 35,300 lb total, 35.3 lb/tube) Figure E4-6.2 gives a conversion of only 53%.

4.5.5 Spherical Packed-Bed Reactors

Let's consider carrying out this reaction in a spherical reactor similar to the one shown in the margin and discussed in detail in the CD-ROM. In a spherical reactor, the cross section varies as we move through the reactor and is greater than in a normal packed-bed reactor. Consequently, the superficial mass velocity $G = \dot{m}/A_C$ will be smaller. From Equation (4-22), we see that a smaller value of G will give a smaller pressure drop and thus a greater conversion. If 40,000 lb of catalyst in the PBR in Example 4-6 had been used in a spherical reactor, 67% conversion would have been achieved instead of 60% conversion. The equations for calculating conversion in spherical reactors along with an example problem are given in the *Professional Reference Shelf* R4.1 for Chapter 4 on the CD-ROM.

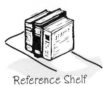

Reference Shelf

4.6 Synthesizing the Design of a Chemical Plant

Synthesizing a chemical plant

Careful study of the various reactions, reactors, and molar flows of the reactants and products used in the example problems in this chapter reveals that they can be arranged to form a chemical plant to produce 200 million pounds of ethylene glycol from a feedstock of 402 million pounds per year of ethane. The flowsheet for the arrangement of the reactors together with the molar flow rates is shown in Figure 4-10. Here 0.425 lb mol/s of ethane is fed to 100 tubular plug-flow reactors connected in parallel; the total volume is 81 ft^3 to produce 0.34 lb mol/s of ethylene (see Example 4-3). The reaction mixture is then fed to a separation unit where 0.04 lb mol/s of ethylene is lost in the separation process in the ethane and hydrogen streams that exit the separator. This process provides a molar flow rate of ethylene of 0.3 lb mol/s, which enters the packed-bed catalytic reactor together with 0.15 lb mol/s of O_2 and 0.564 lb mol/s of N_2. There are 0.18 lb mol/s of ethylene oxide (see Example 4-6) produced in the 1000 pipes arranged in parallel and packed with silver-coated catalyst pellets. There is 60% conversion achieved in each pipe and the total catalyst weight in all the pipes is 44,500 lb. The effluent stream is passed to a separator where 0.03 lb mol/s of ethylene oxide is lost. The ethylene oxide stream is then contacted with water in a gas absorber to produce a 1-lb mol/ft^3 solution of ethylene oxide in water. In the absorption process, 0.022 lb mol/s of ethylene oxide is lost. The ethylene oxide solution is fed to a 197-ft^3 CSTR together with a stream of 0.9 wt % H_2SO_4 solution to produce ethylene glycol at a rate of 0.102 lb mol/s (see Example 4-2). This rate is equivalent to approximately 200 million pounds of ethylene glycol per year.

Always challenge the assumptions, constraints, and boundaries of the problem.

The profit from a chemical plant will be the difference between income from sales and the cost to produce the chemicals. An approximate formula might be

$$\text{Profit} = \text{Value of products} - \text{Cost of reactants}$$

$$- \text{Operating cost} - \text{Separation costs}$$

$$$$

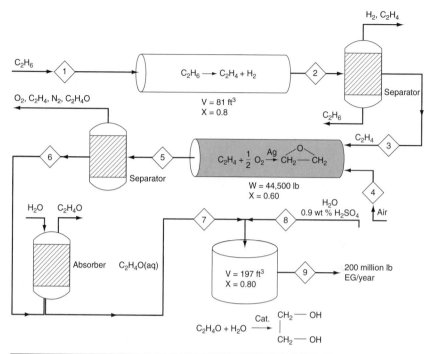

Figure 4-10   Production of ethylene glycol.

Stream	Component[a]	Flow rate (lb mol/s)	Stream	Component[a]	Flow rate (lb mol/s)
1	C_2H_6	0.425	6	EO	0.150
2	C_2H_4	0.340	7	EO	0.128
3	C_2H_4	0.300	8	H_2O	0.443
4	Air	0.714	9	EG	0.102
5	EO	0.180			

[a]EG, ethylene glycol; EO, ethylene oxide.

The operating costs include such costs as energy, labor, overhead, and depreciation of equipment. You will learn more about these costs in your senior design course. While most, if not all, of the streams from the separators could be recycled, lets consider what the profit might be if the streams were to go unrecovered. Also, let's conservatively estimate the operating and other expenses to be $8 million per year and calculate the profit. Your design instructor might give you a better number. The prices of ethane, sulfuric acid, and ethylene glycol are $0.04, $0.043, and $0.38 per pound, respectively. See *www.chemweek.com/* for current prices.

For an ethane feed of 400 million pounds per year and a production rate of 200 million pounds of ethylene glycol per year:

$$
\begin{aligned}
\text{Profit} = \Bigg[&\left(\overbrace{\frac{\$0.38}{lb} \times 2 \times 10^8 \frac{lb}{year}}^{\text{Ethylene glycol cost}} \right) - \left(\overbrace{\frac{\$0.04}{lb} \times 4 \times 10^8 \frac{lb}{year}}^{\text{Ethane cost}} \right) \\
&- \left(\overbrace{\frac{\$0.043}{lb} \times 2.26 \times 10^6 \frac{lb}{year}}^{\text{Sulfuric acid cost}} \right) - \overbrace{\$8,000,000}^{\text{Operating cost}} \Bigg]
\end{aligned}
$$

$$
\begin{aligned}
&= \$76,000,000 - \$16,000,000 - \$54,000 - \$8,000,000 \\
&\cong \$52 \text{ million}
\end{aligned}
$$

Using $52 million a year as a rough estimate of the profit, you can now make different approximations about the conversion, separations, recycle streams, and operating costs to learn how they affect the profit.

PART 2   Mole Balances Written in Terms of Concentration and Molar Flow Rates

Used for:
• Multiple rxns
• Membranes
• Unsteady state

There are many instances when it is much more convenient to work in terms of the number of moles (N_A, N_B) or molar flow rates (F_A, F_B, etc.) rather than conversion. Membrane reactors and multiple reactions taking place in the gas phase are two such cases where molar flow rates are preferred rather than conversion. We now modify our algorithm by using concentrations for liquids and molar flow rates for gases as our dependent variables. The main difference between the conversion algorithm and the molar flow rate/concentration algorithm is that, in the conversion algorithm, we needed to write a mole balance on only *one species*, whereas in the molar flow rate and concentration algorithm, we must write a mole balance on *each and every species*. This algorithm is shown in Figure 4-11. First we write the mole balances on all species present as shown in Step ①. Next we write the rate law, Step ②, and then we relate the mole balances to one another through the relative rates of reaction as shown in Step ③. Steps ④ and ⑤ are used to relate the concentrations in the rate law to the molar flow rates. In Step ⑥, all the steps are combined by the ODE solver (e.g., Polymath).

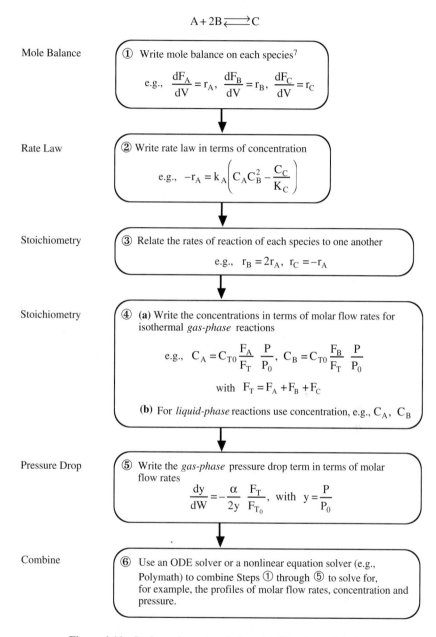

$$A + 2B \rightleftharpoons C$$

Mole Balance

① Write mole balance on each species[7]

e.g., $\dfrac{dF_A}{dV} = r_A,\ \dfrac{dF_B}{dV} = r_B,\ \dfrac{dF_C}{dV} = r_C$

Rate Law

② Write rate law in terms of concentration

e.g., $-r_A = k_A\left(C_A C_B^2 - \dfrac{C_C}{K_C}\right)$

Stoichiometry

③ Relate the rates of reaction of each species to one another

e.g., $r_B = 2r_A,\ r_C = -r_A$

Stoichiometry

④ **(a)** Write the concentrations in terms of molar flow rates for isothermal *gas-phase* reactions

e.g., $C_A = C_{T0}\dfrac{F_A}{F_T}\dfrac{P}{P_0},\ C_B = C_{T0}\dfrac{F_B}{F_T}\dfrac{P}{P_0}$

with $F_T = F_A + F_B + F_C$

(b) For *liquid-phase* reactions use concentration, e.g., $C_A,\ C_B$

Pressure Drop

⑤ Write the *gas-phase* pressure drop term in terms of molar flow rates

$\dfrac{dy}{dW} = -\dfrac{\alpha}{2y}\dfrac{F_T}{F_{T_0}}$, with $y = \dfrac{P}{P_0}$

Combine

⑥ Use an ODE solver or a nonlinear equation solver (e.g., Polymath) to combine Steps ① through ⑤ to solve for, for example, the profiles of molar flow rates, concentration and pressure.

Figure 4-11   Isothermal reaction design algorithm for mole balances.

[7] For a PBR, use $\dfrac{dF_A}{dW} = r_A',\ \dfrac{dF_B}{dW} = r_B',\ \dfrac{dF_C}{dW} = r_C'$.

4.7 Mole Balances on CSTRs, PFRs, PBRs, and Batch Reactors

4.7.1 Liquid Phase

For liquid-phase reactions in which there is no volume change, concentration is the preferred variable. The mole balances for the generic reaction

$$aA + bB \longrightarrow cC + dD \tag{2-1}$$

are shown in Table 4-5 in terms of concentration for the four reactor types we have been discussing. We see from Table 4-5 that we have only to specify the parameter values for the system (C_{A0}, v_0, etc.) and for the rate law parameters (e.g., k_A, α, β) to solve the coupled ordinary differential equations for either PFR, PBR, or batch reactors or to solve the coupled algebraic equations for a CSTR.

$$A + \frac{b}{a}B \longrightarrow \frac{c}{a}C + \frac{d}{a}D \tag{2-2}$$

TABLE 4-5.   MOLE BALANCES FOR LIQUID-PHASE REACTIONS

LIQUIDS

Batch	$\dfrac{dC_A}{dt} = r_A$	and	$\dfrac{dC_B}{dt} = \dfrac{b}{a} r_A$
CSTR	$V = \dfrac{v_0(C_{A0} - C_A)}{-r_A}$	and	$V = \dfrac{v_0(C_{B0} - C_B)}{-(b/a)r_A}$
PFR	$v_0 \dfrac{dC_A}{dV} = r_A$	and	$v_0 \dfrac{dC_B}{dV} = \dfrac{b}{a} r_A$
PBR	$v_0 \dfrac{dC_A}{dW} = r'_A$	and	$v_0 \dfrac{dC_B}{dW} = \dfrac{b}{a} r'_A$

4.7.2 Gas Phase

The mole balances for gas-phase reactions are given in Table 4-6 in terms of number moles (batch) or molar flow rates for the generic rate law for the generic reaction Equation (2-1). The molar flow rates for each species F_j are obtained from a mole balance on each species, as given in Table 4-6. For example, for a plug-flow reactor

Must write a mole balance on each species

$$\frac{dF_j}{dV} = r_j \qquad (1\text{-}11)$$

The generic power law rate law is

Rate law

$$-r_A = k_A C_A^{\alpha} C_B^{\beta}$$

To relate concentrations to molar flow rates, recall Equation (3-42), with $y = P/P_0$

Stoichiometry

$$C_j = C_{T0} \frac{F_j}{F_T} \frac{T_0}{T} y \qquad (3\text{-}42)$$

The pressure drop equation, Equation (4-28), for isothermal operation ($T = T_0$) is

$$\frac{dy}{dW} = \frac{-\alpha}{2y} \frac{F_T}{F_{T0}} \qquad (4\text{-}28)$$

The total molar flow rate is given as the sum of the flow rates of the individual species:

$$F_T = \sum_{j=1}^{n} F_j$$

when species A, B, C, D, and I are the only ones present. Then

$$F_T = F_A + F_B + F_C + F_D + F_I$$

We now combine all the preceding information as shown in Table 4-6.

4.8 Microreactors

Microreactors are emerging as a new technology in CRE. Microreactors are characterized by their high surface area-to-volume ratios in their microstructured regions that contain tubes or channels. A typical channel width might be 100 μm with a length of 20,000 μm (2 cm). The resulting high surface area-to-volume ratio (ca. 10,000 m²/m³) reduces or even eliminates heat and mass

TABLE 4-6.   ALGORITHM FOR GAS-PHASE REACTIONS

$$aA + bB \longrightarrow cC + dD$$

1. Mole balances:

Batch	CSTR	PFR
$\dfrac{dN_A}{dt} = r_A V$	$V = \dfrac{F_{A0} - F_A}{-r_A}$	$\dfrac{dF_A}{dV} = r_A$
$\dfrac{dN_B}{dt} = r_B V$	$V = \dfrac{F_{B0} - F_B}{-r_B}$	$\dfrac{dF_B}{dV} = r_B$
$\dfrac{dN_C}{dt} = r_C V$	$V = \dfrac{F_{C0} - F_C}{-r_C}$	$\dfrac{dF_C}{dV} = r_C$
$\dfrac{dN_D}{dt} = r_D V$	$V = \dfrac{F_{D0} - F_D}{-r_D}$	$\dfrac{dF_D}{dV} = r_D$

2. Rate Law:

$$-r_A = k_A C_A^\alpha C_B^\beta$$

3. Stoichiometry:

Relative rates of reaction:

$$\frac{r_A}{-a} = \frac{r_B}{-b} = \frac{r_C}{c} = \frac{r_D}{d}$$

Gas phase

then

$$r_B = \frac{b}{a} r_A \qquad r_C = -\frac{c}{a} r_A \qquad r_D = -\frac{d}{a} r_A$$

Concentrations:

$$C_A = C_{T0} \frac{F_A T_0}{F_T T} y \qquad C_B = C_{T0} \frac{F_B T_0}{F_T T} y$$

$$C_C = C_{T0} \frac{F_C T_0}{F_T T} y \qquad C_D = C_{T0} \frac{F_D T_0}{F_T T} y$$

$$\frac{dy}{dW} = \frac{-\alpha}{2y} \frac{F_T}{F_{T0}} \frac{T}{T_0}, \quad y = \frac{P}{P_0}$$

Total molar flow rate: $F_T = F_A + F_B + F_C + F_D + F_I$

4. Combine: For an isothermal operation of a PBR with no ΔP

$$\frac{dF_A}{dV} = -k_A C_{T0}^{\alpha+\beta} \left(\frac{F_A}{F_T}\right)^\alpha \left(\frac{F_B}{F_T}\right)^\beta \qquad \frac{dF_B}{dV} = -\frac{b}{a} k_A C_{T0}^{\alpha+\beta} \left(\frac{F_A}{F_T}\right)^\alpha \left(\frac{F_B}{F_T}\right)^\beta$$

$$\frac{dF_C}{dV} = \frac{c}{a} k_A C_{T0}^{\alpha+\beta} \left(\frac{F_A}{F_T}\right)^\alpha \left(\frac{F_B}{F_T}\right)^\beta \qquad \frac{dF_D}{dV} = \frac{d}{a} k_A C_{T0}^{\alpha+\beta} \left(\frac{F_A}{F_T}\right)^\alpha \left(\frac{F_B}{F_T}\right)^\beta$$

1. Specify parameter values: $k_A, C_{T0}, \alpha, \beta, T_0, a, b, c, d$

2. Specify entering numbers: $F_{A0}, F_{B0}, F_{C0}, F_{D0}$ and final values: V_{final}

5. Use an ODE solver.

transfer resistances often found in larger reactors. Consequently, surface-catalyzed reactions can be greatly facilitated, hot spots in highly exothermic reactions can be eliminated, and in many cases highly exothermic reactions can be carried out isothermally. These features provide the opportunity for microreactors to be used to study the intrinsic kinetics of reactions. Another advantage of microreactors is their use in the production of toxic or explosive intermediates where a leak or microexplosion for a single unit will do minimal damage because of the small quantities of material involved. Other advantages include shorter residence times and narrower residence time distributions.

Advantages of microreactors

Figure 4-12 shows (a) a microreactor with heat exchanger and (b) a microplant with reactor, valves, and mixers. Heat, Q, is added or taken away by the fluid flowing perpendicular to the reaction channels as shown in Figure 4-12(a). Production in microreactor systems can be increased simply by adding more units in parallel. For example, the catalyzed reaction

$$R-CH_2OH + \tfrac{1}{2}O_2 \xrightarrow{\text{Ag}} R-CHO+H_2O$$

required only 32 microreaction systems in parallel to produce 2000 tons/yr of acetate!

(b)

(a)

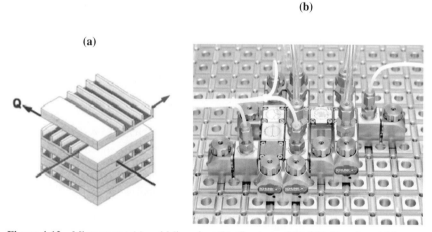

Figure 4-12   Microreactor (a) and Microplant (b). Courtesy of Ehrfeld, Hessel, and Löwe, *Microreactors: New Technology for Modern Chemistry* (Wiley-VCH, 2000).

Microreactors are also used for the production of specialty chemicals, combinatorial chemical screening, lab-on-a-chip, and chemical sensors. In modeling microreactors, we will assume they are either in plug flow for which the mole balance is

$$\frac{dF_A}{dV} = r_A \qquad (1\text{-}12)$$

or in laminar flow, in which case we will use the segregation model discussed in Chapter 13. For the plug-flow case, the algorithm is described in Figure 4-11.

Example 4–7 Gas-Phase Reaction in a Microreactor—Molar Flow Rates

The gas-phase reaction

$$2NOCl \longrightarrow 2NO + Cl_2$$

is carried out at 425°C and 1641 kPa (16.2 atm). Pure NOCl is to be fed, and the reaction follows an elementary rate law.[8] It is desired to produce 20 tons of NO per year in a microreactor system using a bank of ten microreactors in parallel. Each microreactor has 100 channels with each channel 0.2 mm square and 250 mm in length.

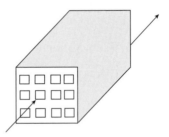

Plot the molar flow rates as a function of volume down the length of the reactor. The volume of each channel is 10^{-5} dm^3.

Additional Information

To produce 20 tons per year of NO at 85% conversion would require a feed rate of 0.0226 mol/s of NOCl, or 2.26×10^{-5} mol/s per channel. The rate constant is

$$k = 0.29 \frac{dm^3}{mol \cdot s} \text{ at 500 K with } E = 24 \frac{kcal}{mol}$$

Solution

For one channel,

Find V.

$$F_{A0} = \frac{22.6 \ \mu mol}{s} \qquad F_B = \frac{19.2 \ \mu mol}{s}, \quad X = 0.85, \quad V = ?$$

---

[8] J. B. Butt, *Reaction Kinetics and Reactor Design*, 2nd ed. (New York: Marcel Dekker, 2001), p. 153.

Although this particular problem could be solved using conversion, we shall illustrate how it can also be solved using molar flow rates as the variable in the mole balance. We first write the reaction in symbolic form and then divide by the stoichiometric coefficient of the limiting reactant, NOCl.

$$2NOCl \rightarrow 2NO + Cl_2$$

$$2A \rightarrow 2B + C$$

$$A \rightarrow B + \tfrac{1}{2}C$$

Following the Algorithm

1. **Mole balances on species A, B, and C:**

$$\frac{dF_A}{dV} = r_A \qquad\qquad\qquad \text{(E4-7.1)}$$

$$\frac{dF_B}{dV} = r_B \qquad\qquad\qquad \text{(E4-7.2)}$$

$$\frac{dF_C}{dV} = r_C \qquad\qquad\qquad \text{(E4-7.3)}$$

2. **Rate law:**

$$-r_A = kC_A^2, \quad k = 0.29 \, \frac{dm^3}{mol \cdot s} \qquad\qquad \text{(E4-7.4)}$$

3. **Stoichiometry:** Gas phase with $T = T_0$ and $P = P_0$, then $v = v_0 \dfrac{F_T}{F_{T0}}$
   a.   Relative rates

$$\frac{r_A}{-1} = \frac{r_B}{1} = \frac{r_C}{\frac{1}{2}}$$

$$r_B = -r_A$$

$$r_C = -\tfrac{1}{2}r_A$$

   b.   Concentration
         Applying Equation (3-42) to species A, B, and C, the concentrations are

$$C_A = C_{T0}\frac{F_A}{F_T}, \quad C_B = C_{T0}\frac{F_B}{F_T}, \quad C_C = C_{T0}\frac{F_C}{F_T} \qquad \text{(E4-7.5)}$$

$$\text{with } F_T = F_A + F_B + F_C$$

4. Combine: the rate law in terms of molar flow rates is

$$-r_A = k_A C_{T0}^2 \left(\frac{F_A}{F_T}\right)^2$$

combining all

$$\frac{dF_A}{dV} = k_A C_{T0}^2 \left(\frac{F_A}{F_T}\right)^2 \tag{E4-7.6}$$

$$\frac{dF_B}{dV} = k_A C_{T0}^2 \left(\frac{F_A}{F_T}\right)^2 \tag{E4-7.7}$$

$$\frac{dF_C}{dV} = \frac{k_A}{2} C_{T0}^2 \left(\frac{F_A}{F_T}\right)^2 \tag{E4-7.8}$$

5. Evaluate:

$$C_{T0} = \frac{P_0}{RT_0} = \frac{(1641 \text{ kPa})}{\left(8.314 \dfrac{\text{kPa} \cdot \text{dm}^3}{\text{mol} \cdot \text{K}}\right) 698 \text{ K}} = 0.286 \frac{\text{mol}}{\text{dm}^3} = \frac{0.286 \text{ mmol}}{\text{cm}^3}$$

When using Polymath or another ODE solver, one does not have to actually combine the mole balances, rate laws, and stoichiometry as was done in the combine step in previous examples in this chapter. The ODE solver will do that for you. Thanks, ODE solver! The Polymath Program and output are shown in Table E4-7.1 and Figure E4-7.1.

TABLE E4-7.1. POLYMATH PROGRAM

ODE REPORT (RKF45)

Differential equations as entered by the user
[1] d(Fa)/d(V) = ra
[2] d(Fb)/d(V) = rb
[3] d(Fc)/d(V) = rc

Explicit equations as entered by the user
[1] T = 698
[2] Cto = 1641/8.314/T
[3] E = 24000
[4] Ft = Fa+Fb+Fc
[5] Ca = Cto*Fa/Ft
[6] k = 0.29*exp(E/1.987*(1/500-1/T))
[7] Fao = 0.0000226
[8] vo = Fao/Cto

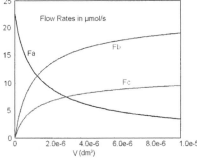

Figure E4-7.1 Profiles of microreactor molar flow rates.

TABLE E4-7.1. POLYMATH PROGRAM (CONTINUED)

ODE REPORT (RKF45) (Continued)

```
[9]  Tau = V/vo
[10] ra = -k*Ca^2
[11] X = 1-Fa/Fao
[12] rb = -ra
[13] rc = -ra/2
[14] rateA = -ra
```

4.9 Membrane Reactors

Membrane reactors can be used to increase conversion when the reaction is thermodynamically limited as well as to increase the selectivity when multiple reactions are occurring. Thermodynamically limited reactions are reactions where the equilibrium lies far to the left (i.e., reactant side) and there is little conversion. If the reaction is exothermic, increasing the temperature will only drive the reaction further to the left, and decreasing the temperature will result in a reaction rate so slow that there is very little conversion. If the reaction is endothermic, increasing the temperature will move the reaction to the right to favor a higher conversion; however, for many reactions these higher temperatures cause the catalyst to become deactivated.

The term *membrane reactor* describes a number of different types of reactor configurations that contain a membrane. The membrane can either provide a barrier to certain components while being permeable to others, prevent certain components such as particulates from contacting the catalyst, or contain reactive sites and be a catalyst in itself. Like reactive distillation, the membrane reactor is another technique for driving reversible reactions to the right toward completion in order to achieve very high conversions. These high conversions can be achieved by having one of the reaction products diffuse out of a semipermeable membrane surrounding the reacting mixture. As a result, the reverse reaction will not be able to take place, and the reaction will continue to proceed to the right toward completion.

By having one of the products pass throughout the membrane, we drive the reaction toward completion.

Two of the main types of catalytic membrane reactors are shown in Figure 4-13. The reactor in Figure 4-13(b) is called an *inert membrane reactor with catalyst pellets on the feed side* (IMRCF). Here the membrane is inert and serves as a barrier to the reactants and some of the products. The reactor in Figure 4-13(c) is a *catalytic membrane reactor* (CMR). The catalyst is deposited directly on the membrane, and only specific reaction products are able to exit the permeate side. For example, in the reversible reaction

$$C_6H_{12} \rightleftharpoons 3H_2 + C_6H_6$$

$$A \rightleftharpoons 3B + C$$

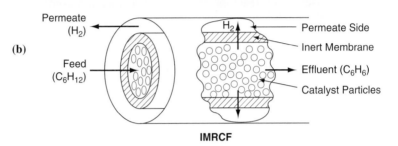

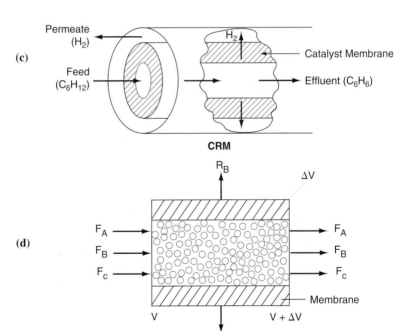

Figure 4-13 Membrane reactors. (Photo courtesy of Coors Ceramics, Golden, Colorado.) (a) Photo of ceramic reactors, (b) cross section of IMRCF, (c) cross section of CRM, (d) schematic of IMRCF for mole balance.

H₂ diffuses through
the membrane
while C₆H₆ does
not.

the hydrogen molecule is small enough to diffuse through the small pores of the membrane while C_6H_{12} and C_6H_6 cannot. Consequently, the reaction continues to proceed to the right even for a small value of the equilibrium constant.

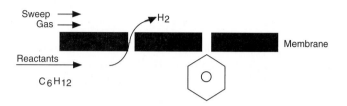

Hydrogen, species B, flows out through the sides of the reactor as it flows down the reactor with the other products, which cannot leave until they exit the reactor.

In analyzing membrane reactors, we only need to make a small change to the algorithm shown in Figure 4-11. We shall choose the reactor volume rather than catalyst weight as our independent variable for this example. The catalyst weight, W, and reactor volume, V, are easily related through the bulk catalyst density, ρ_b (i.e., $W = \rho_b V$). The mole balances on the chemical species that stay *within* the reactor, namely A and C, are shown in Figure 4-11 and also in Table 4-6.

$$V = \frac{W}{\rho_b}$$

$$\frac{dF_A}{dV} = r_A \qquad (1\text{-}11)$$

The mole balance on C is carried out in an identical manner to A, and the resulting equation is

$$\frac{dF_C}{dV} = r_C \qquad (4\text{-}40)$$

However, the mole balance on B (H_2) must be modified because hydrogen leaves through both the sides of the reactor and at the end of the reactor.

First we shall perform mole balances on the volume element ΔV shown in Figure 4-13(c). The mole balance on hydrogen (B) is over a differential volume ΔV shown in Figure 4-13(d) and it yields

Balance on B in the catalytic bed:

$$\begin{bmatrix} \text{In} \\ \text{by flow} \end{bmatrix} - \begin{bmatrix} \text{Out} \\ \text{by flow} \end{bmatrix} - \begin{bmatrix} \text{Out} \\ \text{by diffusion} \end{bmatrix} + [\text{Generation}] = [\text{Accumulation}]$$

Now there are two "OUT" terms for species B.

$$F_B\big|_V \quad - \quad F_B\big|_{V+\Delta V} \quad - \quad R_B \Delta V \quad + \quad r_B \Delta V \quad = 0$$

where R_B is the molar rate of B leaving through the sides the reactor per unit volume of reactor (mol/dm^3 · s). Dividing by ΔV and taking the limit as $\Delta V \to 0$ gives

$$\boxed{\frac{dF_B}{dV} = r_B - R_B}$$
(4-41)

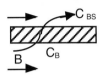

The rate of transport B out through the membrane R_B is the product of the molar flux of B, W_B, and a, the surface area, per unit volume of reactor. The molar flux of B, W_B in (mol/m^2 · s) out of the reactor is a mass transfer coefficient times the concentration driving force across the membrane.

$$W_B = k_C'(C_B - C_{BS})$$
(4-42)

Where k_C' is the overall mass transfer coefficient in m/s and C_{BS} is the concentration of B in the sweep gas channel (mol/dm^3). The overall mass transfer coefficient accounts for all resistances to transport: the tube side resistance of the membrane, the membrane itself, and on the shell (sweep gas) side resistance. Further elaboration of the mass transfer coefficient and its correlations can be found in the literature and in Chapter 11. In general, this coefficient can be a function of the membrane and fluid properties, the fluid velocity, and the tube diameters.

To obtain the rate of removal of B, we need to multiply the flux through the membrane by the surface area of membrane in the reactor. The rate at which B is removed per unit volume of reactor, R_B, is just the flux W_B times the surface area membrane per volume of reactor, a (m^2/m^3); that is,

$$R_B = W_B a = k_C' a(C_B - C_{BS})$$
(4-43)

The membrane surface area per unit volume of reactor is

$$a = \frac{\text{Area}}{\text{Volume}} = \frac{\pi D L}{\frac{\pi D^2}{4} L} = \frac{4}{D}$$

Letting $k_C = k_C'\, a$ and assuming the concentration in the sweep gas is essentially zero (i.e., $C_{BS} \approx 0$), we obtain

Rate of B out through the sides.

$$\boxed{R_B = k_C C_B}$$

where the units of k_C are s^{-1}.

More detailed modeling of the transport and reaction steps in membrane reactors is beyond the scope of this text but can be found in *Membrane Reactor Technology*.[9] The salient features, however, can be illustrated by the following example. When analyzing membrane reactors, it is much more convenient to use molar flow rates than conversion.

[9] R. Govind, and N. Itoh, eds., *Membrane Reactor Technology,* AIChE Symposium Series No. 268, Vol. 85 (1989). T. Sun and S. Khang, *Ind. Eng. Chem. Res., 27,* 1136 (1988).

According to the
DOE, 10 trillion
BTU/yr could be
saved by using
membrane reactors.

Example 4–8   Membrane Reactor

According to The Department of Energy (DOE), an energy saving of 10 trillion BTU per year could result from the use of catalytic membrane reactors as replacements for conventional reactors for dehydrogenation reactions such as the dehydrogenation of ethylbenzene to styrene:

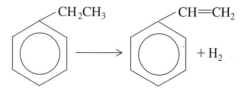

and of butane to butene:

$$C_4H_{10} \longrightarrow C_4H_8 + H_2$$

The dehydrogenation of propane is another reaction that has proven successful with a membrane reactor.[10]

$$C_3H_8 \longrightarrow C_3H_6 + H_2$$

All the preceding dehydrogenation reactions above can be represented symbolically as

$$A \rightleftarrows B + C$$

and will take place on the catalyst side of an IMRCF. The equilibrium constant for this reaction is quite small at 227°C (e.g., $K_C = 0.05$ mol/dm^3). The membrane is permeable to B (e.g., H_2) but not to A and C. Pure gaseous A enters the reactor at 8.2 atm and 227°C at a rate of 10 mol/min.

   We will take the rate of diffusion of B out of the reactor per unit volume of reactor, R_B, to be proportional to the concentration of B (i.e., $R_B = k_C C_B$).

  (a) Perform differential mole balances on A, B, and C to arrive at a set of coupled differential equations to solve.

  (b) Plot the molar flow rates of each species as a function of space time.

  (c) Calculate the conversion.

Additional information: Even though this reaction is a gas–solid catalytic reaction, we will use the bulk catalyst density in order to write our balances in terms of reactor volume rather than catalyst weight (recall $-r_A = -r'_A \rho_b$). For the bulk catalyst density of $\rho_b = 1.5$ g/cm^3 and a 2-cm inside diameter of the tube containing the catalyst pellets, the specific reaction rate, k, and the transport coefficient, k_C, are $k = 0.7$ min^{-1} and $k_C = 0.2$ min^{-1}, respectively.

Solution

We shall choose reactor volume rather than catalyst weight as our independent variable for this example. The catalyst weight, W, and reactor volume, V, are easily

_____

[10] *J. Membrane Sci.*, 77, 221 (1993).

related through the bulk catalyst density, ρ_b, (i.e., $W = \rho_b V$). First, we shall perform mole balances on the volume element ΔV shown in Figure 4-13(d).

1. **Mole balances:**

    Balance on A in the catalytic bed:

$$\left[\begin{array}{c}\text{In} \\ \text{by flow}\end{array}\right] - \left[\begin{array}{c}\text{Out} \\ \text{by flow}\end{array}\right] + \left[\begin{array}{c}\text{Generation}\end{array}\right] = \left[\begin{array}{c}\text{Accumulation}\end{array}\right]$$

$$\overbrace{F_A\big|_V} \quad - \quad \overbrace{F_A\big|_{V+\Delta V}} \quad + \quad \overbrace{r_A \Delta V} \quad = \quad 0$$

Dividing by ΔV and taking the limit as $\Delta V \to 0$ gives

$$\boxed{\frac{dF_A}{dV} = r_A} \tag{E4-8.1}$$

Balance on B in the catalytic bed:
The balance on B is given by Equation (4-41).

$$\boxed{\frac{dF_B}{dV} = r_B - R_B} \tag{E4-8.2}$$

where R_B is the molar flow of B out through the membrane per unit volume of reactor.

    The mole balance on C is carried out in an identical manner to A and the resulting equation is

$$\boxed{\frac{dF_C}{dV} = r_C} \tag{E4-8.3}$$

2. **Rate law:**

$$\boxed{-r_A = k\left(C_A - \frac{C_B C_C}{K_C}\right)} \tag{E4-8.4}$$

3. **Transport out of the reactor.** We apply Equation (4-42) for the case in which the concentration of B of the sweep side is zero, $C_{BS} = 0$, to obtain

$$\boxed{R_B = k_C C_B} \tag{E4-8.5}$$

where k_C is a transport coefficient. In this example, we shall assume that the resistance to species B out of the membrane is a constant and, consequently, k_C is a constant.

4. **Stoichiometry.** Recalling Equation (3-42) for the case of constant temperature and pressure, we have for isothermal operation and no pressure drop ($T = T_0$, $P = P_0$),

Concentrations:

$$C_A = C_{T0} \frac{F_A}{F_T} \tag{E4-8.6}$$

$$C_B = C_{T0} \frac{F_B}{F_T} \tag{E4-8.7}$$

$$C_C = C_{T0} \frac{F_C}{F_T} \tag{E4-8.8}$$

$$F_T = F_A + F_B + F_C \tag{E4-8.9}$$

Relative rates:

$$\frac{r_A}{-1} = \frac{r_B}{1} = \frac{r_C}{1}$$

$$r_B = -r_A \tag{E4-8.10}$$

$$r_C = r_A$$

5. **Combining and summarizing:**

$$\frac{dF_A}{dV} = r_A$$

$$\frac{dF_B}{dV} = -r_A - k_C C_{T0} \left(\frac{F_B}{F_T} \right)$$

Summary of equations describing flow and reaction in a membrane reactor

$$\frac{dF_C}{dV} = -r_A$$

$$-r_A = k C_{T0} \left[\left(\frac{F_A}{F_T} \right) - \frac{C_{T0}}{K_C} \left(\frac{F_B}{F_T} \right) \left(\frac{F_C}{F_T} \right) \right]$$

$$F_T = F_A + F_B + F_C$$

6. **Parameter evaluation:**

$$C_{T0} = \frac{P_0}{RT_0} = \frac{830.6 \text{ kPa}}{8.314 \text{ k Pa} \cdot \text{dm}^3/(\text{mol} \cdot \text{K}) (500 \text{ K})} = 0.2 \frac{\text{mol}}{\text{dm}^3}$$

$$k = 0.7 \text{ min}^{-1}, \, K_C = 0.05 \text{ mol/dm}^3, \, k_C = 0.2 \text{ min}^{-1}$$

$$F_{A0} = 10 \text{ mol/min}$$

$$F_{B0} = F_{C0} = 0$$

7. **Numerical solution.** Equations (E4-8.1) through (E4-8.10) were solved using Polymath and MATLAB, another ODE solver. The profiles of the molar flow rates are shown here. Table E4-8.1 shows the Polymath programs,

and Figure E4-8.1 shows the results of the numerical solution of the initial (entering) conditions.

$$V = 0: \quad F_A = F_{A0}, \quad F_B = 0, \quad F_C = 0$$

TABLE E4-8.1 POLYMATH PROGRAM

ODE REPORT (RKF45)

Differential equations as entered by the user
[1] d(Fa)/d(V) = ra
[2] d(Fb)/d(V) = -ra-kc*Cto*(Fb/Ft)
[3] d(Fc)/d(V) = -ra

Explicit equations as entered by the user
[1] kc = 0.2
[2] Cto = 0.2
[3] Ft = Fa+Fb+Fc
[4] k = 0.7
[5] Kc = 0.05
[6] ra = -k*Cto*((Fa/Ft)-Cto/Kc*(Fb/Ft)*(Fc/Ft))

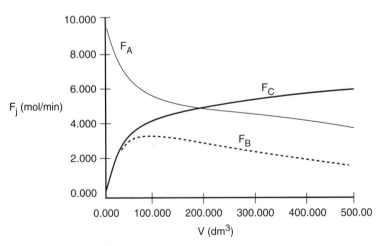

Figure E4-8.1 Polymath solution.

(c)   From Figure E4-8.1 we see the exit molar flow rate of A is 4 mol/min, for which the corresponding conversion is

$$X = \frac{F_{A0} - F_A}{F_{A0}} = \frac{10 - 4}{10} = 0.6$$

Use of Membrane Reactors to Enhance Selectivity. In addition to species leaving through the sides of the membrane reactor, species can also be fed to the reactor through the sides of the membrane. For example, for the reaction

$$A + B \rightarrow C + D$$

A could be fed only to the entrance, and B could be fed only through the membrane as shown here.

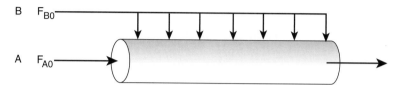

As we will see in Chapter 6, this arrangement is often used to improve selectivity when multiple reactions take place. Here B is usually fed uniformly through the membrane along the length of the reactor. The balance on B is

$$\frac{dF_B}{dV} = r_B + R_B \tag{4-44}$$

where $R_B = F_{B0}/V_t$ with F_{B0} the molar feed rate of B through sides and V_t the total reactor volume. The feed rate of B can be controlled by controlling the pressure drop across the reactor membrane.[11]

4.10 Unsteady-State Operation of Stirred Reactors

In this chapter, we have already discussed the unsteady operation of one type of reactor, the batch reactor. In this section, we discuss two other aspects of unsteady operation: startup of a CSTR and semibatch reactors. First, the startup of a CSTR is examined to determine the time necessary to reach steady-state operation [see Figure 4-14(a)], and then semibatch reactors are discussed. In each of these cases, we are interested in predicting the concentration and conversion as a function of time. Closed-form analytical solutions to the differential equations arising from the mole balance of these reaction types can be obtained only for zero- and first-order reactions. ODE solvers must be used for other reaction orders.

---

[11] The velocity of B through the membrane, U_B, is given by Darcy's law

$$U_B = K(P_s - P_r)$$

where K is the membrane permeability and P_s is the shell-side pressure and P_r the reactor side pressure.

$$F_{B0} = \overbrace{C_{B0} a U_B}^{R_B} V_t = R_B V_t$$

where, as before, a is the membrane surface area per unit volume, C_{B0} is the entering concentration of B, and V_t is the total reactor volume.

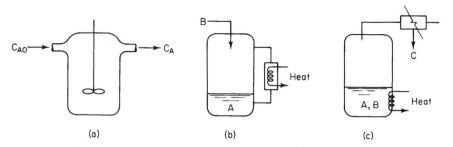

Figure 4-14   Semibatch reactors. (a) Reactor startup, (b) semibatch with cooling, and (c) reactive distillation. [Excerpted by special permission from *Chem. Eng.*, *63*(10) 211 (Oct. 1956). Copyright © 1956 by McGraw-Hill, Inc., New York, NY 10020.]

There are two basic types of semibatch operations. In one type, one of the reactants in the reaction

$$A + B \rightarrow C + D$$

(e.g., B) is slowly fed to a reactor containing the other reactant (e.g., A), which has already been charged to a reactor such as that shown in Figure 4-14(b). This type of reactor is generally used when unwanted side reactions occur at high concentrations of B (Chapter 6) or when the reaction is highly exothermic (Chapter 8). In some reactions, the reactant B is a gas and is bubbled continuously through liquid reactant A. Examples of reactions used in this type of semibatch reactor operation include *ammonolysis*, *chlorination*, and *hydrolysis*. The other type of semibatch reactor is reactive distillation and is shown schematically in Figure 4-14(c). Here reactants A and B are charged simultaneously and one of the products is vaporized and withdrawn continuously. Removal of one of the products in this manner (e.g., C) shifts the equilibrium toward the right, increasing the final conversion above that which would be achieved had C not been removed. In addition, removal of one of the products further concentrates the reactant, thereby producing an increased rate of reaction and decreased processing time. This type of reaction operation is called *reactive distillation*. Examples of reactions carried out in this type of reactor include *acetylation reactions* and *esterification reactions* in which water is removed.

An expanded version of this section can be found on the CD-ROM.

4.10.1 Startup of a CSTR

Summary Notes

The startup of a fixed volume CSTR under isothermal conditions is rare, but it does occur occasionally. We can, however, carry out an analysis to estimate the time necessary to reach steady-state operation. For the case when the reactor is well mixed and as a result there are no spatial variations in r_A, we begin with the general mole balance equation applied to Figure 4-14(a):

$$F_{A0} - F_A + r_A V = \frac{dN_A}{dt} \tag{4-45}$$

Conversion does not have any meaning in startup because one cannot separate the moles reacted from the moles accumulated in the CSTR. Consequently, we *must* use concentration rather than conversion as our variable in the balance equation. For liquid-phase ($v = v_0$) reactions with constant overflow ($V = V_0$), using $\tau = V_0/v_0$, we can transform Equation (4-45) to

$$C_{A0} - C_A + r_A \tau = \tau \frac{dC_A}{dt} \tag{4-46}$$

First-order reaction

For a first-order reaction ($-r_A = kC_A$) Equation (4-46) then becomes

$$\frac{dC_A}{dt} + \frac{1 + \tau k}{\tau} C_A = \frac{C_{A0}}{\tau}$$

which, for the initial conditions $C_A = 0$ at $t = 0$ solves to

$$C_A = \frac{C_{A0}}{1 + \tau k} \left\{ 1 - exp \left[-(1 + \tau k)\frac{t}{\tau} \right] \right\} \tag{4-47}$$

Letting t_s be the time necessary to reach 99% of the steady-state concentration, C_{AS}:

$$C_{AS} = \frac{C_{A0}}{1 + \tau k}$$

Rearranging Equation (4-47) for $C_A = 0.99 C_{AS}$ yields

$$t_s = 4.6 \frac{\tau}{1 + \tau k} \tag{4-48}$$

For slow reactions with small k ($1 \gg \tau k$):

$$\boxed{t_s = 4.6\, \tau} \tag{4-49}$$

For rapid reactions with large k ($\tau k \gg 1$):

Time to reach steady state in an isothermal CSTR

$$\boxed{t_s = \frac{4.6}{k}} \tag{4-50}$$

For most first-order systems, steady state is achieved in three to four space times.

4.10.2 Semibatch Reactors

Motivation

One of the best reasons to use semibatch reactors is to enhance selectivity in liquid-phase reactions. For example, consider the following two simultaneous reactions. One reaction produces the desired product D

$$A + B \xrightarrow{k_D} D$$

with the rate law

$$r_D = kC_A^2 C_B$$

and the other produces an undesired product U

$$A + B \xrightarrow{k_U} U$$

with the rate law

$$r_U = k_U C_A C_B^2$$

The instantaneous selectivity $S_{D/U}$ is the ratio of the relative rates

We want $S_{D/U}$ as large as possible.

$$S_{D/U} = \frac{r_D}{r_U} = \frac{k_D C_A^2 C_B}{k_U C_A C_B^2} = \frac{k_D}{k_U} \frac{C_A}{C_B}$$

and guides us how to produce the most of our desired product and least of our undesired product (see Section 6.1). We see from the instantaneous selectivity that we can increase the formation of D and decrease the formation of U by keeping the concentration of A high and the concentration of B low. This result can be achieved through the use of the semibatch reactor, which is charged with Pure A and to which B is fed slowly to A in the vat.

Of the two types of semibatch reactors, we focus attention primarily on the one with constant molar feed. A schematic diagram of this semibatch reactor is shown in Figure 4-15. We shall consider the elementary liquid-phase reaction

$$A + B \rightarrow C$$

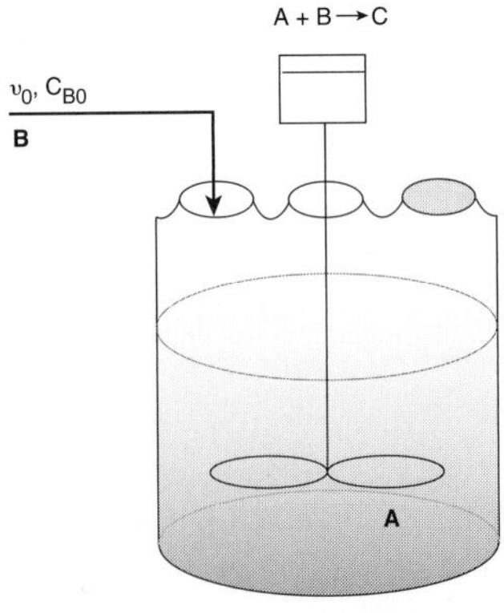

Figure 4-15   Semibatch reactor.

in which reactant B is slowly added to a well-mixed vat containing reactant A.

A mole balance on species A yields

$$\begin{bmatrix} \text{Rate} \\ \text{in} \end{bmatrix} - \begin{bmatrix} \text{Rate} \\ \text{out} \end{bmatrix} + \begin{bmatrix} \text{Rate of} \\ \text{generation} \end{bmatrix} = \begin{bmatrix} \text{Rate of} \\ \text{accumulation} \end{bmatrix} \qquad (4\text{-}51)$$

$$\overbrace{0}^{} \quad - \quad \overbrace{0}^{} \quad + \quad \overbrace{r_A V(t)}^{} \quad = \quad \overbrace{\dfrac{dN_A}{dt}}^{}$$

Three variables can be used to formulate and solve semibatch reactor problems: the concentrations, C_j, the number of moles, N_j, and the conversion, X.

4.10.3 Writing the Semibatch Reactor Equations in Terms of Concentrations

Recalling that the number of moles of A, N_A, is just the product of concentration of A, C_A, and the volume, V, we can rewrite Equation (4-51) as

$$r_A V = \frac{d(C_A V)}{dt} = \frac{V dC_A}{dt} + C_A \frac{dV}{dt} \qquad (4\text{-}52)$$

We note that since the reactor is being filled, the volume, V, varies with time. The reactor volume at any time t can be found from an **overall mass balance** of all species:

Overall mass balance

$$\begin{bmatrix} \text{Rate} \\ \text{in} \end{bmatrix} - \begin{bmatrix} \text{Rate} \\ \text{out} \end{bmatrix} + \begin{bmatrix} \text{Rate of} \\ \text{generation} \end{bmatrix} = \begin{bmatrix} \text{Rate of} \\ \text{accumulation} \end{bmatrix} \qquad (4\text{-}53)$$

$$\overbrace{\rho_0 v_0}^{} \quad - \quad \overbrace{0}^{} \quad + \quad \overbrace{0}^{} \quad = \quad \overbrace{\dfrac{d(\rho V)}{dt}}^{}$$

For a constant-density system, $\rho_0 = \rho$, and

$$\frac{dV}{dt} = v_0 \qquad (4\text{-}54)$$

with the initial condition $V = V_0$ at $t = 0$, integrating for the case of constant volumetric flow rate v_0 yields

Semibatch reactor volume as a function of time

$$\boxed{V = V_0 + v_0 t} \qquad (4\text{-}55)$$

Substituting Equation (4-54) into the right-hand side of Equation (4-52) and rearranging gives us

$$-v_0 C_A + V r_A = \frac{V dC_A}{dt}$$

The **balance on A** [i.e., Equation (4-52)] can be rewritten as

Mole balance on A

$$\frac{dC_A}{dt} = r_A - \frac{v_0}{V}C_A \qquad (4\text{-}56)$$

A **mole balance on B** that is fed to the reactor at a rate F_{B0} is

$$\text{In} \quad + \quad \text{Out} \quad + \quad \text{Generation} \quad = \quad \text{Accumulation}$$

$$\overbrace{F_{B0}} \quad - \quad \overbrace{0} \quad + \quad \overbrace{r_B V} \quad = \quad \overbrace{\frac{dN_B}{dt}}$$

Rearranging

$$\frac{dN_B}{dt} = r_B V + F_{B0} \qquad (4\text{-}57)$$

$$\frac{dVC_B}{dt} = \frac{dV}{dt}C_B + \frac{VdC_B}{dt} = r_B V + F_{B0}$$

Substituting Equation (4-55) in terms of V and differentiating, the mole balance on B becomes

Mole balance on B

$$\frac{dC_B}{dt} = r_B + \frac{v_0(C_{B0} - C_B)}{V} \qquad (4\text{-}58)$$

At time $t = 0$, the initial concentration of B in the vat is zero, $C_{Bi} = 0$. The concentration of B in the feed is C_{B0}. If the reaction order is other than zero- or first-order, or if the reaction is nonisothermal, we must use numerical techniques to determine the conversion as a function of time. Equations (4-56) and (4-58) are easily solved with an ODE solver.

Example 4–9 Isothermal Semibatch Reactor with Second-Order Reaction

The production of methyl bromide is an irreversible liquid-phase reaction that follows an elementary rate law. The reaction

$$CNBr + CH_3NH_2 \rightarrow CH_3Br + NCNH_2$$

is carried out isothermally in a semibatch reactor. An aqueous solution of methyl amine (B) at a concentration of 0.025 mol/dm³ is to be fed at a rate of 0.05 dm³/s to an aqueous solution of bromine cyanide (A) contained in a glass-lined reactor. The initial volume of fluid in a vat is to be 5 dm³ with a bromine cyanide concentration of 0.05 mol/dm³. The specific reaction rate constant is

$$k = 2.2 \text{ dm}^3/\text{s} \cdot \text{mol}$$

Solve for the concentrations of bromine cyanide and methyl bromide and the rate of reaction as a function of time.

Solution

Symbolically, we write the reaction as

$$A + B \rightarrow C + D$$

The reaction is elementary; therefore, the rate law is

Rate Law

$$\boxed{-r_A = kC_A C_B}$$
(E4-9.1)

Substituting the rate law in Equations (4-56) and (4-58) gives

Combined mole
balances and
rate laws on A, B,
C, and D

$$\boxed{\frac{dC_A}{dt} = -kC_A C_B - \frac{v_0}{V}C_A}$$
(E4-9.2)

Polymath will com-
bine for you. Thank
you, Polymath!

$$\boxed{\frac{dC_B}{dt} = -kC_A C_B + \frac{v_0}{V}(C_{B0} - C_B)}$$
(E4-9.3)

$$\boxed{V = V_0 + v_0 t}$$
(E4-9.4)

Similarly for C and D we have

$$\frac{dN_C}{dt} = r_C V = -r_A V$$
(E4-9.5)

$$\frac{dN_C}{dt} = \frac{d(C_C V)}{dt} = V\frac{dC_C}{dt} + C_C\frac{dV}{dt} = V\frac{dC_C}{dt} + v_0 C_C$$
(E4-9.6)

Then

$$\boxed{\frac{dC_C}{dt} = kC_A C_B - \frac{v_0 C_C}{V}}$$
(E4-9.7)

and

$$\boxed{\frac{dC_D}{dt} = kC_A C_B - \frac{v_0 C_D}{V}}$$
(E4-9.8)

We could also calculate the conversion of A.

$$X = \frac{N_{A0} - N_A}{N_{A0}}$$
(E4-9.9)

$$\boxed{X = \frac{C_{A0}V_0 - C_A V}{C_{A0}V_0}}$$
(E4-9.10)

The initial conditions are $t = 0$: $C_{A0} = 0.05$ mol/dm³, $C_B = C_C, = C_D = 0$, and $V_0 = 5$ dm³.

Equations (E4-9.2) through (E4-9.10) are easily solved with the aid of an ODE solver such as Polymath (Table E4-9.1).

TABLE E4-9.1 POLYMATH PROGRAM

ODE REPORT (RKF45)

Differential equations as entered by the user
[1]   d(Ca)/d(t) = -k*Ca*Cb-vo*Ca/V
[2]   d(Cb)/d(t) = -k*Ca*Cb+vo*(Cbo-Cb)/V
[3]   d(Cc)/d(t) = k*Ca*Cb-vo*Cc/V
[4]   d(Cd)/d(t) = k*Ca*Cb-vo*Cd/V

Explicit equations as entered by the user
[1]   k = 2.2
[2]   vo = 0.05
[3]   Cbo = 0.025
[4]   Vo = 5
[5]   Cao = 0.05
[6]   rate = k*Ca*Cb
[7]   V = Vo+vo*t
[8]   X = (Cao*Vo-Ca*V)/(Cao*Vo)

Living Example Problem

The concentrations of bromine cyanide (A) and methyl amine are shown as a function of time in Figure E4-9.1, and the rate is shown in Figure E4-9.2.

Why does the concentration of CH_3Br (C) go through a maximum wrt time?

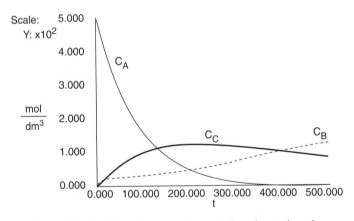

Figure E4-9.1 Polymath output: Concentration–time trajectories.

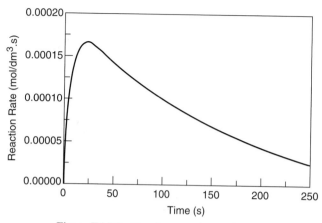

Figure E4-9.2   Reaction rate–time trajectory.

4.10.4 Writing the Semibatch Reactor Equations in Terms of Conversion

Consider the reaction

$$A + B \underset{\longleftarrow}{\longrightarrow} C + D$$

in which B is fed to a vat containing only A initially. The reaction is first-order in A and first-order in B. The number of moles of A remaining at any time, t, is

> The limiting reactant is the one in the vat.

$$\begin{bmatrix} \text{Number of moles} \\ \text{of A in the vat} \\ \text{at time } t \end{bmatrix} = \begin{bmatrix} \text{Number of moles} \\ \text{of A in the vat} \\ \text{initially} \end{bmatrix} - \begin{bmatrix} \text{Number of moles} \\ \text{of A reacted up} \\ \text{to time } t \end{bmatrix} \quad (4\text{-}59)$$

$$\overbrace{N_A} \qquad\qquad = \qquad\qquad \overbrace{N_{A0}} \qquad\qquad - \qquad\qquad \overbrace{N_{A0}X}$$

where X is the moles of A reacted per mole of A initially in the vat. Similarly, for species B,

$$\begin{bmatrix} \text{Number of} \\ \text{moles of B in} \\ \text{the vat at time } t \end{bmatrix} = \begin{bmatrix} \text{Number of} \\ \text{moles of B in} \\ \text{the vat initially} \end{bmatrix} + \begin{bmatrix} \text{Number of} \\ \text{moles of B} \\ \text{added to the vat} \end{bmatrix} - \begin{bmatrix} \text{Number of moles} \\ \text{of B reacted} \\ \text{up to time } t \end{bmatrix} \quad (4\text{-}60)$$

$$\overbrace{N_B} \qquad = \qquad \overbrace{N_{Bi}} \qquad + \qquad \overbrace{\int_0^t F_{B0}\, dt} \qquad - \qquad \overbrace{N_{A0}X}$$

For a constant molar feed rate and no B initially in the vat,

$$N_B = F_{B0}t - N_{A0}X \qquad (4\text{-}61)$$

A **mole balance** on species A gives

$$r_A V = \frac{dN_A}{dt} = -N_{A0}\frac{dX}{dt} \qquad (4\text{-}62)$$

The number of moles of C and D can be taken directly from the stoichiometric table; for example,

$$N_C = N_{Ci} + N_{A0}X \qquad (4\text{-}63)$$

For a reversible second-order reaction $A + B \rightleftarrows C + D$ for which the **rate law** is

$$-r_A = k\left(C_A C_B - \frac{C_C C_D}{K_C}\right) \qquad (4\text{-}64)$$

Recalling Equation (4-55), the **concentrations of A, B, C, and D are**

Concentration of reactants as a function of conversion and time

$$C_A = \frac{N_A}{V} = \frac{N_{A0}(1-X)}{V_0 + v_0 t} \qquad C_C = \frac{N_{A0}X}{V_0 + v_0 t}$$

$$C_B = \frac{N_B}{V} = \frac{N_{Bi} + F_{B0}t - N_{A0}X}{V_0 + v_0 t} \qquad C_D = \frac{N_{A0}X}{V_0 + v_0 t} \qquad (4\text{-}65)$$

Combining equations (4-62), (4-64), and (4-65), substituting for the concentrations, and dividing by N_{A0}, we obtain

$$\boxed{\frac{dX}{dt} = \frac{k[(1-X)(N_{Bi} + F_{B0}t - N_{A0}X) - (N_{A0}X^2/K_C)]}{V_0 + v_0 t}} \qquad (4\text{-}66)$$

Equation (4-66) needs to be solved numerically to determine the conversion as a function of time.

The third variable, in addition to concentration and conversion, we can use to analyze semibatch reactors is number of mole N_A, N_B, etc. This method is discussed in the Summary Notes on the CD-ROM.

Summary Notes

Equilibrium Conversion. For reversible reactions carried out in a semibatch reactor, the maximum attainable conversion (i.e., the equilibrium conversion) will change as the reaction proceeds because more reactant is continuously added to the reactor. This addition shifts the equilibrium continually to the right toward more product.

If the reaction $A + B \rightleftharpoons C + D$ were allowed to reach equilibrium after feeding species B for a time t, the equilibrium conversion could be calculated as follows at equilibrium [see Appendix C]:

$$K_C = \frac{C_{Ce} C_{De}}{C_{Ae} C_{Be}} = \frac{\left(\dfrac{N_{Ce}}{V}\right)\left(\dfrac{N_{De}}{V}\right)}{\left(\dfrac{N_{Ae}}{V}\right)\left(\dfrac{N_{Be}}{V}\right)} \qquad (4\text{-}67)$$

$$= \frac{N_{Ce} N_{De}}{N_{Ae} N_{Be}}$$

The relationship between conversion and number of moles of each species is the same as shown in Table 3-1 except for species B, for which the number of moles is given by Equation (4-61). Thus

$$K_C = \frac{(N_{A0} X_e)(N_{A0} X_e)}{N_{A0}(1 - X_e)(F_{B0} t - N_{A0} X_e)}$$

$$= \frac{N_{A0} X_e^2}{(1 - X_e)(F_{B0} t - N_{A0} X_e)} \qquad (4\text{-}68)$$

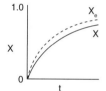

Rearranging yields

$$t = \frac{N_{A0}}{K_C F_{B0}}\left(K_C X_e + \frac{X_e^2}{1 - X_e}\right) \qquad (4\text{-}69)$$

or

Equilibrium
conversion
in a semibatch
reactor

$$X_e = \frac{K_C\left(1 + \dfrac{F_{B0} t}{N_{A0}}\right) - \sqrt{\left[K_C\left(1 + \dfrac{F_{B0} t}{N_{A0}}\right)\right]^2 - 4(K_C - 1) K_C \dfrac{t F_{B0}}{N_{A0}}}}{2(K_C - 1)} \qquad (4\text{-}70)$$

Reactive distillation is used with thermodynamically limited reversible liquid-phase reactions and is particularly attractive when one of the products has a lower boiling point than the reactants. For reversible reactions of this type,

$$A(,) + B(,) \rightleftharpoons C(,) + D(g,,)$$

the equilibrium lies far to the left, and little product is formed. However, if one or more of the products (e.g., D) is removed by vaporization, as shown in Figure 4-16,

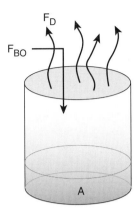

Figure 4-16   Reactive distillation with B fed to a vat containing A and D vaporizing.

the reaction will continue toward completion. The equilibrium constraint is removed, and more product will be formed. The fundamentals of reactive distillation are given on the CD-ROM web module.

4.11 The Practical Side

The material presented in this chapter has been for isothermal ideal reactors. We will build on the concepts developed in this chapter when we discuss nonideal reactors in Chapters 13 and 14. A number of practical guidelines for the operation of chemical reactors have been presented over the years, and tables and some of these descriptions are summarized and presented on the CD-ROM and web. The articles are listed in Table 4-7.

TABLE 4-7   LITERATURE THAT GIVE PRACTICAL GUIDELINES FOR REACTOR OPERATION

D. Mukesh, *Chem. Eng.*, 46 (January 2002). S. Dutta and R. Gualy, *CEP*, 37 (October 2000); *C&EN*, 8 (January 10, 2000). S. Jayakumar, R. G. Squires, G. V. Reklaitis, P. K. Andersen, and L. R. Partin, *Chem. Eng. Educ.*, 136 (Spring 1993). R. W. Cusack, *Chem. Eng.*, 88 (February 2000). A. Bakker, A. H. Haidari, and E. M. Marshall, *CEP*, 30 (December 2001). P. Trambouze, *CEP*, 23 (February 1990). G. Scholwsky and B. Loftus-Koch, *Chem. Eng.*, 96 (February 2000). J. H. Worstell, *CEP*, 55 (June 2000). J. H. Worstell, *CEP*, 68 (March 2001). S. Dutta and R. Gualy, *Chem. Eng.*, 72 (June 2000). A. Abu-Khalaf, *Chem. Eng. Educ.*, 48 (Winter 1994).

- For example, Mukesh gives relationships between the CSTR tank diameter, *T*, impeller size diameter, *D*, tank height, *H*, and the liquid level, ℓ. To scale up a pilot plant (1) to a full scale plant (2), the following guidelines are given

$$\frac{D_2}{D_1} = \frac{T_2}{T_1} = \frac{\ell_2}{\ell_1} = \frac{H_2}{H_1} = R$$

And the rotational speed, N_2, is

$$N_2 = N_1 R^{-n}$$

where values of n for different pumping capacities and Froude numbers are given in Mukesh's article.

Closure. This chapter presents the heart of chemical reaction engineering for isothermal reactors. After completing this chapter, the reader should be able to apply the algorithm building blocks

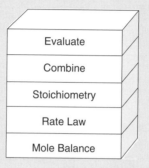

to any of the reactors discussed in this chapter: batch reactor, CSTR, PFR, PBF, membrane reactor, and semibatch reactor. The reader should be able to account for pressure drop and describe the effects of the system variables such as particle size on the conversion and explain why there is an optimum in the conversion when the catalyst particle size is varied. The reader should be able to use either conversions (Part 1) or concentration and molar flow rates (Part 2) to solve chemical reaction engineering problems. Finally, after completing this chapter, the reader should be able to work the California Professional Engineering Exam Problems in approximately 30 minutes [cf. P4-11$_B$ through P4-15$_B$] and to diagnose and troubleshoot malfunctioning reactors [cf. P4-8$_B$].

SUMMARY

1. Solution algorithm—**Conversion**
   a. **Design equations** (Batch, CSTR, PFR, PBR):

$$N_{A0}\frac{dX}{dt} = -r_A V, \quad V = \frac{F_{A0}X}{-r_A}, \quad F_{A0}\frac{dX}{dV} = -r_A, \quad F_{A0}\frac{dX}{dW} = -r_A' \quad \text{(S4-1)}$$

b. **Rate law**: For example,

$$-r_A = kC_A^2 \tag{S4-2}$$

c. **Stoichiometry**:
$$A + \frac{b}{a}B \rightarrow \frac{c}{a}C + \frac{d}{a}D$$

(1) *Gas phase,* $v = v_0(1 + \varepsilon X)\left(\dfrac{P_0}{P}\right)\left(\dfrac{T}{T_0}\right)$

$$C_A = \frac{F_A}{v} = \frac{F_{A0}(1-X)}{v} = \frac{F_{A0}(1-X)}{v_0(1+\varepsilon X)}\left(\frac{P}{P_0}\right)\frac{T_0}{T} = C_{A0}\left(\frac{1-X}{1+\varepsilon X}\right)y\frac{T_0}{T} \tag{S4-3}$$

$$\frac{dy}{dW} = -\frac{\alpha(1+\varepsilon X)}{2y}\left(\frac{T}{T_0}\right) \tag{S4-4}$$

For a packed bed

$$\alpha = \frac{2\beta_0}{A_c(1-\phi)\rho_c P_0} \quad \text{and} \quad \beta_0 = \frac{G(1-\phi)}{\rho_0 g_c D_p \phi^3}\left[\frac{150(1-\phi)\mu}{D_p} + 1.75G\right]$$

(2) *Liquid phase*: $\qquad v = v_0 \tag{S4-5}$

$$C_A = C_{A0}(1-X)$$

d. **Combining** for isothermal operation

Gas: $\quad -r_A' = kC_{A0}^2\dfrac{(1-X)^2}{(1+\varepsilon X)^2}y^2 \tag{S4-6}$

Liquid: $\quad -r_A' = kC_{A0}^2(1-X)^2$

e. Solution techniques:
   (1) Numerical integration—Simpson's rule $\tag{S4-7}$
   (2) Table of integrals
   (3) Software packages
      (a) Polymath
      (b) MATLAB
2. Solution algorithm—**Measures other than conversion**
   When using measures other than conversion for reactor design, the mole balances are written for each species in the reacting mixture:

Mole balances $\qquad\qquad \dfrac{dF_A}{dV} = r_A, \quad \dfrac{dF_B}{dV} = r_B, \quad \dfrac{dF_C}{dV} = r_C, \quad \dfrac{dF_D}{dV} = r_D \tag{S4-8}$

The mole balances are then coupled through their relative rates of reaction. If

Rate law

$$-r_A = kC_A^\alpha C_B^\beta \tag{S4-9}$$

for $aA + bB \rightarrow cC + dD$, then

Stoichiometry

$$r_B = \frac{b}{a} r_A, \quad r_C = -\frac{c}{a} r_A, \quad r_D = -\frac{d}{a} r_A \tag{S4-10}$$

Concentration can also be expressed in terms of the number of moles (batch) and in molar flow rates (flow).

$$\textit{Gas:} \quad C_A = C_{T0} \frac{F_A}{F_T} \frac{P}{P_0} \frac{T_0}{T} = C_{T0} \frac{F_A T_0}{F_T T} y \tag{S4-11}$$

$$C_B = C_{T0} \frac{F_B}{F_T} \frac{T_0}{T} y \tag{S4-12}$$

$$F_T = F_A + F_B + F_C + F_D + F_I \tag{S4-13}$$

$$\frac{dy}{dW} = \frac{-\alpha}{2y} \left(\frac{F_T}{F_{T0}}\right) \left(\frac{T}{T_0}\right) \tag{S4-14}$$

$$\textit{Liquid:} \quad C_A = \frac{F_A}{v_0} \tag{S4-15}$$

Combine

3. An ODE solver (e.g., Polymath) will combine all the eqations for you.

Variable density with $\varepsilon = 0$ or $\varepsilon X \ll 1$ and isothermal:

IFF $\varepsilon = 0$
$$\frac{P}{P_0} = (1 - \alpha W)^{1/2} \tag{S4-16}$$

4. For **membrane reactors** the mole balances for the reaction

$$A \rightleftharpoons B + C$$

when reactant A and product C do not diffuse out the membrane

$$\frac{dF_A}{dV} = r_A, \quad \frac{dF_B}{dV} = r_B - R_B, \quad \text{and} \quad \frac{dF_C}{dV} = r_C \tag{S4-17}$$

with

$$R_B = k_c C_B \tag{S4-18}$$

and k_c is the overall mass transfer coefficient.

5. For **semibatch reactors**, reactant B is fed continuously to a vat initially containing only A:

$$A + B \rightleftharpoons C + D$$

The combined mole balance, rate law, and stoichiometry in terms of conversion is

$$\frac{dX}{dt} = \frac{k[(1-X)(N_{Bi} + F_{B0}t - N_{A0}X) - (N_{A0}X^2/K_C)]}{V_0 + v_0 t} \tag{S4-19}$$

ODE SOLVER ALGORITHM

When using an ordinary differential equation (ODE) solver such as Polymath or MATLAB, it is usually easier to leave the mole balances, rate laws, and concentrations as separate equations rather than combining them into a single equation as we did to obtain an analytical solution. Writing the equations separately leaves it to the computer to combine them and produce a solution. The formulations for a packed-bed reactor with pressure drop and a semibatch reactor are given below for two elementary reactions carried out isothermally.

Gas Phase	Liquid Phase
$A + B \rightarrow 3C$	$A + B \rightarrow 2C$
Packed-Bed Reactor	*Semibatch Reactor*

Gas Phase — *Packed-Bed Reactor*

$$\frac{dX}{dW} = \frac{-r_A}{F_{A0}}$$

$$r_A = -kC_A C_B$$

$$C_A = C_{A0}\frac{1-X}{1+\varepsilon X}y$$

$$C_B = C_{A0}\frac{\theta_B - X}{1+\varepsilon X}y$$

$$\frac{dy}{dW} = -\frac{\alpha(1+\varepsilon X)}{2y}$$

(where $y = P/P_0$)

Liquid Phase — *Semibatch Reactor*

$$\frac{dC_A}{dt} = r_A - \frac{v_0 C_A}{V}$$

$$\frac{dC_B}{dt} = r_A + \frac{v_0(C_{B0} - C_B)}{V}$$

$$\frac{dC_C}{dt} = -2r_A - \frac{v_0 C_C}{V}$$

$$r_A = -kC_A C_B$$

$$V = V_0 + v_0 t$$

Gas Phase	Liquid Phase
$k = 10.0$	$k = 0.15$
$\alpha = 0.01$	$K_C = 4.0$
$\varepsilon = 0.33$	$V_0 = 10.0$
$\theta_B = 2.0$	$v_0 = 0.1$
$C_{A0} = 0.01$	$C_{B0} = 0.1$
$F_{A0} = 15.0$	$C_{Ai} = 0.02$
$W_{final} = 80$	$t_{final} = 200$

CD-ROM MATERIAL

• Learning Resources

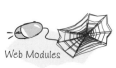

1. *Summary Notes*
2. *Web Modules*

A. Wetlands

B. Reactive Distillation

3. *Interactive Computer Modules*

A. Murder Mystery

B. Tic-Tac

C. Reactor Lab Modules

The following reactor Lab Modules have been developed by Professor Richard Herz in the Chemical Engineering Department at the University of California, San Diego. They are copyrighted by UCSD and Professor Hertz and are used here with their permission.

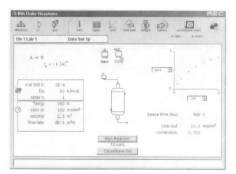

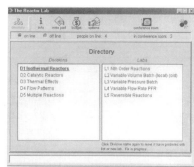

Solved Problems

Living Example Problem

Green engineering

4. *Solved Problems*
   A. CDP4-A$_B$ A Sinister Gentleman Messing with a Batch Reactor
   B. Solution to California Registration Exam Problem
   C. Ten Types of Home Problems: 20 Solved Problems
5. *Analogy of CRE Algorithms to a Menu in a Fine French Restaurant*
6. *Algorithm for Gas Phas Reaction*

• **Living Example Problems**
   Example 4-6 Calculating X in a Reactor with Pressure Drop
   Example 4-7 Gas-Phase Reaction in Microreactor—Molar Flow Rate
   Example 4-8 Membrane Reactor
   Example CDR4.1 Spherical Reactor
   Example 4.3.1 Aerosol Reactor
   Example 4-9 Isothermal Semibatch Reactor

• **Professional Reference Shelf**
   R4.1. *Spherical Packed-Bed Reactors*
       When small catalyst pellets are required, the pressure drop can be significant. One type of reactor that minimizes pressure drop and is also inexpensive to build is the spherical reactor, shown here. In this reactor, called an ultraformer, dehydrogenation reactions such as

$$\text{Paraffin} \longrightarrow \text{Aromatic} + 3H_2$$

   are carried out.

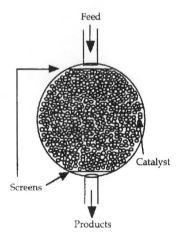

Spherical ultraformer reactor. (Courtesy of Amoco Petroleum Products.) This reactor is one in a series of six used by Amoco for reforming petroleum naphtha. Photo by K. R. Renicker, Sr.

   Analysis of a spherical reactor equation along with an example problem are carried out on the CD-ROM.

   R4.2 *Recycle Reactors*
       Recycle reactors are used (1) when conversion of unwanted (toxic) products is required and they are recycled to extinction, (2) the reaction is autocatalytic or (3) it is necessary to maintain isothermal operation. To design recycle reactors, one simply follows the procedure developed in this chapter and then adds a little additional bookkeeping.

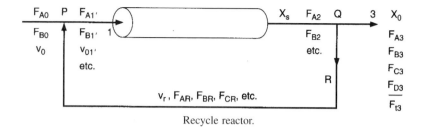

Recycle reactor.

As shown in the CD-ROM, two conversions are usually associated with recycle reactors: the overall conversion, X_0, and the conversion per pass, X_S.

R4.3. *Aerosol Reactors*

Aerosol reactors are used to synthesize nano-size particles. Owing to their size, shape, and high specific surface area, nanoparticles can be used in a number of applications such as in pigments in cosmetics, membranes, photocatalytic reactors, catalysts and ceramics, and catalytic reactors.

We use the production of aluminum particles as an example of an aerosol plug-flow reactor (APFR) operation. A stream of argon gas saturated with Al vapor is cooled.

Nanoparticles

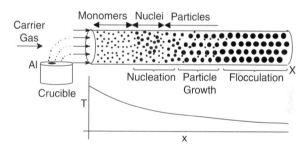

Aerosol reactor and temperature profile.

As the gas is cooled, it becomes supersaturated, leading to the nucleation of particles. This nucleation is a result of molecules colliding and agglomerating until a critical nucleus size is reached and a particle is formed. As these particles move down, the supersaturated gas molecules condense on the particles causing them to grow in size and then to flocculate. In the development on the CD-ROM, we will model the formation and growth of aluminum nanoparticles in an APFR.

R4.4 *Critiquing Journal Articles*

After graduation, your textbooks will be, in part, the professional journals that you read. As you read the journals, it is important that you study them with a critical eye. You need to learn if the author's conclusion is supported by the data, if the article is new or novel, if it advances our understanding, and if the analysis is current. To develop this technique, one of the major assignments used in the graduate course in chemical reaction engineering at the University of Michigan for the past 25 years has been an in-depth analysis and critique of a journal article related to the course material. Significant effort is made to ensure that a cursory or superficial review is not carried out. The CD-ROM gives an example and some guidelines about critiquing journal articles.

QUESTIONS AND PROBLEMS

Homework Problems

The subscript to each of the problem numbers indicates the level of difficulty: A, least difficult; D, most difficult.

$$A = \bullet \quad B = \blacksquare \quad C = \blacklozenge \quad D = \blacklozenge\blacklozenge$$

In each of the following questions and problems, rather than just drawing a box around your answer, write a sentence or two describing how you solved the problem, the assumptions you made, the reasonableness of your answer, what you learned, and any other facts that you want to include. You may wish to refer to W. Strunk and E. B. White, *The Elements of Style*, 4th ed. (New York: Macmillan, 2000) and Joseph M. Williams, *Style: Ten Lessons in Clarity & Grace,* 6th ed. (Glenview, Ill.: Scott, Foresman, 1999) to enhance the quality of your sentences. See the Preface for additional generic parts (x), (y), (z) to the home problems.

P4-1$_A$ Read through all the problems at the end of this chapter. Make up and solve an *original* problem based on the material in this chapter. **(a)** Use real data and reactions. **(b)** Make up a reaction and data. **(c)** Use an example from everyday life (e.g., making toast or cooking spaghetti). In preparing your original problem, first list the principles you want to get across and why the problem is important. Ask yourself how your example will be different from those in the text or lecture. Other things for you to consider when choosing a problem are relevance, interest, impact of the solution, time required to obtain a solution, and degree of difficulty. Look through some of the journals for data or to get some ideas for industrially important reactions or for novel applications of reaction engineering principles (the environment, food processing, etc.). At the end of the problem and solution describe the creative process used to generate the idea for the problem. **(d)** Write a question based on the material in this chapter that requires critical thinking. Explain why your question requires critical thinking. [*Hint:* See Preface, Section B.2] **(e)** Listen to the audios on the CD 🎧 Lecture Notes, pick one, and describe how you might explain it differently.

P4-2$_B$ **What if...** you were asked to explore the example problems in this chapter to learn the effects of varying the different parameters? This sensitivity analysis can be carried out by either downloading the examples from the web or by loading the programs from the CD-ROM supplied with the text. For each of the example problems you investigate, write a paragraph describing your findings.

Before solving the problems, state or sketch qualitatively the expected results or trends.

    (a)   **What if** you were asked to give examples of the material in this book that are found in everyday life? What would you say?

    (b)   **Example 4-1.** What would be the error in k if the batch reactor were only 80% filled with the same composition of reactants instead of being completely filled as in the example? What generalizations can you draw from this example?

(c) **Example 4-2.** How would your reactor volume change if you only needed 50% conversion to produce the 200 million pounds per year required? What generalizations can you draw from this example?

(d) **Example 4-3.** What would be the reactor volume for $X = 0.8$ if the pressure were increased by a factor of 10 assuming everything else remains the same? What generalizations can you draw from this example?

(e) **Example 4-4.** How would the pressure drop change if the particle diameter were reduced by 25%? What generalizations can you draw from this example?

(f) **Example 4-5.** What would be the conversion with and without pressure drop if the entering pressure were increased by a factor of 10? Would the optimum diameter change? If so, how? What would the conversion be if the reactor diameter were decreased by a factor of 2 for the same mass flow rate?

(g) **Example 4-6.** Load the *Living Example Problem 4-6* from the CD-ROM. How much would the catalyst weight change if the pressure was increased by a factor of 5 and the particle size decreased by a factor of 5? (Recall α is also a function of P_0)? Use plots and figures to describe what you find.

(h) **Example 4-7.** Load the *Living Example Problem 4-7* from the CD-ROM. How would the results change if the pressure were doubled and the temperature was decreased 20°C?

(i) **Example 4-8.** Load the *Living Example Problem 4-8* from the CD-ROM. Vary parameters (e.g., k_C), and ratios of parameters (k/k_C), ($k\tau C_{A0}/K_e$), etc., and write a paragraph describing what you find. What ratio of parameters has the greatest effect on the conversion $X = (F_{A0} - F_A)/F_{A0}$?

(j) **Example 4-9.** Load the *Living Example Problem 4-9* from the CD-ROM. The temperature is to be lowered by 35°C so that the reaction rate constant is now (1/10) its original value. (i) If the concentration of B is to be maintained at 0.01 mol/dm³ or below, what is the maximum feed rate of B? (ii) How would your answer change if the concentration of A were tripled?

(k) *Web Module on Wetlands* from the CD-ROM. Load the *Polymath program* and vary a number of parameters such as rainfall, evaporation rate, atrazine concentration, and liquid flow rate, and write a paragraph describing what you find. This topic is a hot Ch.E. research area.

(l) *Web Module on Reactive Distillation* from the CD-ROM. Load the *Polymath program* and vary the parameters such as feed rate, and evaporation rate, and write a paragraph describing what you find.

(m) *Web Module on Aerosol Reactors* from the CD-ROM. Load the *Polymath program* and (1) vary the parameters such as cooling rate and flow rate, and describe their effect on each of the regimes nucleation, growth and flocculation. Write a paragraph describing what you find. (2) It is proposed to replace the carrier gas by helium

   (i)   Compare your plots (He versus Ar) of the number of Al particles as a function of time. Explain the shape of the plots.

   (ii) How does the final value of d_p compare with that when the carrier gas was argon? Explain.

   (iii) Compare the time at which the rate of nucleation reaches a peak in the two cases [carrier gas = Ar and He]. Discuss the comparison.

<u>Data for a He molecule:</u> Mass = 6.64×10^{-27} kg, Volume = 1.33×10^{-29} m^3, Surface area = 2.72×10^{-19} m^2, Bulk density = 0.164 kg/m^3, at normal temperature (25°C) and pressure (1 atm).

(n) Vary some of the operating costs, conversions, and separations in Figure 4-10 to learn how the profit changes. Ethylene oxide, used to make ethylene glycol, sells for $0.56/lb while ethylene glycol sells for $0.38/lb. Is this a money-losing proposition? Explain.

(o) What should you do if some of the ethylene glycol splashed out of the reactor onto your face and clothing? (*Hint*: Recall *www.siri.org/*.)

(p) What safety precautions should you take with the ethylene oxide formation discussed in Example 4-6? With the bromine cyanide discussed in Example 4-9?

(q) **Load reactor lab** on to your computer and call up *D1 Isothermal Reactors*. Detailed instructions with screen shots are given in Chapter 4 of the Summary Notes. **(1)** For **L1** Nth Order Reactions. Vary the parameters n, E, T for a batch, CSTR, and PFR. Write a paragraph discussing the trends (e.g., first order versus second order) and describe what you find. **(2)** Next choose the "Quiz" on membrane at the top of the screen, and find the reaction order **(3)** and turn in your performance number.

Performance number: _____

(r) **The Work Self Tests on the Web.** Write a question for this problem that involves critical thinking and explaining why it involves critical thinking. See examples on the Web Summary Note for Chapter 4.

Interactive

Computer Modules

P4-3$_B$ Load the Interactive Computer Modules (ICM) from the CD-ROM. Run the modules and then record your performance number, which indicates your mastery of the material. Your instructor has the key to decode your performance number.

(a) ICM—Mystery Theater—A real "Who done it?", see *Pulp and Paper*, 25 (January 1993) and also *Pulp and Paper*, 9 (July 1993). The outcome of the murder trial is summarized in the December 1995 issue of *Papermaker*, page 12. You will use fundamental chemical engineering from Sections 4.1 to 4.3 to identify the victim and the murderer.

Performance number: _____

(b) ICM—Tic Tac—Knowledge of all sections is necessary to pit your wit against the computer adversary in playing a game of Tic-Tac-Toe.

Performance number: _____

P4-4$_A$ If it takes 11 minutes to cook spaghetti in Ann Arbor, Michigan, and 14 minutes in Boulder, Colorado, how long would it take in Cuzco, Peru? Discuss ways to make the spaghetti more tasty. If you prefer to make a creative spaghetti dinner for family or friends rather than answering this question, that's OK, too; you'll get full credit—but **only if** you turn in your receipt and bring your instructor a taste. (*Ans. t* = 21 min)

Application Pending for Problem Hall of Fame

P4-5$_A$ The liquid-phase reaction

$$A + B \longrightarrow C$$

follows an elementary rate law and is carried out isothermally in a flow system. The concentrations of the A and B feed streams are 2 *M* before mixing. The volumetric flow rate of each stream is 5 dm^3/min, and the entering temperature is 300 K. The streams are mixed immediately before entering.

Two reactors are available. One is a gray 200.0-dm³ CSTR that can be heated to 77°C or cooled to 0°C, and the other is a white 800.0-dm³ PFR operated at 300 K that cannot be heated or cooled but can be painted red or black. Note $k = 0.07$ dm³/mol·min at 300 K and $E = 20$ kcal/mol.

(a) Which reactor and what conditions do you recommend? Explain the reason for your choice (e.g., color, cost, space available, weather conditions). Back up your reasoning with the appropriate calculations.

(b) How long would it take to achieve 90% conversion in a 200-dm³ batch reactor with $C_{A0} = C_{B0} = 1$ M after mixing at a temperature of 77°C?

(c) What would your answer to part (b) be if the reactor were cooled to 0°C? (*Ans.* 2.5 days)

(d) What conversion would be obtained if the CSTR and PFR were operated at 300 K and connected in series? In parallel with 5 mol/min to each?

(e) Keeping Table 4-1 in mind, what batch reactor volume would be necessary to process the same amount of species A per day as the flow reactors while achieving 90% conversion? Referring to Table 1-1, estimate the cost of the batch reactor.

(f) Write a couple of sentences describing what you learned from the problem and what you believe to be the point of the problem.

P4-6$_B$ Dibutyl phthalate (DBP), a plasticizer, has a potential market of 12 million lb/yr (*AIChE Student Contest Problem*) and is to be produced by reaction of *n*-butanol with monobutyl phthalate (MBP). The reaction follows an elementary rate law and is catalyzed by H_2SO_4 (Figure P4-6). A stream containing MBP and butanol is to be mixed with the H_2SO_4 catalyst immediately before the stream enters the reactor. The concentration of MBP in the stream entering the reactor is 0.2 lb mol/ft³, and the molar feed rate of butanol is five times that of MBP. The specific reaction rate at 100°F is 1.2 ft³/lb mol·h . There is a 1000-gallon CSTR and associated peripheral equipment available for use on this project for 30 days a year (operating 24 h/day).

Hall of Fame

National AICHE
Contest Problem

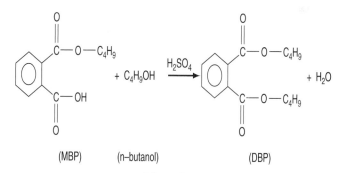

(MBP)   (n–butanol)   (DBP)

Figure P4-6

(a) Determine the exit conversion in the available 1000-gallon reactor if you were to produce 33% of the share (i.e., 4 million lb/yr) of the predicted market. (*Ans.: X* = 0.33)

(b) How might you increase the conversion for the same F_{AO}? For example, what conversion would be achieved if a second 1000-gal CSTR were placed either in series or in parallel with the CSTR? [$X_2 = 0.55$ (series)]

(c) For the same temperature as part (a), what CSTR volume would be necessary to achieve a conversion of 85% for a molar feed rate of MBP of 1 lb mol/min?

(d) *If possible*, calculate the tubular reactor volume necessary to achieve 85% conversion, when the reactor is oblong rather than cylindrical, with a major-to-minor axis ratio of 1.3 : 1.0. There are no radial gradients in either concentration or velocity. If it is not possible to calculate V_{PRF}, explain.

(e) How would your results for parts (a) and (b) change if the temperature were raised to 150°F where k is now 5.0 ft³/lb mol · h but the reaction is reversible with $K_C = 0.3$?

(f) Keeping in mind the times given in Table 4-1 for filling, and other operations, how many 1000-gallon reactors operated in the batch mode would be necessary to meet the required production of 4 million pounds in a 30-day period? Estimate the cost of the reactors in the system. *Note*: Present in the feed stream may be some trace impurities, which you may lump as hexanol. The activation energy is believed to be somewhere around 25 kcal/mol. *Hint*: Plot number of reactors as a function of conversion. (*An Ans.*: 5 reactors)

(g) What generalizations can you make about what you learned in this problem that would apply to other problems?

(h) Write a question that requires critical thinking and then explain why your question requires critical thinking. [*Hint*: See Preface, Section B.2]

P4-7$_A$ The elementary gas-phase reaction

$$(CH_3)_3COOC(CH_3)_3 \rightarrow C_2H_6 + 2CH_3COCH_3$$

is carried out isothermally in a flow reactor with no pressure drop. The specific reaction rate at 50°C is 10^{-4} min^{-1} (from pericosity data) and the activation energy is 85 kJ/mol. Pure di-*tert*-butyl peroxide enters the reactor at 10 atm and 127°C and a molar flow rate of 2.5 mol/min. Calculate the reactor volume and space time to achieve 90% conversion in:

(a) a PFR (*Ans.*: 967 dm³)

(b) a CSTR (*Ans.*: 4700 dm³)

(c) **Pressure drop.** Plot X, y, as a function of the PFR volume when $\alpha = 0.001$ dm^{-3}. What are X, and y at $V = 500$ dm³?

(d) Write a question that requires critical thinking, and explain why it involves critical thinking.

(e) If this reaction is to be carried out isothermally at 127°C and an initial pressure of 10 atm in a constant-volume batch mode with 90% conversion, what reactor size and cost would be required to process (2.5 mol/min × 60 min/h × 24 h/day) 3600 mol of di-*tert*-butyl peroxide per day? (*Hint*: Recall Table 4-1.)

(f) Assume that the reaction is reversible with $K_C = 0.025$ mol^2/dm^6, and calculate the equilibrium conversion; then redo (a) through (c) to achieve a conversion that is 90% of the equilibrium conversion.

(g) **Membrane reactor.** Repeat Part (f) for the case when C_2H_6 flows out through the sides of the reactor and the transport coefficient is $k_C = 0.08$ s^{-1}.

Creative Thinking

P4-8$_B$ **Troubleshooting**

(a) A liquid-phase isomerization A $\longrightarrow$ B is carried out in a 1000-gal CSTR that has a single impeller located halfway down the reactor. The liquid enters at the top of the reactor and exits at the bottom. The reaction is second order. Experimental data taken in a batch reactor predicted the CSTR conversion should be 50%. However, the conversion measured in the actual CSTR was 57%. Suggest reasons for the discrepancy and suggest something that would give closer agreement between the predicted and measured conversions. Back your suggestions with calculations. P.S. It was raining that day.

(b) The first-order gas-phase isomerization reaction

$$A \xrightarrow{\ k\ } B \ \ \text{with } k = 5 \ \text{min}^{-1}$$

Member

Hall of Fame

is to be carried out in a tubular reactor. For a feed of pure A of 5 dm³/min, the expected conversion in a PFR is 63.2%. However, when the reactor was put in operation, the conversion was only 58.6%. We should note that the straight tubular reactor would not fit in the available space. One engineer suggested that the reactor be cut in half and the two reactors be put side by side with equal feed to each. However, the chief engineer overrode this suggestion saying the tubular reactor had to be one piece so he bent the reactor in a U shape. The bend was not a good one. Brainstorm and make a list of things that could cause this off-design specification. Choose the most logical explanation/model, and carry out a calculation to show quantitatively that with your model the conversion is 58.6%. (*An Ans:* 57% of the total)

(c) The liquid-phase reaction

$$A \longrightarrow B$$

was carried out in a CSTR. For an entering concentration of 2 mol/dm³, the conversion was 40%. For the same reactor volume and entering conditions as the CSTR, the expected PFR conversion is 48.6%. However, the PFR conversion was amazingly 52.6% exactly. Brainstorm reasons for the disparity. Quantitatively show how these conversions came about (i.e., the expected conversion and the actual conversion).

(d) The gas-phase reaction

$$A + B \longrightarrow C + D$$

is carried out in a packed bed reactor. When the particle size was decreased by 15%, the conversion remained unchanged. When the particle size was decreased by 20%, the conversion decreased. When the original particle size was increased by 15%, the conversion also decreased. In all cases, the temperature, the total catalyst weight, and all other conditions remained unchanged. What's going on here?

P4-9$_B$ A reversible liquid-phase isomerization A $\rightleftharpoons$ B is carried out *isothermally* in a 1000-gal CSTR. The reaction is second order in both the forward and reverse directions. The liquid enters at the top of the reactor and exits at the bottom. Experimental data taken in a batch reactor shows the CSTR conversion to be 40%. The reaction is reversible with $K_C = 3.0$ at 300 K, and

$\Delta H_{Rx} = -25{,}000$ cal/mol. Assuming that the batch data taken at 300 K are accurate and that $E = 15{,}000$ cal/mol, what CSTR temperature do you recommend to obtain maximum conversion? *Hint:* Read Appendix C and assume $\Delta C_P = 0$ in the appendix Equation (C-8):

$$K_C(T) = K_C(T_0)\exp\left[\frac{\Delta H_{Rx}}{R}\left(\frac{1}{T_0} - \frac{1}{T}\right)\right]$$

Use Polymath to make a plot of X versus T. Does it go through a maximum? If so, explain why.

P4-10   The growth of bacteria to form a product, P, is carried out in a 25 dm³ CSTR (chemostat). The bacteria (e.g., *Zymononas*) consumes the nutrient substrate (e.g., to generate more cells and the desired product—ethanol)

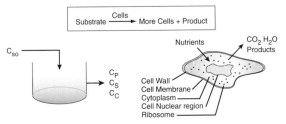

The CSTR was initially inoculated with bacteria and now has reached steady state. Only substrate (nutrient) is fed to the reactor at a volumetric rate of 5 dm³/h and a concentration of 30 g/dm³. The growth law r_g (g/hr dm³) is

$$r_g = \frac{\mu_{max}C_S C_C}{K_m + C_S}$$

and the rate of substrate consumption is related to growth rate by

$$-r_S = Y_{S/C}r_g$$

with the stoichiometric relationship

$$C_C = Y_{C/S}[C_{S0} - C_S]$$

(a) Write a mass balance on the cells and the substrate concentration in the CSTR operated at steady state.
(b) Solve the cell mass balance for the substrate concentration and calculate C_S.
(c) Calculate the cell concentration, C_C.
(d) How would your answers to (b) and (c) change if the volumetric feed rate were cut in half?
(e) How would your answers to (b) and (c) change if the CSTR volume were reduced by a factor of three?
(f) The reaction is now carried out in a 10 dm³ batch reactor with initial concentrations of substrate $C_{S0} = 30$ g/dm³ and cells of $C_{C0} = 0.1$ g/dm³. Plot C_S, C_C, r_g, and $-r_S$ as a function of time.
(g) Repeat (f) for a 100 dm³ reactor.
Additional Information:

$\mu_{max} = 0.5$hr^{-1}, $K_m = 5$ g/dm³
$Y_{C/S} = 0.8$ g cell formed/g substrate consumed $= 1/Y_{S/C}$

Try to work the California problems in 30 minutes.

P4-11$_B$ The gaseous reaction A ⟶ B has a unimolecular reaction rate constant of 0.0015 min^{-1} at 80°F. This reaction is to be carried out in *parallel tubes* 10 ft long and 1 in. inside diameter under a pressure of 132 psig at 260°F. A production rate of 1000 lb/h of B is required. Assuming an activation energy of 25,000 cal/mol, how many tubes are needed if the conversion of A is to be 90%? Assume perfect gas laws. A and B each have molecular weights of 58. (From California Professional Engineers Exam.)

P4-12$_B$ **(a)** The irreversible elementary reaction 2A ⟶ B takes place in the gas phase in an *isothermal tubular (plug-flow) reactor*. Reactant A and a diluent C are fed in equimolar ratio, and conversion of A is 80%. If the molar feed rate of A is cut in half, what is the conversion of A assuming that the feed rate of C is left unchanged? Assume ideal behavior and that the reactor temperature remains unchanged. What was the point of this problem? (From California Professional Engineers Exam.)

(b) Write a question that requires critical thinking, and explain why it involves critical thinking.

P4-13$_B$ Compound A undergoes a reversible isomerization reaction, A ⇌ B, over a supported metal catalyst. Under pertinent conditions, A and B are liquid, miscible, and of nearly identical density; the equilibrium constant for the reaction (in concentration units) is 5.8. In a *fixed-bed isothermal flow reactor* in which backmixing is negligible (i.e., plug flow), a feed of pure A undergoes a net conversion to B of 55%. The reaction is elementary. If a second, identical flow reactor at the same temperature is placed downstream from the first, what overall conversion of A would you expect if:

(a) The reactors are directly connected in series? (*Ans.: X = 0.74.*)

(b) The products from the first reactor are separated by appropriate processing and only the unconverted A is fed to the second reactor?

(From California Professional Engineers Exam.)

P4-14$_C$ A total of 2500 gal/h of metaxylene is being isomerized to a mixture of orthoxylene, metaxylene, and paraxylene in a reactor containing 1000 ft^3 of catalyst. The reaction is being carried out at 750°F and 300 psig. Under these conditions, 37% of the metaxylene fed to the reactor is isomerized. At a flow rate of 1667 gal/h, 50% of the metaxylene is isomerized at the same temperature and pressure. Energy changes are negligible.

It is now proposed that a second plant be built to process 5500 gal/h of metaxylene at the same temperature and pressure as described earlier. What size reactor (i.e., what volume of catalyst) is required if conversion in the new plant is to be 46% instead of 37%? Justify any assumptions made for the scale-up calculation. (*Ans.: 2931 ft^3 of catalyst.*) (From California Professional Engineers Exam.) Make a list of the things you learned from this problem.

P4-15$_A$ It is desired to carry out the gaseous reaction A ⟶ B in an existing *tubular reactor* consisting of 50 parallel tubes 40 ft long with a 0.75-in. inside diameter. Bench-scale experiments have given the reaction rate constant for this first-order reaction as 0.00152 s^{-1} at 200°F and 0.0740 s^{-1} at 300°F. At what temperature should the reactor be operated to give a conversion of A of 80% with a feed rate of 500 lb/h of pure A and an operating pressure of 100 psig? A has a molecular weight of 73. Departures from perfect gas behavior may be neglected, and the reverse reaction is insignificant at these conditions. (*Ans.: T = 275°F.*) (From California Professional Engineers Exam.)

P4-16$_A$ The reversible isomerization

$$m \text{ Xylene} \; \underset{\longleftarrow}{\overset{\longrightarrow}{}} \; \text{para-Xylene}$$

follows an elementary rate law. If X_e is the equilibrium conversion,

(a) Show for a batch and a PFR: $t = \tau_{PFR} = \dfrac{X_e}{k} \ln \dfrac{X_e}{X_e - X}$

(b) Show for a CSTR: $\tau_{PFR} = \dfrac{X_e}{k}\left(\dfrac{X}{X_e - X}\right)$

(c) Show that the volume efficiency is

$$\frac{V_{PFR}}{V_{CSTR}} = \frac{(X_e - X) \ln \left(\dfrac{X_e}{X_e - X}\right)}{X}$$

and then plot the volume efficiency as a function of the ratio (X/X_e) from 0 to 1.

(d) What would be the volume efficiency for two CSTRs in series with the sum of the two CSTR volumes being the same as the PFR volume?

P4-17$_B$ The gas-phase dimerization

$$2A \longrightarrow B$$

follows an elementary rate law and takes place isothermally in a PBR charged with 1.0 kg of catalyst. The feed, consisting of pure A, enters the PBR at a pressure of 20 atm. The conversion exiting the PBR is 0.3, and the pressure at the exit of the PBR is 5 atm.

(a) If the PBR were replaced by a "fluidized" CSTR with 1 kg of catalyst, what will be the conversion at the exit of the CSTR? You may assume that there is no pressure drop in the CSTR. (*Ans.:* $X = 0.4$.)

(b) What would be the conversion in the PBR if the mass flow rate were decreased by a factor of 4 and particle size were doubled? Assume turbulent flow. (*Final exam*)

(c) Discuss what you learned from this problem as well as the strengths and weaknesses of using this as a final exam problem.

(d) Write a question on critical thinking and explain why it involves critical thinking.

P4-18$_B$ The irreversible first-order (wrt partial pressure of A) gas-phase reaction

$$A \rightarrow B$$

Fluidized CSTR

is carried out isothermally in a "fluidized" catalyic CSTR containing 50 kg of catalyst (Figure in margin).

Currently 50% conversion is realized for pure A entering at pressure of 20 atm. There is virtually no pressure drop in the CSTR. It is proposed to put a PBR containing the same catalyst weight in series with the CSTR. The pressure drop parameter for the PBR α, given by Equation (4-33) is $\alpha = 0.018$ kg^{-1}. The particle size is 0.2 mm, the bed porosity is 40%, and the viscosity is the same as that of air at 200°C.

(a) Should the PBR be placed upstream or downstream of the CSTR in order to achieve the highest conversion? Explain qualitatively using concepts you learned in Chapter 2.

(b) What is the conversion exiting the last reactor?

(c) What is the pressure at the exit of the packed bed?

(d) How would your answers change if the catalyst diameter were decreased by a factor of 2 and the PBR diameter were increased by 50% assuming turbulent flow?

P4-19   A microreactor similar to the one shown in Figure P4-19 from the MIT group is used to produce phosgene in the gas phase.

$$CO + Cl_2 \rightarrow COCl_2$$

$$A + B \rightarrow C$$

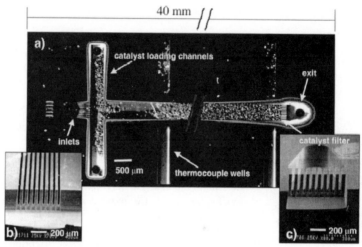

Figure P4-19   Microreactor (Courtesy of S. K. Ajmera, M. W. Losey, K. F. Jensen, and M. A. Schmidt, *AIChE J. 47,* 1639 (2001).

The microreactor is 20 mm long, 500 μm in diameter, and packed with catalyst particles 35 μm in diameter. The entering pressure is 830 kPa (8.2 atm), and the entering flow to each microreactor is equimolar. The molar flow rate of CO is 2×10^{-5} mol/s and the volumetric flow is 2.83×10^{-7} m³/s. The weight of catalyst in one microreactor: $W = 3.5 \times 10^{-6}$ kg. The reactor is kept isothermal at 120°C. Because the catalyst is also slightly different than the one in Figure P4-19, the rate law is different as well:

$$-r'_A = k_A C_A C_B$$

(a) Plot the molar flow rates F_A, F_B, and F_C, the conversion X, and pressure ratio y along the length (i.e., catalyst weight, W) of the reactor.

(b) Calculate the number of microreactors in parallel to produce 10,000 kg/year phosgene.

(c) Repeat part (a) for the case when the catalyst weight remains the same but the particle diameter is cut in half. If possible compare your answer with part (a) and describe what you find, noting anything unusual.

(d) How would your answers to part (a) change if the reaction were reversible with $K_C = 0.4$ dm³/mol? Describe what you find.

(e) What are the advantages and disadvantages of using an array of micro-reactors over using one conventional packed bed reactor that provides the same yield and conversion?

(f) Write a question that involves critical thinking, and explain why it involves critical thinking.

(g) Discuss what you learned from this problem and what you believe to be the point of the problem.

Additional information:

$\alpha = 3.55 \times 10^5$/kg catalyst (based on properties of air and $\phi = 0.4$)

$k = 0.004$ m^6/mol $\cdot$ s $\cdot$ kg catalyst at 120°C

$\upsilon_0 = 2.83 \cdot 10^{-7}$ m^3/s, $\rho = 7$ kg/m^3, $\mu = 1.94 \cdot 10^{-5}$ kg/m $\cdot$ s,

$A_c = 1.96 \cdot 10^{-7}$ m^2, $G = 10.1$ kg/m^2 $\cdot$ s

P4-20$_C$ The elementary gas-phase reaction

$$A + B \longrightarrow C + D$$

is carried out in a packed-bed reactor. Currently, catalyst particles 1 mm in diameter are packed into 4-in. schedule 40 pipe ($A_C = 0.82126$ dm^2). The value of β_0 in the pressure drop equation is 0.001 atm/dm. A stoichiometric mixture of A and B enters the reactor at a total molar flow rate of 10 mol/min, a temperature of 590 K, and a pressure of 20 atm. Flow is turbulent throughout the bed. Currently, only 12% conversion is achieved with 100 kg of catalyst.

Member

Hall of Fame

It is suggested that conversion could be increased by changing the catalyst particle diameter. Use the following data to correlate the specific reaction rate as a function of particle diameter. Then use this correlation to determine the catalyst size that gives the highest conversion. As you will see in Chapter 12, k' for first-order reaction is expected to vary according to the following relationship

$$k' = \eta k = \frac{3}{\Phi^2}(\Phi \coth\Phi - 1) \, k \qquad (P4\text{-}20.1)$$

where Φ varies directly with particle diameter, $\Phi = cD_p$. Although the reaction is not first order, one notes from Figure 12-5 the functionality for a second-order reaction is similar to Equation (P4-20.1).

(a) Show that when the flow is turbulent

$$\alpha(D_P) = \alpha_0\left(\frac{D_{P0}}{D_P}\right)$$

and that $\alpha_0 = 0.8 \times 10^{-4}$ atm/kg and also show that $c = 75$ min^{-1}.

(b) Plot the specific reaction rate k' as a function of D_P, and compare with Figure 12-5.

(c) Make a plot of conversion as a function of catalyst size.

(d) Discuss how your answer would change if you had used the effectiveness factor for a second-order reaction rather than a first-order reaction.

(e) How would your answer to (b) change if both the particle diameter and pipe diameter were increased by 50% when
  (1) the flow is laminar.
  (2) the flow is turbulent.

(f) Write a few sentences describing and explaining what would happen if the pressure drop parameter α is varied.

(g) What generalizations can you make about what you learned in this problem that would apply to other problems?

(h) Discuss what you learned from this problem and what you believe to be the point of the problem.

Additional information:
Void fraction = 0.35   Solid catalyst density = 2.35 kg/dm³
    Bulk density: $\rho_B = (1 - \phi) \rho_C = (0.35)(2.35) = 0.822$

Catalyst Diameter, d_p (mm)	2	1	0.4	0.1	0.02	0.002
k' (dm⁶/mol · min · kg cat)	0.06	0.12	0.30	1.2	2.64	3.00

[*Hint:* You could use Equation (P4.20-1), which would include D_P and an unknown proportionality constant that you could evaluate from the data. For very small values of the Thiele modulus we know $\eta = 1$, and for very large values of the Thiele modulus we know that $\eta = 3/\Phi = 3/cD_p$.]

P4-21$_A$ Nutrition is an important part of ready-to-eat cereal. To make cereal healthier, many nutrients are added. Unfortunately, nutrients degrade over time, making it necessary to add more than the declared amount to assure enough for the life of the cereal. Vitamin V_1 is declared at a level of 20% of the Recommended Daily Allowance per serving size (serving size = 30 g). The Recommended Daily Allowance is 6500 IU (1.7×10^6 IU = 1 g). It has been found that the degradation of this nutrient is first order in the amount of nutrients. Accelerated storage tests have been conducted on this cereal, with the following results:

Temperature (°C)	45	55	65
k (week^{-1})	0.0061	0.0097	0.0185

(a) Given this information and the fact that the cereal needs to have a vitamin level above the declared value of 6500 IU for 1 year at 25°C, what IU should be present in the cereal at the time it is manufactured? Your answer may also be reported in percent overuse: (*Ans.* 12%)

$$\%OU = \frac{C\,(t=0) - C\,(t=1\text{ yr})}{C\,(t=1\text{ yr})} \times 100$$

(b) At what percent of declared value of 6500 IU must you apply the vitamin? If 10,000,000 lb/yr of the cereal is made and the nutrient cost is $100 per pound, how much will this overuse cost?

(c) If this were your factory, what percent overuse would you actually apply and why?

(d) How would your answers change if you stored the material in a Bangkok warehouse for 6 months, where the daily temperature is 40°C, before moving it to the supermarket? (Table of results of accelerated storage tests on cereal; and Problem of vitamin level of cereal after storage courtesy of General Mills, Minneapolis, MN.)

4-22$_A$ A very proprietary industrial waste reaction, which we'll code as A→B+S is to be carried out in a 10-dm³ CSTR followed by 10-dm³ PFR. The reaction is elementary, but A, which enters at a concentration of 0.001 mol/dm³ and a molar flow rate of 20 mol/min, has trouble decomposing. The specific reaction rate at 42°C (i.e., room temperature in the Mojave desert) is 0.0001 s⁻¹.

However, we don't know the activation energy; therefore, we cannot carry out this reaction in the winter in Michigan. Consequently this reaction, while important, is not worth your time to study. Therefore, perhaps you want to take a break and go watch a movie such as *Dances with Wolves* (a favorite of the author), *Bride and Prejudice*, or *Finding Neverland.*

P4-23$_B$ The production of ethylene glycol from ethylene chlorohydrin and sodium bicarbonate

$$CH_2OHCH_2Cl + NaHCO_3 \rightarrow (CH_2OH)_2 + NaCl + CO_2$$

is carried out in a semibatch reactor. A 1.5 molar solution of ethylene chlorohydrin is fed at a rate 0.1 mole/minute to 1500 dm^3 of a 0.75 molar solution of sodium bicarbonate. The reaction is elementary and carried out isothermally at 30°C where the specific reaction rate is 5.1 dm^3/mol·h. Higher temperatures produce unwanted side reactions. The reactor can hold a maximum of 2500 dm^3 of liquid. Assume constant density.

(a) Plot the conversion, reaction rate, concentration of reactants and products, and number of moles of glycol formed as a function of time.

(b) Suppose you could vary the flow rate between 0.01 and 2 mol/min, what flow a rate would and holding time you choose to make the greatest number of moles of ethylene glycol in 24 hours keeping in mind the downtimes for cleaning, filling, etc., shown in Table 4-1.

(c) Suppose the ethylene chlorohydrin is fed at a rate of 0.15 mol/min until the reactor is full and then shut in. Plot the conversion as a function of time.

(d) Discuss what you learned from this problem and what you believe to be the point of this problem.

P4-24$_C$ The following reaction is to be carried out in the liquid phase

$$NaOH + CH_3COOC_2H_5 \longrightarrow CH_3COO^- Na^+ + C_2H_5OH$$

The initial concentrations are 0.2 *M* in NaOH and 0.25 *M* in CH$_3$COOC$_2$H$_5$ with $k = 5.2 \times 10^{-5}$ m^3/mol·s at 20°C with $E = 42,810$ J/mol. Design a set of operating conditions to produce 200 mol/day of ethanol in a semibatch reactor and not operate above 35°C and below a concentration of NaOH of 0.02 molar.[12] The semibatch reactor you have available is 1.5 m in diameter and 2.5 m tall.

P4-25$_C$ *(Membrane reactor)* The first-order, reversible reaction

$$A \rightleftharpoons B + 2C$$

is taking place in a membrane reactor. Pure A enters the reactor, and B diffuses through the membrane. Unfortunately, some of the reactant A also diffuses through the membrane.

(a) Plot the flow rates of A, B, and C down the reactor, as well as the flow rates of A and B through the membrane.

(b) Compare the conversion profiles of a conventional PFR with those of an IMRCF. What generalizations can you make?

---

[12] Manual of Chemical Engineering Laboratory, University of Nancy, Nancy, France, 1994. *eric@ist.uni-stuttgart.de. www.sysbio.del/AICHE*

(c) Would the conversion of A be greater or smaller if C were difusing out instead of B?

(d) Discuss how your curves would change if the temperature were increased significantly or decreased significantly for an exothermic reaction and for an endothermic reaction.

Additional information:

$$k = 10 \text{ min}^{-1} \qquad F_{A0} = 100 \text{ mol/min}$$
$$K_C = 0.01 \text{ mol}^2/\text{dm}^6 \qquad v_0 = 100 \text{ dm}^3/\text{min}$$
$$k_{CA} = 1 \text{ min}^{-1} \qquad V_{\text{reactor}} = 20 \text{ dm}^3$$
$$k_{CB} = 40 \text{ min}^{-1}$$

P4-26$_B$ As we move toward a hydrogen-based energy economy for use in fuel cells. The use of fuel cells to operate appliances ranging from computers to automobiles is rapidly becoming a reality. In the immediate future, fuel cells will use hydrogen to produce electricity, which some have said will lead to a hydrogen-based economy instead of a petroleum-based economy. A large component in the processing train for fuel cells is the water gas shift membrane reactor. (M. Gummala, N. Gupla, B. Olsomer, and Z. Dardas, **Paper 103c**, 2003, AIChE National Meeting, New Orleans, LA.)

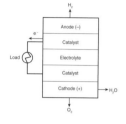

$$CO + H_2O \xrightleftharpoons{\quad} CO_2 + H_2$$

Here CO and water are fed to the membrane reactor containing the catalyst. Hydrogen can diffuse out the sides of the membrane while CO, H_2O, and CO_2 cannot. Based on the following information, plot the concentrations and molar flow rates of each of the reacting species down the length of the membrane reactor. Assume the following. The volumetric feed is 10 dm^3/min at 10 atm, and the equil molar feed of CO and water vapor with $C_{T0} = 0.4$ mol/dm³. The equilibrium constant is $K_e = 1.44$. The k specific reaction rate constant is 1.37 dm^6/mol kg cat · min, and the mass transfer coefficient for hydrogen, $k_{CH_2} = 0.1$ dm³/kg cat · min. What is the reactor volume necessary to achieve 85% conversion of CO? Compare with a PFR. For that same reactor volume, what would be the conversion if the feed rate were doubled?

P4-27 Go to Professor Herz's **Reactor Lab** on the CD-ROM or on the web at *www.reactorlab.net.* Load Division 2, Lab 2 of The Reactor Lab concerning a packed-bed reactor (labeled PFR) in which a gas with the physical properties of air flows over spherical catalyst pellets. Perform experiments here to get a feeling for how pressure drop varies with input parameters such as reactor diameter, pellet diameter, gas flow rate, and temperature. In order to get significant pressure drop, you may need to change some of the input values substantially from those shown when you enter the lab. If you get a notice that you can't get the desired flow, then you need to increase the inlet pressure. In Chapters 10–12, you will learn how to analyze the conversion results in such a reactor.

Good Alternatives on the CD and on the Web

The following problems are either similar to the ones already presented but use different reactions or have a number of figures that would require a lot of text space. Consequently, the full problem statements are on the CD-ROM.

P4-28 Pressure drop in a PBR with a first-order reaction using real data: **What if** questions asked. [3rd Ed. P4-18]

P4-29 *Good Troubleshooting Problem.* Inspector Sergeant Ambercromby investigates possible fraud at Worthless Chemical. [3rd Ed. P4-9]

P4-30 The first-order reaction

$$C_6H_5CH(CH_3)_2 \longrightarrow C_6H_6 + C_3H_6$$

is to be carried out in a packed bed reactor with pressure drop where the rate constant varies inversely with $k \sim (1/D_P)$. One can also choose from various pipe sizes to get the maximum conversion. Similar to Problems P4-22 and P4-23. [3rd Ed. P4-20]

P4-31 Pressure drop in a packed bed reactor to make alkylated cyclohexanols. [3rd Ed. P4-22]

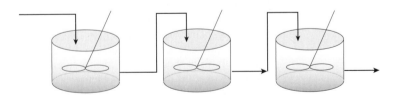

Figure P4-31

P4-32 A semibatch reactor is used to carry out the reaction

$$CH_3COOC_2H_5 + C_4H_9 \rightleftharpoons CH_3COOC_4H_9 + C_2H_5OH$$

Similar to problems 4-26 and 4-27. [3rd Ed. P4-26]

P4-33 A CSTR with two impellers is modeled as three CSTRs in series. [3rd Ed. p4-29]

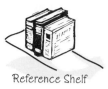

SOME THOUGHTS ON CRITIQUING WHAT YOU READ

Your textbooks after your graduation will be, in part, the professional journals that you read. As you read the journals, it is important that you study them with a critical eye. You need to learn if the author's conclusion is supported by the data, if the article is new or novel, if it advances our understanding, and to learn if the analysis is current. To develop this technique, one of the major assignments used in the graduate course in chemical reaction engineering at the University of Michigan for the past 25 years has been an in-depth analysis and critique of a journal article related to the course material. Significant effort is made to ensure that a cursory or superficial review is not carried out. Students are asked to analyze and critique ideas rather than ask questions such as: Was the pressure measured accurately? They have been told that they are not required to find an error or inconsistency in the article to receive a good grade, but if they do find such things, it just makes the assignment that much more enjoyable. Beginning with Chapter 4, a number of the problems at the end of each chapter in this book are based on students' analyses and critiques of journal articles and are designated with a C (e.g., P4C-1). These problems involve the analysis of journal articles that may have minor or major inconsistencies. A discussion on critiquing journal articles can be found in Professional Reference Shelf R4.4 on the CD-ROM.

JOURNAL CRITIQUE PROBLEMS

P4C-1   In *Water Research, 33* (9), 2130 (1999), is there a disparity in the rate law obtained by batch experiments and continuous flow experiments?

P4C-2   In the article describing the liquid reaction of isoprene and maleic anhydride under pressure [*AIChE J., 16*(5), 766 (1970)], the authors show the reaction rate to be greatly accelerated by the application of pressure. For an equimolar feed they write the second-order reaction rate expression in terms of the mole fraction y:

$$\frac{dy}{dt} = -k_y y^2$$

and then show the effect of pressure on k_y (s^{-1}). Derive this expression from first principles and suggest a possible logical explanation for the increase in the true specific reaction rate constant k ($dm^3/mol \cdot s$) with pressure that is different from the author's. Make a quick check to verify your challenge.

P4C-3   The reduction of NO by char was carried out in a fixed bed between 500 and 845°C [*Int. Chem. Eng., 20*(2), 239 (1980)]. It was concluded that the reaction is first order with respect to the concentration of NO feed (300 to 1000 ppm) over the temperature range studied. It was also found that activation energy begins to increase at about 680°C. Is first order the true reaction order? If there were discrepancies in this article, what might be the reasons for them?

P4C-4   In the article describing vapor phase esterification of acetic acid with ethanol to form ethyl acetate and water [*Ind. Eng. Chem. Res., 26*(2), 198 (1987)], the pressure drop in the reactor was accounted for in a most unusual manner [i.e., $P = P_0(1 - fX)$, where f is a constant].

(a) Using the Ergun equation along with estimating some of the parameter values (e.g., $\phi = 0.4$), calculate the value of α in the packed-bed reactor (2 cm i.d. by 67 cm long).

(b) Using the value of α, redo part (a) accounting for pressure drop along the lines described in this chapter.

(c) Finally, if possible, estimate the value of f used in these equations.

- **Additional Homework Problems**

Solved Problems

CDP4-A$_B$ A sinister looking gentlemen is interested in producing methyl perchlorate in a batch reactor. The reactor has a strange and unsettling rate law. [2nd Ed. P4-28] (Solution Included)

Bioreactors and Reactions

CDP4-B$_C$ (*Ecological Engineering*) A much more complicated version of Problem 4-17 uses actual pond (CSTR) sizes and flow rates in modeling the site with CSTRs for the Des Plaines River experimental wetlands site (EW3) in order to degrade atrazine. [See Web Module on CD or WWW]

CDP4-C$_B$ The rate of binding ligands to receptors is studied in this application of reaction kinetics to *bioengineering*. The time to bind 50% of the ligands to the receptors is required. [2nd Ed. P4-34] J. Lindemann, University of Michigan

Batch Reactors

CDP4-D$_B$ A *batch reactor* is used for the bromination of *p*-chlorophenyl isopropyl ether. Calculate the batch reaction time. [2nd Ed. P4-29]

CDP4-E$_B$ California Professional Engineers Exam Problem, in which the reaction

$$B + H_2 \longrightarrow A$$

is carried out in a batch reactor. [2nd Ed. P4-15]

CDP4-F$_A$ Verify that the liquid-phase reaction of 5,6-benzoquinoline with hydrogen is psuedo first order. [2nd Ed. P4-7]

Flow Reactors

CDP4-G$_B$ Radial flow reactors can be used to good advantage for exothermic reactions with large heats of reaction. The radical velocity is

$$U = \frac{U_0 R_0}{r} (1 + \varepsilon X) \frac{P_0}{P} \frac{T}{T_0}$$

Vary the parameters and plot X as a function of r. [2nd Ed. P4-31]

CDP4-H$_A$ Designed to reinforce the basic CRE principles through very straightforward calculations of CSTR and PFR volumes and batch reactor time. This problem was one of the most frequently assigned problems from the 2nd Edition. [2nd Ed. P4-4]

CDP4-I$_B$ Formation of diphenyl in a batch, CSTR, and PFR. [3rd Ed. P4-10]

Packed Bed Reactors

CDP4-J$_B$ n-Pentane $\longrightarrow$ i-pentane in a packed bed reactor. [3rd Ed. P4-21]
CDP4-K$_C$ Packed bed spherical reactor. [3rd Ed. P4-20]
CDP4-L$_B$ The reaction of A $\rightleftarrows$ B is carried out in a membrane reactor where B diffuses out.

Recycle Reactors

CDP4-M$_B$ The overall conversion is required in a *packed-bed reactor with recycle*. [2nd Ed. P4-22]
CDP4-N$_C$ Excellent reversible reaction with recycle. Good problem by Professor H.S. Shankar, IIT—Bombay. [3rd Ed. P4-28]

Really Difficult Problems

CDP-O$_C$ DQE April 1999
A $\longrightarrow$ B in a PFR and CSTR with unknown order. (30 minutes to solve)
CDP-P$_D$ A Dr. Probjot Singh Problem
A $\longrightarrow$ B $\rightleftarrows$ C Species C starts and ends at the same concentration.

Green Engineering   ## New Problems on the Web

CDP4-New   From time to time new problems relating Chapter 4 material to everyday interests or emerging technologies will be placed on the web. Solutions to these problems can be obtained by e-mailing the author. Also, one can go on the web site, *www.rowan.edu/greenengineering*, and work the home problem specific to this chapter.

These Problems Were on CD-ROM for 3rd Edition but Not in Book for 3rd Edition

CDP4-Q$_A$ The gas-phase reaction $A + 2B \longrightarrow 2D$ has the rate law $-r_A = 2.5 C_A^{0.5} C_B$. Reactor volumes of PFRs and CSTRs are required in this multipart problem. [2nd Ed. P4-8]
CDP4-R$_B$ What type and arrangement of flow reactors should you use for a decomposition reaction with the rate law $-r_A = k_1 C_A^{0.5}/(1 + k_2 C_A)$? [1st Ed. P4-14]
CDP4-S$_B$ The liquid-phase reaction $2A + B \Leftrightarrow C + D$ is carried out in a semibatch reactor. Plot the conversion, volume, and species concentrations as a function of time. Reactive distillation is also considered in part (e). [2nd Ed. P4-27]
CDP4-T$_B$ The growth of a bacterium is to be carried out in excess nutrient.

$$\text{Nutrient} + \text{Cells} \rightarrow \text{More cells} + \text{Product}$$

The growth rate law is $r_B = \mu_m C_B\left(1 - \dfrac{C_B}{C_{B\max}}\right)$ [2nd Ed. P4-35]

CDP4-U$_B$ California Registration Examination Problem. Second-order reaction in different CSTR and PFR arrangements. [2nd Ed. P4-11]

CDP4-V$_B$ An unremarkable semibatch reactor problem, but it does require assessing which equation to use.

SUPPLEMENTARY READING

DAVIS, M. E., and R. J. DAVIS, *Fundamentals of Chemical Reaction Engineering*. New York: McGraw Hill, 2003.

HILL, C. G., *An Introduction to Chemical Engineering Kinetics and Reactor Design*. New York: Wiley, 1977, Chap. 8.

LEVENSPIEL, O., *Chemical Reaction Engineering*, 3rd ed. New York: Wiley, 1998, Chaps. 4 and 5.

SMITH, J. M., *Chemical Engineering Kinetics*, 3rd ed. New York: McGraw-Hill, 1981.

ULRICH, G. D., *A Guide to Chemical Engineering Reactor Design and Kinetics*. Printed and bound by Braun-Brumfield, Inc., Ann Arbor, Mich., 1993.

WALAS, S. M., *Reaction Kinetics for Chemical Engineers*. New York: McGraw-Hill, 1970.

Recent information on reactor design can usually be found in the following journals: *Chemical Engineering Science, Chemical Engineering Communications, Industrial and Engineering Chemistry Research, Canadian Journal of Chemical Engineering, AIChE Journal, Chemical Engineering Progress.*

Collection and Analysis of Rate Data 5

You can observe a lot just by watching.

Yogi Berra, New York Yankees

Overview. In Chapter 4 we have shown that once the rate law is known, it can be substituted into the appropriate design equation, and through the use of the appropriate stoichiometric relationships, we can apply the **CRE** algorithm to size any isothermal reaction system. In this chapter we focus on ways of obtaining and analyzing reaction rate data to obtain the rate law for a specific reaction. In particular, we discuss two common types of reactors for obtaining rate data: the batch reactor, which is used primarily for homogeneous reactions, and the differential reactor, which is used for solid–fluid heterogeneous reactions. In batch reactor experiments, concentration, pressure, and/or volume are usually measured and recorded at different times during the course of the reaction. Data are collected from the batch reactor during transient operation, whereas measurements on the differential reactor are made during steady-state operation. In experiments with a differential reactor, the product concentration is usually monitored for different sets of feed conditions.

Two techniques of data acquisition are presented: concentration-time measurements in a batch reactor and concentration measurements in a differential reactor. Six different methods of analyzing the data collected are used: the differential method, the integral method, the method of half-lives, method of initial rates, and linear and nonlinear regression (least-squares analysis). The differential and integral methods are used primarily in analyzing batch reactor data. Because a number of software packages (e.g., Polymath, MATLAB) are now available to analyze data, a rather extensive discussion of nonlinear regression is included. We close the chapter with a discussion of experimental planning and of laboratory reactors (CD-ROM).

5.1 The Algorithm for Data Analysis

For batch systems, the usual procedure is to collect concentration time data, which we then use to determine the rate law. Table 5-1 gives the procedure we will emphasize in analyzing reaction engineering data.

Data for homogeneous reactions is most often obtained in a batch reactor. After postulating a rate law and combining it with a mole balance, we next use any or all of the methods in Step 5 to process the data and arrive at the reaction orders and specific reaction rate constants.

Analysis of heterogeneous reactions is shown in Step 6. For gas–solid heterogeneous reactions, we need to have an understanding of the reaction and possible mechanisms in order to postulate the rate law in Step 6B. After studying Chapter 10 on heterogeneous reactions, one will be able to postulate different rate laws and then use Polymath nonlinear regression to choose the "best" rate law and reaction rate parameters.

The procedure we should use to delineate the rate law and rate law parameter is given in Table 5-1.

TABLE 5-1. STEPS IN ANALYZING RATE DATA

1. **Postulate a rate law.**
   A. Power law models for homogeneous reactions

   $$-r_A = kC_A^\alpha \,, \quad -r_A = kC_A^\alpha \, C_B^\beta$$

   B. Langmuir-Hinshelwood models for heterogeneous reactions

   $$-r_A' = \frac{kP_A}{1 + K_A P_A} \,, \quad -r_A' = \frac{kP_A P_B}{(1 + K_A P_A + P_B)^2}$$

2. **Select reactor type and corresponding mole balance.**
   A. If batch reactor (Section 5.2), use mole balance on Reactant A

   $$-r_A = -\frac{dC_A}{dt} \qquad\qquad (TE5\text{-}1.1)$$

   B. If differential PBR (Section 5.5), use mole balance on Product P (A $\rightarrow$ P)

   $$-r_A' = \frac{F_P}{\Delta W} = C_p v_0 / \Delta W \qquad\qquad (TE5\text{-}1.2)$$

3. **Process your data in terms of measured variable** (e.g., N_A, C_A, or P_A). If necessary, rewrite your mole balance in terms of the measured variable (e.g., P_A).
4. **Look for simplifications.** For example, if one of the reactants in excess, assume its concentration is constant. If the gas phase mole fraction of reactant is small, set $\varepsilon \approx 0$.
5. **For a batch reactor, calculate $-r_A$ as a function of concentration C_A to determine reaction order.**
   A. Differential analysis
       Combine the mole balance (TE5-1.1) and power law model (TE5-1.3).

   $$-r_A = kC_A^\alpha \qquad\qquad (TE5\text{-}1.3)$$

TABLE 5-1.   STEPS IN ANALYZING RATE DATA (CONTINUED)

$$-\frac{dC_A}{dt} = kC_A^{\alpha} \tag{TE5-1.4}$$

and then take the natural log.

$$\ln\left(-\frac{dC_A}{dt}\right) = \ln(-r_A) = \ln k + \alpha \ln C_A \tag{TE5-1.5}$$

(1) Find $-\dfrac{dC_A}{dt}$ from C_A versus t data by

   (a) Graphical method
   (b) Finite differential method
   (c) Polynominal

(2) Plot $\ln -\dfrac{dC_A}{dt}$ versus $\ln C_A$ and find reaction order α, which is the slope of

   the line fit to the data.
(3) Find k.

B.   Integral method
   For $-r_A = kC_A^{\alpha}$, the combined mole balance and rate law is

$$-\frac{dC_A}{dt} = kC_A^{\alpha} \tag{TE5-1.4}$$

(1) Guess α and integrate Equation (TE5-1.4). Rearrange your equation to obtain the appropriate function of C_A, which when plotted as a function of time should be linear. If it is linear, then the guessed value of α is correct and the slope is specific reaction rate, k. If it is not linear, guess again for α. If you guess $\alpha = 0$, 1, and 2 and none of theses orders fit the data, proceed to nonlinear reression.

(2) Nonlinear regression (Polymath)
   Integrate Equation (TE5-1.4) to obtain

$$t = \frac{1}{k}\left[\frac{C_{A0}^{(1-\alpha)} - C_A^{(1-\alpha)}}{(1-\alpha)}\right] \text{ for } \alpha \neq 1 \tag{TE5-1.6}$$

   Use Polymath regression to find α and k. A Polymath tutorial on regression with screen shots is shown in the Chapter 5 *Summary Notes* on the CD-ROM and web.

6. For differential PBR calculate $-r'_A$ as a function of C_A or P_A

Summary Notes

A.   Calculate $-r'_A = \dfrac{v_0 C_P}{\Delta W}$ as a function of reactant concentration, C_A.

B.   Choose model (see Chapter 10), e.g.,

$$-r'_A = \frac{kP_A}{1 + K_A P_A}$$

C.   Use nonlinear regression to find the best model and model parameters. See example on the CD-ROM *Summary Notes* using data form heterogeneous catalysis, Chapter 10.

7. Analyze your rate law model for "goodness of fit." Calculate a correlation coefficient.

5.2 Batch Reactor Data

Batch reactors are used primarily to determine rate law parameters for homogeneous reactions. This determination is usually achieved by measuring concentration as a function of time and then using either the differential, integral, or nonlinear regression method of data analysis to determine the reaction order, α, and specific reaction rate constant, k. If some reaction parameter other than concentration is monitored, such as pressure, the mole balance must be rewritten in terms of the measured variable (e.g., pressure as shown in the example in *Solved Problems* on the CD).

Process data in terms of the measured variable

When a reaction is *irreversible*, it is possible in many cases to determine the reaction order α and the specific rate constant by either nonlinear regression or by numerically differentiating *concentration versus time data*. This latter method is most applicable when reaction conditions are such that the rate is essentially a function of the concentration of only one reactant; for example, if, for the decomposition reaction,

$$A \rightarrow \text{Products}$$

Assume that the rate law is of the form $-r_A = k_A C_A^\alpha$

$$-r_A = k_A C_A^\alpha \tag{5-1}$$

then the differential method may be used.

However, by utilizing the method of excess, it is also possible to determine the relationship between $-r_A$ and the concentration of other reactants. That is, for the irreversible reaction

$$A + B \rightarrow \text{Products}$$

with the rate law

$$-r_A = k_A C_A^\alpha C_B^\beta \tag{5-2}$$

where α and β are both unknown, the reaction could first be run in an excess of B so that C_B remains essentially unchanged during the course of the reaction and

$$-r_A = k' C_A^\alpha \tag{5-3}$$

where

Method of excess

$$k' = k_A C_B^\beta \approx k_A C_{B0}^\beta$$

After determining α, the reaction is carried out in an excess of A, for which the rate law is approximated as

$$-r_A = k'' C_B^\beta \tag{5-4}$$

where $k'' = k_A C_A^\alpha \approx k_A C_{A0}^\alpha$

Once α and β are determined, k_A can be calculated from the measurement of $-r_A$ at known concentrations of A and B:

$$k_A = \frac{-r_A}{C_A^\alpha C_B^\beta} = (dm^3/mol)^{\alpha+\beta-1}/s \tag{5-5}$$

Both α and β can be determined by using the method of excess, coupled with a differential analysis of data for batch systems.

5.2.1 Differential Method of Analysis

To outline the procedure used in the differential method of analysis, we consider a reaction carried out isothermally in a constant-volume batch reactor and the concentration recorded as a function of time. *By combining the mole balance with the rate law given by Equation (5-1), we obtain*

Constant-volume
batch reactor

$$-\frac{dC_A}{dt} = k_A C_A^\alpha \tag{5-6}$$

After taking the natural logarithm of both sides of Equation (5-6),

$$\boxed{\ln\left(-\frac{dC_A}{dt}\right) = \ln k_A + \alpha \ln C_A} \tag{5-7}$$

observe that the slope of a plot of $\ln(-dC_A/dt)$ as a function of $(\ln C_A)$ is the reaction order, α (Figure 5-1).

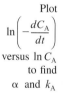

Plot
$$\ln\left(-\frac{dC_A}{dt}\right)$$
versus $\ln C_A$
to find
α and k_A

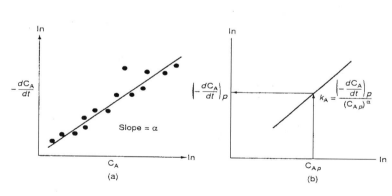

Figure 5-1   Differential method to determine reaction order.

Figure 5-1(a) shows a plot of $[-(dC_A/dt)]$ versus $[C_A]$ on log-log paper (or use Excel to make the plot) where the slope is equal to the reaction order α. The specific reaction rate, k_A, can be found by first choosing a concentration in the plot, say C_{Ap}, and then finding the corresponding value of $[-(dC_A/dt)]$ as shown in Figure 5-1(b). After raising C_{Ap} to the α power, we divide it into $[-(dC_A/dt)_p]$ to determine k_A:

$$k_A = \frac{-(dC_A/dt)_p}{(C_{Ap})^\alpha}$$

To obtain the derivative $-dC_A/dt$ used in this plot, we must differentiate the concentration-time data either numerically or graphically. We describe three methods to determine the derivative from data giving the concentration as a function of time. These methods are:

Methods for finding $-\dfrac{dC_A}{dt}$ from concentration-time data

- Graphical differentiation
- Numerical differentiation formulas
- Differentiation of a polynomial fit to the data

5.2.1A Graphical Method

With this method, disparities in the data are easily seen. Consequently, it is advantageous to use this technique to analyze the data before planning the next set of experiments. As explained in Appendix A.2, the graphical method involves plotting $-\Delta C_A/\Delta t$ as a function of t and then using equal-area differentiation to obtain $-dC_A/dt$. An illustrative example is also given in Appendix A.2.

See Appendix A.2.

In addition to the graphical technique used to differentiate the data, two other methods are commonly used: differentiation formulas and polynomial fitting.

5.2.1B Numerical Method

Numerical differentiation formulas can be used when the data points in the independent variable are *equally spaced*, such as $t_1 - t_0 = t_2 - t_1 = \Delta t$:

Time (min)	t_0	t_1	t_2	t_3	t_4	t_5
Concentration (mol/dm³)	C_{A0}	C_{A1}	C_{A2}	C_{A3}	C_{A4}	C_{A5}

The three-point differentiation formulas[1]

Initial point:

$$\left(\frac{dC_A}{dt}\right)_{t_0} = \frac{-3C_{A0} + 4C_{A1} - C_{A2}}{2\Delta t} \tag{5-8}$$

---

[1] B. Carnahan, H. A. Luther, and J. O. Wilkes, *Applied Numerical Methods* (New York: Wiley, 1969), p. 129.

Interior points:
$$\left(\frac{dC_A}{dt}\right)_{t_i} = \frac{1}{2\Delta t}[(C_{A(i+1)} - C_{A(i-1)})]$$
(5-9)

$$\left[\text{e.g.,}\ \left(\frac{dC_A}{dt}\right)_{t_3} = \frac{1}{2\Delta t}[C_{A4} - C_{A2}]\right]$$

Last point:
$$\left(\frac{dC_A}{dt}\right)_{t_5} = \frac{1}{2\Delta t}[C_{A3} - 4C_{A4} + 3C_{A5}]$$
(5-10)

can be used to calculate dC_A/dt. Equations (5-8) and (5-10) are used for the first and last data points, respectively, while Equation (5-9) is used for all intermediate data points.

5.2.1C Polynomial Fit

Another technique to differentiate the data is to first fit the concentration-time data to an nth-order polynomial:

$$C_A = a_0 + a_1 t + a_2 t^2 + \cdots + a_n t^n$$
(5-11)

Many personal computer software packages contain programs that will calculate the best values for the constants a_i. One has only to enter the concentration-time data and choose the order of the polynomial. After determining the constants, a_i, one has only to differentiate Equation (5-11) with respect to time:

$$\frac{dC_A}{dt} = a_1 + 2a_2 t + 3a_3 t^2 + \cdots + na_n t^{n-1}$$
(5-12)

Thus concentration and the time rate of change of concentration are both known at any time t.

Care must be taken in choosing the order of the polynomial. If the order is too low, the polynomial fit will not capture the trends in the data and not go through many of the points. If too large an order is chosen, the fitted curve can have peaks and valleys as it goes through most all of the data points, thereby producing significant errors when the derivatives, dC_A/dt, are generated at the various points. An example of this higher order fit is shown in Figure 5-2, where the same concentration–time data fit to a third-order polynomial (a) and to a fifth-order polynomial (b). Observe how the derivative for the fifth order changes from a positive value at 15 minutes to a negative value at 20 minutes.

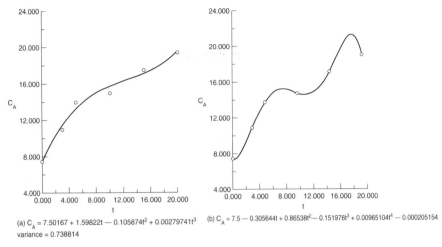

(a) $C_A = 7.50167 + 1.59822t - 0.105874t^2 + 0.00279741t^3$ (b) $C_A = 7.5 - 0.305644t + 0.86538t^2 - 0.151976t^3 + 0.00965104t^4 - 0.000205154$
variance = 0.738814

Figure 5-2   Polynomial fit of concentration–time data.

5.2.1D Finding the Rate Law Parameters

Now, using either the graphical method, differentiation formulas, or the polynomial derivative, the following table can be set up:

Time	t_0	t_1	t_2	t_3
Concentration	C_{A0}	C_{A1}	C_{A2}	C_{A3}
Derivative	$\left(-\dfrac{dC_A}{dt}\right)_0$	$\left(-\dfrac{dC_A}{dt}\right)_1$	$\left(-\dfrac{dC_A}{dt}\right)_2$	$\left(-\dfrac{dC_A}{dt}\right)_3$

The reaction order can now be found from a plot of $\ln(-dC_A/dt)$ as a function of $\ln C_A$, as shown in Figure 5-1(a), since

$$\ln\left(-\frac{dC_A}{dt}\right) = \ln k_A + \alpha \ln C_A \tag{5-7}$$

Before solving an example problem review the steps to determine the reaction rate law from a set of data points (Table 5-1).

Example 5–1 Determining the Rate Law

The reaction of triphenyl methyl chloride (trityl) (A) and methanol (B)

$$(C_6H_5)_3CCl + CH_3OH \rightarrow (C_6H_5)_3\overset{\overset{\displaystyle O}{\displaystyle \|}}{C}CH_3 + HCl$$

$$A \quad + \quad B \quad \rightarrow \qquad C \qquad + \ D$$

was carried out in a solution of benzene and pyridine at 25°C. Pyridine reacts with HCl that then precipitates as pyridine hydrochloride thereby making the reaction irreversible.

The concentration-time data in Table E5-1.1 was obtained in a batch reactor

TABLE E5-1.1.   RAW DATA

Time (min)	0	50	100	150	200	250	300
Concentration of A (mol/dm³) × 10³	50	38	30.6	25.6	22.2	19.5	17.4

(At $t = 0$, $C_A = 0.05\ M$)

The initial concentration of methanol was 0.5 mol/dm³.

Part (1)   Determine the reaction order with respect to triphenyl methyl chloride.

Part (2)   In a separate set of experiments, the reaction order wrt methanol was found to be first order. Determine the specific reaction rate constant.

Solution

Part (1) Find reaction order wrt trityl.

Step 1   **Postulate a rate law.**

$$-r_A = kC_A^\alpha C_B^\beta \tag{E5-1.1}$$

Step 2   **Process your data in terms of the measured variable,** which in this case is C_A.

Step 3   **Look for simplifications.** Because concentration of methanol is 10 times the initial concentration of triphenyl methyl chloride, its concentration is essentially constant

$$C_B = C_{B0} \tag{E5-1.2}$$

Substituting for C_B in Equation (E5-1.1)

$$-r_A = \underbrace{kC_{B0}^\beta}_{k'} C_A^\alpha$$

$$-r_A = k'C_A^\alpha \tag{E5-1.3}$$

Step 4   **Apply the CRE algorithm**
Mole Balance

$$\frac{dN_A}{dt} = r_A V \tag{E5-1.4}$$

Rate Law

$$-r_A = k'C_A^\alpha \tag{E5-1.3}$$

Stoichiometry: Liquid

$$V = V_0 \tag{E5-1.4}$$

$$C_A = \frac{N_A}{V_0}$$

Following the Algorithm

Combine: Mole balance, rate law, and stoichiometry

$$-\frac{dC_A}{dt} = k'C_A^\alpha \qquad \text{(E5-1.5)}$$

Taking the natural log of both sides of Equation (E5-1.5)

$$\ln\left[-\frac{dC_A}{dt}\right] = \ln k' + \alpha \ln C_A \qquad \text{(E5-1.6)}$$

The slope of a plot of $\ln\left[-\dfrac{dC_A}{dt}\right]$ versus $\ln C_A$ will yield the reaction order α with respect to triphenyl methyl chloride (A).

Step 5 Find $\left[-\dfrac{dC_A}{dt}\right]$ as a function of C_A from concentration-time data.

We will find $\left(-\dfrac{dC_A}{dt}\right)$ by each of the three methods just discussed, the graphical, finite difference, and polynomial methods.

Step 5A.1a *Graphical Method.* We now construct Table E5-1.2.

TABLE E5–1.2 PROCESSED DATA

t (min)	$C_A \times 10^3$ (mol/dm^3)	$-\dfrac{\Delta C_A}{\Delta t} \times 10^4$ (mol/dm$^3 \cdot$ min)	$-\dfrac{dC_A}{dt} \times 10^4$ (mol/dm$^3 \cdot$ min)
0	50		3.0
		2.40[a]	
50	38		1.86
		1.48	
100	30.6		1.2
		1.00	
150	25.6		0.8
		0.68	
200	22.2		0.5
		0.54	
250	19.5		0.47
		0.42	
300	17.4		

a
$$-\frac{\Delta C_A}{\Delta t} = -\frac{C_{A2} - C_{A1}}{t_2 - t_1} = -\left(\frac{38-50}{50-0}\right) \times 10^{-3} = 0.24 \times 10^{-3} = 2.4 \times 10^{-4} (\text{mol/dm}^3 \cdot \text{min})$$

The derivative $-dC_A/dt$ is determined by calculating and plotting $(-\Delta C_A/\Delta t)$ as a function of time, t, and then using the equal-area differentiation technique (Appendix A.2) to determine $(-dC_A/dt)$ as a function of C_A. First, we calculate the ratio $(-\Delta C_A/\Delta t)$ from the first two columns of Table E5-1.2; the result is written in the third column. Next we use Table E5-1.2 to plot the third column as a function of the first column in Figure E5-1.1 [i.e., $(-\Delta C_A/\Delta t)$ versus t]. Using equal-area differentiation, the value of $(-dC_A/dt)$ is read off the figure (represented by the arrows); then it is used to complete the fourth column of Table E5-1.2.

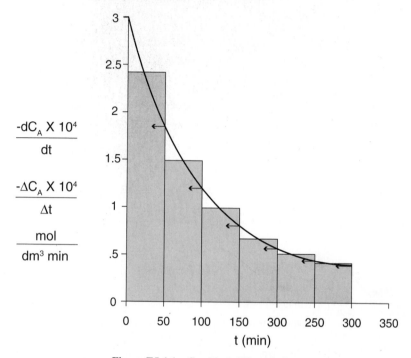

Figure E5-1.1   Graphical differentiation.

Step 5A.1b Finite Difference Method. We now calculate (dC_A/dt) using the finite difference formulas [i.e., Equations (5-8) through (5-10)].

$$t = 0 \quad \left(\frac{dC_A}{dt}\right)_{t=0} = \frac{-3C_{A0} + 4C_{A1} - C_{A2}}{2\Delta t}$$

$$= \frac{[-3(50) + 4(38) - 30.6] \times 10^{-3}}{100}$$

$$= -2.86 \times 10^{-4} \text{ mol/dm}^3 \cdot \text{min}$$

$$-\frac{dC_A}{dt} \times 10^4 = 2.86 \text{ mol/dm}^3 / \text{min}$$

$$t = 50 \quad \left(\frac{dC_A}{dt}\right)_1 = \frac{C_{A2} - C_{A0}}{2\Delta t} = \frac{(30.6 - 50) \times 10^{-3}}{100}$$

$$= -1.94 \times 10^{-4} \text{ mol/dm}^3 \cdot \text{min}$$

$$t = 100 \quad \left(\frac{dC_A}{dt}\right)_2 = \frac{C_{A3} - C_{A1}}{2\Delta t} = \frac{(25.6 - 38) \times 10^{-3}}{100}$$

$$= -1.24 \times 10^{-4} \text{ mol/dm}^3 \cdot \text{min}$$

$$t = 150 \quad \left(\frac{dC_A}{dt}\right)_3 = \frac{C_{A4} - C_{A2}}{2\Delta t} = \frac{(22.2 - 30.6) \times 10^{-3}}{100}$$

$$= -0.84 \times 10^{-4} \text{ mol/dm}^3 \cdot \text{min}$$

$$t = 200 \quad \left(\frac{dC_A}{dt}\right)_4 = \frac{C_{A5} - C_{A3}}{2\Delta t} = \frac{(19.5 - 25.6) \times 10^{-3}}{100}$$

$$= -0.61 \times 10^{-4} \text{ mol/dm}^3 \cdot \text{min}$$

$$t = 250 \quad \left(\frac{dC_A}{dt}\right)_5 = \frac{C_{A6} - C_{A4}}{2\Delta t} = \frac{(17.4 - 22.2) \times 10^{-3}}{100}$$

$$= -0.48 \times 10^{-4} \text{ mol/dm}^3 \cdot \text{min}$$

$$t = 300 \quad \left(\frac{dC_A}{dt}\right)_6 = \frac{C_{A4} - 4C_{A5} + 3C_{A6}}{2\Delta t} = \frac{[22.2 - 4(19.5) + 3(17.4)] \times 10^{-3}}{100}$$

$$= -0.36 \times 10^{-4} \text{ mol/dm}^3 \cdot \text{min}$$

Step 5A.1c Polynomial Method. Another method to determine (dC_A/dt) is to fit the concentration of A to a polynomial in time and then to differentiate the resulting polynomial.

We will use the Polymath software package to express concentration as a function of time. Here we first choose the polynomial degree (in this case, fourth degree) and then type in the values of C_A at various times t to obtain

$$C_A = 0.04999 - 2.978 \times 10^{-4}t + 1.343 \times 10^{-6}t^2 - 3.485 \times 10^{-9}t^3 + 3.697 \times 10^{-12}t^4$$

$$\text{(E5-1.7)}$$

C_A is in (mol/dm^3) and t is in minutes. A plot of C_A versus t and the corresponding fourth-degree polynomial fit are shown in Figure E5-1.2.

TABLE E5–1.3 POLYMATH OUTPUT

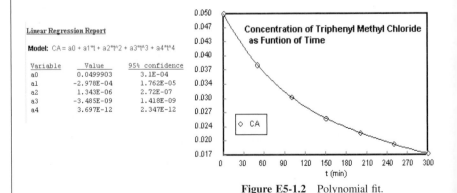

Figure E5-1.2 Polynomial fit.

Differentiating Equation (E5-1.7) yields

$$\frac{dC_A}{dt} \times 10^3 = -0.2987 + 0.002687t - 1.045 \times 10^{-5}t^2 + 1.479 \times 10^{-8}t^3 \qquad \text{(E5-1.8)}$$

Note: You can also obtain Equation (E5-1.9) directly from Polymath.

To find the derivative at various times, we substitute the appropriate time into Equation (E5-1.8) to arrive at the fourth column in Table E5-1.4 and multiply by (−1). We can see that there is quite a close agreement between the graphical technique, finite difference, and the polynomial methods.

TABLE E5-1.4.   SUMMARY OF PROCESSED DATA

	Graphical	*Finite Difference*	*Polynominal*	
	$-\dfrac{dC_A}{dt} \times 10{,}000$	$-\dfrac{dC_A}{dt} \times 10{,}000$	$-\dfrac{dC_A}{dt} \times 10{,}000$	$C_A \times 1{,}000$
t (min)	(mol/dm^3 · min)	(mol/dm^3 · min)	(mol/dm^3 · min)	(mol/dm^3)
0	3.0	2.86	2.98	50
50	1.86	1.94	1.88	38
100	1.20	1.24	1.19	30.6
150	0.80	0.84	0.80	25.6
200	0.68	0.61	0.60	22.2
250	0.54	0.48	0.48	19.5
300	0.42	0.36	0.33	17.4

We will now plot columns 2, 3, and 4 $\left(-\dfrac{dC_A}{dt} \times 10{,}000\right)$ as a function of column 5 ($C_A \times 1{,}000$) on log-log paper as shown in Figure E5-1.3. We could also substitute the parameter values in Table E5-1.4 into Excel to find α and k'. Note that most all of the points for all methods fall virtually on top of one another.

From Figure E5-1.3, we found the slope to be 2.05 so that the reaction is said to be second order wrt triphenyl methyl chloride. To evaluate k', we can evaluate the derivative and $C_{Ap} = 20 \times 10^{-3}$ mol/dm^3, which is

$$\left(-\frac{dC_A}{dt}\right)_p = 0.5 \times 10^{-4} \text{ mol/dm}^3 \cdot \text{min} \qquad \text{(E5-1.9)}$$

then

$$k' = \frac{\left(-\dfrac{dC_A}{dt}\right)_p}{C_{Ap}^2} \qquad \text{(E5-1.10)}$$

$$= \frac{0.5 \times 10^{-4}\,\text{mol/dm}^3 \cdot \text{min}}{(20 \times 10^{-3}\,\text{mol/dm}^3)^2} = 0.125 \text{ dm}^3 / \text{mol} \cdot \text{min}$$

As will be shown in Section 5.1.3, we could also use nonlinear regression on Equation (E5-1.7) to find k':

$$k' = 0.122 \text{ dm}^3/\text{mol} \cdot \text{min} \qquad \text{(E5-1.11)}$$

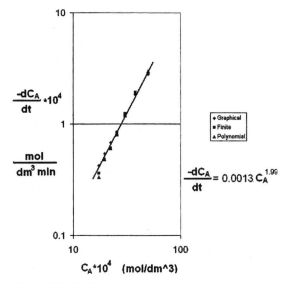

Figure E5-1.3 Excel plot to determine α and k.

Summary Notes

The Excel graph shown in Figure E5-1.3 gives $\alpha = 1.99$ and $k' = 0.13$ dm³/mol · min. We could set $\alpha = 2$ and regress again to find $k' = 0.122$ dm³/mol · min.

ODE Regression. There are techniques and software becoming available whereby an ODE solver can be combined with a regression program to solve differential equations, such as

$$-\frac{dC_A}{dt} = k_A' C_A^\alpha \tag{E5.1.5}$$

to find k_A and α from concentration–time date.

Part (2) The reaction was said to be first order wrt methanol, $\beta = 1$,

$$k' = C_{B0}^\beta k = C_{B0} k \tag{E5-1.12}$$

Assuming C_{B0} is constant at 0.5 mol/dm³ and solving for k yields

$$k = \frac{k'}{C_{B0}} = \frac{0.122 \dfrac{dm^3}{mol \cdot min}}{0.5 \dfrac{mol}{dm^3}}$$

$$k = 0.244 \ (dm^3/mol)^2 \ / \ min$$

The rate law is

$$\boxed{-r_A = [0.244(dm^3/mol)^2/min]C_A^2 C_B} \tag{E5-1.13}$$

5.2.2 Integral Method

To determine the reaction order by the integral method, we guess the reaction order and integrate the differential equation used to model the batch system. If the order we assume is correct, the appropriate plot (determined from this integration) of the concentration–time data should be linear. The integral method is used most often when the reaction order is known and it is desired to evaluate the specific reaction rate constants at different temperatures to determine the activation energy.

The integral method uses a trial-and-error procedure to find reaction order.

In the integral method of analysis of rate data, we are looking for the appropriate function of concentration corresponding to a particular rate law that is linear with time. You should be thoroughly familiar with the methods of obtaining these linear plots for reactions of zero, first, and second order.

For the reaction

$$A \rightarrow Products$$

It is important to know how to generate linear plots of functions of C_A versus t for zero-, first-, and second-order reactions.

carried out in a constant-volume batch reactor, the mole balance is

$$\frac{dC_A}{dt} = r_A$$

For a zero-order reaction, $r_A = -k$, and the combined rate law and mole balance is

$$\frac{dC_A}{dt} = -k \tag{5-13}$$

Integrating with $C_A = C_{A0}$ at $t = 0$, we have

Zero order

$$\boxed{C_A = C_{A0} - kt} \tag{5-14}$$

A plot of the concentration of A as a function of time will be linear (Figure 5-3) with slope $(-k)$ for a zero-order reaction carried out in a constant-volume batch reactor.

If the reaction is first order (Figure 5-4), integration of the combined *mole balance and the rate law*

$$-\frac{dC_A}{dt} = kC_A$$

with the limit $C_A = C_{A0}$ at $t = 0$ gives

First order

$$\boxed{\ln \frac{C_{A0}}{C_A} = kt} \tag{5-15}$$

Consequently, we see that the slope of a plot of $[\ln (C_{A0}/C_A)]$ as a function of time is linear with slope k.

If the reaction is second order (Figure 5-5), then

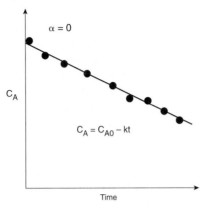

Figure 5-3   Zero-order reaction.

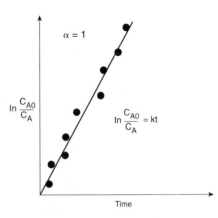

Figure 5-4   First-order reaction.

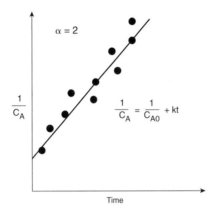

Figure 5-5   Second-order reaction.

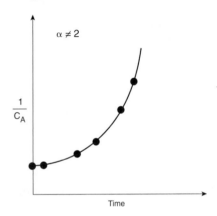

Figure 5-6   Plot of reciprocal concentration as a function of time.

$$-\frac{dC_A}{dt} = kC_A^2$$

Integrating, with $C_A = C_{A0}$ initially, yields

Second order

$$\boxed{\frac{1}{C_A} - \frac{1}{C_{A0}} = kt}$$

(5-16)

We see that for a second-order reaction a plot of $(1/C_A)$ as a function of time should be linear with slope k.

In Figures 5-3, 5-4, and 5-5, we saw that when we plotted the appropriate function of concentration (i.e., C_A, $\ln C_A$, or $1/C_A$) versus time, the plots were linear, and we concluded that the reactions were zero, first, or second order, respectively. However, if the plots of concentration data versus time had turned out **not to be linear**, such as shown in Figure 5-6, we would say that the proposed reaction order did not fit the data. In the case of Figure 5-6, we would conclude the reaction is not second order.

The idea is to arrange the data so that a linear relationship is obtained.

It is important to restate that, given a reaction rate law, you should be able to choose quickly the appropriate function of concentration or conversion that yields a straight line when plotted against time or space time.

Example 5–2   Integral Method of CRE Data Analysis

Use the integral method to confirm that the reaction is second order wrt triphenyl methyl chloride as described in Example 5-1 and to calculate the specific reaction rate k'

$$\text{Trityl (A) + Methanol (B)} \rightarrow \text{Products}$$

Solution

Substituting for $\alpha = 2$ in Equation (E5-1.5)

$$-\frac{dC_A}{dt} = k'C_A^{\alpha} \tag{E5-1.5}$$

we obtain

$$-\frac{dC_A}{dt} = k'C_A^2 \tag{E5-2.1}$$

Integrating with $C_A = C_{A0}$ at $t = 0$

$$t = \frac{1}{k'}\left[\frac{1}{C_A} - \frac{1}{C_{A0}}\right] \tag{E5-2.2}$$

Rearranging

$$\frac{1}{C_A} = \frac{1}{C_{A0}} + k't \tag{E5-2.3}$$

We see if the reaction is indeed second order then a plot of $(1/C_A)$ versus t should be linear. The data in Table E5-1.1 in Example 5-1 will be used to construct Table E5-2.1.

TABLE E5-2.1.   PROCESSED DATA

t (min)	0	50	100	150	200	250	300
C_A (mol/dm^3)	0.05	0.038	0.0306	0.0256	0.0222	0.0195	0.0174
$1/C_A$ (dm^3/mol)	20	26.3	32.7	39.1	45	51.3	57.5

In a graphical solution, the data in Table E5-2.1 can be used to construct a plot of $1/C_A$ as a function of t, which will yield the specific reaction rate k'. This plot is shown in Figure E5-2.1. Again, one could use Excel or Polymath to find k' from the data in Table E5-2.1. The slope of the line is the specific reaction rate k'

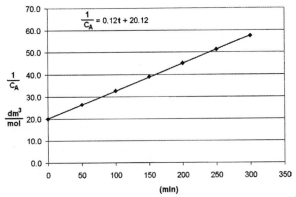

Figure E5-2.1   Plot reciprocal of C_A versut t for a second-order reaction.

We see from the Excel anaysis and plot that the slope of the line is 0.12 dm³/mol/min.

$$k' = 0.12 \ \frac{\text{dm}^3}{\text{mol} \cdot \text{min}}$$

Calculating k,

$$k = \frac{k'}{C_{B0}} = \frac{0.12}{0.5} \frac{\text{dm}^3/\text{mol/min}}{\text{mol/dm}^3} = 0.24 \left(\frac{\text{dm}^3}{\text{mol}} \right)^2 /\text{min}$$

The rate law is

$$-r_A = \left[0.24 \left(\frac{\text{dm}^3}{\text{mol}} \right)^2 /\text{min} \right] C_A^2 C_B$$

We note the integral method tends to smooth the data.

Polymath

An alternate computer solution would be to regress $\dfrac{1}{C_A}$ versus t with a software package such as Polymath.

$$\frac{1}{C_A} = \frac{1}{C_{A0}} + k't \tag{E5-2.4}$$

Let CA inverse $= \dfrac{1}{C_A}$, $a_0 = \dfrac{1}{C_{A0}}$, and $a_1 = k'$ and then enter the data in Table E5-2.1.

Linear Regression Report

Model: CAinverse = a0 + a1*t

Variable	Value	95% confidence
a0	20.117525	0.225264
a1	0.124794	0.0012495

From the Polymath output, we obtain $k' = 0.125$ dm³/mol/min, which yields $k = 0.25$ dm³/mol/min. We shall discuss regression in Example 5-3.

By comparing the methods of analysis of the rate data presented in Example 5-2, we note that the differential method tends to accentuate the uncertainties in the data, while the integral method tends to smooth the data, thereby disguising the uncertainties in it. In most analyses, it is imperative that the engineer know the limits and uncertainties in the data. This prior knowledge is necessary to provide for a safety factor when scaling up a process from laboratory experiments to design either a pilot plant or full-scale industrial plant.

5.2.3 Nonlinear Regression

In nonlinear regression analysis, we search for those parameter values that minimize the sum of the squares of the differences between the measured values and the calculated values for all the data points.[2] Not only can nonlinear regression find the best estimates of parameter values, it can also be used to discriminate between different rate law models, such as the Langmuir–Hinshelwood models discussed in Chapter 10. Many software programs are available to find these parameter values so that all one has to do is enter the data. The Polymath software will be used to illustrate this technique. In order to carry out the search efficiently, in some cases one has to enter initial estimates of the parameter values close to the actual values. These estimates can be obtained using the linear-least-squares technique discussed on the CD-ROM Professional Reference Shelf.

We will now apply nonlinear least-squares analysis to reaction rate data to determine the rate law parameters. Here we make estimates of the parameter values (e.g., reaction order, specific rate constants) in order to calculate the rate of reaction, r_c. We then search for those values that will minimize the sum of the squared differences of the measured reaction rates, r_m, and the calculated reaction rates, r_c. That is, we want the sum of $(r_m - r_c)^2$ for all data points to be minimum. If we carried out N experiments, we would want to find the parameter values (e.g., E, activation energy, reaction orders) that would minimize the quantity

$$\sigma^2 = \frac{s^2}{N-K} = \sum_{i=1}^{N} \frac{(r_{im} - r_{ic})^2}{N-K} \tag{5-34}$$

where

$$s^2 = \sum (r_{im} - r_{ie})^2$$

N = number of runs

K = number of parameters to be determined

r_{im} = measured reaction rate for run i (i.e., $-r_{Aim}$)

r_{ic} = calculated reaction rate for run i (i.e., $-r_{Aic}$)

[2] See also R. Mezakiki and J. R. Kittrell, *AIChE J.*, *14*, 513 (1968), and J. R. Kittrell, *Ind. Eng. Chem.*, *61* (5), 76–78 (1969).

To illustrate this technique, let's consider the first-order reaction

$$A \longrightarrow Product$$

for which we want to learn the reaction order, α, and the specific reaction rate, k,

$$r = kC_A^\alpha$$

The reaction rate will be measured at a number of different concentrations. We now choose values of k and α and calculate the rate of reaction (r_{ic}) at each concentration at which an experimental point was taken. We then subtract the calculated value (r_{ic}) from the measured value (r_{im}), square the result, and sum the squares for all the runs for the values of k and α we have chosen.

This procedure is continued by further varying α and k until we find their best values, that is, those values that minimize the sum of the squares. Many well-known searching techniques are available to obtain the minimum value σ_{min}^2.[3] Figure 5-7 shows a hypothetical plot of the sum of the squares as a function of the parameters α and k:

$$\sigma^2 = f(k, \alpha) \tag{5-17}$$

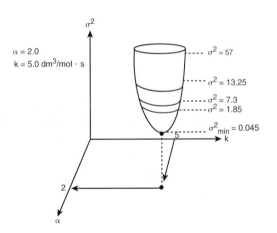

Figure 5-7 Minimum sum of squares.

In searching to find the parameter values that give the minimum of the sum of squares σ^2, one can use a number of optimization techniques or software packages. The searching procedure begins by guessing parameter values and then calculating r_c and then σ^2 for these values. Next a few sets of parameters are chosen around the initial guess, and σ^2 is calculated for these sets as well. The search technique looks for the smallest value of σ^2 in the vicinity of

[3] (a) B. Carnahan and J. O. Wilkes, *Digital Computing and Numerical Methods* (New York: Wiley, 1973), p. 405. (b) D. J. Wilde and C. S. Beightler, *Foundations of Optimization* 2nd ed. (Upper Saddle River, N.J.: Prentice Hall, 1979). (c) D. Miller and M. Frenklach, *Int. J. Chem. Kinet.*, *15*, 677 (1983).

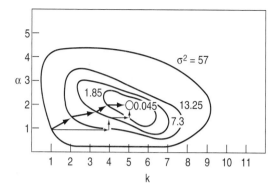

Figure 5-8   Trajectory to find the best values of k and α.

α = 2
k = 5 dm³/mol · s

the initial guess and then proceeds along a trajectory in the direction of decreasing σ^2 to choose different parameter values and determine the corresponding σ^2. The trajectory is continually adjusted so as always to proceed in the direction of decreasing σ^2 until the minimum value of σ^2 is reached. A schematic of this procedure is shown in Figure 5-8, where the parameter values at the minimum are $\alpha = 2$ and $k = 5$ s⁻¹. If the equations are highly nonlinear, the initial guess is extremely important. In some cases it is useful to try different initial guesses of the parameter to make sure that the software program converges on the same minimum for the different initial guesses. The dark lines and heavy arrows represent a computer trajectory, and the light lines and arrows represent the hand calculations. In extreme cases, one can use *linear least squares* (see CD-ROM) to obtain initial estimates of the parameter values.

Vary the initial guesses of parameters to make sure you find the true minimum.

A number of software packages are available to carry out the procedure to determine the best estimates of the parameter values and the corresponding confidence limits. All one has to do is to type the experimental values in the computer, specify the model, enter the initial guesses of the parameters, and then push the computer button, and the best estimates of the parameter values along with 95% confidence limits appear. If the confidence limits for a given parameter are larger than the parameter itself, the parameter is probably not significant and should be dropped from the model. After the appropriate model parameters are eliminated, the software is run again to determine the best fit with the new model equation.

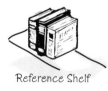

Reference Shelf

Concentration-Time Data.   We will now use nonlinear regression to determine the rate law parameters from concentration–time data obtained in batch experiments. We recall that the combined rate law-stoichiometry-mole balance for a constant-volume batch reactor is

$$\frac{dC_A}{dt} = -kC_A^\alpha \qquad (5\text{-}6)$$

We now integrate Equation (5-6) to give

$$C_{A0}^{1-\alpha} - C_A^{1-\alpha} = (1-\alpha)kt$$

Rearranging to obtain the concentration as a function of time, we obtain

$$C_A = [C_{A0}^{1-\alpha} - (1-\alpha)kt]^{1/(1-\alpha)} \tag{5-18}$$

Now we could use Polymath or MATLAB to find the values of α and k that would minimize the sum of squares of the differences between the measured and calculated concentrations. That is, for N data points,

$$s^2 = \sum_{i=1}^{N}(C_{Ami} - C_{Aci})^2 = \sum_{i=1}^{N}\left[C_{Ami} - [C_{A0}^{1-\alpha} - (1-\alpha)kt_i]^{1/(1-\alpha)}\right]^2 \tag{5-19}$$

we want the values of α and k that will make s^2 a minimum.

If Polymath is used, one should use the absolute value for the term in brackets in Equation (5-19), that is,

$$s^2 = \sum_{i=1}^{n}\left[C_{Ami} - \{(abs[C_{A0}^{1-\alpha} - (1-\alpha)kt_i]\}^{1/(1-\alpha)}\right]^2 \tag{5-20}$$

Another way to solve for the parameter values is to use time rather than concentrations:

$$t_c = \frac{C_{A0}^{1-\alpha} - C_A^{1-\alpha}}{k(1-\alpha)} \tag{5-21}$$

That is, we find the values of k and α that minimize

$$s^2 = \sum_{i=1}^{N}(t_{mi} - t_{ci})^2 = \sum_{i=1}^{N}\left[t_{mi} - \frac{C_{A0}^{1-\alpha} - C_{Ai}^{1-\alpha}}{k(1-\alpha)}\right]^2 \tag{5-22}$$

Reference Shelf

Finally, a discussion of *weighted least squares* as applied to a first-order reaction is provided in the *Professional Reference Shelf R5.2* on the CD-ROM.

Example 5–3   Use of Regression to Find the Rate Law Parameters

We shall use the reaction and data in Example 5-1 to illustrate how to use regression to find α and k'.

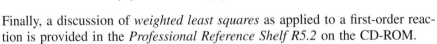

$$(C_6H_5)_3CCl + CH_3OH \rightarrow (C_6H_5)_3\overset{\overset{O}{\|}}{C}CH_3 + HCl$$

$$A \quad + \quad B \quad \rightarrow \quad C \quad + \quad D$$

The Polymath regression program is included on the CD-ROM. Recalling Equation (E5-1.5)

$$-\frac{dC_A}{dt} = k'C_A^\alpha \tag{E5-1.5}$$

and integrating with the initial condition when $t = 0$ and $C_A = C_{A0}$ for $\alpha \ne 1.0$

$$t = \frac{1}{k'}\frac{C_{A0}^{(1-\alpha)} - C_A^{(1-\alpha)}}{(1-\alpha)} \tag{E5-3.1}$$

Substituting for the initial concentration $C_{A0} = 0.05$ mol/dm³

$$t = \frac{1}{k'}\frac{(0.05)^{(1-\alpha)} - C_A^{(1-\alpha)}}{(1-\alpha)} \tag{E5-3.2}$$

Let's do a few calculations by hand to illustrate regression. We will first assume a value of α and k and then calculate t for the concentrations of A given in Table E5-1.1. We will then calculate the sum of the squares of the difference between the measured times t_m and the calculated times (i.e., s^2). For N measurements,

$$s^2 = \sum_{i=1}^{N}(t_{mi} - t_{ci})^2 = \sum_{i=1}^{N}\left(t_{mi} - \frac{C_{A0}^{1-\alpha} - C_A^{1-\alpha}}{k'(1-\alpha)}\right)^2$$

Our first guess is going to be $\alpha = 3$ and $k' = 5$, with $C_{A0} = 0.05$. Equation (E5-3.2) becomes

$$t_c = \frac{1}{2k'}\left[\frac{1}{C_A^2} - \frac{1}{C_{A0}^2}\right] = \frac{1}{10}\left[\frac{1}{C_A^2} - 400\right] \tag{E5-3.3}$$

We now make the calculations for each measurement of concentration and fill in columns 3 and 4 of Table E5-3.1. For example, when $C_A = 0.038$ mol/dm³ then

$$t_{c1} = \frac{1}{10}\left[\frac{1}{(0.038)^2} - 400\right] = 29.2 \text{ min}$$

which is shown in Table E5-3.1 on line 2 for guess 1. We next calculate the squares of difference $(t_{m1} - t_{c1})^2 = (50 - 29.2)^2 = 433$. We continue in this manner for points 2, 3, and 4 to calculate the sum $s^2 = 2916$.

After calculating s^2 for $\alpha = 3$ and $k = 5$, we make a second guess for α and k'. For our second guess we choose $\alpha = 2$ and $k = 5$; Equation (E5-3.2) becomes

$$t_c = \frac{1}{k'}\left[\frac{1}{C_A} - \frac{1}{C_{A0}}\right] = \frac{1}{5}\left[\frac{1}{C_A} - 20\right] \tag{E5-3.4}$$

We now proceed with our second guess to find the sum of $(t_m - t_c)^2$ to be $s^2 = 49{,}895$, which is far worse than our first guess. So we continue to make more guesses of α and k and find s^2. Let's stop and take a look at t_c for guesses 3 and 4.

We shall only use four points for this illustration.

TABLE E5-3.1.   REGRESSION OF DATA

	Original Data		Guess 1 $\alpha = 3$ $k' = 5$		Guess 2 $\alpha = 2$ $k' = 5$		Guess 3 $\alpha = 2$ $k' = 0.2$		Guess 4 $\alpha = 2$ $k' = 0.1$	
	t (min)	$C_A \times 10^3$ (mol/dm³)	t_C	$(t_m - t_C)^2$	t_C	$(t_m - t_C)^2$	t_C	$(t_m - t_C)^2$	t_C	$(t_m - t_C)^2$
1	0	50	0	0	0	0	0	0	0	0
2	50	38	29.2	433	1.26	2,375	31.6	339	63.2	174
3	100	30.6	66.7	1,109	2.5	9,499	63.4	1,340	126.8	718
4	200	22.2	163	1,375	5.0	38,622	125.2	5,591	250	2,540
				$s^2 = 2916$		$s^2 = 49,895$		$s^2 = 7270$		$s^2 = 3432$

Summary Notes

A Polymath tutorial on regression is given on the CD-ROM.

We see that ($k' = 0.2$ dm³/mol · min) underpredicts the time (e.g., 31.6 min versus 50 minutes), while ($k' = 0.1$ dm³/mol · min) overpredicts the time (e.g., 63 min versus 50 minutes). We could continue in this manner by choosing k' between $0.1 < k' < 0.2$, but why bother to go to all the trouble? Nobody has that much time on their hands. Why don't we just let the Polymath regression program find the values of k' and α that will minimize s^2?

The Polymath tutorial on the CD-ROM shows screen shots of how to enter the raw data in Table E5-1.1 and to carry out a nonlinear regression on Equation (E5-3.1). For $C_{A0} = 0.05$ mol/dm³, that is, Equation (E5-3.1) becomes

$$t_c = \frac{1}{k'} \frac{(0.05)^{(1-\alpha)} - C_A^{(1-\alpha)}}{(1-\alpha)}$$

We want to minimize the sum to give α and k'

$$s^2 = \sum_{i=1}^{N} (t_{mi} - t_{ci})^2 = \sum_{i=1}^{N} \left[t_{mi} - \frac{0.05^{1-\alpha} - C_{Ai}^{1-\alpha}}{k(1-\alpha)} \right]^2 \tag{5-22}$$

TABLE E5-3.2.   RESULTS OF 1ST REGRESSION

POLYMATH Results

Example 5-3 Use of Regression to Find Rate Law Parameters 08-05-2004

Nonlinear regression (L-M)

Model: t = (.05^(1-a)-Ca^(1-a))/(k*(1-a))

Variable	Ini guess	Value	95% confidence
a	3	2.04472	0.0317031
k	0.1	0.1467193	0.0164118

Nonlinear regression settings
Max # iterations = 64

Precision
R^2   = 0.9999717
R^2adj = 0.999966
Rmsd   = 0.2011604
Variance = 0.3965618

TABLE E5-3.3.   RESULTS OF 2ND REGRESSION

POLYMATH Results

Example 5-3 Use of Regression to Find Rate Law Parameters 08-05-2004

Nonlinear regression (L-M)

Model: t = (.05^(1-2)-Ca^(1-2))/(k*(1-2))

Variable	Ini guess	Value	95% confidence
k	0.1	0.1253404	7.022E-04

Nonlinear regression settings
Max # iterations = 64

Precision
R^2   = 0.9998978
R^2adj = 0.9998978
Rmsd   = 0.3821581
Variance = 1.1926993

The results shown are

$\alpha = 2.04$
$k' = 0.147 \text{ dm}^3/\text{mol} \cdot \text{min}$

$\alpha = 2.0$
$k' = 0.125 \text{ dm}^3/\text{mol} \cdot \text{min}$

We shall round off α to make the reaction second order, (i.e., $\alpha = 2.00$). Now having fixed α at 2.0, we must do another regression [cf. Table E5-3.3] on k' because the k' given in Table E.5-3.1 is for $\alpha = 2.0447$. We now regress the equation

$$ t = \frac{1}{k'}\left[\frac{1}{C_A} - \frac{1}{C_{A0}}\right] $$

and find $k' = 0.125$ dm³/mol · min.

$$ k = \frac{k'}{C_{A0}} = 0.25 \left(\frac{\text{dm}^3}{\text{mol}}\right)^2 /\text{min} $$

We note that the reaction order is the same as that in Examples 5-1 and 5-2; however, the value of k is about 8% larger.

Model Discrimination.   One can also determine which model or equation best fits the experimental data by comparing the sums of the squares for each model and then choosing the equation with a smaller sum of squares and/or carrying out an F-test. Alternatively, we can compare the residual plots for each model. These plots show the error associated with each data point, and one looks to see if the error is randomly distributed or if there is a trend in the error. When the error is randomly distributed, this is an additional indication that the correct rate law has been chosen. An example of model discrimination using nonlinear regression is given on the CD-ROM in Chapter 10 of the *Summary Notes*.

Links

5.3 Method of Initial Rates

Used when reactions are reversible

The use of the differential method of data analysis to determine reaction orders and specific reaction rates is clearly one of the easiest, since it requires only one experiment. However, other effects, such as the presence of a significant *reverse* reaction, could render the differential method ineffective. In these cases, the method of initial rates could be used to determine the reaction order and the specific rate constant. Here, a series of experiments is carried out at different initial concentrations, C_{A0}, and the initial rate of reaction, $-r_{A0}$, is determined for each run. The initial rate, $-r_{A0}$, can be found by differentiating the data and extrapolating to zero time. For example, in the Trityl–Methanol reaction shown in Example 5-1, the initial rate was found to be 0.00028 mol/dm³ · min. By various plotting or numerical analysis techniques relating $-r_{A0}$ to C_{A0}, we can obtain the appropriate rate law. If the rate law is in the form

$$ -r_{A0} = kC_{A0}^{\alpha} $$

the slope of the plot of $\ln(-r_{A0})$ versus $\ln C_{A0}$ will give the reaction order α.

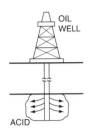

OIL WELL

ACID

Example 5–4   Method of Initial Rates in Solid–Liquid Dissolution Kinetics

The dissolution of dolomite, calcium magnesium carbonate, in hydrochloric acid is a reaction of particular importance in the acid stimulation of dolomite oil reservoirs.[4] The oil is contained in pore space of the carbonate material and must flow through the small pores to reach the well bore. In matrix stimulation, HCl is injected into a well bore to dissolve the porous carbonate matrix. By dissolving the solid carbonate, the pores will increase in size, and the oil and gas will be able to flow out at faster rates, thereby increasing the productivity of the well.[5] The dissolution reaction is

An important reaction for enhancement of oil flow in carbonate reservoirs

$$4HCl + CaMg(CO_3)_2 \rightarrow Mg^{2+} + Ca^{2+} + 4Cl^- + 2CO_2 + 2H_2O$$

The concentration of HCl at various times was determined from atomic absorption spectrophotometer measurements of the calcium and magnesium ions.

Determine the reaction order with respect to HCl from the data presented in Figure E5-4.1 for this batch reaction. Assume that the rate law is in the form given by Equation (5-1) and that the combined rate law and mole balance for HCl can be given by Equation (5-6).

Wormholes etched by acid

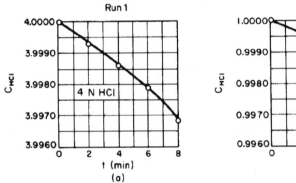
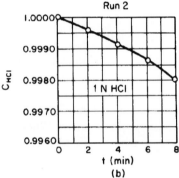

Figure E5-4.1   Concentration–time data.

Solution

Evaluating the mole balance on a constant-volume batch reactor at time $t = 0$ gives

$$\left(-\frac{dC_{HCl}}{dt} \right)_0 = -(r_{HCl})_0 = kC_{HCl,0}^{\alpha} \tag{E5-4.1}$$

Taking the log of both sides of Equation (E5-4.1), we have

$$\ln\left(-\frac{dC_{HCl}}{dt} \right)_0 = \ln k + \alpha \ln C_{HCl,0} \tag{E5-4.2}$$

---

[4] K. Lund, H. S. Fogler, and C. C. McCune, *Chem. Eng. Sci.*, *28*, 691 (1973).

[5] M. Hoefner and H. S. Fogler, *AIChE Journal*, *34*(1), 45 (1988).

The derivative at time $t = 0$ can be found from the slope of the plot of concentration versus time evaluated at $t = 0$. Figure E5-4.1(a) and (b) give

4 N HCl solution	1 N HCl solution
$-r_{HCl,0} = -\dfrac{3.9982 - 4.0000}{5 - 0}$	$-r_{HCl,0} = -\dfrac{0.9987 - 1.0000}{6 - 0}$
$-r_{HCl,0} = 3.6 \times 10^{-4} \text{ mol/dm}^3 \cdot \text{min}$	$-r_{HCl,0} = 2.2 \times 10^{-4} \text{ mol/dm}^3 \cdot \text{min}$

Converting to a rate per unit area, $-r_A''$, and to seconds (30 cm² of solid per liter of solution), the rates at 1 N and 4 N become 1.2×10^{-7} mol/cm²·s and 2.0×10^{-7} mol/cm²·s, respectively. We also could have used either Polymath or the differentiation formulas to find the derivative at $t = 0$.

If we were to continue in this manner, we would generate the following data set in Table E5-4.1.[6]

TABLE E5-4.1.

$C_{HCl,0}$ (mol/dm³)	1.0	4.0	2.0	0.1	0.5
$-r_{HCl,0}''$ (mol/cm²·s) × 10⁷	1.2	2.0	1.36	0.36	0.74

These data are plotted on Figure E5-4.2. The slope of this ln-ln plot of $-r_{HCl,0}''$ versus $C_{HCl,0}$ shown in Figure E5-4.2 gives a reaction order of 0.44. The rate law is

$$-r_{HCl,0}'' = kC_{HCl}^{0.44} \qquad (E5\text{-}4.3)$$

For this dissolution of dolomite in HCl, the reaction order was also found to vary with temperature.

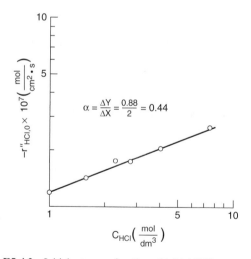

$$\alpha = \frac{\Delta Y}{\Delta X} = \frac{0.88}{2} = 0.44$$

Figure E5-4.2   Initial rate as a function of initial HCl concentration.

[6] K. Lund, H. S. Fogler, and C. C. McCune, *Chem. Eng. Sci., 28*, 691 (1973). C. Fredd and H. S. Fogler, *AIChE J., 44*, p1933 (1998).

5.4 Method of Half-Lives

The method of half-lives requires many experiments

The half-life of a reaction, $t_{1/2}$, is defined as the time it takes for the concentration of the reactant to fall to half of its initial value. By determining the half-life of a reaction as a function of the initial concentration, the reaction order and specific reaction rate can be determined. If two reactants are involved in the chemical reaction, the experimenter will use the method of excess in conjunction with the method of half-lives to arrange the rate law in the form

$$-r_A = kC_A^\alpha \tag{5-1}$$

For the irreversible reaction

$$A \longrightarrow \text{Products}$$

The concept of a half life, $t_{1/2}$, is very important in drug medication.

a mole balance on species A in a constant-volume batch reaction system combined with the rate law results in the following expression:

$$-\frac{dC_A}{dt} = -r_A = kC_A^\alpha \tag{5-6}$$

Integrating with the initial condition $C_A = C_{A0}$ when $t = 0$, we find that

$$t = \frac{1}{k(\alpha - 1)}\left(\frac{1}{C_A^{\alpha-1}} - \frac{1}{C_{A0}^{\alpha-1}}\right)$$

$$= \frac{1}{kC_{A0}^{\alpha-1}(\alpha - 1)}\left[\left(\frac{C_{A0}}{C_A}\right)^{\alpha-1} - 1\right] \tag{5-23}$$

The half-life is defined as the time required for the concentration to drop to half of its initial value; that is,

$$t = t_{1/2} \quad \text{when} \quad C_A = \tfrac{1}{2}C_{A0}$$

Substituting for C_A in Equation (5-23) gives us

$$t_{1/2} = \frac{2^{\alpha-1} - 1}{k(\alpha - 1)}\left(\frac{1}{C_{A0}^{\alpha-1}}\right) \tag{5-24}$$

There is nothing special about using the time required for the concentration to drop to one-half of its initial value. We could just as well use the time required for the concentration to fall to $1/n$ of the initial value, in which case

$$t_{1/n} = \frac{n^{\alpha-1} - 1}{k(\alpha - 1)}\left(\frac{1}{C_{A0}^{\alpha-1}}\right) \tag{5-25}$$

For the method of half-lives, taking the natural log of both sides of Equation (5-24),

$$\ln t_{1/2} = \ln\frac{2^{\alpha-1} - 1}{(\alpha - 1)k} + (1 - \alpha)\ln C_{A0}$$

we see that the slope of the plot of $\ln t_{1/2}$ as a function of $\ln C_{A0}$ is equal to 1 minus the reaction order [i.e., slope = $(1 - \alpha)$]:

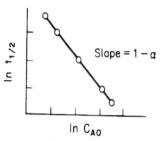

Figure 5-9   Method of half-lives.

Rearranging:

$\alpha = 1 - \text{slope}$

For the plot shown in Figure 5-9, the slope is -1:

$\alpha = 1 - (-1) = 2$

The corresponding rate law is

$$-r_A = kC_A^2$$

Note: We also could have used nonlinear regression on the half-life data. However, by plotting the data, one often gets a "better feel" for the accuracy and precision of the data.

5.5 Differential Reactors

Data acquisition using the method of initial rates and a differential reactor are similar in that the rate of reaction is determined for a specified number of pre-determined initial or entering reactant concentrations. A differential reactor is normally used to determine the rate of reaction as a function of either concen-

tration or partial pressure. It consists of a tube containing a very small amount of catalyst usually arranged in the form of a thin wafer or disk. A typical arrangement is shown schematically in Figure 5-10. The criterion for a reactor being differential is that the conversion of the reactants in the bed is extremely small, as is the change in temperature and reactant concentration through the bed. As a result, the reactant concentration through the reactor is essentially constant and approximately equal to the inlet concentration. That is, the reactor is considered to be gradientless,[7] and the reaction rate is considered spatially uniform within the bed.

The differential reactor is relatively easy to construct at a low cost. Owing to the low conversion achieved in this reactor, the heat release per unit volume will be small (or can be made small by diluting the bed with inert solids) so that the reactor operates essentially in an isothermal manner. When operating this reactor, precautions must be taken so that the reactant gas or liquid does not bypass or channel through the packed catalyst, but instead flows uniformly

across the catalyst. If the catalyst under investigation decays rapidly, the differential reactor is not a good choice because the reaction rate parameters at the

---

[7] B. Anderson, ed., *Experimental Methods in Catalytic Research* (San Diego, Calif.: Academic Press, 1976).

start of a run will be different from those at the end of the run. In some cases sampling and analysis of the product stream may be difficult for small conversions in multicomponent systems.

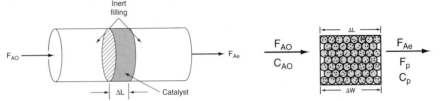

Figure 5-10   Differential reactor.   Figure 5-11   Differential catalyst bed.

$$A \rightarrow P$$

The volumetric flow rate through the catalyst bed is monitored, as are the entering and exiting concentrations (Figure 5-11). Therefore, if the weight of catalyst, ΔW, is known, the rate of reaction per unit mass of catalyst, $-r'_A$, can be calculated. Since the differential reactor is assumed to be *gradientless*, the design equation will be similar to the CSTR design equation. A steady-state mole balance on reactant A gives

$$
\begin{bmatrix} \text{Flow} \\ \text{rate} \\ \text{in} \end{bmatrix} - \begin{bmatrix} \text{Flow} \\ \text{rate} \\ \text{out} \end{bmatrix} + \begin{bmatrix} \text{Rate of} \\ \text{generation} \end{bmatrix} = \begin{bmatrix} \text{Rate of} \\ \text{accumulation} \end{bmatrix}
$$

$$
[F_{A0}] - [F_{Ae}] + \left[\left(\frac{\text{Rate of reaction}}{\text{Mass of cat}} \right)(\text{Mass of cat}) \right] = 0
$$

$$
F_{A0} - F_{Ae} + (r'_A)(\Delta W) = 0
$$

The subscript e refers to the exit of the reactor. Solving for $-r'_A$, we have

$$
-r'_A = \frac{F_{A0} - F_{Ae}}{\Delta W} \tag{5-26}
$$

The mole balance equation can also be written in terms of the concentration

Differential reactor
design equation

$$
\boxed{-r'_A = \frac{v_0 C_{A0} - v C_{Ae}}{\Delta W}} \tag{5-27}
$$

or in terms of the conversion or product flow rate F_P:

$$
\boxed{-r'_A = \frac{F_{A0} X}{\Delta W} = \frac{F_P}{\Delta W}} \tag{5-28}
$$

The term $F_{A0}X$ gives the rate of formation of the product, F_P, when the stoichiometric coefficients of A and of P are identical.

For constant volumetric flow, Equation (5-27) reduces to

$$-r'_A = \frac{v_0(C_{A0} - C_{Ae})}{\Delta W} = \frac{v_0 C_P}{\Delta W}$$ (5-29)

Consequently, we see that the reaction rate, $-r'_A$, can be determined by measuring the product concentration, C_P.

By using very little catalyst and large volumetric flow rates, the concentration difference, $(C_{A0} - C_{Ae})$, can be made quite small. The rate of reaction determined from Equation (5-29) can be obtained as a function of the reactant concentration in the catalyst bed, C_{Ab}:

$$-r'_A = -r'_A(C_{Ab})$$ (5-30)

by varying the inlet concentration. One approximation of the concentration of A within the bed, C_{Ab}, would be the arithmetic mean of the inlet and outlet concentrations:

$$C_{Ab} = \frac{C_{A0} + C_{Ae}}{2}$$ (5-31)

However, since very little reaction takes place within the bed, the bed concentration is essentially equal to the inlet concentration,

$$C_{Ab} \approx C_{A0}$$

so $-r'_A$ is a function of C_{A0}:

$$-r'_A = -r'_A(C_{A0})$$ (5-32)

As with the method of initial rates, various numerical and graphical techniques can be used to determine the appropriate algebraic equation for the rate law. When collecting data for fluid-solid reacting systems, care must be taken that we use high flow rates through the differential reactor and small catalyst particle sizes in order to avoid mass transfer limitations. If data show the reaction to be first order with a low activation energy, say 8 kcal/moles, one should suspect the data is being collected in the mass transfer limited regime. We will expand on mass transfer limitations and how to avoid them in Chapters 10, 11, and 12.

Example 5–5 Differential Reactor

The formation of methane from carbon monoxide and hydrogen using a nickel catalyst was studied by Pursley.[8] The reaction

$$3H_2 + CO \rightarrow CH_4 + 2H_2O$$

was carried out at 500°F in a differential reactor where the effluent concentration of methane was measured.

(a) Relate the rate of reaction to the exit methane concentration.

(b) The reaction rate law is assumed to be the product of a function of the partial pressure of CO, $f(CO)$, and a function of the partial pressure of H_2, $g(H_2)$:

$$r'_{CH_4} = f(CO) \cdot g(H_2) \tag{E5-5.1}$$

Determine the reaction order with respect to carbon monoxide, using the data in Table E5-5.1. Assume that the functional dependence of r'_{CH_4} on P_{CO} is of the form

$$r'_{CH_4} \sim P_{CO}^\alpha \tag{E5-5.2}$$

TABLE E5-5.1.   RAW DATA

Run	P_{CO} (atm)	P_{H_2} (atm)	C_{CH_4} (mol/dm³)
1	1	1.0	1.73×10^{-4}
2	1.8	1.0	4.40×10^{-4}
3	4.08	1.0	10.0×10^{-4}
4	1.0	0.1	1.65×10^{-4}
5	1.0	0.5	2.47×10^{-4}
6	· 1.0	4.0	1.75×10^{-4}

(margin note) P_{H_2} is constant in Runs 1, 2, 3
P_{CO} is constant in Runs 4, 5, 6

The exit volumetric flow rate from a differential packed bed containing 10 g of catalyst was maintained at 300 dm³/min for each run. The partial pressures of H_2 and CO were determined at the entrance to the reactor, and the methane concentration was measured at the reactor exit.

Solution

(a) In this example the product composition, rather than the reactant concentration, is being monitored. $-r'_{CO}$ can be written in terms of the flow rate of methane from the reaction,

$$-r'_{CO} = r'_{CH_4} = \frac{F_{CH_4}}{\Delta W}$$

---

[8] J. A. Pursley, An Investigation of the Reaction between Carbon Monoxide and Hydrogen on a Nickel Catalyst above One Atmosphere, Ph.D. thesis, University of Michigan.

Substituting for F_{CH_4} in terms of the volumetric flow rate and the concentration of methane gives

$$-r'_{CO} = \frac{v_0 C_{CH_4}}{\Delta W} \qquad (E5\text{-}5.3)$$

Since v_0, C_{CH_4}, and ΔW are known for each run, we can calculate the rate of reaction. For run 1:

$$-r'_{CO} = \left(\frac{300 \text{ dm}^3}{\text{min}}\right) \frac{1.73 \times 10^{-4}}{10 \text{ g cat}} \text{ mol/dm}^3 = 5.2 \times 10^{-3} \frac{\text{mol CH}_4}{\text{g cat} \times \text{min}}$$

The rate for runs 2 through 6 can be calculated in a similar manner (Table E5-5.2).

TABLE E5-5.2.   RAW AND CALCULATED DATA

Run	P_{CO} (atm)	P_{H_2} (atm)	C_{CH_4} (mol/dm³)	$r'_{CH_4}\left(\dfrac{\text{mol CH}_4}{\text{cat} \times \text{min}}\right)$
1	1.0	1.0	1.73×10^{-4}	5.2×10^{-3}
2	1.8	1.0	4.40×10^{-4}	13.2×10^{-3}
3	4.08	1.0	10.0×10^{-4}	30.0×10^{-3}
4	1.0	0.1	1.65×10^{-4}	4.95×10^{-3}
5	1.0	0.5	2.47×10^{-4}	7.42×10^{-3}
6	1.0	4.0	1.75×10^{-4}	5.25×10^{-3}

Determining the Rate Law Dependence in CO

For constant hydrogen concentration, the rate law

$$r'_{CH_4} = k P_{CO}^{\alpha} \cdot g(P_{H_2})$$

can be written as

$$r'_{CH_4} = k' P_{CO}^{\alpha} \qquad (E5\text{-}5.4)$$

Taking the log of Equation (E5-5.4) gives us

$$\ln(r'_{CH_4}) = \ln k' + \alpha \ln P_{CO}$$

We now plot $\ln(r'_{CH_4})$ versus $\ln P_{CO}$ for runs 1, 2, and 3.

(b) Runs 1, 2, and 3, for which the H_2 concentration is constant, are plotted in Figure E5-5.1. We see from the Excel plot that $\alpha = 1.22$. Had we included more points we would have found that the reaction is essentially first order with $\alpha = 1$, that is,

$$-r'_{CO} = k' P_{CO} \qquad (E5\text{-}5.5)$$

From the first three data points where the partial pressure of H_2 is constant, we see the rate is linear in partial pressure of CO.

$$r'_{CH_4} = k'' P_{CO} f(P_{H_2})$$

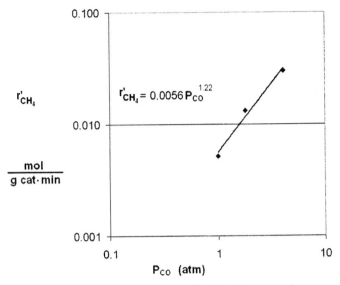

Figure E5-5.1   Reaction rate as a function of concentration.

Now let's look at the hydrogen dependence.

Determining the Rate Law Dependence on H_2

From Table E5-5.2 it appears that the dependence of r'_{CH_4} on P_{H_2} cannot be represented by a power law. Comparing run 4 with run 5 and run 1 with run 6, we see that the reaction rate first increases with increasing partial pressure of hydrogen, and subsequently decreases with increasing P_{H_2}. That is, there appears to be a concentration of hydrogen at which the rate is maximum. One set of rate laws that is consistent with these observations is:

1. At low H_2 concentrations where r'_{CH_4} increases as P_{H_2} increases, the rate law may be of the form

$$r'_{CH_4} \sim P_{H_2}^{\beta_1} \tag{E5-5.6}$$

2. At high H_2 concentrations where r'_{CH_4} decreases as P_{H_2} increases,

$$r'_{CH_4} \sim \frac{1}{P_{H_2}^{\beta_2}} \tag{E5-5.7}$$

We would like to find one rate law that is consistent with reaction rate data at both high and low hydrogen concentrations. Application of Chapter 10 material suggests Equations (E5-5.6) and (E5-5.7) can be combined into the form

$$r'_{CH_4} \sim \frac{P_{H_2}^{\beta_1}}{1 + bP_{H_2}^{\beta_2}} \tag{E5-5.8}$$

We will see in Chapter 10 that this combination and similar rate laws which have reactant concentrations (or partial pressures) in the numerator and denominator are common in *heterogeneous catalysis*.

Let's see if the resulting rate law (E5-5.8) is qualitatively consistent with the rate observed.

1. *For condition 1*: At low P_{H_2}, $(b(P_{H_2})^{\beta_2} \ll 1)$ and Equation (E5-5.8) reduces to

$$r'_{CH_4} \sim P_{H_2}^{\beta_1} \tag{E5-5.9}$$

Equation (E5-5.9) is consistent with the trend in comparing runs 4 and 5.

2. *For condition 2*: At high P_{H_2}, $b((P_{H_2})^{\beta_2} \gg 1)$ and Equation (E5-5.8) reduces to

$$r'_{CH_4} \sim \frac{(P_{H_2})^{\beta_1}}{(P_{H_2})^{\beta_2}} \sim \frac{1}{(P_{H_2})^{\beta_2 - \beta_1}} \tag{E5-5.10}$$

where $\beta_2 > \beta_1$. Equation (E5-5.10) is consistent with the trends in comparing runs 5 and 6.

Combining Equations (E5-5.8) and (E5-5.5)

Typical form of the
rate law for
heterogeneous
catalysis

$$r'_{CH_4} = \frac{aP_{CO}P_{H_2}^{\beta_1}}{1 + bP_{H_2}^{\beta_2}} \tag{E5-5.11}$$

We now use the Polymath regression program to find the parameter values a, b, β_1, and β_2. The results are shown in Table E5-5.3.

TABLE 5-5.3.   FIRST REGRESSION

Summary Notes

Polymath regres-
sion tutorial is in the
Chapter 5 *Summary
Notes*.

POLYMATH Results
No Title 01-31-2004

Nonlinear regression (L-M)

Model: Rate = a*Pco*Ph2^beta1/(1+b*Ph2^beta2)

Variable	Ini guess	Value	95% confidence
a	1	0.0252715	0.4917749
beta1	1	0.6166542	6.9023286
b	1	2.4872569	68.002944
beta2	1	1.0262047	3.2344414

Nonlinear regression settings
Max # iterations = 64

The corresponding rate law is

$$r_{CH_4} = \frac{0.025P_{CO}P_{H_2}^{0.61}}{1 + 2.49P_{H_2}} \tag{E5-5.12}$$

We could use the rate law as given by Equation (E5-5.12) as is, but there are only six data points, and we should be concerned about extrapolating the rate law over a wider range of partial pressures. We could take more data, and/or we could carry out a theoretical analysis of the type discussed in Chapter 10 for heterogeneous reactions. If we assume hydrogen undergoes dissociative adsorption on the catalyst surface one would expect a dependence of hydrogen to the ½ power. Because 0.61 is close to 0.5, we are going to regress the data again setting $\beta_1 = (½)$ and $\beta_2 = 1.0$. The results are shown in Table E5-5.4.

TABLE 5-5.4.   SECOND REGRESSION

POLYMATH Results
No Title 01-31-2004

Nonlinear regression (L-M)

Model: Rate = a*Pco*Ph2^0.5/(1+b*Ph2)

Variable	Ini guess	Value	95% confidence
a	1	0.018059	0.0106293
b	1	1.4898245	1.4787491

The rate law is now

$$r'_{CH_4} = \frac{0.018 P_{CO} P_{H_2}^{1/2}}{1 + 1.49 P_{H_2}}$$

where r'_{CH_4} is in mol/gcat · s and the partial pressure is in atm.

We could also have set $\beta_1 = ½$ and $\beta_2 = 1.0$ and rearranged Equation (E5-5.11) in the form

Linearizing the rate law to determine the rate law parameters

$$\frac{P_{CO} P_{H_2}^{1/2}}{r'_{CH_4}} = \frac{1}{a} + \frac{b}{a} P_{H_2} \qquad (E5\text{-}5.13)$$

A plot of $P_{CO} P_{H_2}^{1/2} / r'_{CH_4}$ as a function of P_{H_2} should be a straight line with an intercept of $1/a$ and a slope of b/a. From the plot in Figure E5-5.2, we see that the rate law is indeed consistent with the rate law data.

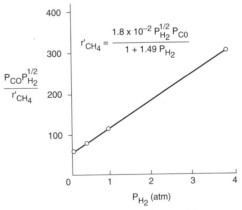

Figure E5-5.2   Linearized plot of data.

5.6 Experimental Planning

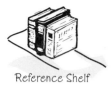

Reference Shelf

> Four to six weeks in the lab can save you an hour in the library.
>
> G. C. Quarderer, Dow Chemical Co.

So far, this chapter has presented various methods of analyzing rate data. It is just as important to know in which circumstances to use each method as it is to know the mechanics of these methods. On the CD-ROM, we give a thumbnail sketch of a heuristic to plan experiments to generate the data necessary for reactor design. However, for a more thorough discussion, the reader is referred to the books and articles by Box and Hunter.[9]

5.7 Evaluation of Laboratory Reactors

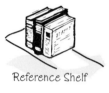

Reference Shelf

The successful design of industrial reactors lies primarily with the *reliability of the experimentally determined parameters used in the scale-up.* Consequently, it is imperative to design equipment and experiments that will generate accurate and meaningful data. Unfortunately, there is usually no single comprehensive laboratory reactor that could be used for all types of reactions and catalysts. In this section, we discuss the various types of reactors that can be chosen to obtain the kinetic parameters for a specific reaction system. We closely follow the excellent strategy presented in the article by V. W. Weekman of Mobil Oil, now ExxonMobil.[10]

5.7.1 Criteria

The criteria used to evaluate various types of laboratory reactors are listed in Table 5-2.

TABLE 5-2.   CRITERIA USED TO EVALUATE LABORATORY REACTORS

1. Ease of sampling and product analysis
2. Degree of isothermality
3. Effectiveness of contact between catalyst and reactant
4. Handling of catalyst decay
5. Reactor cost and ease of construction

Each type of reactor is examined with respect to these criteria and given a rating of good (G), fair (F), or poor (P). What follows is a brief description of each of the laboratory reactors. The reasons for rating each reactor for each of the criteria are given in *Professional Reference Shelf R5.4* on the CD-ROM.

---

[9] G. E. P. Box, W. G. Hunter, and J. S. Hunter, *Statistics for Experimenters: An Introduction to Design, Data Analysis, and Model Building* (New York: Wiley, 1978).

[10] V. W. Weekman, *AIChE J., 20*, 833 (1974).

5.7.2 Types of Reactors

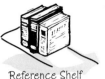

The criteria in Table 5-2 is applied to each of the reactors shown in Figure 5-12 and are also discussed on the CD-ROM in *Professional Reference Shelf R5.4.*

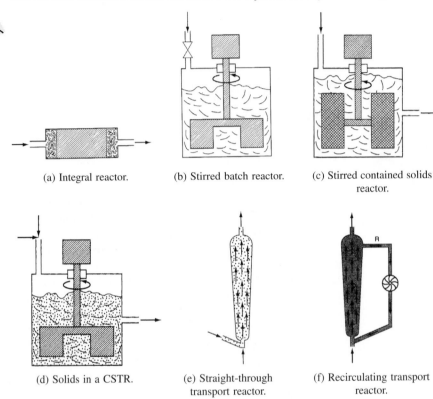

(a) Integral reactor.　　(b) Stirred batch reactor.　　(c) Stirred contained solids reactor.

(d) Solids in a CSTR.　　(e) Straight-through transport reactor.　　(f) Recirculating transport reactor.

Figure 5-12　[From V. Weekman, *AIChe J.*, 20, 833 (1974) with permission of the AIChE. Copyright © 1974 AIChE. All rights reserved.]

5.7.3 Summary of Reactor Ratings

The ratings of the various reactors are summarized in Table 5-3. From this table one notes that the CSTR and recirculating transport reactor appear to be the best choices because they are satisfactory in every category except for construction. However, if the catalyst under study does not decay, the stirred batch and contained solids reactors appear to be the best choices. If the system is not limited by internal diffusion in the catalyst pellet, larger pellets could be used, and the stirred-contained solids is the best choice. If the catalyst is nondecaying and heat effects are negligible, the fixed-bed (integral) reactor would be the top choice, owing to its ease of construction and operation. However, in practice, usually *more than one* reactor type is used in determining the reaction rate law parameters.

TABLE 5-3.   SUMMARY OF REACTOR RATINGS: GAS–LIQUID, POWDERED CATALYST, DECAYING CATALYST SYSTEM[a]

Reactor Type	Sampling and Analysis	Isothermality	Fluid–Solid Contact	Decaying Catalyst	Ease of Construction
Differential	P–F	F–G	F	P	G
Fixed bed	G	P–F	F	P	G
Stirred batch	F	G	G	P	G
Stirred-contained solids	G	G	F–G	P	F–G
Continuous-stirred tank	F	G	F–G	F–G	P-F
Straight-through transport	F–G	P–F	F–G	G	F–G
Recirculating transport	F–G	G	G	F–G	P-F
Pulse	G	F–G	P	F–G	G

[a]G, good; F, fair; P, poor.

Closure. After reading this chapter, the reader should be able to analyze data to determine the rate law and rate law parameters using the graphical and numerical techniques as well as software packages. Nonlinear regression is the easiest method to analyze rate-concentration data to determine the parameters, but the other techniques such as graphical differentiation help one get a feel for the disparities in the data. The reader should be able to describe the care that needs to be taken in using nonlinear regression to ensure you do not arrive on a false minimum for σ^2. Consequently, it is advisable to use more than one method to analyze the data. Finally, the reader should be able to carry out a meaningful discussion on reactor selection to determine the reaction kinetics along with how to efficiently plan experiments.

SUMMARY

1. *Differential method for constant-volume systems*

$$-\frac{dC_A}{dt} = kC_A^\alpha \tag{S5-1}$$

   a. Plot $-\Delta C_A/\Delta t$ as a function of t.
   b. Determine $-dC_A/dt$ from this plot.
   c. Take the ln of both sides of (S5-1) to get

$$\ln\left(-\frac{dC_A}{dt}\right) = \ln k + \alpha \ln C_A \tag{S5-2}$$

Plot $\ln(-dC_A/dt)$ versus $\ln C_A$. The slope will be the reaction order α. We could use finite-difference formulas or software packages to evaluate $(-dC_A/dt)$ as a function of time and concentration.

2. *Integral method*
   a. Guess the reaction order and integrate the mole balance equation.
   b. Calculate the resulting function of concentration for the data and plot it as a function of time. If the resulting plot is linear, you have probably guessed the correct reaction order.
   c. If the plot is not linear, guess another order and repeat the procedure.

3. *Nonlinear regression*: Search for the parameters of the rate law that will minimize the sum of the squares of the difference between the measured rate of reaction and the rate of reaction calculated from the parameter values chosen. For N experimental runs and K parameters to be determined, use Polymath.

$$\sigma_{min}^2 = \sum_{i=1}^{N} \frac{[r_i(\text{measured}) - r_i(\text{calculated})]^2}{N - K} \tag{S5-3}$$

$$s^2 = \sum_{i=1}^{N} (t_{mi} - t_{ci})^2 = \sum_{i=1}^{N} \left[t_{mi} - \frac{C_{A0}^{1-\alpha} - C_{Ai}^{1-\alpha}}{k(1-\alpha)} \right]^2 \tag{S5-4}$$

4. *Method of initial rates:*
   In this method of analysis of rate data, the slope of a plot of $\ln(-r_{A0})$ versus $\ln C_{A0}$ will be the reaction order.

5. *Modeling the differential reactor:*
   The rate of reaction is calculated from the equation

$$-r_A' = \frac{F_{A0}X}{W} = \frac{F_P}{W} = \frac{v_0(C_{A0} - C_{Ae})}{W} = \frac{C_P v_0}{W} \tag{S5-5}$$

In calculating the reaction order, α,

$$-r_A' = kC_A^{\alpha}$$

the concentration of A is evaluated either at the entrance conditions or at a mean value between C_{A0} and C_{Ae}.

CD-ROM MATERIAL

Summary Notes

Interactive

Computer Modules

- **Learning Resources**
    1. *Summary Notes*
    3. *Interactive Computer Modules*
        A. Ecology

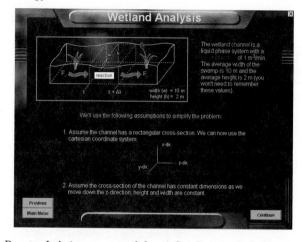

B. Reactor Lab (*www.reactorlab.net*) See Reactor Lab Chapter 4 and P5-3$_B$.
    4. *Solved Problems*
        A. Example Differential Method of Analysis of Pressure-Time Data
        B. Example Integral Method of Analysis of Pressure-Time Data
        C. Example Oxygenating Blood
- **Living Example Problems**
    1. *Example 5-3 Use of Regression to Find the Rate Law Parameters*
- **FAQ [Frequently Asked Questions]—In Updates/FAQ icon section**
- **Professional Reference Shelf**
    R5.1 *Least Squares Analysis of the Linearized Rate Law*
          The CD-ROM describes how the rate law

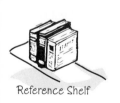

Solved Problems

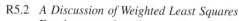

Living Example Problem

Reference Shelf

$$-r_A = kC_A^\alpha \, C_B^\beta$$

is linearized

$$\ln(-r_A) = \ln k + \alpha \ln C_A + \beta \ln C_B$$

and put in the form

$$Y = a_0 + \alpha X_1 + \beta X_2$$

and used to solve for α, β, and k. The etching of a semiconductor, MnO_2, is used as an example to illustrate this technique.

R5.2 *A Discussion of Weighted Least Squares*
      For the case when the error in measurement is not constant, we must use a weighted least squares.
R5.3 *Experimental Planning*
      A. Why perform the experiment?
      B. Are you choosing the correct parameters?
      C. What is the range of your experimental variables?

D. Can you repeat the measurement? (Precision)

E. Milk your data for all it's worth.

F. We don't believe an experiment until it's proven by theory.

G. Tell someone about your result.

R5.4 *Evaluation or Laboratory Reactors (see Table 5-3)*

QUESTIONS AND PROBLEMS

Home Work Problems

The subscript to each of the problem numbers indicates the level of difficulty: A, least difficult; D, most difficult.

$$A = \bullet \quad B = \blacksquare \quad C = \blacklozenge \quad D = \blacklozenge\blacklozenge$$

P5-1$_A$ (a) Compare Table 5-3 on laboratory reactors with a similar table on page 269 of Bisio and Kabel (see Supplementary Reading, listing 1). What are the similarities and differences?

(b) Which of the ICM's for Chapters 4 and 5 was the most fun?

(c) Choose a FAQ from Chapters 4 and 5 and say why it was the most helpful.

(d) Listen to the audios 🎧 on the CD and pick one and say why it could be eliminated.

(e) Create an original problem based on Chapter 5 material.

Creative Thinking

(f) Design an experiment for the undergraduate laboratory that demonstrates the principles of chemical reaction engineering and will cost less than $500 in purchased parts to build. (From 1998 AIChE National Student Chapter Competition) Rules are provided on the CD-ROM.

(g) Plant a number of seeds in different pots (corn works well). The plant and soil of each pot will be subjected to different conditions. Measure the height of the plant as a function of time and fertilizer concentration. Other variables might include lighting, pH, and room temperature. (Great Grade School or High School Science Project)

(h) **Example 5-1.** Discuss the differences for finding $\left[-\dfrac{dC_A}{dt} \right]$ shown in Table E5-3.1 by the three techniques.

(i) **Example 5-2.** Construct a table and plot similar to Table E5-2.1 and Figure E5-2.1, assuming a zero-order and a first-order reaction. Looking at the plots, can either of these orders possibly explain the data?

(j) **Example 5-3.** Explain why the regression had to be carried out twice to find k' and k.

(k) **Example 5-4.** Use regression to analyze the data in Table E5-4.1. What do you find for the reaction order?

(l) **Example 5-5.** Regress the data to fit the rate law

$$r_{CH_4} = kP_{CO}^{\alpha} P_{H_2}^{\beta}$$

What is the difference in the correlation and sums-of-squares compared with those given in Example 5-5? Why was it necessary to regress the data twice, once to obtain Table E5-5.3 and once to obtain Table E5-5.4?

Interactive

P5-2$_A$ Load the Interactive Computer Module (ICM) from the CD-ROM. Run the module and then record your performance number for the module which indicates your mastering of the material. Your professor has the key to decode your performance number.

ICM Ecology Performance # _____.

Computer Modules

Visit Reactor Lab

Links

P5-3$_A$ Go to Professor Herz's **Reactor Lab** on the CD-ROM or on the web at *www.reactorlab.net*. Do (a) one quiz, or (b) two quizzes from Division 1. When you first enter a lab, you see all input values and can vary them. In a lab, click on the Quiz button in the navigation bar to enter the quiz for that lab. In a quiz, you cannot see some of the input values: you need to find those with "???" hiding the values. In the quiz, perform experiments and analyze your data in order to determine the unknown values. See the bottom of the Example Quiz page at *www.reactorlab.net* for equations that relate E and k. Click on the "???" next to an input and supply your value. Your answer will be accepted if is within ±20% of the correct value. Scoring is done with imaginary dollars to emphasize that you should design your experimental study rather than do random experiments. Each time you enter a quiz, new unknown values are assigned. To reenter an unfinished quiz at the same stage you left, click the [i] info button in the Directory for instructions. Turn in copies of your data, your analysis work, and the Budget Report.

P5-4$_B$ When arterial blood enters a tissue capillary, it exchanges oxygen and carbon dioxide with its environment, as shown in this diagram.

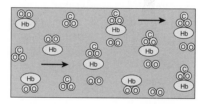

The kinetics of this deoxygenation of hemoglobin in blood was studied with the aid of a **tubular reactor** by Nakamura and Staub [*J. Physiol., 173*, 161].

$$HbO_2 \underset{k_{-1}}{\overset{k_1}{\rightleftharpoons}} Hb + O_2$$

Although this is a reversible reaction, measurements were made in the initial phases of the decomposition so that the reverse reaction could be neglected. Consider a system similar to the one used by Nakamura and Staub: the solution enters a tubular reactor (0.158 cm in diameter) that has oxygen electrodes placed at 5-cm intervals down the tube. The solution flow rate into the reactor is 19.6 cm^3/s with $C_{A0} = 2.33 \times 10^{-6}$ mol/dm^3.

Electrode Position	1	2	3	4	5	6	7
Percent Decomposition of HbO$_2$	0.00	1.93	3.82	5.68	7.48	9.25	11.00

(a) Using the method of differential analysis of rate data, determine the reaction order and the forward specific reaction rate constant k for the deoxygenation of hemoglobin.

(b) Repeat using regression.

P5-5$_A$ The liquid-phase irreversible reaction

$$A \rightarrow B + C$$

is carried out in a CSTR. To learn the rate law the volumetric flow rate, v_0, (hence $\tau = V/v_0$) is varied and the effluent concentrations of species A recorded as a function of the space time τ. Pure A enters the reactor at a concentration of $2\ mol/dm^3$. Steady-state conditions exist when the measurements are recorded.

Run	1	2	3	4	5
τ (min)	15	38	100	300	1200
C_A (mol/dm³)	1.5	1.25	1.0	0.75	0.5

(a) Determine the reaction order and specific reaction rate.
(b) If you were to repeat this experiment to determine the kinetics, what would you do differently? Would you run at a higher, lower, or the same temperature? If you were to take more data, where would you place the measurements (e.g., τ)?
(c) It is believed that the technician may have made a dilution factor-of-10 error in one of the concentration measurements. What do you think? How do your answers compare using regression (Polymath or other software) with those obtained by graphical methods?

Note: All measurements were taken at steady-state conditions.

P5-6$_B$ The reaction

$$A \rightarrow B + C$$

was carried out in a constant-volume batch reactor where the following concentration measurements were recorded as a function of time.

t (min)	0	5	9	15	22	30	40	60
C_A (mol/dm³)	2	1.6	1.35	1.1	0.87	0.70	0.53	0.35

(a) Use nonlinear least squares (i.e., regression) and one other method to determine the reaction order α and the specific reaction rate.
(b) If you were to take more data, where would you place the points? Why?
(c) If you were to repeat this experiment to determine the kinetics, what would you do differently? Would you run at a higher, lower, or the same temperature? Take different data points? Explain.
(d) It is believed that the technician made a dilution error in the concentration measured at 60 min. What do you think? How do your answers compare using regression (Polymath or other software) with those obtained by graphical methods?

P5-7$_B$ The liquid-phase reaction of methanol and triphenyl takes place in a batch reactor at 25°C

$$CH_3OH + (C_6H_5)_3\ CCl \rightarrow (C_6H_5)_3COCH_3 + HCl$$

$$A\quad +\quad B\quad \rightarrow\quad C\quad +\ D$$

For an equal molar feed the following concentration–time data was obtained for methanol:

C_A (mol/dm^3)	0.1	0.95	0.816	0.707	0.50	0.370
t (h)	0	0.278	1.389	2.78	8.33	16.66

The following concentration time data was carried out for an initial methanol concentration 0.01 and an initial triphenyl of 0.1:

C_A (mol/dm^3)	0.01	0.0847	0.0735	0.0526	0.0357
t (h)	0	1	2	5	10

(a) Determine the rate law and rate law parameters.
(b) If you were to take more data points, what would be the reasonable settings (e.g., C_{A0}, C_{B0})? Why?

P5-8$_B$ The following data were reported [C. N. Hinshelwood and P. J. Ackey, *Proc. R. Soc. (Lond)., A115*, 215] for a gas-phase constant-volume decomposition of dimethyl ether at 504°C in a *batch reactor*. Initially, only $(CH_3)_2O$ was present.

Time (s)	390	777	1195	3155	∞
Total Pressure (mmHg)	408	488	562	799	931

(a) Why do you think the total pressure measurement at $t = 0$ is missing? Can you estimate it?
(b) Assuming that the reaction

$$(CH_3)_2O \rightarrow CH_4 + H_2 + CO$$

is irreversible and goes to completion, determine the reaction order and specific reaction rate k.
(c) What experimental conditions would you suggest if you were to obtain more data?
(d) How would the data and your answers change if the reaction were run at a higher or lower temperature?

P5-9$_B$ In order to study the photochemical decay of aqueous bromine in bright sunlight, a small quantity of liquid bromine was dissolved in water contained in a glass battery jar and placed in direct sunlight. The following data were obtained at 25°C:

Time (min)	10	20	30	40	50	60
ppm Br_2	2.45	1.74	1.23	0.88	0.62	0.44

(a) Determine whether the reaction rate is zero, first, or second order in bromine, and calculate the reaction rate constant in units of your choice.
(b) Assuming identical exposure conditions, calculate the required hourly rate of injection of bromine (in pounds) into a sunlit body of water, 25,000 gal in volume, in order to maintain a sterilizing level of bromine of 1.0 ppm. (*Ans.*: 0.43 lb/h)
(c) What experimental conditions would you suggest if you were to obtain more data?

(*Note*: ppm = parts of bromine per million parts of brominated water by weight. In dilute aqueous solutions, 1 ppm ≡ 1 milligram per liter.) (From California Professional Engineers Exam.)

P5-10$_C$ The gas-phase decomposition

$$A \longrightarrow B + 2C$$

is carried out in a *constant-volume batch reactor*. Runs 1 through 5 were carried out at 100°C, while run 6 was carried out at 110°C.

(a) From the data in Table P5-10, determine the reaction order and specific reaction rate.

(b) What is the activation energy for this reaction?

TABLE P5-10.   RAW DATA

Run	Initial Concentration, C_{A0} (g mol/L)	Half-Life, $t_{1/2}$ (min)
1	0.0250	4.1
2	0.0133	7.7
3	0.010	9.8
4	0.05	1.96
5	0.075	1.3
6	0.025	2.0

P5-11$_C$ The reactions of ozone were studied in the presence of alkenes [R. Atkinson et al., *Int. J. Chem. Kinet., 15(8)*, 721 (1983)]. The data in Table P5-11 are for one of the alkenes studied, *cis*-2-butene. The reaction was carried out isothermally at 297 K. Determine the rate law and the values of the rate law parameters.

TABLE P5-11.   RAW DATA

Run	Ozone Rate (mol/s · dm³ × 10⁷)	Ozone Concentration (mol/dm³)	Butene Concentration (mol/dm³)
1	1.5	0.01	10^{-12}
2	3.2	0.02	10^{-11}
3	3.5	0.015	10^{-10}
4	5.0	0.005	10^{-9}
5	8.8	0.001	10^{-8}
6	4.7	0.018	10^{-9}

(*Hint*: Ozone also decomposes by collision with the wall.)

P5-12$_A$ Tests were run on a small experimental reactor used for decomposing nitrogen oxides in an automobile exhaust stream. In one series of tests, a nitrogen stream containing various concentrations of NO_2 was fed to a reactor, and the kinetic data obtained are shown in Figure P5-12. Each point represents one complete run. The reactor operates essentially as an *isothermal backmix reactor (CSTR)*. What can you deduce about the apparent order of the reaction over the temperature range studied?

The plot gives the fractional decomposition of NO_2 fed versus the ratio of reactor volume V (in cm^3) to the NO_2 feed rate, $F_{NO_{2,0}}$ (g mol/h), at different feed concentrations of NO_2 (in parts per million by weight).

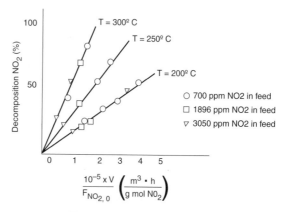

Figure P5-12   Auto exhaust data.

P5-13$_B$ *Microelectronic devices* are formed by first forming SiO_2 on a silicon wafer by chemical vapor deposition (Figure P5-13). This procedure is followed by coating the SiO_2 with a polymer called a photoresist. The pattern of the electronic circuit is then placed on the polymer and the sample is irradiated with ultraviolet light. If the polymer is a positive photoresist, the sections that were irradiated will dissolve in the appropriate solvent, and those sections not irradiated will protect the SiO_2 from further treatment. The wafer is then exposed to strong acids, such as HF, which etch (i.e., dissolve) the exposed SiO_2. It is extremely important to know the kinetics of the reaction so that the proper depth of the channel can be achieved. The dissolution reaction is

$$SiO_2 + 6HF \rightarrow H_2SiF_6 + 2H_2O$$

From the following initial rate data, determine the rate law.

Etching Rate (nm/min)	60	200	600	1000	1400
HF (wt %)	8	20	33	40	48

A total of 1000 thin wafer chips are to be placed in 0.5 dm^3 of 20% HF. If a spiral channel 10 μm wide and 10 m in length were to be etched to a depth of 50 μm on both sides of each wafer, how long should the chips be left in the solution? Assume that the solution is well mixed. (*Ans.*: 330 min)

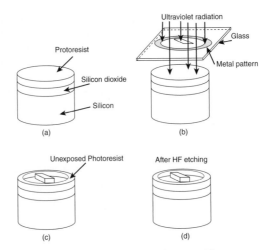

Figure P5-13 Semiconductor etching.

P5-14$_D$ The following reaction

$$C_2H_4Br_2 + 3KI \rightarrow C_2H_4 + 2KBr + KI_3$$

$$(KI_3 \rightleftharpoons I_2 + KI)$$

$$A \quad + \quad 3B \rightarrow \quad C \quad + \quad 2D \quad + \quad E$$

is carried out in a differential reactor, and the flow rate of ethylene is recorded as a function of temperature and entering concentrations.

T(K)	323	333	343	353	363	363
C_2H_4Br (mol/dm³)	0.1	0.1	0.05	0.1	0.2	0.01
KI (mol/dm³)	0.1	0.1	0.1	0.05	0.01	0.01
C_2H_4 (mol/dm³)	0.002	0.006	0.008	0.02	0.02	0.01

The space time for the differential reactor is 2 minutes.

(a)   Determine the rate law and rate law parameters.

(b)   If you were to take more data points, what would be the reasonable settings (e.g., C_{A0})? Why?

P5-15$_B$ The following data as obtained in a batch reactor for the yeast *Saccharomyces cerevisicae*

Yeast Budding

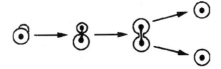

t (h)	0	1	2	3	4	6	8
C_C (g/dm³)	1	1.39	1.93	2.66	3.7	7.12	13.7
C_S (g/dm³)	250	245	238	229	216	197	94.4
C_P (g/dm³)	0	2.17	5.22	9.3	15.3	34	71
$\left(\dfrac{dC_C}{dt}\right)$ (g/dm³ · h)	0.30	0.45	0.63	0.87	1.21	2.32	4.42

Hint: See Chapter 7, section 7.4.

(a) Determine the rate law parameters μ_{max} and K_S, assuming the data can be described by Monod Equation $\dfrac{dC_C}{dt} = r_g = \dfrac{\mu_{max}C_S C_C}{K_S + C_S}$

[*Hint*: It might be best to regress your data taking the reciprocal of the Monod equation in the form $(C_S C_C/r_g)$ vs. C_S. See Chapter 7, section 7.4.]

What is the residual sums of squares?

(b) Determine the rate parameters μ_{max} and k, assuming the data can be fit by the Tessier Equation

$$r_g = \mu_{max}\left[1 - \exp\left(-\frac{C_S}{k}\right)\right]C_C$$

What is the residual sums of squares?

(c) Determine the rate law parameters μ_{max}, k, and λ, assuming the data can be fit by the Moser Equation

$$r_g = \frac{\mu_{max}C_C}{1 + kC_S^{-\lambda}}$$

What is the residual sums of squares?

P5-16$_C$ The thermal decomposition of isopropyl isocyanate was studied in a *differential packed-bed reactor*. From the data in Table P5-16, determine the reaction rate law parameters.

TABLE P5-16.   RAW DATA

Run	Rate (mol/s · dm³)	Concentration (mol/dm³)	Temperature (K)
1	4.9×10^{-4}	0.2	700
2	1.1×10^{-4}	0.02	750
3	2.4×10^{-3}	0.05	800
4	2.2×10^{-2}	0.08	850
5	1.18×10^{-1}	0.1	900
6	1.82×10^{-2}	0.06	950

JOURNAL CRITIQUE PROBLEMS

P5C-1 A *packed-bed reactor* was used to study the reduction of nitric oxide with ethylene on a copper–silica catalyst [*Ind. Eng. Chem. Process Des. Dev., 9*, 455 (1970)]. Develop the integral design equation in terms of the conversion at various initial pressures and temperatures. Is there a significant discrepancy between the experimental results shown in Figures 2 and 3 in the article and the calculated results based on the proposed rate law? If so, what is the possible source of this deviation?

P5C-2 Equation (3) in the article [*J. Chem. Technol. Biotechnol., 31*, 273 (1981)] is the rate of reaction and is incorporated into design equation (2). Rederive the design equation in terms of conversion. Determine the rate dependence on H_2 based on this new equation. How does the order obtained compare with that found by the authors?

P5C-3 See "Kinetics of catalytic esterification of terephthalic acid with methanol vapour" [*Chem. Eng. Sci., 28*, 337 (1973)]. When one observes the data points in Figure 2 of this paper for large times, it is noted that the last data point always falls significantly off the straight-line interpretation. Is it possible to reanalyze these data to determine if the chosen reaction order is indeed correct? Substituting your new rate law into equation (3), derive a new form of equation (10) in the paper relating time and particle radius.

P5C-4 The selective oxidation of toluene and methanol over vanadium pentoxide-supported alkali metal sulfate catalysts was studied [*AIChE J., 27*(1), 41 (1981)]. Examine the experimental technique used (equipment, variables, etc.) in light of the mechanism proposed. Comment on the shortcomings of the analysis and compare with another study of this system presented in *AIChE J., 28*(5), 855 (1982).

- **Additional Homework Problems**

CDP5-A$_B$ The reaction of penicillin G with NH_2OH is carried out in a batch reactor. A colorimeter was used to measure the absorbency as a function of time. [1st Ed. P5-10]

CDP5-B$_B$ The isomerization $A \rightarrow B$ is carried out in a batch reactor. Find α and k. [3rd Ed. P5-3$_B$]

CDP5-C$_C$ The ethane hydrogenolysis over a commercial nickel catalyst was studied in a stirred contained solids reactor.

$$H_2 + C_2H_6 \rightarrow 2CH_4$$

[*Ans: k* = 0.48 mol · atm / kg · h] [3rd ed. P5-14]

CDP5-D$_B$ The half-life of one of the pollutants, NO, in autoexhaust is required. [1st Ed. P5-11]

CDP5-E$_B$ The kinetics of a gas-phase reaction $A_2 \rightarrow 2A$ were studied in a constant-pressure batch reactor in which the volume was measured as a function of time. [1st Ed. P5-6]

CDP5-F$_B$ Reaction kinetics in a tubular reactor.

$$Et_3In(g) + ASH_3 \rightleftharpoons Adduct$$

GaInAs in fiber optics [3rd Ed. P5-6]

CDP5-G$_B$ Differential reactor data for the reaction

$$CH_3CH = CH_2 + O_2 \rightarrow CH_2 = CHCHO + H_2O$$

[3rd Ed. P5-13]

CDP5-H$_B$ Lumping of species for hydrocarbon mixtures. [3rd Ed. P5-16]

CDP5-I$_B$ Prepare an experimental plan to find the rate law. [3rd Ed. P5-17]

CDP5-J$_B$ Batch data on the liquid phase reaction

$$A + B \rightarrow C$$

[3rd Ed. P5-18]

CDP5-K$_B$ Analyze data to see if it fits an elementary reaction

$$2A + B \rightarrow 2C$$

[3rd Ed. P5-21$_B$]

Green Engineering

New Problems on the Web

CDP5-New   From time to time new problems relating Chapter 5 material to everyday interests or emerging technologies will be placed on the web. Solutions to these problems can be obtained by emailing the author. Also, one can go to the web site, *www.rowan.edu/greenengineering*, and work the home problem specific to this chapter.

Links

SUPPLEMENTARY READING

1. A wide variety of techniques for measuring the concentrations of the reacting species may be found in

    BISIO, A., and R. L. KAPEL, *Scaleup of Chemical Processes.* New York: Wiley-Interscience, 1985.

    ROBINSON, J. W., *Undergraduate Instrumental Analysis*, 5th ed. New York: Marcel Dekker, 1995.

    SKOOG, DOUGLAS A., F. JAMES HOLLER, and TIMOTHY A. NIEMAN, *Principles of Instrumental Analysis,* 5th ed. Philadelphia: Saunders College Publishers, Harcourt Brace College Publishers, 1998.

2. A discussion on the methods of interpretation of batch reaction data can be found in

    CRYNES, B. L., and H. S. FOGLER, eds., *AIChE Modular Instruction Series E: Kinetics*, Vol. 2. New York: American Institute of Chemical Engineers, 1981, pp. 51–74.

3. The interpretation of data obtained from flow reactors is also discussed in

    CHURCHILL, S. W., *The Interpretation and Use of Rate Data.* New York: McGraw-Hill, 1974.

4. The design of laboratory catalytic reactors for obtaining rate data is presented in

    RASE, H. F., *Chemical Reactor Design for Process Plants*, Vol. 1. New York: Wiley, 1983, Chap. 5.

5. Model building and current statistical methods applied to interpretation of rate data are presented in

> FROMENT, G. F., and K. B. BISCHOFF, *Chemical Reactor Analysis and Design,* 2nd ed. New York: Wiley, 1990.

> MARKERT, B. A., *Instrumental Element and Multi-Element Analysis of Plant Samples: Methods and Applications.* New York: Wiley, 1996.

6. The sequential design of experiments and parameter estimation is covered in

> BOX, G. E. P., W. G. HUNTER, and J. S. HUNTER, *Statistics for Experimenters: An Introduction to Design, Data Analysis, and Model Building.* New York: Wiley, 1978.

Multiple Reactions **6**

The breakfast of champions is not cereal, it's your opposition.
                                                                    Nick Seitz

Overview. Seldom is the reaction of interest the *only one* that occurs in a chemical reactor. Typically, multiple reactions will occur, some desired and some undesired. One of the key factors in the economic success of a chemical plant is the minimization of undesired side reactions that occur along with the desired reaction.

In this chapter, we discuss reactor selection and general mole balances for multiple reactions. First, we describe the four basic types of multiple reactions: series, parallel, independent, and complex. Next, we define the selectivity parameter and discuss how it can be used to minimize unwanted side reactions by proper choice of operating conditions and reactor selection. We then develop the algorithm that can be used to solve reaction engineering problems when multiple reactions are involved. Finally, a number of examples are given that show how the algorithm is applied to a number of real reactions.

6.1 Definitions

6.1.1 Types of Reactions

There are four basic types of multiple reactions: series, parallel, complex, and independent. These types of multiple reactions can occur by themselves, in pairs, or all together. When there is a combination of parallel and series reactions, they are often referred to as complex reactions.

Parallel **reactions** (also called *competing reactions*) are reactions where the reactant is consumed by two different reaction pathways to form different products:

Parallel reactions

An example of an industrially significant parallel reaction is the oxidation of ethylene to ethylene oxide while avoiding complete combustion to carbon dioxide and water.

Serious chemistry

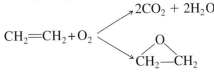

Series **reactions** (also called *consecutive reactions*) are reactions where the reactant forms an intermediate product, which reacts further to form another product:

Series reactions

$$A \xrightarrow{\ k_1\ } B \xrightarrow{\ k_2\ } C$$

An example of a series reaction is the reaction of ethylene oxide (EO) with ammonia to form mono-, di-, and triethanolamine:

$$\overset{O}{\overset{\diagup\,\diagdown}{CH_2-CH_2}} + NH_3 \longrightarrow HOCH_2CH_2NH_2$$

$$\xrightarrow{\ EO\ } (HOCH_2CH_2)_2NH \xrightarrow{\ EO\ } (HOCH_2CH_2)_3N$$

In recent years the shift has been toward the production of diethanolamine as the *desired* product rather than triethanolamine.

Complex **reactions** are multiple reactions that involve a combination of both series and parallel reactions, such as

$$A+B \longrightarrow C+D$$
$$A+C \longrightarrow E$$

An example of a combination of parallel and series reactions is the formation of butadiene from ethanol:

Complex reactions: coupled simulta- neous series and parallel reactions

$$C_2H_5OH \longrightarrow C_2H_4+H_2O$$
$$C_2H_5OH \longrightarrow CH_3CHO+H_2$$
$$C_2H_4+CH_3CHO \longrightarrow C_4H_6+H_2O$$

Independent **reactions** are reactions that occur at the same time but neither the products nor reactants react with themselves or one another.

Independent reactions

$$A \longrightarrow B+C$$
$$D \longrightarrow E+F$$

An example is the cracking of crude oil to form gasoline where two of the many reactions occurring are

$$C_{15}H_{32} \longrightarrow C_{12}C_{26}+C_3H_6$$
$$C_8H_{18} \longrightarrow C_6H_{14}+C_2H_4$$

Desired and Undesired Reactions. Of particular interest are reactants that are consumed in the formation of a *desired product*, D, and the formation of an *undesired product*, U, in a competing or side reaction. In the parallel reaction sequence

$$A \xrightarrow{k_D} D$$
$$A \xrightarrow{k_U} U$$

or in the series sequence

$$A \xrightarrow{k_D} D \xrightarrow{k_U} U$$

We want to minimize the formation of U and maximize the formation of D because the greater the amount of undesired product formed, the greater the cost of separating the undesired product U from the desired product D (Figure 6-1).

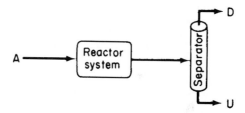

Figure 6-1   Reaction-separation system producing both desired and undesired products.

In a highly efficient and costly reactor scheme in which very little of undesired product U is formed in the reactor, the cost of the separation process could be quite low. On the other hand, even if a reactor scheme is inexpensive and inefficient resulting in the formation of substantial amounts of U, the cost of the separation system could be quite high. Normally, as the cost of a reactor system increases in an attempt to minimize U, the cost of separating species U from D decreases (Figure 6-2).

The economic incentive

Selectivity tells us how one product is favored over another when we have multiple reactions. We can quantify the formation of D with respect to U by defining the selectivity and yield of the system. The **instantaneous selectivity** of D with respect to U is the ratio of the rate of formation of D to the rate of formation of U.

Instantaneous selectivity

$$S_{D/U} = \frac{r_D}{r_U} = \frac{\text{rate of formation of D}}{\text{rate of formation of U}} \qquad (6\text{-}1)$$

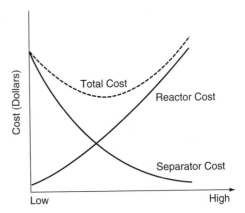

Figure 6-2   Efficiency of a reactor system.

In the next section, we will see how evaluating $S_{D/U}$ will guide us in the design and selection of our reaction system to maximize the selectivity.

Another definition of selectivity used in the current literature, $\tilde{S}_{D/U}$, is given in terms of the flow rates leaving the reactor. $\tilde{S}_{D/U}$ is the **overall selectivity**.

Overall selectivity

$$\boxed{\tilde{S}_{D/U} = \frac{F_D}{F_U} = \frac{\text{Exit molar flow rate of desired product}}{\text{Exit molar flow rate of undesired product}}} \qquad (6\text{-}2)$$

For a batch reactor, the overall selectivity is given in terms of the number of moles of D and U at the end of the reaction time:

$$\tilde{S}_{D/U} = \frac{N_D}{N_U}$$

Example 6–1   Comparing the Overall and Instantaneous Selectives, $\tilde{S}_{D/U}$ and $S_{D/U}$ for a CSTR

Develop a relationship between $S_{D/U}$ and $\tilde{S}_{D/U}$ for a CSTR.

Solution

Consider the instantaneous selectivity for the two parallel reactions just discussed:

$$A \xrightarrow{\ k_D\ } D \qquad\qquad (E6\text{-}1.1)$$

$$A \xrightarrow{\ k_U\ } U \qquad\qquad (E6\text{-}1.2)$$

$$S_{D/U} = \frac{r_D}{r_U} \qquad\qquad (E6\text{-}1.3)$$

The overall selectivity is

$$\tilde{S}_{D/U} = \frac{F_D}{F_U}$$ (E6-1.4)

A mole balance on D for a CSTR yields

$$F_D = r_D V$$ (E6-1.5)

and a mole balance on U yields

$$F_U = r_U V$$ (E6-1.6)

Substituting for F_D and F_U in the overall selectivity Equation (E6-1.4) and canceling the volume, V

$$\tilde{S}_{D/U} = \frac{F_D}{F_U} = \frac{r_D V}{r_U V} = \frac{r_D}{r_U} = S_{D/U}$$ (E6-1.7)

Consequently for a CSTR the overall and instantaneous selectivities are equal; that is,

$$\boxed{\tilde{S}_{D/U} = S_{D/U}}\qquad \text{QED}$$ (E6-1.8)

One can carry out a similar analysis of the series reaction

$$A \to D \to U$$

Two definitions for selectivity and yield are found in the literature.

Reaction yield, like the selectivity, has two definitions: one based on the ratio of reaction rates and one based on the ratio of molar flow rates. In the first case, the yield at a point can be defined as the ratio of the reaction rate of a given product to the reaction rate of the key reactant A. This is sometimes referred to as the instantaneous yield.[1]

Instantaneous yield based on reaction rates

$$\boxed{Y_D = \frac{r_D}{-r_A}}$$ (6-3)

In the case of reaction yield based on molar flow rates, the overall yield, $\tilde{Y}_D$, is defined as the ratio of moles of product formed at the end of the reaction to the number of moles of the key reactant, A, that have been consumed.

For a batch system:

Overall yield based on moles

$$\boxed{\tilde{Y}_D = \frac{N_D}{N_{A0} - N_A}}$$ (6-4)

For a flow system:

Overall yield based on molar flow rates

$$\boxed{\tilde{Y}_D = \frac{F_D}{F_{A0} - F_A}}$$ (6-5)

[1] J. J. Carberry, in *Applied Kinetics and Chemical Reaction Engineering*, R. L. Gorring and V. W. Weekman, eds. (Washington, D.C.: American Chemical Society, 1967), p. 89.

As with selectivity, the instantaneous yield and the overall yield are identical for a CSTR (i.e., $\tilde{Y}_D = Y_D$). Yield and yield coefficients are used extensively in biochemical and biomass reactors. [See P3-14 and Chapter 7.] Yet other definitions of yield even include the stoichiometric coefficients. As a consequence of the different definitions for selectivity and yield, when reading literature dealing with multiple reactions, check carefully to ascertain the definition intended by the author. From an economic standpoint, the *overall* selectivities, $\tilde{S}$, and yields, $\tilde{Y}$, are important in determining profits. However, the rate-based selectivities give insights in choosing reactors and reaction schemes that will help maximize the profit. There often is a conflict between selectivity and conversion (yield) because you want to make a lot of your desired product (D) and at the same time minimize the undesired product (U). However, in many instances, the greater conversion you achieve, not only do you make more D, but you also form more U.

6.2 Parallel Reactions

In this section, we discuss various means of minimizing the undesired product, U, through the selection of reactor type and conditions. We also discuss the development of efficient reactor schemes.

For the competing reactions

$$A \xrightarrow{k_D} D \qquad \text{(desired)}$$

$$A \xrightarrow{k_U} U \qquad \text{(undesired)}$$

the rate laws are

Rate laws for formation of desired and undesired products

$$r_D = k_D C_A^{\alpha_1} \tag{6-6}$$

$$r_U = k_U C_A^{\alpha_2} \tag{6-7}$$

The rate of disappearance of A for this reaction sequence is the sum of the rates of formation of U and D:

$$-r_A = r_D + r_U \tag{6-8}$$

$$-r_A = k_D C_A^{\alpha_1} + k_U C_A^{\alpha_2} \tag{6-9}$$

where α_1 and α_2 are positive reaction orders. We want the rate of formation of D, r_D, to be high with respect to the rate of formation of U, r_U. Taking the ratio of these rates [i.e., Equation (6-6) to Equation (6-7)], we obtain a *rate selectivity parameter*, $S_{D/U}$, which is to be maximized:

Instantaneous selectivity

$$S_{D/U} = \frac{r_D}{r_U} = \frac{k_D}{k_U} C_A^{\alpha_1 - \alpha_2} \tag{6-10}$$

6.2.1 Maximizing the Desired Product for One Reactant

Maximize the rate
selectivity
parameter.

In this section, we examine ways to maximize the instantaneous selectivity, $S_{D/U}$, for different reaction orders of the desired and undesired products.

α_1 is the order of
the desired reaction;
α_2, of the undesired
reaction.

Case 1: $\alpha_1 > \alpha_2$ For the case where the reaction order of the desired product is greater than the reaction order of the undesired product, let a be a positive number that is the difference between these reaction orders ($a > 0$):

$$\alpha_1 - \alpha_2 = a$$

Then, upon substitution into Equation (6-10), we obtain

$$S_{D/U} = \frac{r_D}{r_U} = \frac{k_D}{k_U}\, C_A^a \qquad (6\text{-}11)$$

For $\alpha_1 > \alpha_2$, make
C_A as large as
possible.

To make this ratio as large as possible, we want to carry out the reaction in a manner that will keep the concentration of reactant A as high as possible during the reaction. If the reaction is carried out in the gas phase, we should run it without inerts and at high pressures to keep C_A high. If the reaction is in the liquid phase, the use of diluents should be kept to a minimum.[2]

A batch or plug-flow reactor should be used in this case because, in these two reactors, the concentration of A starts at a high value and drops progressively during the course of the reaction. In a *perfectly mixed* CSTR, the concentration of reactant within the reactor is always at its lowest value (i.e., that of the outlet concentration) and therefore the CSTR should not be chosen under these circumstances.

Case 2: $\alpha_2 > \alpha_1$ When the reaction order of the undesired product is greater than that of the desired product,

$$A \xrightarrow{\;k_D\;} D$$

$$A \xrightarrow{\;k_U\;} U$$

Let $b = \alpha_2 - \alpha_1$, where b is a positive number; then

$$S_{D/U} = \frac{r_D}{r_U} = \frac{k_D C_A^{\alpha_1}}{k_U C_A^{\alpha_2}} = \frac{k_D}{k_U C_A^{\alpha_2 - \alpha_1}} = \frac{k_D}{k_U C_A^b} \qquad (6\text{-}12)$$

For the ratio r_D/r_U to be high, the concentration of A should be as low as possible.

For $\alpha_2 > \alpha_1$ use
a CSTR and dilute
the feed stream.

This low concentration may be accomplished by diluting the feed with inerts and running the reactor at low concentrations of A. A CSTR should be used because the concentrations of reactants are maintained at a low level. A recycle reactor in which the product stream acts as a diluent could be used to maintain the entering concentrations of A at a low value.

Because the activation energies of the two reactions in cases 1 and 2 are not given, it cannot be determined whether the reaction should be run at high

---

[2] For a number of liquid-phase reactions, the proper choice of a solvent can enhance selectivity. See, for example, *Ind. Eng. Chem.*, **62**(9), 16 (1970). In gas-phase heterogeneous catalytic reactions, selectivity is an important parameter of any particular catalyst.

or low temperatures. The sensitivity of the rate selectivity parameter to temperature can be determined from the ratio of the specific reaction rates,

Effect of
temperature on
selectivity

$$S_{D/U} \sim \frac{k_D}{k_U} = \frac{A_D}{A_U} e^{-[(E_D - E_U)/RT]} \tag{6-13}$$

where A is the frequency factor, E the activation energy, and the subscripts D and U refer to desired and undesired product, respectively.

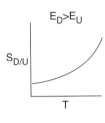

Case 3: $E_D > E_U$ In this case, the specific reaction rate of the desired reaction k_D (and therefore the overall rate r_D) increases more rapidly with increasing temperature than does the specific rate of the undesired reaction k_U. Consequently, the reaction system should be operated at the highest possible temperature to maximize $S_{D/U}$.

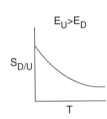

Case 4: $E_U > E_D$ In this case, the reaction should be carried out at a low temperature to maximize $S_{D/U}$, but not so low that the desired reaction does not proceed to any significant extent.

Example 6–2   Maximizing the Selectivity for the Trambouze Reactions

Reactant A decomposes by three simultaneous reactions to form three products, one that is desired, B, and two that are undesired, X and Y. These gas-phase reactions, along with the appropriate rate laws, are called the Trambouze reactions [*AIChE J.* 5, 384 (1959)].

1) $A \xrightarrow{\ k_1\ } X \quad -r_{1A} = r_X = k_1 \quad = 0.0001 \dfrac{\text{mol}}{\text{dm}^3 \cdot \text{s}}$

2) $A \xrightarrow{\ k_2\ } B \quad -r_{2A} = r_B = k_2 C_A = (0.0015\ s^{-1}) C_A$

3) $A \xrightarrow{\ k_3\ } Y \quad -r_{3A} = r_Y = k_3 C_A^2 = 0.008 \dfrac{\text{dm}^3}{\text{mol} \cdot \text{s}} C_A^2$

The specific reaction rates are given at 300 K and the activation energies for reactions (1), (2), and (3) are $E_1 = 10,000$ kcal/mole, $E_2 = 15,000$ kcal/mole, and $E_3 = 20,000$ kcal/mole. How and under what conditions (e.g., reactor type(s), temperature, concentrations) should the reaction be carried out to maximize the selectivity of B for an entering concentration of A of 0.4M and a volumetic flow rate of 2.0 dm³/s.

Solution

The selectivity with respect to B is

$$\boxed{S_{B/XY} = \frac{r_B}{r_X + r_Y} = \frac{k_2 C_A}{k_1 + k_3 C_A^2}} \tag{E6-2.1}$$

Plotting $S_{B/XY}$ vs. C_A, we see there is a maximum as shown in Figure E6-2.1.

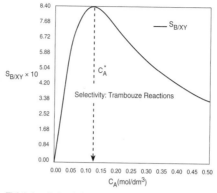

Figure E6-2.1   Selectivity as a function of the concentration of A.

As we can see, the selectivity reaches a maximum at a concentration C_A^p. Because the concentration changes down the length of a PFR, we cannot operate at this maximum. Consequently, we will use a CSTR and design it to operate at this maximum. To find the maximum, C_A^p, we differentiate $S_{B/XY}$ wrt C_A, set the derivative to zero, and solve for C_A^p. That is,

$$\frac{dS_{B/XY}}{dC_A} = 0 = \frac{k_2[k_1 + k_3 C_A^{*2}] - k_2 C_A^*[2k_3 C_A^*]}{[k_1 + k_3 C_A^{*2}]^2} \qquad \text{(E6-2.2)}$$

Solving for C_A^p

$$C_A^* = \sqrt{\frac{k_1}{k_3}} = \sqrt{\frac{0.0001\,(\text{mol/dm}^3 \cdot \text{s})}{0.008\,(\text{dm}^3/\text{mol}) \cdot \text{s}}} = 0.112\,\text{mol/dm}^3 \qquad \text{(E6.2.3)}$$

We see from Figure E6-2.1 the selectivity is indeed a maximum at $C_A^* = 0.112\,\text{mol/dm}^3$

$$\boxed{C_A^* = \sqrt{\frac{k_1}{k_3}} = 0.112\,\text{dm}^3/\text{mol}}$$

We therefore want to carry out our reaction in such a manner that C_A^p is always at this value. Consequently, we will use a CSTR operated at C_A^p. The corresponding selectivity at C_A^p is

$$S_{B/XY} = \frac{k_2 C_A^*}{k_1 + k_3 C_A^{*2}} = \frac{k_2 \sqrt{\dfrac{k_1}{k_3}}}{k_1 + k_1} = \frac{k_2}{2\sqrt{k_1 k_3}} = \frac{0.0015}{2[(0.0001)(0.008)]^{1/2}} \qquad \text{(E6-2.4)}$$

$$\boxed{S_{B/XY} = 0.84}$$

Operate at this CSTR reactant concentration:
$C_A^* = 0.112\,\text{mol/dm}^3$.

We now calculate this CSTR volume and conversion. The net rate of formation of A from reactions (1), (2), and (3) is

$$r_A = r_{1A} + r_{2A} + r_{3A} = -k_{1A} - k_{2A}C_A - k_{3A}C_A^2 \qquad \text{(E6-2.5)}$$

$$-r_A = k_1 + k_2 C_A + k_3 C_A^2$$

Using Equation (E6-2.5) in the mole balance on a CSTR for this liquid-phase reaction ($\upsilon = \upsilon_0$),

$$V = \frac{\upsilon_0 [C_{A0} - C_A^*]}{-r_A} = \frac{\upsilon_0 [C_{A0} - C_A^*]}{[k_1 + k_2 C_A^* + k_3 C_A^{*2}]} \qquad \text{(E6-2.6)}$$

$$\tau = \frac{V}{\upsilon_0} = \frac{C_{A0} - C_A^*}{-r_A^*} = \frac{(C_{A0} - C_A^*)}{k_1 + k_2 C_A^* + k_3 C_A^{*2}} \qquad \text{(E6-2.7)}$$

$$\tau = \frac{(0.4 - 0.112)}{(0.0001) + (0.0015)(0.112) + 0.008(0.112)^2} = \frac{(0.4 - 0.112)}{0.0001 + 0.00168 + 0.0001} = 783 \text{ s}$$

CSTR volume to
maximize selectivity
$\tilde{S}_{B/XY} = S_{B/XY}$

$$V = \upsilon_0 \tau = (2 \text{ dm}^3/\text{s})(783 \text{ s})$$

$$\boxed{V = 1566 \text{ dm}^3}$$

Maximize the selectivity wrt temperature

$$S_{B/XY} = \frac{k_2 C_A^*}{k_1 + k_3 C_A^{*2}} = \frac{k_2 \sqrt{\dfrac{k_1}{k_3}}}{k_1 + k_1} = \frac{k_2}{2\sqrt{k_1 k_3}} \qquad \text{(E6-2.4)}$$

At what temperature should we operate the CSTR?

$$S_{B/XY} = \frac{A_2}{2\sqrt{A_1 A_3}} \exp \left[\frac{\dfrac{E_1 + E_3}{2} - E_2}{RT} \right] \qquad \text{(E6-2.8)}$$

Case 1: If $\dfrac{E_1 + E_3}{2} < E_2$ $\left\{ \begin{array}{l} \text{Run at as high a temperature as possible with existing} \\ \text{equipment and watch out for other side reactions that} \\ \text{might occur at higher temperatures.} \end{array} \right.$

Case 2: If $\dfrac{E_1 + E_3}{2} > E_2$ $\left\{ \begin{array}{l} \text{Run at low temperatures but not so low that a significant} \\ \text{conversion is not achieved.} \end{array} \right.$

For the activation energies given above

$$\frac{E_1 + E_3}{2} - E_2 = \frac{10,000 + 20,000}{2} - 15,000 = 0$$

So the selectivity for this combination of activation energies is independent of temperature!

What is the conversion of A in the CSTR?

$$X^* = \frac{C_{A0} - C_A^*}{C_{A0}} = \frac{0.4 - 0.112}{0.4} = 0.72$$

If greater than 72% conversion of A is required, then the CSTR operated with a reactor concentration of 0.112 mol/dm³ should be followed by a PFR because the concentration and hence selectivity will decrease continuously from C_A^p as we move down the PFR to an exit concentration C_{Af}. Hence the system

How can we
increase the conver-
sion and still have a
high selectivity
$S_{B/XY}$?

$$\left[CSTR\Big|_{C_A^*} + PFR\Big|_{C_A^*}^{C_{Af}} \right]$$

would give the highest selectivity while forming more of the desired product B, beyond what was formed at C_A^p in a CSTR.

Optimum CSTR followed by a PFR.

The exit concentrations of X, Y, and B can now be found from the CSTR mole balances

Species X

$$V = \frac{v_0 C_X}{r_X} = \frac{v_0 C_X}{k_1} \tag{E6-2.9}$$

Rearranging yields

$$C_X = \frac{k_1 V}{v_0} = k_1 \tau = 0.0001 (\text{ mol/dm}^3 \cdot \text{s})(783 \text{ s})$$

$$\boxed{\begin{array}{l} C_X = 0.0783 \text{ mol/dm}^3 \\ F_X = v_0 C_X = 0.156 \text{ mol/s} \end{array}}$$

Species B

$$V = \frac{v_0 C_B}{r_B} = \frac{v_0 C_B}{k_2 C_A^*} \tag{E6-2.10}$$

Rearranging yields

$$C_B = \frac{k_2 V C_A^*}{v_0} = k_2 \tau C_A^* = (0.0015 \text{ s}^{-1})(783 \text{ s})(0.112 \text{ mol/dm}^3) = 0.132 \text{ mol/dm}^3$$

$$\boxed{\begin{array}{l} C_B = 0.132 \text{ mol/dm}^3 \\ F_B = 0.264 \text{ mol/s} \end{array}}$$

Species Y

$$C_Y = \frac{r_Y V}{v_0} = k_3 C_A^{*2} \tau = (0.008 \text{dm}^3/\text{mol} \cdot \text{s})(0.112 \text{mol/dm}^3)^2 (783 \text{ s}) = 0.0786 \text{ mol/dm}^3$$

$$\boxed{\begin{array}{l} C_Y = 0.0786 \text{ mol/dm}^3 \\ F_Y = 0.157 \text{ mol/s} \end{array}}$$

Let's check to make sure the sum of all the species in solution equals the initial concentration $C_{A0} = 0.4$.

$$C_A + C_X + C_B + C_Y = 0.112 + 0.0783 + 0.132 + 0.0786 = 0.4 \quad \text{QED}$$

The reason we want to use a PFR after we reach the maximum selectivity, $S_{B/XY}$, is that the PFR will continue to gradually reduce C_A. Thus, more B will be formed than if another CSTR were to follow. If 90% conversion were required then the exit concentration would be $C_{Af} = (1 - 0.9)(0.4 \text{ mol/dm}^3) = 0.04 \text{ mol/dm}^3$.

The PFR mole balances for this liquid-phase reaction ($v = v_0$) are

$$v_0 \frac{dC_A}{dV} = \frac{dC_A}{d\tau} = r_A, \quad \frac{dC_X}{d\tau} = r_X, \quad \frac{dC_B}{d\tau} = r_B, \quad \frac{dC_Y}{d\tau} = r_Y \qquad \text{(E6.2-11)}$$

Dividing v_0 into V to form τ and then combining the previous mole balances with their respective rate laws, we obtain

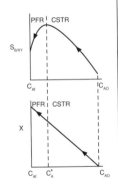

$\dfrac{dC_A}{d\tau} = -k_1 - k_2 C_A - k_3 C_A^2$ (at $\tau = 0$, then $C_A = 0.112$ mol/dm^3)	(E6.2-12)
$\dfrac{dC_X}{d\tau} = k_1$ (at $\tau = 0$, then $C_X = 0.0783$ mol/dm^3)	(E6.2-13)
$\dfrac{dC_B}{d\tau} = k_2 C_A$ (at $\tau = 0$, then $C_B = 0.132$ mol/dm^3)	(E6.2-14)
$\dfrac{dC_Y}{d\tau} = k_3 C_A^2$ (at $\tau = 0$, then $C_Y = 0.0786$ mol/dm^3)	(E6.2-15)

The conversion can be calculated from Equation (E6.2-16)

$$X = \frac{C_{A0} - C_A}{C_{A0}} \qquad \text{(E6.2-16)}$$

We note that at $\tau = 0$, the values of the entering concentrations to the PFR are the exit concentrations from the CSTR. We will use Polymath to plot the exit concentrations as a function of τ and then determine the volume ($V = v_0\tau$) for 90% conversion ($C_A = 0.04 \text{ mol/dm}^3$) and then find C_X, C_B, and C_Y at this volume. This volume turns out to be approximately 600 dm^3. The Polymath program along with the exit concentrations and selectivity are shown in Table E6-2.1.

At the exit of the PFR, $C_A = 0.037$, $C_X = 0.11$, $C_B = 0.16$, and $C_Y = 0.09$ all

An economic decision

in mol/dm^3. The corresponding molar flow rates are $F_X = 0.22$ mol/s, $F_B = 0.32$ mol/s, and $F_Y = 0.18$ mol/s. One now has to make a decision as to whether adding the PFR to increase the conversion of A from 0.72 to 0.9 and the molar flow rate of B from 0.26 to 0.32 mol/s is worth not only the added cost of the PFR, but also the decrease in selectivity. This reaction was carried out isothermally; nonisothermal multiple reactions will be discussed in Chapter 8.

Living Example Problem

TABLE E6-2.1.   POLYMATH PROGRAM FOR PFR FOLLOWING CSTR

POLYMATH Results

Example 6-2 Maximizing the Selectivity for the Trambouze Reactions 08-12-2004, Rev5.1.232

ODE Report (RKF45)   Calculated values of the DEQ variables

Differential equations as entered by the user

[1] d(Ca)/d(tau) = -k1-k2*Ca-k3*Ca^2
[2] d(Cx)/d(tau) = k1
[3] d(Cb)/d(tau) = k2*Ca
[4] d(Cy)/d(tau) = k3*Ca^2

Explicit equations as entered by the user

[1] Cao = 0.4
[2] X = 1-Ca/Cao
[3] k1 = 0.0001
[4] k2 = 0.0015
[5] k3 = 0.008
[6] Sbxy = Cb/(Cx+Cy)

Variable	initial value	minimal value	maximal value	final value
tau	0	0	300	300
Ca	0.112	0.0376688	0.112	0.0376688
Cx	0.0783	0.0783	0.1083	0.1083
Cb	0.132	0.132	0.1634944	0.1634944
Cy	0.0786	0.0786	0.0914368	0.0914368
Cao	0.4	0.4	0.4	0.4
X	0.72	0.72	0.905828	0.905828
k1	1.0E-04	1.0E-04	1.0E-04	1.0E-04
k2	0.0015	0.0015	0.0015	0.0015
k3	0.008	0.008	0.008	0.008
Sbxy	0.8413002	0.8185493	0.8413002	0.8185493

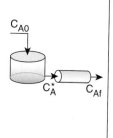

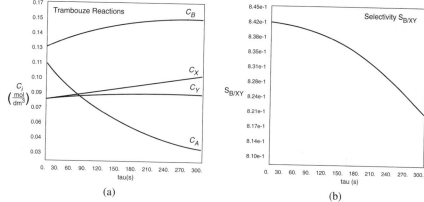

Figure E6-2.2 (a) PFR concentration profiles; (b) PFR Selectivity profile.

6.2.2 Reactor Selection and Operating Conditions

Next consider two simultaneous reactions in which two reactants, A and B, are being consumed to produce a desired product, D, and an unwanted product, U, resulting from a side reaction. The rate laws for the reactions

$$A + B \xrightarrow{k_1} D$$
$$A + B \xrightarrow{k_2} U$$

are

$$r_D = k_1 C_A^{\alpha_1} C_B^{\beta_1} \tag{6-14}$$

$$r_U = k_2 C_A^{\alpha_2} C_B^{\beta_2} \qquad (6\text{-}15)$$

The rate selectivity parameter

Instantaneous
selectivity

$$S_{D/U} = \frac{r_D}{r_U} = \frac{k_1}{k_2} C_A^{\alpha_1 - \alpha_2} C_B^{\beta_1 - \beta_2} \qquad (6\text{-}16)$$

is to be maximized. Shown in Figure 6-3 are various reactor schemes and conditions that might be used to maximize $S_{D/U}$.

Reactor Selection

Criteria:
- Selectivity
- Yield
- Temperature control
- Safety
- Cost

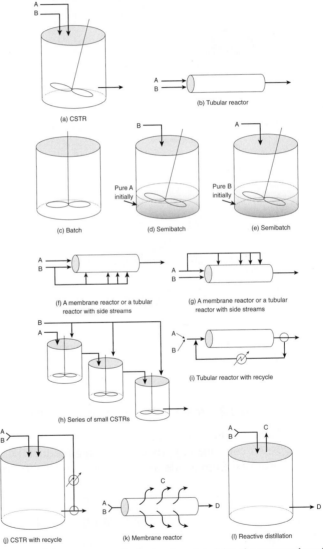

Figure 6.3   Different reactors and schemes for minimizing the unwanted product.

The two reactors with recycle shown in (i) and (j) can be used for highly exothermic reactions. Here the recycle stream is cooled and returned to the reactor to dilute and cool the inlet stream thereby avoiding hot spots and runaway reactions. The PFR with recycle is used for gas-phase reactions, and the CSTR is used for liquid-phase reactions. The last two reactors, (k) and (l), are used for thermodynamically limited reactions where the equilibrium lies far to the left (reactant side)

$$A + B \; \rightleftarrows \; C + D$$

and one of the products must be removed (e.g., C) for the reaction to continue to completion. The membrane reactor (k) is used for thermodynamically limited gas-phase reactions, while reactive distillation (l) is used for liquid-phase reactions when one of the products has a higher volatility (e.g., C) than the other species in the reactor.

Example 6–3   Choice of Reactor and Conditions to Minimize Unwanted Products

For the parallel reactions

$$A + B \longrightarrow D: \qquad r_D = k_1 C_A^{\alpha_1} C_B^{\beta_1}$$

$$A + B \longrightarrow U: \qquad r_U = k_2 C_A^{\alpha_2} C_B^{\beta_2}$$

consider all possible combinations of reaction orders and select the reaction scheme that will maximize $S_{D/U}$.

Solution

Case I: $\alpha_1 > \alpha_2$, $\beta_1 > \beta_2$. Let $a = \alpha_1 - \alpha_2$ and $b = \beta_1 - \beta_2$, where a and b are positive constants. Using these definitions, we can write Equation (6-16) in the form

$$S_{D/U} = \frac{r_D}{r_U} = \frac{k_1}{k_2} C_A^a C_B^b \qquad\qquad \text{(E6-3.1)}$$

To maximize the ratio r_D/r_U, maintain the concentrations of both A and B as high as possible. To do this, use

- A tubular reactor [Figure 6.3(b)]
- A batch reactor [Figure 6.3(c)]
- High pressures (if gas phase), and reduce inerts

Case II: $\alpha_1 > \alpha_2$, $\beta_1 < \beta_2$. Let $a = \alpha_1 - \alpha_2$ and $b = \beta_2 - \beta_1$, where a and b are positive constants. Using these definitions we can write Equation (6.16) in the form

$$S_{D/U} = \frac{r_D}{r_U} = \frac{k_1 C_A^a}{k_2 C_B^b} \qquad\qquad \text{(E6-3.2)}$$

To make $S_{D/U}$ as large as possible, we want to make the concentration of A high and the concentration of B low. To achieve this result, use

- A semibatch reactor in which B is fed slowly into a large amount of A [Figure 6.3(d)]
- A membrane reactor or a tubular reactor with side streams of B continually fed to the reactor [Figure 6.3(f)]
- A series of small CSTRs with A fed only to the first reactor and small amounts of B fed to each reactor. In this way B is mostly consumed before the CSTR exit stream flows into the next reactor [Figure 6.3(h)]

Case III: $\alpha_1 < \alpha_2$, $\beta_1 < \beta_2$. Let $a = \alpha_2 - \alpha_1$ and $b = \beta_2 - \beta_1$, where a and b are positive constants. Using these definitions we can write Equation (6.16) in the form

$$S_{D/U} = \frac{r_D}{r_U} = \frac{k_1}{k_2 C_A^a C_B^b} \tag{E6-3.3}$$

To make $S_{D/U}$ as large as possible, the reaction should be carried out at low concentrations of A and of B. Use

- A CSTR [Figure 6.3(a)]
- A tubular reactor in which there is a large recycle ratio [Figure 6.3(i)]
- A feed diluted with inerts
- Low pressure (if gas phase)

Case IV: $\alpha_1 < \alpha_2$, $\beta_1 > \beta_2$. Let $a = \alpha_2 - \alpha_1$ and $b = \beta_1 - \beta_2$, where a and b are positive constants. Using these definitions we can write Equation (6-16) in the form

$$S_{D/U} = \frac{r_D}{r_U} = \frac{k_1 C_B^b}{k_2 C_A^a} \tag{E6-3.4}$$

To maximize $S_{D/U}$, run the reaction at high concentrations of B and low concentrations of A. Use

- A semibatch reactor with A slowly fed to a large amount of B [Figure 6-3(e)]
- A membrane reactor or a tubular reactor with side streams of A [Figure 6-3(g)]
- A series of small CSTRs with fresh A fed to each reactor [Figure 6-3(h)]

6.3 Maximizing the Desired Product in Series Reactions

In Section 6.1, we saw that the undesired product could be minimized by adjusting the reaction conditions (e.g., concentration) and by choosing the proper reactor. For series of consecutive reactions, the most important variable is time: space-time for a flow reactor and real-time for a batch reactor. To illustrate the importance of the time factor, we consider the sequence

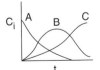

$$A \xrightarrow{k_1} B \xrightarrow{k_2} C$$

in which species B is the desired product.

If the first reaction is slow and the second reaction is fast, it will be extremely difficult to produce species B. If the first reaction (formation of B) is fast and the reaction to form C is slow, a large yield of B can be achieved. However, if the reaction is allowed to proceed for a long time in a batch reactor, or if the tubular flow reactor is too long, the desired product B will be converted to the undesired product C. In no other type of reaction is exactness in the calculation of the time needed to carry out the reaction more important than in series reactions.

Example 6–4 *Maximizing the Yield of the Intermediate Product*

The oxidation of ethanol to form acetaldehyde is carried out on a catalyst of 4 wt % Cu–2 wt % Cr on Al_2O_3.[3] Unfortunately, acetaldehyde is also oxidized on this catalyst to form carbon dioxide. The reaction is carried out in a threefold excess of oxygen and in dilute concentrations (ca. 0.1% ethanol, 1% O_2, and 98.9% N_2). Consequently, the volume change with the reaction can be neglected. Determine the concentration of acetaldehyde as a function of space-time,

$$CH_3CH_2OH(g) \xrightarrow[-H_2O]{+\frac{1}{2}O_2} CH_3CHO \xrightarrow[-2H_2O]{+\frac{5}{2}O_2} 2CO_2$$

The reactions are irreversible and first order in ethanol and acetaldehyde, respectively.

Solution

Because O_2 is in excess, we can write the preceding series reaction as

$$A \xrightarrow{k_1} B \xrightarrow{k_2} C$$

The preceding series reaction can be written as

$$\text{Reaction (1)} \quad A \xrightarrow{k_1} B$$

$$\text{Reaction (2)} \quad B \xrightarrow{k_2} C$$

1. **Mole balance on A:**

$$\frac{dF_A}{dW} = r'_A \tag{E6-4.1}$$

   a. **Rate law:**

$$-r'_A = k_1 C_A$$

   b. **Stoichiometry** (Dilute concentrations: $y_{A0} = 0.001$) $\therefore v = v_0$

$$F_A = C_A v_0$$

---

[3] R. W. McCabe and P. J. Mitchell, *Ind. Eng. Chem. Process Res. Dev.*, 22, 212 (1983).

c. **Combining**, we have

$$v_0 \frac{dC_A}{dW} = -k_1 C_A \tag{E6-4.2}$$

Let $\tau' = W/v_0 = \rho_b V/v_0 = \rho_b \tau$, where ρ_b is the bulk density of the catalyst.

d. Integrating with $C_A = C_{A0}$ at $W = 0$ gives us

$$\boxed{C_A = C_{A0} e^{-k_1 \tau'}} \tag{E6-4.3}$$

2. **Mole balance on B:**

$$\frac{dF_B}{dW} = r'_{B_{net}} \tag{E6-4.4}$$

a. **Rate law** (net):

$$r'_{B_{net}} = r'_{B_{rxn1}} + r'_{B_{rxn2}}$$
$$r'_{B_{net}} = k_1 C_A - k_2 C_B \tag{E6-4.5}$$

b. **Stoichiometry:**

$$F_B = v_0 C_B$$

c. **Combining** yields

$$v_0 \frac{dC_B}{dW} = k_1 C_A - k_2 C_B \tag{E6-4.6}$$

Substituting for C_A, dividing v_0 into W and rearranging, we have

$$\frac{dC_B}{d\tau'} + k_2 C_B = k_1 C_{A0} e^{-k_1 \tau'}$$

$\tau' = \rho_b \tau$
$\rho_b = $ bulk density

d. **Using** the integrating factor gives us

$$\frac{d(C_B e^{+k_2 \tau'})}{d\tau'} = k_1 C_{A0} e^{(k_2 - k_1)\tau'}$$

Links

There is a tutorial on the integrating factor in Appendix A.3 *and* on the web.

At the entrance to the reactor, $W = 0$, $\tau' = W/v_0 = 0$, and $C_B = 0$. Integrating, we get

$$\boxed{C_B = k_1 C_{A0} \left(\frac{e^{-k_1 \tau'} - e^{-k_2 \tau'}}{k_2 - k_1} \right)} \tag{E6-4.7}$$

The concentrations of A, B, and C are shown in Figure E6-4.1.

3. **Optimum yield.** The concentration of B goes through a maximum at a point along the reactor. Consequently, to find the optimum reactor length, we need to differentiate Equation (E6-4.7):

$$\frac{dC_B}{d\tau'} = 0 = \frac{k_1 C_{A0}}{k_2 - k_1} (-k_1 e^{-k_1 \tau'} + k_2 e^{-k_2 \tau'}) \tag{E6-4.8}$$

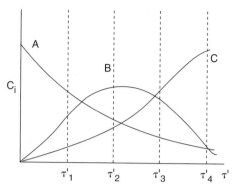

Figure E6-4.1 Concentration profiles down a PBR in terms of space time $\tau' = W/v_0$. [*Note:* $\tau' = \rho_b(W/\rho_b)/v_0 = \rho_b(V/v_0) = \rho_b\tau$.]

Solving for τ'_{opt} gives

$$\tau'_{opt} = \frac{1}{k_1 - k_2} \ln \frac{k_1}{k_2}$$

(E6-4.9)

$$W_{opt} = \frac{v_0}{k_1 - k_2} \ln \frac{k_1}{k_2}$$

(E6-4.10)

The corresponding conversion of A at the maximum C_B is

$$X_{opt} = \frac{C_{A0} - C_A}{C_{A0}} = 1 - e^{-k_1 \tau'_{opt}}$$

Using Equation (E6-4.9) to substitute for τ'_{opt}

$$X_{opt} = 1 - \exp\left[-\ln\left(\frac{k_1}{k_2}\right)^{k_1/(k_1 - k_2)} \right] = 1 - \left(\frac{k_1}{k_2}\right)^{k_1/(k_2 - k_1)}$$

(E6-4.11)

4. Mole Balance on C:

$$\frac{dC_C}{d\tau'} = r'_C = k_2 C_B = \frac{k_1 k_2 C_{A0}}{k_2 - k_1}[e^{-k_1 \tau'} - e^{-k_2 \tau'}]$$

(E6.4-12)

At the entrance to the reactor, no C is present, so the boundary condition is

$$\tau' = 0 \quad C_C = 0$$

$$C_C = \frac{C_{A0}}{k_2 - k_1}[k_2[1 - e^{-k_1 \tau'}] - k_1[1 - e^{-k_2 \tau'}]]$$

(E6.4-13)

Note as $t \rightarrow \infty$ then $C_C = C_{A0}$ as expected

We note that the concentration of C, C_C, could have also been obtained from the overall mole balance

$$C_C = C_{A0} - C_A - C_B$$

(E6.4-14)

We know that the preceding solutions are not valid for $k_1 = k_2$. What would be the equivalent solution for τ_{opt}, W_{opt}, and X_{opt} when $k_1 = k_2$? See P6-1(c).

The yield has been defined as

$$\tilde{Y}_A = \frac{\text{Moles of acetaldehyde in exit}}{\text{Moles of ethanol fed}}$$

and is shown as a function of conversion in Figure E6-4.2.

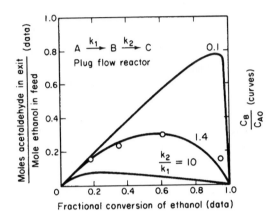

Figure E6-4.2   Yield of acetaldehyde as a function of ethanol conversion. Data were obtained at 518 K. Data points (in order of increasing ethanol conversion) were obtained at space velocities of 26,000, 52,000, 104,000, and 208,000 h^{-1}. The curves were calculated for a first-order series reaction in a plug-flow reactor and show yield of the intermediate species B as a function of the conversion of reactant for various ratios of rate constants k_2 and k_1. [Reprinted with permission from *Ind. Eng. Chem. Prod. Res. Dev.*, 22, 212 (1983). Copyright © 1983 American Chemical Society.]

Another technique is often used to follow the progress for two reactions in series. The concentrations of A, B, and C are plotted as a singular point at different space times (e.g., τ_1', τ_2') on a triangular diagram (see Figure 6-4). The vertices correspond to pure A, B, and C.

For $(k_1/k_2) \gg 1$, a large quantity of B can be obtained.

For $(k_1/k_2) \ll 1$, very little B can be obtained.

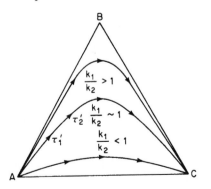

Figure 6-4   Reaction paths for different values of the specific rates.

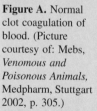

Cut
↓
A + B
↓
C
↓
D
↓
E
↓
F
↓
Clot

Side note: Blood Clotting

Many metabolic reactions involve a large number of sequential reactions, such as those that occur in the coagulation of blood.

$$\text{Cut} \rightarrow \text{Blood} \rightarrow \text{Clotting}$$

Blood coagulation is part of an important host defense mechanism called hemostasis, which causes the cessation of blood loss from a damaged vessel. The clotting process is initiated when a non-enzymatic lipoprotein (called the tissue factor) contacts blood plasma because of cell damage. The tissue factor (TF) normally remains out of contact with the plasma (see Figure B) because of an intact endothelium. The rupture (e.g., cut) of the endothelium exposes the plasma to TF and a cascade of series reactions proceeds (Figure C). These series reactions ultimately result in the conversion of fibrinogen (soluble) to fibrin (insoluble), which produces the clot. Later, as wound healing occurs, mechanisms that restrict formation of fibrin clots, necessary to maintain the fluidity of the blood, start working.

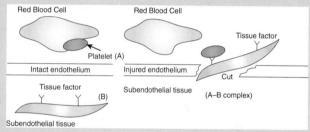

Figure A. Normal clot coagulation of blood. (Picture courtesy of: Mebs, *Venomous and Poisonous Animals*, Medpharm, Stuttgart 2002, p. 305.)

Figure B. Schematic of separation of TF (A) and plasma (B) before cut occurs.

Figure C. Cut allows contact of plasma to initiate coagulation. (A + B → Cascade)

*Platelets provide procoagulant phospholipids-equivalent surfaces upon which the complex-dependent reactions of the blood coagulation cascade are localized.

Solved Problems

An abbreviated form (1) of the initiation and following cascade metabolic reactions that can capture the clotting process is

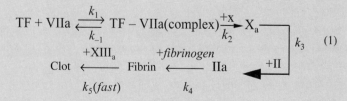

In order to maintain the fluidity of the blood, the clotting sequence (2) must be moderated. The reactions that attenuate the clotting process are

$$\text{ATIII} + \text{Xa} \xrightarrow{k_6} \text{Xa}_{\text{inactive}}$$

$$\text{ATIII} + \text{IIa} \xrightarrow{k_7} \text{IIa}_{\text{inactive}} \tag{2}$$

$$\text{ATIII} + \text{TF} - \text{VIIa} \xrightarrow{k_8} \text{TF} - \text{VIIa}_{\text{inactive}}$$

where TF = tissue factor, VIIa = factor novoseven, X = Stuart Prower factor, Xa = Stuart Prower factor activated, II = prothrombin, IIa = thrombin, ATIII = antithrombin, and XIIIa = factor XIIIa.

One can model the clotting process in a manner identical to the series reactions by writing a mole balance and rate law for each species such as

$$\frac{dC_{\text{TF}}}{dt} = -k_1 \cdot C_{\text{TF}} \cdot C_{\text{VIIa}} + k_{-1} \cdot C_{\text{TF-VIIa}}$$

$$\frac{dC_{\text{VIIa}}}{dt} = -k_1 \cdot C_{\text{TF}} \cdot C_{\text{VIIa}} + k_{-1} \cdot C_{\text{TF-VIIa}}$$

etc.

and then use Polymath to solve the coupled equations to predict the thrombin (shown in Figure D) and other species concentration as a function of time as well as to determine the clotting time. Laboratory data are also shown below for a TF concentration of 5 pM. **One notes that when the complete set of equations are used, the Polymath output is identical to Figure E.** The complete set of equations along with the *Polymath Living Example Problem* code is given on the *Solved Problems* on the CD-ROM. You can load the program directly and vary some of the parameters.

Living Example Problem

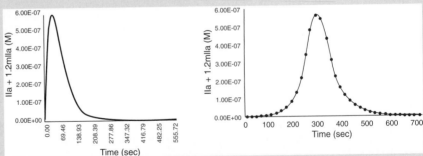

Figure D. Total thrombin as a function of time with an initiating TF concentration of 25 pM (after running Polymath) for the abbreviated blood clotting cascade.

Figure E. Total thrombin as a function of time with an initiating TF concentration of 25 pM (figure courtesy of M. F. Hockin et al., "A Model for the Stoichiometric Regulation of Blood Coagulation," *The Journal of Biological Chemistry,* 277[21], pp. 18322–18333 [2002]). Full blood clotting cascade.

6.4 Algorithm for Solution of Complex Reactions

In complex reaction systems consisting of combinations of parallel and series reactions, the availability of software packages (ODE solvers) makes it much easier to solve problems using moles N_j or molar flow rates F_j rather than conversion. For liquid systems, concentration is usually the preferred variable used in the mole balance equations. The resulting coupled differential mole balance equations can be easily solved using an ODE solver. In fact, this section has been developed to take advantage of the vast number of computational techniques now available on personal computers (Polymath). For gas systems, the molar flow rates are usually the preferred variable in the mole balance equation.

6.4.1 Mole Balances

We begin by writing mole balance equations in variables other than conversion, N_i, C_i, F_i. Table 6-1 gives the forms of the mole balance equation we shall use for complex reactions where r_A and r_B are the **net** rates of formation of A and B.

TABLE 6-1.   MOLE BALANCES FOR MULTIPLE REACTIONS

Mole Balance

$$\frac{dN_A}{dt} = F_{A0} - F_A + \int^V r_A dV$$

These are the forms of the mole balances we will use for multiple reactions.

	Molar Quantities *(Gas or Liquid)*	**Concentration** *(Liquid)*
Batch	$\dfrac{dN_A}{dt} = r_A V$	$\dfrac{dC_A}{dt} = r_A$
	$\dfrac{dN_B}{dt} = r_B V$	$\dfrac{dC_B}{dt} = r_B$
PFR/PBR	$\dfrac{dF_A}{dV} = r_A$	$\dfrac{dC_A}{dV} = \dfrac{r_A}{v_0}$
	$\dfrac{dF_B}{dV} = r_B$	$\dfrac{dC_B}{dV} = \dfrac{r_B}{v_0}$
CSTR	$V = \dfrac{F_{A0} - F_A}{-r_A}$	$V = \dfrac{v_0[C_{A0} - C_A]}{-r_A}$
	$V = \dfrac{F_{B0} - F_B}{-r_B}$	$V = \dfrac{v_0[C_{B0} - C_B]}{-r_B}$
Semibatch B added to **A**	$\dfrac{dN_A}{dt} = r_A V$	$\dfrac{dC_A}{dt} = r_A - \dfrac{v_0 C_A}{V}$
	$\dfrac{dN_B}{dt} = F_{B0} + r_B V$	$\dfrac{dC_B}{dt} = r_B + \dfrac{v_0[C_{B0} - C_B]}{V}$

The algorithm for solving complex reactions is applied to a gas-phase reaction in Figure 6-5. This algorithm is very similar to the one given in Chapter 4 for writing the mole balances in terms of molar flow rates and concentrations (i.e.,

Figure 4-11). After numbering each reaction, we write a mole balance on each species similar to those in Figure 4-11. The major difference between the two algorithms is in rate law step. Here we have four steps (③, ④, ⑤, and ⑥) to find the net rate of reaction for each species in terms of the concentration of the reacting species in order to combine them with their respective mole balances. The remaining steps are analogous to those in Figure 4-8.

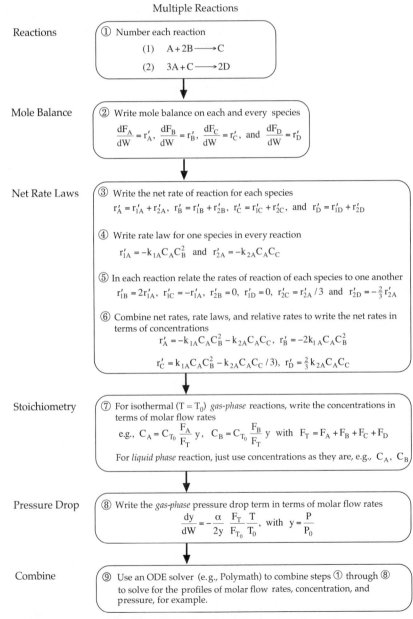

Figure 6-5   Isothermal reaction design algorithm for multiple reactions.

6.4.2 Net Rates of Reaction

6.4.2A Writing the Net Rate of Formation for Each Species

Having written the mole balances, the key point for multiple reactions is to write the **net** rate of formation of each species (e.g., A, B). That is, we have to sum up the rates of formation for each reaction in order to obtain the net rate of formation (e.g., r_A). If q reactions are taking place

$$\text{Reaction 1:} \quad A+B \xrightarrow{k_{1A}} 3C+D$$

$$\text{Reaction 2:} \quad A+2C \xrightarrow{k_{2A}} 3E$$

$$\text{Reaction 3:} \quad 2B+3E \xrightarrow{k_{3B}} 4F$$

$$\vdots$$

$$\text{Reaction } q: \quad E+2F \xrightarrow{k_{qE}} G$$

We note the reaction rate constants, k, in Reactions 1 and 2 are defined with respect to A, while k in Reaction 3 is defined with respect to B.

The net rates of reaction of A and B are found by summing up the rates of formation of A and B for every reaction that species A and B occur.

Net rates of reaction

$$r_A = r_{1A} + r_{2A} + r_{3A} + \cdots + r_{qA} = \sum_{i=1}^{q} r_{iA}$$

$$r_B = r_{1B} + r_{2B} + r_{3B} + \cdots + r_{qB} = \sum_{i=1}^{q} r_{iB}$$

When we sum the rates of the individual reaction for a species, we note that for those reactions in which a species (e.g., A, B) does not appear, the rate is zero. For the first three reactions above, $r_{3A} = 0$, $r_{2B} = 0$, and $r_{2D} = 0$.

In general the net rate of reaction for species j is the sum of all rates of the reactions in which species j appears. For q reactions taking place, the net rate of formation of species j.

r_{ij}
└ species
└ reaction number

$$\boxed{r_j = \sum_{i=1}^{q} r_{ij}} \tag{6-17}$$

6.4.2B Rate Laws

The rate laws for each of the individual reactions are expressed in terms of concentrations, C_j, of the reacting species. A **rate law is needed for one species in each reaction** (i.e., for species j in reaction number i)

$$r_{ij} = k_{ij} f_i(C_A, C_B \cdots C_j \cdots C_n)$$

Here the reaction rate is shown to be dependent on the concentrations of n species. For example, if Reaction 1

Reaction (1): $A + B \xrightarrow{k_{1A}} 3C + D$

followed an elementary rate law, then the rate of disappearance of A in Reaction 1 would be

$$-r_{1A} = k_{1A}C_A C_B$$

and in Reaction 2

Reaction (2): $\dot{A} + 2C \xrightarrow{k_{2A}} 3E$

it would be

$$-r_{2A} = k_{2A}C_A C_C^2$$

or in terms of the rates of formation of A

In Reaction 1: $r_{1A} = -k_{1A}C_A C_B$

In Reaction 2: $r_{2A} = -k_{2A}C_A C_C^2$

Applying Equation (6-17), the net rate of formation of A for these two reactions is

Net rate of reaction
of species A

$$r_A = r_{1A} + r_{2A} = -k_{1A}C_A C_B - k_{2A}C_A C_C^2$$

For a general reaction (Figure 6-5), the rate law for the rate of formation of reactant species A_j in reaction i might depend on the concentration of species A_k and species A_j, for example,

r_{ij}
└ species
└ reaction number

$$r_{ij} = -k_{ij} C_k^2 C_j^3$$

We need to determine the rate law for at least one species in each reaction.

6.4.2C Stoichiometry: Relative Rates of Reaction

The next step is to relate the rate law for a particular reaction and species to other species participating in that reaction. To achieve this relationship, we simply recall the generic reaction from Chapters 2 and 3,

$$aA + bB \longrightarrow cC + dD \tag{2-1}$$

and use Equation (3-1) to relate the rates of disappearance of A and B to the rates of formation of C and D:

$$\frac{-r_A}{a} = \frac{-r_B}{b} = \frac{r_C}{c} = \frac{r_D}{d} \tag{3-1}$$

In working with multiple reactions, it is usually more advantageous to relate the *rates of formation* of each species to one another. This relationship can be achieved by rewriting Equation (3-1) for reaction i in the form

Relative rates of
reaction

$$\boxed{\frac{r_{iA}}{-a_i} = \frac{r_{iB}}{-b_i} = \frac{r_{iC}}{c_i} = \frac{r_{iD}}{d_i}} \tag{6-18}$$

We now will apply Equation (6-18) to reactions 1 and 2

Reaction (1): $A + B \xrightarrow{k_{1A}} 3C + D$ Given: $r_{1A} = -k_{1A}C_A C_B$

We need to relate the rates of formation of the other species in Reaction 1 to the given rate law.

$$\frac{r_{1A}}{-1} = \frac{r_{1B}}{-1} = \frac{r_{1C}}{3} = \frac{r_{1D}}{1}$$

$$r_{1B} = r_{1A} = -k_{1A}C_A C_B$$

$$r_{1C} = 3(-r_{1A}) = 3k_{1A}C_A C_B$$

$$r_{1D} = -r_{1A} = k_{1A}C_A C_B$$

Similarly for Reaction 2,

Reaction (2): $A + 2C \xrightarrow{k_{2A}} 3E$ Given: $r_{2A} = -k_{2A}C_A C_C^2$

$$\frac{r_{2A}}{-1} = \frac{r_{2C}}{-2} = \frac{r_{2E}}{3}$$

the rate of formation of species E in reaction 2, r_{2E}, is

$$r_{2E} = \frac{3}{-1}(r_{2A}) = -3(-k_{2A}C_A C_C^2) = 3k_{2A}C_A C_C^2$$

and the rate of formation of C in reaction 2 is

$$r_{2C} = \frac{-2}{-1}r_{2A} = -2k_{2A}C_A C_C^2$$

r_{2E}
species
reaction
number

6.4.2D Combine Individual Rate Laws to Find the Net Rate

We now substitute the rate laws for each species in each reaction to obtain the net rate of reaction for that species. Again, considering only Reactions 1 and 2

Summary: Rates
- **Relative Rates**
- **Rate Laws**
- **Net Rates**

$$(1) \qquad A + B \xrightarrow{\;k_{1A}\;} 3C + D$$

$$(2) \qquad A + 2C \xrightarrow{\;k_{2A}\;} 3E$$

the net rates of reaction for species A, B, C, D, and E are

$$r_A = r_{1A} + r_{2A} = -k_{1A}C_A C_B - k_{2A}C_A C_C^2$$

$$r_B = r_{1B} = -k_{1A}C_A C_B$$

$$r_C = r_{1C} + r_{2C} = 3k_{1A}C_A C_B - 2k_{2A}C_A C_C^2$$

$$r_D = r_{1D} = k_{1A}C_A C_B$$

$$r_E = r_{2E} = 3k_{2A}C_A C_C^2$$

Now that we have expanded step two of our algorithm, let's consider an example with real reactions.

Example 6–5 Stoichiometry and Rate Laws for Multiple Reactions

Consider the following set of reactions:

Rate Laws[4]

$$\text{Reaction 1:} \quad 4NH_3 + 6NO \longrightarrow 5N_2 + 6H_2O \qquad -r_{1NO} = k_{1NO}C_{NH_3}C_{NO}^{1.5} \qquad (E6\text{-}5.1)$$

$$\text{Reaction 2:} \quad 2NO \longrightarrow N_2 + O_2 \qquad\qquad r_{2N_2} = k_{2N_2}C_{NO}^{2} \qquad\qquad (E6\text{-}5.2)$$

$$\text{Reaction 3:} \quad N_2 + 2O_2 \longrightarrow 2NO_2 \qquad\qquad -r_{3O_2} = k_{3O_2}C_{N_2}C_{O_2}^{2} \qquad (E6\text{-}5.3)$$

Write the rate law for each species in each reaction and then write the net rates of formation of NO, O_2, and N_2.

Solution

A. Net Rates of Reaction

In writing the net rates of reaction, we set the rates to zero for those species that are not in a given reaction. For example, H_2O is not involved in Reactions 2 and 3; therefore, $r_{2H_2O} = 0$ and $r_{3H_2O} = 0$.

The net rates are

$$r_{NO} = r_{1NO} + r_{2NO} + 0 \qquad\qquad (E6\text{-}5.4)$$

$$r_{NH_3} = r_{1NH_3} \qquad\qquad (E6\text{-}5.5)$$

---

[4] From tortusimetry data (11/2/2006).

$$r_{N_2} = r_{1N_2} + r_{2N_2} + r_{3N_2} \tag{E6-5.6}$$

$$r_{H_2O} = r_{1H_2O} \tag{E6-5.7}$$

$$r_{O_2} = r_{2O_2} + r_{3O_2} \tag{E6-5.8}$$

$$r_{NO_2} = r_{3NO_2} \tag{E6-5.9}$$

B.   Relative Rates of Reaction

The rate laws for Reactions 1, 2, and 3 are given in terms of species NO, N_2, and O_2, respectively. Therefore, we need to relate the rates of reactions of other species in a chosen reaction to the given rate laws.

Recalling Equation (6-18), the corresponding rate laws are related by

$$\frac{r_{iA}}{-a_i} = \frac{r_{iB}}{-b_i} = \frac{r_{iC}}{c_i} = \frac{r_{iD}}{d_i} \tag{6-18}$$

Reaction 1: The rate law wrt NO is

$$-r_{1NO} = k_{1NO} C_{NH_3} C_{NO}^{1.5} \tag{E6-5.1}$$

The relative rates are

Multiple reaction stoichiometry

$$\boxed{\frac{r_{1NO}}{-6} = \frac{r_{1NH_3}}{-4} = \frac{r_{1N_2}}{5} = \frac{r_{1H_2O}}{6}} \tag{E6-5.10}$$

Then the rate of disappearance of NH_3 is

Net rate NH_3

$$\boxed{-r_{NH_3} = -r_{1NH_3} = \frac{2}{3}(-r_{1NO}) = \frac{2}{3} k_{1NO} C_{NH_3} C_{NO}^{1.5}} \tag{E6-5.11}$$

$$r_{1N_2} = \frac{5}{6}(-r_{1NO}) = \frac{5}{6} k_{1NO} C_{NH_3} C_{NO}^{1.5} \tag{E6-5.12}$$

$$\boxed{r_{H_2O} = r_{1H_2O} = -r_{1NO} = k_{1NO} C_{NH_3} C_{NO}^{1.5}} \tag{E6-5.13}$$

Reaction 2: $$2NO \rightarrow N_2 + O_2$$

$$\boxed{\frac{r_{2NO}}{-2} = \frac{r_{2N_2}}{1} = \frac{r_{2O_2}}{1}} \tag{E6-5.14}$$

r_{2N_2} is given (i.e., $r_{2N_2} = k_{2N_2} C_{NO}^2$); therefore,

$$-r_{2NO} = 2r_{2N_2} = 2k_{2N_2} C_{NO}^2$$

$$r_{2O_2} = r_{2N_2}$$

Reaction 3: $2O_2 + N_2 \rightarrow 2NO_2$

$$\boxed{\frac{r_{3O_2}}{-2} = \frac{r_{3N_2}}{-1} = \frac{r_{3NO_2}}{2}} \tag{E6-5.15}$$

r_{3O_2} is given (i.e., $r_{3NO_2} = -k_{3O_2}C_{N_2}C_{O_2}^2$; therefore,

$$r_{3NO_2} = -r_{3O_2} = k_{3O_2}C_{N_2}C_{O_2}^2 \tag{E6-5.16}$$

$$-r_{3N_2} = \frac{1}{2}(-r_{3O_2}) = \frac{1}{2}k_{3O_2}C_{N_2}C_{O_2}^2$$

$$\boxed{r_{NO_2} = r_{3NO_2} = -r_{3O_2} = k_{3O_2}C_{N_2}C_{O_2}^2} \tag{E6-5.17}$$

We now combine the individual rates and the rate laws for each reaction to find the net rate of reaction.

Next, let us examine the *net* rate of formations. The net rate of formation of NO is

$$r_{NO} = \sum_{i=1}^{3} r_{iNO} = r_{1NO} + r_{2NO} + 0 \tag{E6-5.18}$$

Net rate NO

$$\boxed{r_{NO} = -k_{1NO}C_{NH_3}C_{NO}^{1.5} - 2k_{2N_2}C_{NO}^2} \tag{E6-5.19}$$

Next consider N_2

$$r_{N_2} = \sum_{i=1}^{3} r_{iN_2} = r_{1N_2} + r_{2N_2} + r_{3N_2} \tag{E6-5.20}$$

Net rate N_2

$$\boxed{r_{N_2} = \frac{5}{6}k_{1NO}C_{NH_3}C_{NO}^{1.5} + k_{2N_2}C_{NO}^2 - \frac{1}{2}k_{3O_2}C_{N_2}C_{O_2}^2} \tag{E6-5.21}$$

Finally O_2

$$r_{O_2} = r_{2O_2} + r_{3O_2} = r_{2N_2} + r_{3O_2} \tag{E6-5.22}$$

Net rate O_2

$$\boxed{r_{O_2} = k_{2N_2}C_{NO}^2 - k_{3O_2}C_{N_2}C_{O_2}^2} \tag{E6-5.23}$$

6.4.3 Stoichiometry: Concentrations

In this step, *if* the reactions are liquid-phase reactions, we can go directly to the combine step. Recall for liquid-phase reactions, $v = v_0$ and

Liquid phase $$C_j = \frac{F_i}{v_0} \tag{6-19}$$

If the reactions are gas-phase reactions, we proceed as follows.

For ideal gases recall Equation (3-42):

Gas phase

$$\boxed{C_j = \frac{F_{T0}}{v_0}\left(\frac{F_j}{F_T}\right)\frac{P}{P_0}\frac{T_0}{T} = C_{T0}\left(\frac{F_j}{F_T}\right)\frac{P}{P_0}\frac{T_0}{T}}$$

(3-42)

where

$$F_T = \sum_{j=1}^{n} F_j$$

(6-20)

and

$$C_{T0} = \frac{P_0}{RT_0}$$

(6-21)

For isothermal systems ($T = T_0$) with no pressure drop ($P = P_0$)

Gas phase

$$C_j = C_{T0}\left(\frac{F_j}{F_T}\right)$$

(6-22)

and we can express the net rates of disappearance of each species (e.g., species 1 and species 2) as a function of the molar flow rates ($F_1, \ldots, F_j$):

$$r_1 = fn_1[C_1, C_2 \ldots C_j] = fn_1\left(C_{T0}\frac{F_1}{F_T}, \ C_{T0}\frac{F_2}{F_T}, \ldots, \ C_{T0}\frac{F_j}{F_T}\right)$$

(6-23)

$$r_2 = fn_2[C_1, C_2 \ldots C_j] = fn_2\left(C_{T0}\frac{F_1}{F_T}, \ldots, \ C_{T0}\frac{F_j}{F_T}\right)$$

(6-24)

where fn represents the functional dependence on concentration of the net rate of formation such as that given in Equation (E6-5.21) for N_2.

6.5 Multiple Reactions in a PFR/PBR

We now insert rate laws written in terms of molar flow rates [e.g., Equation (3-42)] into the mole balances (Table 6-1). After performing this operation for each species, we arrive at a coupled set of first-order ordinary differential equations to be solved for the molar flow rates as a function of reactor volume (i.e., distance along the length of the reactor). In liquid-phase reactions, incorporating and solving for total molar flow rate is not necessary at each step along the solution pathway because there is no volume change with reaction.

Combining mole balance, rate laws, and stoichiometry for species 1 through species j in the gas phase and for isothermal operation with no pressure drop gives us

Coupled ODEs

$$\frac{dF_1}{dV} = r_1 = \sum_{i=1}^{m} r_{i1} = \text{fn}_1\left(C_{T0}\frac{F_1}{F_T}, \ldots, C_{T0}\frac{F_j}{F_T}\right) \qquad (6\text{-}25)$$

$$\vdots$$

$$\frac{dF_j}{dV} = r_j = \sum_{i=1}^{q} r_{ij} = \text{fn}_j\left(C_{T0}\frac{F_1}{F_T}, \ldots, C_{T0}\frac{F_j}{F_T}\right) \qquad (6\text{-}26)$$

For constant-pressure batch systems we would simply substitute N_i for F_i in the preceding equations. For constant-volume batch systems we would use concentrations:

$$C_j = N_j/V_0 \qquad (6\text{-}27)$$

We see that we have j coupled ordinary differential equations that must be solved simultaneously with either a numerical package or by writing an ODE solver. In fact, this procedure has been developed to take advantage of the vast number of computation techniques now available on personal computers (Polymath, MATLAB).

Example 6–6 Combining Mole Balances, Rate Laws, and Stoichiometry for Multiple Reactions

Consider again the reaction in Example 6-5. Write the mole balances on a PFR in terms of molar flow rates for each species.

Reaction 1: $4NH_3 + 6NO \longrightarrow 5N_2 + 6H_2O$ $-r_{1NO} = k_{1NO}C_{NH_3}C_{NO}^{1.5}$ (E6-5.1)

Reaction 2: $2NO \longrightarrow N_2 + O_2$ $r_{2N_2} = k_{2N_2}C_{NO}^2$ (E6-5.2)

Reaction 3: $N_2 + 2O_2 \longrightarrow 2NO_2$ $-r_{3O_2} = k_{3O_2}C_{N_2}C_{O_2}^2$ (E6-5.3)

Solution

For gas-phase reactions, the concentration of species j is

$$C_j = C_{T0}\frac{F_j}{F_T}\frac{P}{P_0}\frac{T_0}{T} \qquad (3\text{-}42)$$

For no pressure drop and isothermal operation,

$$C_j = C_{T0}\frac{F_j}{F_T} \qquad (E6\text{-}6.1)$$

In combining the mole balance, rate laws, and stoichiometry, we will use our results from Example 6-5. The total molar flow rate of all the gases is

$$F_T = F_{NO} + F_{NH_3} + F_{N_2} + F_{H_2O} + F_{O_2} + F_{NO_2} \tag{E6-6.2}$$

We now rewrite mole balances on each species in the total molar flow rate.

Using the results of Example 6-5

(1) Mole balance on NO:

$$\frac{dF_{NO}}{dV} = r_{NO} = -k_{1NO}C_{NH_3}C_{NO}^{1.5} - 2k_{2N_2}C_{NO}^2$$

$$\frac{dF_{NO}}{dV} = -k_{1NO}C_{T0}^{2.5}\left(\frac{F_{NH_3}}{F_T}\right)\left(\frac{F_{NO}}{F_T}\right)^{1.5} - 2k_{2N_2}C_{T0}^2\left(\frac{F_{NO}}{F_T}\right)^2 \tag{E6-6.3}$$

(2) Mole balance on NH_3:

$$\frac{dF_{NH_3}}{dV} = r_{NH_3} = r_{1NH_3} = \frac{2}{3}r_{1NO} = -\frac{2}{3}k_{1NO}C_{NH_3}C_{NO}^{1.5}$$

$$\frac{dF_{NH_3}}{dV} = -\frac{2}{3}k_{1NO}C_{T0}^{2.5}\left(\frac{F_{NH_3}}{F_T}\right)\left(\frac{F_{NO}}{F_T}\right)^{1.5} \tag{E6-6.4}$$

Combined
• Mole balance
• Rate law
• Stoichiometry

(3) Mole balance on H_2O:

$$\frac{dF_{H_2O}}{dV} = r_{H_2O} = r_{1H_2O} = -r_{1NO} = k_{1NO}C_{NH_3}C_{NO}^{1.5}$$

$$\frac{dF_{H_2O}}{dV} = k_{1NO}C_{T0}^{2.5}\left(\frac{F_{NH_3}}{F_T}\right)\left(\frac{F_{NO}}{F_T}\right)^{1.5} \tag{E6-6.5}$$

(4) Mole balance on N_2:

$$\frac{dF_{N_2}}{dV} = r_{N_2} = \frac{5}{6}k_{1NO}C_{NH_3}C_{NO}^{1.5} + k_{2N_2}C_{NO}^2 - \frac{1}{2}k_{3O_2}C_{N_2}C_{O_2}^2$$

$$\frac{dF_{N_2}}{dV} = \frac{5}{6}k_{1NO}C_{T0}^{2.5}\left(\frac{F_{NH_3}}{F_T}\right)\left(\frac{F_{NO}}{F_T}\right)^{1.5} + k_{2N_2}C_{T0}^2\left(\frac{F_{NO}}{F_T}\right)^2 - \frac{1}{2}k_{3O_2}C_{T0}^3\left(\frac{F_{N_2}}{F_T}\right)\left(\frac{F_{O_2}}{F_T}\right)^2$$

$$\tag{E6-6.6}$$

Do we need the
Combine step when
we use Polymath or
another ODE
solver?
(*Answer*: No) (See
Table E6-6.1)

(5) Mole balance on O_2:

$$\frac{dF_{O_2}}{dV} = r_{O_2} = r_{2O_2} + r_{3O_2} = r_{2N_2} + r_{3O_2}$$

$$\boxed{\frac{dF_{O_2}}{dV} = k_{2N_2} C_{T0}^2 \left(\frac{F_{NO}}{F_T}\right)^2 - k_{3O_2} C_{T0}^3 \left(\frac{F_{N_2}}{F_T}\right)\left(\frac{F_{O_2}}{F_T}\right)^2}$$

(E6-6.7)

(6) Mole balance on NO_2:

$$\boxed{\frac{dF_{NO_2}}{dV} = r_{NO_2} = r_{3NO_2} = -r_{3O_2} = k_{3O_2} C_{T0}^3 \left(\frac{F_{N_2}}{F_T}\right)\left(\frac{F_{O_2}}{F_T}\right)^2}$$

(E6-6.8)

The entering molar flow rates, F_{j0}, along with the entering temperature, T_0, and pressure, P_0, $(C_{T0} = P_0/RT_0 = 0.2 \text{ mol/dm}^3)$, are specified as are the specific reaction rates k_{ij} [e.g., $k_{1NO} = 0.43 \text{ (dm}^3/\text{mol)}^{1.5}/\text{s}$, $k_{2N_2} = 2.7 \text{ dm}^3/\text{mol} \cdot \text{s}$, etc.]. Consequently, Equations (E6-6.1) through (E6-6.8) can be solved simultaneously with an ODE solver (e.g., Polymath, MATLAB). In fact, with almost all ODE solvers, the combine step can be eliminated as the ODE solver will do the work. In this case, the ODE solver algorithm is shown in Table E6-6.1.

TABLE E6-6.1.   ODE SOLVER ALGORITHM FOR MULTIPLE REACTIONS

Note: Polymath will
do all the substitu-
tions for you.
Thank you,
Polymath!

(1) $\dfrac{dF_{NO}}{dV} = r_{NO}$	(11) $r_{1N_2} = -\dfrac{5}{6} r_{1NO}$
(2) $\dfrac{dF_{NH_3}}{dV} = r_{NH_3}$	(12) $r_{1H_2O} = -r_{1NO}$
(3) $\dfrac{dF_{N_2}}{dV} = r_{N_2}$	(13) $r_{2NO} = -2r_{2N_2}$
(4) $\dfrac{dF_{O_2}}{dV} = r_{O_2}$	(14) $r_{2O_2} = r_{2N_2}$
(5) $\dfrac{dF_{H_2O}}{dV} = r_{H_2O}$	(15) $r_{3N_2} = \dfrac{1}{2} r_{3O_2}$
(6) $\dfrac{dF_{NO_2}}{dV} = r_{NO_2}$	(16) $r_{3NO_2} = -r_{3O_2}$
(7) $r_{1NO} = -k_{1NO} C_{NH_3} C_{NO}^{1.5}$	(17) $r_{NO} = r_{1NO} + r_{2NO}$
(8) $r_{2N_2} = k_{2N_2} C_{NO}^2$	(18) $r_{NH_3} = r_{1NH_3}$
(9) $r_{3O_2} = -k_{3O_2} C_{N_2} C_{O_2}^2$	(19) $r_{N_2} = r_{1N_2} + r_{2N_2} + r_{3N_2}$
(10) $r_{1NH_3} = \dfrac{2}{3} r_{1NO}$	(20) $r_{O_2} = r_{2O_2} + r_{3O_2}$

TABLE E6-6.1.   ODE SOLVER ALGORITHM FOR MULTIPLE REACTIONS (CONTINUED)

(21) $r_{H_2O} = r_{1H_2O}$	(27) $F_T = F_{NO} + F_{NH_3} + F_{N_2} + F_{O_2} + F_{H_2O} + F_{NO_2}$
(22) $r_{NO_2} = r_{3NO_2}$	(28) $C_{T0} = 0.2$
(23) $C_{NO} = C_{T0}\dfrac{F_{NO}}{F_T}$	(29) $k_{1NO} = 0.43$
(24) $C_{N_2} = C_{T0}\dfrac{F_{N_2}}{F_T}$	(30) $k_{2N_2} = 2.7$
(25) $C_{NH_3} = C_{T0}\dfrac{F_{NH_3}}{F_T}$	(31) $k_{3O_2} = 5.8$
(26) $C_{O_2} = C_{T0}\dfrac{F_{O_2}}{F_T}$	

Summarizing to this point, we show in Table 6-2 the equations for species j and reaction i that are to be combined when we have q reactions and n species.

TABLE 6-2.   SUMMARY OF RELATIONSHIP
FOR MULTIPLE REACTIONS OCCURRING IN A PFR

Mole balances:
$$\frac{dF_j}{dV} = r_j \qquad (6\text{-}26)$$

Rates:

    Relative rates:
$$\frac{r_{iA}}{-a_i} = \frac{r_{iB}}{-b_i} = \frac{r_{iC}}{c_i} = \frac{r_{iD}}{d_i} \qquad (6\text{-}16)$$

    Rate laws:
$$r_{ij} = k_{ij} f_i(C_1, C_j, C_n)$$

The basic equations

    Net laws:
$$r_j = \sum_{i=1}^{q} r_{ij} \qquad (6\text{-}17)$$

Stoichiometry:

    (gas phase)
$$C_j = C_{T0} \frac{F_j}{F_T} \frac{P}{P_0} \frac{T_0}{T} \qquad (3\text{-}42)$$

$$F_T = \sum_{j=1}^{n} F_j \qquad (6\text{-}20)$$

    (liquid phase)
$$C_j = \frac{F_j}{v_0} \qquad (6\text{-}19)$$

Example 6–7 Hydrodealkylation of Mesitylene in a PFR

The production of m-xylene by the hydrodealkylation of mesitylene over a Houdry Detrol catalyst[5] involves the following reactions:

$$\text{[mesitylene structure]} + H_2 \longrightarrow \text{[m-xylene structure]} + CH_4 \qquad \text{(E6-7.1)}$$

m-Xylene can also undergo hydrodealkylation to form toluene:

$$\text{[m-xylene structure]} + H_2 \longrightarrow \text{[toluene structure]} + CH_4 \qquad \text{(E6-7.2)}$$

A significant
economic incentive

The second reaction is undesirable, because m-xylene sells for a higher price than toluene ($1.32/lb vs. $0.30/lb).[6] Thus we see that there is a significant incentive to maximize the production of m-xylene.

The hydrodealkylation of mesitylene is to be carried out isothermally at 1500°R and 35 atm in a packed-bed reactor in which the feed is 66.7 mol% hydrogen and 33.3 mol% mesitylene. The volumetric feed rate is 476 ft³/h and the reactor volume (i.e., $V = W/\rho_b$) is 238 ft³.

The rate laws for reactions 1 and 2 are, respectively,

$$-r_{1M} = k_1 C_M C_H^{0.5} \qquad \text{(E6-7.3)}$$

$$r_{2T} = k_2 C_X C_H^{0.5} \qquad \text{(E6-7.4)}$$

where the subscripts are: M = mesitylene, X = m-xylene, T = toluene, Me = methane, and H = hydrogen (H_2).

At 1500°R, the specific reaction rates are

$$\text{Reaction 1: } k_1 = 55.20 \text{ (ft}^3\text{/lb mol)}^{0.5}\text{/h}$$

$$\text{Reaction 2: } k_2 = 30.20 \text{ (ft}^3\text{/lb mol)}^{0.5}\text{/h}$$

The bulk density of the catalyst has been included in the specific reaction rate (i.e., $k_1 = k_1' \rho_b$).

Plot the concentrations of hydrogen, mesitylene, and xylene as a function of space time. Calculate the space time where the production of xylene is a maximum (i.e., τ_{opt}).

---

[5] *Ind. Eng. Chem. Process Des. Dev.*, 4, 92 (1965); 5, 146 (1966).

[6] November 2004 prices, from *Chemical Market Reporter* (Schnell Publishing Co.), 265, 23 (May 17, 2004). Also see *www.chemweek.com/* and *www.icislor.com*

Solution

$$\text{Reaction 1:} \quad M + H \longrightarrow X + Me \tag{E6-7.1}$$

$$\text{Reaction 2:} \quad X + H \longrightarrow T + Me \tag{E6-7.2}$$

Mole balance on each and every species

1. **Mole balances:**

Hydrogen:
$$\frac{dF_H}{dV} = r_H \tag{E6-7.5}$$

Mesitylene:
$$\frac{dF_M}{dV} = r_M \tag{E6-7.6}$$

Xylene:
$$\frac{dF_X}{dV} = r_X \tag{E6-7.7}$$

Toluene:
$$\frac{dF_T}{dV} = r_T \tag{E6-7.8}$$

Methane:
$$\frac{dF_{Me}}{dV} = r_{Me} \tag{E6-7.9}$$

2. **Rate laws and net rates:** Given

$$\text{Reaction 1:} \quad -r_{1M} = k_1 C_H^{1/2} C_M \tag{E6-7.3}$$

$$\text{Reaction 2:} \quad r_{2T} = k_2 C_H^{1/2} C_X \tag{E6-7.4}$$

Relative rates:

(1) $$\qquad -r_{1H} = -r_{1M} = r_{1Me} = r_{1X} \tag{E6-7.10}$$

(2) $$\qquad r_{2T} = r_{2Me} = -r_{2H} = -r_{2X} \tag{E6-7.11}$$

Net rates:

$$r_M = r_{1M} = -k_1 C_H^{1/2} C_M \tag{E6-7.12}$$

$$r_H = r_{1H} + r_{2H} = r_{1H} - r_{2T} = -k_1 C_H^{1/2} C_M - k_2 C_H^{1/2} C_X \tag{E6-7.13}$$

$$r_X = r_{1X} + r_{2X} = -r_{1H} - r_{2T} = k_1 C_H^{1/2} C_M - k_2 C_H^{1/2} C_X \tag{E6-7.14}$$

$$r_{Me} = r_{1Me} + r_{2Me} = -r_{1H} + r_{2T} = k_1 C_H^{1/2} C_M + k_2 C_H^{1/2} C_X \tag{E6-7.15}$$

$$r_T = r_{2T} = k_2 C_H^{1/2} C_X \tag{E6-7.16}$$

3. **Stoichiometry**

The volumetric flow rate is

$$v = v_0 \frac{F_T}{F_{T0}} \frac{P_0}{P} \frac{T}{T_0}$$

Because there is no pressure drop $P = P_0$ (i.e., $y = 1$), the reaction is carried out isothermally, $T = T_0$, and there is no change in the total number of moles; consequently,

$$v = v_0$$

Flow rates:

$$F_H = v_0 C_H \tag{E6-7.17}$$

$$F_M = v_0 C_M \tag{E6-7.18}$$

$$F_X = v_0 C_X \tag{E6-7.19}$$

$$F_{Me} = v_0 C_{Me} = F_{H0} - F_H = v_0 (C_{H0} - C_H) \tag{E6-7.20}$$

$$F_T = F_{M0} - F_M - F_X = v_0 (C_{M0} - C_M - C_X) \tag{E6-7.21}$$

4. **Combining** and substituting in terms of the space-time yields

$$\tau = \frac{V}{v_0}$$

If we know C_M, C_H, and C_X, then C_{Me} and C_T can be calculated from the reaction stoichiometry. Consequently, we need only to solve the following three equations:

$$\frac{dC_H}{d\tau} = -k_1 C_H^{1/2} C_M - k_2 C_X C_H^{1/2} \tag{E6-7.22}$$

The emergence of user-friendly *ODE solvers* favors this approach over fractional conversion.

$$\frac{dC_M}{d\tau} = -k_1 C_M C_H^{1/2} \tag{E6-7.23}$$

$$\frac{dC_X}{d\tau} = k_1 C_M C_H^{1/2} - k_2 C_X C_H^{1/2} \tag{E6-7.24}$$

5. **Parameter evaluation:**

At $T_0 = 1{,}500°$ R and $P_0 = 35$ atm, the total concentration is

$$C_{T0} = \frac{P_0}{RT_0} = \frac{35 \text{ atm}}{\left(0.73\dfrac{\text{atm ft}^3}{\text{lb mol} \cdot °R}\right)(1{,}500°R)} = 0.032 \text{ lb mol/ft}^3$$

$$C_{H0} = y_{H0} C_{T0} = (0.667)(0.032 \text{ kmol/ft}^3)$$
$$= 0.021 \text{ lb mol/ft}^3$$

$$C_{M0} = \frac{1}{2} C_{H0} = 0.0105 \text{ lb mol/ft}^3$$

$$C_{X0} = 0$$

$$\tau = \frac{V}{v_0} = \frac{238 \text{ ft}^3}{476 \text{ ft}^3/\text{h}} = 0.5 \text{ h}$$

$$F_{T0} = C_{T0}v_0 = \left(0.032\frac{\text{lb mol}}{\text{ft}^3}\right)\left(476\frac{\text{ft}^3}{\text{h}}\right)$$

$$F_{T0} = 15.23 \text{ mol/h}$$

We now solve these three equations, (E6-7.22) to (E6-7.24), simultaneously using Polymath. The program and output in graphical form are shown in Table E6-7.1 and Figure E6-7.1, respectively. However, I hasten to point out that these equations can be solved analytically and the solution was given in the first edition of this text.

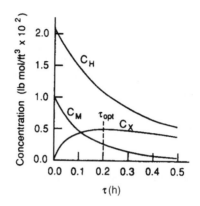

Figure E6-7.1   Concentration profiles in a PFR.

TABLE E6-7.1.   POLYMATH PROGRAM

ODE Report (RKF45)

Differential equations as entered by the user
[1] d(Ch)/d(tau) = r1h+r2h
[2] d(Cm)/d(tau) = r1m
[3] d(Cx)/d(tau) = r1x+r2x

Explicit equations as entered by the user
[1] k1 = 55.2
[2] k2 = 30.2
[3] r1m = -k1* Cm*(Ch^.5)
[4] r2t = k2*Cx*(Ch^.5)
[5] r1h = r1m
[6] r2m = -r2t
[7] r1x = -r1m
[8] r2x = -r2t
[9] r2h = -r2t

Living Example Problem

6.6 Multiple Reactions in a CSTR

For a CSTR, a coupled set of algebraic equations analogous to PFR differential equations must be solved.

$$V = \frac{F_{j0}-F_j}{-r_j} \qquad (6\text{-}28)$$

Rearranging yields

$$F_{j0} - F_j = -r_j V \tag{6-29}$$

Recall that r_j in Equation (6-17) is a function (f_j) of the species concentrations

$$r_j = \sum_{i=1}^{q} r_{ij} = f_j(C_1, C_2, \dots, C_N) \tag{6-17}$$

After writing a mole balance on each species in the reaction set, we substitute for concentrations in the respective rate laws. If there is no volume change with reaction, we use concentrations, C_j, as variables. If the reactions are gas phase and there is volume change, we use molar flow rates, F_j, as variables. The total molar flow rate for n species is

$$F_T = \sum_{j=1}^{n} F_j \tag{6-30}$$

For q reactions occurring in the gas phase, where N different species are present, we have the following set of algebraic equations:

$$F_{10} - F_1 = -r_1 V = V \sum_{i=1}^{q} -r_{i1} = V \cdot f_1 \left(\frac{F_1}{F_T} C_{T0}, \dots, \frac{F_N}{F_T} C_{T0} \right) \tag{6-31}$$

$$F_{j0} - F_j = -r_j V = V \cdot f_j \left(\frac{F_1}{F_T} C_{T0}, \dots, \frac{F_N}{F_T} C_{T0} \right) \tag{6-32}$$

$$F_{N0} - F_N = -r_N V = V \cdot f_N \left(\frac{F_1}{F_T} C_{T0}, \dots, \frac{F_N}{F_T} C_{T0} \right) \tag{6-33}$$

We can use an equation solver in Polymath or a similar program to solve Equations (6-31) through (6-33).

Example 6–8 Hydrodealkylation of Mesitylene in a CSTR

For the multiple reactions and conditions described in Example 6-7, calculate the conversion of hydrogen and mesitylene along with the exiting concentrations of mesitylene, hydrogen, and xylene in a CSTR.

Solution

As in Example 6-7, we assume $v = v_0$; for example,

$$F_A = v C_A = v_0 C_A, \text{ etc.}$$

1. **CSTR Mole Balances:**

Hydrogen: $\qquad v_0 C_{H0} - v_0 C_H = -r_H V$ $\qquad\qquad$ (E6-8.1)

Mesitylene: $\qquad v_0 C_{M0} - v_0 C_M = -r_M V$ $\qquad\qquad$ (E6-8.2)

Xylene: $\qquad\qquad v_0 C_X = r_X V$ $\qquad\qquad$ (E6-8.3)

Toluene: $\qquad\qquad v_0 C_T = r_T V$ $\qquad\qquad$ (E6-8.4)

Methane: $\qquad\qquad v_0 C_{Me} = r_{Me} V$ $\qquad\qquad$ (E6-8.5)

2. **Net Rates**

The rate laws and net rates of reaction for these reactions were given by Equations (E6-7.12) through (E6-7.16) in Example 6-7.

3. **Stoichiometry:** Same as in Example 6-7.

Combining Equations (E6-7.12) through (E6-7.16) with Equations (E6-8.1) through (E6-8.3) and after dividing by v_0, ($\tau = V/v_0$), yields

$$C_{H0} - C_H = (k_1 C_H^{1/2} C_M + k_2 C_H^{1/2} C_X) \tau \qquad\qquad \text{(E6-8.6)}$$

$$C_{M0} - C_M = (k_1 C_H^{1/2} C_M) \tau \qquad\qquad \text{(E6-8.7)}$$

$$C_X = (k_1 C_H^{1/2} C_M - k_2 C_H^{1/2} C_X) \tau \qquad\qquad \text{(E6-8.8)}$$

Next, we put these equations in a form such that they can be readily solved using Polymath.

$$f(C_H) = 0 = C_H - C_{H0} + (k_1 C_H^{1/2} C_M + k_2 C_H^{1/2} C_X) \tau \qquad\qquad \text{(E6-8.9)}$$

$$f(C_M) = 0 = C_M - C_{M0} + k_1 C_H^{1/2} C_M \tau \qquad\qquad \text{(E6-8.10)}$$

$$f(C_X) = 0 = (k_1 C_H^{1/2} C_M - k_2 C_H^{1/2} C_X) \tau - C_X \qquad\qquad \text{(E6-8.11)}$$

The Polymath program and solution to equations (E6-8.9), (E6-8.10), and (E6-8.11) are shown in Table E6-8.1. The problem was solved for different values of τ and the results are plotted in Figure E6-8.1.

For a space time of $\tau = 0.5$, the exiting concentrations are $C_H = 0.0089$, $C_M = 0.0029$, and $C_X = 0.0033$. The overall conversion is

Hydrogen: $\qquad X_H = \dfrac{F_{H0} - F_H}{F_{H0}} = \dfrac{C_{H0} - C_H}{C_{H0}} = \dfrac{0.021 - 0.0089}{0.021} = 0.58$

Overall conversion

Mesitylene: $\qquad X_M = \dfrac{F_{M0} - F_M}{F_{M0}} = \dfrac{C_{M0} - C_M}{C_{M0}} = \dfrac{0.0105 - 0.0029}{0.0105} = 0.72$

TABLE E6-8.1.   POLYMATH PROGRAM AND SOLUTION

NLES Solution

Variable	Value	f(x)	Ini Guess
Ch	0.0089436	1.995E-10	0.006
Cm	0.0029085	7.834E-12	0.0033
Cx	0.0031266	-1.839E-10	0.005
tau	0.5		

NLES Report (safenewt)

Nonlinear equations
[1] f(Ch) = Ch-.021+(55.2*Cm*Ch^.5+30.2*Cx*Ch^.5)*tau = 0
[2] f(Cm) = Cm-.0105+(55.2*Cm*Ch^.5)*tau = 0
[3] f(Cx) = (55.2*Cm*Ch^.5-30.2*Cx*Ch^.5)*tau-Cx = 0

Explicit equations
[1] tau = 0.5

To vary τ_{CSTR}, one
can vary either v_0
for a fixed V or vary
V for a fixed v_0.

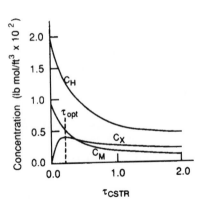

Figure E6-8.1   Concentrations as a function of space time.

We resolve Equations (E6-8.6) through (E6-8.11) for different values of τ to arrive at Figure E6-8.1.

The moles of hydrogen consumed in reaction 1 are equal to the moles of mesitylene consumed. Therefore, the conversion of hydrogen in reaction 1 is

$$X_{1H} = \frac{C_{M0}-C_M}{C_{H0}} = \frac{0.0105-0.0029}{.021} = X_{1H} = 0.36$$

The conversion of hydrogen in reaction 2 is just the overall conversion minus the conversion in reaction 1; that is,

$$X_{2H} = X_H - X_{1H} = 0.58 - 0.36 = 0.22$$

The yield of xylene from mesitylene based on molar flow rates exiting the CSTR for $\tau = 0.5$ is

Overall selectivity,
$\tilde{S}$, and yield, $\tilde{Y}$.

$$\tilde{Y}_{M/X} = \frac{F_X}{F_{M0}-F_M} = \frac{C_X}{C_{M0}-C_M} = \frac{0.00313}{0.0105-0.0029}$$

$$\boxed{\tilde{Y}_{M/X} = \frac{0.41 \text{ mol xylene produced}}{\text{mole mesitylene reacted}}}$$

The overall selectivity of xylene relative to toluene is

$$\tilde{S}_{X/T} = \frac{F_X}{F_T} = \frac{C_X}{C_T} = \frac{C_X}{C_{M0}-C_M-C_X} = \frac{0.00313}{0.0105-0.0029-0.00313}$$

$$\boxed{\tilde{S}_{X/T} = \frac{0.7 \text{ mol xylene produced}}{\text{mole toluene produced}}}$$

Recall that for a CSTR the overall selectivity and yield are identical with the instantaneous selectivity and yield.

6.7 Membrane Reactors to Improve Selectivity in Multiple Reactions

In addition to using membrane reactors to remove a reaction product in order to shift the equilibrium toward completion, we can use membrane reactors to increase selectivity in multiple reactions. This increase can be achieved by injecting one of the reactants along the length of the reactor. It is particularly effective in partial oxidation of hydrocarbons, chlorination, ethoxylation, hydrogenation, nitration, and sulfunation reactions to name a few.[7]

$$C_2H_4 + \frac{1}{2}O_2 \longrightarrow C_2H_4O \xrightarrow{+\frac{5}{2}O_2} 2CO_2 + 2H_2O$$

---

[7] W. J. Asher, D. C. Bomberger, and D. L. Huestis, *Evaluation of SRI's Novel Reactor Process Permix™* (New York: AIChE).

In the top two reactions, the desired product is the intermediate (e.g., C_2H_4O). However, because there is oxygen present, the reactants and intermediates can be completely oxidized to form undesired products CO_2 and water. The desired product in the bottom reaction is xylene. By keeping one of the reactants at a low concentration, we can enhance selectivity. By feeding a reactant through the sides of a membrane reactor, we can keep its concentration low.

Solved Problems

In the solved example problem on the CD-ROM, we have used a membrane reactor (MR) to continue the hydrodealkylation of mesitylene reaction in Examples 6-7 and 6-8. In some ways, this CD example parallels the use of MRs for partial oxidation reactions. We will now do an example for a different reaction to illustrate the advantages of an MR for certain types of reactions.

Example 6–9 Membrane Reactor to Improve Selectivity in Multiple Reactions

The reactions

$$(1)\ A + B \longrightarrow D \qquad -r_{1A} = k_{1A}C_A^2C_B, \quad k_{1A} = 2\ dm^6/mol^2 \cdot s$$

$$(2)\ A + B \longrightarrow U \qquad -r_{2A} = k_{2A}C_AC_B^2, \quad k_{2A} = 3\ dm^6/mol^2 \cdot s$$

take place in the gas phase. The overall selectivities, $\tilde{S}_{D/U}$, are to be compared for a membrane reactor (MR) and a conventional PFR. First, we use the instantaneous selectivity to determine which species should be fed through the membrane

$$S_{D/U} = \frac{k_1 C_A^2 C_B}{k_2 C_B^2 C_A} = \frac{k_1 C_A}{k_2 C_B}$$

We see that to maximize $S_{D/U}$ we need to keep the concentration of A high and the concentration of B low; therefore, we feed B through the membrane. The molar flow rate of A entering the reactor is 4 mol/s and that of B entering through the membrane is 4 mol/s as shown in Figure E6-9.1. For the PFR, B enters along with A.

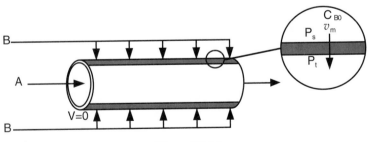

Figure E6-9.1

The reactor volume is 50 dm³ and the entering total concentration is 0.8 mol/dm³. **Plot** the molar flow rates and the overall selectivity, $\tilde{S}_{D/U}$, as a function of reactor volume for both the MR and PFR.

Solution

Mole Balances for both the PFR and the MR

	PFR		**MR**	

Species A: (1) $\dfrac{dF_A}{dV} = r_A$ (E6-9.1[a]) $\dfrac{dF_A}{dV} = r_A$ (E6-9.1[b])

Species B: (2) $\dfrac{dF_B}{dV} = r_B$ (E6-9.2[a]) $\dfrac{dF_B}{dV} = r_B + R_B$ (E6-9.2[b])

Species C: (3) $\dfrac{dF_D}{dV} = r_D$ (E6-9.3[a]) $\dfrac{dF_D}{dV} = r_D$ (E6-9.3[b])

Species D: (4) $\dfrac{dF_U}{dV} = r_U$ (E6-9.4[a]) $\dfrac{dF_U}{dV} = r_U$ (E6-9.4[b])

Net Rates and Rate Laws

$$r_A = r_{1A} + r_{2A} = -k_{1A}C_A^2 C_B - k_{2A}C_A C_B^2 \tag{E6-9.5}$$

$$r_B = r_{1B} + r_{2B} = -k_{1A}C_A^2 C_B - k_{2A}C_A C_B^2 \tag{E6-9.6}$$

$$r_D = r_{1D} = k_{1A}C_A^2 C_B \tag{E6-9.7}$$

$$r_U = r_{2U} = k_{2A}C_A C_B^2 \tag{E6-9.8}$$

Transport Law

The volumetric flow rate through the membrane is given by Darcy's Law (see Chapter 4):

$$v_m = K[P_s - P_t]A_t \tag{E6-9.9}$$

where K is the membrane permeability (m/s · kPa) and P_s (kPa) and P_t (kPa) are the shell side and tube side pressures, and A_t is the membrane surface area m². The flow rate through the membrane can be controlled by pressure drop across the membrane $(P_s - P_t)$. Recall from Equation (4-43) that "a" is the membrane surface area per unit volume of reactor,

$$A_t = aV_t \tag{E6-9.10}$$

The total molar flow rate of B through the sides of the reactor is

$$F_{B0} = C_{B0}v_m = \underbrace{C_{B0}K[P_s - P_t]a}_{R_B} \cdot V_t = R_B V_t \tag{E6-9.11}$$

The molar flow rate of B per unit volume of reactor is

$$R_B = \frac{F_{B0}}{V_t} \tag{E6-9.12}$$

Stoichiometry: Isothermal ($T = T_0$) and neglect pressure drop down the length of the reactor ($P = P_0$, $y = 1.0$)

For both the PFR and MR for no pressure drop down the length of the rector and isothermal operation, the concentrations are

Here $T = T_0$ and
$\Delta P = 0$

$$C_A = C_{T0}\frac{F_A}{F_T} \quad \text{(E6-9.13)} \qquad\qquad C_D = C_{T0}\frac{F_D}{F_T} \qquad \text{(E6-9.15)}$$

$$C_B = C_{T0}\frac{F_B}{F_T} \quad \text{(E6-9.14)} \qquad\qquad C_U = C_{T0}\frac{F_U}{F_T} \qquad \text{(E6-9.16)}$$

Combine

The Polymath Program will combine the mole balance, net rates, and stoichiometric equations to solve for the molar flow rate and selectivity profiles for both the conventional PFR and the MR and also the selectivity profile.

A note of caution on calculating the overall selectivity

$$\tilde{S}_{D/U} = \frac{F_D}{F_U} \qquad\qquad\qquad\qquad \text{(E6-9.17)}$$

Fool Polymath!

We have to fool Polymath because at the entrance of the reactor $F_U = 0$. Polymath will look at Equation (E6-9.17) and will not run because it will say you are dividing by zero. Therefore, we need to add a very small number to the denominator, say 0.0001; that is,

$$\tilde{S}_{D/U} = \frac{F_D}{F_U + 0.0001} \qquad\qquad\qquad \text{(E6-9.18)}$$

Sketch the trends or results you expect **before** working out the details of the problem.

Table E6-9.1 shows the Polymath Program and report sheet.

TABLE E6-9.1.   POLYMATH PROGRAM

POLYMATH Results

Example 6-9 Membrane Reactor to Improve Selectivity in Multiple Reactions   08-18-2004, Rev5.1.232

ODE Report (RKF45)   Calculated values of the DEQ variables

Living Example Problem

Differential equations as entered by the user	Variable	initial value	minimal value	maximal value	final value
[1] d(Fa)/d(V) = ra	V	0	0	50	50
[2] d(Fb)/d(V) = rb+Rb	Fa	4	1.3513875	4	1.3513875
[3] d(Fd)/d(V) = rd	Fb	0	0	1.3513875	1.3513875
[4] d(Fu)/d(V) = ru	Fd	0	0	1.9099789	1.9099789
	Fu	0	0	0.7386336	0.7386336
Explicit equations as entered by the user	Ft	4	4	5.3513875	5.3513875
[1] Ft = Fa+Fb+Fd+Fu	Ct0	0.8	0.8	0.8	0.8
[2] Ct0 = 0.8	k1a	2	2	2	2
[3] k1a = 2	k2a	3	3	3	3
[4] k2a = 3	Cb	0	0	0.2020242	0.2020242
[5] Cb = Ct0*Fb/Ft	Ca	0.8	0.2020242	0.8	0.2020242
[6] Ca = Ct0*Fa/Ft	ra	0	-0.0635265	0	-0.0412269
[7] ra = -k1a*Ca^2*Cb-k2a*Ca*Cb^2	rb	0	-0.0635265	0	-0.0412269
[8] rb = ra	Cd	0	0	0.2855303	0.2855303
[9] Cd = Ct0*Fd/Ft	Cu	0	0	0.1104213	0.1104213
[10] Cu = Ct0*Fu/Ft	rd	0	0	0.0544175	0.0164908
[11] rd = k1a*Ca^2*Cb	ru	0	0	0.0247361	0.0247361
[12] ru = k2a*Ca*Cb^2	Vt	50	50	50	50
[13] Vt = 50	Fbo	4	4	4	4
[14] Fbo = 4	Rb	0.08	0.08	0.08	0.08
[15] Rb = Fbo/Vt	Sdu	0	0	34.315047	2.5854772
[16] Sdu = Fd/(Fu+.0001)					

We can easily modify the program, Table E6-9.1, for the PFR simply by setting R_B equal to zero ($R_B = 0$) and the initial condition for B to be 4.0.

Figures E6-9.2(a) and E6-9.2(b) show the molar flow rate profiles for the conventional PFR and MR, respectively.

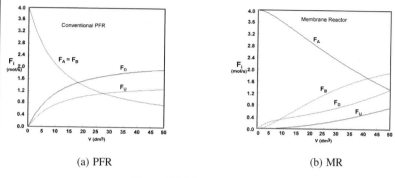

(a) PFR   (b) MR

Figure E6-9.2 Molar flow rates.

Selectivities at
V = 5 dm³
MR: $S_{D/U} = 14$
PFR: $S_{D/U} = 0.65$

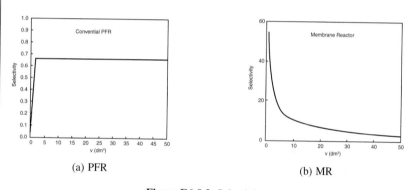

(a) PFR   (b) MR

Figure E6-9.3 Selectivity.

Figures E6-9.3(a) and E6-9.3(b) show the selectivity for the PFR and MR. One notices the enormous enhancement in selectivity the MR has over the PFR.

Be sure to load this living example problem and play with the reactions and reactors. With minor modifications, you can explore reactions analogous to partial oxidations.

$$A + B \longrightarrow D \quad r_D = k_1 C_A C_B \tag{E6-9.19}$$

$$B + D \longrightarrow U \quad r_U = k_2 C_B C_D \tag{E6-9.20}$$

where oxygen (B) is fed through the membrane. See Problems P6-9 and P6-19.

6.8 Complex Reactions of Ammonia Oxidation

In the two preceding examples, there was no volume change with reaction; consequently, we could use concentration as our dependent variable. We now consider a gas-phase reaction with volume change taking place in a PFR. Under these conditions, we must use the molar flow rates as our dependent variables.

Example 6–10 Calculating Concentrations as Functions
of Position for NH₃ Oxidation in a PFR

The following gas-phase reactions take place simultaneously on a metal oxide–supported catalyst:

1. $4NH_3 + 5O_2 \longrightarrow 4NO + 6H_2O$

2. $2NH_3 + 1.5O_2 \longrightarrow N_2 + 3H_2O$

3. $2NO + O_2 \longrightarrow 2NO_2$

4. $4NH_3 + 6NO \longrightarrow 5N_2 + 6H_2O$

Writing these equations in terms of symbols yields

Reaction 1: $4A + 5B \longrightarrow 4C + 6D$ $-r_{1A} = k_{1A}C_A C_B^2$ (E6-10.1)

Reaction 2: $2A + 1.5B \longrightarrow E + 3D$ $-r_{2A} = k_{2A}C_A C_B$ (E6-10.2)

Reaction 3: $2C + B \longrightarrow 2F$ $-r_{3B} = k_{3B}C_C^2 C_B$ (E6-10.3)

Reaction 4: $4A + 6C \longrightarrow 5E + 6D$ $-r_{4C} = k_{4C}C_C C_A^{2/3}$ (E6-10.4)

with[8] $k_{1A} = 5.0 \ (m^3/kmol)^2/min$ $k_{2A} = 2.0 \ m^3/kmol \cdot min$

$k_{3B} = 10.0 \ (m^3/kmol)^2/min$ $k_{4C} = 5.0 \ (m^3/kmol)^{2/3}/min$

Note: We have converted the specific reaction rates to a per unit volume basis by multiplying the k' on a per mass of catalyst basis by the bulk density of the packed bed (i.e., $k = k' \rho_B$).

Determine the concentrations as a function of position (i.e., volume) in a PFR.

Additional information: Feed rate = 10 dm³/min; volume of reactor = 10 dm³; and
$$C_{A0} = C_{B0} = 1.0 \ mol/dm^3, \ C_{T0} = 2.0 \ mol/dm^3$$

Solution

Mole balances:

Species A: $\dfrac{dF_A}{dV} = r_A$ (E6-10.5)

Species B: $\dfrac{dF_B}{dV} = r_B$ (E6-10.6)

Species C: $\dfrac{dF_C}{dV} = r_C$ (E6-10.7)

---

[8] Reaction orders and rate constants were estimated from periscosity measurements for a bulk catalyst density of 1.2 kg/dm³.

Solutions to these
equations are most
easily obtained with
an ODE solver

Species D: $\dfrac{dF_D}{dV} = r_D$ (E6-10.8)

Species E: $\dfrac{dF_E}{dV} = r_E$ (E6-10.9)

Species F: $\dfrac{dF_F}{dV} = r_F$ (E6-10.10)

Total: $F_T = F_A + F_B + F_C + F_D + F_E + F_F$ (E6-10.11)

Rate laws: See above for r_{1A}, r_{2A}, r_{3B}, and r_{4C}.

Stoichiometry:

    A.   *Relative rates*

Reaction 1: $\dfrac{r_{1A}}{-4} = \dfrac{r_{1B}}{-5} = \dfrac{r_{1C}}{4} = \dfrac{r_{1D}}{6}$ (E6-10.12)

Reaction 2: $\dfrac{r_{2A}}{-2} = \dfrac{r_{2B}}{-1.5} = \dfrac{r_{3E}}{1} = \dfrac{r_{2D}}{3}$ (E6-10.13)

Reaction 3: $\dfrac{r_{3C}}{-2} = \dfrac{r_{3B}}{-1} = \dfrac{r_{3F}}{2}$ (E6-10.14)

Reaction 4: $\dfrac{r_{4A}}{-4} = \dfrac{r_{4C}}{-6} = \dfrac{r_{4E}}{5} = \dfrac{r_{4D}}{6}$ (E6-10.15)

    B.   *Concentrations*: For isothermal operation and no pressure drop, the concentrations are given in terms of the molar flow rates by

$$C_j = \frac{F_j}{F_T} C_{T0}$$

Next substitute for the concentration of each species in the rate laws. Writing the rate law for species A in reaction 1 in terms of the rate of formation, r_{1A}, and molar flow rates, F_A and F_B, we obtain

$$r_{1A} = -k_1 C_A C_B^2 = -k_{1A}\left(C_{T0}\frac{F_A}{F_T}\right)\left(C_{T0}\frac{F_B}{F_T}\right)^2$$

Thus

$$r_{1A} = -k_{1A} C_{T0}^3 \frac{F_A F_B^2}{F_T^3}$$ (E6-10.16)

Similarly for the other reactions,

$$r_{2A} = -k_{2A} C_{T0}^2 \frac{F_A F_B}{F_T^2}$$ (E6-10.17)

$$r_{3B} = -k_{3B} C_{T0}^3 \frac{F_C^2 F_B}{F_T^3} \qquad \text{(E6-10.18)}$$

$$r_{4C} = -k_{4C} C_{T0}^{5/3} \frac{F_C F_A^{2/3}}{F_T^{5/3}} \qquad \text{(E6-10.19)}$$

Next, we determine the *net* rate of reaction for each species by using the appropriate stoichiometric coefficients and then summing the rates of the individual reactions.

Net rates of formation:

Species A: $r_A = r_{1A} + r_{2A} + \frac{2}{3} r_{4C}$ (E6-10.20)

Species B: $r_B = 1.25\, r_{1A} + 0.75\, r_{2A} + r_{3B}$ (E6-10.21)

Species C: $r_C = -r_{1A} + 2\, r_{3B} + r_{4C}$ (E6-10.22)

Species D: $r_D = -1.5\, r_{1A} - 1.5\, r_{2A} - r_{4C}$ (E6-10.23)

Species E: $r_E = -\dfrac{r_{2A}}{2} - \dfrac{5}{6} r_{4C}$ (E6-10.24)

Species F: $r_F = -2 r_{3B}$ (E6-10.25)

Combining: Rather than combining the concentrations, rate laws, and mole balances to write everything in terms of the molar flow rate as we did in the past, it is more convenient here to write our computer solution (either Polymath or our own program) using equations for r_{1A}, F_A, and so on. Consequently, we shall write Equations (E6-10.16) through (E6-10.19) and (E6-10.5) through (E6-10.11) as individual lines and let the computer combine them to obtain a solution.

The corresponding Polymath program written for this problem is shown in Table E6-10.1 and a plot of the output is shown in Figure E6-10.1. One notes that there is a maximum in the concentration of NO (i.e., C) at approximately 1.5 dm³.

However, there is one fly in the ointment here: It may not be possible to determine the rate laws for each of the reactions. In this case, it may be necessary to work with the minimum number of reactions and hope that a rate law can be found for each reaction. That is, you need to find the number of linearly independent reactions in your reaction set. In Example 6-10, there are four reactions given [(E6-10.5) through (E6-10.8)]. However, only three of these reactions are independent, as the fourth can be formed from a linear combination of the other three. Techniques for determining the number of independent reactions are given by Aris.[9]

---

[9] R. Aris, *Elementary Chemical Reactor Analysis* (Upper Saddle River, N.J.: Prentice Hall, 1969).

Living Example Problem

TABLE E6-10.1.   POLYMATH PROGRAM

POLYMATH Results
Example 6-10 Calculating Concentrations as a Function of Position for NH3 Oxidation in a PFR

Calculated values of the DEQ variables

Variable	initial value	minimal value	maximal value	final value
V	0	0	10	10
FA	10	1.504099	10	1.504099
FB	10	2.4000779	10	2.4000779
FC	0	0	1.6519497	0.6038017
FD	0	0	12.743851	12.743851
FE	0	0	3.4830019	3.4830019
FF	0	0	0.9260955	0.9260955
Ft	20	20	21.660927	21.660927
r1A	-5	-5	-0.0341001	-0.0341001
r2A	-2	-2	-0.0615514	-0.0615514
r4C	0	-0.5619376	0	-0.0747591
r3B	0	-0.1448551	0	-0.0068877
CA	1	0.1388767	1	0.1388767
rA	-7	-7	-0.1454909	-0.1454909
rB	-7.75	-7.75	-0.0956764	-0.0956764
rC	5	-0.2008343	5	-0.0544343
rD	10.5	0.2182363	10.5	0.2182363
rE	1	0.0930749	1.0317775	0.0930749
rF	0	0	0.2897102	0.0137754

ODE Report (RKF45)

Differential equations as entered by the user
[1] d(FA)/d(V) = rA
[2] d(FB)/d(V) = rB
[3] d(FC)/d(V) = rC
[4] d(FD)/d(V) = rD
[5] d(FE)/d(V) = rE
[6] d(FF)/d(V) = rF

Explicit equations as entered by the user
[1] Ft = FA+FB+FC+FD+FE+FF
[2] r1A = -5*8*(FA/Ft)*(FB/Ft)^2
[3] r2A = -2*4*(FA/Ft)*(FB/Ft)
[4] r4C = -5*3.175*(FC/Ft)*(FA/Ft)^(2/3)
[5] r3B = -10*8*(FC/Ft)^2*(FB/Ft)
[6] CA = 2*FA/Ft
[7] rA = r1A+r2A+2*r4C/3
[8] rB = 1.25*r1A+.75*r2A+r3B
[9] rC = -r1A+2*r3B+r4C
[10] rD = -1.5*r1A-1.5*r2A-r4C
[11] rE = -.5*r2A-5*r4C/6
[12] rF = -2*r3B

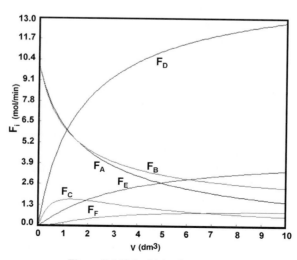

Figure E6-10.1   Molar flow rates profiles.

6.9 Sorting It All Out

In Example 6-9 we were given the rate laws and asked to calculate the product distribution. The inverse of the problem described in Example 6-9 must frequently be solved. Specifically, the rate laws often must be determined from the variation in the product distribution generated by changing the feed concentrations. In some instances this determination may not be possible without carrying out independent experiments on some of the reactions in the sequence. The best strategy to use to *sort out* all of the rate law parameters will vary from reaction sequence to reaction sequence. Consequently, the strategy developed for one system may not be the best approach for other multiple-reaction systems. One general rule is to start an analysis by looking for species produced in only one reaction; next, study the species involved in only two reactions, then three, and so on.

Nonlinear least-squares

When some of the intermediate products are free radicals, it may not be possible to perform independent experiments to determine the rate law parameters. Consequently, we must deduce the rate law parameters from changes in the distribution of reaction products with feed conditions. Under these circumstances, the analysis turns into an optimization problem to estimate the best values of the parameters that will minimize the sums of the squares between the calculated variables and measured variables. This process is basically the same as that described in Section 5.2.3, but more complex, owing to the larger number of parameters to be determined. We begin by estimating the parameter values using some of the methods just discussed. Next, we use our estimates to use nonlinear regression techniques to determine the best estimates of our parameter values from the data for all of the experiments.[10] Software packages are becoming available for an analysis such as this one.

6.10 The Fun Part

I'm not talking about fun you can have at an amusement park, but CRE fun. Now that we have an understanding on how to solve for the exit concentrations of multiple reactions in a CSTR and how to plot the species concentrations down the length of a PFR or PBR, we can address one of the most important and fun areas of chemical reaction engineering. This area, discussed in Section 6.2, is learning how to maximize the desired product and minimize the undesired product. It is this area that can make or break a chemical process financially. It is also an area that requires creativity in designing the reactor schemes and feed conditions that will maximize profits. Here you can mix and match reactors, feed streams, and side streams as well as vary the ratios of feed concentration in order to maximize or minimize the selectivity of a particular species. Problems of this type are what I call *digital-age problems*[11] because we normally need to use ODE solvers along with critical and creative thinking skills to find the best answer. A number of problems at the end of this chapter

---

[10]See, for example, Y. Bard, *Nonlinear Parameter Estimation* (San Diego, Calif.: Academic Press, 1974).

[11]H. Scott Fogler, *Teaching Critical Thinking, Creative Thinking, and Problem Solving in the Digital Age*, Phillips Lecture (Stillwater, Okla.: OSU Press, 1997).

will allow you to practice these critical and creative thinking skills. These problems offer opportunity to explore many different solution alternatives to enhance selectivity and have fun doing it.

However, to carry CRE to the next level and to have a lot more fun solving multiple reaction problems, we will have to be patient a little longer. The reason is that in this chapter we consider only isothermal multiple reactions, and it is nonisothermal multiple reactions where things really get interesting. Consequently, we will have to wait to carry out schemes to maximize the desired product in nonisothermal multiple reactions until we study heat effects in Chapters 8 and 9. After studying these chapters, we will add a new dimension to multiple reactions, as we now have another variable, temperature, that we may or may not be able to use to affect selectivity and yield. In one particularly interesting **problem (P8-26)**, we will study is the production of styrene from ethylbenzene in which two side reactions, one endothermic, and one exothermic, must be taken into account. Here we may vary a whole slew of variables, such as entering temperature, diluent rate, and observe optima, in the production of styrene. However, we will have to delay gratification of the styrene study until we have mastered Chapter 8.

Multiple Reactions with heat effects is unique to this book

Closure. After completing this chapter the reader should be able to describe the different types of multiple reactions (series, parallel, complex, and independent) and to select a reaction system that maximizes the selectivity. The reader should be able to write down and use the algorithm for solving CRE problems with multiple reactions. The reader should also be able to point out the major differences in the CRE algorithm for the multiple reactions from that for the single reactions, and then discuss why care must be taken when writing the rate law and stoichiometric steps to account for the rate laws for each reaction, the relative rates, and the net rates of reaction.

Finally, the readers should feel a sense of accomplishment by knowing they have now reached a level they can solve realistic CRE problems with complex kinetics.

SUMMARY

1. For the competing reactions

$$\text{Reaction 1:} \quad A + B \xrightarrow{k_D} D \quad r_D = A_D e^{-E_D/RT} C_A^{\alpha_1} C_B^{\beta_1} \quad \text{(S6-1)}$$

$$\text{Reaction 2:} \quad A + B \xrightarrow{k_U} U \quad r_U = A_U e^{-E_U/RT} C_A^{\alpha_2} C_B^{\beta_2} \quad \text{(S6-2)}$$

the instantaneous selectivity parameter is defined as

$$S_{D/U} = \frac{r_D}{r_U} = \frac{A_D}{A_U} \exp\left(-\frac{(E_D - E_U)}{RT}\right) C_A^{\alpha_1 - \alpha_2} C_B^{\beta_1 - \beta_2} \quad \text{(S6-3)}$$

a. If $E_D > E_U$, the selectivity parameter $S_{D/U}$ will increase with increasing temperature.

b. If $\alpha_1 > \alpha_2$ and $\beta_2 > \beta_1$, the reaction should be carried out at high concentrations of A and low concentrations of B to maintain the selectivity parameter $S_{D/U}$ at a high value. Use a semibatch reactor with pure A initially or a tubular reactor in which B is fed at different locations down the reactor. Other cases discussed in the text are ($\alpha_2 > \alpha_1$, $\beta_1 > \beta_2$), ($\alpha_2 > \alpha_1$, $\beta_2 > \beta_1$), and ($\alpha_1 > \alpha_2$, $\beta_1 > \beta_2$).

The overall selectivity, based on molar flow rates leaving the reactor, for the reactions given by Equations (S6-1) and (S6-2) is

$$\tilde{S}_{D/U} = \frac{F_D}{F_U} \quad \text{(S6-4)}$$

2. The *overall yield* is the ratio of the number of moles of a product at the end of a reaction to the number of moles of the key reactant that have been consumed:

$$\tilde{Y}_D = \frac{F_D}{F_{A0} - F_A} \quad \text{(S6-5)}$$

1. The algorithm:

Mole balances:

PFR
$$\frac{dF_j}{dV} = r_j \quad \text{(S6-6)}$$

CSTR
$$F_{j0} - F_j = -r_j V \quad \text{(S6-7)}$$

Batch
$$\frac{dN_j}{dt} = r_j V \quad \text{(S6-8)}$$

Membrane ("i" diffuses in)
$$\frac{dF_i}{dW} = r_i + R_i \quad \text{(S6-9)}$$

Liquid-semibatch
$$\frac{dC_j}{dt} = r_j + \frac{v_0(C_{j0} - C_j)}{V} \quad \text{(S6-10)}$$

Rate laws and net rates:

Laws
$$r_{ij} = k_{ij} f_i(C_1, C_j, C_n) \quad \text{(S6-11)}$$

| Net rates | $$r_j = \sum_{i=1}^{q} r_{ij}$$ | (S6-12) |

| Relative rates | $$\frac{r_{iA}}{-a_i} = \frac{r_{iB}}{-b_i} = \frac{r_{iC}}{c_i} = \frac{r_{iD}}{d_i}$$ | (S6-13) |

Stoichiometry:

| *Gas phase* | $$C_j = C_{T0}\frac{F_j}{F_T}\frac{P}{P_0}\frac{T_0}{T} = C_{T0}\frac{F_j}{F_T}\frac{T_0}{T}y$$ | (S6-14) |

$$F_T = \sum_{j=1}^{n} F_j \qquad \text{(S6-15)}$$

$$\frac{dy}{dW} = -\frac{\alpha}{2y}\left(\frac{F_T}{F_{T0}}\right)\frac{T}{T_0} \qquad \text{(S6-16)}$$

Liquid phase $$v = v_0$$

$$C_A, C_B, \ldots$$

CD-ROM MATERIAL

- **Learning Resources**

    1. *Summary Notes*
    2. *Web Modules*
        A. Cobra Bites   B. Oscillating Reactions

Summary Notes

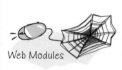

Web Modules

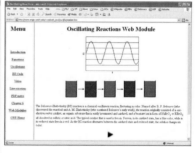

Computer Modules

3. *Interactive Computer Modules (ICM)*
   The Great Race

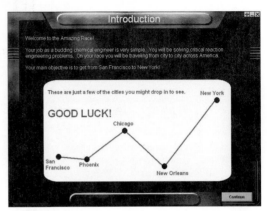

Links

4. *Reactor Lab. See Learning Resources at the end of Chapter 4 for description of these interactive computer oversizes.*
5. *Solved Problems*
   A. Blood Coagulation
   B. Hydrodealkylation of Mesitylene in a Membrane Reactor
   C. All You Wanted to Know About Making Malic Anhydride and More
6. *Clarification: PFR with feed streams along the length of the reactor.*

Solved Problems

- **Living Example Problems**
  1. *Example 6-2 Trambouze Reactions*
  2. *Example 6-7 Hydrodealkylation of Mesitylene in a PFR*
  3. *Example 6-8 Hydrodealkylation of Mesitylene in a CSTR*
  4. *Example 6-9 Membrane Reactor to Improve Selectivity in Multiple Reactions*
  5. *Example 6-10 Calculating Concentrations as a Function of Position for NH_3 Oxidation in a PFR*
  6. *Example web — Cobra Bite Problem*
  7. *Example web — Oscillating Reactions Problem*
  8. *Example CD Solved Problems — Hydrodealkylation of Mesitylene in a Membrane Reactor*
  9. *Example CD Solved Problems — Blood Coagulation*

Living Example Problem

- **FAQ [Frequently Asked Questions] — In Updates/FAQ icon section**
- **Professional Reference Shelf**
  R6.1 *Attainable Region Analysis (ARA)*
  The ARA allows one to find the optimum reaction system for certain types of rate laws. The example used in one of modified **van de Vusse kinetics**

$$A \underset{\longleftarrow}{\overset{\longrightarrow}{\rule{1cm}{0pt}}} B \to C$$
$$2A \to D$$

Reference Shelf

to find the optimum wrt B using a combination of PFRs and CSTRs

Links

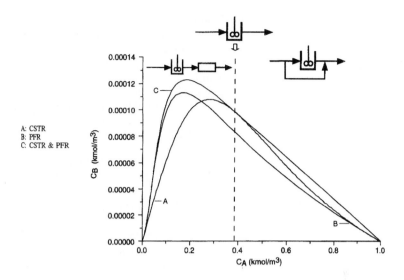

QUESTIONS AND PROBLEMS

The subscript to each of the problem numbers indicates the level of difficulty: A, least difficult; D, most difficult.

$$A = \bullet \quad B = \blacksquare \quad C = \blacklozenge \quad D = \blacklozenge\blacklozenge$$

In each of the following questions and problems, rather than just drawing a box around your answer, write a sentence or two describing how you solved the problem, the assumptions you made, the reasonableness of your answer, what you learned, and any other facts that you want to include.

Homework Problems

P6-1   (a) Make up and solve an original problem to illustrate the principles of this chapter. See Problem P4-1 for guidelines.

(b) Write a question based on the material in this chapter that requires critical thinking. Explain why your question requires critical thinking. [*Hint:* See Preface section B.2.]

(c) Choose a FAQ from Chapter 6 to be eliminated and say why it should be eliminated.

(d) Listen to the audios 🎧 on the CD and then pick one and say why it was the most helpful.

(e) Which example on the CD-ROM Lecture Notes for Chapter 6 was least helpful?

P6-2$_A$ (a) **Example 6-2**. (1) What would have been the selectivity $S_{B/XY}$ and conversion, X, if the reaction had been carried out in a single PFR with the same volume as the CSTR? (2) How would your answers change if the pressure were increased by a factor of 100?

(b) **Example 6-3**. Make a table/list for each reactor shown in Figure 6-3 identifying all the types of reactions that would be best carried out in this reactor. For example, Figure 6-3(d) Semibatch: used for (1) highly exothermic reactions and (2) selectivity, for example, to maintain concentration A high and B low and (3) to control conversion of A or B.

(c) **Example 6-4**. (1) How would τ'_{opt} change if $k_1 = k_2 = 0.25$ m³/s/kg at 300 K? (2) How would τ'_{opt} change for a CSTR? (3) What CSTR (with $\tau' = 0.5$ kg/m³/s) operating temperature would you recommend to maximize B for $C_{A0} = 5$ mol/dm³, $k_1 = 0.4$ m³/kg · s, and $k_2 = 0.01$ m³/kg · s with $E_1 = 10$ kcal/mol and $E_2 = 20$ kcal/mol (plot C_B versus T).

(d) **Example 6-5**. How would your answers change if Reaction 2 were reversible and followed an elementary rate law?

$$2NO \rightleftharpoons N_2 + O_2 \text{ with } K_C \equiv K_e = (C_{N_{2e}})(C_{O_{2e}})/(C^2_{NO_{2e}}) = 0.25$$

(e) **Example 6-6**. How would Equations (E6-6.3) and (E6-6.8) change if the reactions were carried out in the liquid phase under high pressure?

(f) **Example 6-7**. (1) How would your answers change if the feed were equal molar in hydrogen and mesitylene? (2) What is the effect of Θ_H on τ_{opt}? On $\tilde{S}_{X/T}$?

(g) **Example 6-8**. Same question as **P6-2(f)**?

(h) **Example 6-9**. Load the *Living Example Problem* from the CD-ROM. (1) How would your answers change if $F_{B0} = 2F_{A0}$? (2) If reaction (1) were A+2B → D with the rate law remaining the same?

(i) **Example 6-10**. Load the *Living Example Problem* from the CD-ROM. (1) How would your answers change if the reactor volume was cut in half and k_{3B} and k_{4C} were decreased by a factor of 4? (2) What reactor schemes and conditions would you choose to maximize $\tilde{S}_{C/F}$? *Hint:* Plot $\tilde{S}_{C/F}$ versus Θ_B as a start. Describe how pressure drop would affect the selectivity.

(j) **Read Solved Problem A, Blood Coagulation**. Load the living example. (1) Plot out some of the other concentrations, such as TF-VIIa and TF-VIIaX. (2) Why do the curves look the way they do? What reaction in the cascade is most likely to be inhibited causing one to bleed to death? (3) What reactions if eliminated could cause one to die of a blood clot? (*Hint:* Look at ATIIII and/or TFPI.)

(k) **Read Solved Problem B, Membrane Reactor**. Load the *Membrane Reactor* from the CD-ROM. How would your answers change if the feed were the same as in Examples 6-7 and 6-8 (i.e., $F_{H0} = F_{M0}$)? Vary y_{M0} and the reactor volume and describe what you find.

Living Example Problem

(l) **Living Example Web Module: Oscillating Reactions**. Load the *Living Example Polymath Program* for oscillating reactions on the CD-ROM. For the (IO⁻) and (I) reactions set $k_1 = 0.0001$/min⁻¹ and for reaction (1) $C_{P0} = 0.01$ mol/dm³. (1) What did you find? Look at the linearized stability analysis on the CD-ROM. (2) What factors affect the frequency and onset of the oscillations? (3) Explore and write paragraph describing what you find. (4) Load the *Living Example Polymath Program* for the BZ reaction. Vary the parameters and write a paragraph describing what you find. (5) See Problem P6-22.

Interactive   **P6-3$_A$**   Load the Interactive Computer Module (ICM) *The Great Race* from the CD-ROM. Run the module and then record your performance number for the module, which indicates your mastering of the material. Your professor has the key to decode your performance number.

Performance # _____.

Computer Modules

Web Modules

P6-4$_C$ Read the cobra **Web Module** on the CD-ROM.
   (a) Determine how many cobra bites are necessary in order that no amount of anti-venom will save the victim.
   (b) Suppose the victim was bitten by a harmless snake and not bitten by a cobra and anti-venom was injected. How much anti-venom would need to be injected to cause death?
   (c) What is the latest possible time and amount that anti-venom can be injected after being bitten such that the victim would not die?
   (d) Ask another question about this problem.
   Hint: The *Living Example* Polymath program is on the CD-ROM.

P6-5$_B$ The following reactions

$$A \underset{k_2}{\overset{k_1}{\rightleftharpoons}} D \qquad\qquad -r_{1A} = k_1[C_A - C_D/K_{1A}]$$

$$A \rightleftharpoons U \qquad\qquad -r_{2A} = k_2[C_A - C_U/K_{2A}]$$

take place in a batch reactor.
   (a) Plot conversion and the concentrations of A, D, and U as a function of time. When would you stop the reaction to maximize the concentration of D?
   (b) When is the maximum concentration of U?
   (c) What are the equilibrium concentrations of A, D, and U?
   (d) What would be the exit concentrations from a CSTR with a space time of 1.0 min? of 10.0 min? of 100 min?
   Additional information:
   $k_1 = 1.0$ min^{-1}, $K_{1A} = 10$
   $k_2 = 100$ min^{-1}, $K_{2A} = 1.5$
   $C_{A0} = 1$ mol/dm^3
   (Adapted from a problem by John Falkner, University of Colorado.)

P6-6$_A$ Consider the following system of gas-phase reactions:

$$A \longrightarrow X \qquad r_X = k_1 C_A^{1/2} \qquad k_1 = 0.004(\text{mol/dm}^3)^{1/2}\cdot\text{min}$$

$$A \longrightarrow B \qquad r_B = k_2 C_A \qquad k_2 = 0.3 \text{ min}^{-1}$$

$$A \longrightarrow Y \qquad r_Y = k_3 C_A^2 \qquad k_3 = 0.25 \text{ dm}^3/\text{mol}\cdot\text{min}$$

B is the desired product, and X and Y are foul pollutants that are expensive to get rid of. The specific reaction rates are at 27°C. The reaction system is to be operated at 27°C and 4 atm. Pure A enters the system at a volumetric flow rate of 10 dm^3/min.
   (a) Sketch the instantaneous selectivities ($S_{B/X}, S_{B/Y}$, and $S_{B/XY} = r_B/(r_X + r_Y)$) as a function of the concentration of C_A.
   (b) Consider a series of reactors. What should be the volume of the first reactor?
   (c) What are the effluent concentrations of A, B, X, and Y from the first reactor.

(d) What is the conversion of A in the first reactor?

(e) If 99% conversion of A is desired, what reaction scheme and reactor sizes should you use to maximize $S_{B/XY}$?

(f) Suppose that $E_1 = 20,000$ cal/mol, $E_2 = 10,000$ cal/mol, and $E_3 = 30,000$ cal/mol. What temperature would you recommend for a single CSTR with a space time of 10 min and an entering concentration of A of 0.1 mol/dm^3?

(g) If you could vary the pressure between 1 and 100 atm, what pressure would you choose?

P6-7$_B$ Pharmacokinetics concerns the ingestion, distribution, reaction, and elimination reaction of drugs in the body. Consider the application of pharmacokinetics to one of the major problems we have in the United States, drinking and driving. Here we shall model how long one must wait to drive after having a tall martini. In most states, the legal intoxication limit is 0.8 g of ethanol per liter of body fluid. (In Sweden it is 0.5 g/L, and in Eastern Europe and Russia it is any value above 0.0 g/L.) The ingestion of ethanol into the bloodstream and subsequent elimination can be modeled as a series reaction. The rate of absorption from the gastrointestinal tract into the bloodstream and body is a first-order reaction with a specific reaction rate constant of 10 h^{-1}. The rate at which ethanol is broken down in the bloodstream is limited by regeneration of a coenzyme. Consequently, the process may be modeled as a zero-order reaction with a specific reaction rate of 0.192 g/h·L of body fluid. How long would a person have to wait **(a)** in the United States; **(b)** in Sweden; and **(c)** in Russia if they drank two tall martinis immediately after arriving at a party? How would your answer change if **(d)** the drinks were taken $\frac{1}{2}$ h apart; **(e)** the two drinks were consumed at a uniform rate during the first hour? **(f)** Suppose that one went to a party, had one and a half tall martinis right away, and then received a phone call saying an emergency had come up and the person needed to drive home immediately. How many minutes would the individual have to reach home before he/she became legally intoxicated, assuming that the person had nothing further to drink? **(g)** How would your answers be different for a thin person? A heavy person? (*Hint:* Base all ethanol concentrations on the volume of body fluid. Plot the concentration of ethanol in the blood as a function of time.) What generalizations can you make? What is the point of this problem? 24601 = Jean <u>V?</u> <u>J?</u> (i.e., who?)

Additional information:

Ethanol in a tall martini: 40 g

Volume of body fluid: 40 L   **(SADD-MADD problem)**

[See Chapter 7 for a more in-depth look at alcohol metabolism.]

P6-8$_B$ (*Pharmacokinetics*) Tarzlon is a liquid antibiotic that is taken orally to treat infections of the spleen. It is effective only if it can maintain a concentration in the bloodstream (based on volume of body fluid) above 0.4 mg per dm^3 of body fluid. Ideally, a concentration of 1.0 mg/dm^3 in the blood would like to be realized. However, if the concentration in the blood exceeds 1.5 mg/dm^3,

Hall of Fame

Hall of Fame

harmful side effects can occur. Once the Tarzlon reaches the stomach, it can proceed in two pathways, both of which are first order: (1) It can be absorbed into the bloodstream through the stomach walls; (2) it can pass out through the gastrointestinal tract and not be absorbed into the blood. Both these processes are first order in Tarzlon concentration in the stomach. Once in the bloodstream, Tarzlon attacks bacterial cells and is subsequently degraded by a zero-order process. Tarzlon can also be removed from the blood and excreted in urine through a first-order process within the kidneys. In the stomach:

Absorption into blood	$k_1 = 0.15 \text{ h}^{-1}$
Elimination through gastrointestine	$k_2 = 0.6 \text{ h}^{-1}$

In the bloodstream:

Degradation of Tarzlon	$k_3 = 0.1 \text{ mg/dm}^3 \cdot \text{h}$
Elimination through urine	$k_4 = 0.2 \text{ h}^{-1}$

(a) Plot the concentration of Tarzlon in the blood as a function of time when 1 dose (i.e., one liquid capsule) of Tarzlon is taken.

(b) How should the Tarzlon be administered (dosage and frequency) over a 48-h period to be most effective?

(c) Comment on the dose concentrations and potential hazards.

(d) How would your answers change if the drug were taken on a full or empty stomach?

One dose of Tarzlon is 250 mg in liquid form: Volume of body fluid = 40 dm^3

P6-9$_C$ (***Reactor selection and operating conditions***) For each of the following sets of reactions describe your reactor system and conditions to maximize the selectivity to D. Make sketches where necessary to support your choices. The rates are in (mol/dm$^3 \cdot$ s), and concentrations are in (mol/dm^3).

(a) (1) $A + B \rightarrow D$ $-r_{1A} = 10 \exp(-8{,}000 \text{ K}/T)C_A C_B$

     (2) $A + B \rightarrow U$ $-r_{2A} = 100 \exp(-1{,}000 \text{ K}/T)C_A^{1/2} C_B^{3/2}$

(b) (1) $A + B \rightarrow D$ $-r_{1A} = 100 \exp(-1{,}000 \text{ K}/T)C_A C_B$

     (2) $A + B \rightarrow U$ $-r_{2A} = 10^6 \exp(-8{,}000 \text{ K}/T)C_A C_B$

(c) (1) $A + B \rightarrow D$ $-r_{1A} = 10 \exp(-1{,}000 \text{ K}/T)C_A C_B$

     (2) $B + D \rightarrow U$ $-r_{2B} = 10^9 \exp(-10{,}000 \text{ K}/T)C_B C_D$

(d) (1) $A \longrightarrow D$ $-r_{1A} = 4280 \exp(-12{,}000 \text{ K}/T)C_A$

     (2) $D \longrightarrow U_1$ $-r_{2D} = 10{,}100 \exp(-15{,}000 \text{ K}/T)C_D$

     (3) $A \longrightarrow U_2$ $-r_{3A} = 26 \exp(-18{,}800 \text{ K}/T)C_A$

(e) (1) $A + B \rightarrow D$ $-r_{1A} = 10^9 \exp(-10{,}000 \text{ K}/T)C_A C_B$

     (2) $D \rightarrow A + B$ $-r_{2D} = 20 \exp(-2{,}000 \text{ K}/T)C_D$

     (3) $A + B \rightarrow U$ $-r_{3A} = 10^3 \exp(-3{,}000 \text{ K}/T)C_A C_B$

Green Engineering

(f) Consider the following parallel reactions[12]

    (1) $A + B \rightarrow D$ $-r_{1A} = 10 \exp(-8{,}000 \text{ K}/T)C_A C_B$

    (2) $A \rightarrow U$ $-r_{2A} = 26 \exp(-10{,}800 \text{K}/T)C_A$

    (3) $U \rightarrow A$ $-r_{3U} = 10{,}000 \exp(-15{,}000 \text{ K}/T)C_U$

(g) For the following reactions (mol/dm³/min)

    (1) $A + B \rightarrow D$ $-r_{1A} = 800\, e^{(-8{,}000 \text{ K}/T)} C_A^{0.5} C_B$

    (2) $A + B \rightarrow U_1$ $-r_{2B} = 10\, e^{(-300 \text{ K}/T)} C_A C_B$

    (3) $D + B \rightarrow U_2$ $-r_{3D} = 10^6\, e^{(-8{,}000 \text{ K}/T)} C_D C_B$

(h) $CH_3OH + \frac{1}{2}O_2 \rightarrow CH_2O + H_2O$

$$-r_{1CH_3OH} = 0.048 \exp\left[-17{,}630 \text{ K}\left(\frac{1}{T} - \frac{1}{485}\right)\right] P_A^{0.5} P_B^{0.1}$$

$$CH_2O + \frac{1}{2}O_2 \rightarrow CO + H_2O$$

$$-r_{2CH_2O} = \frac{2.9 \exp\left[-8710 \text{ K}\left(\frac{1}{T} - \frac{1}{485}\right)\right] P_B P_C}{\left[1 + 3.9 \exp\left[-5290 \text{ K}\left(\frac{1}{T} - \frac{1}{485}\right)\right] P_C\right]^2}$$

$A = CH_3OH$, $B = O_2$, $C = CH_2O$

For part (h), plot the instantaneous selectivity of CH_2O as a function of the concentration of O_2 at $T = 300$ K and at $T = 1000$ K [*CES, 58,* 801 (2003)].

MR and PFR Comparisons. For parts (h), (i), and (j), compare the PFR with a membrane reactor where one of the reactants is only fed through the sides and the other through the entrance with $v_0 = 10$ dm³/s and $C_{T0} = 0.4$ mol/dm³. The maximum and minimum temperatures are 600 and 275 K, respectively. Plot the molar flow rate of A, B, D, and U and $S_{D/U}$ as a function of volume for both the PFR and the MR.

(i) For the reaction in (a).

(j) For the reaction in (b).

(k) For the reaction in (c).

(l) For the reaction in (g).

P6-10$_B$ The elementary liquid-phase-series reaction

$$A \xrightarrow{k_1} B \xrightarrow{k_2} C$$

Green Engineering

—————————————

[12]Assume the reversible reactions are very fast. Techniques of minimizing the waste U are discussed in *Green Engineering* by D. Allen and D. Shonard (Upper Saddle River, N.J.: Prentice Hall, 2000).

is carried out in a 500-dm³ batch reactor. The initial concentration of A is 1.6 mol/dm³. The desired product is B, and separation of the undesired product C is very difficult and costly. Because the reaction is carried out at a relatively high temperature, the reaction is easily quenched.

$$k_1 = 0.4 \text{ h}^{-1}$$
$$k_2 = 0.01 \text{ h}^{-1} \qquad \text{at } 100°C$$

(a) Assuming that each reaction is irreversible, plot the concentrations of A, B, and C as a function of time.

(b) For a CSTR space time of 0.5 h, what temperature would you recommend to maximize B? ($E_1 = 10,000$ cal/mol, $E_2 = 20,000$ cal/mol)

(c) Assume that the first reaction is reversible with $k_{-1} = 0.3$ h⁻¹. Plot the concentrations of A, B, and C as a function of time.

(d) Plot the concentrations of A, B, and C as a function of time for the case where both reactions are reversible with $k_{-2} = 0.005$ h⁻¹.

(e) Vary k_1, k_2, k_{-1}, and k_{-2}. Explain the consequence of $k_1 > 100$ and $k_2 < 0.1$ with $k_{-1} = k_{-2} = 0$ and with $k_{-2} = 1$, $k_{-1} = 0$, and $k_{-2} = 0.25$.

Note: This problem is extended to include the economics (profit) in CDP6-B.

P6-11$_B$ Terephthalic acid (TPA) finds extensive use in the manufacture of synthetic fibers (e.g., Dacron) and as an intermediate for polyester films (e.g., Mylar). The formation of potassium terephthalate from potassium benzoate was studied using a tubular reactor [*Ind. Eng. Chem. Res., 26,* 1691 (1987)].

It was found that the intermediates (primarily K-phthalates) formed from the dissociation of K-benzoate over a CdCl$_2$ catalyst reacted with K-terephthalate in an autocatalytic reaction step.

$$A \xrightarrow{k_1} R \xrightarrow{k_2} S \qquad \text{Series}$$

$$R + S \xrightarrow{k_3} 2S \qquad \text{Autocatalytic}$$

where A = K-benzoate, R = lumped intermediates (K-phthalatis, K-isophthalates, and K-benzenecarboxylates), and S = K-terephthalate. Pure A is charged to the reactor at a pressure of 110 kPa. The specific reaction rates at 410°C are $k_1 = 1.08 \times 10^{-3}$ s⁻¹ with $E_1 = 42.6$ kcal/mol, $k_2 = 1.19 \times 10^{-3}$ s⁻¹ with $E_2 = 48.6$ kcal/mol, $k_3 = 1.59 \times 10^{-3}$ dm³/mol · s with $E_3 = 32$ kcal/mol.

(a) Plot the concentrations of A, R, and S as a function of time in a batch reactor at 410°C noting when the maximum in R occurs.

(b) Repeat (a) for temperatures of 430°C and 390°C.

(c) What would be the exit concentrations from a CSTR operated at 410°C and a space time of 1200 s.

P6-12$_A$ The following liquid-phase reactions were carried out in a CSTR at 325 K.

$$3A \longrightarrow B + C \qquad -r_{1A} = k_{1A}C_A \qquad k_{1A} = 7.0 \text{ min}^{-1}$$

$$2C + A \longrightarrow 3D \qquad r_{2D} = k_{2D}C_C^2 C_A \qquad k_{2D} = 3.0 \frac{\text{dm}^6}{\text{mol}^2 \cdot \text{min}}$$

$$4D + 3C \longrightarrow 3E \qquad r_{3E} = k_{3E}C_D C_C \qquad k_{3E} = 2.0 \frac{\text{dm}^3}{\text{mol} \cdot \text{min}}$$

Sketch the trends or
results you expect
before working out
the details of the
problem.

The concentrations measured *inside* the reactor were $C_A = 0.10$, $C_B = 0.93$, $C_C = 0.51$, and $C_D = 0.049$ all in mol/dm^3.

(a) What are r_{1A}, r_{2A}, and r_{3A}? ($r_{1A} = -0.7$ mol/dm^3·min)

(b) What are r_{1B}, r_{2B}, and r_{3B}?

(c) What are r_{1C}, r_{2C}, and r_{3C}? ($r_{1C} = 0.23$ mol/dm^3·min)

(d) What are r_{1D}, r_{2D}, and r_{3D}?

(e) What are r_{1E}, r_{2E}, and r_{3E}?

(f) What are the net rates of formation of A, B, C, D, and E?

(g) The entering volumetric flow rate is 100 dm^3/min and, the entering concentration of A is 3 M. What is the CSTR reactor volume? (*Ans.:* 4000 dm^3.)

(h) Write a Polymath program to calculate the exit concentrations when the volume is given as 600 dm^3.

(i) **PFR**. Now assume the reactions take place in the gas phase. Use the preceding data to plot the molar flow rates as a function of PFR volume. The pressure drop parameter is 0.001 dm^{-3}, the total concentration entering the reactor is 0.2 mol/dm^3, and $v_0 = 100$ dm^3/min. What are $\tilde{S}_{D/E}$ and $\tilde{S}_{C/D}$?

(j) **Membrane Reactor**. Repeat (i) when species C diffuses out of membrane reactor and the transport coefficient, k_C, is 10 min^{-1}. Compare your results with part (i).

P6-13$_B$ Calculating the space time for parallel reactions. *m*-Xylene is reacted over a ZSM-5 zeolite catalyst. The following *parallel* elementary reactions were found to occur [*Ind. Eng. Chem. Res., 27,* 942 (1988)]:

$$m\text{-Xylene} \xrightarrow{k_1} \text{Benzene} + \text{Methane}$$

$$m\text{-Xylene} \xrightarrow{k_2} p\text{-Xylene}$$

(a) Calculate the PFR volume to achieve 85% conversion of *m*-xylene in a packed-bed reactor. Plot the overall selectivity and yields as a function of τ. The specific reaction rates are $k_1 = 0.22$ s^{-1} and $k_2 = 0.71$ s^{-1} at 673°C. A mixture of 75% *m*-xylene and 25% inerts is fed to a tubular reactor at volumetric flow rate of 200 dm^3/s and a total concentration of 0.05 mol/dm^3. As a first approximation, neglect any other reactions such as the reverse reactions and isomerization to *o*-xylene.

(b) Suppose that $E_1 = 20,000$ cal/mol and $E_2 = 10,000$ cal/mol, what temperature would you recommend to maximize the formation of *p*-xylene in a 2000-dm^3 CSTR?

P6-14$_B$ The following reactions are carried out isothermally in a 50 dm^3 PFR:

$$A + 2B \longrightarrow C + D \qquad r_{D1} = k_{D1}C_A C_B^2$$

$$2D + 3A \longrightarrow C + E \qquad r_{E2} = k_{E2}C_A C_D$$

$$B + 2C \longrightarrow D + F \qquad r_{F3} = k_{F3}C_B C_C^2$$

Additional information: Liquid phase

$$k_{D1} = 0.25 \text{ dm}^6/\text{mol}^2\cdot\text{min} \qquad\qquad v_0 = 10 \text{ dm}^3/\text{min}$$

$$k_{E2} = 0.1 \text{ dm}^3/\text{mol}\cdot\text{min} \qquad\qquad C_{A0} = 1.5 \text{ mol/dm}^3$$

$$k_{F3} = 5.0 \text{ dm}^6/\text{mol}^2\cdot\text{min} \qquad\qquad C_{B0} = 2.0 \text{ mol/dm}^3$$

(a) Plot the species concentrations and the conversion of A as a function of the distance (i.e., volume) down a 50-dm³ PFR. Note any maxima.

(b) Determine the effluent concentrations and conversion from a 50-dm³ CSTR. (*Ans.:* $C_A = 0.61$, $C_B = 0.79$, $C_F = 0.25$, and $C_D = 0.45$ mol/dm³.)

(c) Plot the species concentrations and the conversion of A as a function of time when the reaction is carried out in a semibatch reactor initially containing 40 dm³ of liquid. Consider two cases: (1) A is fed to B, and (2) B is fed to A. What differences do you observe for these two cases?

(d) Vary the ratio of B to A ($1 < \Theta_B < 10$) in the feed to the PFR and describe what you find. What generalizations can you make from this problem?

(e) Rework this problem for the case when the reaction is a gas-phase reaction. We will keep the constants the same so you won't have to make too many changes in your Polymath program, but we will make $v_0 = 100$ dm³/min, $C_{T0} = 0.4$ mol/dm³, $V = 500$ dm³ and equal molar feed of A and B. Plot the molar flow rates and $S_{C/D}$ and $S_{E/F}$ down a PFR.

(f) Repeat (e) when D diffuses out through the sides of a membrane reactor where the mass transfer coefficient, k_{CD}, can be varied between 0.1 min⁻¹ and 10 min⁻¹. What trends do you find?

(g) Repeat (e) when B is fed through the sides of a membrane reactor.

P6-15$_B$ Review the oxidation of formaldehyde to formic acid reactions over a vanadium titanium oxide catalyst [*Ind. Eng. Chem. Res., 28,* 387 (1989)] shown in the ODE solver algorithm in the Summary Notes on the CD-ROM.

(a) Plot the species concentrations as a function of distance down the PFR for an entering flow rate of 100 dm³/min at 5 atm and 140°C. The feed is 66.7% HCHO and 33.3% O₂. Note any maximum in species concentrations.

(b) Plot the yield of overall HCOOH yield and overall selectivity of HCOH to CO, of HCOOCH₃ to CH₃OH and of HCOOH to HCOOCH₃ as a function of the Θ_{O_2}. Suggest some conditions to best produce formic acid. Write a paragraph describing what you find.

(c) Compare your plot in part (a) with a similar plot when pressure drop is taken into account with $\alpha = 0.002$ dm⁻³.

(d) Suppose that $E_1 = 10{,}000$ cal/mol, $E_2 = 30{,}000$ cal/mol, $E_3 = 20{,}000$ cal/mol, and $E_4 = 10{,}000$ cal/mol, what temperature would you recommend for a 1000-dm³ PFR?

P6-16$_B$ The liquefaction of Kentucky Coal No. 9 was carried out in a slurry reactor [D. D. Gertenbach, R. M. Baldwin, and R. L. Bain, *Ind. Eng. Chem. Process Des. Dev., 21,* 490 (1982)]. The coal particles, which were less than 200 mesh, were dissolved in a ~250°C vacuum cut of recycle oil saturated with hydrogen at 400°C. Consider the reaction sequence

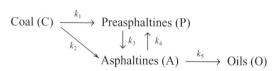

which is a modified version of the one given by Gertenbach et al. All reactions are first order. Calculate the molar flow rate of each species as a function of space time in

(a) A plug-flow reactor.

(b) A 3-m³ CSTR.

(c) What is the point of this problem?

Additional information:

> Entering concentration of coal: 2 kmol/m^3
> Entering flow rate: 10 dm^3/min

At 400°C, $k_1 = 0.12$ min^{-1}, $k_2 = 0.046$ min^{-1}, $k_3 = 0.020$ min^{-1}, $k_4 = 0.034$ min^{-1}, $k_5 = 0.04$ min^{-1}.

P6-17$_B$ The production of acetylene is described by R. W. Wansbaugh [*Chem. Eng., 92*(16), 95 (1985)]. Using the reaction and data in this article, develop a problem and solution.

P6-18$_B$ Read the blood coagulation solved problem on the CD-ROM/Web. Load the blood coagulation *Living Example Problem* from the CD-ROM. Use part (b) for the complete set of coagulation equations. First verify that you can obtain the curve shown in the shaded **Side note** on page 325.

(a) Plot some of the other species as a function of time, specifically TFVIIa, TFVIIaX, TFVIIaXa, and TFVIIaIX. Do you notice anything unusual about these curves, such as two maxima?

(b) How does the full solution compare with the solution for the abbreviated reactions? What reactions can you eliminate and still get a *reasonable* approximation to the thrombin in time curve?

P6-19$_C$ The ethylene epoxydation is to be carried out using a cesium-doped silver catalyst in a packed bed reactor.

$$(1)\quad C_2H_4 + \frac{1}{2}O_2 \rightarrow C_2H_4O \qquad -r_{1E} = \frac{k_{1E}P_EP_O^{0.58}}{(1+K_{1E}P_E)^2}$$

Along with the desired reaction, the complete combustion of ethylene also occurs

$$(2)\quad C_2H_4 + 3O_2 \rightarrow 2CO_2 + 2H_2O \qquad -r_{2E} = \frac{k_{2E}P_EP_O^{0.3}}{(1+K_{2E}P_E)^2}$$

[M. Al-Juaied, D. Lafarga, and A. Varma, *Chem. Eng. Sci. 56*, 395 (2001)]. It is proposed to replace the conventional PBR with a membrane reactor in order to improve the selectivity. As a rule of thumb, a 1% increase in the selectivity to ethylene oxide translates to an increase in profit of about $2 million/yr. The feed consists of 12% (mole) oxygen, 6% ethylene, and the remainder nitrogen at a temperature of 250°C and a pressure of 2 atm. The total molar flow rate is 0.0093 mol/s and to a reactor containing 2 kg of catalyst.

(a) What conversion and selectivity, $\tilde{S}$, are expected in a conventional PFR?

(b) What would be the conversion and selectivity if the total molar flow rate were divided and the 12% oxygen stream (no ethylene) were uniformly fed through the sides of the membrane reactor and 6% ethylene (no oxygen) were fed at the entrance?

(c) Repeat (b) for a case when ethylene is fed uniformly through the sides and oxygen is fed at the entrance. Compare with parts (a) and (b).

(d) Repeat (b) and (c) for the methanol reaction given in problem 6-9 (h) with P_i in Bar and $-r^P_{1CH_2OH}$ and $-r^P_{2CH_2O}$ are in Bar/s.

Additional information:

$$k_{1E} = 0.15 \frac{mol}{kg \cdot s \ atm^{1.58}} \ \text{at 523 K with } E_1 = 60.7 \ kJ/mol$$

$$k_{2E} = 0.0888 \frac{mol}{kg \cdot s \ atm^{1.3}} \ \text{at 523 K with } E_2 = 73.2 \ kJ/mol$$

$$K_{1E} = 6.50 \ atm^{-1}, K_{2E} = 4.33 \ atm^{-1}$$

P6-20$_C$ For the van de Vusse elementary reactions

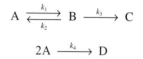

$$2A \xrightarrow{k_4} D$$

Web Hint

determine the reactor or combination of reactors that maximize the amount of B formed. See PRS R6.1.

Links

Additional information:

$$k_1 = 0.01 \ s^{-1} \qquad k_2 = 0.05 \ s^{-1} \qquad k_3 = 10 \ s^{-1} \qquad k_4 = 100 \ m^3/kmol \cdot s$$

$$C_{A0} = 2 \ kmol/m^3 \ \text{and } v_0 = 0.2 \ m^3/s$$

Repeat for $k_2 = 0.002 \ s^{-1}$.

P6-21$_B$ The gas-phase reactions take place isothermally in a membrane reactor packed with catalyst. Pure A enters the reactor at 24.6 atm and 500 K and a flow rate of A of 10 mol/min

$$A \rightleftharpoons B+C \qquad r'_{1C} = k_{1C}\left[C_A - \frac{C_B C_C}{K_{1C}}\right]$$

$$A \longrightarrow D \qquad r'_{2D} = k_{2D} C_A$$

$$2C+D \longrightarrow 2E \qquad r'_{3E} = k_{3E} C_C^2 C_D$$

Only species B diffuses out of the reactor through the membrane.
(a) Plot the concentrations down the length of the reactor.
(b) Explain why your curves look the way they do.
(c) Describe the major differences you observe when C diffuses out instead of B, with the same mass transfer coefficient.
(d) Vary some of the parameters (e.g., k_B, k_{1C}, K_{1C}) and write a paragraph describing what you find.
Additional Information:
Overall mass transfer coefficient $k_B = 1.0 \ dm^3 / kg \ cat \cdot min$
$\quad k_{1C} = 2 \ dm^3 / kg \ cat \cdot min$
$\quad K_{1C} = 0.2 \ mol / dm^3$
$\quad k_{2D} = 0.4 \ dm^3 / kg \ cat \cdot min$
$\quad k_{3E} = 400 \ dm^3 / mol^2 \cdot kg \ cat \cdot min$
$\quad W_f = 100 \ kg$
$\quad \alpha = 0.008 \ kg^{-1}$

Web Modules

P6-22$_C$ Read over the *oscillating reaction Web Module*. For the four reactions involving I^- and IO^-_3 :

(a) What factors influence the amplitude and frequency of the oscillation reaction? What causes these oscillations? (In other words: What makes this reaction different than others we have studied so far in Chapter 6?)

(b) Why do you think that the oscillations eventually cease (in the original experiment by Belousov, they lasted about 50 minutes)?

(c) A 10°C increase in temperature produced the following observations. The dimensionless times at which the oscillations began and ended decreased. The dimensionless period of the oscillation at the start of the oscillation increased while the dimensionless period near the end of the oscillation decreased. What conclusions can you draw about the reactions? Explain your reasoning. Feel free to use plots/sketches or equations if you wish.

(d) What if ... play around with the *Living Example* Polymath Program on the CD-ROM—what are the effects of changing the values of k_0, k_u, k_1, and k_2? Can you make the oscillations damped or unstable?

P6-23$_A$ Go to Professor Herz's **Reactor Lab** on the CD-ROM or on the web at *www.reactorlab.net*.

(a) Load Division 5, Lab 2 of The Reactor Lab from the CD for the selective oxidation of ethylene to ethylene oxide. Click the [i] info button to get information about the system. Perform experiments and develop rate equations for the reactions. Write a technical memo that reports your results and includes plots and statistical measurements of how well your kinetic model fits experimental data.

(b) Load Division 5, Labs 3 and 4 of The Reactor Lab for batch reactors in which parallel and series reactions, respectively, can be carried out. Investigate how dilution with solvent affects the selectivity for different reaction orders, and write a memo describing your findings.

JOURNAL CRITIQUE PROBLEMS

P6C-1 In *J. of Hazardous Materials, B89*, 197 (2002), is there an algebraic error in the equation to calculate the specific reaction rate k_1? If so, what are the ramifications later on in the analysis?

P6C-2 In *Int. J. Chem. Kinet., 35*, 555 (2003), does the model overpredict the ortho diethyl benzene concentration?

P6C-3 Is it possible to extrapolate the curves on Figure 2 [*AIChE J., 17*, 856 (1971)] to obtain the initial rate of reaction? Use the Wiesz–Prater criterion to determine if there are any diffusion limitations in this reaction. Determine the partial pressure of the products on the surface based on a selectivity for ethylene oxide ranging between 51 and 65% with conversions between 2.3 and 3.5%.

P6C-4 Equation 5 [*Chem. Eng. Sci., 35*, 619 (1980)] is written to express the formation rate of C (olefins). As described in equation 2, there is no change in the concentration of C in the third reaction of the series:

$$
\left.\begin{array}{l}
A \xrightarrow{\ k_1\ } B \\
A+B \xrightarrow{\ k_2\ } C \\
B+C \xrightarrow{\ k_3\ } C \\
C \xrightarrow{\ k_4\ } D
\end{array}\right\} \text{ equation 2}
$$

(a) Determine if the rate law given in equation 5 is correct.

(b) Can equations 8, 9, and 12 be derived from equation 5?

(c) Is equation 14 correct?

(d) Are the adsorption coefficients b_i and b_j calculated correctly?

Good alternative problems on the CD and web.

CDP6-24$_B$ The production of maleic anhydride by oxidation with air can be carried out over a vanadium catalyst in (a) a "fluidized" CSTR and (b) a PBR at different temperatures.

$$C_6H_6 + \tfrac{9}{2}O_2 \rightarrow C_4H_2O_3 + 2CO_2 + 2H_2O$$

$$C_4H_2O_3 + 3O_2 \rightarrow 4CO_3 + 3H_2O$$

$$C_6H_6 + \tfrac{15}{2}O_2 \rightarrow 6CO_2 + 3H_2O$$

In excess air these reactions can be represented by

$$A \rightarrow B \rightarrow C$$
$$\searrow$$
$$D$$

[3rd Ed. P6-14$_B$]

CDP6-25$_B$ Rework Problems (a) P6-6, (b) P6-7, (c) P6-8 when the toluene formed in reaction (E6-6.2) also undergoes hydrodealkylation to form benzene

$$H + T \rightarrow B + Me \ , \ r_B = k_3 C_T C_H^{1/2}$$

[3rd Ed. P6-15]

CDP6-26$_B$ A series of five hydrodealkylation reactions beginning with

$$H_2 + C_6H(CH_3)_5 \xrightarrow{\ k_1\ } C_6H_2(CH_3) + CH_4 \qquad -r_{C11} = k_1 C_{H_2}^{1/2} C_{1_1}$$

and ending with

$$H_2 + C_6H_5CH_3 \longrightarrow C_6H_6 + CH_4 \quad -r_{C7} = k_5 C_H^{1/2} C_7$$

are carried out in a PFR. [3rd Ed. P6-16].

CDP6-27$_B$ The hydrogenation of benzene

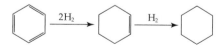

is carried out in a CSTR slurry reactor where the desired product is cyclohexene. [3rd Ed. P6-23]

CDP6-28$_C$ Industrial data for methanol synthesis reactions is provided for the complex reactions

$$CO_2 + 2H_2 \rightleftharpoons CH_3OH$$

$$CO + H_2O \rightleftharpoons CO_2 + H_2$$

$$CH_3 + OH \longrightarrow CH_3O + H_2$$

The reaction is carried out in a PFR. Find the best operating conditions to produce methanol. [3rd Ed. P6-24]

CDP6-29$_C$ Load the *Living Example Problem* for the hydrodealkylation of mesitylene carried out in a membrane reactor. Optimize the parameters to obtain the maximum profit.

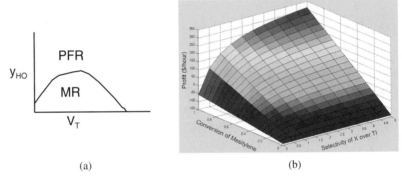

(a)   (b)

Figure P6-30   (a) Comparison (b) cost function.

- **Additional Homework Problems**

- **Oldies, but goodies—problems from previous editions.**

Series Reactions

CDP6-A$_B$ Chlorination of benzene to monochlorobenzene and dichlorobenzene in a CSTR. [1st Ed. P9-14]

CDP6-B$_B$ The reaction sequence A $\longrightarrow$ B, B $\longrightarrow$ C, B $\longrightarrow$ D is carried out in a batch reactor and in a CSTR. [2nd Ed. P9-12]

CDP6-C$_B$ Isobutylene is oxidized to methacrolein, CO, and CO$_2$. [1st Ed. P9-16]

Parallel Reactions

CDP6-D$_B$ Calculate conversion in two independent reactions. [3rd Ed. P6-4(d)]

CDP6-E$_B$ Calculate equilibrium concentrations in reaction $A + B \rightleftharpoons C + D$
$C + B \rightleftharpoons X + Y$ [3rd Ed. P6-5]

CDP6-F$_B$ Find the profit for $A \rightarrow D$ and $A \rightarrow U$. [3rd Ed. P6-10]

Complex Reactions

Ibuprofen

CDP6-G$_C$ Production of ibuprofen intermediate. [3rd Ed. P6-26]

CDP6-H$_B$ Hydrogenation of *o*-cresol. [3rd Ed. P6-3]

CDP6-I$_C$ Design a reaction system to maximize the selectivity of *p*-xylene from methanol and toluene over a HZSM-8 zeolite catalyst. [2nd Ed. P9-17]

CDP6-J$_C$ Oxidation of propylene to acrolein [*Chem. Eng. Sci. 51*, 2189 (1996)].

CDP6-K$_B$ Complex reactions. [Old Exam]

CDP6-L$_C$ California problem. [3rd Ed. P6-19$_B$]

CDP6-M$_B$ Fluidized bed. [3rd Ed. P6-20]

CDP6-N$_D$ Flame retardants. [3rd Ed. P6-21]

CDP6-O$_C$ Oleic acid production. [3rd Ed. P6-25]

New Problems on the Web

CDP1-New   From time to time new problems relating Chapter 6 material to everyday interests or emerging technologies will be placed on the web. Solutions to these problems can be obtained by e-mailing the author.

Also, one can go to the web site, *www.rowan.edu/greenengineering*, and work the home problem on green engineering specific to this chapter.

Green Engineering

SUPPLEMENTARY READING

1. Selectivity, reactor schemes, and staging for multiple reactions, together with evaluation of the corresponding design equations, are presented in

> DENBIGH, K. G., and J. C. R. TURNER, *Chemical Reactor Theory*, 2nd ed. Cambridge: Cambridge University Press, 1971, Chap. 6.
>
> LEVENSPIEL, O., *Chemical Reaction Engineering*, 2nd ed. New York: Wiley, 1972, Chap. 7.

Some example problems on reactor design for multiple reactions are presented in

> HOUGEN, O. A., and K. M. WATSON, *Chemical Process Principles*, Part 3: *Kinetics and Catalysis*. New York: Wiley, 1947, Chap. XVIII.
>
> SMITH, J. M., *Chemical Engineering Kinetics*, 3rd ed. New York: McGraw-Hill, 1980, Chap. 4.

2. Books that have many analytical solutions for parallel, series, and combination reactions are

> CAPELLOS, C., and B. H. J. BIELSKI, *Kinetic Systems*. New York: Wiley, 1972.
>
> WALAS, S. M., *Chemical Reaction Engineering Handbook of Solved Problems*. Newark, N.J.: Gordon and Breach, 1995.

3. A brief discussion of a number of pertinent references on parallel and series reactions is given in

> ARIS, R., *Elementary Chemical Reactor Analysis*. Upper Saddle River, N.J.: Prentice Hall, 1969, Chap. 5.

4. An excellent example of the determination of the specific reaction rates, k_i, in multiple reactions is given in

> BOUDART, M., and G. DJEGA-MARIADASSOU, *Kinetics of Heterogeneous Catalytic Reactions*. Princeton, N.J.: Princeton University Press, 1984.

Reaction Mechanisms, Pathways, Bioreactions, and Bioreactors

7

The next best thing to knowing something is knowing where to find it.

Samuel Johnson (1709–1784)

Overview. One of the main threads that ties this chapter together is the pseudo-steady-state-hypothesis (PSSH) and the concept of active intermediates. We shall use it to develop rate laws for both chemical and biological reactions. We begin by discussing reactions which do not follow elementary rate laws and are not zero, first, or second order. We then show how reactions of this type involve a number of reaction steps, each of which is elementary. After finding the net rates of reaction for each species, we invoke the PSSH to arrive at a rate law that is consistent with experimental observation. After discussing gas-phase reactions, we apply the PSSH to biological reactions, with a focus on enzymatic reactions. Next, the concepts of enzymatic reactions are extended to organisms. Here organism growth kinetics are used in modeling both batch reactors and CSTRs (chemostats). Finally, a physiological-based-pharmacokinetic approach to modeling of the human body is coupled with the enzymatic reactions to develop concentration-time trajectories for the injection of both toxic and nontoxic substances.

7.1 Active Intermediates and Nonelementary Rate Laws

In Chapter 3 a number of simple power law models, that is,

$$-r_A = kC_A^n$$

were presented where n was an integer of 0, 1, or 2 corresponding to a zero-, first-, and second-order reaction. However, for a large number of reactions, the orders are either noninteger such as the decomposition of acetaldehyde at 500°C

$$CH_3CHO \rightarrow CH_4 + CO$$

where the rate law is

$$-r_{CH_3CHO} = kC_{CH_3CHO}^{3/2}$$

or of a form where there are concentration terms in both the numerator and denominator such as the formation of HBr from hydrogen and bromine

$$H_2 + Br_2 \rightarrow 2HBr$$

with

$$r_{HBr} = \frac{k_1 C_{H_2} C_{Br_2}^{3/2}}{C_{HBr} + k_2 C_{Br_2}}$$

Rate laws of this form usually involve a number of elementary reactions and at least one active intermediate. An *active intermediate* is a high-energy molecule that reacts virtually as fast as it is formed. As a result, it is present in very small concentrations. Active intermediates (e.g., A^*) can be formed by collision or interaction with other molecules.

$$A + M \rightarrow A^* + M$$

Properties of an active intermediate A^*

Here the activation occurs when translational kinetic energy is transferred into energy stored in internal degrees of freedom, particularly vibrational degrees of freedom.[1] An unstable molecule (i.e., active intermediate) is not formed solely as a consequence of the molecule moving at a high velocity (high translational kinetic energy). The energy must be absorbed into the chemical bonds where high-amplitude oscillations will lead to bond ruptures, molecular rearrangement, and decomposition. In the absence of photochemical effects or similar phenomena, the transfer of translational energy to vibrational energy to produce an active intermediate can occur only as a consequence of molecular collision or interaction. Collision theory is discussed in the *Professional Reference Shelf* in Chapter 3. Other types of active intermediates that can be formed are *free radicals* (one or more unpaired electrons, e.g., $CH_3\cdot$), ionic intermediates (e.g., carbonium ion), and enzyme-substrate complexes, to mention a few.

The idea of an active intermediate was first postulated in 1922 by F. A. Lindermann[2] who used it to explain changes in reaction order with changes in reactant concentrations. Because the active intermediates were so short lived

---

[1] W. J. Moore, *Physical Chemistry*, (Reading, Mass.: Longman Publishing Group, 1998).

[2] F. A. Lindemann, *Trans. Faraday. Soc.*, *17*, 598 (1922).

and present in such low concentrations, their existence was not really definitively seen until the work of Ahmed Zewail who received the Nobel Prize in 1999 for femtosecond spectroscopy.[3] His work on cyclobutane showed the reaction to form two ethylene molecules did not proceed directly, as shown in Figure 7-1(a), but formed the active intermediate shown in the small trough at the top of the energy reaction coordinate diagram in Figure 7-1(b). As discussed in Chapter 3, an estimation of the barrier height, E, can be obtained using computational software packages such as Spartan, Cerius², or Gaussian as discussed in the *Molecular Modeling Web Module* in Chapter 3.

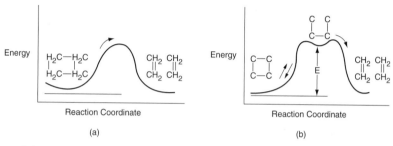

Figure 7-1   Reaction coordinate. Courtesy *Science News, 156*, 247 (1999).

7.1.1 Pseudo-Steady-State Hypothesis (PSSH)

In the theory of active intermediates, decomposition of the intermediate does not occur instantaneously after internal activation of the molecule; rather, there is a time lag, although infinitesimally small, during which the species remains activated. Zewail's work was the first definitive proof of a gas-phase active intermediate that exists for an infinitesimally short time. Because a reactive intermediate reacts virtually as fast as it is formed, the net rate of formation of an active intermediate (e.g., A*) is zero, i.e.,

$$r_{A*} \equiv 0 \tag{7-1}$$

This condition is also referred to as the Pseudo-Steady-State Hypothesis (PSSH). If the active intermediate appears in n reactions, then

$$\boxed{r_{A*} = \sum_{i-1}^{n} r_{iA*} = 0} \tag{7-2}$$

To illustrate how rate laws of this type are formed, we shall first consider the gas-phase decomposition of azomethane, AZO, to give ethane and nitrogen:

$$(CH_3)_2N_2 \longrightarrow C_2H_6 + N_2$$

---

[3] J. Peterson, *Science News, 156*, 247 (1999).

Experimental observations[4] show that the rate of formation of ethane is first order with respect to AZO at pressures greater than 1 atm (relatively high concentrations)

$$r_{C_2H_6} \propto C_{AZO}$$

and second order at pressures below 50 mmHg (low concentrations):

$$r_{C_2H_6} \propto C_{AZO}^2$$

To explain this first and second order depending on the concentration of AZO we shall propose the following mechanism consisting of three elementary reactions.

Mechanism
$$\begin{cases}
\text{Reaction 1:} & (CH_3)_2N_2 + (CH_3)_2N_2 \xrightarrow{k_{1AZO*}} (CH_3)_2N_2 + [(CH_3)_2N_2]^* \\
\text{Reaction 2:} & [(CH_3)_2N_2]^* + (CH_3)_2N_2 \xrightarrow{k_{2AZO*}} (CH_3)_2N_2 + (CH_3)_2N_2 \\
\text{Reaction 3:} & [(CH_3)_2N_2]^* \xrightarrow{k_{3AZO*}} C_2H_6 + N_2
\end{cases}$$

In *reaction 1*, two AZO molecules collide and the kinetic energy of one AZO molecule is transferred to internal rotational and vibrational energies of the other AZO molecule, and it becomes activated and highly reactive (i.e., AZO*). In *reaction 2*, the activated molecule (AZO*) is deactivated through collision with another AZO by transferring its internal energy to increase the kinetic energy of the molecules with which AZO* collides. In *reaction 3*, this high activated AZO* molecule, which is wildly vibrating, spontaneously decomposes into ethane and nitrogen. Because each of the reaction steps is elementary, the corresponding rate laws for the active intermediate AZO* in reactions (1), (2), and (3) are

Note: The specific reaction rates, k, are all defined wrt the active intermediate AZO*.

(1) $$r_{1AZO*} = k_{1AZO*}C_{AZO}^2$$ (7-3)

(2) $$r_{2AZO*} = -k_{2AZO*}C_{AZO*}C_{AZO}$$ (7-4)

(3) $$r_{3AZO*} = -k_{3AZO*}C_{AZO*}$$ (7-5)

These rate laws [Equations (7-3) through (7-5)] are pretty much useless in the design of any reaction system because the concentration of the active intermediate AZO* is not readily measurable. Consequently, we will use the Pseudo-Steady-State-Hypothesis (PSSH) to obtain a rate law in terms of measurable concentrations.

We first write the rate of formation of product (with $k_3 \equiv k_{3AZO*}$)

$$\boxed{r_{C_2H_6} = k_3 C_{AZO*}}$$ (7-6)

---

[4] H. C. Ramsperger, *J. Am. Chem. Soc.*, 49, 912 (1927).

To find the concentration of the active intermediate AZO*, we set the net rate of AZO* equal to zero,[5] $r_{AZO*} \equiv 0$.

$$r_{AZO*} = r_{1AZO*} + r_{2AZO*} + r_{3AZO*} = 0$$

$$= k_1 C_{AZO}^2 - k_2 C_{AZO*} C_{AZO} - k_3 C_{AZO*} = 0 \qquad (7\text{-}7)$$

Solving for C_{AZO*}

$$C_{AZO*} = \frac{k_1 C_{AZO}^2}{k_2 C_{AZO} + k_3} \qquad (7\text{-}8)$$

Substituting Equation (7-8) into Equation (7-6)

$$\boxed{r_{C_2H_6} = \frac{k_1 k_3 C_{AZO}^2}{k_2 C_{AZO} + k_3}} \qquad (7\text{-}9)$$

At low AZO concentrations,

$$k_2 C_{AZO} \ll k_3$$

for which case we obtain the following second-order rate law:

$$r_{C_2H_6} = k_1 C_{AZO}^2$$

At high concentrations

$$k_2 C_{AZO} \gg k_3$$

in which case the rate expression follows first-order kinetics,

$$r_{C_2H_6} = \frac{k_1 k_3}{k_2} C_{AZO} = k C_{AZO}$$

In describing reaction orders for this equation, one would say the reaction is *apparent first order* at high azomethane concentrations and *apparent second order* at low azomethane concentrations.

The PSSH can also explain why one observes so many first-order reactions such as

$$(CH_3)_2O \rightarrow CH_4 + H_2 + CO$$

---

[5] For further elaboration on this section, see R. Aris, *Am. Sci., 58,* 419 (1970).

Symbolically this reaction will be represented as A going to product P, that is,

$$A \rightarrow P$$

with

$$-r_A = kC_A$$

The reaction is first order but the reaction is not elementary. The reaction proceeds by first forming an active intermediate, A*, from the collision of the reactant molecule and an inert molecule of M. Either this wildly oscillating active intermediate is deactivated by collision with inert M, or it decomposes to form product.

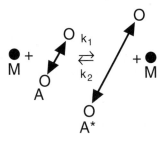

Reaction pathways

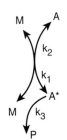

The mechanism consists of the three elementary reactions:

Activation   (1) $A + M \xrightarrow{k_1} A^* + M$

Deactivation   (2) $A^* + M \xrightarrow{k_2} A + M$

Decomposition   (3) $A^* \xrightarrow{k_3} P$

Writing the rate of formation of product

$$r_P = k_3 C_{A^*}$$

and using the PSSH to find the concentrations of A* in a manner similar to the azomethane decomposition described earlier, the rate law can be shown to be

$$r_P = -r_A = \frac{k_3 k_1 C_A C_M}{k_2 C_M + k_3} \tag{7-10}$$

Because the concentration of the inert M is constant, we let

$$k = \frac{k_1 k_3 C_M}{k_2 C_M + k_3} \tag{7-11}$$

to obtain the first-order rate law

$$-r_A = kC_A$$

First-order rate law
for a nonelementary
reaction

Consequently, we see the reaction

$$A \rightarrow P$$

follows an elementary rate law but is not an elementary reaction.

7.1.2 Searching for a Mechanism

In many instances the rate data are correlated before a mechanism is found. It is a normal procedure to reduce the additive constant in the denominator to 1. We therefore divide the numerator and denominator of Equation (7-9) by k_3 to obtain

$$\boxed{r_{C_2H_6} = \frac{k_1 C_{AZO}^2}{1 + k' C_{AZO}}} \tag{7-12}$$

General Considerations. The rules of thumb listed in Table 7-1 may be of some help in the development of a mechanism that is consistent with the experimental rate law. Upon application of Table 7-1 to the azomethane example just discussed, we see the following from rate equation (7-12):

1. The active intermediate, AZO^*, collides with azomethane, AZO [Reaction 2], resulting in the concentration of AZO in the denominator.
2. AZO^* decomposes spontaneously [Reaction 3], resulting in a constant in the denominator of the rate expression.
3. The appearance of AZO in the numerator suggests that the active intermediate AZO^* is formed from AZO. Referring to Reaction 1, we see that this case is indeed true.

TABLE 7-1.   RULES OF THUMB FOR DEVELOPMENT OF A MECHANISM

---

1. Species having the concentration(s) appearing in the *denominator* of the rate law probably collide with the active intermediate, for example,

$$A + A^* \longrightarrow [\text{Collision products}]$$

2. If a constant appears in the *denominator*, one of the reaction steps is probably the spontaneous decomposition of the active intermediate, for example,

$$A^* \longrightarrow [\text{Decomposition products}]$$

3. Species having the concentration(s) appearing in the *numerator* of the rate law probably produce the active intermediate in one of the reaction steps, for example,

$$[\text{reactant}] \longrightarrow A^* + [\text{Other products}]$$

Finding the Reaction Mechanism. Now that a rate law has been synthesized from the experimental data, we shall try to propose a mechanism that is consistent with this rate law. The method of attack will be as given in Table 7-2.

TABLE 7-2.   STEPS TO DEDUCE A RATE LAW

1. Assume an active intermediate(s).

2. Postulate a mechanism, utilizing the rate law obtained from experimental data, if possible.

3. Model each reaction in the mechanism sequence as an elementary reaction.

4. After writing rate laws for the rate of formation of desired product, write the rate laws for each of the active intermediates.

5. Use the PSSH.

6. Eliminate the concentration of the intermediate species in the rate laws by solving the simultaneous equations developed in Steps 4 and 5.

7. If the derived rate law does not agree with experimental observation, assume a new mechanism and/or intermediates and go to Step 3. A strong background in organic and inorganic chemistry is helpful in predicting the activated intermediates for the reaction under consideration.

Once the rate law is found, the search for the mechanism begins.

Example 7–1   The Stern–Volmer Equation

Light is given off when a high-intensity ultrasonic wave is applied to water.[6] This light results from microsize gas bubbles (0.1 mm) being formed by the ultrasonic wave and then being compressed by it. During the compression stage of the wave, the contents of the bubble (e.g., water and whatever else is dissolved in the water, e.g., CS_2, O_2, N_2) are compressed adiabatically.

   This compression gives rise to high temperatures and kinetic energies of the gas molecules, which through molecular collisions generate active intermediates and cause chemical reactions to occur in the bubble.

Collapsing cavitation microbubble

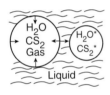

$$M + H_2O \longrightarrow H_2^*O + M$$

The intensity of the light given off, I, is proportional to the rate of deactivation of an activated water molecule that has been formed in the microbubble.

$$H_2O^* \xrightarrow{\ k\ } H_2O + h\nu$$

$$\text{Light intensity (I)} \propto (-r_{H_2O^*}) = k C_{H_2O^*}$$

   An order-of-magnitude increase in the intensity of sonoluminescence is observed when either carbon disulfide or carbon tetrachloride is added to the water. The intensity of luminescence, I, for the reaction

$$CS_2^* \xrightarrow{\ k_4\ } CS_2 + h\nu$$

is

$$I \propto (-r_{CS_2^*}) = k_4 C_{CS_2^*}$$

A similar result exists for CCl_4.

---

[6] P. K. Chendke and H. S. Fogler, *J. Phys. Chem.*, *87*, 1362 (1983).

However, when an aliphatic alcohol, X, is added to the solution, the intensity decreases with increasing concentration of alcohol. The data are usually reported in terms of a Stern–Volmer plot in which relative intensity is given as a function of alcohol concentration, C_X. (See Figure E7-1.1, where I_0 is the sonoluminescence intensity in the absence of alcohol and I is the sonoluminescence intensity in the presence of alcohol.) Suggest a mechanism consistent with experimental observation.

Stern–Volmer plot

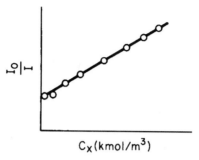

Figure E7-1.1 Ratio of luminescence intensities as a function of Scavenger concentration.

Solution

From the linear plot we know that

$$\frac{I_0}{I} = A + B C_X \equiv A + B(X) \tag{E7-1.1}$$

where $C_X \equiv (X)$. Inverting yields

$$\frac{I}{I_0} = \frac{1}{A + B(X)} \tag{E7-1.2}$$

From rule 1 of Table 7-1, the denominator suggests that alcohol (X) collides with the active intermediate:

$$X + \text{Intermediate} \longrightarrow \text{Deactivation products} \tag{E7-1.3}$$

Reaction Pathways

The alcohol acts as what is called a scavenger to deactivate the active intermediate. The fact that the addition of CCl_4 or CS_2 increases the intensity of the luminescence,

$$I \propto (CS_2) \tag{E7-1.4}$$

leads us to postulate (rule 3 of Table 7-1) that the active intermediate was probably formed from CS_2:

$$M + CS_2 \longrightarrow CS_2^* + M \tag{E7-1.5}$$

where M is a third body (CS_2, H_2O, N_2, etc.).

We also know that deactivation can occur by the reverse of Reaction (E7-1.5). Combining this information, we have as our mechanism:

The mechanism

$$\text{Activation:} \qquad \text{M} + \text{CS}_2 \xrightarrow{k_1} \text{CS}_2^* + \text{M} \qquad\qquad \text{(E7-1.5)}$$

$$\text{Deactivation:} \quad \text{M} + \text{CS}_2^* \xrightarrow{k_2} \text{CS}_2 + \text{M} \qquad\qquad \text{(E7-1.6)}$$

$$\text{Deactivation:} \quad \text{X} + \text{CS}_2^* \xrightarrow{k_3} \text{CS}_2 + \text{X} \qquad\qquad \text{(E7-1.3)}$$

$$\text{Luminescence:} \qquad \text{CS}_2^* \xrightarrow{k_4} \text{CS}_2 + h\nu \qquad\qquad \text{(E7-1.7)}$$

$$I = k_4(\text{CS}_2^*) \qquad\qquad \text{(E7-1.8)}$$

Using the PSSH on CS_2^* yields

$$r_{\text{CS}_2^*} = 0 = k_1(\text{CS}_2)(\text{M}) - k_2(\text{CS}_2^*)(\text{M}) - k_3(\text{X})(\text{CS}_2^*) - k_4(\text{CS}_2^*)$$

Solving for CS_2^* and substituting into Equation (E7-1.8) gives us

$$I = \frac{k_4 k_1 (\text{CS}_2)(\text{M})}{k_2(\text{M}) + k_3(\text{X}) + k_4} \qquad\qquad \text{(E7-1.9)}$$

In the absence of alcohol,

$$I_0 = \frac{k_4 k_1 (\text{CS}_2)(\text{M})}{k_2(\text{M}) + k_4} \qquad\qquad \text{(E7-1.10)}$$

For constant concentrations of CS_2 and the third body, M, we take a ratio of Equation (E7-1.10) to (E7-1.9):

Web Modules

$$\frac{I_0}{I} = 1 + \frac{k_3}{k_2(\text{M}) + k_4}(\text{X}) = 1 + k'(\text{X}) \qquad\qquad \text{(E7-1.11)}$$

which is of the same form as that suggested by Figure E7-1.1. Equation (E7-1.11) and similar equations involving scavengers are called *Stern–Volmer equations.*

A discussion of luminescence is continued on the **CD-ROM Web Module, Glow Sticks.** Here, the PSSH is applied to glow sticks. First, a mechanism for the reactions and luminescence is developed. Next, mole balance equations are written on each species and coupled with rate law obtained using the PSSH and the resulting equations are solved and compared with experimental data.

Glow sticks

7.1.3 Chain Reactions

Now, let us proceed to some slightly more complex examples involving chain reactions. A chain reaction consists of the following sequence:

1. *Initiation:* formation of an active intermediate.

Steps in a chain reaction

2. *Propagation or chain transfer:* interaction of an active intermediate with the reactant or product to produce another active intermediate.
3. *Termination:* deactivation of the active intermediate to form products.

Example 7–2 PSSH Applied to Thermal Cracking of Ethane

The thermal decomposition of ethane to ethylene, methane, butane, and hydrogen is believed to proceed in the following sequence:

Initiation:

(1) $\qquad C_2H_6 \xrightarrow{k_{C_2H_6}} 2CH_3\bullet \qquad\qquad r_{1C_2H_6} = -k_{1C_2H_6}[C_2H_6]$

$$\text{Let } k_1 = k_{1C_2H_6}$$

Propagation:

(2) $CH_3\bullet + C_2H_6 \xrightarrow{k_2} CH_4 + C_2H_5\bullet \qquad r_{2C_2H_6} = -k_2[CH_3\bullet][C_2H_6]$

(3) $\qquad C_2H_5\bullet \xrightarrow{k_3} C_2H_4 + H\bullet \qquad\qquad r_{3C_2H_4} = k_3[C_2H_5\bullet]$

(4) $\quad H\bullet + C_2H_6 \xrightarrow{k_4} C_2H_5\bullet + H_2 \qquad r_{4C_2H_6} = -k_4[H\bullet][C_2H_6]$

Termination:

(5) $\qquad 2C_2H_5\bullet \xrightarrow{k_5} C_4H_{10} \qquad\qquad r_{5C_2H_5\bullet} = -k_{5C_2H_5\bullet}[C_2H_5\bullet]^2$

$$\text{Let } k_5 \equiv k_{5C_2H_5\bullet}$$

(a) Use the PSSH to derive a rate law for the rate of formation of ethylene.
(b) Compare the PSSH solution in Part (a) to that obtained by solving the complete set of ODE mole balances.

Solution

Part (a) *Developing the Rate Law*

The rate of formation of ethylene (Reaction 3) is

$$\boxed{r_{3C_2H_4} = k_3[C_2H_5\bullet]} \qquad\qquad (E7\text{-}2.1)$$

Given the following reaction sequence:
    For the active intermediates: $CH_3\bullet$, $C_2H_5\bullet$, $H\bullet$ the net rates of reaction are

$[\mathbf{C_2H_5\bullet}]$: $\quad r_{C_2H_5\bullet} = r_{2C_2H_5\bullet} + r_{3C_2H_5\bullet} + r_{4C_2H_5\bullet} + r_{5C_2H_5\bullet} = 0$

From reaction stoichiometry we have

$$r_{2C_2H_5\bullet} = -r_{2C_2H_6}, \ r_{3C_2H_5\bullet} = -r_{3C_2H_4} \text{ and } r_{4C_2H_5\bullet} = -r_{4C_2H_6}$$

then $\qquad r_{C_2H_5\bullet} = -r_{2C_2H_6} - r_{3C_2H_4} - r_{4C_2H_6} + r_{5C_2H_5\bullet} = 0$

$$(E7\text{-}2.2)$$

$[\mathbf{H\bullet}]$: $\qquad r_{H\bullet} = r_{3H\bullet} + r_{4H\bullet} = r_{3C_2H_4} + r_{4C_2H_6} = 0 \qquad (E7\text{-}2.3)$

$[\mathbf{CH_3\bullet}]$: $\quad r_{CH_3\bullet} = r_{1CH_3\bullet} + r_{2CH_3\bullet} = -2r_{1C_2H_6} + r_{2C_2H_6} = 0 \quad (E7\text{-}2.4)$

Substituting the concentrations into the elementary Equation (E7-2.4) gives

$$2k_1[C_2H_6] - k_2[CH_3 \bullet][C_2H_6] = 0 \qquad (E7\text{-}2.5)$$

Solving for the concentration of the free radical $[CH_3 \bullet]$,

$$[CH_3 \bullet] = \frac{2k_1}{k_2} \qquad (E7\text{-}2.6)$$

Adding Equations (E7-2.2) and (E7-2.3) yields

$$-r_{2C_2H_6} + r_{5C_2H_5} \bullet = 0$$

Substituting for concentrations in the rate laws

$$k_2[CH_3 \bullet][C_2H_6] - k_5[C_2H_5 \bullet]^2 = 0 \qquad (E7\text{-}2.7)$$

PSSH solution

Solving for $[C_2H_5 \bullet]$ gives us

$$[C_2H_5 \bullet] = \left\{ \frac{k_2}{k_5}[CH_3 \bullet][C_2H_6] \right\}^{1/2} = \left\{ \frac{2k_1k_2}{k_2k_5}[C_2H_6] \right\}^{1/2}$$

$$= \left\{ \frac{2k_1}{k_5}[C_2H_6] \right\}^{1/2} \qquad (E7\text{-}2.8)$$

Substituting for $C_2H_5 \bullet$ in Equation (E7-2.1) yields the rate of formation of ethylene

$$\boxed{r_{C_2H_4} = k_3[C_2H_5 \bullet] = k_3\left(\frac{2k_1}{k_5}\right)^{1/2}[C_2H_6]^{1/2}} \qquad (E7\text{-}2.9)$$

Next we write the net rate of $H \bullet$ formation in Equation (E7-2.3) in terms of concentration

$$k_3[C_2H_5 \bullet] - k_4[H \bullet][C_2H_6] = 0$$

Using Equation (E7-2.8) to substitute for $(C_2H_5 \bullet)$ gives the concentration of the hydrogen radical

$$[H \bullet] = \frac{k_3}{k_4}\left(\frac{2k_1}{k_5}\right)^{1/2}[C_2H_6]^{-1/2} \qquad (E7\text{-}2.10)$$

The rate of disappearance of ethane is

$$r_{C_2H_6} = -k_1[C_2H_6] - k_2[CH_3 \bullet][C_2H_6] - k_4[H \bullet][C_2H_6] \qquad (E7\text{-}2.11)$$

Substituting for the concentration of free radicals, the rate law of disappearance of ethane is

$$\boxed{-r_{C_2H_6} = (k_1 + 2k_1)(C_2H_6) + k_3\left(\frac{2k_1}{k_5}\right)^{1/2}C_2H_6^{1/2}} \qquad (E7\text{-}2.12)$$

For a constant-volume batch reactor, the combined mole balances and rate laws for disappearance of ethane ($P1$) and the formation of ethylene ($P5$) are

$$\frac{dC_{P1}}{dt} = -\left[(3k_1 C_{P1}) + k_3 \left(\frac{2k_1}{k_5}\right)^{1/2} C_{P1}^{1/2} \right] \qquad \text{(E7-2.13)}$$

Combined mole balance and rate law using the PSSH

$$\frac{dC_{P5}}{dt} = k_3 \left(\frac{2k_1}{k_5}\right)^{1/2} C_{P1}^{1/2} \qquad \text{(E7-2.14)}$$

The P in $P1$ (i.e., C_{P1}) and $P5$ (i.e., C_{P5}) is to remind us that we have used the PSSH in arriving at these balances.

At 1000 K the specific reaction rates are $k_1 = 1.5 \times 10^{-3}$ s^{-1}, $k_2 = 2.3 \times 10^6$ dm^3/mol·s, $k_3 = 5.71 \times 10^4$ s^{-1}, $k_4 = 9.53 \times 10^8$ dm^3/mol·s, and $k_5 = 3.98 \times 10^9$ dm^3/mol·s.

For an entering ethane concentration of 0.1 mol/dm^3 and a temperature of 1000 K, Equations (E7-2.13) and (E7-2.14) were solved and the concentrations of ethane, C_{P1}, and ethylene, C_{P5}, are shown as a function of time in Figures E7-2.2 and E7-2.3.

In developing this concentration–time relationship, we used PSSH. However, we can now utilize the techniques described in Chapter 6 to solve the full set of equations for ethane cracking and then compare these results with the much simpler PSSH solutions.

Part (b)   *Testing the PSSH for Ethane Cracking*

The thermal cracking of ethane is believed to occur by the reaction sequence given in Part (a). The specific reaction rates are given as a function of temperature:

$$k_1 = 10e^{(87,500/R)(1/1250 - 1/T)} \text{s}^{-1} \qquad k_2 = 8.45 \times 10^6 e^{(13,000/R)(1/1250 - 1/T)} \text{dm}^3/\text{mol·s}$$

$$k_3 = 3.2 \times 10^6 e^{(40,000/R)(1/1250 - 1/T)} \text{s}^{-1} \quad k_4 = 2.53 \times 10^9 e^{(9700/R)(1/1250 - 1/T)} \text{dm}^3/\text{mol·s}$$

$$k_5 = 3.98 \times 10^9 \text{ dm}^3/\text{mol·s} \qquad E = 0$$

Part (b): Carry out mole balance on every species, solve, and then plot the concentrations of ethane and ethylene as a function of time and compare with the PSSH concentration–time measurements. The initial concentration of ethane is 0.1 mol/dm^3 and the temperature is 1000 K.

Solution Part (b)

Let $1 = C_2H_6$, $2 = CH_3\bullet$, $3 = CH_4$, $4 = C_2H_5\bullet$, $5 = C_2H_4$, $6 = H\bullet$, $7 = H_2$, and $8 = C_4H_{10}$. The combined mole balances and rate laws become

$$(\mathbf{C_2H_6}): \quad \frac{dC_1}{dt} = -k_1 C_1 - k_2 C_1 C_2 - k_4 C_1 C_6 \qquad \text{(E7-2.15)}$$

Full numerical solution

$$(\mathbf{CH_3\bullet}): \quad \frac{dC_2}{dt} = 2k_1 C_1 - k_2 C_2 C_1 \qquad \text{(E7-2.16)}$$

$$(\mathbf{CH_4}): \quad \frac{dC_3}{dt} = k_2 C_1 C_2 \qquad\qquad\qquad (E7\text{-}2.17)$$

$$(\mathbf{C_2H_5 \bullet}): \quad \frac{dC_4}{dt} = k_2 C_1 C_2 - k_3 C_4 + k_4 C_1 C_6 - k_5 C_4^2 \qquad (E7\text{-}2.18)$$

$$(\mathbf{C_2H_4}): \quad \frac{dC_5}{dt} = k_3 C_4 \qquad\qquad\qquad (E7\text{-}2.19)$$

$$(\mathbf{H \bullet}): \quad \frac{dC_6}{dt} = k_3 C_4 - k_4 C_1 C_6 \qquad\qquad (E7\text{-}2.20)$$

$$(\mathbf{H_2}): \quad \frac{dC_7}{dt} = k_4 C_1 C_6 \qquad\qquad\qquad (E7\text{-}2.21)$$

$$(\mathbf{C_4H_{10}}): \quad \frac{dC_8}{dt} = \frac{1}{2} k_5 C_4^2 \qquad\qquad\qquad (E7\text{-}2.22)$$

The Polymath program is given in Table E7-2.1.

TABLE E7-2.1.   POLYMATH PROGRAM

POLYMATH Results
Example 7-2 PSSH Applied to Thermal Cracking of Ethane 08-18-2004, Rev5.1.232
ODE Report (STIFF)

Differential equations as entered by the user
[1] d(C1)/d(t) = -k1*C1-k2*C1*C2-k4*C1*C6
[2] d(C2)/d(t) = 2*k1*C1-k2*C1*C2
[3] d(C6)/d(t) = k3*C4-k4*C6*C1
[4] d(C4)/d(t) = k2*C1*C2-k3*C4+k4*C6*C1-k5*C4^2
[5] d(C7)/d(t) = k4*C1*C6
[6] d(C3)/d(t) = k2*C1*C2
[7] d(C5)/d(t) = k3*C4
[8] d(C8)/d(t) = 0.5*k5*C4^2
[9] d(CP5)/d(t) = k3*(2*k1/k5)^0.5*CP1^0.5
[10] d(CP1)/d(t) = -k1*CP1-2*k1*CP1-(k3*(2*k1/k5)^0.5)*(CP1^0.5)

Explicit equations as entered by the user
[1] k5 = 3980000000
[2] T = 1000
[3] k1 = 10*exp((87500/1.987)*(1/1250-1/T))
[4] k2 = 8450000*exp((13000/1.987)*(1/1250-1/T))
[5] k4 = 2530000000*exp((9700/1.987)*(1/1250-1/T))
[6] k3 = 3200000*exp((40000/1.987)*(1/1250-1/T))

Living Example Problem

Figure E7-2.1 shows the concentration time trajectory for $CH_3 \bullet$ (i.e., C_2). One notes a flat plateau where the PSSH is valid. Figure E7-2.2 shows a comparison of the concentration–time trajectory for ethane calculated from the PSSH (C_{P1}) with the ethane trajectory (C_1) calculated from solving the mole balance Equations (E7-2.13) through (E7-2.22). Figure E7-2.3 shows a similar comparison for ethylene (C_{P5}) and (C_5). One notes that the curves are identical, indicating the validity of the PSSH under these conditions. Figure E7-2.4 shows a comparison of the concentration–time trajectories for methane (C_3) and butane (C_8). Problem P7-2(a) explores the temperature for which the PSSH is valid for the cracking of ethane.

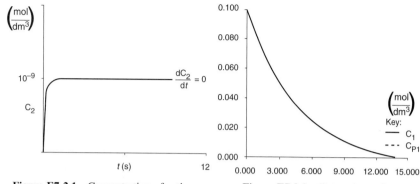

Note: Curves for
C_1 and C_{P1} are
virtually identical.

Figure E7-2.1   Concentration of active
intermediate $CH_3\cdot$ as a function of time.

Figure E7-2.2   Comparison of
concentration–time trajectories for ethane.

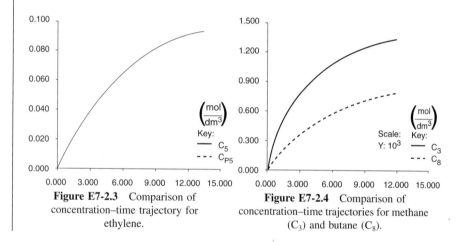

Note: Curves for
C_5 and C_{P5} are
virtually identical.

Figure E7-2.3   Comparison of
concentration–time trajectory for
ethylene.

Figure E7-2.4   Comparison of
concentration–time trajectories for methane
(C_3) and butane (C_8).

7.1.4 Reaction Pathways

Reaction pathways help see the connection of all interacting species for multiple reactions. We have already seen two relatively simple reaction pathways, one to explain the first-order rate law, $-r_A = kC_A$, $(M + A \rightarrow A^* + M)$ and one for the sonoluminescence of CS_2 in Example 7-1. We now will develop reaction pathways for ethane cracking and for smog generation.

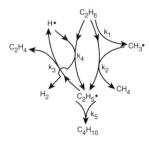

Figure 7-2   Pathway of ethane cracking.

Ethane Cracking. With the increase in computing power, more and more analyses involving free-radical reactions as intermediates are carried out using the coupled sets of differential equations (cf. Example 7-2). The key in any such analyses is to identify which intermediate reactions are important in the overall sequence in predicting the end products. Once the key reactions are identified, one can sketch the pathways in a manner similar to that shown for the ethane cracking in Example 7-2 where Reactions 1 through 5 are shown in Figure 7-2.

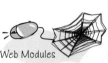

Web Modules

Smog Formation. In Chapter 1, Problem P1-14, in the **CD-ROM Smog Web Module**, we discussed a very simple model for smog removal in the L.A. basin by a Santa Ana wind. We will now look a little deeper into the chemistry of smog formation. Nitrogen and oxygen react to form nitric oxide in the cylinder of automobile engines. The NO from automobile exhaust is oxidized to NO_2 in the presence of peroxide radicals.

$$R\dot{O}O + NO \xrightarrow{k_1} \dot{R}O + NO_2 \qquad (R1)$$

Nitrogen dioxide is then decomposed photochemically to give nascent oxygen

$$NO_2 + h\nu \xrightarrow{k_2} NO + O \qquad (R2)$$

which reacts to form ozone

$$O + O_2 \xrightarrow{k_3} O_3 \qquad (R3)$$

The ozone then becomes involved in a whole series of reactions with hydrocarbons in the atmosphere to form aldehydes, various free radicals, and other intermediates, which react further to produce undesirable products in air pollution:

$$\text{Ozone} + \text{Olefin} \longrightarrow \text{Aldehydes} + \text{Free radicals}$$

$$O_3 + RCH{=}CHR \xrightarrow{k_4} RCHO + \dot{R}O + H\dot{C}O \qquad (R4)$$

$$\xrightarrow[h\nu]{k_5} \dot{R} + H\dot{C}O \qquad (R5)$$

One specific example is the reaction of ozone with 1,3-butadiene to form acrolein and formaldehyde, which are *severe eye irritants*.

Eye irritants

$$\frac{2}{3}O_3 + CH_2{=}CHCH{=}CH_2 \xrightarrow[h\nu]{k_6} CH_2{=}CHCHO + HCHO \qquad (R6)$$

By regenerating NO_2, more ozone can be formed and the cycle continued. This regeneration may be accomplished through the reaction of NO with the free radicals in the atmosphere Reaction (R1). For example, the free radical formed in Reaction (R4) can react with O_2 to give the peroxy free radical,

$$\dot{R} + O_2 \xrightarrow{\ k_7\ } R\dot{O}\dot{O} \qquad\qquad (R7)$$

The coupling of the preceding reactions is shown schematically in Figure 7-3.

   We see that the cycle has been completed and that with a relatively small amount of nitrogen oxides, a large amount of pollutants can be produced. Of course, many other reactions are taking place, so do not be misled by the brevity of the preceding discussion; it does, however, serve to present, in rough outline, the role of nitrogen oxides in air pollution.

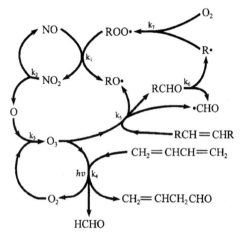

Figure 7-3   Reaction pathways in smog formation.

Metabolic Pathways.   Reaction pathways find their greatest use in metabolic pathways where the various steps are catalyzed by enzymes. The metabolism of alcohol is catalyzed by a different enzyme in each step.

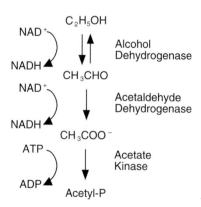

This pathway is discussed in Section 7.5 Pharmacokinetics. However, we first need to discuss enzymes and enzyme kinetics, which we will do in Section 7.2. Check the CD-ROM and the web for future updates on metabolic reaction.

7.2 Enzymatic Reaction Fundamentals

An enzyme is a high-molecular-weight protein or protein-like substance that acts on a substrate (reactant molecule) to transform it chemically at a greatly accelerated rate, usually 10^3 to 10^{17} times faster than the uncatalyzed rate. Without enzymes, essential biological reactions would not take place at a rate necessary to sustain life. Enzymes are usually present in small quantities and are not consumed during the course of the reaction nor do they affect the chemical reaction equilibrium. Enzymes provide an alternate pathway for the reaction to occur thereby requiring a lower activation energy. Figure 7-4 shows the reaction coordinate for the uncatalyzed reaction of a reactant molecule called a substrate (S) to form a product (P)

$$S \rightarrow P$$

The figure also shows the catalyzed reaction pathway that proceeds through an *active intermediate* (E • S), called the enzyme–substrate complex, that is,

$$\boxed{S + E \rightleftharpoons E{\bullet}S \rightarrow E + P}$$

Because enzymatic pathways have lower activation energies, enhancements in reaction rates can be enormous, as in degradation of urea by urease where the degradation rate is on the order of 10^{14} higher than without urease.

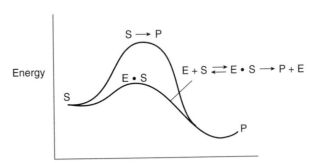

Figure 7-4   Reaction coordinate for enzyme catalysis.

An important property of enzymes is that they are specific; that is, *one* enzyme can usually catalyze only *one* type of reaction. For example, a protease hydrolyzes *only* bonds between specific amino acids in proteins, an amylase works on bonds between glucose molecules in starch, and lipase attacks fats, degrading them to fatty acids and glycerol. Consequently, unwanted products are easily controlled in enzyme-catalyzed reactions. Enzymes are produced only by living organisms, and commercial enzymes are generally produced by bacteria. Enzymes usually work (i.e., catalyze reactions) under

Red Pop

mild conditions: pH 4 to 9 and temperatures 75 to 160°F. Most enzymes are named in terms of the reactions they catalyze. It is a customary practice to add the suffix *-ase* to a major part of the name of the substrate on which the enzyme acts. For example, the enzyme that catalyzes the decomposition of urea is urease and the enzyme that attacks tyrosine is tyrosinase. However, there are exceptions to the naming convention, such as α-amylase. The enzyme α-amylase catalyzes starch in the first step in the production of the soft drink (e.g., Red Pop) sweetener high-fructose corn syrup (HFCS) from corn starch, which is a $4 billion per year business

$$\text{Corn starch} \xrightarrow{\text{α-amylase}} \text{Thinned starch} \xrightarrow[\text{amylase}]{\text{gluco-}} \text{Glucose} \xrightarrow[\text{isomerase}]{\text{Glucose}} \text{HFCS}$$

7.2.1 Enzyme–Substrate Complex

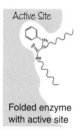

Active Site

Folded enzyme
with active site

The key factor that sets enzymatic reactions apart from other catalyzed reactions is the formation of an enzyme–substrate complex, E•S. Here substrate binds with a specific *active site* of the enzyme to form this complex.[7] Figure 7-5 shows the schematic of the enzyme chymotrypsin (MW = 25,000 Daltons), which catalyzes the hydrolytic cleavage of polypeptide bonds. In many cases the enzyme's active catalytic sites are found where the various folds or loops interact. For chymotrypsin the catalytic sites are noted by the amino acid numbers 57, 102, and 195 in Figure 7-5. Much of the catalytic power is attributed

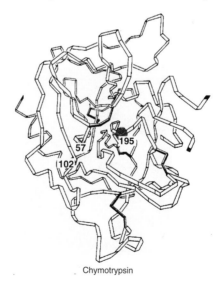

Chymotrypsin

Figure 7-5   Enzyme chymotrypsin. [From *Biochemistry*, 3/E by Stryer © 1988 by Lubert Stryer. Used with permission of W. H. Freeman and Company.]

---

[7] M. L. Shuler and F. Kargi, *Bioprocess Engineering Basic Concepts*, 2nd ed. (Upper Saddle River, N.J.: Prentice Hall, 2002).

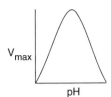

to the binding energy of the substrate to the enzyme through multiple bonds with the specific functional groups on the enzyme (amino side chains, metal ions). The interactions that stabilize the enzyme–substrate complex are hydrogen bonding and hydrophobic, ionic, and London van der Waals forces. If the enzyme is exposed to extreme temperatures or pH environments (i.e., both high and low pH values), it may unfold losing its active sites. When this occurs, the enzyme is said to be *denatured*. See Problem P7-15$_B$.

There are two models for substrate-enzyme interactions: the *lock and key model* and the *induced fit model*, both of which are shown in Figure 7-6. For many years the lock and key model was preferred because of the sterospecific effects of one enzyme acting on one substrate. However, the induced fit model is the more useful model. In the induced fit model both the enzyme molecule and the substrate molecules are distorted. These changes in conformation distort one or more of the substrate bonds, thereby stressing and weakening the bond to make the molecule more susceptible to rearrangement or attachment.

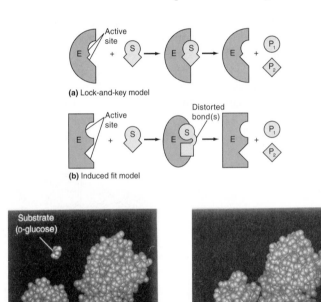

There are six classes of enzymes and only six:

1. Oxidoreductases	$AH_2 + B + E \rightarrow A + BH_2 + E$
2. Transferases	$AB + C + E \rightarrow AC + B + E$
3. Hydrolases	$AB + H_2O + E \rightarrow AH + BOH + E$
4. Isomerases	$A + E \rightarrow isoA + E$
5. Lyases	$AB + E \rightarrow A + B + E$
6. Ligases	$A + B + E \rightarrow AB + E$

Links

More information about enzymes can be found on the following two web sites:

http://us.expasy.org/enzyme/ and *www.chem.qmw.ac.uk/iubmb/enzyme*

These sites also give information about enzymatic reactions in general.

7.2.2 Mechanisms

In developing some of the elementary principles of the kinetics of enzyme reactions, we shall discuss an enzymatic reaction that has been suggested by Levine and LaCourse as part of a system that would reduce the size of an artificial kidney.[8] The desired result is the production of an artificial kidney that could be worn by the patient and would incorporate a replaceable unit for the elimination of the nitrogenous waste products such as uric acid and creatinine. In the microencapsulation scheme proposed by Levine and LaCourse, the enzyme urease would be used in the removal of urea from the bloodstream. Here, the catalytic action of urease would cause urea to decompose into ammonia and carbon dioxide. The mechanism of the reaction is believed to proceed by the following sequence of elementary reactions:

1. The enzyme urease (E) reacts with the substrate urea (S) to form an enzyme–substrate complex (E • S).

The reaction mechanism

$$NH_2CONH_2 + \text{Urease} \xrightarrow{k_1} [NH_2CONH_2 \bullet \text{Urease}]^* \qquad (7\text{-}13)$$

2. This complex (E • S) can decompose back to urea (S) and urease (E):

$$[NH_2CONH_2 \bullet \text{Urease}]^* \xrightarrow{k_2} \text{Urease} + NH_2CONH_2 \qquad (7\text{-}14)$$

3. Or it can react with water (W) to give the products (P) ammonia and carbon dioxide, and recover the enzyme urease (E).

$$[NH_2CONH_2 \bullet \text{Urease}]^* + H_2O \xrightarrow{k_3} 2NH_3 + CO_2 + \text{Urease} \quad (7\text{-}15)$$

We see that some of the enzyme added to the solution binds to the urea, and some remains unbound. Although we can easily measure the total concentration of enzyme, (E_t), it is difficult to measure the concentration of free enzyme, (E).

---

[8] N. Levine and W. C. LaCourse, *J. Biomed. Mater. Res.*, *1*, 275 (1967).

Letting E, S, W, E·S, and P represent the enzyme, substrate, water, the enzyme–substrate complex, and the reaction products, respectively, we can write Reactions (7-13), (7-14), and (7-15) symbolically in the forms

$$S + E \xrightarrow{k_1} E \cdot S \tag{7-16}$$

$$E \cdot S \xrightarrow{k_2} E + S \tag{7-17}$$

$$E \cdot S + W \xrightarrow{k_3} P + E \tag{7-18}$$

Here $P = 2NH_3 + CO_2$.

The corresponding rate laws for Reaction (7-16), (7-17), and (7-18) are

$$r_{1S} = -k_1 (E)(S) \tag{7-16A}$$

$$r_{2S} = k_2 (E \cdot S) \tag{7-17A}$$

$$r_{3P} = k_3 (E \cdot S) (W) \tag{7-18A}$$

The net rate of disappearance of the substrate, $-r_S$, is

$$-r_S = k_1(E)(S) - k_2(E \cdot S) \tag{7-19}$$

This rate law is of not much use to us in making reaction engineering calculations because we cannot measure the concentration of enzyme substrate complex (E • S). We will use the PSSH to express (E • S) in terms of measured variables. The net rate of formation of the enzyme–substrate complex is

$$r_{E \cdot S} = k_1(E)(S) - k_2(E \cdot S) - k_3(W)(E \cdot S) \tag{7-20}$$

Using the PSSH, $r_{E \cdot S} = 0$, we solve equation (7-20) for (E • S)

$$(E \cdot S) = \frac{k_1(E)(S)}{k_2 + k_3(W)} \tag{7-21}$$

and substitute for (E • S) into [Equation (7-19)]

$$-r_S = k_1(E)(S) - k_2 \frac{k_1(E)(S)}{k_2 + k_3(W)}$$

Simplifying

$$-r_S = \frac{k_1 k_3(E)(S)(W)}{k_2 + k_3(W)} \tag{7-22}$$

We need to replace unbound enzyme concentration (E) in the rate law.

We *still* cannot use this rate law because we cannot measure the unbound enzyme concentration (E); however, we can measure the total enzyme concentration, E_t.

In the absence of enzyme denaturization, the total concentration of the enzyme in the system, (E_t), is constant and equal to the sum of the concentrations of the free or unbonded enzyme, (E), and the enzyme–substrate complex, $(E \cdot S)$:

$$(E_t) = (E) + (E \cdot S) \tag{7-23}$$

Total enzyme concentration = Bound + Free enzyme concentration.

Substituting for $(E \cdot S)$

$$(E_t) = (E) + \frac{k_1(E)(S)}{k_2 + k_3(W)}$$

solving for (E)

$$(E) = \frac{(E_t)(k_2 + k_3(W))}{k_2 + k_3(W) + k_1(S)}$$

substituting for (E) in Equation (7-22), the rate law for substrate consumption is

The final form of the rate law

$$\boxed{-r_S = \frac{k_1 k_3(W)(E_t)(S)}{k_1(S) + k_2 + k_3(W)}} \tag{7-24}$$

Note: Throughout, $E_t \equiv (E_t)$ = total concentration of enzyme with typical units $(kmol/m^3$ or $g/dm^3)$.

7.2.3 Michaelis-Menten Equation

Because the reaction of urea and urease is carried out in aqueous solution, water is, of course, in excess, and the concentration of water is therefore considered constant. Let

$$k_{cat} = k_3(W) \quad \text{and} \quad K_M = \frac{k_{cat} + k_2}{k_1}$$

Dividing the numerator and denominator of Equation (7-24) by k_1, we obtain a form of the *Michaelis–Menten equation*:

$$-r_S = \frac{k_{cat}(E_t)(S)}{(S) + K_M} \tag{7-25}$$

The parameter k_{cat} is also referred to as the *turnover number*. It is the number of substrate molecules converted to product in a given time on a single-enzyme molecule when the enzyme is saturated with substrate (i.e., all the active sites on the enzyme are occupied, $S \gg K_M$). For example, turnover number for the decomposition H_2O_2 by the enzyme catalase is $40 \times 10^6 \ s^{-1}$. That is, 40 million molecules of H_2O_2 are decomposed every second on a single-enzyme molecule saturated with H_2O_2. The constant K_M (mol/dm^3) is called the Michaelis constant and for simple systems is a measure of the

Turnover number k_{cat}

Michaelis
constant K_M

attraction of the enzyme for its substrate, so it's also called the *affinity constant*. The Michaelis constant, K_M, for the decomposition of H_2O_2 discussed earlier is 1.1 M while that for chymotrypsin is 0.1 M.[9]

If, in addition, we let V_{max} represent the maximum rate of reaction for a given total enzyme concentration,

$$V_{max} = k_{cat}(E_t)$$

the Michaelis–Menten equation takes the familiar form

Michaelis–Menten
equation

$$-r_S = \frac{V_{max}(S)}{K_M + (S)} \qquad (7\text{-}26)$$

For a given enzyme concentration, a sketch of the rate of disappearance of the substrate is shown as a function of the substrate concentration in Figure 7-7.

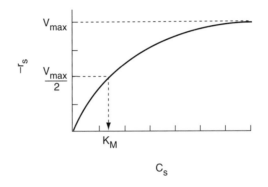

Figure 7-6   Michaelis–Menten plot identifying the parameters V_{max} and K_M.

A plot of this type is sometimes called a Michaelis–Menten plot. At low substrate concentration, $K_M \gg (S)$,

$$-r_S \cong \frac{V_{max}(S)}{K_M}$$

and the reaction is apparent first order in the substrate concentration. At high substrate concentrations,

$$(S) \gg K_M$$

and the reaction is apparent zero order

$$-r_S \cong V_{max}$$

---

[9] D. L. Nelson and M. M. Cox, *Lehninger Principles of Biochemistry*, 3rd ed. (New York: Worth Publishers, 2000).

Consider the case when the substrate concentration is such that the reaction rate is equal to one-half the maximum rate,

$$-r_S = \frac{V_{max}}{2}$$

then

$$\frac{V_{max}}{2} = \frac{V_{max}(S_{1/2})}{K_M + (S_{1/2})} \tag{7-27}$$

Solving Equation (7-27) for the Michaelis constant yields

Interpretation of
Michaelis constant

$$\boxed{K_M = (S_{1/2})} \tag{7-28}$$

The Michaelis constant is equal to the substrate concentration at which the rate of reaction is equal to one-half the maximum rate.

The parameters V_{max} and K_M characterize the enzymatic reactions that are described by Michaelis–Menten kinetics. V_{max} is dependent on total enzyme concentration, whereas K_M is not.

Two enzymes may have the same values for k_{cat} but have different reaction rates because of different values of K_M. One way to compare the catalytic efficiencies of different enzymes is to compare the ratio k_{cat}/K_M. When this ratio approaches 10^8 to 10^9 (dm³/mol/s) the reaction rate approaches becoming diffusion-limited. That is, it takes a long time for the enzyme and substrate to find each other, but once they do they react immediately. We will discuss diffusion-limited reactions in Chapters 11 and 12.

Example 7–3 Evaluation of Michaelis–Menten Parameters V_{max} and K_M

Determine the Michaelis–Menten parameters V_{max} and K_M for the reaction

$S + E \underset{\longleftarrow}{\longrightarrow} E \cdot S \xrightarrow{H_2O} P + E$

$$\text{Urea} + \text{Urease} \underset{k_2}{\overset{k_1}{\rightleftharpoons}} [\text{Urea} \cdot \text{Urease}]^* \xrightarrow[-H_2O]{k_3} 2NH_3 + CO_2 + \text{Urease}$$

The rate of reaction is given as a function of urea concentration in this table.

C_{urea} (kmol/m³)	0.2	0.02	0.01	0.005	0.002
$-r_{urea}$ (kmol/m³·s)	1.08	0.55	0.38	0.2	0.09

Solution

Inverting Equation (7-26) gives us

$$\frac{1}{-r_s} = \frac{(S) + K_M}{V_{max}(S)} = \frac{1}{V_{max}} + \frac{K_M}{V_{max}} \frac{1}{(S)} \tag{E7-3.1}$$

or

$$\boxed{\frac{1}{-r_{\text{urea}}} = \frac{1}{V_{\text{max}}} + \frac{K_{\text{M}}}{V_{\text{max}}}\left(\frac{1}{C_{\text{urea}}}\right)} \qquad \text{(E7-3.2)}$$

A plot of the reciprocal reaction rate versus the reciprocal urea concentration should be a straight line with an intercept $1/V_{\text{max}}$ and slope $K_{\text{M}}/V_{\text{max}}$. This type of plot is called a *Lineweaver–Burk plot*. The data in Table E7-3.1 are presented in Figure E7-3.1 in the form of a Lineweaver–Burk plot. The intercept is 0.75, so

$$\frac{1}{V_{\text{max}}} = 0.75 \text{ m}^3 \cdot \text{s/kmol}$$

TABLE E7-3.1.   RAW AND PROCESSED DATA

C_{urea} (kmol/m^3)	$-r_{\text{urea}}$ (kmol/m$^3 \cdot$s)	$1/C_{\text{urea}}$ (m^3/kmol)	$1/-r_{\text{urea}}$ (m$^3 \cdot$s/kmol)
0.20	1.08	5.0	0.93
0.02	0.55	50.0	1.82
0.01	0.38	100.0	2.63
0.005	0.20	200.0	5.00
0.002	0.09	500.0	11.11

Lineweaver–Burk
plot

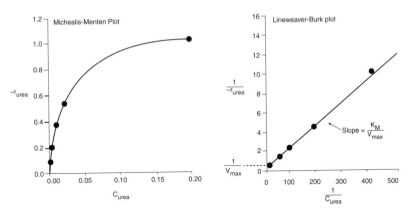

Figure E7-3.1   (a)Michaelis–Menten plot; (b) Lineweaver–Burk plot.

Therefore, the maximum rate of reaction is

$$V_{\text{max}} = 1.33 \text{ kmol/m}^3 \cdot \text{s} = 1.33 \text{ mol/dm}^3 \cdot \text{s}$$

From the slope, which is 0.02 s, we can calculate the Michaelis constant, K_{M}:

For enzymatic reactions, the two key rate-law parameters are V_{max} and K_M.

$$\frac{K_{\text{M}}}{V_{\text{max}}} = \text{slope} = 0.02 \text{ s}$$

$$K_{\text{M}} = 0.0266 \text{ kmol/m}^3$$

Substituting K_M and V_{max} into Equation (7-26) gives us

$$-r_s = \frac{1.33\,C_{urea}}{0.0266 + C_{urea}} \qquad\text{(E7-3.3)}$$

where C_{urea} has units of kmol/m^3 and $-r_s$ has units of $\text{kmol/m}^3\cdot\text{s}$. Levine and LaCourse suggest that the total concentration of urease, (E_t), corresponding to the value of V_{max} above is approximately $5\ \text{g/dm}^3$.

In addition to the Lineweaver–Burk plot, one can also use a Hanes–Woolf plot or an Eadie–Hofstee plot. Here $S \equiv C_{urea}$, and $-r_S \equiv -r_{urea}$. Equation (7-26)

$$-r_S = \frac{V_{max}(S)}{K_M + (S)} \qquad\text{(7-26)}$$

can be rearranged in the following forms. For the Eadie–Hofstee form,

$$-r_S = V_{max} - K_M\!\left(\frac{-r_S}{(S)}\right) \qquad\text{(E7-3.4)}$$

For the Hanes–Woolf form, we have

$$\frac{(S)}{-r_S} = \frac{K_M}{V_{max}} + \frac{1}{V_{max}}(S) \qquad\text{(E7-3.5)}$$

For the Eadie–Hofstee model we plot $-r_S$ as a function of $(-r_S/S)$ and for the Hanes–Woolf model, we plot $[(S)/-r_S]$ as a function of (S). The Eadie–Hofstee plot does not bias the points at low substrate concentrations, while the Hanes–Woolf plot gives a more accurate evaluation of V_{max}. In Table E7-3.2, we add two columns to Table E7-3.1 to generate these plots ($C_{urea} \equiv S$).

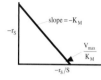

Eadie–Hofstee plot

Hanes–Woolf plot

TABLE E7-3.2.   RAW AND PROCESSED DATA

S (kmol/m³)	$-r_S$ (kmol/m³ • s)	$1/S$ (m³/kmol)	$1/-r_S$ (m³ • s/kmol)	$S/-r_s$ (s)	$-r_S/S$ (1/s)
0.20	1.08	5.0	0.93	0.185	5.4
0.02	0.55	50.0	1.82	0.0364	27.5
0.01	0.38	100.0	2.63	0.0263	38
0.005	0.20	200.0	5.00	0.0250	40
0.002	0.09	500.0	11.11	0.0222	45

Plotting the data in Table E7-3.2, we arrive at Figures E7-3.2 and E7-3.3.

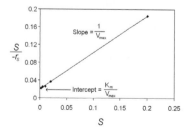

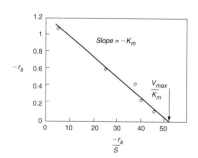

Figure E7-3.2 Hanes–Woolf plot.   **Figure E7-3.3** Eadie-Hofstee plot.

Regression

Equation (7-26) was used in the regression program of Polymath with the following results for V_{max} and K_M.

Nonlinear regression (L-M)

Model: rate = Vmax*Curea/(Km+Curea)

Variable	Ini guess	Value	95% confidence
Vmax	1	1.2057502	0.0598303
Km	0.02	0.0233322	0.003295

Nonlinear regression settings $V_{max} = 1.2$ mol/dm^3 • s
Max # iterations = 64 $K_M = 0.0233$ mol/dm^3

Precision
R^2   = 0.9990611
R^2adj   = 0.9987481
Rmsd   = 0.0047604
Variance = 1.888E-04

The Product-Enzyme Complex

In many reactions the enzyme and product complex (E • P) is formed directly from the enzyme substrate complex (E • S) according to the sequence

$$E + S \rightleftarrows E \bullet S \rightleftarrows P + E$$

Applying the PSSH, we obtain

Briggs–Haldane
Rate Law
$$-r_S = \frac{V_{max}(C_S - C_P/K_C)}{C_S + K_{max} + K_P C_P} \tag{7-29}$$

which is often referred to as the Briggs–Haldane Equation (see Problem P7-10) and the application of the PSSH to enzyme kinetics often called the Briggs–Haldane approximation.

7.2.4 Batch Reactor Calculations for Enzyme Reactions

A mole balance on urea in the batch reactor gives

Mole balance
$$-\frac{dN_{urea}}{dt} = -r_{urea}V$$

Because this reaction is liquid phase, the mole balance can be put in the following form:

$$-\frac{dC_{urea}}{dt} = -r_{urea} \tag{7-30}$$

The rate law for urea decomposition is

Rate law
$$-r_{\text{urea}} = \frac{V_{\text{max}} C_{\text{urea}}}{K_{\text{M}} + C_{\text{urea}}} \tag{7-31}$$

Substituting Equation (7-31) into Equation (7-30) and then rearranging and integrating, we get

Combine
$$t = \int_{C_{\text{urea}}}^{C_{\text{urea0}}} \frac{dC_{\text{urea}}}{-r_{\text{urea}}} = \int_{C_{\text{urea}}}^{C_{\text{urea0}}} \frac{K_{\text{M}} + C_{\text{urea}}}{V_{\text{max}} C_{\text{urea}}} dC_{\text{urea}}$$

Integrate
$$t = \frac{K_{\text{M}}}{V_{\text{max}}} \ln \frac{C_{\text{urea0}}}{C_{\text{urea}}} + \frac{C_{\text{urea0}} - C_{\text{urea}}}{V_{\text{max}}} \tag{7-32}$$

We can write Equation (7-32) in terms of conversion as

$$C_{\text{urea}} = C_{\text{urea0}}(1 - X)$$

Time to achieve a conversion X in a batch enzymatic reaction
$$\boxed{t = \frac{K_{\text{M}}}{V_{\text{max}}} \ln \frac{1}{1-X} + \frac{C_{\text{urea0}} X}{V_{\text{max}}}} \tag{7-32}$$

The parameters K_{M} and V_{max} can readily be determined from batch reactor data by using the integral method of analysis. Dividing both sides of Equation (7-32) by $t K_{\text{M}}/V_{\text{max}}$ and rearranging yields

$$\frac{1}{t} \ln \frac{1}{1-X} = \frac{V_{\text{max}}}{K_{\text{M}}} - \frac{C_{\text{urea0}} X}{K_{\text{M}} t}$$

We see that K_{M} and V_{max} can be determined from the slope and intercept of a plot of $1/t \ln[1/(1-X)]$ versus X/t. We could also express the Michaelis–Menten equation in terms of the substrate concentration S:

$$\boxed{\frac{1}{t} \ln \frac{S_0}{S} = \frac{V_{\text{max}}}{K_{\text{M}}} - \frac{S_0 - S}{K_{\text{M}} t}} \tag{7-33}$$

where S_0 is the initial concentration of substrate. In cases similar to Equation (7-33) where there is no possibility of confusion, we shall not bother to enclose the substrate or other species in parentheses to represent concentration [i.e., $C_S \equiv (S) \equiv S$]. The corresponding plot in terms of substrate concentration is shown in Figure 7-8.

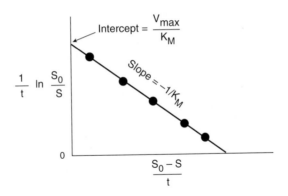

Figure 7-7 Evaluating V_{max} and K_M.

Example 7–4 Batch Enzymatic Reactors

Calculate the time needed to convert 99% of the urea to ammonia and carbon dioxide in a 0.5-dm^3 batch reactor. The initial concentration of urea is 0.1 mol/dm^3, and the urease concentration is 0.001 g/dm^3. The reaction is to be carried out isothermally at the same temperature at which the data in Table E7-3.2 were obtained.

Solution

We can use Equation (7-32),

$$t = \frac{K_M}{V_{max}} \ln \frac{1}{1-X} + \frac{C_{urea0} X}{V_{max}} \tag{7-32}$$

where $K_M = 0.0266$ mol/dm^3, $X = 0.99$, and $C_{urea0} = 0.1$ mol/dm^3, V_{max} was 1.33 $mol/dm^3 \cdot s$. However, for the conditions in the batch reactor, the enzyme concentration is only 0.001 g/dm^3 compared with 5 g in Example 7-3. Because $V_{max} = E_t \cdot k_3$, V_{max} for the second enzyme concentration is

$$V_{max2} = \frac{E_{t2}}{E_{t1}} V_{max1} = \frac{0.001}{5} \times 1.33 = 2.66 \times 10^{-4} \text{ mol/s} \cdot dm^3$$

$$K_M = 0.0266 \text{ mol/dm}^3 \quad \text{and} \quad X = 0.99$$

Substituting into Equation (7-32)

$$t = \frac{2.66 \times 10^{-2} \text{ mol/dm}^3}{2.66 \times 10^{-4} \text{ mol/dm}^3/s} \ln \left(\frac{1}{0.01} \right) + \frac{(0.1 \text{ mol/dm}^3)(0.99)}{2.66 \times 10^{-4} \text{ mol/dm}^3/s} \text{ s}$$

$$= 460 \text{ s} + 380 \text{ s}$$

$$= 840 \text{ s} \ (14 \text{ minutes})$$

Effect of Temperature

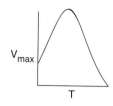

The effect of temperature on enzymatic reactions is very complex. If the enzyme structure would remain unchanged as the temperature is increased, the rate would probably follow the Arrhenius temperature dependence. However, as the temperature increases, the enzyme can unfold and/or become denatured and lose its catalytic activity. Consequently, as the temperature increases, the reaction rate, $-r_S$, increases up to a maximum with increasing temperature and then decreases as the temperature is increased further. The descending part of this curve is called temperature inactivation or thermal denaturizing.[10] Figure 7-8 shows an example of this optimum in enzyme activity.[11]

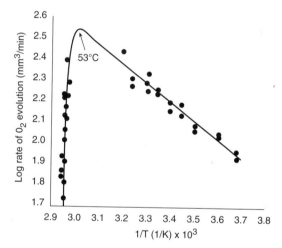

Figure 7-8   Catalytic breakdown rate of H_2O_2 depending on temperature. Courtesy of S. Aiba, A. E. Humphrey, and N. F. Mills, *Biochemical Engineering*, Academic Press (1973).

---

[10]M. L. Shuler and F. Kargi, *Bioprocess Engineering Basic Concepts*, 2nd ed. (Upper Saddle River, N.J.: Prentice Hall, 2002), p. 77.

[11]S. Aiba, A. E. Humphrey, and N. F. Mills, *Biochemical Engineering* (New York: Academic Press, 1973), p. 47.

Side note: Lab-on-a-chip. Enzyme-catalyzed polymerization of nucleotides is a key step in DNA identification. The microfluidic device shown in Figure SN7.1 is used to identify DNA strands. It was developed by Professor Mark Burns's group at the University of Michigan.

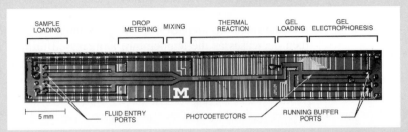

Figure SN7.1   Microfluidic device to identify DNA. Courtesy of *Science, 282*, 484 (1998).

In order to identify the DNA, its concentration must be raised to a level that can be easily quantified. This increase is typically accomplished by replicating the DNA in the following manner. After a biological sample (e.g., purified saliva, blood) is injected into the micro device, it is heated and the hydrogen bonds connecting the DNA strands are broken. After breaking, a primer attaches to the DNA to form a DNA primer complex, DNA*. An enzyme Ⓔ then attaches to this pair forming the DNA* enzyme complex, DNA* • E. Once this complex is formed a polymerization reaction occurs as nucleotides (dNTPs—dATP, dGTP, dCTP, and dTTP—N) attach to the primer one molecule at a time as shown in Figure SN7.2. The enzyme interacts with the DNA strand to add the proper nucleotide in the proper order. The addition continues as the enzyme moves down the strand attaching the nucleotides until the other end of the DNA strand is reached. At this point the enzyme drops off the strand and a duplicate, double-stranded DNA molecule is formed. The reaction sequence is

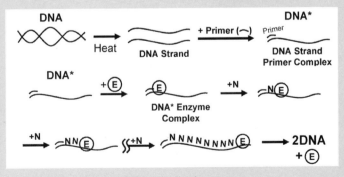

Figure SN7.2   Replication sequence.

The schematic in Figure SN7.2 can be written in terms of single-step reactions where N is one of the four nucleotides.

Complex Formation:

$$DNA + Primer \rightarrow DNA*$$

$$DNA* + E \rightleftharpoons DNA* \cdot E$$

Nucleotide addition/polymerization

$$DNA* \cdot E + N \rightarrow DNA* \cdot N_1 \cdot E$$

$$DNA* \cdot N_1 \cdot E + N \rightarrow DNA* \cdot N_2 \cdot E$$

The process then continues much like a zipper as the enzyme moves along the strand to add more nucleotides to extend the primer. The addition of the last nucleotide is

$$DNA* \cdot N_{i-1} \cdot E + N \rightarrow DNA* \cdot N_i \cdot E$$

where i is the number of nucleotide molecules on the original DNA minus the nucleotides in the primer. Once a complete double-stranded DNA is formed, the polymerization stops, the enzyme drops off, and separation occurs.

$$DNA* \cdot N_i \cdot E \rightarrow 2DNA + E$$

Here 2DNA strands really represents one double-stranded DNA helix. Once replicated in the device, the length of the DNA molecules can be analyzed by electrophoresis to indicate relevant genetic information.

7.3 Inhibition of Enzyme Reactions

In addition to temperature and solution pH, another factor that greatly influences the rates of enzyme-catalyzed reactions is the presence of an inhibitor. Inhibitors are species that interact with enzymes and render the enzyme ineffective to catalyze its specific reaction. The most dramatic consequences of enzyme inhibition are found in living organisms where the inhibition of any particular enzyme involved in a primary metabolic pathway will render the entire pathway inoperative, resulting in either serious damage or death of the organism. For example, the inhibition of a single enzyme, cytochrome oxidase, by cyanide will cause the aerobic oxidation process to stop; death occurs in a very few minutes. There are also beneficial inhibitors such as the ones used in the treatment of leukemia and other neoplastic diseases. Aspirin inhibits the enzyme that catalyzes the synthesis of prostaglandin involved in the pain-producing process.

The three most common types of reversible inhibition occurring in enzymatic reactions are *competitive*, *uncompetitive*, and *noncompetitive*. The enzyme molecule is analogous to a heterogeneous catalytic surface in that it contains active sites. When *competitive* inhibition occurs, the substrate and inhibitor are usually similar molecules that compete for the same site on the enzyme. *Uncompetitive* inhibition occurs when the inhibitor deactivates the enzyme-substrate complex, sometimes by attaching itself to both the substrate and enzyme molecules of the complex. *Noncompetitive* inhibition occurs with enzymes containing at least two different types of sites. The substrate attaches only to one type of site, and the inhibitor attaches only to the other to render the enzyme inactive.

7.3.1 Competitive Inhibition

Competitive inhibition is of particular importance in pharmacokinetics (drug therapy). If a patient were administered two or more drugs that react simultaneously within the body with a common enzyme, cofactor, or active species, this interaction could lead to competitive inhibition in the formation of the respective metabolites and produce serious consequences.

In competitive inhibition another substance, I, competes with the substrate for the enzyme molecules to form an inhibitor–enzyme complex, as shown here.

Reaction Steps   Competitive Inhibition Pathway

Competitive
inhibition pathway

$E + S \rightleftharpoons E \cdot S \longrightarrow E + P$
$+$
I

$\big\downarrow K_I$

$E \cdot I$

(1) $E + S \xrightarrow{\;k_1\;} E \bullet S$

(2) $E \bullet S \xrightarrow{\;k_2\;} E + S$

(3) $E \bullet S \xrightarrow{\;k_3\;} P + E$

(4) $I + E \xrightarrow{\;k_4\;} E \bullet I$ (inactive)

(5) $E \bullet I \xrightarrow{\;k_5\;} E + I$

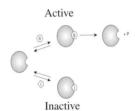

Active

Inactive

(a) Competitive inhibition. Courtesy of D. L. Nelson and M. M. Cox, *Lehninger Principles of Biochemistry*, 3rd ed. (New York: Worth Publishers, 2000), p. 266.

In addition to the three Michaelis–Menten reaction steps, there are two additional steps as the inhibitor reversely ties up the enzyme as shown in reaction steps 4 and 5.

The rate law for the formation of product is the same [cf. Equation (7-18A)] as it was before in the absence of inhibitor

$$r_P = k_3 (E \bullet S) \tag{7-34}$$

Applying the PSSH, the net rate of reaction of the enzyme–substrate complex is

$$r_{E \cdot S} = 0 = k_1(E)(S) - k_2(E \cdot S) - k_3(E \cdot S) \qquad (7\text{-}35)$$

The net rate of reaction of inhibitor-substrate complex is also zero

$$r_{E \cdot I} = 0 = k_4(E)(I) - k_5(E \cdot I) \qquad (7\text{-}36)$$

The total enzyme concentration is the sum of the bound and unbound enzyme concentrations

$$E_t = E + (E \cdot S) + (E \cdot I) \qquad (7\text{-}37)$$

Combining Equations (7-35), (7-36), and (7-37) and solving for $(E \cdot S)$ and substituting in Equation (7-34) and simplifying

Rate law for competitive inhibition

$$\boxed{r_P = -r_S = \frac{V_{max}(S)}{S + K_M\left(1 + \dfrac{I}{K_I}\right)}} \qquad (7\text{-}38)$$

V_{max} and K_M are the same as before when no inhibitor is present, that is,

$$V_{max} = k_3 E_t \text{ and } K_M = \frac{k_2 + k_3}{k_1}$$

and the inhibition constant K_I (mol/dm^3) is

$$K_I = \frac{k_5}{k_4}$$

By letting $K_M' = K_M(1 + I/K_I)$, we can see that the effect of a competitive inhibition is to increase the "apparent" Michaelis constant, K_M'. A consequence of the larger "apparent" Michaelis constant K_M' is that a larger substrate concentration is needed for the rate of substrate decomposition, $-r_S$, to reach half its maximum rate.

Rearranging in order to generate a Lineweaver–Burk plot,

$$\boxed{\frac{1}{-r_S} = \frac{1}{V_{max}} + \frac{1}{(S)} \frac{K_M}{V_{max}}\left(1 + \frac{(I)}{K_I}\right)} \qquad (7\text{-}39)$$

From the Lineweaver–Burk plot (Figure 7-10), we see that as the inhibitor (I) concentration is increased the slope increases (i.e., the rate decreases) while the intercept remains fixed.

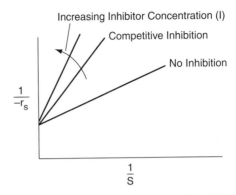

Figure 7-10 Lineweaver–Burk plot for competitive inhibition.

Side note: Methanol Poisoning. An interesting and important example of *competitive substrate inhibition* is the enzyme alcohol dehydrogenase (ADH) in the presence of ethanol and methanol. If a person ingests methanol, ADH will convert it to formaldehyde and then formate, which causes blindness. Consequently, the treatment involves intravenously injecting ethanol (which is metabolized at a slower rate than methanol) at a controlled rate to tie up ADH to slow the metabolism of methanol-to-formaldehyde-to-formate so that the kidneys have time to filter out the methanol which is then excreted in the urine. With this treatment, blindness is avoided. For more on the methanol/ethanol competitive inhibition, see Problem P7-25$_C$.

7.3.2 Uncompetitive Inhibition

Here the inhibitor has no affinity for the enzyme itself and thus does not compete with the substrate for the enzyme; instead it ties up the enzyme–substrate complex by forming an inhibitor–enzyme–substrate complex, (I • E • S) which is inactive. In uncompetitive inhibition, the inhibitor reversibly ties up enzyme–substrate complex *after* it has been formed.

As with competitive inhibition, two additional reaction steps are added to the Michaelis–Menten kinetics for uncompetitive inhibition as shown in reaction steps 4 and 5.

Reaction Steps   **Uncompetitive Pathway**

Uncompetitive
inhibition pathway

$E + S \Longleftrightarrow E \cdot S \longrightarrow E + P$ (1) $E + S \xrightarrow{\;k_1\;} E \bullet S$

$+$
I (2) $E \bullet S \xrightarrow{\;k_2\;} E + S$

$\Big\downarrow K_I$ (3) $E \bullet S \xrightarrow{\;k_3\;} P + E$

$E \cdot S \cdot I$ (4) $I + E \bullet S \xrightarrow{\;k_4\;} I \bullet E \bullet S$ (inactive)

(5) $I \bullet E \bullet S \xrightarrow{\;k_5\;} I \bullet E \bullet S$

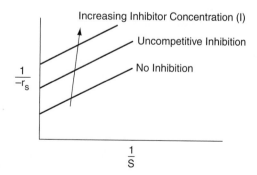

Rate law for
uncompetitive
inhibition

Starting with equation for rate of formation of product, Equation (7-34), and then applying the pseudo-steady-state hypothesis to the intermediate $(I \bullet E \bullet S)$, we arrive at the rate law for uncompetitive inhibition

$$-r_s = r_p = \frac{V_{\max}(S)}{K_M + (S)\left(1 + \dfrac{(I)}{K_I}\right)} \quad \text{where } K_I = \frac{k_5}{k_4} \tag{7-40}$$

Rearranging

$$\frac{1}{-r_s} = \frac{1}{(S)}\frac{K_M}{V_{\max}} + \frac{1}{V_{\max}}\left(1 + \frac{(I)}{K_I}\right) \tag{7-41}$$

The Lineweaver–Burk plot is shown in Figure 7-11 for different inhibitor concentrations. The slope $(K_M/V_{\max})$ remains the same as the inhibition (I) concentration is increased, while the intercept $(1 + (I)/K_I)$ increases.

Figure 7-11   Lineweaver–Burk plot for uncompetitive inhibition.

7.3.3 Noncompetitive Inhibition (Mixed Inhibition)[†]

In noncompetitive inhibition, also called mixed inhibition, the substrate and inhibitor molecules react with different types of sites on the enzyme molecule. Whenever the inhibitor is attached to the enzyme it is inactive and cannot form products. Consequently, the deactivating complex (I • E • S) can be formed by two reversible reaction paths.

1. After a substrate molecule attaches to the enzyme molecule at the substrate site, the inhibitor molecule attaches to the enzyme at the inhibitor site.
2. After an inhibitor molecule attaches to the enzyme molecule at the inhibitor site, the substrate molecule attaches to the enzyme at the substrate site.

These paths, along with the formation of the product, P, are shown here. In noncompetitive inhibition, the enzyme can be tied up in its inactive form either *before or after* forming the enzyme substrate complex as shown in steps 2, 3, and 4.

Reaction Steps   Noncompetitive Pathway

Mixed inhibition

(1) $E + S \rightleftarrows E \cdot S$

(2) $E + I \rightleftarrows I \cdot E$ (inactive)

(3) $I + E \cdot S \rightleftarrows I \cdot E \cdot S$ (inactive)

(4) $S + I \cdot E \rightleftarrows I \cdot E \cdot S$ (inactive)

(5) $E \cdot S \longrightarrow P$

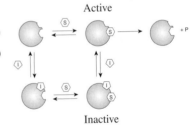

Active

Inactive

Summary Notes

Rate law for noncompetitive inhibition

Again starting with the rate law for the rate of formation of product and then applying the PSSH to the complexes (I • E) and (I • E • S) we arrive at the rate law for the noncompetitive inhibition

$$-r_s = \frac{V_{max}(S)}{((S) + K_M)\left(1 + \dfrac{(I)}{K_I}\right)} \tag{7-42}$$

The derivation of the rate law is given in the *Summary Notes* on the web and CD-ROM. Equation (7-42) is in the form of the rate law that is given for an enzymatic reaction exhibiting noncompetitive inhibition. Heavy metal ions such as Pb^{2+}, Ag^+, and Hg^{2+}, as well as inhibitors that react with the enzyme to form chemical derivatives, are typical examples of noncompetitive inhibitors.

[†] In some texts, mixed inhibition is a combination of competitive and uncompetitive inhibition.

Rearranging

$$\boxed{\frac{1}{-r_s} = \frac{1}{V_{max}}\left(1 + \frac{(I)}{K_I}\right) + \frac{1}{(S)}\frac{K_M}{V_{max}}\left(1 + \frac{(I)}{K_I}\right)}$$

(7-43)

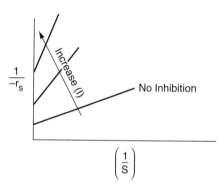

Figure 7-12   Lineweaver–Burk plot for noncompetitive enzyme inhibition.

For noncompetitive inhibition, we see in Figure 7-12 that both the slope $\left(\frac{K_M}{V_{max}}\left[1 + \frac{(I)}{K_I}\right]\right)$ and intercept $\left(\frac{1}{V_{max}}\left[1 + \frac{(I)}{K_I}\right]\right)$ increase with increasing inhibitor concentration. In practice, *uncompetitive inhibition* and *mixed inhibition* are observed only for enzymes with two or more substrates, S_1 and S_2.

The three types of inhibition are compared with a reaction in which no inhibitors are present on the Lineweaver–Burk plot shown in Figure 7-13.

Summary plot of types of inhibition

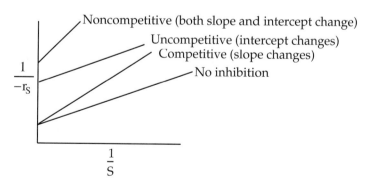

Figure 7-13   Lineweaver–Burk plots for three types of enzyme inhibition.

In summary we observe the following trends and relationships:

1. In *competitive inhibition* the slope increases with increasing inhibitor concentration, while the intercept remains fixed.
2. In *uncompetitive inhibition*, the y-intercept increases with increasing inhibitor concentration while the slope remains fixed.
3. In *noncompetitive inhibition (mixed inhibition)*, both the y–intercept and slope will increase with increasing inhibitor concentration.

Problem P7-14 asks you to find the type of inhibition for the enzyme catalyzed reaction of starch.

7.3.4 Substrate Inhibition

In a number of cases, the substrate itself can act as a inhibitor. In the case of uncompetitive inhibition, the inactive molecules (S • E • S) is formed by the reaction

$$S + E \bullet S \longrightarrow S \bullet E \bullet S \quad \text{(inactive)}$$

Consequently we see that by replacing (I) by (S) in Equation (7-40) the rate law for $-r_s$ is

$$-r_s = \frac{V_{max}S}{K_M + S + \dfrac{S^2}{K_I}} \tag{7-44}$$

We see that at low substrate concentrations

$$K_M \gg \left(S + \frac{S^2}{K_I}\right) \tag{7-45}$$

then

$$-r_s \sim \frac{V_{max}S}{K_M} \tag{7-46}$$

and the rate increases linearly with increasing substrate concentration. At high substrate concentrations $(S^2 / K_I) \gg (K_M + S)$, then

$$-r_S = \frac{V_{max}K_I}{S} \tag{7-47}$$

and we see that the rate decreases as the substrate concentration increases. Consequently, the rate of reaction gives through a maximum in the substrate concentration as shown in Figure 7-14. We also see there is an optimum substrate concentration at which to operate. This maximum is found by taking the derivative of Equation (7-44) wrt S, to obtain

$$\boxed{S_{max} = \sqrt{K_M K_I}} \tag{7-48}$$

Substrate inhibition

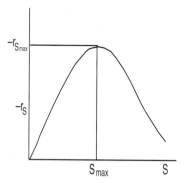

Figure 7-14   Substrate reaction rate as a function of substrate concentration for substrate inhibition.

When substrate inhibition is possible, a semibatch reactor called a *fed batch* is often used as a CSTR to maximize the reaction rate and conversion.

7.3.5 Multiple Enzyme and Substrate Systems

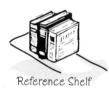

Reference Shelf

In the preceding section, we discussed how the addition of a second substrate, I, to enzyme-catalyzed reactions could deactivate the enzyme and greatly inhibit the reaction. In the present section, we look not only at systems in which the addition of a second substrate is necessary to activate the enzyme, but also at other multiple-enzyme and multiple-substrate systems in which cyclic regeneration of the activated enzyme occurs. Cell growth on multiple substrates is given in the *Professional Reference Shelf R7.4.1*.

Enzyme Regeneration.   The first example considered is the oxidation of glucose (S_r) with the aid of the enzyme glucose oxidase (represented as either G.O. or [E_o]) to give δ-gluconolactone (*P*):

$$\text{Glucose} + \text{G.O.} \; \rightleftharpoons \; (\text{Glucose} \cdot \text{G.O.}) \; \rightleftharpoons \; (\delta\text{-gluconolactone} \cdot \text{G.O.H}_2)$$

$$\rightleftharpoons \; \delta\text{-gluconolactone} + \text{G.O.H}_2$$

In this reaction, the reduced form of glucose oxidase (G.O.H$_2$), which will be represented by E_r, cannot catalyze further reactions until it is oxidized back to E_o. This oxidation is usually carried out by adding molecular oxygen to the system so that glucose oxidase, E_o, is regenerated. Hydrogen peroxide is also produced in this oxidation regeneration step:

$$\text{G.O.H}_2 + \text{O}_2 \longrightarrow \text{G.O.} + \text{H}_2\text{O}_2$$

Overall, the reaction is written

$$\text{Glucose} + \text{O}_2 \; \xrightarrow{\frac{\text{glucose}}{\text{oxidase}}} \; \text{H}_2\text{O}_2 + \delta\text{-Gluconolactone}$$

In biochemistry texts, reactions of this type involving regeneration are usually written in the form

$$\text{Glucose}(S_r) \atop \delta\text{-gluconolactone}(P_1) \Big\rangle \Big\langle {\text{G.O.}(E_o) \atop \text{G.O.H}_2(E_r)} \Big\rangle \Big\langle {\text{H}_2\text{O}_2(P_2) \atop \text{O}_2(S_2)}$$

Derivation of the rate laws for this reaction sequence is given on the CD-ROM.

Enzyme Cofactors. In many enzymatic reactions, and in particular biological reactions, a second substrate (i.e., species) must be introduced to activate the enzyme. This substrate, which is referred to as a *cofactor* or *coenzyme* even though it is not an enzyme as such, attaches to the enzyme and is most often either reduced or oxidized during the course of the reaction. The enzyme–cofactor complex is referred to as a *holoenzyme*. The inactive form of the enzyme–cofactor complex for a specific reaction and reaction direction is called an *apoenzyme*. An example of the type of system in which a cofactor is used is the formation of ethanol from acetaldehyde in the presence of the enzyme alcohol dehydrogenase (ADH) and the cofactor nicotinamide adenine dinucleotide (NAD):

alcohol dehydrogenase

$$\text{acetaldehyde }(S_1) \atop \text{ethanol }(P_1) \Big\rangle \Big| \Big| \Big({\text{NADH }(S_2) \atop \text{H}^+} \atop \text{NAD}^+ (S_2^*)$$

Derivation of the rate laws for this reaction sequence is given in PRS 7.4.

Reference Shelf

7.4   Bioreactors

A bioreactor is a reactor that sustains and supports life for cells and tissue cultures. Virtually all cellular reactions necessary to maintain life are mediated by enzymes as they catalyze various aspects of cell metabolism such as the transformation of chemical energy and the construction, breakdown, and digestion of cellular components. Because enzymatic reactions are involved in the growth of microorganisms, we now proceed to study microbial growth and bioreactors. Not surprisingly, the Monod equation, which describes the growth law for a number of bacteria, is similar to the Michaelis–Menten equation. Consequently, even though bioreactors are not truly homogeneous because of the presence of living cells, we include them in this chapter as a logical progression from enzymatic reactions.

Nutrients

Cell   Products

The growth of biotechnology

$16 billion

The use of living cells to produce marketable chemical products is becoming increasingly important. The number of chemicals, agricultural products and food products produced by biosynthesis has risen dramatically. In 2003, companies in this sector raised over $16 billion of new financing.[12] Both

---

[12] *C & E News*, January 12, 2004, p. 7.

microorganisms and mammalian cells are being used to produce a variety of products, such as insulin, most antibiotics, and polymers. It is expected that in the future a number of organic chemicals currently derived from petroleum will be produced by living cells. The advantages of bioconversions are mild reaction conditions; high yields (e.g., 100% conversion of glucose to gluconic acid with *Aspergillus niger*); the fact that organisms contain several enzymes that can catalyze successive steps in a reaction and, most important, act as stereospecific catalysts. A common example of specificity in bioconversion production of a *single* desired isomer that when produced chemically yields a mixture of isomers is the conversion of *cis*-proenylphosphonic acid to the antibiotic (−) *cis*-1,2-epoxypropyl-phosphonic acid. Bacteria can also be modified and turned into living chemical factories. For example, using recombinant DNA, Biotechnic International engineered a bacteria to produce fertilizer by turning nitrogen into nitrates.[13]

In biosynthesis, the cells, also referred to as the *biomass*, consume nutrients to grow and produce more cells and important products. Internally, a cell uses its nutrients to produce energy and more cells. This transformation of nutrients to energy and bioproducts is accomplished through a cell's use of a number of different enzymes in a series of reactions to produce metabolic products. These products can either remain in the cell (intracellular) or be secreted from the cells (extracellular). In the former case the cells must be lysed (ruptured) and the product purified from the whole broth (reaction mixture). A schematic of a cell is shown in Figure 7-15.

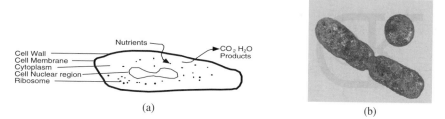

Cell Wall
Cell Membrane
Cytoplasm
Cell Nuclear region
Ribosome

Nutrients

CO_2 H_2O
Products

(a)   (b)

Figure 7-15   (a) Schematic of cell   (b) Photo of cell dividing *E. coli*. Courtesy of D. L. Nelson and M. M. Cox, *Lehninger Principles of Biochemistry*, 3rd ed. (New York: Worth Publishers, 2000).

The cell consists of a cell wall and an outer membrane that encloses cytoplasm containing a nuclear region and ribosomes. The cell wall protects the cell from external influences. The cell membrane provides for selective transport of materials into and out of the cell. Other substances can attach to the cell membrane to carry out important cell functions. The cytoplasm contains the ribosomes that contain ribonucleic acid (RNA), which are important in the synthesis of proteins. The nuclear region contains deoxyribonucleic acid

---

[13]*Chem. Eng. Progr.*, August 1988, p. 18.

(DNA) which provides the genetic information for the production of proteins and other cellular substances and structures.[14]

The reactions in the cell all take place simultaneously and are classified as either class (I) nutrient degradation (fueling reactions), class (II) synthesis of small molecules (amino acids), or class (III) synthesis of large molecules (polymerization, e.g., RNA, DNA). A rough overview with only a fraction of the reactions and metabolic pathways is shown in Figure 7-16. A more detailed model is given in Figures 5.1 and 6.14 of Shuler and Kargi.[15] In the Class I reactions, Adenosine triphosphate (ATP) participates in the degradation of the nutrients to form products to be used in the biosynthesis reactions (Class II) of small molecules (e.g., amino acids), which are then polymerized to form RNA and DNA (Class III). ATP also transfers the energy it releases when it loses a phosphonate group to form adenosine diphosphate (ADP)

$$ATP + H_2O \rightarrow ADT + P + H_2O + Energy$$

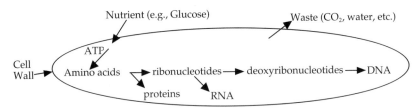

Figure 7-16   Examples of reactions occurring in the cell.

Cell Growth and Division

The cell growth and division typical of mammalian cells is shown schematically in Figure 7-17. The four phases of cell division are called G1, S, G2, and M, and are also described in Figure 7-17.

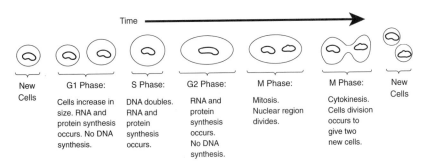

Figure 7-17   Phases of cell division.

---

[14]M. L. Shuler and F. Kargi, *Bioprocess Engineering Basic Concepts*, 2nd ed. (Upper Saddle River, N.J.: Prentice Hall, 2002).

[15]M. L. Shuler and F. Kargi, *Bioprocess Engineering Basic Concepts*, 2nd ed. (Upper Saddle River, N.J.: Prentice Hall, 2002), pp.135, 185.

In general, the growth of an aerobic organism follows the equation

$$[\text{Cells}] + \begin{bmatrix} \text{Carbon} \\ \text{source} \end{bmatrix} + \begin{bmatrix} \text{Nitrogen} \\ \text{source} \end{bmatrix} + \begin{bmatrix} \text{Oxygen} \\ \text{source} \end{bmatrix} + \begin{bmatrix} \text{Phosphate} \\ \text{source} \end{bmatrix} + \cdots$$

Cell multiplication $[CO_2] + [H_2O] + [\text{Products}] + \begin{bmatrix} \text{More} \\ \text{cells} \end{bmatrix} \xleftarrow{\begin{array}{c} \text{Culture media} \\ \text{conditions} \\ \text{(pH, temperature, etc.)} \end{array}}$

(7-49)

A more abbreviated form of Equation (7-49) generally used is

$$\text{Substrate} \xrightarrow{\text{Cells}} \text{More cells} + \text{Product} \qquad (7\text{-}50)$$

The products in Equation (7-50) include CO_2, water, proteins, and other species specific to the particular reaction. An excellent discussion of the stoichiometry (atom and mole balances) of Equation (7-49) can be found in Shuler and Kargi[16] and in Bailey and Ollis.[17] The substrate culture medium contains all the nutrients (carbon, nitrogen, etc.) along with other chemicals necessary for growth. Because, as we will soon see, the rate of this reaction is proportional to the cell concentration, the reaction is autocatalytic. A rough schematic of a simple batch biochemical reactor and the growth of two types of microorganisms, cocci (i.e., spherical) bacteria and yeast, is shown in Figure 7-18.

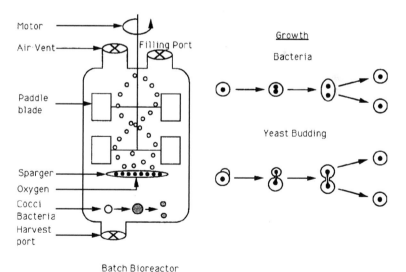

Figure 7-18   Batch bioreactor.

[16]M. L. Shuler and F. Kargi, *Bioprocess Engineering Basic Concepts*, 2nd ed. (Upper Saddle River, N.J.: Prentice Hall, 2002).

[17]J. E. Bailey and D. F. Ollis, *Biochemical Engineering*, 2nd ed. (New York: McGraw-Hill, 1987).

7.4.1 Cell Growth

Stages of cell growth in a batch reactor are shown schematically in Figures 7-19 and 7-20. Initially, a small number of cells is inoculated into (i.e., added to) the batch reactor containing the nutrients and the growth process begins as shown in Figure 7-19. In Figure 7-20, the number of living cells is shown as a function of time.

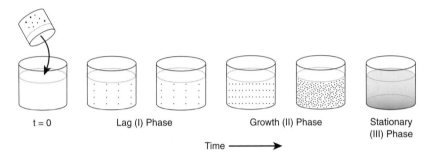

Figure 7-19 Increase in cell concentration.

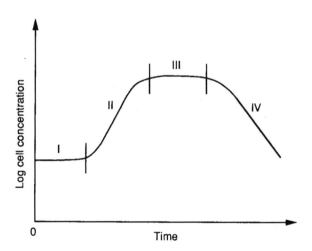

Figure 7-20 Phases of bacteria cell growth.

Lag phase   In phase I, called the lag phase, there is little increase in cell concentration. During the lag phase the cells are adjusting to their new environment, synthesizing enzymes, and getting ready to begin reproducing. During this time the cells carry out such functions as synthesizing transport proteins for moving the substrate into the cell, synthesizing enzymes for utilizing the new substrate, and beginning the work for replicating the cells' genetic material. The duration of the lag phase depends upon the growth medium from which the inoculum was taken relative to the reaction medium in which it is placed. If the inoculum is similar to the medium of the batch reactor, the lag phase will be almost nonexistent. If, however, the inoculum were placed in a medium with a different nutrient or other contents, or if the inoculum culture were in the stationary or death phase, the cells would have to readjust their metabolic path to allow them to consume the nutrients in their new environment.[18]

Exponential growth phase   Phase II is called the exponential growth phase owing to the fact that the cell's growth rate is proportional to the cell concentration. In this phase the cells are dividing at the maximum rate because all of the enzyme's pathways for metabolizing the substrate are in place (as a result of the lag phase) and the cells are able to use the nutrients most efficiently.

Phase III is the stationary phase, during which the cells reach a minimum biological space where the lack of one or more nutrients limits cell growth. During the stationary phase, the net growth rate is zero as a result of the depletion of nutrients and essential metabolites. Many important fermentation products, including most antibiotics, are produced in the stationary phase. For Antibiotics produced during the stationary phase   example, penicillin produced commercially using the fungus *Penicillium chrysogenum* is formed only after cell growth has ceased. Cell growth is also slowed by the buildup of organic acids and toxic materials generated during the growth phase.

Death phase   The final phase, Phase IV, is the death phase where a decrease in live cell concentration occurs. This decline is a result of the toxic by-products, harsh environments, and/or depletion of nutrient supply.

7.4.2 Rate Laws

While many laws exist for the cell growth rate of new cells, that is,

$$\text{Cells + Substrate} \longrightarrow \text{More cells + Product}$$

the most commonly used expression is the *Monod* equation for exponential growth:

$$r_g = \mu C_c \tag{7-51}$$

where r_g = cell growth rate, g/dm³·s
  C_c = cell concentration, g/dm³
  μ = specific growth rate, s⁻¹

The specific cell growth rate can be expressed as

$$\mu = \mu_{max} \frac{C_s}{K_s + C_s} \quad \text{s}^{-1} \tag{7-52}$$

[18]B. Wolf and H. S. Fogler, "Alteration of the Growth Rate and Lag Time of *Leuconostoc mesenteroides* NRRL-B523," *Biotechnology and Bioengineering, 72* (6), 603 (2001). B. Wolf and H. S. Fogler, "Growth of *Leuconostoc mesenteroides* NRRL-B523, in Alkaline Medium," *Biotechnology and Bioengineering, 89* (1), 96 (2005).

where μ_{max} = a maximum specific growth reaction rate, s^{-1}
  K_s = the *Monod* constant, g/dm^3
  C_s = substrate (i.e., nutrient) concentration, g/dm^3

Representative values of μ_{max} and K_s are 1.3 h^{-1} and 2.2×10^{-5} mol/dm^3, respectively, which are the parameter values for the *E. coli* growth on glucose. Combining Equations (7-51) and (7-52), we arrive at the Monod equation for bacterial cell growth rate

Monod equation

$$r_g = \frac{\mu_{max} C_s C_c}{K_s + C_s} \qquad (7\text{-}53)$$

For a number of different bacteria, the constant K_s is small, in which case the rate law reduces to

$$r_g = \mu_{max} C_c \qquad (7\text{-}54)$$

The growth rate, r_g, often depends on more than one nutrient concentration; however, the nutrient that is limiting is usually the one used in Equation (7-53).

In many systems the product inhibits the rate of growth. A classic example of this inhibition is in wine-making, where the fermentation of glucose to produce ethanol is inhibited by the product ethanol. There are a number of different equations to account for inhibition; one such rate law takes the empirical form

$$r_g = k_{obs} \frac{\mu_{max} C_s C_c}{K_s + C_s} \qquad (7\text{-}55)$$

where

Empirical form of Monod equation for product inhibition

$$k_{obs} = \left(1 - \frac{C_p}{C_p^*}\right)^n \qquad (7\text{-}56)$$

with

C_p^* = product concentration at which all metabolism ceases, g/dm^3
n = empirical constant

For the glucose-to-ethanol fermentation, typical inhibition parameters are

$$n = 0.5 \quad \text{and} \quad C_p^* = 93 \ g/dm^3$$

In addition to the Monod equation, two other equations are also commonly used to describe the cell growth rate; they are the *Tessier* equation,

$$r_g = \mu_{max} \left[1 - \exp\left(-\frac{C_s}{k}\right)\right] C_c \qquad (7\text{-}57)$$

and the *Moser* equation,

$$r_g = \frac{\mu_{max}C_c}{(1 + kC_s^{-\lambda})} \tag{7-58}$$

where λ and k are empirical constants determined by a best fit of the data.

The Moser and Tessier growth laws are often used because they have been found to better fit experimental data at the beginning or end of fermentation. Other growth equations can be found in Dean.[19]

The cell death rate is a result of harsh environments, mixing shear forces, local depletion of nutrients and the presence of toxic substances. The rate law is

$$r_d = (k_d + k_t C_t)C_c \tag{7-59}$$

where C_t is the concentration of a substance toxic to the cell. The specific death rate constants k_d and k_t refer to the natural death and death due to a toxic substance, respectively. Representative values of k_d range from 0.1 h^{-1} to less than 0.0005 h^{-1}. The value of k_t depends on the nature of the toxin.

Doubling times   Microbial growth rates are measured in terms of *doubling times*. Doubling time is the time required for a mass of an organism to double. Typical doubling times for bacteria range from 45 minutes to 1 hour but can be as fast as 15 minutes. Doubling times for simple eukaryotes, such as yeast, range from 1.5 to 2 hours but may be as fast as 45 minutes.

Effect of Temperature.   As with enzymes (cf. Figure 7-9), there is an optimum in growth rate with temperature owing to the competition of increased rates with increasing temperature and denaturizing the enzyme at high temperatures. An empirical law that describes this functionality is given in Aiba et al.[20] and is of the form

$$\mu(T) = \mu(T_m)I'$$

$$I' = \frac{aTe^{-E_1/RT}}{1 + be^{-E_2/RT}} \tag{7-60}$$

where I' is the fraction of the maximum growth rate, T_m is the temperature at which the maximum growth occurs, and $\mu(T_m)$ the growth at this temperature. For the rate of oxygen uptake of *Rhizobium trifollic*, the equation takes the form

$$I' = \frac{0.0038 Te^{[21.6 - 6700/T]}}{1 + e^{[153 - 48,000/T]}} \tag{7-61}$$

The maximum growth occurs at 310K.

[19]A. R. C. Dean, *Growth, Function, and Regulation in Bacterial Cells* (London: Oxford University Press, 1964).

[20]S. Aiba, A. E. Humphrey, and N. F. Millis, *Biochemical Engineering* (New York: Academic Press, 1973), p. 407.

7.4.3 Stoichiometry

The stoichiometry for cell growth is very complex and varies with microorganism/nutrient system and environmental conditions such as pH, temperature, and redox potential. This complexity is especially true when more than one nutrient contributes to cell growth, as is usually the case. We shall focus our discussion on a simplified version for cell growth, one that is limited by only one nutrient in the medium. In general, we have

$$\text{Cells} + \text{Substrate} \longrightarrow \text{More cells} + \text{Product}$$

In order to relate the substrate consumed, new cells formed, and product generated, we introduce the *yield* coefficients. The yield coefficient for cells and substrate is

$$Y_{c/s} = \frac{\text{Mass of new cells formed}}{\text{Mass of substrate consumed}} = -\frac{\Delta C_C}{\Delta C_S} \qquad (7\text{-}62)$$

with

$$Y_{c/s} = \frac{1}{Y_{s/c}}$$

A representative value of $Y_{c/s}$ might be 0.4 (g/g). See Chapter 3, Problem P3-14$_B$ where the value of $Y_{c/s}$ was calculated.

Product formation can take place during different phases of the cell growth cycle. When product formation only occurs during the exponential growth phase, the rate of product formation is

Growth associated
product formation

$$r_p = Y_{p/c} r_g = Y_{p/c} \mu C_C = Y_{p/c} \frac{\mu_{\max} C_C C_s}{K_s + C_s} \qquad (7\text{-}63)$$

where

$$Y_{p/c} = \frac{\text{Mass of product formed}}{\text{Mass of new cells formed}} = \frac{\Delta C_p}{\Delta C_c} \qquad (7\text{-}64)$$

The product of $Y_{p/c}$ and μ, that is, $(q_P = Y_{p/c}\,\mu)$ is often called the specific rate of product formation, q_P, (mass product/volume/time). When the product is formed during the stationary phase where no cell growth occurs, we can relate the rate of product formation to substrate consumption by

Nongrowth
associated product
formation

$$r_p = Y_{p/s}\,(-r_s) \qquad (7\text{-}65)$$

The substrate in this case is usually a secondary nutrient, which we discuss in more detail later.

The stoichiometric yield coefficient that relates the amount of product formed per mass of substrate consumed is

$$Y_{p/s} = \frac{\text{Mass of product formed}}{\text{Mass of substrate consumed}} = -\frac{\Delta C_p}{\Delta C_s} \qquad (7\text{-}66)$$

In addition to consuming substrate to produce new cells, part of the substrate must be used just to maintain a cell's daily activities. The corresponding maintenance utilization term is

Cell maintenance

$$m = \frac{\text{Mass of substrate consumed for maintenance}}{\text{Mass of cells} \cdot \text{Time}}$$

A typical value is

$$m = 0.05\ \frac{\text{g substrate}}{\text{g dry weight}}\ \frac{1}{\text{h}} = 0.05\ \text{h}^{-1}$$

The rate of substrate consumption for maintenance whether or not the cells are growing is

$$r_{sm} = mC_c \qquad (7\text{-}67)$$

When maintenance can be neglected, we can relate the concentration of cells formed to the amount of substrate consumed by the equation

Neglecting cell
maintenance

$$C_c = Y_{c/s}[C_{s0} - C_s] \qquad (7\text{-}68)$$

This equation can be used for both batch and continuous flow reactors.

If it is possible to sort out the substrate (S) that is consumed in the presence of cells to form new cells (C) from the substrate that is consumed to form product (P), that is,

$$S \xrightarrow{\text{cells}} Y'_{c/s}C + Y'_{p/s}P$$

the yield coefficients can be written as

$$Y'_{c/s} = \frac{\text{Mass of substrate consumed to form new cells}}{\text{Mass of new cells formed}} \qquad (7\text{-}69\text{A})$$

$$Y'_{s/p} = \frac{\text{Mass of substrate consumed to form product}}{\text{Mass of product formed}} \qquad (7\text{-}69\text{B})$$

Substrate Utilization. We now come to the task of relating the rate of nutrient consumption, $-r_s$, to the rates of cell growth, product generation, and cell maintenance. In general, we can write

Substrate accounting

$$
\begin{bmatrix}
\text{Net rate of} \\
\text{substrate} \\
\text{consumption}
\end{bmatrix}
=
\begin{bmatrix}
\text{Rate} \\
\text{consumed} \\
\text{by cells}
\end{bmatrix}
+
\begin{bmatrix}
\text{Rate} \\
\text{consumed to} \\
\text{form product}
\end{bmatrix}
+
\begin{bmatrix}
\text{Rate} \\
\text{consumed for} \\
\text{maintenance}
\end{bmatrix}
$$

$$
-r_s \quad = \quad Y'_{s/c}r_g \quad + \quad Y'_{s/p}r_p \quad + \quad mC_c
$$

In a number of cases extra attention must be paid to the substrate balance. If product is produced during the growth phase, it may not be possible to separate out the amount of substrate consumed for cell growth from that consumed to produce the product. Under these circumstances all the substrate consumed is lumped into the stoichiometric coefficient, $Y_{s/c}$, and the rate of substrate disappearance is

$$
\boxed{-r_s = Y_{s/c}r_g + mC_c} \tag{7-70}
$$

The corresponding rate of product formation is

Growth-associated product formation in the growth phase

$$
\boxed{r_p = r_g Y_{p/c}} \tag{7-63}
$$

Because there is no growth during the stationary phase, it is clear that Equation (7-70) cannot be used to account for substrate consumption, nor can the rate of product formation be related to the growth rate [e.g., Equation (7-63)]. Many antibiotics, such as penicillin, are produced in the stationary phase. In this phase, the nutrient required for growth becomes virtually exhausted, and a different nutrient, called the secondary nutrient, is used for cell maintenance and to produce the desired product. Usually, the rate law for product formation during the stationary phase is similar in form to the Monod equation, that is,

Nongrowth-associated product formation in the stationary phase

$$
\boxed{r_p = \frac{k_p C_{sn} C_c}{K_{sn} + C_{sn}}} \tag{7-71}
$$

where k_p = specific rate constant with respect to product, $(\text{dm}^3/\text{g} \cdot \text{s})$
C_{sn} = concentration of the secondary nutrient, g/dm^3
C_c = cell concentration, g/dm^3 $(\text{g} \equiv \text{gdw} = \text{gram dry weight})$
K_{sn} = Monod constant, g/dm^3
r_p = $Y_{p/sn}(-r_{sn})$ $(\text{g}/\text{dm}^3 \cdot \text{s})$

The net rate of secondary nutrient consumption during the stationary phase is

In the stationary phase, the concentration of live cells is constant.

$$
-r_{sn} = mC_c + Y_{sn/p}r_p
$$

$$
= mC_c + \frac{Y_{sn/p}k_p C_{sn} C_c}{K_{sn} + C_{sn}} \tag{7-72}
$$

Because the desired product can be produced when there is no cell growth, it is always best to relate the product concentration to the change in secondary nutrient concentration. For a batch system the concentration of product, C_p, formed after a time t in the stationary phase can be related to the substrate concentration, C_s, at that time.

Neglects cell maintenance

$$C_p = Y_{p/s}(C_{sn0} - C_{sn})   \qquad (7\text{-}73)$$

We have considered two limiting situations for relating substrate consumption to cell growth and product formation; product formation only during the growth phase and product formation only during the stationary phase. An example where neither of these situations applies is fermentation using lactobacillus, where lactic acid is produced during both the logarithmic growth and stationary phase.

The specific rate of product formation is often given in terms of the Luedeking–Piret equation, which has two parameters α (growth) and β (non-growth)

Luedeking–Piret equation for the rate of product formation

$$q_p = \alpha\mu_g + \beta   \qquad (7\text{-}74)$$

with

$$r_p = q_p C_C$$

The assumption here in using the β-parameter is that the secondary nutrient is in excess.

Example 7–5   Estimate the Yield Coefficients

The following data was determined in a batch reactor for the yeast *Saccharomyces cerevisiae*

TABLE E7-5.1.   RAW DATA

$$\text{Glucose} \xrightarrow{\text{cells}} \text{More cells} + \text{Ethanol}$$

Time, t (hr)	Cells, C_C (g/dm³)	Glucose, C_S (g/dm³)	Ethanol, C_P (g/dm³)
0	1	250	0
1	1.37	245	2.14
2	1.87	238.7	5.03
3	2.55	229.8	8.96

Determine $Y_{s/c}$, $Y_{c/s}$, $Y_{s/p}$, $Y_{p/s}$, $Y_{p/c}$, μ_{max}, and K_S. Assume no lag and neglect maintenance at the start of the growth where there are just a few cells.

Solution

(a)   Calculate the substrate and cell *yield coefficients*, $Y_{s/c}$ and $Y_{c/s}$.
Between $t = 0$ and $t = 1$ h

$$Y_{s/c} = \frac{-\Delta C_s}{\Delta C_c} = -\frac{245 - 250}{1.37 - 1} = 13.51 \text{ g/g} \tag{E7-5.1}$$

Between $t = 2$ and $t = 3$ h

$$Y_{s/c} = -\frac{229.8 - 238.7}{2.55 - 1.87} = \frac{8.9}{0.68} = 13.1 \tag{E7-5.2}$$

Taking an average

$$Y_{s/c} = 13.3 \text{ g/g}$$

We could also have used Polymath regression to obtain

$$Y_{c/s} = \frac{1}{Y_{s/c}} = \frac{1}{13.3 \text{ g/g}} = 0.075 \text{ g/g} \tag{E7-5.3}$$

(b)   Similarly for the substrate and product *yield coefficients*

$$Y_{s/p} = -\frac{\Delta C_S}{\Delta C_P} = -\frac{238.7 - 245}{5.03 - 2.14} = \frac{6.3}{2.89} = 2.18 \text{ g/g} \tag{E7-5.4}$$

$$Y_{p/s} = \frac{1}{Y_{s/p}} = \frac{1}{2.12 \text{ g/g}} = 0.459 \text{ g/g} \tag{E7-5.5}$$

(c)   The product/cell *yield coefficient* is

$$Y_{p/c} = \frac{\Delta C_p}{\Delta C_c} = \frac{5.03 - 2.14}{1.87 - 1.37} = 5.78 \text{ g/g} \tag{E7-5.6}$$

$$Y_{c/p} = \frac{1}{Y_{p/c}} = \frac{1}{5.78 \text{ g/g}} = 0.173 \text{ g/g} \tag{E7-5.7}$$

We now need to determine the rate law parameters μ_{max} and K_S in the Monod equation

$$r_g = \frac{\mu_{max} C_c C_s}{K_s + C_s} \tag{7-53}$$

For a batch system

$$r_g = \frac{dC_c}{dt} \tag{E7-5.8}$$

How to regress
the Monod equation
for μ_{max} and K_s

     To find the rate law parameters μ_{max} and K_s, we first apply the differential formulas in Chapter 5 to columns 1 and 2 of Table E7-5.1 to find r_g. Because $C_s \gg K_s$ initially, it is best to regress the data using the Hanes-Woolf form of the Monod equation

$$\frac{C_c}{r_g} = \frac{K_s}{\mu_{max}}\left(\frac{1}{C_s}\right) + \frac{1}{\mu_{max}} \tag{E7-5.9}$$

Using Polymath's nonlinear regression and more data points, we find $\mu_{max} = 0.33 \text{ h}^{-1}$ and $K_s = 1.7 \text{g/dm}^3$.

7.4.4 Mass Balances

There are two ways that we could account for the growth of microorganisms. One is to account for the number of living cells, and the other is to account for the mass of the living cells. We shall use the latter. A mass balance on the microorganism in a CSTR (chemostat) (shown in Figure 7-21) of constant volume is

Cell Balance

$$\left[\begin{array}{c} \text{Rate of} \\ \text{accumulation} \\ \text{of cells,} \\ \text{g/s} \end{array}\right] = \left[\begin{array}{c} \text{Rate of} \\ \text{cells} \\ \text{entering,} \\ \text{g/s} \end{array}\right] - \left[\begin{array}{c} \text{Rate of} \\ \text{cells} \\ \text{leaving,} \\ \text{g/s} \end{array}\right] + \left[\begin{array}{c} \text{Net rate of} \\ \text{generation} \\ \text{of live cells,} \\ \text{g/s} \end{array}\right] \quad (7\text{-}75)$$

$$V\frac{dC_c}{dt} = v_0 C_{c0} - v C_c + (r_g - r_d)V$$

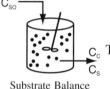

The corresponding substrate balance is

Substrate Balance

$$\left[\begin{array}{c} \text{Rate of} \\ \text{accumulation} \\ \text{of substrate,} \\ \text{g/s} \end{array}\right] = \left[\begin{array}{c} \text{Rate of} \\ \text{substrate} \\ \text{entering,} \\ \text{g/s} \end{array}\right] - \left[\begin{array}{c} \text{Rate of} \\ \text{substrate} \\ \text{leaving,} \\ \text{g/s} \end{array}\right] + \left[\begin{array}{c} \text{Rate of} \\ \text{substrate} \\ \text{generation,} \\ \text{g/s} \end{array}\right] \quad (7\text{-}76)$$

$$V\frac{dC_s}{dt} = v_0 C_{s0} - v C_s + r_s V$$

In most systems the entering microorganism concentration C_{c0} is zero.

Batch Operation

For a batch system $v = v_0 = 0$ and the mass balances are as follows:

Cell

The mass balances

$$V\frac{dC_c}{dt} = r_g V - r_d V$$

Dividing by the reactor volume V gives

$$\frac{dC_c}{dt} = r_g - r_d \quad (7\text{-}77)$$

Substrate

The rate of disappearance of substrate, $-r_s$, results from substrate used for cell growth and substrate used for cell maintenance,

$$V\frac{dC_s}{dt} = r_s V = Y_{s/c}(-r_g)V - mC_c V \quad (7\text{-}78)$$

Dividing by V yields the substrate balance for the growth phase

Growth phase

$$\frac{dC_s}{dt} = Y_{s/c}(-r_g) - mC_c$$

For cells in the stationary phase, where there is no growth, cell maintenance and product formation are the only reactions to consume the substrate. Under these conditions the substrate balance, Equation (7-76), reduces to

Stationary phase

$$V\frac{dC_s}{dt} = -mC_cV + Y_{s/p}(-r_p)V \qquad (7\text{-}79)$$

Typically, r_p will have the same form of the rate law as r_g [e.g., Equation (7-71)]. Of course, Equation (7-79) only applies for substrate concentrations greater than zero.

Product

The rate of product formation, r_p, can be related to the rate of substrate consumption through the following balance:

Batch stationary growth phase

$$V\frac{dC_p}{dt} = r_pV = Y_{p/s}(-r_s)V \qquad (7\text{-}80)$$

During the growth phase we could also relate the rate of formation of product, r_p, to the cell growth rate, r_g. The coupled first-order ordinary differential equations above can be solved by a variety of numerical techniques.

Example 7–6 Bacteria Growth in a Batch Reactor

Glucose-to-ethanol fermentation is to be carried out in a batch reactor using an organism such as *Saccharomyces cerevisiae*. Plot the concentrations of cells, substrate, and product and growth rates as functions of time. The initial cell concentration is 1.0 g/dm³, and the substrate (glucose) concentration is 250 g/dm³.

Additional data [partial source: R. Miller and M. Melick, *Chem. Eng.*, Feb. 16, p. 113 (1987)]:

$$C_p^* = 93 \text{ g/dm}^3 \qquad Y_{c/s} = 0.08 \text{ g/g}$$

$$n = 0.52 \qquad Y_{p/s} = 0.45 \text{ g/g (est.)}$$

$$\mu_{max} = 0.33 \text{ h}^{-1} \qquad Y_{p/c} = 5.6 \text{ g/g (est.)}$$

$$K_s = 1.7 \text{ g/dm}^3 \qquad k_d = 0.01 \text{ h}^{-1}$$

$$m = 0.03 \text{ (g substrate)}/(\text{g cells}\cdot\text{h})$$

Solution

1. **Mass balances:**

Cells: $$V\frac{dC_c}{dt} = (r_g - r_d)V \qquad (E7\text{-}6.1)$$

The algorithm

Substrate: $$V\frac{dC_s}{dt} = Y_{s/c}(-r_g)V - r_{sm}V \qquad (E7\text{-}6.2)$$

Product: $$V\frac{dC_p}{dt} = Y_{p/c}(r_g V) \qquad (E7\text{-}6.3)$$

2. **Rate laws:**

$$r_g = \mu_{max}\left(1 - \frac{C_p}{C_p^*}\right)^{0.52}\frac{C_c C_s}{K_s + C_s} \qquad (E7\text{-}6.4)$$

$$r_d = k_d C_c \qquad (E7\text{-}6.5)$$

$$r_{sm} = mC_c \qquad (7\text{-}67)$$

3. **Stoichiometry:**

$$r_p = Y_{p/c}r_g \qquad (E7\text{-}6.6)$$

4. **Combining gives**

$$\frac{dC_c}{dt} = \mu_{max}\left(1 - \frac{C_p}{C_p^*}\right)^{0.52}\frac{C_c C_s}{K_s + C_s} - k_d C_c \qquad (E7\text{-}6.7)$$

Cells
Substrate
Product

$$\frac{dC_s}{dt} = -Y_{s/c}\,\mu_{max}\left(1 - \frac{C_p}{C_p^*}\right)^{0.52}\frac{C_c C_s}{K_m + C_s} - mC_c \qquad (E7\text{-}6.8)$$

$$\frac{dC_p}{dt} = Y_{p/c}r_g$$

These equations were solved on an ODE equation solver (see Table E7-6.1). The results are shown in Figure E7-6.1 for the parameter values given in the problem statement.

TABLE E7-6.1.   POLYMATH PROGRAM

Living Example Problem

POLYMATH Results

Example 7-6 Bacteria Growth in a Batch Reactor   08-18-2004, Rev5.1.232

ODE Report (RKF45)

Differential equations as entered by the user
[1] d(Cc)/d(t) = rg-rd
[2] d(Cs)/d(t) = Ysc*(-rg)-rsm
[3] d(Cp)/d(t) = rg*Ypc

Explicit equations as entered by the user
[1] rd = Cc*.01
[2] Ysc = 1/.08
[3] Ypc = 5.6
[4] Ks = 1.7
[5] m = .03
[6] umax = .33
[7] rsm = m*Cc
[8] kobs = (umax*(1-Cp/93)^.52)
[9] rg = kobs*Cc*Cs/(Ks+Cs)

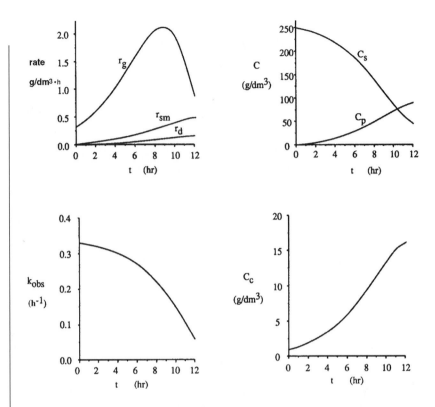

Figure E7-6.1   Concentrations and rates as a function of time.

The substrate concentration C_s can never be less than zero. However, we note that when the substrate is completely consumed, the first term on the right-hand side of Equation (E7-6.8) (and line 3 of the Polymath program) will be zero but the second term for maintenance, mC_c, will not. Consequently, if the integration is carried further in time, the integration program will predict a negative value of C_s! This inconsistency can be addressed in a number of ways such as including an **if** statement in the Polymath program (e.g., if C_s is less than or equal to zero, then $m = 0$).

7.4.5 Chemostats

Chemostats are essentially CSTRs that contain microorganisms. A typical chemostat is shown in Figure 7-21, along with the associated monitoring equipment and pH controller. One of the most important features of the chemostat is that it allows the operator to control the cell growth rate. This control of the growth rate is achieved by adjusting the volumetric feed rate (dilution rate).

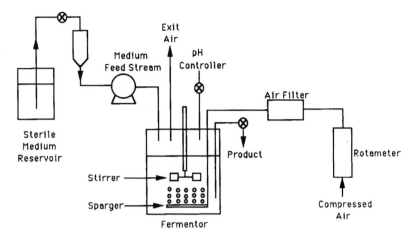

Figure 7-21 Chemostat system.

7.4.6 Design Equations

CSTR

In this section we return to mass equations on the cells [Equation (7-75)] and substrate [Equation (7-76)] and consider the case where the volumetric flow rates in and out are the same and that no live (i.e., viable) cells enter the chemostat. We next define a parameter common to bioreactors called the dilution rate, D. The dilution rate is

$$D = \frac{v_0}{V}$$

and is simply the reciprocal of the space time τ. Dividing Equations (7-75) and (7-76) by V and using the definition of the dilution rate, we have

$$\text{Accumulation} = \text{In} \quad - \text{Out} + \text{Generation}$$

CSTR mass
balances

$$\text{Cell:} \qquad \frac{dC_c}{dt} = 0 \quad - DC_c + (r_g - r_d) \qquad (7\text{-}81)$$

$$\text{Substrate:} \qquad \frac{dC_s}{dt} = DC_{s0} - DC_s + r_s \qquad (7\text{-}82)$$

Using the Monod equation, the growth rate is determined to be

$$r_g = \mu C_c = \frac{\mu_{\max} C_s C_c}{K_s + C_s} \qquad (7\text{-}53)$$

For steady-state operation we have

$$DC_c = r_g - r_d \qquad (7\text{-}83)$$

and

$$D(C_{s0} - C_s) = -r_s \qquad (7\text{-}84)$$

We now neglect the death rate, r_d, and combine Equations (7-51) and (7-83) for steady-state operation to obtain the mass flow rate of cells out of the system, F_c.

$$F_c = C_c v_0 = r_g V = \mu C_c V \tag{7-85}$$

After we divide by $C_c V$,

Dilution rate

$$\boxed{D = \mu} \tag{7-86}$$

How to control cell growth

An inspection of Equation (7-86) reveals that the specific growth rate of the cells *can be controlled* by the operator by controlling the dilution rate D. Using Equation (7-52) to substitute for μ in terms of the substrate concentration and then solving for the steady-state substrate concentration yields

$$C_s = \frac{DK_s}{\mu_{max} - D} \tag{7-87}$$

Assuming that a single nutrient is limiting, cell growth is the only process contributing to substrate utilization, and that cell maintenance can be neglected, the stoichiometry is

$$-r_s = r_g Y_{s/c} \tag{7-88}$$

$$C_c = Y_{c/s}(C_{s0} - C_s) \tag{7-68}$$

Substituting for C_s using Equation (7-87) and rearranging, we obtain

$$\boxed{C_c = Y_{c/s}\left[C_{s0} - \frac{DK_s}{\mu_{max} - D}\right]} \tag{7-89}$$

7.4.7 Wash-out

To learn the effect of increasing the dilution rate, we combine Equations (7-81) and (7-54) and set $r_d = 0$ to get

$$\frac{dC_c}{dt} = (\mu - D)C_c \tag{7-90}$$

We see that if $D > \mu$, then dC_c/dt will be negative, and the cell concentration will continue to decrease until we reach a point where all cells will be washed out:

$$C_c = 0$$

The dilution rate at which wash-out will occur is obtained from Equation (7-89) by setting $C_c = 0$.

Flow rate at which wash-out occurs

$$\boxed{D_{max} = \frac{\mu_{max} C_{s0}}{K_s + C_{s0}}} \tag{7-91}$$

We next want to determine the other extreme for the dilution rate, which is the rate of maximum cell production. The cell production rate per unit volume of reactor is the mass flow rate of cells out of the reactor (i.e., $\dot{m}_c = C_c v_0$) divided by the volume V, or

$$\frac{v_0 C_c}{V} = DC_c$$

Using Equation (7-89) to substitute for C_c yields

$$DC_c = DY_{c/s}\left(C_{s0} - \frac{DK_s}{\mu_{max} - D}\right) \tag{7-92}$$

Figure 7-22 shows production rate, cell concentration, and substrate concentration as functions of dilution rate. We observe a maximum in the production rate, and this maximum can be found by differentiating the production rate, Equation (7-92), with respect to the dilution rate D:

$$\frac{d(DC_c)}{dD} = 0 \tag{7-93}$$

Maximum rate of
cell production
(DC_c)

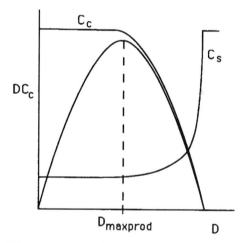

Figure 7-22   Cell concentration and production rate as a function of dilution rate.

Then

Maximum rate of
cell production

$$D_{maxprod} = \mu_{max}\left(1 - \sqrt{\frac{K_s}{K_s + C_{s0}}}\right) \tag{7-94}$$

The organism *Streptomyces aureofaciens* was studied in a 10 dm³ chemostat using sucrose as a substrate. The cell concentration, C_c (mg/ml), the substrate concentration, C_s (mg/ml), and the production rate, DC_c (mg/ml/h), were measured at steady state for different dilution rates. The data are shown in Figure 7-23.[21]

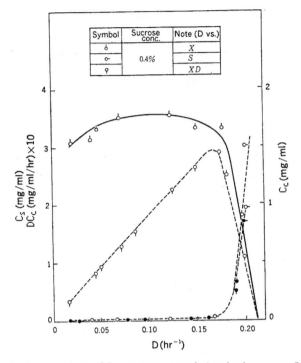

Figure 7-23 Continuous culture of *Streptomyces aureofaciens* in chemostats. (Note: $X \equiv C_c$) Courtesy of S. Aiba, A. E. Humphrey, and N. F. Millis, *Biochemical Engineering*, 2nd Ed. (New York: Academic Press, 1973).

Note that the data follow the same trends as those discussed in Figure 7-22.

7.4.8 Oxygen-Limited Growth

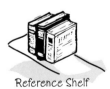

Reference Shelf

Oxygen is necessary for all aerobic growth (by definition). Maintaining the appropriate concentration of dissolved oxygen in the bioreactor is important for efficient operation of a bioreactor. For oxygen-limited systems, it is necessary to design a bioreactor to maximize the oxygen transfer between the injected air bubbles and the cells. Typically, a bioreactor contains a gas sparger, heat transfer surfaces, and an impeller. A chemostat has a similar configuration with the addition of inlet and outlet streams similar to that shown in Figure 7-18. The oxygen transfer rate (OTR) is related to the cell concentration by

$$OTR = Q_{O_2} C_c \qquad (7-95)$$

---

[21]B. Sikyta, J. Slezak, and M. Herold, *Appl. Microbiol.*, *9*, 233 (1961).

where Q_{O_2} is the microbial respiration rate or specific oxygen uptake rate and usually follows Michaelis–Menten kinetics (Monod growth, e.g., $Q_{O_2} = Y_{O_{2/c}} r_g$). (See Problem P7-13$_B$.) The CD-ROM discusses the transport steps from the bulk liquid to and within the microorganism. A series of mass transfer correlations are also given.

7.4.9 Scale-up

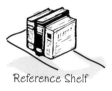

Reference Shelf

Scale-up for the growth of microorganisms is usually based on maintaining a constant dissolved oxygen concentration in the liquid (broth), independent of reactor size. Guidelines for scaling from a pilot-plant bioreactor to a commercial plant reactor are given on the CD-ROM. One key to a scale-up is to have the speed of the end (tip) of the impeller equal to the velocity in both the laboratory pilot reactor and the full-scale plant reactor. If the impeller speed is too rapid, it can lyse the bacteria; if the speed is too slow, the reactor contents will not be well mixed. Typical tip speeds range from 5 to 7 m/s.

7.5 Physiologically Based Pharmacokinetic (PBPK) Models

We now apply the material we have been discussing on enzyme kinetics to modeling reactions in living systems. Physiologically based pharmacokinetic models are used to predict the distribution and concentration-time trajectories of medications, toxins, poisons, alcohol, and drugs in the body. The approach is to model the body components (e.g., liver, muscle) as compartments consisting of PFRs and CSTRs connected to one another with in-flow and out-flow to each organ compartment as shown in Figure 7-24.

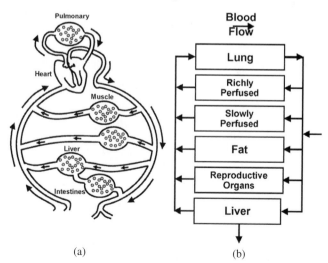

(a)   (b)

Figure 7-24   (a) Compartment model of human body. (b) Generic structure of PBPK models. Courtesy of *Chem. Engrg. Progress*, 100 (6) 38 (June 2004).

Associated with each organ is a certain tissue water volume, TWV, which we will designate as the organ compartment. The TWV for the different organs along with the blood flow into and out of the different organ compartments (called the perfusion rate) can be found in the literature. In the models discussed here, the organ compartments will be modeled as unsteady well-mixed CSTRs with the exception of the liver, which will be modeled as an unsteady PFR. We will apply the chemical reaction engineering algorithm (mole balance, rate law, stoichiometry) to the unsteady operation of each compartment. Some compartments with similar fluid residence times are modeled to consist of several body parts (skin, lungs, etc.) lumped into one compartment, such as the central compartment. The interchange of material between compartments is primarily through blood flow to the various components. The drug/medication concentrations are based on the tissue water volume of a given compartment. Consequently, an important parameter in the systems approach is the perfusion rate for each organ. If we know the perfusion rate, we can determine the exchange of material between the bloodstream and that organ. For example, if organs are connected in series on one in parallel by blood flow as shown in Figure 7-25,

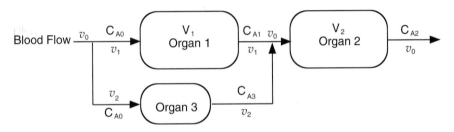

Figure 7-25 Physiologically based model.

then the balance equations on species A in the TWVs of the organs V_1, V_2, and V_3 are

$$V_1\frac{dC_{A1}}{dt} = v_1(C_{A0} - C_{A1}) + r_{A1}V_1 \tag{7-96}$$

$$V_2\frac{dC_{A2}}{dt} = v_1(C_{A1} - C_{A2}) + v_2(C_{A3} - C_{A2}) + r_{A2}V_2 \tag{7-97}$$

$$V_3\frac{dC_{A3}}{dt} = v_2(C_{A0} - C_{A3}) + r_{A3}V_3 \tag{7-98}$$

where r_{A1}, r_{A2}, and r_{A3} are the metabolism rates of species A in organs 1, 2, and 3, respectively, and C_{Aj} is the concentration of species A being metabolized in each of the organ compartments $j = 1$, 2, and 3.

Pharmacokinetic models for drug delivery are given in Professional Reference Shelf R7.5.

Example 7–7   Alcohol Metabolism in the Body[22]

E7-7.A. General

We are going to model the metabolism of ethanol in the human body using fundamental reaction kinetics along with five compartments to represent the human body. Alcohol (Ac) and acetaldehyde (De) will flow between these compartments, but the alcohol and aldehyde will only be metabolized in the liver compartment. Alcohol and acetaldehyde are metabolized in the liver by the following series reactions.

Living Example Problem

$$C_2H_5OH \underset{ADH}{\xrightarrow{\hspace{1cm}}} CH_3CHO \underset{ALDH}{\xrightarrow{\hspace{1cm}}} CH_3COOH$$

The first reaction is catalyzed by the enzyme alcohol dehydrogenease (ADH) and the second reaction is catalyzed by aldehyde dehydrogenease (ADLH).

The reversible enzyme ADH reaction is catalyzed reaction in the presence of a cofactor, nicotinamide adenime dinucleotide (NAD[+])

$$C_2H_5OH + NAD^+ \underset{ADH}{\xrightarrow{\hspace{1cm}}} CH_3CHO + H^+ + NADH$$

The rate law for the disappearance of ethanol follows Michaelis–Menton kinetics and is

$$-r_{AC} = \frac{[V_{max\ ADH}C_{Ac} - V_{rev\ ADH}C_{De}]}{K_M + C_{Ac} + K_{rev\ ADH}C_{De}} \tag{E7-7.1}$$

where V_{max} and K_M are the Michaelis–Menten parameters discussed in Section 7.2, and C_{Ac} and C_{De} are the concentrations of ethanol and acetadehyde, respectively. For the metabolism of acetaldehyde in the presence of acetaldehyde dehydrogenase, and NAD[+]

$$NAD^+ + CH_3CHO + H_2O \xrightarrow{ALDH} CH_3COOH + NADH + H^+$$

the enzymatic rate law is

$$-r_{De} = \frac{V_{max\ ALDH}C_{De}}{K_{MALDH} + C_{De}} \tag{E7-7.2}$$

The parameter values for the rate laws are V_{maxADH} = 2.2 mM/(min • kg liver), K_{MADH} = 0.4 mM, V_{revADH} = 32.6 mmol/(min • kg liver), K_{revADH} = 1 mM, $V_{maxALDH}$ = 2.7 mmol/(min • kg liver), and K_{MALDH} = 1.2 μM (see Summary Notes).

The concentration time trajectories for alcohol concentration in the central compartment are shown in Figure E7-7.1.

E7-7.B The Model System

We are going to use as an example a five-organ compartment model for the metabolism of ethanol in humans. We will apply the CRE algorithm to the tissue water volume in each organ. The TWVs are lumped according to their perfusion rates and

Summary Notes

[22]D. M. Umulis, M. N. Gurmen, P. Singh, and H. S. Fogler, *Alcohol 35* (1), 2005. The complete paper is presented in the Summary Notes on the CD-ROM.

residence times. That is, those compartments receiving only small amounts of blood flow will be lumped together (e.g., fat and muscle) as will those receiving large blood flows (e.g., lungs, kidneys, etc.). The following organs will be modeled as single unsteady CSTRs: stomach, gastrointestinal tract, central system, and muscle and fat. The metabolism of ethanol occurs primarily in the liver, which is modeled as a PFR. A number of unsteady CSTRs in series   approximate the PFR. Figure E7-7.1 gives a diagram showing the connection blood flow (perfusion), and mean residence, τ.

The physiologically
based model

TWV

Muscle & Fat = 25.76 dm³

Central = 11.56 dm³

Liver = 1.1 dm³

G.I. = 2.4 dm³

Residence Time

Muscle & Fat = 27 min

Central = 0.9 min

Liver = 2.4 min

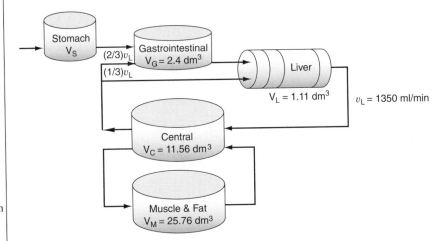

Figure E7-7.1 Compartment model of human body.

The residence times for each organ were obtained from the individual perfusion rates and are shown in the margin note next to Figure E7.7-1.

We will now discuss the balance equation on the tissue water volume of each of the organs/compartments.

Stomach

As a first approximation, we shall neglect the 10% of the total alcohol ingested that is absorbed in the stomach because the majority of the alcohol (90%) is absorbed at the entrance to the gastrointestinal (G.I.) tract. The contents of the stomach are emptied into the G.I. tract at a rate proportional to the volume of the contents in the stomach.

$$\frac{dV_S}{dt} = -k_S V_S \tag{E7-7.3}$$

where V_S, is the volume of the contents of the stomach and k_S is the rate constant

$$k_S = \frac{k_{S\,max}}{1 + a(D)^2}$$

The flow of ethanol from the stomach into the G.I. tract, where it is absorbed virtually instantaneously, is

$$F_{Ac} = k_S V_S C_{S_{Ac}} \tag{E7-7.4}$$

where $C_{S_{Ac}}$ is the ethanol concentration in the stomach, k_{Smax} is the maximum emptying rate, D is the dose of ethanol in the stomach in (mmol), and a is the emptying parameter in $(mmol)^{-2}$.

Gastrointestinal (G.I.) Tract Component

Ethanol is absorbed virtually instantaneously in the duodenum at the entrance of the G.I. tract. In addition, the blood flow to the G.I. compartment from the central compartment to the G.I. tract is two-thirds of the total blood flow with the other third by-passing the G.I. tract to the liver, as shown in Figure E7-7.1. A mole mass balance on ethanol in the G.I. tract tissue water volume (TWV) V_G, gives

$$\begin{bmatrix} In \\ from \\ Stomach \end{bmatrix} + \begin{bmatrix} In \\ from \\ Central \end{bmatrix} - \begin{bmatrix} Out \\ from \\ G.I. \end{bmatrix} + \begin{bmatrix} Generation \\ in \\ G.I. \end{bmatrix} = \begin{bmatrix} Accumulation \\ in \\ G.I. \end{bmatrix}$$

$$k_s V_s C_{S_{Ac}} + \frac{2}{3} v_L C_{C_{Ac}} - \frac{2}{3} v_L C_{G_{Ac}} + \qquad 0 \qquad = \frac{d(V_G C_{G_{Ac}})}{dt} \qquad (E7\text{-}7.5)$$

where $C_{G_{Ac}}$ is the concentration of alcohol in the G.I. compartment. Because the TWV remains constant, the mass balance becomes

Ethanol: $\qquad V_G \dfrac{dC_{G_{Ac}}}{dt} = \dfrac{2}{3} v_L (C_{C_{Ac}} - C_{G_{Ac}}) + k_S V_S C_{S_{Ac}} \qquad (E7\text{-}7.6)$

A similar balance on acetaldehyde gives

Acetaldehyde: $\qquad V_G \dfrac{dC_{G_{De}}}{dt} = \dfrac{2}{3} v_L (C_{C_{De}} - C_{G_{De}}) \qquad (E7\text{-}7.7)$

Central Compartment

The central volume has the largest TWV. Material enters the central compartment from the liver and the muscle/fat compartments. A balance on ethanol in this compartment is

$$[Accumulation] = \begin{bmatrix} In \\ from \\ Liver \end{bmatrix} + \begin{bmatrix} In \\ from \\ Muscle \end{bmatrix} - \begin{bmatrix} Out\ to \\ Liver\ and \\ G.I. \end{bmatrix} - \begin{bmatrix} Out \\ to \\ Muscle \end{bmatrix} + [Generation]$$

Ethanol: $V_C \dfrac{dC_{C_{Ac}}}{dt} = v_L C_{L_{Ac}} + v_M C_{M_{Ac}} - v_L C_{C_{Ac}} - v_M C_{C_{Ac}} + 0 \quad (E7\text{-}7.8)$

Similarly the acetaldehyde balance is

Acetaldehyde: $\qquad V_C \dfrac{dC_{C_{De}}}{dt} = v_L (C_{L_{De}} - C_{C_{De}}) + v_M (C_{M_{De}} - C_{C_{De}}) \qquad (E7\text{-}7.9)$

Muscle/Fat Compartment

Very little material profuses in and out of the muscle and fat compartments compared to the other compartments. The muscle compartment mass balances on ethanol and acetaldehyde are

Ethanol:
$$V_M \frac{dC_{M_{Ac}}}{dt} = \upsilon_M (C_{C_{Ac}} - C_{M_{Ac}}) \tag{E7-7.10}$$

Acetaldehyde:
$$V_M \frac{dC_{M_{De}}}{dt} = \upsilon_M (C_{C_{De}} - C_{M_{De}}) \tag{E7-7.11}$$

Liver Compartment

The liver will be modeled as a number of CSTRs in series to approximate a PFR with a volume of 1.1 dm³. Approximating a PFR with a number of CSTRs in series was discussed in Chapter 2. The total volume of the liver is divided into four CSTRs.

Death by alcohol poisoning can occur when the central compartment concentration reaches 2 g/dm³.

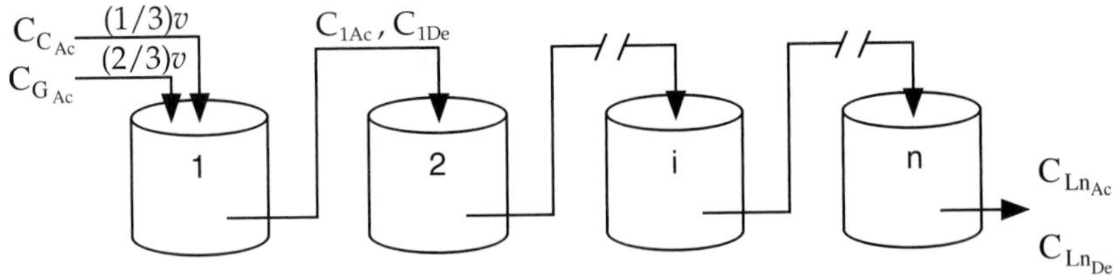

Figure E7-7.2 Liver modeled as a number of CSTRs in series.

Because the first CSTR receives in-flow from the central compartment (1/3 υ) and from the G.I. compartment, it is treated separately. The balance on the first CSTR is

First reactor

Ethanol:
$$\Delta V_L \frac{dC_{L_{Ac}}}{dt} = \upsilon_L \left(\frac{1}{3} C_{C_{Ac}} + \frac{2}{3} C_{G_{Ac}} - C_{L_{Ac}} \right) + r_{L_{Ac}}(C_{L_{Ac}}, C_{L_{De}}) \Delta V_L \tag{E7-7.12}$$

Acetaldehyde:
$$\Delta V_L \frac{dC_{L_{De}}}{dt} = \upsilon_L \left(\frac{1}{3} C_{C_{De}} + \frac{2}{3} C_{G_{De}} - C_{L_{De}} \right) - r_{L_{Ac}}(C_{L_{De}}, C_{L_{Ac}}) \Delta V_L + r_{L_{De}}(C_{L_{De}}) \Delta V_L \tag{E7-7.13}$$

where $C_{L_{Ac}}$ is the concentration of alcohol leaving the first CSTR. A balance on reactor i gives

Later reactors

Ethanol:
$$\Delta V_L \frac{dC_{i_{Ac}}}{dt} = \upsilon_L [C_{(i-1)_{Ac}} - C_{i_{Ac}}] + r_{i_{Ac}}(C_{i_{Ac}}, C_{i_{De}}) \Delta V_L \tag{E7-7.14}$$

Acetaldehyde:

$$\Delta V_{\mathrm{L}} \frac{dC_{i_{De}}}{dt} = v_{\mathrm{L}}[C_{(i-1)_{De}} - C_{i_{De}}] - r_{Ac_i}(C_{Ac_i}, C_{De_i})\Delta V_{\mathrm{L}} + r_{i_{De}}(C_{i_{De}})\Delta V_{\mathrm{L}}$$

(E7-7.15)

The concentrations exiting the last CSTR are $C_{\mathrm{Ln}_{Ac}}$ and $C_{\mathrm{Ln}_{De}}$. Equations (E7-7.1) through (E7-7.15) along with the parameter values are given on the CD-ROM summary notes and the Polymath living example problem. **The Polymath program can be loaded directly from the CD-ROM so that the reader can vary the model parameters.**[†] You can print or view the complete Polymath program and read the complete paper [*Alcohol* **35** (1), p.10, 2005] in the *Summary Notes* on the CD-ROM.

Summary Notes

In the near future, E. R. physicians will go to and interact with the computer to run simulations to help treat patients with drug overdoses and drug interaction.

Living Example Problem

TABLE E7-7.1.   POLYMATH CODE

POLYMATH Results
Example 7-7 Alcohol Metabolism 08-18-2004, Rev5.1.232

Calculated values of the DEQ variables

Variable	initial value	minimal value	maximal value	final value
t	0	0	180	180
Vs	0.15	2.873E-05	0.1877742	2.873E-05
Cc	0	0	11.556606	4.2826187
Cca	0	0	0.0045311	0.0038191
Ct	0	0	10.504887	5.1906634
Cta	0	0	0.0044825	0.004054
CL1	0	0	20.98741	4.2199324
CL2	0	0	20.816457	4.0627945
CL3	0	0	20.645089	3.9062615
CL4	0	0	20.473294	3.7503741
CL5	0	0	20.301064	3.5951778
CL6	0	0	20.128388	3.4407227
CL7	0	0	19.955256	3.2870645
CL8	0	0	19.78166	3.1342649
CL9	0	0	19.60759	2.9823931
CL10	0	0	19.433034	2.8315261
CLa1	0	0	0.0051831	0.0039515
CLa2	0	0	0.0054428	0.0039121
CLa3	0	0	0.0054745	0.0038595
CLa4	0	0	0.0054744	0.0038035
CLa5	0	0	0.0054702	0.0037447
CLa6	0	0	0.0054654	0.0036832
CLa7	0	0	0.0054605	0.0036186
CLa8	0	0	0.0054555	0.0035507
CLa9	0	0	0.0054503	0.0034794
CLa10	0	0	0.0054451	0.0034044
Cg	0	0	27.5776	4.4251559
Cga	0	0	0.0045303	0.0038649
kd	0.013	0	0.013	0
vl	1.35	1.35	1.35	1.35
vt	1.5	1.5	1.5	1.5
Vc	11.56	11.56	11.56	11.56
Vt	25.8	25.8	25.8	25.8
Vg	2.4	2.4	2.4	2.4
Vl	1.1	1.1	1.1	1.1
dVl	0.11	0.11	0.11	0.11
VmAL	2.2	2.2	2.2	2.2
Vrev	29.1	29.1	29.1	29.1
KmAL	0.3898	0.3898	0.3898	0.3898
Krev	1	1	1	1
VmaxAc	2.74	2.74	2.74	2.74
KmAc	0.0015	0.0015	0.0015	0.0015
Cso	2170	2170	2170	2170
a1	1.5	1.5	1.5	1.5
a2	0.06	0.06	0.06	0.06
Vsl	0.15	0.15	0.15	0.15
Ds	325.5	325.5	325.5	325.5
ks	0.0517721	0.0517721	0.0517721	0.0517721

ODE Report (STIFF)

Differential equations as entered by the user
```
[1]  d(Vs)/d(t) = -ks*Vs + kd
[2]  d(Cc)/d(t) = (-vl*(Cc-CL10)-vt*(Cc-Ct))/Vc
[3]  d(Cca)/d(t) = (-vl*(Cca-CLa10)-vt*(Cca-Cta))/Vc
[4]  d(Ct)/d(t) = (vt*(Cc-Ct))/Vt
[5]  d(Cta)/d(t) = (vt*(Cca-Cta))/Vt
[6]  d(CL1)/d(t) = (vl*((1/3)*Cc+(2/3)*Cg-CL1)+(-VmAL*CL1+Vrev*CLa1)/(KmAL+CL1+Krev*CLa1)*dVl)/dVl
[7]  d(CL2)/d(t) = (vl*(CL1-CL2)+(-VmAL*CL2+Vrev*CLa2)/(KmAL+CL2+Krev*CLa2)*dVl)/dVl
[8]  d(CL3)/d(t) = (vl*(CL2-CL3)+(-VmAL*CL3+Vrev*CLa3)/(KmAL+CL3+Krev*CLa3)*dVl)/dVl
[9]  d(CL4)/d(t) = (vl*(CL3-CL4)+(-VmAL*CL4+Vrev*CLa4)/(KmAL+CL4+Krev*CLa4)*dVl)/dVl
[10] d(CL5)/d(t) = (vl*(CL4-CL5)+(-VmAL*CL5+Vrev*CLa5)/(KmAL+CL5+Krev*CLa5)*dVl)/dVl
```

[†] The Polymath program on the CD-ROM was written for Polymath Version 5.1. If you use Version 6.0 or higher, reduce the number of liver compartments from 10 to 9 to avoid exceeding the maximum number of equations allowed in Version 6.0.

Results

Figure E7-7.3 gives the predicted blood ethanol concentration trajectories and experimentally measured trajectories. The different curves are for different initial doses of ethanol. Note that the highest initial dose of ethanol reaches a maximum concentration of 16.5 mM of alcohol and that it takes between 5 and 6 hours to reach a level where it is safe to drive. A comparison of the model and experimental data of Jones et al. for the acetaldehyde concentration is shown in Figure E7-7.4. Because the acetaldehyde concentrations are three orders of magnitude smaller and more difficult to measure, there is a wide range of error bars. The model can predict both the alcohol and acetaldehyde concentration trajectories without adjusting any parameters.

In summary, physiologically based pharamacokinetic models can be used to predict concentration-time trajectories in the TWV of various organs in the body. These models find ever-increasing application of drug delivery to targeted organs and regions. A thorough discussion of the following data and other trends is given in the paper (Ulmulis, Gurmen, Singh, and Fogler cited on page 441) on the CD-ROM.

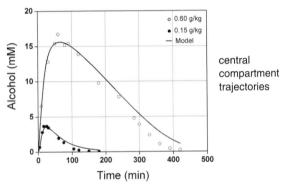

central compartment trajectories

Comparison of model with experimental data

Figure E7-7.3. Blood alcohol–time trajectories from data of Wilkinson et al.[23]

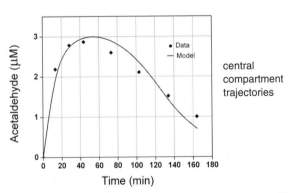

central compartment trajectories

Figure E7-7.4. Blood alcohol–time trajectories from data of Jones et al.[24]

[23]P. K. Wilkinson, et al., "Pharmacokinetics of Ethanol After Oral Administration in the Fasting State," *J. Pharmacoket. Biopharm.*, 5(3):207–24 (1977).

[24]A. W. Jones, J. Neiman, and M. Hillbom, "Concentration-time Profiles of Ethanol and Acetaldehyde in Human Volunteers Treated with the Alcohol-sensitizing Drug, Calcium Carbimide," *Br. J. Clin. Pharmacol.*, 25, 213–21 (1988).

Closure. The theme running through most of this chapter is the pseudo-steady-state hypothesis (PSSH) as it applies to gas-phase reactions and enzymatic reactions. The reader should be able to apply the PSSH to reactions in such problems as P7-7 and P7-12 in order to develop rate laws. Reaction pathways were discussed in order to visualize the various interactions of the reacting species. After completing this chapter the reader should be able to describe and analyze enzymatic reactions and the different types of inhibition as displayed on a Lineweaver–Burk plot. The reader should be able to explain the use of microorganisms to produce chemical products along with the stages of cell growth and how the Monod equation for cell growth is coupled with mass balances on the substrate, cells, and product to obtain the concentration-time trajectories in a batch reactor. The reader should be able to apply the growth laws and balance equations to a chemostat (CSTR) to predict the maximum product rate and the wash-out rate. Finally, the reader should be able to discuss the application the enzyme kinetics to a physiologically based pharmacokinetic (PBPK) model of the human body to describe ethanol metabolism.

SUMMARY

1. In the PSSH, we set the rate of formation of the active intermediates equal to zero. If the active intermediate A^* is involved in m different reactions, we set it to

$$r_{A^*,\text{net}} \equiv \sum_{i=1}^{m} r_{A^*i} = 0 \qquad (S7\text{-}1)$$

This approximation is justified when the active intermediate is highly reactive and present in low concentrations.

2. The azomethane (AZO) decomposition mechanism is

$$2\text{AZO} \underset{k_2}{\overset{k_1}{\rightleftharpoons}} \text{AZO} + \text{AZO}^*$$

$$\text{AZO}^* \xrightarrow{k_3} \text{N}_2 + \text{ethane} \qquad (S7\text{-}2)$$

$$r_{\text{N}_2} = \frac{k(\text{AZO})^2}{1 + k'(\text{AZO})} \qquad (S7\text{-}3)$$

By applying the PSSH to AZO*, we show the rate law, which exhibits first-order dependence with respect to AZO at high AZO concentrations and second-order dependence with respect to AZO at low AZO concentrations.

3. Enzyme Kinetics: Enzymatic reactions follow the sequence

$$\text{E} + \text{S} \underset{k_2}{\overset{k_1}{\rightleftharpoons}} \text{E} \bullet \text{S} \xrightarrow{k_3} \text{E} + \text{P}$$

Using the PSSH for (E • S) and a balance on the total enzyme, E_t, which includes both the bound (E • S) and unbound enzyme (E) concentrations

$$E_t = (E) + (E \cdot S)$$

we arrive at the Michaelis–Menten equation

$$-r_s = \frac{V_{max}(S)}{K_M + (S)} \qquad (S7\text{-}4)$$

where V_{max} is the maximum reaction rate at large substrate concentrations $(S \gg K_M)$ and K_M is the Michaelis constant. K_M is the substrate concentration at which the rate is half the maximum rate $(S_{1/2} = K_M)$.

4. The three different types of inhibition—competitive, uncompetitive, and non-competitive (mixed) inhibition—are shown on the Lineweaver–Burk plot:

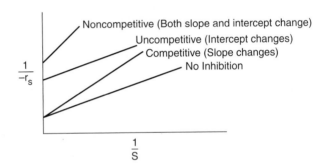

5. Bioreactors:

$$\text{Cells} + \text{Substrate} \longrightarrow \text{More cells} + \text{Product}$$

(a) Phases of bacteria growth:

     I. Lag   II. Exponential   III. Stationary   IV. Death

(b) Monod growth rate law:

$$r_g = \mu_{max} \frac{C_c C_s}{K_s + C_s} \qquad (S7\text{-}5)$$

(c) Stoichiometry:

$$Y_{c/s} = \frac{\text{Mass of new cells formed}}{\text{Substrate consumed}}$$

(d) Unsteady-state mass balance on a chemostat:

$$\frac{dC_c}{dt} = D(C_{c0} - C_c) + r_g - r_d \qquad (S7\text{-}6)$$

$$\frac{dC_s}{dt} = D(C_{s0} - C_s) + r_s \qquad \text{(S7-7)}$$

$$-r_s = Y'_{s/p}\, r_{cg} + Y'_{s/p} r_p + mC_c = Y_{s/c} r_g + mC_c \qquad \text{(S7-8)}$$

6. Physiologically based pharmacokinetic models

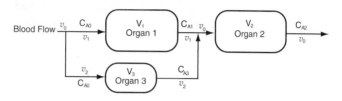

$$V_1 \frac{dC_{A1}}{dt} = v_1(C_{A0} - C_{A1}) + r_{A1} V_1 \qquad \text{(S7-9)}$$

CD-ROM MATERIAL

- **Learning Resources**
    1. *Summary Notes*
    2. *Web Modules*
        A. Ozone Layer   B. Glow Sticks

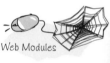

Web Modules

Earth Probe TOMs Total Ozone
September 8, 2000

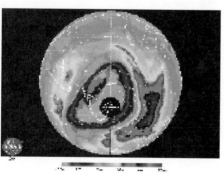

Ozone (Dotson Units)

Photo courtesy of Goddard Space Flight Center (NASA).
See CD-ROM for color pictures of the ozone layer and
the glow sticks.

C.   Russell's Viper

D.   Fer-de-Lance

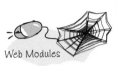

Web Modules

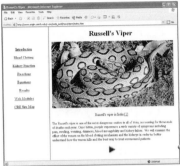

Photo courtesy of Murthy, "Venomous Snakes of Medical Importance in India (Part A)." In *Snakes of Medical Importance (Asia-Pacific Region)* p. 29 (1990). P. Gopalaknishnakone and L. M. Chou, eds.

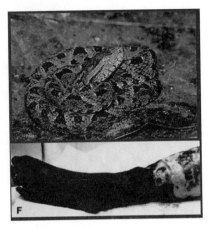

Photo courtesy of
www.jrsphoto.com/reptile gallery.htm
(1st picture).
www.widherps.com/species/B.asper.html
(2nd picture).

3. Interactive Computer Modules
    A.   Enzyme Man

Interactive

Computer Modules

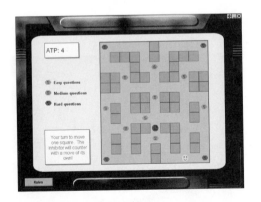

• **Living Example Problems**
    1. Example 7-2 PSSH Applied to Thermal Cracking of Ethane
    2. Example 7-6 Bacteria Growth in a Batch Reactor
    3. Example 7-7 Alcohol Metabolism
    4. Example web module: Ozone
    5. Example web module: Glowsticks
    6. Example web module: Russell's Viper
    7. Example web module: Fer-de-Lance
    8. Example R7.4 Receptor Endocytosis

Living Example Problem

- **Professional Reference Shelf**
  P7.1. *Polymerization*
  A. Step Polymerization
      Mechanism

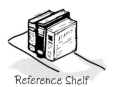

Reference Shelf

$$ARB + ARB \rightarrow AR_2B + AB$$

$$ARB + AR_2B \rightarrow AR_3B + AB$$

$$ARB + AR_3B \rightarrow AR_4B + AB$$

$$AR_2B + AR_2B \rightarrow AR_4B + AB$$

Rate Law

$$r_j = k \sum_{i=1}^{j-1} P_i P_{j-1} - 2kP_j M$$

Concentration

$$P_j = M_o \left(\frac{1}{1 + M_o kt} \right)^2 \left(\frac{M_o kt}{1 + M_o kt} \right)^{j-1}$$

Example R7-1 Determining the Concentration of Polymers for Step
Polymerization

B. Chain Polymerizations
    Free-Radical Polymerization

Initiation $I_2 \rightarrow 2I$

$$I + M \rightarrow R_1$$

Propagation $R_j + M \rightarrow R_{j+1}$

Termination

Addition $R_j + R_k \rightarrow P_{j+k}$

Disproportionation $R_j + R_k \rightarrow P_j + P_k$

$$-r_M = k_p M \sqrt{\frac{2k_0(I_2)f}{k_t}}$$

Mole Fraction of Polymer of Chain Length

$$y_n = (1 - p)p^{n-1}$$

Example R7-2   Parameters of MW Distribution

C. Anionic Polymerization
   Initiation by An Ion

   Initiation $\qquad$ $AB \rightleftharpoons A^- + B^+$

   $\qquad A^- + M \rightarrow R_1$

   Propagation $\qquad R_1 + M \rightarrow R_2$

   $\qquad R_j + M \rightarrow R_{j+1}$

   Transfer to Monomer $\quad R_j + M \rightarrow P_j + R_1$

$$\left.\begin{array}{l} R_1 = I_o e^{-\Theta} \\[2mm] R_j = \dfrac{I_o \Theta^{j-1} e^{-\Theta}}{(j-1)!} \end{array}\right\} \Theta = \int_0^t k_p M \, dt$$

Example 7PRS-3 Calculating the Distribution Parameters from Analytic
Expressions for Anionic Polymerization

Example 7PRS-4 Determination of Dead Polymer Distribution When
Transfer to Monomer Is the Primary Termination Step

R7.2. *Oxygen-Limited Fermentation Scale Up*

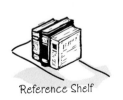

Reference Shelf

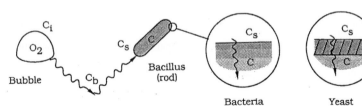

Oxygen Transport to Microorganisms

R7.3. *Receptor Kinetics*
   A. Kinetics of signaling

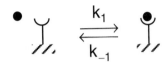

B. Endocytosis

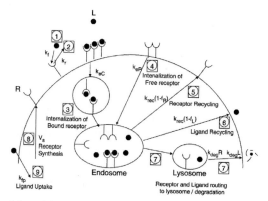

Adapted from D. A. Lauffenburger and J. J. Linerman,
Receptors (New York: Oxford University Press, 1993).

R7.4. *Multiple Enzyme and Substrate Systems*
   A. Enzyme Regeneration
      Example R7.4-1   Construct a Lineweaver–Burk Plot for Different Oxygen
      Concentration
   B. Enzyme Cofactors
      (1) *Example 7.4-2* Derive a Rate Law for Alcohol Dehydrogenase
      (2) *Example 7.4-3* Derive a Rate Law a Multiple Substrate System
      (3) *Example 7.4-4* Calculate the Initial Rate of Formation of Ethanol in
      the Presence of Propanediol
R7.5. *Pharmacokinetics in Drug Delivery*
   Pharmacokinetic models of drug delivery for medication administered either
   orally or intravenously are developed and analyzed.

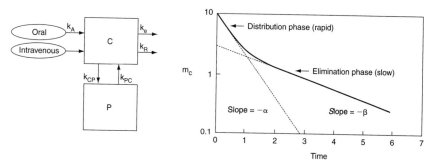

Figure A.   Two-compartment model.   **Figure B.**   Drug response curve.

QUESTIONS AND PROBLEMS

In each of the following questions and problems, rather than just drawing a box around your answer, write a sentence or two describing how you solved the problem, the assumptions you made, the reasonableness of your answer, what you learned, and any other facts that you want to include.

You may wish to refer to W. Strunk and E. B. White, *The Elements of Style, 4th ed.* (New York: Macmillan, 2000) and Joseph M. Williams, *Style: Ten Lessons in Clarity & Grace, 6th ed.* (Glenview, Ill.: Scott, Foresman, 1999) to enhance the quality of your sentences.

See the Preface for additional generic parts (x), (y), (z) to the home problems.

P7-1   (a) **Example 7-1**. How would the results change if the concentration of CS_2 and M were increased?

(b) **Example 7-2**. Over what range of time is the PSSH not valid? Load the *Living Example Problem*. Vary the temperature ($800 < T < 1600$). What temperature gives the greatest disparity with the PSSH results? Specifically compare the PSSH solution with the full numerical solution.

(c) **Example 7-3**. (1) The following additional runs were carried out when an inhibitor was present.

$C_{urea}(kmol/m^3)$	$C_{inhibitor}(kmol/m^3)$	$-r_{urea}(kmol/m^3 \cdot s)$
0.01	0.1	0.125
0.005	0.1	0.066

What type of inhibition is taking place? (2) Sketch the curves for no inhibition, competitive, uncompetitive, noncompetitive (mixed) inhibition, and substrate inhibition on a Woolf–Hanes plot and on an Eadie–Hofstee plot.

(d) **Example 7-4**. (1) What would the conversion be after 10 minutes if the initial concentration of urea were decreased by a factor of 100? (2) What would be the conversion in a CSTR with the same residence time, τ, as the batch reactor? (3) A PFR?

(e) **Example 7-5**. What is the total mass of substrate consumed in grams per mass of cells plus what is consumed to form product? Is there disparity here?

(f) **Example 7-6**. Load the *Living Example Problem*. (1) Plot the concentration up to a time of 24 hours. Did you observe anything unusual? If so, what? (2) Modify the code to carry out the fermentation in a fed-batch (semibatch reactor) in which the substrate is fed at a rate of 0.5 dm³/h and at concentration of 5 g/dm³ to an initial liquid volume of 1.0 dm³ containing a cell mass with an initial concentration of $C_{ci} = 0.2$ mg/dm³ and an initial substrate concentration of $C_{si} = 0.5$ mg/dm³. Plot the concentration of cells, substrate, and product as a function of time along with the mass of product up to 24 hours. Compare your results with (1) above. (3) Repeat (2) when the growth uncompetitively inhibited by the substrate with $K_I = 0.7$ g/dm³. (4) Set $C_P^* = 10,000$ g/dm³, and compare your results with the base case.

(g) **Example 7-7**. This problem is a **gold mine** for things to be learned about the effect of alcohol on the human body. Load the *Polymath Living Example Program* from the CD-ROM. (1) Start by varying the initial doses of alcohol. (2) Next consider individuals who are ALDH enzyme

Explore the blood-
alcohol simulation
on the CD-ROM
Living Example
problem.

Living Example Problem

Interactive

Computer Modules

Enzyme Man

deficient, which includes about 40% to 50% of Asians and Native Americans. Set V_{max} for acetaldehydes between 10% and 50% of its normal value and compare the concentration-time trajectories with the base cases. Hint: Read the journal article in the *Summary Notes* [*Alcohol* **35**, p.1 (2005)]

(h) Load the *Ozone Polymath Living Example Program* from the CD-ROM. Vary the halogen concentrations and describe what you find. Where does PSSH break down? Vary the rate constants and other species concentrations.

(i) Load the *Glowsticks Living Example Problem* from the CD-ROM. Vary the rate constants to learn how you can make the luminescence last longer. Last shorter.

(j) Load the *Russell's Viper Polymath Living Example Program* from the CD-ROM. Describe what would happen if the victim received more than one bite. In the cobra problem in Chapter 6 we saw that after 10 bites, no amount of antivenom would save the victim. What would happen if a victim received 10 bites from a Russell's viper? Replot the concentration-time trajectories for venom, FDP, and other appropriate species. Next, inject different amounts of antivenom to learn if it is possible to negate 10 bites by the viper. What is the number of bites by which no amount of antivenom will save the victim?

(k) Load the *Fer-de-Lance Polymath Living Example Program* from the CD-ROM. Repeat 7-1(j) for the Fer-de-Lance.

(l) Load the *Receptor Endocytosis Living Example Problem* from the CD-ROM. Vary k_{rec}, f_R, and f_L over the ranges in Table R7.31. Describe what you find. When will acute renal failure occur?

(m) List ways you can work this problem incorrectly.

(n) How could you make this problem more difficult?

P7-2 **ICM Enzyme Man.** Load the ICM on your computer and carry out the exercise. Performance number = _____.

(a) List ways you can work this problem incorrectly.

(b) How could you make this problem more difficult?

P7-3$_B$ (*Flame retardants*) Hydrogen radicals are important to sustaining combustion reactions. Consequently, if chemical compounds that can scavange the hydrogen radicals are introduced, the flames can be extinguished. While many reactions occur during the combustion process, we shall choose CO flames as a model system to illustrate the process [S. Senkan et al., *Combustion and Flame*, **69**, 113 (1987)]. In the absence of inhibitors

$$O_2 \longrightarrow O\cdot + O\cdot \tag{P7-3.1}$$

$$H_2O + O\cdot \longrightarrow 2OH\cdot \tag{P7-3.2}$$

$$CO + OH\cdot \longrightarrow CO_2 + H\cdot \tag{P7-3.3}$$

$$H\cdot + O_2 \longrightarrow OH\cdot + O\cdot \tag{P7-3.4}$$

The last two reactions are rapid compared to the first two. When HCl is introduced to the flame, the following additional reactions occur:

$$H\cdot + HCl \longrightarrow H_2 + Cl\cdot$$

$$H\cdot + Cl\cdot \longrightarrow HCl$$

Assume that all reactions are elementary and that the PSSH holds for the $O\cdot$, $OH\cdot$, and $Cl\cdot$ radicals.

(a) Derive a rate law for the consumption of CO when no retardant is present.

(b) Derive an equation for the concentration of $H\cdot$ as a function of time assuming constant concentration of O_2, CO, and H_2O for both uninhibited combustion and combustion with HCl present. Sketch $H\cdot$ versus time for both cases.

(c) Sketch a reaction pathway diagram for this reaction.

(d) List ways you can work this problem incorrectly.

(e) How could you make this problem more difficult?

More elaborate forms of this problem can be found in Chapter 6, where the PSSH is not invoked.

P7-4$_A$ The pyrolysis of acetaldehyde is believed to take place according to the following sequence:

$$CH_3CHO \xrightarrow{k_1} CH_3\cdot + CHO\cdot$$

$$CH_3\cdot + CH_3CHO \xrightarrow{k_2} CH_3\cdot + CO + CH_4$$

$$CHO\cdot + CH_3CHO \xrightarrow{k_3} CH_3\cdot + 2CO + H_2$$

$$2CH_3\cdot \xrightarrow{k_4} C_2H_6$$

(a) Derive the rate expression for the rate of disappearance of acetaldehyde, $-r_{Ac}$.

(b) Under what conditions does it reduce the equation at the top of page 378?

(c) Sketch a reaction pathway diagram for this reaction.

(d) List ways you can work this problem incorrectly.

(e) How could you make this problem more difficult?

P7-5$_B$ **(a)** The gas-phase homogeneous oxidation of nitrogen monoxide (NO) to dioxide (NO_2),

$$2NO + O_2 \xrightarrow{k} 2NO_2$$

is known to have a form of third-order kinetics, which suggests that the reaction is elementary as written, at least for low partial pressures of the nitrogen oxides. However, the rate constant k actually *decreases* with increasing absolute temperature, indicating an apparently *negative* activation energy. Because the activation energy of any elementary reaction must be positive, some explanation is in order.

Provide an explanation, starting from the fact that an active intermediate species, NO_3, is a participant in some other known reactions that involve oxides of nitrogen. Draw the reaction pathway.

(b) The rate law for formation of phosgene, $COCl_2$, from chlorine, Cl_2, and carbon monoxide, CO, has the rate law

$$r_{COCl_2} = kC_{CO}C_{Cl_2}^{3/2}$$

Suggest a mechanism for this reaction that is consistent with this rate law and draw the reaction pathway. [*Hint:* Cl formed from the dissociation of Cl_2 is one of the two active intermediates.]

(c) List ways you can work this problem incorrectly.

(d) How could you make this problem more difficult?

P7-6$_B$ Ozone is a reactive gas that has been associated with respiratory illness and decreased lung function. The following reactions are involved in ozone formation [D. Allen and D. Shonnard, *Green Engineering* (Upper Saddle River, N.J.: Prentice Hall, 2002)].

Green engineering

$$NO_2 + hv \xrightarrow{k_1} NO + O$$

$$O_2 + O + M \xrightarrow{k_2} O_3 + M$$

$$O_3 + NO \xrightarrow{k_3} NO_2 + O_2$$

NO_2 is primarily generated by combustion in the automobile engine.

(a) Show that the steady-state concentration of ozone is directly proportional to NO_2 and inversely proportional to NO.

(b) Drive an equation for the concentration of ozone in solely in terms of the initial concentrations $C_{NO,0}$, $C_{NO_2,0}$, and $O_{3,0}$ and the rate law parameters.

(c) In the absence of NO and NO_2, the rate law for ozone generation is

$$-r_{O_3} = \frac{k(O_2)(M)}{(O_2)(M) + k'(O_3)}$$

Suggest a mechanism.

(d) List ways you can work this problem incorrectly.

(e) How could you make this problem more difficult?

P7-7$_C$ (*Tribology*) One of the major reasons for engine oil degradation is the oxidation of the motor oil. To retard the degradation process, most oils contain an antioxidant [see *Ind. Eng. Chem. 26*, 902 (1987)]. Without an inhibitor to oxidation present, the suggested mechanism at low temperatures is

$$I_2 \xrightarrow{k_0} 2I\cdot$$

Why you need to change the motor oil in your car?

$$I\cdot + RH \xrightarrow{k_i} R\cdot + HI$$

$$R\cdot + O_2 \xrightarrow{k_{p1}} RO_2^{\cdot}$$

$$RO_2^{\cdot} + RH \xrightarrow{k_{p2}} ROOH + R\cdot$$

$$2RO_2^{\cdot} \xrightarrow{k_t} inactive$$

where I_2 is an initiator and RH is the hydrocarbon in the oil.

When an antioxidant is added to retard degradation at low temperatures, the following additional termination steps occur:

$$RO_2\cdot + AH \xrightarrow{k_{A1}} ROOH + A\cdot$$

$$A\cdot + RO_2^{\cdot} \xrightarrow{k_{A2}} inactive$$

Go Blue
Motor Oil

$$\left(\text{for example, AH} = \begin{array}{c} OH \\ \bigcirc \\ CH_3 \end{array} \quad,\quad \text{inactive} = \begin{array}{c} OH \\ \bigcirc \\ CH_2OOR \end{array}\right)$$

(a) Derive a rate law for the degradation of the motor oil in the absence of an antioxidant at low temperatures.

(b) Derive a rate law for the rate of degradation of the motor oil in the presence of an antioxidant for low temperatures.

(c) How would your answer to part (a) change if the radicals $I\cdot$ were produced at a constant rate in the engine and then found their way into the oil?

(d) Sketch a reaction pathway diagram for both high and low temperatures, with and without antioxidant.

(e) See the open-ended problems on the CD-ROM for more on this problem.

(f) List ways you can work this problem incorrectly.

(g) How could you make this problem more difficult?

P7-8$_A$ Consider the application of the PSSH to epidemiology. We shall treat each of the following steps as elementary in that the rate will be proportional to the number of people in a particular state of health. A healthy person, H, can become ill, I, spontaneously, such as by contracting smallpox spores:

$$H \xrightarrow{k_1} I \tag{P7-8.1}$$

or he may become ill through contact with another ill person:

$$I + H \xrightarrow{k_2} 2I \tag{P7-8.2}$$

The ill person may become healthy:

$$I \xrightarrow{k_3} H \tag{P7-8.3}$$

or he may expire:

$$I \xrightarrow{k_4} D \tag{P7-8.4}$$

The reaction given in Equation (P7-8.4) is normally considered completely irreversible, although the reverse reaction has been reported to occur.

(a) Derive an equation for the death rate.

(b) At what concentration of healthy people does the death rate become critical? [*Ans.:* When $[H] = (k_3 + k_4)/k_2$.]

(c) Comment on the validity of the PSSH under the conditions of part (b).

(d) If $k_1 = 10^{-8}\ h^{-1}$, $k_2 = 10^{-16}$ (people·h)$^{-1}$, $k_3 = 5 \times 10^{-10}\ h$, $k_4 = 10^{-11}$ h, and $H_o = 10^9$ people. Use Polymath to plot H, I, and D versus time. Vary k_i and describe what you find. Check with your local *disease control center* or search the WWW to modify the model and/or substitute appropriate values of k_i. Extend the model taking into account what you learn from other sources (e.g., WWW).

(e) List ways you can work this problem incorrectly.

(f) How could you make this problem more difficult?

Chemical Reaction
Engineering in the
Food Industry

P7-9$_C$ (*Postacidification in yogurt*) Yogurt is produced by adding two strains of bacteria (*Lactobacillus bulgaricus* and *Streptococcus thermophilus*) to pasteurized milk. At temperatures of 110°F, the bacteria grow and produce lactic acid. The acid contributes flavor and causes the proteins to coagulate, giving the characteristic properties of yogurt. When sufficient acid has been produced (about 0.90%), the yogurt is cooled and stored until eaten by consumers. A lactic acid level of 1.10% is the limit of acceptability. One limit on the shelf life of yogurt is "postacidification," or continued production of acid by the yogurt cultures during storage. The table that follows shows acid production (% lactic acid) in yogurt versus time at four different temperatures.

Time (days)	35°F	40°F	45°F	50°F
1	1.02	1.02	1.02	1.02
14	1.03	1.05	1.14	1.19
28	1.05	1.06	1.15	1.24
35	1.09	1.10	1.22	1.26
42	1.09	1.12	1.22	1.31
49	1.10	1.12	1.22	1.32
56	1.09	1.13	1.24	1.32
63	1.10	1.14	1.25	1.32
70	1.10	1.16	1.26	1.34

Acid production by yogurt cultures is a complex biochemical process. For the purpose of this problem, assume that acid production follows first-order kinetics with respect to the consumption of lactose in the yogurt to produce lactic acid. At the start of acid production the lactose concentration is about 1.5%, the bacteria concentration is 10^{11} cells/dm^3, and the acid concentration at which all metabolic activity ceases is 1.4% lactic acid.

(a) Determine the activation energy for the reaction.

(b) How long would it take to reach 1.10% acid at 38°F?

(c) If you left yogurt out at room temperature, 77°F, how long would it take to reach 1.10% lactic acid?

(d) Assuming that the lactic acid is produced in the stationary state, do the data fit any of the modules developed in this chapter?

[Problem developed by General Mills, Minneapolis, Minnesota]

(e) List ways you can work this problem incorrectly.

(f) How could you make this problem more difficult?

P7-10$_B$ Derive the rate laws for the following enzymatic reactions and sketch and compare with the plot shown in Figure E7-3.1(a).

(a) $E + S \rightleftarrows E \bullet S \rightleftarrows P + E$	(f) $E + S \rightleftarrows E \bullet S \rightarrow P$
(b) $E + S \rightleftarrows E \bullet S \rightleftarrows E \bullet P \rightarrow P + E$	$E + P \rightleftarrows E \bullet P$
(c) $E + S \rightleftarrows E \bullet S \rightleftarrows E \bullet P \rightleftarrows P + E$	$E \bullet S + P \rightleftarrows E \bullet S \bullet P$
(d) $E + S_1 \rightleftarrows E \bullet S_1$	$E \bullet P + S \rightleftarrows E \bullet S \bullet P$
$\quad E \bullet S_1 + S_2 \rightleftarrows E \bullet S_1 S_2$	(g) $E + S \rightleftarrows E \bullet S \rightarrow P$
$\quad E \bullet S_1 S_2 \rightarrow P + E$	$\quad P + E \rightleftarrows E \bullet P$
(e) $E + S \rightleftarrows (E \bullet S)_1 \rightleftarrows (E \bullet S)_2 + P_1$	
$\quad (E \bullet S)_2 \rightarrow P_2 + E$	

(h) Two products

$$E_0 + S \text{ (Glucose)} \rightleftarrows E_0 \bullet S \rightarrow E_r + P_1 \text{ (s-Lactone)}$$

$$O_2 + E_r \rightarrow E_0 P_1 \rightarrow E_0 + P_2 (H_2 O_2)$$

(i) Cofactor (C) activation

$$E + S \rightleftarrows E \bullet S \rightarrow P$$

$$E + C \rightleftarrows E \bullet C$$

$$E \bullet C + S \rightarrow P + E \bullet C$$

(j) Using the PSSH develop rate laws for each of the six types of enzyme reactions.

(k) Which of the reactions (a) through (j), if any, lend themselves to analysis by a Lineweaver–Burk plot?

P7-11$_B$ Beef catalase has been used to accelerate the decomposition of hydrogen peroxide to yield water and oxygen [*Chem. Eng. Educ.*, 5, 141 (1971)]. The concentration of hydrogen peroxide is given as a function of time for a reaction mixture with a pH of 6.76 maintained at 30°C.

t (min)	0	10	20	50	100
$C_{H_2 O_2}$ (mol/L)	0.02	0.01775	0.0158	0.0106	0.005

(a) Determine the Michaelis–Menten parameters $V_{\max}$ and K_M.

(b) If the total enzyme concentration is tripled, what will the substrate concentration be after 20 minutes?

(c) How could you make this problem more difficult?

(d) List ways you can work this problem incorrectly.

P7-12$_B$ It has been observed that substrate inhibition occurs in the following enzymatic reaction:

$$E + S \rightleftarrows P + E$$

(a) Show that the rate law for substrate inhibition is consistent with the plot in Figure P7-12 of $-r_s$ (mmol/L·min) versus the substrate concentration S (mmol/L).

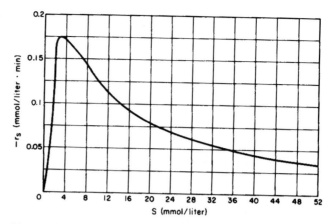

Figure P7-12   Michealis–Menten plot for substrate inhibition.

(b) If this reaction is carried out in a CSTR that has a volume of 1000 dm³, to which the volumetric flow rate is 3.2 dm³/min, determine the three possible steady states, noting, if possible, which are stable. The entrance concentration of the substrate is 50 mmol/dm³. What is the highest conversion?

(c) What would be the effluent substrate concentration if the total enzyme concentration is reduced by 33%?

(d) List ways you can work this problem incorrectly.

(e) How could you make this problem more difficult?

P7-13$_B$ The following data on bakers' yeast in a particular medium at 23.4°C and various oxygen partial pressures were obtained:

P_{O_2}	Q_{O_2} (no sulfanilamide)	Q_{O_2} (20 mg sulfanilamide/mL added to medium)
0.0	0.0	0.0
0.5	23.5	17.4
1.0	33.0	25.6
1.5	37.5	30.8
2.5	42.0	36.4
3.5	43.0	39.6
5.0	43.0	40.0

P_{O_2} = oxygen partial pressure, mmHg; Q_{O_2} = oxygen uptake rate, μL of O_2 per hour per mg of cells.

(a) Calculate the Q_{O_2} maximum (V_{max}), and the Michaelis–Menten constant K_M. (*Ans.*: $V_{max} \doteq 52.63$ μL O_2/h·mg cells.)

(b) Using the Lineweaver–Burk plot, determine the type of inhibition sulfanilamide that causes O_2 to uptake.

(c) List ways you can work this problem incorrectly.

(d) How could you make this problem more difficult?

P7-14$_B$ The enzymatic hydrolysis of starch was carried out with and without maltose and α-dextrin added. [Adapted from S. Aiba, A. E. Humphrey, and N.F. Mills, *Biochemical Engineering* (New York: Academic Press, 1973).

$$\text{Starch} \rightarrow \alpha\text{-dextrin} \rightarrow \text{Limit dextrin} \rightarrow \text{Maltose}$$

No Inhibition

C_S (g/dm^3)	12.5	9.0	4.25	1.0
$-r_S$ (relative)	100	92	70	29

Maltose (I = 12.7 mg/dm^3)

C_S (g/dm^3)	10	5.25	2.0	1.67
$-r_S$ (relative)	77	62	38	34

α-dextrin (I = 3.34 mg/dm^3)

C_S (g/dm^3)	33	10	3.6	1.6
$-r_S$ (relative)	116	85	55	32

Determine the types of inhibition for maltose and for α-dextrin.

P7-15$_B$ The hydrogen ion, H$^+$, binds with the enzyme (E$^-$) to activate it in the form EH. H$^+$ also binds with EH to deactivate it by forming EH$_2^+$

$$H^+ + E^- \rightleftharpoons EH \qquad K_1 = \frac{(EH)}{(H^+)(E^-)}$$

$$H^+ + EH \rightleftharpoons EH_2^+ \qquad K_2 = \frac{(EH_2^+)}{(H^+)(EH)}$$

$$EH + S \overset{K_M}{\rightleftharpoons} EHS \longrightarrow EH + P, \ K_M = \frac{(EHS)}{(EH)(S)}$$

Figure P7-15 Enzyme pH dependence.

where E$^-$ and EH$_2^+$ are inactive. Determine if the preceding sequence can explain the optimum in enzyme activity with pH shown in Figure P7-15.
(a) List ways you can work this problem incorrectly.
(b) How could you make this problem more difficult?

P7-16$_B$ The production of a product P from a particular gram negative bacteria follows the Monod growth law

$$-r_s = \frac{\mu_{max} C_s C_c}{K_M + C_s}$$

with $\mu_{max} = 1$ h^{-1} $K_M = 0.25$ g/dm^3, and $Y_{c/s} = 0.5$ g/g.
(a) The reaction is to be carried out in a batch reactor with the initial cell concentration of $C_{c0} = 0.1$ g/dm^3 and substrate concentration of $C_{s0} = 20$ g/dm^3.

$$C_c = C_{c0} + Y_{c/s}(C_{s0} - C_s)$$

Plot $-r_s$, $-r_c$, C_s, and C_c as a function of time.
(b) Redo part (a) and use a **logistic growth** law.

$$r_g = \mu_{max}\left(1 - \frac{C_c}{C_\infty}\right)C_c$$

and plot C_c and r_c as a function of time. The term C_∞ is the maximum cell mass and is called the carrying capacity and is equal to $C_\infty = 1.0$ g/dm^3. Can you find an analytical solution for the batch reactor? Compare with part (a) for $C_\infty = Y_{c/s} C_{s0} + C_{c0}$.

(c) The reaction is now to be carried out in a CSTR with $C_{s0} = 20$ g/dm³ and $C_{c0} = 0$. What is the dilution rate at which washout occurs?

(d) For the conditions in part (c), what is the dilution rate that will give the maximum product rate (g/h) if $Y_{p/c} = 0.15$ g/g? What are the concentrations C_c, C_s, C_p, and $-r_s$ at this value of D?

(e) How would your answers to (c) and (d) change if cell death could not be neglected with $k_d = 0.02$ h⁻¹?

(f) How would your answers to (c) and (d) change if maintenance could not be neglected with $m = 0.2$ g/h/dm³?

(g) List ways you can work this problem incorrectly.

(h) How could you make this problem more difficult?

P7-17$_B$ Redo Problem P7-16 (a), (c), and (d) using the Tessier equation

$$r_g = \mu_{max}[1 - e^{-C_s/k}]C_c$$

with $\mu_{max} = 1$ and $k = 8$ g/dm³.

(a) List ways you can work this problem incorrectly.

(b) How could you make this problem more difficult?

P7-18$_B$ The bacteria X-II can be described by a simple Monod equation with $\mu_{max} = 0.8$ h⁻¹ and $K_M = 4$, $Y_{p/c} = 0.2$ g/g, and $Y_{s/c} = 2$ g/g. The process is carried out in a CSTR in which the feed rate is 1000 dm³/h at a substrate concentration of 10 g/dm³.

(a) What size fermentor is needed to achieve 90% conversion of the substrate? What is the exiting cell concentration?

(b) How would your answer to (a) change if all the cells were filtered out and returned to the feed stream?

(c) Consider now two 5000 dm³ CSTRs connect in series. What are the exiting concentrations C_s, C_c, and C_p from each of the reactors?

(d) Determine, if possible, the volumetric flow rate at which wash-out occurs and also the flow rate at which the cell production rate ($C_c\ v_0$) in grams per day is a maximum.

(e) Suppose you could use the two 5000-dm³ reactors as batch reactors that take two hours to empty, clean, and fill. What would your production rate be in (grams per day) if your initial cell concentration is 0.5 g/dm³? How many 500-dm³ reactors would you need to match the CSTR production rate?

(f) List ways you can work this problem incorrectly.

(g) How could you make this problem more difficult?

P7-19$_B$ Lactic acid is produced by a *Lactobacillus* species cultured in a CSTR. To increase the cell concentration and production rate, most of the cells in the reactor outlet are recycled to the CSTR, such that the cell concentration in the product stream is 10% of cell concentration in the reactor. Find the optimum dilution rate that will maximize the rate of lactic acid production in the reactor. *How* does this optimum dilution rate change if the exit cell concentration fraction is changed? [$r_p = (\alpha\mu + \beta)C_c$]

$$\mu_{max} = 0.5 \text{ h}^{-1}, \ K_s = 2.0 \text{ g/dm}^3, \alpha = 0.2 \text{ g/g}, \quad C_{SO} = 50 \text{ g/dm}^3$$
$$\beta = 0.1 \text{ g/g} \cdot \text{h}, \ Y_{x/s} = 0.2 \text{ g/g}, \ Y_{p/s} = 0.3 \text{ g/g}$$

(Contributed by Professor D. S. Kompala, University of Colorado)

(a) List ways you can work this problem incorrectly.

(b) How could you make this problem more difficult?

P7-20   (Adapted from Figure 4-20 of Aiba et al.) Mixed cultures of bacteria can develop into predictor prey relationships. Such a system might be cultures of *Alcaligenes faecolis* (prey) and *Colpidium compylum* (predator). Consider two different cells X_1 and X_2 in a chemostat. Cell X_1 only feeds on the substrate and cell X_2 only feeds on cell X_1.

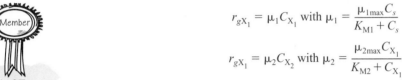

$$X_1 + \text{Substrate} \xrightarrow{\text{Cells } X_1} \text{More } X_1\text{Cells} + \text{Product 1}$$

$$X_2 + X_1 \xrightarrow{\text{Cells } X_2} \text{More } X_2\text{Cells} + \text{Product 2}$$

The growth laws are

Member

Hall of Fame

$$r_{gX_1} = \mu_1 C_{X_1} \text{ with } \mu_1 = \frac{\mu_{1max} C_s}{K_{M1} + C_s}$$

$$r_{gX_1} = \mu_2 C_{X_2} \text{ with } \mu_2 = \frac{\mu_{2max} C_{X_1}}{K_{M2} + C_{X_1}}$$

(a)   For a dilution rate of $D = 0.04$ h^{-1}, an entering substrate concentration of 250 mg/dm^3, initial concentrations of cells of $X_{1i} = 25$ mg/dm^3, $X_{2i} = 7$ mg/dm^3 and of substrate $C_{si} = 10$ mg/dm^3, plot C_{X_1}, C_{X_2}, r_{gX_1}, r_{gX_2}, and C_s as a function of time.

(b)   Vary D between 0.01 h^{-1} and 0.1 h^{-1}, and describe what you find.

(c)   Vary C_{s0} and C_{si} from the base case and describe what you find.

(d)   List ways you can work this problem incorrectly.

(e)   How could you make this problem more difficult?

Additional Information

$\mu_{max1} = 0.5$ h^{-1}, $\mu_{max2} = 0.11$ h^{-1}, $K_{M1} = K_{M2} = 10$ mg/dm^3, $Y_1 = Y_{X_1/s} = 0.14$, and $Y_2 = Y_{X_2/X_1} = 0.5$.

P7-21$_B$ The following data were obtained for *Pyrodictium occultum* at 98°C. Run 1 was carried out in the absence of yeast extract and run 2 with yeast extract. Both runs initially contained Na$_2$S. The vol % of the growth product H$_2$S collected above the broth was reported as a function of time. [*Ann. N.Y. Acad. Sci.*, *506*, 51 (1987).]

Run 1:

Time (h)	0	10	15	20	30	40	50	60	70
Cell Density (cells/mL) $\times 10^{-4}$	2.7	2.8	15	70	400	600	775	600	525
% H$_2$S	0.5	0.8	1.0	1.2	6.8	4.7	7.5	8.0	8.2

Run 2:

Time (h)	0	5	10	15	20	30	40	50	60
Cell Density (cells/mL) $\times 10^{-4}$	2.7	7	11	80	250	350	350	250	—
% H$_2$S	0.1	0.7	0.7	0.8	1.2	4.3	7.5	11.0	12.3

(a) What is the lag time with and without the yeast extract?

(b) What is the difference in the specific growth rates, μ_{max}, of the bacteria with and without the yeast extract?

(c) How long is the stationary phase?

(d) During which phase does the majority production of H_2S occur?

(e) The liquid reactor volume in which these batch experiments were carried out was 0.2 dm³. If this reactor were converted to a continuous-flow reactor, what would be the corresponding wash-out rate?

(f) List ways you can work this problem incorrectly.

(g) How could you make this problem more difficult?

P7-22$_C$ Cell growth with uncompetitive substrate inhibition is taking place in a CSTR. The cell growth rate law for this system is

$$ r_g = \frac{\mu_{max} C_s C_c}{K_s + C_s \left(1 + C_s / K_I\right)} $$

with μ_{max} = 1.5 h⁻¹, K_s = 1 g/dm³, K_I = 50 g/dm³, C_{s0} = 30 g/dm³, $Y_{c/s}$ = 0.08, C_{c0} = 0.5 g/dm³, V = 500 dm³, and D = 0.75 h⁻¹.

(a) Make a plot of the steady-state cell concentration C_c as a function of D. What is the volumetric flow rate (dm³/h) for which the cell production rate is a maximum?

(b) What would be the wash-out rate if C_{c0} = 0? What is the maximum cell production rate and how does it compare with that in part (a)?

(c) Plot C_s as a function of D on the same graph as C_c vs. D? What do you observe?

(d) It is proposed to use the 500-dm³ batch reactor with C_{s0} = 30 g/dm³ and C_{c0} = 0.5 g/dm³. Plot C_c, C_s, r_g, and $-r_s$ as a function of time? Describe what you find.

(e) It is proposed to operate the reactor in the fed-batch mode. A 10-dm³ solution is placed in the 500-dm³ reactor with C_{s0} = 2.0/dm³ and C_{c0} = 0.5 g/dm³. Substrate is fed at a rate of 50 dm³/h and a concentration of 30 g/dm³. Plot C_s, C_c, r_g, r_s, and (VC_c) as a function of time. Can you suggest a better volumetric feed rate? How do your results compare with part (d)?

(f) List ways you can work this problem incorrectly.

(g) How could you make this problem more difficult?

P7-23$_A$ A CSTR is being operated at steady state. The cell growth follows the Monod growth law without inhibition. The exiting substrate and cell concentrations are measured as a function of the volumetric flow rate (represented as the dilution rate), and the results are shown below. Of course, measurements are not taken until steady state is achieved after each change in the flow rate. Neglect substrate consumption for maintenance and the death rate, and assume that $Y_{p/c}$ is zero. For run 4, the entering substrate concentration was 50 g/dm³ and the volumetric flow rate of the substrate was 2 dm³/s.

Run	C_s (g/dm³)	D (day⁻¹)	C_c (g/dm³)
1	1	1	0.9
2	3	1.5	0.7
3	4	1.6	0.6
4	10	1.8	4

(a) Determine the Monod growth parameters μ_{max} and K_s.

(b) Estimate the stoichiometric coefficients, $Y_{c/s}$ and $Y_{s/c}$.

(c) List ways you can work this problem incorrectly.

(d) How could you make this problem more difficult?

P7-24$_C$ In biotechnology industry, *E. coli* is grown aerobically to highest possible concentrations in batch or fed-batch reactors to maximize production of an intracellular protein product. To avoid substrate inhibition, glucose concentration in the initial culture medium is restricted to 100 g/dm^3 in the initial charge of 80-dm^3 culture medium in a 100 dm^3 capacity bioreactor. After much of this glucose is consumed, a concentrated glucose feed (500 g/dm^3) will be fed into the reactor at a constant volumetric feed rate of 1.0 dm^3/h. When the dissolved oxygen concentration in the culture medium falls below a critical value of 0.5 mg/dm^3, acetic acid is produced in a growth-associated mode with a yield coefficient of 0.1 g acetic acid/g cell mass. The product, acetic acid inhibits cell growth linearly, with the toxic concentration (no cell growth) at C_P^* 10 g/dm^3. Find the optimum volumetric flow rate that will maximize the overall rate cell mass production when the bioreactor is filled up and if the feed is turned on after glucose falls below 10 g/dm^3. Inoculum concentration is 1 g cells/dm^3.

Additional parameters:

$$\mu_{max} = 1.2 \text{h}^{-1}, \ K_G = 1.0 \text{ g/dm}^3, \ K_{O_2} = 1 \text{ mg/dm}^3,$$

$$Y_{x/s} = 0.5 \text{ g/g}, \ Y_{p/s} = 0.3 \text{ g/g}, \ q_{O_2/c} = 1.0 \text{ g oxygen/g cell}$$

Oxygen mass transfer rate, $k_L a = 500 \text{ h}^{-1}$; saturation oxygen concentration =

$$C_{O_2}^* = 7.5 \text{ mg/dm}^3 \text{ and } r_g = \left(\frac{\mu_{max} C_s}{K_s + C_s}\right)\left(\frac{C_{O_2}}{K_{O_2} + C_{O_2}}\right)\left(1 - \frac{C_p}{C_p^*}\right)C_c \ ;$$

$$\frac{dC_{O_2}}{dt} = k_L a(C_{O_2}^* - C_{O_2}) - q_{O_2/c} r_g$$

Increase the value of mass transfer rate (up to 1000) or the saturation oxygen concentration (up to 40 mg/dm^3) to see if higher cell densities can be obtained in the fed-batch reactor.

(a) List ways you can work this problem incorrectly.

(b) How could you make this problem more difficult?

(Contributed by Professor D. S. Kompala, University of Colorado.)

P7-25$_C$ (Open-ended problem) You may have to look up/guess/vary some of the constants. If methanol is ingested, it can be metabolized to formaldehyde, which can cause blindness if the formaldehyde reaches a concentration of 0.16 g/dm^3 of fluid in the body. A concentration of 0.75 g/dm^3 will be lethal. After all the methanol has been removed from the stomach, the primary treatment is to inject ethanol intravenously to tie up (competitive inhibition) the enzyme alcohol dehydrogenase (ADH) so that methanol is not converted to formaldehyde and is eliminated from the body through the kidney and bladder (k_7). We will assume as a first approximation that the body is a well-mixed CSTR of 40 dm^3 (total body fluid). In Section 7.5, we applied a more rigorous model.

Web Hint

The following reaction scheme can be applied to the body.

$$\text{Ethanol (E)} \xrightarrow[\text{Dehydrogenase}]{\text{Alcohol}} \text{Acetaldehyde (P}_1\text{) + Water} \qquad (1)$$

$$\text{Methanol (M)} \xrightarrow[\text{Dehydrogenase}]{\text{Alcohol}} \text{Formaldehyde (P}_2\text{) + Water} \qquad (2)$$

$$\left.\begin{array}{l} \text{Ethanol} \\ \text{Methanol} \\ \text{Formaldehyde} \end{array}\right\} \xrightarrow[\text{Excretion}]{k_7} \qquad (2)$$

The complete data set for this reaction is given on the CD-ROM, P7-25 and the open-ended problem H.10. After running the base case, vary the parameters and describe what you find.

P7-26$_B$ **Pharmacokinetics.** The following sets of data give the concentration of different drugs that have been administered either intravenously or orally.

(a) For Giseofulvin and Ampicilin determine the pharmacokinetic parameters for a two-compartment model. Plot the drug concentration in each compartment as a function of time. You may assume that $V_P = V_C$.

(b) Evaluate the model parameters for Aminophylline using a two-compartment model for the intravenous injection and a three-compartment model for oral administration.

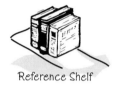

Reference Shelf

(c) If the distribution phase of a drug has a half-life of 1.2 hours and an elimination half-life of 5 hours, plot the concentration in both the peripheral compartment and the central compartment as a function of time

(1) *Giseofulvin* 122-mg intravenous dose (used for fungal infections)

t (h)	0.25	0.5	1	2	3	4	5	10	20	40
C (µg/ml)	2.25	1.85	1.65	1.2	1.05	1.0	1.98	0.8	0.58	0.3

(2) *2-pyridinium adoximine methochloride* 10-mg/kg intravenous dose

t (h)	0.05	0.1	0.2	0.3	0.5	0.7	1	1.5	2.0	3.0
C (µg/ml)	30	25	18	14	8	6	4	2.8	2	1.2

(3) *Ampicilin* 500-mg intravenous dose (used for bacterial infections)

t (h)	0.2	0.5	1	2	4	5	6	7
C (µg/ml)	400	200	100	35	10	5.8	3	1.8

(4) *Diazepam* 105-mg oral dose (sleeping pills)

t (h)	1	2	3	4	5	6	8	12	18	24
C (µg/ml)	2.5	1.9	1.3	1.15	1.0	0.9	0.76	0.65	0.52	0.4

(5) *Aminophylline* (used for bronchial allergies)

t (h)	0.1	1	2	3	4	6	8
Oral C (µmol/dm^3)	0	5	11	14	15	12	6.5
Intravenous C (µmol/dm^3)	26	22	17.5	14	12.5	7	5

JOURNAL CRITIQUE PROBLEMS

P7C-1   Could the mechanism

$$NO + O_2 \rightarrow NO_3$$

$$NO_3 + SO_3 \rightarrow \frac{1}{2}N_2O_2 + SO_2 + \frac{3}{2}O_2$$

$$N_2O_2 + SO_2 \rightarrow N_2O + SO_3$$

$$NO_3 + NO \rightarrow 2NO_2$$

also explain the results in the article *AIChE J., 49*(1), 277 (2003)?

P7C-2   In the article "Behavior Modeling: The Use of Chemical Reaction Kinetics to Investigate Lordosis in Female Rats," *J. Theo. Biol., 174*, 355 (1995), what would be the consequences of product inhibition of the enzyme?

P7C-3   Does the article *Ind. Eng. Chem. Res., 39*, 2837 (2000), provide conclusive evidence that no potassium *o*-phthalate decomposition follows a first-order rate law?

P7C-4   Compare the theoretical curve with actual data points in Figure 5b [*Biotechnol. Bioeng., 24*, 329 (1982)], a normalized residence–time curve. Note that the two curves do not coincide at higher conversions. First, rederive the rate equation and the normalized residence–time equations used by the authors, and then, using the values for kinetic constants and lactase concentration cited by the authors, see if the theoretical curve can be duplicated. Linearize the normalized residence–time equation and replot the data, the theoretical curve in Figure 5b, and a theoretical curve that is obtained by using the constants given in the paper. What is the simplest explanation for the results observed?

P7C-5   In Figure 3 [*Biotechnol. Bioeng., 23*, 361 (1981)], $1/V$ was plotted against $(1/S)(1/PGM)$ at three constant 7-ADCA concentrations, with an attempt to extract V_{max} for the reaction. Does the V_{max} obtained in this way conform to the true value? How is the experimental V_{max} affected by the level of PGM in the medium?

- **Additonal Homework Problems**

- **Complete Data Set**

Hospital E. R.   **P7-25**   Methanol poisoning data and questions for the hospital emergency room (E.R.).

- **New Problems**

    CDP7-A   Anaerobic fermentation of glucose to produce acetic acid [From Prof. D. S. Kim, Chemical Engineering Department, University of Toledo.]

- **Oldies, But Goodies—Problems from Previous Editions**

PSSH

    CDP7-B   Suggest a mechanism for the reaction

$$I^- + OCl^- \longrightarrow OI^- + Cl^-$$

[2nd Ed. P7-8$_B$]

Enzymes

CDP7-C   Determine the diffusion rate in an oxygen fermentor. [2nd Ed. P12-12$_B$].

Bioreactors

CDP7-D$_C$ Plan the scale-up of an oxygen fermentor. [2nd Ed. P12-16$_C$]

CDP7-E$_B$ Assess the effectiveness of bacteria used for denitrification in a batch reactor. [2nd Ed. P12-18$_B$]

CDP7-F$_A$ An understanding of bacteria transport is vital to the efficient operation of water flooding of petroleum reservoirs. [R. Lappan and H. S. Fogler, *SPE Production Eng.*, *7*(2), 167 (1992)]. Analyze the cell concentration time data. [3rd Ed. P7-28$_A$]

CDP7-G$_B$ Design a reactor using *bacillus flavan* to process 10 m^3/day of 2 *M* fumaric acid. [3rd Ed. P7-32$_B$]

CDP7-H$_C$ Find the inconsistencies in the design of the hydrolization of fish oil reactor using lipase. [3rd Ed. P7-10$_C$]

Polymerization

CDP7-I$_B$ Anionic polymerization. Calculate radial concentration as a function of time. [3rd Ed. P7-18]

CDP7-J$_B$ Polymerization. Plot distribution of molecular weight using Flory statistics. [3rd Ed. P7-15]

CDP7-K$_B$ Determine the number average degree of polymerization for free radical polymerization. [3rd Ed. P7-16]

CDP7-L$_B$ Free radical polymerization in a PFR and a CSTR. [3rd Ed. P7-17]

CDP7-M$_B$ Anionic polymerization. Calculate radial concentration as a function of time. [3rd Ed. P7-18]

CDP7-N$_B$ Anionic polymerization carried out in a CSTR. [3rd Ed. P7-19]

CDP7-O$_B$ Anionic polymerization. Comprehensive problem. [3rd Ed. P7-20]

CDP7-P$_B$ Use Flory statistics for molecular weight distribution. [3rd Ed. P7-21]

CDP7-Q$_B$ Anionic polymerization when initiator is slow to dissociate. [3rd Ed. P7-22]

CDP7-R$_B$ Rework CDP7-P$_B$ for a CSTR. [3rd Ed P7-23$_B$]

New Problems on the Web

Green Engineering
Problem

CDP1-New   From time to time new problems relating Chapter 7 material to everyday interests or emerging technologies will be placed on the web. Solutions to these problems can be obtained by e-mailing the author. Also one can go to the web site, *www.rowan.edu/greenengineering*, and work the home problem on green engineering specific to this chapter.

SUPPLEMENTARY READING

Web

Links

Review the following web sites:
   www.cells.com
   www.enzymes.com
   www.pharmacokinetics.com

Text

1. A discussion of complex reactions involving active intermediates is given in

> FROST, A. A., and R. G. PEARSON, *Kinetics and Mechanism*, 2nd ed. New York: Wiley, 1961, Chap. 10.
>
> LAIDLER, K. J., *Chemical Kinetics*, 3rd ed. New York: HarperCollins, 1987.
>
> PILLING, M. J., *Reaction Kinetics*, New York: Oxford University Press, 1995.
>
> SEINFELD, J. H., and S. N. PANDIS, *Atmospheric Chemistry and Physics*, New York: Wiley, 1998.
>
> TEMKIN, O. N., *Chemical Reaction Networks: A Graph-Theoretical Approach*, Boca Raton, Fla.: CRC Press, 1996.

2. Further discussion of enzymatic reactions is presented in

> Just about everything you want to know about enzyme kinetics can be found in SEGEL, I. H. *Enzyme Kinetics*. New York: Wiley-Interscience, 1975.
>
> An excellent description of parameter estimation, biological feedback, and reaction pathways can be found in VOIT, E. O. *Computational Analysis of Biochemical Systems*. Cambridge, UK: Cambridge University Press, 2000.
>
> BLANCH, H. W. and D. S. CLARK, *Biochemical Engineering*. New York: Marcel Dekker, 1996.
>
> CORNISH-BOWDEN, A., *Analysis of Enzyme Kinetic Data*. New York: Oxford University Press, 1995.
>
> LEE, J. M., *Biochemical Engineering*. Upper Saddle River, N.J.: Prentice Hall, 1992.
>
> NELSON, D. L., and M. M. COX, *Lehninger Principles of Biochemistry*, 3rd ed. New York: Worth Publishers, 2000.
>
> SHULER, M. L., and F. KARGI, *Bioprocess Engineering Principles*, 2nd ed. Upper Saddle River, N.J.: Prentice Hall, 2002.
>
> STEPHANOPOULOS, G. N., A. A. ARISTIDOU, and J. NIELSEN, *Metabolic Engineering*. New York: Academic Press, 1998.

3. Material on bioreactors can be found in

> AIBA, S., A. E. HUMPHREY, N. F. MILLIS, *Biochemical Engineering*, 2nd ed. San Diego, Calif.: Academic Press, 1973.
>
> BAILEY, T. J., and D. OLLIS, *Biochemical Engineering*, 2nd ed. New York: McGraw-Hill, 1987.
>
> BLANCH, H. W., and D. S. CLARK, *Biochemical Engineering*. New York: Marcel Dekker, 1996.
>
> CRUEGER, W., and A. CRUEGER, *Biotechnology: A Textbook of Industrial Microbiology*. Madison, Wisc.: Science Tech., 1982.
>
> LEE, J. M., *Biochemical Engineering*. Upper Saddle River, N.J.: Prentice Hall, 1992.
>
> NELSON, D. L., and M. M. COX, *Lehninger Principles of Biochemistry*, 3rd ed. New York: Worth Publishers, 2000.
>
> SCRAGG, A. H., ed., *Biotechnology for Engineers*. New York: Wiley, 1988.
>
> SHULER, M. L., and F. KARGI, *Bioprocess Engineering Principles*, 2nd ed. Upper Saddle River, N.J.: Prentice Hall, 2002.

Steady-State Nonisothermal Reactor Design **8**

If you can't stand the heat, get out of the kitchen.

Harry S Truman

Overview. Because most reactions are not carried out isothermally, we now focus our attention on heat effects in chemical reactors. The basic design equations, rate laws, and stoichiometric relationships derived and used in Chapter 4 for isothermal reactor design are still valid for the design of nonisothermal reactors. The major difference lies in the method of evaluating the design equation when temperature varies along the length of a PFR or when heat is removed from a CSTR. In Section 8.1, we show why we need the energy balance and how it will be used to solve reactor design problems. In Section 8.2, we develop the energy balance to a point where it can be applied to different types of reactors and then give the end result relating temperature and conversion or reaction rate for the main types of reactors we have been studying. Section 8.3 shows how the energy balance is easily applied to design adiabatic reactors, while Section 8.4 develops the energy balance on PFRs/PBRs with heat exchange. In Section 8.5, the chemical equilibrium limitation on conversion is treated along with a strategy for staging reactors to overcome this limitation. Sections 8.6 and 8.7 describe the algorithm for a CSTR with heat effects and CSTRs with multiple steady states, respectively. Section 8.8 describes one of the most important topics of the entire text, multiple reactions with heat effects, which is unique to this textbook. We close the chapter in Section 8.9 by considering both axial and radial concentrations and temperature gradients. The *Professional Reference Shelf* R8.4 on the CD-ROM describes a typical nonisothermal industrial reactor and reaction, the SO_2 oxidation, and gives many practical details.

8.1 Rationale

To identify the additional information necessary to design nonisothermal reactors, we consider the following example, in which a highly exothermic reaction is carried out adiabatically in a plug-flow reactor.

Example 8–1   What Additional Information Is Required?

Calculate the reactor volume necessary for 70% conversion.

$$A \longrightarrow B$$

The reaction is exothermic and the reactor is operated adiabatically. As a result, the temperature will increase with conversion down the length of the reactor.

Solution

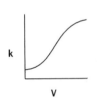

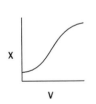

1. Mole Balance (design equation):

$$\frac{dX}{dV} = \frac{-r_A}{F_{A0}} \qquad (E8\text{-}1.1)$$

2. Rate Law: $\qquad -r_A = kC_A \qquad (E8\text{-}1.2)$

Recalling the Arrhenius equation,

$$k = k_1 \exp\left[\frac{E}{R}\left(\frac{1}{T_1} - \frac{1}{T}\right)\right] \qquad (E8\text{-}1.3)$$

we know that k is a function of temperature, T.

3. Stoichiometry (liquid phase): $v = v_0$

$$F_A = C_A v_0$$

$$F_{A0} = C_{A0} v_0$$

$$C_A = C_{A0}(1 - X) \qquad (E8\text{-}1.4)$$

4. Combining:

$$-r_A = k_1 \exp\left[\frac{E}{R}\left(\frac{1}{T_1} - \frac{1}{T}\right)\right] C_{A0}(1 - X) \qquad (E8\text{-}1.5)$$

Combining Equations (E8-1.1), (E8-1.2), and (E8-1.4) and canceling the entering concentration, C_{A0}, yields

$$\frac{dX}{dV} = \frac{k(1 - X)}{v_0} \qquad (E8\text{-}1.6)$$

Because T varies along the length of the reactor, k will also vary, which was not the case for isothermal plug-flow reactors. Combining Equations (E8-1.3) and (E8-1.6) gives us

Why we need the
energy balance

$$\frac{dX}{dV} = k_1 \, \exp\left[\frac{E}{R}\left(\frac{1}{T_1} - \frac{1}{T}\right)\right]\frac{1 - X}{v_0} \qquad (E8\text{-}1.7)$$

We see that we need another relationship relating X and T or T and V to solve this equation. *The energy balance will provide us with this relationship.* So we add another step to our algorithm, this step is the energy balance.

5. Energy Balance:

T_0 = Entering
Temperature

ΔH_{Rx} = Heat of
Reaction

C_{P_A} = Heat Capacity

In this step, we will find the appropriate energy balance to relate temperature and conversion or reaction rate. For example, if the reaction is adiabatic, we will show the temperature-conversion relationship can be written in a form such as

$$T = T_0 + \frac{-\Delta H_{Rx}}{C_{P_A}} X \qquad (E8\text{-}1.8)$$

We now have all the equations we need to solve for the conversion and temperature profiles.

8.2   The Energy Balance

8.2.1   First Law of Thermodynamics

We begin with the application of the first law of thermodynamics first to a closed system and then to an open system. A system is any bounded portion of the universe, moving or stationary, which is chosen for the application of the various thermodynamic equations. For a closed system, in which no mass crosses the system boundaries, the change in total energy of the system, $d\hat{E}$, is equal to the heat flow **to** the system, δQ, minus the work done **by** the system **on** the surroundings, δW. For a *closed system*, the energy balance is

$$d\hat{E} = \delta Q - \delta W \qquad (8\text{-}1)$$

The δ's signify that δQ and δW are not exact differentials of a state function.

The continuous-flow reactors we have been discussing are *open systems* in that mass crosses the system boundary. We shall carry out an energy balance on the open system shown in Figure 8-1. For an open system in which some of the energy exchange is brought about by the flow of mass across the system boundaries, the energy balance for the case of *only one* species entering and leaving becomes

Energy balance on
an open system

$$
\begin{bmatrix}
\text{Rate of} \\
\text{accumulation} \\
\text{of energy} \\
\textit{within} \text{ the} \\
\text{system}
\end{bmatrix}
=
\begin{bmatrix}
\text{Rate of flow} \\
\text{of heat } \textit{to} \\
\text{the system} \\
\textit{from} \text{ the} \\
\text{surroundings}
\end{bmatrix}
-
\begin{bmatrix}
\text{Rate of work} \\
\textit{done by} \\
\text{the system} \\
\textit{on} \text{ the} \\
\text{surroundings}
\end{bmatrix}
+
\begin{bmatrix}
\text{Rate of energy} \\
\text{added to the} \\
\text{system by mass} \\
\text{flow } \textit{into} \text{ the} \\
\text{system}
\end{bmatrix}
-
\begin{bmatrix}
\text{Rate of} \\
\text{energy leaving} \\
\text{system by mass} \\
\text{flow } \textit{out of} \\
\text{the system}
\end{bmatrix}
$$

$$\frac{d\hat{E}_{sys}}{dt} \quad = \quad \dot{Q} \quad - \quad \dot{W} \quad + \quad F_{in}E_{in} \quad - \quad F_{out}E_{out}$$

$$(8\text{-}2)$$

Typical units for each term in Equation (8-2) are (Joule/s).

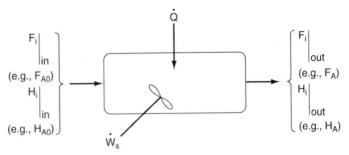

Figure 8-1   Energy balance on a well-mixed open system: schematic.

We will assume the contents of the system volume are well mixed, an assumption that we could relax but that would require a couple pages of text to develop, and the end result would be the same! The unsteady-state energy balance for an open well-mixed system that has n species, each entering and leaving the system at their respective molar flow rates F_i (moles of i per time) and with their respective energy E_i (joules per mole of i), is

The starting point

$$\frac{d\hat{E}_{sys}}{dt} = \dot{Q} - \dot{W} + \sum_{i=1}^{n} E_i F_i \Bigg|_{in} - \sum_{i=1}^{n} E_i F_i \Bigg|_{out} \qquad (8\text{-}3)$$

We will now discuss each of the terms in Equation (8-3).

8.2.2 Evaluating the Work Term

It is customary to separate the work term, $\dot{W}$, into *flow work* and *other work*, $\dot{W}_s$. The term $\dot{W}_s$, often referred to as the *shaft work*, could be produced from such things as a stirrer in a CSTR or a turbine in a PFR. *Flow work* is work that is necessary to get the mass *into* and *out of* the system. For example, when shear stresses are absent, we write

[Rate of flow work]

Flow work and
shaft work

$$\dot{W} = \overbrace{- \sum_{i=1}^{n} F_i P \tilde{V}_i \Bigg|_{in} + \sum_{i=1}^{n} F_i P \tilde{V}_i \Bigg|_{out}} + \dot{W}_s \qquad (8\text{-}4)$$

where P is the pressure (Pa) [1 Pa = 1 Newton/m^2 = 1 kg $\cdot$ m/s^2/m^2] and $\tilde{V}_i$ is the specific molar volume of species i (m^3/mol of i).

Let's look at the units of the flow work term, which is

$$F_i \cdot P \cdot \tilde{V}_i$$

where F_i is in mol/s, P is Pa (1 Pa = 1 Newton/m^2), and $\tilde{V}_i$ is m^3/mol.

$$F_i \cdot P \cdot \tilde{V}_i \; [=] \; \frac{mol}{s} \cdot \frac{Newton}{m^2} \cdot \frac{m^3}{mol} = (Newton \cdot m) \cdot \frac{1}{s} = Joules/s = Watts$$

We see that the units for flow work are consistent with the other terms in Equation (8-2), i.e., J/s.

In most instances, the flow work term is combined with those terms in the energy balance that represent the energy exchange by mass flow across the system boundaries. Substituting Equation (8-4) into (8-3) and grouping terms, we have

$$\frac{d\hat{E}_{sys}}{dt} = \dot{Q} - \dot{W}_s + \sum_{i=1}^{n} F_i(E_i + P\tilde{V}_i)\bigg|_{in} - \sum_{i=1}^{n} F_i(E_i + P\tilde{V}_i)\bigg|_{out} \qquad (8\text{-}5)$$

The energy E_i is the sum of the internal energy (U_i), the kinetic energy $(u_i^2/2)$, the potential energy (gz_i), and any other energies, such as electric or magnetic energy or light:

$$E_i = U_i + \frac{u_i^2}{2} + gz_i + \text{other} \qquad (8\text{-}6)$$

In almost all chemical reactor situations, the kinetic, potential, and "other" energy terms are negligible in comparison with the enthalpy, heat transfer, and work terms, and hence will be omitted; that is,

$$E_i = U_i \qquad (8\text{-}7)$$

We recall that the enthalpy, H_i (J/mol), is defined in terms of the internal energy U_i (J/mol), and the product $P\tilde{V}_i$ (1 Pa·m³/mol = 1 J/mol):

Enthalpy

$$H_i = U_i + P\tilde{V}_i \qquad (8\text{-}8)$$

Typical units of H_i are

$$(H_i) = \frac{J}{\text{mol } i} \text{ or } \frac{\text{Btu}}{\text{lb mol } i} \text{ or } \frac{\text{cal}}{\text{mol } i}$$

Enthalpy carried into (or out of) the system can be expressed as the sum of the net internal energy carried into (or out of) the system by mass flow plus the flow work:

$$F_i H_i = F_i(U_i + P\tilde{V}_i)$$

Combining Equations (8-5), (8-7), and (8-8), we can now write the energy balance in the form

$$\frac{d\hat{E}_{sys}}{dt} = \dot{Q} - \dot{W}_s + \sum_{i=1}^{n} F_i H_i\bigg|_{in} - \sum_{i=1}^{n} F_i H_i\bigg|_{out}$$

The energy of the system at any instant in time, $\hat{E}_{sys}$, is the sum of the products of the number of moles of each species in the system multiplied by their respective energies. This term will be discussed in more detail when unsteady-state reactor operation is considered in Chapter 9.

We shall let the subscript "0" represent the inlet conditions. Unsubscripted variables represent the conditions at the outlet of the chosen system volume.

in

$F_{i0}H_{i0}, \dot{Q}$ $F_iH_i, \dot{W}_s$

out

$$\dot{Q} - \dot{W}_s + \sum_{i=1}^{n} F_{i0}H_{i0} - \sum_{i=1}^{n} F_i H_i = \frac{d\hat{E}_{sys}}{dt} \qquad (8\text{-}9)$$

In Section 8.1, we discussed that in order to solve reaction engineering problems with heat effects, we needed to relate temperature, conversion, and rate of reaction. The energy balance as given in Equation (8-9) is the most convenient starting point as we proceed to develop this relationship.

8.2.3 Overview of Energy Balances

What is the Plan? In the following pages we manipulate Equation (8-9) in order to apply it to each of the reactor types we have been discussing, batch, PFR, PBR, and CSTR. The end result of the application of the energy balance to each type of reactor is shown in Table 8-1. The equations are used in **Step 5** of the algorithm discussed in Example E8-1. The equations in Table 8-1 relate temperature to conversion and molar flow rates and to the system parameters, such as the overall heat-transfer coefficient and area, Ua, and corresponding ambient temperature, T_a, and the heat of reaction, ΔH_{Rx}.

TABLE 8-1.   ENERGY BALANCES OF COMMON REACTORS

1. **Adiabatic** $(\dot{Q} \equiv 0)$ CSTR, PFR, Batch, or PBR. The relationship between conversion, X_{EB}, and temperature for $\dot{W}_s = 0$, constant C_{P_i}, and $\Delta C_P = 0$, is

End results of manipulating the energy balance Sections 8.2.4 to 8.4, 8.6, and 8.8.

$$X_{EB} = \frac{\Sigma \Theta_i C_{P_i}(T - T_0)}{-\Delta H_{Rx}} \qquad (T8\text{-}1.A)$$

$$T = T_0 + \frac{(-\Delta H_{Rx})X}{\Sigma \Theta_i C_{P_i}} \qquad (T8\text{-}1.B)$$

For an exothermic reaction $(-\Delta H_{Rx}) > 0$

X_{EB}

T_0 T

2. **CSTR with heat exchanger**, UA $(T_a - T)$, and large coolant flow rate.

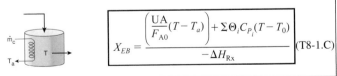

$\dot{m}_c$

T

T_a

$$X_{EB} = \frac{\left(\dfrac{UA}{F_{A0}}(T - T_a)\right) + \Sigma \Theta_i C_{P_i}(T - T_0)}{-\Delta H_{Rx}} \qquad (T8\text{-}1.C)$$

TABLE 8-1.   ENERGY BALANCES OF COMMON REACTORS (CONTINUED)

3. **PFR/PBR with heat exchange**

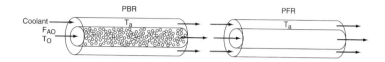

3A. **PBR in terms of conversion**

$$\frac{dT}{dW} = \frac{\dfrac{Ua}{\rho_b}(T_a - T) + r'_A\, \Delta H_{Rx}(T)}{F_{A0}(\Sigma\, \Theta_i C_{P_i} + \Delta C_P X)}$$

(T8-1.D)

3B. **PFR in terms of conversion**

$$\frac{dT}{dV} = \frac{Ua(T_a - T) + r_A \Delta H_{Rx}(T)}{F_{A0}(\Sigma\, \Theta_i C_P + \Delta C_P X)}$$

(T8-1.E)

3C. **PBR in terms of molar flow rates**

$$\frac{dT}{dW} = \frac{\dfrac{Ua}{\rho_b}(T_a - T) + r'_A\, \Delta H_{Rx}(T)}{\Sigma\, F_i C_{P_i}}$$

(T8-1.F)

3D. **PFR in terms of molar flow rates**

$$\frac{dT}{dV} = \frac{Ua(T_a - T) + r_A \Delta H_{Rx}(T)}{\Sigma\, F_i C_{P_i}}$$

(T8-1.G)

4. **Batch**

$$\frac{dT}{dt} = \frac{(-r_A V)(-\Delta H_{Rx}) - UA(T - T_a)}{\Sigma\, N_i C_{P_i}}$$

(T8-1.H)

5. **For Semibatch or unsteady CSTR**

$$\frac{dT}{dt} = \frac{\dot{Q} - \dot{W}_s - \displaystyle\sum_{i=1}^{n} F_{i0}(C_{P_i}(T - T_{i0}) + [-\Delta H_{Rx}(T)](-r_A V))}{\displaystyle\sum_{i=1}^{n} N_i C_{P_i}}$$

(T8-1.I)

6. **For multiple reactions in a PFR**

$$\frac{dT}{dV} = \frac{Ua(T_a - T) + \Sigma\, r_{ij}\Delta H_{Rxij}}{\Sigma\, F_i C_{P_i}}$$

(T8-1.J)

TABLE 8-1.   ENERGY BALANCES OF COMMON REACTORS (CONTINUED)

7. **For a variable coolant temperature, T_a**

$$\boxed{\frac{dT_a}{dV} = \frac{Ua(T - T_a)}{\dot{m}_c C_{P_c}}}$$ (T8-1.K)

These are the equations that we will use to solve reaction engineering problems with heat effects.

– –

[**Nomenclature:** U = overall heat-transfer coefficient, (J/m² • s • K); A = CSTR heat-exchange area, (m²); a = PFR heat-exchange area per volume of reactor, (m²/m³); C_{P_i} = mean heat capacity of species i, (J/mol/K); C_{P_c} = the heat capacity of the coolant, (J/kg/K), $\dot{m}_c$ = coolant flow rate, (kg/s); ΔH_{Rx} = heat of reaction, (J/mol);

$$\Delta H_{Rx}^{\circ} = \left(\frac{d}{a}H_D^{\circ} + \frac{c}{a}H_C^{\circ} - \frac{b}{a}H_B^{\circ} - H_A^{\circ}\right) \text{J/molA}; \; \Delta H_{Rxij} = \text{heat of reaction wrt}$$

species j in reaction i, (J/mol); $\dot{Q}$ = heat added to the reactor, (J/s); and

$$\Delta C_P = \left(\frac{d}{a}C_{P_D} + \frac{c}{a}C_{P_C} - \frac{b}{a}C_{P_B} - C_{P_A}\right) \text{(J/molA • K). All other symbols are as defined in}$$

Chapter 3.]

Examples on How to Use Table 8-1. We now couple the energy balance equations in Table 8-1 with the appropriate reactor mole balance, rate law, stoichiometry algorithm to solve reaction engineering problems with heat effects. For example, recall rate law for a first-order reaction, Equation (E8-1.5) in Example 8-1.

$$-r_A = k_1 \exp\left[\frac{E}{R}\left(\frac{1}{T_1} - \frac{1}{T}\right)\right] C_{A0}(1 - X)$$ (E8-1.5)

If the reaction is carried out adiabatically, then we use Equation (T8-1.B) for the reaction A $\longrightarrow$ B in Example 8-1 to obtain

Adiabatic

$$T = T_0 + \frac{-\Delta H_{Rx} X}{C_{P_A}}$$ (T8-1.B)

Consequently, we can now obtain $-r_A$ as a function of X alone by first choosing X, then calculating T from Equation (T8-1.B), then calculating k from Equation (E8-1.3), and then finally calculating $(-r_A)$ from Equation (E8-1.5).

Choose $X \rightarrow$ calculate $T \rightarrow$ calculate $k \rightarrow$ calculate $-r_A \rightarrow$ calculate $\dfrac{F_{A0}}{-r_A}$

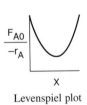

Levenspiel plot

We can use this sequence to prepare a table of $(F_{A0}/-r_A)$ as a function of X. We can then proceed to size PFRs and CSTRs. In the absolute worst case scenario, we could use the techniques in Chapter 2 (e.g., Levenspiel plots or the quadrature formulas in Appendix A).

However, instead of using a Levenspiel plot, we will most likely use Polymath to solve our coupled differential energy and mole balance equations.

If there is cooling along the length of a PFR, we could then apply Equation (T8-1.E) to this reaction to arrive at two coupled differential equations.

Non-adiabatic PFR

$$\frac{dX}{dV} = k_1 \exp\left[\frac{E}{R}\left(\frac{1}{T_1} - \frac{1}{T}\right)\right] C_{A0}(1 - X)$$

$$\frac{dT}{dV} = \frac{Ua(T_a - T) + r_A \Delta H_{Rx}(T)}{F_{A0} C_{P_A}}$$

which are easily solved using an ODE solver such as Polymath.

Similarly, for the case of the reaction A → B carried out in a CSTR, we could use Polymath or MATLAB to solve two nonlinear equations in X and T. These two equations are combined mole balance

Non-adiabatic CSTR

$$V = \frac{F_{A0}X}{k_1 \exp\left[\frac{E}{R}\left(\frac{1}{T_1} - \frac{1}{T}\right)\right] C_{A0}(1 - X)}$$

and the application of Equation (T8-1.C), which is rearranged in the form

$$T = \frac{F_{A0}X(-\Delta H_{Rx}) + UAT_a + F_{A0}C_{P_A}T_0}{UA + C_{P_A}F_{A0}}$$

From these three cases, (1) adiabatic PFR and CSTR, (2) PFR and PBR with heat effects, and (3) CSTR with heat effects, one can see how one couples the energy balances and mole balances. In principle, one could simply use Table 8-1 to apply to different reactors and reaction systems without further discussion. However, understanding the derivation of these equations will greatly facilitate their proper application and evaluation to various reactors and reaction systems. Consequently, the following Sections 8.2, 8.3, 8.4, 8.6, and 8.8 will derive the equations given in Table 8-1.

Why bother? Here is why!! Why bother to derive the equations in Table 8-1? Because I have found that students can *apply* these equations *much* more accurately to solve reaction engineering problems with heat effects if they have gone through the derivation to understand the assumptions and manipulations used in arriving at the equations in Table 8.1.

8.2.4 Dissecting the Steady-State Molar Flow Rates to Obtain the Heat of Reaction

To begin our journey, we start with the energy balance equation (8-9) and then proceed to finally arrive at the equations given in Table 8-1 by first dissecting two terms.

1. The molar flow rates, F_i and F_{i0}
2. The molar enthalpies, H_i, $H_{i0}[H_i \equiv H_i(T)$, and $H_{i0} \equiv H_i(T_0)]$

An animated version of what follows for the derivation of the energy balance can be found in the reaction engineering modules "Heat Effects 1" and "Heat Effects 2" on the CD-ROM. Here equations move around the screen making substitutions and approximations to arrive at the equations shown in Table 8-1. Visual learners find these two ICMs a useful resource.

We will now consider flow systems that are operated at steady state. The steady-state energy balance is obtained by setting $(d\hat{E}_{\text{sys}}/dt)$ equal to zero in Equation (8-9) in order to yield

$$\boxed{\dot{Q} - \dot{W}_s + \sum_{i=1}^{n} F_{i0}H_{i0} - \sum_{i=1}^{n} F_i H_i = 0} \qquad (8\text{-}10)$$

To carry out the manipulations to write Equation (8-10) in terms of the heat of reaction, we shall use the generalized reaction

$$A + \frac{b}{a}B \longrightarrow \frac{c}{a}C + \frac{d}{a}D \qquad (2\text{-}2)$$

The inlet and outlet summation terms in Equation (8-10) are expanded, respectively, to

In: $\quad \Sigma H_{i0}F_{i0} = H_{A0}F_{A0} + H_{B0}F_{B0} + H_{C0}F_{C0} + H_{D0}F_{D0} + H_{I0}F_{I0}$ $\qquad (8\text{-}11)$

and

Out: $\quad \Sigma H_i F_i = H_A F_A + H_B F_B + H_C F_C + H_D F_D + H_I F_I$ $\qquad (8\text{-}12)$

We first express the molar flow rates in terms of conversion.

In general, the molar flow rate of species i for the case of no accumulation and a stoichiometric coefficient v_i is

$$F_i = F_{A0}(\Theta_i + v_i X)$$

Specifically, for Reaction (2-2), $A + \dfrac{b}{a}B \longrightarrow \dfrac{c}{a}C + \dfrac{d}{a}D$, we have

$$F_A = F_{A0}(1 - X)$$

$$\left. \begin{aligned} F_B &= F_{A0}\left(\Theta_B - \frac{b}{a}X\right) \\[6pt] F_C &= F_{A0}\left(\Theta_C + \frac{c}{a}X\right) \\[6pt] F_D &= F_{A0}\left(\Theta_D + \frac{d}{a}X\right) \\[6pt] F_I &= \Theta_I F_{A0} \end{aligned} \right\} \quad \text{where } \Theta_i = \frac{F_{i0}}{F_{A0}}$$

We can substitute these symbols for the molar flow rates into Equations (8-11) and (8-12), then subtract Equation (8-12) from (8-11) to give

$$\sum_{i=1}^{n} H_{i0}F_{i0} - \sum_{i=1}^{n} F_i H_i = F_{A0}[(H_{A0} - H_A) + (H_{B0} - H_B)\Theta_B$$

$$+ (H_{C0} - H_C)\Theta_C + (H_{D0} - H_D)\Theta_D + (H_{I0} - H_I)\Theta_I]$$

$$- \underbrace{\left(\frac{d}{a}H_D + \frac{c}{a}H_C - \frac{b}{a}H_B - H_A\right)}_{\Delta H_{Rx}} F_{A0}X \qquad (8\text{-}13)$$

The term in parentheses that is multiplied by $F_{A0}X$ is called the **heat of reaction** at temperature T and is designated ΔH_{Rx}.

Heat of reaction at
temperature T

$$\boxed{\Delta H_{Rx}(T) = \frac{d}{a}H_D(T) + \frac{c}{a}H_C(T) - \frac{b}{a}H_B(T) - H_A(T)} \qquad (8\text{-}14)$$

All of the enthalpies (e.g., H_A, H_B) are evaluated at the temperature at the outlet of the system volume, and, consequently, $[\Delta H_{Rx}(T)]$ is the heat of reaction at the specific temperature T. The heat of reaction is always given per mole of the species that is the basis of calculation [i.e., species A (joules per mole of A reacted)].

Substituting Equation (8-14) into (8-13) and reverting to summation notation for the species, Equation (8-13) becomes

$$\sum_{i=1}^{n} F_{i0}H_{i0} - \sum_{i=1}^{n} F_i H_i = F_{A0} \sum_{i=1}^{n} \Theta_i (H_{i0} - H_i) - \Delta H_{Rx}(T)F_{A0}X \quad (8\text{-}15)$$

Combining Equations (8-10) and (8-15), we can now write the *steady-state* [i.e., $(d\hat{E}_{sys}/dt = 0)$] energy balance in a more usable form:

One can use this
form of the steady-
state energy
balance if the
enthalpies are
available.

$$\boxed{\dot{Q} - \dot{W}_s + F_{A0} \sum_{i=1}^{n} \Theta_i (H_{i0} - H_i) - \Delta H_{Rx}(T)F_{A0}X = 0} \qquad (8\text{-}16)$$

If a *phase change* takes place during the course of a reaction, this form of the energy balance [i.e., Equation (8-16)] *must* be used (e.g., Problem 8-4$_c$).

8.2.5 Dissecting the Enthalpies

We are neglecting any enthalpy changes resulting from mixing so that the partial molal enthalpies are equal to the molal enthalpies of the pure components. The molal enthalpy of species i at a particular temperature and pressure, H_i, is usually expressed in terms of an *enthalpy of formation* of species i at some reference temperature T_R, $H_i^\circ(T_R)$, plus the change in enthalpy ΔH_{Qi}, that results when the temperature is raised from the reference temperature, T_R, to some temperature T:

$$H_i = H_i^\circ(T_R) + \Delta H_{Qi}$$

For example, if the enthalpy of formation is given at a reference temperature where the species is a solid, then the enthalpy, $H(T)$, of a gas at temperature T is

Calculating the enthalpy when phase changes are involved

$$
\begin{bmatrix} \text{Enthalpy of} \\ \text{species} \\ i \text{ in the gas} \\ \text{at } T \end{bmatrix} = \begin{bmatrix} \text{Enthalpy of} \\ \text{formation} \\ \text{of species} \\ i \text{ in the solid} \\ \text{phase} \\ \text{at } T_R \end{bmatrix} + \begin{bmatrix} \Delta H_Q \text{ in heating} \\ \text{solid from} \\ T_R \text{ to } T_m \end{bmatrix} + \begin{bmatrix} \text{Heat of} \\ \text{melting} \\ \text{at } T_m \end{bmatrix}
$$

$$
+ \begin{bmatrix} \Delta H_Q \text{ in heating} \\ \text{liquid from} \\ T_m \text{ to } T_b \end{bmatrix} + \begin{bmatrix} \text{Heat of} \\ \text{vaporation} \\ \text{at } T_b \end{bmatrix} + \begin{bmatrix} \Delta H_Q \text{ in heating} \\ \text{gas from} \\ T_b \text{ to } T \end{bmatrix}
$$

$$H_i(T) = H_i^\circ(T_R) + \int_{T_R}^{T_m} C_{Ps_i} dT + \Delta H_{mi}(T_m) + \int_{T_m}^{T_b} C_{Pl_i} dT + \Delta H_{vi}(T_b) + \int_{T_b}^{T} C_{Pg_i} dT \quad (8\text{-}17)$$

Here, in addition to the increase in the enthalpies of the solid, liquid, and gas from the temperature increase, one must include the heat of melting at the melting point, $\Delta H_{mi}(T_m)$, and the heat of vaporization at the boiling point, $\Delta H_{vi}(T_b)$. (See Problems P8-4$_C$ and P9-4$_B$.)

The reference temperature at which H_i° is given is usually 25°C. For any substance i that is being heated from T_1 to T_2 in the *absence* of phase change,

No phase change

$$\Delta H_{Qi} = \int_{T_1}^{T_2} C_{P_i} dT \quad (8\text{-}18)$$

Typical units of the heat capacity, C_{P_i}, are

$$(C_{P_i}) = \frac{J}{(\text{mol of } i)(K)} \quad \text{or} \quad \frac{\text{Btu}}{(\text{lb mol of } i)(°R)} \quad \text{or} \quad \frac{\text{cal}}{(\text{mol of } i)(K)}$$

A large number of chemical reactions carried out in industry do not involve phase change. Consequently, we shall further refine our energy balance to apply to *single-phase* chemical reactions. Under these conditions, the enthalpy of species i at temperature T is related to the enthalpy of formation at the reference temperature T_R by

$$H_i = H_i^\circ(T_R) + \int_{T_R}^{T} C_{P_i} dT \quad (8\text{-}19)$$

If phase changes do take place in going from the temperature for which the enthalpy of formation is given and the reaction temperature T, Equation (8-17) must be used instead of Equation (8-19).

The heat capacity at temperature T is frequently expressed as a quadratic function of temperature, that is,

$$C_{P_i} = \alpha_i + \beta_i T + \gamma_i T^2 \tag{8-20}$$

However, while the text will consider only **constant heat capacities**, the PRS R8.3 on the CD-ROM has examples with variable heat capacities.

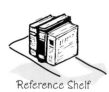

Reference Shelf

To calculate the change in enthalpy $(H_i - H_{i0})$ when the reacting fluid is heated without phase change from its entrance temperature, T_{i0}, to a temperature T, we integrate Equation (8-19) for constant C_{Pi} to write

$$H_i - H_{i0} = \left[H_i^\circ(T_R) + \int_{T_R}^{T} C_{P_i}\, dT \right] - \left[H_i^\circ(T_R) + \int_{T_R}^{T_{i0}} C_{P_i}\, dT \right]$$

$$= \int_{T_{i0}}^{T} C_{P_i}\, dT = C_{P_i}\, [T - T_{i0}] \tag{8-21}$$

Substituting for H_i and H_{i0} in Equation (8-16) yields

Result of dissecting the enthalpies

$$\boxed{\dot{Q} - \dot{W}_s - F_{A0} \sum_{i=1}^{n} \Theta_i C_{P_i}\, [T - T_{i0}] - \Delta H_{Rx}(T) F_{A0} X = 0} \tag{8-22}$$

8.2.6 Relating $\Delta H_{Rx}(T)$, $\Delta H_{Rx}^\circ(T_R)$, and ΔC_p

The heat of reaction at temperature T is given in terms of the enthalpy of each species at temperature T, that is,

$$\Delta H_{Rx}(T) = \frac{d}{a} H_D(T) + \frac{c}{a} H_C(T) - \frac{b}{a} H_B(T) - H_A(T) \tag{8-14}$$

where the enthalpy of each species is given by

$$H_i = H_i^\circ(T_R) + \int_{T_R}^{T} C_{P_i}\, dT = H_i^\circ(T_R) + C_{P_i}(T - T_R) \tag{8-19}$$

If we now substitute for the enthalpy of each species, we have

$A + \frac{b}{a}B \rightarrow \frac{c}{a}C + \frac{d}{a}D$

$$\boxed{\begin{aligned} \Delta H_{Rx}(T) &= \left[\frac{d}{a} H_D^\circ(T_R) + \frac{c}{a} H_C^\circ(T_R) - \frac{b}{a} H_B^\circ(T_R) - H_A^\circ(T_R) \right] \\ &\quad + \left[\frac{d}{a} C_{P_D} + \frac{c}{a} C_{P_C} - \frac{b}{a} C_{P_B} - C_{P_A} \right](T - T_R) \end{aligned}} \tag{8-23}$$

The first set of terms on the right-hand side of Equation (8-23) is the heat of reaction at the reference temperature T_R,

$$\boxed{\Delta H_{Rx}^\circ(T_R) = \frac{d}{a} H_D^\circ(T_R) + \frac{c}{a} H_C^\circ(T_R) - \frac{b}{a} H_B^\circ(T_R) - H_A^\circ(T_R)} \tag{8-24}$$

One can look up the heats of formation at T_R, then calculate the heat of reaction at this reference temperature.

The enthalpies of formation of many compounds, $H_i^\circ(T_R)$, are usually tabulated at 25°C and can readily be found in the *Handbook of Chemistry and Physics*[1] and similar handbooks. For other substances, the heat of combustion (also available in these handbooks) can be used to determine the enthalpy of formation. The method of calculation is described in these handbooks. From these values of the standard heat of formation, $H_i^\circ(T_R)$, we can calculate the heat of reaction at the reference temperature T_R from Equation (8-24).

The second term in brackets on the right-hand side of Equation (8-23) is the overall change in the heat capacity per mole of A reacted, ΔC_P,

$$\Delta C_P = \frac{d}{a}C_{P_D} + \frac{c}{a}C_{P_C} - \frac{b}{a}C_{P_B} - C_{P_A} \qquad (8\text{-}25)$$

Combining Equations (8-25), (8-24), and (8-23) gives us

Heat of reaction at temperature T

$$\Delta H_{Rx}(T) = \Delta H_{Rx}^\circ(T_R) + \Delta C_P(T - T_R) \qquad (8\text{-}26)$$

Equation (8-26) gives the heat of reaction at any temperature T in terms of the heat of reaction at a reference temperature (usually 298 K) and the ΔC_P term. Techniques for determining the heat of reaction at pressures above atmospheric can be found in Chen.[2] For the reaction of hydrogen and nitrogen at 400°C, it was shown that the heat of reaction increased by only 6% as the pressure was raised from 1 atm to 200 atm!

Example 8–2 Heat of Reaction

Calculate the heat of reaction for the synthesis of ammonia from hydrogen and nitrogen at 150°C in kcal/mol of N_2 reacted *and also* in kJ/mol of H_2 reacted.

Solution

$$N_2 + 3H_2 \longrightarrow 2NH_3$$

Calculate the heat of reaction at the reference temperature using the heats of formation of the reacting species obtained from *Perry's Handbook*[3] or the *Handbook of Chemistry and Physics*.

$$\Delta H_{Rx}^\circ(T_R) = 2H_{NH_3}^\circ(T_R) - 3H_{H_2}^\circ(T_R) - H_{N_2}^\circ(T_R) \qquad (E8\text{-}2.1)$$

Note: The heats of formation of the elements (H_2, N_2) are **zero** at 25°C.

---

[1] *CRC Handbook of Chemistry and Physics* (Boca Raton, Fla.: CRC Press, 2003).

[2] N. H. Chen, *Process Reactor Design* (Needham Heights, Mass.: Allyn and Bacon, 1983), p. 26.

[3] R. Perry, D. W. Green, and D. Green, eds., *Perry's Chemical Engineers' Handbook*, 7th ed. (New York: McGraw-Hill, 1999).

$$\Delta H^\circ_{Rx}(T_R) = 2H^\circ_{NH_3}(T_R) - 3(0) - 0 = 2H^\circ_{NH_3}$$

$$= 2(-11{,}020) \frac{cal}{mol \; N_2}$$

$$= -22{,}040 \; cal/mol \; N_2 \; reacted$$

or

$$\Delta H^\circ_{Rx}(298 \; K) = -22.04 \; kcal/mol \; N_2 \; reacted$$

$$= -92.22 \; kJ/mol \; N_2 \; reacted$$

The minus sign indicates the reaction is *exothermic*. If the heat capacities are constant or if the mean heat capacities over the range 25 to 150°C are readily available, the determination of ΔH_{Rx} at 150°C is quite simple.

Exothermic reaction

$$C_{P_{H_2}} = 6.992 \; cal/mol \; H_2 \cdot K$$

$$C_{P_{N_2}} = 6.984 \; cal/mol \; N_2 \cdot K$$

$$C_{P_{NH_3}} = 8.92 \; cal/mol \; NH_3 \cdot K$$

$$\Delta C_P = 2C_{P_{NH_3}} - 3C_{P_{H_2}} - C_{P_{N_2}} \qquad \text{(E8-2.2)}$$

$$= 2(8.92) - 3(6.992) - 6.984$$

$$= -10.12 \; cal/mol \; N_2 \; reacted \cdot K$$

$$\Delta H_{Rx}(T) = \Delta H^\circ_{Rx}(T_R) + \Delta C_P(T - T_R)$$

$$\Delta H_{Rx}(423 \; K) = -22{,}040 + (-10.12)(423 - 298) \qquad \text{(8-26)}$$

$$= -23{,}310 \; cal/mol \; N_2 = -23.31 \; kcal/mol \; N_2$$

$$= -23.3 \; kcal/mol \; N_2 \times 4.184 \; kJ/kcal$$

$$= -97.5 \; kJ/mol \; N_2$$

(Recall: 1 kcal = 4.184 kJ)

The heat of reaction based on the moles of H_2 reacted is

$$\Delta H_{Rx}(423 \; K) = \frac{1 \; mol \; N_2}{3 \; mol \; H_2}\left(-97.53 \; \frac{kJ}{mol \; N_2}\right)$$

$$= -32.51 \; \frac{kJ}{mol \; H_2} \; at \; 423 \; K$$

Now that we see that we can calculate the heat of reaction at any temperature, let's substitute Equation (8-22) in terms of $\Delta H_R(T_R)$ and ΔC_P [i.e., Equation (8-26)]. The steady-state energy balance is now

Energy balance in terms of mean or constant heat capacities

$$\boxed{\dot{Q} - \dot{W}_s - F_{A0}\sum_{i=1}^{n} \Theta_i C_{P_i}(T - T_{i0}) - [\Delta H^\circ_{Rx}(T_R) + \Delta C_P(T - T_R)]F_{A0}X = 0} \qquad \text{(8-27)}$$

From here on, for the sake of brevity we will let

$$\Sigma = \sum_{i=1}^{n}$$

unless otherwise specified.

In most systems, the work term, $\dot{W}_s$, can be neglected (note the exception in the California Registration Exam Problem P8-5$_B$ at the end of this chapter) and the energy balance becomes

$$\boxed{\dot{Q} - F_{A0}\Sigma\Theta_i C_{P_i}(T - T_{i0}) - [\Delta \overset{\circ}{H}_{Rx}(T_R) + \Delta C_P(T - T_R)]F_{A0}X = 0} \quad (8\text{-}28)$$

In almost all of the systems we will study, the reactants will be entering the system at the same temperature; therefore, $T_{i0} = T_0$.

We can use Equation (8-28) to relate temperature and conversion and then proceed to evaluate the algorithm described in Example 8-1. However, unless the reaction is carried out adiabatically, Equation (8-28) is still difficult to evaluate because in nonadiabatic reactors, the heat added to or removed from the system varies along the length of the reactor. This problem does not occur in adiabatic reactors, which are frequently found in industry. Therefore, the adiabatic tubular reactor will be analyzed first.

8.3 Adiabatic Operation

Reactions in industry are frequently carried out adiabatically with heating or cooling provided either upstream or downstream. Consequently, analyzing and sizing adiabatic reactors is an important task.

8.3.1 Adiabatic Energy Balance

In the previous section, we derived Equation (8-28), which relates conversion to temperature and the heat added to the reactor, $\dot{Q}$. Let's stop a minute and consider a system with the special set of conditions of no work, $\dot{W}_s = 0$, adiabatic operation $\dot{Q} = 0$, and then rearrange (8-27) into the form

For adiabatic operation, Example 8.1 can now be solved!

$$\boxed{X = \frac{\Sigma\, \Theta_i C_{P_i}(T - T_{i0})}{-[\Delta H_{Rx}^{\circ}(T_R) + \Delta C_P(T - T_R)]}} \quad (8\text{-}29)$$

In many instances, the $\Delta C_P(T - T_R)$ term in the denominator of Equation (8-29) is negligible with respect to the ΔH_{Rx}° term, so that a plot of X vs. T will usually be linear, as shown in Figure 8-2. To remind us that the conversion in this plot was obtained from the energy balance rather than the mole balance, it is given the subscript EB (i.e., X_{EB}) in Figure 8-2. Equation (8-29) applies to a CSTR, PFR, PBR, and also to a batch (as will be shown in Chapter 9). For $\dot{Q} = 0$ and $\dot{W}_s = 0$, Equation (8-29) gives us the explicit relationship between X and T needed to be used in conjunction with the mole balance to solve reaction engineering problems as discussed in Section 8.1.

Relationship between X and T for *adiabatic* exothermic reactions

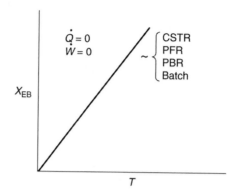

Figure 8-2   Adiabatic temperature–conversion relationship.

8.3.2 Adiabatic Tubular Reactor

We can rearrange Equation (8-29) to solve for temperature as a function of conversion; that is

Energy balance for adiabatic operation of PFR

$$T = \frac{X[-\Delta H_{Rx}(T_R)] + \Sigma \Theta_i C_{P_i} T_0 + X \Delta C_P T_R}{\Sigma \Theta_i C_{P_i} + X \Delta C_P} \qquad (8\text{-}30)$$

This equation will be coupled with the differential mole balance

$$F_{A0} \frac{dX}{dV} = -r_A(X, T)$$

to obtain the temperature, conversion, and concentration profiles along the length of the reactor. One way of analyzing this combination is to use Equation (8-30) to construct a table of T as a function of X. Once we have T as a function of X, we can obtain $k(T)$ as a function of X and hence $-r_A$ as a function of X alone. We could then use the procedures detailed in Chapter 2 to size the different types of reactors; however, software packages such as Polymath and MATLAB can be used to solve the coupled energy balance and mole balance differential equations more easily.

The algorithm for solving PFRs and PBRs operated adiabatically is shown in Table 8-2.

TABLE 8-2A.   ADIABATIC PFR/PBR ALGORITHM

The elementary reversible gas-phase reaction

$$A \underset{\longleftarrow}{\longrightarrow} B$$

is carried out in a PFR in which pressure drop is neglected and pure A enters the reactor.

Mole Balance:
$$\frac{dX}{dV} = \frac{-r_A}{F_{A0}} \tag{T8-2.1}$$

Rate Law:
$$-r_A = k\left(C_A - \frac{C_B}{K_C}\right) \tag{T8-2.2}$$

with
$$k = k_1 \exp\left[\frac{E}{R}\left(\frac{1}{T_1} - \frac{1}{T}\right)\right] \tag{T8-2.3}$$

$$K_C = K(T_2) \exp\left[\frac{\Delta H^{\circ}_{Rx}}{R}\left(\frac{1}{T_2} - \frac{1}{T}\right)\right] \tag{T8-2.4}$$

Stoichiometry: Gas, $\varepsilon = 0$, $P = P_0$

$$C_A = C_{A0}(1 - X)\frac{T_0}{T} \tag{T8-2.5}$$

$$C_B = C_{A0}X\frac{T_0}{T} \tag{T8-2.6}$$

Combine:
$$-r_A = kC_{A0}\left[(1 - X) - \frac{X}{K_C}\right]\frac{T_0}{T} \tag{T8-2.7}$$

Energy Balance:

To relate temperature and conversion, we apply the energy balance to an adiabatic PFR. If all species enter at the same temperature, $T_{i0} = T_0$.

Solving Equation (8-29), with $\dot{Q} = 0$, $\dot{W}_s = 0$, to obtain T as a function of conversion yields

$$T = \frac{X[-\Delta H^{\circ}_{Rx}(T_R)] + \Sigma\,\Theta_i\,C_{P_i}T_0 + X\,\Delta C_P T_R}{\Sigma\,\Theta_i C_{P_i} + X\,\Delta C_P} \tag{T8-2.8}$$

If pure A enters and **iff** $\Delta C_P = 0$, then

$$T = T_0 + \frac{X[-\Delta H^{\circ}_{Rx}(T_R)]}{C_{P_A}} \tag{T8-2.9}$$

Equations (T8-2.1) through (T8-2.9) can easily be solved using either Simpson's rule or an ODE solver.

TABLE 8-2B.   SOLUTION PROCEDURES FOR ADIABATIC PFR/PBR REACTOR

The numerical technique is presented to provide insight to how the variables (k, K_c, etc.) change as we move down the reactor from $V = 0$ and $X = 0$ to V_f and X_f.

A. Numerical Technique

Integrating the PFR mole balance,

$$V = \int_0^{X_3} \frac{F_{A0}}{-r_A} \, dX \qquad (T8-2.10)$$

1. Set $X = 0$.
2. Calculate T using Equation (T8-2.9).
3. Calculate k using Equation (T8-2.3).
4. Calculate K_C using Equation (T8-2.4).
5. Calculate T_0/T (gas phase).
6. Calculate $-r_A$ using Equation (T8-2.7).
7. Calculate $(F_{A0}/-r_A)$.
8. If X is less than the X_3 specified, increment X (i.e., $X_{i+1} = X_i + \Delta X$) and go to Step 2.
9. Prepare table of X vs. $(F_{A0}/-r_A)$.
10. Use numerical integration formulas given in Appendix A, for example,

Use evaluation techniques discussed in Chapter 2.

$$V = \int_0^{X_3} \frac{F_{A0}}{-r_A} \, dX = \frac{3}{8} h \left[\frac{F_{A0}}{-r_A(X=0)} + 3\frac{F_{A0}}{-r_A(X_1)} + 3\frac{F_{A0}}{-r_A(X_2)} + \frac{F_{A0}}{-r_A(X_3)} \right]$$

$$(T8-2.11)$$

with $h = \dfrac{X_3}{3}$

B. Ordinary Differential Equation (ODE) Solver

1. $$\frac{dX}{dV} = \frac{kC_{A0}}{F_{A0}} \left[(1-X) - \frac{X}{K_C} \right] \frac{T_0}{T} \qquad (T8-2.12)$$

Almost always we will use an ODE solver.

2. $$k = k_1(T_1) \exp\left[\frac{E}{R} \left(\frac{1}{T_1} - \frac{1}{T} \right) \right] \qquad (T8-2.13)$$

3. $$K_C = K_{C2}(T_2) \exp\left[\frac{\Delta H_{Rx}}{R} \left(\frac{1}{T_2} - \frac{1}{T} \right) \right] \qquad (T8-2.14)$$

4. $$T = T_0 + \frac{X[-\Delta H_{Rx}(T_R)]}{C_{P_A}} \qquad (T8-2.15)$$

5. Enter parameter values k_1, E, R, K_{C2}, $\Delta H_{Rx}(T_R)$, C_{P_A}, C_{A0}, T_0, T_1, T_2.
6. Enter in intial values $X = 0$, $V = 0$ and final value reactor volume, $V = V_f$.

Living Example Problem

Example 8–3 Liquid-Phase Isomerization of Normal Butane

Normal butane, C_4H_{10}, is to be isomerized to isobutane in a plug-flow reactor. Isobutane is a valuable product that is used in the manufacture of gasoline additives. For example, isobutane can be further reacted to form iso-octane. The 2004 selling price of *n*-butane was 72 cents per gallon, while the price of isobutane was 89 cents per gallon.

The reaction is to be carried out adiabatically in the liquid phase under high pressure using essentially trace amounts of a liquid catalyst which gives a specific reaction rate of 31.1 h^{-1} at 360 K. Calculate the PFR and CSTR volumes necessary to process 100,000 gal/day (163 kmol/h) at 70% conversion of a mixture 90 mol % *n*-butane and 10 mol % *i*-pentane, which is considered an inert. The feed enters at 330 K.

Additional information:

The economic
incentive
$ = 89¢/gal
vs. 72¢/gal

$$\Delta H_{Rx} = -6900 \text{ J/mol} \cdot \text{butane}, \quad \text{Activation energy} = 65.7 \text{ kJ/mol}$$

$$K_C = 3.03 \text{ at } 60°C, \quad C_{A0} = 9.3 \text{ kmol/dm}^3 = 9.3 \text{ kmol/m}^3$$

Butane   *i*-Pentane

$$C_{P_{n-B}} = 141 \text{ J/mol} \cdot \text{K} \qquad\qquad C_{P_{i-P}} = 161 \text{ J/mol} \cdot \text{K}$$

$$C_{P_{i-B}} = 141 \text{ J/mol} \cdot \text{K} = 141 \text{ kJ/kmol} \cdot K$$

Solution

$$n\text{-}C_4H_{10} \rightleftharpoons i\text{-}C_4H_{10}$$

$$A \rightleftharpoons B$$

Mole Balance: $$F_{A0} \frac{dX}{dV} = -r_A \qquad (E8\text{-}3.1)$$

The algorithm

Rate Law: $$-r_A = k\left(C_A - \frac{C_B}{K_C}\right) \qquad (E8\text{-}3.2)$$

with $$k = k(T_1)e^{\left[\frac{E}{R}\left(\frac{1}{T_1} - \frac{1}{T}\right)\right]} \qquad (E8\text{-}3.3)$$

$$K_C = K_C(T_2)e^{\left[\frac{\Delta H_{Rx}}{R}\left(\frac{1}{T_2} - \frac{1}{T}\right)\right]} \qquad (E8\text{-}3.4)$$

Stoichiometry (liquid phase, $v = v_0$):

$$C_A = C_{A0}(1 - X) \qquad (E8\text{-}3.5)$$

$$C_B = C_{A0}X \qquad (E8\text{-}3.6)$$

Combine:

$$-r_A = kC_{A0}\left[1 - \left(1 + \frac{1}{K_C}\right)X\right] \qquad (E8\text{-}3.7)$$

Following the Algorithm

Integrating Equation (E8-3.1) yields

$$V = \int_0^X \frac{F_{A0}}{-r_A} dX \qquad (E8\text{-}3.8)$$

Energy Balance: Recalling Equation (8-27), we have

$$\dot{Q} - \dot{W}_s - F_{A0} \Sigma (\Theta_i C_{P_i}(T - T_0)) - F_{A0}X[\Delta H^\circ_{Rx}(T_R) + \Delta C_P(T - T_R)] = 0 \qquad (8\text{-}27)$$

From the problem statement

$$\text{Adiabatic:} \quad \dot{Q} = 0$$

$$\text{No work:} \quad \dot{W} = 0$$

$$\Delta C_P = C_{P_B} - C_{P_A} = 141 - 141 = 0$$

Applying the preceding conditions to Equation (8-29) and rearranging gives

Nomanclature Note

$\Delta H_{Rx}(T) \equiv \Delta H_{Rx}$

$\Delta H_{Rx}(T_R) \equiv \Delta H^\circ_{Rx}$

$\Delta H_{Rx}=$

$\Delta H^\circ_{Rx} + \Delta C_P(T - T_R)$

$$T = T_0 + \frac{(-\Delta H^\circ_{Rx})X}{\Sigma \Theta_i C_{P_i}} \qquad (E8\text{-}3.9)$$

Parameter Evaluation

$$\Sigma \Theta_i C_{P_i} = C_{P_A} + \Theta_1 C_{P_1} = \left(141 + \frac{0.1}{0.9} 161\right) \text{J/mol} \cdot \text{K}$$

$$= 159 \text{ J/mol} \cdot \text{K}$$

$$T = 330 + \frac{-(-6900)}{159}X$$

$$\boxed{T = 330 + 43.4X} \qquad (E8\text{-}3.10)$$

where T is in degrees Kelvin.

Substituting for the activation energy, T_1, and k_1 in Equation (E8-3.3), we obtain

$$k = 31.1 \exp\left[\frac{65,700}{8.31}\left(\frac{1}{360} - \frac{1}{T}\right)\right](h^{-1})$$

$$\boxed{k = 31.1 \exp\left[7906\left(\frac{T - 360}{360T}\right)\right](h^{-1})} \qquad (E8\text{-}3.11)$$

Substituting for ΔH_{Rx}, T_2, and $K_C(T_2)$ in Equation (E8-3.4) yields

$$K_C = 3.03 \exp\left[\frac{-6900}{8.31}\left(\frac{1}{333} - \frac{1}{T}\right)\right]$$

$$\boxed{K_C = 3.03 \exp\left[-830.3\left(\frac{T - 333}{333T}\right)\right]} \qquad (E8\text{-}3.12)$$

Recalling the rate law gives us

$$-r_A = kC_{A0}\left[1 - \left(1 + \frac{1}{K_C}\right)X\right] \qquad (E8\text{-}3.7)$$

Equilibrium Conversion

At equilibrium

$$-r_A \equiv 0$$

and therefore we can solve Equation (E8-3.7) for the equilibrium conversion

$$X_e = \frac{K_C}{1 + K_C} \tag{E8-3.13}$$

Because we know $K_C(T)$, we can find X_e as a function of temperature.

PFR Solution

It's risky business to ask for 70% conversion in a reversible reaction.

Find the PFR volume necessary to achieve 70% conversion and plot X, X_e, $-r_A$, and T down the length (volume) of the reactor. This problem statement is risky. Why? Because the adiabatic equilibrium conversion may be less than 70%! Fortunately, it's not for the conditions here $0.7 < X_e$. In general, we should ask for the reactor volume to obtain 95% of the equilibrium conversion, $X_f = 0.95 \, X_e$.

    We will solve the preceding set of equations to find the PFR reactor volume using both hand calculations and an ODE computer solution. We carry out the hand calculation to help give an intuitive understanding of how the parameters X_e and $-r_A$ vary with conversation and temperature. The computer solution allows us to readily plot the reaction variables along the length of the reactor and also to study the reaction and reactor by varying the system parameters such as C_{A0} and T_0.

Solution by Hand Calculation to perhaps give greater insight and to build on techniques in Chapter 2.

We will now integrate Equation (E8-3.8) using Simpson's rule after forming a table (E8-3.1) to calculate $(F_{A0}/-r_A)$ as a function of X. This procedure is similar to that described in Chapter 2. We now carry out a sample calculation to show how Table E8-3.1 was constructed. For example, at $X = 0.2$.

(a) $T = 330 + 43.4(0.2) = 338.6$ K

(b) $k = 31.1 \exp\left[7906\left(\dfrac{338.6 - 360}{(360)(338.6)}\right)\right] = 31.1 \exp(-1.388) = 7.76$ h^{-1}

Sample calculation for Table E8-3.1

(c) $K_C = 3.03 \exp\left[-830.3\left(\dfrac{338.6 - 333}{(333)(338.6)}\right)\right] = 3.03 e^{-0.0412} = 2.9$

(d) $X_e = \dfrac{2.9}{1 + 2.9} = 0.74$

(e) $-r_A = \left(\dfrac{7.76}{h}\right)(9.3)\dfrac{mol}{dm^3}\left[1 - \left(1 + \dfrac{1}{2.9}\right)(0.2)\right] = 52.8 \dfrac{mol}{dm^3 \cdot h} = 52.8 \dfrac{kmol}{m^3 \cdot h}$

(f) $\dfrac{F_{A0}}{-r_A} = \dfrac{(0.9 \text{ mol butane/mol total})(163. \text{ kmol total/h})}{52.8 \dfrac{kmol}{m^3 \cdot h}} = 2.78$ m^3

Continuing in this manner for other conversions we can complete Table E8-3.1.

TABLE E8-3.1   HAND CALCULATION

Make a Levenspiel
plot as in Chapter 2.

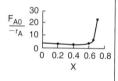

X	T (K)	k (h^{-1})	K_C	X_e	$-r_A$(kmol/m$^3\cdot$h)	$\dfrac{F_{A0}}{-r_A}$ (m^3)
0	330	4.22	3.1	0.76	39.2	3.74
0.2	338.7	7.76	2.9	0.74	52.8	2.78
0.4	347.3	14.02	2.73	0.73	58.6	2.50
0.6	356.0	24.27	2.57	0.72	37.7	3.88
0.65	358.1	27.74	2.54	0.718	24.5	5.99
0.7	360.3	31.67	2.5	0.715	6.2	23.29

In order to construct a Levenspiel plot, the data from Table E8-3.1 ($F_{A0}/-r_A$ vs. X) was used in Example 2-7 to size reactors in series. The reactor volume for 70% will be evaluated using the quadrature formulas. Because ($F_{A0}/-r_A$) increases rapidly as we approach the adiabatic equilibrium conversion, 0.71, we will break the integral into two parts.

$$V = \int_0^{0.7} \frac{F_{A0}}{-r_A}\, dX = \int_0^{0.6} \frac{F_{A0}}{-r_A}\, dX + \int_{0.6}^{0.7} \frac{F_{A0}}{-r_A}\, dX \qquad \text{(E8-3.14)}$$

Using Equations (A-24) and (A-22) in Appendix A, we obtain

Why are we
doing this hand
calculation? If it
isn't helpful, send
me an email and you
won't see this again.

$$V = \frac{3}{8}\times\frac{0.6}{3}[3.74 + 3\times 2.78 + 3\times 2.50 + 3.88]\text{m}^3 \; +\frac{1}{3}\times\frac{0.1}{2}[3.88 + 4\times 5.99 + 23.29]\text{m}^3$$

$$V = 1.75\ m^3 + 0.85\ m^3$$

$$\boxed{V = 2.60\ m^3}$$

You probably will never ever carry out a hand calculation similar to above. So why did we do it? Hopefully, we have given the reader a more intuitive feel of magnitude of each of the terms and how they change as one moves down the reactor (i.e., what the computer solution is doing), as well as to show how the Levenspiel Plots of ($F_{A0}/-r_A$) vs. X in Chapter 2 were constructed. At exit, $V = 2.6$m^3, $X = 0.7$, $X_e = 0.715$, and $T = 360$ K.

Computer Solution

PFR

We could have also solved this problem using Polymath or some other ODE solver. The Polymath program using Equations (E8-3.1), (E8-3.10), (E8-3.7), (E8-3.11), (E8-3.12), and (E8-3.13) is shown in Table E8-3.2.

TABLE E8-3.2.   POLYMATH PROGRAM ADIABATIC ISOMERIZATION

Living Example Problem

Differential equations as entered by the user
[1] d(X)/d(v) = -ra/Fa0

Explicit equations as entered by the user
[1] Ca0 = 9.3
[2] Fa0 = .9*163
[3] T = 330+43.3*X
[4] Kc = 3.03*exp(-830.3*((T-333)/(T*333)))
[5] k = 31.1*exp(7906*(T-360)/(T*360))
[6] Xe = Kc/(1+Kc)
[7] ra = -k*Ca0*(1-(1+1/Kc)*X)
[8] rate = -ra

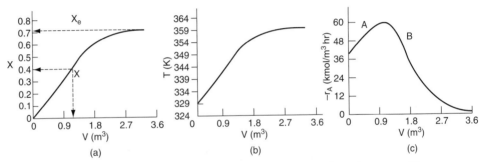

Figure E8-3.1   Conversion, temperature, and reaction rate profiles.

The graphical output is shown in Figure E8-3.1. We see from Figure E8-3.1(a) that 1.15 m³ is required for 40% conversion. The temperature and reaction rate profiles are also shown. One observes that the rate of reaction

Look at the shape of the curves in Figure E8-3.1. Why do they look the way they do?

$$-r_{A} = \underbrace{kC_{A0}}_{A} \underbrace{\left[1 - \left(1 + \frac{1}{K_C}\right)X\right]}_{B} \tag{E8-3.15}$$

goes through a maximum. Near the entrance to the reactor, T increases as does k, causing term A to increase more rapidly than term B decreases, and thus the rate increases. Near the end of the reactor, term B is decreasing more rapidly than term A is increasing. Consequently, because of these two competing effects, we have a maximum in the rate of reaction.

CSTR Solution

Let's calculate the adiabatic CSTR volume necessary to achieve 40% conversion. Do you think the CSTR will be larger or smaller than the PFR? The mole balance is

$$V = \frac{F_{A0}X}{-r_A}$$

Using Equation (E8-3.7) in the mole balance, we obtain

*Is
$V_{PFR} > V_{CSTR}$
or
$V_{PFR} < V_{CSTR}$?*

$$V = \frac{F_{A0}X}{kC_{A0}\left[1 - \left(1 + \frac{1}{K_C}\right)\right]X} \tag{E8-3.16}$$

From the energy balance, we have Equation (E8-3.10):

For 40% conversion
$$T = 330 + 43.4X$$
$$T = 330 + 43.4(0.4) = 347.3$$

Using Equations (E8-3.11) and (E8-3.12) or from Table E8-3.1,

$$k = 14.02 \ \mathrm{h}^{-1}$$
$$K_C = 2.73$$

Then

The adiabatic CSTR
volume is *less*
than the PFR
volume.

$$-r_A = 58.6 \text{ kmol/m}^3 \cdot \text{h}$$

$$V = \frac{(146.7 \text{ kmol butane/h})(0.4)}{58.6 \text{ kmol/m}^3 \cdot \text{h}}$$

$$V = 1.0 \text{ m}^3$$

We see that the CSTR volume (1 m^3) to achieve 40% conversion in this adiabatic reaction is less than the PFR volume (1.15 m^3).

One can readily see why the reactor volume for 40% conversion is smaller for a CSTR than a PFR by recalling the Levenspiel plots from Chapter 2. Plotting ($F_{A0}/-r_A$) as a function of X from the data in Table E8-3.1 is shown here.

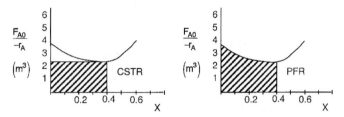

The PFR area (volume) is greater than the CSTR area (volume).

8.4 Steady-State Tubular Reactor with Heat Exchange

In this section, we consider a tubular reactor in which heat is either added or removed through the cylindrical walls of the reactor (Figure 8-3). In modeling the reactor, we shall assume that there are no radial gradients in the reactor and that the heat flux through the wall per unit volume of reactor is as shown in Figure 8-3.

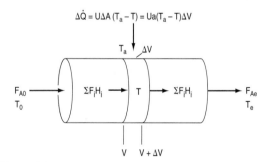

Figure 8-3   Tubular reactor with heat gain or loss.

8.4.1 Deriving the Energy Balance for a PFR

We will carry out an energy balance on the volume ΔV with $\dot{W}_s = 0$, Equation (8-10) becomes

$$\Delta\dot{Q} + \Sigma F_i H_i\big|_v - \Sigma F_i H_i\big|_{v+\Delta v} = 0 \qquad (8\text{-}31)$$

The heat flow to the reactor, $\Delta\dot{Q}$, is given in terms of the overall heat transfer coefficient, U, the heat exchange area, ΔA, and the difference between the ambient temperature T_a and the reactor temperature T.

$$\Delta\dot{Q} = U\Delta A(T_a - T) = Ua\Delta V(T_a - T)$$

where a is the heat exchange area per unit volume of reactor. For the tubular reactor

$$a = \frac{A}{V} = \frac{\pi DL}{\frac{\pi D^2 L}{4}} = \frac{4}{D}$$

where D is the reactor diameter. Substituting for $\Delta\dot{Q}$ in Equation (8-31), dividing Equation (8-31) by ΔV, and taking the limit as $\Delta V \to 0$, we get

$$Ua(T_a - T) - \frac{d\Sigma(F_i H_i)}{dV} = 0$$

Expanding

$$Ua(T_a - T) - \Sigma\frac{dF_i}{dV}H_i - \Sigma F_i\frac{dH_i}{dV} = 0 \qquad (8\text{-}32)$$

From a mole balance on species i, we have

$$\frac{dF_i}{dV} = r_i = v_i(-r_A) \qquad (8\text{-}33)$$

Differentiating the enthalpy Equation (8-19) with respect to V

$$\frac{dH_i}{dV} = C_{P_i}\frac{dT}{dV} \qquad (8\text{-}34)$$

Substituting Equations (8-33) and (8-34) into Equation (8-32), we obtain

$$Ua(T_a - T) - \underbrace{\Sigma v_i H_i}_{\Delta H_{\text{Rx}}}(-r_A) - \Sigma F_i C_{P_i}\frac{dT}{dV} = 0$$

Rearranging, we arrive at

This form of the energy balance will also be applied to multiple reactions.

$$\boxed{\frac{dT}{dV} = \frac{\overbrace{r_A\Delta H_{\text{Rx}}}^{\substack{\text{Heat} \\ \text{"Generated"}}} - \overbrace{Ua(T - T_a)}^{\substack{\text{Heat} \\ \text{Removed}}}}{\Sigma F_i C_{P_i}}} \qquad (8\text{-}35)$$

which is Equation (T8-1G) in Table 8-1. This equation is coupled with the mole balances on each species [Equation (8-33)]. Next we express r_A as a function of either the concentrations for liquid systems or molar flow rates for gas systems as described in Section 4.7.

We will use this form of the energy balance for membrane reactors and also extend this form to multiple reactions.

We could also write Equation (8-35) in terms of conversion by recalling $F_i = F_{A0}(\Theta_i + v_i X)$ and substituting this expression into the denominator of Equation (8-35).

PFR energy balance

$$\frac{dT}{dV} = \frac{Ua(T_a - T) + (r_A)(\Delta H_{Rx})}{F_{A0}(\Sigma \Theta_i C_{P_i} + \Delta C_P X)}$$

(8-36)

For a packed-bed reactor $dW = \rho_b\, dV$ where ρ_b is the bulk density,

PBR energy balance

$$\frac{dT}{dW} = \frac{\dfrac{Ua}{\rho_b}(T_a - T) + (r'_A)(\Delta H_{Rx})}{\Sigma F_i C_{P_i}}$$

(8-37)

Equations (8-36) and (8-37) are also given in Table 8-1 as Equations (T8-1E) and (T8-1F). As noted earlier, having gone through the derivation to these equations it will be easier to apply them accurately to CRE problems with heat effects. The differential equation describing the change of temperature with volume (i.e., distance) down the reactor,

Energy balance

Numerical integration of two coupled differential equations is required.

$$\frac{dT}{dV} = g(X, T)$$

(A)

must be coupled with the mole balance,

Mole balance

$$\frac{dX}{dV} = \frac{-r_A}{F_{A0}} = f(X, T)$$

(B)

and solved simultaneously. If the coolant temperature varies down the reactor we must add the coolant balance, which is

$$\frac{dT_a}{dV} = \frac{U_a(T - T_a)}{\dot{m}_c C_{P_c}}$$

(C)

A variety of numerical schemes can be used to solve the coupled differential equations, (A), (B), and (C).

Example 8–4 Butane Isomerization Continued—OOPS!

When we checked the vapor pressure at the exit to the adiabatic reactor in Example 8-3 where the temperature is 360 K, we found the vapor pressure to be about 1.5 MPa for isobutene, which is greater than the rupture pressure of the glass vessel being used. Fortunately, there is a bank of ten partially insulated (Ua = 5000 kJ/h · m³ · K) tubular reactors each 6 m³ over in the storage shed available for use. We are also going to lower the entering temperature to 310 K. The reactors are cooled by natural convection where average ambient temperature in this tropical location is assumed to be 37°C. The temperature in any of the reactors cannot rise above 325 K. Plot X, X_e, T, and the reaction rate along the length of the reactor. Does the temperature rise above 325 K?

Solution

For ten reactors in parallel

$$F_{A0} = (0.9)(163 \text{ kmol/h}) \times \frac{1}{10} = 14.7 \ \frac{\text{kmol A}}{\text{h}}$$

The mole balance, rate law, and stoichiometry are the same as in the adiabatic case previously discussed in Example 8-3; that is,

Same as Example 8-3

Mole Balance:
$$\boxed{\frac{dX}{dV} = \frac{-r_A}{F_{A0}}}$$
(E8-3.1)

Rate Law and Stoichiometry:
$$\boxed{r_A = -kC_{A0}\left[1 - \left(1 + \frac{1}{K_C}\right)X\right]}$$
(E8-3.7)

with

$$k = 31.1 \exp\left[7906\left(\frac{T - 360}{360T}\right)\right] \text{ h}^{-1}$$
(E8-3.11)

$$K_C = 3.03 \exp\left[-830.3\left(\frac{T - 333}{333T}\right)\right]$$
(E8-3.12)

At equilibrium

$$X_e = \frac{K_C}{1 + K_C}$$
(E8-3.13)

$Q_g = r_A\Delta H_{Rx}$
$Q_r = Ua(T - T_a)$
$\dfrac{dT}{dV} = \dfrac{Q_g - Q_r}{F_{A0}C_{PA}}$

Recalling $\Delta C_P = 0$, Equation (8-36) for the partially insulated reactor can be written as

$$\begin{array}{c} \text{Heat} \qquad \text{Heat} \\ \text{``Generated''} \quad \text{Removed} \\ \frac{dT}{dV} = \frac{\overbrace{r_A\Delta H_{Rx}} - \overbrace{Ua(T - T_a)}}{F_{A0}C_{P_A}} \end{array}$$
(E8-4.1)

where and $C_{P_0} = \Sigma\Theta_i C_{P_i} = 159 \cdot \text{kJ/kmol} \cdot \text{K}$, Ua = 5000 kJ/m³ · h · K, T_a = 310 K, and ΔH_{Rx} = –6900 kJ/kmol. These equations are now solved using Polymath. The Polymath program and profiles of X, X_e, T, and $-r_A$ are shown here.

Living Example Problem

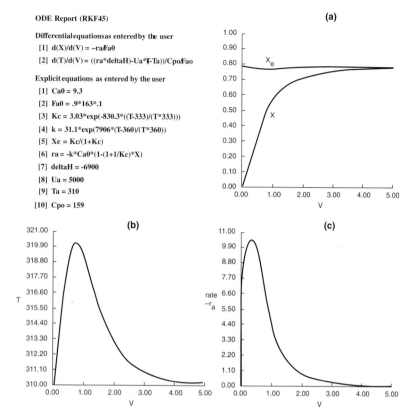

ODE Report (RKF45)   **(a)**

Differential equations as entered by the user
[1] d(X)/d(V) = −ra/Fa0
[2] d(T)/d(V) = ((ra*deltaH)-Ua*(T-Ta))/Cpo/Fao

Explicit equations as entered by the user
[1] Ca0 = 9.3
[2] Fa0 = .9*163*.1
[3] Kc = 3.03*exp(-830.3*((T-333)/(T*333)))
[4] k = 31.1*exp(7906*(T-360)/(T*360))
[5] Xe = Kc/(1+Kc)
[6] ra = -k*Ca0*(1-(1+1/Kc)*X)
[7] deltaH = -6900
[8] Ua = 5000
[9] Ta = 310
[10] Cpo = 159

Figure E8-4.1   (a) Conversion profiles, (b) temperature profile, and (c) reaction rate profile.

We see that the temperature did not rise above 325 K.

8.4.2 Balance on the Coolant Heat Transfer Fluid

The heat transfer fluid will be a coolant for exothermic reactions and a heating medium for endothermic reactions. If the flow rate of the heat transfer fluid is sufficiently high with respect to the heat released (or adsorbed) by the reacting mixture, then the heat transfer fluid temperature will be constant along the reactor.

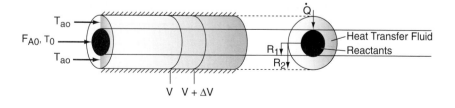

In the material that follows we develop the basic equations for a coolant to remove heat from exothermic reactions, however these same equations apply to endothermic reactions where a heating medium is used to supply heat.

By convention $\dot{Q}$ is the heat **added to** the system. We now carry the balance on the coolant in the annulus between R_1 and R_2 and between V and $V + \Delta V$. The mass flow rate of coolant is $\dot{m}_c$. We will consider the case when the outer radius of the coolant channel R_2 is insulated.

Case A Co-Current Flow

The reactant and the coolant flow in the same direction

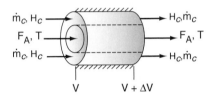

The energy balance on the coolant in the volume between V and $(V + \Delta V)$ is

$$
\begin{bmatrix} \text{Rate of energy} \\ \text{in at } V \end{bmatrix} - \begin{bmatrix} \text{Rate of energy} \\ \text{out at } V + \Delta V \end{bmatrix} + \begin{bmatrix} \text{Rate of heat added} \\ \text{by conduction through} \\ \text{the inner wall} \end{bmatrix} = 0
$$

$$
\dot{m}_c H_c|_V \quad - \quad \dot{m}_c H_c|_{V+\Delta V} \quad + \quad Ua(T - T_a)\Delta V \quad = 0
$$

where T_a is the coolant temperature, and T is the temperature of the reacting mixture in the inner tube.

Dividing by ΔV and taking limit as $\Delta V \to 0$

$$
- \dot{m}_c \frac{dH_c}{dV} + Ua(T - T_a) = 0 \tag{8-38}
$$

The change in enthalpy of the coolant can be written as

$$
\frac{dH_c}{dV} = C_{P_c} \frac{dT_a}{dV} \tag{8-39}
$$

the variation of coolant temperature T_a down the length of reactor is

$$
\boxed{\frac{dT_a}{dV} = \frac{Ua(T - T_a)}{\dot{m}_c C_{P_c}}} \tag{8-40}
$$

Typical heat transfer fluid temperature profiles are shown here for both exothermic and endothermic reactions

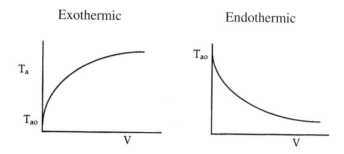

Exothermic   Endothermic

Heat Transfer Fluid Temperature Profiles

Case B Counter Current Flow

Here the reacting mixture and coolant flow in opposite directions for counter current flow of coolant and reactants. At the reactor entrance, $V = 0$, the reactants enter at temperature T_0, and the coolant exits at temperature T_{a2}. At the end of the reactor, the reactants and products exit at temperature T while the coolant enters at T_{a0}.

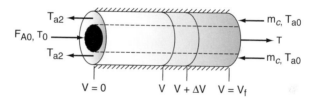

Again we balance over a differential reactor volume to arrive at reactor volume.

$$\boxed{\frac{dT_a}{dV} = \frac{Ua(T_a - T)}{\dot{m}_c C_{P_c}}} \tag{8-41}$$

At the entrance $V = 0$ $\therefore X = 0$ and $T_a = T_{a2}$.
At exit $V = V_f$ $\therefore T_a = T_{a0}$.

We note that the only difference between Equations (8-40) and (8-41) is a minus sign $\boxed{[(T_a - T) = -[T_a - T]]}$.

Solution to this counter current flow problem to find the exit conversion and temperature requires a trial-and-error procedure.

1. Consider an exothermic reaction where the coolant stream enters at the end of the reactor ($V = V_f$) at a temperature T_{a0}, say 300 K. We have to carry out a trial-and-error procedure to find the temperature of coolant exiting the reactor.

2. Assume a coolant temperature at the feed entrance ($X = 0$, $V = 0$) to the reactor to be $T_{a2} = 340$ K as shown in (a).

3. Use an ODE solver to calculate X, T, and T_a as a function of V.

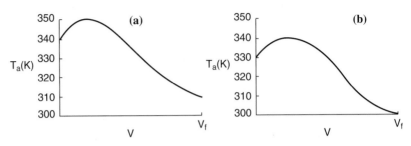

We see from Figure **(a)** that our guess of 340 K for T_{a2} at the feed entrance ($V = 0$ and $X = 0$) gives an entering temperature of the coolant of 310 K ($V = V_f$), which does not match the actual entering coolant temperature of 300 K.

4. Now guess a coolant temperature at $V = 0$ and $X = 0$ of 330 K. We see from Figure **(b)** that an exit coolant temperature of $T_{a2} = 330$ K will give a coolant temperature at V_f of 300 K, which matches the acutal T_{a0}.

We now have all the tools to solve reaction engineering problems involving heat effects in PFR for the cases of both constant and variable coolant temperatures.

Table 8-3 gives the algorithm for the design of PFRs and PBRs with heat exchange for case A: conversion as the reaction variable and case B: molar flow rates as the reaction variable. The procedure in case B must be used when multiple reactions are present.

TABLE 8-3.   PFR/PBR ALGORITHM FOR HEAT EFFECTS

Living Example Problem

A. Conversion as the reaction variable

$$A + B \rightleftharpoons 2C$$

1. Mole Balance:

$$\frac{dX}{dV} = \frac{-r_A}{F_{A0}} \tag{T8-3.1}$$

2. Rate Law:

$$-r_A = k_1\left(C_A C_B - \frac{C_C^2}{K_C}\right) \tag{T8-3.2}$$

$$k = k_1(T_1)\exp\left[\frac{E}{R}\left(\frac{1}{T_1} - \frac{1}{T}\right)\right] \tag{T8-3.3}$$

for $\Delta C_p \cong 0$.

$$K_C = K_{C2}(T_2)\exp\left[\frac{\Delta H_{Rx}^\circ}{R}\left(\frac{1}{T_2} - \frac{1}{T}\right)\right] \tag{T8-3.4}$$

3. Stoichiometry (gas phase, no ΔP):

$$C_A = C_{A0}(1 - X)\frac{T_0}{T} \tag{T8-3.5}$$

$$C_B = C_{A0}(\Theta_B - X)\frac{T_0}{T} \tag{T8-3.6}$$

$$C_C = 2C_{A0}X\frac{T_0}{T} \tag{T8-3.7}$$

TABLE 8-3.   PFR/PBR ALGORITHM FOR HEAT EFFECTS (CONTINUED)

4. Energy Balances:

Reactants: $\dfrac{dT}{dV} = \dfrac{Ua(T_a - T) + (-r_A)(-\Delta H_{Rx})}{F_{A0}[C_{P_A} + \Theta_B C_{P_B} + X\,\Delta C_P]}$ (T8-3.8)

Coolant: $\dfrac{dT_a}{dV} = \dfrac{Ua(T - T_a)}{\dot{m}_c C_{P_c}}$ (T8-3.9)

B. Molar flow rates as the reaction variable

    1. Mole Balances:

$$\dfrac{dF_A}{dV} = r_A \qquad \text{(T8-3.10)}$$

$$\dfrac{dF_B}{dV} = r_B \qquad \text{(T8-3.11)}$$

$$\dfrac{dF_C}{dV} = r_C \qquad \text{(T8-3.12)}$$

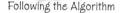

Following the Algorithm   **2. Rate Law:**

$$-r_A = k_1\left(C_A C_B - \dfrac{C_C^2}{K_C}\right) \qquad \text{(T8-3.2)}$$

$$k = k_1(T_1)\exp\left[\dfrac{E}{R}\left(\dfrac{1}{T_1} - \dfrac{1}{T}\right)\right] \qquad \text{(T8-3.3)}$$

$$K_C = K_{C2}(T_2)\exp\left[\dfrac{\Delta H_{Rx}^\circ}{R}\left(\dfrac{1}{T_2} - \dfrac{1}{T}\right)\right] \qquad \text{(T8-3.4)}$$

Living Example Problem
Pb T-8.3

    3. Stoichiometry (gas phase, no ΔP):

$$r_B = r_A \qquad \text{(T8-3.13)}$$

$$r_C = 2r_A \qquad \text{(T8-3.14)}$$

$$C_A = C_{T0}\dfrac{F_A}{F_T}\dfrac{T_0}{T} \qquad \text{(T8-3.15)}$$

$$C_B = C_{T0}\dfrac{F_B}{F_T}\dfrac{T_0}{T} \qquad \text{(T8-3.16)}$$

$$C_C = C_{T0}\dfrac{F_C}{F_T}\dfrac{T_0}{T} \qquad \text{(T8-3.17)}$$

    4. Energy Balance:

Reactor: $\dfrac{dT}{dV} = \dfrac{Ua(T_a - T) + (-r_A)(-\Delta H_{Rx})}{F_A C_{P_A} + F_B C_{P_B} + F_C C_{P_C}}$ (T8-3.18)

Variable coolant temperature   If the coolant temperature, T_a, is not constant, then the energy balance on the coolant fluid gives

Coolant: $\dfrac{dT_a}{dV} = \dfrac{Ua(T - T_a)}{m_c C_{P_c}}$ (T8-3.19)

Summary Notes   where $\dot{m}_c$ is the mass flow rate of the coolant (e.g., kg/s, and C_{P_c} is the heat capacity of the coolant (e.g., kJ/kg·K). (See the CD-ROM for the examples in the Chapter 8 Summary Notes and the Polymath Library for the case when the ambient temperature is not constant.)

TABLE 8-3.   PFR/PBR ALGORITHM FOR HEAT EFFECTS (CONTINUED)

Case A: Conversion as the Independent Variable

$$k_1, E, R, C_{T0}, T_a, T_0, T_1, T_2, K_{C2}, \Theta_B, \Delta H_{Rx}^{\circ}, C_{P_A}, C_{P_B}, C_{P_C}, Ua$$

with initial values T_0 and $X = 0$ at $V = 0$ and final values: $V_f = $ _____

Case B: Molar Flow Rates as the Independent Variables

Same as Case A except the initial values are F_{A0}, and F_{B0} are specified instead of X at $V = 0$.

Summary Notes

Note: The equations in this table have been applied directly to a PBR (recall that we simply use $W = \rho_b V$) using the values for E and ΔH_{Rx} given in Problem P8-2 (m) for the *Living Example Problem* 8-T8-3 on the CD-ROM. Load this *Living Example Problem* from the CD-ROM and vary the cooling rate, flow rate, entering temperature, and other parameters to get an intuitive feel of what happens in flow reactors with heat effects. After carrying out this exercise, go to the **WORKBOOK** in the Chapter 8 Summary Notes on the web/CD-ROM and answer the questions.

The following figures show representative profiles that would result from solving the above equations. The reader is encouraged to load the Living Example Problem 8-T8-3 and vary a number of parameters as discussed in P8 -2 (m). Be sure you can explain why these curves look the way they do.

Be sure you can explain why these curves look the way they do.

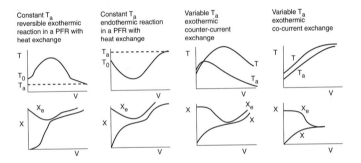

Example 8–5   Production of Acetic Anhydride

Jeffreys,[4] in a treatment of the design of an acetic anhydride manufacturing facility, states that one of the key steps is the vapor-phase cracking of acetone to ketene and methane:

$$CH_3COCH_3 \rightarrow CH_2CO + CH_4$$

He states further that this reaction is first-order with respect to acetone and that the specific reaction rate can be expressed by

$$\ln k = 34.34 - \frac{34,222}{T} \tag{E8-5.1}$$

where k is in reciprocal seconds and T is in kelvin. In this design it is desired to feed 7850 kg of acetone per hour to a tubular reactor. The reactor consists of a bank of 1000 1-inch schedule 40 tubes. We will consider three cases:

---

[4] G. V. Jeffreys, *A Problem in Chemical Engineering Design: The Manufacture of Acetic Anhydride*, 2nd ed. (London: Institution of Chemical Engineers, 1964).

Gas-phase
endothermic
reaction example
1. Adiabatic
2. Heat exchange
T_a is constant
3. Heat exchange
T_a is variable

A. CASE 1   The reactor is operated *adiabatically*.

B. CASE 2   The reactor is surrounded by a *heat exchanger* where the heat-transfer coefficient is 110 $J/m^2 \cdot s \cdot K$, and the temperature of the heating medium, T_a, is constant at 1150 K.

C. CASE 3   The heat exchanger in Case 2 now has a variable heating medium temperature.

The inlet temperature and pressure are the same for both cases at 1035 K and 162 kPa (1.6 atm), respectively. Plot the conversion and temperature along the length of the reactor.

Solution

Let A = CH_3COCH_3, B = CH_2CO, and C = CH_4. Rewriting the reaction symbolically gives us
$$A \rightarrow B + C$$

1. Mole Balance:
$$\frac{dX}{dV} = -\frac{r_A}{F_{A0}} \qquad \text{(E8-5.2)}$$

2. Rate Law:
$$-r_A = kC_A \qquad \text{(E8-5.3)}$$

3. Stoichiometry (gas-phase reaction with no pressure drop):

$$C_A = \frac{C_{A0}(1-X)T_0}{(1+\varepsilon X)T} \qquad \text{(E8-5.4)}$$

$$\varepsilon = y_{A0}\delta = 1(1+1-1) = 1$$

4. Combining yields

$$-r_A = \frac{kC_{A0}(1-X)}{1+X}\frac{T_0}{T} \qquad \text{(E8-5.5)}$$

$$\frac{dX}{dV} = \frac{-r_A}{F_{A0}} = \frac{k}{v_0}\left(\frac{1-X}{1+X}\right)\frac{T_0}{T} \qquad \text{(E8-5.6)}$$

To solve this differential equation, it is first necessary to use the energy balance to determine T as a function of X.

5. Energy Balance:

CASE 1. ADIABATIC OPERATION

For no work done on the system, $\dot{W}_s = 0$, and adiabatic operation, $\dot{Q} = 0$ (i.e., $U \equiv 0$), Equation (8-36) becomes

$$\frac{dT}{dV} = \frac{(-r_A)(-[\Delta H^\circ_{Rx}(T_R) + \Delta C_P(T-T_R)])}{F_{A0}(\Sigma \Theta_i C_{P_i} + X \Delta C_P)} \qquad \text{(E8-5.7)}$$

Because only A enters,

$$\Sigma \Theta_i C_{P_i} = C_{P_A}$$

$\dot{Q} = 0$

and Equation (E8-5.7) becomes

$$\frac{dT}{dV} = \frac{(-r_A)\{-[\Delta H^\circ_{Rx}(T_R) + \Delta C_P(T-T_R)]\}}{F_{A0}(C_{P_A} + X \Delta C_P)} \qquad \text{(E8-5.8)}$$

6. Calculation of Mole Balance Parameters on a Per Tube Basis:

$$F_{A0} = \frac{7,850 \text{ kg/h}}{58 \text{ g/mol}} \times \frac{1}{1,000 \text{ Tubes}} = 0.135 \text{ kmol/h} = 0.0376 \text{ mol/s}$$

$$C_{A0} = \frac{P_{A0}}{RT} = \frac{162 \text{ kPa}}{8.31 \dfrac{\text{kPa} \cdot \text{m}^3}{\text{kmol} \cdot \text{K}} (1035 \text{ K})} = 0.0188 \frac{\text{kmol}}{\text{m}^3} = 18.8 \text{ mol/m}^3$$

$$v_0 = \frac{F_{A0}}{C_{A0}} = 2.037 \text{ dm}^3/\text{s} \quad V = \frac{5 \text{ m}^3}{1000 \text{ tubes}} = 5 \text{ dm}^3$$

7. Calculation of Energy Balance Parameters:

a. $\Delta H_{Rx}^\circ(T_R)$: At 298 K, the standard heats of formation are

$$H_{Rx}^\circ(T_R)_{\text{acetone}} = -216.67 \text{ kJ/mol}$$

$$H_{Rx}^\circ(T_R)_{\text{ketene}} = -61.09 \text{ kJ/mol}$$

$$H_{Rx}^\circ(T_R)_{\text{methane}} = -74.81 \text{ kJ/mol}$$

$$\Delta H_{Rx}^\circ(T_R) = H_B^\circ(T_R) + H_C^\circ(T_R) - H_A^\circ(T_R)$$

$$= (-61.09) + (-74.81) - (-216.67) \text{ kJ/mol}$$

$$= 80.77 \text{ kJ/mol}$$

b. ΔC_P : The mean heat capacities are:

$$\text{CH}_3\text{COCH}_3: \quad C_{P_A} = 163 \text{ J/mol} \cdot \text{K}$$

$$\text{CH}_2\text{CO}: \qquad C_{P_B} = 83 \text{ J/mol} \cdot \text{K}$$

$$\text{CH}_4: \qquad\quad C_{P_C} = 71 \text{ J/mol} \cdot \text{K}$$

$$\Delta C_P = C_{P_B} + C_{P_C} - C_{P_A} = (83 + 71 - 163) \text{ J/mol} \cdot \text{K}$$

$$\Delta C_P = -9 \text{ J/mol} \cdot \text{K}$$

See Table E8-5.1 for a summary of the calculations and Table E8-5.2 and Figure E8-5.1 for the Polymath program and its graphical output. For adiabatic operation, it doesn't matter whether or not we feed everything to one tube with $V = 5 \text{ m}^3$ or distribute the flow to the 1000 tubes each with $V = 5 \text{ dm}^3$. The temperature and conversion profiles are identical because there is no heat exchange.

TABLE E8-5.1.   SUMMARY ADIABATIC OPERATION

Adiabatic PFR

$$\frac{dX}{dV} = \frac{-r_A}{F_{A0}} \tag{E8-5.2}$$

$$r_A = \frac{kC_{A0}(1-X)}{(1+X)} \frac{T_0}{T} \tag{E8-5.5}$$

$$\frac{dT}{dV} = \frac{(-r_A)(\Delta H_{Rx})}{F_{A0}(C_{P_A} + X\Delta C_P)} \tag{E8-5.8}$$

$$k = 8.2 \times 10^{14} \exp\left(\frac{-34,222}{T}\right) \text{s}^{-1} = 3.58 \ \exp\left[34,222\left(\frac{1}{T_0} - \frac{1}{T}\right)\right] \text{s}^{-1} \tag{E8-5.9}$$

$$\Delta H_{Rx} = \Delta H_{Rx}(T_R) + \Delta C_P(T - T_R) \tag{E8-5.10}$$

TABLE E8-5.1.   SUMMARY ADIABATIC OPERATION (CONTINUED)

Parameter Values		
$\Delta C_P = -9$ J/mol • K	ΔH°_{Rx} $(T_R) = 80.77$ J/mol	$T_0 = 1035$ K
$C_{P_A} = 163$ J/mol/A/K	$C_{A0} = 18.8$ mol/m^3	$T_R = 298$ K
$V_f = 5$ dm^3	$F_{A0} = 0.376$ mol/s	

TABLE E8-5.2.   POLYMATH PROGRAM ADIABATIC OPERATION

ODE Report (RKF45)

Differential equations as entered by the user
[1] d(X)/d(V) = -ra/Fao
[2] d(T)/d(V) = -ra*(-deltaH)/(Fao*(Cpa+X*delCp))

Explicit equations as entered by the user
[1] Fao = .0376
[2] Cpa = 163
[3] delCp = -9
[4] Cao = 18.8
[5] To = 1035
[6] deltaH = 80770+delCp*(T-298)
[7] ra = -Cao*3.58*exp(34222*(1/To-1/T))*(1-X)*(To/T)/(1+X)

Living Example Problem

Adiabatic
endothermic
reaction in a PFR

RIP

Reaction
Rate

Figure E8-5.1   Conversion and temperature (a) and reaction rate (b) profiles.

As temperature drops, so does k and hence the rate, $-r_A$, drops to an insignificant value.

Death of a reaction

Note that for this adiabatic endothermic reaction, the reaction virtually *dies out* after 3.5 m^3, owing to the large drop in temperature, and very little conversion is achieved beyond this point. One way to increase the conversion would be to add a diluent such as nitrogen, which could supply the sensible heat for this endothermic reaction. However, if too much diluent is added, the concentration and rate will be quite low. On the other hand, if too little diluent is added, the temperature will drop and virtually extinguish the reaction. How much diluent to add is left as an exercise [see Problem P8-2(e)].

A bank of 1000 1-in. schedule 40 tubes 1.79 m in length corresponds to 1.0 m³ and gives 20% conversion. Ketene is unstable and tends to explode, which is a good reason to keep the conversion low. However, the pipe material and schedule size should be checked to learn if they are suitable for these temperatures and pressures.

CASE 2. HEAT EXCHANGE WITH CONSTANT HEATING MEDIUM TEMPERATURE

Let's now see what happens as we add heat to the reacting mixture. See Figure E8-5.2.

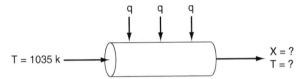

Figure E8-5.2   PFR with heat exchange.

1. Mole Balance:

$$\text{(Step 1)} \qquad \frac{dX}{dV} = \frac{-r_A}{F_{A0}} \qquad \text{(E8-5.2)}$$

Using the algorithm: (Step 2) the **rate law** (E8-5.3) and (Step 3) **stoichiometry** (E8-5.4) for the adiabatic case discussed previously in Case I, we (Step 4) **combine** to obtain the reaction rate as

$$\text{(Step 4)} \qquad -r_A = kC_{A0}\left(\frac{1-X}{1+X}\right)\frac{T_0}{T} \qquad \text{(E8-5.5)}$$

5. Energy Balance:

$$\frac{dT}{dV} = \frac{Ua(T_a - T) + (-r_A)[-\Delta H_{Rx}(T)]}{F_{A0}(\Sigma\,\Theta_i C_{P_i} + X\,\Delta C_P)} \qquad \text{(8-36)}$$

For the acetone reaction system,

$$\frac{dT}{dV} = \frac{Ua(T_a - T) + (r_A)[\Delta H_{Rx}(T)]}{F_{A0}(C_{P_A} + X\,\Delta C_P)} \qquad \text{(E8-5.10)}$$

PFR with heat exchange

6. Parameter Evaluation:

a. *Mole balance.* On a per tube basis, $v_0 = 0.002$ m³/s. The concentration of acetone is 18.8 mol/m³, so the entering molar flow rate is

$$F_{A0} = C_{A0}v_0 = \left(18.8\,\frac{\text{mol}}{\text{m}^3}\right)\left(2\times10^{-3}\,\frac{\text{m}^3}{\text{s}}\right) = 0.0376\,\frac{\text{mol}}{\text{s}}$$

The value of k at 1035 K is 3.58 s^{-1}; consequently, we have

$$k(T) = 3.58 \exp\left[34,222\left(\frac{1}{1035} - \frac{1}{T}\right)\right] \text{s}^{-1} \qquad \text{(E8-5.11)}$$

b. *Energy balance.* From the adiabatic case in Case I, we already have $\Delta C_P, C_{P_A}$. The heat-transfer area per unit volume of pipe is

$$a = \frac{\pi D L}{(\pi D^2/4)L} = \frac{4}{D} = \frac{4}{0.0266 \text{ m}} = 150 \text{ m}^{-1}$$

$$U = 110 \text{ J/m}^2 \cdot \text{s} \cdot \text{K}$$

Combining the overall heat-transfer coefficient with the area yields

$$Ua = 16,500 \text{ J/m}^3 \cdot \text{s} \cdot \text{K}$$

We now use Equations (E8-5.1) through (E8-5.6), and Equations (E8-5.10) and (E8-5.11) along with Equation E8-5.12 in Table E8-5.3 in the Polymath program (Table E8-5.4), to determine the conversion and temperature profiles shown in Figure E8-5.3.

$$V_f = 0.001 \text{ m}^3 = 1 \text{ dm}^3$$

TABLE E8-5.3.   SUMMARY WITH CONSTANT T_a HEAT EXCHANGE

We now apply Equation (8-36) to this example to arrive at Equation (E8-5.12), which we then use to replace Equation (E8-5.8) in Summary Table E8-5.1.

$$\frac{dT}{dV} = \frac{Ua(T_a - T) + (r_A)[\Delta H_{Rx}]}{F_{A0}(C_{P_A} + X \Delta C_P)} \qquad \text{(E8-5.12)}$$

with

$$Ua = 16,500 \text{ J/m}^3/\text{s/K}$$

$$T_a = 1150 \text{ K} \qquad \text{(E8-5.13)}$$

Per tube basis

$$v_0 = 0.002 \text{ m}^3/\text{s}$$

$$F_{A0} = 0.0376 \text{ mol/s}$$

$$V_f = 0.001 \text{ m}^3 = 1 \text{ dm}^3$$

Everything else is the same as shown in Table E8-5.1.

One notes that the reactor temperature goes through a minimum along the length of the reactor. At the front of the reactor, the reaction takes place very rapidly, drawing energy from the sensible heat of the gas causing the gas temperature to drop because the heat exchanger cannot supply energy at an equal or greater rate. This drop in temperature, coupled with the consumption of reactants, slows the reaction rate as we move down the reactor. Because of this slower reaction rate, the heat exchanger supplies energy at a rate greater than reaction draws energy from the gases and as a result the temperature increases.

TABLE E8-5.4.   POLYMATH PROGRAM FOR PFR WITH HEAT EXCHANGE

POLYMATH Results

Example 8-5 Production of Acetic Anhydride with Heat Exchange (Constant Ta) 08-16-2004, Rev5.1.232

ODE Report (RKF45)

Differential equations as entered by the user
[1] d(X)/d(V) = -ra/Fao
[2] d(T)/d(V) = (Ua*(Ta-T)+ra*deltaH)/(Fao*(Cpa+X*delCp))

Explicit equations as entered by the user
[1] Fao = .0376
[2] Cpa = 163
[3] delCp = -9
[4] Cao = 18.8
[5] To = 1035
[6] deltaH = 80770+delCp*(T-298)
[7] ra = -Cao*3.58*exp(34222*(1/To-1/T))*(1-X)*(To/T)/(1+X)
[8] Ua = 16500
[9] Ta = 1150

Living Example Problem

PFR with heat exchange constant heating medium temperature T_a

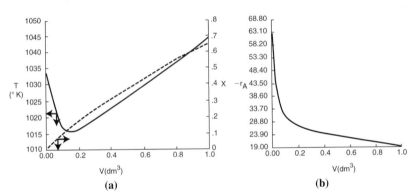

Figure E8-5.3   Temperature and conversion profiles in PFR. Temperature and conversion (a) and reaction rate (b) profiles in a PFR with constant heating medium temperature, T_a.

CASE 3. HEAT EXCHANGE WITH VARIABLE HEATING MEDIUM TEMPERATURE

Air will also be used as a heating stream in a co-current direction entering at a temperature of 1250 K and at molar rate of (0.11 mol/s). The heat capacity of the air is 34.5 J/mol · K.

Solution

For co-current flow

$$\frac{dT_a}{dV} = \frac{Ua(T - T_a)}{\dot{m}_c C_{p_c}}$$ (E8-5.7)

$$\dot{m}_c C_{p_c} = \left(0.11 \frac{mol}{s}\right)\left(34.5 \frac{J}{mol \cdot K}\right) = 3.83 \ J/s/K$$

The Polymath code is modified by replacing $T_a = 1150K$ in Tables E8-5.3 and E8-5.4 by Equation (E8-5.7), and adding the numerical values for $\dot{m}_c$ and C_{P_c}.

TABLE E8-5.5.   SUMMARY HEAT EXCHANGE WITH VARIABLE T_a

POLYMATH Results

Example 8-5 Production of Acetic Anhydride with Heat Exchange (Variable Ta) 08-16-2004, Rev5.1.232

Differential equations as entered by the user
[1] d(X)/d(V) = -ra/Fao
[2] d(T)/d(V) = (Ua*(Ta-T)+ra*deltaH)/(Fao*(Cpa+X*delCp))
[3] d(Ta)/d(V) = Ua*(T-Ta)/mc/Cpc

Explicit equations as entered by the user
[1]   Fao = .0376
[2]   Cpa = 163
[3]   delCp = -9
[4]   Cao = 18.8
[5]   To = 1035
[6]   deltaH = 80770+delCp*(T-298)
[7]   ra = -Cao*3.58*exp(34222*(1/To-1/T))*(1-X)*(To/T)/(1+X)
[8]   Ua = 16500
[9]   mc = .111
[10] Cpc = 34.5

PFR with
heat exchange
variable ambient
temperature T_a

The temperature and conversion profiles are shown in Figure E8-5.4.

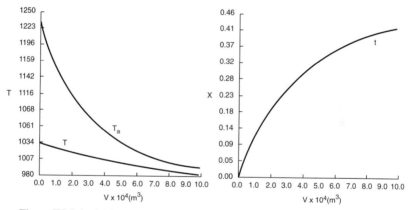

Figure E8-5.4   (a) Temperature and (b) conversion profiles in PFR with a variable heating medium temperature, T_a

8.5 Equilibrium Conversion

For reversible
reactions, the
equilibrium
conversion, X_e, is
usually calculated
first.

The highest conversion that can be achieved in reversible reactions is the equilibrium conversion. For endothermic reactions, the equilibrium conversion increases with increasing temperature up to a maximum of 1.0. For exothermic reactions, the equilibrium conversion decreases with increasing temperature.

8.5.1 Adiabatic Temperature and Equilibrium Conversion

Exothermic Reactions.   Figure 8-4(a) shows the variation of the concentration equilibrium constant as a function of temperature for an exothermic reaction (see Appendix C), and Figure 8-4(b) shows the corresponding equilibrium
conversion X_e as a function of temperature. In Example 8-3, we saw that for a
first-order reaction the equilibrium conversion could be calculated using Equation (E8-3.13)

First-order reversible
reaction

$$X_e = \frac{K_C}{1 + K_C} \qquad \text{(E8-3.13)}$$

Consequently, X_e can be calculated directly using Figure 8-4(a).

For exothermic
reactions,
equilibrium
conversion
decreases with
increasing
temperature.

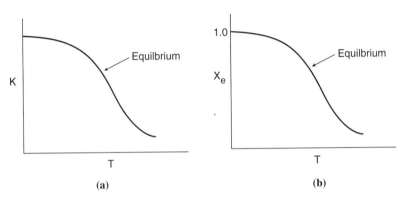

Figure 8-4   Variation of equilibrium constant and conversion with temperature for
an exothermic reaction.

To determine the maximum conversion that can be achieved in an exothermic
reaction carried out adiabatically, we find the intersection of the equilibrium conversion as a function of temperature [Figure 8-4(b)] with temperature–conversion
relationships from the energy balance (Figure 8-2) as shown in Figure 8-5.

$$X_{EB} = \frac{\Sigma \Theta_i C_{P_i}(T - T_0)}{-\Delta H_{Rx}(T)} \qquad \text{(8-29)}$$

If the entering temperature is increased from T_0 to T_{01}, the energy balance line will be shifted to the right and will be parallel to the original line, as
shown by the dashed line. Note that as the inlet temperature increases, the adiabatic equilibrium conversion decreases.

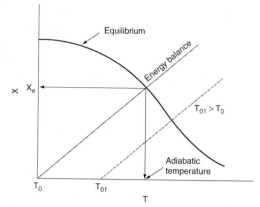

Figure 8-5   Graphical solution of equilibrium and energy balance equations to obtain the adiabatic temperature and the adiabatic equilibrium conversion X_e.

Adiabatic equilibrium conversion for exothermic reactions

Example 8–6   Calculating the Adiabatic Equilibrium Temperature

For the elementary solid-catayzed liquid-phase reaction

$$A \xrightleftharpoons{} B$$

make a plot of equilibrium conversion as a function of temperature. Determine the adiabatic equilibrium temperature and conversion when pure A is fed to the reactor at a temperature of 300 K.

Additional information:

$$H_A^\circ(298\ K) = -40{,}000\ cal/mol \qquad H_B^\circ(298\ K) = -60{,}000\ cal/mol$$

$$C_{P_A} = 50\ cal/mol \cdot K \qquad C_{P_B} = 50\ cal/mol \cdot K$$

$$K_e = 100{,}000\ at\ 298\ K$$

Solution

1. Rate Law:

$$-r_A = k\left(C_A - \frac{C_B}{K_e}\right) \tag{E8-6.1}$$

2. Equilibrium: $-r_A = 0$; so

$$C_{Ae} = \frac{C_{Be}}{K_e}$$

Following the Algorithm

3. Stoichiometry: $(v = v_0)$ yields

$$C_{A0}(1 - X_e) = \frac{C_{A0}X_e}{K_e}$$

Solving for X_e gives

$$X_e = \frac{K_e(T)}{1 + K_e(T)} \tag{E8-6.2}$$

4. Equilibrium Constant: Calculate ΔC_P, then $K_e(T)$

$$\Delta C_P = C_{P_B} - C_{P_A} = 50 - 50 = 0 \text{ cal/mol} \cdot \text{K}$$

For $\Delta C_p = 0$, the equilibrium constant varies with temperature according to the relation

$$K_e(T) = K_e(T_1) \exp\left[\frac{\Delta H^\circ_{Rx}}{R}\left(\frac{1}{T_1} - \frac{1}{T}\right)\right] \tag{E8-6.3}$$

$$\Delta H^\circ_{Rx} = H^\circ_B - H^\circ_A = -20{,}000 \text{ cal/mol}$$

$$K_e(T) = 100{,}000 \exp\left[\frac{-20{,}000}{1.987}\left(\frac{1}{298} - \frac{1}{T}\right)\right]$$

$$K_e = 100{,}000 \exp\left[-33.78\left(\frac{T - 298}{T}\right)\right] \tag{E8-6.4}$$

Substituting Equation (E8-6.4) into (E8-6.2), we can calculate equilibrium conversion as a function of temperature:

5. Equilibrium Conversion from Thermodynamics

$$X_e = \frac{100{,}000 \exp\left[-33.78(T - 298)/T\right]}{1 + 100{,}000 \exp\left[-33.78(T - 298)/T\right]} \tag{E8-6.5}$$

The calculations are shown in Table E8-6.1.

TABLE E8-6.1.   EQUILIBRIUM CONVERSION
AS A FUNCTION OF TEMPERATURE

T	K_e	X_e
298	100,000.00	1.00
350	661.60	1.00
400	18.17	0.95
425	4.14	0.80
450	1.11	0.53
475	0.34	0.25
500	0.12	0.11

6. Energy Balance

For a reaction carried out adiabatically, the energy balance reduces to

$$X_{EB} = \frac{\Sigma \Theta_i C_{p_i}(T - T_0)}{-\Delta H_{Rx}} = \frac{C_{P_A}(T - T_0)}{-\Delta H_{Rx}} \tag{E8-6.6}$$

Conversion
calculated from
energy balance

$$X_{EB} = \frac{50(T-300)}{20,000} = 2.5 \times 10^{-3}(T-300) \qquad \text{(E8-6.7)}$$

Data from Table E8-6.1 and the following data are plotted in Figure E8-6.1.

T (K)	300	400	500	600
X_{EB}	0	0.25	0.50	0.75

$$X_e = 0.42 \quad T_e = 465 \text{ K}$$

For a feed temperature of 300 K, the adiabatic equilibrium temperature is 465 K and the corresponding adiabatic equilibrium conversion is only 0.42.

Adiabatic
equilibrium
conversion and
temperature

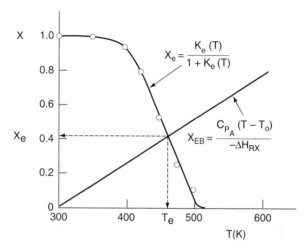

Figure E8-6.1   Finding the adiabatic equilibrium temperature (T_e) and conversion (X_e).

Reactor Staging with Interstage Cooling or Heating

Higher conversions than those shown in Figure E8-6.1 can be achieved for adiabatic operations by connecting reactors in series with interstage cooling:

The conversion–temperature plot for this scheme is shown in Figure 8-6. We see that with three interstage coolers 90% conversion can be achieved compared to an equilibrium conversion of 40% for no interstage cooling.

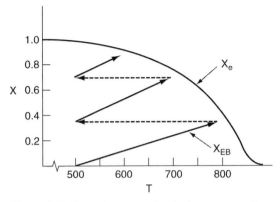

Interstage cooling
used for exothermic
reversible reactions

Figure 8-6 Increasing conversion by interstage cooling.

Typical values
for gasoline
composition

Gasoline	
C_5	10%
C_6	10%
C_7	20%
C_8	25%
C_9	20%
C_{10}	10%
C_{11}-C_{12}	5%

Endothermic Reactions. Another example of the need for interstage heat transfer in a series of reactors can be found when upgrading the octane number of gasoline. The more compact the hydrocarbon molecule for a given number of carbon atoms is, the higher the octane rating is. Consequently, it is desirable to convert straight-chain hydrocarbons to branched isomers, naphthenes, and aromatics. The reaction sequence is

Straight   Naphthenes   Aromatics
Chain

The first reaction step (k_1) is slow compared to the second step, and each step is highly endothermic. The allowable temperature range for which this reaction can be carried out is quite narrow: Above 530°C undesirable side reactions occur, and below 430°C the reaction virtually does not take place. A typical feed stock might consist of 75% straight chains, 15% naphthas, and 10% aromatics.

One arrangement currently used to carry out these reactions is shown in Figure 8-7. Note that the reactors are not all the same size. Typical sizes are on the order of 10 to 20 m high and 2 to 5 m in diameter. A typical feed rate of gasoline is approximately 200 m³/h at 2 atm. Hydrogen is usually separated from the product stream and recycled.

Spring 2005
$2.20/gal for octane
number (ON)
ON = 89

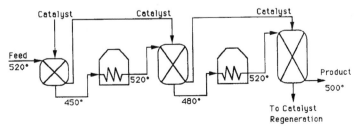

Figure 8-7 Interstage heating for gasoline production in moving-bed reactors.

Because the reaction is endothermic, equilibrium conversion increases with increasing temperature. A typical equilibrium curve and temperature conversion trajectory for the reactor sequence are shown in Figure 8-8.

Interstage heating

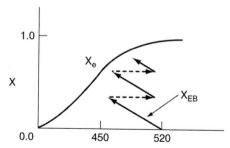

Figure 8-8   Temperature–conversion trajectory for interstage heating of an endothermic reaction analogous to Figure 8-6.

Example 8–7   Interstage Cooling for Highly Exothermic Reactions

What conversion could be achieved in Example 8-6 if two interstage coolers that had the capacity to cool the exit stream to 350 K were available? Also determine the heat duty of each exchanger for a molar feed rate of A of 40 mol/s. Assume that 95% of equilibrium conversion is achieved in each reactor. The feed temperature to the first reactor is 300 K.

Solution

1. Calculate Exit Temperature
We saw in Example 8-6

$$A \rightleftharpoons B$$

that for an entering temperature of 300 K the adiabatic equilibrium conversion was 0.42. For 95% of equilibrium conversion ($X_e = 0.42$), the conversion exiting the first reactor is 0.4. The exit temperature is found from a rearrangement of Equation (E8-6.7):

$$T = 300 + 400X = 300 + (400)(0.4) \tag{A}$$
$$T_1 = 460 \text{ K}$$

We now cool the gas stream exiting the reactor at 460 K down to 350 K in a heat exchanger (Figure E8-7.2).

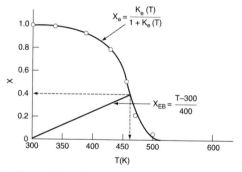

Figure E8-7.1   Determining exit conversion and temperature in the first stage.

2. Calculate the Heat Load

There is no work done on the reaction gas mixture in the exchanger, and the reaction does not take place in the exchanger. Under these conditions ($F_{i|in} = F_{i|out}$), the energy balance given by Equation (8-10)

$$\dot{Q} - \dot{W}_s + \Sigma F_{i0} H_{i0} - \Sigma F_i H_i = 0 \tag{8-10}$$

for $\dot{W}_s = 0$ becomes

Energy balance on the reaction gas mixture **in** the heat exchanger

$$\dot{Q} = \Sigma F_i H_i - \Sigma F_{i0} H_{i0} = \Sigma F_{i0}(H_i - H_{i0}) \tag{E8-7.1}$$

$$= \Sigma F_i C_{P_i}(T_2 - T_1) = (F_A C_{P_A} + F_B C_{P_B})(T_2 - T_1) \tag{E8-7.2}$$

But $C_{P_A} = C_{P_B}$,

$$\dot{Q} = (F_A + F_B)(C_{P_A})(T_2 - T_1) \tag{E8-7.3}$$

Also, $F_{A0} = F_A + F_B$,

$$\dot{Q} = F_{A0} C_{P_A}(T_2 - T_1)$$

$$= \frac{40 \text{ mol}}{\text{s}} \cdot \frac{50 \text{ cal}}{\text{mol} \cdot \text{K}} (350 - 460) \text{ K}$$

$$= -220 \frac{\text{kcal}}{\text{s}} \tag{E8-7.4}$$

That is, 220 kcal/s must be removed to cool the reacting mixture from 460 K to 350 K for a feed rate of 40 mol/s.

3. Calculate the Coolant Flow Rate

We see that 220 kcal/s is removed from the reaction system mixture. The rate at which energy must be absorbed by the coolant stream in the exchanger is

$$\dot{Q} = \dot{m}_c C_{P_c}(T_{out} - T_{in}) \tag{E8-7.5}$$

We consider the case where the coolant is available at 270 K but cannot be heated above 400 K and calculate the coolant flow rate necessary to remove 220 kcal/s from the reaction mixture. Rearranging Equation (E8-7.5) and noting that this coolant has a heat capacity of 18 cal/mol·K gives

Sizing the interstage heat exchanger and coolant flow rate

$$\dot{m}_c = \frac{\dot{Q}}{C_{P_c}(T_{out} - T_{in})} = \frac{220,000 \text{ cal/s}}{18 \frac{\text{cal}}{\text{mol} \cdot \text{K}} (400 - 270) \text{ K}} \tag{E8-7.6}$$

$$= 94 \text{ mol/s} = 1692 \text{ g/s} = 1.69 \text{ kg/s}$$

The necessary coolant flow rate is 1.69 kg/s.

4. Calculate the Heat Exchanger Area

Let's next determine the counter current heat exchanger area. The exchanger inlet and outlet temperatures are shown in Figure E8-7.2. The rate of heat transfer in a counter current heat exchanger is given by the equation[5]

---

[5] See page 268 of C. J. Geankoplis, *Transport Processes and Unit Operations* (Upper Saddle River, N.J.: Prentice Hall, 1993).

Bonding with unit
operations

$$\dot{Q} = UA \frac{[(T_{h2} - T_{c2}) - (T_{h1} - T_{c1})]}{\ln\left(\dfrac{T_{h2} - T_{c2}}{T_{h1} - T_{c1}}\right)} \qquad \text{(E8-7.7)}$$

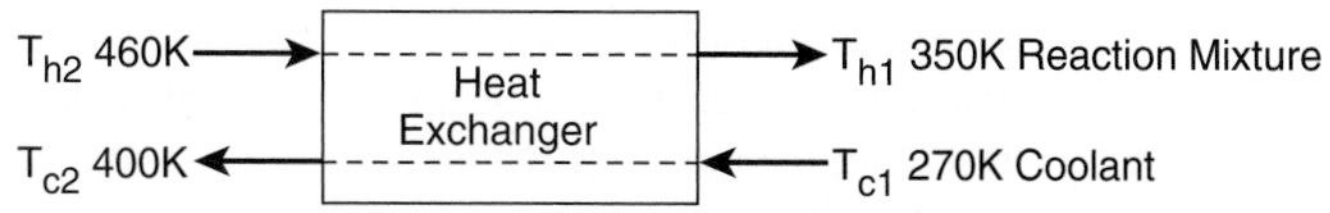

T_{h2} 460K ──→ ┄┄┄┄┄┄ ──→ T_{h1} 350K Reaction Mixture

Heat
Exchanger

T_{c2} 400K ←── ┄┄┄┄┄┄ ←── T_{c1} 270K Coolant

Figure E8-7.2   Counter current heat exchanger.

Rearranging Equation (E8-7.7), assuming a value of U of 100 cal/s·m²·K, and then substituting the appropriate values gives

Sizing the heat
exchanger

$$A = \frac{\dot{Q}\ln\left(\dfrac{T_{h2} - T_{c2}}{T_{h1} - T_{c1}}\right)}{U[(T_{h2} - T_{c2}) - (T_{h1} - T_{c1})]} = \frac{220{,}000\,\dfrac{\text{cal}}{\text{s}}\ln\left(\dfrac{460 - 400}{350 - 270}\right)}{100\,\dfrac{\text{cal}}{\text{s}\cdot\text{m}^2\cdot\text{K}}[(460 - 400) - (350 - 270)]\,\text{K}}$$

$$= \frac{2{,}200\,\ln{(0.75)}}{-20}\,\text{m}^2$$

$$= 31.6\,\text{m}^2$$

The heat-exchanger surface area required to accomplish this rate of heat transfer is 31.6 m².

5. Second Reactor

Now let's return to determine the conversion in the second reactor. The conditions entering the second reactor are $T = 350$ K and $X = 0.4$. The energy balance starting from this point is shown in Figure E8-7.3. The corresponding adiabatic equilibrium conversion is 0.63. Ninety-five percent of the equilibrium conversion is 60% and the corresponding exit temperature is $T = 350 + (0.6 - 0.4)400 = 430$ K.

Rearranging
Equation (E8-6.7)
for the second
reactor

The heat-exchange duty to cool the reacting mixture from 430 K back to 350 K can again be calculated from Equation (E8-7.4):

$$T_2 = T_{20} + \Delta X\left(\frac{-\Delta H_{\text{Rx}}}{C_{P_A}}\right)$$

$$= 350 + 400\Delta X$$

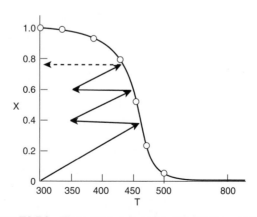

Figure E8-7.3   Three reactors in series with interstage cooling.

$$\dot{Q} = F_{A0}C_{P_A}(350 - 430) = \left(\frac{40 \text{ mol}}{\text{s}}\right)\left(\frac{50 \text{ cal}}{\text{mol} \cdot \text{K}}\right)(-80)$$

$$= -160 \frac{\text{kcal}}{\text{s}}$$

6. Subsequent Reactors

For the final reactor we begin at $T_0 = 350$ K and $X = 0.6$ and follow the line representing the equation for the energy balance along to the point of intersection with the equilibrium conversion, which is $X = 0.8$. Consequently, the final conversion achieved with three reactors and two interstage coolers is $(0.95)(0.8) = 0.76$.

8.5.2 Optimum Feed Temperature

We now consider an adiabatic reactor of fixed size or catalyst weight and investigate what happens as the feed temperature is varied. The reaction is reversible and exothermic. At one extreme, using a very high feed temperature, the specific reaction rate will be large and the reaction will proceed rapidly, but the equilibrium conversion will be close to zero. Consequently, very little product will be formed. At the other extreme of low feed temperatures, little product will be formed because the reaction rate is so low. A plot of the equilibrium conversion and the conversion calculated from the adiabatic energy balance, is shown in Figure 8-9. We see that for an entering temperature of 600 K the adiabatic equilibrium conversion is 0.15. The corresponding conversion profiles down the length of the reactor are shown in Figure 8-10. The equilibrium conversion, which can be calculated from an equation similar to Equation (E8-6.2), also varies along the length of the reactor as shown by the dashed line in Figure 8-10. We also see that because of the high entering temperature, the rate is very rapid and equilibrium is achieved very near the reactor entrance.

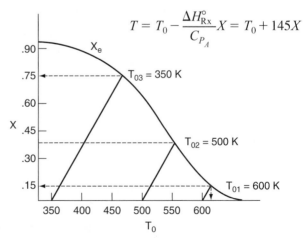

Figure 8-9 Equilibrium conversion for different feed temperatures.

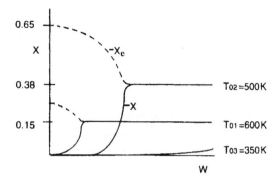

Figure 8-10   Adiabatic conversion profiles for different feed temperatures.

We notice that the conversion and temperature increase very rapidly over a short distance (i.e., a small amount of catalyst). This sharp increase is sometimes referred to as the "point" or temperature at which the reaction ignites. If the inlet temperature were lowered to 500 K, the corresponding equilibrium conversion is increased to 0.38; however, the reaction rate is slower at this lower temperature so that this conversion is not achieved until closer to the end of the reactor. If the entering temperature were lowered further to 350 K, the corresponding equilibrium conversion is 0.75, but the rate is so slow that a conversion of 0.05 is achieved for the specified catalyst weight in the reactor. At a very low feed temperature, the specific reaction rate will be so small that virtually all of the reactant will pass through the reactor without reacting. It is apparent that with conversions close to zero for both high and low feed temperatures there must be an optimum feed temperature that maximizes conversion. As the feed temperature is increased from a very low value, the specific reaction rate will increase, as will the conversion. The conversion will continue to increase with increasing feed temperature until the equilibrium conversion is approached in the reaction. Further increases in feed temperature for this exothermic reaction will only decrease the conversion due to the decreasing equilibrium conversion. This optimum inlet temperature is shown in Figure 8-11.

Optimum inlet
temperature

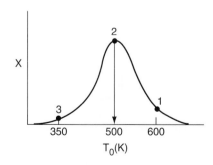

Figure 8-11   Finding the optimum feed temperature.

8.6 CSTR with Heat Effects

In this section we apply the general energy balance [Equation (8-22)] to the CSTR at steady state. We then present example problems showing how the mole and energy balances are combined to size reactors operating adiabatically and non-adiabatically.

Substituting Equation (8-26) into Equation (8-22), the steady-state energy balance becomes

$$\dot{Q} - \dot{W}_s - F_{A0}\Sigma\Theta_i C_{P_i}(T - T_{i0}) - [\Delta H_{Rx}^{\circ}(T_R) + \Delta C_P(T - T_R)]F_{A0}X = 0 \quad (8\text{-}27)$$

These are the forms of the steady-state balance we will use.

[Note: In many calculations the CSTR mole balance ($F_{A0}X = -r_A V$) will be used to replace the term following the brackets in Equation (8-27), that is, ($F_{A0}X$) will be replaced by ($-r_A V$).] Rearranging yields the steady-state balance

$$\dot{Q} - \dot{W}_s - F_{A0}\Sigma\Theta_i C_{P_i}(T - T_{i0}) + (r_A V)(\Delta H_{Rx}) = 0 \quad (8\text{-}42)$$

Although the CSTR is well mixed and the temperature is uniform throughout the reaction vessel, these conditions do not mean that the reaction is carried out isothermally. Isothermal operation occurs when the feed temperature is identical to the temperature of the fluid inside the CSTR.

The $\dot{Q}$ Term in the CSTR

8.6.1 Heat Added to the Reactor, $\dot{Q}$

Figure 8-12 shows schematics of a CSTR with a heat exchanger. The heat transfer fluid enters the exchanger at a mass flow rate $\dot{m}_c$ (e.g., kg/s) at a temperature T_{a1} and leaves at a temperature T_{a2}. The rate of heat transfer *from the* exchanger *to* the reactor is[6]

For exothermic reactions ($T > T_{a2} > T_{a1}$)

$$\dot{Q} = \frac{UA(T_{a1} - T_{a2})}{\ln[(T - T_{a1})/(T - T_{a2})]} \quad (8\text{-}43)$$

For endothermic reactions ($T_{1a} > T_{2a} > T$)

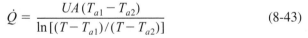

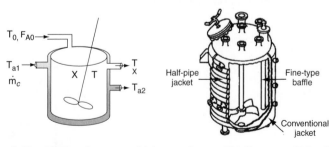

Figure 8-12   CSTR tank reactor with heat exchanger. [(b) Courtesy of Pfaudler, Inc.]

---

[6] Information on the overall heat-transfer coefficient may be found in C. J. Geankoplis, *Transport Processes and Unit Operations*, 3rd ed. Englewood Cliffs, N.J., Prentice Hall (2003), p. 268.

The following derivations, based on a coolant (exothermic reaction) apply also to heating mediums (endothermic reaction). As a first approximation, we assume a quasi-steady state for the coolant flow and neglect the accumulation term (i.e., $dT_a/dt = 0$). An energy balance on the coolant fluid entering and leaving the exchanger is

Energy balance on heat exchanger

$$\begin{bmatrix} \text{Rate of} \\ \text{energy} \\ in \\ \text{by flow} \end{bmatrix} - \begin{bmatrix} \text{Rate of} \\ \text{energy} \\ out \\ \text{by flow} \end{bmatrix} - \begin{bmatrix} \text{Rate of} \\ \text{heat transfer} \\ from \text{ exchanger} \\ to \text{ reactor} \end{bmatrix} = 0 \quad (8\text{-}44)$$

$$\dot{m}_c C_{p_c}(T_{a1} - T_R) - \dot{m}_c C_{p_c}(T_{a2} - T_R) - \frac{UA(T_{a1} - T_{a2})}{\ln(T - T_{a1})/(T - T_{a2})} = 0 \quad (8\text{-}45)$$

where C_{p_C} is the heat capacity of the coolant fluid and T_R is the reference temperature. Simplifying gives us

$$\dot{Q} = \dot{m}_c C_{p_c}(T_{a1} - T_{a2}) = \frac{UA(T_{a1} - T_{a2})}{\ln(T - T_{a1})/(T - T_{a2})} \quad (8\text{-}46)$$

Solving Equation (8-46) for the exit temperature of the coolant fluid yields

$$T_{a2} = T - (T - T_{a1}) \exp\left(\frac{-UA}{\dot{m}_c C_{p_c}}\right) \quad (8\text{-}47)$$

From Equation (8-46)

$$\dot{Q} = \dot{m}_c C_{p_c}(T_{a1} - T_{a2}) \quad (8\text{-}48)$$

Substituting for T_{a2} in Equation (8-48), we obtain

Heat transfer to a CSTR

$$\boxed{\dot{Q} = \dot{m}_c C_{p_c}\left\{(T_{a1} - T)\left[1 - \exp\left(\frac{-UA}{\dot{m}_c C_{p_c}}\right)\right]\right\}} \quad (8\text{-}49)$$

For large values of the coolant flow rate, the exponent will be small and can be expanded in a Taylor series ($e^{-x} = 1 - x + \cdots$) where second-order terms are neglected in order to give

$$\dot{Q} = \dot{m}_c C_{p_c}(T_{a1} - T)\left[1 - \left(1 - \frac{UA}{\dot{m}_c C_{p_c}}\right)\right]$$

Then

Valid only for large coolant flow rates!!

$$\boxed{\dot{Q} = UA(T_a - T)} \quad (8\text{-}50)$$

where $T_{a1} \cong T_{a2} = T_a$.

With the exception of processes involving highly viscous materials such as Problem P8-4$_C$, the California P.E exam problem, the work done by the stirrer can usually be neglected. Setting $\dot{W}_s$ in (8-27) to zero, neglecting ΔC_P, substituting for $\dot{Q}$ and rearranging, we have the following relationship between conversion and temperature in a CSTR.

$$\frac{UA}{F_{A0}}(T_a - T) - \Sigma\Theta_i C_p(T - T_0) - \Delta H^\circ_{\text{Rx}} X = 0 \tag{8-51}$$

Solving for X

$$X = \frac{\dfrac{UA}{F_{A0}}(T - T_a) + \Sigma\Theta_i C_{p_i}(T - T_0)}{[-\Delta H^\circ_{\text{Rx}}(T_R)]} \tag{8-52}$$

Equation (8-52) is coupled with the mole balance equation

$$V = \frac{F_{A0}X}{-r_A(X,T)} \tag{8-53}$$

to size CSTRs.

We now will further rearrange Equation (8-51) after letting $\Sigma\Theta_i C_{P_i} = C_{P_0}$

$$\boxed{\Sigma\Theta_i C_{P_i} = C_{P_0}}$$

$$C_{P_0}\left(\frac{UA}{F_{A0}C_{P_0}}\right)T_a + C_{P_0}T_0 - C_{P_0}\left(\frac{UA}{F_{A0}C_{P_0}} + 1\right)T - \Delta H^\circ_{\text{Rx}} X = 0$$

Let

$$\kappa = \frac{UA}{F_{A0}C_{P_0}} \quad \text{and} \quad T_c = \frac{\kappa T_a + T_0}{1 + \kappa}$$

Then

$$-X\Delta H^\circ_{\text{Rx}} = C_{P_0}(1 + \kappa)(T - T_c) \tag{8-54}$$

The parameters κ and T_c are used to simplify the equations for *non*-adiabatic operation. Solving Equation (8-54) for conversion

$$X = \frac{C_{P_0}(1 + \kappa)(T - T_c)}{-\Delta H^\circ_{\text{Rx}}} \tag{8-55}$$

Forms of the energy balance for a CSTR with heat exchange

Solving Equation (8-54) for the reactor temperature

$$T = T_c + \frac{(-\Delta H^\circ_{\text{Rx}})(X)}{C_{P_0}(1 + \kappa)} \tag{8-56}$$

Figure 8-13 and Table 8-4 show three ways to specify the sizing of a CSTR. This procedure for nonisothermal CSTR design can be illustrated by considering a first-order irreversible liquid-phase reaction. The algorithm for working through either case A (X specified), B (T specified), or C (V specified) is shown in Table 8-4. Its application is illustrated in the following example.

Algorithm

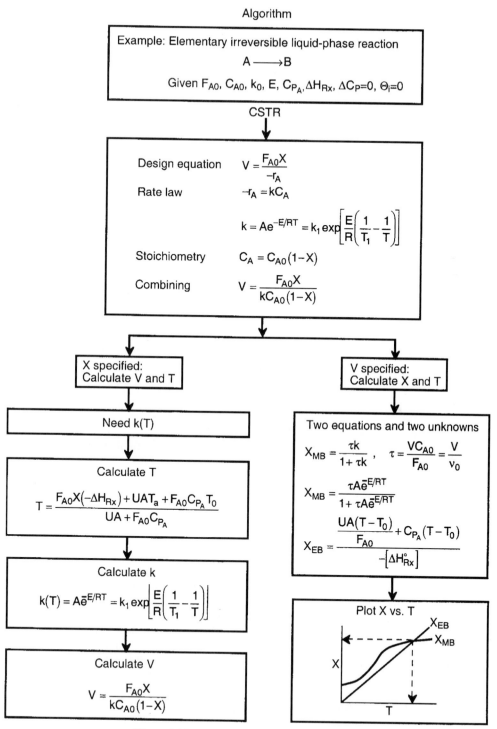

Figure 8-13 Algorithm for adiabatic CSTR design.

TABLE 8-4.   WAYS TO SPECIFY THE SIZING A CSTR

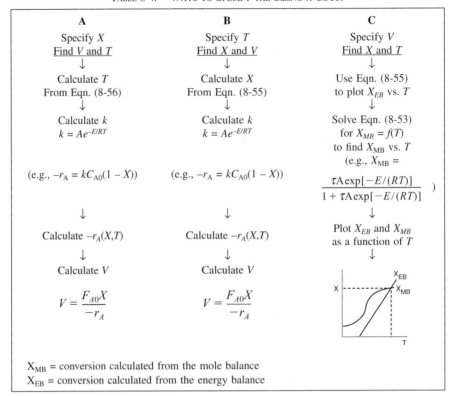

X_{MB} = conversion calculated from the mole balance
X_{EB} = conversion calculated from the energy balance

Example 8–8   Production of Propylene Glycol in an Adiabatic CSTR

Propylene glycol is produced by the hydrolysis of propylene oxide:

$$CH_2—CH—CH_3 + H_2O \xrightarrow{\text{H}_2\text{SO}_4} CH_2—CH—CH_3$$

O   OH   OH

Production, uses,
and economics

Over 800 million pounds of propylene glycol were produced in 2004 and the selling price was approximately $0.68 per pound. Propylene glycol makes up about 25% of the major derivatives of propylene oxide. The reaction takes place readily at room temperature when catalyzed by sulfuric acid.

You are the engineer in charge of an adiabatic CSTR producing propylene glycol by this method. Unfortunately, the reactor is beginning to leak, and you must replace it. (You told your boss several times that sulfuric acid was corrosive and that mild steel was a poor material for construction.) There is a nice-looking overflow CSTR of 300-gal capacity standing idle; it is glass-lined, and you would like to use it.

You are feeding 2500 lb/h (43.04 lb mol/h) of propylene oxide (P.O.) to the reactor. The feed stream consists of (1) an equivolumetric mixture of propylene oxide (46.62 ft^3/h) and methanol (46.62 ft^3/h), and (2) water containing 0.1 wt % H_2SO_4. The volumetric flow rate of water is 233.1 ft^3/h, which is 2.5 times the methanol–P.O. flow rate. The corresponding molar feed rates of methanol and water are 71.87 and 802.8 lb mol/h, respectively. The water–propylene oxide–methanol mixture undergoes a slight decrease in volume upon mixing (approximately 3%), but you neglect this decrease in your calculations. The temperature of both feed streams is 58°F prior to mixing, but there is an immediate 17°F temperature rise upon mixing of the two feed streams caused by the heat of mixing. The entering temperature of all feed streams is thus taken to be 75°F (Figure E8-8.1).

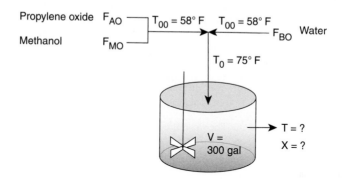

Figure E8-8.1

Furusawa et al.[7] state that under conditions similar to those at which you are operating, the reaction is first-order in propylene oxide concentration and apparent zero-order in excess of water with the specific reaction rate

$$k = Ae^{-E/RT} = 16.96 \times 10^{12} (e^{-32,400/RT}) \ \text{h}^{-1}$$

The units of E are Btu/lb mol.

There is an important constraint on your operation. Propylene oxide is a rather low-boiling substance. With the mixture you are using, you feel that you cannot exceed an operating temperature of 125°F, or you will lose too much oxide by vaporization through the vent system.

Can you use the idle CSTR as a replacement for the leaking one if it will be operated adiabatically? If so, what will be the conversion of oxide to glycol?

Solution

(All data used in this problem were obtained from the *Handbook of Chemistry and Physics* unless otherwise noted.) Let the reaction be represented by

$$A + B \longrightarrow C$$

where

A is propylene oxide ($C_{P_A} = 35$ Btu/lb mol · °F)[8]

B is water ($C_{P_B} = 18$ Btu/lb mol · °F)

---

[7] T. Furusawa, H. Nishimura, and T. Miyauchi, *J. Chem. Eng. Jpn.*, **2**, 95 (1969).

[8] C_{P_A} and C_{P_C} are estimated from the observation that the great majority of low-molecular-weight oxygen-containing organic liquids have a mass heat capacity of 0.6 cal/g · °C ±15%.

C is propylene glycol ($C_{P_C} = 46$ Btu/lb mol $\cdot$ °F)

M is methanol ($C_{P_M} = 19.5$ Btu/lb mol $\cdot$ °F)

In this problem neither the exit conversion nor the temperature of the adiabatic reactor is given. By application of the material and energy balances we can solve two equations with two unknowns (X and T). Solving these coupled equations, we determine the exit conversion and temperature for the glass-lined reactor to see if it can be used to replace the present reactor.

1. Mole Balance and design equation:

$$F_{A0} - F_A + r_A V = 0$$

The design equation in terms of X is

$$V = \frac{F_{A0} X}{-r_A} \tag{E8-8.1}$$

2. Rate Law:

$$-r_A = k C_A \tag{E8-8.2}$$

3. Stoichiometry (liquid phase, $v = v_0$):

$$C_A = C_{A0}(1 - X) \tag{E8-8.3}$$

4. Combining yields

Following the Algorithm

$$V = \frac{F_{A0} X}{k C_{A0}(1 - X)} = \frac{v_0 X}{k(1 - X)} \tag{E8-8.4}$$

Solving for X as a function of T and recalling that $\tau = V/v_0$ gives

$$X_{MB} = \frac{\tau k}{1 + \tau k} = \frac{\tau A e^{-E/RT}}{1 + \tau A e^{-E/RT}} \tag{E8-8.5}$$

This equation relates temperature and conversion through the **mole balance**.

Two equations, two unknown

5. The **energy balance** for this adiabatic reaction in which there is negligible energy input provided by the stirrer is

$$X_{EB} = \frac{\Sigma \, \Theta_i C_{P_i}(T - T_{i0})}{-[\Delta H_{Rx}^\circ(T_R) + \Delta C_P(T - T_R)]} \tag{8-29}$$

This equation relates X and T through the energy balance. We see that two equations [Equations (E8-8.5) and (E8-8.6)] must be solved with $X_{MB} = X_{EB}$ for the two unknowns, X and T.

6. Calculations:

a. *Heat of reaction at temperature T:*[9]

$$\Delta H_{Rx}(T) = \Delta H_{Rx}^\circ(T_R) + \Delta C_p(T - T_R) \tag{8-26}$$

$$H_A^\circ(68°F) : -66{,}600 \text{ Btu/lb mol}$$

[9] H_A° and H_C° are calculated from heat-of-combustion data.

Calculating the
parameter values

ΔH_{Rx}
$\Delta \hat{C}_P$
v_o
τ
C_{A0}
θ_M
θ_B
$\displaystyle\sum_{i=1}^{n} C_{P_i} \theta_i$

$$H_B^\circ(68°F) : -123{,}000 \text{ Btu/lb mol}$$

$$H_C^\circ(68°F) : -226{,}000 \text{ Btu/lb mol}$$

$$\Delta H_{Rx}^\circ(68°F) = -226{,}000 - (-123{,}000) - (-66{,}600) \qquad \text{(E8-8.7)}$$

$$= -36{,}400 \text{ Btu/lb mol propylene oxide}$$

$$\Delta C_P = C_{P_C} - C_{P_B} - C_{P_A}$$

$$= 46 - 18 - 35 = -7 \text{ Btu/lb mol} \cdot °F$$

$$\Delta H_{Rx}^\circ(T) = -36{,}400 - (7)(T - 528) \qquad T \text{ is in } °R$$

b. *Stoichiometry* (C_{A0}, Θ_i, τ): The total liquid volumetric flow rate entering the reactor is

$$v_0 = v_{A0} + v_{M0} + v_{B0}$$

$$= 46.62 + 46.62 + 233.1 = 326.3 \text{ ft}^3/\text{h} \qquad \text{(E8-8.8)}$$

$$V = 300 \text{ gal} = 40.1 \text{ ft}^3$$

$$\tau = \frac{V}{v_0} = \frac{40.1 \text{ ft}^3}{326.3 \text{ ft}^3/\text{h}} = 0.123 \text{ h}$$

$$C_{A0} = \frac{F_{A0}}{v_0} = \frac{43.0 \text{ lb mol/h}}{326.3 \text{ ft}^3/\text{h}}$$

$$= 0.132 \text{ lb mol/ft}^3 \qquad \text{(E8-8.9)}$$

$$\text{For methanol:} \quad \Theta_M = \frac{F_{M0}}{F_{A0}} = \frac{71.87 \text{ lb mol/h}}{43.0 \text{ lb mol/h}} = 1.67$$

$$\text{For water:} \quad \Theta_B = \frac{F_{B0}}{F_{A0}} = \frac{802.8 \text{ lb mol/h}}{43.0 \text{ lb mol/h}} = 18.65$$

c. *Evaluate mole balance terms:* The conversion calculated from the mole balance, X_{MB}, is found from Equation (E8-8.5).

Plot X_{MB} as a
function of
temperature.

$$X_{MB} = \frac{(16.96 \times 10^{12} \text{ h}^{-1})(0.1229 \text{ h}) \exp(-32{,}400/1.987T)}{1 + (16.96 \times 10^{12} \text{ h}^{-1})(0.1229 \text{ h}) \exp(-32{,}400/1.987T)}$$

$$\text{(E8-8.10)}$$

$$X_{MB} = \frac{(2.084 \times 10^{12}) \exp(-16{,}306/T)}{1 + (2.084 \times 10^{12}) \exp(-16{,}306/T)}, \quad T \text{ is in } °R$$

d. *Evaluate energy balance terms:*

$$\Sigma \Theta_i C_{P_i} = C_{P_A} + \Theta_B C_{P_B} + \Theta_M C_{P_M}$$

$$= 35 + (18.65)(18) + (1.67)(19.5)$$

$$= 403.3 \text{ Btu/lb mol} \cdot °F$$

$$T_0 = T_{00} + \Delta T_{mix} = 58°F + 17°F = 75°F$$

$$= 535°R \qquad \text{(E8-8.11)}$$

$$T_R = 68°F = 528°R$$

The conversion calculated from the energy balance, X_{EB}, for an adiabatic reaction is given by Equation (8-29):

$$X_{EB} = -\frac{\Sigma\, \Theta_i C_{P_i}(T - T_{i0})}{\Delta H^{\circ}_{Rx}(T_R) + \Delta C_P(T - T_R)} \tag{8-29}$$

Substituting all the known quantities into the energy balance gives us

$$X_{EB} = \frac{(403.3\ \text{Btu/lb mol}\cdot{}^{\circ}\text{F})(T - 535){}^{\circ}\text{F}}{-[-36,400 - 7(T - 528)]\ \text{Btu/lb mol}}$$

$$\boxed{X_{EB} = \frac{403.3(T - 535)}{36,400 + 7(T - 528)}} \tag{E8-8.12}$$

7. **Solving.** There are a number of different ways to solve these two simultaneous equations. The easiest way is to use the Polymath nonlinear equation solver. However, to give insight into the functional relationship between X and T for the mole and energy balances, we shall obtain a graphical solution. Here X is plotted as a function of T for the mole and energy balances, and the intersection of the two curves gives the solution where both the mole and energy balance solutions are satisfied. In addition, by plotting these two curves we can learn if there is more than one intersection (i.e., multiple steady states) for which both the energy balance and mole balance are satisfied. If numerical root-finding techniques were used to solve for X and T, it would be quite possible to obtain only one root when there are actually more than one. If Polymath were used, you could learn if multiple roots exist by changing your initial guesses in the nonlinear equation solver. We shall discuss multiple steady states further in Section 8.7. We choose T and then calculate X (Table E8-8.1). The calculations are plotted in Figure E8-8.2. The virtually straight line corresponds to the energy balance [Equation (E8-8.12)] and the curved line corresponds to the mole balance [Equation (E8-8.10)]. We observe from this plot that the only intersection point is at 85% conversion and 613°R. At this point, both the energy balance and mole balance are satisfied. Because the temperature must remain below 125°F (585°R), we cannot use the 300-gal reactor as it is now.

TABLE E8-8.1

T (°R)	X_{MB} [Eq. (E8-8.10)]	X_{EB} [Eq. (E8-8.12)]
535	0.108	0.000
550	0.217	0.166
565	0.379	0.330
575	0.500	0.440
585	0.620	0.550
595	0.723	0.656
605	0.800	0.764
615	0.860	0.872
625	0.900	0.980

Don't give up! Head back to the storage shed to check out the heat exchange equipment!

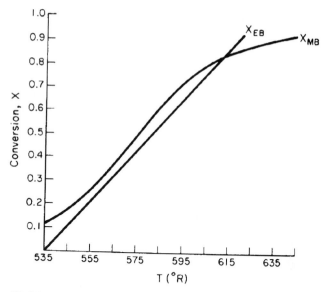

The reactor cannot be used because it will exceed the specified maximum temperature of 585°R.

Figure E8-8.2   The conversions X_{EB} and X_{MB} as a function of temperature.

Example 8–9   CSTR with a Cooling Coil

A cooling coil has been located in equipment storage for use in the hydration of propylene oxide discussed in Example 8-8. The cooling coil has 40 ft² of cooling surface and the cooling water flow rate inside the coil is sufficiently large that a constant coolant temperature of 85°F can be maintained. A typical overall heat-transfer coefficient for such a coil is 100 Btu/h·ft²·°F. Will the reactor satisfy the previous constraint of 125°F maximum temperature if the cooling coil is used?

Solution

If we assume that the cooling coil takes up negligible reactor volume, the conversion calculated as a function of temperature from the mole balance is the same as that in Example 8-8 [Equation (E8-8.10)].

1. **Combining the mole balance, stoichiometry, and rate law**, we have, from Example 8-8,

$$X_{MB} = \frac{\tau k}{1 + \tau k} = \frac{(2.084 \times 10^{12}) \exp(-16{,}306/T)}{1 + (2.084 \times 10^{12}) \exp(-16{,}306/T)} \qquad \text{(E8-8.10)}$$

T is in °R.

2. **Energy balance**. Neglecting the work by the stirrer, we combine Equations (8-27) and (8-50) to write

$$\frac{UA(T_a - T)}{F_{A0}} - X[\Delta H^\circ_{Rx}(T_R) + \Delta C_P(T - T_R)] = \Sigma \Theta_i C_{P_i}(T - T_0) \qquad \text{(E8-9.1)}$$

Solving the energy balance for X_{EB} yields

$$X_{EB} = \frac{\Sigma\Theta_i C_{P_i}(T-T_0) + [UA(T-T_a)/F_{A0}]}{-[\Delta H_{Rx}^\circ(T_R) + \Delta C_P(T-T_R)]} \qquad \text{(E8-9.2)}$$

The cooling coil term in Equation (E8-9.2) is

$$\frac{UA}{F_{A0}} = \left(100\ \frac{\text{Btu}}{\text{h}\cdot\text{ft}^2\cdot{}^\circ\text{F}}\right)\frac{(40\ \text{ft}^2)}{(43.04\ \text{lb mol/h})} = \frac{92.9\ \text{Btu}}{\text{lb mol}\cdot{}^\circ\text{F}} \qquad \text{(E8-9.3)}$$

Recall that the cooling temperature is

$$T_a = 85^\circ\text{F} = 545^\circ\text{R}$$

The numerical values of all other terms of Equation (E8-9.2) are identical to those given in Equation (E8-8.12) but with the addition of the heat exchange term.

$$X_{EB} = \frac{403.3(T-535) + 92.9(T-545)}{36,400 + 7(T-528)} \qquad \text{(E8-9.4)}$$

We now have two equations [(E8-8.10) and (E8-9.4)] and two unknowns, X and T.

The POLYMATH program and solution to these two Equations (E8-8.10), X_{MB}, and (E8-9.4), X_{EB}, are given in Tables E8-9.1 and E8-9.2. The exiting temperature and conversion are 103.7°F (563.7°R) and 36.4%, respectively, i.e.,

$$\boxed{T = 564^\circ\text{R and } X = 0.36}$$

TABLE E8-9.1.   POLYMATH: CSTR WITH HEAT EXCHANGE

Equations:

Nonlinear equations

[1] f(X) = X-(403.3*(T-535)+92.9*(T-545))/(36400+7*(T-528)) = 0
[2] f(T) = X-tau*k/(1+tau*k) = 0

Explicit equations

[1] tau = 0.1229
[2] A = 16.96*10^12
[3] E = 32400
[4] R = 1.987
[5] k = A*exp(-E/(R*T))

Solution output to Polymath program in Table E8-9.1 is shown in Table E8-9.2.

TABLE E8-9.2.   EXAMPLE 8-8 CSTR WITH HEAT EXCHANGE

Variable	Value	f(x)	Ini Guess
X	0.3636087	2.243E-11	0.367
T	563.72893	-5.411E-10	564
tau	0.1229		
A	1.696E+13		
E	3.24E+04		
R	1.987		
k	4.6489843		

8.7 Multiple Steady States

In this section we consider the steady-state operation of a CSTR in which a first-order reaction is taking place. We begin by recalling the hydrolysis of propylene oxide, Example 8-8.

If one were to examine Figure E8-8.2, one would observe that if a parameter were changed slightly, the X_{EB} line could move slightly to the left and there might be more than one intersection of the energy and mole balance curves. When more than one intersection occurs, there is more than one set of conditions that satisfy *both* the energy balance and mole balance; consequently, there will be multiple steady states at which the reactor may operate.

We begin by recalling Equation (8-54), which applies when one neglects shaft work and ΔC_P (i.e., $\Delta C_P = 0$ and therefore $\Delta H_{Rx} = \Delta H_{Rx}^\circ$).

$$-X\Delta H_{Rx}^\circ = C_{P0}(1+\kappa)(T-T_c) \tag{8-54}$$

where

$$\boxed{C_{P0} = \Sigma\,\Theta_i C_{P_i}}$$

$$\boxed{\kappa = \frac{UA}{C_{P0}F_{A0}}}$$

and

$$\boxed{T_c = \frac{T_0 F_{A0} C_{P0} + UAT_a}{UA + C_{P0}F_{A0}} = \frac{\kappa T_a + T_0}{1+\kappa}} \tag{8-57}$$

Using the CSTR mole balance $X = \dfrac{-r_A V}{F_{A0}}$, Equation (8-54) may be rewritten as

$$(-r_A V/F_{A0})(-\Delta H_{Rx}^\circ) = C_{P0}(1+\kappa)(T-T_c) \tag{8-58}$$

The left-hand side is referred to as the *heat-generated term*:

$G(T) =$ Heat-generated term

$$\boxed{G(T) = (-\Delta H_{Rx}^\circ)(-r_A V/F_{A0})} \tag{8-59}$$

The right-hand side of Equation (8-58) is referred to as the *heat-removed term* (by flow and heat exchange) $R(T)$:

$R(T) =$ Heat-removed term

$$\boxed{R(T) = C_{P0}(1+\kappa)(T-T_c)} \tag{8-60}$$

To study the multiplicity of steady states, we shall plot both $R(T)$ and $G(T)$ as a function of temperature on the same graph and analyze the circumstances under which we will obtain multiple intersections of $R(T)$ and $G(T)$.

8.7.1 Heat-Removed Term, $R(T)$

Vary Entering Temperature. From Equation (8-60) we see that $R(T)$ increases linearly with temperature, with slope $C_{P0}(1+\kappa)$. As the entering temperature T_0 is increased, the line retains the same slope but shifts to the right as shown in Figure 8-14.

Heat-removed curve
$R(T)$

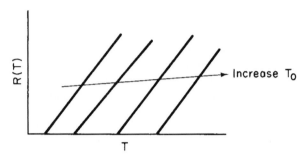

Figure 8-14   Variation of heat removal line with inlet temperature.

Vary Non-adiabatic Parameter κ. If one increases κ by either decreasing the molar flow rate F_{A0} or increasing the heat-exchange area, the slope increases and the ordinate intercept moves to the left as shown in Figure 8-15, for conditions of $T_a < T_0$:

$$\kappa = 0 \quad T_c = T_0$$
$$\kappa = \infty \quad T_c = T_a$$

If $T_a > T_0$, the intercept will move to the right as κ increases.

$$\kappa = \frac{UA}{C_{P0}F_{A0}}$$

$$T_c = \frac{T_0 + \kappa T_a}{1 + \kappa}$$

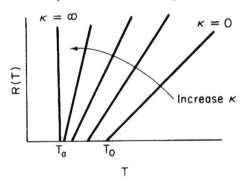

Figure 8-15   Variation of heat removal line with κ ($\kappa = UA/C_{P0}F_{A0}$).

8.7.2 Heat of Generation, $G(T)$

The heat-generated term, Equation (8-59), can be written in terms of conversion. (Recall: $X = -r_A V/F_{A0}$.)

$$G(T) = (-\Delta H^{\circ}_{\text{Rx}})X \tag{8-61}$$

To obtain a plot of heat generated, $G(T)$, as a function of temperature, we must solve for X as a function of T using the CSTR mole balance, the rate law, and stoichiometry. For example, for a first-order liquid-phase reaction, the CSTR mole balance becomes

$$V = \frac{F_{A0}X}{kC_A} = \frac{v_0 C_{A0}X}{kC_{A0}(1-X)}$$

Solving for X yields

1st-order reaction

$$X = \frac{\tau k}{1 + \tau k}$$

Substituting for X in Equation (8-61), we obtain

$$G(T) = \frac{-\Delta H^\circ_{Rx}\tau k}{1 + \tau k} \tag{8-62}$$

Finally, substituting for k in terms of the Arrhenius equation, we obtain

$$\boxed{G(T) = \frac{-\Delta H^\circ_{Rx}\tau A e^{-E/RT}}{1 + \tau A e^{-E/RT}}} \tag{8-63}$$

Note that equations analogous to Equation (8-63) for $G(T)$ can be derived for other reaction orders and for reversible reactions simply by solving the CSTR mole balance for X. For example, for the second-order liquid-phase reaction

2nd-order reaction

$$X = \frac{(2\tau k C_{A0} + 1) - \sqrt{4\tau k C_{A0} + 1}}{2\tau k C_{A0}}$$

the corresponding heat generated is

$$\boxed{G(T) = \frac{-\Delta H^\circ_{Rx}[(2\tau C_{A0}A e^{-E/RT} + 1) - \sqrt{4\tau C_{A0}A e^{-E/RT} + 1}]}{2\tau C_{A0}A e^{-E/RT}}} \tag{8-64}$$

At very low temperatures, the second term in the denominator of Equation (8-63) for the first-order reaction can be neglected so that $G(T)$ varies as

Low T

$$G(T) = -\Delta H^\circ_{Rx}\tau A e^{-E/RT}$$

(Recall that ΔH°_{Rx} means the heat of reaction is evaluated at T_R.)

At very high temperatures, the second term in the denominator dominates, and $G(T)$ is reduced to

High T

$$G(T) = -\Delta H^\circ_{Rx}$$

$G(T)$ is shown as a function of T for two different activation energies, E, in Figure 8-16. If the flow rate is decreased or the reactor volume increased so as to increase τ, the heat of generation term, $G(T)$, changes as shown in Figure 8-17.

Heat-generated
curves, $G(T)$

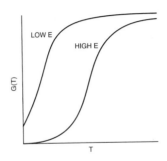

Figure 8-16   Heat generation curve for
identical frequency factors, A.

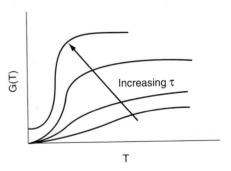

Figure 8-17   Variation of heat
generation curve with space-time.

8.7.3 Ignition-Extinction Curve

The points of intersection of $R(T)$ and $G(T)$ give us the temperature at which the reactor can operate at steady state. Suppose that we begin to feed our reactor at some relatively low temperature, T_{01}. If we construct our $G(T)$ and $R(T)$ curves, illustrated by curves y and a, respectively, in Figure 8-18, we see that there will be only one point of intersection, point 1. From this point of intersection, one can find the steady-state temperature in the reactor, T_{s1}, by following a vertical line down to the T-axis and reading off the temperature as shown in Figure 8-18.

If one were now to increase the entering temperature to T_{02}, the $G(T)$ curve, y, would remain unchanged, but the $R(T)$ curve would move to the right, as shown by line b in Figure 8-18, and will now intersect the $G(T)$ at point 2 and be tangent at point 3. Consequently, we see from Figure 8-18 that there are two steady-state temperatures, T_{s2} and T_{s3}, that can be realized in the CSTR for an entering temperature T_{02}. If the entering temperature is increased to T_{03}, the $R(T)$ curve, line c (Figure 8-19), intersects the $G(T)$ three times and there are three steady-state temperatures. As we continue to increase T_0, we finally reach line e, in which there are only two steady-state temperatures. By further increasing T_0 we reach line f, corresponding to T_{06}, in which we have only one temperature that will satisfy both the mole and energy balances. For the six entering temperatures, we can form Table 8-5, relating the entering temperature to the possible reactor operating temperatures. By plotting T_s as a function of T_0, we obtain the well-known *ignition-extinction curve* shown in Figure 8-20. From this figure we see that as the entering temperature is increased, the steady-state temperature increases along the bottom line until T_{05} is reached. Any fraction of a degree increase in temperature beyond T_{05} and the steady-state reactor temperature will jump up to T_{s11}, as shown in Figure 8-20. The temperature at which

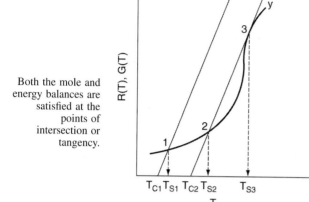

Both the mole and
energy balances are
satisfied at the
points of
intersection or
tangency.

Figure 8-18   Finding multiple steady
states with T_0 varied.

Figure 8-19   Finding multiple steady
states with T_0 varied.

TABLE 8-5.   MULTIPLE STEADY-STATE TEMPERATURES

Entering Temperature	Reactor Temperatures		
T_{01}		T_{s1}	
T_{02}		T_{s2}	T_{s3}
T_{03}	T_{s4}	T_{s5}	T_{s6}
T_{04}	T_{s7}	T_{s8}	T_{s9}
T_{05}		T_{s10}	T_{s11}
T_{06}		T_{s12}	

We must exceed a
certain feed
temperature to
operate at the
upper steady state
where the
temperature and
conversion are
higher.

this jump occurs is called the *ignition temperature*. If a reactor were operating at T_{s12} and we began to cool the entering temperature down from T_{06}, the steady-state reactor temperature T_{s3} would eventually be reached, corresponding to an entering temperature T_{02}. Any slight decrease below T_{02} would drop the steady-state reactor temperature to T_{s2}. Consequently, T_{02} is called the *extinction temperature*.

The middle points 5 and 8 in Figures 8-19 and 8-20 represent unstable steady-state temperatures. Consider the heat removal line d in Figure 8-19 along with the heat-generated curve which is replotted in Figure 8-21.

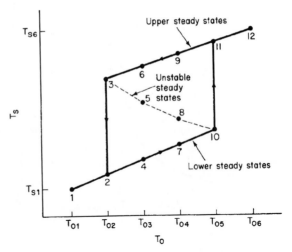

Figure 8-20   Temperature ignition-extinction curve.

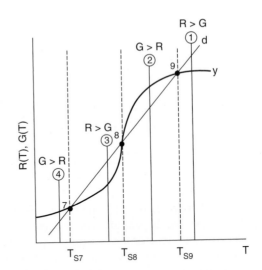

Figure 8-21   Stability on multiple state temperatures.

If we were operating at T_{s8}, for example, and a pulse increase in reactor temperature occurred, we would find ourselves at the temperature shown by vertical line ② between points 8 and 9. We see along this vertical line ② the heat-generated curve, G, is greater than the heat-removed line R ($G > R$). Consequently, the temperature in the reactor would continue to increase until point 9 is reached at the upper steady state. On the other hand, if we had a pulse decrease in temperature from point 8, we would find ourselves a vertical line ③ between points 7 and 8. Here we see the heat-removed curve is greater than the heat-generated curve so the temperature will continue to decrease until the lower steady state is reached. That is a small change in temperature either above or below the middle steady-state temperature, T_{s8} will cause the reactor temperature to move away from this middle steady state. Steady states that behave in the manner are said to be unstable.

In contrast to these unstable operating points, there are stable operating points. Consider what would happen if a reactor operating at T_{s9} were subjected to a pulse increase in reactor temperature indicated by line ① in Figure 8-21. We see that the heat-removed line (d) is greater than the heat-generated curve (y), so that the reactor temperature will decrease and return to T_{s9}. On the other hand, if there is a sudden drop in temperature below T_{s9} as indicated by line ②, we see the heat-generated curve (y) is greater than the heat-removed line (d) and the reactor temperature will increase and return to the upper steady state at T_{s9}.

Next let's look at what happens when the lower steady-state temperature at T_{s7} is subjected to pulse increase to the temperature shown as line ③ in Figure 8-21. Here we again see that the heat removed, R, is greater than the heat generated, G, so that the reactor temperature will drop and return to T_{s7}. If there is a sudden decrease in temperature below T_{s7} to the temperature indicated by line ④, we see that the heat generated is greater than the heat removed, $G > R$, and that the reactor temperature will increase until it returns to T_{s7}. A similar analysis could be carried out for temperature T_{s1}, T_{s2}, T_{s4}, T_{s6}, T_{s11}, and T_{s12} and one would find that reactor temperatures would always return to *local steady-state values*, when subjected to both positive and negative fluctuations.

While these points are locally stable, they are not necessarily globally stable. That is, a perturbation in temperature or concentration, while small, may be sufficient to cause the reactor to fall from the upper steady state (corresponding to high conversion and temperature such as point 9 in Figure 8-21) to the lower steady state (corresponding to low temperature and conversion, point 7). We will examine this case in detail in Section 9.4 and in Problem P9-16$_B$.

An excellent experimental investigation that demonstrates the multiplicity of steady states was carried out by Vejtasa and Schmitz (Figure 8-22). They studied the reaction between sodium thiosulfate and hydrogen peroxide:

$$2Na_2S_2O_3 + 4H_2O_2 \rightarrow Na_2S_3O_6 + Na_2SO_4 + 4H_2O$$

in a CSTR operated adiabatically. The multiple steady-state temperatures were examined by varying the flow rate over a range of space times, τ, as shown in Figure 8-23. One observes from this figure that at a space-time of 12 s, steady-state reaction temperatures of 4, 33, and 80°C are possible. If one were operating on the higher steady-state temperature line and the volumetric flow rates were steadily increased (i.e., the space-time decreased), one notes that if the space velocity dropped below about 7 s, the reaction temperature would drop from 70°C to 2°C. The flow rate at which this drop occurs is referred to as the *blowout velocity*.

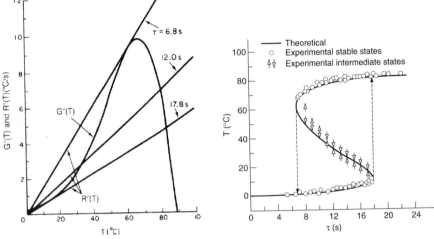

Actual experiments
showing existence of
multiple steady
states

Figure 8-22   Heat generation and removal   **Figure 8-23**   Multiple steady states.
functions for feed mixture of 0.8 M Na$_2$S$_2$O$_3$
and 1.2 M H$_2$O$_2$ at 0°C.

By S. A. Vejtasa and R. A. Schmitz, *AIChE J., 16* (3), 415 (1970). (Reproduced by
permission of the American Institute of Chemical Engineers. Copyright © 1970
AIChE. All right reserved.) See Journal Critique Problem P8C-4.

8.7.4 Runaway Reactions in a CSTR

In many reacting systems, the temperature of the upper steady state may be
sufficiently high that it is undesirable or even dangerous to operate at this condition. For example, at the higher temperatures, secondary reactions can take
place, or as in the case of propylene glycol in Examples 8-8 and 8-9, evaporation of the reacting materials can occur.

We saw in Figure 8-20, that we operated at the upper steady state after
we exceeded the ignition temperature. For a CSTR, we shall consider runaway
(ignition) to occur when we move from the lower steady state to the upper
steady state. The ignition temperature occurs at the point of tangency of the
heat removed curve to the heat-generated curve. If we move slightly off this
point of tangency as shown in Figure 8-24, then runaway is said to have
occurred.

At this point of tangency, T^*, we have not only

$$R(T^*) = G(T^*)$$

$$C_{P_o}(1 + \kappa)(T^* - T_C) = (-\Delta H_{Rx})X^* = (-\Delta H_{Rx})\frac{(-r_A^* V)}{F_{A0}} \qquad (8\text{-}65)$$

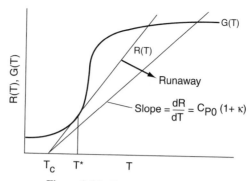

Figure 8-24   Runaway in a CSTR.

but also the slopes of the $R(T)$ and $G(T)$ curves are also equal. For the heat-removed curve, the slope is

$$\left.\frac{dR(T)}{dT}\right|_{T^*} = C_{P_0}(1+\kappa) \tag{8-66}$$

and for the heat-generated curve, the slope S^* is

$$S^* = \left.\frac{dG(T)}{dT}\right|_{T^*} = \frac{d\left[(-\Delta H_{Rx})\dfrac{-r_A V}{F_{A0}}\right]}{dT}\Bigg|_{T^*} = \left(\frac{(-\Delta H_{Rx})V}{F_{A0}}\right)\frac{d(-r_A)}{dT}\Bigg|_{T^*} \tag{8-67}$$

We shall use S^* and Equation (8-67) to develop our stability diagram. As an example we will consider an irrevesible reaction that follows a power law model and the reacting species are a *very* weak function of temperature.[†]

$$-r_A = (Ae^{-E/RT})\, fn(C_i) \tag{8-68}$$

then

$$\left.\frac{d(-r_A)}{dT}\right|_{T^*} = \frac{E}{RT^{*2}}Ae^{-E/RT^*}fn(C_i) = \frac{E}{RT^{*2}}(-r_A^*)$$

Substituting for the derivative of $(-r_A)$ wrt T in Equation (8-67)

$$S^* = \left.\frac{dG(T)}{dT}\right|_{T^*} = \overbrace{\left(\frac{(-\Delta H_{Rx})}{F_{A0}}(-r_A^* V)\right)}^{G(T^*)}\frac{E}{RT^{*2}} = G(T^*)\frac{E}{RT^{*2}} \tag{8-69}$$

$$\left.\frac{dG(T)}{dT}\right|_{T^*} = S^*$$

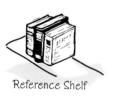

Reference Shelf

[†] Strictly speaking this development is only true for a zero order reaction. A development for other reaction orders is given in Professional Reference Shelf R8.1A.

where for a zero order reaction

$$S^* = G(T^*)\frac{E}{RT^{*2}}$$

Equating Equations (8-66) and (8-69) yields

$$C_{P0}(1 + \kappa) = \overbrace{\frac{E}{RT^{*2}}(-r_A^* V)\frac{(-\Delta H^\circ_{Rx})}{F_{A0}}}^{G(T^*)} = S^* \qquad (8\text{-}70)$$

Next, we divide Equation (8-65) by Equation (8-70) to obtain the following ΔT value for a CSTR operating at $T = T^*$:

$$\Delta T_{rc} = T^* - T_c = \frac{RT^{*2}}{E} \qquad (8\text{-}71)$$

If this difference between the reactor temperature and T_c, ΔT_{rc}, is exceeded, transition to the upper steady state will occur. For many industrial reactions, E/RT is typically between 16 and 24, and the reaction temperatures may be between 300 to 500 K. Consequently, this critical temperature difference ΔT_{rc} will be somewhere around 15 to 30°C.

Stability Diagram. We now want to develop a stability diagram that will show regions of stable operation and unstable operation. One such diagram would be a plot of S^* as a function of T_c. To construct this plot, we first solve Equation (8-71) for T^*, the reactor temperature at the point of tangency,

$$T^* = \frac{E}{2R}\left[1 - \sqrt{1 - 4\frac{T_c R}{E}}\right] \qquad (8\text{-}72)$$

and recalling

$$T_c = \frac{\kappa T_a + T_0}{1 + \kappa}$$

We can now vary T_c, then calculate T^* [Equation (8-72)], calculate k^* ($k^* = Ae^{-E/RT^*}$), calculate $-r_A^*$ at T^* from rate law, calculate $G(T^*)$ [Equation (8-59)], and then finally calculate S^* to make a plot of S^* as a function of T_c as shown in Figure 8-25. We see that any deviation to the right or below the intersection of $C_{P0}(1 + \kappa)$ and S^* will result in runaway.

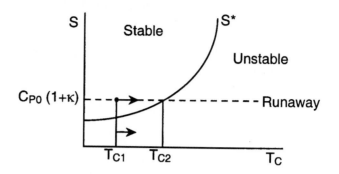

Figure 8-25   CSTR stability diagram.

As shown in PRS R8.1A, the equation for S^* for a first order reaction is

$$S^* = \frac{\tau A e^{-E/RT^*}}{\left(1 + \tau A e^{-E/RT^*}\right)^2}(-\Delta H_{Rx})\frac{E}{RT^{*2}}\qquad(8\text{-}73)$$

Reference Shelf

We simply combine Equation (8-72) and the equation for T_c and then substitute the result into Equation (8-73) and plot S^* as a function of T_0. From Figure 8-25, we see that for a given value of $[C_{P0}(1+\kappa)]$, if we were to increase the entering temperature T_0 from some low-value T_{01} (T_{c1}) to a higher entering temperature value T_{02} (T_{c2}), we would reach a point at which runaway would occur. Further discussions are given on the CD-ROM professional reference shelf R8.2. Referring to Equation (8-70), we can infer

$$C_{P0}(1+\kappa) > S^* \qquad(8\text{-}74)$$

we will not move to the upper steady state, and runaway will not occur. However, if

$$C_{P0}(1+\kappa) < S^* \qquad(8\text{-}75)$$

runaway will occur.

8.8 Nonisothermal Multiple Chemical Reactions

Most reacting systems involve more than one reaction and do not operate isothermally. **This section is one of the most important sections of the book.** It ties together all the previous chapters to analyze multiple reactions that do not take place isothermally.

8.8.1 Energy Balance for Multiple Reactions in Plug-Flow Reactors

In this section we give the energy balance for multiple reactions. We begin by recalling the energy balance for a single reaction taking place in a PFR which is given by Equation (8-35),

$$\frac{dT}{dV} = \frac{Ua(T_a - T) + (-r_A)[-\Delta H_{Rx}(T)]}{\sum_{i=1}^{m} F_i C_{P_i}}$$
(8-35)

When q multiple reactions are taking place in the PFR and there are m species, it is easily shown that Equation (8-35) can be generalized to

Energy balance for multiple reactions

$$\boxed{\frac{dT}{dV} = \frac{Ua(T_a - T) + \sum_{i=1}^{q} (-r_{ij})[-\Delta H_{Rxij}(T)]}{\sum_{j=1}^{m} F_j C_{P_j}}}$$
(8-76)

i = Reaction number
j = Species

The heat of reaction for reaction i must be referenced to the same species in the rate, r_{ij}, by which ΔH_{Rxij} is multiplied, that is,

$$[-r_{ij}][-\Delta H_{Rxij}] = \begin{bmatrix} \dfrac{\text{Moles of } j \text{ reacted in reaction } i}{\text{Volume} \cdot \text{time}} \end{bmatrix} \times \begin{bmatrix} \dfrac{\text{Joules "released" in reaction } i}{\text{Moles of } j \text{ reacted in reaction } i} \end{bmatrix}$$

$$= \begin{bmatrix} \dfrac{\text{Joules "released" in reaction } i}{\text{Volume} \cdot \text{time}} \end{bmatrix}$$

where the subscript j refers to the species, the subscript i refers to the particular reaction, q is the number of **independent** reactions, and m is the number of species.

Consider the following reaction sequence carried out in a PFR:

Reaction 1: $\qquad A \xrightarrow{k_1} B$ (8-77)

Reaction 2: $\qquad B \xrightarrow{k_2} C$ (8-78)

The PFR energy balance becomes

$$\frac{dT}{dV} = \frac{Ua(T_a - T) + (-r_{1A})(-\Delta H_{Rx1A}) + (-r_{2B})(-\Delta H_{Rx2B})}{F_A C_{PA} + F_B C_{PB} + F_C C_{PC}}$$
(8-79)

where ΔH_{Rx1A} = [kJ/mol of A reacted in reaction 1] and
ΔH_{Rx2B} = [kJ/mol of B reacted in reaction 2].

Example 8–10 Parallel Reactions in a PFR with Heat Effects

The following gas-phase reactions occur in a PFR:

$$\text{Reaction 1:} \quad A \xrightarrow{k_1} B \qquad -r_{1A} = k_{1A}C_A \tag{E8-10.1}$$

$$\text{Reaction 2:} \quad 2A \xrightarrow{k_2} C \qquad -r_{2A} = k_{2A}C_A^2 \tag{E8-10.2}$$

Pure A is fed at a rate of 100 mol/s, a temperature of 150°C, and a concentration of 0.1 mol/dm^3. Determine the temperature and flow rate profiles down the reactor.

Living Example Problem

Additional information:

$$\Delta H_{Rx1A} = -20,000 \text{ J/(mol of A reacted in reaction 1)}$$

$$\Delta H_{Rx2A} = -60,000 \text{ J/(mol of A reacted in reaction 2)}$$

$$C_{P_A} = 90 \text{ J/mol} \cdot °C \qquad\qquad k_{1A} = 10 \exp\left[\frac{E_1}{R}\left(\frac{1}{300} - \frac{1}{T}\right)\right] \text{ s}^{-1}$$

$$C_{P_B} = 90 \text{ J/mol} \cdot °C \qquad\qquad E_1/R = 4000 \text{ K}$$

$$C_{P_C} = 180 \text{ J/mol} \cdot °C \qquad\quad k_{2A} = 0.09 \exp\left[\frac{E_2}{R}\left(\frac{1}{300} - \frac{1}{T}\right)\right]\frac{\text{dm}^3}{\text{mol} \cdot \text{s}}$$

$$Ua = 4000 \text{ J/m}^3 \cdot \text{s} \cdot °C \qquad E_2/R = 9000 \text{ K}$$

$$T_a = 100°C$$

Solution

The PFR energy balance becomes [cf. Equation (8-76)]

$$\frac{dT}{dV} = \frac{Ua(T_a - T) + (-r_{1A})(-\Delta H_{Rx1A}) + (-r_{2A})(-\Delta H_{Rx2A})}{F_A C_{PA} + F_B C_{PB} + F_C C_{PC}} \tag{E8-10.3}$$

Mole balances:

$$\frac{dF_A}{dV} = r_A \tag{E8-10.4}$$

$$\frac{dF_B}{dV} = r_B \tag{E8-10.5}$$

$$\frac{dF_C}{dV} = r_C \tag{E8-10.6}$$

Rate laws, relative rates, and net rates:

Rate laws

$$r_{1A} = -k_{1A}C_A \tag{E8-10.1}$$

$$r_{2A} = -k_{2A}C_A^2 \tag{E8-10.2}$$

One of the major goals of this text is that the reader will be able to solve multiple reactions with heat effects.

Relative rates

Reaction 1: $\quad\quad\quad \dfrac{r_{1A}}{-1} = \dfrac{r_{1B}}{1}; \quad r_{1B} = -r_{1A} = k_{1A}C_A$

Reaction 2: $\quad\quad\quad \dfrac{r_{2A}}{-2} = \dfrac{r_{2C}}{1}; \quad r_{2C} = -\dfrac{1}{2}\, r_{2A} = \dfrac{k_{2A}}{2}C_A^2$

Net rates:

$$r_A = r_{1A} + r_{2A} = -k_{1A}C_A - k_{2A}C_A^2 \qquad (E8\text{-}10.7)$$

$$r_B = r_{1B} = k_{1A}C_A \qquad (E8\text{-}10.8)$$

$$r_C = r_{2C} = \dfrac{1}{2}k_{2A}C_A^2 \qquad (E8\text{-}10.9)$$

Stoichiometry (gas phase $\Delta P = 0$):

$$C_A = C_{T0}\left(\dfrac{F_A}{F_T}\right)\left(\dfrac{T_0}{T}\right) \qquad (E8\text{-}10.10)$$

$$C_B = C_{T0}\left(\dfrac{F_B}{F_T}\right)\left(\dfrac{T_0}{T}\right) \qquad (E8\text{-}10.11)$$

The algorithm for multiple reactions with heat effects

$$C_C = C_{T0}\left(\dfrac{F_C}{F_T}\right)\left(\dfrac{T_0}{T}\right) \qquad (E8\text{-}10.12)$$

$$F_T = F_A + F_B + F_C \qquad (E8\text{-}10.13)$$

$$k_{1A} = 10 \exp\left[3000\left(\dfrac{1}{300} - \dfrac{1}{T}\right)\right]\text{s}^{-1}$$

$$(T \text{ in K})$$

$$k_{2A} = 0.09 \exp\left[9000\left(\dfrac{1}{300} - \dfrac{1}{T}\right)\right]\dfrac{\text{dm}^3}{\text{mol}\cdot\text{s}}$$

Energy balance:

$$\dfrac{dT}{dV} = \dfrac{4000(373 - T) + (-r_{1A})(20{,}000) + (-r_{2A})(60{,}000)}{90F_A + 90F_B + 180F_C} \qquad (E8\text{-}10.14)$$

The Polymath program and its graphical outputs are shown in Table E8-10.1 and Figures E8-10.1 and E8-10.2.

Living Example Problem

TABLE E8-10.1. POLYMATH PROGRAM

Equations:

POLYMATH Results

Example 8-10 Parallel Reaction in a PFR with Heat Effects 08-13-2004, Rev5.1.232

Calculated values of the DEQ variables

Variable	initial value	minimal value	maximal value	final value
V	0	0	1	1
Fa	100	2.738E-06	100	2.738E-06
Fb	0	0	55.04326	55.04326
Fc	0	0	22.478369	22.478369
T	423	423	812.19122	722.08816
k1a	482.8247	482.8247	4.484E+04	2.426E+04
k2a	553.05566	553.05566	1.48E+07	3.716E+06
Cto	0.1	0.1	0.1	0.1
Ft	100	77.521631	100	77.521631
To	423	423	423	423
Ca	0.1	2.069E-09	0.1	2.069E-09
Cb	0	0	0.0415941	0.0415941
Cc	0	0	0.016986	0.016986
r1a	-48.28247	-373.39077	-5.019E-05	-5.019E-05
r2a	-5.5305566	-848.11153	-1.591E-11	-1.591E-11

ODE Report (RKF45)

Differential equations as entered by the user
[1] d(Fa)/d(V) = r1a+r2a
[2] d(Fb)/d(V) = -r1a
[3] d(Fc)/d(V) = -r2a/2
[4] d(T)/d(V) = (4000*(373-T)+(-r1a)*20000+(-r2a)*60000)/(90*Fa+90*Fb+180*Fc)

Explicit equations as entered by the user
[1] k1a = 10*exp(4000*(1/300-1/T))
[2] k2a = 0.09*exp(9000*(1/300-1/T))
[3] Cto = 0.1
[4] Ft = Fa+Fb+Fc
[5] To = 423
[6] Ca = Cto*(Fa/Ft)*(To/T)
[7] Cb = Cto*(Fb/Ft)*(To/T)
[8] Cc = Cto*(Fc/Ft)*(To/T)
[9] r1a = -k1a*Ca
[10] r2a = -k2a*Ca^2

Why does the temperature go through a maximum value?

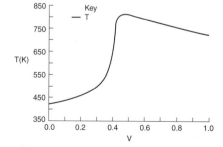

Figure E8-10.1   Temperature profile.

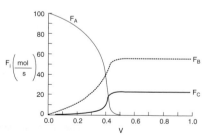

Figure E8-10.2   Profile of molar flow rates $F_A, F_B,$ and F_C.

8.8.2 Energy Balance for Multiple Reactions in CSTR

Recall that $-F_{A0}X = r_A V$ for a CSTR and that $\Delta H_{Rx}(T) = \Delta H_{Rx}^{\circ} + \Delta C_P(T - T_R)$ so that Equation (8-27) for the steady-state energy balance for a single reaction may be written as

$$\dot{Q} - \dot{W}_s - F_{A0} \Sigma \Theta_i C_{P_i}(T - T_0) + [\Delta H_{Rx}(T)][r_A V] = 0 \tag{8-27}$$

For q multiple reactions and m species, the CSTR energy balance becomes

$$\boxed{\dot{Q} - \dot{W}_s - F_{A0} \sum_{i=1}^{m} \Theta_i C_{P_i}(T - T_0) + V \sum_{i=1}^{q} r_{ij} \Delta H_{Rxij}(T) = 0} \tag{8-80}$$

> **Energy balance for multiple reactions in a CSTR**

Substituting Equation (8-50) for $\dot{Q}$, neglecting the work term, and assuming constant heat capacities, Equation (8-80) becomes

$$\boxed{UA(T_a - T) - F_{A0} \sum_{i=1}^{m} C_{P_i}\Theta_i(T - T_0) + V \sum_{i=1}^{q} r_{ij} \Delta H_{Rxij}(T) = 0} \tag{8-81}$$

For the two parallel reactions described in Example 8-10, the CSTR energy balance is

$$UA(T_a - T) - F_{A0} \sum_{i=1}^{m} \Theta_i C_{P_i}(T - T_0) + Vr_{1A}\Delta H_{Rx1A}(T) + Vr_{2A}\Delta H_{Rx2A}(T) = 0$$

$$\tag{8-82}$$

> **Major goal of CRE**

One of the **major goals** of this text is to have the reader solve problems involving multiple reactions with heat effects (cf. Problem P8-26$_C$).

Example 8–11 Multiple Reactions in a CSTR

The elementary liquid-phase reactions

$$A \xrightarrow{k_1} B \xrightarrow{k_2} C$$

take place in a 10-dm³ CSTR. What are the effluent concentrations for a volumetric feed rate of 1000 dm³/min at a concentration of A of 0.3 mol/dm³?

The inlet temperature is 283 K.

Additional information:

$$C_{P_A} = C_{P_B} = C_{P_C} = 200 \text{ J/mol} \cdot \text{K}$$

$k_1 = 3.3 \text{ min}^{-1}$ at 300 K, with $E_1 = 9900 \text{ cal/mol}$

$k_2 = 4.58 \text{ min}^{-1}$ at 500 K, with $E_2 = 27{,}000 \text{ cal/mol}$

$$\Delta H_{Rx1A} = -55,000 \text{ J/mol A} \qquad UA = 40,000 \text{ J/min} \cdot \text{K with } T_a = 57°C$$

$$\Delta H_{Rx2B} = -71,500 \text{ J/mol B}$$

Solution

The reactions follow elementary rate laws

$$r_{1A} = -k_{1A}C_A$$
$$r_{2B} = -k_{2B}C_B$$

1. Mole Balance on Every Species

A: Combined mole balance and rate law for A:

$$V = \frac{F_{A0} - F_A}{-r_A} = \frac{v_0[C_{A0} - C_A]}{-r_{1A}} = \frac{v_0[C_{A0} - C_A]}{k_1 C_A} \qquad (E8\text{-}11.1)$$

Solving for C_A gives us

$$C_A = \frac{C_{A0}}{1 + \tau k_1} \qquad (E8\text{-}11.2)$$

B: Combined mole balance and rate law for B:

$$V = \frac{0 - C_B v_0}{-r_B} = \frac{C_B v_0}{(r_{1B} + r_{2B})} = \frac{C_B v_0}{k_1 C_A - k_2 C_B} \qquad (E8\text{-}11.3)$$

Solving for C_B yields

$$C_B = \frac{\tau k_1 C_A}{1 + \tau k_2} = \frac{\tau k_1 C_{A0}}{(1 + \tau k_1)(1 + \tau k_2)} \qquad (E8\text{-}11.4)$$

2. Rate Laws:

$$\boxed{-r_{1A} = k_1 C_A = \frac{k_1 C_{A0}}{1 + \tau k_1}} \qquad (E8\text{-}11.5)$$

$$\boxed{-r_{2B} = k_2 C_B = \frac{k_2 \tau k_1 C_{A0}}{(1 + \tau k_1)(1 + \tau k_2)}} \qquad (E8\text{-}11.6)$$

3. Energy Balances:

Applying Equation (8-82) to this system gives

$$UA(T_a - T) - F_{A0} C_{P_A}(T - T_0) + V[r_{1A} \Delta H_{Rx1A} + r_{2B} \Delta H_{Rx2B}] = 0 \quad (E8\text{-}11.7)$$

Substituting for r_{1A} and r_{2B} and rearranging, we have

$$\underbrace{\left[-\frac{\Delta H_{Rx1A} \tau k_1}{1 + \tau k_1} - \frac{\tau k_1 \tau k_2 \Delta H_{Rx2B}}{(1 + \tau k_1)(1 + \tau k_2)} \right]}_{G(T)} = \overbrace{C_P(1 + \kappa)[T - T_c]}^{R(T)} \qquad (E8\text{-}11.8)$$

$$\kappa = \frac{UA}{F_{A0} C_{P_A}} = \frac{40,000 \text{ J/min} \cdot \text{K}}{(0.3 \text{ mol/dm}^3)(1000 \text{ dm}^3/\text{min}) \, 200 \text{ J/mol} \cdot \text{K}} = 0.667$$

$$T_c = \frac{T_0 + \kappa T_a}{1 + \kappa} = \frac{283 + (0.666)(330)}{1 + 0.667} = 301.8 \text{ K} \qquad \text{(E8-11.9)}$$

$$G(T) = \left[-\frac{\Delta H_{Rx1A}\tau k_1}{1 + \tau k_1} - \frac{\tau k_1 \tau k_2 \, \Delta H_{Rx2B}}{(1 + \tau k_1)(1 + \tau k_2)} \right] \qquad \text{(E8-11.10)}$$

$$R(T) = C_P(1 + \kappa)[T - T_c] \qquad \text{(E8-11.11)}$$

We are now going to write a Polymath program to increment temperature to obtain $G(T)$ and $R(T)$. The Polymath program to plot $R(T)$ and $G(T)$ vs. T is shown in Table E8-11.1, and the resulting graph is shown in Figure E8-11.1.

TABLE E8-11.1.   POLYMATH

Equations:

POLYMATH Results
Example 8-11 Multiple Reactions in a CSTR 08-13-2004, Rev5.1.232
ODE Report (RKF45)

Differential equations as entered by the user
[1] d(T)/d(t) = 2

Explicit equations as entered by the user
[1]   Cp = 200
[2]   Cao = 0.3
[3]   To = 283
[4]   tau = .01
[5]   DH1 = -55000
[6]   DH2 = -71500
[7]   vo = 1000
[8]   E2 = 27000
[9]   E1 = 9900
[10] UA = 40000
[11] Ta = 330
[12] k2 = 4.58*exp((E2/1.987)*(1/500-1/T))
[13] k1 = 3.3*exp((E1/1.987)*(1/300-1/T))
[14] Ca = Cao/(1+tau*k1)
[15] kappa = UA/(vo*Cao)/Cp
[16] G = -tau*k1/(1+k1*tau)*DH1-k1*tau*k2*tau*DH2/((1+tau*k1)*(1+tau*k2))
[17] Tc = (To+kappa*Ta)/(1+kappa)
[18] Cb = tau*k1*Ca/(1+k2*tau)
[19] R = Cp*(1+kappa)*(T-Tc)
[20] Cc = Cao-Ca-Cb
[21] F = G-R

Incrementing temperature in this manner is an easy way to generate R(T) and G(T) plots

Living Example Problem

When F = 0 G(T) = R(T) and the steady states can be found.

We see that five steady states (SS) exist. The exit concentrations and temperatures listed in Table E8-11.2 were interpreted from the tabular output of the Polymath program.

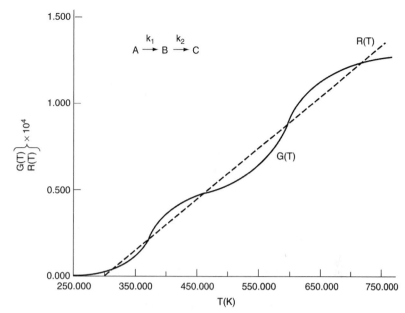

Wow! Five (5)
multiple steady
states!

Figure E8-11.1   Heat-removed and heat-generated curves.

We note there are five steady states (SS) whose values are given in Table E8-11.2. What do you think of the value of tau? Is it a realistic number?

TABLE E8-11.2.   EFFLUENT CONCENTRATIONS AND TEMPERATURES

SS	T	C_A	C_B	C_C
1	310	0.285	0.015	0
2	363	0.189	0.111	0.0
3	449	0.033	0.265	0.002
4	558	0.004	0.163	0.132
5	677	0.001	0.005	0.294

8.9 Radial and Axial Variations in a Tubular Reactor

COMSOL
application

Web Modules

In the previous sections we have assumed that there were no radial variations in velocity, concentration, temperature or reaction rate in the tubular and packed bed reactors. As a result the axial profiles could be determined using an ordinary differential equation (ODE) solver. In this section we will consider the case where we have both axial and radial variations in the system variables in which case will require a partial differential (PDE) solver. A PDE solver such as COMSOL, will allow us to solve tubular reactor problems for both the axial and radial profiles, as shown on the web module.

We are going to carry out differential mole and energy balances on the differential cylindrical annulus shown in Figure 8-26.

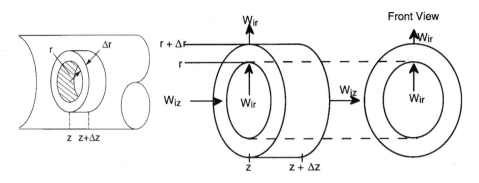

Figure 8-26 Cylindrical shell of thickness Δr, length Δz, and volume $2\pi r \Delta r \Delta z$.

Molar Flux

In order to derive the governing equations we need to define a couple of terms. The first is the molar flux of species i, W_i (mol/m² • s). The molar flux has two components, the radial component W_{ir}, and the axial component, W_{iz}. The molar flow rates are just the product of the molar fluxes and the cross-sectional areas normal to their direction of flow A_{cz}. For example, for species i flowing in the axial (i.e., z) direction

$$F_{iz} = W_{iz} A_{cz}$$

where W_{iz} is the molar flux in the z direction (mol/m²/s), and A_{cz} (m²) is the cross-sectional area of the tubular reactor.

In Chapter 11 we discuss the molar fluxes in some detail, but for now let us just say they consist of a diffusional component, $-D_e(\partial C_i/\partial z)$, and a convective flow component, $U_z C_i$

$$W_{iz} = -D_e \frac{\partial C_i}{\partial z} + U_z C_i \qquad (8\text{-}83)$$

where D_e is the effective diffusivity (or dispersion coefficient) (m²/s), and U_z is the axial molar average velocity (m/s). Similarly, the flux in the radial direction is

Radial Direction

$$W_{ir} = -D_e \frac{\partial C_i}{\partial r} + U_r C_i \qquad (8\text{-}84)$$

where U_r (m/s) is the average velocity in the radial direction. For now we will neglect the velocity in the radial direction, i.e., $U_r = 0$. A mole balance on a cylindrical system volume of length Δz and thickness Δr as shown in Figure 8-26 gives

Mole Balances on Species A

$$\begin{pmatrix} \text{Moles of A} \\ \text{in at } r \end{pmatrix} = W_{Ar} \bullet \begin{pmatrix} \text{Cross-sectional area} \\ \text{normal to radial flux} \end{pmatrix} = W_{Ar} \bullet 2\pi r \Delta z$$

$$\begin{pmatrix} \text{Moles of A} \\ \text{in at } z \end{pmatrix} = W_{Az} \bullet \begin{pmatrix} \text{Cross-sectional area} \\ \text{normal to axial flux} \end{pmatrix} = W_{Az} \bullet 2\pi r \Delta r$$

$$\begin{pmatrix} \text{Moles of A} \\ \text{in at } r \end{pmatrix} - \begin{pmatrix} \text{Moles of A} \\ \text{out at } (r + \Delta r) \end{pmatrix} + \begin{pmatrix} \text{Moles of A} \\ \text{in at } z \end{pmatrix} - \begin{pmatrix} \text{Moles of A} \\ \text{out at } (z + \Delta z) \end{pmatrix}$$

$$+ \begin{pmatrix} \text{Moles of A} \\ \text{formed} \end{pmatrix} = \begin{pmatrix} \text{Moles of A} \\ \text{Accumulated} \end{pmatrix}$$

$$W_{Ar} 2\pi r \Delta z|_r - W_{Ar} 2\pi r \Delta z|_{r+\Delta r} + W_{Az} 2\pi r \Delta r|_z - W_{Az} 2\pi r \Delta r|_{z+\Delta z}$$

$$+ r_A 2\pi r \Delta r \Delta z = \frac{\partial C_A (2\pi r \Delta r \Delta z)}{\partial t}$$

Dividing by $2\pi r \Delta r \Delta z$ and taking the limit as Δr and $\Delta z \to 0$

$$-\frac{1}{r}\frac{\partial(rW_{Ar})}{\partial r} - \frac{\partial W_{Az}}{\partial z} + r_A = \frac{\partial C_A}{\partial t}$$

Similarly, for any species i and steady-state conditions,

$$-\frac{1}{r}\frac{\partial(rW_{ir})}{\partial r} - \frac{\partial W_{iz}}{\partial z} + r_i = 0 \qquad (8\text{-}85)$$

Using Equation (8-83) and (8-84) to substitute for W_{iz} and W_{ir} in Equation (8-85) and then setting the radial velocity to zero, $U_r = 0$, we obtain

$$-\frac{1}{r}\frac{\partial}{\partial r}\left[\left(-D_e \frac{\partial C_i}{\partial r} r\right)\right] - \frac{\partial}{\partial z}\left[-D_e \frac{\partial C_i}{\partial z} + U_z C_i\right] + r_i = 0$$

This equation will also be discussed further in Chapter 14.

For steady-state conditions and assuming U_z does not vary in the axial direction,

$$D_e \frac{\partial^2 C_i}{\partial r^2} + \frac{D_e}{r}\frac{\partial C_i}{\partial r} + D_e \frac{\partial^2 C_i}{\partial z^2} - U_z \frac{\partial C_i}{\partial z} + r_i = 0 \qquad (8\text{-}86)$$

Energy Flux

When we applied the first law of thermodynamics to a reactor to relate either temperature and conversion or molar flow rates and concentration, we arrived at Equation (8-9). Neglecting the work term we have for steady-state conditions

$$
\overset{\text{Conduction}}{\widetilde{\dot{Q}}} + \overset{\text{Convection}}{\overbrace{\sum_{i=1}^{n} F_{i0}H_{i0} - \sum_{i=1}^{n} F_i H_i}} = 0 \qquad (8\text{-}87)
$$

In terms of the molar fluxes and the cross-sectional area and $(\mathbf{q} = \dot{Q}/A_c)$

$$
A_c[\mathbf{q} + (\Sigma \mathbf{W}_{i0}H_{i0} - \Sigma \mathbf{W}_i H_i)] = 0 \qquad (8\text{-}88)
$$

The $\mathbf{q}$ term is the heat added to the system and almost always includes a conduction component of some form. We now define an **energy flux vector, e,** $(\text{J/m}^2 \cdot \text{s})$, to include both the conduction and convection of energy.

e = energy flux J/s·m²

$$
\mathbf{e} = \text{Conduction} + \text{Convection}
$$

$$
\boxed{\mathbf{e} = \mathbf{q} + \Sigma \mathbf{W}_i H_i} \qquad (8\text{-}89)
$$

where the conduction term $\mathbf{q}$ $(\text{kJ/m}^2 \cdot \text{s})$ is given by Fourier's law. For axial and radial conduction Fourier's laws are

$$
q_z = -k_e \frac{\partial T}{\partial z} \qquad \text{and} \qquad q_r = -k_e \frac{\partial T}{\partial r}
$$

where k_e is the thermal conductivity (J/m·s·K). The energy transfer (flow) is the vector flux times the cross-sectional area, A_c, normal to the energy flux

$$
\text{Energy flow} = \mathbf{e} \cdot A_c
$$

Energy Balance

Using the energy flux, e, to carry out an energy balance on our annulus (Figure 8-26) with system volume $2\pi r \Delta r \Delta z$, we have

$$
(\text{Energy flow in at } r) = e_r A_{cr} = e_r \cdot 2\pi r \Delta z
$$

$$
(\text{Energy flow in at } z) = e_z A_{cz} = e_z \cdot 2\pi r \Delta r
$$

$$
\begin{pmatrix} \text{Energy Flow} \\ \text{in at } r \end{pmatrix} - \begin{pmatrix} \text{Energy Flow} \\ \text{out at } r+\Delta r \end{pmatrix} + \begin{pmatrix} \text{Energy Flow} \\ \text{in at } z \end{pmatrix} - \begin{pmatrix} \text{Energy Flow} \\ \text{out at } z+\Delta z \end{pmatrix} = \begin{pmatrix} \text{Accumulation} \\ \text{of Energy in} \\ \text{Volume } (2\pi r \Delta r \Delta z) \end{pmatrix}
$$

$$
(e_r 2\pi r \Delta z)\big|_r - (e_r 2\pi r \Delta z)\big|_{r+\Delta r} + e_z 2\pi r \Delta r\big|_z - e_z 2\pi r \Delta r\big|_{z+\Delta z} = 0 \qquad (8\text{-}90)
$$

Dividing by $2\pi r \Delta r \Delta z$ and taking the limit as Δr and $\Delta z \to 0$,

$$\boxed{-\frac{1}{r}\frac{\partial(re_r)}{\partial r} - \frac{\partial e_z}{\partial z} = 0} \qquad (8\text{-}91)$$

The radial and axial energy fluxes are

$$e_r = q_r + \Sigma W_{ir}H_i$$

$$e_z = q_z + \Sigma W_{iz}H_i$$

Substituting for the energy fluxes, e_r and e_z,

$$-\frac{1}{r}\frac{\partial[r[q_r + \Sigma W_{ir}H_i]]}{\partial r} - \frac{\partial[q_z + \Sigma W_{iz}H_i]}{\partial z} = 0 \qquad (8\text{-}92)$$

and expanding the convective energy fluxes, $\Sigma W_i H_i$,

Radial: $\quad \dfrac{1}{r}\dfrac{\partial}{\partial r}(r\Sigma W_{ir}H_i) = \dfrac{1}{r}\Sigma H_i\dfrac{\partial(rW_{ir})}{\partial r} + \overset{\displaystyle\nearrow\text{Neglect}}{\cancel{\Sigma W_{ir}\dfrac{\partial H_i}{\partial r}}} \qquad (8\text{-}93)$

Axial: $\quad \dfrac{\partial(\Sigma W_{iz}H_i)}{\partial z} = \Sigma H_i\dfrac{\partial W_{iz}}{\partial z} + \Sigma W_{iz}\dfrac{\partial H_i}{\partial z} \qquad (8\text{-}94)$

Substituting Equations (8-93) and (8-94) into Equation (8-92), we obtain upon rearrangement

$$-\frac{1}{r}\frac{\partial(rq_r)}{\partial r} - \frac{\partial q_z}{\partial z} - \Sigma H_i \overbrace{\left(\frac{1}{r}\frac{\partial(rW_{ir})}{\partial r} + \frac{\partial W_{iz}}{\partial z}\right)}^{r_i} - \Sigma W_{iz}\frac{\partial H_i}{\partial z} = 0$$

Recognizing that the term in brackets is related to Equation (8-85) for steady-state conditions and is just the rate of formation of species i, r_i, we have

$$\boxed{-\frac{1}{r}\frac{\partial}{\partial r}(rq_r) - \frac{\partial q_z}{\partial z} - \Sigma H_i r_i - \Sigma W_{iz}\frac{\partial H_i}{\partial z} = 0} \qquad (8\text{-}95)$$

Recalling

$$q_r = -k_e\frac{\partial T}{\partial r}, \quad q_z = -k_e\frac{\partial T}{\partial z}, \quad \frac{\partial H_i}{\partial z} = C_{P_i}\frac{\partial T}{\partial z},$$

and

$$r_i = \nu_i(-r_A)$$

$$\Sigma r_i H_i = \Sigma \nu_i H_i(-r_A) = -\Delta H_{Rx}r_A$$

we have the energy in the form

$$\frac{k_e}{r}\left[\frac{\partial}{\partial r}\left(\frac{r\partial T}{\partial r}\right)\right] + k_e\frac{\partial^2 T}{\partial z^2} + \Delta H_{Rx}r_A - (\Sigma W_{iz}C_{P_i})\frac{\partial T}{\partial z} = 0 \qquad (8\text{-}96)$$

Some Initial Approximations

Assumption 1. Neglect the diffusive term wrt the convective term in the expression involving heat capacities

$$\Sigma C_{P_i}W_{iz} = \Sigma C_{P_i}(0 + U_zC_i) = \Sigma C_{P_i}C_iU_z$$

With this assumption Equation (8-96) becomes

$$\frac{k_e}{r}\frac{\partial}{\partial r}\left(\frac{r\partial T}{\partial r}\right) + k_e\frac{\partial^2 T}{\partial z^2} + \Delta H_{Rx}r_A - (U_z\Sigma C_{P_i}C_i)\frac{\partial T}{\partial z} = 0 \qquad (8\text{-}97)$$

For laminar flow, the velocity profile is

$$U_z = 2U_0\left[1 - \left(\frac{r}{R}\right)^2\right] \qquad (8\text{-}98)$$

where U_0 is the average velocity inside the reactor.

Energy balance
with radial and
axial gradients *Assumption 2.* Assume that the sum $C_{P_m} = \Sigma C_{P_i}C_i = C_{A0}\Sigma\Theta_iC_{P_i}$ is constant. The energy balance now becomes

$$k_e\frac{\partial^2 T}{\partial z^2} + \frac{k_e}{r}\frac{\partial}{\partial r}\left(r\frac{\partial T}{\partial r}\right) + \Delta H_{Rx}r_A - U_zC_{P_m}\frac{\partial T}{\partial z} = 0 \qquad (8\text{-}99)$$

Equation (8-98) is the form we will use in our COMSOL problem. In many instances, the term C_{P_m} is just the product of the solution density and the heat capacity of the solution (kJ/kg • K).

Coolant Balance

We also recall that a balance on the coolant gives the variation of coolant temperature with axial distance where U_{ht} is the overall heat transfer coefficient and R is the reactor wall radius

$$\dot{m}_cC_{P_c}\frac{\partial T_a}{\partial z} = U_{ht}2\pi R[T(z) - T_a] \qquad (8\text{-}100)$$

Boundary and initial conditions

A. Initial conditions *if* other than steady state

$$t = 0, \quad C_i = 0, \qquad T = T_0, \quad \text{for } z > 0 \text{ all } r$$

B. Boundary condition

1) Radial

(a) At $r = 0$, we have symmetry $\partial T/\partial r = 0$ and $\partial C_i/\partial r = 0$.

(b) At the tube wall $r = R$, the temperature flux to the wall on the reaction side equals the convective flux out of the reactor into the shell side of the heat exchanger.

$$-k_e \frac{\partial T}{\partial r}\bigg|_R = U(T_R(z) - T_a)$$

(c) There is no mass flow through the tube walls $\partial C_i/\partial r = 0$ at $r = R$.

(d) At the entrance to the reactor $z = 0$,

$$T = T_0 \text{ and } C_i = C_{i0}$$

(e) At the exit of the reactor $z = L$,

$$\frac{\partial T}{\partial z} = 0 \text{ and } \frac{\partial C_i}{\partial z} = 0$$

The following examples will solve the preceding equations using COMSOL. For the exothermic reaction with cooling, the expected profiles are

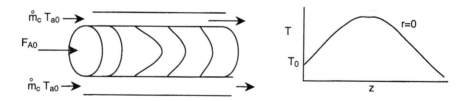

Example 8–12 Radial Effects in Tubular Reactor

Web Modules

This example will highlight the radial effects in a tubular reactor, which up until now have been neglected to simplify the calculations. Now, the effects of parameters such as inlet temperature and flow rate will be studied using the software program COMSOL. Follow the step-by-step procedure in the Web Module on the CD-ROM.

We continue Example 8-8, which discussed the reaction of propylene oxide (A) with water (B) to form propylene glycol (C). The hydrolysis of propylene oxide takes place readily at room temperature when catalyzed by sulfuric acid.

$$A + B \longrightarrow C$$

This exothermic reaction is approximated as a first-order reaction given that the reaction takes place in an excess of water.

The CSTR from Example 8-8 has been replaced by a tubular reactor 1.0 m in length and 0.2 m in diameter.

The feed to the reactor consists of two streams. One stream is an equivolumetric mixture of propylene oxide and methanol, and the other stream is water containing 0.1 wt % sulfuric acid. The water is fed at a volumetric rate 2.5 times larger than the propylene oxide–methanol feed. The molar flow rate of propylene oxide fed to the tubular reactor is 0.1 mol/s.

There is an immediate temperature rise upon mixing the two feed streams caused by the heat of mixing. In these calculations, this temperature rise is already accounted for, and the inlet temperature of both streams is set to 312 K.

The reaction rate law is

$$-r_A = kC_A$$

with

$$k = Ae^{-E/RT}$$

where $E = 75362$ J/mol and A = 16.96×10^{12} h^{-1}, which can also be put in the form

$$k(T) = k_0(T_0) \exp\left[\frac{E}{R}\left(\frac{1}{T_0} - \frac{1}{T}\right)\right]$$

With $k_0 = 1.28$ h^{-1} at 300 K. The thermal conductivity k_e of the reaction mixture and the diffusivity D_e are 0.599 W/m/K and 10^{-9} m^2/s, respectively, and are assumed to be constant throughout the reactor. In the case where there is a heat exchange between the reactor and its surroundings, the overall heat-transfer coefficient is 1300 W/m^2/K and the temperature of the cooling jacket is assumed to be constant and is set to 273 K. The other property data are shown in Table E8-12.1.

TABLE E8–12.1 PHYSICAL PROPERTY DATA

	Propylene Oxide	Methanol	Water	Propylene Glycol
Molar weight (g/mol)	58.095	32.042	18	76.095
Density (kg/m³)	830	791.3	1000	1040
Heat capacity (J/mol • K)	146.54	81.095	75.36	192.59
Heat of formation (J/mol)	−154911.6		−286098	−525676

Solution

Mole Balances: Recalling Equation (8-86) and applying it to species A

A:
$$D_e\frac{\partial^2 C_A}{\partial r^2} + \frac{1}{r}D_e\frac{\partial C_A}{\partial r} + D_e\frac{\partial^2 C_A}{\partial z^2} - U_z\frac{\partial C_A}{\partial z} + r_A = 0 \qquad \text{(E8-12.1)}$$

Rate Law:

$$-r_A = k(T_1)\,\exp\left[\frac{E}{R}\left(\frac{1}{T_1} - \frac{1}{T}\right)\right]C_A \qquad \text{(E8-12.2)}$$

Stoichiometry: The conversion along a streamline (r) at a distance z

$$X(r, z) = 1 - C_A(r, z)/C_{A0} \qquad \text{(E8-12.3)}$$

The overall conversion is

$$\overline{X}(z) = 1 - \frac{2\pi\displaystyle\int_0^R C_A(r,z)U_z r\,dr}{F_{A0}} \qquad \text{(E8-12.4)}$$

The mean concentration at any distance z

$$\overline{C_A}(z) = \frac{2\pi\displaystyle\int_0^R C_A(r,z)U_z r\,dr}{\pi R^2 U_0} \qquad \text{(E8-12.5)}$$

For plug flow the velocity profile is

$$U_z = U_0 \qquad \text{(E8-12.6)}$$

The laminar flow velocity profile is

$$U_z = 2U_0\left[1 - \left(\frac{r}{R}\right)^2\right] \qquad \text{(E8-12.7)}$$

Recalling the Energy Balance

$$\boxed{k_e\frac{\partial^2 T}{\partial z^2} + \frac{k_e}{r}\frac{\partial}{\partial r}\left(r\frac{\partial T}{\partial r}\right) + \Delta H_{Rx}r_A - U_z C_{P_m}\frac{\partial T}{\partial z} = 0} \qquad \text{(8-98)}$$

Living Example Problem

Assumptions

1. U_r is zero.
2. Neglect axial diffusion/dispersion flux wrt convective flux when summing the heat capacity times their fluxes.
3. Steady state.

Cooling jacket

$$\dot{m}C_{P_i}\frac{\partial T_a}{\partial z} = 2\pi R U_{ht}(T_R(z) - T_a) \tag{8-99}$$

Boundary conditions

$$\text{At } r = 0, \text{ then } \frac{\partial C_i}{\partial r} = 0 \text{ and } \frac{\partial T}{\partial r} = 0 \tag{E8-12.8}$$

$$\text{At } r = R, \text{ then } \frac{\partial C_i}{\partial r} = 0 \text{ and } -k_e\frac{\partial T}{\partial r} = U_{ht}(T_R(z) - T_a) \tag{E8-12.9}$$

$$\text{At } z = 0, \text{ then } C_i = C_{i0} \text{ and } T = T_0 \tag{E8-12.10}$$

These equations were solved using COMSOL for a number of cases including adiabatic and non-adiabatic plug flow and laminar flow; they were also solved with and without axial and radial dispersion. A detailed accounting on how to change the parameter values in the COMSOL program can be found in the COMSOL Instructions section on the web in screen shots similar to Figure E8-12.1. Figure E8-12.1 gives the data set in SI units used for the COMSOL example on the CD-ROM.

Note: There is a step-by-step COMSOL tutorial using screen shots for this example on the CD-ROM.

Constants

Name	Expression	Value
Diff	1e-9	1e-9
E	75362	75362
A	16.96e12/3600	4.711111e9
R	8.314	8.314
T0	312	312
v0	4.07365e-5	4.07365e-5
cA0	2454	2454
cB0	47728	47728
Ra	0.1	0.1
rhoCat	1500	1500
dHrx	-84666	-84666
Keq0	1000	1000
ke	0.559	0.559
rho	1173	1173
Cp	3667	3667

OK   Cancel   Apply

Define expression

Figure E8-12.1 COMSOL screen shot of Data Set for Second Order Reaction.

Color surfaces are used to show the concentration and temperature profiles, similar to the black and white figures shown in Figure E8-12.2. Use the COMSOL program on the CD-ROM to develop temperature concentration profiles similar to the ones shown here. Read through the COMSOL web module entitled "Radial and Axial Temperature Gradients" before running the program. One notes in Figure E8-12.2 that the conversion is lower near the wall because of the cooler fluid temperature. These same profiles can be found in color on the web and CD-ROM in the web modules. One notes the maximum and minimum in these profiles. Near the wall, the temperature of the mixture is lower because of the cold wall temperature. Consequently, the rate will be lower, and thus the conversion will be lower. However, right next to the wall, the velocity through the reactor is almost zero so the reactants spend a long time in the reactor; therefore, a greater conversion is achieved as noted by the upturn right next to the wall.

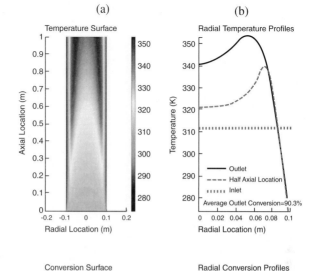

(a)   (b)

Results of the
COMSOL
simulation

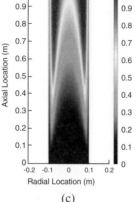
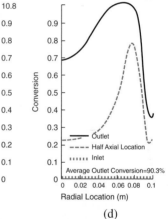

(c)   (d)

Figure E8-12.2 (a) Temperature surface, (b) temperature surface profiles,
(c) conversion surface, and (d) radial conversion profile.

8.10   The Practical Side

Scaling up exothermic chemical reactions can be very tricky. Tables 8.6 and
8.7 give reactions that have resulted in accidents and their causes, respectively.[10]

---

[10]Courtesy of J. Singh, *Chemical Engineering*, 92 (1997) and B. Venugopal, *Chemical
Engineering*, 54 (2002).

TABLE 8-6.   INCIDENCE OF BATCH-PROCESS ACCIDENTS

Process Type	Number of Incidents in U.K., 1962–1987
Polymerization	64
Nitration	15
Sulfurization	13
Hydrolysis	10
Salt formation	8
Halogenation	8
Alkylation (Friedel-Crafts)	5
Amination	4
Diazolization	4
Oxiation	2
Esterification	1
Total:	134

[Source: U.K. Health and Safety Executive]

TABLE 8-7.   CAUSES OF BATCH REACTOR ACCIDENTS IN TABLE 8.6

Cause	Contribution, %
Lack of knowledge of reaction chemistry	20
Problems with material quality	9
Temperature-control problems	19
Agitation problems	10
Mis-charging of reactants or catalyst	21
Poor maintenance	15
Operator error	5

Summary Notes

More information is given in the Summary Notes and Professional Reference Shelf on the web. The use of the ARSST to detect potential problems will be discussed in Chapter 9.

Closure. Virtually all reactions that are carried out in industry involve heat effects. This chapter provides the basis to design reactors that operate at steady state and involve heat effects. To model these reactors, we simply add another step to our algorithm; this step is the energy balance. Here it is important to understand how the energy balance was applied to each reaction type so that you will be able to describe what would happen if you changed some of the operating conditions (e.g., T_0). The living example problems (especially *8T-8-3*) and the ICM module will help you achieve a high level of understanding. Another major goal after studying this chapter is to be able to design reactors that have multiple reactions taking place under nonisothermal conditions. Try working Problem 8-26 to be sure you have achieved this goal. An industrial example that provides a number of practical details is included as an appendix to this chapter. The last example of the chapter considers a tubular reactor that has both axial and radial gradients. As with the other living example problems, one should vary a number of the operating parameters to get a feel of how the reactor behaves and the sensitivity of the parameters for safe operation.

SUMMARY

For the reaction

$$A + \frac{b}{a}B \rightarrow \frac{c}{a}C + \frac{d}{a}D$$

1. The heat of reaction at temperature T, per mole of A, is

$$\Delta H_{Rx}(T) = \frac{c}{a}H_C(T) + \frac{d}{a}H_D(T) - \frac{b}{a}H_B(T) - H_A(T) \quad \text{(S8-1)}$$

2. The mean heat capacity difference, ΔC_p, per mole of A is

$$\Delta C_p = \frac{c}{a}C_{pC} + \frac{d}{a}C_{pD} - \frac{b}{a}C_{pB} - C_{pA} \quad \text{(S8-2)}$$

   where C_{P_i} is the mean heat capacity of species i between temperatures T_R and T.

3. When there are no phase changes, the heat of reaction at temperature T is related to the heat of reaction at the standard reference temperature T_R by

$$\Delta H_{Rx}(T) = H^\circ_{Rx}(T_R) + \Delta C_P(T - T_R) \quad \text{(S8-3)}$$

4. Neglecting changes in potential energy, kinetic energy, and viscous dissipation, and for the case of no work done on or by the system and all species entering at the same temperature, the steady state CSTR energy balance is

$$\frac{UA}{F_{A0}}(T_a - T) - X[\Delta H^\circ_{Rx}(T_R) + \Delta C_P(T - T_R)] = \Sigma\, \Theta_i C_{P_i}(T - T_{i0}) \quad \text{(S8-4)}$$

5. For adiabatic operation of a PFR, PBR, CSTR, or batch reactor, the temperature conversion relationship is

$$X = \frac{\Sigma\, \Theta_i C_{P_i}(T - T_0)}{\Delta H^\circ_{Rx}(T_R) + \Delta C_P(T - T_R)} \quad \text{(S8-5)}$$

   Solving for the temperature, T.

$$T = \frac{X[-\Delta H^\circ_{Rx}(T_R)] + \Sigma\, \Theta_i C_{P_i} T_0 + X\,\Delta C_p T_R}{[\Sigma\, \Theta_i C_{P_i} + X\,\Delta C_p]} \quad \text{(S8-6)}$$

6. The energy balance on a PFR/PBR

$$\frac{dT}{dV} = \frac{Ua(T_a - T) + (-r_A)[-\Delta H_{Rx}(T)]}{\displaystyle\sum_{i=1}^{m} F_i C_{P_i}} \quad \text{(S8-7)}$$

In terms of conversion,

$$\frac{dT}{dV} = \frac{Ua(T_a - T) + (-r_A)[-\Delta H_{Rx}(T)]}{F_{A0}(\Sigma \Theta_i C_{P_i} + X \Delta C_P)} \tag{S8-8}$$

7. The temperature dependence of the specific reaction rate is given in the form

$$k(T) = k_1(T_1) \exp\left[\frac{E}{R}\left(\frac{T - T_1}{TT_1}\right)\right] \tag{S8-9}$$

8. The temperature dependence of the equilibrium constant is given by van't Hoff's equation for $\Delta C_P = 0$,

$$K_P(T) = K_P(T_2) \exp\left[\frac{\Delta H_{Rx}}{R}\left(\frac{1}{T_2} - \frac{1}{T}\right)\right] \tag{S8-10}$$

9. Multiple steady states:

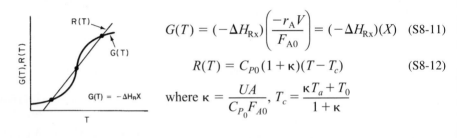

$$G(T) = (-\Delta H_{Rx})\left(\frac{-r_A V}{F_{A0}}\right) = (-\Delta H_{Rx})(X) \tag{S8-11}$$

$$R(T) = C_{P0}(1 + \kappa)(T - T_c) \tag{S8-12}$$

$$\text{where } \kappa = \frac{UA}{C_{P_0}F_{A0}}, \quad T_c = \frac{\kappa T_a + T_0}{1 + \kappa}$$

10. The criteria for *Runaway Reactions* occurs when $(T_r - T_c) > RT_r^2/E$, where T_r is the reactor temperature and $T_c = (T_o + \kappa T_a)/(1 + \kappa)$.

$$S^* = G(T^*)\frac{E}{RT^{*2}}$$

S* ... Runaway ... T_0

11. When q multiple reactions are taking place and there are m species,

$$\frac{dT}{dV} = \frac{Ua(T_a - T) + \sum\limits_{i=1}^{q}(-r_{ij})[-\Delta H_{Rxij}(T)]}{\sum\limits_{j=1}^{m}F_j C_{Pj}} \tag{S8-13}$$

12. Axial or radial temperature and concentration gradients. The following coupled partial differential equations were solved using COMSOL:

$$D_e \frac{\partial^2 C_i}{\partial r^2} + \frac{D_e}{r}\frac{\partial C_i}{\partial r} + D_e \frac{\partial^2 C_i}{\partial z^2} - U_z \frac{\partial C_i}{\partial z} + r_i = 0 \qquad \text{(S8-14)}$$

and

$$k_e \frac{\partial^2 T}{\partial z^2} + \frac{k_e}{r}\frac{\partial}{\partial r}\left(r\frac{\partial T}{\partial r}\right) + \Delta H_{Rx} r_A - U_z C_{P_m} \frac{\partial T}{\partial z} = 0 \qquad \text{(S8-15)}$$

ODE SOLVER ALGORITHM

Packed-Bed Reactor with Heat Exchange and Pressure Drop

$$2A \;\rightleftharpoons\; C$$

Pure gaseous A enters at 5 mol/min at 450 K.

$$\frac{dX}{dW} = \frac{-r'_A}{F_{A0}}$$

$$\frac{dT}{dW} = \frac{UA/\rho_c (T_a - T) + (r'_A)(\Delta H^\circ_{Rx})}{C_{p_A} F_{A0}}$$

$$\frac{dy}{dW} = -\frac{\alpha}{2y}(1 - 0.5X)(T/T_0)$$

$$r'_A = -k[C_A^2 - C_C/K_C]$$

$$C_A = C_{A0}[(1 - X)/(1 - 0.5X)](T_0/T)(y)$$

$$C_C = \tfrac{1}{2}C_{A0}X(T_0/T)y/(1 - 0.5X)$$

$$k = 0.5 \, \exp[5032((1/450) - (1/T))]$$

$$K_C = 25{,}000 \, \exp\left[-10{,}065\left[\left(\frac{1}{450}\right) - \left(\frac{1}{T}\right)\right]\right]$$

$\alpha = 0.019/\text{kg cat.}$

$C_{A0} = 0.25 \text{ mol/dm}^3$

$UA/\rho_b = 0.8 \text{ J/kg cat.} \cdot \text{s} \cdot \text{K}$

$T_a = 500 \text{ K}$

$\Delta H^\circ_{Rx} = -20{,}000 \text{ J/mol}$

$C_{P_A} = 40 \text{ J/mol} \cdot \text{K}$

$F_{A0} = 5.0 \text{ mol/s}$

$T_0 = 450 \text{ K}$

$W_f = 90 \text{ kg cat.}$

CD-ROM MATERIAL

- **Learning Resources**

  1. Summary Notes

  2. Web Module COMSOL Radial and Axial Gradients

  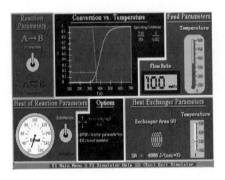
  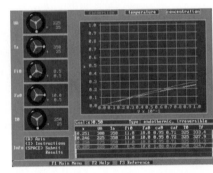

  3. Interactive Computer Modules

  A. Heat Effects I   B. Heat Effects II

  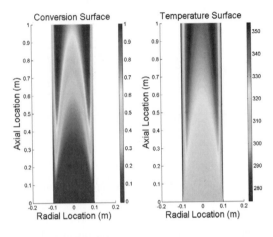

  4. Solved Problems

  A. Example CD8-1 $\Delta H_{Rx}(T)$ for Heat Capacities Expressed as Quadratic Functions of Temperature

  B. Example CD8-2   Second-Order Reaction Carried Out Adiabatically in a CSTR

  5. PFR/PBR Solution Procedure for a Reversible Gas-Phase Reaction

- **Living Example Problems**

  1. Example 8-3 Adiabatic Isomerization of Normal Butane

  2. Example 8-4 Isomerization of Normal Butane with Heat Exchange

  3. Example 8-5 Production of Acetic Anhydride

  4. Example 8-9 CSTR with Cooling Coil

  5. Example 8-10 Parallel Reaction in a PFR with Heat Effects

  6. Example 8-11 Multiple Reactions in a CSTR

Living Example Problem

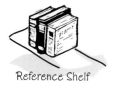

Reference Shelf

7. *Example 8-12 COMSOL Axial and Radial Gradients*
8. *Example R8.2-1 Runaway Reactions in a PFR*
9. *Example R8.4-1 Industrial Oxidation of SO$_2$*
10. *Example 8-T8-3 PBR with Variable Coolant Temperature, T$_a$*

• **Professional Reference Shelf**

R8.1. *Runaway in CSTRs and Plug Flow Reactors*

Phase Plane Plots. We transform the temperature and concentration profiles into a phase plane.

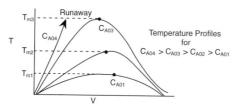

Temperature profiles.

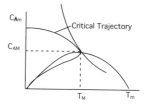

Critical trajectory on the
$C_{Am}-T_m$ phase plane plot.

The trajectory going through the maximum of the "maxima curve" is considered to be *critical* and therefore is the locus of the *critical* inlet conditions for C_A and T corresponding to a given wall temperature.

R8.2. *Steady-State Bifurcation Analysis.* In reactor dynamics, it is particularly important to find out if multiple stationary points exist or if sustained oscillations can arise. We apply bifurcation analysis to learn whether or not multiple steady states are possible. Both CSTR energy and mole balances are of the form

$$F(y) = \alpha y - \beta - G(y)$$

The conditions for uniqueness are then shown to be those that satisfy the relationship

$$\max\left(\frac{\partial G}{\partial y}\right) < \alpha$$

Specifically, the conditions for which multiple steady states exist must satisfy the following set of equations:

$$\left.\frac{dF}{dy}\right|_{y^p} = 0 = \alpha - \left.\frac{dG}{dy}\right|_{y^p} \tag{1}$$

$$F(y^*) = 0 = \alpha y^* - \beta - G(y^*) \tag{2}$$

R8.3. *Variable Heat Capacities.* Next we want to arrive at a form of the energy balance for the case where heat capacities are strong functions of temperature over a wide temperature range. Under these conditions, the mean values of the heat capacity may not be adequate for the relationship between conversion and temperature. Combining heat reaction with the quadratic form of the heat capacity,

$$C_{P_i} = \alpha_i + \beta_i T + \gamma_i T^2$$

Heat capacity as
a function of
temperature

we find that

$$\Delta H_{Rx}(T) = \Delta \overset{\circ}{H}_{Rx}(T_R) + \Delta\alpha(T - T_R) + \frac{\Delta\beta}{2}(T^2 - T_R^2) + \frac{\Delta\gamma}{3}T^3(-T_R^3)$$

Example 8-5 is reworked for the case of variable heat capacities.

Reference Shelf

R8.4. *Manufacture of Sulfuric Acid.* The details of the industrial oxidation of SO_2 are described. Here the catalys quantities reactor configuration, operating conditions, are discussed along with a model to predict the conversion and temperature profiles.

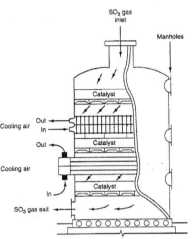
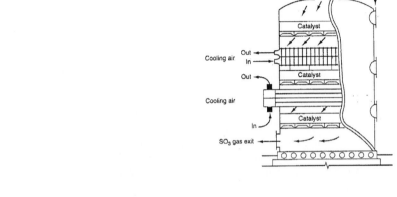
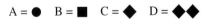

QUESTIONS AND PROBLEMS

Homework Problems

The subscript to each of the problem numbers indicates the level of difficulty: A, least difficult; D, most difficult.

$$A = \bullet \quad B = \blacksquare \quad C = \blacklozenge \quad D = \blacklozenge\blacklozenge$$

In each of the questions and problems, rather than just drawing a box around your answer, write a sentence or two describing how you solved the problem, the assumptions you made, the reasonableness of your answer, what you learned, and any other facts that you want to include. See the Preface for additional generic parts (**x**), (**y**), (**z**) to the home problems.

P8-1$_A$ Read over the problems at the end of this chapter. Make up an original problem that uses the concepts presented in this chapter. To obtain a solution:
(a) Make up your data and reaction.
(b) Use a real reaction and real data. See Problem P4-1 for guidelines.
(c) Prepare a list of safety considerations for designing and operating chemical reactors. (See *www.siri.org/graphics*.)

Creative Problems

Before solving the problems, state or sketch qualitatively the expected results or trends.

See R. M. Felder, *Chem. Eng. Educ.,* *19*(4), 176 (1985). The August 1985 issue of *Chemical Engineering Progress* may be useful for part (c).

(d) Choose a FAQ from Chapter 8 and say why it was most helpful.

(e) Listen to the audios 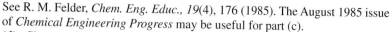 on the CD and pick one and say why it could be eliminated.

(f) Read through the Self Tests and Self Assessments for Chapter 8 on the CD-ROM, and pick one that was most helpful.

(g) Which example on the CD-ROM Lecture Notes for Chapter 8 was the most helpful?

(h) What if you were asked to prepare a list of safety considerations of redesigning and operating a chemical reactor, what would be the first four items on your list?

(i) What if you were asked to give an everyday example that demonstrates the principles discussed in this chapter? (Would sipping a teaspoon of Tabasco or other hot sauce be one?)

Hall of Fame

P8-2$_A$ Load the following Polymath/MATLAB/COMSOL programs from the CD-ROM where appropriate:

(a) Example 8-1. How would this example change if a CSTR were used instead of a PFR?

(b) Example 8-2. What would the heat of reaction be if 50% inerts (e.g., helium) were added to the system? What would be the % error if the ΔC_P term were neglected?

(c) Example 8-3. What if the butane reaction were carried out in a 0.8-m^3 PFR that can be pressurized to very high pressures? What inlet temperature would you recommend? Is there an optimum temperature? How would your answer change for a 2-m^3 CSTR?

(d) Example 8-4. (1) How would the answers change if the reactor were in a counter current exchanger where the coolant temperature was not constant along the length of the reactor? The mass flow rate and heat capacity of the coolant are 50 kg/h and 75 kJ/kg/K, respectively, and the entering coolant temperature is 310 K. Vary the coolant rate, $\dot{m}_c$, make a plot of X versus $\dot{m}_c$. (2) Repeat (1) but change the parameters K_C, E, $1,000 < Ua < 15,000 (\text{J/h/m}^3/\text{K})$, and ΔH_{Rx}. Write a paragraph describing what you find, noting any generalization.

(e) Example 8-5. (1) How would your answer change if the coolant flow was counter current? (2) Make a plot of conversion as a function of F_{A0} for each of the three cases. (3) Make a plot of conversion as a function of coolant rate and coolant temperature. (4) Make a plot of the exit conversion and temperature as a function of reactor diameter but for the same total volume.

(f) Example 8-6. How would the result change if the reaction were second order and reversible 2A $\leftrightarrow$ 2B with K_C remaining the same?

(g) Example 8-7. How would your answers change if the heat of reaction were three times that given in the problem statement?

(h) Example 8-8. Describe how your answers would change if the molar flow of methanol were increased by a factor of 4.

(i) Example 8-9. Other data show $\Delta H_{Rx} = -38,700$ BTU/lbmol and $C_{P_A} = 29$ BTU/lbmol/°F. How would these values change your results? Make a plot of conversion as a function of heat exchanger area. [0 < A < 200 ft^2].

(j) **Example 8-10.** How would your results change if there is (1) a pressure drop with $\alpha = 1.05$ dm^{-3}, (2) Reaction (1) is reversible with $K_C = 10$ at 450 K. (3) How would the selectivity change if Ua is increased? Decreased?

(k) **Example 8-11.** (1) How would the results (e.g., $\tilde{S}_{B/C}$) change if the UA term were varied ($700 < UA < 70,000$ J/m$^3 \cdot$ s $\cdot$ k)? (2) If τ were varied [($0.1 > \tau > 0.00001$ min)]?

(l) **CD Example R8.4-1.** SO$_2$ oxidation. How would your results change if (1) the catalyst particle diameter were cut in half? (2) the pressure were doubled? At what particle size does pressure drop become important for the same catalyst weight assuming the porosity doesn't change? (3) you vary the initial temperature and the coolant temperature? Write a paragraph describing what you find.

(m) **Example T8-3.** Load the Polymath problem from the CD-ROM for this exothermic reversible reaction with a variable coolant temperature. The elementary reaction

$$A + B \underset{\longleftarrow}{\overset{\longrightarrow}{\rule{0pt}{0pt}\qquad}} 2C$$

has the following parameter values for the **base case**.

$E = 25$ kcal/mol $C_{P_A} = C_{P_B} = C_{P_C} = 20$ cal/mol/K

$\Delta H_{Rx} = -20$ kcal/mol $C_{P_I} = 40$ cal/mol/K

$k = \dfrac{0.004 \text{ dm}^6}{\text{mol} \cdot \text{kg} \cdot \text{s}}$ @ 310 K $\dfrac{Ua}{\rho_B} = 0.5 \dfrac{\text{cal}}{\text{kg} \cdot \text{s} \cdot K}$

$K_c = 1000$ @ 303 K $T_a = 320$ K

$\alpha = 0.0002$ / kg $\dot{m}_c = 1,000$ g/s

$F_{a0} = 5$ mol/s $C_{p_c} = 18$ cal/g/K

$C_{T0} = 0.3$ mol/dm^3 $\Theta_I = 1$

Vary the following parameters and write a paragraph describing the trends you find for each parameter variation and why they work the way they do. Use the base case for parameters not varied. *Hint*: See Selftests and Workbook in the *Summary Notes* on the CD-ROM.

Summary Notes

(a) F_{A0}: $1 \le F_{A0} \le 8$ mol/s
(b) Θ_I: $0.5 \le \Theta_I \le 4$
    *Note: The program gives $\Theta_I = 1.0$. Therefore, when you vary Θ_I, you will need to account for the corresponding increase or decrease of C_{A0} because the total concentration, C_{T0}, is constant.

(c) $\dfrac{Ua}{\rho_b}$: $0.1 \le \dfrac{Ua}{\rho_b} \le 0.8 \dfrac{\text{cal}}{\text{kg} \cdot \text{s} \cdot K}$
(d) T_0:   310 K $\le T_0 \le$ 350 K
(e) T_a: 300 K $\le T_a \le$ 340 K

(f) $\dot{m}_c$: $1 \leq m_c \leq 1000$ g/s

$$\left[T_0 = 350 \text{ K}, \frac{Ua}{\rho_b} = 0.5 \frac{\text{cal}}{\text{kg} \cdot \text{s} \cdot \text{K}}, T_{a0} = 320 \text{ K}, F_{A0} = 5 \text{ mol/s}, \Theta_I = 1 \right]$$

(g) Repeat (f) for counter current coolant flow.

(h) Determine the conversion in a 5,000 kg fluidized CSTR where $UA = 500$ cal/s·K with $T_a = 320$ K and $\rho_b = 2$ kg/m^3

(i) Repeat (a), (b), and (d) if the reaction were endothermic with $K_c = 0.01$ at 303 and $\Delta H_{Rx} = +20$ kcal/mol.

(n) Example 8-12. *Instructions:* If you have not installed COMSOL 3.2 ECRE, load the COMSOL 3.2 ECRE CD-ROM and follow the installation instructions.

Double-click on the COMSOL 3.2 ECRE icon on your desktop. In the Model Navigator, select model denoted "4-Non-Isothermal Reactor II" and press "Documentation". This will open your web browser and display the documentation of this specific model. You can also review the detailed documentation for the models listed on the left-hand side in the web browser window. Click on Chapter 2 radial effects in the left hand margin. This section gives an overview of all the model equations. Next, click on non-isothermal reactor with non-isothermal cooling jacket and follow the step by step instructions and screen shots. Select the Model Navigator and press "OK" to open the model.

(1) Why is the concentration of A near the wall lower than the concentration near the center? (2) Where in the reactor do you find the maximum and minimum reaction rates? Why? *Instructions:* Click the "Plot Parameters" button and select the "Surface" tab. Type "–rA" (replace "cA") in the "Expression" edit field to plot the absolute rate of consumption of A (moles m^{-3} s^{-1}). (3) Increase the activation energy of the reaction by 5%. How do the concentration profiles change? Decrease. *Instructions:* Select the "Constants" menu item in the "Options" menu. Multiply the value of "E" in the constants list by 1.05 (just type "*1.05" behind the existing value to increase or multiply by 0.95 to decrease). Press "Apply". Press the "Restart" button in the main toolbar (equal sign with an arrow on top). (4) Change the activation energy back to the original value. *Instructions:* Remove the factor "0.95" in the constant list and press "Apply". (5) Increase the thermal conductivity, k_e, by a factor of 10 and explain how this change affects the temperature profiles. At what radial position do you find the highest conversion? *Instructions:* Multiply the value of "ke" in the constants lists by 10. Press "Apply". Press "Restart". (6) Increase the coolant flow rate by a factor of 10 and explain how this change affects the conversion. (7) In two or three sentences, describe your findings when you varied the parameters (for all parts). (8) What would be your recommendation to maximize the average outlet conversion? (9) Review Figure E8-12.2 and explain why the temperature profile goes through a maximum and why the conversion profile goes through a maximum *and* a minimum. (10) See other problems in the web module.

(o) Example R8.2-1 Runaway Reactions. Load the *Living Example Problem* on runaway trajectories. Vary some of the parameters, such as P_0 and T_0 along with the activation energy and heat of reaction. Write a paragraph describing what you found and what generalizations you can make.

Interactive

Computer Modules

P8-3$_A$ Load the Interactive Computer Module (ICM) from the CD-ROM. Run the module, and then record your performance number for the module, which indicates your mastery of the material. Your professor has the key to decode your performance number.
(a) ICM Heat Effects Basketball 1 Performance # _____.
(b) ICM Heat Effects Simulation 2 Performance # _____.

P8-4$_C$ The following is an excerpt from *The Morning News*, Wilmington, Delaware (August 3, 1977): "Investigators sift through the debris from blast in quest for the cause [that destroyed the new nitrous oxide plant]. A company spokesman said it appears more likely that the [fatal] blast was caused by another gas—ammonium nitrate—used to produce nitrous oxide." An 83% (wt) ammonium nitrate and 17% water solution is fed at 200°F to the CSTR operated at a temperature of about 510°F. Molten ammonium nitrate decomposes directly to produce gaseous nitrous oxide and steam. It is believed that pressure fluctuations were observed in the system and as a result the molten ammonium nitrate feed to the reactor may have been shut off approximately 4 min prior to the explosion. **(a)** Can you explain the cause of the blast? [*Hint:* See Problem P9-3 and Equation (8-75).] **(b)** If the feed rate to the reactor just before shutoff was 310 lb of solution per hour, what was the exact temperature in the reactor just prior to shutdown? **(c)** How would you start up or shut down and control such a reaction? **(d)** What do you learn when you apply the runaway reaction criteria?

Assume that at the time the feed to the CSTR stopped, there was 500 lb of ammonium nitrate in the reactor. The conversion in the reactor is believed to be virtually complete at about 99.99%.

Additional information (approximate but close to the real case):

$$\Delta H^\circ_{Rx} = -336 \text{ Btu/lb ammonium nitrate at } 500°F \text{ (constant)}$$

$$C_P = 0.38 \text{ Btu/lb ammonium nitrate} \cdot °F$$

$$C_P = 0.47 \text{ Btu/lb of steam} \cdot °F$$

$$-r_A V = kC_A V = k\frac{M}{V}V = kM(\text{lb/h})$$

where M is the mass of ammonium nitrate in the CSTR (lb) and k is given by the relationship below.

T (°F)	510	560
k (h^{-1})	0.307	2.912

The enthalpies of water and steam are

$$H_w(200°F) = 168 \text{ Btu/lb}$$

$$H_g(500°F) = 1202 \text{ Btu/lb}$$

(e) Explore this problem and describe what you find. [For example, can you plot a form of $R(T)$ versus $G(T)$?] **(f)** Discuss what you believe to be the point of the problem. The idea for this problem originated from an article by Ben Horowitz.

P8-5$_B$ The endothermic liquid-phase elementary reaction

$$A + B \rightarrow 2C$$

proceeds, substantially, to completion in a single steam-jacketed, continuous-stirred reactor (Table P8-5). From the following data, calculate the steady-state reactor temperature:

>   Reactor volume: 125 gal
>   Steam jacket area: 10 ft²
>   Jacket steam: 150 psig (365.9°F saturation temperature)
>   Overall heat-transfer coefficient of jacket, U: 150 Btu/h · ft² · °F
>   Agitator shaft horsepower: 25 hp
>   Heat of reaction, $\Delta H_{Rx}^{\circ} = +20,000$ Btu/lb mol of A (independent of temperature)

TABLE P8-5

	Component		
	A	B	C
Feed (lbmol/hr)	10.0	10.0	0
Feed temperature (°F)	80	80	—
Specific heat (Btu/lb mol · °F)*	51.0	44.0	47.5
Molecular weight	128	94	222
Density (lb/ft³)	63.0	67.2	65.0

*Independent of temperature.
(*Ans: T* = 199°F)
(Courtesy of the California Board of Registration for Professional & Land Surveyors.)

P8-6$_A$ The elementary irreversible organic liquid-phase reaction.

$$A + B \rightarrow C$$

is carried out adiabatically in a flow reactor. An equal molar feed in A and B enters at 27°C, and the volumetric flow rate is 2 dm³/s and $C_{A0} = 0.1$ kmol/m³.
  (a) Calculate the PFR and CSTR volumes necessary to achieve 85% conversion. What are the reasons for the differences?
  (b) What is the maximum inlet temperature one could have so that the boiling point of the liquid (550 K) would not be exceeded even for complete conversion?
  (c) Plot the conversion and temperature as a function of PFR volume (i.e., distance down the reactor).
  (d) Calculate the conversion that can be achieved in one 500-dm³ CSTR and in two 250-dm³ CSTRs in series.
  (e) Ask another question or suggest another calculation for this reaction.

Additional information:

$$H_A^\circ(273) = -20 \text{ kcal/mol}, \quad H_B^\circ(273) = -15 \text{ kcal/mol}, \quad H_C^\circ(273) = -41 \text{ kcal/mol}$$

$$C_{p_A} = C_{p_B} = 15 \text{ cal/mol} \cdot \text{K} \qquad C_{p_C} = 30 \text{ cal/mol} \cdot \text{K}$$

$$k = 0.01 \frac{\text{dm}^3}{\text{mol} \cdot \text{s}} \text{ at 300 K} \qquad E = 10{,}000 \text{ cal/mol}$$

P8-7$_B$　Use the data and reaction in Problem 8-6 for the following exercices.
  (a) Plot the conversion and temperature of the PFR profiles up to a reactor volume of 10 dm³ for the case when the reaction is reversible with $K_C = 10$ m³/kmol at 450 K. Plot the equilibrium conversion profile.
  (b) Repeat (a) when a heat exchanger is added, $Ua = 20$ cal/m³/s/K, and the coolant temperature is constant at $T_a = 450$ K.
  (c) Repeat (b) for a co-current heat exchanger for a coolant flow rate of 50 g/s and $C_{p_c} = 1$ cal/g • K, and in inlet coolant temperature of $T_{a0} = 450$ K. Vary the coolant rate $(1 < \dot{m}_c < 1{,}000 \text{ g/s})$.
  (d) Repeat (c) for counter current coolant flow.
  (e) Compare your answers to (a) through (d) and describe what you find. What generalizations can you make?
  (f) Repeat (c) and (d) when the reaction is irreversible but endothermic with $\Delta H_{Rx} = 6{,}000$ cal/mol.
  (g) Discuss the application of runaway criteria for the irreversible reaction occurring in a CSTR. What value of T_o would you recommend to prevent runaway if $\kappa = 3$ and $T_a = 450$ K?

P8-8$_A$　The elementary irreversible gas-phase reaction

$$A \rightarrow B + C$$

is carried out adiabatically in a PFR packed with a catalyst. Pure A enters the reactor at a volumetric flow rate of 20 dm³/s at a pressure of 10 atm and a temperature of 450 K.
  (a) Plot the conversion and temperature down the plug-flow reactor until an 80% conversion (if possible) is reached. (The maximum catalyst weight that can be packed into the PFR is 50 kg.) Assume that $\Delta P = 0.0$.
  (b) What catalyst weight is necessary to achieve 80% conversion in a CSTR?
  (c) Write a question that requires critical thinking and then explain why your question requires critical thinking. [*Hint*: See Preface Section B.2.]
  (d) Now take the pressure drop into account in the PFR.

$$\frac{dP}{dW} = -\frac{\alpha}{2}\left(\frac{T}{T_0}\right)\frac{P_0}{(P/P_0)}(1 + \varepsilon X)$$

The reactor can be packed with one of two particle sizes. Choose one.

$$\alpha = 0.019/\text{kg cat. for particle diameter } D_1$$

$$\alpha = 0.0075/\text{kg cat. for particle diameter } D_2$$

Plot the temperature, conversion, and pressure along the length of the reactor. Vary the parameters α and P_0 to learn the ranges of values in which they dramatically affect the conversion.

Additional information:

$C_{P_A} = 40$ J/mol·K $C_{P_B} = 25$ J/mol·K $C_{P_C} = 15$ J/mol·K

$H_A^\circ = -70$ kJ/mol $H_B^\circ = -50$ kJ/mol $H_C^\circ = -40$ kJ/mol

All heats of formation are referenced to 273 K.

$$k = 0.133 \, \exp\left[\frac{E}{R}\left(\frac{1}{450} - \frac{1}{T}\right)\right] \frac{\text{dm}^3}{\text{kg} \cdot \text{cat} \cdot \text{s}} \text{ with } E = 31.4 \text{ kJ/mol}$$

P8-9$_B$ Use the data in Problem 8-8 for the case when heat is removed by a heat exchanger jacketing the reactor. The flow rate of coolant through the jacket is sufficiently high that the ambient exchanger temperature is contant at $T_a = 50°C$.

(a) Plot the temperature and conversion profiles for a PBR with

$$\frac{Ua}{\rho_b} = 0.08 \, \frac{\text{J}}{\text{s} \cdot \text{kg cat.} \cdot \text{K}}$$

where

ρ_b = bulk density of the catalyst (kg/m³)

a = heat-exchange area per unit volume of reactor (m²/m³)

U = overall heat-transfer coefficient (J/s·m²·K)

How would the profiles change if Ua/ρ_b were increased by a factor of 3000?

(b) Repeat part (a) for both co-current and counter current flow with $\dot{m}_c = 0.2$ kg/s, $C_{P_c} = 5,000$ J/kg K and an entering coolant temperature of 50°C.

(c) Find X and T for a "fluidized" CSTR [see margin] with 80 kg of catalyst.

$$UA = 500 \, \frac{\text{J}}{\text{s} \cdot \text{K}}, \qquad \rho_b = 1 \text{ kg/m}^3$$

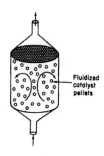

Fluidized
catalyst
pellets

(d) Repeat parts (a) and (b) for W = 80.0 kg assuming a reversible reaction with a reverse specific reaction rate of

$$k_r = 0.2 \, \exp\left[\frac{E_r}{R}\left(\frac{1}{450} - \frac{1}{T}\right)\right]\left(\frac{\text{dm}^6}{\text{kg cat.} \cdot \text{mol} \cdot \text{s}}\right); \qquad E_r = 51.4 \text{ kJ/mol}$$

Vary the entering temperature, T_0, and describe what you find.

(e) Use or modify the data in this problem to suggest another question or calculation. Explain why your question requires either critical thinking or creative thinking. See Preface B.2 and B.3.

P8-10$_B$ The irreversible endothermic vapor-phase reaction follows an elementary rate law

$$CH_3COCH_3 \rightarrow CH_2CO + CH_4$$
$$A \rightarrow B + C$$

and is carried out adiabatically in a 500-dm³ PFR. Species A is fed to the reactor at a rate of 10 mol/min and a pressure of 2 atm. An inert stream is also fed to the reactor at 2 atm, as shown in Figure P8-10. The entrance temperature of both streams is 1100 K.

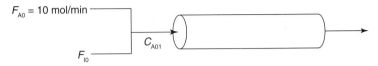

Figure P8-10   Adiabatic PFR with inerts.

(a) First derive an expression for C_{A01} as a function of C_{A0} and Θ_I.

(b) Sketch the conversion and temperature profiles for the case when no inerts are present. Using a dashed line, sketch the profiles when a moderate amount of inerts are added. Using a dotted line, sketch the profiles when a large amount of inerts are added. Sketch or plot the exit conversion as a function of Θ_I. Qualitative sketches are fine.

(c) Is there a ratio of inerts to the entering molar flow rate of A (i.e., $\Theta_I = F_{I0}/F_{A0}$) at which the conversion is at a maximum? Explain why there "is" or "is not" a maximum.

(d) Repeat parts (b) and (c) for an exothermic reaction ($\Delta H_{Rx} = -80$ kJ/mol).

(e) Repeat parts (b) and (c) for a second-order endothermic reaction.

(f) Repeat parts (b) and (c) for an exothermic reversible reaction ($K_C = 2$ dm³/mol at 1100 K).

(g) Repeat (b) through (f) when the total volumetric flow rate v_0 is held constant and the mole fractions are varied.

(h) Sketch or plot F_B for parts (d) through (g).

Additional information:

$$k = \exp\left(34.34 - 34{,}222/T\right)\ \text{dm}^3/\text{mol} \cdot \text{min} \qquad C_{P_I} = 200\ \text{J/mol} \cdot \text{K}$$

(*T* in degrees Kelvin)

$C_{P_A} = 170\ \text{J/mol} \cdot \text{K}$ \qquad\qquad\qquad\qquad $C_{P_B} = 90\ \text{J/mol} \cdot \text{K}$

$C_{P_C} = 80\ \text{J/mol} \cdot \text{K}$ \qquad\qquad\qquad\qquad $\Delta H_{Rx} = 80{,}000\ \text{J/mol}$

P8-11₍C₎ Derive the energy balance for a packed bed membrane reactor. Apply the balance to the reaction in Problem P8-8₍B₎ for the case when it is reversible with $K_c = 0.01\ mol/dm^3$ at 300 K. Species C diffuses out of the membrane.

(a) Plot the concentration profiles or different values of k_c when the reaction is carried out adiabatically.

(b) Repeat part (a) when the heat transfer coefficient is the same as that given in P8-9(a). All other conditions are the same as those in Problem P8-9₍B₎.

P8-12₍B₎   The liquid-phase reaction

$$A + B \rightarrow C$$

Member

Hall of Fame

follows an elementary rate law and takes place in a 1-m³ CSTR, to which the volumetric flow rate is 0.5 m³/min and the entering concentration of A is 1 *M*. The reaction takes place isothermally at 300 K. For an equal molar feed of A and B, the conversion is 20%. When the reaction is carried out adiabatically, the exit temperature is 350K and the conversion is 40%. The heat capacities of A, B, and C are 25, 35, and 60 kJ/mol • K, respectively. It is proposed to add a second reactor of the same size downstream in series with the first CSTR. There is a heat exchanger attached to the second CSTR with $UA = 4.0$ kJ/min $\cdot$ K, and the coolant fluid enters and exits this reactor at virtually the same temperature the coolant feed enters 350 K.

(a) What is the rate of heat removal necessary for isothermal operation?

(b) What is the conversion exiting the second reactor?

(c) What would be the conversion if the second CSTR were replaced with a 1-m^3 PFR with $Ua = 10$ kJ/m^3 $\cdot$ min and $T_a = 300$ K?

(d) A chemist suggests that at temperatures above 380 K the reverse reaction cannot be neglected. From thermodynamics, we know that at 350 K, $K_c = 2$ dm^3/mol. What conversion can be achieved if the entering temperature to the PFR in part (c) is 350 K?

(e) Write an in-depth question that extends this problem and involves critical thinking, and explain why it involves critical thinking.

(f) Repeat part (c) assuming the reaction takes place entirely in the gas phase (same constants for reaction) with $C_{T0} = 0.2$ mol/dm^3.

P8-13$_A$ The reaction

$$A + B \rightleftarrows C + D$$

is carried out adiabatically in a series of staged packed-bed reactors with interstage cooling. The lowest temperature to which the reactant stream may be cooled is 27°C. The feed is equal molar in A and B and the catalyst weight in each reactor is sufficient to achieve 99.9% of the equilibrium conversion. The feed enters at 27°C and the reaction is carried out adiabatically. If four reactors and three coolers are available, what conversion may be achieved?

Additional information:

$$\Delta H_{Rx} = -30{,}000 \text{ cal/mol A} \qquad C_{P_A} = C_{P_B} = C_{P_C} = C_{P_D} = 25 \text{ cal/g mol} \cdot \text{K}$$

$$K_e(50°C) = 500{,}000 \qquad F_{A0} = 10 \text{ mol A/min}$$

First prepare a plot of equilibrium conversion as a function of temperature. [*Partial ans.:* $T = 360$ K, $X_e = 0.984$; $T = 520$ K, $X_e = 0.09$; $T = 540$ K, $X_e = 0.057$]

P8-14$_A$ Figure 8-8 shows the temperature-conversion trajectory for a train of reactors with interstage heating. Now consider replacing the interstage heating with injection of the feed stream in three equal portions as shown here:

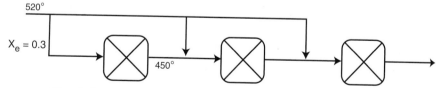

Sketch the temperature-conversion trajectories for **(a)** an endothermic reaction with entering temperatures as shown, and **(b)** an exothermic reaction with the temperatures to and from the first reactor reversed, i.e., $T_0 = 450$°C.

P8-15 The biomass reaction

$$\text{Substrate} \xrightarrow{\text{Cells}} \text{More Cells} + \text{Product}$$

is carried out in a 6 dm^3 chemostat with a heat exchanger.

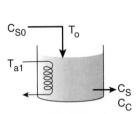
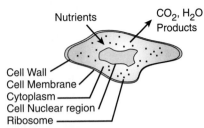

The volumetric flow rate is 1 dm^3/h, the entering substrate concentration and temperature are 100 g/dm^3 and 280 K, respectively. The temperature dependence of the growth rate follows that given by Aibe et al., Equation (7-63)

$$r_g = \mu C_C$$

and

$$-r_S = r_g / Y_{C/S}$$

$$\mu(T) = \mu(310 \text{ K})I' = \mu_{1max}\left[\frac{0.0038 \cdot T \cdot \exp[21.6 - 6700/T]}{1 + \exp[153 - 48000/T]}\right]\frac{C_S}{K_m + C_S}$$

(P8-16.1)

(a) Plot $G(T)$ and $R(T)$ for both adiabatic and non-adiabatic operation assuming a very large coolant rate (i.e., $\dot{Q} = UA(T_a - T)$ with $A = 1.1$ m^2 and $T_a = 290$ K).

(b) What is the heat exchanger area that should be used to maximize the exiting cell concentration for an entering temperature of 288 K? Cooling water is available at 290 K and up to a maximum flow rate of 1 kg/minute.

(c) Identify any multiple steady states and discuss them in light of what you learned in this chapter. *Hint:* Plot T_s vs. T_0 from Part (a).

(d) Vary T_o, $\dot{m}_c$, and T_a and describe what you find.

Additional Information:

$Y_{C/S} = 0.8$ g cell/g substrate
$K_S = 5.0$ g/dm^3
$\mu_{1max} = 0.5h^{-1}$ (note $\mu = \mu$max at 310 K)
C_{P_S} = Heat capacity of substrate solution including all cells = 74 J/g/K

m_s = Mass of substrate solution in chemostat = 6.0 kg
$\Delta H_{Rx} = -20,000$ J/g cells
$U = 50,000$ $J/h/Km^2$
C_{P_c} = Heat capacity of cooling water 74 J/g/K

$\dot{m}_c$ = coolant flow rate (up to 60,000 kg/h)
ρ_S = solution density = 1 kg / dm^3

$$\dot{Q} = \dot{m}_c C_{P_c}[T - T_a]\left[1 - e^{-\frac{UA}{\dot{m}_c C_{P_c}}}\right]$$

P8-16$_B$ The first-order irreversible exothermic liquid-phase reaction

$$A \rightarrow B$$

is to be carried out in a jacketed CSTR. Species A and an inert I are fed to the reactor in equimolar amounts. The molar feed rate of A is 80 mol/min.

(a) What is the reactor temperature for a feed temperature of 450 K?

(b) Plot the reactor temperature as a function of the feed temperature.

(c) To what inlet temperature must the fluid be preheated for the reactor to operate at a high conversion? What are the corresponding temperature and conversion of the fluid in the CSTR at this inlet temperature?

(d) Suppose that the fluid is now heated 5°C above the temperature in part (c) and then cooled 20°C, where it remains. What will be the conversion?

(e) What is the inlet extinction temperature for this reaction system? (*Ans.: $T_0 = 87°C$.*)

Additional information:

Heat capacity of the inert: 30 cal/g mol·°C $\tau = 100$ min

Heat capacity of A and B: 20 cal/g mol·°C $\Delta H_{Rx} = -7500$ cal/mol

UA: 8000 cal/min·°C $k = 6.6 \times 10^{-3}$ min^{-1} at 350 K

Ambient temperature, T_a: 300 K $E = 40,000$ cal/mol·K

P8-17$_C$ The zero-order exothermic liquid-phase reaction

$$A \rightarrow B$$

is carried out at 85°C in a jacketed 0.2-m³ CSTR. The coolant temperature in the reactor is 32°F. The heat-transfer coefficient is 120 W/m²·K. Determine the critical value of the heat-transfer area below which the reactor will run away and explode [*Chem. Eng., 91*(10), 54 (1984)].

Additional information:

$$k = 1.127 \text{ kmol/m}^3 \cdot \text{min at } 40°C$$
$$k = 1.421 \text{ kmol/m}^3 \cdot \text{min at } 50°C$$

The heat capacity of the solution is 4 J/°C/g. The solution density is 0.90 kg/dm³. The heat of reaction is −500 J/g. The feed temperature is 40°C and the feed rate is 90 kg/min. MW of A = 90 g/mol.

P8-18$_B$ The elementary reversible liquid-phase reaction

$$A \rightleftharpoons B$$

takes place in a CSTR with a heat exchanger. Pure A enters the reactor.

(a) Derive an expression (or set of expressions) to calculate $G(T)$ as a function of heat of reaction, equilibrium constant, temperature, and so on. Show a sample calculation for $G(T)$ at $T = 400$ K.

(b) What are the steady-state temperatures? (*Ans.: 310, 377, 418 K.*)

(c) Which steady states are locally stable?

(d) What is the conversion corresponding to the upper steady state?

(e) Vary the ambient temperature T_a and make a plot of the reactor temperature as a function of T_a, identifying the ignition and extinction temperatures.

(f) If the heat exchanger in the reactor suddenly fails (i.e., $UA = 0$), what would be the conversion and the reactor temperature when the new upper steady state is reached? (*Ans.:* 431 K)

(g) What heat exchanger product, UA, will give the maximum conversion?

(h) Write a question that requires critical thinking and then explain why your question requires critical thinking. [*Hint:* See Preface Section B.2.]

(i) What is the adiabatic blowout flow rate, v_0.

(j) Suppose that you want to operate at the lower steady state. What parameter values would you suggest to prevent runaway?

Additional information:

$$UA = 3600 \text{ cal/min} \cdot \text{K} \qquad E/R = 20,000 \text{ K}$$

$$C_{P_A} = C_{P_B} = 40 \text{ cal/mol} \cdot \text{K} \qquad V = 10 \text{ dm}^3$$

$$\Delta H_{Rx} = -80,000 \text{ cal/mol A} \qquad v_0 = 1 \text{ dm}^3/\text{min}$$

$$K_{eq} = 100 \text{ at } 400 \text{ K} \qquad F_{A0} = 10 \text{ mol/min}$$

$$k = 1 \text{ min}^{-1} \text{ at } 400 \text{ K}$$

Ambient temperature, $T_a = 37°C$ \qquad Feed temperature, $T_0 = 37°C$

P8-19$_C$ The first-order irreversible liquid-phase reaction

$$A \rightarrow B$$

is to be carried out in a jacketed CSTR. Pure A is fed to the reactor at a rate of 0.5 g mol/min. The heat-generation curve for this reaction and reactor system,

$$G(T) = \frac{-\Delta H_{Rx}^\circ}{1 + 1/(\tau k)}$$

is shown in Figure P8-19.

(a) To what inlet temperature must the fluid be preheated for the reactor to operate at a high conversion? (*Ans.:* $T_0 \geq 214°C$.)

(b) What is the corresponding temperature of the fluid in the CSTR at this inlet temperature? (*Ans.:* $T_s = 164°C, 184°C$.)

(c) Suppose that the fluid is now heated 5°C above the temperature in part (a) and then cooled 10°C, where it remains. What will be the conversion? (*Ans.:* $X = 0.9$.)

(d) What is the extinction temperature for this reaction system? (*Ans.:* $T_0 = 200°C$.)

(e) Write a question that requires critical thinking and then explain why your question requires critical thinking. [*Hint:* See Preface Section B.2.]

Additional information:

Heat of reaction (constant): -100 cal/g mol A

Heat capacity of A and B: 2 cal/g mol $\cdot$ °C

UA: 1 cal/min $\cdot$ °C , Ambient temperature, T_a : 100°C

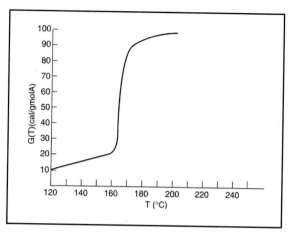

Figure P8-19 $G(T)$ curve.

P8-20$_C$ Troubleshooting. The following reactor system is used to carry out the reversible catalytic reaction

$$A + B \rightleftarrows C + D$$

The feed is equal molar in A and B at a temperature T_1 of 300 K.

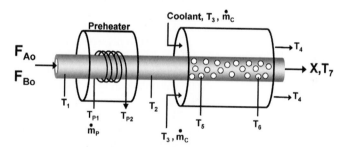

Troubleshoot the reaction system to deduce the problems for an exothermic and an endothermic reaction. Next, suggest measures to correct the problem. You can change $\dot{m}_p$, $\dot{m}_c$, and F_{A0} along with T_2 and T_3.

Troubleshoot what temperatures are normal and what are different and what the distinction is. Explain your reasoning in each of the cases below.

(a) Exothermic reaction. The expected conversion and the exit temperature are $X = 0.75$ and $T = 400$ K. Unfortunately, here is what was found in six different cases.

Case 1 at the exit $X = 0.01$, $T_7 = 305$ K
Case 2 at the exit $X = 0.10$, $T_7 = 550$ K
Case 3 at the exit $X = 0.20$, $T_7 = 350$ K
Case 4 at the exit $X = 0.5$, $T_7 = 450$ K
Case 5 at the exit $X = 0.01$, $T_7 = 400$ K
Case 6 at the exit $X = 0.3$, $T_7 = 500$ K

(b) Endothermic reaction. The expected conversion and the exit temperature are $X = 0.75$ and $T_7 = 350$ K. Here is what was found.

Case 1 at the exit $X = 0.4$, $T_7 = 320$ K
Case 2 at the exit $X = 0.02$, $T_7 = 349$ K
Case 3 at the exit $X = 0.002$, $T_7 = 298$ K
Case 4 at the exit $X = 0.2$, $T_7 = 350$ K

Web Modules

P8-21 If you have not installed COMSOL 3.2 ECRE, load the COMSOL 3.2 CD-ROM and follow the installation instructions discussed in **P8-2(n)**.

(a) Before running the program, sketch the radial temperature profile down a PFR for (1) an exothermic reaction for a PFR with a cooling jacket and (2) an endothermic reaction for a PFR with a heating jacket.

(b) Run the COMSOL 3.2 ECRE program and compare with your results in (a). Double-click on the COMSOL 3.2 ECRE icon on your desktop. In the Model Navigator, select the model denoted "3-Non-Isothermal I" and press OK. You can use this model to compare your results in (1) and (2) above. You can select "Documentation" in the "Help" menu in order to review the instructions for this model and other models in COMSOL ECRE.

Change the velocity profile from laminar parabolic to plug flow. Select "Scalar Expressions" in the "Expression" menu item in the "Options" menu. Change the expression for uz (the velocity) to "u0" (replace the expression "2*u0*(1–(r/Ra)²)", which describes the parabolic velocity profile). Press "Apply".

You can now continue to vary the input data and change the exothermic reaction to an endothermic one. (Hint: Select the "Constants" menu item in the "Options" menu. Do not forget "T_a0" the jacket temperature at the end of the list). Write a paragraph describing your findings.

(c) The thermal conductivity in the reactor, denoted "ke" in Figure E8-12.1, is the molecular thermal conductivity for the solution. In a plug flow reactor, the flow is turbulent. In such a reactor, the apparent thermal conductivity is substantially larger than the molecular thermal conductivity of the fluid. Vary the value of the thermal conductivity "ke" to learn it's influence on the temperature and concentration profile in the reactor.

(d) In turbulent flow, the apparent diffusivity is substantially larger than the molecular diffusivity. Increase the molecular diffusivity in the PFR to reflect turbulent conditions and study the influence on the temperature and concentration profiles. Here you can go to the extremes. Find something interesting to turn in to your instructor. See other problems in the web module.

(e) See other problems in the web module.

P8-22_C A reaction is to be carried out in the packed-bed reactor shown in Figure P8-22.

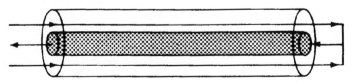

Figure P8-22

The reactants enter in the annular space between an outer insulated tube and an inner tube containing the catalyst. No reaction takes place in the annular region. Heat transfer between the gas in this packed-bed reactor and the gas flowing counter currently in the annular space occurs along the length of the reactor. The overall heat-transfer coefficient is $5 \text{ W}/\text{m}^2 \cdot \text{K}$. Plot the conversion and temperature as a function of reactor length for the data given in

(a) Problem P8-6.

(b) Problem P8-9(d).

P8-23$_B$ The irreversible liquid-phase reactions

Reaction (1)	$A + B \rightarrow 2C$	$r_{1C} = k_{1C} C_A C_B$
Reaction (2)	$2B + C \rightarrow D$	$r_{2D} = k_{2D} C_B C_C$

are carried out in a PFR with heat exchange. The following temperature profile was obtained for the reaction and the coolant stream.

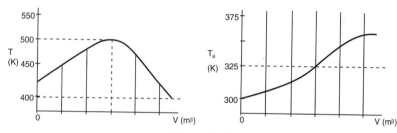

Figure P8-23

The concentrations of A, B, C, and D were measured at the point down the reactor where the liquid temperature, T, reached a maximum, and they were found to be $C_A = 0.1$, $C_B = 0.2$, $C_C = 0.5$, and $C_D = 1.5$ all in mol/dm³. The product of the overall heat-transfer coefficient and the heat-exchanger area per unit volume, Ua, is 10 cal/s • dm³ • K. The feed is equal molar in A and B, and the entering molar flow rate of A is 10 mol/s. What is the activation energy for Reaction (1)? $E = ??$cal/mol.

Additional Information

$C_{P_A} = C_{P_B} = C_{P_C} = 30 \text{ cal/mol/K}$

$C_{P_D} = 90 \text{ cal/mol/K}, \quad C_{P_I} = 100 \text{ cal/mol/K}$

$\Delta H_{Rx1A} = -50,000 \text{ cal/molA}$ $k_{1C} = 0.043 \dfrac{\text{dm}^3}{\text{mol} \cdot \text{s}} \text{ at 400 K}$

$\Delta H_{Rx2B} = +5000 \text{ cal/molB}$ $k_{2D} = 0.4 \dfrac{\text{dm}^3}{\text{mol} \cdot \text{s}} e^{5000 \text{ K}\left[\frac{1}{500} - \frac{1}{T}\right]}$

P8-24   *(Comprehensive Term Problem)* T-Amyl Methyl Ether (TAME) is an oxygenated additive for green gasolines. Besides its use as an octane enhancer, it also improves the combustion of gasoline and reduces the CO and HC (and, in a smaller extent, the NO_x) automobile exhaust emissions. Due to the environmental concerns related to those emissions, this and other ethers (MTBE, ETBE, TAEE) have been lately studied intensively. TAME is currently catalytically produced in the liquid phase by the reaction of methanol (MeOH) and the isoamylenes 2-methyl-1-butene (2M1B) and 2-methyl-2-butene (2M2B). There are three simultaneous equilibrium reactions in the formation and splitting of TAME (the two etherification reactions and the isomerization between the isoamylenes):

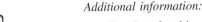

$$2M1B + MeOH \Leftrightarrow TAME \qquad (1)$$

$$2M2B + MeOH \Leftrightarrow TAME \qquad (2)$$

$$2M1B \Leftrightarrow 2M2B \qquad (3)$$

These reactions are to be carried out in a plug-flow reactor and a membrane reactor in which MeOH is fed uniformly through the sides. For isothermal operation:

(a) Plot the concentration profiles for a 10-m^3 PFR.

(b) Vary the entering temperature, T_0, and plot the exit concentrations as a function of T_0.

For a reactor with heat exchange ($U = 10 \text{ J} \cdot \text{m}^{-2} \cdot \text{s}^{-1} \text{ K}^{-1}$):

(c) Plot the temperature and concentration profiles for an entering temperature of 353 K.

(d) Repeat (a) through (c) for a membrane reactor.

Additional information:

The data for this problem is found at the end of the Additional Homework Problems for Chapter 8 on the CD-ROM and on the web. [Problem by M. M. Vilarenho Ferreira, J. M. Loueiro, and D. R. Frias, University of Porto, Portugal.]

P8-25c   *(Multiple reactions with heat effects)* Xylene has three major isomers, *m*-xylene, *o*-xylene, and *p*-xylene. When *o*-xylene is passed over a Cryotite catalyst, the following elementary reactions are observed:

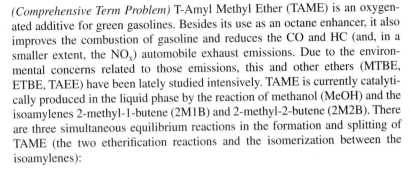

The feed to the reactor is equal molar in in both *m*-xylene and *o*-xylene (species A and B). For a total feed rate of 2 mol/min and the reaction conditions below, plot the temperature and the molar flow rates of each species as a function of catalyst weight up to a weight of 100 kg.

(a) Find the lowest concentration of *o*-xylene achieved in the reactor.
(b) Find the highest concentration of *m*-xylene achieved in the reactor.
(c) Find the maximum concentration of *o*-xylene in the reactor.
(d) Repeat parts (a) to (c) for a pure feed of *o*-xylene.
(e) Vary some of the system parameters and describe what you learn.
(f) What do you believe to be the point of this problem?

Additional information:[11]

All heat capacities are virtually the same at 100 J/mol·K.

$$C_{T0} = 2 \text{ mol/dm}^3$$

$$\Delta H_{Rx1O} = -1800 \text{ J/mol } o\text{-xylene}^{11}$$

$$\Delta H_{Rx3O} = -1100 \text{ J/mol } o\text{-xylene}$$

$$k_1 = 0.5 \exp[2(1 - 320/T)] \text{ dm}^3/\text{kg cat.} \cdot \text{min}$$

$$k_2 = k_1/K_C$$

$$k_3 = 0.005 \exp\{[4.6(1 - (460/T))]\} \text{ dm}^3/\text{kg cat.} \cdot \text{min}$$

$$K_C = 10 \exp[4.8(430/T - 1.5)]$$

$$T_0 = 330 \text{ K}$$

$$T_a = 500 \text{ K}$$

$$Ua/\rho_b = 16 \text{ J/kg cat.} \cdot \text{min} \cdot ^\circ\text{C}$$

$$W = 100 \text{ kg}$$

P8-26$_C$ *(Comprehensive Problem on multiple reactions with heat effects)* Styrene can be produced from ethylbenzene by the following reaction:

$$\text{ethylbenzene} \longleftrightarrow \text{styrene} + H_2 \qquad (1)$$

However, several irreversible side reactions also occur:

$$\text{ethylbenzene} \longrightarrow \text{benzene} + \text{ethylene} \qquad (2)$$

$$\text{ethylbenzene} + H_2 \longrightarrow \text{toluene} + \text{methane} \qquad (3)$$

[J. Snyder and B. Subramaniam, *Chem. Eng. Sci.*, *49*, 5585 (1994)]. Ethylbenzene is fed at a rate of 0.00344 kmol/s to a 10.0-m³ PFR (PBR) along with inert steam at a total pressure of 2.4 atm. The steam/ethylbenzene molar ratio is initially [i.e., parts (a) to (c)] 14.5:1 but can be varied. Given the following data, find the exiting molar flow rates of styrene, benzene, and toluene along with $\tilde{S}_{S/BT}$ for the following inlet temperatures when the reactor is operated adiabatically.

(a) $T_0 = 800$ K
(b) $T_0 = 930$ K
(c) $T_0 = 1100$ K

---

[11] Obtained from inviscid pericosity measurements.

(d) Find the ideal inlet temperature for the production of styrene for a steam/ethylbenzene ratio of 58:1. (*Hint:* Plot the molar flow rate of styrene versus T_0. Explain why your curve looks the way it does.)

(e) Find the ideal steam/ethylbenzene ratio for the production of styrene at 900 K. [*Hint:* See part (d).]

(f) It is proposed to add a counter current heat exchanger with $Ua = 100 \text{ kJ/min/K}$ where T_a is virtually constant at 1000 K. For an entering stream to ethylbenzene ratio of 2, what would you suggest as an entering temperature? Plot the molar flow rates and $\tilde{S}_{St/BT}$.

(g) What do you believe to be the points of this problem?

(h) Ask another question or suggest another calculation that can be made for this problem.

Additional information:

Heat capacities

Methane	68 J/mol·K	Styrene	273 J/mol·K
Ethylene	90 J/mol·K	Ethylbenzene	299 J/mol·K
Benzene	201 J/mol·K	Hydrogen	30 J/mol·K
Toluene	249 J/mol·K	Steam	40 J/mol·K

$$\rho = 2137 \text{ kg/m}^3 \text{ of pellet}$$

$$\phi = 0.4$$

$$\Delta H_{Rx1EB} = 118{,}000 \text{ kJ/kmol ethylbenzene}$$

$$\Delta H_{Rx2EB} = 105{,}200 \text{ kJ/kmol ethylbenzene}$$

$$\Delta H_{Rx3EB} = -53{,}900 \text{ kJ/kmol ethylbenzene}$$

$$K_{p1} = \exp\left\{ b_1 + \frac{b_2}{T} + b_3 \ln(T) + [(b_4 T + b_5)T + b_6]T \right\} \text{ atm}$$

$b_1 = -17.34$	$b_4 = -2.314 \times 10^{-10}$
$b_2 = -1.302 \times 10^4$	$b_5 = 1.302 \times 10^{-6}$
$b_3 = 5.051$	$b_6 = -4.931 \times 10^{-3}$

The kinetic rate laws for the formation of styrene (St), benzene (B), and toluene (T), respectively, are as follows. (EB = ethylbenzene)

$$r_{1St} = \rho(1-\phi)\exp\left(-0.08539 - \frac{10{,}925}{T}\right)\left(P_{EB} - \frac{P_{St}P_{H_2}}{K_{p1}}\right) \quad (\text{kmol/m}^3 \cdot \text{s})$$

$$r_{2B} = \rho(1-\phi)\exp\left(13.2392 - \frac{25{,}000}{T}\right)(P_{EB}) \quad (\text{kmol/m}^3 \cdot \text{s})$$

$$r_{3T} = \rho(1-\phi)\exp\left(0.2961 - \frac{11{,}000}{T}\right)(P_{EB}P_{H_2}) \quad (\text{kmol/m}^3 \cdot \text{s})$$

The temperature T is in kelvin.

P8-27$_B$ Compare the profiles in Figure E8-3.1, E8-4.1, E8-5.1, E8-5.3, E8-5.4, 8-10, E8-10.1, and E8-10.2.
   (a) If you were to classify the profiles into groups, what would they be? What are the common characteristics of each group?
   (b) What are the similarities and differences in the profiles in the various groups and in the various figures?
   (c) Describe why Figure E8-5.3 and Figure E8-3.1 look the way they do. What are the similarities and differences? Describe qualitatively how they would change if inerts were added.
   (d) Repeat (c) for Figure E8-10.1 and for Figure E8-10.2.

TAME

Hall of Fame

- **CD-ROM Complete Data Set**

P8-24$_C$ The TAME data set is given on the CD-ROM. This problem is a very comprehensive problem; perhaps can be used as a term (semester) problem.

- **Good Alternative Problems on CD-ROM Similar to Above Problems**

P8-28$_B$ Industrial data for the reaction

$$2 \text{ vinyl acetylene} \rightarrow \text{styrene}$$

are given. You are asked to make PFR calculations similar to those in Problems P8-6$_B$ to P8-9$_B$. [3rd Ed. P8-9$_B$]

P8-29$_B$ Reactor staging with interstage cooling. Similar to P8-13$_B$, but shorter because X_e versus T is given. [3rd Ed. P8-15$_B$]

P8-30$_B$ Use the data in Problems P8-6 and P8-8 to carry out reactions in a radial flow reactor. [3rd Ed. P8-18$_A$]

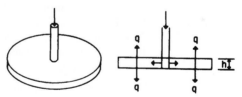

P8-31$_B$ The reactions

$$A \rightarrow B \rightarrow C$$

are carried out in a CSTR with a heat exchanger. [3rd Ed. P8-28$_x$]

P8-32$_B$ Elementary irreversible reaction

$$A \rightarrow 2B$$

is carried out in a PFR with heat exchange and pressure drop. [3rd Ed., P8-12]

P8-33$_B$ Liquid-phase reactions

$$A + B \rightarrow D$$

$$A + B \rightarrow U$$

in a CSTR. Maximize D. [3rd Ed. P8-31]

- **Oldies, But Goodies—Problems from Previous Editions**

  CDP8-A$_C$ The exothermic reaction

  $$A \rightleftharpoons 2B$$

  is carried out in both a plug-flow reactor and a CSTR with heat exchange. You are requested to plot conversion as a function of reactor length for both adiabatic and nonadiabatic operation as well as to size a CSTR. [2nd Ed. P8-16]

  ### Bifurcation Problems

  CDP8-B$_B$ Use bifurcation theory (R8.2 on the CD-ROM) to determine the possible regions for multiple steady states for the gas reaction with the rate law

  $$-r''_A = \frac{k_1 C_A}{(k_2 + k_3 C_A)^2}$$

  [2nd Ed., P8-26]

  CDP8-C$_B$ In this problem bifurcation theory (CD-ROM R8.2) is used to determine if multiple steady states are possible for each of three types of catalyst. [2nd Ed., P8-27$_B$]

  ### SO$_2$ Oxidation Design Problems for R8.4

  CDP8-D$_C$ This problem concerns the SO$_2$ reaction with heat losses. [2nd Ed., P8-33]

  CDP8-E$_C$ This problem concerns the use of interstage cooling in SO$_2$ oxidation. [2nd Ed., P8-34a]

  CDP8-F$_B$ This problem is a continuation of the SO$_2$ oxidation example problem. Reactor costs are considered in the analysis cooling. [2nd Ed., P8-34(b) and (c)]

  ### Multiple Reactions

  CDP8-G$_B$ Parallel reactions take place in a CSTR with heat effects. [1st Ed., P9-21]

  CDP8-H$_B$ Multiple reactions

  $$A \longrightarrow 2B$$
  $$2A + B \longrightarrow C$$

  are carried out adiabatically in a PFR.

  ### Multiple Steady States

  CDP8-I$_D$ In the multiple steady state for
  (Tough Problem)

  $$A \longrightarrow B$$

  The phase plane of C_A vs. T shows a separatrix. [2nd Ed., P8-22]

  CDP8-J$_B$ A second-order reaction with multiple steady states is carried out in different solvents.

Other Problems

CDP8-K$_B$ Extension of COMSOL Example E8-12.

JOURNAL CRITIQUE PROBLEMS

Living Example Problem

P8C-1   Equation (8) in an article in *J. Chem. Technol. Biotechnol., 31,* 273 (1981) is the kinetic model of the system proposed by the authors. Starting from Equation (2), derive the equation that describes the system. Does it agree with Equation (8)? By using the data in Figure 1, determine the new reaction order. The data in Table 2 show the effect of temperature. Figure 2 illustrates this effect. Use Equation (8) and Table 2 to obtain Figure 2. Does it agree with the article's results? Now use Table 2 and your equation. How does the figure obtained compare with Figure 2? What is your new E_{ac}?

P8C-2   The kinetics of the reaction of sodium hypochlorite and sodium sulfate were studied by the flow thermal method in *Ind. Eng. Chem. Fundam., 19,* 207 (1980). What is the flow thermal method? Can the energy balance developed in this article be applied to a plug-flow reactor, and if not, what is the proper energy balance? Under what conditions are the author's equations valid?

P8C-3   In an article on the kinetics of sucrose inversion by invertase with multiple steady states in a CSTR [*Chem. Eng. Commun., 6,* 151 (1980)], consider the following challenges: Are the equations for K_i and K_m correct? If not, what are the correct equations for these variables? Can an analysis be applied to this system to deduce regions of multiple steady states?

P8C-4   Review the article in *AIChE J., 16,* 415 (1970). How was the $G(T)$ curve generated? Is it valid for a CSTR? Should $G(T)$ change when the space-time changes? Critique the article in light of these questions.

SUPPLEMENTARY READING

1. An excellent development of the energy balance is presented in

> ARIS, R., *Elementary Chemical Reactor Analysis.* Upper Saddle River, N.J.: Prentice Hall, 1969, Chaps. 3 and 6.

> HIMMELBLAU, D. M., *Basic Principles and Calculations in Chemical Engineering,* 7th ed. Upper Saddle River, N.J.: Prentice Hall, 2003, Chaps. 4 and 6.

A number of example problems dealing with nonisothermal reactors can be found in

> FROMENT, G. F., AND K. B. BISCHOFF, *Chemical Reactor Analysis and Design,* 2nd ed. New York: Wiley, 1990.

> WALAS, S. M., *Chemical Reaction Engineering Handbook of Solved Problems.* Amsterdam: Gordon and Breach, 1995. See the following solved problems: Problem 4.10.1, page 444; Problem 4.10.08, page 450; Problem 4.10.09, page 451; Problem 4.10.13, page 454; Problem 4.11.02, page 456; Problem 4.11.09, page 462; Problem 4.11.03, page 459; Problem 4.10.11, page 463.

For a thorough discussion on the heat of reaction and equilibrium constant, one might also consult

DENBIGH, K. G., *Principles of Chemical Equilibrium*, 4th ed. Cambridge: Cambridge University Press, 1981.

2. A review of the multiplicity of the steady state and reactor stability is discussed by

FROMENT, G. F., AND K. B. BISCHOFF, *Chemical Reactor Analysis and Design*, 2nd ed. New York: Wiley, 1990.

PERLMUTTER, D. D., *Stability of Chemical Reactors*. Upper Saddle River, N.J.: Prentice Hall, 1972.

3. Partial differential equations describing axial and radial variations in temperature and concentration in chemical reactors are developed in

FROMENT, G. F., AND K. B. BISCHOFF, *Chemical Reactor Analysis and Design*, 2nd ed. New York: Wiley, 1990.

WALAS, S. M., *Reaction Kinetics for Chemical Engineers*. New York: McGraw-Hill, 1959, Chap. 8.

4. The heats of formation, $H_i(T)$, Gibbs free energies, $G_i(T_R)$, and the heat capacities of various compounds can be found in

PERRY, R. H., D. W. GREEN, and J. O. MALONEY, eds., *Chemical Engineers' Handbook*, 6th ed. New York: McGraw-Hill, 1984.

REID, R. C., J. M. PRAUSNITZ, and T. K. SHERWOOD, *The Properties of Gases and Liquids*, 3rd ed. New York: McGraw-Hill, 1977.

WEAST, R. C., ed., *CRC Handbook of Chemistry and Physics*, 66th ed. Boca Raton, Fla.: CRC Press, 1985.

Unsteady-State Nonisothermal Reactor Design

9

Chemical Engineers are not gentle people, they like high temperatures and high pressures.

Steve LeBlanc

Overview. Up to now we have focused on the steady-state operation of nonisothermal reactors. In this section the unsteady-state energy balance will be developed and then applied to CSTRs, as well as well-mixed batch and semibatch reactors. In Section 9.1 we arrange the general energy balance (Equation 8-9) in a more simplified form that can be easily applied to batch and semibatch reactors. In Section 9.2 we discuss the application of the energy balance to the operation of batch reactors and discuss reactor safety and the reasons for the explosion of an industrial batch reactor. This section is followed by a description of the advanced reactor system screening tool (ARSST) and how it is used to determine heats of reaction, activation energies, rate constants, and the size of relief valves in order to make reactors safer. In Section 9.3 we apply the energy balance to a semibatch reactor with a variable ambient temperature. Section 9.4 discusses startup of a CSTR and how to avoid exceeding the practical stability limit. We close the chapter (Section 9.5) with multiple reactions in batch reactors.

9.1 The Unsteady-State Energy Balance

We begin by recalling the unsteady-state form of the energy balance developed in Chapter 8.

$$\dot{Q} - \dot{W}_s + \sum_{i=1}^{n} F_i H_i \big|_{\text{in}} - \sum_{i=1}^{n} F_i H_i \big|_{\text{out}} = \left(\frac{\partial \hat{E}_{\text{sys}}}{\partial t} \right) \qquad (8\text{-}9)$$

591

We shall first concentrate on evaluating the change in the total energy of the system wrt time, $d\hat{E}_{sys}/dt$. The total energy of the system is the sum of the products of specific energies, E_i, of the various species in the system volume and the number of moles of that species:

$$\hat{E}_{sys} = \sum_{i=1}^{n} N_i E_i = N_A E_A + N_B E_B + N_C E_C + N_D E_D + N_I E_I \qquad (9\text{-}1)$$

In evaluating $\hat{E}_{sys}$, we shall neglect changes in the potential and kinetic energies, and substitute for the internal energy U_i in terms of the enthalpy H_i:

$$\hat{E}_{sys} = \sum_{i=1}^{n} N_i E_i = \sum_{i=1}^{n} N_i U_i = \left[\sum_{i=1}^{n} N_i(H_i - PV_i)\right]_{sys} = \sum_{i=1}^{n} N_i H_i - P\underbrace{\sum_{i=1}^{n} N_i \tilde{V}_i}_{V}$$

$$(9\text{-}2)$$

We note the last term on the right-hand side of Equation (9-2) is just the total pressure times the total volume, i.e., PV. For brevity we shall write these sums as

$$\Sigma = \sum_{i=1}^{n}$$

unless otherwise stated.

When no spatial variations are present in the system volume, and time variations in product of the total pressure and volume (PV) are neglected, the energy balance, substitution of Equation (9-2) into (8-9), gives

$$\dot{Q} - \dot{W}_s + \Sigma F_{i0} H_{i0}\big|_{in} - \Sigma F_i H_i\big|_{out} = \left[\Sigma N_i \frac{dH_i}{dt} + \Sigma H_i \frac{dN_i}{dt}\right]_{sys} \qquad (9\text{-}3)$$

Recalling Equation (8-19),

$$H_i = H°(T_R) + \int_{T_R}^{T} C_{pi}\, dT \qquad (8\text{-}19)$$

and differentiating with respect to time, we obtain

$$\frac{dH_i}{dt} = C_{P_i} \frac{dT}{dt} \qquad (9\text{-}4)$$

Then substituting Equation (9-4) into (9-3) gives

$$\dot{Q} - \dot{W}_s + \Sigma F_{i0} H_{i0} - \Sigma F_i H_i = \Sigma N_i C_{pi} \frac{dT}{dt} + \Sigma H_i \frac{dN_i}{dt} \qquad (9\text{-}5)$$

The mole balance on species i is

$$\frac{dN_i}{dt} = -v_i\, r_A V + F_{i0} - F_i \tag{9-6}$$

Using Equation (9-6) to substitute for dN_i/dt, Equation (9-5) becomes

$$\dot{Q} - \dot{W}_s + \Sigma F_{i0} H_{i0} - \Sigma F_i H_i$$

$$= \Sigma N_i C_{P_i} \frac{dT}{dt} + \Sigma v_i H_i(-r_A V) + \Sigma F_{i0} H_i - \Sigma F_i H_i \tag{9-7}$$

Rearranging, and recalling $\sum v_i H_i = \Delta H_{Rx}$, we have

This form of the energy balance should be used when there is a phase change.

$$\boxed{\frac{dT}{dt} = \frac{\dot{Q} - \dot{W}_s - \Sigma F_{i0}(H_i - H_{i0}) + (-\Delta H_{Rx})(-r_A V)}{\Sigma N_i C_{P_i}}} \tag{9-8}$$

Substituting for H_i and H_{i0} for the case of no phase change gives us

Energy balance on a transient CSTR or semibatch reactor.

$$\boxed{\frac{dT}{dt} = \frac{\dot{Q} - \dot{W}_s - \sum F_{i0} C_{P_i}(T - T_{i0}) + [-\Delta H_{Rx}(T)](-r_A V)}{\sum N_i C_{P_i}}} \tag{9-9}$$

Equation (9-9) applies to a semibatch reactor as well as unsteady-state operation of a CSTR.

For liquid-phase reactions where ΔC_p is small and can be neglected, the following approximation is often made:

$$\Sigma N_i C_{P_i} \cong \Sigma N_{i0} C_{P_i} = N_{A0} \overbrace{\Sigma \Theta_i C_{P_i}}^{C_{P_s}} = N_{A0} C_{P_s}$$

where C_{P_s} is the heat capacity of the solution. The units of $N_{A0} C_{P_s}$ are (cal/K) or (Btu/°R) and

$$\Sigma F_{i0} C_{P_i} = F_{A0} C_{P_s}$$

where the units of $F_{A0} C_{P_s}$ are $(\text{cal/s} \cdot \text{K})$ or $(\text{Btu/h} \cdot °\text{R})$.[1] With this approximation and assuming that every species enters the reactor at temperature T_0, we have

$$\frac{dT}{dt} = \frac{\dot{Q} - \dot{W}_s - F_{A0} C_{P_s}(T - T_0) + [-\Delta H_{Rx}(T)](-r_A V)}{N_{A0} C_{P_s}} \tag{9-10}$$

---

[1] We see that if heat capacity were given in terms of mass (i.e., $C_{P_{sm}} = \text{cal/g} \cdot \text{K}$) then both F_{A0} and N_{A0} would have to be converted to mass:

$$m_{A0} C_{P_{sm}} = N_{A0} C_{P_s}$$

and

$$\dot{m}_{A0} C_{P_{sm}} = F_{A0} C_{P_s}$$

but the units of the products would still be the same (cal/K) and $(\text{cal/s} \cdot \text{K})$, respectively.

9.2 Energy Balance on Batch Reactors

A batch reactor is usually well mixed, so that we may neglect spatial variations in the temperature and species concentration. The energy balance on batch reactors is found by setting F_{A0} equal to zero in Equation (9-10) yielding

$$\boxed{\frac{dT}{dt} = \frac{\dot{Q} - \dot{W}_s + (-\Delta H_{Rx})(-r_A V)}{\sum N_i C_{P_i}}} \tag{9-11}$$

Equation (9-11) is the preferred form of the energy balance when the number of moles, N_i, is used in the mole balance rather than the conversion, X. The number of moles of species i at any X is

$$N_i = N_{A0}(\Theta_i + \nu_i X)$$

Consequently, in terms of conversion, the energy balance becomes

$$\boxed{\frac{dT}{dt} = \frac{\dot{Q} - \dot{W}_s + (-\Delta H_{Rx})(-r_A V)}{N_{A0}(\sum \Theta_i C_{P_i} + \Delta C_p X)}} \tag{9-12}$$

Batch reactor energy
and mole balances

Equation (9-12) must be coupled with the mole balance

$$\boxed{N_{A0} \frac{dX}{dt} = -r_A V} \tag{2-6}$$

and the rate law and then solved numerically.

9.2.1 Adiabatic Operation of a Batch Reactor

Batch reactors operated adiabatically are often used to determine the reaction orders, activation energies, and specific reaction rates of exothermic reactions by monitoring the temperature–time trajectories for different initial conditions. In the steps that follow, we will derive the temperature-conversion relationship for adiabatic operation.

For adiabatic operation ($\dot{Q} = 0$) of a batch reactor ($F_{i0} \equiv 0$) and when the work done by the stirrer can be neglected ($\dot{W}_s \cong 0$), Equation (9-11) can be written as

$$\frac{dT}{dt} = \frac{(-\Delta H_{Rx})(-r_A V)}{\sum N_i C_{P_i}}$$

Reference Shelf

rearranging and expanding the summation term

$$-\Delta H_{\mathrm{Rx}}(T)(-r_A V) = N_{A0}(C_{P_s} + \Delta C_P X)\frac{dT}{dt} \tag{9-13}$$

where as before

$$C_{P_s} = \Sigma\, \Theta_i C_{P_i} \tag{9-14}$$

From the mole balance on a batch reactor we have

$$N_{A0}\frac{dX}{dt} = -r_A V \tag{2-6}$$

We combine Equations (9-13) and (2-6) to obtain

$$-[\Delta H_{\mathrm{Rx}}^\circ + \Delta C_P(T - T_R)]\frac{dX}{dt} = [C_{P_s} + \Delta C_P X]\frac{dT}{dt} \tag{9-15}$$

Canceling dt, separating variables, integrating, and rearranging gives (see CD-ROM *Summary Notes* for intermediate steps)

$$\boxed{X = \frac{C_{P_s}(T - T_0)}{-\Delta H_{\mathrm{Rx}}(T)} = \frac{\Sigma\, \Theta_i C_{P_i}(T - T_0)}{-\Delta H_{\mathrm{Rx}}(T)}} \tag{9-16}$$

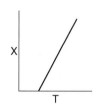

Temperature conversion relationship for an adiabatic batch reactor (or *any* reactor operated adiabatically for that matter)

$$\boxed{T = T_0 + \frac{[-\Delta H_{\mathrm{Rx}}(T_0)]X}{C_{P_s} + X\,\Delta C_P} = T_0 + \frac{[-\Delta H_{\mathrm{Rx}}(T_0)]X}{\displaystyle\sum_{i=1}^{n}\Theta_i C_{P_i} + X\,\Delta C_P}} \tag{9-17}$$

We note that for adiabatic conditions the relationship between temperature and conversion is the same for batch reactors, CSTRs, PBRs, and PFRs. Once we have T as a function of X for a batch reactor, we can construct a table similar to Table E8-3.1 and use techniques analogous to those discussed in Section 8.3.2 to evaluate the following design equation to determine the time necessary to achieve a specified conversion.

$$t = N_{A0}\int_0^x \frac{dX}{-r_A V} \tag{2-9}$$

Example 9–1 Adiabatic Batch Reactor

Although you were hoping for a transfer to the Bahamas, you are still the engineer of the CSTR of Example 8-8, in charge of the production of propylene glycol. You are considering the installation of a new glass-lined 175-gal CSTR, and you decide to make a quick check of the reaction kinetics. You have an insulated instrumented 10-gal stirred batch reactor available. You charge this reactor with 1 gal of methanol and 5 gal of water containing 0.1 wt % H_2SO_4. For safety reasons, the reactor is

located in a storage shed on the banks of Lake Wobegon (you don't want the entire plant to be destroyed if the reactor explodes). At this time of year, the initial temperature of all materials is 38°F.

How many minutes should it take the mixture inside the reactor to reach a conversion of 51.5% if the reaction rate law given in Example 8-8 is correct? What would be the temperature? Use the data presented in Example 8-8.

Solution

1. **Design Equation**:

$$N_{A0} \frac{dX}{dt} = -r_A V \qquad (2\text{-}6)$$

Because there is a negligible change in density during the course of this reaction, the volume V is assumed to be constant.

2. **Rate Law**:

$$-r_A = kC_A \qquad (E9\text{-}1.1)$$

3. **Stoichiometry**:

$$C_A = \left(\frac{N_{A0}}{V}\right)(1 - X) \qquad (E9\text{-}1.2)$$

4. **Combining** Equations (E9-1.1), (E9-1.2), and (2-6), we have

$$\frac{dX}{dt} = k(1 - X) \qquad (E9\text{-}1.3)$$

From the data in Example 8-8,

$$k = (4.71 \times 10^9) \exp\left[\frac{-32,400}{(1.987)(T)}\right] \quad s^{-1}$$

or

$$k = (2.73 \times 10^{-4}) \exp\left[\frac{32,400}{1.987}\left(\frac{1}{535} - \frac{1}{T}\right)\right] \quad s^{-1} \qquad (E9\text{-}1.4)$$

5. **Energy Balance**. Using Equation (9-17), the relationship between X and T for an adiabatic reaction is given by

$$T = T_0 + \frac{[-\Delta H_{Rx}(T_0)]X}{C_{P_s} + \Delta C_P X} \qquad (E9\text{-}1.5)$$

6. **Evaluating the parameters** in the energy balance gives us the heat capacity of the solution:

$$C_{P_s} = \Sigma\, \Theta_i C_{P_i} = \Theta_A C_{P_A} + \Theta_B C_{P_B} + \Theta_C C_{P_C} + \Theta_I C_{P_I}$$

$$= (1)(35) + (18.65)(18) + 0 + (1.670)(19.5)$$

$$= 403 \text{ Btu/lb mol A} \cdot {}^{\circ}F$$

From Example 8-8, $\Delta C_P = -7$ Btu/lb mol·°F and consequently, the second term on the right-hand side of the expression for the heat of reaction,

$$\Delta H_{\text{Rx}}(T) = \Delta H_{\text{Rx}}^{\circ}(T_R) + \Delta C_P(T - T_R)$$

$$= -36{,}400 - 7(T - 528) \tag{E8-8.7}$$

is very small compared with the first term [less than 2% at 51.5% conversion (from Example 8-8)].

Taking the heat of reaction at the initial temperature of 515°R,

$$\Delta H_{\text{Rx}}(T_0) = -36{,}400 - (7)(515 - 528)$$

$$= -36{,}309 \text{ Btu/lb mol}$$

Because terms containing ΔC_P are very small, it can be assumed that

$$\Delta C_P \simeq 0$$

In calculating the initial temperature, we must include the temperature rise from the heat of mixing the two solutions:

$$T_0 = (460 + 38) + 17$$

$$= 515°R$$

$$T = T_0 - \frac{X[\Delta H_{\text{Rx}}(T_0)]}{C_{P_s}} = 515 - \frac{-36{,}309X}{403}$$

$$= 515 + 90.1\,X \tag{E9-1.6}$$

A summary of the heat and mole balance equations is given in Table E9-1.1.

TABLE E9-1.1.   SUMMARY FOR FIRST ORDER ADIABATIC BATCH REACTION

$$\frac{dX}{dt} = k(1 - X)$$

$$k = 2.73 \times 10^{-4} \exp\left[\frac{32{,}400}{1.987}\left(\frac{1}{535} - \frac{1}{T}\right)\right]$$

$$T = 515 + 90.1X$$

where T is in °R and t is in seconds.

A table similar to that used in Example 8-3 can now be constructed.

A software package (e.g., Polymath) was also used to combine Equations (E9-1.3), (E9-1.4), and (E9-1.6) to determine conversion and temperature as a function of time. Table E9-1.2 shows the program, and Figures E9-1.1 and E9-1.2 show the solution results.

TABLE E9-1.2.   POLYMATH PROGRAM

POLYMATH Results

Example 9-1 Adiabatic Batch Reactor 04-14-2005, Rev5.1.233

Calculated values of the DEQ variables

Variable	initial value	minimal value	maximal value	final value
t	0	0	4000	4000
X	0	0	0.9999651	0.9999651
T	515	515	605.09685	605.09685
k	8.358E-05	8.358E-05	0.0093229	0.0093229

ODE Report (RKF45)

Differential equations as entered by the user
[1] d(X)/d(t) = k*(1-X)

Explicit equations as entered by the user
[1] T = 515+90.1*X
[2] k = 0.000273*exp(16306*((1/535)-(1/T)))

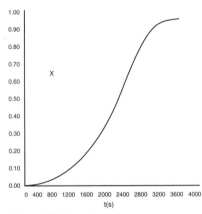

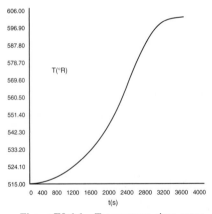

Figure E9-1.1   Temperature–time curve.   **Figure E9-1.2**   Conversion–time curve.

It is of interest to compare this residence time with the residence time in the 175-gal CSTR to attain the same conversion at the same final temperature of 582°R ($k = 0.0032$ s^{-1}):

$$\tau = \frac{V}{v_0} = \frac{X}{k(1-X)} = \frac{0.515}{(0.0032)(0.485)} = 332 \text{ s} = 5.53 \text{ minutes}$$

This occasion is one when the increase in the reaction rate constant caused by the increase in temperature more than compensates for the decrease in rate caused by the decrease in concentration, so the residence time in the CSTR for this conversion is less than it would be in a batch or tubular plug-flow reactor.

9.2.2 Batch Reactor with Interrupted Isothermal Operation

In Chapter 4 we discussed the design of reactors operating isothermally. This operation can be achieved by efficient control of a heat exchanger. The following example shows what can happen when the heat exchanger suddenly fails.

Example 9–2 Safety in Chemical Plants with Exothermic Reactions [2]

A serious accident occurred at Monsanto plant in Sauget, Illinois, on August 8 at 12:18 A.M. (see Figure E9-2.1). The blast was heard as far as 10 miles away in Belleville, Illinois, where people were awakened from their sleep. The explosion occurred in a batch reactor that was used to produce nitroanaline from ammonia and *o*-nitrochlorobenzene (ONCB):

Living Example Problem

Figure E9-2.1 Aftermath of the explosion. (St. Louis Globe Democrat photo by Roy Cook. Courtesy of St. Louis Mercantile Library.)

Was a *Potential Problem Analysis* carried out?

This reaction is normally carried out isothermally at 175°C and about 500 psi. The ambient temperature of the cooling water in the heat exchanger is 25°C. By adjusting the coolant rate the reactor temperature could be maintained at 175°C. At the maximum coolant rate the ambient temperature is 25°C throughout the heat exchanger.

---

[2] Adapted from the problem by Ronald Willey, *Seminar on a Nitroanaline Reactor Rupture*. Prepared for SACHE, Center for Chemical Process Safety, American Institute of Chemical Engineers, New York (1994). Also see *Process Safety Progress*, vol. 20, no. 2 (2001), pp. 123–129. The values of ΔH_{Rx} and UA were estimated in the plant data of the temperature-time trajectory in the article by G. C. Vincent, *Loss Prevention*, 5, 46–52.

Let me tell you something about the operation of this reactor. Over the years the heat exchanger would fail from time to time, but the technicians would be "Johnny on the Spot" and run out and get it up and running in 10 minutes or so, and there was never any problem. One day it is hypothesized that someone looked at the reactor and said, "It looks as if your reactor is only a third full and you still have room to add more reactants and to make more product. How about filling it up to the top so we could triple production?" They did, and you can see what happened in Figure E9-2.1.

On the day of the accident, two changes in normal operation occurred.

A decision was made to triple production.

1. The reactor was charged with 9.044 kmol of ONCB, 33.0 kmol of NH₃, and 103.7 kmol of H₂O. Normally, the reactor is charged with 3.17 kmol of ONCB, 103.6 kmol of H₂O, and 43 kmol of NH₃.

2. The reaction is normally carried out isothermally at 175°C over a 24-h period. Approximately 45 min after the reaction was started, cooling to the reactor failed, but only for 10 min. Cooling may have been halted for 10 min or so on previous occasions when the normal charge of 3.17 kmol of ONCB was used and no ill effects occurred.

The reactor had a rupture disk designed to burst when the pressure exceeded approximately 700 psi. If the disk would have ruptured, the pressure in the reactor would have dropped, causing the water to vaporize, and the reaction would have been cooled (quenched) by the latent heat of vaporization.

Plot the temperature–time trajectory up to a period of 120 min after the reactants were mixed and brought up to 175°C. Show that the following three conditions had to have been present for the explosion to occur: (1) increased ONCB charge, (2) reactor stopped for 10 min, and (3) relief system failure.

Additional Information: The rate law is

$$-r_{\text{ONCB}} = kC_{\text{ONCB}}C_{\text{NH}_3} \quad \text{with} \quad k = 0.00017 \, \frac{m^3}{\text{kmol} \cdot \text{min}} \text{ at } 188°C$$

The reaction volume for the charge of 9.044 kmol of ONCB:

$$V = 3.265 \, m^3 \text{ ONCB/NH}_3 + 1.854 \, m^3 \text{ H}_2\text{O} = 5.119 \, m^3$$

The reaction volume for the charge of 3.17 kmol of ONCB:

$$V = 3.26 \, m^3$$
$$\Delta H_{\text{Rx}} = -5.9 \times 10^5 \text{ kcal/kmol}$$
$$E = 11,273 \text{ cal/mol}$$
$$C_{P_{\text{ONCB}}} = C_{P_A} = 40 \text{ cal/mol} \cdot \text{K}$$
$$C_{P_{\text{H}_2\text{O}}} = C_{P_W} = 18 \text{ cal/mol} \cdot \text{K} \qquad C_{\text{NH}_3} = C_{P_B} = 8.38 \text{ cal/mol} \cdot \text{K}$$

Assume that $\Delta C_P \approx 0$:

$$UA = \frac{35.85 \text{ kcal}}{\text{min} \, °C} \text{ with } T_a = 298 \text{ K}$$

Solution

$$A + 2B \longrightarrow C + D$$

Mole Balance:

$$\frac{dX}{dt} = -r_A \frac{V}{N_{A0}} \tag{E9-2.1}$$

Rate Law:

$$-r_A = k C_A C_B \tag{E9-2.2}$$

Stoichiometry (liquid phase):

$$C_A = C_{A0}(1 - X) \tag{E9-2.3}$$

with

$$C_B = C_{A0}(\Theta_B - 2X) \tag{E9-2.4}$$

$$\Theta_B = \frac{N_{B0}}{N_{A0}}$$

Combine:

$$-r_A = k C_{A0}^2 (1 - X)(\Theta_B - 2X) \tag{E9-2.5}$$

$$k = 0.00017 \exp\left[\frac{11273}{1.987}\left(\frac{1}{461} - \frac{1}{T}\right)\right] \frac{m^3}{kmol \cdot min}$$

Energy Balance:

$$\frac{dT}{dt} = \frac{UA(T_a - T) + (r_A V)(\Delta H_{Rx})}{\sum N_i C_{P_i}} \tag{E9-2.6}$$

For $\Delta C_P = 0$,

$$NC_P = \sum N_i C_{P_i} = N_{A0} C_{P_A} + N_{B0} C_{P_B} + N_W C_{P_W}$$

Let Q_g be the heat generated [i.e., $Q_g = (r_A V)(\Delta H_{Rx})$] and let Q_r be the heat removed [i.e., $Q_r = UA(T - T_a)$]:

$$\frac{dT}{dt} = \frac{\overbrace{UA(T_a - T)}^{-Q_r} + \overbrace{(r_A V)(\Delta H_{Rx})}^{Q_g}}{N_{A0} C_{P_A} + N_{B0} C_{P_B} + N_W C_{P_W}} \tag{E9-2.7}$$

$Q_g = (r_A V)(\Delta H_{Rx})$
$Q_r = UA(T - T_a)$

Then

$$\boxed{\frac{dT}{dt} = \frac{Q_g - Q_r}{NC_P}} \tag{E9-2.8}$$

Parameter evaluation for day of explosion:

$$NC_P = (9.0448)(40) + (103.7)(18) + (33)(8.38)$$

$$\boxed{NC_P = 2504 \text{ kcal/K}}$$

A. Isothermal Operation Up to 45 Minutes

We will first carry out the reaction isothermally at 175°C up to the time the cooling was turned off at 45 min. Combining and canceling yields

$$\frac{dX}{dt} = kC_{A0}(1-X)(\Theta_B - 2X) \tag{E9-2.9}$$

$$\Theta_B = \frac{33}{9.04} = 3.64$$

At 175°C = 448 K, $k = 0.0001167 \, m^3/kmol \cdot min$. Integrating Equation (E9-2.9) gives us

$$t = \left[\frac{V}{kN_{A0}}\right]\left(\frac{1}{\Theta_B - 2}\right)\ln\frac{\Theta_B - 2X}{\Theta_B(1-X)} \tag{E9-2.10}$$

Substituting the parameter values

The calculation and results can also be obtained from the Polymath output on the CD-ROM

Living Example Problem

$$45 \text{ min} = \left[\frac{5.119 \text{ m}^3}{0.0001167 \text{ m}^3/kmol \cdot min \, (9.044 \text{ kmol})}\right]\times\left(\frac{1}{1.64}\right)\ln\frac{3.64 - 2X}{3.64(1-X)}$$

Solving for X, we find that at $t = 45$ min, then $X = 0.033$.
We will calculate the rate of generation Q_g at this temperature and conversion and compare it with the maximum rate of heat removal Q_r. The rate of generation Q_g is

$$Q_g = r_A V \Delta H_{Rx} = k\frac{N_{A0}(1-X)N_{A0}[(N_{B0}/N_{A0}) - 2X]\,V(-\Delta H_{Rx})}{V^2} \tag{E9-2.11}$$

At this time (i.e., $t = 45$ min, $X = 0.033$, $T = 175$°C) we calculate k, then Q_r and Q_g. At 175°C, $k = 0.0001167 \, m^3/min \cdot kmol$.

$$Q_g = (0.0001167)\frac{(9.0448)^2(1-0.033)}{5.119}\left[\frac{33}{(9.0448)} - 2(0.033)\right]5.9\times10^5$$

$$= 3830 \text{ kcal/min}$$

The corresponding maximum cooling rate is

$$Q_r = UA(T - 298)$$
$$= 35.85(448 - 298) \tag{E9-2.12}$$
$$= 5378 \text{ kcal/min}$$

Therefore

$$\boxed{Q_r > Q_g} \tag{E9-2.13}$$

Everything is OK.

The reaction can be controlled. There would have been no explosion had the cooling not failed.

B. Adiabatic Operation for 10 Minutes

The cooling was off for 45 to 55 min. We will now use the conditions at the end of the period of isothermal operation as our initial conditions for adiabatic operation period between 45 and 55 min:

$$t = 45 \text{ min} \quad X = 0.033 \quad T = 448°C$$

Between $t = 45$ and $t = 55$ min, $Q_r = 0$. The Polymath program was modified to account for the time of adiabatic operation by using an *"if statement"* for Q_r in the program, i.e., $Q_r = $ if (t > 45 and t < 55) then (0) else (UA(T − 298)). A similar *"if statement"* is used for isothermal operation, i.e., $(dT/dt) = 0$.

For the 45- to 55-min period without cooling, the temperature rose from 448 K to 468 K, and the conversion increased from 0.033 to 0.0424. Using this temperature and conversion in Equation (E9-2.11), we calculate the rate of generation Q_g at 55 min as

$$Q_g = 6591 \text{ kcal/min}$$

The maximum rate of cooling at this reactor temperature is found from Equation (E9-2.12) to be

$$Q_r = 6093 \text{ kcal/min}$$

Here we see that

The point of no return

$$\boxed{Q_g > Q_r} \tag{E9-2.14}$$

and the temperature will continue to increase. Therefore, the **point of no return** has been passed and the temperature will continue to increase, as will the rate of reaction until the explosion occurs.

C. Batch Operation with Heat Exchange

Return of the cooling occurs at 55 min. The values at the end of the period of adiabatic operation ($T = 468$ K, $X = 0.0423$) become the initial conditions for the period of operation with heat exchange. The cooling is turned on at its maximum capacity, $Q = UA(298 − T)$, at 55 min. Table E9-2.1 gives the Polymath program to determine the temperature–time trajectory. Note that one can change N_{A0} and N_{B0} to 3.17 and 43 kmol in the program and show that if the cooling is shut off for 10 min, at the end of that 10 min Q_r will still be greater than Q_g and no explosion will occur.

Interruptions in the cooling system have happened before with no ill effects.

The complete temperature–time trajectory is shown in Figure E9-2.2. One notes the long plateau after the cooling is turned back on. Using the values of Q_g and Q_r at 55 min and substituting into Equation (E9-2.8), we find that

$$\frac{dT}{dt} = \frac{(6591 \text{ kcal/min}) − (6093 \text{ kcal/min})}{2504 \text{ kcal/°C}} = 0.2°C/\text{min}$$

Living Example Problem

TABLE E9-2.1. POLYMATH PROGRAM

ODE Report (RKF45)

Differential equations as entered by the user
  [1] d(T)/d(t) = if (t<45) then (0) else ((Qg-Qr)/NCp)
  [2] d(X)/d(t) = -ra*V/Nao

Explicit equations as entered by the user
  [1] V = 3.265+1.854
  [2] Nao = 9.045
  [3] UA = 35.83
  [4] k = 0.00017*exp((11273/1.987)*(1/461-1/T))
  [5] ThetaB = 33/9.045
  [6] T1 = T-273
  [7] ra = -k*Nao^2*(1-X)*(ThetaB-2*X)/V^2
  [8] Qr = if(t>45 and t<55) then(0) else (UA*(T-298))
  [9] DeltaHrx = -590000
  [10] Qg = ra*V*DeltaHrx
  [11] NCp = 2504

The explosion occurred shortly after midnight.

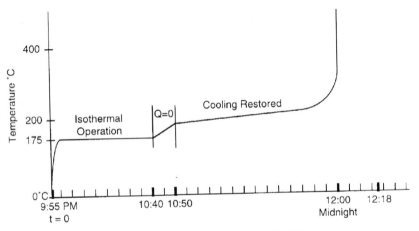

Figure E9-2.2 Temperature–time trajectory.

Consequently, even though dT/dt is positive, the temperature increases very slowly at first, 0.2°C/min. By 11:45, the temperature has reached 240°C and is beginning to increase more rapidly. One observes that 119 min after the batch was started the temperature increases sharply and the reactor explodes at approximately midnight. If the mass and heat capacity of the stirrer and reaction vessel had been included, the NC_p term would have increased by about 5% and extended the time until the explosion occurred by 15 or so minutes, which would predict the actual time the explosion occurred, at 12:18 A.M.

    When the temperature reached 300°C, a secondary reaction, the decomposition of nitroaniline to noncondensable gases such as CO, N_2, and NO_2, occurred, releasing even more energy. The total energy released was estimated to be 6.8×10^9 J, which is enough energy to lift the entire 2500-ton building 300 m (the length of three football fields) straight up.

D. Disk Rupture

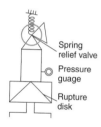

Spring
relief valve

Pressure
guage

Rupture
disk

We note that the pressure relief disk should have ruptured when the temperature reached 265°C (ca. 700 psi) but did not and the temperature continued to rise. If it had ruptured and all the water had vaporized, 10^6 kcal would have been drawn from the reacting solution, thereby lowering its temperature and quenching it.

If the disk had ruptured at 265°C (700 psi), the maximum mass flow rate, $\dot{m}_{vap}$, out of the 2-in. orifice to the atmosphere (1 atm) would have been 830 kg/min at the time of rupture.

$$Q_r = \dot{m}_{vap}\, \Delta H_{vap} + UA(T - T_a)$$

$$= 830\,\frac{kg}{min} \times 540\,\frac{kcal}{kg} + 35.83\,\frac{kcal}{K}\,(538 - 298)K$$

$$= 4.48 \times 10^5\,\frac{kcal}{min} + 8604\,\frac{kcal}{min}$$

$$= 4.49 \times 10^5\,\frac{kcal}{min}$$

This value of Q_r is much greater than Q_g ($Q_g = 27{,}460$ kcal/min), so that the reaction could easily be quenched.

In summary, if any one of the following three things had *not* occurred the explosion would not have happened.

1. Tripled production
2. Heat exchanger failure for 10 minutes
3. Failure of the relieving device (rupture disk)

In other words, all the above had to happen to cause the explosion. If the relief had operated properly, it would not have prevented reaction runaway but it could have prevented the explosion. In addition to using rupture disks as relieving devices, one can also use pressure relief valves. In many cases sufficient care is not taken to obtain data for the reaction at hand and to use it to properly size the relief device. This data can be obtained using a batch reactor called the ARSST.

9.2.3 Reactor Safety: The Use of the ARSST to Find ΔH_{Rx}, E and to Size Pressure Relief Valves

Use ARSST to find:
E, activation energy
A, frequency factor
ΔH_{Rx} heat of reaction

The Advanced Reaction System Screening Tool (ARSST) is a calorimeter that is used routinely in industry experiments to determine activation energies, E; frequency factors, A; heats of reaction, ΔH_{Rx}; and to size vent relief valves for runaway exothermic reactions [*Chemical Engineering Progress, 96* (2), 17 (2000)]. The basic idea is that reactants are placed and sealed in the calorimeter which is then electrically heated as the temperature and pressure in the calorimeter are monitored. As the temperature continues to rise, the rate of reaction also increases to a point where the temperature increases more rapidly from the heat generated by the reaction (called the self-heating rate, $\dot{T}_S$) than the temperature increase by electrical heating. The temperature at which this change in relative heating rates occurs is called the *onset* temperature. A schematic of the calorimeter is shown in Figure 9-1.

Experiments to
obtain data to design
safer reactors

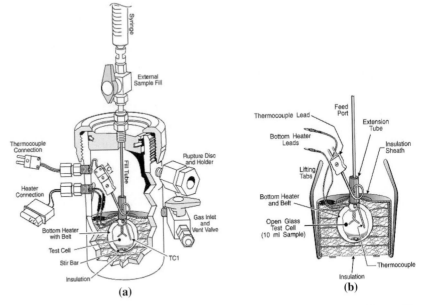

(a)

(b)

Figure 9-1   ARSST (a) Schematic of containment vessel and internals. (b) Test cell assembly.
[Courtesy of Fauske & Associates.]

We shall take as our system the reactants, products and inerts inside the spherical container as well as the spherical container itself because the mass of the container may adsorb a little of the energy given off by the reaction. This system is well insulated and does not lose much heat to the surroundings. Neglecting ΔC_P, the energy balance on the ARSST, Equation (9-12), becomes

$$\frac{dT}{dt} = \frac{\dot{Q} + (-\Delta H_{Rx})(-r_A V)}{\Sigma\, N_i C_{P_i}} = \left(\frac{\dot{Q}}{\Sigma\, N_i C_{P_i}}\right) + \left(\frac{(\Delta H_{Rx})(-r_A V)}{\Sigma\, N_i C_{P_i}}\right) \quad (9\text{-}18)$$

The total heat added, $\dot{Q}$, is the sum of the electrical heat added, $\dot{Q}_E$, and the convective heat added term, $\dot{Q}_C = (UA[T_a - T])$

$$\dot{Q} = \dot{Q}_E \; + \; \underbrace{\dot{Q}_C}_{UA(T_a - T)}$$

Because the system is well insulated, we shall neglect the heat loss from the calorimeter to the surroundings, $\dot{Q}_C$, and further define

$$\dot{T}_E = \frac{\dot{Q}_E}{\Sigma\, N_i C_{P_i}} \tag{9-19}$$

$\dot{T}_E$ is called the *electrical heating rate*. The second term in Equation (9-18) is called the *self-heating rate*, $\dot{T}_S$, that is,

$$\dot{T}_S = \frac{(-\Delta H_{Rx})(-r_A V)}{\Sigma\, N_i C_{P_i}} \tag{9-20}$$

Reference Shelf

The self-heating rate, which is determined from the experiment, is what is used to calculate the vent size of the relief valve, A_V, of the reactor in order to prevent runaway reactions from exploding (see PRS R9.4 on the CD-ROM).

The electrical heating rate is controlled such that the temperature rise, $\dot{T}_E$ (typically 0.5–2°C/min), is maintained constant up to the temperature when the self-heating rate becomes greater than the electrical heating rate

$$\dot{T}_S > \dot{T}_E$$

Again, this temperature is called the *onset* temperature, T_{onset}. A typical thermal history of data collected by the ARSST is shown in Figure 9-2 in terms of the temperature-time trajectory.

Relatively little conversion is achieved for $T < T_{onset}$

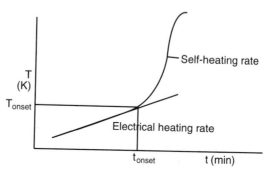

Figure 9-2   Typical temperature history for thermal scan with the ARSST.

The self-heating rate, $\dot{T}_S$, can be easily found by differentiating the temperature-time trajectory, or $\dot{T}_S$ can be determined directly from the output instrumentation and software associated with the ARSST.

We can rewrite Equation (9-18) in the form

$$\frac{dT}{dt} = \dot{T}_E + \dot{T}_S \tag{9-18A}$$

The following example uses data obtained in the senior unit operations laboratory of the University of Michigan.

Safety Valve

Side note: A relief valve is an instrument on the top of the reactor that releases the pressure and contents of the reactor before temperature and pressure builds up to runaway and explosive conditions. The safety relief valve is similar to the rupture disk in that once the reactor pressure exceeds the set pressure P_S, the contents are allowed to flow out through the vent. If the pressure falls below the vapor pressure of the reactor contents, the latent heat of vaporization will cool the reactor. The self-heating rate may be used directly to calculate the vent size necessary to release all the contents of the reactor. The necessary vent area, A_V, is given by the equation for a vapor system and two-phase flashing flow.

$$A_V = 1.5 \cdot 10^{-5} \frac{m_S \dot{T}_S}{F P_S} \text{(in m}^2\text{)}$$

where F is a reduction factor for an ideal nozzle, P_S is the relief set pressure (psia), and m_S the mass of the sample in kg and $\dot{T}_s$ is the self-heating rate discussed in the sample calculation in the *Professional Reference Shelf* R9.1 on the CD-ROM.

Example 9–3   Use of the ARSST

We shall use the ARSST to study the reaction between acetic anhydride and water to form acetic acid

$$(CH_3CO)_2O + H_2O \longrightarrow 2CH_3COOH$$

$$A + B \longrightarrow 2C$$

Living Example Problem

Acetic anhydride is placed in the ARSST to form a 6.7 molar solution of acetic anhydride and a 20.1 M solution of water. The sample volume is 10 ml. The electrical heating is started, and the temperature and its derivative, $\dot{T}_S$, are recorded as a function of time by the ARSST system and computer. Analyze the data to find the heat of reaction ΔH_{Rx}, the activation energy E, and the frequency factor A, and then to compare theoretical and experimental temperature–time trajectories.

The temperature–time trajectory is obtained directly from the computer linked to the ARSST as is shown in Figure E9-3.1.

Physical Properties

Chemical	Density (g/ml)	Heat capacity (J/g°C)	MW	Heat capacity (J/mol°C)	Θ_i
Acetic anhydride	1.0800	1.860	102	189.7	1
Water	1.0000	4.187	18	75.4	3
Glass cell (bomb)	0.1474	0.837	—	0.84 J/g · K	—

Total volume	10 ml with
Water	3.64 g
Acetic anhydride	6.84 g
	10.48 g (10 ml)

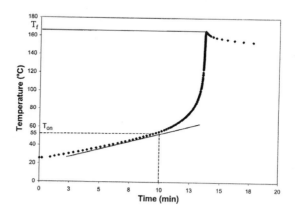

Figure E9-3.1 Temperature–time trajectory for hydrolysis of acetic anhydride.

Solution

Because we are taking our system as the contents inside the bomb as well as the bomb itself, the term $\Sigma N_i C_{P_i}$ needs to be modified to account for the heat absorbed by the bomb calorimeter that holds the reactants. Thus, we include terms for the mass of the calorimeter, m_b, and the heat capacity of the calorimeter, C_{P_b}, in the sum $\Sigma N_i C_{P_i}$, that is,

$$\Sigma\, N_i C_{P_i} = N_A C_{P_A} + N_B C_{P_B} + N_C C_{P_C} + N_I C_{P_I} + m_b C_{P_b}$$

However, for this example: $N_I = 0$ and we neglect ΔC_P and $m_b C_{P_b}$ to obtain

$$\Sigma\, N_i C_{P_i} = N_{A0} \Sigma \Theta_i C_{P_i} \qquad \text{(E9-3.1)}$$

We now apply our algorithm to analyzing the ARSST for the reaction

$$A + B \longrightarrow 2C$$

TABLE E9-3.1.   BALANCE EQUATIONS

Mole balance:

$$\frac{dN_A}{dt} = r_A V \tag{E9-3.2}$$

Rate law:

$$r_A = -k' C_A C_B \tag{E9-3.3}$$

$$k' = A e^{-E/RT} \tag{E9-3.4}$$

Stoichiometry:

$$V = V_0$$

$$C_A = C_{A0}(1 - X) \tag{E9-3.5}$$

$$C_B = C_{A0}(\Theta_B - X) \tag{E9-3.6}$$

$$C_C = 2 C_{A0} X \tag{E9-3.7}$$

For $\Theta_B = 3$, as a first approximation we take

$$C_B \approx C_{A0} \Theta_B = C_{B0}$$

(*Note:* In Problem P9-10$_C$ we do not assume C_{B0} is constant.)

$$r_A = -k' C_{B0} C_A = k C_A \tag{E9-3.8}$$

Combine:

$$k = k' C_{B0}$$

$$\frac{dC_A}{dt} = -k C_A \tag{E9-3.9}$$

Following the Algorithm

Energy balance: $\tag{E9-3.10}$

$$\frac{dT}{dt} = \dot{T}_E + \dot{T}_S$$

$$\tag{9-19}$$

$$\dot{T}_E = \frac{\dot{Q}_E}{N_{A0} \Sigma \Theta_i C_{P_i}}$$

Recalling $N_i = V C_i$ we can substitute Equations (E9-3.5) and (E9-3.6) into Equation (E9-3.1) and then

$$N_{A0} \Sigma \Theta_i C_{P_i} = N_{A0} C_{P_A} + N_{B0} C_{P_A} + \cancel{\Delta C_P N_{A0} X + N_I C_{P_I} + m_b C_{P_b}} \tag{E9-3.11}$$
$$\searrow \text{Neglect}$$

$$= N_{A0}(C_{P_A} + \Theta_B C_{P_B}) \tag{E9-3.12}$$

Usually $N_I = 0$, $\Delta C_P = 0$, and $m_b C_{P_b}$ are negligible with respect to the other terms.

The self-heating rate is

$$\dot{T}_S = \frac{(-\Delta H_{Rx})(-r_A V)}{N_{A0} \Sigma \Theta_i C_{P_i}} = \frac{(-\Delta H_{Rx})(C_{B0} A e^{-E/RT} C_A V)}{N_{A0} \Sigma \Theta_i C_{P_i}} \tag{9-20}$$

Calculating the Heat of Reaction

The heat of reaction for adiabatic operation can be found from the adiabatic temperature rise for complete conversion $X = 1$. The onset temperature is taken as the point at which the self-heating rate is greater than the electrical heating rate. From Figure E9-3.1 we see the onset temperature is 55°C and the maximum temperature is 165°C.

Recalling Equation (8-29) with $\Delta C_P = 0$

$$X = \Sigma\Theta_i C_{P_i}[(T - T_0)/(-\Delta H^\circ_{Rx})]$$

For $X = 1$, $T = T_f$

$$-\Delta H^\circ_{Rx} = (C_{P_A} + \Theta_B C_{P_B})(T_f - T_0)$$

$$\Sigma\Theta_i C_{P_i} = [189.7 + 3(75.4)] = 415\frac{J}{mol \cdot K}$$

$$-\Delta H^\circ_{Rx} = \left(415\frac{J}{mol \cdot °C}\right)(165 - 55)°C = 45,650 \text{ J/mol}$$

$$\boxed{\Delta H^\circ_{Rx} = -(45,650)\frac{J}{mol}}$$

Determining the Activation Energy

The self-heating rate $\dot{T}_S$ is shown as a function of temperature in Figure (E9-3.2). The ARSST software will differentiate the data shown in Figure E9-3.1 directly to give $\dot{T}_S$ **or** (dT/dt) can be found from T versus t data using any of the differentiational techniques discussed in Chapter 5. One notes that the self-heating rate goes to zero at $T = 138°C$ as a result of the reactants having been consumed at this point.

The electrical heating rate, $\dot{T}_E$, either is shut off or becomes negligible wrt $\dot{T}_S$ after the onset temperature T_{onset} is reached. Applying Equation (9-20) to our reaction, we obtain

$$\frac{dT}{dt} \cong \dot{T}_S = \left(\frac{-\Delta H^\circ_{Rx} V C_{B0}}{N_{A0}\Sigma\Theta_i C_{P_i}}\right)Ae^{-E/RT}C_A \qquad \text{(E9.3-13)}$$

We now take the log of the self-heating rate, $\dot{T}_S$, using Equation (E9.3-13) to obtain

$$\ln \dot{T}_S = \ln\frac{(-\Delta H^\circ_{Rx})C_{B0}V}{N_{A0}\Sigma\Theta_i C_{P_i}} + \ln A + \ln C_A - \frac{E}{RT} \qquad \text{(E9.3-14)}$$

After the onset temperature is reached, we can obtain the activation energy from the slope of a plot of $[\ln(dT/dt)]$ as a function of $(1/T)$, neglecting changes in $\ln C_A$. The slope of the line will be $(-E/R)$, as shown in Figure E9-3.3.

From the slope of the plot we find the activation to be

$$E = -R \cdot \text{Slope} = 1.987\frac{cal}{mol \cdot K} \times (-7,750 \text{ K})$$

$$\boxed{E = 15.4\frac{kcal}{mol}}$$

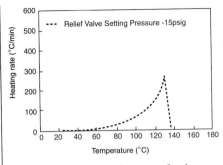

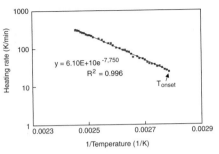

Figure E9-3.2 Self-heating curve for the hydrolysis of acetic anhydride.

Figure E9-3.3 Arrhenius plot of self-heating rate for acetic anhydride.

Calculating the Frequency Factor, A

We will now make an approximate calculation to determine the frequency factor. The intricate details of a more accurate calculation are given in *Professional Reference Shelf* R9.1 where the conversion obtained during the electrical heating phase is taken into account. Neglecting the conversion during electrical heating, then $C_A = C_{A0}$, and Equation (E9-3.13) can be arranged in the form

$$A = \left[\dot{T}_S\big|_{onset} \cdot \left[\frac{N_{A0}\Sigma\Theta_i C_{P_i}}{-\Delta H^{\circ}_{Rx} C_{A0} C_{B0} V} \right] \right] e^{E/RT_{onset}} \qquad \text{(E9-3.15)}$$

At the *onset*, $T_{onset} = 55°C$; at 10 minutes the self-heat rate can be estimated from the slope of the plot of T versus t shown in Figure E9-3.1 to be $\dot{T}_{S\,onset} = 5.2$ K/min. Evaluating the parameters,

$$N_{A0}\Sigma\Theta_i C_{P_i} = \left(6.7\frac{mol}{dm^3}\right)(0.01\ dm^3)\left[415\frac{J}{mol \cdot C}\right]$$

$$= 28\ J/K$$

We now calculate the frequency factor to be

$$A = \frac{5.2\ K}{min}\left[\frac{28\ J/K}{\left(45,650\frac{J}{mol}\right)\left(6.7\frac{mol}{dm^3}\right)\left(\left(20.2\frac{mol}{dm^3}\right) \cdot 0.01\ dm^3\right)} \right]\exp\left[\frac{15,400\ cal/mol}{(1,987cal/mol)(330\ K)} \right]$$

$$\boxed{A = 3.73 \times 10^7\ dm^3/mol/min}$$

Now that we have the activation energy E and the Arrhenius factor A, we can use Polymath to simulate the equations in Table E9-3.1 and compare with the experimental results. The Polymath program is shown in Table E9-3.2, and the corresponding output is shown in Figure E9-3.4.

TABLE E9-3.2.   POLYMATH PROGRAM

ODE Report (RKF45)

Differential equations as entered by the user
[1] d(CA)/d(t) = rA
[2] d(T)/d(t) = Tedot+Tsdot

Explicit equations as entered by the user
[1] CB0 = 20.2
[2] V = 0.01
[3] SUMNA0THEiCpi = 28
[4] dHrx = -45650
[5] A = 3.734e7
[6] E = 15400
[7] R = 1.987
[8] Tedot = if (T>55+273) then 0 else 2
[9] rA = -A*exp(-E/R/T)*CA*CB0
[10] Tsdot = (-dHrx)*(-rA*V)/SUMNA0THEiCpi

Comments
[1] d(CA)/d(t) = rA
    Mole balance on Acetic Anhydride
[2] d(T)/d(t) = Tedot+Tsdot
    Energy Balance
[3] rA = -A*exp(-E/R/T)*CA*CB0
    Rate of the reaction-mol/l.min
[4] V = 0.01
    Volume of the reactive solution-l
[5] SUMNA0THEiCpi = 28
    J/mol.C
[6] dHrx = -45650
    Heat of reaction-J/mol
[7] A = 3.734e7
    rate constant- 1/min
[8] E =   15400
    cal/mol
[9] R = 1.987
    cal/mol.K
[10] Tedot = if (T>55+273) then 0 else 2
    (oK/min) After the onset point, electrical heating is only to compensate for heat loss
[11] Tsdot = (-dHrx)*(-rA*V)/SUMNA0THEiCpi
    Self-heating rate (oK/min)

Living Example Problem

Where SUMNA0THEiCpi = $N_{A0} \sum \Theta_i C_{P_i} = 28$

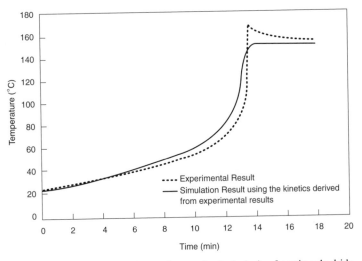

Figure E9-3.4 Temperature–time trajectory for hydrolysis of acetic anhydride.

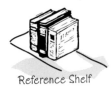

As one can see there is excellent agreement between the simulation and the experiment. The decrease in temperature at 13.5 minutes in the experimental data is a result of a small heat loss to the surroundings which was not accounted for in the simulation. The CD-ROM (R9.1) describes how to size the vent size from this data.

When using the ARSST in the laboratory to actually size the relief vent, follow the procedure in the Professional Reference Shelf, R9.1, which accounts for the conversion during the electrical heating and also taking the onset temperature at a point where $\dot{T}_S \gg \dot{T}_E$. Also see the Fauske web site: *www.fauske.com.*

9.3 Semibatch Reactors with a Heat Exchanger

In our past discussions of reactors with heat exchanges, we assumed that the ambient temperature T_a was spatially uniform throughout the exchanger. This assumption is true if the system is a tubular reactor with the external pipe surface exposed to the atmosphere or if the system is a CSTR or batch where the coolant flow rate through exchanger is so rapid that the coolant temperatures entering and leaving the exchanger are virtually the same.

Oops! The practical stability limit was exceeded.

We now consider the case where the coolant temperature varies along the length of the exchanger while the temperature in the reactor is spatially uniform. The coolant enters the exchanger at a mass flow rate $\dot{m}_c$ at a temperature T_{a1} and leaves at a temperature T_{a2} (see Figure 9-3). As a first approximation, we assume a quasi-steady state for the coolant flow and neglect the accumulation term (i.e., $dT_a/dt = 0$). As a result, Equation (8-49) will give the rate of heat transfer *from* the exchanger *to* the reactor:

$$\dot{Q} = \dot{m}_c C_{P_c}(T_{a1} - T)[1 - \exp - UA/(\dot{m}_c C_{P_c})] \tag{8-49}$$

Using Equation (8-49) to substitute for $\dot{Q}$ in Equation (9-9), we obtain

$$\frac{dT}{dt} = \frac{\dot{m}_c C_{P_c}(T_{a1} - T)[1 - \exp(-UA/\dot{m}_C C_{P_c})] + (r_A V)(\Delta H_{Rx}) - \sum F_{i0} C_{P_i}(T - T_0)}{\sum N_i C_{P_i}}$$

(9-21)

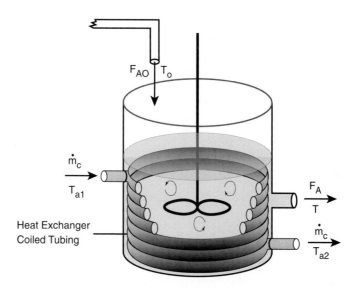

Figure 9-3   Tank reactor with heat exchanger.

At steady state $(dT/dt = 0)$ Equation (9-21) can be solved for the conversion X as a function of reaction temperature by recalling that

$$F_{A0}X = -r_A V$$

and

$$\sum F_{i0} C_{P_i}(T - T_0) = F_{A0} \sum \Theta_i C_{P_i}(T - T_0)$$

and neglecting ΔC_P and then rearranging Equation (9-21) to obtain

Steady-state energy balance

$$X = \frac{\dot{m}_c C_{P_c}(T_{a1} - T)[1 - \exp(-UA/\dot{m}_C C_{P_c})] - F_{A0} \sum \Theta_i C_{P_i}(T - T_0)}{F_{A0}(\Delta H_{Rx}^{\circ})}$$

(9-22)

We are assuming that there is virtually no accumulation of energy in the coolant fluid, that is,

$$\frac{dT_a}{dt} \approx 0$$

Example 9–4 Heat Effects in a Semibatch Reactor

The second-order saponification of ethyl acetate is to be carried out in a semibatch reactor shown schematically in Figure E9-4.1.

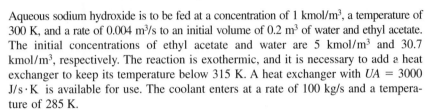

$$C_2H_5(CH_3COO)(aq) + NaOH(aq) \rightleftharpoons Na(CH_3COO)(aq) + C_2H_5OH(aq)$$

$$A \quad + \quad B \quad \rightleftharpoons \quad C \quad + \quad D$$

Living Example Problem

Aqueous sodium hydroxide is to be fed at a concentration of 1 kmol/m³, a temperature of 300 K, and a rate of 0.004 m³/s to an initial volume of 0.2 m³ of water and ethyl acetate. The initial concentrations of ethyl acetate and water are 5 kmol/m³ and 30.7 kmol/m³, respectively. The reaction is exothermic, and it is necessary to add a heat exchanger to keep its temperature below 315 K. A heat exchanger with $UA = 3000$ J/s·K is available for use. The coolant enters at a rate of 100 kg/s and a temperature of 285 K.

Is the heat exchanger and coolant flow rate adequate to keep the reactor temperature below 315 K? Plot temperature, C_A, C_B, and C_C as a function of time.

Additional information:[3]

$$k = 0.39175 \exp\left[5472.7\left(\frac{1}{273} - \frac{1}{T}\right)\right] \text{ m}^3/\text{kmol}\cdot\text{s}$$

$$K_C = 10^{3885.44/T}$$

$$\Delta H_{Rx}^\circ = -79{,}076 \text{ kJ/kmol}$$

$$C_{P_A} = 170.7 \text{ J/mol/K}$$

$$C_{P_B} = C_{P_C} = C_{P_D} \cong C_{P_W} = C_P = 75.24 \text{ J/mol}\cdot\text{K}$$

Feed: $C_{W0} = 55 \text{ kmol/m}^3$ $C_{B0} = 1.0 \text{ kmol/m}^3$

Initially: $C_{Wi} = 30.7 \text{ kmol/m}^3$ $C_{Ai} = 5 \text{ kmol/m}^3$ $C_{Bi} = 0$

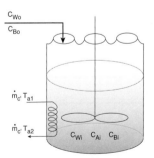

Figure E9-4.1 Semibatch reactor with heat exchange.

---

[3] k from J. M. Smith, *Chemical Engineering Kinetics*, 3rd ed. (New York: McGraw-Hill, 1981), p. 205. ΔH_{Rx} and K_C calculated from values given in *Perry's Chemical Engineers' Handbook*, 6th ed. (New York: McGraw-Hill, 1984), pp. 3–147.

Solution

Mole Balances: (See Section 4.10.2.)

$$\frac{dC_A}{dt} = r_A - \frac{v_0 C_A}{V} \tag{E9-4.1}$$

$$\frac{dC_B}{dt} = r_B + \frac{v_0(C_{B0} - C_B)}{V} \tag{E9-4.2}$$

$$\frac{dC_C}{dt} = r_C - \frac{C_C v_0}{V} \tag{E9-4.3}$$

$$C_D = C_C$$

$$\frac{dN_W}{dt} = C_{Wi0} v_0 \tag{E9-4.4}$$

Initially,

$$N_{Wi} = V_i C_{Wi} = (0.2)(30.7) = 6.14 \text{ kmol}$$

Rate Law:

$$-r_A = k\left(C_A C_B - \frac{C_C C_D}{K_C}\right) \tag{E9-4.5}$$

Stoichiometry:

$$-r_A = -r_B = r_C = r_D \tag{E9-4.6}$$

$$N_A = C_A V \tag{E9-4.7}$$

$$V = V_0 + v_0 t \tag{E9-4.8}$$

Energy Balance: Next we replace $\sum\limits_{i=1}^{n} F_{i0} C_{P_i}$ in Equation (9-9). Because only B and water continually flow into the reactor

$$\sum_{i=1}^{n} F_{i0} C_{P_i} = F_{B0} C_{P_B} + F_{W0} C_W = F_{B0}\left(C_{P_B} + \frac{F_{W0}}{F_{B0}} C_{P_W}\right)$$

However, $C_{P_B} = C_{P_W}$:

$$\sum_{i=1}^{n} F_{i0} C_{P_i} = F_{B0} C_{P_B}(1 + \Theta_W)$$

where

$$\Theta_W = \frac{F_{W0}}{F_{B0}} = \frac{C_{W0}}{C_{B0}} = \frac{55}{1} = 55$$

$$\frac{dT}{dt} = \frac{\dot{Q} - F_{B0} C_{P_B}(1 + \Theta_W)(T - T_0) + (r_A V) \Delta H_{Rx}}{N_A C_{P_A} + N_B C_{P_B} + N_C C_{P_C} + N_D C_{P_D} + N_W C_{P_W}} \tag{E9-4.9}$$

$$\dot{Q} = \dot{m}_c C_{P_c}(T_{a1} - T)[1 - \exp(-UA/\dot{m}_c C_{P_c})] \qquad (8\text{-}49)$$

$$\frac{dT}{dt} = \frac{\dot{m}_c C_{P_c}(T_{a1} - T)[1 - \exp(-UA/\dot{m}_c C_{P_c})] - F_{B0} C_P (1 + \Theta_W)(T - T_0) + (r_A V)\, \Delta H_{Rx}}{C_P (N_B + N_C + N_D + N_W) + C_{P_A} N_A}$$

$$(E9\text{-}4.10)$$

Recalling Equation (8-47) for the outlet temperature of the fluid in the heat exchanger

$$T_{a2} = T - (T - T_{a1})\exp\left[-\frac{UA}{\dot{m}_c C_{P_c}} \right] \qquad (8\text{-}47)$$

The Polymath program is given in Table E9-4.1. The solution results are shown in Figures E9-4.2 and E9-4.3.

TABLE E9-4.1. POLYMATH PROGRAM FOR SEMIBATCH REACTOR

ODE Report (RKF45)

Differential equations as entered by the user
```
[1] d(Ca)/d(t) = ra-(v0*Ca)/V
[2] d(Cb)/d(t) = rb+(v0*(Cb0-Cb)/V)
[3] d(Cc)/d(t) = rc-(Cc*v0)/V
[4] d(T)/d(t) = (Qr-Fb0*cp*(1+55)*(T-T0)+ra*V*dh)/NCp
[5] d(Nw)/d(t) = v0*Cw0
```

Explicit equations as entered by the user
```
[1]  v0 = 0.004
[2]  Cb0 = 1
[3]  UA = 3000
[4]  Ta = 290
[5]  cp = 75240
[6]  T0 = 300
[7]  dh = -7.9076e7
[8]  Cw0 = 55
[9]  k = 0.39175*exp(5472.7*((1/273)-(1/T)))
[10] Cd = Cc
[11] Vi = 0.2
[12] Kc = 10^(3885.44/T)
[13] cpa = 170700
[14] V = Vi+v0*t
[15] Fb0 = Cb0*v0
[16] ra = -k*((Ca*Cb)-((Cc*Cd)/Kc))
[17] Na = V*Ca
[18] Nb = V*Cb
[19] Nc = V*Cc
[20] rb = ra
[21] rc = -ra
[22] Nd = V*Cd
[23] rate = -ra
[24] NCp = cp*(Nb+Nc+Nd+Nw)+cpa*Na
[25] Cpc = 18
[26] Ta1 = 285
[27] mc = 100
[28] Qr = mc*Cpc*(Ta1-T)*(1-exp(-UA/mc/Cpc))
[29] Ta2 = T-(T-Ta1)*exp(-UA/mc/Cpc)
```

Living Example Problem

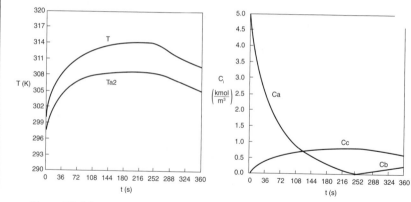

Figure E9-4.2   Temperature–time
trajectory in a semibatch reactor

Figure E9-4.3   Concentration–time
trajectories in a semibatch reactor.

9.4 Unsteady Operation of a CSTR

9.4.1 Startup

Startup of a CSTR   In reactor startup it is often very important *how* temperature and concentrations approach their steady-state values. For example, a significant overshoot in temperature may cause a reactant or product to degrade, or the overshoot may be unacceptable for safe operation. If either case were to occur, we would say that the system exceeded its *practical stability limit*. Although we can solve the unsteady temperature–time and concentration–time equations numerically to see if such a limit is exceeded, it is often more insightful to study the approach to steady state by using the *temperature–concentration phase plane*. To illustrate these concepts we shall confine our analysis to a liquid-phase reaction carried out in a CSTR.

A qualitative discussion of how a CSTR approaches steady state is given in PRS R9.4. This analysis, summarized in Figure S-1 in the Summary for this chapter, is developed to show the four different regions into which the phase plane is divided and how they allow one to sketch the approach to the steady state.

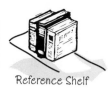

Reference Shelf

Example 9–5   Startup of a CSTR

Again we consider the production of propylene glycol (C) in a CSTR with a heat exchanger in Example 8-8. Initially there is only water at 75°F and 0.1 wt % H_2SO_4 in the 500-gallon reactor. The feed stream consists of 80 lb mol/h of propylene oxide (A), 1000 lb mol/h of water (B) containing 0.1 wt % H_2SO_4, and 100 lb mol/h of methanol (M). Plot the temperature and concentration of propylene oxide as a function of time, and a concentration vs. temperature graph for different entering temperatures and initial concentrations of A in the reactor.

The water coolant flows through the heat exchanger at a rate of 5 lb/s (1000 lb mol/h). The molar densities of pure propylene oxide (A), water (B), and methanol (M) are $\rho_{A0} = 0.932$ lb mol/ft³, $\rho_{B0} = 3.45$ lb mol/ft³, and $\rho_{M0} = 1.54$ lb mol/ft³, respectively.

$$UA = 16,000 \ \frac{\text{Btu}}{\text{h} \cdot °\text{F}} \ \text{with} \ T_{a1} = 60°\text{F}, \ \dot{m}_W = 1000 \ \text{lb mol/h with} \ C_{P_W} = 18 \ \text{Btu/lb mol} \cdot °\text{F}$$

$$C_{P_A} = 35 \ \text{Btu/lb mol} \cdot °\text{F}, \ C_{P_B} = 18 \ \text{Btu/lb mol} \cdot °\text{F},$$
$$C_{P_C} = 46 \ \text{Btu/lb mol} \cdot °\text{F}, \ C_{P_M} = 19.5 \ \text{Btu/lb mol} \cdot °\text{F}$$

Solution

$$A + B \longrightarrow C$$

Mole Balances:

Initial Conditions

A: $\quad \dfrac{dC_A}{dt} = r_A + \dfrac{(C_{A0} - C_A)v_0}{V}$ $\qquad\qquad 0 \qquad\qquad$ (E9-5.1)

B: $\quad \dfrac{dC_B}{dt} = r_B + \dfrac{(C_{B0} - C_B)v_0}{V}$ $\qquad C_{Bi} = 3.45 \ \dfrac{\text{lb mol}}{\text{ft}^3} \qquad$ (E9-5.2)

C: $\quad \dfrac{dC_C}{dt} = r_C + \dfrac{-C_C v_0}{V}$ $\qquad\qquad 0 \qquad\qquad$ (E9-5.3)

M: $\quad \dfrac{dC_M}{dt} = \dfrac{v_0(C_{M0} - C_M)}{V}$ $\qquad\qquad 0 \qquad\qquad$ (E9-5.4)

Rate Law: $\qquad\qquad\qquad -r_A = kC_A$ $\qquad\qquad$ (E9-5.5)

Stoichiometry: $\qquad\qquad\quad -r_A = -r_B = r_C$ $\qquad\qquad$ (E9-5.6)

Energy Balance:

$$\frac{dT}{dt} = \frac{\dot{Q} - F_{A0} \sum \Theta_i C_{P_i}(T - T_0) + (\Delta H_{Rx})(r_A V)}{\sum N_i C_{P_i}}$$ (E9-5.7)

with

$$\dot{Q} = \dot{m}_W C_{P_W}(T_{a1} - T_{a2}) = \dot{m}_W C_{P_W}(T_{a1} - T)\left[1 - \exp\left(-\frac{UA}{\dot{m}_W C_{P_W}}\right)\right]$$ (E9-5.8)

and

$$T_{a2} = T - (T - T_{a1}) \exp\left(-\frac{UA}{\dot{m}_W C_{P_W}}\right)$$

Evaluation of parameters:

$$\Sigma\, N_i C_{P_i} = C_{P_A} N_A + C_{P_B} N_B + C_{P_C} N_C + C_{P_M} N_M$$

$$= 35(C_A V) + 18(C_B V) + 46(C_C V) + 19.5(C_M V)$$

$$\Sigma\, \Theta_i C_{P_i} = C_{P_A} + \frac{F_{B0}}{F_{A0}} C_{P_B} + \frac{F_{M0}}{F_{A0}} C_{P_M}$$

$$= 35 + 18\,\frac{F_{B0}}{F_{A0}} + 19.5\,\frac{F_{M0}}{F_{A0}}$$

$$v_0 = \frac{F_{A0}}{\rho_{A0}} + \frac{F_{B0}}{\rho_{B0}} + \frac{F_{M0}}{\rho_{M0}} = \left(\frac{F_{A0}}{0.923} + \frac{F_{B0}}{3.45} + \frac{F_{M0}}{1.54}\right)\frac{\text{ft}^3}{\text{h}}$$

Neglecting ΔC_P because it changes the heat of reaction insignificantly over the temperature range of the reaction, the heat of reaction is assumed constant at

$$\Delta H_{Rx} = -36{,}000\ \frac{\text{Btu}}{\text{lb mol A}}$$

The Polymath program is shown in Table E9-5.1.

TABLE E9-5.1.   POLYMATH PROGRAM FOR CSTR STARTUP

ODE Report (RKF45)

Differential equations as entered by the user
```
[1]  d(Ca)/d(t) = 1/tau*(Ca0-Ca)+ra
[2]  d(Cb)/d(t) = 1/tau*(Cb0-Cb)+rb
[3]  d(Cc)/d(t) = 1/tau*(0-Cc)+rc
[4]  d(Cm)/d(t) = 1/tau*(Cm0-Cm)
[5]  d(T)/d(t) = (Q-Fa0*ThetaCp*(T-T0)+(-36000)*ra*V)/NCp
```

Explicit equations as entered by the user
```
[1]   Fa0 = 80
[2]   T0 = 75
[3]   V = (1/7.484)*500
[4]   UA = 16000
[5]   Ta1 = 60
[6]   k = 16.96e12*exp(-32400/1.987/(T+460))
[7]   Fb0 = 1000
[8]   Fm0 = 100
[9]   mc = 1000
[10]  ra = -k*Ca
[11]  rb = -k*Ca
[12]  rc = k*Ca
[13]  Nm = Cm*V
[14]  Na = Ca*V
[15]  Nb = Cb*V
[16]  Nc = Cc*V
[17]  ThetaCp = 35+Fb0/Fa0*18+Fm0/Fa0*19.5
[18]  v0 = Fa0/0.923+Fb0/3.45+Fm0/1.54
[19]  Ta2 = T-(T-Ta1)*exp(-UA/(18*mc))
[20]  Ca0 = Fa0/v0
[21]  Cb0 = Fb0/v0
[22]  Cm0 = Fm0/v0
[23]  Q = mc*18*(Ta1-Ta2)
[24]  tau = V/v0
[25]  NCp = Na*35+Nb*18+Nc*46+Nm*19.5
```

Living Example Problem

Figures (E9-5.1) and (E9-5.2) show the reactor concentration and temperature of propylene oxide as a function of time, respectively, for an initial temperature of 75°F and only water in the tank (i.e., $C_{Ai} = 0$). One observes, both the temperature and concentration oscillate around their steady-state values ($T = 138°F$, $C_A = 0.039$ lb mol/ft^3). Figure (E9-5.3) shows the phase plane of temperature and propylene oxide concentration for three different sets of initial conditions ($T_i = 75°F$, $C_{Ai} = 0$; $T_i = 150°F$, $C_{Ai} = 0$; and $T_i = 160°F$, $C_{Ai} = 0.14$ lb mol/ft^3), keeping T_0 constant.

Unacceptable
startup

An upper limit of 180°F should not be exceeded in the tank. This temperature is the *practical stability limit*. The practical stability limit represent a temperature above which it is undesirable to operate because of unwanted side reactions, safety considerations, or damage to equipment. Consequently, we see if we started at an initial temperature of 160°F and an initial concentration of 0.14 mol/dm^3, the practical stability limit of 180°F would be exceeded as the reactor approached its steady-state temperature of 138°F. See the concentration–temperature trajectory in Figure E9-5.4.

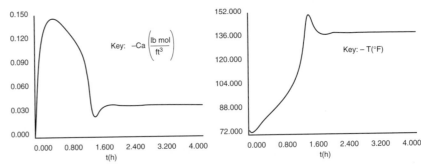

Figure E9-5.1   Propylene oxide concentration as a function of time.

Figure E9-5.2   Temperature–time trajectory for CSTR startup.

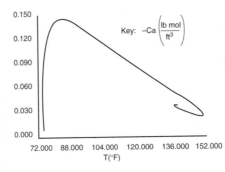

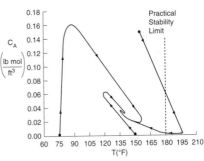

Oops! The practical
stability limit was
exceeded.

Figure E9-5.3   Concentration–temperature phase-plane trajectory.

Figure E9-5.4   Concentration–temperature phase plane.

After about 1.6 h the reactor is operating at steady state with the following values:

$$C_A = 0.0379 \text{ lb mol/ft}^3 \quad C_C = 0.143 \text{ lb mol/ft}^3$$

$$C_B = 2.12 \text{ lb mol/ft}^3 \quad C_M = 0.2265 \text{ lb mol/ft}^3$$

$$T = 138.5°F$$

9.4.2 Falling Off the Steady State

We now consider what can happen to a CSTR operating at an upper steady state when an upset occurs in either the ambient temperature, the entering temperature, the flow rate, the reactor temperature, or some other variable. To illustrate, let's reconsider the production of propylene glycol in a CSTR, which we just discussed.

Example 9–6 Falling Off the Upper Steady State

In Example 9-5 we saw how a 500-gal CSTR used for the production of propylene glycol approached steady state. For the flow rates and conditions (e.g., $T_0 = 75°F$, $T_{a1} = 60°F$), the steady-state temperature was 138°F, and the corresponding conversion was 75.5%. Determine the steady-state temperature and conversion that would result if the entering temperature were to drop from 75°F to 70°F, assuming that all other conditions remain the same. First, sketch the steady-state conversions calculated from the mole and energy balances as a function of temperature before and after the drop in entering temperature occurred. Next, plot the "conversion," concentration of A, and the temperature in the reactor as a function of time after the entering temperature drops from 75°F to 70°F.

Solution

The steady-state conversions can be calculated from the mole balance,

$$X_{MB} = \frac{\tau A e^{-E/RT}}{1 + \tau A e^{-E/RT}} \qquad (E8\text{-}5.5)$$

and from the energy balance,

$$X_{EB} = \frac{\sum \Theta_i C_{p_i}(T - T_0) + [\dot{Q}/F_{A0}]}{-[\Delta H_{Rx}(T_R)]} \qquad (E8\text{-}5.6)$$

before ($T_0 = 75°F$) and after ($T_0 = 70°F$) the upset occurred. We shall use the parameter values given in Example 9-5 (e.g., $F_{A0} = 80$ lb mol/h, $UA = 16,000$ Btu/h · °F) to obtain a sketch of these conversions as a function of temperature, as shown in Figure E9-6.1.

We see that for $T_0 = 70°F$ the reactor has dropped below the extinction temperature and can no longer operate at the upper steady state. In Problem P9-16, we will see it is not always necessary for the temperature to drop below the extinction temperature in order to fall to the lower steady state. The equations describing the dynamic drop from the upper steady state to the lower steady state are identical to those given in Example 9-5; only the initial conditions and entering temperature are different. Consequently, the same Polymath and MATLAB programs can be used with these modifications. (See *Living Example* 9-6 on the CD-ROM.)

Living Example Problem

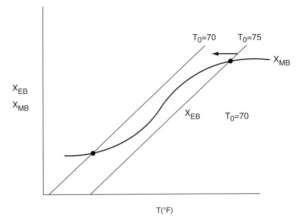

Figure E9-6.1   Conversion from mole and energy balances as a function
of temperature.

Initial conditions are taken from the final steady-state values given in Example 9-5.

$$
\begin{array}{ll}
C_{Ai} = 0.039 \text{ lb mol/ft}^3 & C_{Ci} = 0.143 \text{ lb mol/ft}^3 \\
C_{Bi} = 2.12 \text{ lb mol/ft}^3 & C_{Mi} = 0.226 \text{ lb mol/ft}^3
\end{array}
$$

$$T_i = 138.5°F$$

Change T_0 to 70°F

Because the system is not at steady state, we cannot rigorously define a conversion in terms of the number of moles reacted because of the accumulation within the reactor. However, we can approximate the conversion by the equation $X = (1 - C_A/C_{A0})$. This equation is valid after the steady state is reached. Plots of the temperature and the conversion as a function of time are shown in Figures E9-6.2 and E9-6.3, respectively. The new steady-state temperature and conversion are $T = 83.6°F$ and $X = 0.19$.

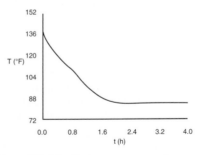

Figure E9-6.2   Temperature versus time.   **Figure E9-6.3**   Conversion versus time.

Reference Shelf

We could now see how we can make adjustments for upsets in the reactor operating conditions (such as we just saw in the drop in the entering temperature) so that we do not fall to the lower steady-state values. We can prevent this drop in conversion by adding a controller to the reactor. The addition of a controller is discussed in the *Professional Reference Shelf* R9.2 on the CD-ROM.

9.5 Nonisothermal Multiple Reactions

For multiple reactions occurring in either a semibatch or batch reactor, Equation (9-21) can be generalized in the same manner as the steady-state energy balance, to give

$$\frac{dT}{dt} = \frac{\dot{m}_c C_{P_c}(T_{a1} - T)[1 - \exp(-UA/\dot{m}_c C_{P_c})] + \sum\limits_{i=1}^{q} r_{ij} V \Delta H_{Rxij}(T) - \sum F_{i0} C_{P_i}(T - T_0)}{\sum N_i C_{P_i}}$$

(9-23)

For large coolant rates Equation (9-23) becomes

$$\frac{dT}{dt} = \frac{UA(T_a - T) - \sum F_{i0} C_{P_i}(T - T_0) + V \sum\limits_{i=1}^{q} r_{ij} \Delta H_{Rxij}}{\sum\limits_{i=1}^{n} N_i C_{P_i}}$$

(9-24)

Example 9–7 Multiple Reactions in a Semibatch Reactor

The series reactions

$$2A \xrightarrow[(1)]{k_{1A}} B \xrightarrow[(2)]{k_{2B}} 3C$$

are catalyzed by H_2SO_4. All reactions are first order in the reactant concentration. The reaction is to be carried out in a semibatch reactor that has a heat exchanger inside with $UA = 35{,}000$ cal/h·K and an exchanger temperature, T_a, of 298 K. Pure A enters at a concentration of 4 mol/dm³, a volumetric flow rate of 240 dm³/h, and a temperature of 305 K. Initially there is a total of 100 dm³ in the reactor, which contains 1.0 mol/dm³ of A and 1.0 mol/dm³ of the catalyst H_2SO_4. The reaction rate is independent of the catalyst concentration. The initial temperature of the reactor is 290 K. Plot the species concentrations and temperature as a function of time.

Living Example Problem

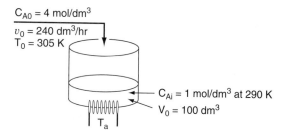

$C_{A0} = 4$ mol/dm³
$v_0 = 240$ dm³/hr
$T_0 = 305$ K

$C_{Ai} = 1$ mol/dm³ at 290 K
$V_0 = 100$ dm³

T_a

Additional information:

$k_{1A} = 1.25\,h^{-1}$ at 320 K with $E_{1A} = 9500$ cal/mol $C_{P_A} = 30$ cal/mol·K

$k_{2B} = 0.08\,h^{-1}$ at 300 K with $E_{2B} = 7000$ cal/mol $C_{P_B} = 60$ cal/mol·K

$\Delta H_{Rx1A} = -6500$ cal/mol A $C_{P_C} = 20$ cal/mol·K

$\Delta H_{Rx2B} = +8000$ cal/mol B $C_{P_{H_2SO_4}} = 35$ cal/mol·K

Solution

Mole Balances:

$$\frac{dC_A}{dt} = r_A + \frac{(C_{A0} - C_A)}{V} v_0 \tag{E9-7.1}$$

$$\frac{dC_B}{dt} = r_B - \frac{C_B}{V} v_0 \tag{E9-7.2}$$

$$\frac{dC_C}{dt} = r_C - \frac{C_C}{V} v_0 \tag{E9-7.3}$$

Rate Laws:

$$-r_{1A} = k_{1A}C_A \tag{E9-7.4}$$
$$-r_{2B} = k_{2B}C_B \tag{E9-7.5}$$

Stoichiometry (liquid phase): Use C_A, C_B, C_C

Relative rates:

$$r_{1B} = -\frac{1}{2}r_{1A} \tag{E9-7.6}$$

$$r_{2C} = -3\,r_{2B} \tag{E9-7.7}$$

Net rates:

$$r_A = r_{1A} = -k_{1A}C_A \tag{E9-7.8}$$

$$r_B = r_{1B} + r_{2B} = \frac{-r_{1A}}{2} + r_{2B} = \frac{k_{1A}C_A}{2} - k_{2B}C_B \tag{E9-7.9}$$

$$r_C = 3\,k_{2B}C_B \tag{E9-7.10}$$

$$N_i = C_i V \tag{E9-7.11}$$

$$V = V_0 + v_0 t \tag{E9-7.12}$$

$$N_{H_2SO_2} = (C_{H_2SO_{4,0}})V_0 = \frac{1\text{ mol}}{dm^3} \times 100\text{ dm}^3 = 100\text{ mol}$$

$$F_{A0} = \frac{4\text{ mol}}{dm^3} \times 240\,\frac{dm^3}{h} = 960\,\frac{mol}{h}$$

Energy Balance:

$$\frac{dT}{dt} = \frac{UA(T_a - T) - \sum F_{i0} C_{P_i}(T - T_0) + V \sum_{i=1}^{q} \Delta H_{Rxij} r_{ij}}{\sum N_i C_{P_i}} \quad (9\text{-}24)$$

$$\frac{dT}{dt} = \frac{UA(T_a - T) - F_{A0} C_{P_A}(T - T_0) + [(\Delta H_{Rx1A})(r_{1A}) + (\Delta H_{Rx2B})(r_{2B})] V}{[C_A C_{P_A} + C_B C_{P_B} + C_C C_{P_C}] V + N_{H_2SO_4} C_{P_{H_2SO_4}}}$$

(E9-7.13)

$$\frac{dT}{dt} = \frac{35,000(298 - T) - (4)(240)(30)(T - 305) + [(-6500)(-k_{1A}C_A) + (+8000)(-k_{2B}C_B)] V}{(30 C_A + 60 C_B + 20 C_C)(100 + 240t) + (100)(35)}$$

(E9-7.14)

Equations (E9-7.1) through (E9-7.3) and (E9-7.8) through (E9-7.12) can be solved simultaneously with Equation (E9-7.14) using an ODE solver. The Polymath program is shown in Table E9-7.1 and the Matlab program is on the CD-ROM. The time graphs are shown in Figures E9-7.1 and E9-7.2.

TABLE E9-7.1.   POLYMATH PROGRAM

ODE Repórt (RKF45)

Differential equations as entered by the user
[1] d(Ca)/d(t) = ra+(Cao-Ca)*vo/V
[2] d(Cb)/d(t) = rb-Cb*vo/V
[3] d(Cc)/d(t) = rc-Cc*vo/V
[4] d(T)/d(t) = (35000*(298-T)-Cao*vo*30*(T-305)+((-6500)*(-k1a*Ca)+(8000)*(-k2b*Cb))*V)/((Ca*30+Cb*60+Cc*20)*V+100*35)

Explicit equations as entered by the user
[1] Cao = 4
[2] vo = 240
[3] k1a = 1.25*exp((9500/1.987)*(1/320-1/T))
[4] k2b = 0.08*exp((7000/1.987)*(1/290-1/T))
[5] ra = -k1a*Ca
[6] V = 100+vo*t
[7] rc = 3*k2b*Cb
[8] rb = k1a*Ca/2-k2b*Cb

Living Example Problem

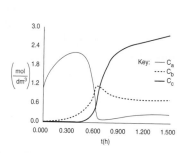

Figure E9-7.1 Concentration–time.

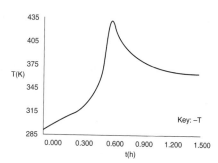

Figure E9-7.2 Temperature (K)–time (h).

9.6 Unsteady Operation of Plug-Flow Reactors

In the CD-ROM, the unsteady energy balance is derived for a PFR. Neglecting changes in total pressure and shaft work, the following equation is derived:

Transient energy
balance on a PFR

$$Ua(T_a - T) - \left(\sum_{i=1}^{n} F_i C_{P_i} \right) \frac{\partial T}{\partial V} + (-r_A)[-\Delta H_{Rx}(T)] = \sum_{i=1}^{n} C_i C_{P_i} \frac{\partial T}{\partial t} \quad (9\text{-}25)$$

This equation must be coupled with the mole balances:

Numerical
solution required for
these three
coupled equations

$$\frac{-\partial F_i}{\partial V} + v_i(-r_A) = \frac{\partial C_i}{\partial t} \quad (9\text{-}26)$$

and the rate law,

$$-r_A = k(T) \cdot \text{fn}(C_i) \quad (9\text{-}27)$$

and solved numerically. A variety of numerical techniques for solving equations of this type can be found in the book *Applied Numerical Methods*.[4]

One can use COMSOL to solve PFR and laminar flow reactors for the time-dependent temperature and concentration profiles. See the COMSOL problems and web module in Chapter 8 and on the COMSOL CD-ROM enclosed with this book. A simpler approach would be to model the PFR as a number of CSTRs in series and then apply Equation (9-9) to each CSTR.

Closure. After completing this chapter, the reader should be able to apply the unsteady-state energy balance to CSTRs, semibatch and batch reactors. The reader should be able to discuss reactor safety using two examples: one a case study of an explosion and the other the use of the ARSST to help prevent explosions. Included in the reader's discussion should be how to start up a reactor so as not to exceed the practical stability limit. After reading these examples, the reader should be able to describe how to operate reactors in a safe manner for both single and multiple reactions.

---

[4] B. Carnahan, H. A. Luther, and J. O. Wilkes, *Applied Numerical Methods* (New York: Wiley, 1969).

SUMMARY

1. Unsteady operation of CSTRs and semibatch reactors

$$\frac{dT}{dt} = \frac{\dot{Q} - \dot{W}_S - \sum\limits_{i=1}^{n} F_{i0} C_{P_i}(T - T_{i0}) + [-\Delta H_{\text{Rx}}(T)](-r_A V)}{\sum\limits_{i=1}^{n} N_i C_{P_i}} \qquad \text{(S9-1)}$$

For large heat-exchanger coolant rates ($T_{a1} = T_{a2}$)

$$\dot{Q} = UA(T_a - T) \qquad \text{(S9-2)}$$

For moderate to low coolant rates

$$\dot{Q} = \dot{m}_c C_{P_c}(T - T_{a1})\left[1 - \exp\left(-\frac{UA}{\dot{m}_c C_{P_c}}\right)\right] \qquad \text{(S9-3)}$$

2. Batch reactors
   a. Nonadiabatic

$$\frac{dT}{dt} = \frac{\dot{Q} - \dot{W}_s + (-\Delta H_{\text{Rx}})(-r_A V)}{N_{\text{A0}}(\sum \Theta_i C_{P_i} + \Delta C_P X)} \qquad \text{(S9-4)}$$

Where $\dot{Q}$ is given by either Equation (S9-2) or (S9-3).

   b. Adiabatic

$$\boxed{X = \frac{C_{P_s}(T - T_0)}{-\Delta H_{\text{Rx}}(T)} = \frac{\sum \Theta_i C_{P_i}(T - T_0)}{-\Delta H_{\text{Rx}}(T)}} \qquad \text{(S9-5)}$$

$$\boxed{T = T_0 + \frac{[-\Delta H_{\text{Rx}}(T_0)]X}{C_{P_s} + X \Delta C_P} = T_0 + \frac{[-\Delta H_{\text{Rx}}(T_0)]X}{\sum\limits_{i=1}^{n} \Theta_i C_{P_i} + X \Delta C_P}} \qquad \text{(S9-6)}$$

3. Startup of a CSTR (Figure S-1) and the approach to the steady state (CD-ROM). By mapping out regions of the concentration–temperature phase plane, one can view the approach to steady state and learn if the practical stability limit is exceeded.

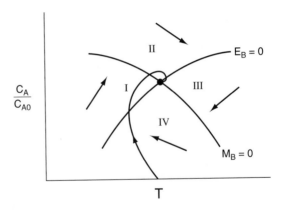

Figure S-1   Startup of a CSTR.

4. Multiple reactions (q reactions and n species)

$$\frac{dT}{dt} = \frac{\dot{m}_c C_{P_c}(T_{a1} - T)[1 - \exp(-UA/\dot{m}_c C_{P_c})] + \sum\limits_{i=1}^{q} r_{ij} V \Delta H_{\text{Rx}ij}(T) - \sum F_{i0} C_{P_i}(T - T_0)}{\sum N_i C_{P_i}}$$

$$(S9\text{-}7)$$

CD-ROM MATERIAL

Summary Notes

Links

Solved Problems

Living Example Problem

- **Learning Resources**
    1. *Summary Notes*
    2. *Web links: SACHE Safety web site www.sache.org.* You will need to get the user name and password from your department chair. The kinetics (i.e., CRE) text, examples, and problems are marked K in the product sections: Safety, Health, and the Environment (S,H, & E).
    3. *Solved Problems*
        Example CD9–1 Startup of a CSTR
        Example CD9–2 Falling Off the Steady State
        Example CD9–3 Proportional-Integral (PI) Control
- **Living Example Problems**
    1. *Example 9–1 Adiabatic Batch Reactor*
    2. *Example 9–2 Safety in Chemical Plants with Exothermic Reactions*
    3. *Example 9–3 Use of the ARSST*
    4. *Example 9–4 Heat Effects in a Semibatch Reactor*
    5. *Example 9–5 Startup of a CSTR*
    6. *Example 9–6 Falling of the Upper Steady State*
    7. *Example 9–7 Multiple Reactions in a Semibatch Reactor*
    8. *Example RE9–1 Integral Control of a CSTR*
    9. *Example RE9–2 Proportion-Integral Control of a CSTR*
    10. *Example RE9–3 Linearized Stability*

- **Professional Reference Shelf**

    R9.1   *The Complete ARSST*

    In this section further details are given to size safety valves to prevent runaway reactions

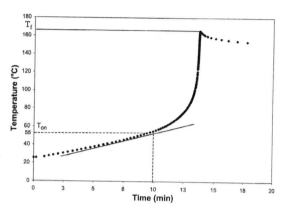

Figure E9-3.1 Temperature-time trajectory for hydrolysis of acetic anhydride.

R9.2.   *Control of a CSTR*

In this section we discuss the use of proportional (P) and integral (I) control of a CSTR. Examples include I and PI control of an exothermic reaction.

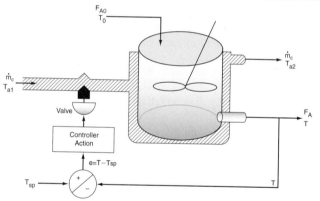

Reactor with control system.

Proportional integral action

$$z = z_0 + k_c(T - T_{SP}) + \frac{k_c}{\tau}\int_0^t (T - T_{SP})dt$$

R9.3.   *Linearized Stability Theory*

In this section we learn if a perturbation will decay in an exponential manner ((**a**) below) in an oscillatory manner (**b**), grown exponentially (**c**), grown exponentially with oscillations (**d**), or just oscillate (**e**).

$$-r_A = -r_{AS} + (C_A - C_{AS})\frac{\partial(-r_A)}{\partial C_A}\bigg|_S + (T - T_S)\frac{\partial(-r_A)}{\partial T}\bigg)_S \quad (R9.3-11)$$

$$\tau\frac{dC_A}{dt} - (C_A - C_{AS}) - \tau k_S(C_A - C_{AS}) - \frac{E}{RT_S^2}(-r_{AS})(T - T_S) \qquad \text{(R9.3-13)}$$

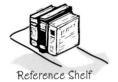

Reference Shelf

TABLE 9C–1 EIGENVALUES OF COUPLED ODES

A.	Tr < 0	Det > 0	_Stable_

$[\text{Tr}^2(\mathbf{M}) - 4\text{Det}(\mathbf{M})] > 0$ **(a)**

$(\text{Tr}^2(\mathbf{M}) - 4\text{Det}(\mathbf{M})) < 0$ **(b)**

B.	Tr > 0	Det > 0	_Unstable_

$[\text{Tr}^2(\mathbf{M}) - 4\text{Det}(\mathbf{M})] > 0$ **(c)**

$[\text{Tr}^2(\mathbf{M}) - 4\text{Det}(\mathbf{M})] < 0$ **(d)**

C.	Tr $(\mathbf{M})= 0$	Det $(\mathbf{M}) > 0$	_Pure Oscillation_

    (e)

R9.4. *Approach to the Steady-State Phase-Plane Plots and Trajectories of Concentration versus Temperature*
Here we learn if the practical stability is exceeded during startup.
Example RE9–4.1 Start Up of a CSTR
Example RE9–4.2 Falling Off the Steady State
Example RE9–4.3 Revisit Example RE9-2.

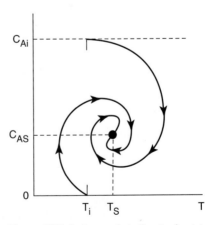

Figure CD9-5 Approach to the steady state.

R9.5. *Adiabatic Operation of a Batch Reactor*
R9.6. *Unsteady Operation of Plug-Flow Reactors*

QUESTIONS AND PROBLEMS

P9-1$_A$ Read over the problems at the end of this chapter. Refer to the guidelines given in Problem 4-1, and make up an original problem that uses the concepts presented in this chapter. To obtain a solution:

(a) Make up your data and reaction.

(b) Use a real reaction and real data.

Also,

(c) Prepare a list of safety considerations for designing and operating chemical reactors.

See R. M. Felder, *Chem. Eng. Educ.*, *19* (4), 176 (1985). The August 1985 issue of *Chemical Engineering Progress* may be useful for part (c).

P9-2$_A$ Review the example problems in this chapter, choose one, and use a software package such as Polymath or MATLAB to carry out a parameter sensitivity analysis.

What if...

(a) Example 9-1. How much time would it take to achieve 90% conversion if the reaction were started on a very cold day where the initial temperature was 20°F? (Methanol won't freeze at this temperature.)

(b) Example 9-2. Explore the ONCB explosion described in Example 9-2. Show that no explosion would have occurred if the cooling was not shut off for the 9.04-kmol charge of ONCB or if the cooling was shut off for 10 min after 45 min of operation for the 3.17-kmol ONCB charge. Show that if the cooling had been shut off for 10 min after 12 h of operation, no explosion would have occurred for the 9.04-kmol charge. Develop a set of guidelines as to when the reaction should be quenched should the cooling fail. Perhaps safe operation could be discussed using a plot of the time after the reaction began at which the cooling failed, t_0, versus the length of the cooling failure period, t_f, for the different charges of ONCB. Parameter values used in this example predict that the reactor will explode at midnight. What parameter values would predict the time the reactor would explode at the actual time of 18 min after midnight? Find a set of parameter values that would cause the explosion to occur at exactly 12:18 A.M. For example, include heat capacities of metal reactor and/or make a new estimate of *UA*. Finally, what if a 1/2-in. rupture disk rated at 800 psi had been installed and did indeed rupture at 800 psi (270°C)? Would the explosion still have occurred? (*Note:* The mass flow rate $\dot{m}$ varies with the cross-sectional area of the disk. Consequently, for the conditions of the reaction the maximum mass flow rate out of the 1/2-in. disk can be found by comparing it with the mass flow rate of 830 kg/min of the 2-in. disk.

(c) Example 9-3. What would be the conversion at the onset temperature if the heating rate were reduced by a factor of 10? Increased by a factor of 10?

(d) Example 9-4. What would the *X* versus *t* and *T* versus t trajectories look like if the coolant rate is decreased by a factor of 10? Increased by a factor of 10?

(e) **Example 9-5.** Load the Living Example Problem for *Startup of a CSTR*, for an entering temperature of 70°F, an initial reactor temperature of 160°F, and an initial concentration of propylene oxide of 0.1 *M*. Try other combinations of T_0, T_i, and C_{Ai}, and report your results in terms of temperature–time trajectories and temperature–concentration phase planes.

(f) **Example 9-6.** Load the Living Example Problem for *Falling Off the Upper Steady State*. Try varying the entering temperature, T_0, to between 80 and 68°F and plot the steady-state conversion as a function of T_0. Vary the coolant rate between 10,000 and 400 mol/h. Plot conversion and reactor temperature as a function of coolant rate.

(g) **Example 9-7.** What happens if you increase the heat transfer coefficient by a factor of 10 and decrease T_a to 280 K? Which trajectories change the most?

T

t

Integral Controller

(h) **Example RE9-1.** Load the Living Example Problem. Vary the gain, k_C, between 0.1 and 500 for the integral controller of the CSTR. Is there a lower value of k_C that will cause the reactor to fall to the lower steady state or an upper value to cause it to become unstable? What would happen if T_0 were to fall to 65°F or 60°F?

(i) **Example RE9-2.** Load the Living Example Problem. Learn the effects of the parameters k_C and τ_I. Which combination of parameter values generates the least and greatest oscillations in temperature? Which values of k_C and τ_I return the reaction to steady state the quickest?

(j) **Reactor Safety.** Enter the SACHE web site, *www.sache.org*. [Note you will need to obtain the user name and password for your school from your department chair or SACHE representative.] After entering hit the current year (e.g., 2004). Go to product: Safety, Health and the Environment (S,H, & E). The problems are for KINETICS (i.e., CRE). There are some example problems marked K and explanations in each of the above S,H, & E selections. Solutions to the problems are in a different section of the site. Specifically look at: *Loss of Cooling Water (K-1) Runaway Reactions* (HT-1), *Design of Relief Values* (D-2), *Temperature Control and Runaway* (K-4) and (K-5), and *Runaway and the Critical Temperature Region* (K-7).

Links

Member

Hall of Fame

P9-3$_B$ The following is an excerpt from *The Morning News*, Wilmington, Delaware (August 3, 1977): "Investigators sift through the debris from blast in quest for the cause [that destroyed the new nitrous oxide plant]. A company spokesman said it appears more likely that the [fatal] blast was caused by another gas—ammonium nitrate—used to produce nitrous oxide." An 83% (wt) ammonium nitrate and 17% water solution is fed at 200°F to the CSTR operated at a temperature of about 520°F. Molten ammonium nitrate decomposes directly to produce gaseous nitrous oxide and steam. It is believed that pressure fluctuations were observed in the system and as a result the molten ammonium nitrate feed to the reactor may have been shut off approximately 4 min prior to the explosion. Can you explain the cause of the blast? If the feed rate to the reactor just before shutoff was 310 lb of solution per hour, what was the exact temperature in the reactor just prior to shutdown? Using the following data, calculate the time it took to explode after the feed was shut off for the reactor. How would you start up or shut down and control such a reaction?

Assume that at the time the feed to the CSTR stopped, there was 500 lb of ammonium nitrate in the reactor at a temperature of 520°F. The conversion in the reactor is virtually complete at about 99.99%. Additional data for this problem are given in Problem 8-3. How would your answer change if 100 lb of solution were in the reactor? 310 lb? 800 lb? What if $T_0 = 100°F$? 500°F?

P9-4$_B$ The first-order irreversible reaction

$$A(l) \longrightarrow B(g) + C(g)$$

is carried out adiabatically in a CSTR into which 100 mol/min of pure liquid A is fed at 400 K. The reaction goes virtually to completion (i.e., the feed rate into the reactor equals the product of reaction rate inside the reactor and the reactor volume).

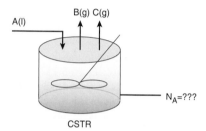

How many moles of liquid A are in the CSTR under steady-state conditions? Plot the temperature and moles of A in the reactor as a function of time after the feed to the reactor has been shut off.

Additional information:

Temperature (K)	400	800	1200
k (min^{-1})	0.19	0.32	2.5
H_A (kJ/mol)	-38	-30	-22
H_B (kJ/mol)	-26	-22	-18
H_C (kJ/mol)	-20	-16	-12

P9-5$_B$ The liquid-phase reaction in Problem P8-5 is to be carried out in a semibatch reactor. There is 500 mol of A initially in the reactor at 25°C. Species B is fed to the reactor at 50°C and a rate of 10 mol/min. The feed to the reactor is stopped after 500 mol of B has been fed.
(a) Plot the temperature and conversion as a function of time when the reaction is carried out adiabatically. Calculate to $t = 2$ h.
(b) Plot the conversion as a function of time when a heat exchanger ($UA = 100$ cal/min·K) is placed in the reactor and the ambient temperature is constant at 50°C. Calculate to $t = 3$ h.
(c) Repeat part (b) for the case where the reverse reaction cannot be neglected.

New parameter values:

$k = 0.01$ (dm^3/mol · min) at 300 K with $E = 10$ kcal/mol
$V_0 = 50$ dm^3, $v_0 = 1$ dm^3/min, $C_{A0} = C_{B0} = 10$ mol/dm^3
For the reverse reaction: $k_r = 10$ s^{-1} at 300 K with $E_r = 16$ kcal/mol

P9-6$_B$ You are operating a batch reactor and the reaction is first-order, liquid-phase, and exothermic. An inert coolant is added to the reaction mixture to control the temperature. The temperature is kept constant by varying the flow rate of the coolant (see Figure P9-6).

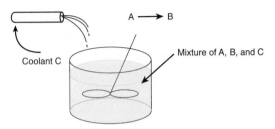

Figure P9-6

(a) Calculate the flow rate of the coolant 2 h after the start of the reaction. (*Ans.*: $F_C = 3.157$ lb/s.)

(b) It is proposed that rather than feeding a coolant to the reactor, a solvent be added that can be easily boiled off, even at moderate temperatures. The solvent has a heat of vaporization of 1000 Btu/lb and initially there are 25 lb mol of A placed in the tank. The initial volume of solvent and reactant is 300 ft³. Determine the solvent evaporation rate as a function of time. What is the rate at the end of 2 h?

Additional information:

Temperature of reaction: 100°F
Value of k at 100°F: 1.2×10^{-4} s^{-1}
Temperature of coolant: 80°F
Heat capacity of all components: 0.5 Btu/lb·°F
Density of all components: 50 lb/ft³
ΔH°_{Rx}: $-25{,}000$ Btu/lb mol
Initially:
Vessel contains only A (no B or C present)
C_{A0}: 0.5 lb mol/ft³
Initial volume: 50 ft³

P9-7$_B$ The reaction

$$A + B \longrightarrow C$$

is carried out adiabatically in a constant-volume batch reactor. The rate law is

$$-r_A = k_1 C_A^{1/2} C_B^{1/2} - k_2 C_C$$

Plot the conversion, temperature, and concentrations of the reacting species as a function of time.

Additional information:

Entering Temperature = 100°C

k_1 (373 K) = 2 × 10⁻³ s⁻¹ E_1 = 100 kJ/mol
k_2 (373 K) = 3 × 10⁻⁵ s⁻¹ E_2 = 150 kJ/mol
          C_{A0} = 0.1 mol/dm³ C_{P_A} = 25 J/mol·K
          C_{B0} = 0.125 mol/dm³ C_{P_B} = 25 J/mol·K
ΔH°_{Rx}(298 K) = −40,000 J/mol A C_{P_C} = 40 J/mol·K

P9-8$_B$ The biomass reaction

$$\text{Substrate} \xrightarrow{\text{Cells}} \text{More cells} + \text{Product}$$

is carried out in a 25 dm³ batch chemostat with a heat exchanger.

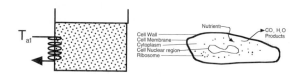

The initial concentration of cells and substrate are 0.1 and 300 g/dm³, respectively. The temperature dependence of the growth rate follows that given by Aiba et al., Equation (7-61)[5]

$$\mu(T) = \mu(310 \text{ K})I' = \mu_{1\max}\left[\frac{0.0038 \cdot T \exp[21.6 - 6700/T]}{1 + \exp[153 - 48,000/T]}\right]\frac{C_S}{K_m + C_S}$$

(7-61)

(a) For adiabatic operation and initial temperature of 278 K, plot T, I', r_g, $-r_S$, C_C, and C_S as a function of time up to 300 hours. Discuss the trends.

(b) Repeat (a) and increase the initial temperature in 10°C increments up to 330 K and describe what you find. Plot the concentration of cells at 24 hours as a function of inlet temperature.

(c) What heat exchange area should be added to maximize the total number of cells at the end of 24 hours? For an initial temperature of 310 K and a constant coolant temperature of 290 K, what would be the cell concentration after 24 hours? (Ans. C_C = _____ g/dm³.)

Additional Information:

$Y_{C/S}$ = 0.8 g cell/g substrate
K_m = 5.0 g/dm³
$\mu_{1\max}$ = 0.5 h⁻¹ (note $\mu = \mu_{\max}$ at 310 K and $C_S \to \infty$)
C_{P_s} = Heat capacity of substrate solution including all cells

C_{P_s} = 74 J/g/K

$\dot{m}_C$ = 100 kg/h

---

[5] S. Abia, A. E. Humphrey, and N. F. Mills, *Biochemical Engineering* (New York: Academic Press, 1973).

ρ = density of solution including cells = 1000
ΔH_{Rx} = –20,000 J/g cells
C_{P_C} = Heat capacity of cooling water 74 J/g/K
U = 50,000 J/h/K/m²

$$Q = \dot{m}_C C_{P_C}[T - T_a]\left[1 - \exp\left[-\frac{UA}{\dot{m}_C C_{P_C}}\right]\right]$$

P9-9$_B$ The first order exothermic liquid-phase reaction

$$A \longrightarrow B$$

is carried out at 85°C in a jacketed 0.2-m³ CSTR. The coolant temperature in the reactor is 32°F. The heat-transfer coefficient is 120 W/m²·K. Determine the critical value of the heat-transfer area below which the reaction will run away and the reactor will explode [*Chem. Eng.*, *91*(10), 54 (1984)].

Additional information:

Specific reaction rate:
  k = 1.1 min⁻¹ at 40°C
  k = 3.4 min⁻¹ at 50°C

The heat capacity of the solution is 20 J/g·K. The solution density is 0.90 kg/dm³. The heat of reaction is −2500 J/g. The feed temperature is 40°C and the feed rate is 90 kg/min. MW of A = 90 g/mol. C_{A0} = 2 *M*.

P9-10$_C$ The ARSST adiabatic bomb calorimeter reactor can also be used to determine the reaction orders. The hydrolysis of acetic anhydride to form acetic acid was carried out adiabatically

$$A + B \longrightarrow 2c$$

The rate law is postulated to be of the form

$$-r_A = kC_A^\alpha \, C_B^\beta$$

The following temperature time data were obtained for two different critical concentrations of acetic anhydride under adiabatic operation. The heating rate was 2°C/min.

$$C_{A0} = 6.7 \ M, \ C_{B0} = 0.2 \ M$$

t (min)	0.0	2.0	4.3	6.2	8.1	10.2	12.0	13.0	13.5	13.6	13.7	13.8	14.0
T (K)	299	303	309	314	321	329	344	361	386	403	439	438	435

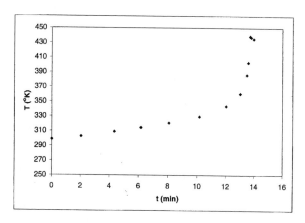

Figure P9-10.1 Data from Undergraduate Laboratory University of Michigan.

(a) Assume $\Delta C_P = 0$ and show for complete conversion, $X = 1$, the difference between the final temperature, T_f, and the initial temperature, T_0,

$$T_f = T_0 + \frac{\Delta H^{\circ}_{\mathrm{Rx}}}{\Sigma \Theta_i C_{p_i}} \qquad\qquad \text{(P9-10.1)}$$

(b) Show that the concentration of A can be written as

$$C_{\mathrm{A}} = C_{\mathrm{A}0} - C_{\mathrm{A}0}\frac{(T - T_0)}{(T_f - T_0)} = C_{\mathrm{A}0}\left[\frac{T_f - T}{T_f - T_0}\right] \qquad \text{(P9-10.2)}$$

and C_{B} as

$$C_{\mathrm{B}} = C_{\mathrm{B}0} - C_{\mathrm{A}0}\frac{(T - T_0)}{(T_f - T_0)} = C_{\mathrm{A}0}\left[\frac{\Theta_{\mathrm{B}}T_f - T + (T_0 - \Theta_{\mathrm{B}}T_0)}{T_f - T_0}\right] \qquad \text{(P9-10.3)}$$

and $-r_{\mathrm{A}}$ as

$$-r_{\mathrm{A}} = \frac{kC_{\mathrm{A}0}^{\alpha+\beta}\;[(T_f - T)^{\alpha}(\Theta_{\mathrm{B}}T_f - T + (T_0 - \Theta_{\mathrm{B}}T_0))^{\beta}]}{[T_f - T_0]^{\alpha+\beta}} \qquad \text{(P9-10.4)}$$

(c) Show the unsteady energy balance can be written as

$$\frac{dT}{dt} = \left[\frac{-\Delta H^{\circ}_{\mathrm{Rx}}}{\Sigma \Theta_i C_{p_i}}\right]\frac{kC_{\mathrm{A}0}^{\alpha+\beta-1}}{[T_f - T_0]^{\alpha+\beta}}[(T_f - T)^{\alpha}(\Theta_{\mathrm{B}}T_f - T + (T_0 - \Theta_{\mathrm{B}}T_0))^{\beta}] \quad \text{(P9-10.5)}$$

(d) Assume first order in A and in B and that $\Theta_{\mathrm{B}} = 3$ then show

$$\frac{dT}{dt} = \left[\frac{-\Delta H^{\circ}_{\mathrm{Rx}}}{\Sigma \Theta_i C_{p_i}}\right]\frac{kC_{\mathrm{A}0}}{(T_f - T_0)^2}[(T_f - T)(3T_f - T - 2T_0)] \qquad \text{(P9-10.6)}$$

(e) Rearrange Equation (P9-10.6) in the form

$$\ln\left(\frac{\frac{dT}{dt}}{[(T_f - T)(3T_f - T - 2T_0)]}\right) = \ln\left(\frac{k_1 C_{A0}}{(T_f - T_0)}\right) + \frac{E}{RT_1} - \frac{E}{RT} \quad \text{(P9-10.7)}$$

(f) Plot the data to obtain the activation energy and the specific reaction rate k_1.

(g) Find the heat of reaction.

Additional information:

Chemical	Density (g/ml)	Heat capacity (J/g•°C)	MW	Heat capacity (J/mol•°C)
Acetic anhydride	1.0800	1.860	102	189.7
Water	1.0000	4.187	18	75.4
Glass cell (bomb)	0.1474	0.837		0.84 J/g/°C

Total volume	10 ml with
Water	3.638 g
Acetic anhydride	6.871 g

$(M_S C_{P_S} = 28.012 \text{ J/°C and } \phi = 1.004 \text{ and } m_S C_{P_S} = \phi M_S C_{P_S})$

P9-11$_B$ The elementary irreversible liquid-phase reaction

$$A + 2B \longrightarrow C$$

is to be carried out in a semibatch reactor in which B is fed to A. The volume of A in the reactor is 10 dm³, the initial concentration of A in the reactor is 5 mol/dm³, and the initial temperature in the reactor is 27°C. Species B is fed at a temperature of 52°C and a concentration of 4 M. It is desired to obtain at least 80% conversion of A in as short a time as possible, but at the same time the temperature of the reactor must not rise above 130°C. You should try to make approximately 120 mol of C in a 24-hour day allowing for 30 minutes to empty and fill the reactor between each batch. The coolant flow rate through the reactor is 2000 mol/min. There is a heat exchanger in the reactor.

(a) What volumetric feed rate (dm³/min) do you recommend?

(b) How would your answer or strategy change if the maximum coolant rate dropped to 200 mol/min? To 20 mol/min?

Additional information:

$\Delta H_{Rx}^{\circ} = -55,000 \text{ cal/mol A}$

$C_{P_A} = 35 \text{ cal/mol·K}, \quad C_{P_B} = 20 \text{ cal/mol·K}, \quad C_{P_C} = 75 \text{ cal/mol·K}$

$k = 0.0005 \dfrac{dm^6}{mol^2 \cdot min} \text{ at } 27°C \text{ with } E = 8000 \text{ cal/mol}$

$UA = 2500 \dfrac{cal}{min \cdot K} \text{ with } T_a = 17°C$

$C_P(\text{coolant}) = 18 \text{ cal/mol·K}$ [Old exam]

P9-12$_B$ Read Section R9.2 on the CD-ROM and then rework Example RE9-1 using
  (a) Only a proportional controller.
  (b) Only an integral controller.
  (c) A combined proportional and integral controller.

P9-13$_B$ Apply the different types of controllers to the reactions in Problem P9-11.

P9-14$_B$ **(a)** Rework Example R9-1 for the case of a 5°F decrease in the outlet temperature when the controlled input variable is the reactant feed rate.
  (b) Consider a 5°F drop in the ambient temperature, T_a, when the controlled variable is the inlet temperature, T_0.
  (c) Use each of the controllers (P with $k_C = 10$, I with $\tau_I = 1.0$ h, D with $\tau_D = 0.1$ h) to keep the reactor temperature at the unstable steady state (i.e., $T = 112.5$°F and $X = 0.3$).

T

t

Integral Controller

P9-15$_B$ Rework Problem P9-3 for the case when a heat exchanger with $UA = 10,000$ Btu/h·ft^2 and a control system are added and the mass flow rate is increased to 310 lb of solution per hour.
  (a) Plot temperature and mass of ammonium nitrate in the tank as a function of time when there is no control system on the reactor. Assume that all the ammonium nitrate reacts, and show that the mass balance is

$$\frac{dM_A}{dt} = \dot{m}_{A0} - kM_A$$

  There is 500 lb of A in the CSTR, and the reactor temperature, T, is 516°F at time $t = 0$.
  (b) Plot T and M_A as a function of time when a proportional controller is added to control T_a in order to keep the reactor temperature at 516°F. The controller gain, k_C, is -5 with T_{a0} set at 975 R.
  (c) Plot T and M_A versus time when a PI controller is added with $\tau_I = 1$.
  (d) Plot T and M_A versus time when two PI controllers are added to the reactor: one to control T and a second to control M by manipulating the feed rate $\dot{m}_{A0}$.

$$\dot{m}_{A0} = \dot{m}_{A00} + \frac{k_{C2}}{\tau_{I2}} I_M + k_{C2}(M - M_{sp})$$

  with $M_{sp} = 500$ lb, $k_{C2} = 25$ h^{-1}, $\tau_{I2} = 1$h.

Member

Hall of Fame

P9-16$_B$ The elementary liquid phase reaction

$$A \longrightarrow B$$

is carried out in a CSTR. Pure A is fed at a rate of 200 lb mol/h at 530 R and a concentration of 0.5 lb mol/ft^3. [M. Shacham, N. Brauner, and M. B. Cutlip, *Chem. Engr. Edu. 28* (1), 30 (Winter 1994).] The mass density of the solution is constant at 50 lb/ft^3.
  (a) Plot $G(T)$ and $R(T)$ as a function of temperature.
  (b) What are the steady-state concentrations and temperatures? [One answer $T = 628.645$ R, $C_A = 0.06843$ lb mol/ft^3.] Which ones are stable? What is the extinction temperature?
  (c) Apply the unsteady-state mole and energy balances to this system. Consider the upper steady state. Use the values you obtained in part (b) as your initial values to plot C_A and T versus time up to 6 hours and then to plot C_A versus T. What did you find? Do you want to change any of your answers to part (b)?

(d) Expand your results for part (c) by varying T_o and T_a. [*Hint:* Try $T_o = 590°R$]. Describe what you find.

(e) What are the parameters in part (d) for the other steady states? Plot T and C_A as a function of time using the steady-state values as the initial conditions at the lower steady state by value of $T_o = 550$ R and $T_o = 560$ R. Start at the lower steady state ($T = 547.1$, $C_A = 0.425$) and make a C_A/T phase-plane plot for the base case in part (a) $T_o = 530$. Now increase T_a to 550 and then to 560, and describe what you find. Vary T_o.

(f) Explore this problem. Write a paragraph describing your results and what you learned from this problem.

(g) Read Section 9.3 on the CD-ROM. Carry out a linearized stability analysis. What were your values of A, B, C, τ, J, L, and M? Find the roots for the upper, lower, and middle steady states you found in part (a).

(h) Normalize x and y by the steady-state values, $x1 = x/C_{As}$ and $y1 = y/T_s$, and plot $x1$ and $y1$ as a function of time and also $x1$ as a function of $y1$ Plot $x1$ and $y1$ as a function of time for each of the three steady states. [*Hint:* First try initial values of x and y of 0.02 and 2, respectively.]

Additional information:

$v = 400 \text{ ft}^3/hr$

$C_{A0} = 0.50 \text{ lb mol/ft}^3$

$V = 48 \text{ ft}^3$

$k \begin{cases} A = 1.416 \times 10^{12} \text{ h}^{-1} \\ E = 30{,}000 \text{ BTU/lb mol} \\ R = 1.987 \text{ BTU/lb mol °R} \\ U = 150 \text{ BTU/h-ft}^2\text{-°R} \end{cases}$

$A = 250 \text{ ft}^2 \text{ (Heat Exchanger)}$

$T_a = 530°R$

$T_o = 530°R$

$\Delta H_{Rx} = -30{,}000 \text{ Btu/lb mol}$

$C_p = 0.75 \text{ Btu/lb °R}$

$\rho = 50 \text{ lbm/ft}^3$

$$C_P = C_P\rho = \left(0.75\frac{BTU}{lb\,R}\right)\left(50\frac{lb}{ft^3}\right) = 37.5\frac{BTU}{ft^3 R}$$

P9-17$_B$ The reaction in Example 8-6 are to be carried out in a 10-dm³ batch reactor. Plot the temperature and the concentrations of A, B, and C as a function of time for the following cases:

(a) Adiabatic operation.

(b) Values of UA of 10,000, 40,000, and 100,000 J/min·K.

(c) Use $UA = 40{,}000$ J/min·K and different initial reactor temperatures.

P9-18$_B$ The reaction in Problem P8-34$_B$ is to be carried out in a semibatch reactor.

(a) How would you carry out this reaction (i.e., T_0, v_0, T_i)? The molar concentrations of pure A and pure B are 5 and 4 mol/dm³ respectively. Plot concentrations, temperatures, and the overall selectivity as a function of time for the conditions you chose.

(b) Vary the reaction orders for each reaction and describe what you find.

(c) Vary the heats of reaction and describe what you find.

P9-19$_B$ The following temperature–time data was taken on the ARSST. Determine the heat of reaction and the activation energy. Heat capacity of the solution $C_P = 18$ cal/mol/K.

P9-20$_D$ The formation of high-molecular-weight olefins, for example,

Time (min)	0	2.1	3.75	6.0	8.2	11.4	15.4	20.4	25.7	28.4	31.0
T (°C)	18.0	21.5	25.5	30.7	36.0	39.5	42.5	46.8	52.53	57.1	62.3
Time (min)	33.2	34.45	35.2	35.8	36.8	37.2	37.8	38.0	38.2	38.0	40.6
T (°C)	68.8	74.2	78.3	82.8	94.0	103.8	125.3	142.3	157.0	166.8	164.6

$$C_2(=) + C_2(=) \xrightarrow{\ k_2\ } C_4(=) \ \frac{+C_2(=)}{k_4} \to C_6(=) \xrightarrow{\ k_6\ }$$

is carried out in the CSTR shown in Figure P9-20(a) where (=) denotes the molecule is an olefin. The reaction is exothermic, and heat is removed from the reactor by heat exchange with a cooling water stream as shown in Figure P9-20(a).

Troubleshooting

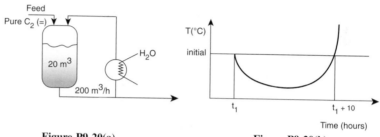

Figure P9-20(a)   Figure P9-20(b)

Initially, there is a temperature controller that regulates the reactor temperature. At $t = t_1$, the controller is set into manual, and a step *increase* in the water flow rate through the heat exchanger was made. The temperature response is shown in Figure P9-20(b). What are the parameters to match this trend?

Explain the observed temperature–time trajectory. To make your analysis simpler, you may assume that the only reaction taking place is

$$C_2(=) + C_2(=) \xrightarrow{\ k_2\ } C_4(=)$$

with

$$\boxed{-r_A = Ae^{E/RT} C_2^2 (=)}$$

Also, you can assume that the exchanger is inside the reactor (to avoid recycle stream calculations). **See P9-20 CD-ROM Complete Data Set.**

- **Additional Homework Problems**

    CDP9-A$_B$ The production of propylene glycol discussed in Examples 8-4, 9-4, 9-5, 9-6, and 9-7 is carried out in a semibatch reactor.

    CDP9-B$_C$ Reconsider Problem P9-14 when a PI controller is added to the coolant stream.

    CDP9-C$_B$ Calculate time to achieve 90% of equilibrium in a batch reactor. [3rd Ed. P9-8$_B$]

    CDP9-D$_B$ Startup of a CSTR. [3rd Ed. P9-10$_B$]

SUPPLEMENTARY READING

1. A number of solved problems for batch and semibatch reactors can be found in

   WALAS, S. M., *Chemical Reaction Engineering Handbook*. Amsterdam: Gordon and Breach, 1995, pp. 386–392, 402, 460–462, and 469.

2. Basic control textbooks

   SEBORG, D. E., T. F. EDGAR, and D. A. MELLICHAMP, *Process Dynamics and Control*, 2nd ed. New York: Wiley, 2004.

   OGUNNAIKE, B. A., and W. H. RAY, *Process Dynamics, Modeling and Control*. Oxford: Oxford University Press, 1994.

3. A nice historical perspective of process control is given in

   EDGAR, T. F., "From the Classical to the Postmodern Era," *Chem. Eng. Educ.*, *31*, 12 (1997).

Links

Links

1. The **SACHE web site** has a great discussion on reactor safety with examples (*www.sache.org*). You will need a user name and password; both can be obtained from your department chair. Hit 2003 Tab. Go to K Problems.

2. The **reactor lab** developed by Professor Herz and discussed in Chapters 4 and 5 could also be used here: *www.reactorlab.net* and also on the CD-ROM.

Catalysis and Catalytic Reactors

10

It isn't that they can't see the solution. It is that they can't see the problem.

G. K. Chesterton

Overview. The objectives of this chapter are to develop an understanding of catalysts, reaction mechanisms, and catalytic reactor design. Specifically, after reading this chapter one should be able to (1) define a catalyst and describe its properties, (2) describe the steps in a catalytic reaction and in chemical vapor deposition (CVD) and apply the concept of a rate-limiting step to derive a rate law, (3) develop a rate law and determine the rate-law parameters from a set of gas–solid reaction rate data, (4) describe the different types of catalyst deactivation, determine an equation for catalytic activity from concentration–time data, define temperature–time trajectories to maintain a constant reaction rate, and (5) calculate the conversion or catalyst weight for packed (fixed) beds, moving beds, well-mixed (CSTR) and straight-through (STTR) fluid-bed reactors for both decaying and nondecaying catalysts. The various sections of this chapter roughly correspond to each of these objectives.

10.1 Catalysts

Catalysts have been used by humankind for over 2000 years.[1] The first observed uses of catalysts were in the making of wine, cheese, and bread. It was found that it was always necessary to add small amounts of the previous batch to make the current batch. However, it wasn't until 1835 that Berzelius

---

[1] S. T. Oyama and G. A. Somorjai, *J. Chem. Educ.*, *65*, 765 (1986).

began to tie together observations of earlier chemists by suggesting that small amounts of a foreign source could greatly affect the course of chemical reactions. This mysterious force attributed to the substance was called catalytic. In 1894, Ostwald expanded Berzelius' explanation by stating that catalysts were substances that accelerate the rate of chemical reactions without being consumed. In over 150 years since Berzelius' work, catalysts have come to play a major economic role in the world market. In the United States alone, sales of process catalysts in 2007 will be over $3.5 billion, the major uses being in petroleum refining and in chemical production.

10.1.1 Definitions

A *catalyst* is a substance that affects the rate of a reaction but emerges from the process unchanged. A catalyst usually changes a reaction rate by promoting a different molecular path ("mechanism") for the reaction. For example, gaseous hydrogen and oxygen are virtually inert at room temperature, but react rapidly when exposed to platinum. The reaction coordinate shown in Figure 10-1 is a measure of the progress along the reaction path as H_2 and O_2 approach each other and pass over the activation energy barrier to form H_2O. A more exact comparison of the pathways, similar to the margin figure, is given in the Summary Notes for Chapter 10. *Catalysis* is the occurrence, study, and use of catalysts and catalytic processes. Commercial chemical catalysts are immensely important. Approximately one third of the material gross national product of the United States involves a catalytic process somewhere between raw material and finished product.[2] The development and use of catalysts is a major part of the constant search for new ways of increasing product yield and selectivity from chemical reactions. Because a catalyst makes it possible to obtain an end product by a different pathway with a lower energy barrier, it can affect both the yield and the selectivity.

Summary Notes

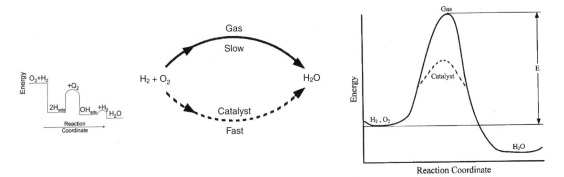

Figure 10-1   Different reaction paths.

Normally when we talk about a catalyst, we mean one that speeds up a reaction, although strictly speaking, a catalyst can either accelerate or slow the

[2] V. Haensel and R. L. Burwell, Jr., *Sci. Am.*, 225(10), 46.

Catalysts can accel-
erate the reaction
rate but can-
not change the
equilibrium.

formation of a particular product species. *A catalyst changes only the rate of a reaction; it does not affect the equilibrium.*

Homogeneous catalysis concerns processes in which a catalyst is in solution with at least one of the reactants. An example of homogeneous catalysis is the industrial Oxo process for manufacturing normal isobutylaldehyde. It has propylene, carbon monoxide, and hydrogen as the reactants and a liquid-phase cobalt complex as the catalyst.

$$CH_3-CH=CH_2 + CO + H_2 \xrightarrow{\ Co\ } \begin{cases} CH_3-CH \begin{smallmatrix} \diagup CHO \\ \diagdown CH_3 \end{smallmatrix} \\ CH_3-CH_2-CH_2-CHO \end{cases}$$

Reactions carried out in supercritical fluids have been found to accelerate the reaction rate greatly.[3] By manipulating the properties of the solvent in which the reaction is taking place, interphase mass transfer limitations can be eliminated. Application of transition states theory discussed in the *Professional Reference Shelf* on the CD-ROM of Chapter 3 has proven useful in the analysis of these reactions.

A *heterogeneous catalytic process* involves more than one phase; usually the catalyst is a solid and the reactants and products are in liquid or gaseous form. Much of the benzene produced in this country today is manufactured from the dehydrogenation of cyclohexane (obtained from the distillation of crude petroleum) using platinum-on-alumina as the catalyst:

$$\bighexagon \xrightarrow[\ Al_2O_3 \times H_2O\ ]{Pt\ on} \bigcirc\!\!\!\!\!\!\bigcirc + 3H_2$$

Cyclohexane   Benzene Hydrogen

Examples of
heterogeneous
catalytic reactions

Sometimes the reacting mixture is in both the liquid and gaseous forms, as in the hydrodesulfurization of heavy petroleum fractions. Of these two types of catalysis, heterogeneous catalysis is the more common type. The simple and complete separation of the fluid product mixture from the solid catalyst makes heterogeneous catalysis economically attractive, especially because many catalysts are quite valuable and their reuse is demanded. Only heterogeneous catalysts will be considered in this chapter.

A heterogeneous catalytic reaction occurs at or very near the fluid–solid interface. The principles that govern heterogeneous catalytic reactions can be applied to both catalytic and noncatalytic fluid–solid reactions. The two other types of heterogeneous reactions involve gas–liquid and gas–liquid–solid systems. Reactions between gases and liquids are usually mass-transfer limited.

---

[3] P. E. Savage, S. Gopalan, T. I. Mizan, C. J. Martino, and E. E. Brock, *AIChE J.*, *41* (7) 1723 (1995). B. Subramanian and M. A. McHugh, *IEC Process Design and Development*, *25*, 1 (1986).

10.1.2 Catalyst Properties

Because a catalytic reaction occurs at the fluid–solid interface, a large interfacial area is almost always essential in attaining a significant reaction rate. In many catalysts, this area is provided by an inner porous structure (i.e., the solid contains many fine pores, and the surface of these pores supplies the area needed for the high rate of reaction). The area possessed by some porous materials is surprisingly large. A typical silica-alumina cracking catalyst has a pore volume of 0.6 cm^3/g and an average pore radius of 4 nm. The corresponding surface area is 300 m^2/g.

Ten grams of this catalyst possess more surface area than a U.S. football field

A catalyst that has a large area resulting from pores is called a *porous catalyst*. Examples of these include the Raney nickel used in the hydrogenation of vegetable and animal oils, the platinum-on-alumina used in the reforming of petroleum naphthas to obtain higher octane ratings, and the promoted iron used in ammonia synthesis. Sometimes pores are so small that they will admit small molecules but prevent large ones from entering. Materials with this type of pore are called *molecular sieves*, and they may be derived from natural substances such as certain clays and zeolites, or be totally synthetic, such as some crystalline aluminosilicates (see Figure 10-2). These sieves can form the basis

Catalyst types:
· Porous
· Molecular sieves
· Monolithic
· Supported
· Unsupported

Typical zeolite catalyst

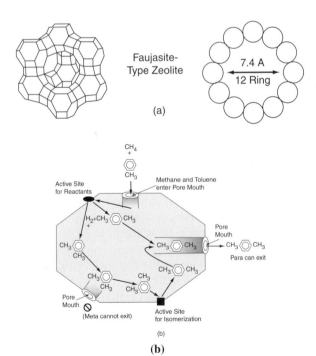

Faujasite-Type Zeolite

7.4 A
12 Ring

(a)

(b)

(b)

Figure 10-2   (a) Framework structures and (b) pore cross sections of two types of zeolites. (a) Faujasite-type zeolite has a three-dimensional channel system with pores at least 7.4 Å in diameter. A pore is formed by 12 oxygen atoms in a ring. (b) Schematic of reaction CH$_4$ and C$_6$H$_5$CH$_3$. (Note the size of the pore mouth and the interior of the zeolite are not to scale.) [(a) from N. Y. Chen and T. F. Degnan, *Chem. Eng. Prog.*, **84**(2), 33 (1988). Reproduced by permission of the American Institute of Chemical Engineers. Copyright © 1988 AIChE. All rights reserved.]

for quite selective catalysts; the pores can control the residence time of various molecules near the catalytically active surface to a degree that essentially allows *only* the desired molecules to react. One example of the high selectivity of zeolite catalysts is the formation of xylene from toluene and methane shown in Figure 10-2(b).[4] Here benzene and toluene enter through the pore of the zeolite and react on the interior surface to form a mixture of ortho, meta, and para xylenes. However, the size of the pore mouth is such that only para-xylene can exit through the pore mouth as meta and ortho xylene with their methyl group on the side cannot fit through the pore mouth. There are interior sites that can isomerize ortho and meta to para-xylene. Hence we have a very high selectivity to form para-xylene. Another example of zeolite specificity is controlled placement of the reacting molecules. After the molecules are inside the zeolite, the configuration of the reacting molecules may be able to be controlled by placement of the catalyst atoms at specific sites in the zeolite. This placement would facilitate cyclization reactions, such as orienting ethane molecules in a ring on the surface of the catalyst so that they form benzene:

High selectivity to para xylene

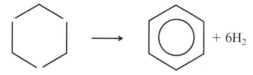

$+ 6H_2$

Monolithic catalysts can be either porous or nonporous.

    Not all catalysts need the extended surface provided by a porous structure, however. Some are sufficiently active so that the effort required to create a porous catalyst would be wasted. For such situations one type of catalyst is the monolithic catalyst. Monolithic catalysts are normally encountered in processes where pressure drop and heat removal are major considerations. Typical examples include the platinum gauze reactor used in the ammonia oxidation portion of nitric acid manufacture and catalytic converters used to oxidize pollutants in automobile exhaust. They can be porous (honeycomb) or nonporous (wire gauze). A photograph of an automotive catalytic converter is shown in PRS Figure R11.1-2. Platinum is a primary catalytic material in the monolith.

    In some cases a catalyst consists of minute particles of an active material dispersed over a less active substance called a *support*. The active material is frequently a pure metal or metal alloy. Such catalysts are called *supported catalysts*, as distinguished from *unsupported catalysts*. Catalysts can also have small amounts of active ingredients added called *promoters*, which increase their activity. Examples of supported catalysts are the packed-bed catalytic converter automobile, the platinum-on-alumina catalyst used in petroleum reforming, and the vanadium pentoxide on silica used to oxidize sulfur dioxide in manufacturing sulfuric acid. On the other hand, the platinum gauze for ammonia oxidation, the promoted iron for ammonia synthesis, and the silica–alumina dehydrogenation catalyst used in butadiene manufacture typify unsupported catalysts.

---

[4] R. J. Masel, *Chemical Kinetics and Catalysis* (New York: Wiley Interscience, 2001), p. 741.

Most catalysts do not maintain their activities at the same levels for indefinite periods. They are subject to *deactivation*, which refers to the decline in a catalyst's activity as time progresses. Catalyst deactivation may be caused by (1) *aging* phenomenon, such as a gradual change in surface crystal structure; (2) by *poisoning*, which is the irreversible deposition of a substance on the active site; or (3) by *fouling* or *coking*, which is the deposit of carbonaceous or other material on the entire surface. Deactivation may occur very fast, as in the catalytic cracking of petroleum naphthas, where *coking* on the catalyst requires that the catalyst be removed after only a couple of minutes in the reaction zone. In other processes poisoning might be very slow, as in automotive exhaust catalysts, which gradually accumulate minute amounts of lead even if unleaded gasoline is used because of residual lead in the gas station storage tanks.

For the moment, let us focus our attention on gas-phase reactions catalyzed by solid surfaces. For a catalytic reaction to occur, at least one and frequently all of the reactants must become attached to the surface. This attachment is known as *adsorption* and takes place by two different processes: physical adsorption and chemisorption. *Physical adsorption* is similar to condensation. The process is exothermic, and the heat of adsorption is relatively small, being on the order of 1 to 15 kcal/mol. The forces of attraction between the gas molecules and the solid surface are weak. These van der Waals forces consist of interaction between permanent dipoles, between a permanent dipole and an induced dipole, and/or between neutral atoms and molecules. The amount of gas physically adsorbed decreases rapidly with increasing temperature, and above its critical temperature only very small amounts of a substance are physically adsorbed.

The type of adsorption that affects the rate of a chemical reaction is *chemisorption*. Here, the adsorbed atoms or molecules are held to the surface by valence forces of the same type as those that occur between bonded atoms in molecules. As a result the electronic structure of the chemisorbed molecule is perturbed significantly, causing it to be extremely reactive. Interaction with the catalyst causes bonds of the adsorbed reactant to be stretched, making them easier to break.

Figure 10-3 shows the bonding from the adsorption of ethylene on a platinum surface to form chemisorbed ethylidyne. Like physical adsorption, chemisorption is an exothermic process, but the heats of adsorption are generally of the same magnitude as the heat of a chemical reaction (i.e., 40 to 400 kJ/mol). If a catalytic reaction involves chemisorption, it must be carried out within the temperature range where chemisorption of the reactants is appreciable.

In a landmark contribution to catalytic theory, Taylor[5] suggested that a reaction is not catalyzed over the entire solid surface but only at certain *active sites* or centers. He visualized these sites as unsaturated atoms in the solids that resulted from surface irregularities, dislocations, edges of crystals, and cracks along grain boundaries. Other investigators have taken exception to this

---

[5] H. S. Taylor, *Proc. R. Soc. London*, *A108*, 105 (1928).

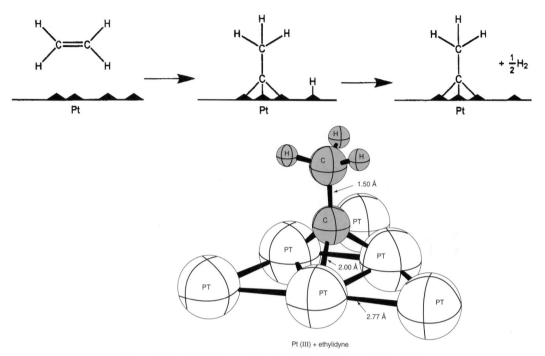

Figure 10-3 Ethylidyne as chemisorbed on platinum. (Adapted from G. A. Somorjai, *Introduction to Surface Chemistry and Catalysis*, Wiley, New York, 1994.)

definition, pointing out that other properties of the solid surface are also important. The active sites can also be thought of as places where highly reactive intermediates (i.e., chemisorbed species) are stabilized long enough to react. This stabilization of a reactive intermediate is key in the design of any catalyst. However, for our purposes we will define an *active site* as *a point on the catalyst surface that can form strong chemical bonds with an adsorbed atom or molecule.*

One parameter used to quantify the activity of a catalyst is the *turnover frequency, (TOF), f.* It is the number of molecules reacting per active site per second at the conditions of the experiment. When a metal catalyst such as platinum is deposited on a support, the metal atoms are considered active sites. The *dispersion, D,* of the catalyst is the fraction of the metal atoms deposited that are on the surface.

Example 10–1   Turnover Frequency in Fischer–Tropsch Synthesis

$$CO + 3H_2 \longrightarrow CH_4 + H_2O$$

The Fischer–Tropsch synthesis was studied using a commercial 0.5 wt % Ru on γ-Al_2O_3.[6] The catalyst dispersion percentage of atoms exposed, determined from

[6] R. S. Dixit and L. L. Tavlarides, *Ind. Eng. Chem. Process Des. Dev.*, 22, 1 (1983).

hydrogen chemisorption, was found to be 49%. At a pressure of 988 kPa and a temperature of 475 K, a turnover frequency, f_{CH_4}, of 0.044 s^{-1} was reported for methane. What is the rate of formation of methane, r'_M, in mol/s·g of catalyst (metal plus support)?

Solution

$$r'_M = f_{CH_4} D \left(\frac{1}{MW_{Ru}} \right) \frac{\% \, Ru}{100}$$

$$= \frac{0.044 \text{ molecules}}{(\text{surface atom Ru}) \cdot s} \times \frac{1 \text{ mol CH}_4}{6.02 \times 10^{23} \text{ molecules}}$$

$$\times \frac{0.49 \text{ surface atoms}}{\text{total atoms Ru}} \times \frac{6.02 \times 10^{23} \text{ atoms Ru}}{\text{g atom (mol) Ru}}$$

$$\times \frac{\text{g atoms Ru}}{101.1 \text{ g Ru}} \times \frac{0.005 \text{ g Ru}}{\text{g total}}$$

$$= 1.07 \times 10^{-6} \text{ mol/s} \cdot \text{g catalyst} \qquad \text{(E10-1.1)}$$

That is, 1.07×10^{-6} moles (6.1×10^{17} molecules) of methane are jumping off one gram of catalyst every second.

Figure 10-4 shows the range of turnover frequencies (molecules/site·s) as a function of temperature and type of reaction. One notes that the turnover frequency in Example 10-1 is in the same range as the frequencies shown in the box for hydrogenation catalysts.

10.1.3 Classification of Catalysts

While platinum can be used for some of the reactions shown in Figure 10-4, we shall also discuss several other classes of reactions and the catalysts.[7]

Alkylation and Dealkylation Reactions. *Alkylation* is the addition of an alkyl group to an organic compound. This type of reaction is commonly carried out in the presence of the Friedel–Crafts catalysts, AlCl$_3$ along with a trace of HCl. One such reaction is

$$C_4H_8 + i\text{-}C_4H_{10} \underset{\longleftarrow}{\overset{AlCl_3}{\longrightarrow}} i\text{-}C_8H_{18}$$

A similar alkylation is the formation of ethyl benzene from benzene and ethylene:

$$C_6H_6 + C_2H_4 \longrightarrow C_6H_5C_2H_5$$

---

[7] J. H. Sinfelt, *Ind. Eng. Chem.*, 62(2), 23 (1970); 62(10), 66 (1970). Also, W. B. Innes, in P. H. Emmertt, Eds., *Catalysis*, Vol. 2 (New York: Reinhold, 1955), p. 1, and R. Masel, *Kinetics* (New York: Wiley, 2003).

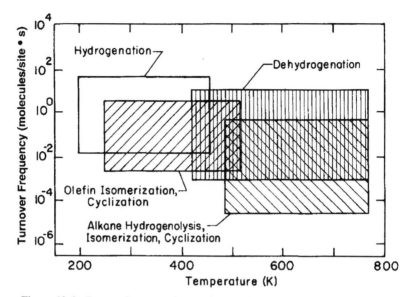

Figure 10-4   Range of turnover frequencies as a function for different reactions and temperatures. (Adapted from G. A. Somorjai, *Introduction to Surface Chemistry and Catalysis*, Wiley, New York, 1994.)

The cracking of petrochemical products is probably the most common dealkylation reaction. Silica-alumina, silica-magnesia, and a clay (montmoril-lonite) are common dealkylation catalysts.

Isomerization Reactions.   In petrochemical production, the conversion of normal hydrocarbon chains to branched chains is important, since the latter has a higher gasoline octane number. When *n*-pentane is isomerized to *i*-pentane, the octane number increases from 62 to 90!! Acid-promoted Al_2O_3 is a catalyst used in such isomerization reactions. Although this and other acid catalysts are used in isomerization reactions, it has been found that the conversion of nor-mal paraffins to isoparaffins is easiest when both acid sites and hydrogenation sites are present, such as in the catalyst Pt on Al_2O_3.

Hydrogenation and Dehydrogenation Reactions.   The bonding strength be-tween hydrogen and metal surfaces increases with an increase in vacant *d*-orbit-als. Maximum catalytic activity will not be realized if the bonding is too strong and the products are not readily desorbed from the surface. Consequently, this maximum in catalytic activity occurs when there is approximately one vacant *d*-orbital per atom. The most active metals for reactions involving hydrogen are generally Co, Ni, Rh, Ru, Os, Pd, Ir, and Pt. On the other hand, V, Cr, Nb, Mo, Ta, and W, each of which has a large number of vacant *d*-orbitals, are relatively inactive as a result of the strong adsorption for the reactants or the products or both. However, the oxides of Mo (MoO_2) and Cr (Cr_2O_3) are quite active for most reactions involving hydrogen. Dehydrogenation reactions are favored at high temperatures (at least 200°C), and hydrogenation reactions are favored at lower temperatures. Industrial butadiene, which has been used to produce synthetic rubber, can be obtained by the dehydrogenation of butenes:

$$CH_3CH{=}CHCH_3 \xrightarrow{\text{catalyst}} CH_2{=}CHCH{=}CH_2 + H_2$$

(possible catalysts: calcium nickel phosphate, Cr_2O_3, etc.)

The same catalysts could also be used in the dehydrogenation of ethyl benzene to form styrene:

$$\phi CH_2CH_3 \xrightarrow{\text{catalyst}} H_2 + \phi CH{=}CH_2$$

An example of cyclization, which may be considered to be a special type of dehydrogenation, is the formation of cyclohexane from *n*-hexane.

Oxidation Reactions.   The transition group elements (group VIII) and subgroup I are used extensively in oxidation reactions. Ag, Cu, Pt, Fe, Ni, and their oxides are generally good oxidation catalysts. In addition, V_2O_5 and MnO_2 are frequently used for oxidation reactions. A few of the principal types of catalytic oxidation reactions are:

1. Oxygen addition:

$$2C_2H_4 + O_2 \xrightarrow{\text{Ag}} 2C_2H_4O$$
$$2SO_2 + O_2 \xrightarrow{V_2O_5} 2SO_3$$
$$2CO + O_2 \xrightarrow{\text{Cu}} 2CO_2$$

2. Oxygenolysis of carbon–hydrogen bonds:

$$2C_2H_5OH + O_2 \xrightarrow{\text{Cu}} 2CH_3CHO + 2H_2O$$
$$2CH_3OH + O_2 \xrightarrow{\text{Ag}} 2HCHO + 2H_2O$$

3. Oxygenation of nitrogen–hydrogen bonds:

$$5O_2 + 4NH_3 \xrightarrow{\text{Pt}} 4NO + 6H_2O$$

4. Complete combustion:

$$2C_2H_6 + 7O_2 \xrightarrow{\text{Ni}} 4CO_2 + 6H_2O$$

Platinum and nickel can be used for both oxidation reactions and hydrogenation reactions.

Hydration and Dehydration Reactions.   Hydration and dehydration catalysts have a strong affinity for water. One such catalyst is Al_2O_3, which is used in the dehydration of alcohols to form olefins. In addition to alumina, silica-alumina gels, clays, phosphoric acid, and phosphoric acid salts on inert carriers have also been used for hydration–dehydration reactions. An example of an industrial catalytic hydration reaction is the synthesis of ethanol from ethylene:

$$CH_2{=}CH_2 + H_2O \longrightarrow CH_3CH_2OH$$

Halogenation and Dehalogenation Reactions. Usually, reactions of this type take place readily without utilizing catalysts. However, when selectivity of the desired product is low or it is necessary to run the reaction at a lower temperature, the use of a catalyst is desirable. Supported copper and silver halides can be used for the halogenation of hydrocarbons. Hydrochlorination reactions can be carried out with mercury copper or zinc halides.

Summary. Table 10-1 gives a summary of the representative reactions and catalysts discussed previously.

TABLE 10-1.   TYPES OF REACTIONS AND REPRESENTATIVE CATALYSTS

Reaction	*Catalysts*
1. Halogenation–dehalogenation	$CuCl_2$, AgCl, Pd
2. Hydration–dehydration	Al_2O_3, MgO
3. Alkylation–dealkylation	$AlCl_3$, Pd, Zeolites
4. Hydrogenation–dehydrogenation	Co, Pt, Cr_2O_3, Ni
5. Oxidation	Cu, Ag, Ni, V_2O_5
6. Isomerization	$AlCl_3$, Pt/Al_2O_3, Zeolites

If, for example, we were to form styrene from an equimolar mixture of ethylene and benzene, we could carry out an alkylation reaction to form ethyl benzene, which is then dehydrogenated to form styrene. We need both an alkylation catalyst and a dehydrogenation catalyst:

$$C_2H_4 + C_6H_6 \xrightarrow[\text{trace HCl}]{AlCl_3} C_6H_5C_2H_5 \xrightarrow{Ni} C_6H_5CH{=}CH_2 + H_2$$

10.2 Steps in a Catalytic Reaction

A photograph of different types and sizes of catalyst is shown in Figure 10-5(a). A schematic diagram of a tubular reactor packed with catalytic pellets is shown in Figure 10-5(b). The overall process by which heterogeneous catalytic reactions proceed can be broken down into the sequence of individual steps shown in Table 10-2 and pictured in Figure 10-6 for an isomerization reaction.

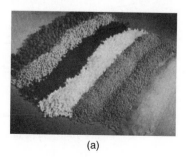

(a)

Figure 10-5   (a) Different shapes and sizes of catalyst. (Courtesy of the Engelhard Corporation.)

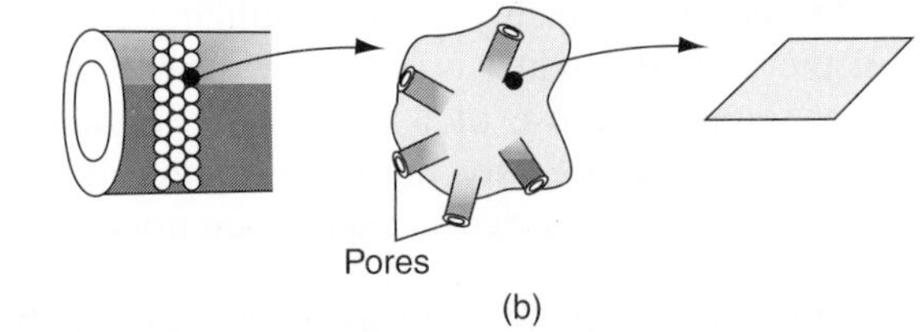

Packed catalyst bed   Catalyst pellet   Catalyst pellet surface

Pores

(b)

Figure 10-5   (b) Catalytic packed-bed reactor—schematic.

Each step in Table 10-2 is shown schematically in Figure 10-6.

TABLE 10-2.   STEPS IN A CATALYTIC REACTION

1. Mass transfer (diffusion) of the reactant(s) (e.g., species A) from the bulk fluid to the external surface of the catalyst pellet
2. Diffusion of the reactant from the pore mouth through the catalyst pores to the immediate vicinity of the internal catalytic surface
3. Adsorption of reactant A onto the catalyst surface
4. Reaction on the surface of the catalyst (e.g., A $\longrightarrow$ B)
5. Desorption of the products (e.g., B) from the surface
6. Diffusion of the products from the interior of the pellet to the pore mouth at the external surface
7. Mass transfer of the products from the external pellet surface to the bulk fluid

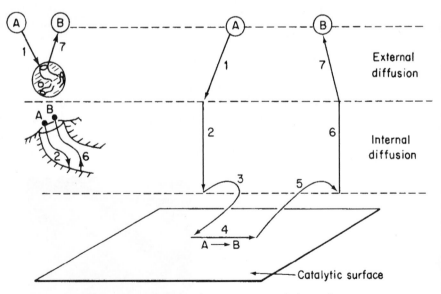

Figure 10-6   Steps in a heterogeneous catalytic reaction.

A reaction takes place *on* the surface, but the species involved in the reaction must get *to* and *from* the surface

The overall rate of reaction is equal to the rate of the slowest step in the mechanism. When the diffusion steps (1, 2, 6, and 7 in Table 10-2) are very fast compared with the reaction steps (3, 4, and 5), the concentrations in the immediate vicinity of the active sites are indistinguishable from those in the bulk fluid. In this situation, the transport or diffusion steps do not affect the overall rate of the reaction. In other situations, if the reaction steps are very fast compared with the diffusion steps, mass transport does affect the reaction rate. In systems where diffusion from the bulk gas or liquid to the catalyst surface or to the mouths of catalyst pores affects the rate, changing the flow conditions past the catalyst should change the overall reaction rate. In porous catalysts, on the other hand, diffusion within the catalyst pores may limit the rate of reaction. Under these circumstances, the overall rate will be unaffected by external flow conditions even though diffusion affects the overall reaction rate.

There are many variations of the situation described in Table 10-2. Sometimes, of course, two reactants are necessary for a reaction to occur, and both of these may undergo the steps listed earlier. Other reactions between two substances may have only one of them adsorbed.

In this chapter we focus on:
3. Adsorption
4. Surface reaction
5. Desorption

With this introduction, we are ready to treat individually the steps involved in catalytic reactions. In this chapter only the steps of adsorption, surface reaction, and desorption will be considered [i.e., it is assumed that the diffusion steps (1, 2, 6, and 7) are very fast, such that the overall reaction rate is not affected by mass transfer in any fashion]. Further treatment of the effects involving diffusion limitations is provided in Chapters 11 and 12.

Where Are We Heading?   As we saw in Chapter 5, one of the tasks of a chemical reaction engineer is to analyze rate data and to develop a rate law that can be used in reactor design. Rate laws in heterogeneous catalysis seldom follow power law models and hence are inherently more difficult to formulate from the data. To develop an in-depth understanding and insight as to how the rate laws are formed from heterogeneous catalytic data, we are going to proceed in somewhat of a reverse manner than what is normally done in industry when one is asked to develop a rate law. That is, we will postulate catalytic mechanisms and *then* derive rate laws for the various mechanisms. The mechanism will typically have an adsorption step, a surface reaction step, and a desorption step, one of which is usually rate-limiting. Suggesting mechanisms and rate-limiting steps is not the first thing we normally do when presented with data. However, by deriving equations for different mechanisms, we will observe the various forms of the rate law one can have in heterogeneous catalysis. Knowing the different forms that catalytic rate equations can take, it will be easier to view the trends in the data and deduce the appropriate rate law. This deduction is usually what is done first in industry before a mechanism is proposed. Knowing the form of the rate law, one can then numerically evaluate the rate law parameters and postulate a reaction mechanism and rate-limiting step that is consistent with the rate data. Finally, we use the rate law to design catalytic reactors. This procedure is shown in Figure 10-7. The dashed lines represent feedback to obtain new data in specific regions (e.g., concentrations,

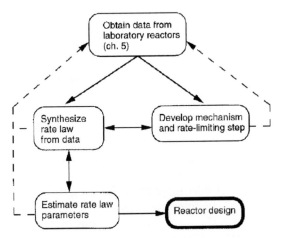

An algorithm

Figure 10-7 Collecting information for catalytic reactor design.

temperature) to evaluate the rate law parameters more precisely or to differentiate between reaction mechanisms.

We will discuss each of the steps shown in Figure 10-6 and Table 10-2. As mentioned earlier, this chapter focuses on Steps 3, 4, and 5 (the adsorption, surface reaction and desorption steps) by assuming Steps 1, 2, 6, and 7 are very rapid. Consequently to understand when this assumption is valid, we shall give a quick overview of Steps 1, 2, 6, and 7. Steps 1 and 2 involve diffusion of the reactants to and within the catalyst pellet. While these steps are covered in detail in Chapters 11 and 12, it is worthwhile to give a brief description of these two mass transfer steps to better understand the entire sequence of steps.

10.2.1 Step 1 Overview: Diffusion from the Bulk to the External Transport

For the moment let's assume the transport of A from the bulk fluid to the external surface of the catalyst is the slowest step in the sequence. We lump all the resistance to transfer from the bulk fluid to the surface in the boundary layer surrounding the pellet. In this step the reactant A at a bulk concentration C_{Ab} must travel through the boundary layer of thickness δ to the external surface of the pellet where the concentration is C_{As} as shown in Figure 10-8. The rate of transfer (and hence rate of reaction, $-r'_A$) for this slowest step is

$$\text{Rate} = k_C (C_{Ab} - C_{As})$$

where the mass transfer coefficient, k_C, is a function of the hydrodynamic conditions, namely fluid velocity, U, and the particle diameter, D_p.

As we will see (Chapter 11) the mass transfer coefficient is inversely proportional to the boundary layer thickness, δ,

$$k_C = \frac{D_{AB}}{\delta}$$

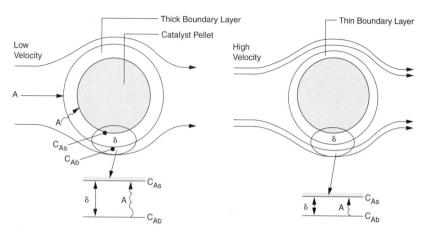

Figure 10-8   Diffusion through the external boundary layer. [Also see Figure E11-1.1.]

and directly proportional to the coefficient diffusion (i.e., the diffusivity D_{AB}). At low velocities of fluid flow over the pellet, the boundary layer across which A and B must diffuse is thick, and it takes a long time for A to travel to the surface, resulting in a small mass transfer coefficient k_C. As a result, mass transfer across the boundary layer is slow and limits the rate of the overall reaction. As the velocity across the pellet is increased, the boundary layer becomes smaller and the mass transfer rate is increased. At very high velocities the boundary layer is so small it no longer offers any resistance to the diffusion across the boundary layer. As a result, external mass transfer no longer limits the rate of reaction. This external resistance also decreases as the particle size is decreased. As the fluid velocity increases and/or the particle diameter decreases, the mass transfer coefficient increases until a plateau is reached, as shown in Figure 10-9. On this plateau, $C_{Ab} \approx C_{As}$, and one of the other steps in the sequence is the slowest step and limits the overall rate. Further details on external mass transfer are discussed in Chapter 11.

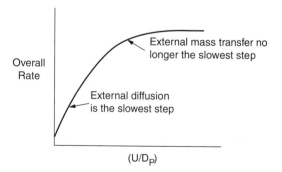

Figure 10-9   Effect of overall rate on particle size and fluid velocity.

10.2.2 Step 2 Overview: Internal Diffusion

Now consider that we are operating at a fluid velocity where external diffusion is no longer the rate-limiting step and that internal diffusion is the slowest step. In Step 2 the reactant A diffuses from the external surface at a concentration C_{As} into the pellet interior where the concentration is C_A. As A diffuses into the interior of the pellet, it reacts with catalyst deposited on the sides of the pore walls.

For large pellets, it takes a long time for the reactant A to diffuse into the interior compared to the time it takes for the reaction to occur on the interior pore surface. Under these circumstances, the reactant is only consumed near the exterior surface of the pellet and the catalyst near the center of the pellet is wasted catalyst. On the other hand, for very small pellets it takes very little time to diffuse into and out of the pellet interior and, as a result, internal diffusion no longer limits the rate of reaction. The rate of reaction can be expressed as

$$\text{Rate} = k_r\, C_{As}$$

where C_{As} is the concentration at the external surface and k_r is an overall rate constant and is a function of particle size. The overall rate constant, k_r, increases as the pellet diameter decreases. In Chapter 12 we show that Figure 12-5 can be combined with Equation (12-34) to arrive at the plot of k_r as a function of D_P shown in Figure 10-10(b).

We see that at small particle sizes internal diffusion is no longer the slow step and that the surface reaction sequence of adsorption, surface reaction, and desorption (Steps 3, 4, and 5) limit the overall rate of reaction. Consider now one more point about internal diffusion and surface reaction. These steps (2 through 6) are **not at all** affected by flow conditions external to the pellet.

In the material that follows, we are going to choose our pellet size and external fluid velocity such that neither external diffusion nor internal diffusion

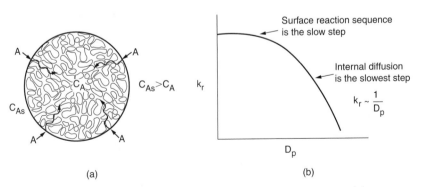

(a)   (b)

Figure 10-10 Effect of particle size on overall reaction rate constant. (a) Branching of a single pore with deposited metal; (b) Decrease in rate constant with increasing particle diameter..

is limiting. Instead, we assume that either Step 3 adsorption, Step 4 surface reaction, or Step 5 desorption, or a combination of these steps, limits the overall rate of reaction.

10.2.3 Adsorption Isotherms

Because chemisorption is usually a necessary part of a catalytic process, we shall discuss it before treating catalytic reaction rates. The letter S will represent an active site; alone it will denote a vacant site, with no atom, molecule, or complex adsorbed on it. The combination of S with another letter (e.g., A·S) will mean that one unit of species A will be adsorbed on the site S. Species A can be an atom, molecule, or some other atomic combination, depending on the circumstances. Consequently, the adsorption of A on a site S is represented by

$$A + S \rightleftharpoons A \cdot S$$

The total molar concentration of active sites per unit mass of catalyst is equal to the number of active sites per unit mass divided by Avogadro's number and will be labeled C_t (mol/g·cat). The molar concentration of vacant sites, C_v, is the number of vacant sites per unit mass of catalyst divided by Avogadro's number. In the absence of catalyst deactivation, we assume that the total concentration of active sites remains constant. Some further definitions include

P_i partial pressure of species i in the gas phase, atm or kPa
$C_{i \cdot S}$ surface concentration of sites occupied by species i, mol/g cat

A conceptual model depicting species A and B on two sites is shown in Figure 10-11.

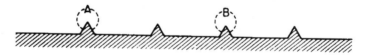

Figure 10-11   Vacant and occupied sites.

For the system shown, the total concentration of sites is

Site balance

$$C_t = C_v + C_{A \cdot S} + C_{B \cdot S} \tag{10-1}$$

This equation is referred to as a *site balance*.

    Now consider the adsorption of a nonreacting gas onto the surface of a catalyst. Adsorption data are frequently reported in the form of adsorption *isotherms*. Isotherms portray the amount of a gas adsorbed on a solid at different pressures but at one temperature.

Postulate models; then see which one(s) fit(s) the data.

    First, a model system is proposed and then the isotherm obtained from the model is compared with the experimental data shown on the curve. If the curve predicted by the model agrees with the experimental one, the model may reasonably describe what is occurring physically in the real system. If the predicted curve does not agree with data obtained experimentally, the model fails to match the physical situation in at least one important characteristic and perhaps more.

Two models will be postulated for the adsorption of carbon monoxide on metal surfaces. In one model, CO is adsorbed as molecules, CO,

$$CO + S \rightleftharpoons CO \cdot S$$

as is the case on nickel

$$
\begin{array}{ccc}
CO & & \overset{C}{\underset{O}{\diagdown}} \\
+ & & \vdots \\
-Ni-Ni-Ni- & \rightleftharpoons & -Ni-Ni-Ni-
\end{array}
$$

In the other, carbon monoxide is adsorbed as oxygen and carbon atoms instead of molecular CO.

$$CO + 2S \rightleftharpoons C \cdot S + O \cdot S$$

as is the case on iron[8]

$$
\begin{array}{ccc}
CO & & C \quad O \\
+ & & \vdots \quad \vdots \\
-Fe-Fe-Fe- & \rightleftharpoons & -Fe-Fe-Fe-
\end{array}
$$

Two models:
1. Adsorption as CO
2. Adsorption as C and O

The former is called *molecular* or *nondissociated adsorption* (e.g., CO) and the latter is called *dissociative adsorption* (e.g., C and O). Whether a molecule adsorbs nondissociatively or dissociatively depends on the surface.

The adsorption of carbon monoxide molecules will be considered first. Because the carbon monoxide does not react further after being adsorbed, we need only to consider the adsorption process:

$$CO + S \rightleftharpoons CO \cdot S \qquad (10\text{-}2)$$

See

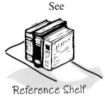

Reference Shelf
For H$_2$ Adsorption

$P_{CO} = C_{CO}RT$

In obtaining a rate law for the rate of adsorption, the reaction in Equation (10-2) can be treated as an *elementary reaction*. The rate of attachment of the carbon monoxide molecules to the active site on the surface is proportional to the number of collisions that these molecules make with a surface active site per second. In other words, a specific fraction of the molecules that strike the surface become adsorbed. The collision rate is, in turn, directly proportional to the carbon monoxide partial pressure, P_{CO}. Because carbon monoxide molecules adsorb only on vacant sites and not on sites already occupied by other carbon monoxide molecules, the rate of attachment is also directly proportional to the concentration of vacant sites, C_v. Combining these two facts means that the

---

[8] R. L. Masel, *Principles of Adsorption and Reaction on Solid Surfaces* (New York: Wiley, 1996).

rate of attachment of carbon monoxide molecules to the surface is directly proportional to the product of the partial pressure of CO and the concentration of vacant sites; that is,

$$\text{Rate of attachment} = k_A P_{CO} C_v$$

The rate of detachment of molecules from the surface can be a first-order process; that is, the detachment of carbon monoxide molecules from the surface is usually directly proportional to the concentration of sites occupied by the adsorbed molecules (e.g., $C_{CO\cdot S}$):

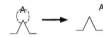

$$\text{rate of detachment} = k_{-A} C_{CO\cdot S}$$

The net rate of adsorption is equal to the rate of molecular attachment to the surface minus the rate of detachment from the surface. If k_A and k_{-A} are the constants of proportionality for the attachment and detachment processes, then

$$r_{AD} = k_A P_{CO} C_v - k_{-A} C_{CO\cdot S} \tag{10-3}$$

The ratio $K_A = k_A/k_{-A}$ is the *adsorption equilibrium constant*. Using it to rearrange Equation (10-3) gives

Adsorption
$$A + S \underset{}{\overset{}{\rightleftharpoons}} A \cdot S$$
$$r_{AD} = k_A \left(P_A C_v - \frac{C_{A\cdot S}}{K_A} \right)$$

$$\boxed{r_{AD} = k_A \left(P_{CO} C_v - \frac{C_{CO\cdot S}}{K_A} \right)} \tag{10-4}$$

The adsorption rate constant, k_A, for molecular adsorption is virtually independent of temperature, while the desorption constant, k_{-A}, increases exponentially with increasing temperature. The equilibrium constant K_A decreases exponentially with increasing temperature.

$$-r_A' = r_{AD} = \left(\frac{\text{mol}}{\text{gcat}\cdot\text{s}} \right)$$

Because carbon monoxide is the only material adsorbed on the catalyst, the site balance gives

$$C_t = C_v + C_{CO\cdot S} \tag{10-5}$$

$$k_A = \left(\frac{1}{\text{atm}\cdot\text{s}} \right)$$

At equilibrium, the net rate of adsorption equals zero. Setting the left-hand side of Equation (10-4) equal to zero and solving for the concentration of CO adsorbed on the surface, we get

$$P_A = (\text{atm})$$

$$C_v = \left(\frac{\text{mol}}{\text{gcat}} \right)$$

$$C_{CO\cdot S} = K_A C_v P_{CO} \tag{10-6}$$

$$K_A = \left(\frac{1}{\text{atm}} \right)$$

Using Equation (10-5) to give C_v in terms of $C_{CO\cdot S}$ and the total number of sites C_t, we can solve for $C_{CO\cdot S}$ in terms of constants and the pressure of carbon monoxide:

$$C_{A\cdot S} = \left(\frac{\text{mol}}{\text{gcat}} \right)$$

$$C_{CO\cdot S} = K_A C_v P_{CO} = K_A P_{CO}(C_t - C_{CO\cdot S})$$

Rearranging gives us

$$C_{\text{CO}\cdot\text{S}} = \frac{K_A P_{\text{CO}} C_t}{1 + K_A P_{\text{CO}}} \qquad (10\text{-}7)$$

This equation thus gives the concentration of carbon monoxide adsorbed on the surface, $C_{\text{CO}\cdot\text{S}}$, as a function of the partial pressure of carbon monoxide, and is an equation for the adsorption isotherm. This particular type of isotherm equation is called a Langmuir *isotherm*.[9] Figure 10-12(a) shows a plot of the amount of CO adsorbed per unit mass of catalyst as a function of the partial pressure of CO.

One method of checking whether a model (e.g., molecular adsorption versus dissociative adsorption) predicts the behavior of the experimental data is to linearize the model's equation and then plot the indicated variables against one another. For example, Equation (10-7) may be arranged in the form

Molecular adsorption

$$\frac{P_{\text{CO}}}{C_{\text{CO}\cdot\text{S}}} = \frac{1}{K_A C_t} + \frac{P_{\text{CO}}}{C_t} \qquad (10\text{-}8)$$

and the linearity of a plot of $P_{\text{CO}}/C_{\text{CO}\cdot\text{S}}$ as a function of P_{CO} will determine if the data conform to a Langmuir single-site isotherm.

Next, the isotherm for carbon monoxide adsorbing as atoms is derived:

Dissociative adsorption

$$\text{CO} + 2\text{S} \; \rightleftharpoons \; \text{C}\cdot\text{S} + \text{O}\cdot\text{S}$$

When the carbon monoxide molecule dissociates upon adsorption, it is referred to as the *dissociative adsorption* of carbon monoxide. As in the case of molecular adsorption, the rate of adsorption is proportional to the pressure of carbon monoxide in the system because this rate is governed by the number of gaseous collisions with the surface. For a molecule to dissociate as it adsorbs, however, two adjacent vacant active sites are required rather than the single site needed when a substance adsorbs in its molecular form. The probability of

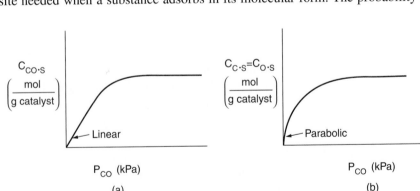

Figure 10-12   Langmuir isotherm for (a) molecular adsorption (b) dissociative adsorption of CO.

<hr />

[9] Named after Irving Langmuir (1881–1957), who first proposed it. He received the Nobel Prize in 1932 for his discoveries in surface chemistry.

two vacant sites occurring adjacent to one another is proportional to the square of the concentration of vacant sites. These two observations mean that the rate of adsorption is proportional to the product of the carbon monoxide partial pressure and the square of the vacant-site concentration, $P_{CO}C_v^2$.

For desorption to occur, two occupied sites must be adjacent, meaning that the rate of desorption is proportional to the product of the occupied-site concentration, $(C \cdot S) \times (O \cdot S)$. The net rate of adsorption can then be expressed as

$$r_{AD} = k_A P_{CO}C_v^2 - k_{-A}C_{O \cdot S}C_{C \cdot S} \tag{10-9}$$

Factoring out k_A, the equation for *dissociative adsorption* is

Rate of dissociative adsorption

$$r_{AD} = k_A \left(P_{CO}C_v^2 - \frac{C_{C \cdot S}C_{O \cdot S}}{K_A} \right)$$

where

$$K_A = \frac{k_A}{k_{-A}}$$

For dissociative adsorption both k_A and k_{-A} increase exponentially with increasing temperature while K_A decreases with increasing temperature.

At equilibrium, $r_{AD} = 0$, and

$$k_A P_{CO}C_v^2 = k_{-A}C_{C \cdot S}C_{O \cdot S}$$

For $C_{C \cdot S} = C_{O \cdot S}$,

$$(K_A P_{CO})^{1/2}C_v = C_{O \cdot S} \tag{10-10}$$

Substituting for $C_{C \cdot S}$ and $C_{O \cdot S}$ in a site balance Equation (10-1),

Site Balance: $C_t = C_v + C_{O \cdot S} + C_{C \cdot S} = C_v + (K_{CO}P_{CO})^{1/2}C_v + (K_{CO}P_{CO})^{1/2}C_v = C_v(1 + 2(K_{CO}P_{CO})^{1/2})$

Solving for C_v

$$C_v = C_t / (1 + 2(K_{CO}P_{CO})^{1/2})$$

This value may be substituted into Equation (10-10) to give an expression that can be solved for $C_{O \cdot S}$. The resulting equation for the isotherm shown in Figure 10-12(b) is

Langmuir isotherm for adsorption as atomic carbon monoxide

$$C_{O \cdot S} = \frac{(K_A P_{CO})^{1/2}C_t}{1 + 2(K_A P_{CO})^{1/2}} \tag{10-11}$$

Dissociative adsorption

Taking the inverse of both sides of the equation, then multiplying through by $(P_{CO})^{1/2}$, yields

$$\frac{(P_{CO})^{1/2}}{C_{O \cdot S}} = \frac{1}{C_t(K_A)^{1/2}} + \frac{2(P_{CO})^{1/2}}{C_t} \tag{10-12}$$

If dissociative adsorption is the correct model, a plot of $(P_{CO}^{1/2}/C_{O \cdot S})$ versus $P_{CO}^{1/2}$ should be linear with slope $(2/C_t)$.

When more than one substance is present, the adsorption isotherm equations are somewhat more complex. The principles are the same, though, and the isotherm equations are easily derived. It is left as an exercise to show that the adsorption isotherm of A in the presence of adsorbate B is given by the relationship

$$C_{A \cdot S} = \frac{K_A P_A C_t}{1 + K_A P_A + K_B P_B} \qquad (10\text{-}13)$$

$C_t = C_v + C_{A \cdot S} + C_{B \cdot S}$

When the adsorption of both A and B are first-order processes, the desorptions are also first order, and both A and B are adsorbed as molecules. The derivations of other Langmuir isotherms are relatively easy and are left as an exercise.

Note assumptions in the model and check their validity.

In obtaining the Langmuir isotherm equations, several aspects of the adsorption system were presupposed in the derivations. The most important of these, and the one that has been subject to the greatest doubt, is that a *uniform* surface is assumed. In other words, any active site has the same attraction for an impinging molecule as does any other site. Isotherms different from the Langmuir types, such as the Freundlich isotherm, may be derived based on various assumptions concerning the adsorption system, including different types of nonuniform surfaces.

10.2.4 Surface Reaction

The rate of adsorption of species A onto a solid surface,

$$A + S \;\rightleftharpoons\; A \cdot S$$

is given by

$$r_{AD} = k_A \left(P_A C_v - \frac{C_{A \cdot S}}{K_A} \right) \qquad (10\text{-}14)$$

Surface reaction models

After a reactant has been adsorbed onto the surface, it is capable of reacting in a number of ways to form the reaction product. Three of these ways are:

1. *Single site.* The surface reaction may be a single-site mechanism in which only the site on which the reactant is adsorbed is involved in the reaction. For example, an adsorbed molecule of A may isomerize (or perhaps decompose) directly on the site to which it is attached, such as

N = n-pentene   I = i-pentene

$$A \cdot S \;\rightleftharpoons\; B \cdot S$$

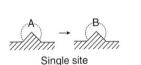

Single site

Because in each step the reaction mechanism is elementary, the surface reaction rate law is

Single Site

$$k_S = \left(\frac{1}{s}\right)$$

K_s = (dimensionless)

$$r_S = k_S\left(C_{A\cdot S} - \frac{C_{B\cdot S}}{K_S}\right) \qquad (10\text{-}15)$$

where K_S is the surface reaction equilibrium constant $K_S = k_S/k_{-S}$

2. *Dual site.* The surface reaction may be a dual-site mechanism in which the adsorbed reactant interacts with another site (either unoccupied or occupied) to form the product.

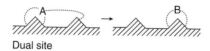

Dual site

For example, adsorbed A may react with an adjacent vacant site to yield a vacant site and a site on which the product is adsorbed, such as the dehydration of butanol.

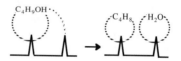

For the generic reaction

$$A\cdot S + S \rightleftharpoons B\cdot S + S$$

Dual Site

$$r_S = \left(\frac{mol}{gcat\cdot s}\right)$$

$$k_S = \left(\frac{gcat}{mol\cdot s}\right)$$

K_S = (dimensionless)

the corresponding surface reaction rate law is

$$r_S = k_S\left(C_{A\cdot S}C_v - \frac{C_{B\cdot S}C_v}{K_S}\right) \qquad (10\text{-}16)$$

Another example of a dual-site mechanism is the reaction between two adsorbed species, such as the reaction of CO with O

For the generic reaction

Dual site

$$A\cdot S + B\cdot S \rightleftharpoons C\cdot S + D\cdot S$$

the corresponding surface reaction rate law is

$$r_S = k_S\left(C_{A\cdot S}C_{B\cdot S} - \frac{C_{C\cdot S}C_{D\cdot S}}{K_S}\right) \qquad (10\text{-}17)$$

A third dual-site mechanism is the reaction of two species adsorbed on different types of sites S and S′, such as reaction of CO with O.

For the generic reaction

$$A \cdot S + B \cdot S' \; \rightleftharpoons \; C \cdot S' + D \cdot S$$

the corresponding surface reaction rate law is

$$r_S = k_S \left(C_{A \cdot S} C_{B \cdot S'} - \frac{C_{C \cdot S'} C_{D \cdot S}}{K_S} \right) \tag{10-18}$$

Langmuir-Hinshelwood kinetics

Reactions involving either single- or dual-site mechanisms, which were described earlier are sometimes referred to as following *Langmuir–Hinshelwood kinetics*.

3. *Eley–Rideal.* A third mechanism is the reaction between an adsorbed molecule and a molecule in the gas phase, such as the reaction of propylene and benzene

Dual site

Eley-Rideal mechanism

For the generic reaction

$$A \cdot S + B(g) \; \rightleftharpoons \; C \cdot S$$

the corresponding surface reaction rate law is

$$r_S = k_S \left(C_{A \cdot S} P_B - \frac{C_{C \cdot S}}{K_S} \right) \tag{10-19}$$

$$k_s = \left(\frac{1}{\text{atm} \cdot \text{s}} \right)$$

This type of mechanism is referred to as an *Eley–Rideal mechanism*.

$$K_S = \left(\frac{1}{\text{atm}} \right)$$

10.2.5 Desorption

In each of the preceding cases, the products of the surface reaction adsorbed on the surface are subsequently desorbed into the gas phase. For the desorption of a species (e.g., C),

$$K_{DC} = (\text{atm})$$

$$C \cdot S \; \rightleftharpoons \; C + S$$

$$k_D = \left(\frac{1}{\text{s}} \right)$$

the rate of desorption of C is

$$r_{DC} = k_D \left(C_{C \cdot S} - \frac{P_C C_v}{K_{DC}} \right) \tag{10-20}$$

where K_{DC} is the desorption equilibrium constant. We note that the desorption step for C is just the reverse of the adsorption step for C and that the rate of desorption of C, r_D, is just opposite in sign to the rate of adsorption of C, r_{ADC}:

$$r_{DC} = -r_{ADC}$$

In addition, we see that the desorption equilibrium constant K_{DC} is just the reciprocal of the adsorption equilibrium constant for C, K_C:

$K_{DC} = (\text{atm})$

$K_C = \left(\dfrac{1}{\text{atm}} \right)$

$$K_{DC} = \frac{1}{K_C}$$

$$r_{DC} = k_D(C_{C \cdot S} - K_C P_C C_v) \tag{10-21}$$

In the material that follows, the form of the equation for the desorption step that we will use will be similar to Equation (10-21).

10.2.6 The Rate-Limiting Step

When heterogeneous reactions are carried out at steady state, the rates of each of the three reaction steps in series (adsorption, surface reaction, and desorption) are equal to one another:

$$\boxed{-r_A' = r_{AD} = r_S = r_D}$$

However, one particular step in the series is usually found to be *rate-limiting* or *rate-controlling*. That is, if we could make this particular step go faster, the entire reaction would proceed at an accelerated rate. Consider the analogy to the electrical circuit shown in Figure 10-13. A given concentration of reactants is analogous to a given driving force or electromotive force (EMF). The current I (with units of Coulombs/s) is analogous to the rate of reaction, $-r_A'$ (mol/s·g cat), and a resistance R_i is associated with each step in the series. Since the resistances are in series, the total resistance is just the sum of the individual resistances, for adsorption (R_{AD}), surface reaction (R_S), and desorption (R_D), and the current, I, for a given voltage, E, is

Who is slowing us down?

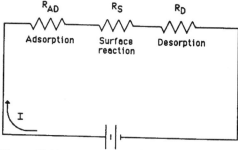

Figure 10-13   Electrical analog to heterogeneous reactions.

$$I = \frac{E}{R_{\text{tot}}} = \frac{E}{R_{\text{AD}} + R_S + R_D}$$

The concept of a
rate-limiting step

Since we observe only the total resistance, R_{tot}, it is our task to find which resistance is much larger (say, 100 Ω) than the other two (say, 0.1 Ω). Thus, if we could lower the largest resistance, the current I (i.e., $-r'_A$), would be larger for a given voltage, E. Analogously, we want to know which step in the adsorption–reaction–desorption series is limiting the overall rate of reaction.

An algorithm to
determine the
rate-limiting step

The approach in determining catalytic and heterogeneous mechanisms is usually termed the *Langmuir–Hinshelwood approach*, since it is derived from ideas proposed by Hinshelwood[10] based on Langmuir's principles for adsorption. The Langmuir–Hinshelwood approach was popularized by Hougen and Watson[11] and occasionally includes their names. It consists of first assuming a sequence of steps in the reaction. In writing this sequence, one must choose among such mechanisms as molecular or atomic adsorption, and single- or dual-site reaction. Next, rate laws are written for the individual steps as shown in the preceding section, assuming that all steps are reversible. Finally, a rate-limiting step is postulated, and steps that are not rate-limiting are used to eliminate all coverage-dependent terms. The most questionable assumption in using this technique to obtain a rate law is the hypothesis that the activity of the surface toward adsorption, desorption, or surface reaction is independent of coverage; that is, the surface is essentially uniform as far as the various steps in the reaction are concerned.

An example of an adsorption-limited reaction is the synthesis of ammonia from hydrogen and nitrogen,

$$3H_2 + N_2 \longrightarrow 2NH_3$$

over an iron catalyst that proceeds by the following mechanism.[12]

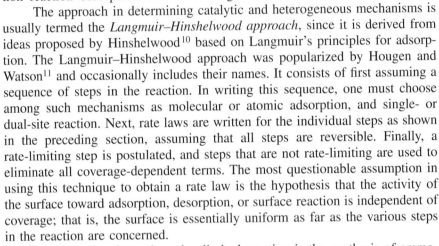

Dissociative
adsorption
of N_2 limits

$$H_2 + 2S \longrightarrow 2H \cdot S \qquad \text{Rapid}$$

$$\left.\begin{array}{l} N_2 + S \rightleftarrows N_2 \cdot S \\ N_2 \cdot S + S \longrightarrow 2N \cdot S \end{array}\right\} \qquad \text{Rate-limiting}$$

$$\left.\begin{array}{l} N \cdot S + H \cdot S \rightleftarrows HN \cdot S + S \\ NH \cdot S + H \cdot S \rightleftarrows H_2N \cdot S + S \\ H_2N \cdot S + H \cdot S \rightleftarrows NH_3 \cdot S + S \\ NH_3 \cdot S \rightleftarrows NH_3 + S \end{array}\right\} \qquad \text{Rapid}$$

---

[10]C. N. Hinshelwood, *The Kinetics of Chemical Change* (Oxford: Clarendon Press, 1940).

[11]O. A. Hougen and K. M. Watson, *Ind. Eng. Chem.*, *35*, 529 (1943).

[12]From the literature cited in G. A. Somorjai, *Introduction to Surface Chemistry and Catalysis* (New York: Wiley, 1994), p. 482.

The rate-limiting step is believed to be the adsorption of the N_2 molecule as an N atom.

   An example of a surface-limited reaction is the reaction of two noxious automobile exhaust products, CO and NO,

$$CO + NO \longrightarrow CO_2 + \tfrac{1}{2} N_2$$

carried out over copper catalyst to form environmentally acceptable products, N_2 and CO_2:

$$\left.\begin{array}{l} CO + S \rightleftarrows CO \cdot S \\ NO + S \rightleftarrows NO \cdot S \end{array}\right\} \qquad \text{Rapid}$$

Surface reaction limits

$$NO \cdot S + CO \cdot S \rightleftarrows CO_2 + N \cdot S + S\} \quad \text{Rate-limiting}$$

$$\left.\begin{array}{l} N \cdot S + N \cdot S \rightleftarrows N_2 \cdot S \\ N_2 \cdot S \longrightarrow N_2 + S \end{array}\right\} \qquad \text{Rapid}$$

Analysis of the rate law suggests that CO_2 and N_2 are weakly adsorbed (see Problem P10-7$_B$).

10.3 Synthesizing a Rate Law, Mechanism, and Rate-Limiting Step

We now wish to develop rate laws for catalytic reactions that are not diffusion-limited. In developing the procedure to obtain a mechanism, a rate-limiting step, and a rate law consistent with experimental observation, we shall discuss a particular catalytic reaction, the decomposition of cumene to form benzene and propylene. The overall reaction is

$$C_6H_5CH(CH_3)_2 \longrightarrow C_6H_6 + C_3H_6$$

   A conceptual model depicting the sequences of steps in this platinum-catalyzed reaction is shown in Figure 10-14. Figure 10-14 is only a schematic representation of the adsorption of cumene; a more realistic model is the formation of a complex of the π orbitals of benzene with the catalytic surface, as shown in Figure 10-15.

· Adsorption
· Surface
  reaction
· Desorption

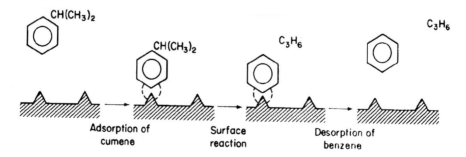

Figure 10-14   Sequence of steps in reaction-limited catalytic reaction.

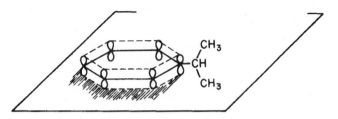

Figure 10-15 π-orbital complex on surface.

The nomenclature in Table 10-3 will be used to denote the various species in this reaction: C = cumene, B = benzene, and P = propylene. The reaction sequence for this decomposition is

TABLE 10-3.   STEPS IN A LANGMUIR-HINSHELWOOD KINETIC MECHANISM

$C + S \xrightleftharpoons[k_{-A}]{k_A} C \cdot S$	Adsorption of cumene on the surface (10-22)
$C \cdot S \xrightleftharpoons[k_{-S}]{k_S} B \cdot S + P$	Surface reaction to form adsorbed (10-23) benzene and propylene in the gas phase
$B \cdot S \xrightleftharpoons[k_{-D}]{k_D} B + S$	Desorption of benzene from surface (10-24)

These three steps represent the mechanism for cumene decomposition

Equations (10-22) through (10-24) represent the mechanism proposed for this reaction.

When writing rate laws for these steps, we treat each step as an elementary reaction; the only difference is that the species concentrations in the gas phase are replaced by their respective partial pressures:

$$C_C \longrightarrow P_C$$

Ideal Gas Law
$P_C = C_C RT$

There is no theoretical reason for this replacement of the concentration, C_C, with the partial pressure, P_C; it is just the convention initiated in the 1930s and used ever since. Fortunately P_C can be calculated directly from C_C using the ideal gas law.

The rate expression for the adsorption of cumene as given in Equation (10-22) is

$C + S \xrightleftharpoons[k_{-A}]{k_A} C \cdot S$

$$r_{AD} = k_A P_C C_v - k_{-A} C_{C \cdot S}$$

Adsorption: $r_{AD} = k_A \left(P_C C_v - \dfrac{C_{C \cdot S}}{K_C} \right)$ (10-25)

If r_{AD} has units of (mol/g cat·s) and $C_{C \cdot S}$ has units of (mol cumene adsorbed/g cat) typical units of k_A, k_{-A}, and K_C would be

$$[k_A] \equiv (kPa \cdot s)^{-1} \text{ or } (atm \cdot h)^{-1}$$

$$[k_{-A}] \equiv h^{-1} \text{ or } s^{-1}$$

$$[K_C] \equiv \left[\frac{k_A}{k_{-A}}\right] \equiv kPa^{-1}$$

The rate law for the surface reaction step producing adsorbed benzene and propylene in the gas phase,

$$C \cdot S \underset{k_{-S}}{\overset{k_S}{\rightleftarrows}} B \cdot S + P(g) \tag{10-23}$$

is

$$r_S = k_S C_{C \cdot S} - k_{-S} P_P C_{B \cdot S}$$

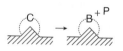

$$\boxed{\text{Surface reaction:} \quad r_S = k_S \left(C_{C \cdot S} - \frac{P_P C_{B \cdot S}}{K_S}\right)} \tag{10-26}$$

with the *surface reaction equilibrium constant* being

$$K_S = \frac{k_S}{k_{-S}}$$

Typical units for k_S and K_S are s^{-1} and kPa, respectively.

Propylene is not adsorbed on the surface. Consequently, its concentration on the surface is zero.

$$C_{P \cdot S} = 0$$

The rate of benzene desorption [see Equation (10-24)] is

$$r_D = k_D C_{B \cdot S} - k_{-D} P_B C_v \tag{10-27}$$

$$\boxed{\text{Desorption:} \quad r_D = k_D \left(C_{B \cdot S} - \frac{P_B C_v}{K_{DB}}\right)} \tag{10-28}$$

Typical units of k_D and K_{DB} are s^{-1} and kPa, respectively. By viewing the desorption of benzene,

$$B \cdot S \rightleftarrows B + S$$

from right to left, we see that desorption is just the reverse of the adsorption of benzene. Consequently, it is easily shown that the benzene adsorption equilibrium constant K_B is just the reciprocal of the benzene desorption constant K_{DB}:

$$K_B = \frac{1}{K_{DB}}$$

and Equation (10-28) can be written as

> Desorption: $r_D = k_D(C_{B \cdot S} - K_B P_B C_v)$ (10-29)

Because there is no accumulation of reacting species on the surface the rates of each step in the sequence are all equal:

$$-r'_C = r_{AD} = r_S = r_D$$ (10-30)

For the mechanism postulated in the sequence given by Equations (10-22) through (10-24), we wish to determine which step is rate-limiting. We first assume one of the steps to be rate-limiting (rate-controlling) and then formulate the reaction rate law in terms of the partial pressures of the species present. From this expression we can determine the variation of the initial reaction rate with the initial total pressure. If the predicted rate varies with pressure in the same manner as the rate observed experimentally, the implication is that the assumed mechanism and rate-limiting step are correct.

10.3.1 Is the Adsorption of Cumene Rate-Limiting?

To answer this question we shall assume that the adsorption of cumene is indeed rate-limiting, derive the corresponding rate law, and then check to see if it is consistent with experimental observation. By assuming that this (or any other) step is rate-limiting, we are considering that the reaction rate constant of this step (in this case k_A) is small with respect to the specific rates of the other steps (in this case k_S and k_D).[13] The rate of adsorption is

Need to express C_v and $C_{C \cdot S}$ in terms of P_C, P_B, and P_P

$$-r'_C = r_{AD} = k_A \left(P_C C_v - \frac{C_{C \cdot S}}{K_C} \right)$$ (10-25)

Because we can measure neither C_v or $C_{C \cdot S}$, we must replace these variables in the rate law with measurable quantities for the equation to be meaningful.

For steady-state operation we have

$$-r'_C = r_{AD} = r_S = r_D$$ (10-30)

---

[13]Strictly speaking one should compare the product $k_A P_C$ with k_S and k_D.

$$r_{AD} = k_{AD} P_C \left[C_v - \frac{C_{C \cdot S}}{K_C P_C} \right]$$

$$\frac{mol}{s \cdot kg\ cat} = \left[\frac{1}{s\ atm} \right] \cdot [atm] \cdot \left[\frac{mol}{kg\ cat} \right] = \left[\frac{1}{s} \right] \frac{mol}{kg\ cat}$$

Dividing by $K_{AD} P_C$ we note $\frac{r_{ADC}}{k_{AD} P_C} = \frac{mol}{kg\ cat}$. The reason is that in order to compare terms,

the ratios $\left(\frac{r_{AD}}{k_{AD} P_C} \right)$, $\left(\frac{r_S}{k_S} \right)$ and $\left(\frac{r_D}{k_D} \right)$ must all have the same units $\left[\frac{mol}{kg \cdot cat} \right]$. The end

result is the same however.

For adsorption-limited reactions, k_A is small and k_S and k_D are large. Consequently, the ratios r_S/k_S and r_D/k_D are very small (approximately zero), whereas the ratio r_{AD}/k_A is relatively large.

The surface reaction rate law is

$$r_S = k_S \left(C_{C\cdot S} - \frac{C_{B\cdot S}P_P}{K_S} \right) \qquad (10\text{-}31)$$

Again, for adsorption-limited reactions the surface specific reaction rate k_S is large by comparison, and we can set

$$\frac{r_S}{k_S} \simeq 0 \qquad (10\text{-}32)$$

and solve Equation (10-31) for $C_{C\cdot S}$:

$$C_{C\cdot S} = \frac{C_{B\cdot S}P_P}{K_S} \qquad (10\text{-}33)$$

To be able to express $C_{C\cdot S}$ solely in terms of the partial pressures of the species present, we must evaluate $C_{B\cdot S}$. The rate of desorption of benzene is

$$r_D = k_D (C_{B\cdot S} - K_B P_B C_v) \qquad (10\text{-}29)$$

> Using
> $$\frac{r_S}{k_S} \simeq 0 \simeq \frac{r_D}{k_D}$$
> to find $C_{B\cdot S}$ and $C_{C\cdot S}$ in terms of partial pressures

However, for adsorption-limited reactions, k_D is large by comparison, and we can set

$$\frac{r_D}{k_D} \simeq 0 \qquad (10\text{-}34)$$

and then solve Equation (10-29) for $C_{B\cdot S}$:

$$C_{B\cdot S} = K_B P_B C_v \qquad (10\text{-}35)$$

After combining Equations (10-33) and (10-35), we have

$$C_{C\cdot S} = K_B \frac{P_B P_P}{K_S} C_v \qquad (10\text{-}36)$$

Replacing $C_{C\cdot S}$ in the rate equation by Equation (10-36) and then factoring C_v, we obtain

$$r_{AD} = k_A \left(P_C - \frac{K_B P_B P_P}{K_S K_C} \right) C_v = k_A \left(P_C - \frac{P_B P_P}{K_P} \right) C_v \qquad (10\text{-}37)$$

Observe that by setting $r_{AD} = 0$, the term $(K_S K_C / K_B)$ is simply the overall partial pressure equilibrium constant, K_P, for the reaction

$$C \xrightleftharpoons{} B + P$$

$$\boxed{\frac{K_S K_C}{K_B} = K_P} \tag{10-38}$$

The equilibrium constant can be determined from thermodynamic data and is related to the change in the Gibbs free energy, $\Delta G°$, by the equation (see Appendix C)

$$\boxed{RT \ln K = -\Delta G°} \tag{10-39}$$

where R is the ideal gas constant and T is the absolute temperature.

The concentration of vacant sites, C_v, can now be eliminated from Equation (10-37) by utilizing the site balance to give the total concentration of sites, C_t, which is assumed constant:[14]

$$\boxed{\text{Total sites} = \text{Vacant sites} + \text{Occupied sites}}$$

Because cumene and benzene are adsorbed on the surface, the concentration of occupied sites is $(C_{C \cdot S} + C_{B \cdot S})$, and the total concentration of sites is

Site balance
$$C_t = C_v + C_{C \cdot S} + C_{B \cdot S} \tag{10-40}$$

Substituting Equations (10-35) and (10-36) into Equation (10-40), we have

$$C_t = C_v + \frac{K_B}{K_S} P_B P_P C_v + K_B P_B C_v$$

Solving for C_v, we have

$$C_v = \frac{C_t}{1 + P_B P_P K_B / K_S + K_B P_B} \tag{10-41}$$

Combining Equations (10-41) and (10-37), we find that the rate law for the catalytic decompositon of cumene, assuming that the adsorption of cumene is the rate-limiting step, is

Cumene reaction
rate law if
adsorption were
the limiting step
$$\boxed{-r'_C = r_{AD} = \frac{C_t k_A (P_C - P_P P_B / K_P)}{1 + K_B P_P P_B / K_S + K_B P_B}} \tag{10-42}$$

---

[14]Some prefer to write the surface reaction rate in terms of the fraction of the surface of sites covered (i.e., f_A) rather than the number of sites $C_{A \cdot S}$ covered, the difference being the multiplication factor of the total site concentration, C_t. In any event, the final form of the rate law is the same because C_t, K_A, k_S, and so on, are all lumped into the reaction rate constant, k.

We now wish to sketch a plot of the initial rate as a function of the partial pressure of cumene, P_{C0}. Initially, no products are present; consequently, $P_P = P_B = 0$. The initial rate is given by

$$-r'_{C0} = C_t k_A P_{C0} = k P_{C0} \qquad (10\text{-}43)$$

If the cumene decompostion is adsorption rate limited, then the initial rate will be linear with the initial partial pressure of cumene as shown in Figure 10-16.

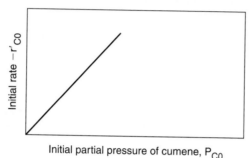

Initial partial pressure of cumene, P_{C0}

Figure 10-16   Uninhibited adsorption-limited reaction.

If adsorption were rate-limiting, the data should show $-r'_0$ increasing linearly with P_{C0}.

Before checking to see if Figure 10-16 is consistent with experimental observation, we shall derive the corresponding rate laws and initial rate plots when the surface reaction is rate-limiting and then when the desorption of benzene is rate-limiting.

10.3.2 Is the Surface Reaction Rate-Limiting?

The rate of surface reaction is

Single-site mechanism

$$r_S = k_S \left(C_{C\cdot S} - \frac{P_P C_{B\cdot S}}{K_S} \right) \qquad (10\text{-}26)$$

Since we cannot readily measure the concentrations of the adsorbed species, we must utilize the adsorption and desorption steps to eliminate $C_{C\cdot S}$ and $C_{B\cdot S}$ from this equation.

From the adsorption rate expression in Equation (10-25) and the condition that k_A and k_D are very large by comparison with k_S when surface reaction is controlling (i.e., $r_{AD}/k_A \approx 0$), we obtain a relationship for the surface concentration for adsorbed cumene:

$$C_{C\cdot S} = K_C P_C C_v$$

In a similar manner, the surface concentration of adsorbed benzene can be evaluated from the desorption rate expression [Equation (10-29)] together with the approximation:

Using
$$\frac{r_{AD}}{k_A} \cong 0 \cong \frac{r_D}{k_D}$$
to find $C_{B\cdot S}$ and $C_{C\cdot S}$ in terms of partial pressures

$$\text{when } \frac{r_D}{k_D} \cong 0 \qquad \text{then } C_{B\cdot S} = K_B P_B C_v$$

Substituting for $C_{B \cdot S}$ and $C_{C \cdot S}$ in Equation (10-26) gives us

$$r_S = k_S \left(P_C K_C - \frac{K_B P_B P_P}{K_S} \right) C_v = k_S K_C \left(P_C - \frac{P_B P_P}{K_P} \right) C_v$$

The only variable left to eliminate is C_v:

Site balance
$$C_t = C_v + C_{B \cdot S} + C_{C \cdot S}$$

Substituting for concentrations of the adsorbed species, $C_{B \cdot S}$, and $C_{C \cdot S}$ yields

$$C_v = \frac{C_t}{1 + K_B P_B + K_C P_C}$$

Cumene rate law
for surface-
reaction-limiting

$$-r'_C = r_S = \frac{\overset{k}{\overbrace{k_S C_t K_C}}(P_C - P_P P_B / K_P)}{1 + P_B K_B + K_C P_C} \tag{10-44}$$

The initial rate is

$$-r'_{C0} = \frac{\overset{k}{\overbrace{k_S C_t K_C}} P_{C0}}{1 + K_C P_{C0}} = \frac{k P_{C0}}{1 + K_C P_{C0}} \tag{10-45}$$

At low partial pressures of cumene

$$1 \gg K_C P_{C0}$$

and we observe that the initial rate will increase linearly with the initial partial pressure of cumene:

$$-r'_{C0} \approx k P_{C0}$$

At high partial pressures

$$K_C P_{C0} \gg 1$$

and Equation (10-45) becomes

$$-r'_{C0} \cong \frac{k P_{C0}}{K_C P_{C0}} = \frac{k}{K_C}$$

and the rate is independent of the partial pressure of cumene. Figure 10-17 shows the initial rate of reaction as a function of initial partial pressure of cumene for the case of surface reaction controlling.

10.3.3 Is the Desorption of Benzene Rate-Limiting?

The rate expression for the desorption of benzene is

$$r_D = k_D (C_{B \cdot S} - K_B P_B C_v) \tag{10-29}$$

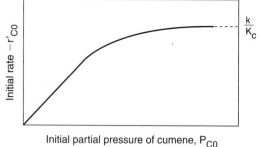

Figure 10-17   Surface-reaction-limited.

If surface reaction were rate-limiting, the data would show this behavior.

For desorption-limited reactions, both k_{AD} and k_S are very large compared with k_D, which is small.

From the rate expression for surface reaction, Equation (10-26), we set

$$\frac{r_S}{k_S} \simeq 0$$

to obtain

$$C_{B \cdot S} = K_S \left(\frac{C_{C \cdot S}}{P_P} \right) \qquad (10\text{-}46)$$

Similarly, for the adsorption step, Equation (10-25), we set

$$\frac{r_{AD}}{k_A} \simeq 0$$

to obtain

$$C_{C \cdot S} = K_C P_C C_v$$

then substitute for $C_{C \cdot S}$ in Equation (10-46):

$$C_{B \cdot S} = \frac{K_C K_S P_C C_v}{P_P} \qquad (10\text{-}47)$$

Combining Equations (10-28) and (10-47) gives us

$$r_D = k_D K_C K_S \left(\frac{P_C}{P_P} - \frac{P_B}{K_P} \right) C_v \qquad (10\text{-}48)$$

where K_C is the cumene adsorption constant, K_S is the surface reaction equilibrium constant, and K_P is the gas-phase equilibrium constant for the reaction. To obtain an expression for C_v, we again perform a site balance:

Site balance: $C_t = C_{C \cdot S} + C_{B \cdot S} + C_v$

After substituting for the respective surface concentrations, we solve the site balance for C_v:

$$C_v = \frac{C_t}{1 + K_C K_S P_C / P_P + K_C P_C} \qquad (10\text{-}49)$$

Replacing C_v in Equation (10-48) by Equation (10-49) and multiplying the numerator and denominator by P_P, we obtain the rate expression for desorption control:

$$-r'_C = r_D = \frac{\overbrace{k_D C_t K_S K_C}^{k} (P_C - P_B P_P / K_P)}{P_P + P_C K_C K_S + K_C P_P P_C} \qquad (10\text{-}50)$$

To determine the dependence of the initial rate on partial pressure of cumene, we again set $P_P = P_B = 0$; and the rate law reduces to

$$-r'_{C0} = k_D C_t$$

with the corresponding plot of $-r'_{C0}$ shown in Figure 10-18. If desorption were controlling, we would see that the initial rate would be independent of the initial partial pressure of cumene.

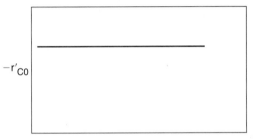

$-r'_{C0}$

Initial partial pressure of cumene, P_{C0}

Figure 10-18   Desorption-limited reaction.

10.3.4 Summary of the Cumene Decomposition

The experimental observations of $-r'_{C0}$ as a function of P_{C0} are shown in Figure 10-19. From the plot in Figure 10-19, we can clearly see that neither adsorption nor desorption is rate-limiting. For the reaction and mechanism given by

$$C + S \;\rightleftarrows\; C \cdot S \qquad (10\text{-}22)$$

$$C \cdot S \;\rightleftarrows\; B \cdot S + P \qquad (10\text{-}23)$$

$$B \cdot S \;\rightleftarrows\; B + S \qquad (10\text{-}24)$$

the rate law derived by assuming that the surface reaction is rate-limiting agrees with the data.

The rate law for the case of no inerts adsorbing on the surface is

$$-r'_C = \frac{k(P_C - P_B P_P / K_P)}{1 + K_B P_B + K_C P_C} \qquad (10\text{-}44)$$

Surface limited
reaction mechanism
is consistent with
experimental data.

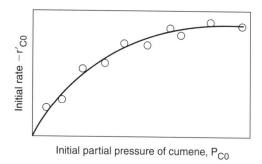

Figure 10-19   Actual initial rate as a function of partial pressure of cumene.

The forward cumene decomposition reaction is a single-site mechanism involving only adsorbed cumene while the reverse reaction of propylene in the gas phase reacting with adsorbed benzene is an Eley–Rideal mechanism.

If we were to have an adsorbing inert in the feed, the inert would not participate in the reaction but would occupy sites on the catalyst surface:

$$I + S \; \rightleftarrows \; I \cdot S$$

Our site balance is now

$$C_t = C_v + C_{C \cdot S} + C_{B \cdot S} + C_{I \cdot S} \qquad (10\text{-}51)$$

Because the adsorption of the inert is at equilibrium, the concentration of sites occupied by the inert is

$$C_{I \cdot S} = K_I P_I C_v \qquad (10\text{-}52)$$

Substituting for the inert sites in the site balance, the rate law for surface reaction control when an adsorbing inert is present is

Adsorbing inerts

$$-r_C' = \frac{k(P_C - P_B P_P / K_P)}{1 + K_C P_C + K_B P_B + K_I P_I} \qquad (10\text{-}53)$$

10.3.5 Reforming Catalysts

We now consider a dual-site mechanism, which is a reforming reaction found in petroleum refining to upgrade the octane number of gasoline.

Side Note: Octane Number. Fuels with low octane numbers can produce spontaneous combustion in the cylinder before the air/fuel mixture is compressed to its desired value and ignited by the spark plug. The following figure shows the desired combustion wave front moving down from the spark plug and the unwanted spontaneous combustion wave in the lower right-hand corner. This spontaneous combustion produces detonation waves which constitute engine knock. The lower the octane number the greater the chance of engine knock.

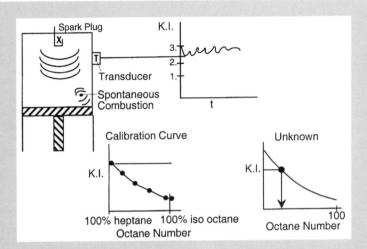

The octane number of a gasoline is determined from a calibration curve relating knock intensity to the % iso-octane in a mixture of iso-octane and heptane. One way to calibrate the octane number is to place a transducer on the side of the cylinder to measure the knock intensity (K.I.) (pressure pulse) for various mixtures of heptane and iso-octane. The octane number is the percentage of iso-octane in this mixture. That is, pure iso-octane has an octane number of 100, 80% iso-octane/20% heptane has an octane number of 80, and so on. The knock intensity is measured for this 80/20 mixture and recorded. The relative percentages of iso-octane and heptane are changed (e.g., 90/10), and the test is repeated. After a series of experiments, a calibration curve is constructed. The gasoline to be calibrated is then used in the test engine, where the standard knock intensity is measured. Knowing the knock intensity, the octane rating of the fuel is read off the calibration curve. A gasoline with an octane rating of 92 means that it matches the performance of a mixture of 92% iso-octane and 8% heptane. Another way to calibrate octane number is to set the knock intensity and increase the compression ratio. A fixed percentage of iso-octane and heptane is placed in a test engine and the compression ratio (CR) is increased continually until spontaneous combustion occurs, producing an engine knock. The compression ratio and the corresponding composition of the mixture are then recorded, and the test is repeated to obtain the calibration curve of compression ratio as a function of % iso-octane. After the calibration curve is obtained, the unknown is placed in the cylinder, and the compression ratio (CR) is increased until the set knock intensity is exceeded. The CR is then matched to the calibration curve to find the octane number.

The more compact the hydrocarbon molecule, the less likely it is to cause spontaneous combustion and engine knock. Consequently, it is desired to isomerize straight-chain hydrocarbon molecules into more compact molecules through a catalytic process called reforming.

One common reforming catalyst is platinum on alumina. Platinum on alumina (Al_2O_3) (see SEM photo below) is a bifunctional catalyst that can be prepared by exposing alumina pellets to a chloroplatinic acid solution, drying, and then heating in air 775 to 875 K for several hours. Next, the material is exposed to hydrogen at temperatures around 725 to 775 K to produce very small clusters of Pt on alumina. These clusters have sizes on the order of 10 Å, while the alumina pore sizes on which the Pt is deposited are on the order of 100 to 10,000 Å (i.e., 10 to 1000 nm).

Catalyst manufacture

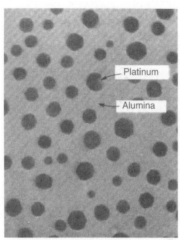

Platinum on alumina. (Figure from R.I. Masel, *Chemical Kinetics and Catalysis*, Wiley, New York, 2001, p. 700).

As an example of catalytic reforming we shall consider the isomerization of *n*-pentane to *i*-pentane:

$$n\text{-pentane} \xrightleftharpoons[Al_2O_3]{0.75 \text{ wt\% Pt}} i\text{-pentane}$$

Gasoline	
C_5	10%
C_6	10%
C_7	20%
C_8	25%
C_9	20%
C_{10}	10%
C_{11-12}	5%

Normal pentane has an octane number of 62, while *iso*-pentane has an octane number of 90! The *n*-pentane adsorbs onto the platinum, where it is dehydrogenated to form *n*-pentene. The *n*-pentene desorbs from the platinum and adsorbs onto the alumina, where it is isomerized to *i*-pentene, which then desorbs and subsequently adsorbs onto platinum, where it is hydrogenated to form *i*-pentane. That is,

$$n\text{-pentane} \xrightleftharpoons[Pt]{-H_2} n\text{-pentene} \xrightleftharpoons{Al_2O_3} i\text{-pentene} \xrightleftharpoons[Pt]{+H_2} i\text{-pentane}$$

We shall focus on the isomerization step to develop the mechanism and the rate law:

$$n\text{-pentene} \xrightleftharpoons{Al_2O_3} i\text{-pentene}$$

$$N \rightleftharpoons I$$

The procedure for formulating a mechanism, rate-limiting step, and corresponding rate law is given in Table 10-4.

Table 10-5 gives the forms of rate laws for different reaction mechanisms that are irreversible and surface-reaction-limited.

TABLE 10-5.   IRREVERSIBLE SURFACE-REACTION-LIMITED RATE LAWS

Single site

$$A \cdot S \longrightarrow B \cdot S \qquad\qquad -r'_A = \frac{kP_A}{1 + K_A P_A + K_B P_B}$$

Dual site

$$A \cdot S + S \longrightarrow B \cdot S + S \qquad\qquad -r'_A = \frac{kP_A}{(1 + K_A P_A + K_B P_B)^2}$$

$$A \cdot S + B \cdot S \longrightarrow C \cdot S + S \qquad\qquad -r'_A = \frac{kP_A P_B}{(1 + K_A P_A + K_B P_B + K_C P_C)^2}$$

Eley–Rideal

$$A \cdot S + B(g) \longrightarrow C \cdot S \qquad\qquad -r'_A = \frac{kP_A P_B}{1 + K_A P_A + K_C P_C}$$

We need a word of caution at this point. Just because the mechanism and rate-limiting step may fit the rate data does not imply that the mechanism is correct.[15] Usually, spectroscopic measurements are needed to confirm a mechanism absolutely. However, the development of various mechanisms and rate-limiting steps can provide insight into the best way to correlate the data and develop a rate law.

10.3.6 Rate Laws Derived from the Pseudo-Steady-State Hypothesis

In Section 7.1 we discussed the PSSH where the net rate of formation of reactive intermediates was assumed to be zero. An alternative way to derive a catalytic rate law rather than setting

$$\frac{r_{AD}}{k_A} \cong 0$$

is to assume that each species adsorbed on the surface is a reactive intermediate. Consequently, the net rate of formation of species i adsorbed on the surface will be zero:

The PSSH should be used when more than one step is limiting

$$r^*_{i \cdot S} = 0 \qquad\qquad\qquad (10\text{-}54)$$

[15]R. I. Masel, *Principles of Adsorption and Reaction on Solid Surfaces* (New York: Wiley, 1996), p. 506. http://www.uiuc.edu/ph/www/r-masel/

TABLE 10-4.   ALGORITHM FOR DETERMINING REACTION MECHANISM AND RATE-LIMITING STEP

Isomerization of n-pentene (N) to i-pentene (I) over alumina

$$N \underset{}{\overset{Al_2O_3}{\rightleftharpoons}} I$$

Reforming reaction to increase octane number of gasoline

1. *Select a mechanism.* (Mechanism Dual Site)

   Adsorption: $N + S \rightleftharpoons N \cdot S$

   Surface reaction: $N \cdot S + S \rightleftharpoons I \cdot S + S$

   Desorption: $I \cdot S \rightleftharpoons I + S$

   Treat each reaction step as an elementary reaction when writing rate laws.

2. *Assume a rate-limiting step.* Choose the surface reaction first, since *more than 75% of all heterogeneous reactions that are not diffusion-limited are surface-reaction-limited.* The rate law for the surface reaction step is

$$-r_N' = r_S = k_S \left(C_v C_{N \cdot S} - \frac{C_{I \cdot S} C_v}{K_S} \right)$$

3. *Find the expression for concentration of the adsorbed species $C_{i \cdot S}$.* Use the other steps that are not limiting to solve for $C_{i \cdot S}$ (e.g., $C_{N \cdot S}$ and $C_{I \cdot S}$). For this reaction,

   From $\dfrac{r_{AD}}{k_{AD}} \approx 0$: $C_{N \cdot S} = P_N K_N C_v$

   From $\dfrac{r_D}{k_D} \approx 0$: $C_{I \cdot S} = \dfrac{P_I C_v}{K_D} = K_I P_I C_v$

4. *Write a site balance.*

$$C_t = C_v + C_{N \cdot S} + C_{I \cdot S}$$

Following the Algorithm

5. *Derive the rate law.* Combine Steps 2, 3, and 4 to arrive at the rate law:

$$-r_N' = r_S = \frac{\overbrace{k_S C_t^2 K_N}^{k} (P_N - P_I / K_P)}{(1 + K_N P_N + K_I P_I)^2}$$

6. *Compare with data.* Compare the rate law derived in Step 5 with experimental data. If they agree, there is a good chance that you have found the correct mechanism and rate-limiting step. If your derived rate law (i.e., model) does not agree with the data:

   a. Assume a different rate-limiting step and repeat Steps 2 through 6.

   b. If, after assuming that each step is rate-limiting, none of the derived rate laws agree with the experimental data, select a different mechanism (e.g., A Single Site Mechanism):

$$N + S \rightleftharpoons N \cdot S$$
$$N \cdot S \rightleftharpoons I \cdot S$$
$$I \cdot S \rightleftharpoons I + S$$

   and then proceed through steps 2 through 6.
   The single-site mechanism turns out to be the correct one. For this mechanism the rate law is

$$-r_N' = \frac{k(P_N - P_I / K_P)}{(1 + K_N P_N + K_I P_I)}$$

   c. If two or more models agree, the statistical tests discussed in Chapter 5 (e.g., comparison of residuals) should be used to discriminate between them (see the Supplementary Reading).

While this method works well for a single rate-limiting step, it also works well when two or more steps are rate-limiting (e.g., adsorption *and* surface reaction). To illustrate how the rate laws are derived using the PSSH, we shall consider the isomerization of normal pentene to *iso*-pentene by the following mechanism, shown as item 6 of Table 10-4:

$$
\begin{array}{c}
\mathrm{N + S} \underset{k_{-N}}{\overset{k_N}{\rightleftharpoons}} \mathrm{N \cdot S} \\[2ex]
\mathrm{N \cdot S} \xrightarrow{k_S} \mathrm{I \cdot S} \\[2ex]
\mathrm{I \cdot S} \underset{k_{-I}}{\overset{k_I}{\rightleftharpoons}} \mathrm{I + S}
\end{array}
$$

The rate law for the irreversible surface reaction is

$$-r'_N = r_S = k_S C_{N \cdot S} \tag{10-55}$$

The net rates of generation of $\mathrm{N \cdot S}$ sites and $\mathrm{I \cdot S}$ sites are

> The net rate of the adsorbed species (i.e., active intermediates) is zero.

$$r^*_{N \cdot S} = k_N P_N C_v - k_{-N} C_{N \cdot S} - k_S C_{N \cdot S} = 0 \tag{10-56}$$

$$r^*_{I \cdot S} = k_S C_{N \cdot S} - k_I C_{I \cdot S} + k_{-I} P_I C_v = 0 \tag{10-57}$$

Solving for $C_{N \cdot S}$ and $C_{I \cdot S}$ gives

$$C_{N \cdot S} = \frac{k_N P_N C_v}{k_{-N} + k_S} \tag{10-58}$$

$$C_{I \cdot S} = \frac{k_S C_{N \cdot S} + k_{-I} P_I C_v}{k_I} = \left(\frac{k_S k_N P_N}{k_I (k_{-N} + k_S)} + \frac{k_{-I}}{k_I} P_I \right) C_v \tag{10-59}$$

and substituting for $C_{N \cdot S}$ in the surface reaction rate law gives

$$-r'_N = \frac{k_N k_S}{k_{-N} + k_S} P_N C_v \tag{10-60}$$

From a site balance we obtain

$$C_t = C_{N \cdot S} + C_{I \cdot S} + C_v \tag{10-61}$$

After substituting for $C_{N \cdot S}$ and $C_{I \cdot S}$, solving for C_v, which we then substitute in the rate law, we find

> Use the PSSH method when
> · Some steps are irreversible.
> · Two or more steps are rate-limiting.

$$-r'_N = \left(\frac{k_N k_S C_t}{k_{-N} + k_S} \right) \frac{P_N}{1 + \dfrac{k_N}{k_{-N} + k_S} \left(1 + \dfrac{k_S}{k_I} \right) P_N + \dfrac{k_{-I}}{k_I} P_I} \tag{10-62}$$

The adsorption constants are just the ratio of their respective rate constants:

$$K_I = \frac{k_{-I}}{k_I} \quad \text{and} \quad K_N = \frac{k_N}{k_{-N}}$$

$$-r'_N = \left(\overbrace{\frac{K_N k_S C_t}{1 + k_S/k_{-N}}}^{k} \right) \frac{P_N}{1 + \dfrac{1}{1 + k_S/k_{-N}} K_N P_N + K_I P_I}$$

We have taken the surface reaction step to be rate-limiting; therefore, the surface specific reaction constant, k_S, is much smaller than the rate constant for the desorption of normal pentene, k_{-N}; that is,

$$1 \gg \frac{k_S}{k_{-N}}$$

and the rate law given by Equation (10-62) becomes

$$-r'_A = \frac{k P_N}{1 + K_N P_N + K_I P_I} \tag{10-63}$$

This rate law [Equation (10-63)] is identical to the one derived assuming that $r_{AD}/k_{AD} \approx 0$ and $r_D/k_{AD} \approx 0$. However, this technique is preferred if two or more steps are rate-limiting or if some of the steps are irreversible or if none of the steps are rate-limiting.

10.3.7 Temperature Dependence of the Rate Law

Consider a surface-reaction-limited irreversible isomerization

$$A \longrightarrow B$$

in which both A and B are adsorbed on the surface, the rate law is

$$-r'_A = \frac{k P_A}{1 + K_A P_A + K_B P_B} \tag{10-64}$$

The specific reaction rate, k, will usually follow an Arrhenius temperature dependence and increase exponentially with temperature. However, the adsorption of all species on the surface is exothermic. Consequently, the higher the temperature, the smaller the adsorption equilibrium constant. That is, as the temperature increases, K_A and K_B decrease resulting in less coverage of the surface by A and B. Therefore, at high temperatures, the denominator of catalytic rate laws approaches 1. That is, at high temperatures (low coverage)

$$1 \gg (P_A K_A + P_B K_B)$$

The rate law could then be approximated as

Neglecting the
adsorbed species at
high temperatures

$$-r'_A \simeq kP_A \qquad (10\text{-}65)$$

or for a reversible isomerization we would have

$$-r'_A \simeq k\left(P_A - \frac{P_B}{K_P}\right) \qquad (10\text{-}66)$$

The algorithm we can use as a start in postulating a reaction mechanism and rate-limiting step is shown in Table 10-4. Again, we can never really prove a mechanism by comparing the derived rate law with experimental data. Independent spectroscopic experiments are usually needed to confirm the mechanism. We can, however, prove that a proposed mechanism is inconsistent with the experimental data by following the algorithm in Table 10-4. Rather than taking all the experimental data and then trying to build a model from the data, Box et al.[16] describe techniques of sequential data taking and model building.

Available strategies
for model building

10.4 Heterogeneous Data Analysis for Reactor Design

<u>Algorithm</u>
Deduce
  Rate law
Find
  Mechanism
Evaluate
  Rate law
  parameters
Design
  PBR
  CSTR

In this section we focus on four operations that reaction engineers need to be able to accomplish: (1) developing an algebraic rate law consistent with experimental observations, (2) analyzing the rate law in such a manner that the rate law parameters (e.g., k, K_A) can readily be extracted from the experimental data, (3) finding a mechanism and rate-limiting step consistent with the experimental data, and (4) designing a catalytic reactor to achieve a specified conversion. We shall use the hydrodemethylation of toluene to illustrate these four operations.

Hydrogen and toluene are reacted over a solid mineral catalyst containing clinoptilolite (a crystalline silica-alumina) to yield methane and benzene:[17]

Following the Algorithm

$$C_6H_5CH_3 + H_2 \longrightarrow C_6H_6 + CH_4$$

We wish to design a packed-bed reactor and a fluidized CSTR to process a feed consisting of 30% toluene, 45% hydrogen, and 25% inerts. Toluene is fed at a rate of 50 mol/min at a temperature of 640°C and a pressure of 40 atm (4052 kPa). To design the PBR, we must first determine the rate law from the differential reactor data presented in Table 10-6. In this table we find the rate of reaction of toluene as a function of the partial pressures of hydrogen (H), toluene (T), benzene (B), and methane (M). In the first two runs, methane was introduced into the feed together with hydrogen and toluene, while the other product, benzene, was fed to the reactor together with the reactants only in runs 3, 4, and 6. In runs 5 and 16 both methane and benzene were introduced in the feed. In the remaining runs, neither of the products was present in the feed-

[16] G. E. P. Box, W. G. Hunter, and J. S. Hunter, *Statistics for Engineers* (New York: Wiley, 1978).

[17] J. Papp, D. Kallo, and G. Schay, *J. Catal.*, **23**, 168 (1971).

TABLE 10-6.   DATA FROM A DIFFERENTIAL REACTOR

Run	$r_T' \times 10^{10}$ $\left(\dfrac{\text{g mol toluene}}{\text{g cat.} \cdot \text{s}}\right)$	Partial Pressure (atm)			
		Toluene, P_T	Hydrogen (H_2),[a] P_{H_2}	Methane, P_M	Benzene, P_B
Set A					
1	71.0	1	1	1	0
2	71.3	1	1	4	0
Set B					
3	41.6	1	1	0	1
4	19.7	1	1	0	4
5	42.0	1	1	1	1
6	17.1	1	1	0	5
Set C					
7	71.8	1	1	0	0
8	142.0	1	2	0	0
9	284.0	1	4	0	0
Set D					
10	47.0	0.5	1	0	0
11	71.3	1	1	0	0
12	117.0	5	1	0	0
13	127.0	10	1	0	0
14	131.0	15	1	0	0
15	133.0	20	1	0	0
16	41.8	1	1	1	1

Unscramble the data to find the rate law

[a] $P_H \equiv P_{H_2}$.

stream. Because the conversion was less than 1% in the differential reactor, the partial pressures of the products, methane and benzene, in these runs were essentially zero, and the reaction rates were equivalent to initial rates of reaction.

10.4.1 Deducing a Rate Law from the Experimental Data

Assuming that the reaction is essentially irreversible (which is reasonable after comparing runs 3 and 5), we ask what qualitative conclusions can be drawn from the data about the dependence of the rate of disappearance of toluene, $-r_T'$, on the partial pressures of toluene, hydrogen, methane, and benzene.

$$T + H_2 \longrightarrow M + B$$

1. *Dependence on the product methane.* If the methane were adsorbed on the surface, the partial pressure of methane would appear in the denominator of the rate expression and the rate would vary inversely with methane concentration:

$$-r_T' \sim \frac{[\cdot]}{1 + K_M P_M + \cdots} \qquad (10\text{-}67)$$

Following the Algorithm

If it is in the
denominator, it is
probably on the
surface.

However, from runs 1 and 2 we observe that a fourfold increase in the pressure of methane has little effect on $-r'_T$. Consequently, we assume that methane is either very weakly adsorbed (i.e., $K_M P_M \ll 1$) or goes directly into the gas phase in a manner similar to propylene in the cumene decomposition previously discussed.

2. *Dependence on the product benzene.* In runs 3 and 4, we observe that for fixed concentrations (partial pressures) of hydrogen and toluene the rate decreases with increasing concentration of benzene. A rate expression in which the benzene partial pressure appears in the denominator could explain this dependency:

$$-r'_T \sim \frac{1}{1 + K_B P_B + \cdots} \tag{10-68}$$

The type of dependence of $-r'_T$ on P_B given by Equation (10-68) suggests that benzene is adsorbed on the clinoptilolite surface.

3. *Dependence on toluene.* At low concentrations of toluene (runs 10 and 11), the rate increases with increasing partial pressure of toluene, while at high toluene concentrations (runs 14 and 15), the rate is essentially independent of the toluene partial pressure. A form of the rate expression that would describe this behavior is

$$-r'_T \sim \frac{P_T}{1 + K_T P_T + \cdots} \tag{10-69}$$

A combination of Equations (10-68) and (10-69) suggests that the rate law may be of the form

$$-r'_T \sim \frac{P_T}{1 + K_T P_T + K_B P_B + \cdots} \tag{10-70}$$

4. *Dependence on hydrogen.* When we examine runs 7, 8, and 9 in Table 10-6, we see that the rate increases linearly with increasing hydrogen concentration, and we conclude that the reaction is first order in H_2. In light of this fact, hydrogen is either not adsorbed on the surface or its coverage of the surface is extremely low ($1 \gg K_{H_2} P_{H_2}$) for the pressures used. If it were adsorbed, $-r'_T$ would have a dependence on P_{H_2} analogous to the dependence of $-r'_T$ on the partial pressure of toluene, P_T [see Equation (10-69)]. For first-order dependence on H_2,

$$-r'_T \sim P_{H_2} \tag{10-71}$$

Combining Equations (10-67) through (10-71), we find that the rate law

$$\boxed{-r'_T = \frac{k P_{H_2} P_T}{1 + K_B P_B + K_T P_T}}$$

is in qualitative agreement with the data shown in Table 10-6.

10.4.2 Finding a Mechanism Consistent
with Experimental Observations

We now propose a mechanism for the hydrodemethylation of toluene. We assume that toluene is adsorbed on the surface and then reacts with hydrogen in the gas phase to produce benzene adsorbed on the surface and methane in the gas phase. Benzene is then desorbed from the surface. Because approximately 75% of all heterogeneous reaction mechanisms are surface-reaction-limited rather than adsorption- or desorption-limited, we begin by assuming the reaction between adsorbed toluene and gaseous hydrogen to be reaction-rate-limited. Symbolically, this mechanism and associated rate laws for each elementary step are

Approximately 75% of all heterogeneous reaction mechanisms are surface-reaction-limited.

Proposed mechanism

$$\text{Adsorption:} \quad T(g) + S \;\rightleftharpoons\; T{\cdot}S$$

$$r_{AD} = k_A \left(C_v P_T - \frac{C_{T{\cdot}S}}{K_T} \right) \tag{10-72}$$

$$\text{Surface reaction:} \quad H_2(g) + T{\cdot}S \;\rightleftharpoons\; B{\cdot}S + M(g)$$

$$r_S = k_S \left(P_{H_2} C_{T{\cdot}S} - \frac{C_{B{\cdot}S} P_M}{K_S} \right) \tag{10-73}$$

$$\text{Desorption:} \quad B{\cdot}S \;\rightleftharpoons\; B(g) + S$$

$$r_D = k_D \left(C_{B{\cdot}S} - K_B P_B C_v \right) \tag{10-74}$$

For surface-reaction-limited mechanisms,

$$r_S = k_S \left(P_{H_2} C_{T{\cdot}S} - \frac{C_{B{\cdot}S} P_M}{K_S} \right) \tag{10-73}$$

we see that we need to replace $C_{T{\cdot}S}$ and $C_{B{\cdot}S}$ in Equation (10-73) by quantities that we can measure.

For surface-reaction-limited mechanisms, we use the adsorption rate Equation (10-72) to obtain $C_{T{\cdot}S}$:

$$\frac{r_{AD}}{k_A} \approx 0$$

Then

$$C_{T{\cdot}S} = K_T P_T C_v \tag{10-75}$$

and we use the desorption rate Equation (10-74) to obtain $C_{B{\cdot}S}$:

$$\frac{r_D}{k_D} \approx 0$$

Then

$$C_{B \cdot S} = K_B P_B C_v \tag{10-76}$$

The total concentration of sites is

Perform a site
balance to obtain C_v.

$$\boxed{C_t = C_v + C_{T \cdot S} + C_{B \cdot S}} \tag{10-77}$$

Substituting Equations (10-75) and (10-76) into Equation (10-77) and re-arranging, we obtain

$$C_v = \frac{C_t}{1 + K_T P_T + K_B P_B} \tag{10-78}$$

Next, substitute for $C_{T \cdot S}$ and $C_{B \cdot S}$ and then substitute for C_v in Equation (10-73) to obtain the rate law for the case of surface-reaction control:

$$-r_T' = \frac{\overbrace{C_t k_S K_T}^{k}(P_{H_2} P_T - P_B P_M / K_P)}{1 + K_T P_T + K_B P_B} \tag{10-79}$$

Neglecting the reverse reaction we have

Rate law for
surface-reaction-
limited mechanism

$$\boxed{-r_T' = \frac{k P_{H_2} P_T}{1 + K_B P_B + K_T P_T}} \tag{10-80}$$

Again we note that the adsorption equilibrium constant of a given species is exactly the reciprocal of the desorption equilibrium constant of that species.

10.4.3 Evaluation of the Rate Law Parameters

In the original work on this reaction by Papp et al.,[18] over 25 models were tested against experimental data, and it was concluded that the preceding mechanism and rate-limiting step (i.e., the surface reaction between adsorbed toluene and H_2 gas) is the correct one. Assuming that the reaction is essentially irreversible, the rate law for the reaction on clinoptilolite is

$$-r_T' = k \frac{P_{H_2} P_T}{1 + K_B P_B + K_T P_T} \tag{10-80}$$

We now wish to determine how best to analyze the data to evaluate the rate law parameters, k, K_T, and K_B. This analysis is referred to as *parameter estimation*.[19] We now rearrange our rate law to obtain a linear relationship between our measured variables. For the rate law given by Equation (10-80),

---

[18] Ibid.

[19] See the Supplementary Reading for a variety of techniques for estimating the rate law parameters.

we see that if both sides of Equation (10-80) are divided by $P_{H_2} P_T$ and the equation is then inverted,

Linearize the rate
equation to extract
the rate law
parameters.

$$\boxed{\frac{P_{H_2} P_T}{-r'_T} = \frac{1}{k} + \frac{K_B P_B}{k} + \frac{K_T P_T}{k}} \tag{10-81}$$

The regression techniques described in Chapter 5 could be used to determine the rate law parameters by using the equation

A linear-least-
squares analysis of
the data shown in
Table 10-6 is
presented on the
CD-ROM.

$$\boxed{Y_j = a_0 + a_1 X_{1j} + a_2 X_{2j}} \tag{R5.1-3}$$

One can use the linearized least-squares analysis to obtain initial estimates of the parameters k, K_T, K_B, in order to obtain convergence in nonlinear regression. However, in many cases it is possible to use a nonlinear regression analysis directly as described in Section 5.2.3 and in Example 10-4.

Living Example Problem

Example 10–2   *Regression Analysis to Determine the Model Parameters k, K_B, and K_T*

Solution

The data from Table 10-6 were entered into the Polymath nonlinear-least-squares program with the following modification. The rates of reaction in column 1 were multiplied by 10^{10}, so that each of the numbers in column 1 was entered directly (i.e., 71.0, 71.3, ...). The model equation was

$$\text{Rate} = \frac{k P_T P_{H_2}}{1 + K_B P_B + K_T P_T} \tag{E10-2.1}$$

Following the step-by-step regression procedure in Chapter 5 and the Summary Notes, we arrive at the following parameter values shown in Table E10-2.1.

TABLE E10-2.1.   PARAMETER VALUES

POLYMATH Results
Example 10-2 04-06-2004

Nonlinear regression (L-M)

Model: RATE = k*PT*PH2/(1+KB*PB+KT*PT)

Variable	Ini guess	Value	95% confidence
k	144	144.7673	1.2403071
KB	1.4	1.3905253	0.0457965
KT	1.03	1.0384105	0.0131585

Nonlinear regression settings
Max # iterations = 64

Precision
R^2   = 0.9999509
R^2adj   = 0.9999434
Rmsd   = 0.1128555
Variance = 0.2508084

Converting the rate law to kilograms of catalyst and minutes,

$$-r'_T = \frac{1.45 \times 10^{-8} P_T P_{H_2}}{1 + 1.39 P_B + 1.038 P_T} \frac{\text{mol T}}{\text{g cat} \cdot \text{s}} \times \frac{1000 \text{ g}}{1 \text{ kg}} \times \frac{60 \text{ s}}{\text{min}} \qquad \text{(E10-2.2)}$$

we have

$$-r'_T = \frac{8.7 \times 10^{-4} P_T P_{H_2}}{1 + 1.39 P_B + 1.038 P_T} \left(\frac{\text{g mol T}}{\text{kg cat.} \cdot \text{min}} \right) \qquad \text{(E10-2.3)}$$

Ratio of sites occupied by toluene to those occupied by benzene

After we have the adsorption constants, K_T and K_B, we can calculate the ratio of sites occupied by the various adsorbed species. For example, the ratio of toluene sites to benzene sites at 40% conversion is

$$\frac{C_{T \cdot S}}{C_{B \cdot S}} = \frac{C_v K_T P_T}{C_v K_B P_B} = \frac{K_T P_T}{K_B P_B} = \frac{K_T P_{A0}(1-X)}{K_B P_{A0} X}$$

$$= \frac{K_T(1-X)}{K_B X} = \frac{1.038(1-0.4)}{1.39(0.4)} = 1.12$$

We see that at 40% conversion there are approximately 12% more sites occupied by toluene than by benzene.

10.4.4 Reactor Design

Our next step is to express the partial pressures P_T, P_B, and P_{H_2} as a function of X, combine the partial pressures with the rate law, $-r'_A$, as a function of conversion, and carry out the integration of the packed-bed design equation

$$\frac{dX}{dW} = \frac{-r'_A}{F_{A0}} \qquad \text{(2-17)}$$

Example 10–3   Fixed (i.e., Packed)-Bed Reactor Design

Living Example Problem

The hydrodemethylation of toluene is to be carried out in a packed-bed reactor. Plot the conversion, the pressure ratio, y, and the partial pressures of toluene, hydrogen, and benzene as a function of catalyst weight. The molar feed rate of toluene to the reactor is 50 mol/min, and the reactor is operated at 40 atm and 640°C. The feed consists of 30% toluene, 45% hydrogen, and 25% inerts. Hydrogen is used in excess to help prevent coking. The pressure drop parameter, α, is 9.8×10^{-5} kg^{-1}. Also determine the catalyst weight in a CSTR with a bulk density of 400 kg/m^3 (0.4 g/cm^3).

$$C_6H_5CH_3 + H_2 \longrightarrow C_6H_6 + CH_4$$

Solution

1. **Design Equation:**

Balance on
toluene (T)

$$\frac{dF_T}{dW} = r'_T$$

$$\frac{dX}{dW} = \frac{-r'_T}{F_{T0}} \qquad \text{(E10-3.1)}$$

2. **Rate Law.** From Equation (E10-2.1) we have

$$-r'_T = \frac{kP_{H_2}P_T}{1 + K_B P_B + K_T P_T} \qquad \text{(E10-3.2)}$$

with $k = 0.00087$ mol/atm^2/kg cat/min and $K_B = 1.39$ atm^{-1} and $K_T = 1.038$ atm^{-1}.

3. **Stoichiometry:**

$$P_T = C_T RT = C_{T0}RT_0\left(\frac{1-X}{1+\varepsilon X}\right)y = P_{T0}\left(\frac{1-X}{1+\varepsilon X}\right)y$$

$$\varepsilon = y_{T0}\delta = 0.3(0) = 0$$

$$P_T = P_{T0}(1-X)y \qquad \text{(E10-3.3)}$$

Relating
Toluene (T)
Benzene (B)
Hydrogen (H$_2$)

$$P_{H_2} = P_{T0}(\Theta_{H_2} - X)y$$

$$\Theta_{H_2} = \frac{0.45}{0.30} = 1.5$$

$$P_{H_2} = P_{T0}(1.5 - X)y \qquad \text{(E10-3.4)}$$

$$P_B = P_{T0}Xy \qquad \text{(E10-3.5)}$$

Because $\varepsilon = 0$, we can use the integrated form of the pressure drop term.

P_0 = total pressure
at the entrance

$$y = \frac{P}{P_0} = (1 - \alpha W)^{1/2} \qquad \text{(4-33)}$$

$$\alpha = 9.8 \times 10^{-5} \text{ kg}^{-1}$$

Note that P_{T0} designates the initial partial pressure of toluene. In this example the initial total pressure is designated P_0 to avoid any confusion. The initial mole fraction of toluene is 0.3 (i.e., $y_{T0} = 0.3$), so that the initial partial pressure of toluene is

Pressure drop in
PBRs is discussed in
Section 4.5

$$P_{T0} = (0.3)(40) = 12 \text{ atm}$$

The maximum catalyst weight we can have and not fall below 1 atm is found from Equation (4-33) for an entering pressure of 40 atm and an exit pressure of 1 atm.

$$\frac{1}{40} = (1 - 9.8 \times 10^{-5}W)^{1/2}$$

$$W = 10{,}197 \text{ kg}$$

Consequently, we will set our final weight at 10,000 kg and determine the conversion as a function of catalyst weight up to this value. Equations (E10-3.1) through

The calculations for no ΔP are given on the CD-ROM.

Reference Shelf

(E10-3.5) are shown in the Polymath program in Table E10-3.1. The conversion is shown as a function of catalyst weight in Figure E10-3.1, and profiles of the partial pressures of toluene, hydrogen, and benzene are shown in Figure E10-3.2. We note that the pressure drop causes the partial pressure of benzene to go through a maximum as one traverses the reactor.

For the case of no pressure drop, the conversion that would have been achieved with 10,000 kg of catalyst weight would have been 79%, compared with 69% when there is pressure drop in the reactor. *For the feed rate given, eliminating or minimizing pressure drop would increase the production of benzene by up to 61 million pounds per year!*

Living Example Problem

TABLE E10-3.1.   POLYMATH PROGRAM

ODE Report (RKF45)

Differential equations as entered by the user
[1] d(X)/d(w) = -rt/FTo

Explicit equations as entered by the user
[1] FTo = 50
[2] k = .00087
[3] KT = 1.038
[4] KB = 1.39
[5] alpha = 0.000098
[6] Po = 40
[7] PTo = 0.3*Po
[8] y = (1-alpha*w)^0.5
[9] P = y*Po
[10] PH2 = PTo*(1.5-X)*y
[11] PB = PTo*X*y
[12] PT = PTo*(1-X)*y
[13] rt = -k*PT*PH2/(1+KB*PB+KT*PT)
[14] RATE = -rt

Conversion profile
down the
packed bed

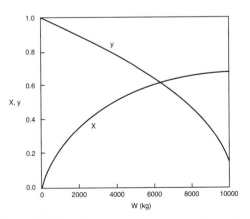

Figure E10-3.1   Conversion and pressure ratio profiles.

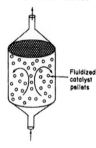

Figure E10-3.2   Partial pressure profiles.

Note the partial pressure of benzene goes through a maximum. Why?

We note in Figure E10-3.2 that the partial pressure of benzene goes through a maximum as a result of the decrease in total pressure because of the pressure drop.

We will now calculate the fluidized CSTR catalyst weight necessary to achieve the same conversion as in the packed-bed reactor at the same operating conditions. The bulk density in the fluidized reactor is 0.4 g/cm³. The design equation is

Fluidized CSTR

In	–	Out	+	Gen	=	Accum
F_{T0}	–	F_T	+	$r_T' W$	=	0

Fluidized catalyst pellets

Rearranging

$$W = \frac{F_{T0}-F_T}{-r_T'} = \frac{F_{T0}X}{-r_T'}$$

Writing Equation (E10-2.3) in terms of conversion and then substituting $X = 0.65$ and $P_{T0} = 12$ atm, we have

$$-r_T' = \frac{8.7\times10^{-4}P_T P_{H_2}}{1+1.39P_B+1.038P_T} = \frac{8.7\times10^{-4}P_{T0}^2(1-X)(1.5-X)}{1+1.39P_{T0}X+1.038P_{T0}(1-X)} = 2.3\times10^{-3}\frac{\text{mol}}{\text{kgcat}\cdot\text{min}}$$

$$W = \frac{F_{T0}X}{-r_T'} = \frac{(50 \text{ mol T/min})(0.65)}{1.043\times10^{-3} \text{ mol T/kg cat}\cdot\text{min}}$$

$$\boxed{W = 3.12\times10^4 \text{ kg of catalyst}}$$

$$V = \frac{3.12\times10^4 \text{ kg}}{400 \text{ kg/m}^3} = 77.9 \text{ m}^3$$

These values of the catalyst weight and reactor volume are quite high, especially for the low feed rates given. *Consequently, the temperature of the reacting mixture should be increased to reduce the catalyst weight, provided that side reactions do not become a problem at higher temperatures.*

How can the weight of catalyst be reduced?

Example 10-3 illustrated the major activities pertinent to catalytic reactor design described earlier in Figure 10-7. In this example the rate law was extracted directly from the data and then a mechanism was found that was consistent with experimental observation. Conversely, developing a feasible mechanism may guide one in the synthesis of the rate law.

10.5 Reaction Engineering in Microelectronic Fabrication

10.5.1 Overview

We now extend the principles of the preceding sections to one of the emerging technologies in chemical engineering. Chemical engineers are now playing an important role in the electronics industry. Specifically, they are becoming more involved in the manufacture of electronic and photonic devices and recording materials.

Surface reactions play an important role in the manufacture of microelectronics devices. One of the single most important developments of this century was the invention of the integrated circuit. Advances in the development of integrated circuitry have led to the production of circuits that can be placed on a single semiconductor chip the size of a pinhead and perform a wide variety of tasks by controlling the electron flow through a vast network of channels. These channels, which are made from semiconductors such as silicon, gallium arsenide, indium phosphide, and germanium, have led to the development of a multitude of novel microelectronic devices. Examples of microelectronic sensing devices manufactured using chemical reaction engineering principles are shown in the left-hand margin.

The manufacture of an integrated circuit requires the fabrication of a network of pathways for electrons. The principal reaction engineering steps of the fabrication process include depositing material on the surface of a material called a substrate (e.g., by chemical vapor deposition), changing the conductivity of regions of the surface (e.g., by boron doping or ion-inplantation), and removing unwanted material (e.g., by etching). By applying these steps systematically, miniature electronic circuits can be fabricated on very small semiconductor chips. The fabrication of microelectronic devices may include as few as 30 or as many as 200 individual steps to produce chips with up to 10^9 elements per chip.

An abbreviated schematic of the steps involved in producing a typical metal-oxide semiconductor field-effect transistor (MOSFET) device is shown in Figure 10-20. Starting from the upper left, we see that single-crystal silicon ingots are grown in a Czochralski crystallizer, sliced into wafers, and chemically and physically polished. These polished wafers serve as starting materials

for a variety of microelectronic devices. A typical fabrication sequence is shown for processing the wafer beginning with the formation of an SiO_2 layer on top of the silicon. The SiO_2 layer may be formed either by oxidizing a silicon layer or by laying down a SiO_2 layer by chemical vapor deposition (CVD). Next, the wafer is masked with a polymer photoresist, a template with the pattern to be etched onto the SiO_2 layer is placed over the photoresist, and the wafer is exposed to ultraviolet irradiation. If the mask is a positive photoresist, the light will cause the exposed areas of the polymer to dissolve when the wafer is placed in the developer. On the other hand, when a negative photoresist mask is exposed to ultraviolet irradiation, cross-linking of the polymer chains occurs, and the *unexposed* areas dissolve in the developer. The undeveloped portion of the photoresist (in either case) will protect the covered areas from etching.

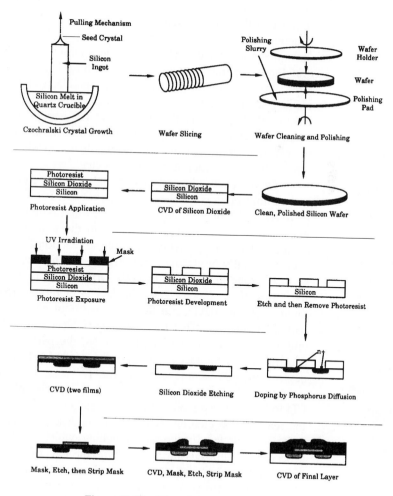

Figure 10-20   Microelectronic fabrication steps.

After the exposed areas of SiO_2 are etched to form trenches [either by wet etching (see Problem P5-12) or by plasma etching], the remaining photoresist is removed. Next, the wafer is placed in a furnace containing gas molecules of the desired dopant, which then diffuse into the exposed silicon. After diffusion of dopant to the desired depth in the wafer, the wafer is removed and covered with SiO_2 by CVD. The sequence of masking, etching, CVD, and metallization continues until the desired device is formed. A schematic of a final chip is shown in the lower right-hand corner of Figure 10-20. In Section 10.5.3 we discuss one of the key processing steps, CVD.

10.5.2 Etching

We have seen in Figure 10-20 that etching (i.e., the dissolution or physical or chemical removal of material) is also an important step in the fabrication process. Etching takes on a priority role in microelectronics manufacturing because of the need to create well-defined structures from an essentially homogeneous material. In integrated circuits, etching is necessary to remove unwanted material that could provide alternative pathways for electrons and thus hinder operation of the circuit. Etching is also of vital importance in the fabrication of micromechanical and optoelectronic devices. By selectively etching semiconductor surfaces, it is possible to fabricate motors and valves, ultrasmall diaphragms that can sense differences in pressure, or cantilever beams that can sense acceleration. In each of these applications, proper etching is crucial to remove material that would either short out a circuit or hinder movement of the micromechanical device.

There are two basic types of etching: wet etching and dry etching. The wet etching process, as described in Problem P5-12$_B$, uses liquids such as HF or KOH to dissolve the layered material that is unprotected by the photoresist mask. Wet etching is used primarily in the manufacture of micromechanical devices. Dry etching involves gas-phase reactions, which form highly reactive species, usually in plasmas, that impinge on the surface either to react with the surface, erode the surface, or both. Dry etching is used almost exclusively for the fabrication of optoelectronic devices. Optoelectronic devices differ from microelectronic devices in that they use light and electrons to carry out their particular function. That function may be detecting light, transmitting light, or emitting light. Etching is used to create the pathways or regions where light can travel and interact to produce the desired effects. Appliances using such devices include remote controls for TV sets, LED displays on clocks and microwave ovens, laser printers, and compact disc players. The material on the CD-ROM gives examples of both dry etching and wet etching. For *dry etching*, the reactive ion etching (RIE) of InP is described. Here the PSSH is used to arrive at a rate law for the rate of etching, which is compared with experimental observation. In discussing *wet etching*, the idea of dissolution catalysis is introduced and rate laws are derived and compared with experimental observation.

Reference Shelf
P10.3

In the formation of microcircuits, electrically interconnected films are laid down by chemical reactions (see Section 12.10). One method by which these films are made is chemical vapor deposition.

10.5.3 Chemical Vapor Deposition

The mechanisms by which CVD occurs are very similar to those of heterogeneous catalysis discussed earlier in this chapter. The reactant(s) adsorbs on the surface and then reacts on the surface to form a new surface. This process may be followed by a desorption step, depending on the particular reaction.

**Ge used in
Solar Cells**

    The growth of a germanium epitaxial film as an interlayer between a gallium arsenide layer and a silicon layer and as a contact layer is receiving increasing attention in the microelectronics industry.[20] Epitaxial germanium is also an important material in the fabrication of tandem solar cells. The growth of germanium films can be accomplished by CVD. A proposed mechanism is

Gas-phase dissociation: $GeCl_4(g) \rightleftharpoons GeCl_2(g) + Cl_2(g)$

Adsorption: $GeCl_2(g) + S \underset{}{\overset{k_A}{\rightleftharpoons}} GeCl_2 \cdot S$

Adsorption: $H_2 + 2S \underset{}{\overset{k_H}{\rightleftharpoons}} 2H \cdot S$

Surface reaction: $GeCl_2 \cdot S + 2H \cdot S \xrightarrow{k_S} Ge(s) + 2HCl(g) + 2S$

At first it may appear that a site has been lost when comparing the right- and left-hand sides of the surface reaction step. However, the newly formed germanium atom on the right-hand side is a site for the future adsorption of $H_2(g)$ or $GeCl_2(g)$, and there are three sites on both the right- and left-hand sides of the surface reaction step. These sites are shown schematically in Figure 10-21.

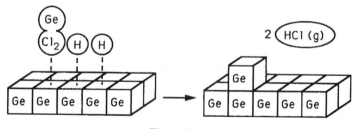

Figure 10-21

    The surface reaction between adsorbed molecular hydrogen and germanium dichloride is believed to be rate-limiting:

**Rate law for
rate-limiting step**

$$r''_{Dep} = k_S f_{GeCl_2} f_H^2 \qquad (10\text{-}82)$$

where r''_{Dep} = deposition rate, nm/s
          k_S = surface specific reaction rate, nm/s

[20]H. Ishii and Y. Takahashi, *J. Electrochem. Soc.*, *135*, p. 1539.

f_{GeCl_2} = fraction on the surface occupied by germanium dichloride

f_{H} = fraction of the surface covered by molecular hydrogen

The deposition rate (film growth rate) is usually expressed in nanometers per second and is easily converted to a molar rate $(\text{mol}/\text{m}^2\cdot\text{s})$ by multiplying by the molar density of solid germanium.

The difference between developing CVD rate laws and rate laws for catalysis is that the site concentration (e.g., C_v) is replaced by the fractional surface area coverage (e.g., the fraction of the surface that is vacant, f_v). The total fraction of surface available for adsorption should, of course, add up to 1.0.

Area balance   Fractional area balance: $\boxed{f_v + f_{\text{GeCl}_2} + f_{\text{H}} = 1}$ (10-83)

We will first focus our attention on the adsorption of GeCl_2. The rate of jumping on to the surface is proportional to the partial pressure of GeCl_2, P_{GeCl_2}, and the fraction of the surface that is vacant, f_v. The net rate of GeCl_2 adsorption is

$$r_{\text{AD}} = k_{\text{A}} \left(f_v P_{\text{GeCl}_2} - \frac{f_{\text{GeCl}_2}}{K_{\text{A}}} \right)$$ (10-84)

Since the surface reaction is rate-limiting, in a manner analogous to catalysis reactions, we have for the adsorption of GeCl_2

**Adsorption of
GeCl_2 not
rate-limiting**

$$\frac{r_{\text{AD}}}{k_{\text{A}}} \approx 0$$

Solving Equation (10-84) for the fractional surface coverage of GeCl_2 gives

$$\boxed{f_{\text{GeCl}_2} = f_v K_{\text{A}} P_{\text{GeCl}_2}}$$ (10-85)

For the dissociative adsorption of hydrogen on the Ge surface, the equation analogous to (10-84) is

$$r_{\text{H}_2} = k_{\text{H}} \left(P_{\text{H}_2} f_v^2 - \frac{f_{\text{H}}^2}{K_{\text{H}}} \right)$$ (10-86)

Since the surface reaction is rate-limiting,

**Adsorption of H_2 is
not rate-limiting**

$$\frac{r_{\text{H}_2}}{k_{\text{H}}} \approx 0$$

Then

$$\boxed{f_{\text{H}} = f_v \sqrt{K_{\text{H}} P_{\text{H}_2}}}$$ (10-87)

Recalling the rate of deposition of germanium, we substitute for f_{GeCl_2} and f_H in Equation (10-82) to obtain

$$r''_{Dep} = f_v^3 k_S K_A P_{GeCl_2} K_H P_{H_2} \tag{10-88}$$

We solve for f_v in an identical manner to that for C_v in heterogeneous catalysis. Substituting Equations (10-85) and (10-87) into Equation (10-83) gives

$$f_v + f_v \sqrt{K_H P_{H_2}} + f_v K_A P_{GeCl_2} = 1$$

Rearranging yields

$$f_v = \frac{1}{1 + K_A P_{GeCl_2} + \sqrt{K_H P_{H_2}}} \tag{10-89}$$

Finally, substituting for f_v in Equation (10-88), we find that

$$r''_{Dep} = \frac{k_S K_H K_A P_{GeCl_2} P_{H_2}}{(1 + K_A P_{GeCl_2} + \sqrt{K_H P_{H_2}})^3}$$

and lumping K_A, K_H, and k_S into a specific reaction rate k' yields

Rate of deposition of Ge

$$\boxed{r''_{Dep} = \frac{k' P_{GeCl_2} P_{H_2}}{(1 + K_A P_{GeCl_2} + \sqrt{K_H P_{H_2}})^3}} \tag{10-90}$$

If we assume that the gas-phase reaction

Equilibrium in gas phase

$$GeCl_4(g) \rightleftharpoons GeCl_2(g) + Cl_2(g)$$

is in equilibrium, we have

$$K_p = \frac{P_{GeCl_2} P_{Cl_2}}{P_{GeCl_4}}$$

$$P_{GeCl_2} = \frac{P_{GeCl_4}}{P_{Cl_2}} \cdot K_p$$

and if hydrogen is weakly adsorbed, $(\sqrt{K_H P_{H_2}} < 1)$ we obtain the rate of deposition as

$$r''_{Dep} = \frac{k P_{GeCl_4} P_{H_2} P_{Cl_2}^2}{(P_{Cl_2} + K P_{GeCl_4})^3} \tag{10-91}$$

It should also be noted that it is possible that $GeCl_2$ may also be formed by the reaction of $GeCl_4$ and a Ge atom on the surface, in which case a different rate law would result.

10.6 Model Discrimination

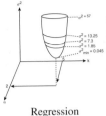

Regression

We have seen that for each mechanism and each rate-limiting step we can derive a rate law. Consequently, if we had three possible mechanisms and three rate-limiting steps for each mechanism, we would have nine possible rate laws to fit to the experimental data. We will use the regression techniques discussed in Chapter 5 to identify which model equation best fits the data by choosing the one with the smaller sums of squares and/or carrying out an F-test. We could also compare the residual plots for each model, which not only show the error associated with each data point but also show if the error is randomly distributed or if there is a trend in the error. If the error is randomly distributed, this result is an additional indication that the correct rate law has been chosen.

We need to raise a caution here about choosing the model with the smallest sums of squares. The caution is that the model parameter values that give the smallest sum must be realistic. In the case of heterogeneous catalysis, all values of the adsorption equilibrium constant must be positive. In addition, if the temperature dependence is given, because adsorption is exothermic, the adsorption equilibrium constant must decrease with increasing temperature. To illustrate these principles, let's look at the following example.

Example 10–4 Hydrogenation of Ethylene to Ethane

The hydrogenation (H) of ethylene (E) of form ethane (EA),

$$H_2 + C_2H_4 \rightarrow C_2H_6$$

is carried out over a cobalt molybdenum catalyst [*Collect. Czech. Chem. Commun.*, *51*, 2760 (1988)]. Carry out a nonlinear regression analysis on the data given in Table E10-4.1, and determine which rate law best describes the data.

TABLE E10-4.1.   DIFFERENTIAL REACTOR DATA

Run Number	Reaction Rate (mol/kg cat. • s)	P_E (atm)	P_{EA} (atm)	P_H (atm)
1	1.04	1	1	1
2	3.13	1	1	3
3	5.21	1	1	5
4	3.82	3	1	3
5	4.19	5	1	3
6	2.391	0.5	1	3
7	3.867	0.5	0.5	5
8	2.199	0.5	3	3
9	0.75	0.5	5	1

Determine which of the following rate laws best describes the data.

(a) $-r'_E = \dfrac{kP_E P_H}{1+K_{EA}P_{EA}+K_E P_E}$

(c) $-r'_E = \dfrac{kP_E P_H}{(1+K_E P_E)^2}$

(b) $-r'_E = \dfrac{kP_E P_H}{1+K_E P_E}$

(d) $-r'_E k P_E^a P_H^b$

Solution

Polymath was chosen as the software package to solve this problem. The data in Table E10-4.1 were typed into the system. A screen-shot by screen-shot set of instructions on how to carry out regression is given on the CD-ROM and the web.

After entering the data and following the step-by-step procedures described in the Summary Notes on the Web/CD-ROM of Chapter 5, the results shown in Table E10-4.2 were obtained.

Summary Notes

Living Example Problem

TABLE E10-4.2.   RESULTS OF THE POLYMATH NONLINEAR REGRESSION

Model (a)

```
Model: RATE = k*Pe*PH2/(1+KEA*Pea+KE*Pe)

Variable     Value      95% confidence
k            3.3478805    0.2922517
KEA          0.0428419    0.0636262
KE           2.2110797    0.2392585

Nonlinear regression settings
Max # iterations = 64

Precision
R^2      = 0.998321
R^2adj   = 0.9977614
Rmsd     = 0.0191217
Variance = 0.0049361
```

Model (b)

```
Model: RATE = k*Pe*PH2/(1+KE*Pe)

Variable     Value      95% confidence
k            3.1867851    0.287998
KE           2.1013363    0.2638835

Nonlinear regression settings
Max # iterations = 64

Precision
R^2      = 0.9975978
R^2adj   = 0.9972547
Rmsd     = 0.022872
Variance = 0.0060534
```

Model (c)

```
Model: RATE = k*Pe*PH2/(1+KE*Pe)^2

Variable     Value      95% confidence
k            2.0087761    0.2661838
KE           0.3616652    0.0623045

Nonlinear regression settings
Max # iterations = 64

Precision
R^2      = 0.9752762
R^2adj   = 0.9717442
Rmsd     = 0.0733772
Variance = 0.0623031
```

Model (d)

```
Model: RATE = k*Pe^a*PH2^b

Variable     Value      95% confidence
k            0.8940237    0.2505474
a            0.2584412    0.0704628
b            1.0615542    0.2041339

Nonlinear regression settings
Max # iterations = 64

Precision
R^2      = 0.9831504
R^2adj   = 0.9775338
Rmsd     = 0.0605757
Variance = 0.0495372
```

Model (a)

From Table E10-4.2 data, we can obtain

$$-r'_E = \frac{3.348\, P_E P_H}{1+0.043P_{EA}+2.21\, P_E}$$

We now examine the sums of squares (variance) and range of variables themselves. While the sums of squares is reasonable and in fact the smallest of all the models at 0.0049. However, let's look at K_{EA}. We note that from the values for the 95% confidence limit of ± 0.0636 is greater than the nominal value of $K_{EA} = 0.043$ atm^{-1} itself (i.e., $K_{EA} = 0.043 \pm 0.0636$). The 95% times means that if the experiment were run 100 times then 95 times it would fall within the range $(-0.021) < K_{EA} < (0.1066)$. Because K_{EA} can never be negative, we are going to reject this model. Consequently, we set $K_{EA} = 0$ and proceed to Model (b).

Model (b)

From Table E10-4.2 we can obtain

$$-r'_E = \frac{3.187\, P_E P_H}{1+2.1\, P_E}$$

The value of the adsorption constant $K_E = 2.1$ atm^{-1} is reasonable and is not negative within the 95% confidence limit. Also the variance is small at $\sigma_B^2 = 0.0061$.

Model (c)

From Table E10-4.2 we can obtain

$$-r'_E = \frac{2.0\, P_E P_{H_2}}{(1+0.036\, P_E)^2}$$

While K_E is small, it never goes negative within the 95% confidence interval. The variance of this model at $\sigma_C^2 = 0.0623$ is much larger than the other models. Comparing the variance of model **(c)** with model **(b)**

$$\frac{\sigma_C^2}{\sigma_B^2} = \frac{0.0623}{0.0061} = 10.2$$

We see that the σ_C^2 is an order of magnitude greater than σ_B^2, and we eliminate model **(c)**.[21]

Model (d)

Similarly for the power law model, we obtain from Table E10-4.2

$$-r'_E = 0.894\, P_E^{0.26} P_{H_2}^{1.06}$$

---

[21]See G. F. Froment and K. B. Bishoff, *Chemical Reaction Analysis and Design*, 2nd ed. (New York: Wiley, 1990), p. 96.

As with model **(c)** the variance is quite large compared to model **(b)**

$$\frac{\sigma_D^2}{\sigma_B^2} = \frac{0.049}{0.0061} = 8.03$$

Consequently, we see that for heterogeneous reactions, Langmuir-Hinshelwood rate laws are preferred over power law models.

Choose the Best Model.

Because all the parameter values are realistic for model **(b)** and the sums of squares are significantly smaller for model **(b)** than for the other models, we choose model **(b)**. We note again that there is a caution we need to point out regarding the use of regression! One cannot simply carry out a regression and then choose the model with the lowest value of the sums of squares. If this were the case, we would have chosen model **(a)**, which had the smallest sums of squares of all the models with σ^2 = 0.0049. However, one must consider the physical realism of the parameters in the model. In model **(a)** the 95% confidence interval was greater than the parameter itself, thereby yielding negative values of the parameter, K_{AE}, which is physically impossible.

10.7 Catalyst Deactivation

In designing fixed and ideal fluidized-bed catalytic reactors, we have assumed up to now that the activity of the catalyst remains constant throughout the catalyst's life. That is, the total concentration of active sites, C_t, accessible to the reaction does not change with time. Unfortunately, Mother Nature is not so kind as to allow this behavior to be the case in most industrially significant catalytic reactions. One of the most insidious problems in catalysis is the loss of catalytic activity that occurs as the reaction takes place on the catalyst. A wide variety of mechanisms have been proposed by Butt and Petersen,[22] to explain and model catalyst deactivation.

Catalytic deactivation adds another level of complexity to sorting out the reaction rate law parameters and pathways. In addition, we need to make adjustments for the decay of the catalysts in the design of catalytic reactors. This adjustment is usually made by a quantitative specification of the catalyst's activity, $a(t)$. In analyzing reactions over decaying catalysts we divide the reactions into two categories: *separable kinetics* and *nonseparable kinetics*. In separable kinetics, we separate the rate law and activity:

[22] J. B. Butt and E. E. Petersen, *Activation, Deactivation and Poisoning of Catalysts* (New York: Academic Press, 1988). See also S. Szépe and O. Levenspiel, *Chem. Eng. Sci.*, 23, 881–894 (1968).

Separable kinetics: $-r_A' = a(\text{Past history}) \times -r_A'$ (Fresh catalyst)

When the kinetics and activity are separable, it is possible to study catalyst decay and reaction kinetics independently. However, nonseparability,

Nonseparable kinetics: $-r_A' = -r_A'$ (Past history, Fresh catalyst)

must be accounted for by assuming the existence of a nonideal surface or by describing deactivation by a mechanism composed of several elementary steps.[23]

In this section we shall consider only *separable kinetics* and define the activity of the catalyst at time t, $a(t)$, as the ratio of the rate of reaction on a catalyst that has been used for a time t to the rate of reaction on a fresh catalyst ($t = 0$):

$a(t)$: catalyst
activity

$$a(t) = \frac{-r_A'(t)}{-r_A'(t = 0)} \tag{10-92}$$

Because of the catalyst decay, the activity decreases with time and a typical curve of the activity as a function of time is shown in Figure 10-22.

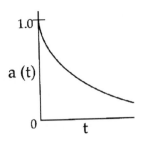

Figure 10-22   Activity as a function of time.

Combining Equations (10-92) and (3-2), the rate of disappearance of reactant A on a catalyst that has been utilized for a time t is

Reaction rate law
accounting for
catalyst activity

$$\boxed{-r_A' = a(t)\,k(T)\,\text{fn}(C_A, C_B, \ldots, C_P)} \tag{10-93}$$

where $a(t)$ = catalytic activity, time-dependent
$k(T)$ = specific reaction rate, temperature-dependent
C_i = gas-phase concentration of reactants, products, or contaminant

The rate of catalyst decay, r_d, can be expressed in a rate law analogous to Equation (10-93):

Catalyst decay
rate law

$$r_d = -\frac{da}{dt} = p[a(t)]k_d(T)h(C_A, C_B, \ldots, C_P) \tag{10-94}$$

[23] D. T. Lynch and G. Emig, *Chem. Eng. Sci.*, *44*(6), 1275–1280 (1989).

where $p[a(t)]$ is some function of the activity, k_d is the specific decay constant, and $h(C_i)$ is the functionality of r_d on the reacting species concentrations. For the cases presented in this chapter, this functionality either will be independent of concentration (i.e., $h = 1$) or will be a linear function of species concentration (i.e., $h = C_i$).

The functionality of the activity term, $p[a(t)]$, in the decay law can take a variety of forms. For example, for a first-order decay,

$$p(a) = a \tag{10-95}$$

and for a second-order decay,

$$p(a) = a^2 \tag{10-96}$$

The particular function, $p(a)$, will vary with the gas catalytic system being used and the reason or mechanism for catalytic decay.

10.7.1 Types of Catalyst Deactivation

- Sintering
- Fouling
- Poisoning

There are three categories into which the loss of catalytic activity can traditionally be divided: sintering or aging, fouling or coking, and poisoning.

Deactivation by Sintering (Aging).[24]   Sintering, also referred to as aging, is the loss of catalytic activity due to a loss of active surface area resulting from the prolonged exposure to high gas-phase temperatures. The active surface area may be lost either by crystal agglomeration and growth of the metals deposited on the support or by narrowing or closing of the pores inside the catalyst pellet. A change in the surface structure may also result from either surface recrystallization or the formation or elimination of surface defects (active sites). The reforming of heptane over platinum on alumina is an example of catalyst deactivation as a result of sintering.

Figure 10-23 shows the loss of surface area resulting from the flow of the solid porous catalyst support at high temperatures to cause pore closure. Figure 10-24 shows the loss of surface by atomic migration and agglomeration of small metal sites deposited on the surface into a larger site where the interior atoms are not accessible to the reaction. Sintering is usually negligible at temperatures below 40% of the melting temperature of the solid.[25]

The catalyst support becomes soft and flows, resulting in pore closure.

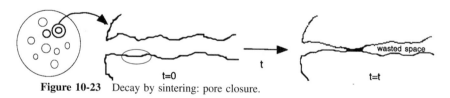

Figure 10-23   Decay by sintering: pore closure.

[24] See G. C. Kuczynski, Ed., *Sintering and Catalysis*, Vol. 10 of *Materials Science Research* (New York: Plenum Press, 1975).

[25] R. Hughes, *Deactivation of Catalysts* (San Diego: Academic Press, 1984).

The atoms move
along the surface
and agglomerate.

Top View

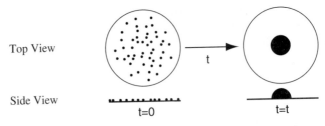

Side View

t=0   t=t

Figure 10-24   Decay by sintering: agglomeration of deposited
metal sites.

Deactivation by sintering may in some cases be a function of the main-stream gas concentration. Although other forms of the sintering decay rate laws exist, one of the most commonly used decay laws is second order with respect to the present activity:

$$r_d = k_d a^2 = -\frac{da}{dt} \tag{10-97}$$

Integrating, with $a = 1$ at time $t = 0$, yields

Sintering: second-order decay

$$\boxed{a(t) = \frac{1}{1+k_d t}} \tag{10-98}$$

The amount of sintering is usually measured in terms of the active surface area of the catalyst S_a:

$$S_a = \frac{S_{a0}}{1+k_d t} \tag{10-99}$$

The sintering decay constant, k_d, follows the Arrhenius equation

$$\boxed{k_d = k_d(T_0) \exp\left[\frac{E_d}{R}\left(\frac{1}{T_0} - \frac{1}{T}\right)\right]} \tag{10-100}$$

Minimizing
sintering

The decay activation energy, E_d, for the reforming of heptane on Pt/Al$_2$O$_3$ is on the order of 70 kcal/mol, which is rather high. As mentioned earlier sintering can be reduced by keeping the temperature below 0.3 to 0.4 times the metal's melting point.

We will now stop and consider reactor design for a fluid–solid system with decaying catalyst. To analyze these reactors, we only add one step to our algorithm, that is, determine the catalyst decay law. The sequence is shown here.

The algorithm

Mole balance $\longrightarrow$ Reaction rate law $\longrightarrow$ *Decay rate law* $\longrightarrow$

Stoichiometry $\longrightarrow$ Combine and solve $\longrightarrow$ Numerical techniques

Example 10–5   Calculating Conversion with Catalyst Decay in Batch Reactors

The first-order isomerization

$$A \longrightarrow B$$

is being carried out isothermally in a batch reactor on a catalyst that is decaying as a result of aging. Derive an equation for conversion as a function of time.

Solution

1. **Design equation:**

$$N_{A0} \frac{dX}{dt} = -r'_A W \qquad (E10\text{-}5.1)$$

2. **Reaction rate law:**

$$-r'_A = k' a(t) C_A \qquad (E10\text{-}5.2)$$

One extra step (number 3) is added to the algorithm.

3. **Decay law.** For second-order decay by sintering:

$$a(t) = \frac{1}{1 + k_d t} \qquad (10\text{-}98)$$

Following the Algorithm

4. **Stoichiometry:**

$$C_A = C_{A0}(1 - X) = \frac{N_{A0}}{V}(1 - X) \qquad (E10\text{-}5.3)$$

5. **Combining** gives us

$$\frac{dX}{dt} = \frac{W}{V} k' a(t)(1 - X) \qquad (E10\text{-}5.4)$$

Let $k = k'W/V$. Then, separating variables, we have

$$\frac{dX}{1 - X} = k a(t)\, dt \qquad (E10\text{-}5.5)$$

Substituting for a and integrating yields

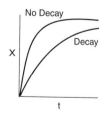

$$\int_0^X \frac{dX}{1 - X} = k \int_0^t \frac{dt}{1 + k_d t} \qquad (E10\text{-}5.6)$$

$$\ln \frac{1}{1 - X} = \frac{k}{k_d} \ln(1 + k_d t) \qquad (E10\text{-}5.7)$$

6. **Solving** for the conversion X at any time t, we find that

$$X = 1 - \frac{1}{(1 + k_d t)^{k/k_d}} \qquad (E10\text{-}5.8)$$

This is the conversion that will be achieved in a batch reactor for a first-order reaction when the catalyst decay law is second order. The purpose of this example was to demonstrate the algorithm for isothermal catalytic reactor design for a decaying catalyst. In problem 10-2(e) you are asked to sketch the temperature–time trajectories for various values of k and k_d.

Deactivation by Coking or Fouling. This mechanism of decay (see Figures 10-25 and 10-26) is common to reactions involving hydrocarbons. It results from a carbonaceous (coke) material being deposited on the surface of a catalyst.

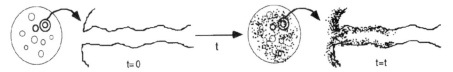

Figure 10-25   Schematic of decay by coking.

(a) Fresh catalyst   (b) Spent catalyst

Figure 10-26   Decay by coking. (Photos courtesy of Engelhard catalyst, copyright by Michael Gaffney Photographer, Mendham, N.J.)

The amount of coke on the surface after a time t has been found to obey the following empirical relationship:

$$\boxed{C_C = At^n}$$

(10-101)

where C_C is the concentration of carbon on the surface (g/m^2) and n and A are fouling parameters, which can be functions of the feed rate. This expression was originally developed by Voorhies[26] and has been found to hold for a wide variety of catalysts and feed streams. Representative values of A and n for the cracking of East Texas light gas oil[27] yield

[26] A. Voorhies, *Ind. Eng. Chem.*, *37*, 318 (1945).

[27] C. O. Prater and R. M. Lago, *Adv. Catal.*, *8*, 293 (1956).

$$\% \text{ coke} = 0.47 \sqrt{t(\text{min})}$$

Different functionalities between the activity and amount of coke on the surface have been observed. One commonly used form is

$$a = \frac{1}{k_{Ck} C_C^p + 1} \tag{10-102}$$

or, in terms of time,

$$a = \frac{1}{k_{Ck} A^p t^{np} + 1} = \frac{1}{1 + k' t^m} \tag{10-103}$$

For light Texas gas oil being cracked at 750°F over a synthetic catalyst for short times, the decay law is

$$a = \frac{1}{1 + 7.6 \, t^{1/2}} \tag{10-104}$$

where t is in seconds.

Activity for deactivation by coking

Other commonly used forms are

$$a = e^{-\alpha_1 C_C} \tag{10-105}$$

and

$$a = \frac{1}{1 + \alpha_2 C_C} \tag{10-106}$$

A dimensionless fouling correlation has been developed by Pacheco and Petersen.[28]

Minimizing coking

When possible, coking can be reduced by running at elevated pressures (2000 to 3000 kPa) and hydrogen-rich streams. A number of other techniques for minimizing fouling are discussed by Bartholomew.[29] Catalysts deactivated by coking can usually be regenerated by burning off the carbon. The use of the shrinking core model to describe regeneration is discussed in Section 11.5.1.

Deactivation by Poisoning. Deactivation by this mechanism occurs when the poisoning molecules become irreversibly chemisorbed to active sites, thereby reducing the number of sites available for the main reaction. The poisoning molecule, P, may be a reactant and/or a product in the main reaction, or it may be an impurity in the feed stream.

---

[28] M. A. Pacheco and E. E. Petersen, *J. Catal.*, *86*, 75 (1984).
[29] C. Bartholomew, *Chem. Eng.*, Sept. 12, 1984, p. 96.

Side Note.   One of the most significant examples of catalyst poisoning occurred at the gasoline pump. Oil companies found that adding lead to the gasoline increased the octane number. The television commercials said "We are going to enhance your gasoline, *but it's going to cost you* for the added tetra-ethyl lead." So for many years they used lead as an antiknock component. As awareness grew about NO, HC, and CO emission from the engine, it was decided to add a catalytic afterburner in the exhaust system to reduce these emissions. Unfortunately, it was found that the lead in the gasoline poisoned the reactive catalytic sites. So, the television commercials now said "We are going to take the lead out of gasoline but to receive the same level of performance as without lead, *but it's going to cost you* because of the added refining costs to raise the octane number." Do you think that financially, the consumer would have been better off if they never put the lead in the gasoline in the first place??

Poison in the Feed.   Many petroleum feed stocks contain trace impurities such as sulfur, lead, and other components which are too costly to remove, yet poison the catalyst slowly over time. For the case of an impurity, P, in the feed stream, such as sulfur, for example, in the reaction sequence

Main
reaction:
$$\left.\begin{array}{l} A+S \rightleftharpoons (A\cdot S) \\ A\cdot S \rightleftharpoons (B\cdot S + C(g)) \\ B\cdot S \rightleftharpoons (B+S) \end{array}\right\} \quad -r'_A = a(t)\,\frac{kC_A}{1+K_A C_A + K_B C_B}$$

Poisoning
reaction:
$$P+S \longrightarrow P\cdot S \qquad r_d = -\frac{da}{dt} = k'_d C_p^m a^q \qquad (10\text{-}107)$$

the surface sites would change with time as shown in Figure 10-27.

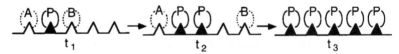

Figure 10-27   Decay by poisoning.

If we assume the rate of removal of the poison, $r_{P\cdot S}$, from the reactant gas stream onto the catalyst sites is proportional to the number of sites that are unpoisoned ($C_{t0} - C_{P\cdot S}$) and the concentration of poison in the gas phase, C_P:

$$r_{P\cdot S} = k_d(C_{t0} - C_{P\cdot S})C_P$$

where $C_{P\cdot S}$ is the concentration of poisoned sites and C_{t0} is the total number of sites initially available. Because every molecule that is adsorbed from the gas phase onto a site is assumed to poison the site, this rate is also equal to the rate of removal of total active sites (C_t) from the surface:

$$-\frac{dC_t}{dt} = \frac{dC_{P\cdot S}}{dt} = r_{P\cdot S} = k_d(C_{t0} - C_{P\cdot S})C_P$$

Dividing through by C_{t0} and letting f be the fraction of the total number of sites that have been poisoned yields

$$\frac{df}{dt} = k_d(1 - f)C_P \tag{10-108}$$

The fraction of sites available for adsorption $(1 - f)$ is essentially the activity $a(t)$. Consequently, Equation (10-108) becomes

$$-\frac{da}{dt} = a(t)k_dC_P \tag{10-109}$$

A number of examples of catalysts with their corresponding catalyst poisons are given by Bartholomew.[30]

Packed-Bed Reactors.   In packed-bed reactors where the poison is removed from the gas phase by being adsorbed on the specific catalytic sites, the deactivation process can move through the packed bed as a wave front. Here, at the start of the operation, only those sites near the entrance to the reactor will be deactivated because the poison (which is usually present in trace amounts) is removed from the gas phase by the adsorption; consequently, the catalyst sites farther down the reactor will not be affected. However, as time continues, the sites near the entrance of the reactor become saturated, and the poison must travel farther downstream before being adsorbed (removed) from the gas phase and attaching to a site to deactivate it. Figure 10-28 shows the corresponding activity profile for this type of poisoning process. We see in Figure 10-28 that by time t_4 the entire bed has become deactivated. The corresponding overall conversion at the exit of the reactor might vary with time as shown in Figure 10-29.

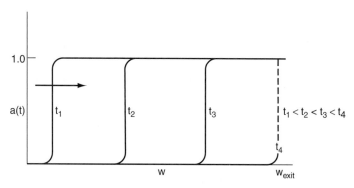

Figure 10-28   Movement of activity front in a packed bed.

Summary Notes

The partial differential equations that describe the movement of the reaction front shown in Figure 10-28 are derived and solved in an example in the CD-ROM/Web Summary Notes for Chapter 10.

---

[30] R. J. Farrauto and C. H. Bartholomew, *Fundamentals of Industrial Catalytic Processes* (New York: Blackie Academic and Professional, 1997). This book is one of the definitive resources on catalyst decay.

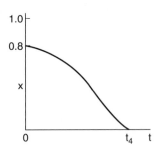

Figure 10-29   Exit conversion as a function of time.

Poisoning by Either Reactants or Products.   For the case where the main reactant also acts as a poison, the rate laws are:

Main reaction: $A + S \longrightarrow B + S$ $-r'_A = k_A C_A^n$

Poisoning reaction: $A + S \longrightarrow A \cdot S$ $r_d = k'_d C_A^m a^q$

An example where one of the reactants acts as a poison is in the reaction of CO and H_2 over ruthenium to form methane, with

$$-r_{CO} = k a(t) C_{CO}$$

$$-\frac{da}{dt} = r_d = k'_d a(t) C_{CO}$$

Similar rate laws can be written for the case when the product B acts as a poison.

For *separable deactivation kinetics* resulting from contacting a poison at a constant concentration C_{P_0} and no spatial variation:

Separable
deactivation
kinetics

$$-\frac{da}{dt} = r_d = k'_d C_{P_0}^n a^n(t) = k_d a^n \qquad (10\text{-}110)$$

The solution to this equation for the case of first-order decay, $n = 1$

$$-\frac{da}{dt} = k_d a \qquad (10\text{-}111)$$

is

$$a = e^{-k_d t} \qquad (10\text{-}112)$$

Empirical Decay Laws.   Table 10-7 gives a number of empirical decay laws along with the reaction systems to which they apply.

Key Resource for
catalyst deactivation

One should also see *Fundamentals of Industrial Catalytic Processes,* by Farrauto and Bartholomew,[31] which contains rate laws similar to those in Table 10-7, and also gives a comprehensive treatment of catalyst deactivation.

---

[31]Ibid.

TABLE 10-7.   DECAY RATE LAWS

Functional Form of Activity	Decay Reaction Order	Differential Form	Integral Form	Examples
Linear	0	$-\dfrac{da}{dt} = \beta_0$	$a = 1 - \beta_0 t$	Conversion of *para*-hydrogen on tungsten when poisoned with oxygen[a]
Exponential	1	$-\dfrac{da}{dt} = \beta_1 a$	$a = e^{-\beta_1 t}$	Ethylene hydrogenation on Cu poisoned with CO[b]
				Paraffin dehydrogenation on $Cr \cdot Al_2O_3$[c]
				Cracking of gas oil[d]
				Vinyl chloride monomer formation[e]
Hyperbolic	2	$-\dfrac{da}{dt} = \beta_2 a^2$	$\dfrac{1}{a} = 1 + \beta_2 t$	Vinyl chloride monomer formation[f]
				Cyclohexane dehydrogenation on Pt/Al_2O_3[g]
				Isobutylene hydrogenation on Ni[h]
Reciprocal power	$\dfrac{\beta_3 + 1}{\beta_3} = \gamma$	$-\dfrac{da}{dt} = \beta_3 a^n A_0^{1/5}$	$a = A_0 t^{-\beta_3}$	Cracking of gas oil and gasoline on clay[i]
	$\dfrac{\beta_4 + 1}{\beta_4} = n$	$-\dfrac{da}{dt} = \beta_4 a^n A_0^{1/5}$	$a = A_0 t^{-\beta_4}$	Cyclohexane aromatization on NiAl[j]

[a] D. D. Eley and E. J. Rideal, *Proc. R. Soc. London*, *A178*, 429 (1941).
[b] R. N. Pease and L. Y. Steward, *J. Am. Chem. Soc.*, *47*, 1235 (1925).
[c] E. F. K. Herington and E. J. Rideal, *Proc. R. Soc. London*, *A184*, 434 (1945).
[d] V. W. Weekman, *Ind. Eng. Chem. Process Des. Dev.*, *7*, 90 (1968).
[e] A. F. Ogunye and W. H. Ray, *Ind. Eng. Chem. Process Des. Dev.*, *9*, 619 (1970).
[f] A. F. Ogunye and W. H. Ray, *Ind. Eng. Chem. Process Des. Dev.*, *10*, 410 (1971).
[g] H. V. Maat and L. Moscou, *Proc. 3rd Int. Congr. Catal.* (Amsterdam: North-Holland, 1965), p. 1277.
[h] A. L. Pozzi and H. F. Rase, *Ind. Eng. Chem.*, *50*, 1075 (1958).
[i] A. Voorhies, Jr., *Ind. Eng. Chem.*, *37*, 318 (1945); E. B. Maxted, *Adv. Catal.*, *3*, 129 (1951).
[j] C. G. Ruderhausen and C. C. Watson, *Chem. Eng. Sci.*, *3*, 110 (1954).
Source: J. B. Butt, Chemical Reactor Engineering–Washington, *Advances in Chemistry Series 109* (Washington, D.C.: American Chemical Society, 1972), p. 259. Also see CES 23, 881(1968)

Examples of reactions with decaying catalysts and their decay laws

Example 10–6   Catalyst Decay in a Fluidized Bed Modeled as a CSTR

The gas-phase cracking reaction[†]

$$Gas\ oil\ (g) \longrightarrow Products\ (g)$$

$$A \longrightarrow B + C$$

[†] For simplicity, gas oil is used to represent the reactive portion of the feed. In actuality, gas oil, distilled from crude, is made up of complex hydrocarbons, which can be cracked, and simple hydrocarbons, which will not crack and are therefore inert in this application.

is carried out in a *fluidized* CSTR reactor. The feed stream contains 80% gas oil (A) and 20% inert I. The gas oil contains sulfur compounds, which poison the catalyst. As a first approximation we will assume that the cracking reaction is first order in the gas oil concentration. The rate of catalyst decay is first order in the present activity, and first order in the reactant concentration. Assuming that the bed can be modeled as a well-mixed CSTR, determine the reactant concentration, activity, and conversion as a function of time. The volumetric feed rate to the reactor is 5000 m³/h. There are 50,000 kg of catalyst in the reactor and the bulk density is 500 kg/m³.

Living Example Problem

Additional information:

$$C_{A0} = 0.8 \text{ mol/dm}^3 \qquad k = \rho_B k' = 45 \text{ h}^{-1}$$

$$C_{T0} = 1.0 \text{ mol/dm}^3 \qquad k_d = 9 \text{ dm}^3/\text{mol} \cdot \text{h}$$

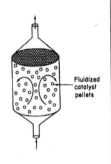

Fluidized catalyst pellets

Solution

1. **Mole Balance** on reactant:

$$v_0 C_{A0} - v C_A + r'_A W = \frac{dN_A}{dt} \tag{E10-6.1}$$

Recalling $N_A = C_A V$ and $r_A V = r'_A W$, then for constant volume we have

$$\boxed{v_0 C_{A0} - v C_A + r_A V = V \frac{dC_A}{dt}} \tag{E10-6.2}$$

2. **Rate Law:**

$$\boxed{-r_A = k a C_A} \tag{E10-6.3}$$

3. **Decay Law:**

$$\boxed{-\frac{da}{dt} = k_d a C_A} \tag{E10-6.4}$$

4. **Stoichiometry** (gas phase, $P = P_0$, $T = T_0$). From Equation (3-41) we have

$$\frac{v}{v_0} = \frac{F_T}{F_{T0}} = (1 + \varepsilon X) \tag{E10-6.5}$$

$$X = 1 - \frac{F_A}{F_{A0}} = 1 - \frac{C_A v}{C_{A0} v_0}$$

$$\frac{v}{v_0} = 1 + \varepsilon - \varepsilon \frac{C_A}{C_{A0}} \frac{v}{v_0}$$

$$\varepsilon = y_{A0} \delta = (1 + 1 - 1) y_{A0} = y_{A0} = \frac{C_{A0}}{C_{T0}}$$

$$\frac{v}{v_0} = 1 + y_{A0} - \frac{C_A}{C_{T0}}\frac{v}{v_0} \tag{E10-6.6}$$

Solving for v yields

$$v = v_0 \frac{1 + y_{A0}}{1 + C_A/C_{T0}} \tag{E10-6.7}$$

5. **Combining** gives us

$$v_0 C_{A0} - \frac{v_0(1+y_{A0})}{1+C_A/C_{T0}}C_A - kaC_A V = V\frac{dC_A}{dt} \tag{E10-6.8}$$

Dividing both sides of Equation (E10-6.8) by the volume and writing the equation in terms of $\tau = V/v_0$, we obtain

$$\frac{dC_A}{dt} = \frac{C_{A0}}{\tau} - \frac{(1+y_{A0})/(1+C_A/C_{T0}) + a\tau k}{\tau}C_A \tag{E10-6.9}$$

As an approximation we assume the conversion to be

$$X = \frac{F_{A0} - F_A}{F_{A0}} = 1 - \frac{vC_A}{v_0 C_{A0}} = 1 - \left(\frac{1+y_{A0}}{1+C_A/C_{T0}}\right)\left(\frac{C_A}{C_{A0}}\right) \tag{E10-6.10}$$

Calculation of reactor volume and space time yields

$$V = \frac{W}{\rho_b} = \frac{50{,}000}{500 \text{ kg/m}^3} = 100 \text{ m}^3$$

$$\tau = \frac{V}{v_0} = \frac{100 \text{ m}^3}{5000 \text{ m}^3/\text{h}} = 0.02 \text{ h}$$

Equations (E10-6.4), (E10-6.9), and (E10-6.10) are solved using Polymath as the ODE solver. The Polymath program is shown in Table E10-6.1. The solution is shown in Figure E10-6.1.

The conversion variable X does not have much meaning in flow systems not at steady state, owing to the accumulation of reactant. However, here the space time is relatively short ($\tau = 0.02$ h) in comparison with the time of decay $t = 0.5$ h. Consequently, we can assume a quasi-steady state and consider the conversion as defined by Equation (E10-6.10) valid. Because the catalyst decays in less than an hour, a fluidized bed would not be a good choice to carry out this reaction.

TABLE E10-6.1.   POLYMATH PROGRAM

Living Example Problem

ODE Report (RKF45)

Differential equations as entered by the user
[1] d(a)/d(t) = -kd*a*Ca
[2] d(Ca)/d(t) = Ca0/tau-((1+yao)/(1+Ca/Ct0)+tau*a*k)*Ca/tau

Explicit equations as entered by the user
[1] kd = 9
[2] Ca0 = .8
[3] tau = .02
[4] Ct0 = 1.
[5] k = 45
[6] yao = Ca0/Ct0
[7] X = 1-(1+yao)/(1+Ca/Ct0)*Ca/Ca0

C_A, X, and a time
trajectories in a
CSTR *not* at
steady state

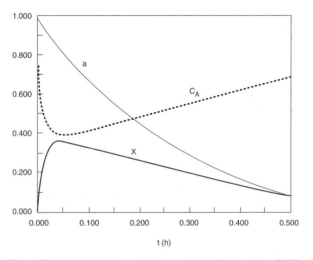

Figure E10-6.1   Variation of C_A, a, and X with time in a CSTR.

We will now consider three reaction systems that can be used to handle systems with decaying catalyst. We will classify these systems as those having slow, moderate, and rapid losses of catalytic activity. To offset the decline in chemical reactivity of decaying catalysts in continuous flow reactors, the following three methods are commonly used:

Matching the reactor
type with speed of
catalyst decay

- Slow decay – *Temperature–Time Trajectories* (10.7.2)
- Moderate decay – *Moving-Bed Reactors* (10.7.3)
- Rapid decay – *Straight-Through Transport Reactors* (10.7.4)

10.7.2 Temperature–Time Trajectories

In many large-scale reactors, such as those used for hydrotreating, and reaction systems where deactivation by poisoning occurs, the catalyst decay is relatively slow. In these continuous flow systems, constant conversion is usually necessary in order that subsequent processing steps (e.g., separation) are not upset. One way to maintain a constant conversion with a decaying catalyst in a packed or fluidized bed is to increase the reaction rate by steadily increasing the feed temperature to the reactor. Operation of a "fluidized" bed in this manner is shown in Figure 10-30.

Slow rate of catalyst decay

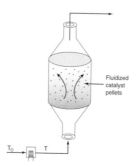

Fluidized
catalyst
pellets

Figure 10-30   Reactor with preheater to increase feed temperature.

We are going to increase the feed temperature in such a manner that the reaction rate remains constant with time:

$$-r_A' (t = 0, T_0) = -r_A' (t, T) = a(t, T)[-r_A' (t = 0, T)]$$

For a first-order reaction we have

$$k(T_0)C_A = a(t, T)k(T)C_A$$

We will neglect any variations in concentration so that the product of the activity (a) and specific reaction rate (k) is constant and equal to the specific reaction rate, k_0 at time $t = 0$ and temperature T_0; that is,

$$k(T)a(t, T) = k_0 \tag{10-113}$$

Gradually raising the temperature can help offset effects of catalyst decay.

The goal is to find *how* the temperature should be increased with time (i.e., the temperature–time trajectory) to maintain constant conversion. Using the Arrhenius equation to substitute for k in terms of the activation energy, E_A, gives

$$k_0 e^{(E_A/R)(1/T_0 - 1/T)} a = k_0 \tag{10-114}$$

Solving for $1/T$ yields

$$\frac{1}{T} = \frac{R}{E_A} \ln a + \frac{1}{T_0} \tag{10-115}$$

The decay law also follows an Arrhenius-type temperature dependence.

$$-\frac{da}{dt} = k_{d0}e^{(E_d/R)(1/T_0 - 1/T)}\,a^n \qquad (10\text{-}116)$$

where k_{d0} = decay constant at temperature T_0, s^{-1}
$\qquad E_A$ = activation energy for the main reaction (e.g., A $\rightarrow$ B), kJ/mol
$\qquad E_d$ = activation energy for catalyst decay, kJ/mol

Substituting Equation (10-115) into (10-116) and rearranging yields

$$-\frac{da}{dt} = k_{d0}\exp\left(-\frac{E_d}{E_A}\ln a\right)a^n = k_{d0}a^{(n-E_d/E_A)} \qquad (10\text{-}117)$$

Integrating with $a = 1$ at $t = 0$ for the case $n \neq (1 + E_d/E_A)$, we obtain

$$t = \frac{1-a^{1-n+E_d/E_A}}{k_{d0}(1-n+E_d/E_A)} \qquad (10\text{-}118)$$

Solving Equation (10-114) for a and substituting in (10-118) gives

$$\boxed{t = \frac{1-\exp\left[\dfrac{E_A-nE_A+E_d}{R}\left(\dfrac{1}{T}-\dfrac{1}{T_0}\right)\right]}{k_{d0}(1-n+E_d/E_A)}} \qquad (10\text{-}119)$$

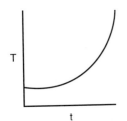

Equation (10-119) tells us how the temperature of the catalytic reactor should be increased with time in order for the reaction rate to remain constant.

In many industrial reactions, the decay rate law changes as temperature increases. In hydrocracking, the temperature–time trajectories are divided into three regimes. Initially, there is fouling of the acidic sites of the catalyst followed by a linear regime due to slow coking, and finally, accelerated coking characterized by an exponential increase in temperature. The temperature–time trajectory for a deactivating hydrocracking catalyst is shown in Figure 10-31.

For a first-order decay, Krishnaswamy and Kittrell's expression [Equation (10-119)] for the temperature–time trajectory reduces to

$$t = \frac{E_A}{k_{d0}E_d}\,[1-e^{(E_d/R)(1/T-1/T_0)}] \qquad (10\text{-}120)$$

10.7.3 Moving-Bed Reactors

Reaction systems with significant catalyst decay require the continual regeneration and/or replacement of the catalyst. Two types of reactors currently in commercial use that accommodate production with decaying catalysts are the moving-bed and straight-through transport reactor. A schematic diagram of a moving-bed reactor (used for catalytic cracking) is shown in Figure 10-32.

Comparing theory
and experiment

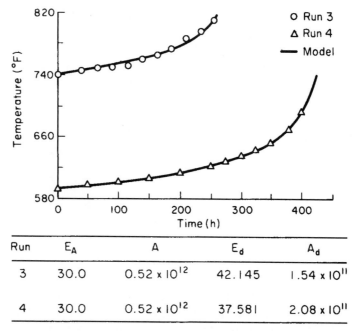

Run	E_A	A	E_d	A_d
3	30.0	0.52×10^{12}	42.145	1.54×10^{11}
4	30.0	0.52×10^{12}	37.581	2.08×10^{11}

Figure 10-31   Temperature–time trajectories for deactivating hydrocracking catalyst, runs 3 and 4. [Reprinted with permission from S. Krishnaswamy and J. R. Kittrell, *Ind. Eng. Chem. Process Des. Dev.*, *18*, 399 (1979). Copyright © 1979 American Chemical Society.]

Moving bed reactor:
Used for reactions
with moderate rate
of catalyst decay.

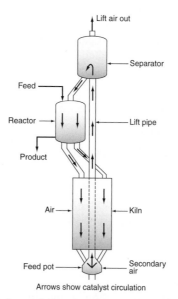

Figure 10-32   Thermofor catalytic cracking (TCC) unit. [From V. Weekman, *AIChE Monogr. Ser.*, *75*(11), 4 (1979). With permission of the AIChE. Copyright © 1979 AIChE. All rights reserved.]

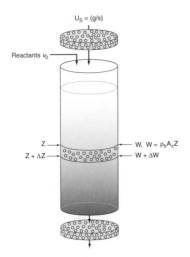

Figure 10-33   Moving-bed reactor–schematic.

The freshly regenerated catalyst enters the top of the reactor and then moves through the reactor as a compact packed bed. The catalyst is coked continually as it moves through the reactor until it exits the reactor into the kiln, where air is used to burn off the carbon. The regenerated catalyst is lifted from the kiln by an airstream and then fed into a separator before it is returned to the reactor. The catalyst pellets are typically between $\frac{1}{8}$ and $\frac{1}{4}$ in. in diameter.

The reactant feed stream enters at the top of the reactor and flows rapidly through the reactor relative to the flow of the catalyst through the reactor (Figure 10-33). If the feed rates of the catalyst and the reactants do not vary with time, the reactor operates at steady state; that is, conditions at any point in the reactor do not change with time. The mole balance on reactant A over ΔW is

$$
\begin{bmatrix} \text{Molar} \\ \text{flow} \\ \text{rate of A in} \end{bmatrix} - \begin{bmatrix} \text{Molar} \\ \text{flow} \\ \text{rate of A out} \end{bmatrix} + \begin{bmatrix} \text{Molar} \\ \text{rate of} \\ \text{generation of A} \end{bmatrix} = \begin{bmatrix} \text{Molar} \\ \text{rate of} \\ \text{accumulation of A} \end{bmatrix}
$$

$$F_A(W) \quad - \quad F_A(W+\Delta W) + \quad r'_A \Delta W \quad = \quad 0 \qquad (10\text{-}121)$$

Dividing by ΔW, letting ΔW approach zero, and expressing the flow rate in terms of conversion gives

$$F_{A0}\frac{dX}{dW} = -r'_A \qquad (2\text{-}17)$$

The rate of reaction at any time t is

$$-r'_A = a(t)k(T)[-r'_A(t=0)] = a(t)[k\,\text{fn}(C_A, C_B, ..., C_P)] \qquad (10\text{-}93)$$

The activity, as before, is a function of the time the catalyst has been in contact with the reacting gas stream. The decay rate law is

$$-\frac{da}{dt} = k_d a^n \tag{10-110}$$

We now need to relate the contact time to the weight of the catalyst. Consider a point z in the reactor, where the reactant gas has passed cocurrently through a catalyst weight W. Since the solid catalyst is moving through the bed at a rate U_s (mass per unit time), the time t that the catalyst has been in contact with the gas when the catalyst reaches a point z is

$$t = \frac{W}{U_s} \tag{10-122}$$

If we now differentiate Equation (10-122)

$$dt = \frac{dW}{U_s} \tag{10-123}$$

and combine it with the decay rate law, we obtain

$$\boxed{-\frac{da}{dW} = \frac{k_d}{U_s} a^n} \tag{10-124}$$

The activity equation is combined with the mole balance:

The design equation
for moving-bed
reactors

$$\boxed{\frac{dX}{dW} = \frac{a[-r'_A(t=0)]}{F_{A0}}} \tag{10-125}$$

Living Example Problem

Example 10–7 Catalytic Cracking in a Moving-Bed Reactor

The catalytic cracking of a gas oil charge, A, to form C_5+ (B) and to form coke and dry gas (C) is to be carried out in a screw-type conveyor moving-bed reactor at 900°F:

$$\text{Gas oil} \begin{cases} \xrightarrow{k_B} C_{5^+} \\ \xrightarrow{k_C} \text{Dry gas, Coke} \end{cases}$$

This reaction can also be written as

$$A \xrightarrow{k_1} \text{Products}$$

While pure hydrocarbons are known to crack according to a first-order rate law, the fact that the gas oil exhibits a wide spectrum of cracking rates gives rise to the fact that the lumped cracking rate is well represented by a second-order rate law (see Problem CDP5-H$_B$) with the following specific reaction rate:[32]

---

[32]Estimated from V. W. Weekman and D. M. Nace, *AIChE J.*, *16*, 397 (1970).

$$-r'_A = 0.60 \frac{(dm)^6}{(\text{g cat})(\text{mol})(\text{min})} C_A^2$$

The catalytic deactivation is independent of gas-phase concentration and follows a first-order decay rate law, with a decay constant of 0.72 reciprocal minutes. The feed stream is diluted with nitrogen so that as a first approximation, volume changes can be neglected with reaction. The reactor contains 22 kg of catalyst that moves through the reactor at a rate of 10 kg/min. The gas oil is fed at a rate of 30 mol/min at a concentration of 0.075 mol/dm³. Determine the conversion that can be achieved in this reactor.

Solution

1. **Design Equation:**

$$F_{A0} \frac{dX}{dW} = a(-r'_A) \tag{E10-7.1}$$

2. **Rate Law:**

$$-r'_A = kC_A^2 \tag{E10-7.2}$$

3. **Decay Law.** First-order decay

$$-\frac{da}{dt} = k_d a$$

Using Equation (10-124), we obtain

$$-\frac{da}{dW} = \frac{k_d}{U_s} a \tag{E10-7.3}$$

Integrating

$$a = e^{-(k_d/U_s)W} \tag{E10-7.4}$$

4. **Stoichiometry.** If $v \approx v_0$ [see Problem P10-2(g)] then

$$C_A = C_{A0}(1 - X) \tag{E10-7.5}$$

5. **Combining,** we have

$$F_{A0} \frac{dX}{dW} = e^{-(k_d/U_s)W} kC_{A0}^2 (1-X)^2 \tag{E10-7.6}$$

6. **Separating and integrating** yields

$$\frac{F_{A0}}{kC_{A0}^2} \int_0^X \frac{dX}{(1-X)^2} = \int_0^W e^{-(k_d/U_s)W} dW \tag{E10-7.7}$$

$$\frac{X}{1-X} = \frac{kC_{A0}^2 U_s}{F_{A0}k_d} (1 - e^{-k_d W/U_s}) \tag{E10-7.8}$$

Moving beds: moderate rate of catalyst decay

7. Numerical evaluation:

$$\frac{X}{1-X} = \frac{0.6 \text{ dm}^6}{\text{mol}\cdot\text{g cat.}\cdot\text{min}} \times \frac{(0.075 \text{ mol/dm}^3)^2}{30 \text{ mol/min}} \frac{10,000 \text{ g cat/min}}{0.72 \text{ min}^{-1}}$$

$$\times \left(1-\exp\left[\frac{(-0.72 \text{ min}^{-1})(22 \text{ kg})}{10 \text{ kg/min}}\right]\right)$$

$$\frac{X}{1-X} = 1.24$$

$$X = 55\%$$

We will now rearrange Equation (E10-7.8) to a form more commonly found in the literature. Let λ be a dimensionless decay time:

$$\lambda = k_d t = \frac{k_d W}{U_s} \tag{10-126}$$

and Da_2 be the Damköhler number for a second-order reaction (*a reaction rate divided by a transport rate*) for a packed-bed reactor:

$$\text{Da}_2 = \frac{(kC_{A0}^2)(W)}{F_{A0}} = \frac{kC_{A0}W}{v_0} \tag{10-127}$$

Through a series of manipulations we arrive at the equation for the conversion in a *moving bed* where a second-order reaction is taking place:[33]

Second-order
reaction in a
moving-bed reactor

$$\boxed{X = \frac{\text{Da}_2(1-e^{-\lambda})}{\lambda + \text{Da}_2(1-e^{-\lambda})}} \tag{10-128}$$

Similar equations are given or can easily be obtained for other reaction orders or decay laws.

Heat Effects in Moving Beds.   We shall consider two cases for modeling the temperature profile in the moving-bed reactor. In one case the temperature of the solid catalyst and the temperature of the gas are different and in the other case they are the same.

Case 1 ($T \neq T_s$).   The rate of heat transfer between the gas at temperature T and the solid catalyst particles at temperature T_S is

$$Q_P = \tilde{h}\tilde{a}_P(T-T_S) \tag{10-129}$$

where $\underset{\sim}{h}$ = heat transfer coefficient, kJ/m$^2\cdot$s$\cdot$K
$\tilde{a}_P$ = solid catalyst surface area per mass of catalyst in the bed, m^2/kg cat
T_S = temperature of the solid, K. Also, T_a = temperature of heat exchange fluid, K

[33]Ibid.

The energy balance on the gas phase is

Energy balance

$$\frac{dT}{dW} = \frac{U\tilde{a}_W(T_a-T)+h\tilde{a}_p(T_S-T)}{\Sigma F_i \tilde{C}_{P_i}} \tag{10-130}$$

If D_P is the pipe diameter (m), ρ_B is the bulk catalyst density (kg/m^3), and $\tilde{a}_w$ is the wall surface area per mass of catalyst (m^2/kg)

$$\tilde{a}_w = \frac{4}{D_P \rho_B} \tag{10-131}$$

The energy balance on the solid catalyst is

Heat exchange
between catalyst
particle and gas

$$\frac{dT_S}{dW} = -\frac{h\tilde{a}_p(T_S-T)+(r'_A)(\Delta H_{Rx})}{U_S C_{P_S}} \tag{10-132}$$

where C_{P_S} (J/kg·K) is the heat capacity of the solids, U_s (kg/s) the catalyst loading, and $\tilde{a}_p$ is the external surface area of the catalyst pellet per unit mass of catalyst bed:

$$\tilde{a}_P = \frac{6}{d_P \rho_b} \tag{10-133}$$

where d_P is the pellet diameter.

Case 2 ($T_s = T$). If the product of the heat transfer coefficient, h, and the surface area, $\tilde{a}_p$, is very large, we can assume that the solid and gas temperatures are identical. Under these circumstances the energy balance becomes

$$\frac{dT}{dW} = \frac{U\tilde{a}_W(T_a-T)+(r'_A)(\Delta H_{Rx})}{U_S C_{P_S}+\Sigma F_i C_{P_i}} \tag{10-134}$$

10.7.4 Straight-Through Transport Reactors (STTR)

This reactor is used for reaction systems in which the catalyst deactivates very rapidly. Commercially, the STTR is used in the production of gasoline from the cracking of heavier petroleum fractions where coking of the catalyst pellets occurs very rapidly. In the STTR, the catalyst pellets and the reactant feed enter together and are transported very rapidly through the reactor. The bulk density of the catalyst particle in the STTR is significantly smaller than in moving-bed reactors, and often the particles are carried through at the same velocity as the gas velocity. In some places the STTR is also referred to as a circulating fluidized bed (CFB). A schematic diagram is shown in Figure 10-34.

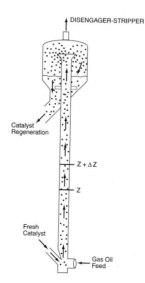

DISENGAGER-STRIPPER

Catalyst
Regeneration

$z + \Delta z$

z

Fresh
Catalyst

Gas Oil
Feed

Figure 10-34 Straight-through transport reactor.

A mole balance on the reactant A over the differential reactor volume

$$\Delta V = A_C \, \Delta z$$

is

$$F_A\big|_z - F_A\big|_{z+\Delta z} + r_A A_C \, \Delta z = 0$$

Dividing by Δz and taking the limit as $\Delta z \to 0$ and recalling that $r_A = \rho_B r_A'$, we obtain

$$\frac{dF_A}{dz} = r_A A_C = r_A' \rho_B A_C \tag{10-135}$$

In terms of conversion and catalyst activity,

$$\frac{dX}{dz} = \left(\frac{\rho_B A_C}{F_{A0}} \right) [-r_A'(t=0)] \, a(t) \tag{10-136}$$

For a catalyst particle traveling through the reactor with a velocity U_P, the time the catalyst pellet has been in the reactor when it reaches a height z is just

$$t = \frac{z}{U_P} \tag{10-137}$$

Substituting for time t in terms of distance z [i.e., $a(t) = a(z/U_P)$], the mole balance now becomes

$$\frac{dX}{dz} = \frac{\rho_B A_C [-r_A'(t=0)] \, a(z/U_P)}{F_{A0}}$$

STTR: Used when catalyst decay (usually coking) is very rapid

The entering molar flow rate, F_{A0}, can be expressed in terms of the gas velocity U_g, C_{A0}, and A_C:

$$F_{A0} = U_o A_C C_{A0}$$

Substituting for F_{A0}, we have

$$\frac{dX}{dz} = \frac{\rho_B a \, (z/U_P) \, [-r_A'(t=0)]}{C_{A0} U_o} \qquad (10\text{-}138)$$

Living Example Problem

A typical cost of the catalyst in the system is $1 million

Example 10–8 Decay in a Straight-Through Transport Reactor

The vapor-phase cracking of a gas oil is to be carried out in a straight-through transport reactor (STTR) that is 10 m high and 1.5 m in diameter. Gas-oil is a mixture of normal and branched paraffins (C_{12}–C_{40}), naphthenes, and aromatics, all of which will be lumped as a single species, A. We shall lump the primary hydrocarbon products according to distillate temperature into two respective groups, dry gas (C–C_4) B and gasoline (C_5–C_{14}) C. The reaction

$$\text{Gas-oil (g)} \longrightarrow \text{Products (g)} + \text{Coke}$$

can be written symbolically as

$$\text{A} \longrightarrow \text{B} + \text{C} + \text{Coke}$$

Both B and C are adsorbed on the surface. The rate law for a gas-oil cracking reaction on fresh catalyst can be approximated by

$$-r_A' = \frac{k' P_A}{1 + K_A P_A + K_B P_B + K_C P_C}$$

with $k' = 0.0014$ kmol/kg cat·s·atm, $K_A = 0.05$ atm^{-1}, $K_B = 0.15$ atm^{-1}, and $K_C = 0.1$ atm^{-1}. The catalyst decays by the deposition of coke, which is produced in most cracking reactions along with the reaction products. The decay law is

$$a = \frac{1}{1 + At^{1/2}} \qquad \text{with } A = 7.6 \text{ s}^{-1/2}$$

Pure gas-oil enters at a pressure of 12 atm and a temperature of 400°C. The bulk density of catalyst in the STTR is 80 kg cat/m^3. Plot the activity and conversion of gas oil up the reactor for entering gas velocity $U_0 = 2.5$ m/s.

Solution

Mole Balance:

$$F_{A0} \frac{dX}{dz} = -r_A A_C$$

$$\boxed{\frac{dX}{dz} = \frac{-r_A}{U_o C_{A0}}} \qquad (E10\text{-}8.1)$$

The height of the catalyst particle at time "t" after entering the STTR is

$$z = \int_0^t U\,dt$$

Differentiating we can find a relation between the time of the catalyst particle has been in the STTR a height of the practice which we can use to find the activity a.

$$\frac{dt}{dz} = \frac{1}{U}$$

Following the Algorithm

Rate Law:

$$-r_A = \rho_B(-r_A')$$ (E10-8.2)

$$-r_A' = a\,[-r_A'(t=0)]$$ (E10-8.3)

On fresh catalyst

$$-r_A'(t=0) = k'\,\frac{P_A}{1+K_A P_A + K_B P_B + K_C P_C}$$ (E10-8.4)

Combining Equations (E10-8.2) through (E10-8.4) gives

$$-r_A = a\left(\rho_B k'\,\frac{P_A}{1+K_A P_A + K_B P_B + K_C P_C}\right)$$ (E10-8.5)

Decay law. Assuming that the catalyst particle and gas travel up the reactor at the velocity $U_P = U_g = U$, we obtain

$$t = \frac{z}{U}$$ (E10-8.6)

$$a = \frac{1}{1 + A\,(z/U)^{1/2}}$$ (E10-8.7)

where $U = v/A_C = v_0(1+\varepsilon X)/A_C$ and $A_C = \pi D^2/4$.

Stoichiometry (gas phase isothermal and no pressure drop):

$$P_A = P_{A0}\,\frac{1-X}{1+\varepsilon X}$$ (E10-8.8)

$$P_B = \frac{P_{A0} X}{1+\varepsilon X}$$ (E10-8.9)

$$P_C = P_B$$ (E10-8.10)

Parameter Evaluation:

$$\varepsilon = y_{A0}\delta = (1+1-1) = 1$$

$$U = U_0(1+\varepsilon X)$$

$$C_{A0} = \frac{P_{A0}}{RT_0} = \frac{12\ \text{atm}}{(0.082\ \text{m}^3 \cdot \text{atm/kmol} \cdot \text{K})(673\ \text{K})} = 0.22\ \frac{\text{kmol}}{\text{m}^3}$$

Equations (E10-8.1), (E10-8.5), (E10-8.7), and (E10-8.8) through (E10-8.10) are now combined and solved using an ODE solver. The Polymath program is shown in Table E10-8.1, and the computer output is shown in Figure E10-8.1.

TABLE E10-8.1.   EQUATIONS FOR THE STTR: LANGMUIR–HINSHELWOOD KINETICS

ODE Report (RKF45)

Differential equations as entered by the user
[1] d(X)/d(z) = -ra/Uo/Cao

Explicit equations as entered by the user
[1] Ka = 0.05
[2] Kb = .15
[3] Pao = 12
[4] eps = 1
[5] A = 7.6
[6] R = 0.082
[7] T = 400+273
[8] rho = 80
[9] kprime = 0.0014
[10] D = 1.5
[11] Uo = 2.5
[12] Kc = 0.1
[13] U = Uo*(1+eps*X)
[14] Pa = Pao*(1-X)/(1+eps*X)
[15] Pb = Pao*X/(1+eps*X)
[16] vo = Uo*3.1416*D*D/4
[17] Cao = Pao/R/T
[18] Kca = Ka*R*T
[19] Pc = Pb
[20] a = 1/(1+A*(z/U)^0.5)
[21] raprime = a*(-kprime*Pa/(1+Ka*Pa+Kb*Pb+Kc*Pc))
[22] ra = rho*raprime

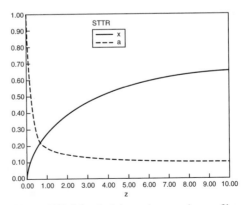

Figure E10-8.1   Activity and conversion profiles.

Closure. After reading this chapter, the reader should be able to discuss the steps in a heterogeneous reaction (adsorption, surface reaction, and desorption) and describe what is meant by a rate-limiting step. The differences between molecular adsorption and dissociated adsorption should be explained by the reader as should the different types of surface reactions (single site, dual site, and Eley-Rideal). Given heterogeneous reaction rate data, the reader should be able to analyze the data and to develop a rate law for Langmuir-Hinshelwood kinetics.

Applications of CRE in the electronics industry were discussed and readers should be able to describe the analog between Langmuir-Hinshelwood kinetics and chemical vapor deposition (CVD) and to derive a rate law for CVD mechanisms. Because under a harsh environment catalysts do not maintain their original activity, the reader needs to define and describe the three basic types of catalyst decay (sintering, fouling, coking, and poisoning). In addition, the reader should be able to carry out calculations that predict conversion in the types of reactors (namely *moving bed* and STTR) used to offset the deactivation of the catalyst.

SUMMARY

1. Types of adsorption:
   a. Chemisorption
   b. Physical adsorption

2. The **Langmuir isotherm** relating the concentration of species A on the surface to the partial pressure of A in the gas phase is

$$C_{A \cdot S} = \frac{K_A C_t P_A}{1 + K_A P_A}$$

(S10-1)

3. The sequence of steps for the solid-catalyzed isomerization

$$A \longrightarrow B$$

(S10-2)

   is:

   a. **Mass transfer of A** from the bulk fluid to the external surface of the pellet
   b. **Diffusion of A** into the interior of the pellet
   c. **Adsorption of A** onto the catalytic surface
   d. **Surface reaction of A** to form **B**
   e. **Desorption of B** from the surface
   f. **Diffusion of B** from the pellet interior to the external surface
   g. **Mass transfer of B** away from the solid surface to the bulk fluid

4. Assuming that mass transfer is not rate-limiting, the rate of adsorption is

$$r_{AD} = k_A \left(C_v P_A - \frac{C_{A \cdot S}}{K_A} \right)$$

(S10-3)

The rate of surface reaction is

$$r_S = k_S \left(C_{A \cdot S} - \frac{C_{B \cdot S}}{K_S} \right) \tag{S10-4}$$

The rate of desorption is

$$r_D = k_D (C_{B \cdot S} - K_B P_B C_v) \tag{S10-5}$$

At steady state,

$$r_{AD} = r_S = r_D \tag{S10-6}$$

If there are no inhibitors present, the total concentration of sites is

$$C_t = C_v + C_{A \cdot S} + C_{B \cdot S} \tag{S10-7}$$

5. If we assume that the surface reaction is rate-limiting, we set

$$\frac{r_D}{k_D} \simeq 0 \qquad \frac{r_{AD}}{k_A} \simeq 0$$

and solve for $C_{A \cdot S}$ and $C_{B \cdot S}$ in terms of P_A and P_B. After substitution of these quantities in Equation (S10-4), the concentration of vacant sites is eliminated with the aid of Equation (S10-7):

$$-r'_A = r_S = \frac{C_t k_S K_A (P_A - P_B / K_P)}{1 + K_A P_A + K_B P_B} \tag{S10-8}$$

Recall that the equilibrium constant for desorption of species B is the reciprocal of the equilibrium constant for the adsorption of species B:

$$K_B = \frac{1}{K_{DB}} \tag{S10-9}$$

and

$$K_P = K_A K_S / K_B \tag{S10-10}$$

6. Chemical vapor deposition:

$$SiH_4 \; \rightleftharpoons \; SiH_2(g) + H_2(g) \tag{S10-11}$$

$$SiH_2 + S \; \longrightarrow \; SiH_2 \cdot S$$

$$SiH_2 \cdot S \; \longrightarrow \; Si(s) + H_2(g) \tag{S10-12}$$

$$r_{Dep} = \frac{k P_{SiH_4}}{P_{H_2} + K P_{SiH_4}}$$

7. **Catalyst deactivation.** The catalyst activity is defined as

$$a(t) = \frac{-r_A'(t)}{-r_A'(t=0)} \tag{S10-13}$$

The rate of reaction at any time t is

$$-r_A' = a(t)k(T) \, fn(C_A, C_B, \ldots, C_P) \tag{S10-14}$$

The rate of catalyst decay is

$$r_d = -\frac{da}{dt} = p[a(t)]k_d(T)g(C_A, C_B, \ldots, C_P) \tag{S10-15}$$

For first-order decay:

$$p(a) = a \tag{S10-16}$$

For second-order decay:

$$p(a) = a^2 \tag{S10-17}$$

8. For slow catalyst decay the idea of a **temperature–time trajectory** is to increase the temperature in such a way that the rate of reaction remains constant.

9. The coupled differential equations to be solved for a **moving-bed reactor** are

$$F_{A0} \frac{dX}{dW} = a(-r_A') \tag{S10-18}$$

For nth-order activity decay and m order in a gas-phase concentration of species i,

$$-\frac{da}{dW} = \frac{k_d a^n C_i^m}{U_s} \tag{S10-19}$$

$$t = \frac{W}{U_s} \tag{S10-20}$$

10. The coupled differential equations to be solved in a **straight-through transport reactor** for the case when the particle and gas velocities, U, are identical are

$$\frac{dX}{dz} = \frac{a(t)[-r_A'(t=0)]}{U_g}\left(\frac{\rho_b A_c}{C_{A0}}\right) \tag{S10-21}$$

$$t = \frac{z}{U_g} \tag{S10-22}$$

For coking

$$a(t) = \frac{1}{1 + At^{1/2}} \tag{S10-23}$$

 O D E   S O L V E R
A L G O R I T H M

The isomerization A $\rightarrow$ B is carried out over a decaying catalyst in a *moving-bed reactor*. Pure A enters the reactor and the catalyst flows through the reactor at a rate of 2.0 kg/s.

$$\frac{dX}{dW} = \frac{-r'_A}{F_{A0}}$$

$k = 0.1$ mol/(kg cat·s·atm)

$$r'_A = \frac{-akP_A}{1 + K_A P_A}$$

$K_A = 1.5$ atm^{-1}

$$\frac{da}{dW} = \frac{-k_d a^2 P_B}{U_s}$$

$$k_d = \frac{0.75}{\text{s·atm}}$$

$$P_A = P_{A0}(1 - X)y$$

$F_{A0} = 10$ mol/s

$$P_B = P_{A0}Xy$$

$P_{A0} = 20$ atm

$$\frac{dy}{dW} = -\frac{\alpha}{2y}$$

$U_s = 2.0$ kg cat/s

$$\alpha = 0.0019 \text{ kg}^{-1}$$

$W_f = 500$ kg cat

C D - R O M   M A T E R I A L

Summary Notes

Interactive

Computer Modules

- **Learning Resources**
    1. *Summary Notes for Chapter 10*
    2. *Interactive Computer Modules*
        A. Heterogeneous Catalysis

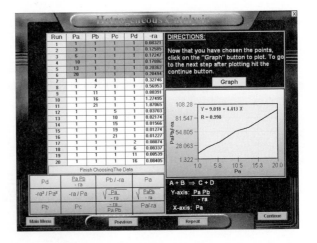

Solved Problems

4. *Solved Problems*
   Example CD10-1 Analysis of a Heterogeneous Reaction [Class Problem University of Michigan]
   Example CD10-2 Least Squares Analysis to Determine the Rate Law Parameters k, k_T, and k_B
   Example CD10-3 Decay in a Straight-Through Reactor
   Example CD10-4 Catalyst Poisoning in a Batch Reactor

- **Living Example Problems**
   1. *Example 10-2 Regression Analysis to Determine Model Parameters*
   2. *Example 10-3 Fixed-Bed Reactor Design*
   3. *Example 10-4 Model /Discrimination*
   4. *Example 10-6 Catalyst Decay in a Fluidized Bed Modeled as a CSTR*
   5. *Example 10-8 Decay in a Straight-Through Transport Reactor*

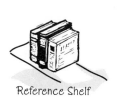

Living Example Problem

- **Professional Reference Shelf**
   R10.1. *Hydrogen Adsorption*
      A. Molecular Adsorption
      B. Dissociative Adsorption

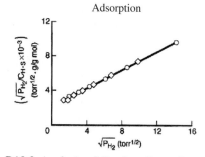

Adsorption

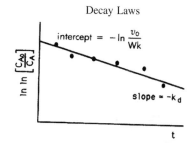

Decay Laws

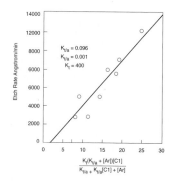

Reference Shelf

   R10.2. *Analysis of Catalyst Decay Laws*
      A. Integral Method
      B. Differential Method
   R10.3. *Etching of Semiconductors*
      A. Dry Etching

$$r_{\text{etch}} = \frac{k_C[K_1 + (Ar)(Cl)]}{K_1 + K_2(Cl) + (Ar)}$$

B. Wet Etching

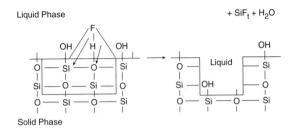

Liquid Phase   + SiF$_t$ + H$_2$O

Solid Phase

C. Dissolution Catalysis

$$H^+ + S \longrightarrow H^+ \bullet S$$

$$HF + H^+ \bullet S \longrightarrow Product$$

> **After Reading Each Page in
> This Book, Ask Yourself a Question
> About What You Read**

QUESTIONS AND PROBLEMS

The subscript to each of the problem numbers indicates the level of difficulty: A, least
difficult; D, most difficult.

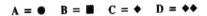

$$A = \bullet \quad B = \blacksquare \quad C = \blacklozenge \quad D = \blacklozenge\blacklozenge$$

Member

Hall of Fame

P10-1$_A$ Read over the problems at the end of this chapter. Make up an original problem that uses the concepts presented in this chapter. See Problem P4-1 for
guidelines. To obtain a solution:
   (a) Create your data and reaction.
   (b) Use a real reaction and real data.
   The journals listed at the end of Chapter 1 may be useful for part (b).
   (c) Choose an FAQ from Chapter 10 and say why it was most helpful.
   (d) Listen to the audios 🎧 on the CD and pick one and say why it was
   most helpful.

P10-2$_B$ **(a)** **Example 10-1.** Combine Table 10-1, Figure 10-4, and Example 10-1 to
   calculate the maximum and minimum rates of reaction in (mol/g cat/s)
   for (1) the isomerization of *n*-pentane, (2) the oxidation of SO$_2$, and (3)
   the hydration of ethylene. Assume the dispersion is 50% in all cases as is
   the amount of catalyst at 1%.
   (b) **Example 10-2.** (1) What is the fraction of vacant sites at 60% conversion? (2) At 80% and 1 atm, what is the fraction of toluene sites? (3)
   How would you linearize the rate law to evaluate the parameters k, K_B,
   and K_T from various linear plots? Explain.
   (c) **Example 10-3.** (1) What if the entering pressure were increased to 80
   atm or reduced 1 atm, how would your answers change? (2) What if the
   molar flow rate were reduced by 50%, how would X and y change? What
   catalyst weight would be required for 60% conversion?
   (d) **Example 10-4.** (1) How would your answers change if the following data
   for Run 10 were incorporated in your regression table?

$$-r_A' = 0.8, \; P_E = 0.5 \text{ atm}, \; P_{EA} = 15 \text{ atm}, \; P_H = 2$$

(1) How do the rate laws (e) and (f)

$$\text{(e)} \quad -r_E' = \frac{kP_EP_H}{(1+K_AP_{EA}+K_EP_E)^2} \qquad \text{(f)} \quad -r_E' = \frac{kP_HP_E}{1+K_AP_{EA}}$$

compare with the other rate laws?

(e) Example 10-5. (1) Sketch X vs. t for various values of k_d and k. Pay particular attention to the ratio k/k_d. (2) Repeat (1) for this example (i.e., the plotting of X vs. t) for a second-order reaction with ($C_{A0} = 1$ mol/dm³) and first-order decay. (3) Repeat (2) for this example for a first-order reaction and first-order decay. (4) Repeat (1) for this example for a second-order reaction ($C_{A0} = 1$ mol/dm³) and a second-order decay.

Member

Hall of Fame

(f) Example 10-6. What if . . . (1) the space time were changed? How would the minimum reactant concentration change? Compare your results with the case when the reactor is full of inerts at time $t = 0$ instead of 80% reactant. Is your catalyst lifetime longer or shorter? (2) What if the temperature were increased so that the specific rate constants increase to $k = 120$ and $k_d = 12$? Would your catalyst lifetime be longer or shorter than at the lower temperature? (3) Describe how the minimum in reactant concentration changes as the space time τ changes? What is the minimum if $\tau = 0.005$ h? If $\tau = 0.01$ h?

(g) Example 10-7. (1) What if the solids and reactants entered from opposite ends of the reactor? How would your answers change? (2) What if the decay in the moving bed were second order? By how much must the catalyst charge, U_s, be increased to obtain the same conversion? (3) What if $\varepsilon = 2$ (e.g., A → 3B) instead of zero, how would the results be affected?

(h) Example 10-8. (1) What if you varied the parameters P_{A0}, U_g, A, and k' in the STTR? What parameter has the greatest effect on either increasing or decreasing the conversion? Ask questions such as: What is the effect of varying the ratio of k to U_g or of k to A on the conversion? Make a plot of conversion versus distance as U_g is varied between 0.5 and 50 m/s. Sketch the activity and conversion profiles for $U_g = 0.025, 0.25, 2.5,$ and 25 m/s. What generalizations can you make? Plot the exit conversion and activity as a function of gas velocity between velocities of 0.02 and 50 m/s. What gas velocity do you suggest operating at? What is the corresponding entering volumetric flow rate? What concerns do you have operating at the velocity you selected? Would you like to choose another velocity? If so, what is it?

(i) What if you were asked to sketch the temperature–time trajectories and to find the catalyst lifetimes for first- and for second-order decay when $E_A = 35$ kcal/mol, $E_d = 10$ kcal/mol, $k_{d0} = 0.01$ day⁻¹, and $T_0 = 400$ K? How would the trajectory of the catalyst lifetime change if $E_A = 10$ kcal/mol and $E_d = 35$ kcal/mol? At what values of k_{d0} and ratios of E_d to E_A would temperature–time trajectories not be effective? What would your temperature–time trajectory look like if $n = 1 + E_d/E_A$?

(j) Write a question for this problem that involves critical thinking and explain why it involves critical thinking.

Interactive

Computer Modules

P10-3 Load the Interactive Computer Module (ICM) from the CD-ROM. Run the module and then record your performance number for the module, which indicates your mastering of the material. Your professor has the key to decode your performance number. ICM Heterogeneous Catalysis Performance # _____.

P10-4$_A$ t-Butyl alcohol (TBA) is an important octane enhancer that is used to replace lead additives in gasoline [*Ind. Eng. Chem. Res.*, 27, 2224 (1988)]. t-Butyl alcohol was produced by the liquid-phase hydration (W) of isobutene (I) over an Amberlyst-15 catalyst. The system is normally a multiphase mixture of hydrocarbon, water, and solid catalysts. However, the use of cosolvents or excess TBA can achieve reasonable miscibility.

The reaction mechanism is believed to be

$$I + S \rightleftharpoons I \cdot S \tag{P10-4.1}$$

$$W + S \rightleftharpoons W \cdot S \tag{P10-4.2}$$

$$W \cdot S + I \cdot S \rightleftharpoons TBA \cdot S + S \tag{P10-4.3}$$

$$TBA \cdot S \rightleftharpoons TBA + S \tag{P10-4.4}$$

Derive a rate law assuming:

(a) The surface reaction is rate-limiting.
(b) The adsorption of isobutene is limiting.
(c) The reaction follows Eley–Rideal kinetics

$$I \cdot S + W \longrightarrow TBA \cdot S \tag{P10-4.5}$$

and that the surface reaction is limiting.

(d) Isobutene (I) and water (W) are adsorbed on different sites

$$I + S_1 \rightleftharpoons I \cdot S_1 \tag{P10-4.6}$$

$$W + S_2 \rightleftharpoons W \cdot S_2 \tag{P10-4.7}$$

TBA is *not* on the surface, and the surface reaction is rate-limiting.

$$\left[Ans.: \ r'_{TBA} = -r'_I = \frac{k[C_I C_W - C_{TBA}/K_c]}{(1 + K_W C_W)(1 + K_I C_I)} \right]$$

(e) What generalization can you make by comparing the rate laws derived in parts (a) through (d)?

The process flow sheet for the commercial production of TBA is shown in Figure P10-4.

(f) What can you learn from this problem and the process flow sheet?

P10-5$_A$ The rate law for the hydrogenation (H) of ethylene (E) to form ethane (A) over a cobalt-molybdenum catalyst [*Collection Czech. Chem. Commun.*, 51, 2760 (1988)] is

$$-r'_E = \frac{k P_E P_H}{1 + K_E P_E}$$

(a) Suggest a mechanism and rate-limiting step consistent with the rate law.
(b) What was the most difficult part in finding the mechanism?

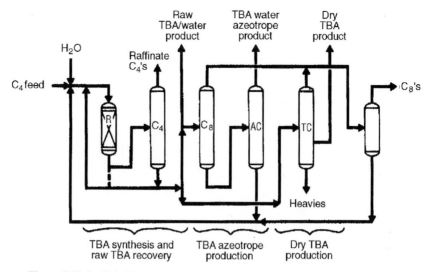

Figure P10-4   Hüls TBA synthesis process. R, reactor; C_4, C_4 column; C_8, C_8 column; AC, azeotrope column; TC, TBA column. (Adapted from R. E. Meyers, Ed., *Handbook of Chemicals Production Processes*, *Chemical Process Technology Handbook Series*, McGraw-Hill, New York, 1983, p. 1.19-3. ISBN 0-67-041 765-2.)

(c) The formation of propanol on a catalytic surface is believed to proceed by the following mechanism

$$O_2 + 2S \rightleftharpoons 2O \bullet S$$

$$C_3H_6 + O \bullet S \rightarrow C_3H_5OH \bullet S$$

$$C_3H_5OH \bullet S \rightleftharpoons C_3H_5OH + S$$

Suggest a rate-limiting step and derive a rate law.

P10-6$_B$ The dehydration of *n*-butyl alcohol (butanol) over an alumina-silica catalyst was investigated by J. F. Maurer (Ph.D. thesis, University of Michigan). The data in Figure P10-6 were obtained at 750°F in a modified differential reactor. The feed consisted of pure butanol.

(a) Suggest a mechanism and rate-controlling step that is consistent with the experimental data.

(b) Evaluate the rate law parameters.

(c) At the point where the initial rate is a maximum, what is the fraction of vacant sites? What is the fraction of occupied sites by both A and B?

(d) What generalizations can you make from studying this problem?

(e) Write a question that requires critical thinking and then explain why your question requires critical thinking. [*Hint:* See Preface Section B.2.]

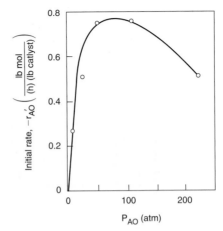

Figure P10-6

P10-7$_B$ The catalytic dehydration of methanol (ME) to form dimethyl ether (DME) and water was carried out over an ion exchange catalyst [K. Klusacek, *Collection Czech. Chem. Commun.*, *49*, 170(1984)]. The packed bed was initially filled with nitrogen and at $t = 0$ a feed of pure methanol vapor entered the reactor at 413 K, 100 kPa, and 0.2 cm³/s. The following partial pressures were recorded at the exit to the differential reactor containing 1.0 g of catalyst in 4.5 cm³ of reactor volume.

					t(s)		
	0	10	50	100	150	200	300
P_{N_2} (kPa)	100	50	10	2	0	0	0
P_{ME} (kPa)	0	2	15	23	25	26	26
P_{H_2O} (kPa)	0	10	15	30	35	37	37
P_{DME} (kPa)	0	38	60	45	40	37	37

Discuss the implications of these data.

P10-8$_B$ In 1981 the U.S. government put forth the following plan for automobile manufacturers to reduce emissions from automobiles over the next few years.

		Year	
	1981	1993	2004
Hydrocarbons	0.41	0.25	0.125
CO	3.4	3.4	1.7
NO	1.0	0.4	0.2

Web Hint

All values are in grams per mile. An automobile emitting 3.74 lb of CO and 0.37 lb of NO on a journey of 1000 miles would meet the current government requirements.

To remove oxides of nitrogen (assumed to be NO) from automobile exhaust, a scheme has been proposed that uses unburned carbon monoxide (CO) in the exhaust to reduce the NO over a solid catalyst, according to the reaction

$$CO + NO \longrightarrow Products\ (N_2, CO_2)$$

Experimental data for a particular solid catalyst indicate that the reaction rate can be well represented over a large range of temperatures by

$$-r'_N = \frac{kP_N P_C}{(1 + K_1 P_N + K_2 P_C)^2}$$

where P_N = gas-phase partial pressure of NO
           P_C = gas-phase partial pressure of CO
    k, K_1, K_2 = coefficients depending only on temperature

(a) Based on your experience with other such systems, you are asked to propose an adsorption-surface reaction-desorption mechanism that will explain the experimentally observed kinetics.
(b) A certain engineer thinks that it would be desirable to operate with a very large stoichiometric excess of CO to minimize catalytic reactor volume. Do you agree or disagree? Explain.
(c) When this reaction is carried out over a supported Rh catalyst [*J. Phys. Chem.*, 92, 389 (1988)], the reaction mechanism is believed to be

$$CO + S \rightleftharpoons CO \cdot S$$
$$NO + S \rightleftharpoons NO \cdot S$$
$$NO \cdot S + S \longrightarrow N \cdot S + O \cdot S$$
$$CO \cdot S + O \cdot S \longrightarrow CO_2 + 2S$$
$$N \cdot S + N \cdot S \longrightarrow N_2 + 2S$$

When the ratio of P_{CO}/P_{NO} is small, the rate law that is consistent with the experimental data is

$$-r'_{CO} = \frac{kP_{CO}}{(1 + K_{CO}P_{CO})^2}$$

What are the conditions for which the rate law and mechanism are consistent?

P10-9$_B$ Methyl ethyl ketone (MEK) is an important industrial solvent that can be produced from the dehydrogenation of butan-2-ol (Bu) over a zinc oxide catalyst [*Ind. Eng. Chem. Res.*, 27, 2050 (1988)]:

$$Bu \rightarrow MEK + H_2$$

The following data giving the reaction rate for MEK were obtained in a differential reactor at 490°C.

P_{Bu} (atm)	2	0.1	0.5	1	2	1
P_{MEK} (atm)	5	0	2	1	0	0
P_{H_2} (atm)	0	0	1	1	0	10
r'_{MEK} (mol/h·g cat.)	0.044	0.040	0.069	0.060	0.043	0.059

(a) Suggest a rate law consistent with the experimental data.
(b) Suggest a reaction mechanism and rate-limiting step consistent with the rate law. (*Hint:* Some species might be weakly adsorbed.)
(c) What do you believe to be the point of this problem?
(d) Plot conversion (up to 90%) and reaction rate as a function of catalyst weight for an entering molar flow rate of pure butan-2-ol of 10 mol/min and an entering pressure $P_0 = 10$ atm. $W_{max} = 23$ kg.
(e) Write a question that requires critical thinking and then explain why your question requires critical thinking. [*Hint:* See Preface Section B.2.]
(f) Repeat part (d) accounting for pressure drop and $\alpha = 0.03$ kg^{-1}. Plot y and X as a function of catalyst weight down the reactor.

P10-10$_C$ The following data for the hydrogenation of *i*-octene to form *i*-octane were obtained using a differential reactor operated at 200°C.

Run	Rate (mol/g·h)	Partial Pressure (atm)		
		Hydrogen	i-Octene	i-Octane
1	0.0362	1	1	0
2	0.0239	1	1	1
3	0.0390	3	1	1
4	0.0351	1	3	1
5	0.0114	1	1	3
6	0.0534	10	1	0
7	0.0280	1	10	0
8	0.0033	1	1	10
9	0.0380	2	2	2
10	0.0090	1	1	4
11	0.0127	0.6	0.6	0.6
12	0.0566	5	5	5

(a) Develop a rate law, and evaluate all the rate law parameters.
(b) Suggest a mechanism consistent with the experimental data.
     Hydrogen and *i*-octene are to be fed in stoichiometric proportions at a total rate of 5 mol/min at 200°C and 3 atm.
(c) Neglecting pressure drop, calculate the catalyst weight necessary to achieve 80% conversion of *i*-octane in a CSTR and in a plug-flow reactor.
(d) If pressure drop is taken into account and the $\frac{1}{8}$-in. catalyst pellets are packed in $1\frac{1}{2}$-in. schedule 80 pipes 35 ft long, what catalyst weight is necessary to achieve 80% conversion? The void fraction is 40% and the density of the catalyst is 2.6 g/cm^3.

P10-11$_B$ Cyclohexanol was passed over a catalyst to form water and cyclohexene:

$$\text{Cyclohexanol} \longrightarrow \text{Water} + \text{Cyclohexene}$$

The following data were obtained.

Run	Reaction Rate (mol/dm³·s) × 10⁵	Partial Pressure of Cyclohexanol	Partial Pressure of Cyclohexene	Partial Pressure of Steam (H₂O)
1	3.3	1	1	1
2	1.05	5	1	1
3	0.565	10	1	1
4	1.826	2	5	1
5	1.49	2	10	1
6	1.36	3	0	5
7	1.08	3	0	10
8	0.862	1	10	10
9	0	0	5	8
10	1.37	3	3	3

It is suspected that the reaction may involve a dual-site mechanism, but it is not known for certain. It is believed that the adsorption equilibrium constant for cyclohexanol is around 1 and is roughly one or two orders of magnitude greater than the adsorption equilibrium constants for the other compounds. Using these data:

(a) Suggest a rate law and mechanism consistent with the data given here.

(b) Determine the constants needed for the rate law.

[*Ind. Eng. Chem. Res.*, *32*, 2626–2632 (1993).]

(c) Why do you think estimates of the rate law parameters were given?

P10-12$_B$ A recent study of the *chemical vapor deposition* of silica from silane (SiH_4) is believed to proceed by the following irreversible two-step mechanism [*J. Electrochem. Soc.*, *139*(9), 2659 (1992)]:

$$SiH_4 + S \xrightarrow{k_1} SiH_2 \cdot S + H_2 \qquad (1)$$

$$SiH_2 \cdot S \xrightarrow{k_2} Si + H_2 \qquad (2)$$

This mechanism is somewhat different in that while SiH_2 is irreversibly adsorbed, it is highly reactive. In fact, adsorbed SiH_2 reacts as fast as it is formed [i.e., $r^*_{SiH_2 \cdot S} = 0$, i.e., PSSH (Chapter 7)], so that it can be assumed to behave as an active intermediate.

(a) Determine if this mechanism is consistent with the following data:

Deposition Rate (mm/min)	0.25	0.5	0.75	0.80
Silane Pressure (mtorr)	5	15	40	60

(b) At what partial pressures of silane would you take the next two data points?

P10-13$_A$ Vanadium oxides are of interest for various sensor applications, owing to the sharp metal–insulator transitions they undergo as a function of temperature, pressure, or stress. Vanadium triisopropoxide (VTIPO) was used to grow vanadium oxide films by *chemical vapor deposition* [*J. Electrochem. Soc.*, *136*, 897 (1989)]. The deposition rate as a function of VTIPO pressure for two different temperatures follows:

$T = 120°C$:

Growth Rate (μm/h)	0.004	0.015	0.025	0.04	0.068	0.08	0.095	0.1
VTIPO Pressure (torr)	0.1	0.2	0.3	0.5	0.8	1.0	1.5	2.0

$T = 200°C$:

Growth Rate (μm/h)	0.028	0.45	1.8	2.8	7.2
VTIPO Pressure (torr)	0.05	0.2	0.4	0.5	0.8

In light of the material presented in this chapter, analyze the data and describe your results. Specify where additional data should be taken.

P10-14$_A$ Titanium dioxide is a wide-bandgap semiconductor that is showing promise as an insulating dielectric in VLSI capacitors and for use in solar cells. Thin films of TiO_2 are to be prepared by *chemical vapor deposition* from gaseous titanium tetraisopropoxide (TTIP). The overall reaction is

$$Ti(OC_3H_7)_4 \longrightarrow TiO_2 + 4C_3H_6 + 2H_2O$$

The reaction mechanism in a CVD reactor is believed to be [K. L. Siefering and G. L. Griffin, *J. Electrochem. Soc.*, 137, 814 (1990)]

$$TTIP(g) + TTIP(g) \rightleftharpoons I + P_1$$

$$I + S \rightleftharpoons I \cdot S$$

$$I \cdot S \longrightarrow TiO_2 + P_2$$

where I is an active intermediate and P_1 is one set of reaction products (e.g., H_2O, C_3H_6) and P_2 is another set. Assuming the homogeneous gas-phase reaction for TTIP is in equilibrium, derive a rate law for the deposition of TiO_2. The experimental results show that at 200°C the reaction is second order at low partial pressures of TTIP and zero order at high partial pressures, while at 300°C the reaction is second order in TTIP over the entire pressure range. Discuss these results in light of the rate law you derived.

P10-15$_B$ The dehydrogenation of methylcyclohexane (M) to produce toluene (T) was carried out over a 0.3% Pt/Al_2O_3 catalyst in a differential catalytic reactor. The reaction is carried out in the presence of hydrogen (H_2) to avoid coking [*J. Phys. Chem.*, 64, 1559 (1960)].

(a) Determine the model parameters for each of the following rate laws.

$$(1) \ -r'_M = kP_M^\alpha P_{H_2}^\beta \qquad (3) \ -r'_M = \frac{kP_M P_{H_2}}{(1 + K_M P_M)^2}$$

$$(2) \ -r'_M = \frac{kP_M}{1 + K_M P_M} \qquad (4) \ -r'_M = \frac{kP_M P_{H_2}}{1 + K_M P_M + K_{H_2} P_{H_2}}$$

Use the data in Table P10-15.

(b) Which rate law best describes the data? (*Hint*: Neither K_{H_2} or K_M can take on negative values.)

(c) Where would you place additional data points?

TABLE P10-15.   DEHYDROGENATION OF METHYLCYCLOHEXANE

P_{H_2} (atm)	P_M (atm)	$r'_T \left(\dfrac{\text{mol toluene}}{\text{s} \cdot \text{kg cat}} \right)$
1	1	1.2
1.5	1	1.25
0.5	1	1.30
0.5	0.5	1.1
1	0.25	0.92
0.5	0.1	0.64
3	3	1.27
1	4	1.28
3	2	1.25
4	1	1.30
0.5	0.25	0.94
2	0.05	0.41

P10-16$_B$ In the production of ammonia

$$NO + \tfrac{5}{2}H_2 \; \rightleftarrows \; H_2O + NH_3 \tag{1}$$

the following side reaction occurs:

$$NO + H_2 \; \rightleftarrows \; H_2O + \tfrac{1}{2}N_2 \tag{2}$$

Ayen and Peters [*Ind. Eng. Chem. Process Des. Dev., 1*, 204 (1962)] studied catalytic reaction of nitric oxide with Girdler G–50 catalyst in a differential reactor at atmospheric pressure. Table P10-16 shows the reaction rate of the side reaction as a function of P_{H_2} and P_{NO} at a temperature of 375°C.

TABLE P10-16.   FORMATION OF WATER

P_{H_2} (atm)	P_{NO} (atm)	Reaction Rate $r_{H_2O} \times 10^5$ (g mol/min · g cat) $T = 375°C,\ W = 2.39$ g
0.00922	0.0500	1.60
0.0136	0.0500	2.56
0.0197	0.0500	3.27
0.0280	0.0500	3.64
0.0291	0.0500	3.48
0.0389	0.0500	4.46
0.0485	0.0500	4.75
0.0500	0.00918	1.47
0.0500	0.0184	2.48
0.0500	0.0298	3.45
0.0500	0.0378	4.06
0.0500	0.0491	4.75

The following rate laws for side reaction (2), based on various catalytic mechanisms, were suggested:

$$r_{H_2O} = \frac{kK_{NO}P_{NO}P_{H_2}}{1+K_{NO}P_{NO}+K_{H_2}P_{H_2}} \tag{3}$$

$$r_{H_2O} = \frac{kK_{H_2}K_{NO}P_{NO}}{1+K_{NO}P_{NO}+K_{H_2}P_{H_2}} \tag{4}$$

$$r_{H_2O} = \frac{k_1K_{H_2}K_{NO}P_{NO}P_{H_2}}{(1+K_{NO}P_{NO}+K_{H_2}P_{H_2})^2} \tag{5}$$

Find the parameter values of the different rate laws and determine which rate law best represents the experimental data.

P10-17$_B$ Rework Example 10-6 when
  (a) The reaction is carried out in a *moving-bed reactor* at a catalyst loading rate of 250,000 kg/h.
  (b) Repeat (a) when the catalyst and feed enter at opposite ends of the bed.
  (c) The reaction is carried out in a *packed-bed reactor* modeled as five CSTRs in series.
  (d) Determine the temperature–time trajectory to keep conversion constant in a CSTR if activation energies for reaction and decay are 30 kcal/mol and 10 kcal/mol, respectively.
  (e) How would your answer to part (d) change if the activation energies were reversed?

P10-18$_A$ Sketch *qualitatively* the reactant, product, and activity profiles as a function of length at various times for a *packed-bed reactor* for each of the following cases. In addition, sketch the effluent concentration of A as a function of time. The reaction is a simple isomerization:

$$A \longrightarrow B$$

  (a) Rate law: $-r_A' = kaC_A$

  Decay law: $r_d = k_d a C_A$

  Case I: $k_d \ll k$, Case II: $k_d = k$, Case III: $k_d \gg k$

  (b) $-r_A' = kaC_A$ and $r_d = k_d a^2$

  (c) $-r_A' = kaC_A$ and $r_d = k_d a C_B$
  (d) Sketch similar profiles for the rate laws in parts (a) and (c) in a *moving-bed reactor* with the solids entering at the same end of the reactor as the reactant.
  (e) Repeat part (d) for the case where the solids and the reactant enter at opposite ends.

P10-19$_B$ The elementary irreversible gas phase catalytic reaction

$$A + B \xrightarrow{k} C + D$$

is to be carried out in a moving-bed reactor at constant temperature. The reactor contains 5 kg of catalyst. The feed is stoichiometric in A and B. The entering concentration of A is 0.2 mol/dm³. The catalyst decay law is zero order with $k_D = 0.2$ s^{-1} and $k = 1.0$ dm⁶/(mol · kg cat · s) and the volumetric flow rate is $v_0 = 1$ dm³/s.

(a) What conversion will be achieved for a catalyst feed rate of 0.5 kg/s?
(b) Sketch the catalyst activity as a function of catalyst weight (i.e., distance) down the reactor length for a catalyst feed rate of 0.5 kg/s.
(c) What is the maximum conversion that could be achieved (i.e., at infinite catalyst loading rate)?
(d) What catalyst loading rate is necessary to achieve 40% conversion?
(e) At what catalyst loading rate (kg/s) will the catalyst activity be exactly zero at the exit of the reactor?
(f) What does an activity of zero mean? Can catalyst activity be less than zero?
(g) How would your answers change in part (a) if the catalyst and reactant were fed at opposite ends? Compare with part (a).
(h) Now consider the reaction to be also zero order with $k = 0.2$ mol/kg cat · min.
   •   The product sells for $160 per gram mole.
   •   The cost of operating the bed is $10 per kilogram of catalyst exiting the bed.

What is the feed rate of solids (kg/min) that will give the maximum profit? (*Ans.: U$_s$ = 4 kg/min.*)

(*Note:* For the purpose of this calculation, ignore all other costs, such as the cost of the reactant, etc.)

The economics

P10-20$_B$ With the increasing demand for xylene in the petrochemical industry, the production of xylene from toluene disproportionation has gained attention in recent years [*Ind. Eng. Chem. Res.*, 26, 1854 (1987)]. This reaction,

$$2\ \text{Toluene} \longrightarrow \text{Benzene} + \text{Xylene}$$

$$2T \xrightarrow{\text{catalyst}} B + X$$

was studied over a hydrogen mordenite catalyst that decays with time. As a first approximation, assume that the catalyst follows second-order decay,

$$r_d = k_d a^2$$

and the rate law for low conversions is

$$-r'_T = k_T P_T a$$

with $k_T = 20$ g mol/h·kg cat·atm and $k_d = 1.6$ h^{-1} at 735 K.

(a) Compare the conversion time curves in a batch reactor containing 5 kg cat at different initial partial pressures (1 atm, 10 atm, etc.). The reaction volume containing pure toluene initially is 1 dm³ and the temperature is 735 K.

(b) What conversion can be achieved in a *moving-bed reactor* containing 50 kg of catalyst with a catalyst feed rate of 2 kg/h? Toluene is fed at a pressure of 2 atm and a rate of 10 mol/min.

(c) Explore the effect of catalyst feed rate on conversion.

(d) Suppose that E_T = 25 kcal/mol and E_d = 10 kcal/mol. What would the temperature–time trajectory look like for a CSTR? What if E_T = 10 kcal/mol and E_d = 25 kcal/mol?

(e) The decay law more closely follows the equation

$$r_d = k_d P_T^2 a^2$$

with k_d = 0.2 atm^{-2} h^{-1}. Redo parts (b) and (c) for these conditions.

P10-21$_B$ The elementary irreversible gas-phase catalytic reaction

$$A \xrightarrow{k_1} B$$

is carried out isothermically in a batch reactor. The catalyst deactivation follows a first-order decay law and is independent of the concentrations of both A and B.

(a) Determine a general expression for catalyst activity as a function of time.

(b) Make a qualitative sketch of catalyst activity as a function of time. Does $a(t)$ ever equal zero for a first-order decay law?

(c) Write out the general algorithm and derive an expression for conversion as a function of time, the reactor parameters, and the catalyst parameters. Fill in the following algorithm

     Mole balance
     Rate law
     Decay law
     Stoichiometry
     Combine

   Solve
   1. Separate
   2. Integrate

$$\left[Ans.: \ X = 1 - \exp\left[-\frac{k_1 W}{k_d V_0}(1 - \exp(-k_d t)) \right] \right]$$

(d) Calculate the conversion and catalyst activity in the reactor after 10 minutes at 300 K.

(e) How would you expect your results in parts (b) and (d) to change if the reaction were run at 400 K? Briefly describe the trends qualitatively.

(f) Calculate the conversion and catalyst activity in the reactor after 10 minutes if the reaction were run at 400 K instead of 300 K. Do your results match the predictions in part (e)?

Additional information:

C_{A0} = 1 mol/dm^3
V_0 = 1 dm^3
W = 1 kg
k_d = 0.1 min^{-1} at 300 K E_d/R = 2000 K
k_1 = 0.2 dm^3/(kg cat • min) at 300 K E_A/R = 500 K

P10-22$_A$ The vapor-phase cracking of gas-oil in Example 10-8 is carried out over a different catalyst, for which the rate law is

$$-r_A' = k'P_A^2 \qquad \text{with } k' = 5 \times 10^{-5} \; \frac{\text{kmol}}{\text{kg cat} \cdot \text{s} \cdot \text{atm}^2}$$

(a) Assuming that you can vary the entering pressure and gas velocity, what operating conditions would you recommend?

(b) What could go wrong with the conditions you chose?
Now assume the decay law is

$$-\frac{da}{dt} = k_D a \, C_{\text{coke}} \qquad \text{with } k_D = 100 \; \frac{\text{dm}^3}{\text{mol} \cdot \text{s}} \text{ at } 400°\text{C}$$

where the concentration, C_{coke}, in mol/dm^3 can be determined from a stoichiometric table.

(c) For a temperature of 400°C and a reactor height of 15 m, what gas velocity do you recommend? Explain. What is the corresponding conversion?

(d) The reaction is now to be carried in an STTR 15 m high and 1.5 m in diameter. The gas velocity is 2.5 m/s. You can operate in the temperature range between 100 and 500°C. What temperature do you choose, and what is the corresponding conversion?

(e) What would the temperature–time trajectory look like for a CSTR?

Additional information:

$$E_R = 3000 \text{ cal/mol}$$
$$E_D = 15{,}000 \text{ cal/mol}$$

P10-23$_C$ When the impurity cumene hydroperoxide is present in trace amounts in a cumene feed stream, it can deactivate the silica-alumina catalyst over which cumene is being cracked to form benzene and propylene. The following data were taken at 1 atm and 420°C in a differential reactor. The feed consists of cumene and a trace (0.08 mol %) of cumene hydroperoxide (CHP).

Benzene in Exit Stream (mol %)	2	1.62	1.31	1.06	0.85	0.56	0.37	0.24
t (s)	0	50	100	150	200	300	400	500

(a) Determine the order of decay and the decay constant. (*Ans.: $k_d = 4.27 \times 10^{-3} \text{ s}^{-1}$.*)

(b) As a first approximation (actually a rather good one), we shall neglect the denominator of the catalytic rate law and consider the reaction to be first order in cumene. Given that the specific reaction rate with respect to cumene is $k = 3.8 \times 10^3$ mol/kg fresh cat ·s·atm, the molar flow rate of cumene (99.92% cumene, 0.08% CHP) is 200 mol/min, the entering concentration is 0.06 kmol/m^3, the catalyst weight is 100 kg, the velocity of solids is 1.0 kg/min, what conversion of cumene will be achieved in a *moving-bed reactor*?

P10-24$_C$ The decomposition of spartanol to wulfrene and CO$_2$ is often carried out at high temperatures [*J. Theor. Exp.*, *15*, 15 (2014)]. Consequently, the denominator of the catalytic rate law is easily approximated as unity, and the reaction is first order with an activation energy of 150 kJ/mol. Fortunately, the reaction is irreversible. Unfortunately, the catalyst over which the reaction occurs decays with time on stream. The following conversion-time data were obtained in a differential reactor:

For $T = 500$ K:

t (days)	0	20	40	60	80	120
X (%)	1	0.7	0.56	0.45	0.38	0.29

For $T = 550$ K:

t (days)	0	5	10	15	20	30	40
X (%)	2	1.2	0.89	0.69	0.57	0.42	0.33

(a) If the initial temperature of the catalyst is 480 K, determine the *temperature–time trajectory* to maintain a constant conversion.
(b) What is the catalyst lifetime?

P10-25$_B$ The hydrogenation of ethylbenzene to ethylcyclohexane over a nickel-mordenite catalyst is zero order in both reactants up to an ethylbenzene conversion of 75% [*Ind. Eng. Chem. Res.*, 28 (3), 260 (1989)]. At 553 K, $k = 5.8$ mol ethylbenzene/(dm^3 of catalyst·h). When a 100 ppm thiophene concentration entered the system, the ethylbenzene conversion began to drop.

Time (h)	0	1	2	4	6	8	12
Conversion	0.92	0.82	0.75	0.50	0.30	0.21	0.10

Hall of Fame

The reaction was carried out at 3 MPa and a molar ratio of H$_2$/ETB = 10. Discuss the catalyst decay. Be quantitative where possible.

P10-26$_C$ (Modified P8-9$_B$) The gas-phase exothermic elementary reaction

$$A \xrightarrow{\ k\ } B + C$$

is carried out in a moving-bed reactor.

$$k = 0.33 \exp\left[\frac{E_r}{R}\left(\frac{1}{450} - \frac{1}{T}\right)\right] \text{s}^{-1} \text{ , with } \frac{E_r}{R} = 3777 \text{ K}$$

Heat is removed by a heat exchanger jacketing the reactor.

$$\frac{Ua}{\rho_b} = \frac{0.8 \text{ J}}{\text{s} \cdot \text{kg cat} \cdot \text{K}}$$

The flow rate of the coolant in the exchanger is sufficiently high that the ambient temperature is constant at 50°C. Pure A enters the reactor at a rate of 5.42 mol/s at a concentration of 0.27 mol/dm^3. Both the solid catalyst and the reactant enter the reactor at a temperature of 450 K, and the heat transfer coefficient between the catalyst and gas is virtually infinite. The heat capacity of the solid catalyst is 100 J/kg cat/K.

The catalyst decay is first order in activity with

$$k_d = 0.01 \, \exp\left[7000 \text{ K} \left(\frac{1}{450} - \frac{1}{T} \right) \right] \text{s}^{-1}$$

There are 50 kg of catalyst in the bed.

(a) What catalyst charge rate (kg/s) will give the highest conversion?

(b) What is the corresponding conversion?

(c) Redo parts (a) and (b) for Case 1 $(T \neq T_S)$ of heat effects in moving beds. Use realistic values of the parameter values h and $\tilde{a}_p$ and $\tilde{a}_w$. Vary the entering temperatures of T_S and T.

Additional information:

$C_{P_A} = 40 \text{ J/mol} \cdot \text{K}$

$C_{P_B} = 25 \text{ J/mol} \cdot \text{K}$

$C_{P_C} = 15 \text{ J/mol} \cdot \text{K}$

$\Delta H_{Rx} = -80 \text{ kJ/mol A}$

JOURNAL CRITIQUE PROBLEMS

P10C-1 See *AIChE J.*, 7 (4), 658 (1961). Determine if the following mechanism can also be used to explain the data in this paper.

$$\text{NO} + \text{S} \rightleftharpoons \text{NO} \cdot \text{S}$$

$$\text{NO} + \text{NO} \cdot \text{S} \rightleftharpoons \text{N}_2 + \text{O}_2 \cdot \text{S}$$

$$\text{O}_2 \cdot \text{S} \rightleftharpoons \text{O}_2 + \text{S}$$

P10C-2 In *J. Catal.*, *63*, 456 (1980), a rate expression is derived by assuming a reaction that is first order with respect to the pressure of hydrogen and first order with respect to the pressure of pyridine [Equation (10)]. Would another reaction order describe the data just as well? Explain and justify. Is the rate law expression derived by the authors correct?

P10C-3 See "The decomposition of nitrous oxide on neodymium oxide, dysprosium oxide and erbium oxide," *J. Catal.*, *28*, 428 (1973). Some investigators have reported the rate of this reaction to be independent of oxygen concentration and first order in nitrous oxide concentration, while others have reported the reaction to be first order in nitrous oxide concentration and negative one-half order in oxygen concentration. Can you propose a mechanism that is consistent with both observations?

P10C-4 The kinetics of self-poisoning of Pd/Al_2O_3 catalysis in the hydrogenolysis of cyclopentane is discussed in *J. Catal.*, *54*, 397 (1978). Is the effective diffusivity that is used realistic? Is the decay homographic? The authors claim that the deactivation of the catalyst is independent of metal dispersion. If one were to determine the specific reaction rate as a function of percent dispersion, would this information support or reject the authors' hypotheses?

P10C-5 A packed-bed reactor was used to study the reduction of nitric oxide with ethylene on a copper-silica catalyst [*Ind. Eng. Chem. Process Des. Dev.*, *9* (3), 455 (1970)]. Develop the integral design equation in terms of the conversion and the initial pressure using the author's proposed rate law. If this equation is solved for conversion at various initial pressures and temperatures, is there a significant discrepancy between the experimental results shown in Figures 2 and 3 and the calculated results based on the proposed rate law? What is the possible source of this deviation?

P10C-6 The thermal degradation of rubber wastes was studied [*Int. J. Chem. Eng.*, *23* (4), 645 (1983)], and it was shown that a sigmoidal-shaped plot of conversion versus time would be obtained for the degradation reaction. Propose a model with physical significance that can explain this sigmoidal-shaped curve rather than merely curve fitting as the authors do. Also, what effect might the particle size distribution of the waste have on these curves? [*Hint:* See O. Levenspiel, *The Chemical Reactor Omnibook* (Corvallis, Ore.: Oregon State University Press, 1979) regarding gas-solid reactions.]

• **Additional Homework Problems**

• **Mechanisms**

CDP10-A_B Suggest a rate law and mechanism for the catalytic oxidation of ethanol over tantalum oxide when adsorption of ethanol and oxygen take place on different sites. [2nd ed. P6-17]

CDP10-B_B Analyze the data for the vapor-phase esterification of acetic acid over a resin catalyst at 118°C.

CDP10-C_B Silicon dioxide is grown by CVD according to the reaction

$$SiH_2Cl_2(g) + 2N_2O(g) \longrightarrow SiO_2(s) + 2N_2(g) + 2HCl(g)$$

Use the rate data to determine the rate law, reaction mechanism, and rate law parameters. [2nd Ed. P6-13]

CDP10-D_B Determine the rate law and rate law parameters for the wet etching of an aluminum silicate.

CDP10-E_B Titanium films are used in decorative coatings as well as wear-resistant tools because of their thermal stability and low electrical resistivity. TiN is produced by CVD from a mixture of $TiCl_4$ and NH_3TiN. Develop a rate law, mechanism, and rate-limiting step and evaluate the rate law parameters.

CDP10-F_B The dehydrogenation of ethylbenzene is carried out over a shell catalyst. From the data provided, find the cost of the catalyst to produce a specified amount of styrene. [2nd Ed. P6-20]

CDP10-G_B The formation of CH_4 from CO of H_2 is studied in a differential reactor.

CDP10-H_A Determine the rate law and mechanism for the reaction A + B $\longrightarrow$ C.

CDP10-I_B Determine the rate law from data where the pressures are varied in such a way that the rate is constant. [2nd Ed. P6-18]

CDP10-J_B Determine the rate law and mechanism for the vapor phase dehydration of ethanol. [2nd Ed. P6-21]

CDP10-K_B Analyze the rate data.

CDP10-L_B Carbonation of allyl chloride whereby the complexes Pd•CO and Pd•CO•NaOH* are formed. [3rd Ed. P10-11]. Find the rate-limiting step.

Octane Number

- **Catalyst Decay**

CDP10-M$_B$ California problem. Isomerization reaction with catalyst decay. [3rd ed. P10-5]. Review the data and make a recommendation.

CDP10-N$_B$ Fluidized bed reactor with catalyst decay as measured by decline in octane number. Real data. [3rd Ed. P10-20]

CDP10-O$_B$ Catalyst decay in a batch reactor. [3rd Ed. P10-23]

CDP10-P$_B$ Deactivation by *coking* in a differential reactor. [3rd Ed. P10-24]

CDP10-Q$_B$ The autocatalytic reaction A + B $\longrightarrow$ 2B is carried out in a moving-bed reactor. The decay law is first order in B. Plot the activity and concentration of A and B as a function of catalyst weight.

CDP10-R$_C$ The decompositon of cumene is carried out over a LaY zeolite catalyst, and deactivation is found to occur by coking. Determine the decay law and rate law and use these to design a STTR.

CDP10-S$_B$ Analyze a second-order reaction over a decaying catalyst takes place in a moving-bed reactor. [Final Exam, Winter 1994] Plot the conversion and activity profiles.

CDP10-T$_B$ Analyze a first-order reaction A $\rightarrow$ B + C takes place in a moving-bed reactor. Plot the profiles of conversion and activity.

CDP10-U$_B$ For the cracking of normal paraffins (P$_n$), the rate has been found to increase with increasing temperature up to a carbon number of 15 (i.e., $n \leq 15$) and to decrease with increasing temperature for a carbon number greater than 16. [J. Wei, *Chem. Eng. Sci., 51,* 2995 (1996)] Provide an explanation.

CDP10-V$_B$ The reaction A + B $\rightarrow$ C + D is carried out in a moving-bed reactor. Plot conversion and activity profiles.

CDP10-W$_B$ Second-order reaction and zero-order decay in a batch reactor. Plot X and a as a function of time.

CDP10-X$_B$ Analyze first-order decay in a moving-bed reactor for the series reaction A $\rightarrow$ B $\rightarrow$ C.

CDP10-Y   New Problem on CD-ROM.

SUPPLEMENTARY READING

1. A terrific discussion of heterogeneous catalytic mechanisms and rate-controlling steps may be found in

> DAVIS, M. E., and R. J. DAVIS, *Fundamentals of Chemical Reaction Engineering.* New York: McGraw-Hill, 2003.

> MASEL, R. I., *Principles of Adsorption and Reaction on Solid Surfaces.* New York: Wiley, 1996.

> SOMORJAI, G. A., *Introduction to Surface Chemistry and Catalysis.* New York: Wiley, 1994.

2. A truly excellent discussion of the types and rates of adsorption together with techniques used in measuring catalytic surface areas is presented in

> MASEL, R. I., *Principles of Adsorption and Reaction on Solid Surfaces.* New York: Wiley, 1996.

3. A discussion of the types of catalysis, methods of catalyst selection, methods of preparation, and classes of catalysts can be found in

> ENGELHARD CORPORATION, *Engelhard Catalysts and Precious Metal Chemicals Catalog*. Newark, N.J.: Engelhard Corp., 1985.
>
> GATES, BRUCE C., *Catalytic Chemistry*. New York: Wiley, 1992.
>
> SCHMIDT, L. D., *The Engineering of Chemical Reactions*. New York: Oxford Press, 1998.
>
> VAN SANTEN, R. A., and J. W. NIEMANTSVERDRIET, *Chemical Kinetics and Catalysis*. New York: Plenum Press, 1995.

4. Heterogeneous catalysis and catalytic reactors can be found in

> HARRIOTT, P., *Chemical Reactor Design*. New York: Marcel Pekker, 2003.
>
> SATERFIELD, C. N., *Heterogeneous Catalysis in Industrial Practice*, 2nd ed. New York: McGraw-Hill, 1991.

and in the following journals: *Advances in Catalysis*, *Journal of Catalysis*, and *Catalysis Reviews*.

5. Techniques for discriminating between mechanisms and models can be found in

> BOX, G. E. P., W. G. HUNTER, and J. S. HUNTER, *Statistics for Experimenters*. New York: Wiley, 1978.
>
> FROMENT, G. F., and K. B. BISCHOFF, *Chemical Reactor Analysis and Design*. New York: Wiley, 1979, Sec. 2.3.

6. A reasonably complete listing of the different decay laws coupled with various types of reactors is given by

> BUTT, J. B., and E. E. PETERSEN, *Activation, Deactivation, and Poisoning of Catalysts*. San Diego, Calif.: Academic Press, 1988.
>
> FARRAUTO, R. J., and C. H. BARTHOLOMEW, *Fundamentals of Industrial Catalytic Processes*. New York: Blackie Academic and Professional, 1997.

7. Examples of applications of catalytic principles to microelectronic manufacturing can be found in

> DOBKIN, D. M., and M. K. ZURAW. The Netherlands: Kluwer Academic Publishers, 2003.
>
> HESS, D. W., and K. F. JENSEN, *Microelectronics Processing*. Washington, D.C.: American Chemical Society, 1989.
>
> JENSEN, K. F., "Modeling of chemical vapor deposition reactors for the fabrication of microelectronic devices," in *Chemical and Catalytic Reactor Modeling*. Washington, D.C.: American Chemical Society, 1984.
>
> LEE, H. H., *Fundamentals of Microelectronics Processing*. New York: McGraw-Hill, 1990.

External **11** *Diffusion Effects on Heterogeneous Reactions*

Giving up is the ultimate tragedy.

Robert J. Donovan

or

It ain't over 'til it's over.

Yogi Berra

Overview In many industrial reactions, the overall rate of reaction is limited by the rate of mass transfer of reactants between the bulk fluid and the catalytic surface. By mass transfer, we mean any process in which diffusion plays a role. In the rate laws and catalytic reaction steps described in Chapter 10 (diffusion, adsorption, surface reaction, desorption, and diffusion), we neglected the diffusion steps by saying we were operating under conditions where these steps are fast when compared to the other steps and thus could be neglected. We now examine the assumption that diffusion can be neglected. In this chapter we consider the external resistance to diffusion, and in the next chapter we consider internal resistance to diffusion.

We begin with presentation of the fundamentals of diffusion, molar flux, and then write the mole balance in terms of the mole fluxes for rectangular and cylindrical coordinates. Using Fick's law, we write the full equations describing flow, reaction, and diffusion. We consider a few simple geometries and solve the mass flux equations to obtain the concentration gradients and rate of mass transfer. We then discuss mass transfer rates in terms of mass transfer coefficients and correlations for the mass transfer coefficients. Here we include two examples that ask "What if . . ." questions about the system variables. We close the chapter with a discussion on dissolving solids and the shrinking core model, which has applications in drug delivery.

11.1 Diffusion Fundamentals

The Algorithm
1. Mole balance
2. Rate law
3. Stoichiometry
4. Combine
5. Evaluate

The first step in our CRE algorithm is the mole balance, which we now need to extend to include the molar flux, W_{Az}, and diffusional effects. The molar flow rate of A in a given direction, such as the z direction down the length of a tubular reactor, is just the product of the flux, W_{Az} (mol/m$^2 \cdot$ s), and the cross-sectional area, A_c (m^2), that is,

$$F_{Az} = A_c W_{Az}$$

In the previous chapters we have only considered plug flow in which case

$$W_{Az} = \frac{C_A \upsilon}{A_C}$$

We now will extend this concept to consider diffusion superimposed on the molar average velocity.

11.1.1 Definitions

Diffusion is the spontaneous intermingling or mixing of atoms or molecules by random thermal motion. It gives rise to motion of the species *relative* to motion of the mixture. In the absence of other gradients (such as temperature, electric potential, or gravitational potential), molecules of a given species within a single phase will always diffuse from regions of higher concentrations to regions of lower concentrations. This gradient results in a molar flux of the species (e.g., A), $\mathbf{W}_A$ (moles/area$\cdot$time), in the direction of the concentration gradient. The flux of A, $\mathbf{W}_A$, is relative to a fixed coordinate (e.g., the lab bench) and is a vector quantity with typical units of mol/m$^2 \cdot$s. In rectangular coordinates

$$\mathbf{W}_A = iW_{Ax} + jW_{Ay} + kW_{Az} \qquad (11\text{-}1)$$

We now apply the mole balance to species A, which flows and reacts in an element of volume $\Delta V = \Delta x \Delta y \Delta z$ to obtain the variation of the molar fluxes in three dimensions.

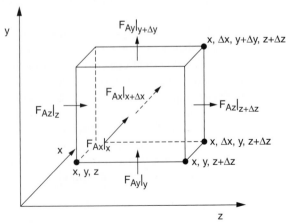

$$F_{Az} = W_{Az}\Delta x\Delta y$$

$$F_{Ay} = W_{Ay}\Delta x\Delta z$$

$$F_{Ax} = W_{Ax}\Delta z\Delta y$$

$$\left[\begin{array}{c} \text{Molar} \\ \text{flow rate} \\ \text{in} \end{array}\right]_z - \left[\begin{array}{c} \text{Molar} \\ \text{flow rate} \\ \text{out} \end{array}\right]_{z+\Delta z} + \left[\begin{array}{c} \text{Molar} \\ \text{flow rate} \\ \text{in} \end{array}\right]_y - \left[\begin{array}{c} \text{Molar} \\ \text{flow rate} \\ \text{out} \end{array}\right]_{y+\Delta y} +$$

Mole Balance

$$\Delta x\Delta y W_{Az}\big|_z - \Delta x\Delta y W_{Az}\big|_{z+\Delta z} + \Delta x\Delta z W_{Ay}\big|_y - \Delta x\Delta z W_{Ay}\big|_{y+\Delta y} +$$

$$\left[\begin{array}{c} \text{Molar} \\ \text{flow rate} \\ \text{in} \end{array}\right]_x - \left[\begin{array}{c} \text{Molar} \\ \text{flow rate} \\ \text{out} \end{array}\right]_{x+\Delta x} + \left[\begin{array}{c} \text{Rate of} \\ \text{generation} \end{array}\right] = \left[\begin{array}{c} \text{Rate of} \\ \text{accumulation} \end{array}\right]$$

$$\Delta z\Delta y W_{Ax}\big|_x - \Delta z\Delta y W_{Ax}\big|_{x+\Delta x} + r_A\Delta x\Delta y\Delta z = \Delta x\Delta y\Delta z\frac{\partial C_A}{\partial t}$$

Dividing by $\Delta x\Delta y\Delta z$ and taking the limit as they go to zero, we obtain the molar flux balance in rectangular coordinates

$$\boxed{-\frac{\partial W_{Ax}}{\partial x} - \frac{\partial W_{Ay}}{\partial y} - \frac{\partial W_{Az}}{\partial z} + r_A = \frac{\partial C_A}{\partial t}} \qquad (11\text{-}2)$$

The corresponding balance in cylindrical coordinates with no variation in the rotation about the z-axis is

COMSOL

$$\boxed{-\frac{1}{r}\frac{\partial}{\partial r}(rW_{Ar}) - \frac{\partial W_{Az}}{\partial z} + r_A = \frac{\partial C_A}{\partial t}} \qquad (11\text{-}3)$$

We will now evaluate the flux terms $\mathbf{W}_A$. We have taken the time to derive the molar flux equations in this form because they are now in a form that is consistent with the partial differential equation (PDE) solver COMSOL, which is included on the CD with this textbook.

11.1.2 Molar Flux

The molar flux of A, $\mathbf{W}_A$, is the result of two contributions: $\mathbf{J}_A$, the molecular diffusion flux relative to the bulk motion of the fluid produced by a concentration gradient, and $\mathbf{B}_A$, the flux resulting from the bulk motion of the fluid:

Total flux =
diffusion + bulk
motion

$$\boxed{\mathbf{W}_A = \mathbf{J}_A + \mathbf{B}_A} \qquad (11\text{-}4)$$

The bulk flow term for species A is the total flux of all molecules relative to a fixed coordinate times the mole fraction of A, y_A; i.e., $\mathbf{B}_A = y_A \sum \mathbf{W}_i$.

The bulk flow term $\mathbf{B}_A$ can also be expressed in terms of the concentration of A and the molar average velocity $\mathbf{V}$:

$$\mathbf{B}_A = C_A \mathbf{V} \qquad (11\text{-}5)$$

$$\frac{\text{mol}}{\text{m}^2 \cdot \text{s}} = \frac{\text{mol}}{\text{m}^3} \cdot \frac{\text{m}}{\text{s}}$$

where the molar average velocity is

Molar average
velocity

$$\mathbf{V} = \sum y_i \mathbf{V}_i$$

Here $\mathbf{V}_i$ is the particle velocity of species i, and y_i is the mole fraction of species i. By particle velocities, we mean the vector-average velocities of millions of A molecules at a point. For a binary mixture of species A and B, we let $\mathbf{V}_A$ and $\mathbf{V}_B$ be the particle velocities of species A and B, respectively. The flux of A with respect to a fixed coordinate system (e.g., the lab bench), $\mathbf{W}_A$, is just the product of the concentration of A and the particle velocity of A:

$$\mathbf{W}_A = C_A \mathbf{V}_A \qquad (11\text{-}6)$$

$$\left(\frac{\text{mol}}{\text{dm}^2 \cdot \text{s}}\right) = \left(\frac{\text{mol}}{\text{dm}^3}\right)\left(\frac{\text{dm}}{\text{s}}\right)$$

The molar average velocity for a binary system is

$$\mathbf{V} = y_A \mathbf{V}_A + y_B \mathbf{V}_B \qquad (11\text{-}7)$$

The total molar flux of A is given by Equation (11-4). $\mathbf{B}_A$ can be expressed either in terms of the concentration of A, in which case

$$\boxed{\mathbf{W}_A = \mathbf{J}_A + C_A \mathbf{V}} \qquad (11\text{-}8)$$

or in terms of the mole fraction of A:

Binary system of
A and B

$$\boxed{\mathbf{W}_A = \mathbf{J}_A + y_A(\mathbf{W}_A + \mathbf{W}_B)} \qquad (11\text{-}9)$$

We now need to evaluate the molar flux of A, $\mathbf{J}_A$, that is superimposed on the molar average velocity $\mathbf{V}$.

11.1.3 Fick's First Law

Our discussion on diffusion will be restricted primarily to binary systems containing only species A and B. We now wish to determine *how* the molar diffusive flux of a species (i.e., $\mathbf{J}_A$) is related to its concentration gradient. As an aid in the discussion of the transport law that is ordinarily used to describe diffusion, recall similar laws from other transport processes. For example, in conductive heat transfer the constitutive equation relating the heat flux $\mathbf{q}$ and the temperature gradient is Fourier's law:

Experimentation
with frog legs led to
Fick's first law

$$\mathbf{q} = -k_t \nabla T \qquad (11\text{-}10)$$

where k_t is the thermal conductivity.

In rectangular coordinates, the gradient is in the form

Constitutive
equations in heat,
momentum, and
mass transfer

$$\nabla = i \frac{\partial}{\partial x} + j \frac{\partial}{\partial y} + k \frac{\partial}{\partial z}$$

(11-11)

The one-dimensional form of Equation (11-10) is

Heat Transfer

$$q_z = -k_t \frac{dT}{dz}$$

(11-12)

In momentum transfer, the constitutive relationship between shear stress, τ, and shear rate for simple planar shear flow is given by Newton's law of viscosity:

Momentum Transfer

$$\tau = -\mu \frac{du}{dz}$$

The mass transfer flux law is analogous to the laws for heat and momentum transport, i.e., for constant total concentration

Mass Transfer

$$\mathbf{J}_{Az} = -D_{AB}\frac{dC_A}{dz}$$

The general 3-dimensional constitutive equation for $\mathbf{J}_A$, the diffusional flux of **A** resulting from a concentration difference, is related to the mole fraction gradient by Fick's first law:

$$\mathbf{J}_A = -cD_{AB}\nabla y_A$$

(11-13)

where c is the total concentration (mol/dm³), D_{AB} is the diffusivity of A in B (dm²/s), and y_A is the mole fraction of A. Combining Equations (11-9) and (11-13), we obtain an expression for the molar flux of A:

Molar flux equation

$$\boxed{\mathbf{W}_A = -cD_{AB}\nabla y_A + y_A(\mathbf{W}_A + \mathbf{W}_B)}$$

(11-14)

In terms of concentration for constant total concentration

Molar flux equation

$$\boxed{\mathbf{W}_A = -D_{AB}\nabla C_A + C_A\mathbf{V}}$$

(11-15)

11.2 Binary Diffusion

Although many systems involve more than two components, the diffusion of each species can be treated as if it were diffusing through another single species rather than through a mixture by defining an effective diffusivity. Methods and examples for calculating this effective diffusivity can be found in Hill.[1]

11.2.1 Evaluating the Molar Flux

The task is to now
evaluate the bulk
flow term.

We now consider four typical conditions that arise in mass transfer problems and show how the molar flux is evaluated in each instance. The first two

---

[1] C. G. Hill, *Chemical Engineering Kinetics and Reactor Design* (New York: Wiley, 1977), p. 480.

Summary Notes

conditions, equal molar counter diffusion (EMCD) and dilute concentration give the same equation for $\mathbf{W}_A$, that is,

$$\mathbf{W}_A = -D_{AB}\nabla C_A$$

The third condition, diffusion through a stagnant film, does not occur as often and is discussed in the summary notes and the solved problems on the CD. The fourth condition is the one we have been discussing up to now for plug flow and the PFR, that is,

$$F_A = vC_A$$

Solved Problems

We will first consider equimolar counter diffusion (EMCD).

11.2.1A Equimolar Counter Diffusion. In equimolar counter diffusion (EMCD), for every mole of A that diffuses in a given direction, one mole of B diffuses in the opposite direction. For example, consider a species A that is diffusing at steady state from the bulk fluid to a catalyst surface, where it isomerizes to form B. Species B then diffuses back into the bulk (see Figure 11-1). For every mole of A that diffuses to the surface, 1 mol of the isomer B diffuses away from the surface. The fluxes of A and B are equal in magnitude and flow counter to each other. Stated mathematically,

$$\mathbf{W}_A = -\mathbf{W}_B \tag{11-16}$$

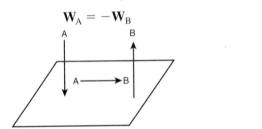

Figure 11-1   EMCD in isomerization reaction.

An expression for $\mathbf{W}_A$ in terms of the concentration of A, C_A, for the case of EMCD can be found by first substituting Equation (11-16) into Equation (11-9):

$$\mathbf{W}_A = \mathbf{J}_A + y_A[\mathbf{W}_A + (-\mathbf{W}_A)] = \mathbf{J}_A + 0$$
$$= \mathbf{J}_A = -cD_{AB}\nabla y_A$$

For constant total concentration

EMCD flux
equation

$$\boxed{\mathbf{W}_A = \mathbf{J}_A = -D_{AB}\nabla C_A} \tag{11-17}$$

11.2.1B Dilute Concentrations. When the mole fraction of the diffusing solute and the bulk motion in the direction of the diffusion are small, the second term on the right-hand side of Equation (11-14) [i.e., $y_A(\mathbf{W}_A + \mathbf{W}_B)$] can usually be neglected compared with the first term, $\mathbf{J}_A$. Under these conditions, together with the condition of constant total concentration, the flux of A is identical to that in Equation (11-16), that is,

Flux at dilute
concentrations

$$\boxed{\mathbf{W}_A \simeq \mathbf{J}_A = -D_{AB}\nabla C_A}$$

(11-18)

This approximation is almost always used for molecules diffusing within aqueous systems when the convective motion is small. For example, the mole fraction of a 1 *M* solution of a solute *A* diffusing in water whose molar concentration, C_W, is

$$C_W = 55.6 \text{ mol/dm}^3$$

would be

$$y_A = \frac{C_A}{C_W + C_A} = \frac{1}{55.6 + 1} = 0.018$$

Consequently, in most liquid systems the concentration of the diffusing solute is small, and Equation (11-18) is used to relate $\mathbf{W}_A$ and the concentration gradient within the boundary layer.

Equation (11-14) also reduces to Equation (11-17) for porous catalyst systems in which the pore radii are very small. Diffusion under these conditions, known as *Knudsen diffusion*, occurs when the mean free path of the molecule is greater than the diameter of the catalyst pore. Here the reacting molecules collide more often with pore walls than with each other, and molecules of different species do not affect each other. The flux of species A for Knudsen diffusion (where bulk flow is neglected) is

$$\mathbf{W}_A = \mathbf{J}_A = -D_K\nabla C_A$$

(11-19)

where D_K is the Knudsen diffusivity.[2]

Summary Notes

11.2.1C Diffusion Through a Stagnant Gas ($W_B = 0$).   Because this condition only effects mass transfer in a limited number of situations, we will discuss this condition in the *Summary Notes* of the CD-ROM.

11.2.1D Forced Convection.   In systems where the flux of A results primarily from forced convection, we assume that the diffusion in the direction of the flow (e.g., axial *z* direction), J_{Az}, *is small in comparison with the bulk flow* contribution in that direction, $B_{Az}(V_z \equiv U)$,

$$J_{Az} \simeq 0$$

---

[2] C. N. Satterfield, *Mass Transfer in Heterogeneous Catalysis* (Cambridge: MIT Press, 1970), pp. 41–42, discusses Knudsen flow in catalysis and gives the expression for calculating D_K.

$$W_{Az} \simeq B_{Az} = C_A V_z \equiv C_A U = \frac{v}{A_c} C_A = \frac{F_A}{A_c}$$

where A_c is the cross-sectional area and v is the volumetric flow rate. Although
the component of the diffusional flux vector of A in the direction of flow, J_{Az},
is neglected, the component of the flux of A in the x direction, J_{Ax}, which is
normal to the direction of flow, may not necessarily be neglected (see Figure 11-2).

Molar flux of species A when axial diffusion effects are negligible

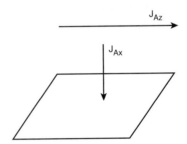

Figure 11-2   Forced axial convection with diffusion to surface.

When diffusional effects can be neglected, F_A can be written as the product of the volumetric flow rate and concentration:

Plug flow

$$F_A = v C_A$$

11.2.1E Diffusion and Convective Transport.   When accounting for diffusional effects, the molar flow rate of species A, F_A, in a specific direction z, is the product of molar flux in that direction, W_{Az}, and the cross-sectional area normal to the direction of flow, A_c:

$$F_{Az} = A_c W_{Az}$$

In terms of concentration the flux is

$$W_{Az} = -D_{AB} \frac{dC_A}{dz} + C_A U_z$$

The molar flow rate is

$$F_{Az} = W_{Az} A_c = \left[-D_{AB} \frac{dC_A}{dz} + C_A U_z \right] A_c \qquad (11\text{-}20)$$

Similar expressions follow for W_{Ax} and W_{Az}. Substituting for the flux W_{Ax},
W_{Ay}, and W_{Az} into Equation (11-2), we obtain

Flow, diffusion, and reaction.

This form is used in Comsol Multiphysics.

$$\boxed{ D_{AB} \left[\frac{\partial^2 C_A}{\partial x^2} + \frac{\partial^2 C_A}{\partial y^2} + \frac{\partial^2 C_A}{\partial z^2} \right] - U_x \frac{\partial C_A}{\partial x} - U_y \frac{\partial C_A}{\partial y} - U_z \frac{\partial C_A}{\partial z} + r_A = \frac{\partial C_A}{\partial t} } \qquad (11\text{-}21)$$

COMSOL   Equation (11-21) is in a user-friendly form to apply to the PDE solver, COM-
SOL. For one-dimension at steady state, Equation (11-21) reduces to

$$\boxed{D_{AB}\frac{d^2 C_A}{dz^2} - U_z\frac{dC_A}{dz} + r_A = 0}$$ (11-22)

In order to solve Equation (11-22) we need to specify the boundary conditions.
In this chapter we will consider some of the simple boundary conditions, and
in Chapter 14 we will consider the more complicated boundary conditions,
such as the Danckwerts' boundary conditions.

We will now use this form of the molar flow rate in our mole balance in
the z direction of a tubular flow reactor

$$\frac{dF_{Az}}{dV} = r_A$$ (1-11)

However, we first have to discuss the boundary conditions in solving the equation.

11.2.2 Boundary Conditions

The most common boundary conditions are presented in Table 11-1.

TABLE 11-1.   TYPES OF BOUNDARY CONDITIONS

1. Specify a concentration at a boundary (e.g., $z = 0$, $C_A = C_{A0}$). For an instantaneous reaction
   at a boundary, the concentration of the reactants at the boundary is taken to be zero (e.g.,
   $C_{As} = 0$). See Chapter 14 for the more exact and complicated Danckwerts' boundary conditions at $z = 0$ and $z = L$.
2. Specify a flux at a boundary.
   a. No mass transfer to a boundary,

$$W_A = 0$$ (11-23)

     For example, at the wall of a nonreacting pipe,

$$\frac{dC_A}{dr} = 0 \qquad \text{at } r = R$$

     That is, because the diffusivity is finite, the only way the flux can be zero is if the concen-
   tration gradient is zero.
   b. Set the molar flux to the surface equal to the rate of reaction on the surface,

$$W_A(\text{surface}) = -r_A''(\text{surface})$$ (11-24)

     c. Set the molar flux to the boundary equal to convective transport across a boundary layer,

$$W_A(\text{boundary}) = k_c(C_{Ab} - C_{As})$$ (11-25)

     where k_c is the mass transfer coefficient and C_{As} and C_{Ab} are the surface and bulk concen-
   trations, respectively.
3. Planes of symmetry. When the concentration profile is symmetrical about a plane, the concentration gradient is zero in that plane of symmetry. For example, in the case of radial diffusion
   in a pipe, at the center of the pipe

$$\frac{dC_A}{dr} = 0 \qquad \text{at } r = 0$$

11.2.3 Modeling Diffusion Without Reaction

In developing mathematical models for chemically reacting systems in which diffusional effects are important, the first steps are:

Step 1: Perform a differential mole balance on a particular species A.
Step 2: Substitute for F_{Az} in terms of W_{Az}.
Step 3: Replace W_{Az} by the appropriate expression for the concentration gradient.
Step 4: State the boundary conditions.
Step 5: Solve for the concentration profile.
Step 6: Solve for the molar flux.

Steps in modeling mass transfer

We are now going to apply this algorithm to one of the most important cases, diffusion through a boundary layer. Here we consider the boundary layer to be a hypothetical "stagnant film" in which all the resistance to mass transfer is lumped.

Example 11–1 Diffusion Through a Film to a Catalyst Particle

Species A, which is present in dilute concentrations, is diffusing at steady state from the bulk fluid through a stagnant film of B of thickness δ to the external surface of the catalyst (Figure E11-1.1). The concentration of A at the external boundary is C_{Ab} and at the external catalyst surface is C_{As}, with $C_{Ab} > C_{As}$. Because the thickness of the "hypothetical stagnant film" next to the surface is small with respect to the diameter of the particle (i.e., $\delta \ll d_p$), we can neglect curvature and represent the diffusion in rectilinear coordinates as shown in Figure E11-1.2.

Determine the concentration profile and the flux of A to the surface using (a) shell balances and (b) the general balance equations.

External Mass Transfer

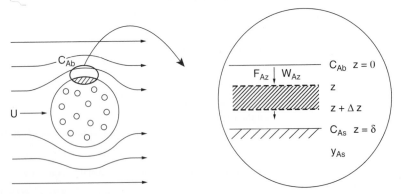

Figure E11-1.1 Transport to a sphere.   **Figure E11-1.2** Boundary layer.

Additional information

$$D_{AB} = 0.01 \text{ cm}^2/\text{s} = 10^{-6} \text{ m}^2/\text{s} \qquad C_{T0} = 0.1 \ kmol/m^3$$

$$y_{Ab} = 0.9$$

$$y_{As} = 0.2$$

Solution

(a) Shell Balances

Our first step is to perform a mole balance on species A over a differential element of width Δz and cross-sectional area A_c and then arrive at a first-order differential equation in W_{Az} [i.e., Equation (E11-1.2)].

Step 1: The general **mole balance** equation is

$$
\begin{bmatrix} \text{Rate} \\ \text{in} \end{bmatrix} - \begin{bmatrix} \text{Rate} \\ \text{out} \end{bmatrix} + \begin{bmatrix} \text{Rate of} \\ \text{generation} \end{bmatrix} = \begin{bmatrix} \text{Rate of} \\ \text{accumulation} \end{bmatrix}
$$

$$
F_{Az}|_z \quad - \quad F_{Az}|_{z+\Delta z} \quad + \quad 0 \quad = \quad 0
$$

An algorithm

- Mole balance
- Bulk flow = ?
- Differential equation
- Boundary conditions
- Concentration profile
- Molar flux

Dividing by $-\Delta z$ gives us

$$
\frac{F_{Az}|_{z+\Delta z} - F_{Az}|_z}{\Delta z} = 0
$$

and taking the limit as $\Delta z \to 0$, we obtain

$$
\frac{dF_{Az}}{dz} = 0 \tag{E11-1.1}
$$

Step 2: Next, **substitute** for F_A in terms of W_{Az} and A_c,

$$
F_{Az} = W_{Az} A_c
$$

Following the Algorithm

Divide by A_c to get

$$
\frac{dW_{Az}}{dz} = 0 \tag{E11-1.2}
$$

Step 3: To **evaluate the bulk flow term**, we now must relate W_{Az} to the concentration gradient utilizing the specification of the problem statement. For diffusion of almost all solutes through a liquid, the concentration of the diffusing species is considered dilute. For a dilute concentration of the diffusing solute, we have for constant total concentration,

$$
W_{Az} = -D_{AB} \frac{dC_A}{dz} \tag{E11-1.3}
$$

Differentiating Equation (E11-1.3) for constant diffusivity yields

$$
\frac{dW_{Az}}{dz} = -D_{AB} \frac{d^2 C_A}{dz^2}
$$

However, Equation (E11-1.2) yielded

$$
\frac{dW_{Az}}{dz} = 0 \tag{E11-1.4}
$$

Therefore, the differential equation describing diffusion through a liquid film reduces to

$$
\frac{d^2 C_A}{dz^2} = 0 \tag{E11-1.5}
$$

Step 4: The **boundary** conditions are:

$$\text{When } z = 0, \qquad C_A = C_{Ab}$$

$$\text{When } z = \delta, \qquad C_A = C_{A\delta}$$

***Step 5:* Solve for the concentration profile.** Equation (E11-1.5) is an elementary differential equation that can be solved directly by integrating twice with respect to z. The first integration yields

$$\frac{dC_A}{dz} = K_1$$

and the second yields

$$C_A = K_1 z + K_2 \qquad\qquad \text{(E11-1.6)}$$

where K_1 and K_2 are arbitrary constants of integration. We now use the boundary conditions to evaluate the constants K_1 and K_2.
At $z = 0$, $C_A = C_{Ab}$; therefore,

$$C_{Ab} = 0 + K_2$$

At $z = \delta$, $C_A = C_{As}$;

$$C_{As} = K_1 \delta + K_2 = K_1 \delta + C_{Ab}$$

Eliminating K_1 and rearranging gives the following concentration profile:

$$\boxed{\frac{C_A - C_{Ab}}{C_{As} - C_{Ab}} = \frac{z}{\delta}} \qquad\qquad \text{(E11-1.7)}$$

Rearranging (E11-1.7), we get the concentration profile shown in Figure E11-1.3.

$$C_A = C_{Ab} + (C_{As} - C_{Ab})\frac{z}{\delta} \qquad\qquad \text{(E11-1.8)}$$

Concentration
profile

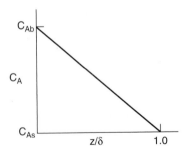

Figure E11-1.3   Concentration profile.

Step 6: The next step is to **determine the molar flux** of A diffusing through the stagnant film. For dilute solute concentrations and constant total concentration,

$$W_{Az} = -D_{AB} \left(\frac{dC_A}{dz} \right) \qquad (E11\text{-}1.3)$$

To determine the flux, differentiate Equation (E11-1.8) with respect to z and then multiply by $-D_{AB}$:

$$W_{Az} = (-D_{AB}) \frac{d}{dz} \left[C_{Ab} + (C_{As} - C_{Ab}) \frac{z}{\delta} \right]$$

$$W_{Az} = -D_{AB} \left(\frac{C_{As} - C_{Ab}}{\delta} \right)$$

$$\boxed{W_{Az} = \frac{D_{AB}}{\delta} (C_{Ab} - C_{As})} \qquad (E11\text{-}1.9)$$

Equation (E11-1.9) could also be written in terms of mole fractions

$$W_{Az} = \frac{D_{AB} C_{T0}}{\delta} (y_{Ab} - y_{As}) \qquad (E11\text{-}1.10)$$

If $D_{AB} = 10^{-6}$ m²/s, $C_{T0} = 0.1$ kmol/m³, and $\delta = 10^{-6}$ m, $y_{Ab} = 0.9$, and $y_{As} = 0.2$, are substituted into Equation (E11-1.9). Then

EMCD or dilute concentration

$$W_{Az} = \frac{(10^{-6} \text{ m}^2/\text{s})(0.1 \text{ kmol/m}^3)(0.9 - 0.2)}{10^{-6} \text{ m}}$$

$$= 0.07 \text{ kmol/m}^2 \cdot \text{s}$$

(b) General Balance Equations (11-2) and (11-21)
Another method used to arrive at the equation describing flow, reaction, and diffusion for a particular geometry and set of conditions is to use the general balance equations (11-2) and (11-21). In this method, we examine each term and then cross out terms that do not apply. In this example, there is no reaction, $-r_A = 0$, no flux in the x-direction, $(W_{Ax} = 0)$ or the y-direction $(W_{Ay} = 0)$, and we are at steady state $\left(\frac{\partial C_A}{\partial t} = 0 \right)$ so that Equation (11-2) reduces to

$$-\frac{dW_A}{dz} = 0 \qquad (E11.1.11)$$

which is the same as Equation (E11-1.4).

Similarly one could apply Equation (11-21) to this example realizing we are at steady state, no reaction, and there is no variation in concentration in either the x-direction or the y-direction

$$\left(\frac{\partial C_A}{\partial x} = 0, \ \frac{\partial^2 C_A}{\partial x^2} = 0 \ \text{ and } \ \frac{\partial C_A}{\partial y} = 0, \ \frac{\partial^2 C_A}{\partial y^2} = 0 \right)$$

so that Equation (11-21) reduces to

$$D_{AB} \frac{d^2 C_A}{dz^2} = 0 \qquad (E11.1.12)$$

Solved Problems

After dividing both sides by the diffusivity, we realize this equation is the same as Equation (E11-1.5).

This problem is reworked for diffusion through a stagnant film in the solved example problems on the CD-ROM/web solved problems.

11.2.4 Temperature and Pressure Dependence of D_{AB}

Before closing this brief discussion on mass transfer fundamentals, further mention should be made of the diffusion coefficient.[3] Equations for predicting gas diffusivities are given by Fuller[4] and are also given in Perry's *Handbook*.[5] The orders of magnitude of the diffusivities for gases, liquids,[6] and solids and the manner in which they vary with temperature and pressure are given in Table 11-2. We note that the Knudsen, liquid, and solid diffusivities are independent of total pressure.

TABLE 11-2.   DIFFUSIVITY RELATIONSHIPS FOR GASES, LIQUIDS, AND SOLIDS

It is important to know the magnitude and the T and P dependence of the diffusivity

Gas:

Liquid:

Phase	Order of Magnitude		Temperature and Pressure Dependences[a]
	cm²/s	m²/s	
Gas			
Bulk	10^{-1}	10^{-5}	$D_{AB}(T_2, P_2) = D_{AB}(T_1, P_1) \dfrac{P_1}{P_2}\left(\dfrac{T_2}{T_1}\right)^{1.75}$
Knudsen	10^{-2}	10^{-6}	$D_A(T_2) = D_A(T_1)\left(\dfrac{T_2}{T_1}\right)^{1/2}$
Liquid	10^{-5}	10^{-9}	$D_{AB}(T_2) = D_{AB}(T_1)\dfrac{\mu_1}{\mu_2}\left(\dfrac{T_2}{T_1}\right)$
Solid	10^{-9}	10^{-13}	$D_{AB}(T_2) = D_{AB}(T_1)\exp\left[\dfrac{E_D}{R}\left(\dfrac{T_2-T_1}{T_1 T_2}\right)\right]$

[a] μ_1, μ_2, liquid viscosities at temperatures T_1 and T_2, respectively; E_D, diffusion activation energy.

---

[3] For further discussion of mass transfer fundamentals, see R. B. Bird, W. E. Stewart, and E. N. Lightfoot, *Transport Phenomena*, 2nd ed. (New York: Wiley, 1960).

[4] E. N. Fuller, P. D. Schettler, and J. C. Giddings, *Ind. Eng. Chem.*, 58(5), 19 (1966). Several other equations for predicting diffusion coefficients can be found in B. E. Polling, J. M. Prausnitz, and J. P. O'Connell, *The Properties of Gases and Liquids*, 5th ed. (New York: McGraw-Hill, 2001).

[5] R. H. Perry and D. W. Green, *Chemical Engineer's Handbook*, 7th ed. (New York: McGraw-Hill, 1999).

[6] To estimate liquid diffusivities for binary systems, see K. A. Reddy and L. K. Doraiswamy, *Ind. Eng. Chem. Fund.*, 6, 77 (1967).

11.2.5 Modeling Diffusion with Chemical Reaction

The method used in solving diffusion problems similar to Example 11-1 is shown in Table 11-3. Also see Cussler.[7]

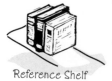

Reference Shelf

TABLE 11-3.   STEPS IN MODELING CHEMICAL SYSTEMS WITH DIFFUSION AND REACTION

1. Define the problem and state the assumptions. (**See Problem Solving on the CD.**)
2. Define the system on which the balances are to be made.
3. Perform a differential mole balance on a particular species.
4. Obtain a differential equation in W_A by rearranging your balance equation properly and taking the limit as the volume of the element goes to zero.
5. Substitute the appropriate expression involving the concentration gradient for W_A from Section 11.2 to obtain a second-order differential equation for the concentration of A.[a]
6. Express the reaction rate r_A (if any) in terms of concentration and substitute into the differential equation.
7. State the appropriate boundary and initial conditions.
8. Put the differential equations and boundary conditions in dimensionless form.
9. Solve the resulting differential equation for the concentration profile.
10. Differentiate this concentration profile to obtain an expression for the molar flux of A.
11. Substitute numerical values for symbols.

Expanding the previous six modeling steps just a bit

[a] In some instances it may be easier to integrate the resulting differential equation in Step 4 before substituting for W_A.

*Move
In ⇄ Out
of the algorithm
(Steps 1 → 6)
to generate
creative solutions.*

The purpose of presenting algorithms (e.g., Table 11-3) to solve reaction engineering problems is to give the readers a starting point or framework with which to work if they were to get stuck. It is expected that once readers are familiar and comfortable using the algorithm/framework, they will be able to move in and out of the framework as they develop creative solutions to nonstandard chemical reaction engineering problems.

11.3 External Resistance to Mass Transfer

11.3.1 The Mass Transfer Coefficient

To begin our discussion on the diffusion of reactants from the bulk fluid to the external surface of a catalyst, we shall focus attention on the flow past a single catalyst pellet. Reaction takes place only on the external catalyst surface and not in the fluid surrounding it. The fluid velocity in the vicinity of the spherical pellet will vary with position around the sphere. The hydrodynamic boundary layer is usually defined as the distance from a solid object to where the fluid velocity is 99% of the bulk velocity U_0. Similarly, the mass transfer boundary layer thickness, δ, is defined as the distance from a solid object to where the concentration of the diffusing species reaches 99% of the bulk concentration.

A reasonable representation of the concentration profile for a reactant A diffusing to the external surface is shown in Figure 11-3. As illustrated, the

[7] E. L. Cussler, *Diffusion Mass Transfer in Fluid Systems*, 2nd ed. (New York: Cambridge University Press, 1997).

Side Note: Transdermal Drug Delivery

The principles of steady state diffusion have been used in a number of drug delivery systems. Specifically, medicated patches are commonly used to attach to the skin to deliver drugs for nicotine withdrawal, birth control, and motion sickness, to name a few. The U.S. transdermal drug delivery market was $1.2 billion in 2001. Equations similar to Equation 11-26 have been used to model the release, diffusion, and absorption of the drug from the patch into the body. Figure SN11.1 shows a drug delivery vehicle (patch) along with the concentration gradient in the epidermis and dermis skin layers.

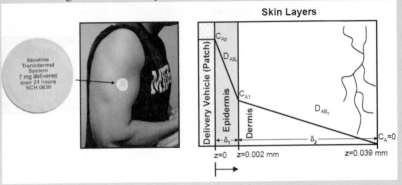

Figure SN 11.1 Transdermal drug delivery schematic.

As a first approximation, the delivery rate can be written as

$$F_A = A_p W_A = \frac{A_p [C_{Ap} - 0]}{R}$$

where

$$R = R_p + \frac{\delta_1}{D_{AB_1}} + \frac{\delta_2}{D_{AB_2}}$$

Where A_p is the area of the patch; C_{Ap}, the concentration of drug in the patch; R, the overall resistance; and R_p, the resistance to release from the patch. There are a number of situations one can consider, such as the patch resistance limits the delivery, diffusion through the epidermis limits delivery, or the concentration of the drug is kept constant in the patch by using solid hydrogels. When diffusion through the epidermis layer limits, the rate of drug delivery rate is

$$F_A = A_p \left[\frac{D_{AB_1}}{\delta_1} \right] C_{Ap}$$

Other models include the use of a quasi-steady analysis to couple the diffusion equation with a balance on the drug in the patch or the zero order dissolution of the hydrogel in the patch.[†] Problem 11-2(f) explores these situations.

[†] For further information see: Y. H. Kalia and R. Guy, *Advanced Drug Delivery Reviews* 48, 159 (2001); B. Muller, M. Kasper, C. Surber, and G. Imanidis, *European Journal of Pharmaceutical Science 20*, 181 (2003); *www.drugdeliverytech.com/cgi-bin/articles.cgi?idArticle=143; www.pharmacy.umaryland.edu/faculty/rdalby/Teaching%20Web%20Pages/Teaching.htm*

change in concentration of A from C_{Ab} to C_{As} takes place in a very narrow fluid layer next to the surface of the sphere. Nearly all of the resistance to mass transfer is found in this layer.

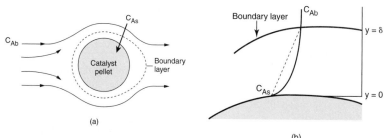

(a)

(b)

Figure 11-3   Boundary layer around the surface of a catalyst pellet.

11.3.2 Mass Transfer Coefficient

The concept of a hypothetical stagnant film within which all the resistance to external mass transfer exists

A useful way of modeling diffusive transport is to treat the fluid layer next to a solid boundary as a stagnant film of thickness δ. We say that *all* the resistance to mass transfer is found (i.e., lumped) within this hypothetical stagnant film of thickness δ, and the properties (i.e., concentration, temperature) of the fluid at the outer edge of the film are identical to those of the bulk fluid. This model can readily be used to solve the differential equation for diffusion through a stagnant film. The dashed line in Figure 11-3b represents the concentration profile predicted by the hypothetical stagnant film model, while the solid line gives the actual profile. If the film thickness is much smaller than the radius of the pellet, curvature effects can be neglected. As a result, only the one-dimensional diffusion equation must be solved, as was shown in Section 11.1 (see also Figure 11-4).

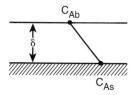

Figure 11-4   Concentration profile for EMCD in stagnant film model.

For either EMCD or dilute concentrations, the solution was shown in Example E11-1 to be in the form

$$W_{Az} = \frac{D_{AB}}{\delta} \cdot [C_{Ab} - C_{As}] \qquad (11\text{-}26)$$

While the boundary layer thickness will vary around the sphere, we will take it to have a mean film thickness δ. The ratio of the diffusivity D_{AB} to the film thickness δ is the mass transfer coefficient, k_c, that is,

The mass transfer coefficient

$$\boxed{k_c = \frac{D_{AB}}{\delta}} \tag{11-27}$$

Combining Equations (11-26) and (11-27), we obtain the average molar flux from the bulk fluid to the surface

Molar flux of A to the surface

$$\boxed{W_{Ar} = k_c(C_{Ab} - C_{As})} \tag{11-28}$$

In this stagnant film model, we consider all the resistance to mass transfer to be lumped into the thickness δ. The reciprocal of the mass transfer coefficient can be thought of as this resistance

$$\boxed{W_{Az} = \text{Flux} = \frac{\text{Driving force}}{\text{Resistance}} = \frac{C_{Ab} - C_{As}}{(1/k_c)}}$$

11.3.3 Correlations for the Mass Transfer Coefficient

The mass transfer coefficient k_c is analogous to the heat transfer coefficient h. The heat flux q from the bulk fluid at a temperature T_0 to a solid surface at T_s is

$$q_r = h(T_0 - T_s) \tag{11-29}$$

For forced convection, the heat transfer coefficient is normally correlated in terms of three dimensionless groups: the Nusselt number, Nu; the Reynolds number, Re; and the Prandtl number, Pr. For the single spherical pellets discussed here, Nu and Re take the following forms:

$$\text{Nu} = \frac{h d_p}{k_t} \tag{11-30}$$

$$\text{Re} = \frac{U \rho d_p}{\mu} \tag{11-31}$$

The Prandtl number is not dependent on the geometry of the system.

The Nusselt, Prandtl, and Reynolds numbers are used in forced convection heat transfer correlations

$$\text{Pr} = \frac{\mu C_p}{k_t} = \frac{\mu}{\rho}\left(\frac{\rho C_p}{k_t}\right) = \frac{\nu}{\alpha_t} \tag{11-32}$$

where $\alpha_t = k_t/\rho C_p$ = thermal diffusivity, m^2/s

$\nu = \dfrac{\mu}{\rho}$ = kinematic viscosity (momentum diffusivity), m^2/s

d_p = diameter of pellet, m

U = free-stream velocity, m/s

k_t = thermal conductivity, J/K·m·s

ρ = fluid density, kg/m^3

h = heat transfer coefficient, J/m^2·s·K or Watts/m^2 K

The other symbols are as defined previously.

The heat transfer correlation relating the Nusselt number to the Prandtl and Reynolds numbers for flow around a sphere is[8]

$$Nu = 2 + 0.6 Re^{1/2} Pr^{1/3} \tag{11-33}$$

Although this correlation can be used over a wide range of Reynolds numbers, it can be shown theoretically that if a sphere is immersed in a stagnant fluid (Re = 0), then

$$Nu = 2 \tag{11-34}$$

and that at higher Reynolds numbers in which the boundary layer remains laminar, the Nusselt number becomes

$$Nu \simeq 0.6 Re^{1/2} Pr^{1/3} \tag{11-35}$$

Although further discussion of heat transfer correlations is no doubt worthwhile, it will not help us to determine the mass transfer coefficient and the mass flux from the bulk fluid to the external pellet surface. However, the preceding discussion on heat transfer was not entirely futile, because, for similar geometries, *the heat and mass transfer correlations are analogous.* If a heat transfer correlation for the Nusselt number exists, the mass transfer coefficient can be estimated by replacing the Nusselt and Prandtl numbers in this correlation by the Sherwood and Schmidt numbers, respectively:

Converting a heat transfer correlation to a mass transfer correlation

$$Sh \longrightarrow Nu$$

$$Sc \longrightarrow Pr$$

The heat and mass transfer coefficients are analogous. The corresponding fluxes are

$$q_z = h(T - T_s) \tag{11-36}$$

$$W_{Az} = k_c(C_A - C_{As})$$

The one-dimensional differential forms of the mass flux for EMCD and the heat flux are, respectively,

For EMCD the heat and molar flux equations are analogous.

$$W_{Az} = -D_{AB} \frac{dC_A}{dz} \tag{E11-1.3}$$

$$q_z = -k_t \frac{dT}{dz} \tag{11-12}$$

If we replace h by k_c and k_t by D_{AB} in Equation (11-30), i.e.,

$$\left. \begin{array}{l} h \longrightarrow k_c \\ k_t \longrightarrow D_{AB} \end{array} \right\} Nu \longrightarrow Sh$$

---

[8] W. E. Ranz and W. R. Marshall, Jr., *Chem. Eng. Prog.*, *48*, 141–146, 173–180 (1952).

we obtain the mass transfer Nusselt number (i.e., the Sherwood number):

Sherwood number

$$\text{Sh} = \frac{k_c d_p}{D_{AB}} = \frac{(\text{m/s})(\text{m})}{\text{m}^2/\text{s}} \quad \text{dimensionless} \tag{11-38}$$

The Prandtl number is the ratio of the kinematic viscosity (i.e., the momentum diffusivity) to the thermal diffusivity. Because the Schmidt number is analogous to the Prandtl number, one would expect that Sc is the ratio of the momentum diffusivity (i.e., the kinematic viscosity), ν, to the mass diffusivity D_{AB}. Indeed, this is true:

$$\alpha_t \longrightarrow D_{AB}$$

The Schmidt number is

Schmidt number

$$\text{Sc} = \frac{\nu}{D_{AB}} = \frac{\text{m}^2/\text{s}}{\text{m}^2/\text{s}} \quad \text{dimensionless} \tag{11-39}$$

The Sherwood, Reynolds, and Schmidt numbers are used in forced convection mass transfer correlations.

Consequently, the correlation for mass transfer for flow around a spherical pellet is analogous to that given for heat transfer [Equation (11-33)], that is,

$$\text{Sh} = 2 + 0.6\text{Re}^{1/2}\text{Sc}^{1/3} \tag{11-40}$$

This relationship is often referred to as the *Frössling correlation.*[9]

11.3.4 Mass Transfer to a Single Particle

In this section we consider two limiting cases of diffusion and reaction on a catalyst particle.[10] In the first case the reaction is so rapid that the rate of diffusion of the reactant to the surface limits the reaction rate. In the second case, the reaction is so slow that virtually no concentration gradient exists in the gas phase (i.e., rapid diffusion with respect to surface reaction).

Example 11–2 Rapid Reaction on a Catalyst Surface

If rapid reaction, then diffusion limits the overall rate.

Calculate the mass flux of reactant A to a single catalyst pellet 1 cm in diameter suspended in a large body of liquid. The reactant is present in dilute concentrations, and the reaction is considered to take place instantaneously at the external pellet surface (i.e., $C_{As} \simeq 0$). The bulk concentration of the reactant is 1.0 M, and the free-system liquid velocity is 0.1 m/s. The kinematic viscosity is 0.5 centistoke (cS; 1 centistoke = 10^{-6} m^2/s), and the liquid diffusivity of A is 10^{-10} m^2/s. T = 300K.

Solution

For dilute concentrations of the solute the radial flux is

$$W_{Ar} = k_c(C_{Ab} - C_{As}) \tag{11-28}$$

---

[9] N. Frössling, *Gerlands Beitr. Geophys.*, 52, 170 (1938).

[10] A comprehensive list of correlations for mass transfer to particles is given by G. A. Hughmark, *Ind. Eng. Chem. Fund.*, 19(2), 198 (1980).

Because reaction is assumed to occur instantaneously on the external surface of the pellet, $C_{As} = 0$. Also, C_{Ab} is given as 1 mol/dm³. The mass transfer coefficient for single spheres is calculated from the Frössling correlation:

$$\text{Sh} = \frac{k_c d_p}{D_{AB}} = 2 + 0.6\text{Re}^{1/2}\text{Sc}^{1/3} \qquad (11\text{-}40)$$

$$\text{Re} = \frac{\rho d_p U}{\mu} = \frac{d_p U}{\nu} = \frac{(0.01\ \text{m})(0.1\ \text{m/s})}{0.5 \times 10^{-6}\ \text{m}^2/\text{s}} = 2000$$

$$\text{Sc} = \frac{\nu}{D_{AB}} = \frac{5 \times 10^{-7}\ \text{m}^2/\text{s}}{10^{-10}\ \text{m}^2/\text{s}} = 5000$$

Substituting these values into Equation (11-40) gives us

$$\text{Sh} = 2 + 0.6(2000)^{0.5}(5000)^{1/3} = 460.7 \qquad (\text{E11-2.1})$$

$$k_c = \frac{D_{AB}}{d_p}\text{Sh} = \frac{10^{-10}\ \text{m}^2/\text{s}}{0.01\ \text{m}} \times 460.7 = 4.61 \times 10^{-6}\ \text{m/s} \qquad (\text{E11-2.2})$$

$$C_{Ab} = 1.0\ \text{mol/dm}^3 = 10^3\ \text{mol/m}^3$$

Substituting for k_c and C_{Ab} in Equation (11-28), the molar flux to the surface is

$$W_{Ar} = (4.61 \times 10^{-6})\ \text{m/s}\ (10^3 - 0)\ \text{mol/m}^3 = 4.61 \times 10^{-3}\ \text{mol/m}^2\cdot\text{s}$$

Because $W_{Ar} = -r''_{As}$, this rate is also the rate of reaction per unit surface area of catalyst.

$$\boxed{-r''_{As} = 0.0046\ \text{mol/m}^2\cdot\text{s} = 0.46\ \text{mol/dm}^2\cdot\text{s}}$$

In Example 11-2, the surface reaction was extremely rapid and the rate of mass transfer to the surface dictated the overall rate of reaction. We now consider a more general case. The isomerization

$$A \longrightarrow B$$

is taking place on the surface of a solid sphere (Figure 11-5). The surface reaction follows a Langmuir–Hinshelwood single-site mechanism for which the rate law is

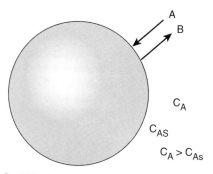

Figure 11-5   Diffusion to, and reaction on, external surface of pellet.

$$-r''_{As} = \frac{k_r C_{As}}{1 + K_A C_{As} + K_B C_{Bs}} \tag{11-41}$$

The temperature is sufficiently high that we need only consider the case of very weak adsorption (i.e., low surface coverage) of A and B; thus

$$(K_B C_{Bs} + K_A C_{As}) \ll 1$$

Therefore,

$$-r''_{As} = k_r C_{As} \tag{11-42}$$

Using boundary conditions 2b and 2c in Table 11-1, we obtain

$$W_A\big|_{surface} = -r''_{As} \tag{11-43}$$

$$W_A = k_c(C_A - C_{As}) = k_r C_{As} \tag{11-44}$$

The concentration C_{As} is not as easily measured as the bulk concentration. Consequently, we need to eliminate C_{As} from the equation for the flux and rate of reaction. Solving Equation (11-44) for C_{As} yields

$$C_{As} = \frac{k_c C_A}{k_r + k_c} \tag{11-45}$$

and the rate of reaction on the surface becomes

Molar flux of A to the surface is equal to the rate of consumption of A on the surface

$$\boxed{W_A = -r''_{As} = \frac{k_c k_r C_A}{k_r + k_c}} \tag{11-46}$$

One will often find the flux to or from the surface as written in terms of an *effective* transport coefficient k_{eff}:

$$W_A = -r''_{As} = k_{eff} C_A \tag{11-47}$$

where

$$k_{eff} = \frac{k_c k_r}{k_c + k_r}$$

Rapid Reaction.   We first consider how the overall rate of reaction may be increased when the rate of mass transfer to the surface limits the overall rate of reaction. Under these circumstances the specific reaction rate constant is much greater than the mass transfer coefficient

$$k_r \gg k_c$$

and

$$\frac{k_c}{k_r} \ll 1$$

$$-r''_A = \frac{k_c C_A}{1 + k_c/k_r} \approx k_c C_A \tag{11-48}$$

To increase the rate of reaction per unit surface area of solid sphere, one must increase C_A and/or k_c. In this gas-phase catalytic reaction example, and for most liquids, the Schmidt number is sufficiently large that the number 2 in Equation (11-40) is negligible with respect to the second term when the Reynolds number is greater than 25. As a result, Equation (11-40) gives

It is important to know how the mass transfer coefficient varies with fluid velocity, particle size, and physical properties.

$$k_c = 0.6 \left(\frac{D_{AB}}{d_p}\right) \mathrm{Re}^{1/2} \mathrm{Sc}^{1/3}$$

$$= 0.6 \left(\frac{D_{AB}}{d_p}\right) \left(\frac{U d_p}{\nu}\right)^{1/2} \left(\frac{\nu}{D_{AB}}\right)^{1/3}$$

$$k_c = 0.6 \times \frac{D_{AB}^{2/3}}{\nu^{1/6}} \times \frac{U^{1/2}}{d_p^{1/2}} \tag{11-49}$$

$$k_c = 0.6 \times (\text{Term 1}) \times (\text{Term 2})$$

Mass Transfer Limited

Term 1 is a function of the physical properties D_{AB} and ν, which depend on temperature and pressure only. The diffusivity always increases with increasing temperature for both gas and liquid systems. However, the kinematic viscosity ν increases with temperature ($\nu \propto T^{3/2}$) for gases and decreases exponentially with temperature for liquids. Term 2 is a function of flow conditions and particle size. Consequently, to increase k_c and thus the overall rate of reaction per unit surface area, one may either decrease the particle size or increase the velocity of the fluid flowing past the particle. For this particular case of flow past a single sphere, we see that if the velocity is doubled, the mass transfer coefficient and consequently the rate of reaction is increased by a factor of

$$(U_2/U_1)^{0.5} = 2^{0.5} = 1.41 \text{ or } 41\%$$

Reaction Rate Limited

Slow Reaction. Here the specific reaction rate constant is small with respect to the mass transfer coefficient:

$$k_r \ll k_c$$

$$-r''_{As} = \frac{k_r C_A}{1 + k_r/k_c} \approx k_r C_A \tag{11-50}$$

Mass transfer effects are not important when the reaction rate is limiting.

The specific reaction rate is independent of the velocity of fluid and for the solid sphere considered here, independent of particle size. *However, for porous catalyst pellets, k_r may depend on particle size for certain situations, as shown in Chapter 12.*

Figure 11-6 shows the variation in reaction rate with Term 2 in Equation (11-49), the ratio of velocity to particle size. At low velocities the mass transfer boundary layer thickness is large and diffusion limits the reaction. As the velocity past the sphere is increased, the boundary layer thickness decreases, and the mass transfer across the boundary layer no longer limits the rate of reaction. One also notes that for a given velocity, reaction-limiting conditions can be achieved by using very small particles. However, the smaller the particle size, the greater the pressure drop in a packed bed. When one is obtaining reaction rate data in the laboratory, one must operate at sufficiently high velocities or sufficiently small particle sizes to ensure that the reaction is not mass transfer–limited.

When collecting rate law data, operate in the reaction-limited region.

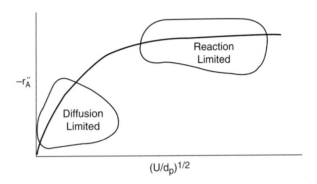

Figure 11-6   Regions of mass transfer–limited and reaction–limited reactions.

11.3.5 Mass Transfer–Limited Reactions in Packed Beds

A number of industrial reactions are potentially mass transfer–limited because they may be carried out at high temperatures without the occurrence of undesirable side reactions. In mass transfer–dominated reactions, the surface reaction is so rapid that the rate of transfer of reactant from the bulk gas or liquid phase to the surface limits the overall rate of reaction. Consequently, mass transfer–limited reactions respond quite differently to changes in temperature and flow conditions than do the rate-limited reactions discussed in previous chapters. In this section the basic equations describing the variation of conversion with the various reactor design parameters (catalyst weight, flow conditions) will be developed. To achieve this goal, we begin by carrying out a mole balance on the following mass transfer–limited reaction:

$$A + \frac{b}{a} B \longrightarrow \frac{c}{a} C + \frac{d}{a} D \tag{2-2}$$

carried out in a packed-bed reactor (Figure 11-7). A steady-state mole balance on reactant A in the reactor segment between z and $z + \Delta z$ is

$$
\begin{bmatrix} \text{Molar} \\ \text{rate in} \end{bmatrix} - \begin{bmatrix} \text{Molar} \\ \text{rate out} \end{bmatrix} + \begin{bmatrix} \text{Molar rate of} \\ \text{generation} \end{bmatrix} = \begin{bmatrix} \text{Molar rate of} \\ \text{accumulation} \end{bmatrix}
$$

$$
F_{Az}|_z \quad - \quad F_{Az}|_{z+\Delta z} \quad + \quad r''_A a_c (A_c \Delta z) \quad = \quad 0 \qquad (11\text{-}51)
$$

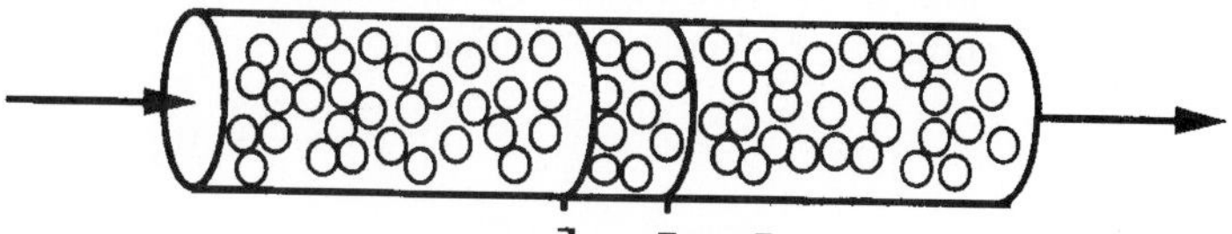

Figure 11-7   Packed-bed reactor.

where r''_A = rate of generation of A per unit catalytic surface area, mol/s·m²
 a_c = external surface area of catalyst per volume of catalytic bed, m²/m³
 $= 6(1 - \phi)/d_p$ for packed beds, m²/m³
 ϕ = porosity of the bed (i.e., void fraction)[11]
 d_p = particle diameter, m
 A_c = cross-sectional area of tube containing the catalyst, m²

Dividing Equation (11-51) by $A_c \Delta z$ and taking the limit as $\Delta z \longrightarrow 0$, we have

$$
-\frac{1}{A_c}\left(\frac{dF_{Az}}{dz}\right) + r''_A a_c = 0 \qquad (11\text{-}52)
$$

We now need to express F_{Az} and r''_A in terms of concentration.
    The molar flow rate of A in the axial direction is

$$
F_{Az} = A_c W_{Az} = (J_{Az} + B_{Az})A_c \qquad (11\text{-}53)
$$

Axial diffusion is neglected. In almost all situations involving flow in packed-bed reactors, the amount of material transported by diffusion or dispersion in the axial direction is negligible compared with that transported by convection (i.e., bulk flow):

$$
J_{Az} \ll B_{Az}
$$

(In Chapter 14 we consider the case when dispersive effects **must** be taken into account.) Neglecting dispersion, Equation (11-20) becomes

$$
F_{Az} = A_c W_{Az} = A_c B_{Az} = U C_A A_c \qquad (11\text{-}54)
$$

where U is the superficial molar average velocity through the bed (m/s). Substituting for F_{Az} in Equation (11-52) gives us

$$
-\frac{d(C_A U)}{dz} + r''_A a_c = 0 \qquad (11\text{-}55)
$$

For the case of constant superficial velocity U,

---

[11]In the nomenclature for Chapter 4, for Ergun Equation for pressure drop.

Differential
equation describing
flow and reaction in
a packed bed

$$-U \frac{dC_A}{dz} + r''_A a_c = 0 \qquad (11\text{-}56)$$

For reactions at steady state, the molar flux of A to the particle surface, W_{Ar} (mol/m$^2\cdot$s) (see Figure 11-8), is equal to the rate of disappearance of A on the surface $-r''_A$ (mol/m$^2\cdot$s); that is,

$$-r''_A = W_{Ar} \qquad (11\text{-}57)$$

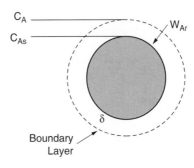

Figure 11-8 Diffusion across stagnant film surrounding catalyst pellet.

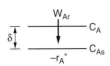

From Table 11-1, the boundary condition at the external surface is

$$-r''_A = W_{Ar} = k_c(C_A - C_{As}) \qquad (11\text{-}58)$$

where k_c = mass transfer coefficient = D_{AB}/δ (s^{-1})
       C_A = bulk concentration (mol/m^3)
      C_{As} = concentration of A at the catalytic surface (mol/m^3)

Substituting for r''_A in Equation (11-56), we have

$$-U \frac{dC_A}{dz} - k_c a_c(C_A - C_{As}) = 0 \qquad (11\text{-}59)$$

In reactions that
are completely mass
transfer–limited, it is
not necessary to
know the rate law.

In most mass transfer–limited reactions, the surface concentration is negligible with respect to the bulk concentration (i.e., $C_A \gg C_{As}$):

$$-U \frac{dC_A}{dz} = k_c a_c C_A \qquad (11\text{-}60)$$

Integrating with the limit, at $z = 0$, $C_A = C_{A0}$:

$$\frac{C_A}{C_{A0}} = \exp\left(-\frac{k_c a_c}{U} z\right) \qquad (11\text{-}61)$$

The corresponding variation of reaction rate along the length of the reactor is

$$-r_A'' = k_c C_{A0} \exp\left(-\frac{k_c a_c}{U} z\right) \tag{11-62}$$

The concentration and conversion profiles down a reactor of length L are shown in Figure 11-9.

Reactor concentration profile for a mass transfer–limited reaction

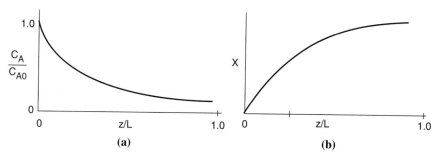

(a)   **(b)**

Figure 11-9   Axial concentration (a) and conversion (b) profiles in a packed bed.

To determine the reactor length L necessary to achieve a conversion X, we combine the definition of conversion,

$$X = \frac{C_{A0} - C_{AL}}{C_{A0}} \tag{11-63}$$

with the evaluation of Equation (11-61) at $z = L$ to obtain

$$\boxed{\ln\frac{1}{1-X} = \frac{k_c a_c}{U} L} \tag{11-64}$$

11.3.6 Robert the Worrier

Robert is an engineer who is always worried. He thinks something bad will happen if we change an operating condition such as flow rate or temperature or an equipment parameter such as particle size. Robert's motto is "If it ain't broke, don't fix it." We can help Robert be a little more adventuresome by analyzing how the important parameters vary as we change operating conditions in order to predict the outcome of such a change. We first look at Equation (11-64) and see that conversion depends upon k_c, a_c, U, and L. We now examine how each of these parameters will change as we change operating conditions. We first consider the effects of temperature and flow rate on conversion.

To learn the effect of flow rate and temperature on conversion, we need to know how these parameters affect the mass transfer coefficient. That is, we must determine the correlation for the mass transfer coefficient for the particular geometry and flow field. For flow through a packed bed, the correlation given by Thoenes and Kramers[12] for $0.25 < \phi < 0.5$, $40 < Re' < 4000$, and $1 < Sc < 4000$ is

$$Sh' = 1.0(Re')^{1/2}Sc^{1/3} \tag{11-65}$$

Thoenes–Kramers correlation for flow through packed beds

$$\left[\frac{k_c d_p}{D_{AB}}\left(\frac{\phi}{1-\phi}\right)\frac{1}{\gamma}\right] = \left[\frac{U d_p \rho}{\mu(1-\phi)\gamma}\right]^{1/2}\left(\frac{\mu}{\rho D_{AB}}\right)^{1/3} \tag{11-66}$$

where $Re' = \dfrac{Re}{(1-\phi)\gamma}$

$Sh' = \dfrac{Sh\,\phi}{(1-\phi)\gamma}$

d_p = particle diameter (equivalent diameter of sphere of the same volume), m

$= [(6/\pi)\,(\text{volume of pellet})]^{1/3}$, m

ϕ = void fraction (porosity) of packed bed

γ = shape factor (external surface area divided by πd_p^2)

U = superficial gas velocity through the bed, m/s

μ = viscosity, kg/m·s

ρ = fluid density, kg/m³

$\nu = \dfrac{\mu}{\rho}$ = kinematic viscosity, m²/s

D_{AB} = gas-phase diffusivity, m²/s

For constant fluid properties and particle diameter:

$$k_c \propto U^{1/2} \tag{11-67}$$

We see that the mass transfer coefficient increases with the square root of the superficial velocity through the bed. Therefore, *for a fixed concentration, C_A,* such as that found in a differential reactor, the rate of reaction should vary with $U^{1/2}$:

For diffusion-limited reactions, reaction rate depends on particle size and fluid velocity.

$$-r''_A \propto k_c C_A \propto U^{1/2}$$

However, if the gas velocity is continually increased, a point is reached where the reaction becomes reaction rate–limited and consequently, is independent of the superficial gas velocity, as shown in Figure 11-6.

Most mass transfer correlations in the literature are reported in terms of the Colburn J factor (i.e., J_D) as a function of the Reynolds number. The relationship between J_D and the numbers we have been discussing is

Colburn J factor

$$J_D = \frac{Sh}{Sc^{1/3}\,Re} \tag{11-68}$$

Figure 11-10 shows data from a number of investigations for the J factor as a function of the Reynolds number for a wide range of particle shapes and

[12]D. Thoenes, Jr. and H. Kramers, *Chem. Eng. Sci.*, 8, 271 (1958).

gas flow conditions. *Note:* There are serious deviations from the Colburn analogy when the concentration gradient and temperature gradient are coupled as shown by Venkatesan and Fogler.[13]

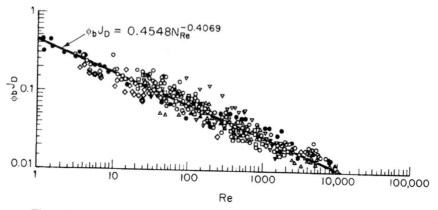

Figure 11-10 Mass transfer correlation for packed beds. $\phi b \equiv \phi$ [Reprinted with permission from P. N. Dwidevi and S. S. Upadhyay, *Ind. Eng. Chem. Process Des. Dev.*, *16*, 157 (1977). Copyright © 1977 American Chemical Society.]

Dwidevi and Upadhyay[14] review a number of mass transfer correlations for both fixed and fluidized beds and arrive at the following correlation, which is valid for both gases (Re > 10) and liquids (Re > 0.01) in either fixed or fluidized beds:

<aside>A correlation for flow through packed beds in terms of the Colburn *J* factor</aside>

$$\phi J_D = \frac{0.765}{Re^{0.82}} + \frac{0.365}{Re^{0.386}} \tag{11-69}$$

For nonspherical particles, the equivalent diameter used in the Reynolds and Sherwood numbers is $d_p = \sqrt{A_p/\pi} = 0.564\sqrt{A_p}$, where A_p is the external surface area of the pellet.

To obtain correlations for mass transfer coefficients for a variety of systems and geometries, see either D. Kunii and O. Levenspiel, *Fluidization Engineering*, 2nd ed. (Butterworth-Heinemann, 1991), Chap. 7, or W. L. McCabe, J. C. Smith, and P. Harriott, *Unit Operations in Chemical Engineering*, 6th ed. (New York: McGraw-Hill, 2000). For other correlations for packed beds with different packing arrangements, see I. Colquhoun-Lee and J. Stepanek, *Chemical Engineer*, 108 (Feb. 1974).

<aside>Actual case history and current application</aside>

Example 11–3 Maneuvering a Space Satellite

Hydrazine has been studied extensively for use in monopropellant thrusters for space flights of long duration. Thrusters are used for altitude control of communication satellites. Here the decomposition of hydrazine over a packed bed of alumina-supported

---

[13]R. Venkatesan and H. S. Fogler, *AIChE J.*, *50*, 1623 (July 2004).

[14]P. N. Dwidevi and S. N. Upadhyay, *Ind. Eng. Chem. Process Des. Dev.*, *16*, 157 (1977).

iridium catalyst is of interest.[15] In a proposed study, a 2% hydrazine in 98% helium mixture is to be passed over a packed bed of cylindrical particles 0.25 cm in diameter and 0.5 cm in length at a gas-phase velocity of 15 m/s and a temperature of 750 K. The kinematic viscosity of helium at this temperature is 4.5×10^{-4} m²/s. The hydrazine decomposition reaction is believed to be externally mass transfer–limited under these conditions. If the packed bed is 0.05 m in length, what conversion can be expected? Assume isothermal operation.

Additional information

$D_{AB} = 0.69 \times 10^{-4}$ m²/s at 298 K
Bed porosity: 30%
Bed fluidicity: 95.7%

Solution

Rearranging Equation (11-64) gives us

$$X = 1 - e^{-(k_c a_c / U)L}$$ (E11-3.1)

(a) Thoenes–Kramers correlation

1. First we find the volume-average particle diameter:

$$d_p = \left(\frac{6V}{\pi} \right)^{1/3} = \left(6 \frac{\pi D^2}{4} \frac{L}{\pi} \right)^{1/3}$$ (E11-3.2)

$$= [1.5(0.0025 \text{ m})^2 (0.005 \text{ m})]^{1/3} = 3.61 \times 10^{-3} \text{ m}$$

2. Surface area per volume of bed:

$$a_c = 6 \left(\frac{1 - \phi}{d_p} \right) = 6 \left(\frac{1 - 0.3}{3.61 \times 10^{-3} \text{ m}} \right) = 1163 \text{ m}^2/\text{m}^3$$ (E11-3.3)

3. Mass transfer coefficient:

$$\text{Re} = \frac{d_p U}{\nu} = \frac{(3.61 \times 10^{-3} \text{ m})(15 \text{ m/s})}{4.5 \times 10^{-4} \text{ m}^2/\text{s}} = 120.3$$

For cylindrical pellets,

$$\gamma = \frac{2\pi r L_p + 2\pi r^2}{\pi d_p^2} = \frac{(2)(0.0025/2)(0.005) + (2)(0.0025/2)^2}{(3.61 \times 10^{-3})^2} = 1.20$$ (E11-3.4)

$$\text{Re}' = \frac{\text{Re}}{(1 - \phi)\gamma} = \frac{120.3}{(0.7)(1.2)} = 143.2$$

Correcting the diffusivity to 750 K using Table 11-2 gives us

$$D_{AB}(750 \text{ K}) = D_{AB}(298 \text{ K}) \times \left(\frac{750}{298} \right)^{1.75} = (0.69 \times 10^{-4} \text{ m}^2/\text{s})(5.03)$$

---

[15]O. I. Smith and W. C. Solomon, *Ind. Eng. Chem. Fund.*, *21*, 374 (1982).

$$D_{AB}\,(750\text{ K}) = 3.47 \times 10^{-4}\text{ m}^2/\text{s} \tag{E11-3.5}$$

$$Sc = \frac{\nu}{D_{AB}} = \frac{4.5 \times 10^{-4}\text{ m}^2/\text{s}}{3.47 \times 10^{-4}\text{ m}^2/\text{s}} = 1.30$$

Substituting Re′ and Sc into Equation (11-65) yields

$$Sh' = (143.2)^{1/2}(1.3)^{1/3} = (11.97)(1.09) = 13.05 \tag{E11-3.6}$$

$$k_c = \frac{D_{AB}(1 - \phi)}{d_p\phi}\,\gamma\,(Sh') = \left(\frac{3.47 \times 10^{-4}\text{ m}^2/\text{s}}{3.61 \times 10^{-3}\text{ m}}\right)\left(\frac{1 - 0.3}{0.3}\right)$$

$$\times\,(1.2)(13.05) = 3.52\text{ m/s} \tag{E11-3.7}$$

The conversion is

$$X = 1 - \exp\left[-(3.52\text{ m/s})\left(\frac{1163\text{ m}^2/\text{m}^3}{15\text{ m/s}}\right)(0.05\text{ m})\right] \tag{E11-3.8}$$

$$= 1 - 1.18 \times 10^{-6} \simeq 1.00 \qquad \text{virtually complete conversion}$$

(b) Colburn J_D factor. Calculate the surface-area-average particle diameter.
For cylindrical pellets the external surface area is

$$A = \pi d L_p + 2\pi\left(\frac{d^2}{4}\right) \tag{E11-3.9}$$

$$d_p = \sqrt{\frac{A}{\pi}} = \sqrt{\frac{\pi d L_p + 2\pi\,(d^2/4)}{\pi}} \tag{E11-3.10}$$

$$= \sqrt{(0.0025)(0.005) + \frac{(0.0025)^2}{2}} = 3.95 \times 10^{-3}\text{ m}$$

$$a = \frac{6(1 - \phi)}{d_p} = 1063\text{ m}^2/\text{m}^3$$

Typical
values

Gas Phase
Re = 130
J_D = 0.23
Sc = 1.3
Sh = 33
k_c = 3 m/s

$$Re = \frac{d_p U}{\nu} = \frac{(3.95 \times 10^{-3}\text{ m})(15\text{ m/s})}{4.5 \times 10^{-4}\text{ m}^2/\text{s}}$$

$$= 131.6$$

$$\phi J_D = \frac{0.765}{Re^{0.82}} + \frac{0.365}{Re^{0.386}} \tag{11-69}$$

$$= \frac{0.765}{(131.6)^{0.82}} + \frac{0.365}{(131.6)^{0.386}} = 0.014 + 0.055 \tag{E11-3.11}$$

$$= 0.069$$

$$J_D = \frac{0.069}{0.3} = 0.23 \tag{E11-3.12}$$

$$\text{Sh} = \text{Sc}^{1/3}\text{Re}(J_D) \qquad\qquad \text{(E11-3.13)}$$

$$= (1.3)^{1/3}(131.6)(0.23) = 33.0$$

$$k_c = \frac{D_{AB}}{d_p}\,\text{Sh} = \frac{3.47 \times 10^{-4}}{3.95 \times 10^{-3}}\,(33) = 2.9 \text{ m/s}$$

Then

$$X = 1 - \exp\left[-(2.9 \text{ m/s})\left(\frac{1063 \text{ m}^2/\text{m}^3}{15 \text{ m/s}}\right)(0.05 \text{ m})\right] \qquad \text{(E11-3.14)}$$

$$= 1 - 0.0000345$$

$$\simeq 1 \qquad \text{again, virtually complete conversion}$$

If there were such a thing as the **bed fluidity**, given in the problem statement, it would be a useless piece of information. Make sure that you know what information you need to solve problems, and go after it. Do not let additional data confuse you or lead you astray with useless information or facts that represent someone else's bias, which are probably not well-founded.

11.4 What If . . . ? (Parameter Sensitivity)

One of the most important skills of an engineer is to be able to predict the effects of changes of system variables on the operation of a process. The engineer needs to determine these effects quickly through approximate but reasonably close calculations, which are sometimes referred to as "back-of-the-envelope calculations."[16] This type of calculation is used to answer such questions as "**What** will happen **if** I decrease the particle size?" "**What if** I triple the flow rate through the reactor?"

J. D. Goddard's

Back of the
Envelope

We will now proceed to show how this type of question can be answered using the packed-bed, mass transfer–limited reactors as a model or example system. Here we want to learn the effect of changes of the various parameters (e.g., temperature, particle size, superficial velocity) on the conversion. We begin with a rearrangement of the mass transfer correlation, Equation (11-49), to yield

$$k_c \propto \left(\frac{D_{AB}^{2/3}}{\nu^{1/6}}\right)\left(\frac{U^{1/2}}{d_p^{1/2}}\right) \qquad\qquad \text{(11-70)}$$

Find out how the mass transfer coefficient varies with changes in physical properties and system properties

The first term on the right-hand side is dependent on physical properties (temperature and pressure), whereas the second term is dependent on system properties (flow rate and particle size). One observes from this equation that the mass transfer coefficient increases as the particle size decreases. The use of sufficiently small particles offers another technique to escape from the mass transfer–limited regime into the reaction rate–limited regime.

---

[16]Prof. J. D. Goddard, University of Michigan, 1963–1976. Currently at University of California, San Diego.

Example 11–4   The Case of Divide and Be Conquered

A mass transfer–limited reaction is being carried out in two reactors of equal volume and packing, connected in series as shown in Figure E11-4.1. Currently, 86.5% conversion is being achieved with this arrangement. It is suggested that the reactors be separated and the flow rate be divided equally among each of the two reactors (Figure E11-4.2) to decrease the pressure drop and hence the pumping requirements. In terms of achieving a higher conversion, Robert is wondering if this is a good idea?

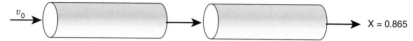

Figure E11-4.1   Series arrangement

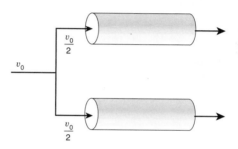

Figure E11-4.2   Parallel arrangement.

Reactors in series versus reactors in parallel

Solution

As a first approximation, we neglect the effects of small changes in temperature and pressure on mass transfer. We recall Equation (11-64), which gives conversion as a function of reactor length. For a mass transfer–limited reaction

$$\ln \frac{1}{1-X} = \frac{k_c a_c}{U} L \tag{11-64}$$

For case 1, the undivided system:

$$\left(\ln \frac{1}{1-X_1} \right) = \frac{k_{c1} a_c}{U_1} L_1 \tag{E11-4.1}$$

$$X_1 = 0.865$$

For case 2, the divided system:

$$\left(\ln \frac{1}{1-X_2} \right) = \frac{k_{c2} a_c}{U_2} L_2 \tag{E11-4.2}$$

$$X_2 = ?$$

We now take the ratio of case 2 (divided system) to case 1 (undivided system):

$$\frac{\ln \dfrac{1}{1-X_2}}{\ln \dfrac{1}{1-X_1}} = \frac{k_{c2}}{k_{c1}} \left(\frac{L_2}{L_1} \right) \frac{U_1}{U_2} \tag{E11-4.3}$$

The surface area per unit volume a_c is the same for both systems.

From the conditions of the problem statement we know that

$$L_2 = \tfrac{1}{2}L_1$$

$$U_2 = \tfrac{1}{2}U_1$$

$$X_1 = 0.865$$

$$X_2 = ?$$

However, we must also consider the effect of the division on the mass transfer coefficient. From Equation (11-70) we know that

$$k_c \propto U^{1/2}$$

Then

$$\frac{k_{c2}}{k_{c1}} = \left(\frac{U_2}{U_1}\right)^{1/2} \tag{E11-4.4}$$

Multiplying by the ratio of superficial velocities yields

$$\frac{U_1}{U_2}\left(\frac{k_{c2}}{k_{c1}}\right) = \left(\frac{U_1}{U_2}\right)^{1/2} \tag{E11-4.5}$$

$$\ln\frac{1}{1-X_2} = \left(\ln\frac{1}{1-X_1}\right)\frac{L_2}{L_1}\left(\frac{U_1}{U_2}\right)^{1/2} \tag{E11-4.6}$$

$$= \left(\ln\frac{1}{1-0.865}\right)\left[\frac{\tfrac{1}{2}L_1}{L_1}\left(\frac{U_1}{\tfrac{1}{2}U_1}\right)^{1/2}\right]$$

$$= 2.00\left(\frac{1}{2}\right)\sqrt{2} = 1.414$$

Solving for X_2 gives us

$$X_2 = 0.76$$

Bad idea!! Robert
was right to worry.

Consequently, we see that although the divided arrangement will have the advantage of a smaller pressure drop across the bed, it is a bad idea in terms of conversion. Recall that if the reaction were reaction rate–limited, both arrangements would give the same conversion.

Example 11–5   The Case of the Overenthusiastic Engineers

The same reaction as that in Example 11-4 is being carried out in the same two reactors in series. A new engineer suggests that the rate of reaction could be increased by a factor of 2^{10} by increasing the reaction temperature from 400°C to 500°C, reasoning that the reaction rate doubles for every 10°C increase in temperature. Another engineer arrives on the scene and berates the new engineer with quotations from Chapter 3 concerning this rule of thumb. She points out that it is valid only for a specific activation energy within a specific temperature range. She then

Robert worries if this temperature increase will be worth the trouble.

suggests that he go ahead with the proposed temperature increase but should only expect an increase on the order of 2^3 or 2^4. What do you think? Who is correct?

Solution

Because almost all surface reaction rates increase more rapidly with temperature than do diffusion rates, increasing the temperature will only increase the degree to which the reaction is mass transfer–limited.

We now consider the following two cases:

Case 1: $T = 400°C$ $X = 0.865$

Case 2: $T = 500°C$ $X = ?$

Taking the ratio of case 2 to case 1 and noting that the reactor length is the same for both cases ($L_1 = L_2$), we obtain

$$\frac{\ln \dfrac{1}{1-X_2}}{\ln \dfrac{1}{1-X_1}} = \frac{k_{c2}}{k_{c1}} \left(\frac{L_2}{L_1}\right) \frac{U_1}{U_2} = \frac{k_{c2}}{k_{c1}} \left(\frac{U_1}{U_2}\right) \tag{E11-5.1}$$

The molar feed rate F_{T0} remains unchanged:

$$F_{T0} = v_{01} \left(\frac{P_{01}}{RT_{01}}\right) = v_{02} \left(\frac{P_{02}}{RT_{02}}\right) \tag{E11-5.2}$$

Because $v = A_c U$, the superficial velocity at temperature T_2 is

$$U_2 = \frac{T_2}{T_1} U_1 \tag{E11-5.3}$$

We now wish to learn the dependence of the mass transfer coefficient on temperature:

$$k_c \propto \left(\frac{U^{1/2}}{d_p^{1/2}}\right) \left(\frac{D_{AB}^{2/3}}{\nu^{1/6}}\right) \tag{E11-5.4}$$

Taking the ratio of case 2 to case 1 and realizing that the particle diameter is the same for both cases gives us

$$\frac{k_{c2}}{k_{c1}} = \left(\frac{U_2}{U_1}\right)^{1/2} \left(\frac{D_{AB2}}{D_{AB1}}\right)^{2/3} \left(\frac{\nu_1}{\nu_2}\right)^{1/6} \tag{E11-5.5}$$

The temperature dependence of the gas-phase diffusivity is (from Table 11-2)

$$D_{AB} \propto T^{1.75} \tag{E11-5.6}$$

For most gases, viscosity increases with increasing temperature according to the relation

$$\mu \propto T^{1/2}$$

From the ideal gas law,

$$\rho \propto T^{-1}$$

Then

$$\nu = \frac{\mu}{\rho} \propto T^{3/2} \tag{E11-5.7}$$

It's really important to know how to do this type of analysis.

$$\frac{\ln\dfrac{1}{1-X_2}}{\ln\dfrac{1}{1-X_1}} = \frac{U_1}{U_2}\left(\frac{k_{c2}}{k_{c1}}\right) = \left(\frac{U_1}{U_2}\right)^{1/2}\left(\frac{D_{AB2}}{D_{AB1}}\right)^{2/3}\left(\frac{\nu_1}{\nu_2}\right)^{1/6} \tag{E11-5.8}$$

$$= \left(\frac{T_1}{T_2}\right)^{1/2}\left[\left(\frac{T_2}{T_1}\right)^{1.75}\right]^{2/3}\left[\left(\frac{T_1}{T_2}\right)^{3/2}\right]^{1/6}$$

$$= \left(\frac{T_1}{T_2}\right)^{1/2}\left(\frac{T_2}{T_1}\right)^{7/6}\left(\frac{T_1}{T_2}\right)^{1/4} = \left(\frac{T_2}{T_1}\right)^{5/12} \tag{E11-5.9}$$

$$= \left(\frac{773}{673}\right)^{5/12} = 1.059$$

$$\ln\frac{1}{1-X_1} = \ln\frac{1}{1-0.865} = 2$$

$$\ln\frac{1}{1-X_2} = 1.059\left(\ln\frac{1}{1-X_1}\right) = 1.059(2) \tag{E11-5.10}$$

$$X_2 = 0.88$$

Bad idea!! Robert was right to worry.

Consequently, we see that increasing the temperature from 400°C to 500°C increases the conversion by only 1.7%. Both engineers would have benefited from a more thorough study of this chapter.

For a packed catalyst bed, the temperature-dependence part of the mass transfer coefficient for a gas-phase reaction can be written as

$$k_c \propto U^{1/2}\ \ (D_{AB}^{2/3}/\nu^{1/6}) \tag{11-71}$$

$$k_c \propto U^{1/2}T^{11/12} \tag{11-72}$$

Important concept

Depending on how one fixes or changes the molar feed rate, F_{T0}, U may also depend on the feed temperature. *As an engineer, it is extremely important that you reason out the effects of changing conditions*, as illustrated in the preceding two examples.

11.5 The Shrinking Core Model

The shrinking core model is used to describe situations in which solid particles are being consumed either by dissolution or reaction and, as a result, the

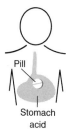

Pill

Stomach
acid

amount of the material being consumed is "shrinking." This model applies to areas ranging from pharmacokinetics (e.g., dissolution of pills in the stomach) to the formation of an ash layer around a burning coal particle, to catalyst regeneration. To design the time release of drugs into the body's system, one must focus on the rate of dissolution of capsules and solid pills injected into the stomach. See PRS11.4. In this section we focus primarily on catalyst regeneration and leave other applications such as drug delivery as exercises at the end of the chapter.

11.5.1 Catalyst Regeneration

Many situations arise in heterogeneous reactions where a gas-phase reactant reacts with a species contained in an inert solid matrix. One of the most common examples is the removal of carbon from catalyst particles that have been deactivated by fouling (see Section 10.7.1). The catalyst regeneration process to reactivate the catalyst by burning off the carbon is shown in Figures 11-11 through 11-13. Figure 11-11 shows a schematic diagram of the removal of carbon from a single porous catalyst pellet as a function of time. Carbon is first removed from the outer edge of the pellet and then in the final stages of the regeneration from the center core of the pellet.

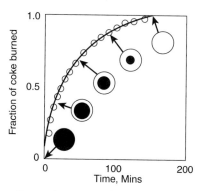

Progressive regeneration of fouled pellet

Figure 11-11   Shell progressive regeneration of fouled pellet. [Reprinted with permission from J. T. Richardson, *Ind. Eng. Chem. Process Des. Dev.*, *11*(1), 8 (1972); copyright American Chemical Society.]

As the carbon continues to be removed from the porous catalyst pellet, the reactant gas must diffuse farther into the material as the reaction proceeds to reach the unreacted solid phase. Note that approximately 3 hours was required to remove all of the carbon from the pellets at these conditions. The regeneration time can be reduced by increasing the gas-phase oxygen concentration and temperature.

To illustrate the principles of the shrinking core model, we shall consider the removal of carbon from the catalyst particle just discussed. In Figure 11-12

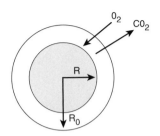

Figure 11-12   Partially regenerated catalyst pellet.

a core of unreacted carbon is contained between $r = 0$ and $r = R$. Carbon has been removed from the porous matrix between $r = R$ and $r = R_0$. Oxygen diffuses from the outer radius R_0 to the radius R, where it reacts with carbon to form carbon dioxide, which then diffuses out of the porous matrix. The reaction

$$C + O_2 \longrightarrow CO_2$$

QSSA
Use steady-state
profiles

at the solid surface is very rapid, so the rate of oxygen diffusion to the surface controls the rate of carbon removal from the core. Although the core of carbon is shrinking with time (an unsteady-state process), we assume the concentration profiles at any instant in time to be the steady-state profiles over the distance $(R_0 - R)$. This assumption is referred to as the *quasi-steady state assumption* (QSSA).

Oxygen must
diffuse through the
porous pellet matrix
until it reaches the
unreacted
carbon core.

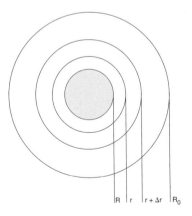

Figure 11-13   Sphere with unreacted carbon core of radius R.

To study how the radius of unreacted carbon changes with time, we must first find the rate of diffusion of oxygen to the carbon surface. Next, we perform a mole balance on the elemental carbon and equate the rate of consumption of carbon to the rate of diffusion of oxygen to the gas carbon interface.

In applying a differential oxygen mole balance over the increment Δr located somewhere between R_0 and R, we recognize that O_2 does not react in this region and reacts only when it reaches the solid carbon interface located at $r = R$. We shall let species A represent O_2.

Step 1: The mole balance on O_2 (i.e., A) between r and $r + \Delta r$ is

$$\begin{bmatrix} \text{Rate} \\ \text{in} \end{bmatrix} - \begin{bmatrix} \text{Rate} \\ \text{out} \end{bmatrix} + \begin{bmatrix} \text{Rate of} \\ \text{generation} \end{bmatrix} = \begin{bmatrix} \text{Rate of} \\ \text{accumulation} \end{bmatrix}$$

$$W_{Ar} 4\pi r^2 |_r - W_{Ar} 4\pi r^2 |_{r+\Delta r} + \qquad 0 \qquad = \qquad 0$$

Dividing through by $-4\pi\Delta r$ and taking the limit gives

Mole balance on oxygen

$$\lim_{\Delta r \to 0} \frac{W_{Ar} r^2 |_{r+\Delta r} - W_{Ar} r^2 |_r}{\Delta r} = \frac{d(W_{Ar} r^2)}{dr} = 0 \qquad (11\text{-}73)$$

Step 2: For every mole of O_2 that diffuses into the spherical pellet, 1 mol of CO_2 diffuses out ($W_{CO_2} = -W_{O_2}$), that is, EMCD. The constitutive equation for constant total concentration becomes

$$W_{Ar} = -D_e \frac{dC_A}{dr} \qquad (11\text{-}74)$$

where D_e is an *effective diffusivity* in the porous catalyst. In Chapter 12 we present an expanded discussion of effective diffusivities in a porous catalyst [cf. Equation (12-1)].

Following the Algorithm

Step 3: Combining Equations (11-73) and (11-74) yields

$$\frac{d}{dr}\left(-D_e \frac{dC_A}{dr} r^2\right) = 0$$

Dividing by $(-D_e)$ gives

$$\boxed{\frac{d}{dr}\left(r^2 \frac{dC_A}{dr}\right) = 0} \qquad (11\text{-}75)$$

Step 4: The boundary conditions are:

At the outer surface of the particle, $r = R_0$: $C_A = C_{A0}$

At the fresh carbon/gas interface, $r = R(t)$: $C_A = 0$

(rapid reaction)

Step 5: Integrating twice yields

$$r^2 \frac{dC_A}{dr} = K_1$$

$$C_A = \frac{-K_1}{r} + K_2$$

Using the boundary conditions to eliminate K_1 and K_2, the concentration profile is given by

$$\boxed{\frac{C_A}{C_{A0}} = \frac{1/R - 1/r}{1/R - 1/R_0}} \tag{11-76}$$

A schematic representation of the profile of O_2 is shown in Figure 11-14 at a time when the inner core is receded to a radius R. The zero on the r axis corresponds to the center of the sphere.

Concentration profile at a given time, t (i.e., core radius, R)

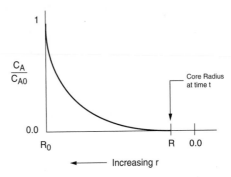

Increasing r

Figure 11-14   Oxygen concentration profile shown from the external radius of the pellet (R_0) to the pellet center. The gas–carbon interface is located at R.

Step 6: The molar flux of O_2 to the gas–carbon interface is

$$W_{Ar} = -D_e \frac{dC_A}{dr} = \frac{-D_e C_{A0}}{(1/R - 1/R_0) r^2} \tag{11-77}$$

Step 7: We now carry out an overall balance on elemental carbon. Elemental carbon does not enter or leave the particle.

$$\begin{bmatrix} \text{Rate} \\ \text{in} \end{bmatrix} - \begin{bmatrix} \text{Rate} \\ \text{out} \end{bmatrix} + \begin{bmatrix} \text{Rate of} \\ \text{generation} \end{bmatrix} = \begin{bmatrix} \text{Rate of} \\ \text{accumulation} \end{bmatrix}$$

Mole balance on shrinking core

$$0 \quad - \quad 0 \quad + \quad r_C'' 4\pi R^2 \quad = \quad \frac{d\left(\frac{4}{3}\pi R^3 \rho_C \phi_C\right)}{dt}$$

where ρ_C is the molar density of the carbon and ϕ_C is the volume fraction of carbon in the porous catalyst. Simplifying gives

$$\frac{dR}{dt} = \frac{r_C''}{\phi_C \rho_C} \tag{11-78}$$

Step 8: The rate of disappearance of carbon is equal to the flux of O_2 to the gas–carbon interface:

$$-r''_C = -W_{Ar}\Big|_{r=R} = \frac{D_e C_{A0}}{R - R^2/R_0} \qquad (11\text{-}79)$$

The minus sign arises with respect to W_{Ar} in Equation (11-79) because O_2 is diffusing in an inward direction [i.e., opposite to the increasing coordinate (r) direction]:

$$-\frac{dR}{dt} = \frac{D_e C_{A0}}{\phi_C \rho_C}\left(\frac{1}{R - R^2/R_0}\right)$$

Step 9: Integrating with limits $R = R_0$ at $t = 0$, the time necessary for the solid carbon interface to recede inward to a radius R is

$$t = \frac{\rho_C R_0^2 \phi_C}{6 D_e C_{A0}}\left[1 - 3\left(\frac{R}{R_0}\right)^2 + 2\left(\frac{R}{R_0}\right)^3\right] \qquad (11\text{-}80)$$

We see that as the reaction proceeds, the reacting gas–solid moves closer to the center of the core. The corresponding oxygen concentration profiles at three different times are shown in Figure 11-15.

Concentration
profiles at different
times at inner
core radii

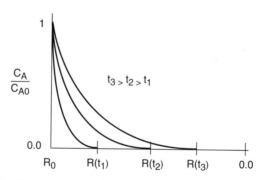

Figure 11-15   Oxygen concentration profile at various times. At t_1, the gas–carbon interface is located at $R(t_1)$; at t_2 it is located at $R(t_2)$.

The time necessary to consume all the carbon in the catalyst pellet is

Time to complete
regeneration
of the particle

$$t_c = \frac{\rho_C R_0^2 \phi_C}{6 D_e C_{A0}} \qquad (11\text{-}81)$$

For a 1-cm diameter pellet with a 0.04 volume fraction of carbon, the regeneration time is the order of 10 s.

Variations on the simple system we have chosen here can be found on page 360 of Levenspiel[17] and in the problems at the end of this chapter.

---

[17]O. Levenspiel, *Chemical Reactor Engineering*, 2nd ed. (New York: Wiley, 1972).

11.5.2 Pharmacokinetics—Dissolution of Monodispersed Solid Particles

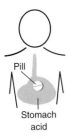

Pill

Stomach acid

We now consider the case where the total particle is being completely consumed. We choose as an example the case where species A must diffuse to the surface to react with solid B at the liquid–solid interface. Reactions of this type are typically zero order in B and first order in A. The rate of mass transfer to the surface is equal to the rate of surface reaction.

$$W_{Ar} = k_c(C_A - C_{As}) = -r''_{As} = k_r C_{As}$$

(Diffusion)   (Surface reaction)

Eliminating C_{As}, we arrive at an equation identical to Equation (11-46) for the radial flux:

$$W_{Ar} = -r''_{As} = \frac{k_c k_r}{k_c + k_r} C_A \tag{11-46}$$

For the case of small particles and negligible shear stress at the fluid boundary, the Frössling equation, Equation (11-40), is approximated by

$$Sh = 2$$

or

$$k_c = \frac{2D_e}{D} \tag{11-82}$$

where D is the diameter of the dissolving particle. Substituting Equation (11-82) into (11-46) and rearranging yields

Diameter at which mass transfer and reaction rate resistances are equal is D^*

$$-r''_{As} = \frac{k_r C_A}{1 + k_r/k_c} = \frac{k_r C_A}{1 + k_r D/2D_e} = \frac{k_r C_A}{1 + D/D^*} \tag{11-83}$$

where $D^* = 2D_e/k_r$ is the diameter at which the resistances to mass transfer and reaction rate are equal.

$D > D^*$	mass transfer controls
$D < D^*$	reaction rate controls

A mole balance on the solid particle yields

$$\text{In} - \text{Out} + \text{Generation} = \text{Accumulation}$$

Balance on the dissolving solid B

$$0 - 0 + r''_{Bs} \pi D^2 = \frac{d(\rho \pi D^3/6)}{dt}$$

where ρ is the molar density of species B. If 1 mol of A dissolves 1 mol of B, then $-r''_{As} = -r''_{Bs}$, and after differentiation and rearrangement we obtain

Ibuprofen

$$\frac{dD}{dt} = -\left[\frac{2\left(-r''_{As}\right)}{\rho}\right] = -\frac{2k_r C_A}{\rho}\left(\frac{1}{1+D/D^*}\right)$$

$$\boxed{\frac{dD}{dt} = -\frac{\alpha}{1+D/D^*}} \qquad (11\text{-}84)$$

where

$$\alpha = \frac{2k_r C_A}{\rho}$$

At time $t = 0$, the initial diameter is $D = D_i$. Integrating Equation (11-84) for the case of excess concentration of reactant A, we obtain the following diameter–time relationship:

Excess A

$$D_i - D + \frac{1}{2D^*}\left(D_i^2 - D^2\right) = \alpha t \qquad (11\text{-}85)$$

The time to complete dissolution of the solid particle is

Reference Shelf

$$t_c = \frac{1}{\alpha}\left(D_i + \frac{D_i^2}{2D^*}\right) \qquad (11\text{-}86)$$

The dissolution of polydisperse particle sizes is analyzed using population balances and is discussed on the CD-ROM.

Closure. After completing this chapter, the reader should be able to define and describe molecular diffusion and how it varies with temperature and pressure, the molar flux, bulk flow, the mass transfer coefficient, the Sherwood and Schmidt numbers, and the correlations for the mass transfer coefficient. The reader should be able to choose the appropriate correlation and calculate the mass transfer coefficient, the molar flux, and the rate of reaction. The reader should be able to describe the regimes and conditions under which mass transfer–limited reactions occur and when reaction rate limited reactions occur and to make calculations of the rates of reaction and mass transfer for each case. One of the most important areas for the reader apply the knowledge of this (and other chapters) is in their ability to ask and answer "What if . . ." questions. Finally, the reader should be able to describe the shrinking core model and apply it to catalyst regeneration and pharmacokinetics.

SUMMARY

1. The molar flux of A in a binary mixture of A and B is

$$\mathbf{W}_A = -cD_{AB}\nabla y_A + y_A(\mathbf{W}_A + \mathbf{W}_B) \qquad (S11\text{-}1)$$

   a. For equimolar counterdiffusion (EMCD) or for dilute concentration of the solute,

$$\mathbf{W}_A = -cD_{AB}\nabla y_A \qquad (S11\text{-}2)$$

   b. For diffusion through a stagnant gas,

$$\mathbf{W}_A = cD_{AB}\nabla\ln(1 - y_A) \qquad (S11\text{-}3)$$

   c. For negligible diffusion,

$$\mathbf{W}_A = y_A\mathbf{W} = C_A\mathbf{V} \qquad (S11\text{-}4)$$

2. The rate of mass transfer from the bulk fluid to a boundary at concentration C_{As} is

$$\mathbf{W}_A = k_c(C_{Ab} - C_{As}) \qquad (S11\text{-}5)$$

where k_c is the mass transfer coefficient.

3. The Sherwood and Schmidt numbers are, respectively,

$$\text{Sh} = \frac{k_c d_p}{D_{AB}} \qquad (S11\text{-}6)$$

$$\text{Sc} = \frac{\nu}{D_{AB}} \qquad (S11\text{-}7)$$

Representative Values
Liquid Phase
$\text{Re} \sim 5000$
$\text{Sc} \sim 4000$
$\text{Sh} \sim 500$
$k_c = 10^{-2}$ m/s
Gas Phase
$\text{Re} \sim 500$
$\text{Sc} \sim 1$
$\text{Sh} \sim 10$
$k_c = 5$ m/s

4. If a heat transfer correlation exists for a given system and geometry, the mass transfer correlation may be found by replacing the Nusselt number by the Sherwood number and the Prandtl number by the Schmidt number in the existing heat transfer correlation.

5. Increasing the gas-phase velocity and decreasing the particle size will increase the overall rate of reaction for reactions that are externally mass transfer–limited.

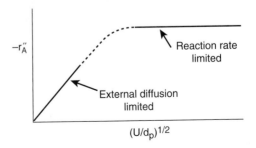

6. The conversion for externally mass transfer–limited reactions can be found from the equation

$$\ln \frac{1}{1-X} = \frac{k_c a_c}{U} L$$ (S11-8)

7. Back-of-the-envelope calculations should be carried out to determine the magnitude and direction that changes in process variables will have on conversion. **What if . . .?**

8. The shrinking core model states that the time to regenerate a coked catalyst particle is

$$t_c = \frac{\rho_C R_0^2 \phi_c}{6 D_e C_{A0}}$$ (S11-9)

CD-ROM MATERIAL

Summary Notes

- **Learning Resources**
    1. *Summary Notes*
       Diffusion Through a Stagnant Film
    4. *Solved Problems*
       Example CD11-1   Calculating Steady State Diffusion
       Example CD11-2   Relative Fluxes W_A, B_A, and J_A
       Example CD11-3   Diffusion Through a Stagnant Gas
       Example CD11-4   Measuring Gas-Phase Diffusivities

Solved Problems

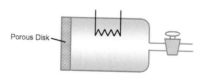

Porous Disk

Example CD11-5   Measuring Liquid-Phase Diffusivities
- **Professional Reference Shelf**
    R11.1. *Mass Transfer-Limited Reactions on Metallic Surfaces*
       A. Catalyst Monoliths
       B. Wire Gauze Reactors

R11.2. *Methods to Experimentally Measure Diffusivities*

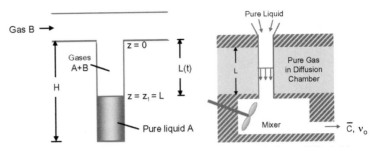

A. Gas-phase diffusivities   B. Liquid-phase diffusivities

R11.3. *Facilitated Heat Transfer*

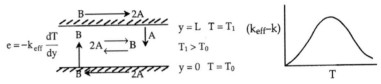

R11.4. *Dissolution of Polydisperse Solids (e.g., Pills in the stomach)*

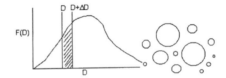

Ibuprofen

Reference Shelf

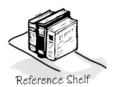

QUESTIONS AND PROBLEMS

The subscript to each of the problem numbers indicates the level of difficulty: A, least difficult; D, most difficult.

$$A = \bullet \quad B = \blacksquare \quad C = \blacklozenge \quad D = \blacklozenge\blacklozenge$$

Creative Thinking

P11-1$_A$ Read over the problems at the end of this chapter. Make up an original problem that uses the concepts presented in this chapter. See Problem 4-1 for the guidelines. To obtain a solution:
(a) Make up your data and reaction.
(b) Use a real reaction and real data.
The journals listed at the end of Chapter 1 may be useful for part (b).

P11-2$_A$ (a) **Example 1-1.** Consider the mass transfer–limited reaction

$$A \longrightarrow 2B$$

What would your concentration (mole fraction) profile look like? Using the same values for D_{AB}, and so on, in Example 11-1, what is the flux of A?
(b) **Example 11-2.** How would your answers change if the temperature was increased by 50°C, the particle diameter was doubled, and fluid velocity was cut in half? Assume properties of water can be used for this system.

(c) **Example 11-3.** How would your answers change if you had a 50–50 mixture of hydrazine and helium? If you increase d_p by a factor of 5?

(d) **Example 11-4.** *What if* you were asked for representative values for Re, Sc, Sh, and k_c for both liquid- and gas-phase systems for a velocity of 10 cm/s and a pipe diameter of 5 cm (or a packed-bed diameter of 0.2 cm)? What numbers would you give?

(e) **Example 11-5.** How would your answers change if the reaction were carried out in the liquid phase where kinetic viscosity varied as

$$v(T_2) = v(T_1)\exp\left[-4000\left(\frac{1}{T_1} - \frac{1}{T_2}\right)\right]?$$

(f) **Side Note.** Derive equations (SN11-1.1) and (SN11.2). Next consider there are no gradients inside the patch and that the equilibrium solubility in the skin immediately adjacent to the skin is $C_{A0} = HC_{AP}$ where H is a form of Henry's law constant. Write the flux as a function of H, δ_1, D_{AB1}, D_{AB2}, δ_2, and C_{AP}. Finally carry out a quasi-steady analysis, i.e.,

$$V_P\frac{dC_{AP}}{dt} = -\frac{A_P}{R}C_{AP}$$

to predict the drug delivery as a function of time. Compare this result with that where the drug in the patch is in a dissolving solid and a hydro-gel and therefore constant with time. Explore this problem using different models and parameter values.

Additional information

$H = 0.1$, $D_{AB1} = 10^{-6}$ cm²/s, $D_{AB2} = 10^{-5}$ cm²/s, $A_p = 5$ cm², $V = 1$ cm³ and $C_{AP} = 10$ mg/dm³

P11-3$_B$ Pure oxygen is being absorbed by xylene in a catalyzed reaction in the experimental apparatus sketched in Figure P11-3. Under constant conditions of temperature and liquid composition the following data were obtained:

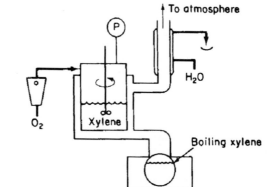

Figure P11-3

Stirrer Speed (rpm)	Rate of Uptake of O_2 (mL/h) for System Pressure (absolute)			
	1.2 atm	1.6 atm	2.0 atm	3.0 atm
400	15	31	75	152
800	20	59	102	205
1200	21	62	105	208
1600	21	61	106	207

No gaseous products were formed by the chemical reaction. What would you conclude about the relative importance of liquid-phase diffusion and about the order of the kinetics of this reaction? (**California Professional Engineers Exam**)

P11-4$_C$ In a diving-chamber experiment, a human subject breathed a mixture of O_2 and He while small areas of his skin were exposed to nitrogen gas. After awhile the exposed areas became blotchy, with small blisters forming on the skin. Model the skin as consisting of two adjacent layers, one of thickness δ_1 and the other of δ_2. If counterdiffusion of He out through the skin occurs at the same time as N_2 diffuses into the skin, at what point in the skin layers is the sum of the partial pressures a maximum? If the saturation partial pressure for the sum of the gases is 101 kPa, can the blisters be a result of the sum of the gas partial pressures exceeding the saturation partial pressure and the gas coming out of the solution (i.e., the skin)?

Before answering any of these questions, derive the concentration profiles for N_2 and He in the skin layers. Hint: See Side Note.

Diffusivity of He and N_2 in the inner skin layer
$$= 5 \times 10^{-7} \text{ cm}^2/\text{s and } 1.5 \times 10^{-7} \text{ cm}^2/\text{s, respectively}$$

Diffusivity of He and N_2 in the outer skin layer
$$= 10^{-5} \text{ cm}^2/\text{s and } 3.3 \times 10^{-4} \text{ cm}^2/\text{s, respectively}$$

	External Skin Boundary Partial Pressure	Internal Skin Boundary Partial Pressure
N_2	101 kPa	0
He	0	81 kPa
δ_1	20 μm	Stratum corneum
δ_2	80 μm	Epidermis

Green engineering

P11-5$_B$ The decomposition of cyclohexane to benzene and hydrogen is mass transfer–limited at high temperatures. The reaction is carried out in a 5-cm-ID pipe 20 m in length packed with cylindrical pellets 0.5 cm in diameter and 0.5 cm in length. The pellets are coated with the catalyst only on the outside. The bed porosity is 40%. The entering volumetric flow rate is 60 dm³/min.

    (a) Calculate the number of pipes necessary to achieve 99.9% conversion of cyclohexane from an entering gas stream of 5% cyclohexane and 95% H_2 at 2 atm and 500°C.

    (b) Plot conversion as a function of length.

(c) How much would your answer change if the pellet diameter and length were each cut in half?

(d) How would your answer to part (a) change if the feed were pure cyclohexane?

(e) What do you believe is the point of this problem?

P11-6 Assume the minimum respiration rate of a chipmunk is 1.5 micromoles of O_2/min. The corresponding volumetric rate of gas intake is 0.05 dm³/min at STP.

(a) What is the deepest a chipmunk can burrow a 3-cm diameter hole beneath the surface in Ann Arbor, Michigan? $D_{AB} = 0.18 \times 10^{-5}$ m/s

(b) In Boulder, Colorado?

(c) How would your answers to (a) and (b) change in the dead of winter when $T = 0°F$?

(d) Critique and extend this problem (e.g., CO_2 poisoning).

P11-7$_C$ Carbon disulphide (A) is evaporating into air (B) at 35°C $P_{v.p.c.} = 510$ mm Hg and 1 atm from the bottom of a 1.0 cm diameter vertical tube. The distance from the CS_2 surface to the open end is 20.0 cm, and this is held constant by continuous addition of liquid CS_2 from below. The experiment is arranged so that the vapor concentration of CS_2 at the open end is zero.

(a) Calculate the molecular diffusivity of CS_2 in air (D_{ca}) and its vapor pressure at 35°C. (*Ans.:* $D_{AB} = 0.12$ cm²/s.)

(b) Find the molar and mass fluxes (W_{Az} and n_c of CS_2) in the tube.

(c) Calculate the following properties at 0.0, 5.0, 10.0, 15.0, 18.0, and 20.0 cm from the CS_2 surface. Arrange columns in the following order on one sheet of paper. (Additional columns may be included for computational purposes if desired.) On a separate sheet give the relations used to obtain each quantity. Try to put each relation into a form involving the minimum computation and the highest accuracy:

(1) y_A and y_B (mole fractions), C_A

(2) V_A, V_B, V*, V (mass velocity)

(3) J_A, J_B

(d) Plot each of the groups of quantities in (c)(1), (2), and (3) on separate graphs. Name all variables and show units. Do *not* plot those parameters in parentheses.

(e) What is the rate of evaporation of CS_2 in cm/day?

(f) Discuss the physical meaning of the value of V_A and J_A at the open end of the tube.

(g) Is *molecular* diffusion of air taking place?

P11-8$_B$ A device for measuring the diffusion coefficient of a gas mixture (Figure P11-8) consists of two chambers connected by a small tube. Initially the chambers contain different proportions of two gases, A and B. The total pressure is the same in each chamber.

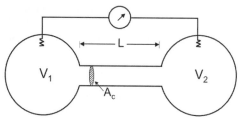

Figure P11-8   Diffusion cell.

(a) Assuming that diffusion may be described by Fick's law, that the concentration in each flask is uniform, and that the concentration gradient in the *tube* is linear show that

$$\ln(C_{A1} - C_{A2}) = -D_{AB}\left(\frac{1}{V_1} + \frac{1}{V_2}\right)t + \text{Constant}$$

State any other assumptions needed.

(b) B. G. Bray (Ph.D. Thesis, University of Michigan, 1960) used a similar device. The concentration of hydrogen in hydrogen-argon mixtures was determined from measurements of an ionizing current in each chamber. The ionizing current is proportional to concentration. The difference in ionizing currents between chambers one and two is measured (ΔIC). Compute the diffusion coefficient, D_{AB}, for the following data. 769 psia, $T = 35°C$, $C_T = 2.033$ mol/dm^3, cell constant,

$$\frac{A}{L}\left(\frac{1}{V_1} + \frac{1}{V_2}\right) = 0.01025 \text{ cm}^{-2}$$

Time, min	10	20	33	50	66	83	100	117	133
ΔIC	36.60	32.82	28.46	23.75	19.83	16.60	13.89	11.67	9.79

P11-9$_A$ A spherical particle is dissolving in a liquid. The rate of dissolution is first order in the solvent concentration, C. Assuming that the solvent is in excess, show that the following conversion time relationships hold.

Rate-Limiting Regime	Conversion Time Relationship
Surface reaction	$1 - (1 - X)^{1/3} = \dfrac{\alpha t}{D_i}$
Mass transfer	$\dfrac{D_i}{2D^*}[1 - (1 - X)^{2/3}] = \dfrac{\alpha t}{D_i}$
Mixed	$[1 - (1 - X)^{1/3}] + \dfrac{D_i}{2D^*}[1 - (1 - X)^{2/3}] = \dfrac{\alpha t}{D_i}$

P11-10$_C$ A powder is to be completely dissolved in an aqueous solution in a large, well-mixed tank. An acid must be added to the solution to render the spherical particle soluble. The particles are sufficiently small that they are unaffected by fluid velocity effects in the tank. For the case of excess acid, $C_0 = 2\ M$, derive an equation for the diameter of the particle as a function of time when

(a) Mass transfer limits the dissolution: $-W_A = k_c C_{A0}$
(b) Reaction limits the dissolution: $-r''_A = k_r C_{A0}$
What is the time for complete dissolution in each case?
(c) Now assume that the acid is not in excess and that mass transfer is limiting the dissolution. One mole of acid is required to dissolve 1 mol of solid. The molar concentration of acid is 0.1 M, the tank is 100 L, and 9.8 mol of solid is added to the tank at time $t = 0$. Derive an expression

for the radius of the particles as a function of time and calculate the time for the particles to dissolve completely.

(d) How could you make the powder dissolve faster? Slower?

Additional information:

$$D_e = 10^{-10} \text{ m}^2/\text{s}, \qquad k = 10^{-18}/\text{s}$$

$$\text{initial diameter} = 10^{-5} \text{ m}$$

P11-11$_B$ The irreversible gas-phase reaction

$$A \xrightarrow{\text{catalyst}} B$$

is carried out adiabatically over a packed bed of solid catalyst particles. The reaction is first order in the concentration of A on the catalyst surface:

$$-r'_{As} = k'C_{As}$$

The feed consists of 50% (mole) A and 50% inerts and enters the bed at a temperature of 300 K. The entering volumetric flow rate is 10 dm³/s (i.e., 10,000 cm³/s). The relationship between the Sherwood number and the Reynolds number is

$$\text{Sh} = 100\, \text{Re}^{1/2}$$

As a first approximation, one may neglect pressure drop. The entering concentration of A is 1.0 *M*. Calculate the catalyst weight necessary to achieve 60% conversion of A for

(a) Isothermal operation.

(b) Adiabatic operation.

(c) What generalizations can you make after comparing parts (a) and (b)?

Additional information:

Kinematic viscosity: $\mu/\rho = 0.02 \text{ cm}^2/\text{s}$
Particle diameter: $d_p = 0.1 \text{ cm}$
Superficial velocity: $U = 10 \text{ cm/s}$
Catalyst surface area/mass of catalyst bed: $a = 60 \text{ cm}^2/\text{g cat.}$
Diffusivity of A: $D_e = 10^{-2} \text{ cm}^2/\text{s}$
Heat of reaction: $\Delta H_{Rx} = -10,000 \text{ cal/g mol A}$
Heat capacities:

$$C_{pA} = C_{pB} = 25 \text{ cal/g mol} \cdot \text{K}$$

$$C_{pS} \text{ (solvent)} = 75 \text{ cal/g mol} \cdot \text{K}$$

$$k' (300 \text{ K}) = 0.01 \text{ cm}^3/\text{s} \cdot \text{g cat with } E = 4000 \text{ cal/mol}$$

P11-12$_C$ *(Pills)* An antibiotic drug is contained in a solid inner core and is surrounded by an outer coating that makes it palatable. The outer coating and the drug are dissolved at different rates in the stomach, owing to their differences in equilibrium solubilities.

(a) If $D_2 = 4$ mm and $D_1 = 3$ mm, calculate the time necessary for the pill to dissolve completely.

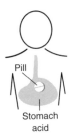

Pill

Stomach
acid

(b) Assuming first-order kinetics ($k_A = 10\ h^{-1}$) for the absorption of the dissolved drug (i.e., in solution in the stomach) into the bloodstream, plot
the concentration in grams of the drug in the blood per gram of body
weight as a function of time when the following three pills are taken
simultaneously:

$$\text{Pill 1:}\quad D_2 = 5\ \text{mm},\qquad D_1 = 3\ \text{mm}$$

$$\text{Pill 2:}\quad D_2 = 4\ \text{mm},\qquad D_1 = 3\ \text{mm}$$

$$\text{Pill 3:}\quad D_2 = 3.5\ \text{mm},\qquad D_1 = 3\ \text{mm}$$

(c) Discuss how you would maintain the drug level in the blood at a constant
level using different-size pills?
(d) How could you arrange a distribution of pill sizes so that the concentration in the blood was constant over a period (e.g., 3 hr) of time?

Additional information:

> Amount of drug in inner core = 500 mg
> Solubility of outer layer at stomach conditions = 1.0 mg/cm^3
> Solubility of inner layer at stomach conditions = 0.4 mg/cm^3
> Volume of fluid in stomach = 1.2 L
> Typical body weight = 75 kg
> Sh = 2., $D_{AB} = 6 \times 10^{-4}$ cm^2/min

Green engineering

P11-13$_B$ If disposal of industrial liquid wastes by incineration is to be a feasible process, it is important that the toxic chemicals be completely decomposed into
harmless substances. One study carried out concerned the atomization and
burning of a liquid stream of "principal" organic hazardous constituents
(POHCs) [*Environ. Prog.*, 8, 152 (1989)]. The following data give the burning
droplet diameter as a function of time (both diameter and time are given in
arbitrary units):

Time	20	40	50	70	90	110
Diameter	9.7	8.8	8.4	7.1	5.6	4.0

What can you learn from these data?

P11-14$_B$ (*Estimating glacial ages*) The following oxygen-18 data were obtained from
soil samples taken at different depths in Ontario, Canada. Assuming that all
the ^{18}O was laid down during the last glacial age and that the transport of ^{18}O
to the surface takes place by molecular diffusion, estimate the number of
years since the last glacial age from the following data. Independent measurements give the diffusivity of ^{18}O in soil as 2.64×10^{-10} m^2/s.

Figure P11-14   Glaciers.

Depth (m)	(surface) 0	3	6	9	12	18
^{18}O Conc. Ratio (C/C_0)	0	0.35	0.65	0.83	0.94	1.0

C_0 is the concentration of ^{18}O at 25 m.

JOURNAL ARTICLE PROBLEM

P11J-1 After reading the article "Designing gas-sparged vessels for mass transfer" [*Chem. Eng.*, *89*(24), p. 61 (1982)], design a gas-sparged vessel to saturate 0.6 m^3/s of water up to an oxygen content of 4×10^{-3} kg/m^3 at 20°C. A liquid holding time of 80 s is required.

JOURNAL CRITIQUE PROBLEMS

P11C-1 The decomposition of nitric oxide on a heated platinum wire is discussed in *Chem. Eng. Sci.*, *30*, 781 (1975). After making some assumptions about the density and the temperatures of the wire and atmosphere, and using a correlation for convective heat transfer, determine if mass transfer limitations are a problem in this reaction.

P11C-2 Given the proposed rate equation on page 296 of the article in *Ind. Eng. Chem. Process Des. Dev.*, *19*, 294 (1980), determine whether or not the concentration dependence on sulfur, C_s, is really second-order. Also, determine if the intrinsic kinetic rate constant, K_{2p}, is indeed only a function of temperature and partial pressure of oxygen and not of some other variables as well.

P11C-3 Read through the article on the oxidation kinetics of oil shale char in *Ind. Eng. Chem. Process Des. Dev.*, *18*, 661 (1979). Are the units for the mass transfer coefficient k_m in Equation (6) consistent with the rate law? Is the mass transfer coefficient dependent on sample size? Would the shrinking core model fit the authors' data as well as the model proposed by the authors?

• **Additional Homework Problems**

CDP11-A$_A$ An isomerization reaction that follows Langmuir–Hinshelwood kinetics is carried out on monolith catalyst. [2nd Ed. P10-11]

CDP11-B$_B$ A parameter sensitivity analysis is required for this problem in which an isomerization is carried out over a 20-mesh gauze screen. [2nd Ed. P10-12]

CDP11-C$_C$ This problem examines the effect on temperature in a catalyst monolith. [2nd Ed. P10-13]

CDP11-D$_D$ A second-order catalytic reaction is carried out in a catalyst monolith. [2nd Ed. P10-14]

CDP11-E$_C$ Fracture acidizing is a technique to increase the productivity of oil wells. Here acid is injected at high pressures to fracture the rock and form a channel that extends out from the well bore. As the acid flows through the channel it etches the sides of the channel to make it larger and thus less resistant to the flow of oil. Derive an equation for the concentration profile of acid and the channel width as a function of distance from the well bore.

CDP11-F$_C$ The solid–gas reaction of silicon to form SiO_2 is an important process in microelectronics fabrication. The oxidation occurs at the Si–SiO$_2$ interface. Derive an equation for the thickness of the SiO_2 layer as a function of time. [2nd Ed. P10-17]

CDP11-G$_B$ Mass transfer limitations in CVD processing to product material with ferroelectric and piezoelectric properties. [2nd Ed. P10-17]

CDP11-H$_B$ Calculate multicomponent diffusivities. [2nd Ed. P10-9]

CDP11-I$_B$ Application of the shrinking core model to FeS_2 rock samples in acid mine drainage. [2nd Ed. P10-18]

CDP11-J$_B$ Removal of chlorine by adsorption from a packed bed reactor: Vary system parameters to predict their effect on the conversion. [3rd Ed. P11-5$_B$]

CDP11-K$_B$ The reversible reaction A $\rightleftharpoons$ B is carried out in a PBR. The rate law is

Green Engineering
Problem

$$-r_A'' = \frac{k_A\left(C_A - \dfrac{1}{K_C}C_B\right)}{\left(1 + \dfrac{k_A}{k_B K_C}\right)}$$

[3rd Ed. P11-6$_B$]

CDP11-L$_B$ Oxidation of ammonia over wire screens. [3rd Ed. P11-7$_B$]

CDP11-M$_B$ Sgt. Ambercromby on the scene again to investigate foul play. [3rd Ed. P11-13$_B$]

CDP11-N$_B$ Additional Green Engineering problems can be found on the web at *www.rowan.edu/greenengineering*.

SUPPLEMENTARY READING

1. The fundamentals of diffusional mass transfer can be found in

   BIRD, R. B., W. E. STEWART, and E. N. LIGHTFOOT, *Transport Phenomena*, 2nd ed. New York: Wiley, 2003, Chaps. 17 and 18.

   CUSSLER, E. L., *Diffusion Mass Transfer in Fluid Systems*, 2nd ed. New York: Cambridge University Press, 1997.

   FAHIEN, R. W., *Fundamentals of Transport Phenomena*. New York: McGraw-Hill, 1983, Chap. 7.

   GEANKOPLIS, C. J., *Transport Processes and Unit Operations*. Upper Saddle River, N.J.: Prentice Hall, 2003.

   HINES, A. L., and R. N. MADDOX, *Mass Transfer: Fundamentals and Applications*. Upper Saddle River, N.J.: Prentice Hall, 1984.

   LEVICH, V. G., *Physiochemical Hydrodynamics*. Upper Saddle River, N.J.: Prentice Hall, 1962, Chaps. 1 and 4.

2. Equations for predicting gas diffusivities are given in Appendix D. Experimental values of the diffusivity can be found in a number of sources, two of which are

PERRY, R. H., D. W. GREEN, and J. O. MALONEY, *Chemical Engineers' Handbook*, 7th ed. New York: McGraw-Hill, 1997.

SHERWOOD, T. K., R. L. PIGFORD, and C. R. WILKE, *Mass Transfer*. New York: McGraw-Hill, 1975.

3. A number of correlations for the mass transfer coefficient can be found in

LYDERSEN, A. L., *Mass Transfer in Engineering Practice*. New York: Wiley-Interscience, 1983, Chap. 1.

MCCABE, W. L., J. C. SMITH, and P. HARRIOTT, *Unit Operations of Chemical Engineering*, 6th ed. New York: McGraw-Hill, 2000, Chap. 17.

TREYBAL, R. E., *Mass Transfer Operations*, 3rd ed. New York: McGraw-Hill, 1980.

Diffusion 12

and Reaction

Research is to see what everybody else sees, and
to think what nobody else has thought.

Albert Szent-Gyorgyi

Overview This chapter presents the principles of diffusion and reaction. While the focus is primarily on catalyst pellets, examples illustrating these principles are also drawn from biomaterials engineering and microelectronics. In our discussion of catalytic reactions in Chapter 10, we assumed each point on the interior of catalyst surface was accessible to the same concentration. However, we know there are many, many situations where this equal accessibility will not be true. For example, when the reactants must diffuse inside the catalyst pellet in order to react, we know the concentration at the pore mouth must be higher than that inside the pore. Consequently, the entire catalytic surface is not accessible to the same concentration; therefore, the rate of reaction throughout the pellet will vary. To account for variations in reaction rate throughout the pellet, we introduce a parameter known as the effectiveness factor, which is the ratio of the overall reaction rate in the pellet to the reaction rate at the external surface of the pellet. In this chapter we will develop models for diffusion and reaction in two-phase systems, which include catalyst pellets, tissue generation, and chemical vapor deposition (CVD). The types of reactors discussed in this chapter will include packed beds, bubbling fluidized beds, slurry reactors, trickle bed reactors, and CVD boat reactors. After studying this chapter, you will be able to describe diffusion and reaction in two- and three-phase systems, determine when internal diffusion limits the overall rate of reaction, describe how to go about eliminating this limitation, and develop models for systems in which both diffusion and reaction play a role (e.g., tissue growth, CVD).

The concentration in the internal surface of the pellet is less than that of the external surface

In a heterogeneous reaction sequence, mass transfer of reactants first takes place from the bulk fluid to the external surface of the pellet. The reactants then diffuse from the external surface into and through the pores within the pellet, with reaction taking place only on the catalytic surface of the pores. A schematic representation of this two-step diffusion process is shown in Figures 10-6 and 12-1.

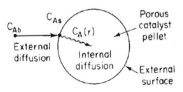

Figure 12-1 Mass transfer and reaction steps for a catalyst pellet.

12.1 Diffusion and Reaction in Spherical Catalyst Pellets

In this section we will develop the internal effectiveness factor for spherical catalyst pellets. The development of models that treat individual pores and pellets of different shapes is undertaken in the problems at the end of this chapter. We will first look at the internal mass transfer resistance to either the products or reactants that occurs between the external pellet surface and the interior of the pellet. To illustrate the salient principles of this model, we consider the irreversible isomerization

$$A \longrightarrow B$$

that occurs on the surface of the pore walls within the spherical pellet of radius R.

12.1.1 Effective Diffusivity

The pores in the pellet are not straight and cylindrical; rather, they are a series of tortuous, interconnecting paths of pore bodies and pore throats with varying cross-sectional areas. It would not be fruitful to describe diffusion within each and every one of the tortuous pathways individually; consequently, we shall define an effective diffusion coefficient so as to describe the average diffusion taking place at any position r in the pellet. We shall consider only radial variations in the concentration; the radial flux W_{Ar} will be based on the total area (voids and solid) normal to diffusion transport (i.e., $4\pi r^2$) rather than void area alone. This basis for W_{Ar} is made possible by proper definition of the effective diffusivity D_e.

The effective diffusivity accounts for the fact that:

1. Not all of the area normal to the direction of the flux is available (i.e., the area occupied by solids) for the molecules to diffuse.

2. The paths are tortuous.
3. The pores are of varying cross-sectional areas.

An equation that relates D_e to either the bulk or the Knudsen diffusivity is

The effective
diffusivity

$$D_e = \frac{D_{AB}\phi_p\sigma_c}{\tilde{\tau}} \tag{12-1}$$

where

$$\tilde{\tau} = \text{tortuosity}[1] = \frac{\text{Actual distance a molecule travels between two points}}{\text{Shortest distance between those two points}}$$

$$\phi_p = \text{pellet porosity} = \frac{\text{Volume of void space}}{\text{Total volume (voids and solids)}}$$

$\sigma_c = $ Constriction factor

The constriction factor, σ_c, accounts for the variation in the cross-sectional area that is normal to diffusion.[2] It is a function of the ratio of maximum to minimum pore areas (Figure 12-2(a)). When the two areas, A_1 and A_2, are equal, the constriction factor is unity, and when $\beta = 10$, the constriction factor is approximately 0.5.

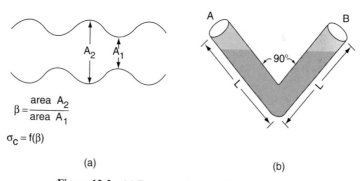

Figure 12-2 — (a) Pore constriction; (b) pore tortuosity.

Example 12–1   Finding the Tortuosity

Calculate the tortuosity for the hypothetical pore of length, L (Figure 12-2(b)), from the definition of $\tilde{\tau}$.

---

[1] Some investigators lump constriction and tortuosity into one factor, called the tortuosity factor, and set it equal to $\tilde{\tau}/\sigma_c$. C. N. Satterfield, *Mass Transfer in Heterogeneous Catalysis* (Cambridge, Mass.: MIT Press, 1970), pp. 33–47, has an excellent discussion on this point.

[2] See E. E. Petersen, *Chemical Reaction Analysis* (Upper Saddle River, N.J.: Prentice Hall, 1965), Chap. 3; C. N. Satterfield and T. K. Sherwood, *The Role of Diffusion in Catalysis* (Reading, Mass.: Addison-Wesley, 1963), Chap. 1.

Solution

$$\tilde{\tau} = \frac{\text{Actual distance molecule travels from } A \text{ to } B}{\text{Shortest distance between } A \text{ and } B}$$

The shortest distance between points A and B is $\sqrt{2L}$. The actual distance the molecule travels from A to B is $2L$.

$$\tilde{\tau} = \frac{2L}{\sqrt{2L}} = \sqrt{2} = 1.414$$

Although this value is reasonable for $\tilde{\tau}$, values for $\tilde{\tau} = 6$ to 10 are not unknown. Typical values of the constriction factor, the tortuosity, and the pellet porosity are, respectively, $\sigma_c = 0.8$, $\tilde{\tau} = 3.0$, and $\phi_p = 0.40$.

12.1.2 Derivation of the Differential Equation Describing Diffusion and Reaction

We now perform a steady-state mole balance on species A as it enters, leaves, and reacts in a spherical shell of inner radius r and outer radius $r + \Delta r$ of the pellet (Figure 12-3). Note that even though A is diffusing inward toward the center of the pellet, the convention of our shell balance dictates that the flux be in the direction of increasing r. We choose the flux of A to be positive in the direction of increasing r (i.e., the outward direction). Because A is actually diffusing inward, the flux of A will have some negative value, such as $-10 \text{ mol/m}^2 \cdot \text{s}$, indicating that the flux is actually in the direction of decreasing r.

First we will derive the concentration profile of reactant A in the pellet.

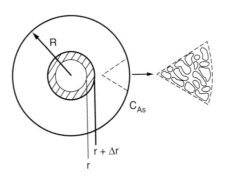

Figure 12-3   Shell balance on a catalyst pellet.

We now proceed to perform our shell balance on A. The area that appears in the balance equation is the total area (voids and solids) *normal* to the direction of the molar flux:

$$\text{Rate of A in at } r = W_{Ar} \cdot \text{Area} = W_{Ar} \times 4\pi r^2 \big|_r \qquad (12\text{-}2)$$

$$\text{Rate of A out at } (r + \Delta r) = W_{Ar} \cdot \text{Area} = W_{Ar} \times 4\pi r^2 \big|_{r+\Delta r} \qquad (12\text{-}3)$$

$$
\begin{bmatrix}
\text{Rate of} \\
\text{generation} \\
\text{of A within a} \\
\text{shell thickness} \\
\text{of } \Delta r
\end{bmatrix}
=
\begin{bmatrix}
\dfrac{\text{Rate of reaction}}{\text{Mass of catalyst}}
\end{bmatrix}
\times
\begin{bmatrix}
\dfrac{\text{Mass catalyst}}{\text{Volume}}
\end{bmatrix}
\times
\begin{bmatrix}
\text{Volume of shell}
\end{bmatrix}
$$

$$
=\qquad r'_A \qquad \times \qquad \rho_c \qquad \times \qquad 4\pi r_m^2 \Delta r
$$

(12-4)

Mole balance for diffusion and reaction inside the catalyst pellet

where r_m is some mean radius between r and $r + \Delta r$ that is used to approximate the volume ΔV of the shell and ρ_c is the density of the pellet.

The mole balance over the shell thickness Δr is

Mole balance

(In at r)   −   **(Out at $r+\Delta r$)**   + **(Generation within Δr)** = 0

$$
(W_{Ar} \times 4\pi r^2 |_r) - (W_{Ar} \times 4\pi r^2 |_{r+\Delta r}) + \quad (r'_A \, \rho_c \times 4\pi r_m^2 \, \Delta r) \quad = 0 \qquad \text{(12-5)}
$$

After dividing by $(-4\pi \Delta r)$ and taking the limit as $\Delta r \to 0$, we obtain the following differential equation:

$$
\frac{d(W_{Ar} r^2)}{dr} - r'_A \rho_c r^2 = 0 \tag{12-6}
$$

Because 1 mol of A reacts under conditions of constant temperature and pressure to form 1 mol of B, we have Equal Molar Counter Diffusion (EMCD) at constant total concentration (Section 11.2.1A), and, therefore,

The flux equation

$$
W_{Ar} = -cD_e \frac{dy_A}{dr} = -D_e \frac{dC_A}{dr} \tag{12-7}
$$

where C_A is the number of moles of A per dm^3 of open pore volume (i.e., volume of gas) as opposed to (mol/vol of gas and solids). In systems where we do not have EMCD in catalyst pores, it may still be possible to use Equation (12-7) if the reactant gases are present in dilute concentrations.

After substituting Equation (12-7) into Equation (12-6), we arrive at the following differential equation describing diffusion with reaction in a catalyst pellet:

$$
\boxed{\frac{d[-D_e(dC_A/dr)r^2]}{dr} - r^2 \rho_c r'_A = 0} \tag{12-8}
$$

We now need to incorporate the rate law. In the past we have based the rate of reaction in terms of either per unit volume,

$$
\boxed{-r_A [=] (\text{mol/dm}^3 \cdot s)}
$$

or per unit mass of catalyst,

$$
\boxed{-r_A [=] (\text{mol/g cat} \cdot s)}
$$

When we study reactions on the internal surface area of catalysts, the rate of reaction and rate law are often based on per unit surface area,

$$-r_A''[=](\text{mol/m}^2 \cdot \text{s})$$

As a result, the surface area of the catalyst per unit mass of catalyst,

$$S_a [=] (\text{m}^2/\text{g cat.})$$

is an important property of the catalyst. The rate of reaction per unit mass of catalyst, $-r_A'$, and the rate of reaction per unit surface area of catalyst are related through the equation

$$-r_A' = -r_A'' S_a$$

A typical value of S_a might be 150 m²/g of catalyst.

As mentioned previously, at high temperatures, the denominator of the catalytic rate law approaches 1. Consequently, for the moment, it is reasonable to assume that the surface reaction is of nth order in the gas-phase concentration of A within the pellet.

$$-r_A'' = k_n'' C_A^n \qquad (12\text{-}9)$$

where

$$-r_A'': k_n'' [=] \left(\frac{\text{m}^3}{\text{kmol}}\right)^{n-1} \frac{\text{m}}{\text{s}}$$

Similarly,

$$-r_A': k_n' = S_a k_n'' [=] \left(\frac{\text{m}^3}{\text{kmol}}\right)^{n-1} \frac{\text{m}^3}{\text{kg} \cdot \text{s}}$$

$$-r_A: k_n = k_n' \rho_c = \rho_c S_a k_n'' [=] \left(\frac{\text{m}^3}{\text{kmol}}\right)^{n-1} \frac{1}{\text{s}}$$

Substituting the rate law equation (12-9) into Equation (12-8) gives

$$\frac{d[r^2(-D_e dC_A/dr)]}{dr} + r^2 \overbrace{k_n'' S_a \rho_c}^{k_n} C_A^n = 0 \qquad (12\text{-}10)$$

By differentiating the first term and dividing through by $-r^2 D_e$, Equation (12-10) becomes

$$\boxed{\frac{d^2 C_A}{dr^2} + \frac{2}{r}\left(\frac{dC_A}{dr}\right) - \frac{k_n}{D_e} C_A^n = 0} \qquad (12\text{-}11)$$

Side notes:

Inside the Pellet
$$-r_A' = S_a(-r_A'')$$
$$-r_A = \rho_c(-r_A')$$
$$-r_A = \rho_c S_a(-r_A'')$$

S_a: 10 grams of catalyst may cover as much surface area as a football field

The rate law

For **first-order** catalytic reaction
per unit Surface Area: $k_1'' = [\text{m/s}]$
per unit Mass of Catalyst: $k_1' = k_1'' S_a = [\text{m}^3/(\text{kg} \cdot \text{s})]$
per unit Volume: $k = k_1'' S_a \rho_c = [\text{s}^{-1}]$

Differential equation and boundary conditions describing diffusion and reaction in a catalyst pellet

The boundary conditions are:

1. The concentration remains finite at the center of the pellet:

$$\boxed{C_A \text{ is finite} \qquad \text{at } r = 0}$$

2. At the external surface of the catalyst pellet, the concentration is C_{As}:

$$\boxed{C_A = C_{As} \qquad \text{at } r = R}$$

12.1.3 Writing the Equation in Dimensionless Form

We now introduce dimensionless variables ψ and λ so that we may arrive at a parameter that is frequently discussed in catalytic reactions, the *Thiele modulus*. Let

$$\psi = \frac{C_A}{C_{As}} \tag{12-12}$$

$$\lambda = \frac{r}{R} \tag{12-13}$$

With the transformation of variables, the boundary condition

$$C_A = C_{As} \qquad \text{at } r = R$$

becomes

$$\psi = \frac{C_A}{C_{As}} = 1 \qquad \text{at } \lambda = 1$$

and the boundary condition

$$C_A \text{ is finite} \qquad \text{at } r = 0$$

becomes

$$\psi \text{ is finite} \qquad \text{at } \lambda = 0$$

We now rewrite the differential equation for the molar flux in terms of our dimensionless variables. Starting with

$$W_{Ar} = -D_e \frac{dC_A}{dr} \tag{11-7}$$

we use the chain rule to write

$$\frac{dC_A}{dr} = \left(\frac{dC_A}{d\lambda}\right)\frac{d\lambda}{dr} = \frac{d\psi}{d\lambda}\left(\frac{dC_A}{d\psi}\right)\frac{d\lambda}{dr} \tag{12-14}$$

Then differentiate Equation (12-12) with respect to ψ and Equation (12-13) with respect to r, and substitute the resulting expressions,

$$\frac{dC_A}{d\psi} = C_{As} \quad \text{and} \quad \frac{d\lambda}{dr} = \frac{1}{R}$$

into the equation for the concentration gradient to obtain

$$\frac{dC_A}{dr} = \frac{d\psi}{d\lambda} \frac{C_{As}}{R} \tag{12-15}$$

The flux of A in terms of the dimensionless variables, ψ and λ, is

The total rate of consumption of A inside the pellet, M_A (mol/s)

$$W_{Ar} = -D_e \frac{dC_A}{dr} = -\frac{D_e C_{As}}{R}\left(\frac{d\psi}{d\lambda}\right) \tag{12-16}$$

At steady state, the net flow of species A that enters into the pellet at the external pellet surface reacts completely within the pellet. The overall rate of reaction is therefore equal to the total molar flow of A into the catalyst pellet. The overall rate of reaction, M_A, can be obtained by multiplying the molar flux into the pellet at the outer surface by the external surface area of the pellet, $4\pi R^2$:

All the reactant that diffuses into the pellet is consumed (a black hole)

$$M_A = -4\pi R^2 W_{Ar}\big|_{r=R} = +4\pi R^2 D_e \frac{dC_A}{dr}\bigg|_{r=R} = 4\pi R D_e C_{As} \frac{d\psi}{d\lambda}\bigg|_{\lambda=1} \tag{12-17}$$

Consequently, to determine the overall rate of reaction, which is given by Equation (12-17), we first solve Equation (12-11) for C_A, differentiate C_A with respect to r, and then substitute the resulting expression into Equation (12-17). Differentiating the concentration gradient, Equation (12-15), yields

$$\frac{d^2C_A}{dr^2} = \frac{d}{dr}\left(\frac{dC_A}{dr}\right) = \frac{d}{d\lambda}\left(\frac{d\psi}{d\lambda}\frac{C_{As}}{R}\right)\frac{d\lambda}{dr} = \frac{d^2\psi}{d\lambda^2}\left(\frac{C_{As}}{R^2}\right) \tag{12-18}$$

After dividing by C_{As}/R^2, the dimensionless form of Equation (12-11) is written as

$$\frac{d^2\psi}{d\lambda^2} + \frac{2}{\lambda}\frac{d\psi}{d\lambda} - \frac{k_n R^2 C_{As}^{n-1}}{D_e}\psi^n = 0$$

Then

Dimensionless form of equations describing diffusion and reaction

$$\boxed{\frac{d^2\psi}{d\lambda^2} + \frac{2}{\lambda}\left(\frac{d\psi}{d\lambda}\right) - \phi_n^2\psi^n = 0} \tag{12-19}$$

where

$$\phi_n^2 = \frac{k_n R^2 C_{As}^{n-1}}{D_e}$$

(12-20)

The square root of the coefficient of ψ^n, (i.e., ϕ_n) is called the Thiele modulus. The Thiele modulus, ϕ_n, will always contain a subscript (e.g., n), which will distinguish this symbol from the symbol for porosity, ϕ, defined in Chapter 4, which has no subscript. The quantity ϕ_n^2 is a measure of the ratio of "a" surface reaction rate to "a" rate of diffusion through the catalyst pellet:

Thiele modulus

$$\phi_n^2 = \frac{k_n R^2 C_{As}^{n-1}}{D_e} = \frac{k_n R C_{As}^n}{D_e[(C_{As}-0)/R]} = \frac{\text{"a" surface reaction rate}}{\text{"a" diffusion rate}}$$

(12-20)

When the Thiele modulus is large, internal diffusion usually limits the overall rate of reaction; when ϕ_n is small, the surface reaction is usually rate-limiting. If for the reaction

$$A \longrightarrow B$$

the surface reaction were rate-limiting with respect to the adsorption of A and the desorption of B, and if species A and B are weakly adsorbed (i.e., low coverage) and present in very dilute concentrations, we can write the apparent first-order rate law

$$-r_A'' \simeq k_1'' C_A$$

(12-21)

The units of k_1'' are m^3/m^2s ($= m/s$).

For a first-order reaction, Equation (12-19) becomes

$$\frac{d^2\psi}{d\lambda^2} + \frac{2}{\lambda}\frac{d\psi}{d\lambda} - \phi_1^2\psi = 0$$

(12-22)

where

$$\phi_1 = R\sqrt{\frac{k_1''\rho_c S_a}{D_e}} = R\sqrt{\frac{k_1}{D_e}}$$

$$k_1 = [k_1''\rho_c S_a] = \left(\frac{m}{s} \cdot \frac{g}{m^3} \cdot \frac{m^2}{g}\right) = 1/s$$

$$\frac{k_1}{D_e} = \left(\frac{1/s}{m^2/s}\right) = \frac{1}{m^2}$$

$$\phi_1 = R\sqrt{\frac{k_1}{D_e}} = m\left(\frac{s^{-1}}{m^2/s}\right)^{1/2} = \frac{1}{1} \quad \text{(Dimensionless)}$$

The boundary conditions are

$$
\boxed{
\begin{array}{lll}
\text{B.C. 1: } \psi = 1 & \text{at } \lambda = 1 & \text{(12-23)} \\
\text{B.C. 2: } \psi \text{ is finite} & \text{at } \lambda = 0 & \text{(12-24)}
\end{array}
}
$$

12.1.4 Solution to the Differential Equation for a First-Order Reaction

Differential equation (12-22) is readily solved with the aid of the transformation $y = \psi\lambda$:

$$
\frac{d\psi}{d\lambda} = \frac{1}{\lambda}\left(\frac{dy}{d\lambda}\right) - \frac{y}{\lambda^2}
$$

$$
\frac{d^2\psi}{d\lambda^2} = \frac{1}{\lambda}\left(\frac{d^2y}{d\lambda^2}\right) - \frac{2}{\lambda^2}\left(\frac{dy}{d\lambda}\right) + \frac{2y}{\lambda^3}
$$

With these transformations, Equation (12-22) reduces to

$$
\frac{d^2y}{d\lambda^2} - \phi_1^2 y = 0 \tag{12-25}
$$

This differential equation has the following solution (Appendix A.3):

$$
y = A_1 \cosh\phi_1\lambda + B_1 \sinh\phi_1\lambda
$$

In terms of ψ,

$$
\psi = \frac{A_1}{\lambda} \cosh\phi_1\lambda + \frac{B_1}{\lambda} \sinh\phi_1\lambda \tag{12-26}
$$

The arbitrary constants A_1 and B_1 can easily be evaluated with the aid of the boundary conditions. At $\lambda = 0$; $\cosh\phi_1\lambda \to 1$, $(1/\lambda) \to \infty$, and $\sinh\phi_1\lambda \to 0$. Because the second boundary condition requires ψ to be finite at the center (i.e., $\lambda = 0$), therefore A_1 must be zero.

The constant B_1 is evaluated from B.C. 1 (i.e., $\psi = 1$, $\lambda = 1$) and the dimensionless concentration profile is

Concentration
profile

$$
\boxed{
\psi = \frac{C_A}{C_{As}} = \frac{1}{\lambda}\left(\frac{\sinh\phi_1\lambda}{\sinh\phi_1}\right)
}
\tag{12-27}
$$

Figure 12-4 shows the concentration profile for three different values of the Thiele modulus, ϕ_1. Small values of the Thiele modulus indicate surface reaction controls and a significant amount of the reactant diffuses well into the pellet interior without reacting. Large values of the Thiele modulus indicate that the surface reaction is rapid and that the reactant is consumed very close to the external pellet surface and very little penetrates into the interior of the pellet. Consequently, if the porous pellet is to be plated with a precious metal catalyst (e.g., Pt), it should only be plated in the immediate vicinity of the external surface when large values of ϕ_n characterize the diffusion and reaction. That is, it would be a waste of the precious metal to plate the entire pellet when internal diffusion is limiting because the reacting gases are consumed near the outer surface. Consequently, the reacting gases would never contact the center portion of the pellet.

For large values of the Thiele modulus, internal diffusion limits the rate of reaction.

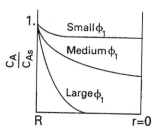

Figure 12-4   Concentration profile in a spherical catalyst pellet.

Example 12–2   Applications of Diffusion and Reaction to Tissue Engineering

The equations describing diffusion and reaction in porous catalysts also can be used to derive rates of tissue growth. One important area of tissue growth is in cartilage tissue in joints such as the knee. Over 200,000 patients per year receive knee joint replacements. Alternative strategies include the growth of cartilage to repair the damaged knee.[3]

One approach currently being researched by Professor Kristi Anseth at the University of Colorado is to deliver cartilage forming cells in a hydrogel to the damaged area such as the one shown in Figure E12-2.1.

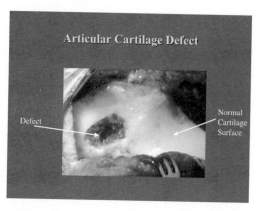

Figure E12-2.1 Damaged cartilage. (Figure courtesy of *Newsweek*, September 3, 2001.)

---

[3] *www.genzymebiosurgery.com/prod/cartilage/gzbx_p_pt_cartilage.asp.*

Here the patient's own cells are obtained from a biopsy and embedded in a hydro gel, which is a cross-linked polymer network that is swollen in water. In order for the cells to survive and grow new tissue, many properties of the gel must be tuned to allow diffusion of important species in and out (e.g., nutrients *in* and cell-secreted extracellular molecules *out*, such as collagen). Because there is no blood flow through the cartilage, oxygen transport to the cartilage cells is primarily by diffusion. Consequently, the design must be that the gel can maintain the necessary rates of diffusion of nutrients (e.g., O_2) into the hydrogel. These rates of exchange in the gel depend on the geometry and the thickness of the gel. To illustrate the application of chemical reaction engineering principles to tissue engineering, we will examine the diffusion and consumption of one of the nutrients, oxygen.

Our examination of diffusion and reaction in catalyst pellets showed that in many cases the reactant concentration near the center of the particle was virtually zero. If this condition were to occur in a hydrogel, the cells at the center would die. Consequently, the gel thickness needs to be designed to allow rapid transport of oxygen.

Let's consider the simple gel geometry shown in Figure E12-2.2.

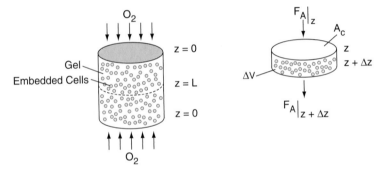

Figure E12-2.2 Schematic of cartilage cell system.

We want to find the gel thickness at which the minimum oxygen consumption rate is 10^{-13} mol/cell/h. The cell density in the gel is 10^{10} cells/dm³, and the bulk concentration of oxygen ($z = 0$) is 2×10^{-4} mol/dm³, and the diffusivity is 10^{-5} cm²/s.

Solution

A mole balance on oxygen, A, in the volume $\Delta V = A_c \Delta z$ is

$$F_A|_z - F_A|_{z+\Delta z} + r_A A_c \Delta z = 0 \qquad (E12\text{-}2.1)$$

Dividing by Δz and taking the limit as $\Delta z \to 0$ gives

$$\frac{1}{A_c}\frac{dF_A}{dz} + r_A = 0 \qquad (E12\text{-}2.2)$$

$$F_A = A_c\left[-D_{AB}\frac{dC_A}{dz} + UC_A\right] \qquad (E12\text{-}2.3)$$

For dilute concentrations we neglect UC_A and combine Equations (E12-2.2) and (E12-2.3) to obtain

$$\boxed{D_{AB}\frac{d^2C_A}{dz^2}+r_A=0}$$

(E12-2.4)

If we assume the O_2 consumption rate is zero order, then

$$D_{AB}\frac{d^2C_A}{dz^2}-k=0$$

(E12-2.5)

Putting our equation in dimensionless form using $\psi=C_A/C_{A0}$ and z/L, we obtain

$$\frac{d^2\psi}{d\lambda^2}-\frac{kL^2}{D_{AB}C_{A0}}=0$$

(E12-2.6)

Recognizing the second term is just the ratio of a reaction rate to a diffusion rate for a zero order reaction, we call this ratio the Thiele modulus, ϕ_0. We divide and multiply by two to facilitate the integration:

$$\boxed{\phi_0=\frac{k}{2D_{AB}C_{A0}}L^2}$$

(E12-2.7)

$$\boxed{\frac{d^2\psi}{d\lambda^2}-2\phi_0=0}$$

(E12-2.8)

The boundary conditions are

> At $\lambda=0$ $\psi=1$ $C_A=C_{A0}$ (E12-2.9)

> At $\lambda=1$ $\dfrac{d\psi}{d\lambda}=0$ Symmetry condition   (E12-2.10)

Recall that at the midplane ($z=L$, $\lambda=1$) we have symmetry so that there is no diffusion across the midplane so the gradient is zero at $\lambda=1$.

Integrating Equation (E12-2.8) once yields

$$\frac{d\psi}{d\lambda}=2\phi_0\lambda+K_1$$

(E12-2.11)

Using the symmetry condition that there is no gradient across the midplane, Equation (E12-2.10), gives $K_1=-2\phi_0$:

$$\frac{d\psi}{d\lambda}=2\phi_0(\lambda-1)$$

(E12-2.12)

Integrating a second time gives

$$\psi=\phi_0\lambda^2-2\phi_0\lambda+K_2$$

Using the boundary condition $\psi = 1$ at $\lambda = 0$, we find $K_2 = 1$. The dimensionless concentration profile is

$$\boxed{\psi = \phi_0 \lambda (\lambda - 2) + 1} \qquad \text{(E12-2.13)}$$

Note: The dimensionless concentration profile given by Equation (E12-2.13) is only valid for values of the Thiele modulus less than or equal to 1. This restriction can be easily seen if we set $\phi_0 = 10$ and then calculate ψ at $\lambda = 0.1$ to find $\psi = -0.9$, which is a negative concentration!! This condition is explored further in Problem P12-10$_B$.

Parameter Evaluation

Evaluating the zero-order rate constant, k, yields

$$k = \frac{10^{10} \text{cells}}{\text{dm}^3} \cdot \frac{10^{-13} \text{mole O}_2}{\text{cell} \cdot \text{h}} = 10^{-3} \text{mole} / \text{dm}^3 \cdot \text{h}$$

and then the ratio

$$\frac{k}{2 C_{A0} D_{AB}} = \frac{10^{-3} \text{mol/dm}^3 \cdot \text{h}}{2 \times 0.2 \times 10^{-3} \text{mol/dm}^3 \cdot 10^{-5} \dfrac{\text{cm}^2}{\text{s}} \times \dfrac{3600 \text{ s}}{\text{h}}} = 70 \text{ cm}^{-2} \qquad \text{(E12-2.14)}$$

The Thiele modulus is

$$\boxed{\phi_0 = 70 \text{ cm}^{-2} L^2} \qquad \text{(E12-2.15)}$$

(a)   Consider the gel to be completely effective such that the concentration of oxygen is reduced to zero by the time it reaches the center of the gel. That is, if $\psi = 0$ at $\lambda = 1$, we solve Equation (E12-2.13) to find that $\phi_0 = 1$

$$\phi_0 = 1 = \frac{70}{\text{cm}^2} L^2 \qquad \text{(E12-2.16)}$$

Solving for the gel half thickness L yields

$$L = 0.12 \text{ cm}$$

Let's critique this answer. We said the oxygen concentration was zero at the center, and the cells can't survive without oxygen. Consequently, we need to redesign so C_{O_2} is not zero at the center.

(b)   Now consider the case where the minimum oxygen concentration for the cells to survive is 0.1 mmol/dm^3, which is one half that at the surface (i.e., $\psi = 0.5$ at $\lambda = 1.0$). Then Equation (E12-2.13) gives

$$\phi_0 = 0.5 = \frac{70 L^2}{\text{cm}^2} \qquad \text{(E12-2.17)}$$

Solving Equation (E12-2.17) for L gives

$$\boxed{L = 0.085 \text{ cm} = 0.85 \text{ mm} = 850 \text{ }\mu\text{m}}$$

Consequently, we see that the maximum thickness of the cartilage gel ($2L$) is the order of 1 mm, and engineering a thicker tissue is challenging.

(c)   One can consider other perturbations to the preceding analysis by considering the reaction kinetics to follow a first-order rate law, $-r_A = k_A C_A$, or Monod kinetics,

$$-r_A = \frac{\mu_{max} C_A}{K_S + C_A}$$ (E12-2.18)

The author notes the similarities to this problem with his research on wax build-up in subsea pipeline gels.[4] Here as the paraffin diffuses into the gel to form and grow wax particles, these particles cause paraffin molecules to take a longer diffusion path, and as a consequence the diffusivity is reduced. An analogous diffusion pathway for oxygen in the hydrogel containing collagen is shown for in Figure E12-2.3.

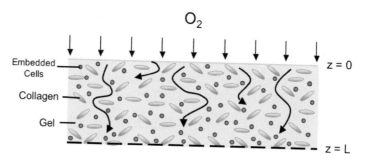

Figure E12-2.3 Diffusion of O_2 around collagen.

$$D_e = \frac{D_{AB}}{1 + \alpha^2 F_w^2/(1 - F_w)}$$ (E12-2.19)

where α and F_w are predetermined parameters that account for diffusion around particles. Specifically, for collagen, α is the aspect ratio of the collagen particle and F_w is weight fraction of "solid" collagen obstructing the diffusion.[4] A similar modification could be made for cartilage growth. These situations are left as an exercise in the end-of-the-chapter problems, e.g., P12-2(b).

12.2 Internal Effectiveness Factor

The magnitude of the effectiveness factor (ranging from 0 to 1) indicates the relative importance of diffusion and reaction limitations. The internal effectiveness factor is defined as

[4] P. Singh, R. Venkatesan, N. Nagarajan, and H. S. Fogler, *AIChE J.*, 46, 1054 (2000).

η is a measure of how far the reactant diffuses into the pellet before reacting.

$$\eta = \frac{\text{Actual overall rate of reaction}}{\text{Rate of reaction that would result if entire interior surface were exposed to the external pellet surface conditions } C_{As}, T_s} \qquad (12\text{-}28)$$

The overall rate, $-r'_A$, is also referred to as the observed rate of reaction $[-r_A(\text{obs})]$. In terms of symbols, the effectiveness factor is

$$\eta = \frac{-r_A}{-r_{As}} = \frac{-r'_A}{-r'_{As}} = \frac{-r''_A}{-r''_{As}}$$

To derive the effectiveness factor for a first-order reaction, it is easiest to work in reaction rates of (moles per unit time), M_A, rather than in moles per unit time per volume of catalyst (i.e., $-r_A$)

$$\eta = \frac{-r_A}{-r_{As}} = \frac{-r_A \times \text{Volume of catalyst particle}}{-r_{As} \times \text{Volume of catalyst particle}} = \frac{M_A}{M_{As}}$$

First we shall consider the denominator, M_{As}. If the entire surface were exposed to the concentration at the external surface of the pellet, C_{As}, the rate for a first-order reaction would be

$$M_{As} = \frac{\text{Rate at external surface}}{\text{Volume}} \times \text{Volume of catalyst}$$

$$= -r_{As} \times \left(\frac{4}{3}\pi R^3\right) = kC_{As}\left(\frac{4}{3}\pi R^3\right) \qquad (12\text{-}29)$$

The subscript s indicates that the rate $-r_{As}$ is evaluated at the conditions present at the **external** surface of the pellet (i.e., $\lambda = 1$).

The actual rate of reaction is the rate at which the reactant diffuses into the pellet at the outer surface. We recall Equation (12-17) for the actual rate of reaction,

The actual rate of reaction

$$M_A = 4\pi RD_e C_{As} \left.\frac{d\psi}{d\lambda}\right|_{\lambda=1} \qquad (12\text{-}17)$$

Differentiating Equation (12-27) and then evaluating the result at $\lambda = 1$ yields

$$\left.\frac{d\psi}{d\lambda}\right|_{\lambda=1} = \left(\frac{\phi_1 \cosh \lambda\phi_1}{\lambda \sinh \phi_1} - \frac{1}{\lambda^2}\frac{\sinh \lambda\phi_1}{\sinh \phi_1}\right)_{\lambda=1} = (\phi_1 \coth \phi_1 - 1) \qquad (12\text{-}30)$$

Substituting Equation (12-30) into (12-17) gives us

$$M_A = 4\pi RD_e C_{As}(\phi_1 \coth \phi_1 - 1) \qquad (12\text{-}31)$$

We now substitute Equations (12-29) and (12-31) into Equation (12-28) to obtain an expression for the effectiveness factor:

$$\eta = \frac{M_A}{M_{As}} = \frac{M_A}{(-r_{As})\left(\dfrac{4}{3}\pi R^3\right)} = \frac{4\pi R D_e C_{As}}{k_1 C_{As}\dfrac{4}{3}\pi R^3}(\phi_1 \coth \phi_1 - 1)$$

$$= 3\,\frac{1}{\underbrace{k_1 R^2/D_e}_{\phi_1^2}}\,(\phi_1 \coth \phi_1 - 1)$$

Internal
effectiveness factor
for a first-order
reaction in a
spherical catalyst
pellet

$$\boxed{\eta = \frac{3}{\phi_1^2}(\phi_1 \coth \phi_1 - 1)} \qquad (12\text{-}32)$$

A plot of the effectiveness factor as a function of the Thiele modulus is shown in Figure 12-5. Figure 12-5(a) shows η as a function of the Thiele modulus ϕ_s for a spherical catalyst pellet for reactions of zero, first, and second order. Figure 12-5(b) corresponds to a first-order reaction occurring in three differently shaped pellets of volume V_p and external surface area A_p, and the Thiele modulus for a first-order reaction, ϕ_1, is defined differently for each shape. When volume change accompanies a reaction (i.e., $\varepsilon \neq \phi$) the corrections shown in Figure 12-6 apply to the effectiveness factor for a first-order reaction.

If $\phi_1 > 2$

then $\eta \approx \dfrac{3}{\phi_1^2}[\phi_1 - 1]$

If $\phi_1 > 20$

then $\eta \approx \dfrac{3}{\phi_1}$

We observe that as the particle diameter becomes very small, ϕ_n decreases, so that the effectiveness factor approaches 1 and the reaction is surface-reaction-limited. On the other hand, when the Thiele modulus ϕ_n is large (~ 30), the internal effectiveness factor η is small (i.e., $\eta \ll 1$), and the reaction is diffusion-limited within the pellet. Consequently, factors influencing the rate of external mass transport will have a negligible effect on the overall reaction rate. For large values of the Thiele modulus, the effectiveness factor can be written as

$$\eta \approx \frac{3}{\phi_1} = \frac{3}{R}\sqrt{\frac{D_e}{k_1}} \qquad (12\text{-}33)$$

To express the overall rate of reaction in terms of the Thiele modulus, we rearrange Equation (12-28) and use the rate law for a first-order reaction in Equation (12-29)

$$-r_A = \left(\frac{\text{Actual reaction rate}}{\text{Reaction rate at } C_{As}}\right) \times (\text{Reaction rate at } C_{As})$$

$$= \eta\,(-r_{As})$$

$$= \eta\,(k_1 C_{As}) \qquad (12\text{-}34)$$

Combining Equations (12-33) and (12-34), the overall rate of reaction for a first-order, internal-diffusion-limited reaction is

$$-r_A = \frac{3}{R}\sqrt{D_e k_1}\,C_{As} = \frac{3}{R}\sqrt{D_e S_a \rho_c k''}\,C_{As}$$

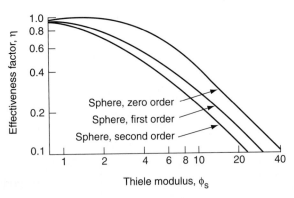

$$\text{Zero order} \qquad \phi_{s0} = R\sqrt{k_0'' S_a \rho_c / D_e C_{A0}} = R\sqrt{k_0 / D_e C_{A0}}$$

$$\text{First order} \qquad \phi_{s1} = R\sqrt{k_1'' S_a \rho_c / D_e} = R\sqrt{k_1 / D_e}$$

$$\text{Second order} \qquad \phi_{s2} = R\sqrt{k_2'' S_a \rho_c C_{A0} / D_e} = R\sqrt{k_2 C_{A0} / D_e}$$

(a)

Internal
effectiveness factor
for different
reaction orders and
catalyst shapes

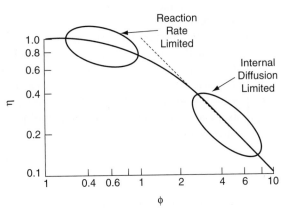

$$\text{Sphere} \qquad \phi_1 = (R/3)\sqrt{k_1'' S_a \rho_c / D_e} = \frac{R}{3}\sqrt{k_1 / D_e}$$

$$\text{Cylinder} \qquad \phi_1 = (R/2)\sqrt{k_1'' S_a \rho_c / D_e} = \frac{R}{2}\sqrt{k_1 / D_e}$$

$$\text{Slab} \qquad \phi_1 = L\sqrt{k_1'' S_a \rho_c / D_e} = L\sqrt{k_1 / D_e}$$

(b)

Figure 12-5   (a) Effectiveness factor plot for nth-order kinetics spherical catalyst
particles (from *Mass Transfer in Heterogeneous Catalysis*, by C. N. Satterfield,
1970; reprint edition: Robert E. Krieger Publishing Co., 1981; reprinted by
permission of the author). (b) First-order reaction in different pellet geometrics
(from R. Aris, *Introduction to the Analysis of Chemical Reactors*, 1965, p. 131;
reprinted by permission of Prentice-Hall, Englewood Cliffs, N.J.).

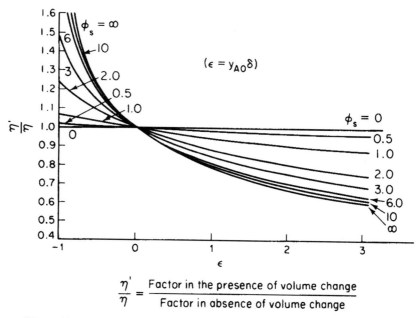

$$\frac{\eta'}{\eta} = \frac{\text{Factor in the presence of volume change}}{\text{Factor in absence of volume change}}$$

Figure 12-6 Effectiveness factor ratios for first-order kinetics on spherical catalyst pellets for various values of the Thiele modulus of a sphere, ϕ_s, as a function of volume change. [From V. W. Weekman and R. L. Goring, *J. Catal.*, 4, 260 (1965).]

How can the rate of reaction be increased?

Therefore, to increase the overall rate of reaction, $-r'_A$: (1) decrease the radius R (make pellets smaller); (2) increase the temperature; (3) increase the concentration; and (4) increase the internal surface area. For reactions of order n, we have, from Equation (12-20),

$$\phi_n^2 = \frac{k_n'' S_a \rho_c R^2 C_{As}^{n-1}}{D_e} = \frac{k_n R^2 C_{As}^{n-1}}{D_e} \tag{12-20}$$

For large values of the Thiele modulus, the effectiveness factor is

$$\eta = \left(\frac{2}{n+1}\right)^{1/2} \frac{3}{\phi_n} = \left(\frac{2}{n+1}\right)^{1/2} \frac{3}{R} \sqrt{\frac{D_e}{k_n}} \, C_{As}^{(1-n)/2} \tag{12-35}$$

Consequently, for reaction orders greater than 1, the effectiveness factor decreases with increasing concentration at the external pellet surface.

The preceding discussion of effectiveness factors is valid only for isothermal conditions. When a reaction is exothermic and nonisothermal, the effectiveness factor can be significantly greater than 1 as shown in Figure 12-7. Values of η greater than 1 occur because the external surface temperature of the pellet is less than the temperature inside the pellet where the exothermic reaction is taking place. Therefore, the rate of reaction inside the pellet is greater than the rate at the surface. Thus, because the effectiveness factor is the

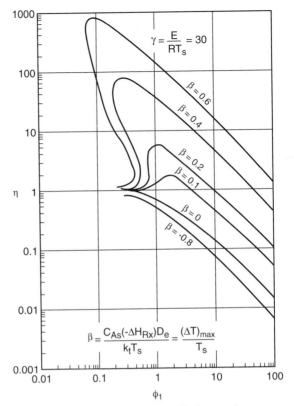

Figure 12-7   Nonisothermal effectiveness factor.

ratio of the actual reaction rate to the rate at surface conditions, the effectiveness factor can be greater than 1, depending on the magnitude of the parameters β and γ. The parameter γ is sometimes referred to as the Arrhenius number, and the parameter β represents the maximum temperature difference that could exist in the pellet relative to the surface temperature T_s.

$$\gamma = \text{Arrhenius number} = \frac{E}{RT_s}$$

$$\beta = \frac{\Delta T_{max}}{T_s} = \frac{T_{max} - T_s}{T_s} = \frac{-\Delta H_{Rx} D_e C_{As}}{k_t T_s}$$

(See Problem P12-13$_B$ for the derivation of β.) The Thiele modulus for a first-order reaction, ϕ_1, is evaluated at the external surface temperature. Typical values of γ for industrial processes range from a value of $\gamma = 6.5$ ($\beta = 0.025$, $\phi_1 = 0.22$) for the synthesis of vinyl chloride from HCl and acetone to a value of $\gamma = 29.4$ ($\beta = 6 \times 10^{-5}$, $\phi_1 = 1.2$) for the synthesis of ammonia.[5] The lower the thermal conductivity k_t and the higher the heat of reaction, the greater the temperature difference (see Problems P12-13$_B$ and P12-14$_B$). We observe

---

[5] H. V. Hlavacek, N. Kubicek, and M. Marek, *J. Catal.*, *15*, 17 (1969).

Can you find regions where multiple solutions (MSS) exist?

Typical parameter values

from Figure 12-7 that multiple steady states can exist for values of the Thiele modulus less than 1 and when β is greater than approximately 0.2. There will

be no multiple steady states when the criterion developed by Luss[6] is fulfilled.

$$4(1+\beta) > \beta\gamma \qquad (12\text{-}36)$$

12.3 Falsified Kinetics

There are circumstances under which the measured reaction order and activation energy are not the true values. Consider the case in which we obtain reaction rate data in a differential reactor, where precautions are taken to virtually eliminate external mass transfer resistance (i.e., $C_{As} = C_{Ab}$). From these data we construct a log-log plot of the measured rate of reaction $-r'_A$ as a function of the gas-phase concentration, C_{As} (Figure 12-8). The slope of this plot is the apparent reaction order n' and the rate law takes the form

$$-r'_A = k'_n C_{As}^{n'} \qquad (12\text{-}37)$$

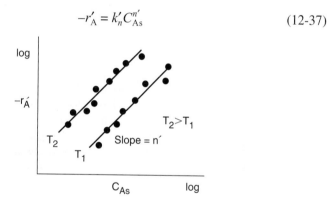

Figure 12-8   Determining the apparent reaction order $(-r_A = \rho_b\,(-r'_A))$.

We will now proceed to relate this measured reaction order n' to the true reaction order n. Using the definition of the effectiveness factor, note that the actual rate $-r'_A$, is the product of η and the rate of reaction evaluated at the external surface, $k_n C_{As}^n$, i.e.,

$$-r'_A = \eta\,(-r'_{As}) = \eta\,(k_n C_{As}^n) \qquad (12\text{-}38)$$

For large values of the Thiele modulus ϕ_n, we can use Equation (12-35) to substitute into Equation (12-38) to obtain

[6] D. Luss, *Chem. Eng. Sci.*, 23, 1249 (1968).

$$-r_A' = \frac{3}{\phi_n}\sqrt{\frac{2}{n+1}} \; k_n C_{As}^n = \frac{3}{R}\sqrt{\frac{D_e}{k_n}C_{As}^{1-n}}\sqrt{\frac{2}{n+1}} \; k_n C_{As}^n$$

$$= \frac{3}{R}\sqrt{\frac{2D_e}{(n+1)}} \; k_n^{1/2} C_{As}^{(n+1)/2} \tag{12-39}$$

We equate the true reaction rate, Equation (12-39), to the measured reaction rate, Equation (12-37), to get

$$-r_A' = \sqrt{\frac{2}{n+1}}\left(\frac{3}{R}\sqrt{D_e} \; k_n^{1/2} C_{As}^{(n+1)/2}\right) = k_n' C_{As}^{n'} \tag{12-40}$$

We now compare Equations (12-39) and (12-40). Because the overall exponent of the concentration, C_{As}, must be the same for both the analytical and measured rates of reaction, the apparent reaction order n' is related to the true reaction order n by

The true and the apparent reaction order

$$\boxed{n' = \frac{1+n}{2}} \tag{12-41}$$

In addition to an apparent reaction order, there is also an apparent activation energy, E_{App}. This value is the activation energy we would calculate using the experimental data, from the slope of a plot of $\ln(-r_A)$ as a function of $1/T$ at a fixed concentration of A. Substituting for the measured and true specific reaction rates in terms of the activation energy gives

$$\underbrace{k_n' = A_{App}e^{-E_{App}/RT}}_{\text{measured}} \qquad \underbrace{k_n = A_T e^{-E_T/RT}}_{\text{true}}$$

into Equation (12-40), we find that

$$\boxed{n_{true} = 2n_{apparent} - 1}$$

$$-r_A' = \left(\frac{3}{R}\sqrt{\frac{2}{n+1}} \; D_e\right) A_T^{1/2}\left[\exp\left(\frac{-E_T}{RT}\right)\right]^{1/2} C_{As}^{(n+1)/2} = A_{App}\left[\exp\left(\frac{-E_{App}}{RT}\right)\right] C_{As}^{n'}$$

Taking the natural log of both sides gives us

$$\ln\left[\frac{3}{R}\sqrt{\frac{2}{n+1}} \; D_e \; A_T^{1/2} C_{As}^{(n+1)/2}\right] - \frac{E_T}{2RT} = \ln(A_{App}C_{As}^{n'}) - \frac{E_{App}}{RT} \tag{12-42}$$

where E_T is the true activation energy.

Comparing the temperature-dependent terms on the right- and left-hand sides of Equation (12-42), we see that the true activation energy is equal to twice the apparent activation energy.

The true activation energy

$$\boxed{E_T = 2E_{App}} \tag{12-43}$$

This measurement of the apparent reaction order and activation energy results primarily when internal diffusion limitations are present and is referred to as *disguised* or *falsified kinetics*. Serious consequences could occur if the laboratory data were taken in the disguised regime and the reactor were operated in a different regime. For example, what if the particle size were reduced so that internal diffusion limitations became negligible? The higher activation energy, E_T, would cause the reaction to be much more temperature-sensitive, and there is the possibility for *runaway reaction conditions* to occur.

12.4 Overall Effectiveness Factor

For first-order reactions we can use an overall effectiveness factor to help us analyze diffusion, flow, and reaction in packed beds. We now consider a situation where external and internal resistance to mass transfer to and within the pellet are of the same order of magnitude (Figure 12-9). At steady state, the transport of the reactant(s) from the bulk fluid to the external surface of the catalyst is equal to the net rate of reaction of the reactant within and on the pellet.

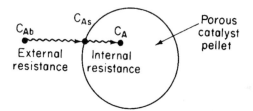

Figure 12-9   Mass transfer and reaction steps.

The molar rate of mass transfer from the bulk fluid to the external surface is

$$\text{Molar rate} = (\text{Molar flux}) \cdot (\text{External surface area})$$
$$M_A = W_{Ar} \cdot (\text{Surface area/Volume})(\text{Reactor volume})$$
$$= W_{Ar} \cdot a_c \Delta V \tag{12-44}$$

where a_c is the external surface area per unit reactor volume (cf. Chapter 11) and ΔV is the volume.

This molar rate of mass transfer to the surface, M_A, is equal to the net (total) rate of reaction *on* **and** *within* the pellet:

$$\boxed{M_A = -r_A'' \ (\text{External area} + \text{Internal area})}$$

$$\text{External area} = \frac{\text{External area}}{\text{Reactor volume}} \times \text{Reactor volume}$$

$$= a_c \, \Delta V$$

$$\text{Internal area} = \frac{\text{Internal area}}{\text{Mass of catalyst}} \times \frac{\text{Mass of catalyst}}{\text{Volume of catalyst}} \times \frac{\text{Volume of catalyst}}{\text{Reactor volume}} \times \text{Reactor volume}$$

$$= S_a \, \overbrace{\rho_c(1-\phi)}^{\rho_b} \, \Delta V$$

$$= S_a \rho_b \, \Delta V$$

(12-45)

$$\boxed{M_A = -r_A''[a_c \, \Delta V + S_a \rho_b \, \Delta V \,]}$$

Combining Equations (12-44) and (12-45) and canceling the volume ΔV, one obtains

$$W_A a_c = -r_A'' \cdot (a_c + S_a \rho_b)$$

For most catalysts the internal surface area is much greater than the external surface area (i.e., $S_a \rho_b \gg a_c$), in which case we have

$$\boxed{W_A a_c = -r_A'' S_a \rho_b} \qquad (12\text{-}46)$$

where $-r_A''$ is the overall rate of reaction within and on the pellet per unit surface area. The relationship for the rate of mass transport is

$$M_A = W_{Ar} a_c \, \Delta V = k_c (C_{Ab} - C_{As}) a_c \, \Delta V \qquad (12\text{-}47)$$

where k_c is the external mass transfer coefficient (m/s). Because internal diffusion resistance is also significant, not all of the interior surface of the pellet is accessible to the concentration at the external surface of the pellet, C_{As}. We have already learned that the effectiveness factor is a measure of this surface accessibility [see Equation (12-38)]:

$$-r_A'' = -r_{As}'' \, \eta$$

Assuming that the surface reaction is first order with respect to A, we can utilize the internal effectiveness factor to write

$$-r_A'' = \eta k_1'' C_{As} \qquad (12\text{-}48)$$

We need to eliminate the surface concentration from any equation involving the rate of reaction or rate of mass transfer, because C_{As} cannot be measured by standard techniques. To accomplish this elimination, first substitute Equation (12-48) into Equation (12-46):

$$W_{Ar}a_c = \eta k_1'' S_a C_{As}\rho_b$$

Then substitute for $W_{Ar}a_c$ using Equation (12-47)

$$k_c a_c(C_{Ab} - C_{As}) = \eta k_1'' S_a \rho_b C_{As} \tag{12-49}$$

Solving for C_{As}, we obtain

Concentration at the pellet surface as a function of bulk gas concentration

$$C_{As} = \frac{k_c a_c C_{Ab}}{k_c a_c + \eta k_1'' S_a \rho_b} \tag{12-50}$$

Substituting for C_{As} in Equation (12-48) gives

$$-r_A'' = \frac{\eta k_1'' k_c a_c C_{Ab}}{k_c a_c + \eta k_1'' S_a \rho_b} \tag{12-51}$$

In discussing the surface accessibility, we defined the internal effectiveness factor η with respect to the concentration at the external surface of the pellet, C_{As}:

$$\boxed{\eta = \frac{\text{Actual overall rate of reaction}}{\begin{array}{c}\text{Rate of reaction that would result if entire interior surface were}\\ \text{exposed to the external pellet surface conditions, } C_{As}, T_s\end{array}}} \tag{12-28}$$

Two different effectiveness factors

We now define an overall effectiveness factor that is based on the bulk concentration:

$$\boxed{\Omega = \frac{\text{Actual overall rate of reaction}}{\begin{array}{c}\text{Rate that would result if the entire surface were}\\ \text{exposed to the bulk conditions, } C_{Ab}, T_b\end{array}}} \tag{12-52}$$

Dividing the numerator and denominator of Equation (12-51) by $k_c a_c$, we obtain the net rate of reaction (total molar flow of A to the surface in terms of the bulk fluid concentration), which is a measurable quantity:

$$-r_A'' = \frac{\eta}{1 + \eta k_1'' S_a \rho_b / k_c a_c} k_1'' C_{Ab} \tag{12-53}$$

Consequently, the overall rate of reaction in terms of the bulk concentration C_{Ab} is

$$\boxed{-r''_A = \Omega(-r''_{Ab}) = \Omega k''_1 C_{Ab}} \tag{12-54}$$

where

$$\boxed{\Omega = \frac{\eta}{1 + \eta k''_1 S_a \rho_b / k_c a_c}} \tag{12-55}$$

The rates of reaction based on surface and bulk concentrations are related by

$$-r''_A = \Omega(-r''_{Ab}) = \eta(-r''_{As}) \tag{12-56}$$

where

$$-r''_{As} = k''_1 C_{As}$$
$$-r''_{Ab} = k''_1 C_{Ab}$$

The actual rate of reaction is related to the reaction rate evaluated at the bulk concentrations. The actual rate can be expressed in terms of the rate per unit volume, $-r_A$, the rate per unit mass, $-r'_A$, and the rate per unit surface area, $-r''_A$, which are related by the equation

$$-r_A = -r'_A \rho_b = -r''_A S_a \rho_b$$

In terms of the overall effectiveness factor for a first-order reaction and the reactant concentration in the bulk

$$-r_A = r_{Ab}\Omega = r'_{Ab}\rho_b\Omega = -r''_{Ab}S_a\rho_b\Omega = k''_1 C_{Ab}S_a\rho_b\Omega \tag{12-57}$$

where again

$$\Omega = \frac{\eta}{1 + \eta k''_1 S_a \rho_b / k_c a_c}$$

Recall that k''_1 is given in terms of the catalyst surface area ($m^3/m^2 \cdot s$).

12.5 Estimation of Diffusion- and Reaction-Limited Regimes

In many instances it is of interest to obtain "quick and dirty" estimates to learn which is the rate-limiting step in a heterogeneous reaction.

12.5.1   Weisz–Prater Criterion for Internal Diffusion

The Weisz–Prater criterion uses measured values of the rate of reaction, $-r_A'$ (obs), to determine if internal diffusion is limiting the reaction. This criterion can be developed intuitively by first rearranging Equation (12-32) in the form

$$\eta \phi_1^2 = 3(\phi_1 \coth \phi_1 - 1) \tag{12-58}$$

Showing where the Weisz–Prater comes from

The left-hand side is the Weisz–Prater parameter:

$$C_{WP} = \eta \times \phi_1^2 \tag{12-59}$$

$$= \frac{\text{Observed (actual) reaction rate}}{\text{Reaction rate evaluated at } C_{As}} \times \frac{\text{Reaction rate evaluated at } C_{As}}{\text{A diffusion rate}}$$

$$= \frac{\text{Actual reaction rate}}{\text{A diffusion rate}}$$

Substituting for

$$\eta = \frac{-r_A'(\text{obs})}{-r_{As}'} \qquad \text{and} \qquad \phi_1^2 = \frac{-r_{As}'' S_a \rho_c R^2}{D_e C_{As}} = \frac{-r_{As}' \rho_c R^2}{D_e C_{As}}$$

in Equation (12-59) we have

$$C_{WP} = \frac{-r_A'(\text{obs})}{-r_{As}'} \left(\frac{-r_{As}' \rho_c R^2}{D_e C_{As}} \right) \tag{12-60}$$

$$\boxed{C_{WP} = \eta \phi_1^2 = \frac{-r_A'(\text{obs}) \rho_c R^2}{D_e C_{As}}} \tag{12-61}$$

Are there any internal diffusion limitations indicated from the Weisz–Prater criterion?

All the terms in Equation (12-61) are either measured or known. Consequently, we can calculate C_{WP}. However, if

$$\boxed{C_{WP} \ll 1}$$

there are no diffusion limitations and consequently no concentration gradient exists within the pellet. However, if

$$\boxed{C_{WP} \gg 1}$$

internal diffusion limits the reaction severely. Ouch!

Example 12–3   Estimating Thiele Modulus and Effectiveness Factor

The first-order reaction

$$A \longrightarrow B$$

was carried out over two different-sized pellets. The pellets were contained in a spinning basket reactor that was operated at sufficiently high rotation speeds that external mass transfer resistance was negligible. The results of two experimental runs made under identical conditions are as given in Table E12-3.1. **(a)** Estimate the Thiele modulus and effectiveness factor for each pellet. **(b)** How small should the pellets be made to virtually eliminate all internal diffusion resistance?

TABLE E12-3.1 DATA FROM A SPINNING BASKET REACTOR[†]

	Measured Rate (obs) (mol/g cat ·s) $\times 10^5$	Pellet Radius (m)
Run 1	3.0	0.01
Run 2	15.0	0.001

[†] See Figure 5-12(c).

These two experiments yield an enormous amount of information.

Solution

(a) Combining Equations (12-58) and (12-61), we obtain

$$\frac{-r_A'(\text{obs})\,R^2\rho_c}{D_eC_{As}} = \eta\phi_1^2 = 3(\phi_1\coth\phi_1 - 1) \tag{E12-3.1}$$

Letting the subscripts 1 and 2 refer to runs 1 and 2, we apply Equation (E12-3.1) to runs 1 and 2 and then take the ratio to obtain

$$\frac{-r_{A2}'\,R_2^2}{-r_{A1}'\,R_1^2} = \frac{\phi_{12}\coth\phi_{12}-1}{\phi_{11}\coth\phi_{11}-1} \tag{E12-3.2}$$

The terms ρ_c, D_e, and C_{As} cancel because the runs were carried out under identical conditions. The Thiele modulus is

$$\phi_1 = R\sqrt{\frac{-r_{As}'\rho_c}{D_eC_{As}}} \tag{E12-3.3}$$

Taking the ratio of the Thiele moduli for runs 1 and 2, we obtain

$$\boxed{\frac{\phi_{11}}{\phi_{12}} = \frac{R_1}{R_2}} \tag{E12-3.4}$$

or

$$\phi_{11} = \frac{R_1}{R_2}\phi_{12} = \frac{0.01\text{ m}}{0.001\text{ m}}\phi_{12} = 10\phi_{12} \tag{E12-3.5}$$

Substituting for ϕ_{11} in Equation (E12-3.2) and evaluating $-r_A'$ and R for runs 1 and 2 gives us

$$\left(\frac{15\times10^{-5}}{3\times10^{-5}}\right)\frac{(0.001)^2}{(0.01)^2} = \frac{\phi_{12}\coth\phi_{12}-1}{10\phi_{12}\coth(10\phi_{12})-1} \tag{E12-3.6}$$

$$0.05 = \frac{\phi_{12} \coth \phi_{12} - 1}{10\phi_{12} \coth (10\phi_{12}) - 1} \qquad \text{(E12-3.7)}$$

We now have one equation and one unknown. Solving Equation (E12-3.7) we find that

$$\phi_{12} = 1.65 \qquad \text{for } R_2 = 0.001 \text{ m}$$

Then

$$\phi_{11} = 10\phi_{12} = 16.5 \qquad \text{for } R_1 = 0.01 \text{ m}$$

The corresponding effectiveness factors are

Given two experimental points, one can predict the particle size where internal mass transfer does not limit the rate of reaction.

$$\text{For } R_2: \quad \eta_2 = \frac{3(\phi_{12} \coth \phi_{12} - 1)}{\phi_{12}^2} = \frac{3(1.65 \coth 1.65 - 1)}{(1.65)^2} = 0.856$$

$$\text{For } R_1: \quad \eta_1 = \frac{3(16.5 \coth 16.5 - 1)}{(16.5)^2} \approx \frac{3}{16.5} = 0.182$$

(b) Next we calculate the particle radius needed to virtually eliminate internal diffusion control (say, $\eta = 0.95$):

$$0.95 = \frac{3(\phi_{13} \coth \phi_{13} - 1)}{\phi_{13}^2} \qquad \text{(E12-3.8)}$$

Solution to Equation (E12-2.8) yields $\phi_{13} = 0.9$:

$$R_3 = R_1 \frac{\phi_{13}}{\phi_{11}} = (0.01)\left(\frac{0.9}{16.5}\right) = 5.5 \times 10^{-4} \text{ m}$$

A particle size of 0.55 mm is necessary to virtually eliminate diffusion control (i.e., $\eta = 0.95$).

12.5.2 Mears' Criterion for External Diffusion

The Mears[7] criterion, like the Weisz–Prater criterion, uses the measured rate of reaction, $-r_A'$, (kmol/kg cat·s) to learn if mass transfer from the bulk gas phase to the catalyst surface can be neglected. Mears proposed that when

Is external diffusion limiting?

$$\frac{-r_A' \rho_b Rn}{k_c C_{Ab}} < 0.15 \qquad \text{(12-62)}$$

external mass transfer effects can be neglected.

---

[7] D. E. Mears, *Ind. Eng. Chem. Process Des. Dev.*, *10*, 541 (1971). Other interphase transport-limiting criteria can be found in *AIChE Symp. Ser. 143* (S. W. Weller, ed.), 70 (1974).

where n = reaction order
       R = catalyst particle radius, m
       ρ_b = bulk density of catalyst bed, kg/m^3
          = $(1 - \phi)\rho_c$ (ϕ = porosity)
       ρ_c = solid density of catalyst pellet, kg/m^3
       C_{Ab} = solid catalyst density, kg/m^3
       k_c = mass transfer coefficient, m/s

The mass transfer coefficient can be calculated from the appropriate correlation, such as that of Thoenes–Kramers, for the flow conditions through the bed. When Equation (12-62) is satisfied, no concentration gradients exist between the bulk gas and external surface of the catalyst pellet.

Mears also proposed that the bulk fluid temperature, T, will be virtually the same as the temperature at the external surface of the pellet when

Is there a temperature gradient?

$$\left| \frac{-\Delta H_{Rx}(-r'_A)\rho_b RE}{h T^2 R_g} \right| < 0.15 \qquad (12\text{-}63)$$

where h = heat transfer coefficient between gas and pellet, kJ/m$^2\cdot$s$\cdot$K
       R_g = gas constant, = 8.31 kJ/mol$\cdot$K
       ΔH_{Rx} = heat of reaction, kJ/mol
       E = activation energy, kJ/kmol

and the other symbols are as in Equation (12-62).

12.6 Mass Transfer and Reaction in a Packed Bed

We now consider the same isomerization taking place in a packed bed of catalyst pellets rather than on one single pellet (see Figure 12-10). The concentration C_A is the bulk gas-phase concentration of A at any point along the length of the bed.

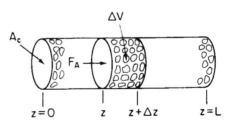

Figure 12-10   Packed-bed reactor.

We shall perform a balance on species A over the volume element ΔV, neglecting any radial variations in concentration and assuming that the bed is operated at steady state. The following symbols will be used in developing our model:

$$A_c = \text{cross-sectional area of the tube, dm}^2$$

$$C_{Ab} = \text{bulk gas concentration of A, mol/dm}^3$$

$$\rho_b = \text{bulk density of the catalyst bed, g/dm}^3$$

$$v_0 = \text{volumetric flow rate, dm}^3/\text{s}$$

$$U = \text{superficial velocity} = v_0/A_c, \text{dm/s}$$

Mole Balance   A mole balance on the volume element $(A_c \Delta z)$ yields

$$[\text{Rate in}] - [\text{Rate out}] + [\text{Rate of formation of A}] = 0$$

$$A_c W_{Az}|_z - A_c W_{Az}|_{z+\Delta z} + \qquad r'_A \rho_b A_c \Delta z \qquad = 0$$

Dividing by $A_c \, \Delta z$ and taking the limit as $\Delta z \longrightarrow 0$ yields

$$-\frac{dW_{Az}}{dz} + r'_A \rho_b = 0 \tag{12-64}$$

Assuming that the total concentration c is constant, Equation (11-14) can be expressed as

$$W_{Az} = -D_{AB}\frac{dC_{Ab}}{dz} + y_{Ab}(W_{Az} + W_{Bz})$$

Also, writing the bulk flow term in the form

$$B_{Az} = y_{Ab}(W_{Az} + W_{Bz}) = y_{Ab}cU = UC_{Ab}$$

Equation (12-64) can be written in the form

$$D_{AB}\frac{d^2C_{Ab}}{dz^2} - U\frac{dC_{Ab}}{dz} + r'_A \rho_b = 0 \tag{12-65}$$

Now we will see how to use η and Ω to calculate conversion in a packed bed.   The term $D_{AB}(d^2C_{Ab}/dz^2)$ is used to represent either diffusion and/or dispersion in the axial direction. Consequently, we shall use the symbol D_a for the dispersion coefficient to represent either or both of these cases. We will come back to this form of the diffusion equation when we discuss dispersion in Chapter 14. The overall reaction rate within the pellet, $-r'_A$, is the overall rate of reaction within and on the catalyst per unit mass of catalyst. It is a function of the reactant concentration within the catalyst. This overall rate can be related to the rate of reaction of A that would exist if the entire surface were exposed to the bulk concentration C_{Ab} through the overall effectiveness factor Ω:

$$-r'_A = -r'_{Ab} \times \Omega \tag{12-57}$$

For the first-order reaction considered here,

$$-r'_{Ab} = -r''_{Ab}S_a = k''S_a C_{Ab} \tag{12-66}$$

Substituting Equation (12-66) into Equation (12-57), we obtain the overall rate of reaction per unit mass of catalyst in terms of the bulk concentration C_{Ab}:

$$-r'_A = \Omega k'' S_a C_{Ab}$$

Substituting this equation for $-r'_A$ into Equation (12-65), we form the differential equation describing diffusion with a first-order reaction in a catalyst bed:

Flow and first-order reaction in a packed bed

$$\boxed{D_a \frac{d^2 C_{Ab}}{dz^2} - U \frac{dC_{Ab}}{dz} - \Omega \rho_b k'' S_a C_{Ab} = 0} \tag{12-67}$$

As an example, we shall solve this equation for the case in which the flow rate through the bed is very large and the axial diffusion can be neglected. Finlayson[8] has shown that axial dispersion can be neglected when

Criterion for neglecting axial dispersion/diffusion

$$\boxed{\left| \frac{-r'_A \rho_b d_p}{U_0 C_{Ab}} \right| \ll \left| \frac{U_0 d_p}{D_a} \right|} \tag{12-68}$$

where U_0 is the superficial velocity, d_p the particle diameter, and D_a is the effective axial dispersion coefficient. In Chapter 14 we will consider solutions to the complete form of Equation (12-67).

Neglecting axial dispersion with respect to forced axial convection,

$$\left| U \frac{dC_{Ab}}{dz} \right| \gg \left| D_a \frac{d^2 C_{Ab}}{dz^2} \right|$$

Equation (12-67) can be arranged in the form

$$\frac{dC_{Ab}}{dz} = - \left(\frac{\Omega \rho_b k'' S_a}{U} \right) C_{Ab} \tag{12-69}$$

With the aid of the boundary condition at the entrance of the reactor,

$$C_{Ab} = C_{Ab0} \qquad \text{at } z = 0$$

Equation (12-69) can be integrated to give

$$C_{Ab} = C_{Ab0} e^{-(\rho_b k'' S_a \Omega z)/U} \tag{12-70}$$

The conversion at the reactor's exit, $z = L$, is

Conversion in a packed-bed reactor

$$\boxed{X = 1 - \frac{C_{Ab}}{C_{Ab0}} = 1 - e^{-(\rho_b k'' S_a \Omega L)/U}} \tag{12-71}$$

---

[8] L. C. Young and B. A. Finlayson, *Ind. Eng. Chem. Fund.*, *12*, 412 (1973).

Example 12–4   Reducing Nitrous Oxides in a Plant Effluent

In Section 7.1.4 we saw the role that nitric oxide plays in smog formation and the incentive we would have for reducing its concentration in the atmosphere. It is proposed to reduce the concentration of NO in an effluent stream from a plant by passing it through a packed bed of spherical porous carbonaceous solid pellets. A 2% NO–98% air mixture flows at a rate of 1×10^{-6} m³/s (0.001 dm³/s) through a 2-in.-ID tube packed with porous solid at a temperature of 1173 K and a pressure of 101.3 kPa. The reaction

$$\text{NO} + \text{C} \longrightarrow \text{CO} + \tfrac{1}{2}\,\text{N}_2$$

is first order in NO, that is,

$$-r'_{\text{NO}} = k'' S_a C_{\text{NO}}$$

and occurs primarily in the pores inside the pellet, where

Green
chemical reaction
engineering

$$S_a = \text{Internal surface area} = 530 \text{ m}^2/\text{g}$$

$$k'' = 4.42 \times 10^{-10} \text{ m}^3/\text{m}^2 \cdot \text{s}$$

Calculate the weight of porous solid necessary to reduce the NO concentration to a level of 0.004%, which is below the Environmental Protection Agency limit.

Additional information:

At 1173 K, the fluid properties are

$$\nu = \text{Kinematic viscosity} = 1.53 \times 10^{-8} \text{ m}^2/\text{s}$$

$$D_{\text{AB}} = \text{Gas-phase diffusivity} = 2.0 \times 10^{-8} \text{ m}^2/\text{s}$$

$$D_e = \text{Effective diffusivity} = 1.82 \times 10^{-8} \text{ m}^2/\text{s}$$

Also see Web site
www.rowan.edu/
greenengineering

The properties of the catalyst and bed are

$$\rho_c = \text{Density of catalyst particle} = 2.8 \text{ g/cm}^3 = 2.8 \times 10^6 \text{ g/m}^3$$

$$\phi = \text{Bed porosity} = 0.5$$

$$\rho_b = \text{Bulk density of bed} = \rho_c(1 - \phi) = 1.4 \times 10^6 \text{ g/m}^3$$

$$R = \text{Pellet radius} = 3 \times 10^{-3} \text{ m}$$

$$\gamma = \text{Sphericity} = 1.0$$

Solution

It is desired to reduce the NO concentration from 2.0% to 0.004%. Neglecting any volume change at these low concentrations gives us

$$X = \frac{C_{\text{Ab0}} - C_{\text{Ab}}}{C_{\text{Ab0}}} = \frac{2 - 0.004}{2} = 0.998$$

where A represents NO.

The variation of NO down the length of the reactor is given by Equation (12-69)

$$\frac{dC_{Ab}}{dz} = -\frac{\Omega k'' S_a \rho_b C_{Ab}}{U} \tag{12-69}$$

Multiplying the numerator and denominator on the right-hand side of Equation (12-69) by the cross-sectional area, A_c, and realizing that the weight of solids up to a point z in the bed is

(Mole balance)
+
(Rate law)
+
(Overall
effectiveness
factor)

$$W = \rho_b A_c z$$

the variation of NO concentration with solids is

$$\frac{dC_{Ab}}{dW} = -\frac{\Omega k'' S_a C_{Ab}}{v} \tag{E12-4.1}$$

Because NO is present in dilute concentrations, we shall take $\varepsilon \ll 1$ and set $v = v_0$. We integrate Equation (E12-4.1) using the boundary condition that when $W = 0$, then $C_{Ab} = C_{Ab0}$:

$$X = 1 - \frac{C_{Ab}}{C_{Ab0}} = 1 - \exp\left(-\frac{\Omega k'' S_a W}{v_0}\right) \tag{E12-4.2}$$

where

$$\Omega = \frac{\eta}{1 + \eta k'' S_a \rho_c / k_c a_c} \tag{12-55}$$

Rearranging, we have

$$W = \frac{v_0}{\Omega k'' S_a} \ln \frac{1}{1-X} \tag{E12-4.3}$$

1. *Calculating the internal effectiveness factor* for spherical pellets in which a first-order reaction is occurring, we obtained

$$\eta = \frac{3}{\phi_1^2} (\phi_1 \coth \phi_1 - 1) \tag{12-32}$$

As a first approximation, we shall neglect any changes in the pellet size resulting from the reactions of NO with the porous carbon. The Thiele modulus for this system is[9]

$$\phi_1 = R \sqrt{\frac{k_1'' \rho_c S_a}{D_e}} \tag{E12-4.4}$$

where

R = pellet radius = 3×10^{-3} m
D_e = effective diffusivity = 1.82×10^{-8} m²/s
ρ_c = 2.8 g/cm³ = 2.8×10^6 g/m³

---

[9] L. K. Chan, A. F. Sarofim, and J. M. Beer, *Combust. Flame*, *52*, 37 (1983).

$$k = \text{specific reaction rate} = 4.42 \times 10^{-10} \text{ m}^3/\text{m}^2 \cdot \text{s}$$

$$\phi_1 = 0.003 \text{ m} \sqrt{\frac{(4.42 \times 10^{-10} \text{ m/s})(530 \text{ m}^2/\text{g})(2.8 \times 10^6 \text{ g/m}^3)}{1.82 \times 10^{-8} \text{ m}^2/\text{s}}}$$

$$\phi_1 = 18$$

Because ϕ_1 is large,

$$\eta \cong \frac{3}{18} = 0.167$$

2. *To calculate the external mass transfer coefficient*, the Thoenes–Kramers correlation is used. From Chapter 11 we recall

$$\text{Sh}' = (\text{Re}')^{1/2} \text{Sc}^{1/3} \tag{11-65}$$

For a 2-in.-ID pipe, $A_c = 2.03 \times 10^{-3} \text{ m}^2$. The superficial velocity is

$$U = \frac{v_0}{A_c} = \frac{10^{-6} \text{ m}^3/\text{s}}{2.03 \times 10^{-3} \text{ m}^2} = 4.93 \times 10^{-4} \text{ m/s}$$

Procedure
Calculate
Re'
Sc
Then
Sh'
Then
k_c''

$$\text{Re}' = \frac{U d_p}{(1-\phi)\nu} = \frac{(4.93 \times 10^{-4} \text{ m/s})(6 \times 10^{-3} \text{ m})}{(1-0.5)(1.53 \times 10^{-8} \text{ m}^2/\text{s})} = 386.7$$

Nomenclature note: ϕ with subscript 1, $\phi_1 = $ Thiele modulus
ϕ without subscript, $\phi = $ porosity

$$\text{Sc} = \frac{\nu}{D_{AB}} = \frac{1.53 \times 10^{-8} \text{ m}^2/\text{s}}{2.0 \times 10^{-8} \text{ m}^2/\text{s}} = 0.765$$

$$\text{Sh}' = (386.7)^{1/2}(0.765)^{1/3} = (19.7)(0.915) = 18.0$$

$$k_c'' = \frac{1-\phi}{\phi}\left(\frac{D_{AB}}{d_p}\right)\text{Sh}' = \frac{0.5}{0.5}\left(\frac{2.0 \times 10^{-8} \text{ (m}^2/\text{s)}}{6.0 \times 10^{-3} \text{ m}}\right)(18.0)$$

$$k_c'' = 6 \times 10^{-5} \text{ m/s}$$

3. *Calculating the external area per unit reactor volume*, we obtain

$$a_c = \frac{6(1-\phi)}{d_p} = \frac{6(1-0.5)}{6 \times 10^{-3} \text{ m}} \tag{E12-4.5}$$

$$= 500 \text{ m}^2/\text{m}^3$$

4. *Evaluating the overall effectiveness factor.* Substituting into Equation (12-55), we have

$$\Omega = \frac{\eta}{1 + \eta k_1'' \, S_a \rho_b / k_c a_c}$$

$$\Omega = \frac{0.167}{1 + \dfrac{(0.167)(4.4 \times 10^{-10} \ \text{m}^3/\text{m}^2 \cdot \text{s})(530 \ \text{m}^2/\text{g})(1.4 \times 10^6 \ \text{g/m}^3)}{(6 \times 10^{-5} \ \text{m/s})(500 \ \text{m}^2/\text{m}^3)}}$$

$$= \frac{0.167}{1 + 1.83} = 0.059$$

In this example we see that both the external and internal resistances to mass transfer are significant.

5. *Calculating the weight of solid necessary to achieve 99.8% conversion.* Substituting into Equation (E12-4.3), we obtain

$$W = \frac{1 \times 10^{-6} \ \text{m}^3/\text{s}}{(0.059)(4.42 \times 10^{-10} \ \text{m}^3/\text{m}^2 \cdot \text{s})(530 \ \text{m}^2/\text{g})} \ \ln \frac{1}{1 - 0.998}$$

$$= 450 \ \text{g}$$

6. *The reactor length is*

$$L = \frac{W}{A_c \rho_b} = \frac{450 \ \text{g}}{(2.03 \times 10^{-3} \ \text{m}^2)(1.4 \times 10^6 \ \text{g/m}^3)}$$

$$= 0.16 \ \text{m}$$

12.7 Determination of Limiting Situations from Reaction Data

For external mass transfer-limited reactions in packed beds, the rate of reaction at a point in the bed is

$$-r_A' = k_c a_c C_A \tag{12-72}$$

Variation of reaction rate with system variables

The correlation for the mass transfer coefficient, Equation (11-66), shows that k_c is directly proportional to the square root of the velocity and inversely proportional to the square root of the particle diameter:

$$k_c \propto \frac{U^{1/2}}{d_p^{1/2}} \tag{12-73}$$

We recall from Equation (E12-4.5), $a_c = 6(1 - \phi)/d_p$, that the variation of external surface area with catalyst particle size is

$$a_c \propto \frac{1}{d_p}$$

Consequently, for external mass transfer-limited reactions, the rate is inversely proportional to the particle diameter to the three-halves power:

$$-r'_A \propto \frac{1}{d_p^{3/2}} \tag{12-74}$$

From Equation (11-72) we see that for gas-phase external mass transfer-limited reactions, the rate increases approximately linearly with temperature.

When internal diffusion limits the rate of reaction, we observe from Equation (12-39) that the rate of reaction varies inversely with particle diameter, is independent of velocity, and exhibits an exponential temperature dependence which is not as strong as that for surface-reaction-controlling reactions. For surface-reaction-limited reactions the rate is independent of particle size and is a strong function of temperature (exponential). Table 12-1 summarizes the dependence of the rate of reaction on the velocity through the bed, particle diameter, and temperature for the three types of limitations that we have been discussing.

TABLE 12-1   LIMITING CONDITIONS

Type of Limitation	Variation of Reaction Rate with:		
	Velocity	Particle Size	Temperature
External diffusion	$U^{1/2}$	$(d_p)^{-3/2}$	$\approx$ Linear
Internal diffusion	Independent	$(d_p)^{-1}$	Exponential
Surface reaction	Independent	Independent	Exponential

The exponential temperature dependence for internal diffusion limitations is usually not as strong a function of temperature as is the dependence for surface reaction limitations. If we would calculate an activation energy between 8 and 24 kJ/mol, chances are that the reaction is strongly diffusion-limited. An activation energy of 200 kJ/mol, however, suggests that the reaction is reaction-rate-limited.

12.8 Multiphase Reactors

Multiphase reactors are reactors in which two or more phases are necessary to carry out the reaction. The majority of multiphase reactors involve gas and liquid phases that contact a solid. In the case of the slurry and trickle bed reactors, the reaction between the gas and the liquid takes place on a solid catalyst surface (see Table 12-2). However, in some reactors the liquid phase is an inert medium for the gas to contact the solid catalyst. The latter situation arises when a large heat sink is required for highly exothermic reactions. In many cases the catalyst life is extended by these milder operating conditions.

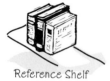

Reference Shelf

The multiphase reactors discussed in this edition of the book are the slurry reactor, fluidized bed, and the trickle bed reactor. The trickle bed reactor, which has reaction and transport steps similar to the slurry reactor, is discussed in the first edition of this book and on the CD-ROM along with the bubbling

TABLE 12-2.   APPLICATIONS OF THREE-PHASE REACTORS

I. *Slurry reactor*
  A. Hydrogenation
       1. of fatty acids over a supported nickel catalyst
       2. of 2-butyne-1,4-diol over a Pd-CaCO$_3$ catalyst
       3. of glucose over a Raney nickel catalyst
  B. Oxidation
       1. of C$_2$H$_4$ in an inert liquid over a PdCl$_2$-carbon catalyst
       2. of SO$_2$ in inert water over an activated carbon catalyst
  C. Hydroformation
       of CO with high-molecular-weight olefins on either a cobalt or ruthenium complex bound
       to polymers
  D. Ethynylation
       Reaction of acetylene with formaldehyde over a CaCl$_2$-supported catalyst
II. *Trickle bed reactors*
  A. Hydrodesulfurization
       Removal of sulfur compounds from crude oil by reaction with hydrogen on Co-Mo on
       alumina
  B. Hydrogenation
       1. of aniline over a Ni-clay catalyst
       2. of 2-butyne-1,4-diol over a supported Cu-Ni catalyst
       3. of benzene, α-CH$_3$ styrene, and crotonaldehyde
       4. of aromatics in napthenic lube oil distillate
  C. Hydrodenitrogenation
       1. of lube oil distillate
       2. of cracked light furnace oil
  D. Oxidation
       1. of cumene over activated carbon
       2. of SO$_2$ over carbon

Source: C. N. Satterfield, *AIChE J.*, *21*, 209 (1975); P. A. Ramachandran and R. V. Chaudhari,
Chem. Eng., *87*(24), 74 (1980); R. V. Chaudhari and P. A. Ramachandran, *AIChE J.*, *26*, 177
(1980).

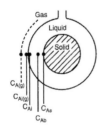

Reference Shelf

fluidized bed. In slurry reactors, the catalyst is suspended in the liquid, and gas
is bubbled through the liquid. A slurry reactor may be operated in either a
semibatch or continuous mode.

12.8.1 Slurry Reactors

A complete description of the slurry reactor and the transport and reaction
steps are given on the CD-ROM, along with the design equations and a number of examples. Methods to determine which of the transport and reaction
steps are rate limiting are included. See Professional Reference Shelf R12.1.

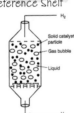

12.8.2 Trickle Bed Reactors

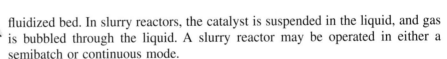

The CD-ROM includes all the material on trickle bed reactors from the first
edition of this book. A comprehensive example problem for trickle bed reactor
design is included. See Professional Reference Shelf R12.2.

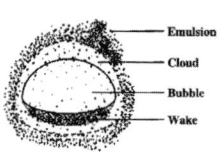

12.9 Fluidized Bed Reactors

The Kunii-Levenspiel model for fluidization is given on the CD-ROM along with a comprehensive example problem. The rate limiting transport steps are also discussed. See Professional Reference Shelf R12.3.

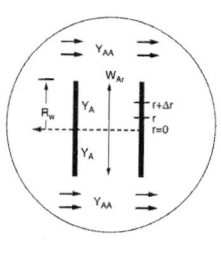

12.10 Chemical Vapor Deposition (CVD)

Chemical Vapor Deposition in boat reactors is discussed and modeled. The equations and parameters which affect wafer thickness and shape are derived and analyzed. This material is taken directly from the second edition of this book. See Professional Reference Shelf R12.4.

> **Closure** After completing this chapter, the reader should be able to derive differential equations describing diffusion and reaction, discuss the meaning of the effectiveness factor and its relationship to the Thiele modulus, and identify the regions of mass transfer control and reaction rate control. The reader should be able to apply the Weisz–Prater and Mears criteria to identify gradients and diffusion limitations. These principles should be able to be applied to catalyst particles as well as biomaterial tissue engineering. The reader should be able to apply the overall effectiveness factor to a packed bed reactor to calculate the conversion at the exit of the reactor. The reader should be able to describe the reaction and transport steps in slurry reactors, trickle bed reactors, fluidized-bed reactors, and CVD boat reactors and to make calculations for each reactor.

SUMMARY

1. The concentration profile for a first-order reaction occurring in a spherical catalyst pellet is

$$\frac{C_A}{C_{As}} = \frac{R}{r}\left[\frac{\sinh(\phi_1 r/R)}{\sinh \phi_1}\right] \tag{S12-1}$$

where ϕ_1 is the Thiele modulus. For a first-order reaction

$$\phi_1^2 = \frac{k_1}{D_e} R^2 \tag{S12-2}$$

2. The effectiveness factors are

$$\begin{array}{c}\text{Internal}\\ \text{effectiveness} = \eta =\\ \text{factor}\end{array} \quad \frac{\text{Actual rate of reaction}}{\begin{array}{c}\text{Reaction rate if entire interior}\\ \text{surface is exposed to concentration}\\ \text{at the external pellet surface}\end{array}}$$

$$
\begin{array}{c}
\text{Overall} \\
\text{effectiveness} = \Omega = \\
\text{factor}
\end{array}
\frac{\text{Actual rate of reaction}}{\begin{array}{c}\text{Reaction rate if entire surface area}\\\text{is exposed to bulk concentration}\end{array}}
$$

3. For large values of the Thiele modulus for an nth order reaction,

$$
\eta = \left(\frac{2}{n+1}\right)^{1/2}\frac{3}{\phi_n} \tag{S12-3}
$$

4. For internal diffusion control, the true reaction order is related to the measured reaction order by

$$
n_{\text{true}} = 2n_{\text{apparent}} - 1 \tag{S12-4}
$$

The true and apparent activation energies are related by

$$
E_{\text{true}} = 2E_{\text{app}} \tag{S12-5}
$$

5. A. The Weisz–Prater Parameter

$$
C_{\text{WP}} = \phi_1^2\eta = \frac{-r_A'(\text{observed})\rho_c R^2}{D_e C_{As}} \tag{S12-6}
$$

The Weisz–Prater criterion dictates that

If $C_{\text{WP}} \ll 1$ no internal diffusion limitations present

If $C_{\text{WP}} \gg 1$ internal diffusion limitations present

B. Mears Criteria for Neglecting External Diffusion and Heat Transfer

$$
\boxed{\frac{-r_A'\rho_b R n}{k_c C_{Ab}} < 0.15} \tag{S12-7}
$$

and

$$
\boxed{\left|\frac{-\Delta H_{Rx}(-r_A')(\rho_b R E)}{h T^2 R_g}\right| < 0.15} \tag{S12-8}
$$

CD-ROM MATERIAL

- **Learning Resources**
  1. *Summary Notes*

- **Professional Reference Shelf**

  R12.1. *Slurry Reactors*

Reference Shelf

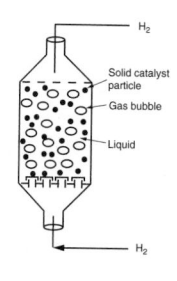

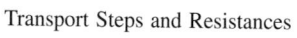

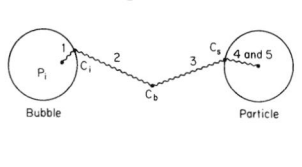

Transport Steps and Resistances

  A. Description of Use of Slurry Reactors

      Example R12-1 Industrial Slurry Reactor

  B. Reaction and Transport Steps in a Step in Slurry Reactor

$$
\frac{C_i}{R_A} = \frac{1}{k_b a_b} + \frac{1}{m}\left(\frac{\overbrace{1}^{r_c}}{k_b a_b} + \frac{\overbrace{1}^{r_r}}{k\eta}\right)
$$

  C. Determining the Rate-Limiting Step

      1. Effect of Loading, Particle Size and Gas Adsorption

      2. Effect of Shear

          Example R12-2 Determining the Controlling Resistance

  D. Slurry Reactor Design

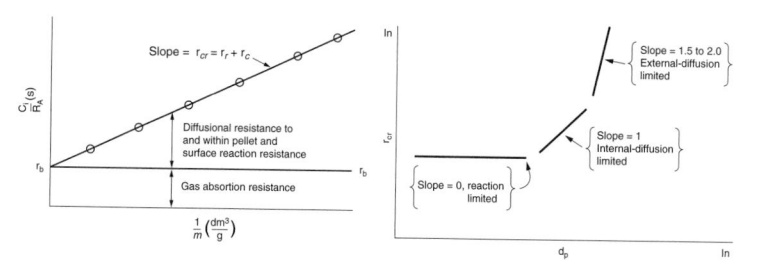

Example R12-3 Slurry Reactor Design

R12.2. *Trickle Bed Reactors*

  A. Fundamentals

Reference Shelf

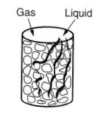

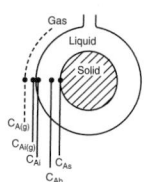

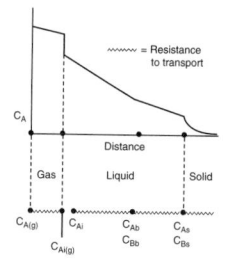

  B. Limiting Situations

  C. Evaluating the Transport Coefficients

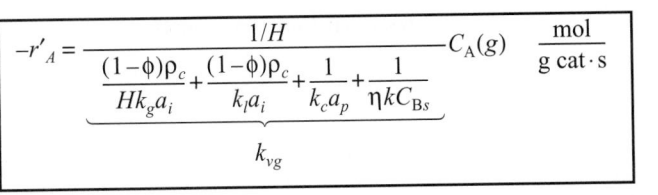

$$-r'_A = \frac{1/H}{\underbrace{\frac{(1-\phi)\rho_c}{Hk_g a_i} + \frac{(1-\phi)\rho_c}{k_l a_i} + \frac{1}{k_c a_p} + \frac{1}{\eta k C_{Bs}}}_{k_{vg}}} C_A(g) \qquad \frac{mol}{g\ cat \cdot s}$$

Reference Shelf

R12.3. *Fluidized-Bed Reactors*
    A. Descriptive Behavior of the Kunii-Levenspiel Bubbling Bed Model

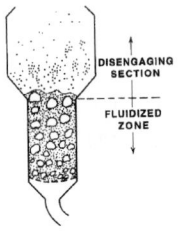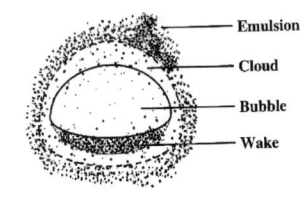

    B. Mechanics of Fluidized Beds
        Example R12-4 Maximum Solids Hold-Up
    C. Mass Transfer in Fluidized Beds
    D. Reaction in a Fluidized Bed
    E. Solution to the Balance Equations for a First Order Reaction

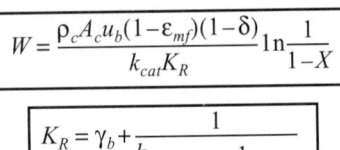

$$W = \frac{\rho_c A_c u_b (1-\varepsilon_{mf})(1-\delta)}{k_{cat} K_R} \ln \frac{1}{1-X}$$

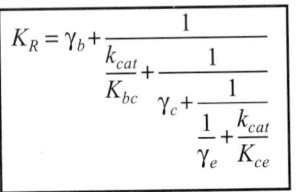

$$K_R = \gamma_b + \cfrac{1}{\cfrac{k_{cat}}{K_{bc}} + \cfrac{1}{\gamma_c + \cfrac{1}{\cfrac{1}{\gamma_e} + \cfrac{k_{cat}}{K_{ce}}}}}$$

        Example R12-5 Catalytic Oxidation of Ammonia
    F. Limiting Situations
        Example R12-6 Calculation of the Resistances
        Example R12-7 Effect of Particle Size on Catalyst Weight for
                a Slow Reaction
        Example R12-8 Effect of Catalyst Weight for a Rapid Reaction
R12.4. *Chemical Vapor Deposition Reactors*

Reference Shelf

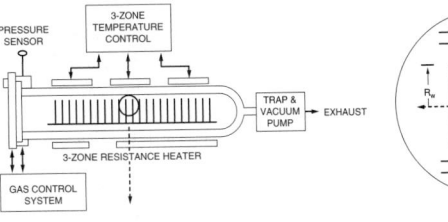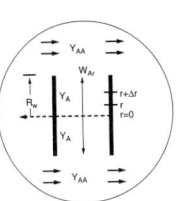

A. Chemical Reaction Engineering in Microelectronic Processing
B. Fundamentals of CVD
C. Effectiveness Factors for Boat Reactors

$$\eta = \frac{2I_1(\phi_1)}{\phi_1 I_o(\phi_1)}$$

Example R12-9 Diffusion Between Wafers
Example R12-10 CVD Boat Reactor

Reference Shelf

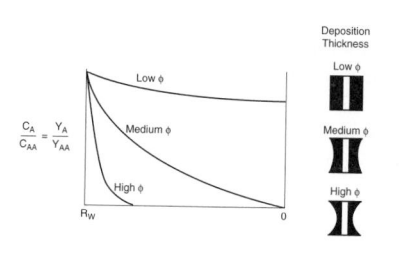

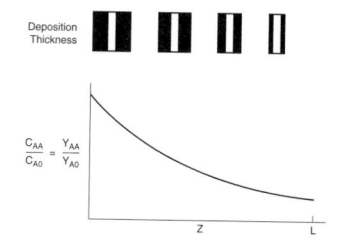

QUESTIONS AND PROBLEMS

The subscript to each of the problem numbers indicates the level of difficulty: A, least difficult; D, most difficult.

A = ●   B = ■   C = ◆   D = ◆◆

Member

Hall of Fame

P12-1$_C$ Make up an original problem using the concepts presented in Section _____ (your instructor will specify the section). Extra credit will be given if you obtain and use real data from the literature. (See Problem P4-1 for the guidelines.)

P12-2$_B$ **(a) Example 12-1.** *Effective Diffusivity.* Make a sketch of a diffusion path for which the tortuosity is 5. How would your effective gas-phase diffusivity change if the absolute pressure were tripled and the temperature were increased by 50%?

(b) Example 12-2. *Tissue Engineering.* How would your answers change if the reaction kinetics were (1) first order in O_2 concentration with $k_1 = 10^{-2}$ h^{-1}? (2) For zero-order kinetics carry out a quasi steady state analysis using Equations (E12-2.19) along with the overall balance

$$\frac{dF_w}{dt} = v_c W_{O_2}\Big|_{z=0} A_c$$

to predict the O_2 flux and collagen build-up as a function of time. Sketch ψ versus λ at different times. Sketch λ_c as a function of time. *Hint:* See P12-10$_c$. *Note:* $V = A_c L$. Assume $\alpha = 10$ and the stoichiometric coefficient for oxygen to collagen, v_c, is 0.05 mass fraction of cell/mol O_2. $A_c = 2$ cm^2.

(c) Example 12-3. (1) What is the percent of the total resistance for internal diffusion and for reaction rate for each of the three particles studied? (2) Apply the Weisz-Prater criteria to a particle 0.005 m in diameter.

(d) Example 12-4. *Overall Effectiveness Factor.* (1) Calculate the percent of the total resistance for external diffusion, internal diffusion, and surface reaction. Qualitatively how would each of your percentages change (2) if the temperature were increased significantly? (3) if the gas velocity were tripled? (4) if the particle size were decreased by a factor of 2? How would the reactor length change in each case? (5) What length would be required to achieve 99.99% conversion of the pollutant NO?

What if...

(e) you applied the Mears and Weisz–Prater criteria to Examples 11-4 and 12-4? What would you find? What would you learn if $\Delta H_{Rx} = -25$ k cal/mol, $h = 100$ Btu/h·ft²·°F and $E = 20$ k cal/mol?

(f) we let $\gamma = 30$, $\beta = 0.4$, and $\phi = 0.4$ in Figure 12-7? What would cause you to go from the upper steady state to the lower steady state and vice versa?

(g) your internal surface area decreased with time because of sintering. How would your effectiveness factor change and the rate of reaction change with time if $k_d = 0.01$ h⁻¹ and $\eta = 0.01$ at $t = 0$? Explain.

(h) someone had used the false kinetics (i.e., wrong E, wrong n)? Would their catalyst weight be overdesigned or underdesigned? What are other positive or negative effects that occur?

Green engineering
Web site
www.rowan.edu/
greenengineering

(i) you were to assume the resistance to gas absorption in CDROM Example R12.1 were the same as in Example R12.3 and that the liquid phase reactor volume in Example R12.3 was 50% of the total, could you estimate the controlling resistance? If so, what is it? What other things could you calculate in Example R12.1 (e.g., selectivity, conversion, molar flow rates in and out)? *Hint:* Some of the other reactions that occur include

$$CO + 3H_2 \longrightarrow CH_4 + H_2O$$

$$H_2O + CO \longrightarrow CO_2 + H_2$$

(j) the temperature in CDROM Example R12.2 were increased? How would the relative resistances in the slurry reactor change?

(k) you were asked for all the things that could go wrong in the operation of a slurry reactor, what would you say?

P12-3$_B$ The catalytic reaction

$$A \longrightarrow B$$

takes place within a fixed bed containing spherical porous catalyst X22. Figure P12-3 shows the overall rates of reaction at a point in the reactor as a function of temperature for various entering total molar flow rates, F_{T0}.

(a) Is the reaction limited by external diffusion?

(b) If your answer to part (a) was "yes," under what conditions [of those shown (i.e., T, F_{T0})] is the reaction limited by external diffusion?

(c) Is the reaction "reaction-rate-limited"?

(d) If your answer to part (c) was "yes," under what conditions [of those shown (i.e., T, F_{T0})] is the reaction limited by the rate of the surface reactions?

(e) Is the reaction limited by internal diffusion?

(f) If your answer to part (e) was "yes," under what conditions [of those shown (i.e., T, F_{T0})] is the reaction limited by the rate of internal diffusion?

(g) For a flow rate of 10 g mol/h, determine (if possible) the overall effectiveness factor, Ω, at 360 K.

(h) Estimate (if possible) the internal effectiveness factor, η, at 367 K.

(i) If the concentration at the external catalyst surface is 0.01 mol/dm³, calculate (if possible) the concentration at $r = R/2$ inside the porous catalyst at 367 K. (Assume a first-order reaction.)

Member

Hall of Fame

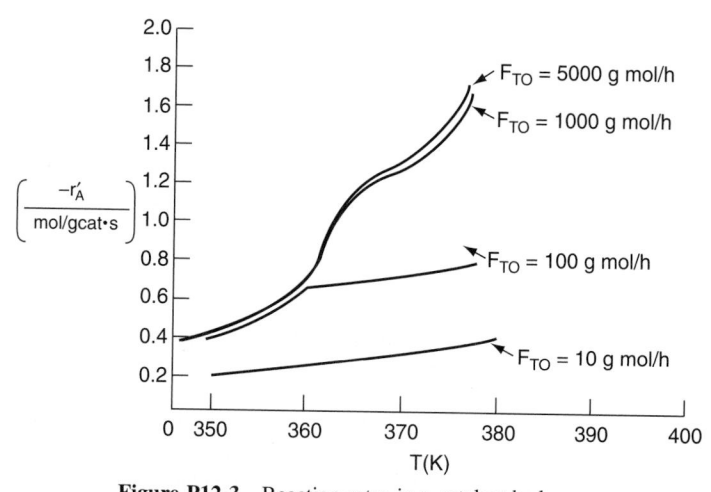

Figure P12-3   Reaction rates in a catalyst bed.

Additional information:

Gas properties:

Diffusivity: 0.1 cm²/s
Density: 0.001 g/cm³
Viscosity: 0.0001 g/cm·s

Bed properties:

Tortuosity of pellet: 1.414
Bed permeability: 1 millidarcy
Porosity = 0.3

Hall of Fame

P12-4$_B$ The reaction

$$A \longrightarrow B$$

is carried out in a differential packed-bed reactor at different temperatures, flow rates, and particle sizes. The results shown in Figure P12-4 were obtained.

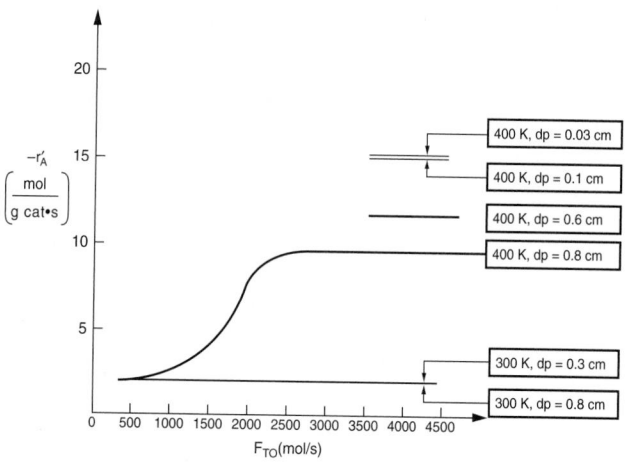

Figure P12-4   Reaction rates in a catalyst bed.

(a) What regions (i.e., conditions d_p, T, F_{T0}) are external mass transfer-limited?
(b) What regions are reaction-rate-limited?
(c) What region is internal-diffusion-controlled?
(d) What is the internal effectiveness factor at $T = 400$ and $d_p = 0.8$ cm?

P12-5$_A$ Curves A, B, and C in Figure P12-5 show the variations in reaction rate for three different reactions catalyzed by solid catalyst pellets. What can you say about each reaction?

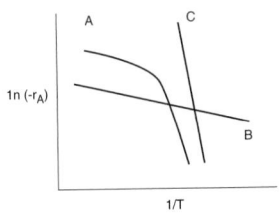

Figure P12-5   Temperature dependence of three reactions.

P12-6$_B$ A first-order heterogeneous irreversible reaction is taking place within a spherical catalyst pellet which is plated with platinum throughout the pellet (see Figure 12-3). The reactant concentration halfway between the external surface and the center of the pellet (i.e., $r = R/2$) is equal to one-tenth the concentration of the pellet's external surface. The concentration at the external surface is 0.001 g mol/dm³, the diameter ($2R$) is 2×10^{-3} cm, and the diffusion coefficient is 0.1 cm²/s.

$$A \longrightarrow B$$

(a) What is the concentration of reactant at a distance of 3×10^{-4} cm in from the external pellet surface? (*Ans.:* $C_A = 2.36 \times 10^{-4}$ mol/dm³.)

(b) To what diameter should the pellet be reduced if the effectiveness factor is to be 0.8? (*Ans.:* $d_p = 6.8 \times 10^{-4}$ cm. Critique this answer!)

(c) If the catalyst support were not yet plated with platinum, how would you suggest that the catalyst support be plated *after* it had been reduced by grinding?

P12-7$_B$ The swimming rate of a small organism [*J. Theoret. Biol.*, 26, 11 (1970)] is related to the energy released by the hydrolysis of adenosine triphosphate (ATP) to adenosine diphosphate (ADP). The rate of hydrolysis is equal to the rate of diffusion of ATP from the midpiece to the tail (see Figure P12-7). The diffusion coefficient of ATP in the midpiece and tail is 3.6×10^{-6} cm²/s. ADP is converted to ATP in the midsection, where its concentration is 4.36×10^{-5} mol/cm³. The cross-sectional area of the tail is 3×10^{-10} cm².

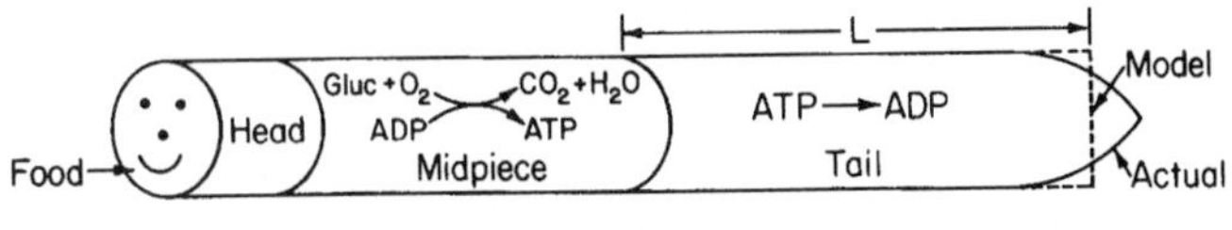

Figure P12-7   Swimming of an organism.

(a) Derive an equation for diffusion and reaction in the tail.

(b) Derive an equation for the effectiveness factor in the tail.

(c) Taking the reaction in the tail to be of zero order, calculate the length of the tail. The rate of reaction in the tail is 23×10^{-18} mol/s.

(d) Compare your answer with the average tail length of 41 μm. What are possible sources of error?

P12-8$_B$ A first-order, heterogeneous, irreversible reaction is taking place within a catalyst pore which is plated with platinum entirely along the length of the pore (Figure P12-8). The reactant concentration at the plane of symmetry (i.e., equal distance from the pore mouths) of the pore is equal to one-tenth the concentration of the pore mouth. The concentration at the pore mouth is 0.001 mol/dm³, the pore length (2L) is 2 × 10⁻³ cm, and the diffusion coefficient is 0.1 cm²/s.

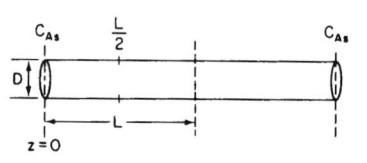

Figure P12-8   Single catalyst pore.

(a) Derive an equation for the effectiveness factor.
(b) What is the concentration of reactant at $L/2$?
(c) To what length should the pore length be reduced if the effectiveness factor is to be 0.8?
(d) If the catalyst support were not yet plated with platinum, how would you suggest the catalyst support be plated *after* the pore length, L, had been reduced by grinding?

P12-9$_A$ A first-order reaction is taking place inside a porous catalyst. Assume dilute concentrations and neglect any variations in the axial (x) direction.

(a) Derive an equation for both the internal and overall effectiveness factors for the rectangular porous slab shown in Figure P12-9.
(b) Repeat part (a) for a cylindrical catalyst pellet where the reactants diffuse inward in the radial direction.

Figure P12-9   Flow over porous catalyst slab.

P12-10$_B$ The irreversible reaction

$$A \longrightarrow B$$

is taking place in the porous catalyst disk shown in Figure P12-9. The reaction is zero order in A.

(a) Show that the concentration profile using the symmetry B.C. is

$$\frac{C_A}{C_{As}} = 1 + \phi_0^2 \left[\left(\frac{z}{L} \right)^2 - 1 \right] \qquad (P10\text{-}10.1)$$

where

$$\phi_0^2 = \frac{kL^2}{2D_eC_{As}}$$ (P10-10.2)

(b) For a Thiele modulus of 1.0, at what point in the disk is the concentration zero? For $\phi_0 = 4$?

(c) What is the concentration you calculate at $z = 0.1\ L$ and $\phi_0 = 10$ using Equation (P12-10.1)? What do you conclude about using this equation?

(d) Plot the dimensionless concentration profile $\psi = C_A/C_{As}$ as a function of $\lambda = z/L$ for $\phi_0 = 0.5$, 1, 5, and 10. *Hint:* there are regions where the concentration is zero. Show that $\lambda_C = 1 - 1/\phi_0$ is the start of this region where the gradient and concentration are both zero. [L. K. Jang, R. L. York, J. Chin, and L. R. Hile, *Inst. Chem. Engr.*, 34, 319 (2003).] Show that $\psi = \phi_0^2\ \lambda^2 - 2\phi_0(\phi_0 - 1)\ \lambda + (\phi_0 - 1)^2$ for $\lambda_C \leq \lambda < 1$.

(e) The effectiveness factor can be written as

$$\eta = \frac{\int_0^L -r_A A_c\ dz}{-r_{As} A_c L} = \frac{\int_0^{z_C} -r_A A_c\ dz + \int_{z_C}^L -r_A A_c\ dz}{-r_{As} A_c L}$$ (P12-10.3)

where z_C (λ_C) is the point where both the concentration gradients and flux go to zero and A_c is the cross-sectional area of the disk. Show for a zero-order reaction that

$$\eta = \begin{cases} 1 & \text{for } \phi_0 \leq 1.0 \\ 1 - \lambda_C = \dfrac{1}{\phi_0} & \text{for } \phi_0 \geq 1 \end{cases}$$ (P10-10.4)

(f) Make a sketch for η versus ϕ_0 similar to the one shown in Figure 12-5.

(g) Repeat parts (a) to (f) for a spherical catalyst pellet.

(h) What do you believe to be the point of this problem?

P12-11$_C$ The second-order decomposition reaction

$$A \longrightarrow B + 2C$$

is carried out in a tubular reactor packed with catalyst pellets 0.4 cm in diameter. The reaction is internal-diffusion-limited. Pure A enters the reactor at a superficial velocity of 3 m/s, a temperature of 250°C, and a pressure of 500 kPa. Experiments carried out on smaller pellets where surface reaction is limiting yielded a specific reaction rate of 0.05 m^6/mol·g cat·s. Calculate the length of bed necessary to achieve 80% conversion. Critique the numerical answer.

Additional information:

    Effective diffusivity: 2.66×10^{-8} m^2/s
    Ineffective diffusivity: 0.00 m^2/s
    Bed porosity: 0.4
    Pellet density: 2×10^6 g/m^3
    Internal surface area: 400 m^2/g

P12-12$_C$ Derive the concentration profile and effectiveness factor for cylindrical pellets 0.2 cm in diameter and 1.5 cm in length. Neglect diffusion through the ends of the pellet.

(a) Assume that the reaction is a first-order isomerization. (*Hint:* Look for a Bessel function.)

(b) Rework Problem P12-11 for these pellets.

P12-13$_C$ Reconsider diffusion and reaction in a spherical catalyst pellet for the case where the reaction is not isothermal. Show that the energy balance can be written as

$$\frac{1}{r^2} \frac{d}{dr} \left(r^2 k_t \frac{dT}{dr} \right) + (-\Delta H_R)(-r_A) = 0 \qquad \text{(P12-13.1)}$$

where k_t is the effective thermal conductivity, cal/s·cm·K of the pellet with $dT/dr = 0$ at $r = 0$ and $T = T_s$ at $r = R$.

(a) Evaluate Equation (12-11) for a first-order reaction and combine with Equation (P12-13.1) to arrive at an equation giving the maximum temperature in the pellet.

$$T_{max} = T_s + \frac{(-\Delta H_{Rx})(D_e C_{As})}{k_t} \qquad \text{(P12-13.2)}$$

Note: At T_{max}, $C_A = 0$.

(b) Choose representative values of the parameters and use a software package to solve Equations (12-11) and (P12-13.1) simultaneously for $T(r)$ and $C_A(r)$ when the reaction is carried out adiabatically. Show that the resulting solution agrees qualitatively with Figure 12-7.

P12-14$_C$ Determine the effectiveness factor for a nonisothermal spherical catalyst pellet in which a first-order isomerization is taking place.

Additional information:

$A_i = 100 \text{ m}^2/\text{m}^3$
$\Delta H_R = -800{,}000 \text{ J/mol}$
$D_e = 8.0 \times 10^{-8} \text{ m}^2/\text{s}$
$C_{As} = 0.01 \text{ kmol/m}^3$
External surface temperature of pellet, $T_s = 400 \text{ K}$
$E = 120{,}000 \text{ J/mol}$
Thermal conductivity of pellet $= 0.004 \text{ J/m·s·K}$
$d_p = 0.005 \text{ m}$
Specific reaction rate $= 10^{-1} \text{ m/s at } 400 \text{ K}$
Density of calf's liver $= 1.1 \text{ g/dm}^3$

How would your answer change if the pellets were 10^{-2}, 10^{-4}, and 10^{-5} m in diameter? What are typical temperature gradients in catalyst pellets?

P12-15$_B$ *Extension of Problem P12-8.* The elementary isomerization reaction

$$A \longrightarrow B$$

is taking place on the walls of a cylindrical catalyst pore. (See Figure P12-8.) In one run a catalyst poison P entered the reactor together with the reactant A. To estimate the effect of poisoning, we assume that the poison renders the catalyst pore walls near the pore mouth ineffective up to a distance z_1, so that no reaction takes place on the walls in this entry region.

(a) Show that before poisoning of the pore occurred, the effectiveness factor was given by

$$\eta = \frac{1}{\phi} \tanh \phi$$

where

$$\phi = L \sqrt{\frac{2k}{rD_e}}$$

with k = reaction rate constant (length/time)
        r = pore radius (length)
        D_e = effective molecular diffusivity (area/time)

(b) Derive an expression for the concentration profile and also for the molar flux of A in the ineffective region $0 < z < z_1$, in terms of z_1, D_{AB}, C_{A1}, and C_{As}. Without solving any further differential equations, obtain the new effectiveness factor η' for the poisoned pore.

P12-16$_B$ *Falsified Kinetics.* The irreversible gas-phase dimerization

$$2A \longrightarrow A_2$$

is carried out at 8.2 atm in a stirred contained-solids reactor to which only pure A is fed. There is 40 g of catalyst in each of the four spinning baskets. The following runs were carried out at 227°C:

Total Molar Feed Rate, F_{T0} (g mol/min)	1	2	4	6	11	20
Mole Fraction A in Exit, y_A	0.21	0.33	0.40	0.57	0.70	0.81

The following experiment was carried out at 237°C:

$$F_{T0} = 9 \text{ g mol/min} \qquad y_A = 0.097$$

(a) What are the apparent reaction order and the apparent activation energy?
(b) Determine the true reaction order, specific reaction rate, and activation energy.
(c) Calculate the Thiele modulus and effectiveness factor.
(d) What diameter of pellets should be used to make the catalyst more effective?
(e) Calculate the rate of reaction on a rotating disk made of the catalytic material when the gas-phase reactant concentration is 0.01 g mol/L and the temperature is 527°C. The disk is flat, nonporous, and 5 cm in diameter.

Additional information:

   Effective diffusivity: 0.23 cm²/s
   Surface area of porous catalyst: 49 m²/g cat
   Density of catalyst pellets: 2.3 g/cm³
   Radius of catalyst pellets: 1 cm
   Color of pellets: blushing peach

P12-17$_B$ Derive Equation (12-35). *Hint:* Multiply both sides of Equation (12-25) for *n*th order reaction, that is,

$$\frac{d^2y}{d\lambda^2} - \phi_n^2 y^n = 0$$

by $2dy/d\lambda$, rearrange to get

$$\frac{d}{d\lambda}\left(\frac{dy}{d\lambda}\right)^2 = \phi_n^2 y^n 2\frac{dy}{d\lambda}$$

and solve using the boundary conditions $dy/d\lambda = 0$ at $\lambda = 0$.

JOURNAL ARTICLE PROBLEMS

P12J-1 The article in *Trans. Int. Chem. Eng.*, *60*, 131 (1982) may be advantageous in answering the following questions.
   (a) Describe the various types of gas–liquid–solid reactors.
   (b) Sketch the concentration profiles for gas absorption with:
      (1) An instantaneous reaction
      (2) A very slow reaction
      (3) An intermediate reaction rate

P12J-2 After reading the journal review by Y. T. Shah et al. [*AIChE J.*, *28*, 353 (1982)], design the following bubble column reactor. One percent carbon dioxide in air is to be removed by bubbling through a solution of sodium hydroxide. The reaction is mass-transfer-limited. Calculate the reactor size (length and diameter) necessary to remove 99.9% of the CO_2. Also specify a type of sparger. The reactor is to operate in the bubbly flow regime and still process 0.5 m³/s of gas. The liquid flow rate through the column is 10^{-3} m³/s.

JOURNAL CRITIQUE PROBLEMS

P12C-1 Use the Weisz–Prater criterion to determine if the reaction discussed in *AIChE J.*, *10*, 568 (1964) is diffusion-rate-limited.

P12C-2 Use the references given in *Ind. Eng. Chem. Prod. Res. Dev.*, *14*, 226 (1975) to define the iodine value, saponification number, acid number, and experimental setup. Use the slurry reactor analysis to evaluate the effects of mass transfer and determine if there are any mass transfer limitations.

• **Additional Homework Problems**

   CDP12-A$_B$ Determine the catalyst size that gives the highest conversion in a packed bed reactor.

   CDP12-B$_B$ Determine importance of concentration and temperature gradients in a packed bed reactor.

   CDP12-C$_B$ Determine concentration profile and effectiveness factor for the first order gas phase reaction

$$A \rightarrow 3B$$

Slurry Reactors

CDP12-D$_B$ Hydrogenation of methyl linoleate-comparing catalyst. [3rd Ed. P12-19]

CDP12-E$_B$ Hydrogenation of methyl linoleate. Find the rate-limiting step. [3rd Ed. P12-20]

CDP12-F$_B$ Hydrogenation of 2-butyne–1,4-diol to butenediol. Calculate percent resistance of total of each for each step and the conversion. [3rd Ed. P12-21]

CVD Boat Reactors

CDP12-G$_D$ Determine the temperature profile to achieve a uniform thickness. [2nd Ed. P11-18]

CDP12-H$_B$ Explain how varying a number of the parameters in the *CVD boat reactor* will affect the wafer shape. [2nd Ed. P11-19]

CDP12-I$_B$ Determine the wafer shape in a CVD boat reactor for a series of operating conditions. [2nd Ed. P11-20]

CDP12-J$_C$ Model the build-up of a silicon wafer on parallel sheets. [2nd Ed. P11-21]

CDP12-K$_C$ Rework CVD boat reactor accounting for the reaction

$$SiH_4 \rightleftharpoons SiH_2 + H_2$$

[2nd ed. P11-22]

Trickle Bed Reactors

CDP12-L$_B$ Hydrogenation of an unsaturated organic is carried out in a *trickle bed reactor*. [2nd Ed. P12-7]

CDP12-M$_B$ The oxidation of ethanol is carried out in a *trickle bed reactor*. [2nd Ed. P12-9]

CDP12-N$_C$ Hydrogenation of aromatics in a *trickle bed reactor* [2nd Ed. P12-8]

Fluidized Bed Reactors

CDP12-O$_C$ Open-ended fluidization problem that requires critical thinking to compare the two-phase fluid models with the three-phase bubbling bed model.

CDP12-P$_A$ Calculate reaction rates at the top and the bottom of the bed for Example R12.3-3.

CDP12-Q$_B$ Calculate the conversion for A $\rightarrow$ B in a bubbling fluidized bed.

CDP12-R$_B$ Calculate the effect of operating parameters on conversion for the reaction limited and transport limited operation.

CDP12-S$_B$ *Excellent Problem* Calculate all the parameters in Example R12-3.3 for a different reaction and different bed.

CDP12-T$_B$ Plot conversion and concentration as a function of bed height in a bubbling fluidized bed.

CDP12-U$_B$ Use RTD studies to compare bubbling bed with a fluidized bed.

CDP12-V$_B$ New problems on web and CD-ROM.

CDP12-W$_B$ Green Engineering, *www.rowan.edu/greenengineering*.

Hall of Fame

Green Engineering

SUPPLEMENTARY READING

1. There are a number of books that discuss internal diffusion in catalyst pellets; however, one of the first books that should be consulted on this and other topics on heterogeneous catalysis is

> LAPIDUS, L., and N. R. AMUNDSON, *Chemical Reactor Theory: A Review.* Upper Saddle River, N.J.: Prentice Hall, 1977.

In addition, see

> ARIS, R., *Elementary Chemical Reactor Analysis.* Upper Saddle River, N.J.: Prentice Hall, 1969, Chap. 6. One should find the references listed at the end of this reading particularly useful.

> LUSS, D., "Diffusion—Reaction Interactions in Catalyst Pellets," p. 239 in *Chemical Reaction and Reactor Engineering.* New York: Marcel Dekker, 1987.

The effects of mass transfer on reactor performance are also discussed in

> DENBIGH, K., and J. C. R. TURNER, *Chemical Reactor Theory,* 3rd ed. Cambridge: Cambridge University Press, 1984, Chap. 7.

> SATERFIELD, C. N., *Heterogeneous Catalysis in Industrial Practice,* 2nd ed. New York: McGraw-Hill, 1991.

2. Diffusion with homogeneous reaction is discussed in

> ASTARITA, G., and R. OCONE, *Special Topics in Transport Phenomena.* New York: Elsevier, 2002.

> DANCKWERTS, P. V., *Gas–Liquid Reactions.* New York: McGraw-Hill, 1970.

Gas-liquid reactor design is also discussed in

> CHARPENTIER, J. C., review article, *Trans. Inst. Chem. Eng., 60,* 131 (1982).

> SHAH, Y. T., *Gas–Liquid–Solid Reactor Design.* New York: McGraw-Hill, 1979.

3. Modeling of CVD reactors is discussed in

> HESS, D. W., K. F. JENSEN, and T. J. ANDERSON, "Chemical Vapor Deposition: A Chemical Engineering Perspective," *Rev. Chem. Eng., 3,* 97, 1985.

> JENSEN, K. F., "Modeling of Chemical Vapor Deposition Reactors for the Fabrication of Microelectronic Devices," *Chemical and Catalytic Reactor Modeling,* ACS Symp. Ser. 237, M. P. Dudokovic, P. L. Mills, eds., Washington, D.C.: American Chemical Society, 1984, p. 197.

> LEE, H. H., *Fundamentals of Microelectronics Processing.* New York: McGraw-Hill, 1990.

4. Multiphase reactors are discussed in

> RAMACHANDRAN, P. A., and R. V. CHAUDHARI, *Three-Phase Catalytic Reactors.* New York: Gordon and Breach, 1983.

> RODRIGUES, A. E., J. M. COLO, and N. H. SWEED, eds., *Multiphase Reactors,* Vol. 1: *Fundamentals.* Alphen aan den Rijn, The Netherlands: Sitjhoff and Noordhoff, 1981.

> RODRIGUES, A. E., J. M. COLO, and N. H. SWEED, eds., *Multiphase Reactors,* Vol. 2: *Design Methods.* Alphen aan den Rijn, The Netherlands: Sitjhoff and Noordhoff, 1981.

SHAH, Y. T., B. G. KELKAR, S. P. GODBOLE, and W. D. DECKWER, "Design Parameters Estimations for Bubble Column Reactors" (journal review), *AIChE J.*, *28*, 353 (1982).

TARHAN, M. O., *Catalytic Reactor Design*. New York: McGraw-Hill, 1983.

YATES, J. G., *Fundamentals of Fluidized-Bed Chemical Processes*, 3rd ed. London: Butterworth, 1983.

The following *Advances in Chemistry Series* volume discusses a number of multiphase reactors:

FOGLER, H. S., ed., *Chemical Reactors*, ACS Symp. Ser. 168. Washington, D.C.: American Chemical Society, 1981, pp. 3–255.

5. Fluidization

In addition to Kunii and Levenspiel's book, many correlations can be found in

DAVIDSON, J. F., R. CLIFF, and D. HARRISON, *Fluidization*, 2nd ed. Orlando: Academic Press, 1985.

A discussion of the different models can be found in

YATES, J. G., *Fluidized Bed Chemical Processes*. London: Butterworth-Heinemann, 1983. Also see GELDART, D. ed. *Gas Fluidization Technology*. Chichester: Wiley-Interscience, 1986.

Distributions $\mathbf{13}$
of Residence Times
for Chemical Reactors

Nothing in life is to be feared. It is only to be understood.

Marie Curie

Overview In this chapter we learn about nonideal reactors, that is, reactors that do not follow the models we have developed for ideal CSTRs, PFRs, and PBRs. In Part 1 we describe how to characterize these nonideal reactors using the residence time distribution function $E(t)$, the mean residence time t_m, the cumulative distribution function $F(t)$, and the variance σ^2. Next we evaluate $E(t)$, $F(t)$, t_m, and σ^2 for ideal reactors, so that we have a reference point as to how far our real (i.e., nonideal) reactor is off the norm from an ideal reactor. The functions $E(t)$ and $F(t)$ will be developed for ideal PPRs, CSTRs and laminar flow reactors. Examples are given for diagnosing problems with real reactors by comparing t_m and $E(t)$ with ideal reactors. We will then use these ideal curves to help diagnose and troubleshoot bypassing and dead volume in real reactors.

In Part 2 we will learn how to use the residence time data and functions to make predictions of conversion and exit concentrations. Because the residence time distribution is not unique for a given reaction system, we must use new models if we want to predict the conversion in our nonideal reactor. We present the five most common models to predict conversion and then close the chapter by applying two of these models, the segregation model and the maximum mixedness model, to single and to multiple reactions.

After studying this chapter the reader will be able to describe the cumulative $F(t)$ and external age $E(t)$ and residence-time distribution functions, and to recognize these functions for PFR, CSTR, and laminar flow reactors. The reader will also be able to apply these functions to calculate the conversion and concentrations exiting a reactor using the segregation model and the maximum mixedness model for both single and multiple reactions.

13.1 General Characteristics

The reactors treated in the book thus far—the perfectly mixed batch, the plug-flow tubular, the packed bed, and the perfectly mixed continuous tank reactors—have been modeled as ideal reactors. Unfortunately, in the real world we often observe behavior very different from that expected from the exemplar; this behavior is true of students, engineers, college professors, and chemical reactors. Just as we must learn to work with people who are not perfect, so the reactor analyst must learn to diagnose and handle chemical reactors whose performance deviates from the ideal. Nonideal reactors and the principles behind their analysis form the subject of this chapter and the next.

> We want to analyze and characterize nonideal reactor behavior.

Part 1 Characterization and Diagnostics

The basic ideas that are used in the distribution of residence times to characterize and model nonideal reactions are really few in number. The two major uses of the residence time distribution to characterize nonideal reactors are

1. To diagnose problems of reactors in operation
2. To predict conversion or effluent concentrations in existing/available reactors when a new reaction is used in the reactor

System 1   In a gas–liquid continuous-stirred tank reactor (Figure 13-1), the gaseous reactant was bubbled into the reactor while the liquid reactant was fed through an inlet tube in the reactor's side. The reaction took place at the gas–liquid interface of the bubbles, and the product was a liquid. The continuous liquid phase could be regarded as perfectly mixed, and the reaction rate was proportional to the total bubble surface area. The surface area of a particular bubble depended on the time it had spent in the reactor. Because of their different sizes, some gas bubbles escaped from the reactor almost immediately, while others spent so much time in the reactor that they were almost com-

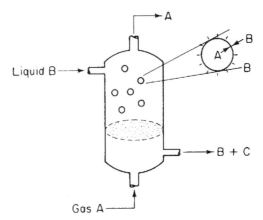

Figure 13-1 Gas–liquid reactor.

pletely consumed. The time the bubble spends in the reactor is termed the *bubble residence time*. What was important in the analysis of this reactor was not the average residence time of the bubbles but rather the residence time of each bubble (i.e., the residence time distribution). The total reaction rate was found by summing over all the bubbles in the reactor. For this sum, the distribution of residence times of the bubbles leaving the reactor was required. An understanding of residence-time distributions (RTDs) and their effects on chemical reactor performance is thus one of the necessities of the technically competent reactor analyst.

Not all molecules are spending the same time in the reactor.

System 2   A packed-bed reactor is shown in Figure 13-2. When a reactor is packed with catalyst, the reacting fluid usually does not flow through the reactor uniformly. Rather, there may be sections in the packed bed that offer little resistance to flow, and as a result a major portion of the fluid may channel through this pathway. Consequently, the molecules following this pathway do not spend as much time in the reactor as those flowing through the regions of high resistance to flow. We see that there is a distribution of times that molecules spend in the reactor in contact with the catalyst.

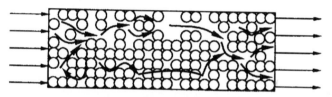

Figure 13-2   Packed-bed reactor.

System 3   In many continuous-stirred tank reactors, the inlet and outlet pipes are close together (Figure 13-3). In one operation it was desired to scale up pilot plant results to a much larger system. It was realized that some short circuiting occurred, so the tanks were modeled as perfectly mixed CSTRs with a bypass stream. In addition to short circuiting, stagnant regions (dead zones) are often encountered. In these regions there is little or no exchange of material with the well-mixed regions, and, consequently, virtually no reaction occurs

We want to find ways of determining the dead volume and amount of bypassing.

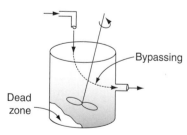

Figure 13-3   CSTR.

there. Experiments were carried out to determine the amount of the material effectively bypassed and the volume of the dead zone. A simple modification of an ideal reactor successfully modeled the essential physical characteristics of the system and the equations were readily solvable.

The three concepts
• RTD
• Mixing
• Model

Three concepts were used to describe nonideal reactors in these examples: *the distribution of residence times in the system, the quality of mixing,* and *the model used to describe the system.* All three of these concepts are considered when describing deviations from the mixing patterns assumed in ideal reactors. The three concepts can be regarded as characteristics of the mixing in nonideal reactors.

One way to order our thinking on nonideal reactors is to consider modeling the flow patterns in our reactors as either CSTRs or PFRs as a *first* approximation. In real reactors, however, nonideal flow patterns exist, resulting in ineffective contacting and lower conversions than in the case of ideal reactors. We must have a method of accounting for this nonideality, and to achieve this goal we use the next-higher level of approximation, which involves the use of *macromixing* information (RTD) (Sections 13.1 to 13.4). The next level uses microscale (*micromixing*) information to make predictions about the conversion in nonideal reactors. We address this third level of approximation in Sections 13.6 to 13.9 and in Chapter 14.

13.1.1 Residence-Time Distribution (RTD) Function

The idea of using the distribution of residence times in the analysis of chemical reactor performance was apparently first proposed in a pioneering paper by MacMullin and Weber.[1] However, the concept did not appear to be used extensively until the early 1950s, when Prof. P. V. Danckwerts[2] gave organizational structure to the subject of RTD by defining most of the distributions of interest. The ever-increasing amount of literature on this topic since then has generally followed the nomenclature of Danckwerts, and this will be done here as well.

In an ideal plug-flow reactor, all the atoms of material leaving the reactor have been inside it for exactly the same amount of time. Similarly, in an ideal batch reactor, all the atoms of materials within the reactor have been inside it for an identical length of time. The time the atoms have spent in the reactor is called the *residence time* of the atoms in the reactor.

The idealized plug-flow and batch reactors are the only two classes of reactors in which all the atoms in the reactors have the same residence time. In all other reactor types, the various atoms in the feed spend different times inside the reactor; that is, there is a distribution of residence times of the material within the reactor. For example, consider the CSTR; the feed introduced into a CSTR at any given time becomes completely mixed with the material already in the reactor. In other words, some of the atoms entering the CSTR

---

[1] R. B. MacMullin and M. Weber, Jr., *Trans. Am. Inst. Chem. Eng., 31,* 409 (1935).

[2] P. V. Danckwerts, *Chem. Eng. Sci., 2,* 1 (1953).

The "RTD": Some
molecules leave
quickly, others
overstay their
welcome.
leave it almost immediately because material is being continuously withdrawn
from the reactor; other atoms remain in the reactor almost forever because all
the material is never removed from the reactor at one time. Many of the atoms,
of course, leave the reactor after spending a period of time somewhere in the
vicinity of the mean residence time. In any reactor, the distribution of residence times can significantly affect its performance.

The *residence-time distribution* (RTD) of a reactor is a characteristic of
the mixing that occurs in the chemical reactor. There is no axial mixing in a
plug-flow reactor, and this omission is reflected in the RTD. The CSTR is thoroughly mixed and possesses a far different kind of RTD than the plug-flow
reactor. As will be illustrated later, not all RTDs are unique to a particular reactor type; markedly different reactors can display identical RTDs. Nevertheless,
the RTD exhibited by a given reactor yields distinctive clues to the type of
mixing occurring within it and is one of the most informative characterizations
of the reactor.

We will use the
RTD to
characterize
nonideal reactors.

13.2 Measurement of the RTD

The RTD is determined experimentally by injecting an inert chemical, molecule, or atom, called a *tracer*, into the reactor at some time $t = 0$ and then
measuring the tracer concentration, C, in the effluent stream as a function of
time. In addition to being a nonreactive species that is easily detectable, the
tracer should have physical properties similar to those of the reacting mixture
and be completely soluble in the mixture. It also should not adsorb on the
walls or other surfaces in the reactor. The latter requirements are needed so
that the tracer's behavior will honestly reflect that of the material flowing
through the reactor. Colored and radioactive materials along with inert gases
are the most common types of tracers. The two most used methods of injection
are *pulse input* and *step input*.

Use of tracers to
determine the RTD

13.2.1 Pulse Input Experiment

In a pulse input, an amount of tracer N_0 is suddenly injected in one shot into
the feedstream entering the reactor in as short a time as possible. The outlet
concentration is then measured as a function of time. Typical concentration–time curves at the inlet and outlet of an arbitrary reactor are shown in Figure 13-4. The effluent concentration–time curve is referred to as the *C* curve in
RTD analysis. We shall analyze the injection of a tracer pulse for a single-input
and single-output system in which *only flow* (i.e., no dispersion) carries the
tracer material across system boundaries. First, we choose an increment of
time Δt sufficiently small that the concentration of tracer, $C(t)$, exiting
between time t and $t + \Delta t$ is essentially the same. The amount of tracer material, ΔN, leaving the reactor between time t and $t + \Delta t$ is then

The *C* curve

$$\Delta N = C(t)v\,\Delta t \tag{13-1}$$

where v is the effluent volumetric flow rate. In other words, ΔN is the amount of material exiting the reactor that has spent an amount of time between t and $t + \Delta t$ in the reactor. If we now divide by the total amount of material that was injected into the reactor, N_0, we obtain

$$\frac{\Delta N}{N_0} = \frac{vC(t)}{N_0} \Delta t \tag{13-2}$$

which represents the fraction of material that has a residence time in the reactor between time t and $t + \Delta t$.

For pulse injection we define

$$E(t) = \frac{vC(t)}{N_0} \tag{13-3}$$

so that

$$\frac{\Delta N}{N_0} = E(t)\,\Delta t \tag{13-4}$$

| Interpretation of $E(t)\,dt$ |

The quantity $E(t)$ is called the **residence-time distribution function**. It is the function that describes in a quantitative manner how much time different fluid elements have spent in the reactor. The quantity $E(t)dt$ is the fraction of fluid exiting the reactor that has spent between time t and $t + dt$ inside the reactor.

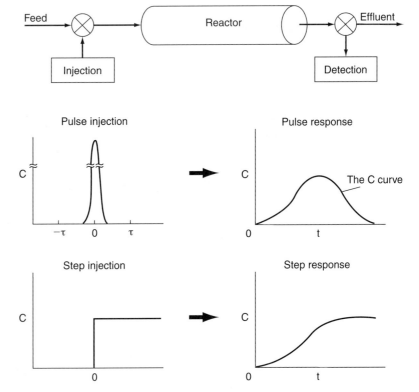

Figure 13-4   RTD measurements.

The C curve

Area = $\int_0^\infty C(t)\,dt$

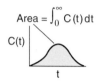

We find the RTD function, **E(t)**, from the tracer concentration **C(t)**

The E curve

Eventually all must leave

Pulse Input

If N_0 is not known directly, it can be obtained from the outlet concentration measurements by summing up all the amounts of materials, ΔN, between time equal to zero and infinity. Writing Equation (13-1) in differential form yields

$$dN = vC(t)\,dt \qquad (13\text{-}5)$$

and then integrating, we obtain

$$N_0 = \int_0^\infty vC(t)\,dt \qquad (13\text{-}6)$$

The volumetric flow rate v is usually constant, so we can define $E(t)$ as

$$E(t) = \frac{C(t)}{\displaystyle\int_0^\infty C(t)\,dt} \qquad (13\text{-}7)$$

The integral in the denominator is the area under the C curve.

An alternative way of interpreting the residence-time function is in its integral form:

$$\left[\begin{array}{c}\text{Fraction of material leaving the reactor}\\ \text{that has resided in the reactor}\\ \text{for times between } t_1 \text{ and } t_2\end{array}\right] = \int_{t_1}^{t_2} E(t)\,dt$$

We know that the fraction of all the material that has resided for a time t in the reactor between $t = 0$ and $t = \infty$ is 1; therefore,

$$\int_0^\infty E(t)\,dt = 1 \qquad (13\text{-}8)$$

The following example will show how we can calculate and interpret $E(t)$ from the effluent concentrations from the response to a pulse tracer input to a real (i.e., nonideal) reactor.

Example 13–1   Constructing the C(t) and E(t) Curves

A sample of the tracer hytane at 320 K was injected as a pulse to a reactor, and the effluent concentration was measured as a function of time, resulting in the data shown in Table E13–1.1.

TABLE E13–1.1   TRACER DATA

t (min)	0	1	2	3	4	5	6	7	8	9	10	12	14
C (g/m³)	0	1	5	8	10	8	6	4	3.0	2.2	1.5	0.6	0

The measurements represent the exact concentrations at the times listed and not average values between the various sampling tests. **(a)** Construct figures showing $C(t)$ and $E(t)$ as functions of time. **(b)** Determine both the fraction of material leaving

the reactor that has spent between 3 and 6 min in the reactor and the fraction of
material leaving that has spent between 7.75 and 8.25 min in the reactor, and **(c)**
determine the fraction of material leaving the reactor that has spent 3 min or less in
the reactor.

Solution

(a) By plotting C as a function of time, using the data in Table E13-1.1, the curve
shown in Figure E13-1.1 is obtained.

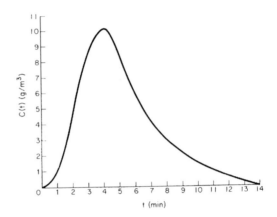

The **C** curve

Figure E13-1.1   The C curve.

To obtain the $E(t)$ curve from the $C(t)$ curve, we just divide $C(t)$ by the integral
$\int_0^\infty C(t)\,dt$, which is just the area under the C curve. Because one quadrature (integration) formula will not suffice over the entire range of data in Table E13-1.1, we
break the data into two regions, 0-10 minutes and 10 to 14 minutes. The area under
the C curve can now be found using the numerical integration formulas (A-21) and
(A-25) in Appendix A.4:

$$\int_0^\infty C(t)\,dt = \int_0^{10} C(t)\,dt + \int_{10}^{14} C(t)\,dt \qquad \text{(E13-1.1)}$$

$$\int_0^{10} C(t)\,dt = \tfrac{1}{3}[1(0) + 4(1) + 2(5) + 4(8) \qquad \text{(A-25)}$$
$$+ 2(10) + 4(8) + 2(6)$$
$$+ 4(4) + 2(3.0) + 4(2.2) + 1(1.5)]$$
$$= 47.4 \text{ g}\cdot\text{min/m}^3$$

$$\int_{10}^{14} C(t)\,dt = \tfrac{2}{3}[1.5 + 4(0.6) + 0] = 2.6 \text{ g}\cdot\text{min/m}^3 \qquad \text{(A-21)}$$

$$\int_0^\infty C(t)\,dt = 50.0 \text{ g}\cdot\text{min/m}^3 \qquad \text{(E13-1.2)}$$

We now calculate

$$E(t) = \frac{C(t)}{\int_0^\infty C(t)\, dt} = \frac{C(t)}{50 \text{ g} \cdot \text{min/m}^3} \tag{E13-1.3}$$

with the following results:

TABLE E13–1.2 $C(t)$ AND $E(t)$

t (min)	1	2	3	4	5	6	7	8	9	10	12	14
$C(t)$ (g/m^3)	1	5	8	10	8	6	4	3	2.2	1.5	0.6	0
$E(t)$ (min^{-1})	0.02	0.1	0.16	0.2	0.16	0.12	0.08	0.06	0.044	0.03	0.012	0

(b) These data are plotted in Figure E13-1.2. The shaded area represents the fraction of material leaving the reactor that has resided in the reactor between 3 and 6 min.

The *E* curve

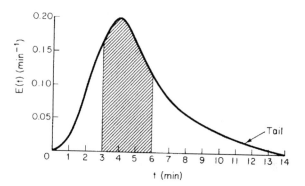

Figure E13-1.2   Analyzing the E curve.

Using Equation (A-22) in Appendix A.4:

$$\int_3^6 E(t)\, dt = \text{shaded area} \tag{A-22}$$

$$= \tfrac{3}{8}\Delta t (f_1 + 3f_2 + 3f_3 + f_4)$$

$$= \tfrac{3}{8}(1)[0.16 + 3(0.2) + 3(0.16) + 0.12] = 0.51$$

Evaluating this area, we find that 51% of the material leaving the reactor spends between 3 and 6 min in the reactor.

Because the time between 7.75 and 8.25 min is very small relative to a time scale of 14 min, we shall use an alternative technique to determine this fraction to reinforce the interpretation of the quantity $E(t)\,dt$. The average value of $E(t)$ between these times is 0.06 min^{-1}:

$$E(t)\, dt = (0.06 \text{ min}^{-1})(0.5 \text{ min}) = 0.03$$

The tail

Consequently, 3.0% of the fluid leaving the reactor has been in the reactor between 7.75 and 8.25 min. The long-time portion of the $E(t)$ curve is called the *tail*. In this example the tail is that portion of the curve between *say* 10 and 14 min.

(c) Finally, we shall consider the fraction of material that has been in the reactor for a time t or less, that is, the fraction that has spent between 0 and t minutes in the reactor. This fraction is just the shaded area under the curve up to $t = t$ minutes. This area is shown in Figure E13-1.3 for $t = 3$ min. Calculating the area under the curve, we see that 20% of the material has spent 3 min *or less* in the reactor.

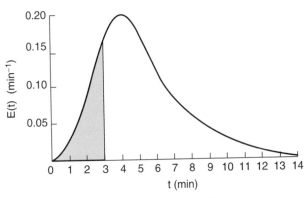

Figure E13-1.3　Analyzing the E curve.

The principal difficulties with the pulse technique lie in the problems connected with obtaining a reasonable pulse at a reactor's entrance. The injection must take place over a period which is very short compared with residence times in various segments of the reactor or reactor system, and there must be a negligible amount of dispersion between the point of injection and the entrance to the reactor system. If these conditions can be fulfilled, this technique represents a simple and direct way of obtaining the RTD.

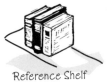

Drawbacks to the pulse injection to obtain the RTD

Reference Shelf

There are problems when the concentration–time curve has a long tail because the analysis can be subject to large inaccuracies. This problem principally affects the denominator of the right-hand side of Equation (13-7) [i.e., the integration of the $C(t)$ curve]. It is desirable to extrapolate the tail and analytically continue the calculation. The tail of the curve may sometimes be approximated as an exponential decay. The inaccuracies introduced by this assumption are very likely to be much less than those resulting from either truncation or numerical imprecision in this region. Methods of fitting the tail are described in the *Professional Reference Shelf* 13 R.1.

13.2.2 Step Tracer Experiment

Now that we have an understanding of the meaning of the RTD curve from a pulse input, we will formulate a more general relationship between a time-varying tracer injection and the corresponding concentration in the effluent. We shall state without development that the output concentration from a vessel is related to the input concentration by the convolution integral:[3]

---

[3] A development can be found in O. Levenspiel, *Chemical Reaction Engineering*, 2nd ed. (New York: Wiley, 1972), p. 263.

$$C_{out}(t) = \int_0^t C_{in}(t - t')E(t')\,dt \qquad (13\text{-}9)$$

The inlet concentration most often takes the form of either a perfect *pulse input* (Dirac delta function), *imperfect pulse injection* (see Figure 13-4), or a *step input*.

Just as the RTD function $E(t)$ can be determined directly from a pulse input, the cumulative distribution $F(t)$ can be determined directly from a step input. We will now analyze a *step input* in the tracer concentration for a system with a constant volumetric flow rate. Consider a constant rate of tracer addition to a feed that is initiated at time $t = 0$. Before this time no tracer was added to the feed. Stated symbolically, we have

Step Input

C_{in}

$$C_0(t) = \begin{cases} 0 & t < 0 \\ (C_0)\ \text{constant} & t \geq 0 \end{cases}$$

C_{out}

The concentration of tracer in the feed to the reactor is kept at this level until the concentration in the effluent is indistinguishable from that in the feed; the test may then be discontinued. A typical outlet concentration curve for this type of input is shown in Figure 13-4.

Because the inlet concentration is a constant with time, C_0, we can take it outside the integral sign, that is,

$$C_{out} = C_0 \int_0^t E(t')\,dt'$$

Dividing by C_0 yields

$$\left[\frac{C_{out}}{C_0}\right]_{step} = \int_0^t E(t')\,dt' = F(t)$$

$$F(t) = \left[\frac{C_{out}}{C_0}\right]_{step} \qquad (13\text{-}10)$$

We differentiate this expression to obtain the RTD function $E(t)$:

$$E(t) = \frac{d}{dt}\left[\frac{C(t)}{C_0}\right]_{step} \qquad (13\text{-}11)$$

The positive step is usually easier to carry out experimentally than the pulse test, and it has the additional advantage that the total amount of tracer in the feed over the period of the test does not have to be known as it does in the pulse test. One possible drawback in this technique is that it is sometimes difficult to maintain a constant tracer concentration in the feed. Obtaining the RTD from this test also involves differentiation of the data and presents an additional and probably more serious drawback to the technique, because differentiation of data can, on occasion, lead to large errors. A third problem lies with the large amount of tracer required for this test. If the tracer is very expensive, a pulse test is almost always used to minimize the cost.

Advantages and drawbacks to the step injection

Other tracer techniques exist, such as negative step (i.e., elution), frequency-response methods, and methods that use inputs other than steps or pulses. These methods are usually much more difficult to carry out than the ones presented and are not encountered as often. For this reason they will not be treated here, and the literature should be consulted for their virtues, defects, and the details of implementing them and analyzing the results. A good source for this information is Wen and Fan.[4]

13.3 Characteristics of the RTD

From $E(t)$ we can learn how long different molecules have been in the reactor.

Sometimes $E(t)$ is called the *exit-age distribution function*. If we regard the "age" of an atom as the time it has resided in the reaction environment, then $E(t)$ concerns the age distribution of the effluent stream. It is the most used of the distribution functions connected with reactor analysis because it characterizes the lengths of time various atoms spend at reaction conditions.

13.3.1 Integral Relationships

The fraction of the exit stream that has resided in the reactor for a period of time shorter than a given value t is equal to the sum over all times less than t of $E(t)\,\Delta t$, or expressed continuously,

The cumulative RTD function $F(t)$

$$\int_0^t E(t)\,dt = \begin{bmatrix} \text{Fraction of effluent} \\ \text{that has been in reactor} \\ \text{for less than time } t \end{bmatrix} = F(t) \qquad (13\text{-}12)$$

Analogously, we have

$$\int_t^\infty E(t)\,dt = \begin{bmatrix} \text{Fraction of effluent} \\ \text{that has been in reactor} \\ \text{for longer than time } t \end{bmatrix} = 1 - F(t) \qquad (13\text{-}13)$$

Because t appears in the integration limits of these two expressions, Equations (13-12) and (13-13) are both functions of time. Danckwerts[5] defined Equation (13-12) as a *cumulative distribution function and called it $F(t)$*. We can calculate $F(t)$ at various times t from the area under the curve of an $E(t)$ versus t plot. For example, in Figure E13-1.3 we saw that $F(t)$ at 3 min was 0.20, meaning that 20% of the molecules spent 3 min or less in the reactor. Similarly, using Figure E13-1.3 we calculate $F(t) = 0.4$ at 4 minutes. We can continue in this manner to construct $F(t)$. The shape of the $F(t)$ curve is shown in Figure 13-5. One notes from this curve that 80% $[F(t)]$ of the molecules spend 8 min or less in the reactor, and 20% of the molecules $[1 - F(t)]$ spend longer than 8 min in the reactor.

[4] C. Y. Wen and L. T. Fan, *Models for Flow Systems and Chemical Reactors* (New York: Marcel Dekker, 1975).

[5] P. V. Danckwerts, *Chem. Eng. Sci.*, 2, 1 (1953).

The *F* curve

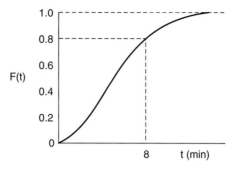

Figure 13-5   Cumulative distribution curve, $F(t)$.

The *F* curve is another function that has been defined as the normalized response to a particular input. Alternatively, Equation (13-12) has been used as a definition of $F(t)$, and it has been stated that as a result it can be obtained as the response to a positive-step tracer test. Sometimes the *F* curve is used in the same manner as the RTD in the modeling of chemical reactors. An excellent example is the study of Wolf and White,[6] who investigated the behavior of screw extruders in polymerization processes.

13.3.2 Mean Residence Time

In previous chapters treating ideal reactors, a parameter frequently used was the space time or average residence time τ, which was defined as being equal to V/v. It will be shown that, in the absence of dispersion, and for constant volumetric flow ($v = v_0$) no matter what RTD exists for a particular reactor, ideal or nonideal, this nominal space time, τ, is equal to the mean residence time, t_m.

As is the case with other variables described by distribution functions, the mean value of the variable is equal to the first moment of the RTD function, $E(t)$. Thus the mean residence time is

The first moment gives the average time the effluent molecules spent in the reactor.

$$t_m = \frac{\displaystyle\int_0^\infty tE(t)\,dt}{\displaystyle\int_0^\infty E(t)\,dt} = \int_0^\infty tE(t)\,dt \tag{13-14}$$

We now wish to show how we can determine the total reactor volume using the cumulative distribution function.

---

[6] D. Wolf and D. H. White, *AIChE J.*, **22**, 122 (1976).

What we are going to do now is prove $t_m = \tau$ for constant volumetric flow, $v = v_0$. You can skip what follows and go directly to Equation (13–21) if you can accept this result.

Consider the following situation: We have a reactor completely filled with maize molecules. At time $t = 0$ we start to inject blue molecules to replace the maize molecules that currently fill the reactor. Initially, the reactor volume V is equal to the volume occupied by the maize molecules. Now, in a time dt, the volume of molecules that will leave the reactor is $(v\,dt)$. The fraction of these molecules that have been in the reactor a time t or greater is $[1 - F(t)]$. Because only the maize molecules have been in the reactor a time t or greater, the volume of maize molecules, dV, leaving the reactor in a time dt is

$$dV = (v\,dt)[1 - F(t)] \tag{13-15}$$

If we now sum up all of the maize molecules that have left the reactor in time $0 < t < \infty$, we have

All we are doing here is proving that the space time and mean residence time are equal.

$$V = \int_0^\infty v\,[1 - F(t)]\,dt \tag{13-16}$$

Because the volumetric flow rate is constant,[†]

$$V = v \int_0^\infty [1 - F(t)]\,dt \tag{13-17}$$

Using the integration-by-parts relationship gives

$$\int x\,dy = xy - \int y\,dx$$

and dividing by the volumetric flow rate gives

$$\frac{V}{v} = t[1 - F(t)]\,\Big|_0^\infty + \int_0^1 t\,dF \tag{13-18}$$

At $t = 0$, $F(t) = 0$; and as $t \to \infty$, then $[1 - F(t)] = 0$. The first term on the right-hand side is zero, and the second term becomes

$$\frac{V}{v} = \tau = \int_0^1 t\,dF \tag{13-19}$$

However, $dF = E(t)\,dt$; therefore,

$$\boxed{\tau = \int_0^\infty tE(t)\,dt} \tag{13-20}$$

The right-hand side is just the mean residence time, and we see that the mean residence time is just the space time τ:

[†] *Note:* For gas-phase reactions at constant temperature and no pressure drop $t_m = \tau/(1 + \varepsilon X)$.

$\tau = t_m$, Q.E.D.

$$\boxed{\tau = t_m}$$ (13-21)

and no change in volumetric flow rate. For gas-phase reactions, this means no pressure drop, isothermal operation, and no change in the total number of moles (i.e., $\varepsilon \equiv 0$, as a result of reaction).

End of proof!

This result is true *only* for a *closed system* (i.e., no dispersion across boundaries; see Chapter 14). The exact reactor volume is determined from the equation

$$\boxed{V = vt_m}$$ (13-22)

13.3.3 Other Moments of the RTD

It is very common to compare RTDs by using their moments instead of trying to compare their entire distributions (e.g., Wen and Fan[7]). For this purpose, three moments are normally used. The first is the mean residence time. The second moment commonly used is taken about the mean and is called the variance, or square of the standard deviation. It is defined by

The second moment about the mean is the variance.

$$\boxed{\sigma^2 = \int_0^\infty (t - t_m)^2 E(t)\, dt}$$ (13-23)

The magnitude of this moment is an indication of the "spread" of the distribution; the greater the value of this moment is, the greater a distribution's spread will be.

The third moment is also taken about the mean and is related to the *skewness*. The skewness is defined by

The two parameters most commonly used to characterize the RTD are τ and σ^2

$$s^3 = \frac{1}{\sigma^{3/2}} \int_0^\infty (t - t_m)^3 E(t)\, dt$$ (13-24)

The magnitude of this moment measures the extent that a distribution is skewed in one direction or another in reference to the mean.

Rigorously, for complete description of a distribution, all moments must be determined. Practically, these three are usually sufficient for a reasonable characterization of an RTD.

Example 13–2   Mean Residence Time and Variance Calculations

Calculate the mean residence time and the variance for the reactor characterized in Example 13-1 by the RTD obtained from a pulse input at 320 K.

Solution

First, the mean residence time will be calculated from Equation (13-14):

[7] C. Y. Wen and L. T. Fan, *Models for Flow Systems and Chemical Reactors* (New York: Decker, 1975), Chap. 11.

$$t_m = \int_0^\infty tE(t)\, dt \qquad \text{(E13-2.1)}$$

The area under the curve of a plot of $tE(t)$ as a function of t will yield t_m. Once the mean residence time is determined, the variance can be calculated from Equation (13-23):

$$\sigma^2 = \int_0^\infty (t - t_m)^2 E(t)\, dt \qquad \text{(E13-2.2)}$$

To calculate t_m and σ^2, Table E13-2.1 was constructed from the data given and interpreted in Example 13-1. One quadrature formula will not suffice over the entire range. Therefore, we break the integral up into two regions, 0 to 10 min and 10 to 14 (minutes), i.e., infinity (∞).

$$t_m = \int_0^\infty tE(t)\, dt = \int_0^{10} tE(t)\, dt + \int_{10}^\infty tE(t)\, dt$$

Starting with Table E13-1.2 in Example 13-1, we can proceed to calculate $tE(t)$, $(t - t_m)$ and $(t - t_m)^2 E(t)$ and $t^2 E(t)$ shown in Table E13-2.1.

TABLE E13-2.1.   CALCULATING $E(t)$, t_m, AND σ^2

t	$C(t)$	$E(t)$	$tE(t)$	$(t - t_m)^a$	$(t - t_m)^2 E(t)^a$	$t^2 E(t)^a$
0	0	0	0	−5.15	0	0
1	1	0.02	0.02	−4.15	0.34	0.02
2	5	0.10	0.20	−3.15	0.992	0.4
3	8	0.16	0.48	−2.15	0.74	1.44
4	10	0.20	0.80	−1.15	0.265	3.2
5	8	0.16	0.80	−0.15	0.004	4.0
6	6	0.12	0.72	0.85	0.087	4.32
7	4	0.08	0.56	1.85	0.274	3.92
8	3	0.06	0.48	2.85	0.487	3.84
9	2.2	0.044	0.40	3.85	0.652	3.56
10	1.5	0.03	0.30	4.85	0.706	3.0
12	0.6	0.012	0.14	6.85	0.563	1.73
14	0	0	0	8.85	0	0

[a] The last two columns are completed after the mean residence time (t_m) is found.

Again, using the numerical integration formulas (A-25) and (A-21) in Appendix A.4, we have

$$t_m = \int_0^h f(x)dx = \frac{h_1}{3}(f_1 + 4f_2 + 2f_3 + 4f_4 + \cdots + 4f_{n-1} + f_n) \qquad \text{(A-25)}$$

$$+ \frac{h_2}{3}(f_{n+1} + 4f_{n+2} + f_{n+3}) \qquad \text{(A-21)}$$

$$t_m = \tfrac{1}{3}[1(0) + 4(0.02) + 2(0.2) + 4(0.48) + 2(0.8) + 4(0.8)$$

$$+ 2(0.72) + 4(0.56) + 2(0.48) + 4(0.40) + 1(0.3)]$$

Numerical integration to find the mean residence time, t_m

$$+ \tfrac{2}{3}[0.3 + 4(0.14) + 0]$$

$$= 4.58 + 0.573 = 5.15 \text{ min}$$

Calculating the
mean residence
time,

$$\tau = t_m = \int_0^\infty tE(t)\,dt$$

Note: One could also use the spreadsheets in Polymath or Excel to formulate Table E13-2.1 and to calculate the mean residence time t_m and variance σ.

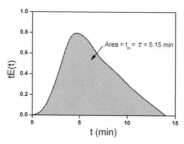

Figure E13-2.1   Calculating the mean residence time.

Plotting $tE(t)$ versus t we obtain Figure E13-2.1. The area under the curve is 5.15 min.

$$\boxed{t_m = 5.15 \text{ min}}$$

Calculating the
variance,

$$\sigma^2 = \int_0^\infty (t - t_m)^2 E(t)\,dt$$

$$\sigma^2 = \int_0^\infty t^2 E(t)\,dt - t_m^2$$

Now that the mean residence time has been determined, we can calculate the variance by calculating the area under the curve of a plot of $(t - t_m)^2 E(t)$ as a function of t (Figure E13-2.2[a]). The area under the curve(s) is 6.11 min².

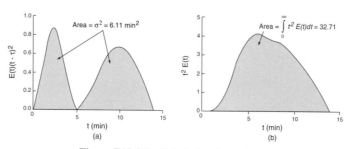

Figure E13-2.2   Calculating the variance.

One could also
use Polymath or
Excel to make
these calculations.

Expanding square term in Equation (13-23)

$$\sigma^2 = \int_0^\infty t^2 E(t)\,dt - 2t_m \int_0^\infty tE(t) + t_m^2 \int_0^\infty E(t)\,dt \qquad \text{(E13-2.2)}$$

$$= \int_0^\infty t^2 E(t)\,dt - 2t_m^2 + t_m^2$$

$$\boxed{\sigma^2 = \int_0^\infty t^2 E(t)\,dt - t_m^2} \qquad \text{(E13-2.3)}$$

We will use quadrature formulas to evaluate the integral using the data (columns 1 and 7) in Table E13-2.1. Integrating between 1 and 10 minutes and 10 and 14 minutes using the same form as Equation (E13-2.3)

$$\int_0^\infty t^2 E(t)\,dt = \int_0^{10} t^2 E(t)\,dt + \int_0^{14} t^2 E(t)\,dt$$

$$= \frac{1}{3}[0 + 4(0.02) + 2(0.4) + 4(1.44) + 2(3.2)$$

$$+4(4.0) + 2(4.32) + 4(3.92) + 2(3.84)$$

$$+4(3.56) + 3.0] + \frac{2}{3}[3.0 + 4(1.73) + 0]\ \text{min}^2$$

$$= 32.71\ \text{min}^2$$

This value is also the shaded area under the curve in Figure E13-2.2(b).

$$\boxed{\sigma^2 = \int_0^\infty t^2 E(t)\,dt - t_m^2 = 32.71\ \text{min}^2 - (5.15\ \text{min})^2 = 6.19\ \text{min}^2}$$

The square of the standard deviation is $\sigma^2 = 6.19\ \text{min}^2$, so $\sigma = 2.49$ min.

13.3.4 Normalized RTD Function, $E(\Theta)$

Frequently, a normalized RTD is used instead of the function $E(t)$. If the parameter Θ is defined as

$$\Theta \equiv \frac{t}{\tau} \qquad (13\text{-}25)$$

a dimensionless function $E(\Theta)$ can be defined as

$$E(\Theta) \equiv \tau E(t) \qquad (13\text{-}26)$$

and plotted as a function of Θ. The quantity Θ represents the number of reactor volumes of fluid based on entrance conditions that have flowed through the reactor in time t.

Why we use a normalized RTD

The purpose of creating this normalized distribution function is that the flow performance inside reactors of different sizes can be compared directly. For example, if the normalized function $E(\Theta)$ is used, *all* perfectly mixed CSTRs have numerically the same RTD. If the simple function $E(t)$ is used, numerical values of $E(t)$ can differ substantially for different CSTRs. As will be shown later, for a perfectly mixed CSTR,

$E(t)$ for a CSTR

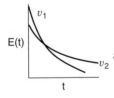

$v_1 > v_2$

$$\boxed{E(t) = \frac{1}{\tau} e^{-t/\tau}} \qquad (13\text{-}27)$$

and therefore

$$\boxed{E(\Theta) = \tau E(t) = e^{-\Theta}} \qquad (13\text{-}28)$$

From these equations it can be seen that the value of $E(t)$ at identical times can be quite different for two different volumetric flow rates, say v_1 and v_2. But for

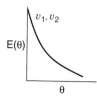

the same value of Θ, the value of $E(\Theta)$ is the same irrespective of the size of a perfectly mixed CSTR.

It is a relatively easy exercise to show that

$$\int_0^\infty E(\Theta)\, d\Theta = 1 \tag{13-29}$$

and is recommended as a 93-s divertissement.

13.3.5 Internal-Age Distribution, $I(\alpha)$

Tombstone jail
How long have you
been here? $I(\alpha)$
When do you
expect to get out?

Although this section is not a prerequisite to the remaining sections, the internal-age distribution is introduced here because of its close analogy to the external-age distribution. We shall let α represent the age of a molecule inside the reactor. The internal-age distribution function $I(\alpha)$ is a function such that $I(\alpha)\Delta\alpha$ is the fraction of material *inside the reactor* that has been inside the reactor for a period of time between α and $\alpha + \Delta\alpha$. It may be contrasted with $E(\alpha)\Delta\alpha$, which is used to represent the material *leaving the reactor* that has spent a time between α and $\alpha + \Delta\alpha$ in the reaction zone; $I(\alpha)$ characterizes the time the material has been (and still is) in the reactor at a particular time. The function $E(\alpha)$ is viewed outside the reactor and $I(\alpha)$ is viewed inside the reactor. In unsteady-state problems it can be important to know what the particular state of a reaction mixture is, and $I(\alpha)$ supplies this information. For example, in a catalytic reaction using a catalyst whose activity decays with time, the internal age distribution of the catalyst in the reactor $I(\alpha)$ is of importance and can be of use in modeling the reactor.

Reference Shelf

The internal-age distribution is discussed further on the Professional Reference Shelf where the following relationships between the cumulative internal age distribution $I(\alpha)$ and the cumulative external age distribution $F(\alpha)$

$$I(\alpha) = (1 - F(\alpha))/\tau \tag{13-30}$$

and between $E(t)$ and $I(t)$

$$E(\alpha) = -\frac{d}{d\alpha}[\tau I(\alpha)] \tag{13-31}$$

are derived. For a CSTR it is shown that the internal age distribution function is

$$I(\alpha) = -\frac{1}{\tau}e^{-\alpha/\tau}$$

13.4 RTD in Ideal Reactors

13.4.1 RTDs in Batch and Plug-Flow Reactors

The RTDs in plug-flow reactors and ideal batch reactors are the simplest to consider. All the atoms leaving such reactors have spent precisely the same amount of time within the reactors. The distribution function in such a case is a spike of infinite height and zero width, whose area is equal to 1; the spike occurs at $t = V/v = \tau$, or $\Theta = 1$. Mathematically, this spike is represented by the Dirac delta function:

$$\boxed{E(t) = \delta(t - \tau)} \tag{13-32}$$

The Dirac delta function has the following properties:

$$\delta(x) = \begin{cases} 0 & \text{when } x \neq 0 \\ \infty & \text{when } x = 0 \end{cases} \tag{13-33}$$

$$\int_{-\infty}^{\infty} \delta(x)\, dx = 1 \tag{13-34}$$

$$\int_{-\infty}^{\infty} g(x)\, \delta(x - \tau)\, dx = g(\tau) \tag{13-35}$$

To calculate τ the mean residence time, we set $g(x) = t$

$$t_m = \int_0^{\infty} tE(t)\, dt = \int t\,\delta(t - \tau)\, dt = \tau \tag{13-36}$$

But we already knew this result. To calculate the variance we set, $g(t) = (t - \tau)^2$, and the variance, σ^2, is

$$\sigma^2 = \int_0^{\infty} (t-\tau)^2\, \delta(t - \tau)\, dt = 0 \tag{13-37}$$

All material spends exactly a time τ in the reactor, there is no variance! The cumulative distribution function $F(t)$ is

$$F(t) = \int_0^t E(t)dt = \int_0^t \delta(t - \tau)dt$$

The $E(t)$ function is shown in Figure 13-6(a), and $F(t)$ is shown in Figure 13-6(b).

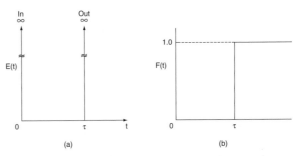

Figure 13-6   Ideal plug-flow response to a pulse tracer input.

13.4.2 Single-CSTR RTD

In an ideal CSTR the concentration of any substance in the effluent stream is identical to the concentration throughout the reactor. Consequently, it is possible to obtain the RTD from conceptual considerations in a fairly straightforward manner. A material balance on an inert tracer that has been injected as a pulse at time $t = 0$ into a CSTR yields for $t > 0$

From a tracer balance we can determine $E(t)$.

$$\text{In} - \text{Out} = \text{Accumulation}$$

$$\overset{\frown}{0} - \overset{\frown}{vC} = \overset{\frown}{V \frac{dC}{dt}} \qquad (13\text{-}38)$$

Because the reactor is perfectly mixed, C in this equation is the concentration of the tracer either in the effluent or within the reactor. Separating the variables and integrating with $C = C_0$ at $t = 0$ yields

$$C(t) = C_0 e^{-t/\tau} \qquad (13\text{-}39)$$

This relationship gives the concentration of tracer in the effluent at any time t.

To find $E(t)$ for an ideal CSTR, we first recall Equation (13-7) and then substitute for $C(t)$ using Equation (13-39). That is,

$$E(t) = \frac{C(t)}{\displaystyle\int_0^\infty C(t)\,dt} = \frac{C_0 e^{-t/\tau}}{\displaystyle\int_0^\infty C_0 e^{-t/\tau}\,dt} = \frac{e^{-t/\tau}}{\tau} \qquad (13\text{-}40)$$

Evaluating the integral in the denominator completes the derivation of the RTD for an ideal CSTR given by Equations (13-27) and (13-28):

$E(t)$ and $E(\Theta)$ for a CSTR

$$\boxed{\begin{aligned} E(t) &= \frac{e^{-t/\tau}}{\tau} \\[2mm] E(\Theta) &= e^{-\Theta} \end{aligned}}$$

$\qquad (13\text{-}27)$

$\qquad (13\text{-}28)$

Recall $\Theta = t/\tau$ and $E(\Theta) = \tau E(t)$.

Response of an ideal CSTR

$\boxed{\begin{aligned} E(\Theta) &= e^{-\Theta} \\ F(\Theta) &= 1 - e^{-\Theta} \end{aligned}}$

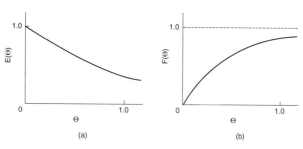

(a)   (b)

Figure 13-7 $E(\Theta)$ and $F(\Theta)$ for an Ideal CSTR.

The cumulative distribution $F(\Theta)$ is

$$F(\Theta) = \int_0^\Theta E(\Theta)d\Theta = 1 - e^{-\Theta}$$

The $E(\Theta)$ and $F(\Theta)$ functions for an Ideal CSTR are shown in Figure 13-7 (a) and (b), respectively.

Earlier it was shown that for a constant volumetric flow rate, the mean residence time in a reactor is equal to V/υ, or τ. This relationship can be shown in a simpler fashion for the CSTR. Applying the definition of a mean residence time to the RTD for a CSTR, we obtain

$$t_m = \int_0^\infty tE(t)\,dt = \int_0^\infty \frac{t}{\tau} e^{-t/\tau}\,dt = \tau \tag{13-20}$$

Thus the nominal holding time (space time) $\tau = V/\upsilon$ is also the mean residence time that the material spends in the reactor.

The second moment about the mean is a measure of the spread of the distribution about the mean. The variance of residence times in a perfectly mixed tank reactor is (let $x = t/\tau$)

For a perfectly mixed CSTR: $t_m = \tau$ and $\sigma = \tau$.

$$\sigma^2 = \int_0^\infty \frac{(t-\tau)^2}{\tau} e^{-t/\tau}\,dt = \tau^2 \int_0^\infty (x-1)^2 e^{-x}\,dx = \tau^2 \tag{13-41}$$

Then $\sigma = \tau$. The standard deviation is the square root of the variance. For a CSTR, the standard deviation of the residence-time distribution is as large as the mean itself!!

13.4.3 Laminar Flow Reactor (LFR)

Before proceeding to show how the RTD can be used to estimate conversion in a reactor, we shall derive $E(t)$ for a laminar flow reactor. For laminar flow in a tubular reactor, the velocity profile is parabolic, with the fluid in the center of the tube spending the shortest time in the reactor. A schematic diagram of the fluid movement after a time t is shown in Figure 13-8. The figure at the left shows how far down the reactor each concentric fluid element has traveled after a time t.

Molecules near the center spend a shorter time in the reactor than those close to the wall.

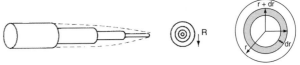

Figure 13-8   Schematic diagram of fluid elements in a laminar flow reactor.

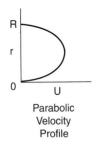

Parabolic Velocity Profile

The velocity profile in a pipe of outer radius R is

$$U = U_{max}\left[1 - \left(\frac{r}{R}\right)^2\right] = 2U_{avg}\left[1 - \left(\frac{r}{R}\right)^2\right] = \frac{2\upsilon_0}{\pi R^2}\left[1 - \left(\frac{r}{R}\right)^2\right] \tag{13-42}$$

where U_{max} is the centerline velocity and U_{avg} is the average velocity through the tube. U_{avg} is just the volumetric flow rate divided by the cross-sectional area.

The time of passage of an element of fluid at a radius r is

$$t(r) = \frac{L}{U(r)} = \frac{\pi R^2 L}{v_0} \frac{1}{2[1-(r/R)^2]}$$

$$= \frac{\tau}{2[1-(r/R)^2]}$$

(13-43)

The volumetric flow rate of fluid out between r and $(r + dr)$, dv, is

$$dv = U(r)\, 2\pi r dr$$

The fraction of total fluid passing between r and $(r + dr)$ is dv/v_0, i.e.

$$\frac{dv}{v_0} = \frac{U(r)2(\pi r dr)}{v_0}$$

(13-44)

The fraction of fluid between r and $(r + dr)$ that has a flow rate between v and $(v + dv)$ spends a time between t and $(t + dt)$ in the reactor is

$$E(t)dt = \frac{dv}{v_0}$$

(13-45)

We now need to relate the fluid fraction [Equation (13-45)] to the fraction of fluid spending between time t and $t + dt$ in the reactor. First we differentiate Equation (13-43):

$$dt = \frac{\tau}{2R^2} \frac{2r\, dr}{[1-(r/R)^2]^2} = \frac{4}{\tau R^2} \left\{ \frac{\tau/2}{[1-(r/R)^2]} \right\}^2 r\, dr$$

and then substitute for t using Equation (13-43) to yield

$$dt = \frac{4t^2}{\tau R^2} r\, dr$$

(13-46)

Combining Equations (13-44) and (13-46), and then using Equation (13-43) for $U(r)$, we now have the fraction of fluid spending between time t and $t + dt$ in the reactor:

$$E(t)dt = \frac{dv}{v_0} = \frac{L}{t} \left(\frac{2\pi r\, dr}{v_0} \right) = \frac{L}{t} \left(\frac{2\pi}{v_0} \right) \frac{\tau R^2}{4t^2} dt = \frac{\tau^2}{2t^3} dt$$

$$E(t) = \frac{\tau^2}{2t^3}$$

The minimum time the fluid may spend in the reactor is

$$t = \frac{L}{U_{max}} = \frac{L}{2U_{avg}}\left(\frac{\pi R^2}{\pi R^2}\right) = \frac{V}{2v_0} = \frac{\tau}{2}$$

Consequently, the complete RTD function for a laminar flow reactor is

E(t) for a laminar flow reactor

$$E(t) = \begin{cases} 0 & t < \dfrac{\tau}{2} \\[2mm] \dfrac{\tau^2}{2t^3} & t \geq \dfrac{\tau}{2} \end{cases}$$

(13-47)

The cumulative distribution function for $t \geq \tau/2$ is

$$F(t) = \int_0^t E(t)dt = 0 + \int_{\tau/2}^t E(t)dt = \int_{\tau/2}^t \frac{\tau^2}{2t^3}\,dt = \frac{\tau^2}{2}\int_{\tau/2}^t \frac{dt}{t^3} = 1 - \frac{\tau^2}{4t^2}$$

(13-48)

The mean residence time t_m is

For LFR $t_m = \tau$

$$t_m = \int_{\tau/2}^{\infty} tE(t)\,dt = \frac{\tau^2}{2}\int_{\tau/2}^{\infty}\frac{dt}{t^2}$$

$$= \frac{\tau^2}{2}\left[-\frac{1}{t}\right]_{\tau/2}^{\infty} = \tau$$

This result was shown previously to be true for any reactor. The mean residence time is just the space time τ.

The dimensionless form of the RTD function is

Normalized RTD function for a laminar flow reactor

$$E(\Theta) = \begin{cases} 0 & \Theta < 0.5 \\[2mm] \dfrac{1}{2\Theta^3} & \Theta \geq 0.5 \end{cases}$$

(13-49)

and is plotted in Figure 13-9.

The dimensionless cumulative distribution, $F(\Theta)$ for $\Theta \geq 1/2$, is

$$F(\Theta) = 0 + \int_{\frac{1}{2}}^{\Theta} E(\Theta)d\Theta = \int_{\frac{1}{2}}^{\Theta} \frac{d\Theta}{2\Theta^3} = \left(1 - \frac{1}{4\Theta^2}\right)$$

$$F(\Theta) = \begin{cases} 0 & \Theta < \dfrac{1}{2} \\[2mm] \left(1 - \dfrac{1}{4\Theta^2}\right) & \Theta \geq \dfrac{1}{2} \end{cases}$$

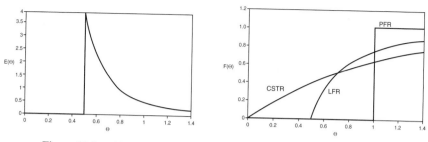

Figure 13-9   (a) $E(\Theta)$ for an LFR; (b) $F(\Theta)$ for a PFR, CSTR, and LFR.

Figure 13-9(a) shows $E(\Theta)$ for a laminar flow reactor (LFR), while Figure 9-13(b) compares $F(\Theta)$ for a PFR, CSTR, and LFR.

Experimentally injecting and measuring the tracer in a laminar flow reactor can be a difficult task if not a nightmare. For example, if one uses as a tracer chemicals that are photo-activated as they enter the reactor, the analysis and interpretation of $E(t)$ from the data become much more involved.[8]

13.5 Diagnostics and Troubleshooting

13.5.1 General Comments

As discussed in Section 13.1, the RTD can be used to diagnose problems in existing reactors. As we will see in further detail in Chapter 14, the RTD functions $E(t)$ and $F(t)$ can be used to model the real reactor as combinations of ideal reactors.

Figure 13-10 illustrates typical RTDs resulting from different nonideal reactor situations. Figures 13-10(a) and (b) correspond to nearly ideal PFRs and CSTRs, respectively. In Figure 13-10(d) one observes that a principal peak occurs at a time smaller than the space time ($\tau = V/v_0$) (i.e., early exit of fluid) and also that some fluid exits at a time greater than space-time τ. This curve could be representative of the RTD for a packed-bed reactor with channeling and dead zones. A schematic of this situation is shown in Figure 13-10(c). Figure 13-10(f) shows the RTD for the nonideal CSTR in Figure 13-10(e), which has dead zones and bypassing. The dead zone serves to reduce the effective reactor volume, so the active reactor volume is smaller than expected.

[8] D. Levenspiel, *Chemical Reaction Engineering,* 3rd ed. (New York: Wiley, 1999), p. 342.

RTDs that are commonly observed

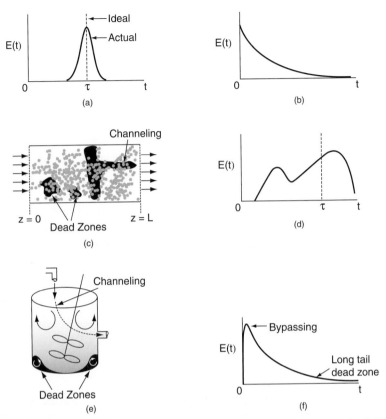

Figure 13-10   (a) RTD for near plug-flow reactor; (b) RTD for near perfectly mixed CSTR; (c) Packed-bed reactor with dead zones and channeling; (d) RTD for packed-bed reactor in (c); (e) tank reactor with short-circuiting flow (bypass); (f) RTD for tank reactor with channeling (bypassing or short circuiting) and a dead zone in which the tracer slowly diffuses in and out.

13.5.2 Simple Diagnostics and Troubleshooting Using the RTD for Ideal Reactors

13.5.2A   The CSTR

We will first consider a CSTR that operates (a) normally, (b) with bypassing, and (c) with a dead volume. For a well-mixed CSTR, the mole (mass) balance on the tracer is

$$\frac{V dC}{dt} = -v_0 C$$

Rearranging, we have

$$\frac{dC}{dt} = -\frac{1}{\tau}C$$

We saw the response to a pulse tracer is

Concentration: $C(t) = C_{T0}e^{-t/\tau}$

RTD Function: $E(t) = \dfrac{1}{\tau}e^{-t/\tau}$

Cumulative Function: $F(t) = 1 - e^{-t/\tau}$

$$\tau = \frac{V}{v_0}$$

where τ is the space time—the case of perfect operation.

a. Perfect Operation (P)

Here we will measure our reactor with a yardstick to find V and our flow rate with a flow meter to find v_0 in order to calculate $\tau = V/v_0$. We can then compare the curves shown below for the perfect operation in Figure 13-11 with the subsequent cases, which are for imperfect operation.

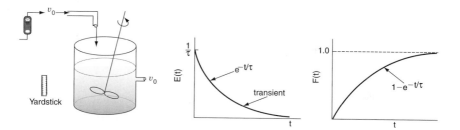

Figure 13-11   Perfect operation of a CSTR.

$$\tau = \frac{V}{v_0}$$

If τ is large, there will be a slow decay of the output transient, $C(t)$, and $E(t)$ for a pulse input. If τ is small, there will be rapid decay of the transient, $C(t)$, and $E(t)$ for a pulse input.

b. Bypassing (BP)

A volumetric flow rate v_b bypasses the reactor while a volumetric flow rate v_{SB} enters the system volume and $(v_0 = v_{SB} + v_b)$. The reactor system volume V_S is the well-mixed portion of the reactor, and the volumetric flow rate entering the system volume is v_{SB}. The subscript SB denotes that part of the flow has bypassed and only v_{SB} enters the system. Because some of the fluid bypasses, the flow passing through the system will be less than the total volumetric rate, $v_{SB} < v_0$, consequently $\tau_{SB} > \tau$. Let's say the volumetric flow rate that bypasses the reactor, v_b, is 25% of the total (e.g., $v_b = 0.25\ v_0$). The volumetric flow rate entering the reactor system, v_{SB} is 75% of the total ($v_{SB} = 0.75\ v_0$) and the corresponding true space time (τ_{SB}) for the system volume with bypassing is

$$\tau_{SB} = \frac{V}{v_{SB}} = \frac{V}{0.75v_0} = 1.33\tau$$

The space time, τ_{SB}, will be greater than that if there were no bypassing. Because τ_{SB} is greater than τ there will be a slower decay of the transients $C(t)$ and $E(t)$ than that of perfect operation.

An example of a corresponding $E(t)$ curve for the case of bypassing is

$$E(t) = \frac{v_b}{v_0}\delta(t-0) + \frac{v_{SB}^2}{Vv_0}e^{-t/\tau_{SB}}$$

The CSTR with bypassing will have RTD curves similar to those in Figure 13-12.

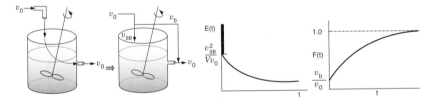

Figure 13-12   Ideal CSTR with bypass.

We see from the $F(t)$ curve that we have an initial jump equal to the fraction by-passed.

c. Dead Volume (DV)

Consider the CSTR in Figure 13-13 without bypassing but instead with a stagnant or dead volume.

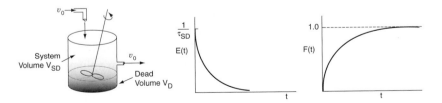

Figure 13-13   Ideal CSTR with dead volume.

The total volume, V, is the same as that for perfect operation, $V = V_D + V_{SD}$.

We see that because there is a dead volume which the fluid does not enter, there is less system volume, V_{SD}, than in the case of perfect operation, $V_{SD} < V$. Consequently, the fluid will pass through the reactor with the dead volume more quickly than that of perfect operation, i.e., $\tau_{SD} < \tau$.

If $\quad V_D = 0.2V$, $V_{SD} = 0.8V$, then $\quad \tau_{SD} = \dfrac{0.8V}{v_0} = 0.8\tau$

Also as a result, the transients $C(t)$ and $E(t)$ will decay more rapidly than that for perfect operation because there is a smaller system volume.

Summary

A summary for ideal CSTR mixing volume is shown in Figure 13-14.

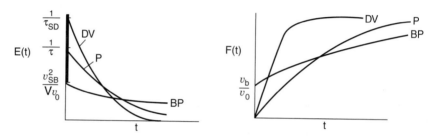

Figure 13-14 Comparison of $E(t)$ and $F(t)$ for CSTR under perfect operation, bypassing, and dead volume. (BP = bypassing, P = perfect, and DV = dead volume).

Knowing the volume V measured with a yardstick and the flow rate v_0 entering the reactor measured with a flow meter, one can calculate and plot $E(t)$ and $F(t)$ for the ideal case (P) and then compare with the measured RTD $E(t)$ to see if the RTD suggests either bypassing (BP) or dead zones (DV).

13.5.2B Tubular Reactor

A similar analysis to that for a CSTR can be carried out on a tubular reactor.
a. **Perfect Operation of PFR (P)**

We again measure the volume V with a yardstick and v_0 with a flow meter. The $E(t)$ and $F(t)$ curves are shown in Figure 13-15. The space time for a perfect PFR is

$$\tau = V/v_0$$

b. **PFR with Channeling (Bypassing, BP)**

Let's consider channeling (bypassing), as shown in Figure 13-16, similar to that shown in Figures 13-2 and 13-10(d). The space time for the reactor system with bypassing (channeling) τ_{SB} is

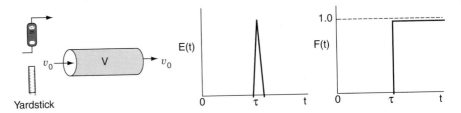

Figure 13-15   Perfect operation of a PFR.

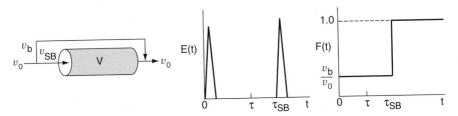

Figure 13-16   PFR with bypassing similar to the CSTR.

$$\tau_{SB} = \frac{V}{v_{SB}}$$

Because $v_{SB} < v_0$, the space time for the case of bypassing is greater when compared to perfect operation, i.e.,

$$\tau_{SB} > \tau$$

If 25% is bypassing (i.e., $v_b = 0.25\ v_0$) and 75% is entering the reactor system (i.e., $v_{SB} = 0.75\ v_0$), then $\tau_{SB} = V/(0.75v_0) = 1.33\tau$. The fluid that does enter the reactor system flows in plug flow. Here we have two spikes in the $E(t)$ curve. One spike at the origin and one spike at τ_{SB} that comes after τ for perfect operation. Because the volumetric flow rate is reduced, the time of the second spike will be greater than τ for perfect operation.

c. **PFR with Dead Volume (DV)**
The dead volume, V_D, could be manifested by internal circulation at the entrance to the reactor as shown in Figure 13-17.

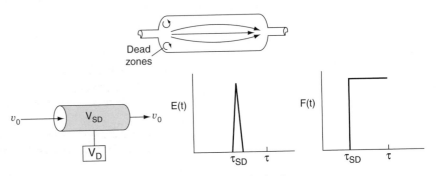

Figure 13-17   PFR with dead volume.

The system V_{SD} is where the reaction takes place and the total reactor volume is $(V = V_{SD} + V_D)$. The space time, τ_{SD}, for the reactor system with only dead volume is

$$\tau_{SD} = \frac{V_{SD}}{v_0}$$

Compared to perfect operation, the space time τ_{SD} is smaller and the tracer spike will occur before τ for perfect operation.

$$\tau_{SD} < \tau$$

Here again, the dead volume takes up space that is not accessible. As a result, the tracer will exit early because the system volume, V_{SD}, through which it must pass is smaller than the perfect operation case.

Summary

Figure 13-18 is a summary of these three cases.

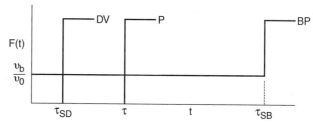

Figure 13-18   Comparison of PFR under perfect operation, bypassing, and dead volume (DV = dead volume, P = perfect PFR, BP = bypassing).

In addition to its use in diagnosis, the RTD can be used to predict conversion in existing reactors when a new reaction is tried in an old reactor. However, as we will see in Section 13.5.3, the RTD is not unique for a given system, and we need to develop models for the RTD to predict conversion.

13.5.3 PFR/CSTR Series RTD

Modeling the real reactor as a CSTR and a PFR in series

In some stirred tank reactors, there is a highly agitated zone in the vicinity of the impeller that can be modeled as a perfectly mixed CSTR. Depending on the location of the inlet and outlet pipes, the reacting mixture may follow a somewhat tortuous path either before entering or after leaving the perfectly mixed zone—or even both. This tortuous path may be modeled as a plug-flow reactor. Thus this type of tank reactor may be modeled as a CSTR in series with a plug-flow reactor, and the PFR may either precede or follow the CSTR. In this section we develop the RTD for this type of reactor arrangement.

First consider the CSTR followed by the PFR (Figure 13-19). The residence time in the CSTR will be denoted by τ_s and the residence time in the PFR by τ_p. If a pulse of tracer is injected into the entrance of the CSTR, the

Side Note: Medical Uses of RTD The application of RTD analysis in biomedical engineering is being used at an increasing rate. For example, Professor Bob Langer's[*] group at MIT used RTD analysis for a novel Taylor-Couette flow device for blood detoxification while Lee et al.[†] used an RTD analysis to study arterial blood flow in the eye. In this later study, sodium fluorescein was injected into the anticubical vein. The cumulative distribution function $F(t)$ is shown schematically in Figure 13.5.N-1. Figure 13.5N-2 shows a laser ophthalmoscope image after injection of the sodium fluorescein. The mean residence time can be calculated for each artery to estimate the mean circulation time (ca. 2.85 s). Changes in the retinal blood flow may provide important decision-making information for sickle-cell disease and retinitis pigmentosa.

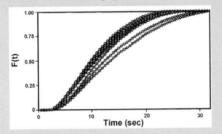

Figure 13.5.N-1   Cumulative RTD function for arterial blood flow in the eye. Courtesy of *Med. Eng. Phys.*[†]

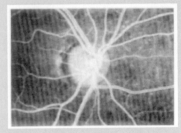

Figure 13.5.N-2   Image of eye after tracer injection. Courtesy of *Med. Eng. Phys.*[†]

---

[*] G. A. Ameer, E. A. Grovender, B. Olradovic, C. L. Clooney, and R. Langer, *AIChE J. 45*, 633 (1999).

[†] E. T. Lee, R. G. Rehkopf, J. W. Warnicki, T. Friberg, D. N. Finegold, and E. G. Cape, *Med. Eng. Phys. 19*, 125 (1997).

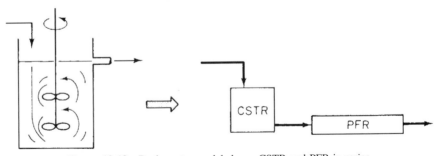

Figure 13-19   Real reactor modeled as a CSTR and PFR in series.

CSTR output concentration as a function of time will be

$$C = C_0 e^{-t/\tau_s}$$

This output will be delayed by a time τ_p at the outlet of the plug-flow section of the reactor system. Thus the RTD of the reactor system is

$$E(t) = \begin{cases} 0 & t < \tau_p \\ \dfrac{e^{-(t-\tau_p)/\tau_s}}{\tau_s} & t \geq \tau_p \end{cases} \qquad (13\text{-}50)$$

See Figure 13-20.

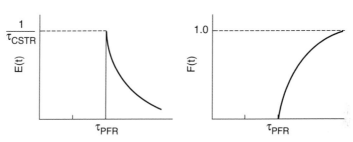

The RTD is not
unique to a
particular reactor
sequence.

Figure 13-20 RTD curves $E(t)$ and $F(t)$ for a CSTR and a PFR in series.

Next the reactor system in which the CSTR is preceded by the PFR will be treated. If the pulse of tracer is introduced into the entrance of the plug-flow section, then the same pulse will appear at the entrance of the perfectly mixed section τ_p seconds later, meaning that the RTD of the reactor system will be

$E(t)$ is the same no
matter which
reactor comes first.

$$E(t) = \begin{cases} 0 & t < \tau_p \\ \dfrac{e^{-(t-\tau_p)/\tau_s}}{\tau_s} & t \geq \tau_p \end{cases} \qquad (13\text{-}51)$$

which is *exactly* the same as when the CSTR was followed by the PFR.

It turns out that no matter where the CSTR occurs within the PFR/CSTR reactor sequence, the same RTD results. Nevertheless, this is not the entire story as we will see in Example 13-3.

Example 13–3 Comparing Second-Order Reaction Systems

Examples of *early*
and late mixing for
a given RTD

Consider a second-order reaction being carried out in a *real* CSTR that can be modeled as two different reactor systems: In the first system an ideal CSTR is followed by an ideal PFR; in the second system the PFR precedes the CSTR. Let τ_s and τ_p each equal 1 min, let the reaction rate constant equal 1.0 m³/kmol·min, and let the initial concentration of liquid reactant, C_{A0}, equal 1 kmol/m³. Find the conversion in each system.

Solution

Again, consider first the CSTR followed by the plug-flow section (Figure E13-3.1). A mole balance on the CSTR section gives

$$v_0(C_{A0} - C_{Ai}) = kC_{Ai}^2 V \qquad (E13\text{-}3.1)$$

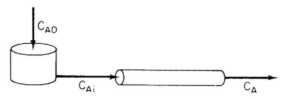

Figure E13-3.1   Early mixing scheme.

Rearranging, we have

$$\tau_s k C_{Ai}^2 + C_{Ai} - C_{A0} = 0$$

Solving for C_{Ai} gives

$$C_{Ai} = \frac{\sqrt{1 + 4\tau_s k C_{A0}} - 1}{2\tau_s k} \qquad (E13\text{-}3.2)$$

Then

$$C_{Ai} = \frac{-1 + \sqrt{1+4}}{2} = 0.618 \text{ kmol/m}^3 \qquad (E13\text{-}3.3)$$

This concentration will be fed into the PFR. The PFR mole balance

$$\frac{dF_A}{dV} = v_0 \frac{dC_A}{dV} = \frac{dC_A}{d\tau_p} = r_A = -kC_A^2 \qquad (E13\text{-}3.4)$$

$$\frac{1}{C_A} - \frac{1}{C_{Ai}} = \tau_p k \qquad (E13\text{-}3.5)$$

Substituting $C_{Ai} = 0.618$, $\tau = 1$, and $k = 1$ in Equation (E13-3.5) yields

$$\frac{1}{C_A} - \frac{1}{0.618} = (1)(1)$$

Solving for C_A gives

CSTR → PFR
X = 0.618

$$C_A = 0.382 \text{ kmol/m}^3$$

as the concentration of reactant in the effluent from the reaction system. Thus, the conversion is 61.8%—i.e., $X = ([1 - 0.382]/1) = 0.618$.

When the perfectly mixed section is preceded by the plug-flow section (Figure E13-3.2) the outlet of the PFR is the inlet to the CSTR, C_{Ai}:

$$\frac{1}{C_{Ai}} - \frac{1}{C_{A0}} = \tau_p k$$

$$\frac{1}{C_{Ai}} - \frac{1}{1} = (1)(1) \qquad (E13\text{-}3.6)$$

$$C_{Ai} = 0.5 \text{ kmol/m}^3$$

and a material balance on the perfectly mixed section (CSTR) gives

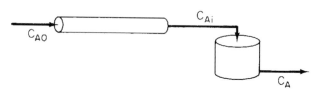

Figure E13-3.2 Late mixing scheme.

PFR → CSTR
$X = 0.634$

$$\tau_s k C_A^2 + C_A - C_{Ai} = 0 \tag{E13-3.7}$$

$$C_A = \frac{\sqrt{1 + 4\tau_s k C_{Ai}} - 1}{2\tau_s k} \tag{E13-3.8}$$

$$= \frac{-1 + \sqrt{1+2}}{2} = 0.366 \text{ kmol/m}^3$$

Early Mixing
$X = 0.618$
Late Mixing
$X = 0.634$

as the concentration of reactant in the effluent from the reaction system. The corresponding conversion is 63.4%.

In the one configuration, a conversion of 61.8% was obtained; in the other, 63.4%. While the difference in the conversions is small for the parameter values chosen, **the point is that there is a difference**.

While $E(t)$ was the same for both reaction systems, the conversion was not.

The conclusion from this example is of extreme importance in reactor analysis: **The RTD is not a complete description of structure for a particular reactor or system of reactors**. The RTD is unique for a particular reactor. However, the reactor or reaction system is not unique for a particular RTD. When analyzing nonideal reactors, the RTD alone is not sufficient to determine its performance, and more information is needed. It will be shown that in addition to the RTD, an adequate model of the nonideal reactor flow pattern and knowledge of the quality of mixing or "degree of segregation" are both required to characterize a reactor properly.

There are many situations where the fluid in a reactor neither is well mixed nor approximates plug flow. The idea is this: We have seen that the RTD can be used to diagnose or interpret the type of mixing, bypassing, etc., that occurs in an existing reactor that is currently on stream and is not yielding the conversion predicted by the ideal reactor models. Now let's envision another use of the RTD. Suppose we have a nonideal reactor either on line or sitting in storage. We have characterized this reactor and obtained the RTD function. What will be the conversion of a reaction with a known rate law that is carried out in a reactor with a known RTD?

The Question

How can we use the RTD to predict conversion in a real reactor?

In Part 2 we show how this question can be answered in a number of ways.

Part 2   Predicting Conversion and Exit Concentration

13.6 Reactor Modeling Using the RTD

Now that we have characterized our reactor and have gone to the lab to take data to determine the reaction kinetics, we need to choose a model to predict conversion in our real reactor.

The Answer

$$\boxed{\text{RTD} + \text{MODEL} + \text{KINETIC DATA} \Rightarrow \left\{ \frac{\text{EXIT CONVERSION and}}{\text{EXIT CONCENTRATION}} \right.}$$

We now present the five models shown in Table 13-1. We shall classify each model according to the number of adjustable parameters. We will discuss the first two in this chapter and the other three in Chapter 14.

TABLE 13-1.   MODELS FOR PREDICTING CONVERSION FROM RTD DATA

Ways we use the RTD data to predict conversion in nonideal reactors

1. **Zero adjustable parameters**

   a. Segregation model
   b. Maximum mixedness model

2. **One adjustable parameter**

   a. Tanks-in-series model
   b. Dispersion model

3. **Two adjustable parameters**

   Real reactors modeled as combinations of ideal reactors

The RTD tells us how long the various fluid elements have been in the reactor, but it does not tell us anything about the exchange of matter between the fluid elements (i.e., *the mixing*). The mixing of reacting species is one of the major factors controlling the behavior of chemical reactors. Fortunately for first-order reactions, knowledge of the length of time each molecule spends in the reactor is all that is needed to predict conversion. For first-order reactions the conversion is independent of concentration (recall Equation E9-1.3):

$$\frac{dX}{dt} = k(1 - X) \tag{E9-1.3}$$

Consequently, mixing with the surrounding molecules is not important. Therefore, once the RTD is determined, we can predict the conversion that will be achieved in the real reactor provided that the specific reaction rate for the first-order reaction is known. However, for reactions other than first order, knowledge of the RTD is not sufficient to predict conversion. In these cases the degree of mixing of molecules must be known in addition to how long each molecule spends in the reactor. Consequently, we must develop models that account for the mixing of molecules inside the reactor.

The more complex models of nonideal reactors necessary to describe reactions other than first order must contain information about *micromixing* in addition to that of *macromixing*. **Macromixing** produces a distribution of residence times *without*, however, specifying how molecules of different ages encounter one another in the reactor. **Micromixing**, on the other hand, describes how molecules of different ages encounter one another in the reactor. There are two extremes of *micromixing*: (1) all molecules of the same age group remain together as they travel through the reactor and are not mixed with any other age until they exit the reactor (i.e., complete segregation); (2) molecules of different age groups are completely mixed at the molecular level as soon as they enter the reactor (complete micromixing). For a given state of macromixing (i.e., a given RTD), these two extremes of micromixing will give the upper and lower limits on conversion in a nonideal reactor. For reaction orders greater than one or less than zero, the segregation model will predict the highest conversion. For reaction orders between zero and one, the maximum mixedness model will predict the highest conversion. This concept is discussed further in Section 13.7.3.

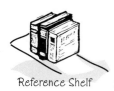
Reference Shelf

We shall define a globule as a fluid particle containing millions of molecules all of the same age. A fluid in which the globules of a given age do not mix with other globules is called a macrofluid. A *macrofluid* could be visualized as a noncoalescent globules where all the molecules in a given globule have the same age. A fluid in which molecules are not constrained to remain in the globule and are free to move everywhere is called a *microfluid*.[9] There are two extremes of mixing of the *macrofluid* globules to form a *microfluid* we shall study—early mixing and late mixing. These two extremes of late and early mixing are shown in Figure 13-21 (a) and (b), respectively. These extremes can also be seen by comparing Figures 13-23 (a) and 13-24 (a). The extremes of late and early mixing are referred to as *complete segregation* and *maximum mixedness*, respectively.

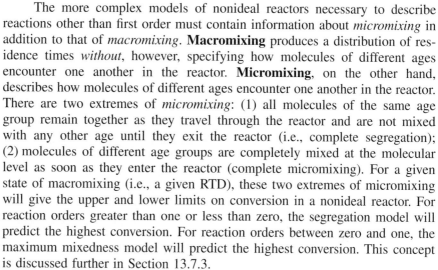

Figure 13-21   (a) Macrofluid; and (b) microfluid mixing on the molecular level.

---

[9] J. Villermaux, *Chemical Reactor Design and Technology* (Boston: Martinus Nijhoff, 1986).

13.7 Zero-Parameter Models

13.7.1 Segregation Model

In a "perfectly mixed" CSTR, the entering fluid is assumed to be distributed immediately and evenly throughout the reacting mixture. This mixing is assumed to take place even on the microscale, and elements of different ages mix together thoroughly to form a completely micromixed fluid. If fluid elements of different ages do not mix together at all, the elements remain segregated from each other, and the fluid is termed *completely segregated*. The extremes of complete micromixing and complete segregation are the limits of the micromixing of a reacting mixture.

In developing the segregated mixing model, we first consider a CSTR because the application of the concepts of mixing quality are illustrated most easily using this reactor type. In the segregated flow model we visualize the flow through the reactor to consist of a continuous series of globules (Figure 13-22).

In the segregation model globules behave as batch reactors operated for different times

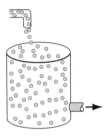

Figure 13-22   Little batch reactors (globules) inside a CSTR.

These globules retain their identity; that is, they do not interchange material with other globules in the fluid during their period of residence in the reaction environment, i.e., they remain segregated. In addition, each globule spends a different amount of time in the reactor. In essence, what we are doing is lumping all the molecules that have exactly the same residence time in the reactor into the same globule. The principles of reactor performance in the presence of completely segregated mixing were first described by Danckwerts[10] and Zwietering.[11]

The segregation model has mixing at the latest possible point.

Another way of looking at the segregation model for a continuous flow system is the PFR shown in Figures 13-23(a) and (b). Because the fluid flows down the reactor in plug flow, each exit stream corresponds to a specific residence time in the reactor. Batches of molecules are removed from the reactor at different locations along the reactor in such a manner as to duplicate the RTD function, $E(t)$. The molecules removed near the entrance to the reactor correspond to those molecules having short residence times in the reactor.

[10]P. V. Danckwerts, *Chem. Eng. Sci.*, 8, 93 (1958).
[11]T. N. Zwietering, *Chem. Eng. Sci.*, 11, 1 (1959).

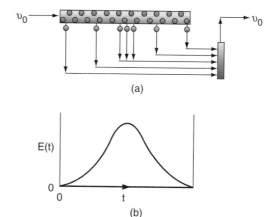

(a)

E(t) matches the
removal of the
batch reactors

E(t)

0

(b)

Figure 13-23   Mixing at the latest possible point.

Little batch reactors

Physically, this effluent would correspond to the molecules that channel rap-idly through the reactor. The farther the molecules travel along the reactor before being removed, the longer their residence time. The points at which the various groups or batches of molecules are removed correspond to the RTD function for the reactor.

Because there is no molecular interchange between globules, each acts essentially as its own batch reactor. The reaction time in any one of these tiny batch reactors is equal to the time that the particular globule spends in the reaction environment. The distribution of residence times among the globules is given by the RTD of the particular reactor.

$$\boxed{\text{RTD} + \text{MODEL} + \text{KINETIC DATA} \quad \Rightarrow \begin{cases} \text{EXIT CONVERSION and} \\ \text{EXIT CONCENTRATION} \end{cases}}$$

To determine the mean conversion in the effluent stream, we must aver-age the conversions of all of the various globules in the exit stream:

$$\begin{bmatrix} \text{Mean} \\ \text{conversion} \\ \text{of those globules} \\ \text{spending between} \\ \text{time } t \text{ and } t + dt \\ \text{in the reactor} \end{bmatrix} = \begin{bmatrix} \text{Conversion} \\ \text{achieved in a globule} \\ \text{after spending a time } t \\ \text{in the reactor} \end{bmatrix} \times \begin{bmatrix} \text{Fraction} \\ \text{of globules that} \\ \text{spend between } t \\ \text{and } t + dt \text{ in the} \\ \text{reactor} \end{bmatrix}$$

then

$$d\overline{X} = X(t) \times E(t)\, dt$$

$$\boxed{\frac{d\bar{X}}{dt} = X(t)E(t)} \tag{13-52}$$

Summing over all globules, the mean conversion is

Mean conversion
for the segregation
model

$$\boxed{\bar{X} = \int_0^\infty X(t)E(t)\,dt} \tag{13-53}$$

Consequently, if we have the batch reactor equation for $X(t)$ and measure the RTD experimentally, we can find the mean conversion in the exit stream. *Thus, if we have the RTD, the reaction rate expression, then for a segregated flow situation (i.e., model), we have sufficient information to calculate the conversion.* An example that may help give additional physical insight to the segregation model is given in the Summary Notes on the CD-ROM, click the More back button just before section 4A.2.

Summary Notes

Consider the following first-order reaction:

$$A \xrightarrow{\ k\ } \text{products}$$

For a batch reactor we have

$$-\frac{dN_A}{dt} = -r_A V$$

For constant volume and with $N_A = N_{A0}(1 - X)$,

$$N_{A0}\frac{dX}{dt} = -r_A V = kC_A V = kN_A = kN_{A0}(1 - X)$$

$$\boxed{\frac{dX}{dt} = k(1 - X)} \tag{13-54}$$

Solving for $X(t)$, we have

$$X(t) = 1 - e^{-kt}$$

Mean conversion
for a first-order
reaction

$$\bar{X} = \int_0^\infty X(t)E(t)\,dt = \int_0^\infty (1 - e^{-kt})E(t)\,dt = \int_0^\infty E(t)\,dt - \int_0^\infty e^{-kt}E(t)\,dt \tag{13-55}$$

$$\bar{X} = 1 - \int_0^\infty e^{-kt}E(t)\,dt \tag{13-56}$$

We will now determine the mean conversion predicted by the segregation model for an ideal PFR, a CSTR, and a laminar flow reactor.

Example 13–4   Mean Conversion Is an Ideal PFR, an Ideal CSTR, and a Laminar Flow Reactor

Derive the equation of a first-order reaction using the segregation model when the RTD is equivalent to (a) an ideal PFR, (b) an ideal CSTR, and (c) a laminar flow reactor. Compare these conversions with those obtained from the design equation.

Solution

(a) For the **PFR**, the RTD function was given by Equation (13-32)

$$E(t) = \delta(t - \tau) \tag{13-32}$$

Recalling Equation (13-55)

$$\bar{X} = \int_0^\infty X(t)E(t)\, dt = 1 - \int_0^\infty e^{-kt}E(t)\, dt \tag{13-55}$$

Substituting for the RTD function for a PFR gives

$$\bar{X} = 1 - \int_0^\infty (e^{-kt})\delta(t - \tau)\, dt \tag{E13-4.1}$$

Using the integral properties of the Dirac delta function, Equation (13-35) we obtain

$$\bar{X} = 1 - e^{-k\tau} = 1 - e^{-Da} \tag{E13-4.2}$$

where for a first-order reaction the Damköhler number is $Da = \tau k$. Recall that for a PFR after combining the mole balance, rate law, and stoichiometric relationships (cf. Chapter 4), we had

$$\frac{dX}{d\tau} = k(1 - X) \tag{E13-4.3}$$

Integrating yields

$$X = 1 - e^{-k\tau} = 1 - e^{-Da} \tag{E13-4.4}$$

which is identical to the conversion predicted by the segregation model $\bar{X}$.

(b) For the **CSTR**, the RTD function is

$$E(t) = \frac{1}{\tau} e^{-t/\tau} \tag{13-27}$$

Recalling Equation (13-56), the mean conversion for a first-order reaction is

$$\bar{X} = 1 - \int_0^\infty e^{-kt}E(t)\, dt \tag{13-56}$$

$$\bar{X} = 1 - \int_0^\infty \frac{e^{-(1/\tau + k)t}}{\tau}\, dt$$

$$\bar{X} = 1 + \frac{1}{k + 1/\tau} \frac{1}{\tau} e^{-(k + 1/\tau)t} \Big|_0^\infty$$

$$\bar{X} = \frac{\tau k}{1+\tau k} = \frac{Da}{1+Da} \qquad (E13\text{-}4.5)$$

Combining the CSTR mole balance, the rate law, and stoichiometry, we have

$$F_{A0}X = -r_A V$$

$$v_0 C_{A0}X = kC_{A0}(1-X)V$$

$$X = \frac{\tau k}{1+\tau k} \qquad (E13\text{-}4.6)$$

As expected, using the $E(t)$ for an ideal PFR and CSTR with the segregation model gives a mean conversion $\bar{X}$ identical to that obtained by using the algorithm in Ch. 4.

which is identical to the conversion predicted by the segregation model $\bar{X}$.

(c)   For a **laminar flow reactor** the RTD function is

$$E(t) = \begin{cases} 0 & \text{for } (t < \tau/2) \\ \dfrac{\tau^2}{2t^3} & \text{for } (t \ge \tau/2) \end{cases} \qquad (13\text{-}47)$$

The dimensionless form is

$$E(\Theta) = \begin{cases} 0 & \text{for } \Theta < 0.5 \\ \dfrac{1}{2\Theta^3} & \text{for } \Theta \ge 0.5 \end{cases} \qquad (13\text{-}49)$$

From Equation (13-15), we have

$$\bar{X} = 1 - \int_0^\infty e^{-kt} E(t)\,dt = 1 - \int_0^\infty e^{-\tau k\Theta} E(\Theta)\,d\Theta \qquad (E13\text{-}4.7)$$

$$\bar{X} = 1 - \int_{0.5}^\infty \frac{e^{-\tau k\Theta}}{2\Theta^3}\,d\Theta \qquad (E13\text{-}4.8)$$

Integrating twice by parts

$$\boxed{\bar{X} = 1 - (1 - 0.5\tau k)e^{-0.5k\tau} - (0.5\tau k)^2 \int_{0.5}^\infty \frac{e^{-\tau k\Theta}}{\Theta}\,d\Theta} \qquad (E13\text{-}4.9)$$

The last integral is the *exponential integral* and can be evaluated from tabulated values. Fortunately, Hilder[12] developed an approximate formula ($\tau k = Da$).

---

[12] M. H. Hilder, *Trans. I. ChemE,* 59, 143 (1979).

$$\bar{X} = 1 - \frac{1}{(1 + 0.25\tau k)e^{0.5\tau k} + 0.25\tau k} \equiv 1 - \frac{1}{(1 + 0.25\,Da)\,e^{0.5Da} + 0.25\,Da}$$

$$\bar{X} = \frac{(4 + Da)e^{0.5Da} + Da - 4}{(4 + Da)e^{0.5Da} + Da} \qquad \text{(E13-4.10)}$$

A comparison of the exact value along with Hilder's approximation is shown in Table E13-4.1 for various values of the Damköhler number, τk, along with the conversion in an ideal PFR and an ideal CSTR.

TABLE E13-4.1.   COMPARISON OF CONVERSION IN PFR, CSTR, AND LAMINAR FLOW REACTOR FOR DIFFERENT DAMKÖHLER NUMBERS FOR A FIRST-ORDER REACTION

$Da = \tau k$	$X_{\text{L.F. Exact}}$	$X_{\text{L.F. Approx.}}$	X_{PFR}	X_{CSTR}
0.1	0.0895	0.093	0.0952	0.091
1	0.557	0.56	0.632	0.501
2	0.781	0.782	0.865	0.667
4	0.940	0.937	0.982	0.80
10	0.9982	0.9981	0.9999	0.90

where $X_{\text{L.F. Exact}}$ = exact solution to Equation (E13-4.9) and $X_{\text{L.F. Approx.}}$ = Equation (E13-4.10).

For large values of the Damköhler number then, there is complete conversion along the streamlines off the center streamline so that the conversion is determined along the pipe axis such that

$$\bar{X} = 1 - \int_{0.5}^{\infty} 4e^{-\tau k\Theta}\,d\Theta = 1 - 4e^{-0.5\tau k}/\tau k \qquad \text{(E13-4.11)}$$

Figure E13-4.1 shows a comparison of the mean conversion of the conversion in an LFR, PFR, and CSTR as a function of the Damköhler number for a first-order reaction.

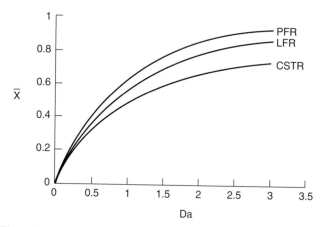

Figure E13-4.1 Conversion in a PFR, LFR, and CSTR as a function of a Damköhler number (Da) for a first-order reaction ($Da = \tau k$).

We have just shown for a first-order reaction that whether you assume complete micromixing [Equation (E13-4.6)] or complete segregation [Equation (E13-4.5)] in a CSTR, the same conversion results. This phenomenon occurs because the rate of change of conversion for a first-order reaction does *not* depend on the concentration of the reacting molecules [Equation (13-54)]; it does not matter what kind of molecule is next to it or colliding with it. Thus the extent of micromixing does not affect a first-order reaction, so the segregated flow model can be used to calculate the conversion. As a result, *only the RTD is necessary to calculate the conversion for a first-order reaction in any type of reactor* (see Problem P13-3$_c$). Knowledge of neither the degree of micromixing nor the reactor flow pattern is necessary. We now proceed to calculate conversion in a real reactor using RTD data.

Example 13–5 Mean Conversion Calculations in a Real Reactor

Calculate the mean conversion in the reactor we have characterized by RTD measurements in Examples 13-1 and 13-2 for a first-order, liquid-phase, irreversible reaction in a completely segregated fluid:

$$A \longrightarrow \text{products}$$

The specific reaction rate is 0.1 min^{-1} at 320 K.

Solution

Because each globule acts as a **batch reactor** of constant volume, we use the batch reactor design equation to arrive at the equation giving conversion as a function of time:

$$X = 1 - e^{-kt} = 1 - e^{-0.1t} \qquad \text{(E13-5.1)}$$

To calculate the mean conversion we need to evaluate the integral:

$$\bar{X} = \int_0^\infty X(t)E(t)\,dt \qquad \text{(13-53)}$$

The RTD function for this reactor was determined previously and given in Table E13-2.1 and is repeated in Table E13-5.1. To evaluate the integral we make a plot of $X(t)E(t)$ as a function of t as shown in Figure E13-5.1 and determine the area under the curve.

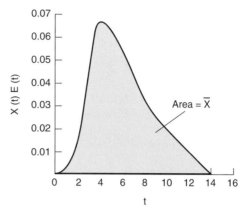

For a given RTD, the segregation model gives the upper bound on conversion for reaction orders less than zero or greater than 1.

Figure E13-5.1   Plot of columns 1 and 4 from the data in Table E13-5.1.

TABLE E13-5.1   PROCESSED DATA TO FIND THE MEAN CONVERSION $\bar{X}$

t (min)	$E(t)$ (min^{-1})	$X(t)$	$X(t)E(t)$ (min^{-1})
0	0.000	0	0
1	0.020	0.095	0.0019
2	0.100	0.181	0.0180
3	0.160	0.259	0.0414
4	0.200	0.330	0.0660
5	0.160	0.393	0.0629
6	0.120	0.451	0.0541
7	0.080	0.503	0.0402
8	0.060	0.551	0.0331
9	0.044	0.593	0.0261
10	0.030	0.632	0.01896
12	0.012	0.699	0.0084
14	0.000	0.75	0

Using the quadature formulas in Appendix A.4

$$\int_0^\infty X(t)E(t)\,dt = \int_0^{10} X(t)E(t)\,dt + \int_{10}^{14} X(t)E(t)\,dt$$

$$= \tfrac{1}{3}[0 + 4(0.0019) + 2(0.018) + 4(0.0414) + 2(0.066)$$
$$+ 4(0.0629) + 2(0.0541) + 4(0.0402) + 2(0.0331)$$
$$+ 4(0.0261) + 0.01896] + \tfrac{2}{3}[0.01896 + 4(0.0084) + 0]$$

$$= (0.350) + (0.035) = 0.385$$

$$\bar{X} = \text{area} = 0.385$$

The mean conversion is 38.5%. Polymath or Excel will easily give $\bar{X}$ after setting up columns 1 and 4 in Table E13-5.1. The area under the curve in Figure E13-5.1 is the mean conversion $\bar{X}$.

As discussed previously, because the reaction is *first order*, the conversion calculated in Example 13-5 would be valid for a reactor with complete mixing, complete segregation, or any degree of mixing between the two. Although early or late mixing does not affect a first-order reaction, micromixing or complete segregation can modify the results of a second-order system significantly.

Example 13–6   Mean Conversion for a Second-Order Reaction in a Laminar Flow Reactor

The liquid-phase reaction between cytidine and acetic anhydride

| cytidine | acetic anhydride | | |
| (A) | (B) | (C) | (D) |

$$A + B \rightarrow C + D$$

is carried out isothermally in an inert solution of *N*-methyl-2-pyrrolidone (NMP) with $\Theta_{NMP} = 28.9$. The reaction follows an elementary rate law. The feed is equal molar in A and B with $C_{A0} = 0.75$ mol/dm^3, a volumetric flow rate of 0.1 dm^3/s and a reactor volume of 100 dm^3. Calculate the conversion in (a) a PFR, (b) a batch reactor, and (c) a laminar flow reactor.

Additional information:[13]

$k = 4.93 \times 10^{-3}$ dm^3/mol · s at 50°C with $E = 13.3$ kcal/mol, $\Delta H_{RX} = -10.5$ kcal/mol

Heat of mixing for $\Theta_{NMP} = \dfrac{F_{NMP}}{F_{A0}} = 28.9$, $\Delta H_{mix} = -0.44$ kcal/mol

Solution

The reaction will be carried out isothermally at 50°C. The space time is

$$\tau = \frac{V}{v_0} = \frac{100 \text{ dm}^3}{0.1 \text{ dm}^3/\text{s}} = 1000 \text{ s}$$

(a)   **For a PFR**
Mole Balance

$$\frac{dX}{dV} = \frac{-r_A}{F_{A0}} \tag{E13-6.1}$$

---

[13]J. J. Shatynski and D. Hanesian, *Ind. Eng. Chem. Res., 32*, 594 (1993).

Rate Law

$$-r_A = kC_A C_B \tag{E13-6.2}$$

Stoichiometry, $\Theta_B = 1$

$$C_A = C_{A0}(1 - X) \tag{E13-6.3}$$

$$C_B = C_A \tag{E13-6.4}$$

Combining

$$\frac{dX}{dV} = \frac{kC_{A0}(1 - X)^2}{v_0} \tag{E13-6.5}$$

PFR Calculation

Solving with $\tau = V/v_0$ and $X = 0$ for $V = 0$ gives

$$X = \frac{\tau k C_{A0}}{1 + \tau k C_{A0}} = \frac{Da_2}{1 + Da_2} \tag{E13-6.6}$$

where Da_2 is the Damköhler number for a second-order reaction.

$$Da_2 = \tau k C_{A0} = (1000\text{s})(4.9 \times 10^{-3}\ \text{dm}^3/\text{s} \cdot \text{mol})(0.75\ \text{mol/dm}^3)$$

$$= 3.7$$

$$X = \frac{3.7}{4.7}$$

$$\boxed{X = 0.787}$$

(b)   **Batch Reactor**

$$\frac{dX}{dt} = \frac{-r_A}{C_{A0}} \tag{E13-6.7}$$

Batch Calculation

$$\frac{dX}{dt} = kC_{A0}(1 - X)^2 \tag{E13-6.8}$$

$$X(t) = \frac{kC_{A0}t}{1 + kC_{A0}t} \tag{E13-6.9}$$

If the batch reaction time is the same time as the space time the batch conversion is the same as the PFR conversion $X = 0.787$.

(c)   **Laminar Flow Reactor**

The differential form for the mean conversion is obtained from Equation (13-52)

$$\frac{d\bar{X}}{dt} = X(t)E(t) \tag{13-52}$$

We use Equation (E13-6.9) to substitute for $X(t)$ in Equation (13-52). Because $E(t)$ for the LFR consists of two parts, we need to incorporate the IF statement in our ODE solver program. For the laminar flow reaction, we write

$$E_1 = 0 \text{ for } t < \tau/2 \tag{E13-6.10}$$

$$E_2 = \frac{\tau^2}{2t^3} \text{ for } t \geq \tau/2 \qquad\qquad \text{(E13-6.11)}$$

Let $t_1 = \tau/2$ so that the IF statement now becomes

LFR Calculation

$$E = \text{If } (t < t_1) \text{ then } (E_1) \text{ else } (E_2) \qquad\qquad \text{(E13-6.12)}$$

One other thing is that the ODE solver will recognize that $E_2 = \infty$ at $t = 0$ and refuse to run. So we must add a very small number to the denominator such as (0.001); for example,

$$E_2 = \frac{\tau^2}{(2t^3 + 0.001)} \qquad\qquad \text{(E13-6.13)}$$

The integration time should be carried out to 10 or more times the reactor space time τ. The Polymath Program for this example is shown below.

POLYMATH Results
Example 13-6 07-24-2004, Rev5.1.225

Calculated values of the DEQ variables

Variable	initial value	minimal value	maximal value	final value
t	0	0	2.0E+04	2.0E+04
Xbar	0	0	0.7413021	0.7413021
k	0.00493	0.00493	0.00493	0.00493
Cao	0.75	0.75	0.75	0.75
X	0	0	0.9866578	0.9866578
tau	1000	1000	1000	1000
t1	500	500	500	500
E2	5.0E+10	6.25E-08	5.0E+10	6.25E-08
E	0	0	0.0034671	6.25E-08

ODE Report (RKF45)

Differential equations as entered by the user
[1] d(Xbar)/d(t) = X*E

Explicit equations as entered by the user
[1] k = .00493
[2] Cao = 0.75
[3] X = k*Cao*t/(1+k*Cao*t)
[4] tau = 1000
[5] t1 = tau/2
[6] E2 = tau^2/2/(t^3+.00001)
[7] E = if(t<t1) then (0)else (E2)

Living Example Problem

We see that the mean conversion Xbar ($\overline{X}$) for the LFR is 74.1%. In summary,

$$\boxed{\begin{array}{l} X_{PRF} = 0.786 \\ X_{LFR} = 0.741 \end{array}}$$

Compare this result with the exact analytical formula[14] for the laminar flow reactor with a second-order reaction

Analytical Solution

$$\boxed{\overline{X} = Da[1 - (Da/2) \ln(1+2/Da)]}$$

where $Da = kC_{A0}\tau$. For $Da = 3.70$ we get $\overline{X} = 0.742$.

---

[14] K. G. Denbigh, *J. Appl. Chem.*, *1*, 227 (1951).

13.7.2 Maximum Mixedness Model

Segregation model
mixing occurs at the
latest possible point.
In a reactor with a segregated fluid, mixing between particles of fluid does not occur until the fluid leaves the reactor. The reactor exit is, of course, the *latest* possible point that mixing can occur, and any effect of mixing is postponed until after all reaction has taken place as shown in Figure 13-23. We can also think of completely segregated flow as being in a state of *minimum mixedness*. We now want to consider the other extreme, that of *maximum mixedness* consistent with a given residence-time distribution.

We return again to the plug-flow reactor with side entrances, only this time the fluid enters the reactor along its length (Figure 13-24). As soon as the fluid enters the reactor, it is completely mixed radially (but not longitudinally) with the other fluid already in the reactor. The entering fluid is fed into the reactor through the side entrances in such a manner that the RTD of the plug-flow reactor with side entrances is identical to the RTD of the real reactor.

Maximum
mixedness:
mixing occurs
at the earliest
possible point

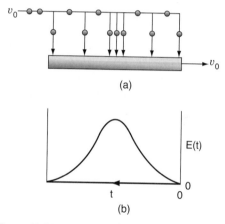

Figure 13-24   Mixing at the earliest possible point.

The globules at the far left of Figure 13-24 correspond to the molecules that spend a long time in the reactor while those at the far right correspond to the molecules that channel through the reactor. In the reactor with side entrances, mixing occurs at the *earliest* possible moment consistent with the RTD. Thus the effect of mixing occurs as early as possible throughout the reactor, and this situation is termed the condition of *maximum mixedness*.[15] The approach to calculating conversion for a reactor in a condition of maximum mixedness will now be developed. In a reactor with side entrances, let λ be the time it takes for the fluid to move from a particular point to the end of the reactor. In other words, λ is the life expectancy of the fluid in the reactor at that point (Figure 13-25).

---

[15]T. N. Zwietering, *Chem. Eng. Sci., 11*, 1 (1959).

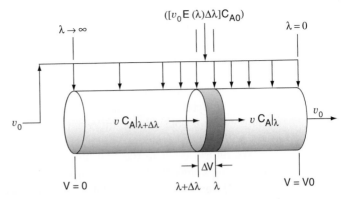

Figure 13-25 Modeling maximum mixedness by a plug-flow reactor with side entrances.

Moving down the reactor from left to right, λ decreases and becomes zero at the exit. At the left end of the reactor, λ approaches infinity or the maximum residence time if it is other than infinite.

Consider the fluid that enters the reactor through the sides of volume ΔV in Figure 13-25. The fluid that enters here will have a life expectancy between λ and $\lambda + \Delta\lambda$. The fraction of fluid that will have this life expectancy between the product of the total volumetric flow rate, v_0, and the fraction of the total that has life expectancy between λ and $\lambda + \Delta\lambda$ is $E(\lambda)\,\Delta\lambda$. That is the volumetric rate of fluid entering through the sides of volume ΔV is $v_0\,E(\lambda)\,\Delta\lambda$.

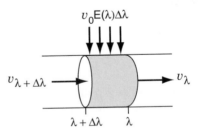

The volumetric flow rate at λ, v_λ, is the flow rate that entered at $\lambda + \Delta\lambda$, $v_{\lambda+\Delta\lambda}$ plus what entered through the sides $v_0\,E(\lambda)\Delta\lambda$, i.e.,

$$v_\lambda = v_{\lambda+\Delta\lambda} + v_0 E(\lambda)\Delta\lambda$$

Rearranging and taking the limit as $\Delta\lambda \to 0$

$$\frac{dv_\lambda}{d\lambda} = -v_0 E(\lambda) \tag{13-57}$$

The volumetric flow rate v_0 at the entrance to the reactor $(X = 0)$ is zero because the fluid only enters through the sides along the length.

Integrating equation (13-57) with limits $v_\lambda = 0$ at $\lambda = \infty$ and $v_\lambda = v_\lambda$ at $\lambda = \lambda$, we obtain

$$\boxed{v_\lambda = v_0 \int_\lambda^\infty E(\lambda)d\lambda = v_0[1 - F(\lambda)]}$$

(13-58)

The volume of fluid with a life expectancy between λ and $\lambda + \Delta\lambda$ is

$$\Delta V = v_0[1 - F(\lambda)]\,\Delta\lambda$$

(13-59)

The rate of generation of the substance A in this volume is

$$r_A \Delta V = r_A v_0[1 - F(\lambda)]\,\Delta\lambda$$

(13-60)

We can now carry out a mole balance on substance A between λ and $\lambda + \Delta\lambda$:

Mole balance

$$\begin{bmatrix} \text{In} \\ \text{at } \lambda + \Delta\lambda \end{bmatrix} + \begin{bmatrix} \text{In} \\ \text{through side} \end{bmatrix} - \begin{bmatrix} \text{Out} \\ \text{at } \lambda \end{bmatrix} + \begin{bmatrix} \text{Generation} \\ \text{by reaction} \end{bmatrix} = 0$$

$$v_0[1 - F(\lambda)]C_A|_{\lambda+\Delta\lambda} + v_0 C_{A0}E(\lambda)\,\Delta\lambda$$

$$- v_0[1 - F(\lambda)]C_A|_\lambda + r_A v_0[1 - F(\lambda)]\,\Delta\lambda = 0 \quad (13\text{-}61)$$

Dividing Equation (13-61) by $v_0\,\Delta\lambda$ and taking the limit as $\Delta\lambda \to 0$ gives

$$E(\lambda)C_{A0} + \frac{d\{[1 - F(\lambda)]C_A(\lambda)\}}{d\lambda} + r_A[1 - F(\lambda)] = 0$$

Taking the derivative of the term in brackets

$$C_{A0}E(\lambda) + [1 - F(\lambda)]\frac{dC_A}{d\lambda} - C_A E(\lambda) + r_A[1 - F(\lambda)] = 0$$

or

$$\boxed{\frac{dC_A}{d\lambda} = -r_A + (C_A - C_{A0})\frac{E(\lambda)}{1 - F(\lambda)}}$$

(13-62)

We can rewrite Equation (13-62) in terms of conversion as

$$-C_{A0}\frac{dX}{d\lambda} = -r_A - C_{A0}X\frac{E(\lambda)}{1 - F(\lambda)}$$

(13-63)

or

$$\boxed{\frac{dX}{d\lambda} = \frac{r_A}{C_{A0}} + \frac{E(\lambda)}{1 - F(\lambda)}(X)}$$

(13-64)

The boundary condition is as $\lambda \to \infty$, then $C_A = C_{A0}$ for Equation (13-62) [or $X = 0$ for Equation (13-64)]. To obtain a solution, the equation is integrated backwards numerically, starting at a very large value of λ and ending with the final conversion at $\lambda = 0$. For a given RTD and reaction orders greater than one, the maximum mixedness model gives the lower bound on conversion.

MM gives the lower bound on X.

Example 13–7 Conversion Bounds for a Nonideal Reactor

The liquid-phase, second-order dimerization

$$2A \longrightarrow B \qquad r_A = -kC_A^2$$

for which $k = 0.01$ dm³/mol·min is carried out at a reaction temperature of 320 K. The feed is pure A with $C_{A0} = 8$ mol/dm³. The reactor is nonideal and perhaps could be modeled as two CSTRs with interchange. The reactor volume is 1000 dm³, and the feed rate for our dimerization is going to be 25 dm³/min. We have run a tracer test on this reactor, and the results are given in columns 1 and 2 of Table E13-7.1. We wish to know the bounds on the conversion for different possible degrees of micromixing for the RTD of this reactor. What are these bounds?

Tracer test on tank reactor: $N_0 = 100$ g, $v = 25$ dm³/min.

Living Example Problem

Columns 3 through 5 are calculated from columns 1 and 2.

TABLE E13-7.1   RAW AND PROCESSED DATA

t (min)	C (mg/dm³)	$E(t)$ (min⁻¹)	$1 - F(t)$	$E(t)/[1 - F(t)]$ (min⁻¹)	λ (min)
0	112	0.0280	1.000	0.0280	0
5	95.8	0.0240	0.871	0.0276	5
10	82.2	0.0206	0.760	0.0271	10
15	70.6	0.0177	0.663	0.0267	15
20	60.9	0.0152	0.584	0.0260	20
30	45.6	0.0114	0.472	0.0242	30
40	34.5	0.00863	0.353	0.0244	40
50	26.3	0.00658	0.278	0.0237	50
70	15.7	0.00393	0.174	0.0226	70
100	7.67	0.00192	0.087	0.0221	100
150	2.55	0.000638	0.024	0.0266	150
200	0.90	0.000225	0.003	0.075	200
1	2	3	4	5	6

Solution

The bounds on the conversion are found by calculating conversions under conditions of complete segregation and maximum mixedness.

Conversion if fluid is completely segregated. The batch reactor equation for a second-order reaction of this type is

$$X = \frac{kC_{A0}t}{1 + kC_{A0}t}$$

Spreadsheets work
quite well here.

The conversion for a completely segregated fluid in a reactor is

$$\bar{X} = \int_0^\infty X(t)E(t)\,dt$$

The calculations for this integration are carried out in Table E13-7.2. The numerical integration uses the simple trapezoid rule. The conversion for this system if the fluid were completely segregated is 0.61 or 61%.

TABLE E13-7.2.   SEGREGATION MODEL

t (min)	$X(t)$	$X(t)E(t)$ (min^{-1})	$X(t)E(t)\,\Delta t$
0	0	0	0
5	0.286	0.00686	0.0172[†]
10	0.444	0.00916	0.0400
15	0.545	0.00965	0.0470
20	0.615	0.00935	0.0475
30	0.706	0.00805	0.0870
40	0.762	0.00658	0.0732
50	0.800	0.00526	0.0592
70	0.848	0.00333	0.0859
100	0.889	0.00171	0.0756
150	0.923	0.000589	0.0575
200	0.941	0.000212	0.0200
			0.610

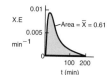

[†]For the first point we have $X(t)E(t)\Delta t = (0 + 0.00686)\,(5/2) = 0.0172$

Conversion for maximum mixedness. The Euler method will be used for numerical integration:

$$X_{i+1} = X_i + (\Delta\lambda)\left[\frac{E(\lambda_i)}{1-F(\lambda_i)}\,X_i - kC_{A0}(1-X_i)^2\right]$$

Integrating this equation presents some interesting results. If the equation is integrated from the exit side of the reactor, starting with $\lambda = 0$, the solution is unstable and soon approaches large negative or positive values, depending on what the starting value of X is. We want to find the conversion at the exit to the reactor $\lambda = 0$. Consequently, we need to integrate backwards.

$$X_{i-1} = X_i - \Delta\lambda\left[\frac{E(\lambda_i)}{1-F(\lambda_i)} - kC_{A0}(1-X_i)^2\right]$$

If integrated from the point where $\lambda \longrightarrow \infty$, oscillations may occur but are damped out, and the equation approaches the same final value no matter what initial value of X between 0 and 1 is used. We shall start the integration at $\lambda = 200$ and let $X = 0$ at this point. If we set $\Delta\lambda$ too large, the solution will blow up, so we will start out with $\Delta\lambda = 25$ and use the average of the measured values of $E(t)/[(1 - F(t)]$ where necessary. We will now use the data in column 5 of Table E13-7.1 to carry out the integration.

At $\lambda = 200$, $X = 0$

$\lambda = 175$:

$$X(\lambda = 175) = X(\lambda = 200) - \Delta\lambda\left[\frac{E(200)X(200)}{1 - F(200)} - kC_{A0}(1 - X(200))^2\right]$$

$$X = 0 - (25)[(0.075)(0) - ((0.01)(8)(1))^2] = 2$$

$\lambda = 150$:

$$X(\lambda = 150) = X(\lambda = 175) - \Delta\lambda\left[\frac{E(175)X(175)}{1 - F(175)} - kC_{A0}(1 - X(175))^2\right]$$

we need to take an average of $E / (1 - F)$ between $\lambda = 200$ and $\lambda = 150$.

$$X = 2 - (25)\left[\left(\frac{0.075 + 0.0266}{2}\right)(2) - (0.01)(8)(1 - 2)^2\right] = 1.46$$

$\lambda = 125$:

$$X = 1.46 - (25)[(0.0266)(1.46) - (0.01)(8)(1 - 1.46)^2] = 0.912$$

$\lambda = 100$:

$$X = 0.912 - (25)\left[\left(\frac{0.0266 + 0.0221}{2}\right)(0.912) - (0.01)(8)(1 - 0.912)^2\right]$$

$$= 0.372$$

$\lambda = 70$:

$$X = 0.372 - (30)[(0.0221)(0.372) - (0.01)(8)(1 - 0.372)^2] = 1.071$$

$\lambda = 50$:

$$X = 1.071 - (20)[(0.0226)(1.071) - (0.01)(8)(1 - 1.071)^2] = 0.595$$

$\lambda = 40$:

$$X = 0.595 - (10)[(0.0237)(0.595) - (0.01)(8)(1 - 0.595)^2] = 0.585$$

Summary

PFR	76%
Segregation	61%
CSTR	58%
Max. mix	56%

Solved Problems

Running down the values of X along the right-hand side of the preceding equation shows that the oscillations have now damped out. Carrying out the remaining calculations down to the end of the reactor completes Table E13-7.3. The conversion for a condition of maximum mixedness in this reactor is 0.56 or 56%. It is interesting to note that there is little difference in the conversions for the two conditions of complete segregation (61%) and maximum mixedness (56%). With bounds this narrow, there may not be much point in modeling the reactor to improve the predictability of conversion.

For comparison it is left for the reader to show that the conversion for a PFR of this size would be 0.76, and the conversion in a perfectly mixed CSTR with complete micromixing would be 0.58.

TABLE E13-7.3.   MAXIMUM MIXEDNESS MODEL

λ (min)	X
200	0.0
175	2.0
150	1.46
125	0.912
100	0.372
70	1.071
50	0.595
40	0.585
30	0.580
20	0.581
10	0.576
5	0.567
0	0.564

Calculate backwards to reactor exit.

The Intensity Function, $\Lambda(t)$ can be thought of as the probability of a particle escaping the system between a time t and $(t + dt)$ provided the particle is still in the system. Equations (13-62) and (13-64) can be written in a slightly more compact form by making use of the **intensity function**.[16] The intensity function $\Lambda(\lambda)$ is the fraction of fluid in the vessel with age λ that will leave between λ and $\lambda + d\lambda$. We can relate $\Lambda(\lambda)$ to $I(\lambda)$ and $E(\lambda)$ in the following manner:

$$
\begin{bmatrix}
\text{Volume of} \\
\text{fluid leaving} \\
\text{between times} \\
\lambda \text{ and } \lambda + d\lambda
\end{bmatrix}
=
\begin{bmatrix}
\text{Volume of fluid} \\
\text{remaining at} \\
\text{time } \lambda
\end{bmatrix}
\begin{bmatrix}
\text{Fraction of} \\
\text{the fluid with} \\
\text{age } \lambda \text{ that} \\
\text{will leave between} \\
\text{time } \lambda \text{ and } \lambda + d\lambda
\end{bmatrix}
$$

$$[v_0 E(\lambda)\, d\lambda] = [V\, I(\lambda)][\Lambda(\lambda)\, d\lambda] \tag{13-65}$$

Then

$$\Lambda(\lambda) = \frac{E(\lambda)}{\tau I(\lambda)} = -\frac{d\ln[\tau I(\lambda)]}{d\lambda} = \frac{E(\lambda)}{1 - F(\lambda)} \tag{13-66}$$

Combining Equations (13-64) and (13-66) gives

$$\frac{dX}{d\lambda} = \frac{r_A(\lambda)}{C_{A0}} + \Lambda(\lambda)X(\lambda) \tag{13-67}$$

We also note that the exit age, t, is just the sum of the internal age, α, and the life expectancy, λ:

$$t = \alpha + \lambda \tag{13-68}$$

[16]D. M. Himmelblau and K. B. Bischoff, *Process Analysis and Simulation* (New York: Wiley, 1968).

In addition to defining *maximum mixedness* discussed above, Zwietering[17] also generalized a measure of micromixing proposed by Danckwerts[18] and defined the **degree of segregation, J,** as

$$J = \frac{\text{variance of ages between fluid "points"}}{\text{variance of ages of all molecules in system}}$$

A fluid "point" contains many molecules but is small compared to the scale of mixing. The two extremes of the *degree of segregation* are

$$J = 1: \qquad \text{complete segregation}$$
$$J = 0: \qquad \text{maximum mixedness}$$

Equations for the variance and J for the intermediate cases can be found in Zwietering.[17]

13.7.3 Comparing Segregation and Maximum Mixedness Predictions

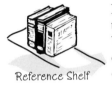

Reference Shelf

In the previous example we saw that the conversion predicted by the segregation model, X_{seg}, was greater than that by the maximum mixedness model X_{max}. Will this always be the case? No. To learn the answer we take the second derivative of the rate law as shown in the Professional Reference Shelf R13.3 on the CD-ROM.

Comparing X_{seg} and X_{mm}

$$\text{If} \quad \left| \frac{\partial^2(-r_A)}{\partial C_A^2} > 0 \right| \qquad \text{then} \qquad X_{seg} > X_{mm}$$

$$\text{If} \quad \left| \frac{\partial^2(-r_A)}{\partial C_A^2} < 0 \right| \qquad \text{then} \qquad X_{mm} > X_{seg}$$

$$\text{If} \quad \left| \frac{\partial^2(-r_A)}{\partial C_A^2} = 0 \right| \qquad \text{then} \qquad X_{mm} = X_{seg}$$

For example, if the rate law is a power law model

$$-r_A = kC_A^n$$

$$\frac{\partial(-r_A)}{\partial C_A} = nkC_A^{n-1}$$

$$\frac{\partial^2(-r_A)}{\partial C_A^2} = n(n-1)kC_A^{n-2}$$

From the product $[(n)(n-1)]$, we see

$$\text{If} \quad n > 1, \quad \text{then} \quad \frac{\partial^2(-r_A)}{\partial C_A^2} > 0 \quad \text{and} \quad X_{seg} > X_{mm}$$

$$\text{If} \quad n < 0, \quad \text{then} \quad \frac{\partial^2(-r_A)}{\partial C_A^2} > 0 \quad \text{and} \quad X_{seg} > X_{mm}$$

---

[17]T. N. Zwietering, *Chem. Eng. Sci., 11,* 1 (1959).
[18]P. V. Danckwerts, *Chem. Eng. Sci., 8,* 93 (1958).

If $0 < n < 1$,   then $\dfrac{\partial^2(-r_A)}{\partial C_A^2} < 0$ and $X_{mm} > X_{seg}$

We note that in some cases X_{seg} is not too different from X_{mm}. However, when one is considering the destruction of toxic waste where $X > 0.99$ is desired, then even a small difference is significant!!

Important point

In this section we have addressed the case where all we have is the RTD and no other knowledge about the flow pattern exists. Perhaps the flow pattern cannot be assumed because of a lack of information or other possible causes. Perhaps we wish to know the extent of possible error from assuming an incorrect flow pattern. We have shown how to obtain the conversion, using only the RTD, for two limiting mixing situations: the earliest possible mixing consistent with the RTD, or maximum mixedness, and mixing only at the reactor exit, or complete segregation. Calculating conversions for these two cases gives bounds on the conversions that might be expected for different flow paths consistent with the observed RTD.

13.8 Using Software Packages

Example 13-7 could have been solved with an ODE solver after fitting $E(t)$ to a polynomial.

Fitting the E(t) Curve to a Polynomial

Some forms of the equation for the conversion as a function of time multiplied by $E(t)$ will not be easily integrated analytically. Consequently, it may be easiest to use ODE software packages. The procedure is straightforward. We recall Equation (13-52)

$$\frac{d\bar{X}}{dt} = X(t)E(t) \tag{13-52}$$

where $\bar{X}$ is the mean conversion and $X(t)$ is the batch reactor conversion at time t. The mean conversion $\bar{X}$ is found by integrating between $t = 0$ and $t = \infty$ or a very large time.

Next we obtain the mole balance on $X(t)$ from a batch reactor

$$\frac{dX}{dt} = -\frac{r_A}{C_{A0}}$$

and would write the rate law in terms of conversion, e.g.,

$$-r_A = kC_{A0}^2(1 - X)^2$$

The ODE solver will combine these equations to obtain $X(t)$ which will be used in Equation (13-56). Finally we have to specify $E(t)$. This equation can be an analytical function such as those for an ideal CSTR,

$$E(t) = \frac{e^{-t/\tau}}{\tau}$$

or it can be polynomial or a combination of polynomials that have been used to fit the experimental RTD data

$$E(t) = a_0 + a_1 t + a_2 t^2 + \cdots \qquad (13-69)$$

or

$$F(t) = b_0 + b_1 t + b_2 t^2 + \cdots \qquad (13-70)$$

We now simply combine Equations (13-52), (13-69), and (13-70) and use an ODE solver. There are three cautions one must be aware of when fitting $E(t)$ to a polynomial. First, you use one polynomial $E_1(t)$ as $E(t)$ increases with time to the top of the curve shown in Figure 13-27. A second polynomial $E_2(t)$ is used from the top as $E(t)$ decreases with time. One needs to match the two curves at the top.

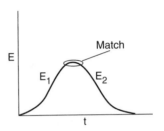

Figure 13-27   Matching $E_1(t)$ and $E_2(t)$.

Summary Notes

Polymath Tutorial

Second, one should be certain that the polynomial used for $E_2(t)$ does not become negative when extrapolated to long times. If it does, then constraints must be placed on the fit using **IF** statements in the fitting program. Finally, one should check that the area under the $E(t)$ curve is virtually one and that the cumulative distribution $F(t)$ at long times is never greater than 1. A tutorial on how to fit the $C(t)$ and $E(t)$ data to a polynomial is given in the *Summary Notes* for Chapter 5 on the CD-ROM and on the web.

Segregation Model

Here we simply use the coupled set of differential equations for the mean or exit conversion, $\overline{X}$, and the conversion $X(t)$ inside a globule at any time, t.

$$\frac{d\overline{X}}{dt} = X(t) E(t) \qquad (13-52)$$

$$\frac{dX}{dt} = \frac{-r_A}{C_{A0}} \qquad (13-71)$$

The rate of reaction is expressed as a function of conversion: for example,

$$-r_A = k_A C_{A0}^2 \frac{(1-X)^2}{(1+X)^2}$$

and the equations are then solved numerically with an ODE solver.

Maximum Mixedness Model

Because most software packages won't integrate backwards, we need to change the variable such that the integration proceeds forward as λ decreases from some large value to zero. We do this by forming a new variable, z, which is the difference between the longest time measured in the $E(t)$ curve, $\bar{T}$, and λ. In the case of Example 13-7, the longest time at which the tracer concentration was measured was 200 minutes (Table E13-7.1). Therefore we will set $\bar{T} = 200$.

$$z = \bar{T} - \lambda = 200 - \lambda$$

$$\lambda = \bar{T} - z = 200 - z$$

Then,

$$\frac{dX}{dz} = -\frac{r_A}{C_{A0}} - \frac{E(\bar{T}-z)}{1-F(T-z)}X \qquad (13\text{-}72)$$

One now integrates between the limit $z = 0$ and $z = 200$ to find the exit conversion at $z = 200$ which corresponds to $\lambda = 0$.

In fitting $E(t)$ to a polynomial, one has to make sure that the polynomial does not become negative at large times. Another concern in the maximum mixedness calculations is that the term $1 - F(\lambda)$ does not go to zero. Setting the maximum value of $F(t)$ at 0.999 rather than 1.0 will eliminate this problem. It can also be circumvented by integrating the polynomial for $E(t)$ to get $F(t)$ and then setting the maximum value of $F(t)$ at 0.999. If $F(t)$ is ever greater than one when fitting a polynomial, the solution will blow up when integrating Equation (13-72) numerically.

Example 13–8   Using Software to Make Maximum Mixedness
Model Calculations

Use an ODE solver to determine the conversion predicted by the **maximum mixedness model** for the $E(t)$ curve given in Example E13-7.

Solution

Because of the nature of the $E(t)$ curve, it is necessary to use two polynomials, a third order and a fourth order, each for a different part of the curve to express the RTD, $E(t)$, as a function of time. The resulting $E(t)$ curve is shown in Figure E13-8.1.

To use Polymath to carry out the integration, we change our variable from λ to z using the largest time measurements that were taken from $E(t)$ in Table E13-7.1, which is 200 min:

First, we fit $E(t)$.

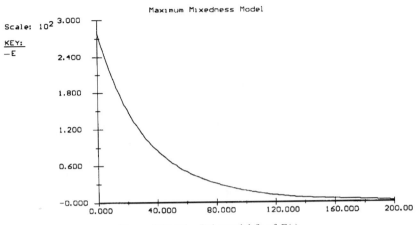

Figure E13-8.1 Polynomial fit of $E(t)$.

$$z = 200 - \lambda$$

The equations to be solved are

$$\lambda = 200 - z \qquad\qquad\qquad \text{(E13-8.1)}$$

Maximum
mixedness model

$$\frac{dX}{dz} = -\frac{r_A}{C_{A0}} - \frac{E(200-z)}{1-F(200-z)} X \qquad \text{(E13-8.2)}$$

For values of λ less than 70, we use the polynomial

$$E_1(\lambda) = 4.447e^{-10}\lambda^4 - 1.180e^{-7}\lambda^3 + 1.353e^{-5}\lambda^2 - 8.657e^{-4}\lambda + 0.028 \quad \text{(E13-8.3)}$$

For values of λ greater than 70, we use the polynomial

$$E_2(\lambda) = -2.640e^{-9}\lambda^3 + 1.3618e^{-6}\lambda^2 - 2.407e^{-4}\lambda + 0.015 \quad \text{(E13-8.4)}$$

$$\frac{dF}{d\lambda} = E(\lambda) \qquad\qquad\qquad \text{(E13-8.5)}$$

with $z = 0$ $(\lambda = 200)$, $X = 0$, $F = 1$ [i.e., $F(\lambda) = 0.999$]. **Caution:** Because $[1 - F(\lambda)]^{-1}$ tends to infinity at $F = 1$, $(z = 0)$, we set the maximum value of F at 0.999 at $z = 0$.

The Polymath equations are shown in Table E13-8.1. The solution is

$$\text{at } z = 200 \qquad X = 0.563$$

The conversion predicted by the maximum mixedness model is 56.3%.

Table E13-8.1.   Polymath Program for Maximum Mixedness Model

ODE Report (RKF45)

Differential equations as entered by the user
[1] d(x)/d(z) = -(ra/cao+E/(1-F)*x)

Explicit equations as entered by the user
[1] cao = 8
[2] k = .01
[3] lam = 200-z
[4] ca = cao*(1-x)
[5] E1 = 4.44658e-10*lam^4-1.1802e-7*lam^3+1.35358e-5*lam^2-.000865652*lam+.028004
[6] E2 = -2.64e-9*lam^3+1.3618e-6*lam^2-.00024069*lam+.015011
[7] F1 = 4.44658e-10/5*lam^5-1.1802e-7/4*lam^4+1.35358e-5/3*lam^3-.000865652/2*lam^2+.028004*lam
[8] F2 = -(-9.30769e-8*lam^3+5.02846e-5*lam^2-.00941*lam+.618231-1)
[9] ra = -k*ca^2
[10] E = if (lam<=70) then (E1) else (E2)
[11] F = if (lam<=70) then (F1) else (F2)
[12] EF = E/(1-F)

Independent variable
variable name : z
initial value : 0
final value : 200

Living Example Problem

Polynomials used to
fit $E(t)$ and $F(t)$

13.8.1 Heat Effects

If tracer tests are carried out isothermally and then used to predict nonisothermal conditions, one must couple the segregation and maximum mixedness models with the energy balance to account for variations in the specific reaction rate. This approach will only be valid for liquid phase reactions because the volumetric flow rate remains constant. For adiabatic operation and $\Delta \hat{C}_P = 0$,

$$T = T_0 + \frac{(-\Delta H_{Rx})}{\Sigma \, \Theta_i C_{P_i}} X \qquad \text{(E8-4.9)}$$

As before, the specific reaction rate is

$$k = k_1 \exp\left[\frac{E}{R}\left(\frac{1}{T_1} - \frac{1}{T}\right)\right] \qquad \text{(T8-2.3)}$$

Assuming that $E(t)$ is unaffected by temperature variations in the reactor, one simply solves the segregation and maximum mixedness models, accounting for the variation of k with temperature [i.e., conversion; see Problem P13-2(i)].

13.9 RTD and Multiple Reactions

As discussed in Chapter 6, when multiple reactions occur in reacting systems, it is best to work in concentrations, moles, or molar flow rates rather than conversion.

13.9.1 Segregation Model

In the **segregation model** we consider each of the globules in the reactor to have different concentrations of reactants, C_A, and products, C_P. These globules are mixed together immediately upon exiting to yield the exit concentration of A, $\overline{C_A}$, which is the average of all the globules exiting:

$$\overline{C_A} = \int_0^\infty C_A(t) E(t)\, dt \tag{13-73}$$

$$\overline{C_B} = \int_0^\infty C_B(t) E(t)\, dt \tag{13-74}$$

The concentrations of the individual species, $C_A(t)$ and $C_B(t)$, in the different globules are determined from batch reactor calculations. For a constant-volume batch reactor, where q reactions are taking place, the coupled mole balance equations are

$$\frac{dC_A}{dt} = r_A = \sum_{i=1}^{i=q} r_{iA} \tag{13-75}$$

$$\frac{dC_B}{dt} = r_B = \sum_{i=1}^{i=q} r_{iB} \tag{13-76}$$

These equations are solved simultaneously with

$$\frac{d\overline{C_A}}{dt} = C_A(t) E(t) \tag{13-77}$$

$$\frac{d\overline{C_B}}{dt} = C_B(t) E(t) \tag{13-78}$$

to give the exit concentration. The RTDs, $E(t)$, in Equations (13-77) and (13-78) are determined from experimental measurements and then fit to a polynomial.

13.9.2 Maximum Mixedness

For the **maximum mixedness model,** we write Equation (13-62) for each species and replace r_A by the net rate of formation

$$\frac{dC_A}{d\lambda} = -\sum r_{iA} + (C_A - C_{A0}) \frac{E(\lambda)}{1 - F(\lambda)} \tag{13-79}$$

$$\frac{dC_B}{d\lambda} = -\sum r_{iB} + (C_B - C_{B0}) \frac{E(\lambda)}{1 - F(\lambda)} \tag{13-80}$$

After substitution for the rate laws for each reaction (e.g., $r_{1A} = k_1 C_A$), these equations are solved numerically by starting at a very large value of λ, say $\overline{T} = 200$, and integrating backwards to $\lambda = 0$ to yield the exit concentrations C_A, C_B, ...

We will now show how different RTDs with the *same* mean residence time can produce different product distributions for multiple reactions.

Example 13–9   RTD and Complex Reactions

Consider the following set of liquid-phase reactions:

$$A + B \xrightarrow{k_1} C$$

$$A \xrightarrow{k_2} D$$

$$B + D \xrightarrow{k_3} E$$

which are occurring in two different reactors with the same mean residence time $t_m = 1.26$ min. However, the RTD is very different for each of the reactors, as can be seen in Figures E13-9.1 and E13-9.2.

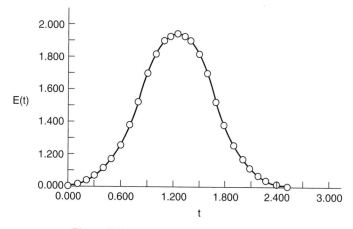

Figure E13-9.1 $E_1(t)$: asymmetric distribution.

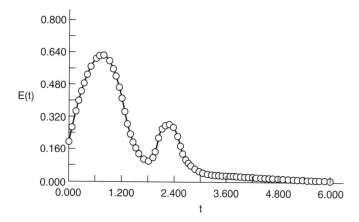

Figure E13-9.2 $E_2(t)$: bimodal distribution.

(a) Fit a polynomial to the RTDs.

(b) Determine the product distribution (e.g., $S_{C/D}$, $S_{D/E}$) for
  1. The segregation model
  2. The maximum mixedness model

Additional Information

$k_1 = k_2 = k_3 = 1$ in appropriate units at 350K.

Solution

Segregation Model

Combining the mole balance and rate laws for a constant-volume batch reactor (i.e., globules), we have

$$\frac{dC_A}{dt} = r_A = r_{1A} + r_{2A} = -k_1 C_A C_B - k_2 C_A \qquad \text{(E13-9.1)}$$

$$\frac{dC_B}{dt} = r_B = r_{1B} + r_{3B} = -k_1 C_A C_B - k_3 C_B C_D \qquad \text{(E13-9.2)}$$

$$\frac{dC_C}{dt} = r_C = r_{1C} = k_1 C_A C_B \qquad \text{(E13-9.3)}$$

$$\frac{dC_D}{dt} = r_D = r_{2D} + r_{3D} = k_2 C_A - k_3 C_B C_D \qquad \text{(E13-9.4)}$$

$$\frac{dC_E}{dt} = r_E = r_{3E} = k_3 C_B C_D \qquad \text{(E13-9.5)}$$

and the concentration for each species exiting the reactor is found by integrating the equation

$$\frac{d\overline{C_i}}{dt} = C_i E(t) \qquad \text{(E13-9.6)}$$

over the life of the $E(t)$ curve. For this example the life of the $E_1(t)$ is 2.42 minutes (Figure E13-9.1), and the life of $E_2(t)$ is 6 minutes (Figure E13-9.2).

The initial conditions are $t = 0$, $C_A = C_B = 1$, and $C_C = C_D = C_E = 0$.

The Polymath program used to solve these equations is shown in Table E13-9.1 for the asymmetric RTD, $E_1(t)$.

With the exception of the polynomial for $E_2(t)$, an identical program to that in Table E13-9.1 for the bimodal distribution is given on the CD-ROM. A comparison of the exit concentration and selectivities of the two RTD curves is shown in Table E13-9.2.

TABLE E13-9.1.   POLYMATH PROGRAM
FOR SEGREGATION MODEL WITH ASYMMETRIC RTD (MULTIPLE REACTIONS)

ODE Report (RKF45)

Differential equations as entered by the user
[1] d(ca)/d(t) = ra
[2] d(cb)/d(t) = rb
[3] d(cc)/d(t) = rc
[4] d(cabar)/d(t) = ca*E
[5] d(cbbar)/d(t) = cb*E
[6] d(ccbar)/d(t) = cc*E
[7] d(cd)/d(t) = rd
[8] d(ce)/d(t) = re
[9] d(cdbar)/d(t) = cd*E
[10] d(cebar)/d(t) = ce*E

Explicit equations as entered by the user
[1] k1 = 1
[2] k2 = 1
[3] k3 = 1
[4] E1 = -2.104*t^4+4.167*t^3-1.596*t^2+0.353*t-0.004
[5] E2 = -2.104*t^4+17.037*t^3-50.247*t^2+62.964*t-27.402
[6] rc = k1*ca*cb
[7] re = k3*cb*cd
[8] ra = -k1*ca*cb-k2*ca
[9] rb = -k1*ca*cb-k3*cb*cd
[10] E = if(t<=1.26)then(E1)else(E2)
[11] rd = k2*ca-k3*cb*cd

Living Example Problem

TABLE E13-9.2.   SEGREGATION MODEL RESULTS

| Asymmetric Distribution | | Bimodal Distribution | |
The solution for $E_1(t)$ is:		The solution for $E_2(t)$ is:	
$\overline{C_A} = 0.151$	$\overline{C_E} = 0.178$	$\overline{C_A} = 0.245$	$\overline{C_E} = 0.162$
$\overline{C_B} = 0.454$	$\overline{X} = 84.9\%$	$\overline{C_B} = 0.510$	$\overline{X} = 75.5\%$
$\overline{C_C} = 0.357$	$S_{C/D} = 1.18$	$\overline{C_C} = 0.321$	$S_{C/D} = 1.21$
$\overline{C_D} = 0.303$	$S_{D/E} = 1.70$	$\overline{C_D} = 0.265$	$S_{D/E} = 1.63$

Maximum Mixedness Model

The equations for each species are

$$\frac{dC_A}{d\lambda} = k_1 C_A C_B + k_2 C_A + (C_A - C_{A0})\frac{E(\lambda)}{1 - F(\lambda)} \tag{E13-9.7}$$

$$\frac{dC_B}{d\lambda} = k_1 C_A C_B + k_3 C_B C_D + (C_B - C_{B0})\frac{E(\lambda)}{1 - F(\lambda)} \tag{E13-9.8}$$

$$\frac{dC_C}{d\lambda} = -k_1 C_A C_B + (C_C - C_{C0})\frac{E(\lambda)}{1 - F(\lambda)} \tag{E13-9.9}$$

$$\frac{dC_D}{d\lambda} = -k_2 C_A + k_3 C_B C_D + (C_D - C_{D0})\frac{E(\lambda)}{1 - F(\lambda)} \tag{E13-9.10}$$

$$\frac{dC_E}{d\lambda} = -k_3 C_B C_D + (C_E - C_{E0}) \frac{E(\lambda)}{1 - F(\lambda)} \tag{E13-9.11}$$

Solved Problems

The Polymath program for the bimodal distribution, $E_2(t)$, is shown in Table E13-9.3. The Polymath program for the asymmetric distribution is identical with the exception of the polynomial fit for $E_1(t)$ and is given on the CD-ROM. A comparison of the exit concentration and selectivities of the two RTD distributions is shown in Table E13-9.4.

Living Example Problem

TABLE E13-9.3.   POLYMATH PROGRAM
FOR MAXIMUM MIXEDNESS MODEL WITH BIMODAL DISTRIBUTION (MULTIPLE REACTIONS)

ODE Report (RKF45)

Differential equations as entered by the user
[1] d(ca)/d(z) = -(-ra+(ca-cao)*EF)
[2] d(cb)/d(z) = -(-rb+(cb-cbo)*EF)
[3] d(cc)/d(z) = -(-rc+(cc-cco)*EF)
[4] d(F)/d(z) = -E
[5] d(cd)/d(z) = -(-rd+(cd-cdo)*EF)
[6] d(ce)/d(z) = -(-re+(ce-ceo)*EF)

Explicit equations as entered by the user
[1] cbo = 1
[2] cao = 1
[3] cco = 0
[4] cdo = 0
[5] ceo = 0
[6] lam = 6-z
[7] k2 = 1
[8] k1 = 1
[9] k3 = 1
[10] rc = k1*ca*cb
[11] re = k3*cb*cd
[12] E1 = 0.47219*lam^4-1.30733*lam^3+0.31723*lam^2+0.85688*lam+0.20909
[13] E2 = 3.83999*lam^6-58.16185*lam^5+366.2097*lam^4-1224.66963*lam^3+2289.84857*lam^2-2265.62125*lam+925.46463
[14] E3 = 0.00410*lam^4-0.07593*lam^3+0.52276*lam^2-1.59457*lam+1.84445
[15] rb = -k1*ca*cb-k3*cb*cd
[16] ra = -k1*ca*cb-k2*ca
[17] rd = k2*ca-k3*cb*cd
[18] E = if(lam<=1.82)then(E1)else(if(lam<=2.8)then(E2)else(E3))
[19] EF = E/(1-F)

TABLE E13-9.4.   MAXIMUM MIXEDNESS MODEL RESULTS

Asymmetric Distribution		*Bimodal Distribution*	
The solution for $E_1(t)$ (1) is:		The solution for $E_2(t)$ (2) is:	
$C_A = 0.161$	$C_E = 0.192$	$C_A = 0.266$	$C_E = 0.190$
$C_B = 0.467$	$X = 83.9\%$	$C_B = 0.535$	$X = 73.4\%$
$C_C = 0.341$	$S_{C/D} = 1.11$	$C_C = 0.275$	$S_{C/D} = 1.02$
$C_D = 0.306$	$S_{D/E} = 1.59$	$C_D = 0.269$	$S_{D/E} = 1.41$

Solved Problems

Calculations similar to those in Example 13-9 are given in an example on the CD-ROM for the series reaction

$$A \xrightarrow{k_1} B \xrightarrow{k_2} C$$

In addition, the effect of the variance of the RTD on the parallel reactions in Example 13-9 and on the series reaction in the CD-ROM is shown on the CD-ROM.

Closure After completing this chapter the reader will use the tracer concentration time data to calculate the external age distribution function $E(t)$, the cumulative distribution function $F(t)$, the mean residence time, t_m, and the variance, σ^2. The reader will be able to sketch $E(t)$ for ideal reactors, and by comparing $E(t)$ from experiment with $E(t)$ for ideal reactors (PFR, PBR, CSTR, laminar flow reactor) the reader will be able to diagnose problems in real reactors. The reader will also be able to couple RTD data with reaction kinetics to predict the conversion and exit concentrations using the segregation and the maximum mixedness models without using any adjustable parameters. By analyzing the second derivative of the reaction rate with respect to concentration, the reader will be able to determine whether the segregation model or maximum mixedness model will give the greater conversion.

SUMMARY

1. The quantity $E(t)\,dt$ is the fraction of material exiting the reactor that has spent between time t and $t + dt$ in the reactor.
2. The mean residence time

$$t_m = \int_0^\infty tE(t)\,dt = \tau \tag{S13-1}$$

is equal to the space time τ for constant volumetric flow, $v = v_0$.
3. The variance about the mean residence time is

$$\sigma^2 = \int_0^\infty (t - t_m)^2 E(t)\,dt \tag{S13-2}$$

4. The cumulative distribution function $F(t)$ gives the fraction of effluent material that has been in the reactor a time t or less:

$$F(t) = \int_0^t E(t)\,dt$$

$1 - F(t) =$ fraction of effluent material that has been in   (S13-3)
the reactor a time t or longer

5. The RTD functions for an ideal reactor are

Plug-flow: $E(t) = \delta(t - \tau)$ (S13-4)

CSTR: $E(t) = \dfrac{e^{-t/\tau}}{\tau}$ (S13-5)

Laminar flow: $E(t) = 0 \qquad t < \dfrac{\tau}{2}$ (S13-6)

$E(t) = \dfrac{\tau^2}{2t^3} \qquad t \geq \dfrac{\tau}{2}$ (S13-7)

6. The dimensionless residence time is

$$\Theta = \frac{t}{\tau} \tag{S13-8}$$

$$E(\Theta) = \tau E(t) \tag{S13-9}$$

7. The internal-age distribution, $[I(\alpha)\, d\alpha]$, gives the fraction of material inside the reactor that has been inside between a time α and a time $(\alpha + d\alpha)$.

8. Segregation model: The conversion is

$$\bar{X} = \int_0^\infty X(t)E(t)\, dt \tag{S13-10}$$

and for multiple reactions

$$\bar{C}_A = \int_0^\infty C_A(t)E(t)\, (dt)$$

9. Maximum mixedness: Conversion can be calculated by solving the following equations:

$$\frac{dX}{d\lambda} = \frac{r_A}{C_{A0}} + \frac{E(\lambda)}{1 - F(\lambda)}(X) \tag{S13-11}$$

and for multiple reactions

$$\frac{dC_A}{d\lambda} = -r_{A_{net}} + (C_A - C_{A0})\frac{E(\lambda)}{1 - F(\lambda)} \tag{S13-12}$$

$$\frac{dC_B}{d\lambda} = -r_{B_{net}} + (C_B - C_{B0})\frac{E(\lambda)}{1 - F(\lambda)} \tag{S13-13}$$

from λ_{max} to $\lambda = 0$. To use an ODE solver let $z = \lambda_{max} - \lambda$.

CD-ROM MATERIAL

Summary Notes

Links

Solved Problems

- **Learning Resources**
  1. *Summary Notes*
  2. *Web Material Links*
     A. The Attainable Region Analysis
        www.engin.umich.edu/~cre/Chapters/ARpages/Intro/intro.htm and
        www.wits.ac.za/fac/engineering/promat/aregion
  4. *Solved Problems*
     A. Example CD13-1 Calculate the exit concentrations series reaction

$$A \longrightarrow B \longrightarrow C$$

     B. Example CD13-2 Determination of the effect of variance on the exit concentrations for the series reaction

$$A \longrightarrow B \longrightarrow C$$

Living Example Problem

- **Living Example Problems**
    1. *Example 13–6 Laminar Flow Reactor*
    2. *Example 13–8 Using Software to Make Maximum Mixedness Model Calculations*
    3. *Example 13–9 RTD and Complex Reactions*
    4. *Example CD13-1 A → B → C Effect of RTD*
    5. *Example CD13-2 A → B → C Effect of Variance*
- **Professional Reference Shelf**
    13R.1. *Fitting the Tail*

    Whenever there are dead zones into which the material diffuses in and out, the C and E curves may exhibit long tails. This section shows how to analytically describe fitting these tails to the curves.

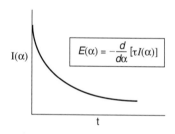

    $$E(t) = ae^{-bt}$$

    $$b = \text{slope of } \ln E \text{ vs. } t$$

    $$a = be^{bt_1}[1 - F(t_1)]$$

    13R.2. *Internal-Age Distribution*

    The internal-age distribution currently in the reactor is given by the distribution of ages with respect to how long the molecules have been in the reactor.

    The equation for the internal-age distribution is derived along and an example is given showing how it is applied to catalyst deactivation in a "fluidized CSTR."

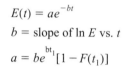

$I(\alpha)$

$$E(\alpha) = -\frac{d}{d\alpha}[\tau I(\alpha)]$$

t

$$E(\alpha) = -\frac{d}{d\alpha}[\tau I(\alpha)]$$

Example 13R2.1 Mean Catalyst Activity in a Fluidized Bed Reactor.

13R.3. *Comparing X_{seg} with X_{mm}*

The derivation of equations using the second derivative criteria

$$\frac{\partial^2(-r_A)}{\partial C_A^2} = ?$$

is carried out.

QUESTIONS AND PROBLEMS

The subscript to each of the problem numbers indicates the level of difficulty: A, least difficult; D, most difficult.

$$A = \bullet \quad B = \blacksquare \quad C = \blacklozenge \quad D = \blacklozenge\blacklozenge$$

P13-1$_A$ Read over the problems of this chapter. Make up an original problem that uses the concepts presented in this chapter. The guidelines are given in Problem P4-1. RTDs from real reactors can be found in *Ind. Eng. Chem.*, *49*, 1000 (1957); *Ind. Eng. Chem. Process Des. Dev.*, *3*, 381 (1964); *Can. J. Chem. Eng.*, *37*, 107 (1959); *Ind. Eng. Chem.*, *44*, 218 (1952); *Chem. Eng. Sci.*, *3*, 26 (1954); and *Ind. Eng. Chem.*, *53*, 381 (1961).

P13-2$_A$ **What if...**

    (a) **Example 13-1.** What fraction of the fluid spends nine minutes or longer in the reactor?

    (b) The combinations of ideal reactors modeled the following real reactors, given $E(\Theta)$, $F(\Theta)$, or $1 - F(\Theta)$.

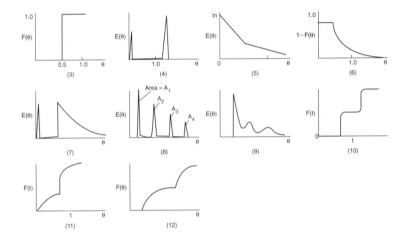

    Suggest a model for each figure.

    (c) **Example 13-3.** How would the $E(t)$ change if τ_{PF} as reduced by 50% and τ_{CSTR} was increased by 50%?

    (d) **Example 13-4.** For 75% conversion, what are the relative sizes of the CSTR, PFR, and LFR?

    (e) **Example 13-5.** How does the mean conversion compare with the conversion calculated with the same, t_m applied to an ideal PFR and CSTR? Can you give examples of $E(t)$ where this calculation would and would not be a good estimate of X?

    (f) **Example 13-6.** Load the *Living Example Problem*. How would your results change if $T = 40°C$? How would your answer change if the reaction was pseudo first order with $kC_{A0} = 4 \times 10^{-3}/s$? What if the reaction were carried out adiabatically where

$$C_{P_A} = C_{P_B} = 20 \ cal/mol/K, \ \Delta H_{Rx} = -10 \ kcal/mol$$
$$k = 0.01 \ dm^3/mol/min \ at \ 25°C \ with \ E = 8 \ kcal/mol$$

(g) **Example 13-7.** Load the *Living Example Problem*. How does the X_{seg} and X_{MM} compare with the conversion calculated for a PFR and a CSTR at the mean residence time?

(h) **Example 13-8.** Load the *Living Example Problem*. How would your results change if the reaction was pseudo first order with $k_1 = C_{A0}k = 0.08$ min^{-1}? If the reaction was third order with $kC_{A0}^2 = 0.08$ min^{-1}? The reaction was half order with $kC_{A0}^{1/2} = 0.08$ min^{-1}. Describe any trends.

(i) **Example 13-9.** Load the *Living Example Problem*. If the activation energies in cal/mol and $E_1 = 5{,}000$, $E_2 = 1{,}000$, and $E_3 = 9{,}000$, how would the selectives and conversion of A change as the temperature was raised or lowered around 350 K?

(j) **Heat Effects.** Redo *Living Example Problems* 13-7 and 13-8 for the case when the reaction is carried out adiabatically with
   (1) Exothermic reaction with

$$T(K) = T_0 + \left(\frac{-\Delta H_{Rx}}{C_P}\right)X = 320 + 150X \qquad (13\text{-}2.j.1)$$

   with k given at 320 K and $E = 10{,}000$ cal/mol.
   (2) Endothermic reaction with

$$T = 320 - 100X \qquad (13\text{-}2.j.2)$$

   and $E = 45$ kJ/mol. How will your answers change?

(k) you were asked to compare the results from **Example 13-9** for the asymmetric and bimodal distributions in Tables E13-9.2 and E13-9.4. What similarities and differences do you observe? What generalizations can you make?

(l) Repeat 13-2(h) above using the RTD in Polymath program E13-8 to predict and compare conversions predicted by the segregation model.

(m) the reaction in **Example 13-5** was half order with $kC_{A0}^{1/2} = 0.08$ min^{-1}? How would your answers change? *Hint:* Modify the Living Example 13-8 program.

(n) you were asked to vary the specific reaction rates k_1 and k_2 in the series reaction A $\xrightarrow{k_1}$ B $\xrightarrow{k_2}$ C given on the Solved Problems CD-ROM? What would you find?

(o) you were asked to vary the isothermal temperature in **Example 13-9** from 300 K, at which the rate constants are given, up to a temperature of 500 K? The activation energies in cal/mol are $E_1 = 5000$, $E_2 = 7000$, and $E_3 = 9000$. How would the selectivity change for each RTD curve?

(p) the reaction in **Example 13-7** were carried out adiabatically with the same parameters as those in Equation [P13-2(j).1]? How would your answers change?

Heat effects

(q) If the reaction in **Examples 13-8** and **13-5** were endothermic and carried out adiabatically with

$$T(K) = 320 - 100X \quad \text{and} \quad E = 45 \text{ kJ/mol} \qquad [P13\text{-}2(j).1]$$

how would your answers change? What generalizations can you make about the effect of temperature on the results (e.g., conversion) predicted from the RTD?

(r) If the reaction in Example 8-12 were carried out in the reactor described by the RTD in Example 13-9 with the exception that RTD is in seconds rather than minutes (i.e., $t_m = 1.26$ s), how would your answers change?

P13-3$_C$ Show that for a first-order reaction

$$A \longrightarrow B$$

the exit concentration maximum mixedness equation

$$\frac{dC_A}{d\lambda} = kC_A + \frac{E(\lambda)}{1-F(\lambda)}(C_A - C_{A0}) \qquad \text{(P13-3.1)}$$

is the same as the exit concentration given by the segregation model

$$C_A = C_{A0}\int_0^\infty E(t)e^{-kt}\,dt \qquad \text{(P13-3.2)}$$

[*Hint:* Verify

$$C_A(\lambda) = \frac{C_{A0}e^{k\lambda}}{1-F(\lambda)}\int_\lambda^\infty E(t)e^{-kt}\,dt \qquad \text{(P13-3.3)}$$

is a solution to Equation (P13-3.1).]

P13-4$_C$ The first-order reaction

$$A \longrightarrow B$$

with $k = 0.8$ min^{-1} is carried out in a real reactor with the following RTD function

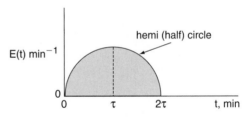

For $2\tau \geq t \geq 0$ then $E(t) = \sqrt{\tau^2 - (t-\tau)^2}$ min^{-1}(hemi circle)
For $t > 2\tau$ then $E(t) = 0$
(a) What is the mean residence time?
(b) What is the variance?
(c) What is the conversion predicted by the segregation model?
(d) What is the conversion predicted by the maximum mixedness model?

P13-5$_B$ A step tracer input was used on a real reactor with the following results:
    For $t \leq 10$ min, then $C_T = 0$
    For $10 \leq t \leq 30$ min, then $C_T = 10$ g/dm^3
    For $t \geq 30$ min, then $C_T = 40$ g/dm^3
The second-order reaction A $\rightarrow$ B with $k = 0.1$ dm^3/mol · min is to be carried out in the real reactor with an entering concentration of A of 1.25 mol/dm^3 at a volumetric flow rate of 10 dm^3/min. Here k is given at 325 K.
(a) What is the mean residence time t_m?
(b) What is the variance σ^2?
(c) What conversions do you expect from an ideal PFR and an ideal CSTR in a real reactor with t_m?

(d) What is the conversion predicted by
    (1) the segregation model?
    (2) the maximum mixedness model?
(e) What conversion is predicted by an ideal laminar flow reactor?
(f) Calculate the conversion using the segregation model assuming
    $T(\text{K}) = 325 - 500X$ and E/R = 5000K.

P13-6$_B$ The following $E(t)$ curves were obtained from a tracer test on two tubular reactors in which dispersion is believed to occur.

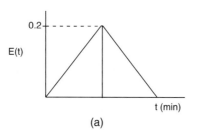
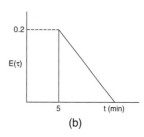

(a)   (b)

Figure P13-6$_B$ (a) RTD Reactor A; (b) RTD Reactor B.

A second-order reaction

$$A \xrightarrow{\ k\ } B \quad \text{with} \quad kC_{A0} = 0.2 \text{ min}^{-1}$$

is to be carried out in this reactor. There is no dispersion occurring either upstream or downstream of the reactor, but there is dispersion inside the reactor. (a) Find the quantities asked for in parts (a) through (e) in problem P13-5$_B$ for reactor A. (b) Repeat for Reactor B.

P13-7$_B$ The irreversible liquid phase reaction

$$A \xrightarrow{\ k_1\ } B$$

is half order in A. The reaction is carried out in a nonideal CSTR, which can be modeled using the segregation model. RTD measurements on the reactor gave values of $\tau = 5$ min and $\sigma = 3$ min. For an entering concentration of pure A of 1.0 mol/dm^3 the mean exit conversion was 10%. Estimate the specific reaction rate constant, k_1. *Hint:* Assume a Gaussian distribution.

P13-8$_B$ The third-order liquid-phase reaction with an entering concentration of 2M.

$$A \xrightarrow{\ k_3\ } B$$

was carried out in a reactor that has the following RTD

$$\begin{aligned}
E(t) &= 0 & \text{for} &\quad t < 1 \text{ min} \\
E(t) &= 1.0 \text{ min}^{-1} & \text{for} &\quad 1 \le t \le 2 \text{ min} \\
E(t) &= 0 & \text{for} &\quad t > 2 \text{ min}
\end{aligned}$$

(a) For isothermal operation, what is the conversion predicted by
    1) a CSTR, a PFR, an LFR, and the segregation model, X_{seg}.
      Hint: Find t_m (i.e., τ) from the data and then use it with $E(t)$ for each
      of the ideal reactors.
    2) the maximum mixedness model, X_{MM}. Plot X vs. z (or λ) and explain
      why the curve looks the way it does.
(b) For isothermal operation, at what temperature is the discrepancy between X_{seg} and X_{MM} the greatest in the range $300 < T < 350$ K.

(c) Suppose the reaction is carried out adiabatically with an entering temperature of 305 K. Calculate X_{seq}.

Additional Information

$k = 0.3$ dm^6/mol^2/min at 300 K	$E/R = 20,000$
$\Delta H_{RX} = -40,000$ cal/mol	$C_{P_A} = C_{P_B} = 25$ cal/mol/K

P13-9$_A$ Consider again the nonideal reactor characterized by the RTD data in Example 13-5. The irreversible gas-phase nonelementary reaction

$$A + B \longrightarrow C + D$$

is first order in A and second order in B and is to be carried out isothermally. Calculate the conversion for:

(a) A PFR, a laminar flow reactor with complete segregation, and a CSTR.

(b) The cases of complete segregation and maximum mixedness.

Also

(c) Plot $I(\alpha)$ and $\Lambda(\lambda)$ as a function of time and then determine the mean age $\bar{\alpha}$ and the mean life expectancy $\bar{\lambda}$.

(d) How would your answers change if the reaction is carried out adiabatically with parameter values given by Equation [P13-2(h).1]?

Additional information:

$$C_{A0} = C_{B0} = 0.0313 \text{ mol/dm}^3, V = 1000 \text{ dm}^3,$$
$$v_0 = 10 \text{ dm}^3/\text{s}, k = 175 \text{ dm}^6/\text{mol}^2\cdot\text{s at 320 K}.$$

P13-10$_B$ An irreversible first-order reaction takes place in a long cylindrical reactor. There is no change in volume, temperature, or viscosity. The use of the simplifying assumption that there is plug flow in the tube leads to an estimated degree of conversion of 86.5%. What would be the actually attained degree of conversion if the real state of flow is laminar, with negligible diffusion?

P13-11$_A$ Consider a PFR, CSTR, and LFR.

(a) Evaluate the first movement about the mean $m_1 = \int_0^\infty (t - \tau) \ E(t)dt$ for a PFR, a CSTR, and a laminar flow reactor.

(b) Calculate the conversion in each of these ideal reactors for a second-order liquid-phase reaction with Da = 1.0 ($\tau = 2$ min and $kC_{A0} = 0.5$ min^{-1}).

P13-12$_B$ For the catalytic reaction

$$A \xrightarrow[\text{cat}]{} C + D$$

the rate law can be written as

$$-r_A = \frac{kC_A}{(1 + K_A C_A)^2}$$

Which will predict the highest conversion, the maximum mixedness model or the segregation model?

Hint: Are the different ranges of the conversion where one model will dominate over the other.

Additional Information

$C_{A0} = 2$ mol/dm^3

$k = 0.01$ dm^3/mol · s

$K_A = 0.25$ dm^3/mol

P13-13$_D$ Too unlucky! Skip!

P13-14$_C$ The second-order liquid-phase reaction

$$2A \xrightarrow{k_{1A}} B$$

is carried out in a nonideal CSTR. At 300 K the specific reaction rate is 0.5 dm^3/mol·min. In a tracer test, the tracer concentration rose linearly up to 1 mg/dm^3 at 1.0 minutes and then decreased linearly to zero at exactly 2.0 minutes. Pure A enters the reactor at a temperature of 300 K.

(a) Calculate the conversion predicted by segregation and maximum mixedness models.

(b) Now consider that a second reaction also takes place

$$A + B \xrightarrow{k_{2C}} C, \qquad k_{2C} = 0.12 \text{ dm}^3/\text{mol} \cdot \text{min}$$

Compare the selectivities $\tilde{S}_{B/C}$ predicted by the segregation and maximum mixedness models.

(c) Repeat (a) for adiabatic operation.

Additional information:

$$C_{P_A} = 50 \text{ J/mol} \cdot \text{K}, \qquad C_{P_B} = 100 \text{ J/mol} \cdot \text{K}, \qquad E = 10 \text{ kcal/mol}$$

$$\Delta H_{Rx1A} = -7500 \text{ J/mol} \qquad C_{A0} = 2 \text{ mol/dm}^3$$

P13-15$_B$ The reactions described in Problem P6-16$_B$ are to be carried out in the reactor whose RTD is described in Problem CDP13-N$_B$.
Determine the exit selectivities

(a) Using the segregation model.

(b) Using the maximum mixedness model.

(c) Compare the selectivities in parts (a) and (b) with those that would be found in an ideal PFR and ideal CSTR in which the space time is equal to the mean residence time.

(d) What would your answers to parts (a) to (c) be if the reactor in Problem P13-7$_B$ were used?

P13-16$_B$ The reactions described in Example 6-10 are to be carried out in the reactor whose RTD is described in Example 13-7 with $C_{A0} = C_{B0} = 0.05$ mol/dm^3.

(a) Determine the exit selectivities using the segregation model.

(b) Determine the exit selectivities using the maximum mixedness model.

(c) Compare the selectivities in parts (a) and (b) with those that would be found in an ideal PFR and ideal CSTR in which the space time is equal to the mean residence time.

(d) What would your answers to parts (a) to (c) be if the RTD curve rose from zero at $t = 0$ to a maximum of 50 mg/dm^3 after 10 min, and then fell linearly to zero at the end of 20 min?

P13-17$_B$ The reactions described in Problem P6-12$_B$ are to be carried out in the reactor whose RTD is described in Example 13-9.

(a) Determine the exit selectivities using the segregation model.

(b) Determine the exit selectivities using the maximum mixedness model.

(c) Compare the selectivities in parts (a) and (b) with those that would be found in an ideal PFR and ideal CSTR in which the space time is equal to the mean residence time.

P13-18$_C$ The reactions described in Problem P6-11$_B$ are to be carried out in the reactor whose RTD is described in Problem CDP13-I$_B$ with C_{A0} = 0.8 mol/dm^3 and C_{B0} = 0.6 mol/dm^3.

(a) Determine the exit selectivities using the segregation model.

(b) Determine the exit selectivities using the maximum mixedness model.

(c) Compare the selectivities in parts (a) and (b) with those that would be found in an ideal PFR and ideal CSTR in which the space time is equal to the mean residence time.

(d) How would your answers to parts (a) and (b) change if the reactor in Problem P13-14$_C$ were used?

P13-19$_B$ The flow through a reactor is 10 dm^3/min. A pulse test gave the following concentration measurements at the outlet:

Member

Hall of Fame

t (min)	$c \times 10^5$	t (min)	$c \times 10^5$
0	0	15	238
0.4	329	20	136
1.0	622	25	77
2	812	30	44
3	831	35	25
4	785	40	14
5	720	45	8
6	650	50	5
8	523	60	1
10	418		

(a) Plot the external age distribution $E(t)$ as a function of time.

(b) Plot the external age cumulative distribution $F(t)$ as a function of time.

(c) What are the mean residence time t_m and the variance, σ^2?

(d) What fraction of the material spends between 2 and 4 min in the reactor?

(e) What fraction of the material spends longer than 6 min in the reactor?

(f) What fraction of the material spends less than 3 min in the reactor?

(g) Plot the normalized distributions $E(\Theta)$ and $F(\Theta)$ as a function of Θ.

(h) What is the reactor volume?

(i) Plot internal age distribution $I(t)$ as a function of time.

(j) What is the mean internal age α_m?

(k) Plot the intensity function, $\Lambda(t)$, as a function of time.

Problem P13-19$_B$
will be continued
in Chapter 14,
P14-13$_B$.

(l) The activity of a "fluidized" CSTR is maintained constant by feeding fresh catalyst and removing spent catalyst at constant rate. Using the preceding RTD data, what is the mean catalytic activity if the catalyst decays according to the rate law

$$-\frac{da}{dt} = k_D a^2$$

with

$$k_D = 0.1 \text{ s}^{-1}?$$

(m) What conversion would be achieved in an ideal PFR for a second-order reaction with $kC_{A0} = 0.1$ min^{-1} and $C_{A0} = 1$ mol/dm^3?

(n) Repeat (m) for a laminar flow reactor.

(o) Repeat (m) for an ideal CSTR.

(p) What would be the conversion for a second-order reaction with $kC_{A0} = 0.1$ min^{-1} and $C_{A0} = 1$ mol/dm^3 using the segregation model?

(q) What would be the conversion for a second-order reaction with $kC_{A0} = 0.1$ min^{-1} and $C_{A0} = 1$ mol/dm^3 using the maximum mixedness model?

- **Additional Homework Problems**

CDP13-A$_C$ After showing that $E(t)$ for two CSTRs in series having different volumes is

$$E(t) = \frac{1}{\tau(2m-1)} \left\{ \exp\left(\frac{-t}{\tau_m}\right) - \exp\left[\frac{-t}{(1-m)\tau}\right] \right\}$$

you are asked to make a number of calculations. [2nd Ed. P13-11]

CDP13-B$_B$ Determine $E(t)$ and from data taken from a pulse test in which the pulse is not perfect and the inlet concentration varies with time. [2nd Ed. P13-15]

CDP13-C$_B$ Derive the $E(t)$ curve for a Bingham plastic flowing in a cylindrical tube. [2nd Ed. P13-16]

CDP13-D$_B$ The order of a CSTR and PFR in series is investigated for a third-order reaction. [2nd Ed. P13-10]

CDP13-E$_B$ Review the Murphree pilot plant data when a second-order reaction occurs in the reactor. [1st Ed. P13-15]

CDP13-F$_A$ Calculate the mean waiting time for gasoline at a service station and in a parking garage. [2nd Ed. P13-3]

CDP13-G$_B$ Apply the RTD given by

$$E(t) = \begin{cases} A - B(t_0 - t)^2 & \text{for } 0 \le t \le 2t_0 \\ 0 & \text{for } 0 > 2t_0 \end{cases}$$

to Examples 13-6 through 13-8. [2nd Ed. P13-2$_B$]

CDP13-H$_B$ The multiple reactions in Problem 6-27 are carried out in a reactor whose RTD is described in Example 13-7.

CDP13-I$_B$ Real RTD data from an industrial packed bed reactor operating under poor operation. [3rd Ed. P13-5]

CDP13-J$_B$ Real RTD data from distribution in a stirred tank. [3rd Ed. P13-7$_B$]

CDP13-K$_B$ Triangle RTD with second-order reaction. [3rd Ed. P13-8$_B$]

CDP13-L$_B$ Derive $E(t)$ for a turbulent flow reactor with 1/7th power law.

CDP13-M$_B$ Good problem—must use numerical techniques. [3rd Ed. P13-12 $_B$]

CDP13-N$_B$ Internal age distribution for a catalyst. [3rd Ed. P13-13 $_B$]
CDP13-DQEA U of M, Doctoral Qualifying Exam (DQE), May, 2000
CDP13-DQEB U of M, Doctoral Qualifying Exam (DQE), April, 1999
CDP13-DQEC U of M, Doctoral Qualifying Exam (DQE), January, 1999
CDP13-DQED U of M, Doctoral Qualifying Exam (DQE), January, 1999
CDP13-DQEE U of M, Doctoral Qualifying Exam (DQE), January, 1998
CDP13-DQEF U of M, Doctoral Qualifying Exam (DQE), January, 1998
CDP13-ExG   U of M, Graduate Class Final Exam
CDP13-New   New Problems will be inserted from time to time on the web.

SUPPLEMENTARY READING

1. Discussions of the measurement and analysis of residence-time distribution can be found in

> CURL, R. L., and M. L. McMILLIN, "Accuracies in residence time measurements," *AIChE J.*, *12*, 819–822 (1966).

> LEVENSPIEL, O., *Chemical Reaction Engineering*, 3rd ed. New York: Wiley, 1999, Chaps. 11–16.

2. An excellent discussion of segregation can be found in

> DOUGLAS, J. M., "The effect of mixing on reactor design," *AIChE Symp. Ser.*, *48*, Vol. 60, p. 1 (1964).

3. Also see

> DUDUKOVIC, M., and R. FELDER, in *CHEMI Modules on Chemical Reaction Engineering*, Vol. 4, ed. B. Crynes and H. S. Fogler. New York: AIChE, 1985.

> NAUMAN, E. B., "Residence time distributions and micromixing," *Chem. Eng. Commun.*, *8*, 53 (1981).

> NAUMAN, E. B., and B. A. BUFFHAM, *Mixing in Continuous Flow Systems*. New York: Wiley, 1983.

> ROBINSON, B. A., and J. W. TESTER, *Chem. Eng. Sci.*, *41*(3), 469–483 (1986).

> VILLERMAUX, J., "Mixing in chemical reactors," in *Chemical Reaction Engineering—Plenary Lectures*, ACS Symposium Series 226. Washington, D.C.: American Chemical Society, 1982.

Models for Nonideal **14**
Reactors

Success is a journey, not a destination.
Ben Sweetland

Overview Not all tank reactors are perfectly mixed nor do all tubular reactors exhibit plug-flow behavior. In these situations, some means must be used to allow for deviations from ideal behavior. Chapter 13 showed how the RTD was sufficient if the reaction was first order *or* if the fluid was either in a state of complete segregation or maximum mixedness. We use the segregation and maximum mixedness models to bound the conversion when no adjustable parameters are used. For non-first-order reactions in a fluid with good micromixing, more than just the RTD is needed. These situations compose a great majority of reactor analysis problems and cannot be ignored. For example, we may have an existing reactor and want to carry out a new reaction in that reactor. To predict conversions and product distributions for such systems, a model of reactor flow patterns is necessary. To model these patterns, we use combinations and/or modifications of ideal reactors to represent real reactors. With this technique, we classify a model as being either a one-parameter model (e.g., tanks-in-series model or dispersion model) or a two-parameter model (e.g., reactor with bypassing and dead volume). The RTD is then used to evaluate the parameter(s) in the model. After completing this chapter, the reader will be able to apply the tanks-in-series model and the dispersion model to tubular reactors. In addition, the reader will be able to suggest combinations of ideal reactors to model a real reactor.

Use the RTD to
evaluate
parameters

14.1 Some Guidelines

The overall goal is to use the following equation

$$\boxed{\text{RTD Data} + \text{Kinetics} + \text{Model} = \text{Prediction}}$$

Conflicting goals

The choice of the particular model to be used depends largely on the engineering judgment of the person carrying out the analysis. It is this person's job to choose the model that best combines the conflicting goals of mathematical simplicity and physical realism. There is a certain amount of art in the development of a model for a particular reactor, and the examples presented here can only point toward a direction that an engineer's thinking might follow.

For a given real reactor, it is not uncommon to use all the models discussed previously to predict conversion and then make a comparison. Usually, the real conversion will be *bounded* by the model calculations.

The following guidelines are suggested when developing models for nonideal reactors:

A Model Must
• Fit the data
• Be able to extrapolate theory and experiment
• Have realistic parameters

1. *The model must be mathematically tractable.* The equations used to describe a chemical reactor should be able to be solved without an inordinate expenditure of human or computer time.
2. *The model must realistically describe the characteristics of the nonideal reactor.* The phenomena occurring in the nonideal reactor must be reasonably described physically, chemically, and mathematically.
3. *The model must not have more than two adjustable parameters.* This constraint is used because an expression with more than two adjustable parameters can be fitted to a great variety of experimental data, and the modeling process in this circumstance is nothing more than an exercise in curve fitting. The statement "Give me four adjustable parameters and I can fit an elephant; give me five and I can include his tail!" is one that I have heard from many colleagues. Unless one is into modern art, a substantially larger number of adjustable parameters is necessary to draw a reasonable-looking elephant.[1] A one-parameter model is, of course, superior to a two-parameter model if the one-parameter model is sufficiently realistic. To be fair, however, in complex systems (e.g., internal diffusion and conduction, mass transfer limitations) where other parameters may be measured *independently*, then more than two parameters are quite acceptable.

Table 14-1 gives some guidelines that will help your analysis and model building of nonideal reaction systems.

---

[1] J. Wei, *Chem. Technol.*, *5*, 128 (1975).

TABLE 14-1.   A PROCEDURE FOR CHOOSING A MODEL
TO PREDICT THE OUTLET CONCENTRATION AND CONVERSION

1. *Look at the reactor.*
   a.   Where are the inlet and outlet streams to and from the reactors? (Is by-passing a possibility?)
   b.   Look at the mixing system. How many impellers are there? (Could there be multiple mixing zones in the reactor?)
   c.   Look at the configuration. (Is internal recirculation possible? Is the packing of the catalyst particles loose so channeling could occur?)
2. *Look at the tracer data.*
   a.   Plot the E(t) and F(t) curves.
   b.   Plot and analyze the shapes of the $E(\Theta)$ and $F(\Theta)$ curves. Is the shape of the curve such that the curve or parts of the curve can be fit by an ideal reactor model? Does the curve have a long tail suggesting a stagnant zone? Does the curve have an early spike indicating bypassing?
   c.   Calculate the mean residence time, t_m, and variance, σ^2. How does the t_m determined from the RTD data compare with τ as measured with a yardstick and flow meter? How large is the variance; is it larger or smaller than τ^2?
3. *Choose a model or perhaps two or three models.*
4. *Use the tracer data to determine the model parameters* (e.g., n, D_a, v_b).
5. *Use the CRE algorithm in Chapter 4.* Calculate the exit concentrations and conversion for the model system you have selected.

14.1.1 One-Parameter Models

Here we use a single parameter to account for the nonideality of our reactor. This parameter is most always evaluated by analyzing the RTD determined from a tracer test. Examples of one-parameter models for nonideal CSTRs include a reactor dead volume V_D, where no reaction takes place, or a fraction f of fluid bypassing the reactor, thereby exiting unreacted. Examples of one-parameter models for tubular reactors include the tanks-in-series model and the dispersion model. For the tanks-in-series model, the parameter is the number of tanks, n, and for the dispersion model, it is the dispersion coefficient, D_a. Knowing the parameter values, we then proceed to determine the conversion and/or effluent concentrations for the reactor.

Nonideal tubular reactors

We first consider nonideal tubular reactors. Tubular reactors may be empty, or they may be packed with some material that acts as a catalyst, heat-transfer medium, or means of promoting interphase contact. Until now when analyzing ideal tubular reactors, it usually has been assumed that the fluid moved through the reactor in piston-like flow (PFR), and every atom spends an identical length of time in the reaction environment. Here, the *velocity profile is flat,* and there is no axial mixing. Both of these assumptions are false to some extent in every tubular reactor; frequently, they are sufficiently false to warrant some modification. Most popular tubular reactor models need to have means to allow for failure of the plug-flow model and insignificant axial mixing assumptions; examples include the unpacked laminar flow tubular reactor, the unpacked turbulent flow, and packed-bed reactors. One of two approaches is usually taken to compensate for failure of either or both of the ideal assumptions. One approach involves modeling the nonideal tubular reactor as a series of identically sized CSTRs. The other approach (the dispersion model) involves a modification of the ideal reactor by imposing axial dispersion on plug flow.

14.1.2 Two-Parameter Models

The premise for the two-parameter model is that we can use a combination of ideal reactors to model the real reactor. For example, consider a packed bed reactor with channeling. Here the response to a pulse tracer input would show two dispersed pulses in the output as shown in Figure 13-10 and Figure 14-1.

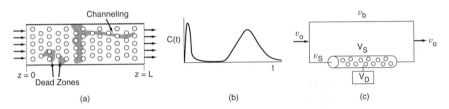

Figure 14-1   (a) Real system; (b) outlet for a pulse input; (c) model system.

Here we could model the real reactor as two ideal PBRs in parallel with the two parameters being the fluid that channels, v_b, and the reactor dead volume, V_D. The real reactor voume is $V = V_D + V_S$ with $v_0 = v_b + v_S$.

14.2 Tanks-in-Series (T-I-S) Model

In this section we discuss the use of the tanks-in-series (T-I-S) model to describe nonideal reactors and calculate conversion. The T-I-S model is a one-parameter model. We will analyze the RTD to determine the number of ideal tanks, n, in series that will give approximately the same RTD as the non-ideal reactor. Next we will apply the reaction engineering algorithm developed in Chapters 1 through 4 to calculate conversion. We are first going to develop the RTD equation for three tanks in series (Figure 14-2) and then generalize to n reactors in series to derive an equation that gives the number of tanks in series that best fits the RTD data.

$n = ?$

In Figure 2-9, we saw how tanks in series could approximate a PFR.

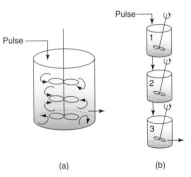

Figure 14-2   Tanks in series: (a) real system, (b) model system.

The RTD will be analyzed from a tracer pulse injected into the first reactor of three equally sized CSTRs in series. Using the definition of the RTD presented in Section 13.2, the fraction of material leaving the system of three reactors (i.e., leaving the third reactor) that has been in the system between time t and $t + \Delta t$ is

$$E(t)\, \Delta t = \frac{v C_3(t)\, \Delta t}{N_0} = \frac{C_3(t)}{\displaystyle\int_0^\infty C_3(t)\, dt}\, \Delta t$$

Then

$$E(t) = \frac{C_3(t)}{\displaystyle\int_0^\infty C_3(t)\, dt} \tag{14-1}$$

In this expression, $C_3(t)$ is the concentration of tracer in the effluent from the third reactor and the other terms are as defined previously.

It is now necessary to obtain the outlet concentration of tracer, $C_3(t)$, as a function of time. As in a single CSTR, a material balance on the first reactor gives

$$V_1 \frac{dC_1}{dt} = -v C_1 \tag{14-2}$$

We perform a tracer balance on each reactor to obtain $C_3(t)$

Integrating gives the expression for the tracer concentration in the effluent from the first reactor:

$$C_1 = C_0 e^{-vt/V_1} = C_0 e^{-t/\tau_1} \tag{14-3}$$

$$C_0 = N_0/V_1 = \frac{v_0 \displaystyle\int_0^\infty C_3(t)\, dt}{V_1}$$

The volumetric flow rate is constant ($v = v_0$) and all the reactor volumes are identical ($V_1 = V_2 = V_i$); therefore, all the space times of the individual reactors are identical ($\tau_1 = \tau_2 = \tau_i$). Because V_i is the volume of a single reactor in the series, τ_i here is the residence time in *one* of the reactors, *not* in the entire reactor system (i.e., $\tau_i = \tau/n$).

A material balance on the tracer in the second reactor gives

$$V_2 \frac{dC_2}{dt} = v C_1 - v C_2$$

Using Equation (14-3) to substitute for C_1, we obtain the first-order ordinary differential equation

$$\frac{dC_2}{dt} + \frac{C_2}{\tau_i} = \frac{C_0}{\tau_i} e^{-t/\tau_i}$$

This equation is readily solved using an integrating factor e^{t/τ_i} along with the initial condition $C_2 = 0$ at $t = 0$, to give

$$C_2 = \frac{C_0 t}{\tau_i} e^{-t/\tau_i} \tag{14-4}$$

The same procedure used for the third reactor gives the expression for the concentration of tracer in the effluent from the third reactor (and therefore from the reactor system),

$$C_3 = \frac{C_0 t^2}{2\tau_i^2} e^{-t/\tau_i} \tag{14-5}$$

Substituting Equation (14-5) into Equation (14-1), we find that

$$E(t) = \frac{C_3(t)}{\displaystyle\int_0^\infty C_3(t)dt} = \frac{C_0 t^2/(2\tau_i^2)e^{-t/\tau_i}}{\displaystyle\int_0^\infty \frac{C_0 t^2 e^{-t/\tau_i}}{2\tau_i^2}dt}$$

$$= \frac{t^2}{2\tau_i^3} e^{-t/\tau_i} \tag{14-6}$$

Generalizing this method to a series of n CSTRs gives the RTD for n CSTRs in series, $E(t)$:

RTD for equal-size
tanks in series

$$\boxed{E(t) = \frac{t^{n-1}}{(n-1)!\tau_i^n} e^{-t/\tau_i}} \tag{14-7}$$

Because the total reactor volume is nV_i, then $\tau_i = \tau/n$, where τ represents the total reactor volume divided by the flow rate, v:

$$E(\Theta) = \tau E(t) = \frac{n(n\Theta)^{n-1}}{(n-1)!} e^{-n\Theta} \tag{14-8}$$

where $\Theta = t/\tau$.

     Figure 14-3 illustrates the RTDs of various numbers of CSTRs in series in a two-dimensional plot (a) and in a three-dimensional plot (b). As the number becomes very large, the behavior of the system approaches that of a plug-flow reactor.

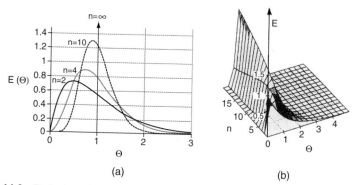

Figure 14-3 Tanks-in-series response to a pulse tracer input for different numbers of tanks.

We can determine the number of tanks in series by calculating the dimensionless variance σ_Θ^2 from a tracer experiment.

$$\sigma_\Theta^2 = \frac{\sigma^2}{\tau^2} = \int_0^\infty (\Theta - 1)^2 E(\Theta)\, d\Theta$$

$$= \int_0^\infty \Theta^2 E(\Theta)\, d\Theta - 2 \int_0^\infty \Theta E(\Theta)\, d\Theta + \int_0^\infty E(\Theta)\, d\Theta \qquad (14\text{-}9)$$

$$\boxed{\sigma_\Theta^2 = \int_0^\infty \Theta^2 E(\Theta)\, d\Theta - 1} \qquad\qquad (14\text{-}10)$$

$$= \int_0^\infty \Theta^2 \frac{n(n\Theta)^{n-1}}{(n-1)!} e^{-n\Theta}\, d\Theta - 1$$

$$\sigma_\Theta^2 = \frac{n^n}{(n-1)!} \int_0^\infty \Theta^{n+1} e^{-n\Theta}\, d\Theta - 1$$

$$= \frac{n^n}{(n-1)!} \left[\frac{(n+1)!}{n^{n+2}} \right] - 1$$

As the number of tanks increases, the variance decreases.

$$= \frac{1}{n} \qquad\qquad (14\text{-}11)$$

The number of tanks in series is

$$\boxed{n = \frac{1}{\sigma_\Theta^2} = \frac{\tau^2}{\sigma^2}} \qquad\qquad (14\text{-}12)$$

This expression represents the number of tanks necessary to model the real reactor as n ideal tanks in series. If the number of reactors, n, turns out to be small, the reactor characteristics turn out to be those of a single CSTR or perhaps two CSTRs in series. At the other extreme, when n turns out to be large, we recall from Chapter 2 the reactor characteristics approach those of a PFR.

If the reaction is first order, we can use Equation (4-11) to calculate the conversion,

$$X = 1 - \frac{1}{(1 + \tau_i k)^n} \tag{4-11}$$

where

$$\tau_i = \frac{V}{v_0 n}$$

It is acceptable (and usual) for the value of n calculated from Equation (14-12) to be a noninteger in Equation (4-11) to calculate the conversion. For reactions other than first order, an integer number of reactors must be used and sequential mole balances on each reactor must be carried out. If, for example, $n = 2.53$, then one could calculate the conversion for two tanks and also for three tanks to bound the conversion. The conversion and effluent concentrations would be solved sequentially using the algorithm developed in Chapter 4. That is, after solving for the effluent from the first tank, it would be used as the input to the second tank and so on as shown in Table 14-2.

TABLE 14-2.   TANKS-IN-SERIES SECOND-ORDER REACTION

Two-Reactor System	Three-Reactor System
For two equally sized reactors	For three equally sized reactors

$$V = V_1 + V_2 \qquad\qquad\qquad V = V_1 + V_2 + V_3$$

$$V_1 = V_2 = \frac{V}{2} \qquad\qquad V_1 = V_2 = V_3 = V/3$$

$$\tau_2 = \frac{V_2}{v_0} = \frac{V/2}{v_0} = \frac{\tau}{2} \qquad\qquad \tau_1 = \tau_2 = \tau_3 = \frac{V/3}{v_0} = \frac{\tau}{3}$$

For a second-order reaction, the combined mole balance, rate law, and stoichiometry for the first reactor gives

$$\tau = \frac{C_{Ain} - C_{Aout}}{k_1 C_{Aout}^2}$$

Solving for C_{Aout}

$$C_{Aout} = \frac{-1 + \sqrt{1 + 4k\tau C_{Ain}}}{2k\tau}$$

Two-Reactor System: $\tau_2 = \dfrac{\tau}{2}$ $\qquad\qquad$ Three-Reactor System: $\tau_3 = \dfrac{\tau}{3}$

Solving for exit concentration from reactor 1 for each reactor system gives

$$C_{A1} = \frac{-1 + \sqrt{1 + 4\tau_2 k C_{A0}}}{2\tau_2 k} \qquad\qquad C'_{A1} = \frac{-1 + \sqrt{1 + 4\tau_3 k C_{A0}}}{2\tau_3 k}$$

The exit concentration from the second reactor for each reactor system gives

TABLE 14-2.   TANKS-IN-SERIES SECOND-ORDER REACTION (CONTINUED)

Two-Reactor System	Three-Reactor System
$C_{A2} = \dfrac{-1 + \sqrt{1 + 4\tau_2 k C_{A1}}}{2\tau_2 k}$	$C'_{A2} = \dfrac{-1 + \sqrt{1 + 4\tau_3 k C'_{A1}}}{2\tau_3 k}$

Balancing on the third reactor for the three-reactor system and solving for its outlet concentration, C'_{A3}, gives

$$C'_{A3} = \frac{-1 + \sqrt{1 + \tau_3 k C'_{A2}}}{2\tau_3 k}$$

The corresponding conversion for the two- and three-reactor systems are

$$X_2 = \frac{C_{A0} - C_{A2}}{C_{A0}} \qquad\qquad X'_3 = \frac{C_{A0} - C'_{A3}}{C_{A0}}$$

For $n = 2.53$, $(X_2 < X < X'_3)$

Tanks-in-Series Versus Segregation for a First-Order Reaction   We have already stated that the segregation and maximum mixedness models are equivalent for a first-order reaction. The proof of this statement was left as an exercise in Problem P13-3$_B$. We now show the tanks-in-series model and the segregation models are equivalent for a first-order reaction.

Example 14–1   Equivalency of Models for a First-Order Reaction

Show that $X_{T–I–S} = X_{MM}$ for a first-order reaction

$$A \xrightarrow{\ k\ } B$$

Solution

For a first-order reaction, we already showed in Problem P13-3 that

$$\boxed{X_{Seg} = X_{MM}}$$

Therefore we only need to show $X_{Seg} = X_{T\text{-}I\text{-}S}$.
For a first-order reaction in a batch reactor the conversion is

$$X = 1 - e^{-kt} \tag{E14-1.1}$$

Segregation Model

$$\overline{X} = \int_0^\infty X(t)E(t)dt = \int_0^\infty (1 - e^{-kt})E(t)dt \tag{E14-1.2}$$

$$= 1 - \int_0^\infty e^{-kt}E(t)dt \tag{E14-1.3}$$

Using Maclaurin's series expansion gives

$$e^{-kt} = 1 - kt + \frac{k^2 t^2}{2} + \text{Error} \tag{E14-1.4}$$

neglecting the error term

$$\overline{X} = 1 - \int_0^\infty \left[1 - kt + \frac{k^2 t^2}{2} \right] E(t)dt \tag{E14-1.5}$$

$$\overline{X} = \tau k - \frac{k^2}{2} \int_0^\infty t^2 E(t)dt \tag{E14-1.6}$$

To evaluate the second term, we first recall Equation (E13-2.5) for the variance

$$\sigma^2 = \int_0^\infty (t - \tau)^2 E(t)dt = \int t^2 E(t)dt - 2\tau \int_0^\infty tE(t)dt + \tau^2 \int_0^\infty E(t)dt \tag{E14-1.7}$$

$$\sigma^2 = \int_0^\infty t^2 E(t)dt - 2\tau^2 + \tau^2 \tag{E14-1.8}$$

Rearranging Equation (E14-1.8)

$$\int_0^\infty t^2 E(t)dt = \sigma^2 + \tau^2 \tag{E14-1.9}$$

Combining Equations (E14-1.6) and (E14-1.9), we find the mean conversion for the segregation model for a first-order reaction is

$$\boxed{\overline{X} = \tau k - \frac{k^2}{2}(\sigma^2 + \tau^2)} \tag{E14-1.10}$$

Tanks in Series

Recall from Chapter 4, for n tanks in series for a first-order reaction, the conversion is

$$X = 1 - \frac{1}{\left(1 + \frac{\tau}{n}k \right)^n} \tag{4-11}$$

Rearranging yields

$$X = 1 - \left(1 + \frac{\tau}{n}k \right)^{-n} \tag{E14-1.11}$$

We now expand in a binomial series

$$X = 1 - \left[1 - n\frac{\tau}{n}k + \frac{n(n+1)}{2}\frac{\tau^2 k^2}{n^2} + \text{Error} \right] \tag{E14-1.12}$$

$$= k\tau - \frac{\tau^2 k^2}{2} - \frac{\tau^2 k^2}{2n} + \text{Error} \tag{E14-1.13}$$

Neglecting the error gives

$$X = k\tau - \frac{k^2}{2}\left[\tau^2 + \frac{\tau^2}{n}\right] \qquad\qquad \text{(E14-1.14)}$$

Rearranging Equation (14-12) in the form

$$\frac{\tau^2}{n} = \sigma^2 \qquad\qquad \text{(14-12)}$$

and substituting in Equation (E14-1.14) the mean conversion for the T-I-S model is

$$\boxed{X = \tau k - \frac{k^2}{2}(\tau^2 + \sigma^2)} \qquad\qquad \text{(E14-1.15)}$$

We see that Equations (E14-1.10) and (E14-1.15) are identical; thus, the conversions are identical, and for a first-order reaction we have

Important Result

$$\boxed{X_{\text{T-I-S}} = X_{\text{Seg}} = X_{\text{MM}}}$$

But this is true only for a first-order reaction.

14.3 Dispersion Model

The dispersion model is also used to describe nonideal tubular reactors. In this model, there is an axial dispersion of the material, which is governed by an analogy to Fick's law of diffusion, superimposed on the flow as shown in Figure 14-4. So in addition to transport by bulk flow, UA_cC, every component in the mixture is transported through any cross section of the reactor at a rate equal to $[-D_a A_c(dC/dz)]$ resulting from molecular and convective diffusion. By convective diffusion (i.e., dispersion) we mean either Aris-Taylor dispersion in laminar flow reactors or turbulent diffusion resulting from turbulent eddies. Radial concentration profiles for plug flow (a) and a representative axial and radial profile for dispersive flow (b) are shown in Figure 14-4. Some molecules will diffuse forward ahead of molar average velocity while others will lag behind.

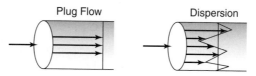

Plug Flow   Dispersion

Figure 14-4   Concentration profiles (a) without and (b) with dispersion.

To illustrate how dispersion affects the concentration profile in a tubular reactor we consider the injection of a perfect tracer pulse. Figure 14-5 shows how dispersion causes the pulse to broaden as it moves down the reactor and becomes less concentrated.

Tracer pulse with dispersion

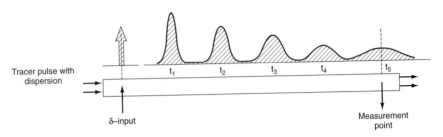

Figure 14-5 Dispersion in a tubular reactor. (From O. Levenspiel, *Chemical Reaction Engineering*, 2nd ed. Copyright © 1972 John Wiley & Sons, Inc. Reprinted by permission of John Wiley & Sons, Inc. All rights reserved.)

Recall Equation (11-20). The molar flow rate of tracer (F_T) by both convection and dispersion is

$$F_T = \left[-D_a \frac{\partial C_T}{\partial z} + U C_T \right] A_c \tag{11-20}$$

In this expression D_a is the effective dispersion coefficient (m²/s) and U (m/s) is the superficial velocity. To better understand how the pulse broadens, we refer to the concentration peaks t_2 and t_3 in Figure 14-6. We see that there is a concentration gradient on both sides of the peak causing molecules to diffuse away from the peak and thus broaden the pulse. The pulse broadens as it moves through the reactor.

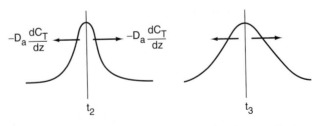

Figure 14-6 Symmetric concentration gradients causing the spreading by dispersion of a pulse input.

Correlations for the dispersion coefficients in both liquid and gas systems may be found in Levenspiel.[2] Some of these correlations are given in Section 14.4.5.

---

[2] O. Levenspiel, *Chemical Reaction Engineering* (New York: Wiley, 1962), pp. 290–293.

A mole balance on the inert tracer T gives

$$-\frac{\partial F_T}{\partial z} = A_c \frac{\partial C_T}{\partial t}$$

Substituting for F_T and dividing by the cross-sectional area A_c, we have

Pulse tracer balance
dispersion

$$D_a \frac{\partial^2 C_T}{\partial z^2} - \frac{\partial (U C_T)}{\partial z} = \frac{\partial C_T}{\partial t} \qquad (14\text{-}13)$$

Once we know the boundary conditions, the solution to Equation (14-13) will give the outlet tracer concentration–time curves. Consequently, we will have to wait to obtain this solution until we discuss the boundary conditions in Section 14.4.2.

The Plan   We are now going to proceed in the following manner. First, we will write the balance equations for dispersion with reaction. We will discuss the two types of boundary conditions: closed-closed and open-open. We will then obtain an analytical solution for the closed-closed system for the conversion for a first-order reaction in terms of the Peclet number (dispersion coefficient) and the Damköhler number. We then will discuss how the dispersion coefficient can be obtained either from correlations or from the analysis of the RTD curve.

14.4 Flow, Reaction, and Dispersion

Now that we have an intuitive feel for how dispersion affects the transport of molecules in a tubular reactor, we shall consider two types of dispersion in a tubular reactor, *laminar* and *turbulent*.

14.4.1 Balance Equations

A mole balance is taken on a particular component of the mixture (say, species A) over a short length Δz of a tubular reactor of cross section A_c in a manner identical to that in Chapter 1, to arrive at

$$-\frac{1}{A_c} \frac{dF_A}{dz} + r_A = 0 \qquad (14\text{-}14)$$

Combining Equations (14-14) and the equation for the molar flux F_A, we can rearrange Equation (11-22) in Chapter 11 as

$$\boxed{\frac{D_a}{U} \frac{d^2 C_A}{dz^2} - \frac{dC_A}{dz} + \frac{r_A}{U} = 0} \qquad (14\text{-}15)$$

This equation is a second-order ordinary differential equation. It is nonlinear when r_A is other than zero or first order.

When the reaction rate r_A is first order, $r_A = -kC_A$, then Equation (14-16)

Flow, reaction, and
dispersion

$$\boxed{\frac{D_a}{U}\frac{d^2C_A}{dz^2} - \frac{dC_A}{dz} - \frac{kC_A}{U} = 0}$$ (14-16)

is amenable to an analytical solution. However, before obtaining a solution, we put our Equation (14-16) describing dispersion and reaction in dimensionless form by letting $\psi = C_A/C_{A0}$ and $\lambda = z/L$:

D_a = Dispersion
coefficient

$$\boxed{\frac{1}{Pe_r}\frac{d^2\psi}{d\lambda^2} - \frac{d\psi}{d\lambda} - \boldsymbol{Da}\cdot\psi = 0}$$ (14-17)

$\boldsymbol{Da}$ = Damköhler
number

The quantity $\boldsymbol{Da}$ appearing in Equation (14-17) is called the *Damköhler number* for convection and physically represents the ratio

Damköhler number
for a first-order
reaction

$$\boldsymbol{Da} = \frac{\text{Rate of consumption of A by reaction}}{\text{Rate of transport of A by convection}} = k\tau$$ (14-18)

The other dimensionless term is the *Peclet number*, Pe,

$$Pe = \frac{\text{Rate of transport by convection}}{\text{Rate of transport by diffusion or dispersion}} = \frac{Ul}{D_a}$$ (14-19)

For open tubes
$Pe_r \sim 10^6$,
$Pe_f \sim 10^4$

in which l is the characteristic length term. There are two different types of Peclet numbers in common use. We can call Pe_r the reactor Peclet number; it uses the reactor length, L, for the characteristic length, so $Pe_r \equiv UL/D_a$. It is Pe_r that appears in Equation (14-17). The reactor Peclet number, Pe_r, for mass dispersion is often referred to in reacting systems as the Bodenstein number, Bo, rather than the Peclet number. The other type of Peclet number can be called the fluid Peclet number, Pe_f; it uses the characteristic length that determines the fluid's mechanical behavior. In a packed bed this length is the particle diameter d_p, and $Pe_f \equiv Ud_p/\phi D_a$. (The term U is the empty tube or superficial velocity. For packed beds we often wish to use the average interstitial velocity, and thus U/ϕ is commonly used for the packed-bed velocity term.) In an empty tube, the fluid behavior is determined by the tube diameter d_t, and $Pe_f = Ud_t/D_a$. The fluid Peclet number, Pe_f, is given in all correlations relating the Peclet number to the Reynolds number because both are directly related to the fluid mechanical behavior. It is, of course, very simple to convert Pe_f to Pe_r: Multiply by the ratio L/d_p or L/d_t. The reciprocal of Pe_r, D_a/UL, is sometimes called the *vessel dispersion number*.

For packed beds
$Pe_r \sim 10^3$,
$Pe_f \sim 10^1$

14.4.2 Boundary Conditions

There are two cases that we need to consider: boundary conditions for closed vessels and open vessels. In the case of closed-closed vessels, we assume that there is no dispersion or radial variation in concentration either upstream (closed) or downstream (closed) of the reaction section; hence this is a closed-closed vessel. In an open vessel, dispersion occurs both upstream (open) and downstream (open) of the reaction section; hence this is an open-open vessel. These two cases are shown in Figure 14-7, where fluctuations in concentration due to dispersion are superimposed on the plug-flow velocity profile. A closed-open vessel boundary condition is one in which there is no dispersion in the entrance section but there is dispersion in the reaction and exit sections.

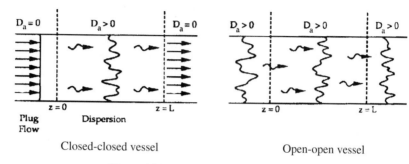

Closed-closed vessel   Open-open vessel

Figure 14-7   Types of boundary conditions.

14.4.2A Closed-Closed Vessel Boundary Condition

For a closed-closed vessel, we have plug flow (no dispersion) to the immediate left of the entrance line ($z = 0^-$) (closed) and to the immediate right of the exit $z = L$ ($z = L^+$) (closed). However, between $z = 0^+$ and $z = L^-$, we have dispersion and reaction. The corresponding entrance boundary condition is

At $z = 0$: $$F_A(0^-) = F_A(0^+)$$

Substituting for F_A yields

$$UA_c C_A(0^-) = -A_c D_a \left(\frac{\partial C_A}{\partial z}\right)_{z=0^+} + UA_c C_A(0^+)$$

Solving for the entering concentration $C_A(0^-) = C_{A0}$:

Concentration boundary conditions at the entrance

$$\boxed{C_{A0} = \frac{-D_a}{U}\left(\frac{\partial C_A}{\partial z}\right)_{z=0^+} + C_A(0^+)}$$ (14-20)

At the exit to the reaction section, the concentration is continuous, and there is no gradient in tracer concentration.

Concentration boundary conditions at the exit

At $z = L$:

$$\boxed{\begin{aligned} C_{A0}(L^-) &= C_A(L^+) \\ \frac{\partial C_A}{\partial z} &= 0 \end{aligned}}$$ (14-21)

These two boundary conditions, Equations (14-19) and (14-20), first stated by Danckwerts,[3] have become known as the famous *Danckwerts boundary conditions*. Bischoff[4] has given a rigorous derivation of them, solving the differential equations governing the dispersion of component A in the entrance and exit sections and taking the limit as D_a in entrance and exit sections approaches zero. From the solutions he obtained boundary conditions on the reaction section identical with those Danckwerts proposed.

The closed-closed concentration boundary condition at the entrance is shown schematically in Figure 14-8. One should not be uncomfortable with the discontinuity in concentration at $z = 0$ because if you recall for an ideal CSTR the concentration drops immediately on entering from C_{A0} to C_{Aexit}. For the other boundary condition at the exit $z = L$, we see the concentration gradient has gone to zero. At steady state, it can be shown that this Danckwerts boundary condition at $z = L$ also applies to the open-open system at steady state.

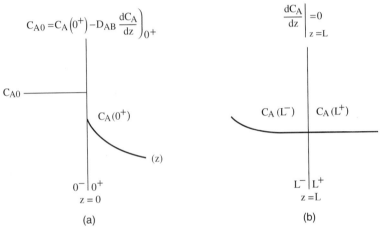

$$C_{A0} = C_A\left(0^+\right) - D_{AB} \left.\frac{dC_A}{dz}\right)_{0^+}$$

$$\left.\frac{dC_A}{dz}\right|_{z=L} = 0$$

Prof. P. V. Danckwerts,
Cambridge University,
U.K.

Figure 14-8   Schematic of Danckwerts boundary conditions. (a) Entrance (b) Exit

14.4.2B Open-Open System

For an open-open system there is continuity of flux at the boundaries at $z = 0$,

$$F_A(0^-) = F_A(0^+)$$

Open-open
boundary condition

$$-D_a\left.\frac{\partial C_A}{\partial z}\right)_{0^-} + UC_A(0^-) = -D_a\left.\frac{\partial C_A}{\partial z}\right)_{0^+} + UC_A(0^+) \qquad (14\text{-}22)$$

[3] P. V. Danckwerts, *Chem. Eng. Sci.*, 2, 1 (1953).

[4] K. B. Bischoff, *Chem. Eng. Sci.*, 16, 131 (1961).

At $z = L$, we have continuity of concentration and

$$\boxed{\frac{dC_A}{\partial z} = 0}$$

(14-23)

14.4.2C Back to the Solution for a Closed-Closed System

We now shall solve the dispersion reaction balance for a first-order reaction

$$\frac{1}{Pe_r}\frac{d^2\psi}{d\lambda^2} - \frac{d^2\psi}{d\lambda} - \mathbf{Da}\psi = 0$$

(14-17)

For the closed-closed system, the Danckwerts boundary conditions in dimensionless form are

$$\text{At } \lambda = 0 \text{ then } 1 = -\frac{1}{Pe_r}\frac{d\psi}{d\lambda}\bigg)_{0^+} + \psi(0^+)$$

(14-24)

$$\text{At } \lambda = 1 \text{ then } \frac{d\psi}{d\lambda} = 0$$

(14-25)

At the end of the reactor, where $\lambda = 1$, the solution to Equation (14-17) is

$$\boxed{\begin{aligned}\psi_L &= \frac{C_{AL}}{C_{A0}} = 1 - X \\ &= \frac{4q\exp(Pe_r/2)}{(1+q)^2 \exp(Pe_r q/2) - (1-q)^2 \exp(-Pe_r q/2)} \\ \text{where } q &= \sqrt{1 + 4\mathbf{Da}/Pe_r}\end{aligned}}$$

(14-26)

This solution was first obtained by Danckwerts[5] and has been published many places (e.g., Levenspiel[6]). With a slight rearrangement of Equation (14-26), we obtain the conversion as a function of $\mathbf{Da}$ and Pe_r.

$$\boxed{X = 1 - \frac{4q \exp(Pe_r/2)}{(1+q)^2 \exp(Pe_r q/2) - (1-q)^2 \exp(-Pe_r q/2)}}$$

(14-27)

Outside the limited case of a first-order reaction, a numerical solution of the equation is required, and because this is a split-boundary-value problem, an iterative technique is required.

To evaluate the exit concentration given by Equation (14-26) or the conversion given by (14-27), we need to know the Damköhler and Peclet numbers. The Damköhler number for a first-order reaction, $\mathbf{Da} = \tau k$, can be found using the techniques in Chapter 5. In the next section, we discuss methods to determine D_a, finding the Peclet number.

[5] P. V. Danckwerts, *Chem. Eng. Sci., 2*, 1 (1953).

[6] Levenspiel, *Chemical Reaction Engineering,* 3rd ed. (New York: Wiley, 1999).

14.4.3 Finding D_a and the Peclet Number

Three ways to
find D_a
There are three ways we can use to find D_a and hence Pe_r
1. Laminar flow with radial and axial molecular diffusion theory
2. Correlations from the literature for pipes and packed beds
3. Experimental tracer data

At first sight, simple models described by Equation (14-13) appear to have the capability of accounting only for axial mixing effects. It will be shown, however, that this approach can compensate not only for problems caused by axial mixing, *but also for those caused by radial mixing and other nonflat velocity profiles.*[7] These fluctuations in concentration can result from different flow velocities and pathways and from molecular and turbulent diffusion.

14.4.4 Dispersion in a Tubular Reactor with Laminar Flow

In a laminar flow reactor, we know that the axial velocity varies in the radial direction according to the Hagen–Poiseuille equation:

$$u(r) = 2U\left[1 - \left(\frac{r}{R}\right)^2\right]$$

where U is the average velocity. For laminar flow, we saw that the RTD function $E(t)$ was given by

$$E(t) = \begin{cases} 0 & \text{for } t < \dfrac{\tau}{2} \quad \left(\tau = \dfrac{L}{U}\right) \\[3mm] \dfrac{\tau^2}{2t^3} & \text{for } t \geq \dfrac{\tau}{2} \end{cases} \qquad (13\text{-}47)$$

In arriving at this distribution $E(t)$, it was assumed that there was no transfer of molecules in the radial direction between streamlines. Consequently, with the aid of Equation (13-43), we know that the molecules on the center streamline ($r = 0$) exited the reactor at a time $t = \tau/2$, and molecules traveling on the streamline at $r = 3R/4$ exited the reactor at time

$$t = \frac{L}{u} = \frac{L}{2U[1 - (r/R)^2]} = \frac{\tau}{2[1 - (3/4)^2]}$$

$$= \frac{8}{7} \cdot \tau$$

---

[7] R. Aris, *Proc. R. Soc. (London)*, A235, 67 (1956).

The question now arises: What would happen if some of the molecules traveling on the streamline at $r = 3R/4$ jumped (i.e., diffused) onto the streamline at $r = 0$? The answer is that they would exit sooner than if they had stayed on the streamline at $r = 3R/4$. Analogously, if some of the molecules from the faster streamline at $r = 0$ jumped (i.e., diffused) on to the streamline at $r = 3R/4$, they would take a longer time to exit (Figure 14-9). In addition to the molecules diffusing between streamlines, they can also move forward or backward relative to the average fluid velocity by molecular diffusion (Fick's law). With both axial and radial diffusion occurring, the question arises as to what will be the distribution of residence times when molecules are transported between and along streamlines by diffusion. To answer this question we will derive an equation for the axial dispersion coefficient, D_a, that accounts for the axial and radial diffusion mechanisms. In deriving D_a, which is referred to as the Aris–Taylor dispersion coefficient, we closely follow the development given by Brenner and Edwards.[8]

Molecules diffusing between streamlines and back and forth along a streamline

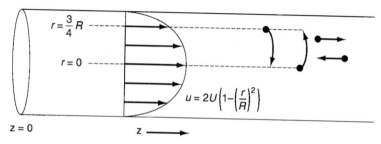

Figure 14-9   Radial diffusion in laminar flow.

The convective–diffusion equation for solute (e.g., tracer) transport in both the axial and radial direction can be obtained by combining Equations (11-3) and (11-15),

$$\frac{\partial c}{\partial t} + u(r)\frac{\partial c}{\partial z} = D_{AB}\left\{\frac{1}{r}\frac{\partial[r(\partial c/\partial r)]}{\partial r} + \frac{\partial^2 c}{\partial z^2}\right\} \qquad (14\text{-}28)$$

where c is the solute concentration at a particular r, z, and t.

We are going to change the variable in the axial direction z to z^*, which corresponds to an observer moving with the fluid

$$z^* = z - Ut \qquad (14\text{-}29)$$

A value of $z^* = 0$ corresponds to an observer moving with the fluid on the center streamline. Using the chain rule, we obtain

[8] H. Brenner and D. A. Edwards, *Macrotransport Processes* (Boston: Butterworth-Heinemann, 1993).

$$\left(\frac{\partial c}{\partial t}\right)_{z^*} + [u(r) - U]\frac{\partial c}{\partial z^*} = D_{AB}\left[\frac{1}{r}\frac{\partial}{\partial r}\left(r\frac{\partial c}{\partial r}\right) + \frac{\partial^2 c}{\partial z^{*2}}\right] \quad (14\text{-}30)$$

Because we want to know the concentrations and conversions at the exit to the reactor, we are really only interested in the average axial concentration $\overline{C}$, which is given by

$$\overline{C}(z, t) = \frac{1}{\pi R^2}\int_0^R c(r, z, t)2\pi r\, dr \quad (14\text{-}31)$$

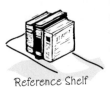

Reference Shelf

Consequently, we are going to solve Equation (14-30) for the solution concentration as a function of r and then substitute the solution c (r, z, t) into Equation (14-31) to find $\overline{C}$ (z, t). All the intermediate steps are given on the CD-ROM R14.1, and the partial differential equation describing the variation of the average axial concentration with time and distance is

$$(14\text{-}32)$$

$$\frac{\partial \overline{C}}{\partial t} + U\frac{\partial \overline{C}}{\partial z^*} = D^*\frac{\partial^2 \overline{C}}{\partial z^{*2}}$$

where D^* is the Aris–Taylor dispersion coefficient:

Aris–Taylor
dispersion
coefficient

$$\boxed{D^* = D_{AB} + \frac{U^2 R^2}{48 D_{AB}}} \quad (14\text{-}33)$$

That is, for laminar flow in a pipe

$$D_a \equiv D^*$$

Figure 14-10 shows the dispersion coefficient D^* in terms of the ratio $D^*/U(2R) = D^*/Ud_t$ as a function of the product of the Reynolds and Schmidt numbers.

14.4.5 Correlations for D_a

14.4.5A Dispersion for Laminar and Turbulent Flow in Pipes

An estimate of the dispersion coefficient, D_a, can be determined from Figure 14-11. Here d_t is the tube diameter and Sc is the Schmidt number discussed in Chapter 11. The flow is laminar (streamline) below 2,100, and we see the ration (D_aU/d_t) increases with increasing Schmidt and Reynolds numbers. Between Reynolds numbers of 2,100 and 30,000, one can put bounds on D_a by calculating the maximum and minimum values at the top and bottom of the shaded region.

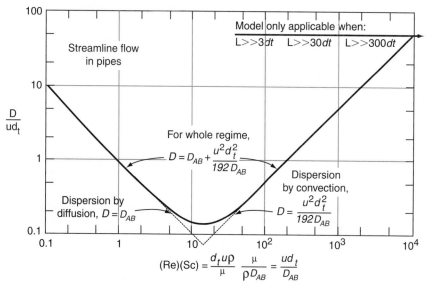

Figure 14-10   Correlation for dispersion for streamline flow in pipes. (From O. Levenspiel, *Chemical Reaction Engineering*, 2nd ed. Copyright © 1972 John Wiley & Sons, Inc. Reprinted by permission of John Wiley & Sons, Inc. All rights reserved.) [Note: $D \equiv D_a$ and $D \equiv D_{AB}$]

Once the Reynolds number is calculated, D_a can be found.

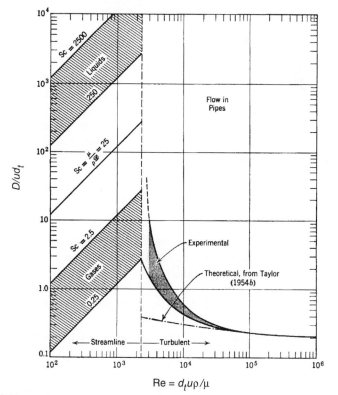

Figure 14-11   Correlation for dispersion of fluids flowing in pipes. (From O. Levenspiel, *Chemical Reaction Engineering*, 2nd ed. Copyright © 1972 John Wiley & Sons, Inc. Reprinted by permission of John Wiley & Sons, Inc. All rights reserved.) [*Note*: $D \equiv D_a$]

14.4.5B Dispersion in Packed Beds

For the case of gas–solid catalytic reactions that take place in packed-bed reactors, the dispersion coefficient, D_a, can be estimated by using Figure 14-12. Here d_p is the particle diameter and ε is the porosity.

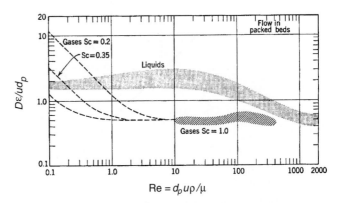

Figure 14-12 Experimental findings on dispersion of fluids flowing with mean axial velocity u in packed beds. (From O. Levenspiel, *Chemical Reaction Engineering*, 2nd ed. Copyright © 1972 John Wiley & Sons, Inc. Reprinted by permission of John Wiley & Sons, Inc. All rights reserved.) [*Note*: $D \equiv D_a$]

14.4.6 Experimental Determination of D_a

The dispersion coefficient can be determined from a pulse tracer experiment. Here, we will use t_m and σ^2 to solve for the dispersion coefficient D_a and then the Peclet number, Pe_r. Here the effluent concentration of the reactor is measured as a function of time. From the effluent concentration data, the mean residence time, t_m, and variance, σ^2, are calculated, and these values are then used to determine D_a. To show how this is accomplished, we will write

$$D_a \frac{\partial^2 C_T}{\partial z^2} - \frac{\partial (U C_T)}{\partial z} = \frac{\partial C_T}{\partial t} \tag{14-13}$$

in dimensionless form, discuss the different types of boundary conditions at the reactor entrance and exit, solve for the exit concentration as a function of dimensionless time ($\Theta = t/\tau$), and then relate D_a, σ^2, and τ.

14.4.6A The Unsteady-State Tracer Balance

The first step is to put Equation (14-13) in dimensionless form to arrive at the dimensionless group(s) that characterize the process. Let

$$\psi = \frac{C_T}{C_{T0}}, \quad \lambda = \frac{z}{L}, \quad \text{and} \quad \Theta = \frac{tU}{L}$$

For a pulse input, C_{T0} is defined as the mass of tracer injected, M, divided by the vessel volume, V. Then

$$\boxed{\frac{1}{\mathrm{Pe}_r}\frac{\partial^2 \psi}{\partial \lambda^2} - \frac{\partial \psi}{\partial \lambda} = \frac{\partial \psi}{\partial \Theta}} \tag{14-34}$$

The initial condition is

Initial condition

$$\text{At } t = 0, \quad z > 0, \quad C_T(0^+,0) = 0, \quad \psi(0^+) = 0 \tag{14-35}$$

The mass of tracer injected, M is

$$M = UA_c \int_0^\infty C_T(0^-, t)\, dt$$

14.4.6B Solution for a Closed-Closed System

In dimensionless form, the Danckwerts boundary conditions are

At $\lambda = 0$:
$$\left(-\frac{1}{\mathrm{Pe}_r}\frac{\partial \psi}{\partial \lambda}\right)_{0^+} + \psi\frac{(0^2)}{0^+} = \frac{C_T(0^-, t)}{C_{T0}} = 1 \tag{14-36}$$

At $\lambda = 1$:
$$\frac{\partial \psi}{\partial \lambda} = 0 \tag{14-37}$$

Equation (14-34) has been solved numerically for a pulse injection, and the resulting dimensionless effluent tracer concentration, ψ_{exit}, is shown as a function of the dimensionless time Θ in Figure 14-13 for various Peclet numbers. Although analytical solutions for ψ can be found, the result is an infinite series. The corresponding equations for the mean residence time, t_m, and the variance, σ^2, are[9]

$$\boxed{t_m = \tau} \tag{14-38}$$

and

$$\frac{\sigma^2}{t_m^2} = \frac{1}{\tau^2}\int_0^\infty (t - \tau)^2 E(t)\, dt$$

which can be used with the solution to Equation (14-34) to obtain

[9] See K. Bischoff and O. Levenspiel, *Adv. Chem. Eng.*, *4*, 95 (1963).

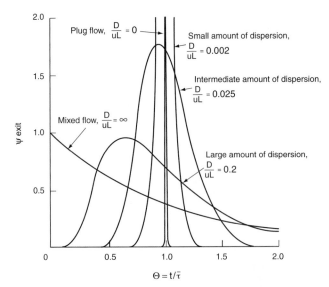

Figure 14-13 *C* curves in closed vessels for various extents of back-mixing as predicted by the dispersion model. (From O. Levenspiel, *Chemical Reaction Engineering*, 2nd ed. Copyright © 1972 John Wiley & Sons, Inc. Reprinted by permission of John Wiley & Sons, Inc. All rights reserved.) [*Note*: $D \equiv D_a$][10]

$$\frac{\sigma^2}{t_m^2} = \frac{2}{Pe_r} - \frac{2}{Pe_r^2}(1 - e^{-Pe_r}) \qquad (14\text{-}39)$$

Consequently, we see that the Peclet number, Pe_r (and hence D_a), can be found experimentally by determining t_m and σ^2 from the RTD data and then solving Equation (14-39) for Pe_r.

14.4.6C Open-Open Vessel Boundary Conditions

When a tracer is injected into a packed bed at a location more than two or three particle diameters downstream from the entrance and measured some distance upstream from the exit, the open-open vessel boundary conditions apply. For an open-open system, an analytical solution to Equation (14-13) can be obtained for a pulse tracer input.

For an open-open system, the boundary conditions at the entrance are

$$F_T(0^-, t) = F_T(0^+, t)$$

---

[10]O. Levenspiel, *Chemical Reaction Engineering*, 2nd ed. (New York: Wiley, 1972), pp. 282–284.

Then for the case when the dispersion coefficient is the same in the entrance and reaction sections:

$$-D_a\left(\frac{\partial C_T}{\partial z}\right)_{z=0^-} + UC_T(0^-,t) = -D_a\left(\frac{\partial C_T}{\partial z}\right)_{z=0^+} + UC_T(0^+,t) \quad (14\text{-}40)$$

Because there are no discontinuities across the boundary at $z = 0$

$$C_T(0^-,t) = C_T(0^+,t) \qquad (14\text{-}41)$$

At the exit

Open at the exit

$$-D_a\left(\frac{\partial C_T}{\partial z}\right)_{z=L^-} + UC_T(L^-,t) = -D_a\left(\frac{\partial C_T}{\partial z}\right)_{z=L^+} + UC_T(L^+,t) \quad (14\text{-}42)$$

$$C_T(L^-,t) = C_T(L^+,t) \qquad (14\text{-}43)$$

There are a number of perturbations of these boundary conditions that can be applied. The dispersion coefficient can take on different values in each of the three regions ($z < 0$, $0 \le z \le L$, and $z > L$), and the tracer can also be injected at some point z_1 rather than at the boundary, $z = 0$. These cases and others can be found in the supplementary readings cited at the end of the chapter. We shall consider the case when there is no variation in the dispersion coefficient for all z and an impulse of tracer is injected at $z = 0$ at $t = 0$.

For long tubes (Pe > 100) in which the concentration gradient at $\pm \infty$ will be zero, and the solution to Equation (14-34) at the exit is[11]

Valid for Pe$_r$ > 100

$$\psi(1,\Theta) = \frac{C_T(L,t)}{C_{T0}} = \frac{1}{2\sqrt{\pi\Theta/\text{Pe}_r}} \exp\left[\frac{-(1-\Theta)^2}{4\Theta/\text{Pe}_r}\right] \qquad (14\text{-}44)$$

The mean residence time for an open-open system is

Calculate τ for an open-open system

$$\boxed{t_m = \left(1 + \frac{2}{\text{Pe}_r}\right)\tau} \qquad (14\text{-}45)$$

where τ is based on the volume between $z = 0$ and $z = L$ (i.e., reactor volume measured with a yardstick). We note that the mean residence time for an open system is greater than that for a closed system. The reason is that the molecules can diffuse back into the reactor after they exit. The variance for an open-open system is

Calculate Pe$_r$ for an open–open system.

$$\boxed{\frac{\sigma^2}{\tau^2} = \frac{2}{\text{Pe}_r} + \frac{8}{\text{Pe}_r^2}} \qquad (14\text{-}46)$$

---

[11]W. Jost, *Diffusion in Solids, Liquids and Gases* (New York: Academic Press, 1960), pp. 17, 47.

We now consider two cases for which we can use Equations (14-39) and (14-46) to determine the system parameters:

Case 1. The space time τ is *known*. That is, V and v_0 are measured independently. Here we can determine the Peclet number by determining t_m and σ^2 from the concentration–time data and then using Equation (14-46) to calculate Pe$_r$. We can also calculate t_m and then use Equation (14-45) as a check, but this is usually less accurate.

Case 2. The space time τ is *unknown*. This situation arises when there are dead or stagnant pockets that exist in the reactor along with the dispersion effects. To analyze this situation we first calculate t_m and σ^2 from the data as in case 1. Then use Equation (14-45) to eliminate τ^2 from Equation (14-46) to arrive at

$$\boxed{\frac{\sigma^2}{t_m^2} = \frac{2\text{Pe} + 8}{\text{Pe}^2 + 4\text{Pe} + 4}} \qquad (14\text{-}47)$$

Finding the effective
reactor voume

We now can solve for the Peclet number in terms of our experimentally determined variables σ^2 and t_m^2. Knowing Pe$_r$, we can solve Equation (14-45) for τ, and hence V. The dead volume is the difference between the measured volume (i.e., with a yardstick) and the effective volume calculated from the RTD.

14.4.7 Sloppy Tracer Inputs

It is not always possible to inject a tracer pulse cleanly as an input to a system because it takes a finite time to inject the tracer. When the injection does not approach a perfect pulse input (Figure 14-14), the differences in the variances between the input and output tracer measurements are used to calculate the Peclet number:

$$\Delta\sigma^2 = \sigma^2_{\text{in}} - \sigma^2_{\text{out}}$$

where σ^2_{in} is the variance of the tracer measured at some point upstream (near the entrance) and σ^2_{out} is the variance measured at some point downstream (near the exit).

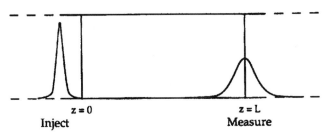

Figure 14-14 Imperfect tracer input.

For an open-open system, it has been shown[12] that the Peclet number can be calculated from the equation

$$\boxed{\frac{\Delta\sigma^2}{t_m^2} = \frac{2}{Pe_r}}$$

(14-48)

Now let's put all the material in Section 14.4 together to determine the conversion in tubular reactor for a first-order reactor.

Example 14–2   Conversion Using Dispersion and Tanks-in-Series Models

The first-order reaction

$$A \longrightarrow B$$

is carried out in a 10-cm-diameter tubular reactor 6.36 m in length. The specific reaction rate is 0.25 min^{-1}. The results of a tracer test carried out on this reactor are shown in Table E14-2.1.

TABLE E14-2.1.   EFFLUENT TRACER CONCENTRATION AS A FUNCTION OF TIME

t (min)	0	1	2	3	4	5	6	7	8	9	10	12	14
C (mg/L)	0	1	5	8	10	8	6	4	3	2.2	1.5	0.6	0

Calculate conversion using (a) the closed vessel dispersion model, (b) PFR, (c) the tanks-in-series model, and (d) a single CSTR.

Solution

(a) We will use Equation (14-27) to calculate the conversion

$$X = 1 - \frac{4q \, \exp(Pe_r/2)}{(1+q)^2 \, \exp(Pe_r q/2) - (1-q)^2 \, \exp(-Pe_r q/2)}$$

(14-27)

where $q = \sqrt{1 + 4Da/Pe_r}$, $Da = \tau k$, and $Pe_r = UL/D_a$. We can calculate Pe_r from Equation (14-39):

$$\frac{\sigma^2}{\tau^2} = \frac{2}{Pe_r} - \frac{2}{Pe_r^2}(1 - e^{-Pe_r})$$

(14-39)

First calculate t_m and σ^2 from RTD data.

However, we must find τ^2 and σ^2 from the tracer concentration data first.

$$\tau = \int_0^\infty tE(t) \, dt = \frac{V}{\upsilon}$$

(E14-2.1)

$$\sigma^2 = \int_0^\infty (t - \tau)^2 E(t) \, dt = \int_0^\infty t^2 E(t) \, dt - \tau^2$$

(E14-2.2)

---

[12]R. Aris, *Chem. Eng. Sci.*, 9, 266 (1959).

Consider the data listed in Table E14-2.2.

TABLE E14-2.2.   CALCULATIONS TO DETERMINE t_m AND σ^2

t	0	1	2	3	4	5	6	7	8	9	10	12	14
$C(t)$	0	1	5	8	10	8	6	4	3	2.2	1.5	0.6	0
$E(t)$	0	0.02	0.1	0.16	0.2	0.16	0.12	0.08	0.06	0.044	0.03	0.012	0
$tE(t)$	0	0.02	0.2	0.48	0.8	0.80	0.72	0.56	0.48	0.40	0.3	0.14	0
$t^2E(t)$	0	0.02	0.4	1.44	3.2	4.0	4.32	3.92	3.84	3.60	3.0	1.68	0

Here again spreadsheets can be used to calculate τ^2 and σ^2.

To find $E(t)$ and then t_m, we first find the area under the C curve, which is

$$\int_0^\infty C(t)\, dt = 50 \text{ g} \cdot \text{min}$$

Then

$$\tau = t_m = \int_0^\infty tE(t)\, dt = 5.15 \text{ min}$$

Calculating the first term on the right-hand side of Equation (E14-2.2), we find

$$\int_0^\infty t^2 E(t)\, dt = (\tfrac{1}{3})[1(0) + 4(0.02) + 2(0.4) + 4(1.44) + 2(3.2) + 4(4.0)$$
$$+ 2(4.32) + 4(3.92) + 2(3.84) + 4(3.6) + 1(3.0)]$$
$$+ (\tfrac{2}{3})[3.0 + 4(1.68) + 0]$$
$$= 32.63 \text{ min}^2$$

Substituting these values to Equation (E14-2.2), we obtain the variance, σ^2.

$$\sigma^2 = 32.63 - (5.15)^2 = 6.10 \text{ min}^2$$

Most people, including the author, would use Polymath or Excel to form Table E14-2.2 and to calculate t_m and σ^2. Dispersion in a closed vessel is represented by

Calculate Pe_r from t_m and σ^2.

$$\frac{\sigma^2}{\tau^2} = \frac{2}{Pe_r^2}(Pe_r - 1 + e^{-Pe_r}) \qquad (14\text{-}39)$$

$$= \frac{6.1}{(5.15)^2} = 0.23 = \frac{2}{Pe_r^2}(Pe_r - 1 + e^{-Pe_r})$$

Solving for Pe_r either by trial and error or using Polymath, we obtain

$$\boxed{Pe_r = 7.5}$$

Next, calculate Da, q, and X.

Next we calculate Da to be

$$\boxed{Da = \tau k = (5.15 \text{ min})(0.25 \text{ min}^{-1}) = 1.29}$$

Using the equations for q and X gives

$$q = \sqrt{1 + \frac{4Da}{Pe_r}} = \sqrt{1 + \frac{4(1.29)}{7.5}} = 1.30$$

Then

$$\frac{Pe_r q}{2} = \frac{(7.5)(1.3)}{2} = 4.87$$

Substitution into Equation (14-40) yields

Dispersion Model

$$X = 1 - \frac{4(1.30)\, e^{(7.5/2)}}{(2.3)^2 \, \exp(4.87) - (-0.3)^2 \, \exp(-4.87)}$$

$$\boxed{X = 0.68} \qquad 68\% \text{ conversion for the dispersion model}$$

When dispersion effects are present in this tubular reactor, 68% conversion is achieved.

(b) If the reactor were operating ideally as a plug-flow reactor, the conversion would be

PFR

$$\boxed{X = 1 - e^{-\tau k} = 1 - e^{-Da} = 1 - e^{-1.29} = 0.725}$$

That is, 72.5% conversion would be achieved in an ideal plug-flow reactor.

Tanks-in-series model

(c) Conversion using the tanks-in-series model: We recall Equation (14-12) to calculate the number of tanks in series:

$$n = \frac{\tau^2}{\sigma^2} = \frac{(5.15)^2}{6.1} = 4.35$$

To calculate the conversion, we recall Equation (4-11). For a first-order reaction for n tanks in series, the conversion is

$$\boxed{\begin{array}{l} X = 1 - \dfrac{1}{(1 + \tau_i k)^n} = 1 - \dfrac{1}{[1 + (\tau/n)\,k]^n} = 1 - \dfrac{1}{(1 + 1.29/4.35)^{4.35}} \\[2mm] X = \textbf{67.7\% for the tanks-in-series model} \end{array}}$$

(d) For a single CSTR,

CSTR

$$X = \frac{\tau k}{1 + \tau k} = \frac{1.29}{2.29} = 0.563$$

So 56.3% conversion would be achieved in a single ideal tank.
Summary:

Summary

$$\boxed{\begin{array}{l} \text{PFR: } X = 72.5\% \\ \text{Dispersion: } X = 68.0\% \\ \text{Tanks in series: } X = 67.7\% \\ \text{Single CSTR: } X = 56.3\% \end{array}}$$

In this example, correction for finite dispersion, whether by a dispersion model or a tanks-in-series model, is significant when compared with a PFR.

14.5 Tanks-in-Series Model Versus Dispersion Model

We have seen that we can apply both of these one-parameter models to tubular reactors using the variance of the RTD. For first-order reactions, the two models can be applied with equal ease. However, the tanks-in-series model is mathematically easier to use to obtain the effluent concentration and conversion for reaction orders other than one and for multiple reactions. However, we need to ask what would be the accuracy of using the tanks-in-series model over the dispersion model. These two models are equivalent when the Peclet–Bodenstein number is related to the number of tanks in series, n, by the equation[13]

$$\text{Bo} = 2(n-1) \qquad (14\text{-}49)$$

Equivalency between models of tanks-in-series and dispersion

or

$$n = \frac{\text{Bo}}{2} + 1 \qquad (14\text{-}50)$$

where $\text{Bo} = UL/D_a$, where U is the superficial velocity, L the reactor length, and D_a the dispersion coefficient.

For the conditions in Example 14-2, we see that the number of tanks calculated from the Bodenstein number, Bo (i.e., Pe_r), Equation (14-50), is 4.75, which is very close to the value of 4.35 calculated from Equation (14-12). Consequently, for reactions other than first order, one would solve successively for the exit concentration and conversion from each tank in series for both a battery of four tanks in series and of five tanks in series in order to bound the expected values.

In addition to the one-parameter models of tanks-in-series and dispersion, many other one-parameter models exist when a combination of ideal reactors is to model the real reactor as shown in Section 13.5 for reactors with bypassing and dead volume. Another example of a one-parameter model would be to model the real reactor as a PFR and a CSTR in series with the one parameter being the fraction of the total volume that behaves as a CSTR. We can dream up many other situations that would alter the behavior of ideal reactors in a way that adequately describes a real reactor. However, it may be that one parameter is not sufficient to yield an adequate comparison between theory and practice. We explore these situations with combinations of ideal reactors in the section on two-parameter models.

The reaction rate parameters are usually known (i.e., *Da*), but the Peclet number is usually not known because it depends on the flow and the vessel. Consequently, we need to find Pe_r using one of the three techniques discussed earlier in the chapter.

---

[13]K. Elgeti, *Chem. Eng. Sci., 51,* 5077 (1996).

14.6 Numerical Solutions to Flows with Dispersion and Reaction

We now consider dispersion and reaction. We first write our mole balance on species A by recalling Equation (14-28) and including the rate of formation of A, r_A. At steady state we obtain

$$D_{AB}\left[\frac{1}{r}\frac{\partial\left(r\frac{\partial C_A}{\partial r}\right)}{\partial r} + \frac{\partial^2 C_A}{\partial z^2}\right] - U(r)\frac{\partial C_A}{\partial z} + r_A = 0 \qquad (14\text{-}51)$$

Analytical solutions to dispersion with reaction can only be obtained for isothermal zero- and first-order reactions. We are now going to use COMSOL to solve the flow with reaction and dispersion with reaction. A COMSOL CD-ROM is included with the text.

We are going to compare two solutions: one which uses the Aris–Taylor approach and one in which we numerically solve for both the axial and radial concentration using COMSOL.

Case A. Aris–Taylor Analysis for Laminar Flow

For the case of an nth-order reaction, Equation (14-15) is

$$\frac{D_a}{U}\frac{d^2 C_A}{dz^2} - \frac{dC_A}{dz} - \frac{kC_A^n}{U} = 0 \qquad (14\text{-}52)$$

If we use the Aris–Taylor analysis, we can use Equation (14-15) with a caveat that $\overline{\psi} = \overline{C}_A/C_{A0}$ where $\overline{C}_A$ is the average concentration from $r = 0$ to $r = R$ as given by

$$\frac{1}{Pe_r}\frac{d^2\overline{\psi}}{d\lambda^2} - \frac{d\overline{\psi}}{d\lambda} - \boldsymbol{Da}\overline{\psi}^n = 0 \qquad (14\text{-}53)$$

where

$$Pe_r = \frac{UL}{D_a} \text{ and } \boldsymbol{Da} = \tau k C_{A0}^{n-1}$$

For the closed-closed boundary conditions we have

$$\text{At}\quad \lambda = 0: \quad -\frac{1}{Pe_r}\frac{d\overline{\psi}}{d\lambda}\bigg|_{0^+} = \overline{\psi}(0^+) = 1 \qquad (14\text{-}54)$$

Danckwerts boundary conditions

$$\text{At}\quad \lambda = 1: \quad \frac{d\overline{\psi}}{d\lambda} = 0$$

For the open-open boundary condition we have

$$\text{At} \quad \lambda = 0: \quad \overline{\psi}(0^-) - \frac{1}{Pe_r}\frac{d\overline{\psi}}{d\lambda}\bigg|_{0^-} = \overline{\psi}(0^+) - \frac{1}{Pe_r}\frac{d\overline{\psi}}{d\lambda}\bigg|_{0^+} \quad (14\text{-}55)$$

$$\text{At} \quad \lambda = 1: \quad \frac{d\overline{\psi}}{d\lambda} = 0$$

Equation (14-53) is a nonlinear second order ODE that is solved on the COMSOL CD-ROM.

Case B. Full Numerical Solution

To obtain axial and radial profile we now solve Equation (14-51)

$$D_{AB}\left[\frac{1}{r}\frac{\partial\left(r\frac{\partial C_A}{\partial r}\right)}{\partial r} + \frac{\partial^2 C_A}{\partial z^2}\right] - U(r)\frac{\partial C_A}{\partial z} + r_A = 0 \quad (14\text{-}51)$$

First we will put the equations in dimensionless form by letting $\psi = C_A/C_{A0}$, $\lambda = z/L$, and $\phi = r/R$. Following our earlier transformation of variables, Equation (14-51) becomes

$$\left(\frac{L}{R}\right)\frac{1}{Pe_r}\left[\frac{1}{\phi}\frac{\partial\left(r\frac{\partial\psi}{\partial\phi}\right)}{\partial\phi}\right] + \frac{1}{Pe_r}\frac{d^2\psi}{d\lambda^2} - 2(1-\phi^2)\frac{d\psi}{d\lambda} - Da\psi^n = 0 \quad (14\text{-}56)$$

Example 14–3 Dispersion with Reaction

(a) First, use COMSOL to solve the dispersion part of Example 14-2 again. How does the COMSOL result compare with the solution to Example 14-2?

(b) Repeat (a) for a second-order reaction with k = 0.5 dm³/mol • min.

(c) Repeat (a) but assume laminar flow and consider radial gradients in concentration. Use D_{AB} for both the radial and axial diffusion coefficients. Plot the axial and radial profiles. Compare your results with part (a).

Additional information:

C_{A0} = 0.5 mol/dm³, U_0 = L/τ = 1.24 m/min, D_a = $U_0 L/Pe_r$ = 1.05 m²/min. D_{AB} = 7.6E-5 m²/min.

Note: For part (a), the two-dimensional model with no radial gradients (plug flow) becomes a one-dimensional model. The inlet boundary condition for part (a) and part (b) is a closed-closed vessel (flux[z = 0⁻] = flux[z = 0⁺] or $U_z \cdot C_{A0}$ = flux) at the inlet boundary. In COMSOL format it is: $-N_i \cdot n$ = U0*CA0. The boundary condition for laminar flow in COMSOL format for part (c) is: $-N_i \cdot n$ = 2*U0*(1–(r/Ra)₂)*CA0.

Solution

(a) Equation (14-52) was used in the COMSOL program along with the rate law

The different types of COMSOL Boundary Conditions are given in Problem P14-19$_c$

$$r_A = -kC_A = -kC_{A0}\,\psi$$

We see that we get the same results as the analytical solution in Example 14-2. With the Aris–Taylor analysis the two-dimensional profile becomes one-dimensional plug flow velocity profile. Figure E14-3.1(a) shows a uniform concentration surface and shows the plug flow behavior of the reactor. Figure E14-3.1(b) shows the corresponding cross-section plots at the inlet, half axial location, and outlet. The average outlet conversion is 67.9%.

The average outlet concentration at an axial distance z is found by integrating across the radius as shown below

$$C_A(z) = \int_0^R \frac{2\pi r C_A(r,z)dr}{\pi R^2}$$

From the average concentrations at the inlet and outlet we can calculate the average conversion as

$$X = \frac{C_{A0} - C_A}{C_{A0}}$$

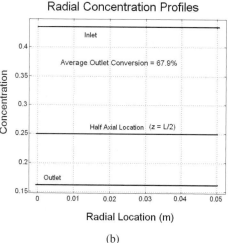

(a)   (b)

Figure E14-3.1 COMSOL results for a plug flow reactor with first-order reaction. (Concentrations in mol/dm³.)

(b)   Now we expand our results to consider the case when the reaction is second order ($-r_A = kC_A^2 = kC_{A0}^2\,\psi^2$) with $k = 0.5$ dm³/mol·min and $C_{A0} = 0.5$ mol/dm³. Let's assume the radial dispersion coefficient is equal to the molecular diffusivity. Keeping everything else constant, the average outlet conversion is 52.3%. However, because the flow inside the reactor is modeled as plug flow the concentration profiles are still flat, as shown in Figure E14-3.2.

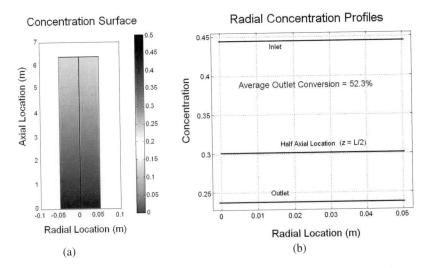

Figure E14-3.2 COMSOL results for a plug flow reactor with second-order reaction. (Concentrations in mol/dm³.)

(c)   Now, we will change the flow assumption from plug flow to laminar flow and solve Equation (14-51) for a first-order reaction.

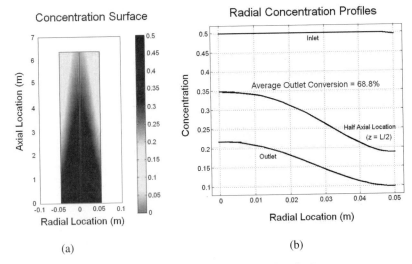

Figure E14-3.3 COMSOL output for laminar flow in the reactor. (Concentrations in mol/dm³.)

The average outlet conversion becomes 68.8%, not much different from the one in part (a) in agreement with the Aris–Taylor analysis. However, due to the laminar flow assumption in the reactor, the radial concentration profiles are very different throughout the reactor.

(d)   As a homework exercise, repeat part (c) for the second-order reaction given in part (b).

14.7 Two-Parameter Models—Modeling Real Reactors with Combinations of Ideal Reactors

Creativity and engineering judgment are necessary for model formulation

We now will see how a real reactor might be modeled by one of two different combinations of ideal reactors. These are but two of an almost unlimited number of combinations that could be made. However, if we limit the number of adjustable parameters to two (e.g., bypass flow rate, v_b, and dead volume, V_D), the situation becomes much more tractable. After reviewing the steps in Table 14-1, choose a model and determine if it is reasonable by qualitatively comparing it with the RTD, and if it is, determine the model parameters. Usually, the simplest means of obtaining the necessary data is some form of tracer test. These tests have been described in Chapter 13, together with their uses in determining the RTD of a reactor system. Tracer tests can be used to determine the RTD, which can then be used in a similar manner to determine the suitability of the model and the value of its parameters.

A tracer experiment is used to evaluate the model parameters.

In determining the suitability of a particular reactor model and the parameter values from tracer tests, it may not be necessary to calculate the RTD function $E(t)$. The model parameters (e.g., V_D) may be acquired directly from measurements of effluent concentration in a tracer test. The theoretical prediction of the particular tracer test in the chosen model system is compared with the tracer measurements from the real reactor. The parameters in the model are chosen so as to obtain the closest possible agreement between the model and experiment. If the agreement is then sufficiently close, the model is deemed reasonable. If not, another model must be chosen.

The quality of the agreement necessary to fulfill the criterion "sufficiently close" again depends on creatively in developing the model and on engineering judgment. The most extreme demands are that the maximum error in the prediction not exceed the estimated error in the tracer test and that there be no observable trends with time in the difference between prediction (the model) and observation (the real reactor). To illustrate how the modeling is carried out, we will now consider two different models for a CSTR.

14.7.1 Real CSTR Modeled Using Bypassing and Dead Space

A real CSTR is believed to be modeled as a combination of an ideal CSTR of volume V_s, a dead zone of volume V_d, and a bypass with a volumetric flow rate v_b (Figure 14-15). We have used a tracer experiment to evaluate the parameters of the model V_s and v_s. Because the total volume and volumetric flow rate are known, once V_s and v_s are found, v_b and V_d can readily be calculated.

14.7.1A Solving the Model System for C_A and X

We shall calculate the conversion for this model for the first-order reaction

$$A \longrightarrow B$$

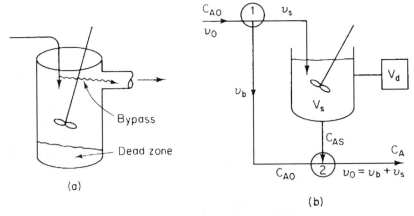

(a)

(b)

Figure 14-15   (a) Real system; (b) model system.

The bypass stream and effluent stream from the reaction volume are mixed at point 2. From a balance on species A around this point,

Balance at junction

$$[\text{In}] = [\text{Out}]$$

$$[C_{A0}v_b + C_{As}v_s] = [C_A(v_b + v_s)] \tag{14-57}$$

We can solve for the concentration of A leaving the reactor,

$$C_A = \frac{v_b C_{A0} + C_{As}v_s}{v_b + v_s} = \frac{v_b C_{A0} + C_{As}v_s}{v_0}$$

Let $\alpha = V_s/V$ and $\beta = v_b/v_0$. Then

$$\boxed{C_A = \beta C_{A0} + (1 - \beta)C_{As}} \tag{14-58}$$

For a first-order reaction, a mole balance on V_s gives

Mole balance on CSTR

$$v_s C_{A0} - v_s C_{As} - kC_{As}V_s = 0 \tag{14-59}$$

or, in terms of α and β,

$$\boxed{C_{As} = \frac{C_{A0}(1 - \beta)v_0}{(1 - \beta)v_0 + \alpha Vk}} \tag{14-60}$$

Substituting Equation (14-60) into (14-58) gives the effluent concentration of species A:

Conversion as a function of model parameters

$$\boxed{\frac{C_A}{C_{A0}} = 1 - X = \beta + \frac{(1 - \beta)^2}{(1 - \beta) + \alpha \tau k}} \tag{14-61}$$

We have used the ideal reactor system shown in Figure 14-15 to predict the conversion in the real reactor. The model has two parameters, α and β. If these parameters are known, we can readily predict the conversion. In the following section, we shall see how we can use tracer experiments and RTD data to evaluate the model parameters.

14.7.1B Using a Tracer to Determine the Model Parameters in CSTR-with-Dead-Space-and-Bypass Model

In Section 14.7.1A, we used the system shown in Figure 14-16, with bypass flow rate v_b and dead volume V_d, to model our real reactor system. We shall inject our tracer, T, as a positive-step input. The unsteady-state balance on the nonreacting tracer T in the reactor volume V_s is

$$\text{In} - \text{out} = \text{accumulation}$$

Tracer balance for
step input

$$\boxed{v_s C_{T0} - v_s C_{Ts} = \frac{dN_{Ts}}{dt} = V_s \frac{dC_{Ts}}{dt}} \qquad (14\text{-}62)$$

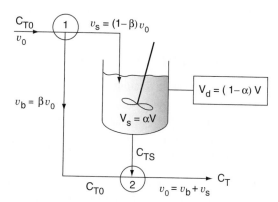

Model system

Figure 14-16   Model system: CSTR with dead volume and bypassing.

The conditions for the positive-step input are

$$\text{At } t < 0 \quad C_T = 0$$

$$\text{At } t \geq 0 \quad C_T = C_{T0}$$

The junction
balance

A balance around junction point 2 gives

$$\boxed{C_T = \frac{v_b C_{T0} + C_{Ts} v_s}{v_0}} \qquad (14\text{-}63)$$

As before,

$$V_s = \alpha V$$

$$v_b = \beta v_0$$

$$\tau = \frac{V}{v_0}$$

Integrating Equation (14-62) and substituting in terms of α and β gives

$$\frac{C_{Ts}}{C_{T0}} = 1 - \exp\left[-\frac{1-\beta}{\alpha}\left(\frac{t}{\tau}\right)\right] \qquad (14\text{-}64)$$

Combining Equations (14-63) and (14-64), the effluent tracer concentration is

$$\frac{C_T}{C_{T0}} = 1 - (1 - \beta)\, \exp\left[-\frac{1-\beta}{\alpha}\left(\frac{t}{\tau}\right)\right] \qquad (14\text{-}65)$$

We now need to rearrange this equation to extract the model parameters, α and β, either by regression (Polymath/MATLAB/Excel) or from the proper plot of the effluent tracer concentrations as a function of time. Rearranging yields

Evaluating the
model parameters

$$\ln\frac{C_{T0}}{C_{T0} - C_T} = \ln\frac{1}{1-\beta} + \left(\frac{1-\beta}{\alpha}\right)\frac{t}{\tau} \qquad (14\text{-}66)$$

Consequently, we plot $\ln[C_{T0}/(C_{T0} - C_T)]$ as a function of t. If our model is correct, a straight line should result with a slope of $(1 - \beta)/\tau\alpha$ and an intercept of $\ln[1/(1 - \beta)]$.

Example 14–4 CSTR with Dead Space and Bypass

The elementary reaction

$$A + B \longrightarrow C + D$$

is to be carried out in the CSTR shown schematically in Figure 14-15. There is both bypassing and a stagnant region in this reactor. The tracer output for this reactor is shown in Table E14-4.1. The measured reactor volume is 1.0 m³ and the flow rate to the reactor is 0.1 m³/min. The reaction rate constant is 0.28 m³/kmol·min. The feed is equimolar in A and B with an entering concentration of A equal to 2.0 kmol/m³. Calculate the conversion that can be expected in this reactor (Figure E14-4.1).

TABLE E14-4.1   TRACER DATA FOR STEP INPUT

C_T (mg/dm³)	1000	1333	1500	1666	1750	1800
t (min)	4	8	10	14	16	18

The entering tracer concentration is $C_{T0} = 2000$ mg/dm³.

Two-parameter
model

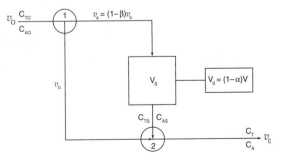

Figure E14-4.1 Schematic of real reactor modeled with dead space (V_d) and bypass (v_b).

Solution

Recalling Equation (14-66)

$$ln\frac{C_{T0}}{C_{T0}-C_T} = ln\frac{1}{1-\beta} + \frac{(1-\beta)}{\alpha}\frac{t}{\tau} \qquad (14\text{-}66)$$

Equation (14-66) suggests that we construct Table E14-4.2 from Table E14-4.1 and plot $C_{T0}/(C_{T0} - C_T)$ as a function of time on semilog paper. Using this table we get Figure E14-4.2.

TABLE E14-4.2.   PROCESSED DATA

t (min)	4	8	10	14	16	18
$\dfrac{C_{T0}}{C_{T0}-C_T}$	2	3	4	6	8	10

We can find α and β from either a semilog plot as shown in Figure E14-4.2 or by regression using Polymath, MATLAB, or Excel.

Evaluating the parameters α and β

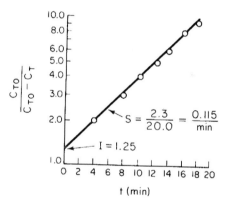

Figure E14-4.2   Response to a step input.

The volumetric flow rate to the well-mixed portion of the reactor, v_s, can be determined from the intercept, I:

$$\frac{1}{1-\beta} = I = 1.25$$

$$\boxed{\beta = \frac{v_b}{v_0} = 0.2}$$

The volume of the well-mixed region, V_s, can be calculated from the slope:

$$\frac{1-\beta}{\alpha\tau} = S = 0.115 \text{ min}^{-1}$$

$$\alpha\tau = \frac{1-0.2}{0.115} = 7 \text{ min}$$

$$\tau = \frac{V}{v_0} = \frac{1 \text{ m}^3}{(0.1 \text{ m}^3/\text{min})} = 10 \text{ min}$$

$$\boxed{\alpha = \frac{7 \text{ min}}{\tau} = 0.7}$$

We now proceed to determine the conversion corresponding to these model parameters.

1. **Balance on reactor volume V_s:**

$$[\text{In}] - \text{Out} + \text{Generation} = \text{Accumulation}$$

$$v_s C_{A0} - v_s C_{As} + r_{As} V_s = 0 \tag{E14-4.1}$$

2. **Rate law:**

$$-r_{AS} = k C_{As} C_{Bs}$$

Equimolar feed $\therefore C_{As} = C_{Bs}$

$$-r_{As} = k C_{As}^2 \tag{E14-4.2}$$

3. **Combining** Equations (E14-4.1) and (E14-4.2) gives

$$v_s C_{A0} - v_s C_{As} - k C_{As}^2 V_s = 0 \tag{E14-4.3}$$

Rearranging, we have

$$\tau_s k C_{As}^2 + C_{As} - C_{A0} = 0 \tag{E14-4.4}$$

Solving for C_{As} yields

$$\boxed{C_{As} = \frac{-1 + \sqrt{1 + 4\tau_s k C_{A0}}}{2\tau_s k}} \tag{E14-4.5}$$

*The Duck Tape Council would like to point out the new wrinkle: **The Junction Balance.***

4. **Balance around junction** point 2:

$$[\text{In}] = [\text{Out}]$$

$$[v_b C_{A0} + v_s C_{As}] = [v_0 C_A] \tag{E14-4.6}$$

Rearranging Equation (E14-4.6) gives us

$$\boxed{C_A = \frac{v_0 - v_s}{v_0} C_{A0} + \frac{v_s}{v_0} C_{As}} \tag{E14-4.7}$$

5. Parameter evaluation:

$$v_s = 0.8\,v_0 = (0.8)(0.1\ \text{m}^3/\text{min}) = 0.08\ \text{m}^3/\text{min}$$

$$V_s = (\alpha\tau)v_0 = (7.0\ \text{min})(0.1\ \text{m}^3/\text{min}) = 0.7\ \text{m}^3$$

$$\tau_s = \frac{V_s}{v_s} = 8.7\ \text{min}$$

$$C_{As} = \frac{\sqrt{1 + 4\tau_s kC_{A0}} - 1}{2\tau_s k} \tag{E14-4.8}$$

$$= \frac{\sqrt{1 + (4)(8.7\ \text{min})(0.28\ \text{m}^3/\text{kmol}\cdot\text{min})(2\ \text{kmol}/\text{m}^3)} - 1}{(2)(8.7\ \text{min})(0.28\ \text{m}^3/\text{kmol}\cdot\text{min})}$$

$$= 0.724\ \text{kmol}/\text{m}^3$$

Substituting into Equation (E14-4.7) yields

Finding the conversion

$$C_A = \frac{0.1 - 0.08}{0.1}\,(2) + (0.8)(0.724) = 0.979$$

$$X = 1 - \frac{0.979}{2.0} = 0.51$$

If the real reactor were acting as an ideal CSTR, the conversion would be

$$\boxed{C_A = \frac{\sqrt{1 + 4\tau kC_{A0}} - 1}{2\tau k}} \tag{E14-4.9}$$

$$C_A = \frac{\sqrt{1 + 4(10)(0.28)(2)} - 1}{2(10)(0.28)} = 0.685$$

$$\boxed{X = 1 - \frac{C_A}{C_{A0}} = 1 - \frac{0.685}{2.0} = 0.66} \tag{E14-4.10}$$

$X_{\text{model}} = 0.51$

$X_{\text{Ideal}} = 0.66$

Other Models. In Section 14.7.1 it was shown how we formulated a model consisting of ideal reactors to represent a real reactor. First, we solved for the exit concentration and conversion for our model system in terms of two parameters α and β. We next evaluated these parameters from data of tracer concentration as a function of time. Finally, we substituted these parameter values into the mole balance, rate law, and stoichiometric equations to predict the conversion in our real reactor.

To reinforce this concept, we will use one more example.

14.7.2 Real CSTR Modeled as Two CSTRs with Interchange

In this particular model there is a highly agitated region in the vicinity of the impeller; outside this region, there is a region with less agitation (Figure 14-17). There is considerable material transfer between the two regions. Both inlet and outlet flow channels connect to the highly agitated region. We shall model the highly agitated region as one CSTR, the quieter region as another CSTR, with material transfer between the two.

The model system

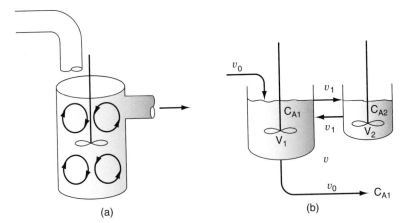

Figure 14-17   (a) Real reaction system; (b) model reaction system.

14.7.2A Solving the Model System for C_A and X

Let β represent that fraction of the total flow that is exchanged between reactors 1 and 2, that is,

$$v_1 = \beta v_0$$

and let α represent that fraction of the total volume V occupied by the highly agitated region:

Two parameters:
α and β

$$V_1 = \alpha V$$

Then

$$V_2 = (1 - \alpha)V$$

The space time is

$$\tau = \frac{V}{v_0}$$

Reference Shelf

As shown on the CD-ROM 14R.2, for a first-order reaction, the exit concentration and conversion are

$$C_{A1} = \frac{C_{A0}}{1 + \beta + \alpha\tau k - \{\beta^2/[\beta + (1-\alpha)\tau k]\}} \tag{14-67}$$

and

Conversion for
two-CSTR model

$$X = 1 - \frac{C_{A1}}{C_{A0}} = \frac{(\beta + \alpha\tau k)[\beta + (1-\alpha)\tau k] - \beta^2}{(1+\beta+\alpha\tau k)[\beta + (1-\alpha)\tau k] - \beta^2} \quad (14\text{-}68)$$

where C_{A1} is the reactor concentration exiting the first reactor in Figure 14-17(b).

14.7.2B Using a Tracer to Determine the Model Parameters in a CSTR with an Exchange Volume

The problem now is to evaluate the parameters α and β using the RTD data. A mole balance on a tracer pulse injected at $t = 0$ for each of the tanks is

$$\text{Accumulation} = \text{Rate in} - \text{Rate out}$$

Unsteady-state
balance of inert
tracer

Reactor 1: $\qquad V_1 \dfrac{dC_{T1}}{dt} = v_1 C_{T2} - (v_0 C_{T1} + v_1 C_{T1}) \qquad (14\text{-}69)$

Reactor 2: $\qquad V_2 \dfrac{dC_{T2}}{dt} = v_1 C_{T1} - v_1 C_{T2} \qquad (14\text{-}70)$

C_{T1} and C_{T2} are the tracer concentrations in reactors 1 and 2, respectively, with initial conditions $C_{T10} = N_{T0}/V_1$ and $C_{T20} = 0$.

Substituting in terms of α, β, and τ, we arrive at two coupled differential equations describing the unsteady behavior of the tracer that must be solved simultaneously.

See Appendix A.3
for method of
solution

$$\tau\alpha \frac{dC_{T1}}{dt} = \beta C_{T2} - (1+\beta)C_{T1} \qquad (14\text{-}71)$$

$$\tau(1-\alpha) \frac{dC_{T2}}{dt} = \beta C_{T1} - \beta C_{T2} \qquad (14\text{-}72)$$

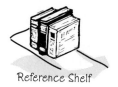

Reference Shelf

Analytical solutions to Equations (14-71) and (14-72) are given in the CD-ROM, in Appendix A.3 and in Equation (14-73), below. However, for more complicated systems, analytical solutions to evaluate the system parameters may not be possible.

$$\left(\frac{C_{T1}}{C_{T10}}\right)_{\text{pulse}} = \frac{(\alpha m_1 + \beta + 1)e^{m_2 t/\tau} - (\alpha m_2 + \beta + 1)e^{m_1 t/\tau}}{\alpha(m_1 - m_2)} \qquad (14\text{-}73)$$

where

$$m_1, m_2 = \left[\frac{1-\alpha+\beta}{2\alpha(1-\alpha)}\right]\left[-1 \pm \sqrt{1 - \frac{4\alpha\beta(1-\alpha)}{(1-\alpha+\beta^2)}}\right]$$

By regression on Equation (14-73) and the data in Table E14-4.2 or by an appropriate semilog plot of C_{T1}/C_{T0} versus time, one can evaluate the model parameters α and β.

14.8 Use of Software Packages to Determine the Model Parameters

If analytical solutions to the model equations are not available to obtain the parameters from RTD data, one could use ODE solvers. Here, the RTD data would first be fit to a polynomial to the effluent concentration–time data and then compared with the model predictions for different parameter values.

Example 14–5 CSTR with Bypass and Dead Volume

(a) Determine parameters α and β that can be used to model two CSTRs with interchange using the tracer concentration data listed in Table E14-5.1.

TABLE E14-5.1.   RTD DATA

t (min)	0.0	20	40	60	80	120	160	200	240
C_{Te} (g/m³)	2000	1050	520	280	160	61	29	16.4	10.0

(b) Determine the conversion of first-order reaction with $k = 0.03\ \text{min}^{-1}$ and $\tau = 40\ \text{min}$.

Solution

First we will use Polymath to fit the RTD to a polynomial. Because of the steepness of the curve, we shall use two polynomials.

For $t \le 80$ min,

$$C_{Te} = 2000 - 59.6t + 0.642t^2 - 0.00146t^3 - 1.04 \times 10^{-5}t^4 \quad \text{(E14-5.1)}$$

For $t > 80$,

$$C_{Te} = 921 - 17.3t + 0.129t^2 - 0.000438t^3 - 5.6 \times 10^{-7}t^4 \quad \text{(E14-5.2)}$$

where C_{Te} is the exit concentration of tracer determined experimentally. Next we would enter the tracer mole (mass) balances Equations (14-71) and (14-72) into an ODE solver. The Polymath program is shown in Table E14-5.2. Finally, we vary the parameters α and β and then compare the calculated effluent concentration C_{T1} with the experimental effluent tracer concentration C_{Te}. After a few trials we converge on the values $\alpha = 0.8$ and $\beta = 0.1$. We see from Figure E14-5.1 and Table E14-5.3 that the agreement between the RTD data and the calculated data are quite good, indicating the validity of our values of α and β. The graphical solution to this problem is given on the CD-ROM and in the 2nd Edition. We now substitute these values in Equation (14-68), and as shown in the CD-ROM, the corresponding conversion is 51% for the model system of two CSTRs with interchange:

Trial and error using software packages

$$\boxed{X = 1 - \frac{C_{A1}}{C_{A0}} = \frac{(\beta + \alpha\tau k)[\beta + (1-\alpha)\tau k] - \beta^2}{(1+\beta+\alpha\tau k)[\beta + (1-\alpha)\tau k] - \beta^2}} \quad \text{(14-68)}$$

$$\tau k = (40 \text{ min})(0.03 \text{ min}^{-1}) = 1.2$$

$$X = \frac{[0.1 + (0.8)(1.2)][0.1 + (1 - 0.8)(1.2)] - (0.1)^2}{[1 + 0.1 + (0.8)(1.2)][0.1 + (1 - 0.8)(1.2) - (0.1)^2]}$$

$$X = 0.51$$

Comparing models, we find

$$(X_{\text{model}} = 0.51) < (X_{\text{CSTR}} = 0.55) < (X_{\text{PFR}} = 0.7)$$

TABLE E14-5.2.   POLYMATH PROGRAM: TWO CSTRS WITH INTERCHANGE

Living Example Problem

ODE Report (RKF45)

Differential equations as entered by the user
[1] d(CT1)/d(t) = (beta*CT2-(1+beta)*CT1)/alpha/tau
[2] d(CT2)/d(t) = (beta*CT1-beta*CT2)/(1-alpha)/tau

Explicit equations as entered by the user
[1] beta = 0.1
[2] alpha = 0.8
[3] tau = 40
[4] CTe1 = 2000-59.6*t+0.64*t^2-0.00146*t^3-1.047*10^(-5)*t^4
[5] CTe2 = 921-17.3*t+0.129*t^2-0.000438*t^3+5.6*10^(-7)*t^4
[6] t1 = t-80
[7] CTe = if(t<80)then(CTe1)else(CTe2)

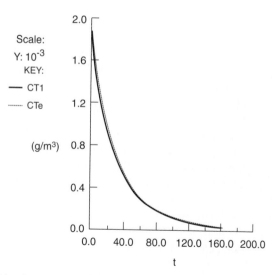

Figure E14-5.1 Comparison of model and experimental exit tracer concentrations.

TABLE E14-5.3.   COMPARING MODEL (CT1) WITH EXPERIMENT (CTe)

Two CSTRs with
interchange

t	CT1	CTe
0	2000	2000
10	1421.1968	1466.4353
20	1014.8151	1050.6448
30	728.9637	740.0993
40	527.4236	519.7568
50	384.9088	372.0625
60	283.7609	276.9488
70	211.6439	211.8353
80	159.9355	161.2816
100	95.43456	99
120	60.6222	61.8576
140	40.92093	40.6576
160	29.10943	28.3536

14.9 Other Models of Nonideal Reactors Using CSTRs and PFRs

Several reactor models have been discussed in the preceding pages. All are based on the physical observation that in almost all agitated tank reactors, there is a well-mixed zone in the vicinity of the agitator. This zone is usually represented by a CSTR. The region outside this well-mixed zone may then be modeled in various fashions. We have already considered the simplest models, which have the main CSTR combined with a dead-space volume; if some short-circuiting of the feed to the outlet is suspected, a bypass stream can be added. The next step is to look at all possible combinations that we can use to model a nonideal reactor using only CSTRs, PFRs, dead volume, and bypassing. The rate of transfer between the two reactors is one of the model parameters. The positions of the inlet and outlet to the model reactor system depend on the physical layout of the real reactor.

Figure 14-18(a) describes a real PFR or PBR with channeling that is modeled as two PFRs/PBRs in parallel. The two parameters are the fraction of flow to the reactors [i.e., β and $(1 - \beta)$] and the fractional volume [i.e., α and $(1 - \alpha)$] of each reactor. Figure 14-18(b) describes a real PFR/PBR that has a backmix region and is modeled as a PFR/PBR in parallel with a CSTR. Figures 14-19(a) and (b) show a real CSTR modeled as two CSTRs with interchange. In one case, the fluid exits from the top CSTR (a) and in the other case the fluid exits from the bottom CSTR. The parameter β represents the interchange volumetric flow rate and α the fractional volume of the top reactor, where the fluid exits the reaction system. We note that the reactor in model 14-19(b) was found to describe extremely well a real reactor used in the production of terephthalic acid.[14] A number of other combinations of ideal reactions can be found in Levenspiel.[15]

A case history for
terephthalic acid

---

[14]Proc. Indian Inst. Chem. Eng. Golden Jubilee, a Congress, Delhi, 1997, p. 323.

[15]Levenspiel, O. *Chemical Reaction Engineering*, 3rd ed. (New York: Wiley, 1999), pp. 284–292.

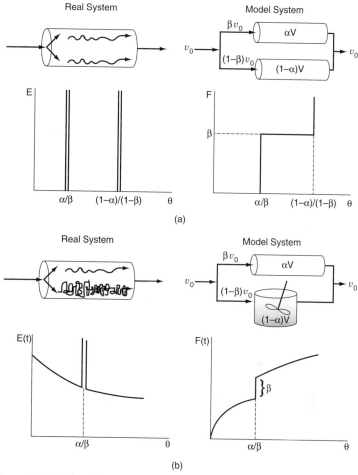

Figure 14-18   Combinations of ideal reactors used to model real tubular reactors. (a) two ideal PFRs in parallel (b) ideal PFR and ideal CSTR in parallel.

14.10 Applications to Pharmacokinetic Modeling

The use combinations of ideal reactors to model metabolism and drug distribution in the human body is becoming commonplace. For example, one of the simplest models for drug adsorption and elimination is similar to that shown in Figure 14-19(a). The drug is injected intravenously into a central compartment containing the blood (the top reactor). The blood distributes the drug back and forth to the tissue compartment (the bottom reactor) before being eliminated (top reactor). This model will give the familiar linear semi-log plot found in pharmacokinetics textbooks. As can be seen in the figure for *Professional Reference Shelf* R7.5 on pharmacokinetics on page 453, there are two different slopes, one for the drug distribution phase and one for the elmination phase. More elaborate models using combinations of ideal reactors to model a real system are described in section 7.5 where alcohol metabolism is discussed.

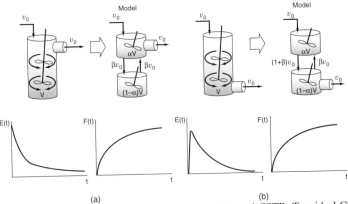

Figure 14-19   Combinations of ideal reactors to model a real CSTR. Two ideal CSTRs with interchange (a) exit from the top of the CSTR (b) exit from the bottom of the CSTR.

Closure

$$\boxed{\text{RTD Data} + \text{Kinetics} + \text{Model} = \text{Prediction}}$$

In this chapter, models were developed for existing reactors to obtain a more precise estimate of the exit conversion and concentration than estimates of the examples given by the zero-order parameter models of segregation and maximum mixedness. After completing this chapter, the reader will use the RTD data and kinetic rate law and reactor model to make predictions of the conversion and exit concentrations using the tank-in-series and dispersion one-parameter models. In addition, the reader should be able to create combinations of ideal reactors that mimic the RTD data and to solve for the exit conversions and concentrations. The choice of a proper model is almost pure art *requiring creativity and engineering judgment*. The flow pattern of the model must possess the most important characteristics of that in the real reactor. Standard models are available that have been used with some success, and these can be used as starting points. Models of real reactors usually consist of combinations of PFRs, perfectly mixed CSTRs, and dead spaces in a configuration that matches the flow patterns in the reactor. For tubular reactors, the simple dispersion model has proven most popular.

The parameters in the model, which with rare exception should not exceed two in number, are obtained from the RTD data. Once the parameters are evaluated, the conversion in the model, and thus in the real reactor, can be calculated. For typical tank-reactor models, this is the conversion in a series–parallel reactor system. For the dispersion model, the second-order differential equation must be solved, usually numerically. Analytical solutions exist for first-order reactions, but as pointed out previously, no model has to be assumed for the first-order system if the RTD is available.

Correlations exist for the amount of dispersion that might be expected in common packed-bed reactors, so these systems can be designed using the dispersion model without obtaining or estimating the RTD. This situation is perhaps the only one where an RTD is not necessary for designing a non-ideal reactor.

S U M M A R Y

The models

1. The models for predicting conversion from RTD data are:
   a. Zero adjustable parameters
      (1) Segregation model
      (2) Maximum mixedness model
   b. One adjustable parameter
      (1) Tanks-in-series model
      (2) Dispersion model
   c. Two adjustable parameters: real reactor modeled as combinations of ideal reactors
2. Tanks-in-series model: Use RTD data to estimate the number of tanks in series,

$$n = \frac{\tau^2}{\sigma^2} \tag{S14-1}$$

For a first-order reaction

$$X = 1 - \frac{1}{(1 + \tau_i k)^n}$$

3. Dispersion model: For a first-order reaction, use the Danckwerts boundary conditions

$$X = 1 - \frac{4q \exp(\text{Pe}_r/2)}{(1 + q)^2 \exp(\text{Pe}_r q/2) - (1 - q)^2 \exp(-\text{Pe}_r q/2)} \tag{S14-2}$$

where

$$q = \sqrt{1 + \frac{4\boldsymbol{Da}}{\text{Pe}_r}} \tag{S14-3}$$

$$\boldsymbol{Da} = \tau k \qquad \text{Pe}_r = \frac{UL}{D_a} \qquad \text{Pe}_f = \frac{Ud_p}{D_a \varepsilon} \tag{S14-4}$$

4. Determine D_a
   a. For laminar flow the dispersion coefficient is

$$D^* = D_{\text{AB}} + \frac{U^2 R^2}{48 D_{\text{AB}}} \tag{S14-5}$$

   b. Correlations. Use Figures 14-10 through 14-12.
   c. Experiment in RTD analysis to find t_m and σ^2.
   For a closed-closed system use Equation (S14-6) to calculate Pe_r from the RTD data:

$$\frac{\sigma^2}{\tau^2} = \frac{2}{\text{Pe}_r} - \frac{2}{\text{Pe}_r^2} (1 - e^{-\text{Pe}_r}) \tag{S14-6}$$

For an open-open system, use

$$\frac{\sigma^2}{t_m^2} = \frac{2Pe + 8}{Pe^2 + 4Pe + 4}$$

(14-47)

5. If a real reactor is modeled as a combination of ideal reactors, the model should have at most two parameters.

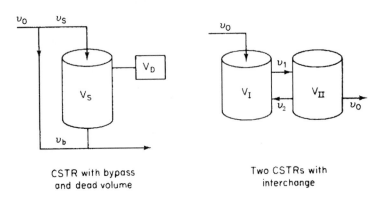

CSTR with bypass   Two CSTRs with
and dead volume   interchange

6. The RTD is used to extract model parameters.
7. Comparison of conversions for a PFR and CSTR with the zero-parameter and two-parameter models. X_{seg} symbolizes the conversion obtained from the segregation model and X_{mm} that from the maximum mixedness model for reaction orders greater than one.

$$X_{PFR} > X_{seg} > X_{mm} > X_{CSTR}$$

$$X_{PFR} > X_{model} \quad \text{with } X_{model} < X_{CSTR} \quad \text{or} \quad X_{model} > X_{CSTR}$$

Cautions: For rate laws with unusual concentration functionalities or for nonisothermal operation, these bounds may not be accurate for certain types of rate laws.

CD-ROM MATERIAL

Summary Notes

Living Example Problem

- **Learning Resources**
  1. *Summary Notes*
  2. *Web Material*
     COMSOL CD-ROM
- **Living Example Problems**
  1. *Example 14-3 Dispersion with Reaction—COMSOL*
  2. *Example 14-5 CSTR with Bypass and Dead Volume*

COMSOL results

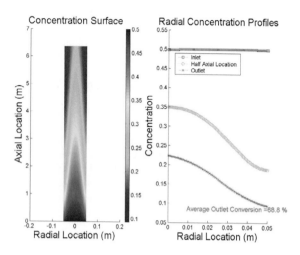

Concentration Surface   Radial Concentration Profiles

Reference Shelf

- **Professional Reference Shelf**

R14.1 *Derivation of Equation for Taylor–Aris Dispersion*

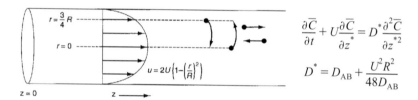

$$\frac{\partial \overline{C}}{\partial t} + U\frac{\partial \overline{C}}{\partial z^*} = D^*\frac{\partial^2 \overline{C}}{\partial z^{*2}}$$

$$D^* = D_{AB} + \frac{U^2 R^2}{48 D_{AB}}$$

R14.2 *Real Reactor Modeled in an Ideal CSTR with Exchange Volume*
   Example R14-1 Two CSTRs with interchange.

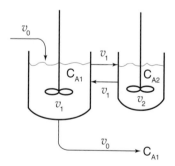

QUESTIONS AND PROBLEMS

The subscript to each of the problem numbers indicates the level of difficulty: A, least difficult; D, most difficult.

$$A = \bullet \quad B = \blacksquare \quad C = \blacklozenge \quad D = \blacklozenge\blacklozenge$$

Creative Thinking

P14-1$_B$ Make up and solve an original problem. The guidelines are given in Problem P4-1. However, make up a problem in reverse by first choosing a model system such as a CSTR in parallel with a CSTR and PFR [with the PFR modeled as four small CSTRs in series; Figure P14-1(a)] or a CSTR with recycle and bypass [Figure P14-1(b)]. Write tracer mass balances and use an ODE solver to predict the effluent concentrations. In fact, you could build up an arsenal of tracer curves for different model systems to compare against real reactor RTD data. In this way you could deduce which model best describes the real reactor.

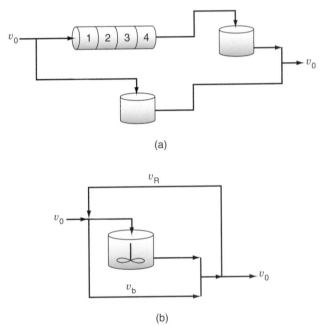

(a)

(b)

Figure P14-1.1   Model systems.

Member

Hall of Fame

P14-2$_B$ **(a)** **Example 14-1.** How large would the error term be in Equation E14-1.4 if $\tau k = 0.1$? $\tau k = 1$? $\tau k = 10$?
(b) **Example 14-2.** Vary D_a, k, U, and L. To what parameters or groups of parameters (e.g., kL^2/D_a) would the conversion be most sensitive? What if the first-order reaction were carried out in tubular reactors of different diameters, but with the space time, τ, remaining constant? The diameters would range from a diameter of 0.1 dm to a diameter of 1 m for kinematic viscosity $v = \mu/\rho = 0.01$ cm^2/s, $U = 0.1$ cm/s, and $D_{AB} = 10^{-5}$ cm^2/s. How would your conversion change? Is there a diameter that would maximize or minimize conversion in this range?

(c) **Example 14-3.** (1) Load the reaction and dispersion program from the COMSOL CD-ROM. Vary the Damköhler number for a second-order reaction using the Aris–Taylor approximation (part (b) in Example 14-3). (2) Vary the Peclet and Damköhler numbers for a second-order reaction in laminar flow. What values of the Peclet number affect the conversion significantly?

(d) **Example 14-4.** How would your answers change if the slope was 4 min^{-1} and the intercept was 2 in Figure E14-4.2?

(e) **Example 14-5.** Load the *Living Example Polymath Program*. Vary α and β and describe what you find. What would be the conversion if $\alpha = 0.75$ and $\beta = 0.15$?

(f) **What if** you were asked to design a tubular vessel that would minimize dispersion? What would be your guidelines? How would you maximize the dispersion? How would your design change for a packed bed?

(g) **What if** someone suggested you could use the solution to the flow-dispersion-reactor equation, Equation (14-27), for a second-order equation by linearizing the rate law by lettering $-r_A = kC_A^2 \cong (kC_{A0}/2)\, C_A = k'C_A$? Under what circumstances might this be a good approximation? Would you divide C_{A0} by something other than 2? What do you think of linearizing other non-first-order reactions and using Equation (14-27)? How could you test your results to learn if the approximation is justified?

(h) **What if** you were asked to explain why physically the shapes of the curves in Figure 14-3 look the way they do, what would you say? What if the first pulse in Figure 14.1(b) broke through at $\Theta = 0.5$ and the second pulse broke through at $\Theta = 1.5$ in a tubular reactor in which a second-order liquid-phase reaction

$$2A \longrightarrow B + C$$

was occurring. What would the conversion be if $\tau = 5$ min, $C_{A0} = 2$ mol/dm^3, and $k = 0.1$ dm^3/mol·min?

P14-3$_B$ The second-order liquid-phase reaction

$$A \longrightarrow B + C$$

is to be carried out isothermally. The entering concentration of A is 1.0 mol/dm^3. The specific reaction rate is 1.0 dm^3/mol·min. A number of used reactors (shown below) are available, each of which has been characterized by an RTD. There are two crimson and white reactors and three maize and blue reactors available.

Reactor	σ(min)	τ(min)	Cost
Maize and blue	2	2	$25,000
Green and white	4	4	50,000
Scarlet and gray	3.05	4	50,000
Orange and blue	2.31	4	50,000
Purple and white	5.17	4	50,000
Silver and black	2.5	4	50,000
Crimson and white	2.5	2	25,000

(a) You have $50,000 available to spend. What is the greatest conversion you can achieve with the available money and reactors?

(b) How would your answer to (a) change if you had $75,000 available to spend?

(c) From which cities do you think the various used reactors came from?

P14-4$_B$ The elementary liquid-phase reaction

$$A \xrightarrow{k_1} B, \qquad k_1 = 1.0 \text{ min}^{-1}$$

is carried out in a packed bed reactor in which dispersion is present. What is the conversion?

Additional information

Porosity = 50%	Reactor length = 0.1 m
Particle size = 0.1 cm	Mean velocity = 1 cm/s
Kinematic viscosity = 0.01 cm²/s	

P14-5$_A$ A gas-phase reaction is being carried out in a 5-cm-diameter tubular reactor that is 2 m in length. The velocity inside the pipe is 2 cm/s. As a very first approximation, the gas properties can be taken as those of air (kinematic viscosity = 0.01 cm²/s), and the diffusivities of the reacting species are approximately 0.005 cm²/s.

(a) How many tanks in series would you suggest to model this reactor?

(b) If the second-order reaction A + B $\longrightarrow$ C + D is carried out for the case of equal molar feed and with $C_{A0} = 0.01$ mol/dm³, what conversion can be expected at a temperature for which $k = 25$ dm³/mol·s?

(c) How would your answers to parts (a) and (b) change if the fluid velocity were reduced to 0.1 cm/s? Increased to 1 m/s?

(d) How would your answers to parts (a) and (b) change if the superficial velocity was 4 cm/s through a packed bed of 0.2-cm-diameter spheres?

(e) How would your answers to parts (a) to (d) change if the fluid were a liquid with properties similar to water instead of a gas, and the diffusivity was 5×10^{-6} cm²/s?

P14-6$_A$ Use the data in Example 13-2 to make the following determinations. (The volumetric feed rate to this reactor was 60 dm³/min.)

(a) Calculate the Peclet numbers for both open and closed systems.

(b) For an open system, determine the space-time τ and then calculate the % dead volume in a reactor for which the manufacturer's specifications give a volume of 420 dm³.

(c) Using the dispersion and tanks-in-series models, calculate the conversion for a closed vessel for the first-order isomerization

$$A \longrightarrow B$$

with $k = 0.18$ min^{-1}.

(d) Compare your results in part (c) with the conversion calculated from the tanks-in-series model, a PFR, and a CSTR.

P14-7$_A$ A tubular reactor has been sized to obtain 98% conversion and to process 0.03 m³/s. The reaction is a first-order irreversible isomerization. The reactor is 3 m long, with a cross-sectional area of 25 dm². After being built, a pulse tracer test on the reactor gave the following data: $t_m = 10$ s and $\sigma^2 = 65$ s². What conversion can be expected in the real reactor?

P14-8$_B$ The following $E(t)$ curve was obtained from a tracer test on a reactor.

$$
\begin{aligned}
E(t) &= 0.25t & 0 < t < 2 \\
&= 1 - 0.25t & 2 < t < 4 \\
&= 0 & t > 4
\end{aligned}
$$

t in minutes, and E(t) in min^{-1}.

The conversion predicted by the tanks-in-series model for the isothermal elementary reaction

$$A \longrightarrow B$$

was 50% at 300 K.

(a) If the temperature is to be raised 10°C ($E = 25{,}000$ cal/mol) and the reaction carried out isothermally, what will be the conversion predicted by the maximum mixedness model? The T-I-S model?

(b) The elementary reactions

$$A \xrightarrow{k_1} B \xrightarrow{k_2} C$$

$$A \xrightarrow{k_3} D$$

$k_1 = k_2 = k_3 = 0.1$ min^{-1} at 300 K, $C_{A0} = 1$ mol/dm^3

were carried out isothermally at 300 K in the same reactor. What is the concentration of B in the exit stream predicted by the maximum mixedness model?

(c) For the multiple reactions given in part (b), what is the conversion of A predicted by the dispersion model in an isothermal closed-closed system?

P14-9$_B$ Revisit Problem P13-4$_C$ where the RTD function is a hemicircle. What is the conversion predicted by

(a) the tanks-in-series model?

(b) the dispersion model?

P14-10$_B$ Revisit Problem P13-5$_B$.

(a) What combination of ideal reactors would you use to model the RTD?

(b) What are the model parameters?

(c) What is the conversion predicted for your model?

P14-11$_B$ Revisit Problem P13-6$_B$.

(a) What conversion is predicted by the tanks-in-series model?

(b) What is the Peclet number?

(c) What conversion is predicted by the dispersion model?

P14-12$_C$ Consider a real tubular reactor in which dispersion is occurring.

(a) For small deviations from plug flow, show that the conversion for a first-order reaction is given approximately as

$$X = 1 - \exp\left[-\tau k + \frac{(\tau k)^2}{\text{Pe}_r}\right] \qquad (P12.1)$$

(b) Show that to achieve the same conversion, the relationship between the volume of a plug-flow reactor, V_P, and volume of a real reactor, V, in which dispersion occurs is

$$\frac{V}{V_P} = 1 - \frac{(k\tau)}{\text{Pe}} = 1 - \frac{kD_e}{U^2} \qquad (P12.2)$$

(c) For a Peclet number of 0.1 based on the PFR length, how much bigger than a PFR must the real reactor be to achieve the 99% conversion predicted by the PFR?

(d) For an nth-order reaction, the ratio of exit concentration for reactors of the same length has been suggested as

$$\frac{C_A}{C_{A_{plug}}} = 1 + \frac{n}{Pe}(\tau k C_{A0}^{n-1})\ln\frac{C_{A0}}{C_{A_{plug}}} \qquad (P12.3)$$

What do you think of this suggestion?

(e) What is the effect of dispersion on zero-order reactions?

P14-13$_B$ Let's continue Problem P13-19$_B$.

(a) What would be the conversion for a second-order reaction with $kC_{A0} = 0.1$ min^{-1} and $C_{A0} = 1$ mol/dm^3 using the segregation model?

(b) What would be the conversion for a second-order reaction with $kC_{A0} = 0.1$ min^{-1} and $C_{A0} = 1$ mol/dm^3 using the maximum mixedness model?

(c) If the reactor is modeled as tanks in series, how many tanks are needed to represent this reactor? What is the conversion for a first-order reaction with $k = 0.1$ min^{-1}?

(d) If the reactor is modeled by a dispersion model, what are the Peclet numbers for an open system and for a closed system? What is the conversion for a first-order reaction with $k = 0.1$ min^{-1} for each case?

(e) Use the dispersion model to estimate the conversion for a second-order reaction with $k = 0.1$ dm^3/mol·s and $C_{A0} = 1$ mol/dm^3.

(f) It is suspected that the reactor might be behaving as shown in Figure P14-13, with *perhaps (?)* $V_1 = V_2$. What is the "backflow" from the second to the first vessel, as a multiple of v_0?

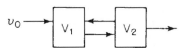

Figure P14-13 Proposed model system

(g) If the model above is correct, what would be the conversion for a second-order reaction with $k = 0.1$ dm^3/mol·min if $C_{A0} = 1.0$ mol/dm^3?

(h) Prepare a table comparing the conversion predicted by each of the models described above.

(i) How would your answer to part (a) change if the reaction were carried out adiabatically with the parameter values given in Problem P13-2(j)?

P14-14$_D$ It is proposed to use the elementary reactions

$$A + B \xrightarrow{\ k_1\ } C + D$$

$$C + B \xrightarrow{\ k_2\ } X + Y$$

to characterize mixing in a real reactor by monitoring the product distribution at different temperatures. The ratio of specific reaction rates (k_2/k_1) at temperatures T_1, T_2, T_3, and T_4 is 5.0, 2.0, 0.5, and 0.1, respectively. The corresponding values of $\tau k_1 C_{A0}$ are 0.2, 2, 20, and 200.

(1) (a) Calculate the product distribution for the CSTR and PFR in series described in Example 13-3 for $\tau_{CSTR} = \tau_{PFR} = 0.5\tau$.

(b) Compare the product distribution at two temperatures using the RTD shown in Examples 13-1 and 13-2 for the complete segregation model and the maximum mixedness model.

(c) Explain how you could use the product distribution as a function of temperature (and perhaps flow rate) to characterize your reactor. For example, could you use the test reactions to determine whether the early mixing scheme or the late mixing scheme in Example 13-3 is more representative of a real reactor? Recall that both schemes have the same RTD.

(d) How should the reactions be carried out (i.e., at high or low temperatures) for the product distribution to best characterize the micromixing in the reactor?

P14-15$_B$ The second-order reaction is to be carried out in a real reactor which gives the following outlet concentration for a step input.

For $0 \le t < 10$ min then $C_T = 10\,(1-e^{-.1t})$

For $10 \le t$ then $C_T = 5+10\,(1-e^{-.1t})$

(a) What model do you propose and what are your model parameters, α and β?

(b) What conversion can be expected in the real reactor?

(c) How would your model change and conversion change if your outlet tracer concentration was

For $t \le 10$ min, then $C_T = 0$

For $t \ge 10$ min, then $C_T = 5+10\,(1-e^{-0.2(t-10)})$

$v_0 = 1$ dm^3/min, $k = 0.1$ dm^3/mol $\cdot$ min, $C_{A0} = 1.25$ mol/dm^3

P14-16$_B$ Suggest combinations of ideal reactors to model real reactors given in Problem 13-2(b) for either $E(\theta),E(t)$, $F(\theta)$, $F(t)$, or $(1 - F(\theta))$.

P14-17$_B$ Below are two COMSOL simulations for a laminar flow reactor with heat effects: Run 1 and Run 2. The figures below show the cross-section plot of concentration for species A at the middle of the reactor. Run 2 shows a minimum on the cross-section plot. This minimum could be the result of (circle all that apply and explain your reasoning for each suggestion (a) through (e))

(a) the thermal conductivity of reaction mixture decreases

(b) overall heat transfer coefficient increases

(c) overall heat transfer coefficient decreases

(d) the coolant flow rate increases

(e) the coolant flow rate decreases

Hint: Explore "Nonisothermal Reactor II" on the COMSOL CD-ROM.

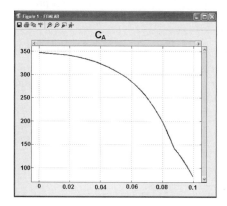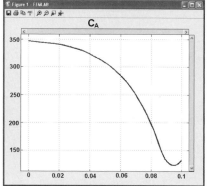

Figure P14-17   COMSOL screen shots

P14-18$_B$ Load the laminar flow with dispersion example on the COMSOL CD-ROM. Keep ***Da*** and L/R constant and vary the reaction order n, $(0.5 \le n \le 5)$ for different Peclet numbers. Are there any combinations of n and Pe where dispersion is more important or less important on the exit concentration? What generalizations can you make? *Hint:* for n < 1 use $r_A = -k \cdot (\text{Abs}(C_A^n))$

COMSOL Problem   **P14-19$_C$** Revisit the COMSOL Example 14-3 for laminar flow with dispersions.

(a) Plot the radial concentration profiles for $z/L = 0.5$ and 1.0 for a second-order reaction with $C_{A0} = 0.5$ mol/dm^3 and $kC_{A0} = 0.7$ min^{-1} using both the closed-vessel and the laminar flow open-vessel boundary conditions at the inlet. Is the average outlet conversion for the open-vessel boundary condition lower than that which uses the closed-vessel boundary condition? In what situation, if any, will the two boundary conditions result in significantly different outlet concentrations? Vary Pe and ***Da*** and describe what you find, i.e., $C_{A0} = 0.5$ mol/dm^3.

(b) Repeat (a) for both a third order with $kC_{A0}^2 = 0.7$ min^{-1} and a half-order reaction with $k = 0.495$ (mol/dm$^3)^{1/2}$ min^{-1}. Compare the radial conversion profiles for a first-, a second-, a third-, and a half-order reaction at different locations down the reactor.

Note in COMSOL:
Open-vessel Boundary (Laminar Flow): $-Ni \cdot n = 2*U0*(1-(r/Ra).\wedge 2)*CA0$
Close-vessel Boundary: $-Ni \cdot n = U0*CA0$
Concentration Boundary Condition CA = CA0
Symmetry/Insulation Condition $n \cdot N = 0$

P14-20$_B$ The F curves for two tubular reactors are shown here, for a closed–closed system.

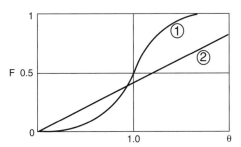

Figure P14-20   F Curves

(a) Which curve has the higher Peclet number? Explain.
(b) Which curve has the higher dispersion coefficient? Explain.
(c) If this F curve is for the tanks-in-series model applied to two different reactors, which curve has the largest number of T-I-S (1) or (2)?

U of M, ChE528 Fall 2000 Exam II

P14-21$_B$ Consider the following system used to model a real reactor:

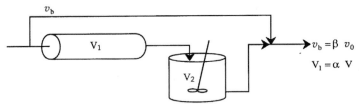

Figure P14-21   Model system

Describe how you would evaluate the parameters α and β.

(a) Draw the F and E curves for this system of ideal reactors used to model a real reactor using $\beta = 0.2$ and $\alpha = 0.4$. Identify the numerical values of the points on the F curve (e.g., t_1) as they relate to τ.

(b) If the reaction $A \rightarrow B$ is second order with $kC_{A0} = 0.5$ min^{-1}, what is the conversion assuming the space time for the real reactor is 2 min?

U of M, ChE528 Fall 2000 Final Exam

P14-22$_B$ There is a 2 m^3 reactor in storage that is to be used to carry out the liquid-phase second-order reaction

$$A + B \longrightarrow C$$

A and B are to be fed in equal molar amounts at a volumetric rate of 1 m^3/min. The entering concentration of A is 2 molar, and the specific reaction rate is 1.5 m^3/kmol $\cdot$ min. A tracer experiment was carried out and reported in terms of F as a function of time in minutes.

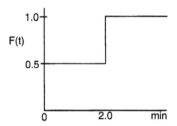

Figure P14-22   F curve for a nonideal reactor

Suggest a two-parameter model consistent with the data; evaluate the model parameters and the expected conversion.

U of M, ChE528 Fall 2001 Final Exam

P14-23$_B$ The following E curve was obtained from a tracer test:

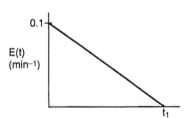

Figure P14-23   E curve for a nonideal reactor

(a) What is the mean residence time?

(b) What is the Peclet number for a closed-closed system?

(c) How many tanks in series are necessary to model this non-ideal reactor?

U of M, Doctoral Qualifying Exam (DQE), May 2001

P14-24$_B$ The A first-order reaction is to be carried out in the reactor with $k = 0.1$ min^{-1}.

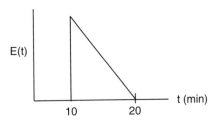

P14-25$_B$ Fill in the following table with the conversion predicted by each type of model/reactor.

Ideal PFR	Ideal CSTR	Ideal laminar flow reactor	Segregation	Maximum mixedness	Dispersion	Tanks in series

P14-24$_B$ The following outlet concentration trajectory was obtained from a step input to a nonideal reactor. The entering concentration was 10 millimolar of tracer.

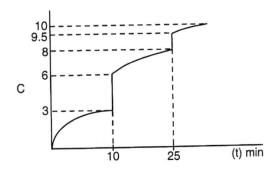

Figure P14-24 C curve for a nonideal reactor

Suggest a model using the collection of ideal reactors to model the nonideal reactor.

U of M, Doctoral Qualifying Exam (DQE), May 2001

- **Additional Homework Problems**

CDP14-A$_C$ A real reactor is modeled as a combination of ideal PFRs and CSTRs. [2nd Ed. P14-5]

CDP14-B$_B$ A real batch reactor is modeled as a combination of two ideal reactors. [2nd Ed. P14-13]

CDP14-C$_C$ Develop a model for a real reactor for RTD obtained from a step input. [2nd Ed. P14-10]

CDP14-D$_B$ Calculate D_a and X from sloppy tracer data. [2nd Ed. P14-6$_A$]

CDP14-E$_B$ Use RTD data from Oak Ridge National Laboratory to calculate the conversion from the tanks-in-series and the dispersion models. [2nd Ed. P14-7$_B$]

CDP14-F$_B$ RTD data from a slurry reactor. [3rd Ed. P14-8]

CDP14-G$_C$ RTD data to calculate conversion for a second-order reaction for all
             models. [3rd Ed. P14-9]
CDP14-H$_B$ RTD data from barge spill on Mississippi River. [3rd Ed. P14-10]
CDP14-I$_B$ RTD data to calculate conversion using all models. [3rd Ed. P14-11]
CDP14-J$_B$ Apply two-parameter model to multiple reactions. [3rd Ed. P14-15]
CDP14-New New problems will be inserted from time to time on the web.

SUPPLEMENTARY READING

1. Excellent discussions of maximum mixedness can be found in

   DOUGLAS, J. M., "The effect of mixing on reactor design," *AIChE Symp. Ser.
      48*, Vol. 60, p. 1 (1964).

   ZWIETERING, TH. N., *Chem. Eng. Sci.*, *11*, 1 (1959).

2. Modeling real reactors with a combination of ideal reactors is discussed together
   with axial dispersion in

   LEVENSPIEL, O., *Chemical Reaction Engineering*, 3rd ed. New York: Wiley,
      1999.

   WEN, C. Y., and L. T. FAN, *Models for Flow Systems and Chemical Reactors*.
      New York: Marcel Dekker, 1975.

3. Mixing and its effects on chemical reactor design have been receiving increasingly
   sophisticated treatment. See, for example:

   BISCHOFF, K. B., "Mixing and contacting in chemical reactors," *Ind. Eng.
      Chem.*, *58*(11), 18 (1966).

   NAUMAN, E. B., "Residence time distributions and micromixing," *Chem. Eng.
      Commun.*, *8*, 53 (1981).

   NAUMAN, E. B., and B. A. BUFFHAM, *Mixing in Continuous Flow Systems*. New
      York: Wiley, 1983.

   PATTERSON, G. K., "Applications of turbulence fundamentals to reactor model-
      ing and scaleup," *Chem. Eng. Commun.*, *8*, 25 (1981).

4. See also

   DUDUKOVIC, M., and R. FELDER, in *CHEMI Modules on Chemical Reaction
      Engineering*, Vol. 4, ed. B. Crynes and H. S. Fogler. New York: AIChE,
      1985.

5. Dispersion. A discussion of the boundary conditions for closed-closed, open-open,
   closed-open, and open-closed vessels can be found in

   ARIS, R., *Chem. Eng. Sci.*, *9*, 266 (1959).

   LEVENSPIEL, O., and K. B. BISCHOFF, *Adv. in Chem. Eng.*, *4*, 95 (1963).

   NAUMAN, E. B., *Chem. Eng. Commun.*, *8*, 53 (1981).

This is not the end.
It is not even the beginning of the end.
But it is the end of the beginning.

Winston Churchill

Appendices

Text Appendices

CD-ROM Appendices

Numerical **A**
Techniques

A.1 Useful Integrals in Reactor Design

Links

Also see *www.integrals.com.*

$$\int_0^x \frac{dx}{1-x} = \ln \frac{1}{1-x} \tag{A-1}$$

$$\int_{x_1}^{x_2} \frac{dx}{(1-x)^2} = \frac{1}{1-x_2} - \frac{1}{1-x_1} \tag{A-2}$$

$$\int_0^x \frac{dx}{(1-x)^2} = \frac{x}{1-x} \tag{A-3}$$

$$\int_0^x \frac{dx}{1+\varepsilon x} = \frac{1}{\varepsilon} \ln(1+\varepsilon x) \tag{A-4}$$

$$\int_0^x \frac{(1+\varepsilon x)dx}{1-x} = (1+\varepsilon)\ln\frac{1}{1-x} - \varepsilon x \tag{A-5}$$

$$\int_0^x \frac{(1+\varepsilon x)dx}{(1-x)^2} = \frac{(1+\varepsilon)x}{1-x} - \varepsilon \ln\frac{1}{1-x} \tag{A-6}$$

$$\int_0^x \frac{(1+\varepsilon x)^2 dx}{(1-x)^2} = 2\varepsilon(1+\varepsilon)\ln(1-x) + \varepsilon^2 x + \frac{(1+\varepsilon)^2 x}{1-x} \tag{A-7}$$

$$\int_0^x \frac{dx}{(1-x)(\Theta_B - x)} = \frac{1}{\Theta_B - 1} \ln \frac{\Theta_B - x}{\Theta_B(1-x)} \qquad \Theta_B \neq 1 \qquad \text{(A-8)}$$

$$\int_0^W (1 - \alpha W)^{1/2} dW = \frac{2}{3\alpha}[1 - (1 - \alpha W)^{3/2}] \qquad \text{(A-9)}$$

$$\int_0^x \frac{dx}{ax^2 + bx + c} = \frac{-2}{2ax + b} + \frac{2}{b} \qquad \text{for } b^2 = 4ac \qquad \text{(A-10)}$$

$$\int_0^x \frac{dx}{ax^2 + bx + c} = \frac{1}{a(p-q)} \ln\left(\frac{q}{p} \cdot \frac{x-p}{x-q}\right) \qquad \text{for } b^2 > 4ac \qquad \text{(A-11)}$$

where p and q are the roots of the equation.

$$ax^2 + bx + c = 0 \qquad \text{i.e., } p, q = \frac{-b \mp \sqrt{b^2 - 4ac}}{2a}$$

$$\int_0^x \frac{a + bx}{c + gx} dx = \frac{bx}{g} + \frac{ag - bc}{g^2} \ln \frac{c + gx}{c} \qquad \text{(A-12)}$$

A.2 Equal-Area Graphical Differentiation

There are many ways of differentiating numerical and graphical data. We shall confine our discussions to the technique of equal-area differentiation. In the procedure delineated here, we want to find the derivative of y with respect to x.

1. Tabulate the (y_i, x_i) observations as shown in Table A-1.
2. For each *interval*, calculate $\Delta x_n = x_n - x_{n-1}$ and $\Delta y_n = y_n - y_{n-1}$.

This method finds use in Chapter 5.

TABLE A-1

x_i	y_i	Δx	Δy	$\dfrac{\Delta y}{\Delta x}$	$\dfrac{dy}{dx}$
x_1	y_1				$\left(\dfrac{dy}{dx}\right)_1$
		$x_2 - x_1$	$y_2 - y_1$	$\left(\dfrac{\Delta y}{\Delta x}\right)_2$	
x_2	y_2				$\left(\dfrac{dy}{dx}\right)_2$
		$x_3 - x_2$	$y_3 - y_2$	$\left(\dfrac{\Delta y}{\Delta x}\right)_3$	
x_3	y_3				$\left(\dfrac{dy}{dx}\right)_3$
x_4	y_4		etc.		

3. Calculate $\Delta y_n / \Delta x_n$ as an estimate of the *average* slope in an interval x_{n-1} to x_n.

4. Plot these values as a histogram versus x_i. The value between x_2 and x_3, for example, is $(y_3 - y_2)/(x_3 - x_2)$. Refer to Figure A-1.

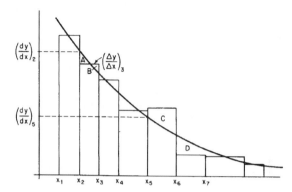

Figure A-1   Equal-area differentiation.

5. Next draw in the *smooth curve* that best approximates the *area* under the histogram. That is, attempt in each interval to balance areas such as those labeled A and B, but when this approximation is not possible, balance out over several intervals (as for the areas labeled C and D). From our definitions of Δx and Δy we know that

$$y_n - y_1 = \sum_{i=2}^{n} \frac{\Delta y}{\Delta x_i} \Delta x_i \tag{A-13}$$

The equal-area method attempts to estimate dy/dx so that

See CD-ROM,
Appendix A, for
worked example

$$y_n - y_1 = \int_{x_1}^{x_n} \frac{dy}{dx} \, dx \tag{A-14}$$

that is, so that the area under $\Delta y / \Delta x$ is the same as that under dy/dx, *everywhere possible*.

6. Read estimates of dy/dx from this curve at the data points $x_1, x_2, \ldots$ and complete the table.

An example illustrating the technique is given in the CD-ROM, Appendix A.
Differentiation is, at best, less accurate than integration. This method also *clearly indicates bad data* and allows for compensation of such data. Differentiation is only valid, however, when the data are presumed to differentiate *smoothly*, as in rate-data analysis and the interpretation of transient diffusion data.

A.3 Solutions to Differential Equations

A.3.A   First-Order Ordinary Differential Equation

Links

See *www.ucl.ac.uk/Mathematics/geomath/level2/deqn/de8.html* and CD-ROM Appendix K.

$$\frac{dy}{dt} + f(t)y = g(t) \tag{A-15}$$

Using integrating factor $= \exp\left(\int f dt\right)$, the solution is

$$y = e^{-\int f dt} \int g(t)\, e^{\int f dt}\, dt + K_1 e^{-\int f dt} \tag{A-16}$$

Example A–1   Integrating Factor for Series Reactions

$$\frac{dy}{dt} + k_2 y = k_1 e^{-k_1 t}$$

$$\text{Integrating factor} = \exp\int k_2 dt = e^{k_2 t}$$

$$\frac{d(y e^{k_2 t})}{dt} = e^{k_2 t} k_1 e^{-k_1 t} = k_1 e^{(k_2 - k_1)t}$$

$$e^{k_2 t} y = k_1 \int e^{(k_2 - k_1 t)} dt = \frac{k_1}{k_2 - k_1} e^{(k_2 - k_1)t} + K_1$$

$$y = \frac{k_1}{k_2 - k_1} e^{-k_1 t} + K_1 e^{-k_2 t}$$

$$t = 0 \quad y = 0$$

$$y = \frac{k_1}{k_2 - k_1}\left[e^{-k_1 t} - e^{-k_2 t} \right]$$

A.3.B   Coupled Differential Equations

Techniques to solve coupled first order linear ODEs such as

$$\frac{dx}{dt} = ax + by$$

$$\frac{dy}{dt} = cx + dy$$

are given in Appendix K in the CD-ROM.

A.3.C   Second-Order Ordinary Differential Equation

Methods of solving differential equations of the type

$$\frac{d^2y}{dx^2} - \beta y = 0 \tag{A-17}$$

can be found in such texts as *Applied Differential Equations* by M. R. Spiegel (Upper Saddle River, N.J.: Prentice Hall, 1958, Chapter 4; a great book even though it's old) or in *Differential Equations* by F. Ayres (New York: Schaum Outline Series, McGraw-Hill, 1952). One method of solution is to determine the characteristic roots of

$$\left(\frac{d^2}{dx^2} - \beta\right)y = (m^2 - \beta)y$$

which are

$$m = \pm\sqrt{\beta}$$

Solutions of this type are required in Chapter 12.

The solution to the differential equation is

$$y = A_1 e^{-\sqrt{\beta}x} + B_1 e^{+\sqrt{\beta}x} \tag{A-18}$$

where A_1 and B_1 are arbitrary constants of integration. It can be verified that Equation (A-18) can be arranged in the form

$$y = A \sinh\sqrt{\beta}x + B \cosh\sqrt{\beta}x \tag{A-19}$$

Equation (A-19) is the more useful form of the solution when it comes to evaluating the constants A and B because $\sinh(0) = 0$ and $\cosh(0) = 1.0$.

A.4 Numerical Evaluation of Integrals

In this section, we discuss techniques for numerically evaluating integrals for solving first-order differential equations.

1. *Trapezoidal rule* (two-point) (Figure A-2). This method is one of the simplest and most approximate, as it uses the integrand evaluated at the limits of integration to evaluate the integral:

$$\int_{X_0}^{X_1} f(X)\, dX = \frac{h}{2}[f(X_0) + f(X_1)] \tag{A-20}$$

when $h = X_1 - X_0$.

2. *Simpson's one-third rule* (three-point) (Figure A-3). A more accurate evaluation of the integral can be found with the application of Simpson's rule:

$$\int_{X_0}^{X_2} f(X)\, dX = \frac{h}{3}[f(X_0) + 4f(X_1) + f(X_2)] \qquad \text{(A-21)}$$

where

$$h = \frac{X_2 - X_0}{2} \qquad X_1 = X_0 + h$$

Methods to solve
$$\int_0^X \frac{F_{A0}}{-r_A}\, dX$$
in Chapters 2, 4, 8, and
$$\int_0^\infty X(t)E(t)\, dt$$
in Chapter 13

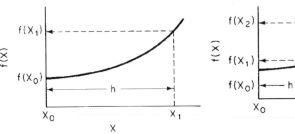

Figure A-2   Trapezoidal rule illustration.

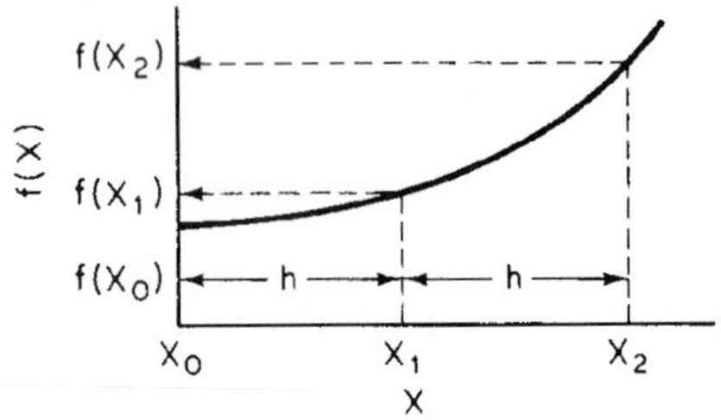

Figure A-3   Simpson's three-point rule illustration.

3. *Simpson's three-eighths rule* (four-point) (Figure A-4). An improved version of Simpson's one-third rule can be made by applying *Simpson's three-eighths rule*:

$$\int_{X_0}^{X_3} f(X)\, dX = \tfrac{3}{8} h\, [f(X_0) + 3f(X_1) + 3f(X_2) + f(X_3)] \qquad \text{(A-22)}$$

where

$$h = \frac{X_3 - X_0}{3} \qquad X_1 = X_0 + h \qquad X_2 = X_0 + 2h$$

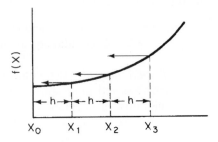

Figure A-4   Simpson's four-point rule illustration.

4. *Five-point quadrature formula.*

$$\int_{X_0}^{X_4} f(X)\, dX = \frac{h}{3}(f_0 + 4f_1 + 2f_2 + 4f_3 + f_4) \qquad \text{(A-23)}$$

where
$$h = \frac{X_4 - X_0}{4}$$

5. For $N + 1$ points, where $(N/3)$ is an integer,

$$\int_{X_0}^{X_N} f(X)\, dX = \tfrac{3}{8} h\, [f_0 + 3f_1 + 3f_2 + 2f_3$$

$$+ 3f_4 + 3f_5 + 2f_6 + \cdots + 3f_{N-1} + f_N] \qquad \text{(A-24)}$$

where $h = \dfrac{X_N - X_0}{N}$

6. For $N + 1$ points, where N is even,

$$\int_{X_0}^{X_N} f(X)\, dX = \frac{h}{3}\,(f_0 + 4f_1 + 2f_2 + 4f_3 + 2f_4 + \cdots + 4f_{N-1} + f_N)$$

$$\text{(A-25)}$$

where
$$h = \frac{X_N - X_0}{N}$$

These formulas are useful in illustrating how the reaction engineering integrals and coupled ODEs (ordinary differential equation(s)) can be solved and also when there is an ODE solver power failure or some other malfunction.

A.5 Software Packages

Links

Instructions on how to use Polymath, MATLAB, COMSOL, and Aspen can be found on the CD-ROM.

For the ordinary differential equation solver (ODE solver), contact:

Polymath
CACHE Corporation
P.O. Box 7939
Austin, Texas 78713-7379
Web site: *www.polymath-software.com/fogler*

Aspen Technology, Inc.
10 Canal Park
Cambridge, Massachusetts
02141-2201
Email: info@aspentech.com
Web site: *//www.aspentech.com*

MATLAB
The Math Works, Inc.
20 North Main Street, Suite 250
Sherborn, Massachusetts 01770

COMSOL Multiphysics
COMSOL, Inc.
8 New England Executive Park, Suite 310
Burlington, Massachusetts 01803
Tel: 781-273-3322; Fax: 781-273-6603
Email: info@comsol.com
Web site: *www.comsol.com*

A critique of some of these software packages (and others) can be found in *Chem. Eng. Educ., XXV* (Winter) 54 (1991).

Ideal Gas Constant B
and Conversion
Factors

See *www.onlineconversion.com*.

Ideal Gas Constant

$$R = \frac{0.73 \ \text{ft}^3 \cdot \text{atm}}{\text{lb mol} \cdot {}^{\circ}\text{R}}$$

$$R = \frac{1.987 \ \text{Btu}}{\text{lb-mol} \cdot {}^{\circ}\text{R}}$$

$$R = \frac{8.314 \ \text{kPa} \cdot \text{dm}^3}{\text{mol} \cdot \text{K}}$$

$$R = \frac{8.3144 \ \text{J}}{\text{mol} \cdot \text{K}}$$

$$R = 0.082 \ \frac{\text{dm}^3 \cdot \text{atm}}{\text{mol} \cdot \text{K}} = \frac{0.082 \ \text{m}^3 \cdot \text{atm}}{\text{kmol} \cdot \text{K}}$$

$$R = \frac{1.987 \text{cal}}{\text{mol} \cdot \text{K}}$$

Boltzmann's constant $k_B = 1.381 \times 10^{-23} \ \dfrac{\text{J}}{\text{molecule} \bullet \text{K}}$

$$= 1.381 \times 10^{-23} \ \text{kg m}^2/\text{s}^2/\text{molecule}/\text{K}$$

Volume of Ideal Gas

1 lb-mol of an ideal gas at 32°F and 1 atm occupies 359 ft³ (0.00279 lbmol/ft³).
1 mol of an ideal gas at 0°C and 1 atm occupies 22.4 dm³ (0.0446 mol/dm³).

$$C_A = \frac{P_A}{RT} = \frac{y_A P}{RT}$$

where C_A = concentration of A, mol/dm³ T = temperature, K
       P = pressure, kPa y_A = mole fraction of A
       R = ideal gas constant, 8.314 kPa·dm³/mol·K

$1M = 1 \ \text{molar} = 1 \ \text{mol/liter} = 1 \ \text{mol/dm}^3 = 1 \ \text{kmol/m}^3 = 0.062 \ \text{lb·mol/ft}^3$

Volume

1 cm^3	= 0.001 dm^3
1 in^3	= 0.0164 dm
1 fluid oz	= 0.0296 dm^3
1 ft^3	= 28.32 dm^3
1 m^3	= 1000 dm^3
1 U.S. gallon	= 3.785 dm^3
1 liter (L)	= 1 dm^3

$$\left(1 \text{ ft}^3 = 28.32 \text{ dm}^3 \times \frac{1 \text{ gal}}{3.785 \text{ dm}^3} = 7.482 \text{ gal} \right)$$

Length

1 Å	= 10^{-8} cm
1 dm	= 10 cm
1 μm	= 10^{-4} cm
1 in.	= 2.54 cm
1 ft	= 30.48 cm
1 m	= 100 cm

Pressure

1 torr (1 mmHg)	= 0.13333 kPa
1 in. H$_2$O	= 0.24886 kPa
1 in. Hg	= 3.3843 kPa
1 atm	= 101.33 kPa
1 psi	= 6.8943 kPa
1 megadyne/cm^2	= 100 kPa

Energy (Work)

1 kg·m^2/s^2	= 1 J
1 Btu	= 1055.06 J
1 cal	= 4.1868 J
1 L·atm	= 101.34 J
1 hp·h	= 2.6806 × 10^6 J
1 kWh	= 3.6 × 10^6 J

Temperature

°F	= 1.8 × °C + 32
R	= °F + 459.69
K	= °C + 273.16
°R	= 1.8 × K
°Réamur	= 1.25 × °C

Mass

1 lb	= 454 g
1 kg	= 1000 g
1 grain	= 0.0648 g
1 oz (avoird.)	= 28.35 g
1 ton	= 908,000 g

Viscosity

1 poise = 1 g/cm/s = 0.1 kg/m/s
1 centipoise = 1 cp = 0.01 poise = 0.1 micro Pascal · second

Force

$$1 \text{ dyne} = 1 \text{ g} \cdot \text{cm/s}^2$$

$$1 \text{ Newton} = 1 \text{ kg} \cdot \text{m/s}^2$$

Pressure

$$1 \text{ Pa} = 1 \text{ Newton/m}^2$$

Work

A. Work = Force × Distance

$$1 \text{ Joule} = 1 \text{ Newton} \cdot \text{meter} = 1 \text{ kg m}^2/\text{s}^2 = 1 \text{ Pa} \cdot \text{m}^3$$

B. Pressure × Volume = Work

$$1 \text{ Newton/ m}^2 \cdot \text{m}^3 = 1 \text{ Newton} \cdot \text{m} = 1 \text{ Joule}$$

Time Rate of Change of Energy with Time

$$1 \text{ watt} = 1 \text{ J/s}$$

$$1 \text{ hp} = 746 \text{ J/s}$$

Gravitational Conversion Factor

Gravitational constant

$$g = 32.2 \text{ ft/s}^2$$

American Engineering System

$$g_c = 32.174 \frac{(\text{ft})(\text{lb}_m)}{(\text{s}^2)(\text{lb}_f)}$$

SI/cgs System

$$g_c = 1 \text{ (Dimensionless)}$$

TABLE B.1 TYPICAL PROPERTY VALUES

	Liquid (water)	Gas (air, 77°C, 101 kPa)	Solid
Density	1000 kg/m^3	1.0 kg/m^3	3000 kg/m^3
Concentration	55.5 mol/dm^3	0.04 mol/dm^3	–
Diffusivity	10^{-8} m^2/s	10^{-5} m^2/s	10^{-11} m^2/s
Viscosity	10^{-3} kg/m/s	1.82×10^{-5} kg/m/s	–
Heat capacity	4.31 J/g/K	40 J/mol/K	0.45 J/g/K
Thermal conductivity	1.0 J/s/m/K	10^{-2} J/s/m/K	100 J/s/m/K
Kinematic viscosity	10^{-6} m^2/s	1.8×10^{-5} m^2/s	–
Prandtl number	7	0.7	–
Schmidt number	200	2	–

TABLE B.1 TYPICAL TRANSPORT VALUES

	Liquid	Gas
Heat Transfer Coefficient, h	1000 W/m^2/K	65 W/m^2/K
Mass Transfer Coefficient, k_c	10^{-2} m/s	3 m/s

Thermodynamic C
Relationships
Involving
the Equilibrium
Constant[1]

For the gas-phase reaction

$$A + \frac{b}{a} B \rightleftharpoons \frac{c}{a} C + \frac{d}{a} D \qquad (2\text{-}2)$$

1. The true (dimensionless) equilibrium constant

$$RT \ln K = -\Delta G$$

$$\boxed{K = \frac{a_C^{c/a}\, a_D^{d/a}}{a_A\, a_B^{b/a}}}$$

where a_i is the activity of species i

$$a_i = \frac{f_i}{f_i^o}$$

where f_i = fugacity of species i

f_i^o = fugacity of the standard state. For gases, the standard state is 1 atm.

$$a_i = \frac{f_i}{f_i^o} = \gamma_i P_i$$

---

[1] For the limitations and for further explanation of these relationships, see, for example, K. Denbigh, *The Principles of Chemical Equilibrium*, 3rd ed. (Cambridge: Cambridge University Press, 1971), p. 138.

where γ_i is the activity coefficient

K = True equilibrium constant
K_γ = Activity equilibrium constant
K_p = Pressure equilibrium constant
K_c = Concentration equilibrium constant

$$K = \underbrace{\frac{\gamma_C^{c/a}\,\gamma_D^{d/a}}{\gamma_A\,\gamma_B^{b/a}}}_{K_\gamma} \cdot \underbrace{\frac{P_C^{c/a}\,P_D^{d/a}}{P_A\,P_B^{b/a}}}_{K_p} = K_\gamma K_p$$

K_γ has units of $[\text{atm}]^{-\left(d/a+c/a-\frac{b}{a}-1\right)} = [\text{atm}]^{-\delta}$

K_p has units of $[\text{atm}]^{\left(d/a+c/a-\frac{b}{a}-1\right)} = [\text{atm}]^{\delta}$

For ideal gases $K_\gamma = 1.0$ atm$^{-\delta}$

2. The pressure equilibrium constant K_P is

$$K_P = \frac{P_C^{c/a}P_D^{d/a}}{P_A P_B^{b/a}}$$

(C-1)

P_i = partial pressure of species i, atm, kPa.

$$P_i = C_i\,RT$$

3. The concentration equilibrium constant K_C is

It is important to be able to relate
K, K_γ, K_c, and K_p.

$$K_C = \frac{C_C^{c/a}C_D^{d/a}}{C_A C_B^{b/a}}$$

(C-2)

4. For ideal gases, K_C and K_P are related by

$$K_P = K_C(RT)^\delta$$

(C-3)

$$\delta = \frac{c}{a} + \frac{d}{a} - \frac{b}{a} - 1$$

(C-4)

5. K_P is a function of temperature only, and the temperature dependence of K_P is given by van't Hoff's equation:

Van't Hoff's equation

$$\boxed{\frac{d\,\ln K_P}{dT} = \frac{\Delta H_{\text{Rx}}(T)}{RT^2}}$$

(C-5)

$$\frac{d\,\ln K_P}{dT} = \frac{\Delta H_{\text{Rx}}^\circ(T_R) + \Delta\hat{C}_p(T - T_R)}{RT^2}$$

(C-6)

6. Integrating, we have

$$\ln\frac{K_P(T)}{K_P(T_1)} = \frac{\Delta H_{\text{Rx}}^\circ(T_R) - T_R\,\Delta\hat{C}_p}{R}\left(\frac{1}{T_1} - \frac{1}{T}\right) + \frac{\Delta\hat{C}_p}{R}\ln\frac{T}{T_1}$$

(C-7)

K_P and K_C are related by

$$\boxed{K_C = \frac{K_P}{(RT)^\delta}}$$

(C-8)

when

$$\delta = \left(\frac{d}{a} + \frac{c}{a} - \frac{b}{a} - 1\right) = 0$$

then

$$K_P = K_C$$

7. K_P neglecting ΔC_P. Given the equilibrium constant at one temperature T_1, $K_P(T_1)$ and the heat of reaction ΔH_{Rx}, the partial pressure equilibrium constant at any temperature T is

$$K_P(T) = K_P(T_1)\exp\left[\frac{\Delta H_{Rx}(T_R)}{R}\left(\frac{1}{T_1} - \frac{1}{T}\right)\right] \tag{C-9}$$

8. From Le Châtelier's principle we know that for exothermic reactions, the equilibrium shifts to the left (i.e., K and X_e decrease) as the temperature increases. Figures C-1 and C-2 show how the equilibrium constant varies with temperature for an exothermic reaction and for an endothermic reaction, respectively.

Variation of equilibrium constant with temperature

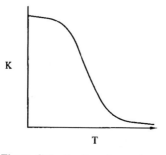

Figure C-1 Exothermic reaction.

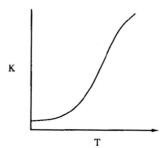

Figure C-2 Endothermic reaction.

9. The equilibrium constant at temperature T can be calculated from the change in the Gibbs free energy using

$$-RT \ln[K(T)] = \Delta G_{Rx}^{\circ}(T) \tag{C-10}$$

$$\Delta G_{Rx}^{\circ} = \frac{c}{a}\,G_C^{\circ} + \frac{d}{a}\,G_D^{\circ} - \frac{b}{a}\,G_B^{\circ} - G_A^{\circ} \tag{C-11}$$

10. Tables that list the standard Gibbs free energy of formation of a given species G_i° are available in the literature.
    1) *www/uic.edu:80/~mansoori/Thermodynamic.Data.and.Property_html*
    2) *webbook.nist.gov*
11. The relationship between the change in Gibbs free energy and enthalpy, H, and entropy, S, is

$$\Delta G = \Delta H - T\,\Delta S \tag{C-12}$$

See *bilbo.chm.uri.edu/CHM112/lectures/lecture31.htm*.

Links

Example C–1   Water-Gas Shift Reaction

The water-gas shift reaction to produce hydrogen,

$$H_2O + CO \; \underset{\longleftarrow}{\overset{\longrightarrow}{}} \; CO_2 + H_2$$

is to be carried out at 1000 K and 10 atm. For an equal-molar mixture of water and carbon monoxide, calculate the equilibrium conversion and concentration of each species.

Data: At 1000 K and 10 atm the Gibbs free energies of formation are $G^{\circ}_{CO} = -47,860$ cal/mol; $G^{\circ}_{CO_2} = -94,630$ cal/mol; $G^{\circ}_{H_2O} = -46,040$ cal/mol; $G^{\circ}_{H_2} = 0$.

Solution

We first calculate the equilibrium constant. The first step in calculating K is to calculate the change in Gibbs free energy for the reaction. Applying Equation (C-10) gives us

Calculate ΔG°_{Rx}

$$\Delta G^{\circ}_{Rx} = G^{\circ}_{H_2} + G^{\circ}_{CO_2} - G^{\circ}_{CO} - G^{\circ}_{H_2O} \tag{EC-1.1}$$

$$= 0 + (-94,630) - (-47,860) - (-46,040)$$

$$= -730 \text{ cal/mol}$$

$$-RT \ln K = \Delta G^{\circ}_{Rx}(T) \tag{C-10}$$

$$\ln K = -\frac{\Delta G^{\circ}_{Rx}(T)}{RT} = \frac{-(-730 \text{ cal/mol})}{1.987 \text{ cal/mol} \cdot K (1000 \text{ K})} \tag{EC-1.2}$$

$$= 0.367$$

then

Calculate K

$$K = 1.44$$

Expressing the equilibrium constant first in terms of activities and then finally in terms of concentration, we have

$$K = \frac{a_{CO_2} a_{H_2}}{a_{CO} a_{H_2O}} = \frac{f_{CO_2} f_{H_2}}{f_{CO} f_{H_2O}} = \frac{\gamma_{CO_2} y_{CO_2} \gamma_{H_2} y_{H_2}}{\gamma_{CO} y_{CO} \gamma_{H_2O} y_{CO_2}} \tag{EC-1.3}$$

where a_i is the activity, f_i is the fugacity, γ_i is the activity coefficient (which we shall take to be 1.0 owing to high temperature and low pressure), and y_i is the mole fraction of species i.[2] Substituting for the mole fractions in terms of partial pressures gives

$$y_i = \frac{P_i}{P_T} = \frac{C_i RT}{P_T} \tag{EC-1.4}$$

$$K = \frac{P_{CO_2} P_{H_2}}{P_{CO} P_{H_2O}} = \frac{C_{CO_2} C_{H_2}}{C_{CO} C_{H_2O}} \tag{EC-1.5}$$

---

[2] See Chapter 9 in J. M. Smith, *Introduction to Chemical Engineering Thermodynamics*, 3rd ed. (New York: McGraw-Hill, 1959), and Chapter 9 in S. I. Sandler, *Chemical and Engineering Thermodynamics*, 2nd ed. (New York: Wiley, 1989), for a discussion of chemical equilibrium including nonideal effects.

In terms of conversion for an equal molar feed, we have

Relate K and X_e

$$K = \frac{C_{CO,0}X_e C_{CO,0}X_e}{C_{CO,0}(1-X_e)C_{CO,0}(1-X_e)} \qquad \text{(EC-1.6)}$$

$$= \frac{X_e^2}{(1-X_e)^2} = 1.44 \qquad \text{(EC-1.7)}$$

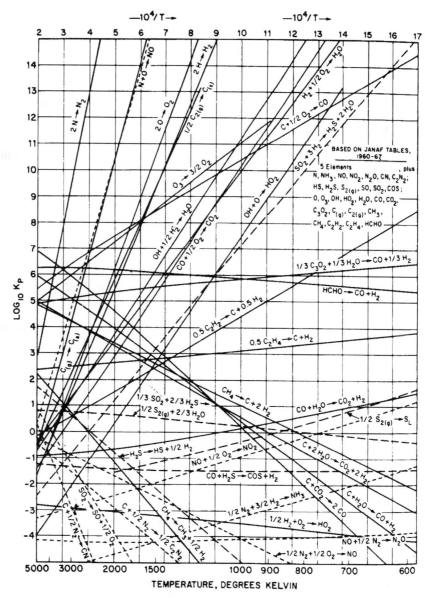

Figure EC-1.1   From M. Modell and R. Reid, *Thermodynamics and Its Applications*, © 1983. Reprinted by permission of Prentice Hall, Inc., Upper Saddle River, N.J.

From Figure EC-1.1 we read at 1000 K that log $K_P = 0.15$; therefore, $K_P = 1.41$, which is close to the calculated value. We note that there is no net change in the number of moles for this reaction (i.e., $\delta = 0$); therefore,

$$K = K_p = K_c \text{ (dimensionless)}$$

Taking the square root of Equation (EC-1.7) yields

Calculate X_e, the Equilibrium Conversion

$$\frac{X_e}{1 - X_e} = (1.44)^{1/2} = 1.2 \qquad \text{(EC-1.8)}$$

Solving for X_e, we obtain

$$\boxed{X_e = \frac{1.2}{2.2} = 0.55}$$

Then

$$C_{CO,0} = \frac{y_{CO,0} P_0}{RT_0}$$

$$= \frac{(0.5)(10 \text{ atm})}{(0.082 \text{ dm}^3 \cdot \text{atm}/\text{mol} \cdot \text{K})(1000 \text{ K})}$$

$$= 0.061 \text{ mol}/\text{dm}^3$$

Calculate $C_{CO,e}$, the Equilibrium Conversion of CO

$$C_{CO,e} = C_{CO,0}(1 - X_e) = (0.061)(1 - 0.55) = 0.0275 \text{ mol}/\text{dm}^3$$

$$C_{H_2O,e} = 0.0275 \text{ mol}/\text{dm}^3$$

$$\boxed{C_{CO,e} = C_{H_2,e} = C_{CO,0}X_e = 0.0335 \text{ mol}/\text{dm}^3}$$

Figure EC-1.1 gives the equilibrium constant as a function of temperature for a number of reactions. Reactions in which the lines increase from left to right are exothermic.

The following links give thermochemical data. (Heats of Formation, C_P, etc.)

Links

1)   *www.uic.edu/~mansoori/Thermodynamic.Data.and.Property_html*
2)   *webbook.nist.gov*

Also see *Chem. Tech.,* 28 (3) (March), 19 (1998).

Measurement of Slopes on Semilog Paper **D**

By plotting data directly on the appropriate log-log or semilog graph paper, a great deal of time may be saved over computing the logs of the data and then plotting them on linear graph paper. In the CD-ROM, we review the various techniques for plotting data and measuring slopes on semilog paper.

Links

Also see Physics Department web site at the University of Guelph, Ontario, Canada, *www.physics.uoguelph.ca/tutorials/GLP.*

Software Packages **E**

Polymath 5.1 and a special version of COMSOL Multiphysics are the software packages included with this book, along with extensive tutorials using screen shots.

E.1 Polymath

E.1.A About Polymath

Polymath is the primary software package used in this textbook. Polymath is an easy-to-use numerical computation package that allows students and professionals to use personal computers to solve the following types of problems:

- Simultaneous linear algebraic equations
- Simultaneous nonlinear algebraic equations
- Simultaneous ordinary differential equations
- Data Regressions (including the following)
  - Curve fitting by polynomials and splines
  - Multiple linear regression with statistics
  - Nonlinear regression with statistics

Polymath is unique in that the problems are entered just like their mathematical equations, and there is a minimal learning curve. Problem solutions are easily found with robust algorithms. This allows very convenient problem solving to be used in chemical reaction engineering and other areas of chemical engineering, leading to an enhanced educational experience for students.

The following special Polymath web site for software use and updating will be maintained for users of this textbook:

Links

www.polymath-software.com/fogler

E.1.B   Polymath Tutorials

Polymath tutorials are given in the Summary Notes. Here screen shots of the various steps are shown for each of the Polymath programs.

Summary Notes

Summary Notes

Chapter 1
    A. Ordinary Differential Equation (ODE) Tutorial
    B. Nonlinear (NLE) Solver Tutorial
Chapter 5
    A. Fitting a Polynomial Tutorial
    B. Nonlinear Regression Tutorial

Note: The copy of Polymath supplied in the CD-ROM can be "opened" up to 200 times. Once Polymath is "opened," any of the options (e.g., ODE solver, regression) can be run as many times as desired. That is, you could open Polymath once, leave it open, and make an infinite number of runs and it would only count as one "opening." After Polymath has been opened 200 times, directions on how to renew Polymath (with all recent updates and revisions and for a small fee) may be found on the web at

www.polymath-software.com

Links

To obtain a discounted fee for Polymath, have the following information available when you sign on the above web site to order Polymath! The ISBN number of the text (0-13-047394-4) and the printing number which can be found on the back of the title page.

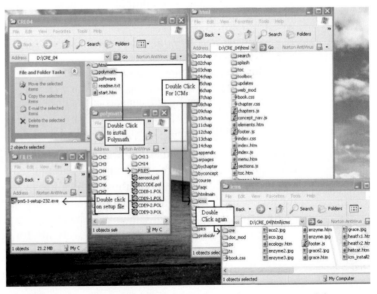

Figure E-1   Install Polymath and the interactive computer modules.

E.2 MATLAB

Sample MATLAB programs are given in CD-ROM, Appendix E. The disadvantage of the MATLAB ODE solver is that it is not particularly user-friendly when trying to determine the variation of secondary parameter values. MATLAB can be used for the same four types of programs as Polymath.

E.3 Aspen

Aspen is a process simulator that is used primarily in senior design courses. It has the steepest learning curve of the software packages used in this text. It has a built-in database of the physical properties of reactants and products. Consequently, one has only to type in the chemicals and the rate law parameters. It is really too powerful to be used for the types of home problems given here.

Aspen Example on
CD-ROM

The Pyrlosis of Benzene using Aspen is given on the CD-ROM using Aspen. Perhaps one home assignment should be devoted to using Aspen to solve a problem with heat effects in order to help familiarize the student with Aspen.

An Aspen tutorial can be accessed directly from the CD-ROM home page.

E.4 COMSOL

COMSOL is a partial differential equation solver (PDE) available commercially from COMSOL, Inc. Included with this text is a special version of COMSOL that has been prepared to solve problems involving tubular reactors. Specifically, one can solve CRE problems with heat effects involving both axial and radial gradients in concentration and temperature simply by loading the COMSOL CD on one's computer and running the program. One can also use it to solve isothermal CRE problems with reaction and diffusion.

Web Modules

A step-by-step COMSOL tutorial with screen shots is given on the CRE CD-ROM web modules.

There are two sections on the COMSOL ECRE CD. The first one is "Heat effects in tubular reactors," and the second section is "Tubular reactors with dispersion." In the first section, the four examples focus on the effects of radial velocity profile and external cooling on the performances of isothermal and nonisothermal tubular reactors. In the second section, two examples examine the dispersion effects in a tubular reactor.

Heat Effects

1. **Isothermal reactor.** This example concerns an elementary, exothermic, second-order reversible liquid-phase reaction in a tubular reactor with a parabolic velocity distribution. Only the mole, rate law, and stoichiometry balance in the tubular reactor are required in this COMSOL chemical engineering module.

2. **Nonisothermal adiabatic reactor.** The isothermal reactor is extended to include heat effects whereby the tubular reactor is treated as an adiabatic reactor. In this model, material and energy balances in the reactor are solved simultaneously in the COMSOL module.

3. **Nonisothermal reactor with isothermal cooling jacket.** A coolant at constant temperature cooling jacket is added to the previous example to examine the performance of a nonisothermal reactor. In this model, the boundary condition for the energy balance at the radial boundary is changed from the thermal insulation boundary condition to a heat flux boundary condition.

4. **Nonisothermal reactor with variable coolant temperature.** This example extends the third example by including the energy balance on the coolant in the cooling jacket as the temperature of the coolant varies along the reactor.

Dispersion and Reaction

1. **One-dimensional model with Danckwerts boundary conditions.** In this example, the mass balance of an arbitrary reaction in a tubular reactor is described by a dimensionless ordinary differential equation in terms of the Péclet number and the Damköhler number. This model uses the Danckwerts boundary conditions when dispersion and reaction take place simultaneously.

2. **One-dimensional model with upstream and downstream sections.** This example uses the open-vessel boundary conditions where an inlet (upstream) section and an outlet (downstream) section are added to a tubular reactor where dispersion occurs but no reaction.

It is suggested that one first use and play with the COMSOL program for solutions to Examples 8-12 and 14-4, before making any changes. One should also review the web module for Chapter 8 on *Axial and radial gradients in tubular reactors* before running the program.

Tutorial

Tutorials with step-by-step explanation and screen shots can be found by opening the documentation section on the COMSOL CD-ROM. Start with the documentation for **1. Isothermal Reactor** and then go to the documentation section for each of the other modules. Tutorial material discussed in this module is built on, but not repeated in modules that follow. Next go to the nonisothermal adiabatic reaction, and read the document there. This tutorial material will be built on but not repeated in the later modules. The same hold for each of the sequential modules.

Nomenclature F

A	Chemical species
A_c	Cross-sectional area (m²)
A_p	Total external surface area of particle (m²)
a_p	External surface area of catalyst per unit bed volume (m²/m³)
a	Area of heat exchange per unit volume of reactor (m⁻¹)
a_c	External surface area per volume of catalyst pellets (m²/m³)
B	Chemical species
$\mathbf{B_A}$	Flux of A resulting from bulk flow (mol/m²·s)
Bo	Bodenstein number
C	Chemical species
C_i	Concentration of species i (mol/dm³)
C_{pi}	Heat capacity of species i at temperature T (cal/gmol·K)
$\overline{C}_{pi}$	Mean heat capacity of species i between temperature T_0 and temperature T (cal/mol·K)
$\hat{C}_{pi}$	Mean heat capacity of species i between temperature T_R and temperature T (cal/mol·K)
c	Total concentration (mol/dm³) (Chapter 11)
D	Chemical species
Da	Damköhler Number (dimensionless)
D_{AB}	Binary diffusion coefficient of A in B (dm²/s)
D_a	Dispersion coefficient (cm²/s)
D_e	Effective diffusivity (dm²/s)
D_K	Knudsen diffusivity (dm²/s)
D^*	Taylor dispersion coefficient
E	Activation energy (cal/gmol)
(E)	Concentration of free (unbound) enzyme (mol/dm³)
F_i	Molar flow rate of species i (mol/s)
F_{i0}	Entering molar flow of species i (mol/s)
G	Superficial mass velocity (g/dm²·s)

G_i	Rate of generation of species i (mol/s)
$G_i^\circ(T)$	Gibbs free energy of species i at temperature T (cal/gmol·K)
$H_i(T)$	Enthalpy of species i at temperature T (cal/mol i)
$H_{i0}(T)$	Enthalpy of species i at temperature T_0 (cal/mol i)
H_i°	Enthalpy of formation of species i at temperature T_R (cal/gmol i)
h	Heat transfer coefficient (cal/m²·s·K)
$\mathbf{J_A}$	Molecular diffusive flux of species A (mol/m²·s)
K_A	Adsorption equilibrium constant
K_c	Concentration equilibrium constant
K_e	Equilibrium constant (dimensionless)
K_P	Partial pressure equilibrium constant
k	Specific reaction rate
k_c	Mass transfer coefficient (m/s)
M_i	Molecular weight of species i (g/mol)
m_i	Mass of species i (g)
N_i	Number of moles of species i (mol)
n	Overall reaction order
Nu	Nusselt Number (dimensionless)
Pe	Peclet number (Chapter 14)
P_i	Partial pressure of species i (atm)
Pr	Prandtl Number (dimensionless)
Q	Heat flow from the surroundings to the system (cal/s)
R	Ideal gas constant
Re	Reynolds number
r	Radial distance (m)
r_A	Rate of generation of species A per unit volume (gmol A/s·dm³)
$-r_A$	Rate of disappearance of species A per unit volume (mol A/s·dm³)
$-r_A'$	Rate of disappearance of species A per mass of catalyst (mol A/g·s)
$-r_A''$	Rate of disappearance of A per unit area of catalytic surface (mol A/m²·s)
S	An active site (Chapter 10)
(S)	Substrate concentration (gmol/dm³) (Chapter 7)
S_a	Surface area per unit mass of catalyst (m²/g)
$S_{D/U}$	Selectivity parameter (instantaneous selectivity) (Chapter 6)
$\tilde{S}_{D/U}$	Overall selectivity of D to U
Sc	Schmidt number (dimensionless) (Chapter 10)
Sh	Sherwood number (dimensionless) (Chapter 10)
SV	Space velocity (s⁻¹)
T	Temperature (K)
t	Time (s)
U	Overall heat transfer coefficient (cal/m²·s·K)
V	Volume of reactor (dm³)
V_0	Initial reactor volume (dm³)
v	Volumetric flow rate (dm³/s)

v_0	Entering volumetric flow rate (dm^3/s)
W	Weight of catalyst (kg)
$\mathbf{W_A}$	Molar flux of species A ($mol/m^2 \cdot s$)
X	Conversion of key reactant, A
Y_i	Instantaneous yield of species i
$\tilde{Y}_i$	Overall yield of species i
y	Pressure ratio P/P_0
y_i	Mole fraction of species i
y_{i0}	Initial mole fraction of species i
Z	Compressibility factor
z	Linear distance (cm)

Subscripts

0	Entering or initial condition
b	Bed (bulk)
c	Catalyst
e	Equilibrium
p	Pellet

Greek Symbols

α	Reaction order (Chapter 3)
α	Pressure drop parameter (Chapter 4)
α_i	Parameter in heat capacity (Chapter 8)
β_i	Parameter in heat capacity
β	Reaction order
γ_i	Parameter in heat capacity
δ	Change in the total number of moles per mole of A reacted
ε	Fraction change in volume per mole of A reacted resulting from the change in total number of moles
η	Internal effectiveness factor
Θ_i	Ratio of the number of moles of species i initially (entering) to the number of moles of A initially (entering)
λ	Dimensionless distance (z/L) (Chapter 12)
λ	Life expectancy (s) (Chapter 13)
μ	Viscosity ($g/cm \cdot s$)
ρ	Density (g/cm^3)
ρ_c	Density of catalyst pellet (g/cm^3 of pellet)
ρ_b	Bulk density of catalyst (g/cm^3 of reactor bed)
τ	Space time (s)
ϕ	Void fraction (Porosity)
ϕ_n	Thiele modulus
ψ	Dimensionless concentration (C_A/C_{As})
Ω	External (overall) effectiveness factor

Rate Law Data   G

Reaction rate laws and data can be obtained from the following web sites:

1. National Institute of Standards and Technology (NIST)
   Chemical Kinetics Database on the Web
   Standard Reference Database 17, Version 7.0 (Web Version), Release 1.2
   This web site provides a compilation of kinetics data on gas-phase reactions.
   kinetics.nist.gov/index.php

2. International Union of Pure and Applied Chemistry (IUPAC)
   This web site provides kinetic and photochemical data for gas kinetic data evaluation.
   www.iupac-kinetic.ch.cam.ac.uk/

3. *NASA/JPL (Jet Propulsion Laboratory: California Institute of Technology)*
   This web site provides chemical kinetics and photochemical data for use in atmospheric studies.
   jpldataeval.jpl.nasa.gov/download.html

4. BRENDA: University of Cologne
   This web site provides enzyme data and metabolic information. BRENDA is maintained and developed at the Institute of Biochemistry at the University of Cologne.
   www.brenda.uni-koeln.de

5. NDRL Radiation Chemistry Data Center: Notre Dame Radiation Laboratory
   This web site provides the reaction rate data for transient radicals, radical ions, and excited states in solution.
   www.rcdc.nd.edu

Open-Ended Problems H

Links

The following are summaries for open-ended problems that have been used as term problems at the University of Michigan. The complete problem statement of the problems can be found in the CD-ROM, Appendix H.

H.1 Design of Reaction Engineering Experiment

The experiment is to be used in the undergraduate laboratory and cost less than $500 to build. The judging criteria are the same as the criteria for the National AIChE Student Chapter Competition. The design is to be displayed on a poster board and explained to a panel of judges. Guidelines for the poster board display are provided by Jack Fishman and are given on the CD-ROM.

H.2 Effective Lubricant Design

Lubricants used in car engines are formulated by blending a base oil with additives to yield a mixture with the desirable physical attributes. In this problem, students examine the degradation of lubricants by oxidation and design an improved lubricant system. The design should include the lubricant system's physical and chemical characteristics, as well as an explanation as to how it is applied to automobiles. Focus: automotive industry, petroleum industry.

H.3 Peach Bottom Nuclear Reactor

The radioactive effluent stream from a newly constructed nuclear power plant must be made to conform with Nuclear Regulatory Commission standards. Students use chemical reaction engineering and creative problem solving to propose solutions for the treatment of the reactor effluent. Focus: problem analysis, safety, ethics.

H.4 Underground Wet Oxidation

You work for a specialty chemicals company, which produces large amounts of aqueous waste. Your chief executive officer (CEO) read in a journal about an emerging technology for reducing hazardous waste, and you must evaluate the system and its feasibility. Focus: waste processing, environmental issues, ethics.

H.5 Hydrodesulfurization Reactor Design

Your supervisor at Kleen Petrochemical wishes to use a hydrodesulfurization reaction to produce ethylbenzene from a process waste stream. You have been assigned the task of designing a reactor for the hydrodesulfurization reaction. Focus: reactor design.

H.6 Continuous Bioprocessing

Most commercial bioreactions are carried out in batch reactors. The design of a continuous bioreactor is desired since it may prove to be more economically rewarding than batch processes. Most desirable is a reactor that can sustain cells that are suspended in the reactor while growth medium is fed in, without allowing the cells to exit the reactor. Focus: mixing modeling, separations, bioprocess kinetics, reactor design.

The complete problem statements are found in the CD-ROM, Appendix H

H.7 Methanol Synthesis

Kinetic models based on experimental data are being used increasingly in the chemical industry for the design of catalytic reactors. However, the modeling process itself can influence the final reactor design and its ultimate performance by incorporating different interpretations of experimental design into the basic kinetic models. In this problem, students are asked to develop kinetic modeling methods/approaches and apply them in the development of a model for the production of methanol from experimental data. Focus: kinetic modeling, reactor design.

H.8 Cajun Seafood Gumbo

Most gourmet foods are prepared by batch processes. In this problem, students are challenged to design a continuous process for the production of gourmet-quality Cajun seafood gumbo from an old family recipe. Focus: reactor design.

Most gourmet foods are prepared by a batch process (actually in a batch reactor). Some of the most difficult gourmet foods to prepare are Louisiana specialties, owing to the delicate balance between spices (hotness) and subtle flavors that must be achieved. In preparing Creole and Cajun food, certain flavors are released only by cooking some of the ingredients in hot oil for a period of time.

We shall focus on one specialty, Cajun seafood gumbo. Develop a continuous-flow reactor system that would produce 5 gal/h of a gourmet-quality seafood gumbo. Prepare a flow sheet of the entire operation. Outline certain experiments and areas of research that would be needed to ensure the success of your project. Discuss how you would begin to research these problems. Make a plan for any experiments to be carried out (see Section 5.6).

Following is an old family formula for Cajun seafood gumbo for batch operation (10 quarts, serves 40):

1 cup flour	4 bay leaves, crushed
$1\frac{1}{2}$ cups olive oil	$\frac{1}{2}$ cup chopped parsley
1 cup chopped celery	3 large Idaho potatoes (diced)
2 large red onions (diced)	1 tablespoon ground pepper
5 qt fish stock	1 tablespoon tomato paste
6 lb fish (combination of cod, red	5 cloves garlic (diced)
snapper, monk fish, and halibut)	$\frac{1}{2}$ tablespoon Tabasco sauce
12 oz crabmeat	1 bottle dry white wine
1 qt medium oysters	1 lb scallops
1 lb medium to large shrimp	

1. Make a roux (i.e., add 1 cup flour to 1 cup of boiling olive oil). Cook until dark brown. Add roux to fish stock.
2. Cook chopped celery and onion in boiling olive oil until onion is translucent. Drain and add to fish stock.
3. Add $\frac{1}{3}$ of the fish (2 lb) and $\frac{1}{3}$ of the crabmeat, liquor from oysters, bay leaves, parsley, potatoes, black pepper, tomato paste, garlic, Tabasco, and $\frac{1}{4}$ cup of the olive oil. Bring to a slow boil and cook 4 h, stirring intermittently.
4. Add 1 qt cold water, remove from the stove, and refrigerate (at least 12 h) until $2\frac{1}{2}$ h before serving.
5. Remove from refrigerator, add $\frac{1}{4}$ cup of the olive oil, wine, and scallops. Bring to a light boil, then simmer for 2 h. Add remaining fish (cut to bite size), crabmeat, and water to bring total volume to 10 qt. Simmer for 2 h, add shrimp, then 10 minutes later, add oysters and serve immediately.

H.9   Alcohol Metabolism

The purpose of this open-ended problem is for the students to apply their knowledge of reaction kinetics to the problem of modeling the metabolism of alcohol in humans. In addition, the students will present their findings in a poster session. The poster presentations will be designed to bring a greater awareness to the university community of the dangers associated with alcohol consumption.

Living Example Problem

Students should choose one of the following four major topics to further investigate:

1. Death caused by acute alcohol overdose
2. Long-term effects of alcohol
3. Interactions of alcohol with common medications
4. Factors affecting metabolism of alcohol

General information regarding each of these topics can be found on the CD-ROM.

The metabolism and model equations are given in section 7.5. One can load the Living Example problem for alcohol metabolism directly from the CD-ROM.

Living Example Problem

H.10 Methanol Poisoning

The emergency room treatment for methanol poisoning is to inject ethanol intravenously to tie up the alcohol dehydrogenase enzyme so that methanol will not be converted to formic acid and formate, which causes blindness. The goal of this open-ended problem is to build on the physiological-based model for ethanol metabolism to predict the ethanol injection rate for methanol poisoning. One can find a start on this problem by reading problem P7-25c.

How to Use **I**
the CD-ROM

The primary purpose of the CD-ROM is to serve as an enrichment resource. The benefits of using the CD-ROM are fourfold:

1. To facilitate different student learning styles
   www.engin.umich.edu/~cre/asyLearn/itresources.htm.
2. To provide the student with the option/opportunity for further study or clarification of a particular concept or topic
3. To provide the opportunity to practice critical thinking, creative thinking, and problem-solving skills
4. To provide additional technical material for the practicing engineer
5. To provide other tutorial information such as additional homework problems and instructions on using computational software in chemical engineering

I.1 CD-ROM Components

There are two types of information on this CD-ROM: information that is organized **by chapter** and information organized **by concept.** Material in the by chapter section on the CD-ROM corresponds to the material found in this book and is further divided into five sections.

- **Objectives.** The objectives page lists what the students will learn from the chapter. When students are finished working on a chapter, they can come back to the objectives to see if you have covered everything in that chapter. Or if students need additional help on a specific topic, they can see if that topic is covered in a chapter from the objectives page.
- **Learning Resources.** These resources give an overview of the material in each chapter and provide extra explanations, examples, and applications to reinforce the basic concepts of chemical reaction engineering. Summary Notes serve as an overview of each chapter and contain a logical flow of derived equations and

additional examples. Web Modules and Interactive Computer Modules (ICM) show how the principles from the text can be applied to nonstandard problems. Solved Problems provide more examples for students to use the knowledge gained from each chapter.

Living Example Problem

- **Living Example Problems.** These problems are usually examples from the text that require computational software to solve. Students can "play" with the problem and ask "what if . . . ?" questions to practice critical and creative thinking skills. Students can change parameter values, such as the reaction rate constants, to learn to deduce trends or predict the behavior of a given reaction system.

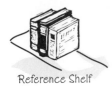

Reference Shelf

- **Professional Reference Shelf.** The Professional Reference Shelf contains two types of information. First, it includes material that is important to the practicing engineer but that is typically not included in the majority of chemical reaction engineering courses. Second, it includes material that gives a more detailed explanation of derivations that were abbreviated in the text. The intermediate steps to these derivations are shown on the CD-ROM.

Homework Problems

- **Additional Homework Problems.** New problems were developed for this edition. They provide a greater opportunity to use today's computing power to solve realistic problems. Instead of omitting some of the more traditional, yet excellent problems of previous editions, these problems were placed on the CD-ROM and can serve as practice problems along with those unassigned problems in the text.

The materials in Learning Resources are further divided into Summary Notes, Web Modules, Interactive Computer Modules, and Solved Problems. Table I-1 shows which enrichment resources can be found in each chapter.

TABLE I-1.   CD-ROM ENRICHMENT RESOURCES

Chapters

Learning Resources:	1	2	3	4	5	6	7	8	9	10	11	12	13	14
Summary Notes	■	■	■	■	■	■	■	■	■	■	■	■	■	■
Web Modules	■	■	■	■		■		■					■	
Interactive Computer Modules	■	■	■	■	■	■		■		■				■
Solved Problems	■	■	■	■	■	■	■	■	■	■		■	■	■
Living Example Problems	■			■	■	■	■	■	■	■			■	■
Professional Reference Shelf	■	■	■	■	■	■	■	■	■	■	■	■	■	■
Additional Homework Problems	■	■	■	■	■	■	■	■	■	■	■	■	■	■

Note: The ICMs are high-memory-use programs. Because of the memory intensive nature of the ICMs, there have been intermittent problems (10–15% of Windows computers) with the modules. You can usually solve the problem by trying the ICM on a different computer. In the Heatfx 2 ICM, only the first three reactors can be solved, and users cannot continue on to part 2 because of a bug currently in the program.

The information that can be accessed in the by concept sections is not specific to a single chapter. Although the material can be accessed from the by chapter sections, the by concept section allows you to access certain material quickly without browsing through chapters.

Interactive

Computer Modules

Web Modules

- **Interactive Modules.** The CD-ROM includes both **Web Modules** and **ICMs**. The Web Modules use a web browser for an interface and give examples of how chemical reaction engineering principles can be applied to a wide range of situations such as modeling cobra bites and cooking a potato. ICMs are modules that use a Windows or DOS-based program for an interface. They test knowledge on different aspects of chemical reaction engineering through a variety of games such as basketball and jeopardy.
- **Problem Solving.** Here students can learn different strategies for problem solving in both closed- and open-ended problems. See the ten different types of home problems and suggestions for approaching them. Extensive information on critical and creative thinking can also be found in this section.
- **Frequently Asked Questions (FAQs).** Over the years that I have taught this course, I have collected a number of questions that the students have asked over and over for years and years. The questions usually ask for clarification or for a different way of explaining the material or for another example of the principle being discussed. The FAQs and answers are arranged by chapter.
- **Syllabi.** Three representative syllabi have been included on the CD-ROM. See a 3-credit course schedule from the University of Illinois, or a couple of 4-credit course schedules from the University of Michigan. You can also practice for midterms, finals, or Doctoral Qualifying Exams with the actual exams given at the University of Michigan in previous years.
- **Credits.** See who was responsible for putting this CD-ROM together.

I.2 Navigation

Students can use the CD-ROM in conjunction with the text in a number of different ways. The CD-ROM provides *enrichment resources*. It is up to each student to determine how to use these resources to generate the greatest benefit. Table I-2 shows some of the clickable buttons found in the Summary Notes within the Learning Resources and a brief description of what the students will see when they click on the buttons.

TABLE I-2.   HOT BUTTONS IN SUMMARY NOTES

Clickable Button	*Where it goes*
Example	Solved example problem
Link	General material that may not be related to the chapter
Tip	Hints and tips for solving problems
Self Test	A test on the material in a section with solutions
Derive	Derivations of equations when not shown in the notes

TABLE I-2.   HOT BUTTONS IN SUMMARY NOTES

Critical	Critical Thinking Question related to the chapter
Module	Web Module related to the chapter
Assess	Chapter objectives
Polymath	Polymath solution of a problem from the Summary Notes
Biography	Biography of the person who developed an equation or principle
More	Chapter insert with more information on a topic
Workbook	Detailed solution of a problem
Plot	Plot of an equation or solution
Side Note	Extra information on a specific topic
wm qt raw	Audio clip

The creators of the CD-ROM tried to make navigating through the resources as easy and logical as possible. A more comprehensive guide to usage and navigation can be found on the CD-ROM. Figure I-1 shows how to access the installation files for Polymath and the ICMs on the CD-ROM. The upper left window (CRE04) in the figure is what appears when the disc is inserted and "Explore the CD" is chosen from the auto-run pop-up window.

I.3 How the CD-ROM/Web Can Help Learning Style

I.3.1   Global vs. Sequential Learners

See *www.engin.umich.edu/~cre/asyLearn/itresources.htm*.

Global

- Use the summary lecture notes to get an overview of each chapter on the CD-ROM and see the big picture
- Review real-world examples and pictures on the CD-ROM
- Look at concepts outlined in the ICMs

Sequential

- Use the Derive hot buttons to go through derivations in lecture notes on the web
- Follow all derivations in the ICMs step by step
- Do all self-tests, audios, and examples in the CD-ROM lecture notes step by step

I.3.2   Active vs. Reflective Learners

Active

- Use all the hot buttons to interact with the material to keep active
- Use self-tests as a good source of practice problems
- Use Living Example Problems to change settings/parameters and see the result
- Review for exams using the ICMs

Reflective

- Self-tests allow you to consider the answer before seeing it
- Use Living Learning Problems to think about topics independently

I.3.3   Sensing vs. Intuitive Learners

Sensing

- Use Web Modules (cobra, hippo, nanoparticles) to see how material is applied to real-world topics
- Relate how Living Example Problems are linked to real-world topics

Intuitive

- Vary parameters in supplied Polymath problems and understand their influence on a problem
- Use the trial-and-error portions of some ICMs to understand "what if . . . " style questions

I.3.4   Visual vs. Verbal Learners

Visual

- Study the examples and self-tests on the CD-ROM summary notes that have graphs and figures showing trends
- Do ICMs to see how each step of a derivation/problem leads to the next
- Use the graphical output from Living Example Problems/Polymath code to obtain a visual understanding of how various parameters affect a system
- Use the Professional Reference Shelf to view pictures of real reactors

Verbal

- Listen to audios on the web to hear information in another way
- Work with a partner to answer questions on the ICMs

Use of Computational **J**
Chemistry Software
Packages

J.1 Computational Chemical Engineering

As a prologue to the future, our profession is evolving to one of molecular chemical engineering. For chemical reaction engineers, computation chemistry and molecular modeling, this could well be our future.

Thermodynamic properties of molecular species that are used in reactor design problems can be readily estimated from thermodynamic data tabulated in standard reference sources such as Perry's Handbook or the JANAF Tables. Thermochemical properties of molecular species not tabulated can usually be estimated using group contribution methods. Estimation of activation energies is, however, much more difficult owing to the lack of reliable information on transition state structures, and the data required to carry out these calculations is not readily available.

Recent advances in computational chemistry and the advent of powerful easy-to-use software tools have made it possible to estimate important reaction rate quantities (such as activation energy) with sufficient accuracy to permit incorporation of these new methods into the reactor design process. Computational chemistry programs are based on theories and equations from quantum mechanics, which until recently, could only be solved for the simplest systems such as the hydrogen atom. With the advent of inexpensive high-speed desktop computers, the use of these programs in both engineering research and industrial practice is increasing rapidly. Molecular properties such as bond length, bond angle, net dipole moment, and electrostatic charge distribution can be calculated. Additionally, reaction energetics can be accurately determined by using quantum chemistry to estimate heats of formation of reactants, products, and also for transition state structures.

Links

Examples of commercially available computational chemistry programs include Spartan developed by Wavefunction, Inc. (*www.wavefun.com*), and Cerius2 from Molecular Simulations, Inc. (*www.accelrys.com*). The following example utilizes Spartan 4.0 to estimate the activation energy for a nucleophilic substitution reaction (SN2). The following calculations were performed on an IBM 43-P RS-6000 UNIX workstation.

An example using Spartan to calculate the activation energy for the reaction

$$C_2H_5Cl + OH^- \longrightarrow C_2H_5OH + Cl^-$$

is given on the CD-ROM.

CDPApp.J-1$_A$ Redo Example Appendix J.1 on the CD-ROM.
   (a) Choose different methods of calculation, such as using a value of 2.0 Å to constrain the C—Cl and C—O bonds.
   (b) Choose different methods to calculate the potential energy surface. Compare the Ab Initio to the semiempirical method.
   (c) Within the semiempirical method, compare the AM1 and PM3 models.

Index

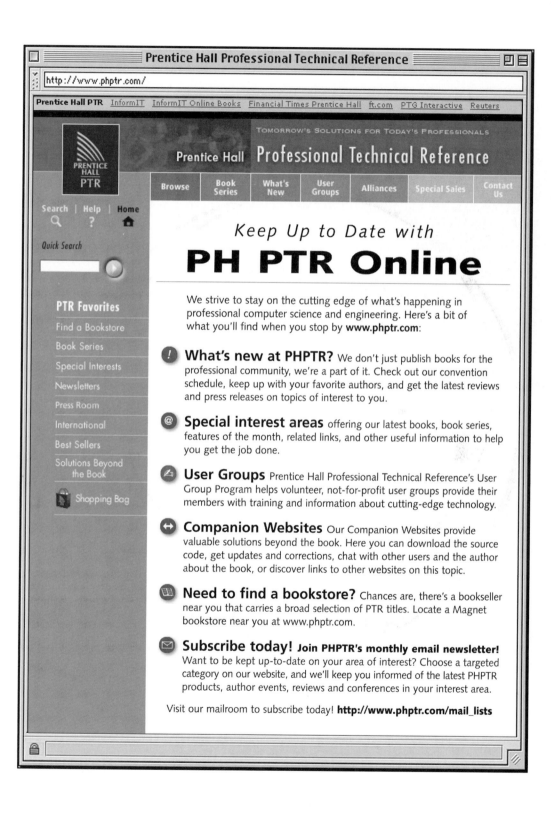

2.   <u>Term of License</u>. This License Agreement is effective until terminated ("Term"). You may terminate this License Agreement at any time by destroying the disk and insuring that you have uninstalled the Content, including all copies in whole or in part in any form. This License Agreement will also terminate automatically if you fail to comply with any Term, including any Third Party Term. Upon such automatic termination, you agree to either (i) destroy the disk and insure that you have uninstalled the Content, including all copies in whole or in part in any form, or (ii) within seven (7) days after demand by us, return at your expense the disk and all copies thereof and all such copies of the Content to us at the address in Section 1 above.

3.   <u>Compliance with Laws</u>. You shall comply fully with all the Terms and all applicable laws and regulations, including without limitation, those relating to import and export of technical data and computer software. Without limiting the foregoing, you will not exercise any of the rights under this Agreement in violation of such laws and regulations, nor may you transfer this Agreement to any party where doing so might result in such a violation.

4.   <u>No Maintenance or Support</u>. We do not provide any maintenance and support for the disk and Content. If you have any problems with the disk, please see your seller.

5.   <u>Limited Warranty; Sole Remedy; Limitation of Liability</u>.

    (a) <u>Disk and Content</u>. Your seller warrants that for a period of thirty (30) days from delivery to you the disk shall be free of defects in material and workmanship under normal use, and we warrant that the Content shall conform in all material respects to its functional specifications in the documentation. ALL OTHER WARRANTIES AND PROMISES, INCLUDING, WITHOUT LIMITATION, THOSE RELATING TO CONDITION, OPERATION, DESIGN, MERCHANTABILITY, NON-INFRINGEMENT, OR FITNESS FOR A PARTICULAR PURPOSE OF THE CONTENT, WHETHER EXPRESS OR IMPLIED, ARE HEREBY DISCLAIMED. All requests for warranty assistance should be directed to your seller.

    (b) <u>Sole Remedy</u>. If the disk or Content is not as warranted, your exclusive remedy and your seller's and our sole liability shall be (i) the correction or workaround of major defects within a reasonable time, or (ii) if that is not practical or possible in our reasonable judgment, we may terminate this Agreement and refund your purchase price on a straight-line three-year depreciated basis provided you have returned the disk and all Content, including all copies in whole or in part, to us at the address specified in Section 1.

    (c) <u>Limitation of Liability</u>. EXCEPT AS EXPRESSLY WARRANTED ABOVE (OR AS IMPLIED BY LAW WHERE THE LAW PROVIDES THAT THE PARTICULAR TERMS IMPLIED CANNOT BE EXCLUDED BY CONTRACT), IN NO EVENT WILL WE BE LIABLE TO YOU, ANY USER OF THE CONTENT, OR TO ANY OTHER PERSON OR ENTITY FOR CLAIMS OR DAMAGES ARISING FROM THE USE OF THE DISK OR CONTENT OR RELATED TO THIS AGREEMENT, INCLUDING BUT NOT LIMITED TO LOST REVENUE OR PROFIT, CONSEQUENTIAL, INCIDENTAL, SPECIAL, PUNITIVE, INDIRECT, OR OTHER DAMAGES, LOSS OF DATA, AND INTERRUPTION OR LOSS OF USE, EVEN IF WE HAVE BEEN ADVISED OF THE POSSIBILITY THEREOF, AND REGARDLESS OF WHETHER SUCH CLAIMS OR DAMAGES ARE BASED ON CONTRACT, TORT (INCLUDING NEGLIGENCE), STRICT LIABILITY, OR OTHER THEORY.

6.   <u>General</u>. If any part of this Agreement should, for any reason, be held invalid or unenforceable in any respect, the remainder of this Agreement shall be enforced to the full extent permitted by law. A court of competent jurisdiction is hereby empowered to modify any invalid or unenforceable provision to make it valid and enforceable as closely as possible to reflect its original intent. The failure of a parity to insist on strict compliance with any Term will not be deemed a waiver of such Term, and any waiver will not be deemed a continuing waiver or a waiver of any other Term. This Agreement and the rights and obligations of the parties will be binding upon and inure to the benefit of the parties and their successors and permitted assigns. You may not assign this Agreement without our prior written consent. This Agreement will be interpreted under the laws of New York, without regard to its conflict of laws principles and without giving any effect to the U.N. Conventional on the International Sale of Goods or to the provisions of UCITA as it may be adopted in any state having jurisdiction or whose law governs this Agreement. Any dispute arising under this Agreement shall be brought in the State and federal courts located in Rochester, New York, and you consent to the exclusive personal jurisdiction of such courts in the event of any such dispute. This Agreement contains the entire understanding of the parties with respect to the disk and Content, and it may not be changed except by a writing signed by both you and us. We expressly reserve all rights not expressly granted to you herein.

About the CD-ROMs

The two enclosed CD-ROMs are companions to the book *Elements of Chemical Reaction Engineering, Fourth Edition,* by H. Scott Fogler. The primary CD contains browser-driven content intended to supplement the information in the book. (Please review the README file on the primary CD-ROM for system requirements.) The COMSOL CD-ROM contains a specially prepared version of the software for use in solving problems in the book.

Included on these CD-ROMs is executable software intended to be installed and run directly from your computer. There are instructions and links to this software on each of the CDs.

The Reactor Lab End User License Agreement is available on the book's primary CD-ROM. The COMSOL End User License is on the COMSOL CD-ROM. Detailed information on the use of POLYMATH and on ordering updates and future versions of the software is available from *www.polymath-software.com.*

Prentice Hall warrants the enclosed CD-ROMs to be free of defects in materials and faulty workmanship under normal use for a period of ninety days after purchase (when purchased new). If a defect is discovered in either of the CD-ROMs during this warranty period, a replacement CD-ROM can be obtained at no charge by sending the defective CD-ROM, postage prepaid, with proof of purchase to:

> Disc Exchange
> Prentice Hall
> Pearson Technology Group
> 75 Arlington Street, Suite 300
> Boston, MA 02116
> Email: AWPro@aw.com

Prentice Hall makes no warranty or representation, either expressed or implied, with respect to this software, its quality, performance, merchantability, or fitness for a particular purpose. In no event will Prentice Hall, its distributors, or dealers be liable for direct, indirect, special, incidental, or consequential damages arising out of the use or inability to use the software. The exclusion of implied warranties is not permitted in some states. Therefore, the above exclusion may not apply to you. This warranty provides you with specific legal rights. There may be other rights that you may have that vary from state to state. The contents of these CD-ROMs are intended for personal use only.

More information and updates are available at:
www.prenhallprofessional.com and *www.engin.umich.edu/~cre*